Intermediate Algebra

SENIOR CONTRIBUTING AUTHOR
LYNN MARECEK, SANTA ANA COLLEGE

OpenStax
Rice University
6100 Main Street MS-375
Houston, Texas 77005

To learn more about OpenStax, visit https://openstax.org.
Individual print copies and bulk orders can be purchased through our website.

PRINT BOOK ISBN-10	0-9986257-2-8
PRINT BOOK ISBN-13	978-0-9986257-2-0
PDF VERSION ISBN-10	1-947172-26-3
PDF VERSION ISBN-13	978-1-947172-26-5
Revision Number	IA-2017-001(06/17)-LC
Original Publication Year	2017

Printed in Indiana, USA

OPENSTAX

OpenStax provides free, peer-reviewed, openly licensed textbooks for introductory college and Advanced Placement® courses and low-cost, personalized courseware that helps students learn. A nonprofit ed tech initiative based at Rice University, we're committed to helping students access the tools they need to complete their courses and meet their educational goals.

RICE UNIVERSITY

OpenStax, OpenStax CNX, and OpenStax Tutor are initiatives of Rice University. As a leading research university with a distinctive commitment to undergraduate education, Rice University aspires to path-breaking research, unsurpassed teaching, and contributions to the betterment of our world. It seeks to fulfill this mission by cultivating a diverse community of learning and discovery that produces leaders across the spectrum of human endeavor.

FOUNDATION SUPPORT

OpenStax is grateful for the tremendous support of our sponsors. Without their strong engagement, the goal of free access to high-quality textbooks would remain just a dream.

Laura and John Arnold Foundation (LJAF) actively seeks opportunities to invest in organizations and thought leaders that have a sincere interest in implementing fundamental changes that not only yield immediate gains, but also repair broken systems for future generations. LJAF currently focuses its strategic investments on education, criminal justice, research integrity, and public accountability.

The William and Flora Hewlett Foundation has been making grants since 1967 to help solve social and environmental problems at home and around the world. The Foundation concentrates its resources on activities in education, the environment, global development and population, performing arts, and philanthropy, and makes grants to support disadvantaged communities in the San Francisco Bay Area.

Calvin K. Kazanjian was the founder and president of Peter Paul (Almond Joy), Inc. He firmly believed that the more people understood about basic economics the happier and more prosperous they would be. Accordingly, he established the Calvin K. Kazanjian Economics Foundation Inc, in 1949 as a philanthropic, nonpolitical educational organization to support efforts that enhanced economic understanding.

Guided by the belief that every life has equal value, the Bill & Melinda Gates Foundation works to help all people lead healthy, productive lives. In developing countries, it focuses on improving people's health with vaccines and other life-saving tools and giving them the chance to lift themselves out of hunger and extreme poverty. In the United States, it seeks to significantly improve education so that all young people have the opportunity to reach their full potential. Based in Seattle, Washington, the foundation is led by CEO Jeff Raikes and Co-chair William H. Gates Sr., under the direction of Bill and Melinda Gates and Warren Buffett.

The Maxfield Foundation supports projects with potential for high impact in science, education, sustainability, and other areas of social importance.

Our mission at The Michelson 20MM Foundation is to grow access and success by eliminating unnecessary hurdles to affordability. We support the creation, sharing, and proliferation of more effective, more affordable educational content by leveraging disruptive technologies, open educational resources, and new models for collaboration between for-profit, nonprofit, and public entities.

The Bill and Stephanie Sick Fund supports innovative projects in the areas of Education, Art, Science and Engineering.

☰ Table of Contents

PREFACE

Welcome to *Intermediate Algebra,* an OpenStax resource. This textbook was written to increase student access to high-quality learning materials, maintaining highest standards of academic rigor at little to no cost.

About OpenStax

OpenStax is a nonprofit based at Rice University, and it's our mission to improve student access to education. Our first openly licensed college textbook was published in 2012, and our library has since scaled to over 25 books for college and AP courses used by hundreds of thousands of students. Our adaptive learning technology, designed to improve learning outcomes through personalized educational paths, is being piloted in college courses throughout the country. Through our partnerships with philanthropic foundations and our alliance with other educational resource organizations, OpenStax is breaking down the most common barriers to learning and empowering students and instructors to succeed.

About OpenStax Resources

Customization

Intermediate Algebra is licensed under a Creative Commons Attribution 4.0 International (CC BY) license, which means that you can distribute, remix, and build upon the content, as long as you provide attribution to OpenStax and its content contributors.

Because our books are openly licensed, you are free to use the entire book or pick and choose the sections that are most relevant to the needs of your course. Feel free to remix the content by assigning your students certain chapters and sections in your syllabus, in the order that you prefer. You can even provide a direct link in your syllabus to the sections in the web view of your book.

Instructors also have the option of creating a customized version of their OpenStax book. The custom version can be made available to students in low-cost print or digital form through their campus bookstore. Visit your book page on openstax.org for more information.

Errata

All OpenStax textbooks undergo a rigorous review process. However, like any professional-grade textbook, errors sometimes occur. Since our books are web based, we can make updates periodically when deemed pedagogically necessary. If you have a correction to suggest, submit it through the link on your book page on openstax.org. Subject matter experts review all errata suggestions. OpenStax is committed to remaining transparent about all updates, so you will also find a list of past errata changes on your book page on openstax.org.

Format

You can access this textbook for free in web view or PDF through openstax.org, and for a low cost in print.

About *Intermediate Algebra*

Intermediate Algebra is designed to meet the scope and sequence requirements of a one-semester Intermediate algebra course. The book's organization makes it easy to adapt to a variety of course syllabi. The text expands on the fundamental concepts of algebra while addressing the needs of students with diverse backgrounds and learning styles. Each topic builds upon previously developed material to demonstrate the cohesiveness and structure of mathematics.

Coverage and Scope

Intermediate Algebra continues the philosophies and pedagogical features of *Prealgebra* and *Elementary Algebra*, by Lynn Marecek and MaryAnne Anthony-Smith. By introducing the concepts and vocabulary of algebra in a nurturing, non-threatening environment while also addressing the needs of students with diverse backgrounds and learning styles, the book helps students gain confidence in their ability to succeed in the course and become successful college students.

The material is presented as a sequence of small, and clear steps to conceptual understanding. The order of topics was carefully planned to emphasize the logical progression throughout the course and to facilitate a thorough understanding of each concept. As new ideas are presented, they are explicitly related to previous topics.

> **Chapter 1: Foundations**
> Chapter 1 reviews arithmetic operations with whole numbers, integers, fractions, decimals and real numbers, to give the student a solid base that will support their study of algebra.
>
> **Chapter 2: Solving Linear Equations and Inequalities**
> In Chapter 2, students learn to solve linear equations using the Properties of Equality and a general strategy. They use a problem-solving strategy to solve number, percent, mixture and uniform motion applications. Solving a formula for a specific variable, and also solving both linear and compound inequalities is presented.
>
> **Chapter 3: Graphs and Functions**
> Chapter 3 covers the rectangular coordinate system where students learn to plot graph linear equations in two variables, graph with intercepts, understand slope of a line, use the slope-intercept form of an equation of a line, find the equation of a line, and create graphs of linear inequalities. The chapter also introduces relations and

functions as well as graphing of functions.

Chapter 4: Systems of Linear Equations
Chapter 4 covers solving systems of equations by graphing, substitution, and elimination; solving applications with systems of equations, solving mixture applications with systems of equations, and graphing systems of linear inequalities. Systems of equations are also solved using matrices and determinants.

Chapter 5: Polynomials and Polynomial Functions
In Chapter 5, students learn how to add and subtract polynomials, use multiplication properties of exponents, multiply polynomials, use special products, divide monomials and polynomials, and understand integer exponents and scientific notation.

Chapter 6: Factoring
In Chapter 6, students learn the process of factoring expressions and see how factoring is used to solve quadratic equations.

Chapter 7: Rational Expressions and Functions
In Chapter 7, students work with rational expressions, solve rational equations and use them to solve problems in a variety of applications, and solve rational inequalities.

Chapter 8: Roots and Radical
In Chapter 8, students simplify radical expressions, rational exponents, perform operations on radical expressions, and solve radical equations. Radical functions and the complex number system are introduced

Chapter 9: Quadratic Equations
In Chapter 9, students use various methods to solve quadratic equations and equations in quadratic form and learn how to use them in applications. Students will graph quadratic functions using their properties and by transformations.

Chapter 10: Exponential and Logarithmic Functions
In Chapter 10, students find composite and inverse functions, evaluate, graph, and solve both exponential and logarithmic functions.

Chapter 11: Conics
In Chapter 11, the properties and graphs of circles, parabolas, ellipses and hyperbolas are presented. Students also solve applications using the conics and solve systems of nonlinear equations.

Chapter 12: Sequences, Series and the Binomial Theorem
In Chapter 12, students are introduced to sequences, arithmetic sequences, geometric sequences and series and the binomial theorem.

All chapters are broken down into multiple sections, the titles of which can be viewed in the **Table of Contents**.

Key Features and Boxes

Examples Each learning objective is supported by one or more worked examples, which demonstrate the problem-solving approaches that students must master. Typically, we include multiple examples for each learning objective to model different approaches to the same type of problem, or to introduce similar problems of increasing complexity.

All examples follow a simple two- or three-part format. First, we pose a problem or question. Next, we demonstrate the solution, spelling out the steps along the way. Finally (for select examples), we show students how to check the solution. Most examples are written in a two-column format, with explanation on the left and math on the right to mimic the way that instructors "talk through" examples as they write on the board in class.

Be Prepared! Each section, beginning with Section 2.1, starts with a few "Be Prepared!" exercises so that students can determine if they have mastered the prerequisite skills for the section. Reference is made to specific Examples from previous sections so students who need further review can easily find explanations. Answers to these exercises can be found in the supplemental resources that accompany this title.

Try It

> **Try it** The Try It feature includes a pair of exercises that immediately follow an Example, providing the student with an immediate opportunity to solve a similar problem with an easy reference to the example. In the Web View version of the text, students can click an Answer link directly below the question to check their understanding. In the PDF, answers to the Try It exercises are located in the Answer Key.

How To

How To Examples use a three column format to demonstrate how to solve an example with a certain procedure. The first column states the formal step, the second column is in words as the teacher would explain the process, and then the third column is the actual math. A How To procedure box follows each of these How To examples and summarizes the series of steps from the example. These procedure boxes provide an easy reference for students.

Media

      Media The "Media" icon appears at the conclusion of each section, just prior to the Self Check. This icon marks a list of links to online video tutorials that reinforce the concepts and skills introduced in the section.

Disclaimer: While we have selected tutorials that closely align to our learning objectives, we did not produce these tutorials, nor were they specifically produced or tailored to accompany *Intermediate Algebra*.

Self Check The Self Check includes the learning objectives for the section so that students can self-assess their mastery and make concrete plans to improve.

Art Program

Intermediate Algebra contains many figures and illustrations. Art throughout the text adheres to a clear, understated style, drawing the eye to the most important information in each figure while minimizing visual distractions.

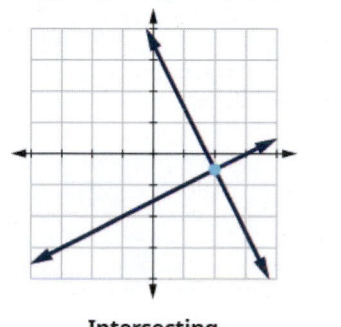
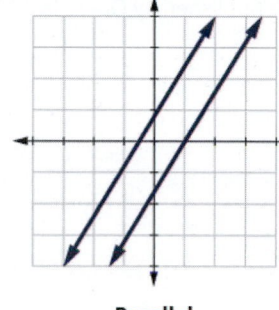
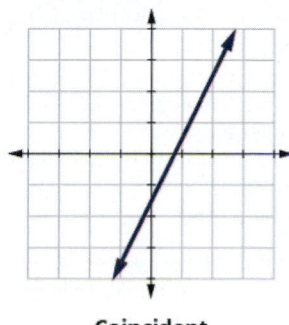

    Intersecting   **Parallel**   **Coincident**

Section Exercises and Chapter Review

Section Exercises Each section of every chapter concludes with a well-rounded set of exercises that can be assigned as homework or used selectively for guided practice. Exercise sets are named *Practice Makes Perfect* to encourage completion of homework assignments.

    Exercises correlate to the learning objectives. This facilitates assignment of personalized study plans based on individual student needs.

    Exercises are carefully sequenced to promote building of skills.

    Values for constants and coefficients were chosen to practice and reinforce arithmetic facts.

    Even and odd-numbered exercises are paired.

    Exercises parallel and extend the text examples and use the same instructions as the examples to help students easily recognize the connection.

    Applications are drawn from many everyday experiences, as well as those traditionally found in college math texts.

    Everyday Math highlights practical situations using the concepts from that particular section

    Writing Exercises are included in every exercise set to encourage conceptual understanding, critical thinking, and literacy.

Chapter review Each chapter concludes with a review of the most important takeaways, as well as additional practice problems that students can use to prepare for exams.

    Key Terms provide a formal definition for each bold-faced term in the chapter.

    Key Concepts summarize the most important ideas introduced in each section, linking back to the relevant Example(s) in case students need to review.

    Chapter Review Exercises include practice problems that recall the most important concepts from each section.

    Practice Test includes additional problems assessing the most important learning objectives from the chapter.

    Answer Key includes the answers to all Try It exercises and every other exercise from the Section Exercises, Chapter Review Exercises, and Practice Test.

Additional Resources
Student and Instructor Resources

We've compiled additional resources for both students and instructors, including Getting Started Guides, manipulative mathematics worksheets, an answer key to the Be Prepared Exercises, and an answer guide to the section review exercises. Instructor resources require a verified instructor account, which can be requested on your openstax.org log-in. Take advantage of these resources to supplement your OpenStax book.

Partner Resources

OpenStax partners are our allies in the mission to make high-quality learning materials affordable and accessible to students and instructors everywhere. Their tools integrate seamlessly with our OpenStax titles at a low cost. To access the partner resources for your text, visit your book page on openstax.org.

About the Authors
Senior Contributing Author

Lynn Marecek, Santa Ana College

Lynn Marecek has been teaching mathematics at Santa Ana College for many years has focused her career on meeting the needs of developmental math students. At Santa Ana College, she has been awarded the Distinguished Faculty Award, Innovation Award, and the Curriculum Development Award four times. She is a Coordinator of the Freshman Experience Program, the Department Facilitator for Redesign, and a member of the Student Success and Equity Committee, and the Basic Skills Initiative Task Force.

She is the coauthor with MaryAnne Anthony-Smith of *Strategies for Success: Study Skills for the College Math Student*, *Prealgebra* published by OpenStax and *Elementary Algebra* published by OpenStax.

Reviewers

Shaun Ault, Valdosta State University
Brandie Biddy, Cecil College
Kimberlyn Brooks, Cuyahoga Community College
Michael Cohen, Hofstra University
Robert Diaz, Fullerton College
Dianne Hendrickson, Becker College
Linda Hunt, Shawnee State University
Stephanie Krehl, Mid-South Community College
Yixia Lu, South Suburban College
Teresa Richards, Butte-Glenn College
Christian Roldán- Johnson, College of Lake County Community College
Yvonne Sandoval, El Camino College
Gowribalan Vamadeva, University of Cincinnati Blue Ash College
Kim Watts, North Lake college
Libby Watts, Tidewater Community College
Matthew Watts, Tidewater Community College

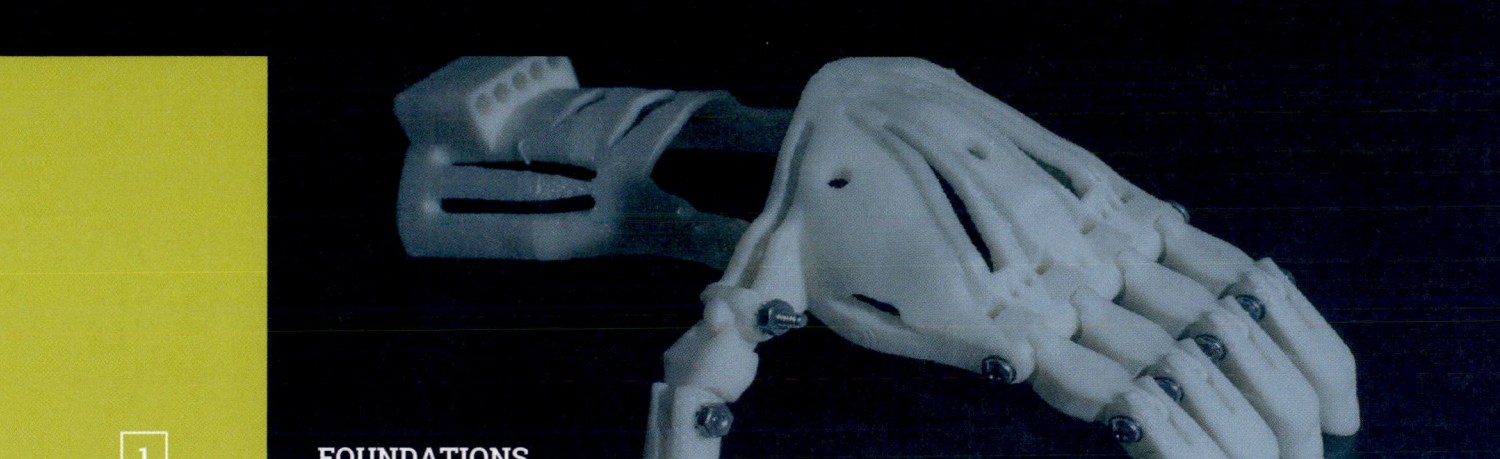

Figure 1.1 This hand may change someone's life. Amazingly, it was created using a special kind of printer known as a 3D printer. (credit: U.S. Food and Drug Administration/Wikimedia Commons)

Chapter Outline

1.1 Use the Language of Algebra

1.2 Integers

1.3 Fractions

1.4 Decimals

1.5 Properties of Real Numbers

Introduction

For years, doctors and engineers have worked to make artificial limbs, such as this hand for people who need them. This particular product is different, however, because it was developed using a 3D printer. As a result, it can be printed much like you print words on a sheet of paper. This makes producing the limb less expensive and faster than conventional methods.

Biomedical engineers are working to develop organs that may one day save lives. Scientists at NASA are designing ways to use 3D printers to build on the moon or Mars. Already, animals are benefitting from 3D-printed parts, including a tortoise shell and a dog leg. Builders have even constructed entire buildings using a 3D printer.

The technology and use of 3D printers depends on the ability to understand the language of algebra. Engineers must be able to translate observations and needs in the natural world to complex mathematical commands that can provide directions to a printer. In this chapter, you will review the language of algebra and take your first steps toward working with algebraic concepts.

1.1 Use the Language of Algebra

Learning Objectives

By the end of this section, you will be able to:

› Find factors, prime factorizations, and least common multiples
› Use variables and algebraic symbols
› Simplify expressions using the order of operations
› Evaluate an expression
› Identify and combine like terms
› Translate an English phrase to an algebraic expression

Be Prepared!

This chapter is intended to be a brief review of concepts that will be needed in an Intermediate Algebra course. A more thorough introduction to the topics covered in this chapter can be found in the *Elementary Algebra* chapter,

Foundations.

Find Factors, Prime Factorizations, and Least Common Multiples

The numbers 2, 4, 6, 8, 10, 12 are called multiples of 2. A **multiple** of 2 can be written as the product of a counting number and 2.

Multiples of 2:

2,   4,   6,   8,   10,   12, ...
2·1 2·2 2·3 2·4 2·5 2·6

Similarly, a multiple of 3 would be the product of a counting number and 3.

Multiples of 3:

3,   6,   9,   12,   15,   18, ...
3·1 3·2 3·3 3·4 3·5 3·6

We could find the multiples of any number by continuing this process.

Counting Number	1	2	3	4	5	6	7	8	9	10	11	12
Multiples of 2	2	4	6	8	10	12	14	16	18	20	22	24
Multiples of 3	3	6	9	12	15	18	21	24	27	30	33	36
Multiples of 4	4	8	12	16	20	24	28	32	36	40	44	48
Multiples of 5	5	10	15	20	25	30	35	40	45	50	55	60
Multiples of 6	6	12	18	24	30	36	42	48	54	60	66	72
Multiples of 7	7	14	21	28	35	42	49	56	63	70	77	84
Multiples of 8	8	16	24	32	40	48	56	64	72	80	88	96
Multiples of 9	9	18	27	36	45	54	63	72	81	90	99	108

Multiple of a Number

A number is a **multiple** of n if it is the product of a counting number and n.

Another way to say that 15 is a multiple of 3 is to say that 15 is **divisible** by 3. That means that when we divide 3 into 15, we get a counting number. In fact, $15 \div 3$ is 5, so 15 is $5 \cdot 3$.

Divisible by a Number

If a number m is a multiple of n, then m is **divisible** by n.

If we were to look for patterns in the multiples of the numbers 2 through 9, we would discover the following divisibility tests:

Divisibility Tests

A number is divisible by:

2 if the last digit is 0, 2, 4, 6, or 8.

3 if the sum of the digits is divisible by 3.

5 if the last digit is 5 or 0.

6 if it is divisible by both 2 and 3.

10 if it ends with 0.

EXAMPLE 1.1

Is 5,625 divisible by ⓐ 2? ⓑ 3? ⓒ 5 or 10? ⓓ 6?

✓ **Solution**

ⓐ
Is 5,625 divisible by 2?

Does it end in 0, 2, 4, 6 or 8?   No.
5,625 is not divisible by 2.

ⓑ
Is 5,625 divisible by 3?

What is the sum of the digits? $5 + 6 + 2 + 5 = 18$
Is the sum divisible by 3?   Yes.
5,625 is divisible by 3.

ⓒ
Is 5,625 divisible by 5 or 10?

What is the last digit? It is 5.   5,625 is divisible by 5 but not by 10.

ⓓ
Is 5,625 divisible by 6?

Is it divisible by both 2 and 3?   No, 5,625 is not divisible by 2, so 5,625 is
not divisible by 6.

> **TRY IT : : 1.1**   Is 4,962 divisible by ⓐ 2? ⓑ 3? ⓒ 5? ⓓ 6? ⓔ 10?

> **TRY IT : : 1.2**   Is 3,765 divisible by ⓐ 2? ⓑ 3? ⓒ 5? ⓓ 6? ⓔ 10?

In mathematics, there are often several ways to talk about the same ideas. So far, we've seen that if m is a multiple of n, we can say that m is divisible by n. For example, since 72 is a multiple of 8, we say 72 is divisible by 8. Since 72 is a multiple of 9, we say 72 is divisible by 9. We can express this still another way.

Since $8 \cdot 9 = 72$, we say that 8 and 9 are **factors** of 72. When we write $72 = 8 \cdot 9$, we say we have factored 72.

$$\underbrace{8 \cdot 9}_{factors} = \underbrace{72}_{product}$$

Other ways to factor 72 are $1 \cdot 72$, $2 \cdot 36$, $3 \cdot 24$, $4 \cdot 18$, and $6 \cdot 12$. The number 72 has many factors: $1, 2, 3, 4, 6, 8, 9, 12, 18, 24, 36$, and 72.

Factors

If $a \cdot b = m$, then a and b are **factors** of m.

Some numbers, such as 72, have many factors. Other numbers have only two factors. A **prime number** is a counting number greater than 1 whose only factors are 1 and itself.

Prime number and Composite number

A **prime number** is a counting number greater than 1 whose only factors are 1 and the number itself.

A **composite number** is a counting number that is not prime. A composite number has factors other than 1 and the number itself.

The counting numbers from 2 to 20 are listed in the table with their factors. Make sure to agree with the "prime" or "composite" label for each!

Number	Factors	Prime or Composite?	Number	Factors	Prime or Composite?
2	1,2	Prime	12	1,2,3,4,6,12	Composite
3	1,3	Prime	13	1,13	Prime
4	1,2,4	Composite	14	1,2,7,14	Composite
5	1,5	Prime	15	1,3,5,15	Composite
6	1,2,3,6	Composite	16	1,2,4,8,16	Composite
7	1,7	Prime	17	1,17	Prime
8	1,2,4,8	Composite	18	1,2,3,6,9,18	Composite
9	1,3,9	Composite	19	1,19	Prime
10	1,2,5,10	Composite	20	1,2,4,5,10,20	Composite
11	1,11	Prime			

The prime numbers less than 20 are 2, 3, 5, 7, 11, 13, 17, and 19. Notice that the only even prime number is 2.

A composite number can be written as a unique product of primes. This is called the **prime factorization** of the number. Finding the prime factorization of a composite number will be useful in many topics in this course.

Prime Factorization

The **prime factorization** of a number is the product of prime numbers that equals the number.

To find the prime factorization of a composite number, find any two factors of the number and use them to create two branches. If a factor is prime, that branch is complete. Circle that prime. Otherwise it is easy to lose track of the prime numbers.

If the factor is not prime, find two factors of the number and continue the process. Once all the branches have circled primes at the end, the factorization is complete. The composite number can now be written as a product of prime numbers.

EXAMPLE 1.2   HOW TO FIND THE PRIME FACTORIZATION OF A COMPOSITE NUMBER

Factor 48.

⊘ **Solution**

Step 1. Find two factors whose product is the given number. Use these numbers to create two branches.	$48 = 2 \cdot 24$	48 / 2 24
Step 2. If a factor is prime, that branch is complete. Circle the prime.	2 is prime. Circle the prime.	48 / ② 24
Step 3. If a factor is not prime, write it as the product of two factors and continue the process.	24 is not prime. Break it into 2 more factors.	48 / ② 24 / 4 6
	4 and 6 are not prime. Break them each into two factors.	48 / ② 24 / 4 6 / ② ② ② ③
	2 and 3 are prime, so circle them.	

Step 4. Write the composite number as the product of all the circled primes.		$48 = 2 \cdot 2 \cdot 2 \cdot 2 \cdot 3$

We say $2 \cdot 2 \cdot 2 \cdot 2 \cdot 3$ is the prime factorization of 48. We generally write the primes in ascending order. Be sure to multiply the factors to verify your answer.

If we first factored 48 in a different way, for example as $6 \cdot 8,$ the result would still be the same. Finish the prime factorization and verify this for yourself.

> **TRY IT :: 1.3**   Find the prime factorization of $80.$

> **TRY IT :: 1.4**   Find the prime factorization of $60.$

 HOW TO :: FIND THE PRIME FACTORIZATION OF A COMPOSITE NUMBER.

Step 1. Find two factors whose product is the given number, and use these numbers to create two branches.

Step 2. If a factor is prime, that branch is complete. Circle the prime, like a leaf on the tree.

Step 3. If a factor is not prime, write it as the product of two factors and continue the process.

Step 4. Write the composite number as the product of all the circled primes.

One of the reasons we look at primes is to use these techniques to find the **least common multiple** of two numbers. This will be useful when we add and subtract fractions with different denominators.

Least Common Multiple

The **least common multiple (LCM)** of two numbers is the smallest number that is a multiple of both numbers.

To find the least common multiple of two numbers we will use the Prime Factors Method. Let's find the LCM of 12 and 18 using their prime factors.

EXAMPLE 1.3   HOW TO FIND THE LEAST COMMON MULTIPLE USING THE PRIME FACTORS METHOD

Find the least common multiple (LCM) of 12 and 18 using the prime factors method.

 Solution

Step 1. Write each number as a product of primes.		(18 and 12 factor trees)
Step 2. List the primes of each number. Match primes vertically when possible.	List the primes of 12. List the primes of 18. Line up with the primes of 12 when possible. If not create a new column.	$12 = 2 \cdot 2 \cdot 3$ $18 = 2 \cdot \quad 3 \cdot 3$
Step 3. Bring down the number from each column.		$12 = 2 \cdot 2 \cdot 3$ $18 = 2 \cdot \quad 3 \cdot 3$ $LCM = 2 \cdot 2 \cdot 3 \cdot 3$

Step 4. Multiply the factors.	LCM = 36

Notice that the prime factors of 12 $(2 \cdot 2 \cdot 3)$ and the prime factors of 18 $(2 \cdot 3 \cdot 3)$ are included in the LCM $(2 \cdot 2 \cdot 3 \cdot 3)$. So 36 is the least common multiple of 12 and 18.

By matching up the common primes, each common prime factor is used only once. This way you are sure that 36 is the *least* common multiple.

> **TRY IT :: 1.5**   Find the LCM of 9 and 12 using the Prime Factors Method.

> **TRY IT :: 1.6**   Find the LCM of 18 and 24 using the Prime Factors Method.

  HOW TO :: FIND THE LEAST COMMON MULTIPLE USING THE PRIME FACTORS METHOD.

Step 1.   Write each number as a product of primes.

Step 2.   List the primes of each number. Match primes vertically when possible.

Step 3.   Bring down the columns.

Step 4.   Multiply the factors.

Use Variables and Algebraic Symbols

In algebra, we use a letter of the alphabet to represent a number whose value may change. We call this a **variable** and letters commonly used for variables are x, y, a, b, c.

Variable

A **variable** is a letter that represents a number whose value may change.

A number whose value always remains the same is called a **constant**.

Constant

A **constant** is a number whose value always stays the same.

To write algebraically, we need some operation symbols as well as numbers and variables. There are several types of symbols we will be using. There are four basic arithmetic operations: addition, subtraction, multiplication, and division. We'll list the symbols used to indicate these operations below.

Operation Symbols

Operation	Notation	Say:	The result is...
Addition	$a + b$	a plus b	the sum of a and b
Subtraction	$a - b$	a minus b	the difference of a and b
Multiplication	$a \cdot b$, ab, $(a)(b)$, $(a)b$, $a(b)$	a times b	the product of a and b
Division	$a \div b$, a/b, $\frac{a}{b}$, $b\overline{)a}$	a divided by b	the quotient of a and b; a is called the dividend, and b is called the divisor

When two quantities have the same value, we say they are equal and connect them with an equal sign.

Equality Symbol

$a = b$ is read "a is equal to b."

The symbol "$=$" is called the equal sign.

On the number line, the numbers get larger as they go from left to right. The number line can be used to explain the symbols "$<$" and "$>$".

Inequality

$a < b$ is read "a is less than b"
a is to the left of b on the number line

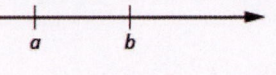

$a > b$ is read "a is greater than b"
a is to the right of b on the number line

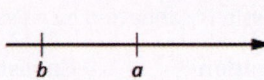

The expressions $a < b$ or $a > b$ can be read from left to right or right to left, though in English we usually read from left to right. In general,

$a < b$ is equivalent to $b > a$. For example, $7 < 11$ is equivalent to $11 > 7$.

$a > b$ is equivalent to $b < a$. For example, $17 > 4$ is equivalent to $4 < 17$.

Inequality Symbols

Inequality Symbols	Words
$a \neq b$	a is *not equal to b.*
$a < b$	a is *less than b.*
$a \leq b$	a is *less than or equal to b.*
$a > b$	a is *greater than b.*
$a \geq b$	a is *greater than or equal to b.*

Grouping symbols in algebra are much like the commas, colons, and other punctuation marks in English. They help identify an **expression**, which can be made up of number, a variable, or a combination of numbers and variables using operation symbols. We will introduce three types of grouping symbols now.

Grouping Symbols

Parentheses   ( )
Brackets   [ ]
Braces   { }

Here are some examples of expressions that include grouping symbols. We will simplify expressions like these later in this section.

$$8(14 - 8) \qquad 21 - 3[2 + 4(9 - 8)] \qquad 24 \div \{13 - 2[1(6 - 5) + 4]\}$$

What is the difference in English between a phrase and a sentence? A phrase expresses a single thought that is incomplete by itself, but a sentence makes a complete statement. A sentence has a subject and a verb. In algebra, we have *expressions* and *equations*.

Expression

An **expression** is a number, a variable, or a combination of numbers and variables using operation symbols.

Expression	Words	English Phrase
$3 + 5$	3 plus 5	the sum of three and five
$n - 1$	n minus one	the difference of n and one
$6 \cdot 7$	6 times 7	the product of six and seven
$\frac{x}{y}$	x divided by y	the quotient of x and y

Notice that the English phrases do not form a complete sentence because the phrase does not have a verb.

An **equation** is two expressions linked by an equal sign. When you read the words the symbols represent in an equation, you have a complete sentence in English. The equal sign gives the verb.

Equation

An **equation** is two expressions connected by an equal sign.

Equation	English Sentence
$3 + 5 = 8$	The sum of three and five is equal to eight.
$n - 1 = 14$	n minus one equals fourteen.
$6 \cdot 7 = 42$	The product of six and seven is equal to forty-two.
$x = 53$	x is equal to fifty-three.
$y + 9 = 2y - 3$	y plus nine is equal to two y minus three.

Suppose we need to multiply 2 nine times. We could write this as $2 \cdot 2 \cdot 2 \cdot 2 \cdot 2 \cdot 2 \cdot 2 \cdot 2 \cdot 2$. This is tedious and it can be hard to keep track of all those 2s, so we use exponents. We write $2 \cdot 2 \cdot 2$ as 2^3 and $2 \cdot 2 \cdot 2 \cdot 2 \cdot 2 \cdot 2 \cdot 2 \cdot 2 \cdot 2$ as 2^9. In expressions such as 2^3, the 2 is called the *base* and the 3 is called the *exponent*. The exponent tells us how many times we need to multiply the base.

$$2^{3} \leftarrow \text{exponent}$$
$$\underset{base}{\uparrow}$$

means multiply 2 by itself, three times, as in $2 \cdot 2 \cdot 2$.

We read 2^3 as "two to the third power" or "two cubed."

Exponential Notation

We say 2^3 is in *exponential notation* and $2 \cdot 2 \cdot 2$ is in *expanded notation*.

a^n means multiply a by itself, n times.

$$\text{base} \rightarrow a^n \leftarrow \text{exponent}$$
$$a^n = \underbrace{a \cdot a \cdot a \cdot \ldots \cdot a}_{n\ factors}$$

The expression a^n is read a to the nth power.

While we read a^n as "a to the nth power", we usually read:

$$a^2 \qquad \text{"}a \text{ squared"}$$
$$a^3 \qquad \text{"}a \text{ cubed"}$$

We'll see later why a^2 and a^3 have special names.

Table 1.1 shows how we read some expressions with exponents.

Expression	In Words	
7^2	7 to the second power or	7 squared
5^3	5 to the third power or	5 cubed
9^4	9 to the fourth power	
12^5	12 to the fifth power	

Table 1.1

Simplify Expressions Using the Order of Operations

To **simplify an expression** means to do all the math possible. For example, to simplify $4 \cdot 2 + 1$ we would first multiply $4 \cdot 2$ to get 8 and then add the 1 to get 9. A good habit to develop is to work down the page, writing each step of the process below the previous step. The example just described would look like this:

$$4 \cdot 2 + 1$$
$$8 + 1$$
$$9$$

By not using an equal sign when you simplify an expression, you may avoid confusing expressions with equations.

Simplify an Expression

To **simplify an expression**, do all operations in the expression.

We've introduced most of the symbols and notation used in algebra, but now we need to clarify the **order of operations**. Otherwise, expressions may have different meanings, and they may result in different values.

For example, consider the expression $4 + 3 \cdot 7$. Some students simplify this getting 49, by adding $4 + 3$ and then multiplying that result by 7. Others get 25, by multiplying $3 \cdot 7$ first and then adding 4.

The same expression should give the same result. So mathematicians established some guidelines that are called the order of operations.

 HOW TO :: USE THE ORDER OF OPERATIONS.

Step 1. Parentheses and Other Grouping Symbols

      ◦ Simplify all expressions inside the parentheses or other grouping symbols, working on the innermost parentheses first.

Step 2. Exponents

      ◦ Simplify all expressions with exponents.

Step 3. Multiplication and Division

      ◦ Perform all multiplication and division in order from left to right. These operations have equal priority.

Step 4. Addition and Subtraction

      ◦ Perform all addition and subtraction in order from left to right. These operations have equal priority.

Students often ask, "How will I remember the order?" Here is a way to help you remember: Take the first letter of each key word and substitute the silly phrase "Please Excuse My Dear Aunt Sally".

Parentheses	**Please**
Exponents	**Excuse**
Multiplication Division	**My Dear**
Addition Subtraction	**Aunt Sally**

It's good that "**My Dear**" goes together, as this reminds us that **m**ultiplication and **d**ivision have equal priority. We do not always do multiplication before division or always do division before multiplication. We do them in order from left to right.

Similarly, "**A**unt **S**ally" goes together and so reminds us that **a**ddition and **s**ubtraction also have equal priority and we do them in order from left to right.

EXAMPLE 1.4

Simplify: $18 \div 6 + 4(5 - 2)$.

✓ **Solution**

	$18 \div 6 + 4(5 - 2)$
Parentheses? Yes, subtract first.	$18 \div 6 + 4(3)$
Exponents? No.	
Multiplication or division? Yes.	
Divide first because we multiply and divide left to right.	$3 + 4(3)$
Any other multiplication or division? Yes.	
Multiply.	$3 + 12$
Any other multiplication of division? No.	
Any addition or subtraction? Yes.	
Add.	15

> **TRY IT : : 1.7**   Simplify: $30 \div 5 + 10(3 - 2)$.

> **TRY IT : : 1.8**   Simplify: $70 \div 10 + 4(6 - 2)$.

When there are multiple grouping symbols, we simplify the innermost parentheses first and work outward.

EXAMPLE 1.5

Simplify: $5 + 2^3 + 3[6 - 3(4 - 2)]$.

✓ **Solution**

	$5 + 2^3 + 3[6 - 3(4 - 2)]$
Are there any parentheses (or other grouping symbols)? Yes.	$5 + 2^3 + 3[6 - 3(4 - 2)]$
Focus on the parentheses that are inside the brackets. Subtract.	$5 + 2^3 + 3[6 - 3(2)]$
Continue inside the brackets and multiply.	$5 + 2^3 + 3[6 - 6]$
Continue inside the brackets and subtract.	$5 + 2^3 + 3[0]$
The expression inside the brackets requires no further simplification.	
Are there any exponents? Yes. Simplify exponents.	$5 + 8 + 3[0]$
Is there any multiplication or division? Yes.	

Multiply.	$5 + 8 + 0$
Is there any addition of subtraction? Yes.	
Add.	$13 + 0$
Add.	13

> **TRY IT :: 1.9**   Simplify: $9 + 5^3 - [4(9 + 3)]$.

> **TRY IT :: 1.10**   Simplify: $7^2 - 2[4(5 + 1)]$.

Evaluate an Expression

In the last few examples, we simplified expressions using the order of operations. Now we'll evaluate some expressions—again following the order of operations. To **evaluate an expression** means to find the value of the expression when the variable is replaced by a given number.

Evaluate an Expression

To **evaluate an expression** means to find the value of the expression when the variable is replaced by a given number.

To evaluate an expression, substitute that number for the variable in the expression and then simplify the expression.

EXAMPLE 1.6

Evaluate when $x = 4$:   ⓐ x^2 ⓑ 3^x ⓒ $2x^2 + 3x + 8$.

⊘ **Solution**

ⓐ

	x^2
Replace x with 4.	4^2
Use definition of exponent.	$4 \cdot 4$
Simplify.	16

ⓑ

	3^x
Replace x with 4.	3^4
Use definition of exponent.	$3 \cdot 3 \cdot 3 \cdot 3$
Simplify.	81

ⓒ

	$2x^2 + 3x + 8$
Replace x with 4.	$2(4)^2 + 3(4) + 8$
Follow the order of operations.	$2(16) + 3(4) + 8$
	$32 + 12 + 8$
	52

> **TRY IT :: 1.11**   Evaluate when $x = 3$, ⓐ x^2 ⓑ 4^x ⓒ $3x^2 + 4x + 1$.

> **TRY IT :: 1.12**   Evaluate when $x = 6$, ⓐ x^3 ⓑ 2^x ⓒ $6x^2 - 4x - 7$.

Identify and Combine Like Terms

Algebraic expressions are made up of terms. A **term** is a constant, or the product of a constant and one or more variables.

> **Term**
>
> A **term** is a constant or the product of a constant and one or more variables.

Examples of terms are 7, y, $5x^2$, $9a$, and b^5.

The constant that multiplies the variable is called the **coefficient**.

> **Coefficient**
>
> The **coefficient** of a term is the constant that multiplies the variable in a term.

Think of the coefficient as the number in front of the variable. The coefficient of the term $3x$ is 3. When we write x, the coefficient is 1, since $x = 1 \cdot x$.

Some terms share common traits. When two terms are constants or have the same variable and exponent, we say they are **like terms**.

Look at the following 6 terms. Which ones seem to have traits in common?

$$5x \quad 7 \quad n^2 \quad 4 \quad 3x \quad 9n^2$$

We say,

7 and 4 are like terms.

$5x$ and $3x$ are like terms.

n^2 and $9n^2$ are like terms.

> **Like Terms**
>
> Terms that are either constants or have the same variables raised to the same powers are called **like terms**.

If there are like terms in an expression, you can simplify the expression by combining the like terms. We add the coefficients and keep the same variable.

Simplify.	$4x + 7x + x$
Add the coefficients.	$12x$

EXAMPLE 1.7   HOW TO COMBINE LIKE TERMS

Simplify: $2x^2 + 3x + 7 + x^2 + 4x + 5$.

⊘ **Solution**

Step 1. Identify the like terms.	$2x^2 + 3x + 7 + x^2 + 4x + 5$
	$2\underline{x^2} + \underline{3x} + \underline{7} + \underline{x^2} + \underline{4x} + \underline{5}$
	$2x^2 + 3x + 7 + x^2 + 4x + 5$
Step 2. Rearrange the expression so the like terms are together.	$2x^2 + x^2 + 3x + 4x + 7 + 5$
	$2x^2 + x^2 + 3x + 4x + 7 + 5$
Step 3. Combine like terms.	$3x^2 + 7x + 12$

> **TRY IT ::** 1.13   Simplify: $3x^2 + 7x + 9 + 7x^2 + 9x + 8$.

> **TRY IT ::** 1.14   Simplify: $4y^2 + 5y + 2 + 8y^2 + 4y + 5$.

 HOW TO :: COMBINE LIKE TERMS.

Step 1.   Identify like terms.

Step 2.   Rearrange the expression so like terms are together.

Step 3.   Add or subtract the coefficients and keep the same variable for each group of like terms.

Translate an English Phrase to an Algebraic Expression

We listed many operation symbols that are used in algebra. Now, we will use them to translate English phrases into algebraic expressions. The symbols and variables we've talked about will help us do that. Table 1.7 summarizes them.

Operation	Phrase	Expression
Addition	a plus b the sum of a and b a increased by b b more than a the total of a and b b added to a	$a + b$
Subtraction	a minus b the difference of a and b a decreased by b b less than a b subtracted from a	$a - b$
Multiplication	a times b the product of a and b twice a	$a \cdot b$, ab, $a(b)$, $(a)(b)$ $2a$
Division	a divided by b the quotient of a and b the ratio of a and b b divided into a	$a \div b$, a/b, $\frac{a}{b}$, $b\overline{)a}$

Table 1.7

Look closely at these phrases using the four operations:

the **sum** *of a and b*

the **difference** *of a and b*

the **product** *of a and b*

the **quotient** *of a and b*

Each phrase tells us to operate on two numbers. Look for the words *of* and *and* to find the numbers.

EXAMPLE 1.8

Translate each English phrase into an algebraic expression:

ⓐ the difference of $14x$ and 9

ⓑ the quotient of $8y^2$ and 3

ⓒ twelve more than y

ⓓ seven less than $49x^2$

⊘ **Solution**

ⓐ The key word is *difference*, which tells us the operation is subtraction. Look for the words *of* and *and* to find the numbers to subtract.

the *difference of* 14*x and* 9

14*x* minus 9

14*x* – 9

ⓑ The key word is *quotient*, which tells us the operation is division.

$$\text{the } \textit{quotient of } 8y^2 \textit{ and } 3$$

$$\text{divide } 8y^2 \text{ by } 3$$

$$8y^2 \div 3$$

$$\text{This can also be written } 8y^2/3 \text{ or } \frac{8y^2}{3}.$$

ⓒ The key words are *more than*. They tell us the operation is addition. *More than* means "added to."

$$\text{twelve more than } y$$

$$\text{twelve added to } y$$

$$y + 12$$

ⓓ The key words are *less than*. They tell us to subtract. *Less than* means "subtracted from."

$$\text{seven less than } 49x^2$$

$$\text{seven subtracted from } 49x^2$$

$$49x^2 - 7$$

> **TRY IT : : 1.15**   Translate the English phrase into an algebraic expression:

ⓐ the difference of $14x^2$ and 13

ⓑ the quotient of $12x$ and 2

ⓒ 13 more than z

ⓓ 18 less than $8x$

> **TRY IT : : 1.16**   Translate the English phrase into an algebraic expression:

ⓐ the sum of $17y^2$ and 19

ⓑ the product of 7 and y

ⓒ Eleven more than x

ⓓ Fourteen less than $11a$

We look carefully at the words to help us distinguish between multiplying a sum and adding a product.

EXAMPLE 1.9

Translate the English phrase into an algebraic expression:

ⓐ eight times the sum of x and y

ⓑ the sum of eight times x and y

⊘ Solution

There are two operation words—*times* tells us to multiply and *sum* tells us to add.

ⓐ Because we are multiplying 8 times the sum, we need parentheses around the sum of x and y, $(x + y)$. This forces us to determine the sum first. (Remember the order of operations.)

$$\text{eight times the sum of } x \text{ and } y$$

$$8(x + y)$$

ⓑ To take a sum, we look for the words *of* and *and* to see what is being added. Here we are taking the sum *of* eight times x and y.

$$\text{the sum } \textit{of} \text{ eight times } x \textit{ and } y$$

$$8x + y$$

> **TRY IT : : 1.17**   Translate the English phrase into an algebraic expression:

ⓐ four times the sum of p and q

ⓑ the sum of four times p and q

> **TRY IT : : 1.18**   Translate the English phrase into an algebraic expression:

ⓐ the difference of two times x and 8

ⓑ two times the difference of x and 8

Later in this course, we'll apply our skills in algebra to solving applications. The first step will be to translate an English phrase to an algebraic expression. We'll see how to do this in the next two examples.

EXAMPLE 1.10

The length of a rectangle is 14 less than the width. Let w represent the width of the rectangle. Write an expression for the length of the rectangle.

⊘ **Solution**

Write a phrase about the length of the rectangle.	14 less than the width
Substitute w for "the width."	w
Rewrite *less than* as *subtracted from*.	14 subtracted from w
Translate the phrase into algebra.	$w - 14$

> **TRY IT : : 1.19**
>
> The length of a rectangle is 7 less than the width. Let w represent the width of the rectangle. Write an expression for the length of the rectangle.

> **TRY IT : : 1.20**
>
> The width of a rectangle is 6 less than the length. Let l represent the length of the rectangle. Write an expression for the width of the rectangle.

The expressions in the next example will be used in the typical coin mixture problems we will see soon.

EXAMPLE 1.11

June has dimes and quarters in her purse. The number of dimes is seven less than four times the number of quarters. Let q represent the number of quarters. Write an expression for the number of dimes.

⊘ **Solution**

Write a phrase about the number of dimes.	seven less than four times the number of quarters
Substitute q for the number of quarters.	7 less than 4 times q
Translate 4 times q.	7 less than $4q$
Translate the phrase into algebra.	$4q - 7$

> **TRY IT : : 1.21**
>
> Geoffrey has dimes and quarters in his pocket. The number of dimes is eight less than four times the number of quarters. Let q represent the number of quarters. Write an expression for the number of dimes.

> **TRY IT : : 1.22**
>
> Lauren has dimes and nickels in her purse. The number of dimes is three more than seven times the number of nickels. Let n represent the number of nickels. Write an expression for the number of dimes.

1.1 EXERCISES

Practice Makes Perfect

Identify Multiples and Factors

In the following exercises, use the divisibility tests to determine whether each number is divisible by 2, by 3, by 5, by 6, and by 10.

1. 84

2. 96

3. 896

4. 942

5. 22,335

6. 39,075

Find Prime Factorizations and Least Common Multiples

In the following exercises, find the prime factorization.

7. 86

8. 78

9. 455

10. 400

11. 432

12. 627

In the following exercises, find the least common multiple of each pair of numbers using the prime factors method.

13. 8, 12

14. 12, 16

15. 28, 40

16. 84, 90

17. 55, 88

18. 60, 72

Simplify Expressions Using the Order of Operations

In the following exercises, simplify each expression.

19. $2^3 - 12 \div (9 - 5)$

20. $3^2 - 18 \div (11 - 5)$

21. $2 + 8(6 + 1)$

22. $4 + 6(3 + 6)$

23. $20 \div 4 + 6(5 - 1)$

24. $33 \div 3 + 4(7 - 2)$

25. $3(1 + 9 \cdot 6) - 4^2$

26. $5(2 + 8 \cdot 4) - 7^2$

27. $2[1 + 3(10 - 2)]$

28. $5[2 + 4(3 - 2)]$

29. $8 + 2[7 - 2(5 - 3)] - 3^2$

30. $10 + 3[6 - 2(4 - 2)] - 2^4$

Evaluate an Expression

In the following exercises, evaluate the following expressions.

31. When $x = 2$,
ⓐ x^6
ⓑ 4^x
ⓒ $2x^2 + 3x - 7$

32. When $x = 3$,
ⓐ x^5
ⓑ 5^x
ⓒ $3x^2 - 4x - 8$

33. When $x = 4$, $y = 1$
$x^2 + 3xy - 7y^2$

34. When $x = 3$, $y = 2$
$6x^2 + 3xy - 9y^2$

35. When $x = 10$, $y = 7$
$(x - y)^2$

36. When $a = 3$, $b = 8$
$a^2 + b^2$

Simplify Expressions by Combining Like Terms

In the following exercises, simplify the following expressions by combining like terms.

37. $7x + 2 + 3x + 4$

38. $8y + 5 + 2y - 4$

39. $10a + 7 + 5a - 2 + 7a - 4$

40. $7c + 4 + 6c - 3 + 9c - 1$

41. $3x^2 + 12x + 11 + 14x^2 + 8x + 5$

42. $5b^2 + 9b + 10 + 2b^2 + 3b - 4$

Translate an English Phrase to an Algebraic Expression

In the following exercises, translate the phrases into algebraic expressions.

43.

ⓐ the difference of $5x^2$ and $6xy$

ⓑ the quotient of $6y^2$ and $5x$

ⓒ Twenty-one more than y^2

ⓓ $6x$ less than $81x^2$

44.

ⓐ the difference of $17x^2$ and $5xy$

ⓑ the quotient of $8y^3$ and $3x$

ⓒ Eighteen more than a^2 ;

ⓓ $11b$ less than $100b^2$

45.

ⓐ the sum of $4ab^2$ and $3a^2b$

ⓑ the product of $4y^2$ and $5x$

ⓒ Fifteen more than m

ⓓ $9x$ less than $121x^2$

46.

ⓐ the sum of $3x^2y$ and $7xy^2$

ⓑ the product of $6xy^2$ and $4z$

ⓒ Twelve more than $3x^2$

ⓓ $7x^2$ less than $63x^3$

47.

ⓐ eight times the difference of y and nine

ⓑ the difference of eight times y and 9

48.

ⓐ seven times the difference of y and one

ⓑ the difference of seven times y and 1

49.

ⓐ five times the sum of $3x$ and y

ⓑ the sum of five times $3x$ and y

50.

ⓐ eleven times the sum of $4x^2$ and $5x$

ⓑ the sum of eleven times $4x^2$ and $5x$

51. Eric has rock and country songs on his playlist. The number of rock songs is 14 more than twice the number of country songs. Let *c* represent the number of country songs. Write an expression for the number of rock songs.

52. The number of women in a Statistics class is 8 more than twice the number of men. Let *m* represent the number of men. Write an expression for the number of women.

53. Greg has nickels and pennies in his pocket. The number of pennies is seven less than three the number of nickels. Let *n* represent the number of nickels. Write an expression for the number of pennies.

54. Jeannette has $5 and $10 bills in her wallet. The number of fives is three more than six times the number of tens. Let *t* represent the number of tens. Write an expression for the number of fives.

Writing Exercises

55. Explain in your own words how to find the prime factorization of a composite number.

56. Why is it important to use the order of operations to simplify an expression?

57. Explain how you identify the like terms in the expression $8a^2 + 4a + 9 - a^2 - 1$.

58. Explain the difference between the phrases "4 times the sum of x and y" and "the sum of 4 times x and y".

Self Check

ⓐ *Use this checklist to evaluate your mastery of the objectives of this section.*

I can...	Confidently	With some help	No-I don't get it!
identify multiples and apply divisibility tests.			
find prime factorizations and least common multiples.			
use variables and algebraic symbols.			
simplify expressions using the order of operations.			
evaluate an expression.			
identify and combine like terms.			
translate English phrases to algebraic expressions.			

ⓑ *If most of your checks were:*

...confidently. *Congratulations! You have achieved the objectives in this section. Reflect on the study skills you used so that you can continue to use them. What did you do to become confident of your ability to do these things? Be specific.*

...with some help. *This must be addressed quickly because topics you do not master become potholes in your road to success. In math every topic builds upon previous work. It is important to make sure you have a strong foundation before you move on. Who can you ask for help? Your fellow classmates and instructor are good resources. Is there a place on campus where math tutors are available? Can your study skills be improved?*

...no - I don't get it! *This is a warning sign and you must not ignore it. You should get help right away or you will quickly be overwhelmed. See your instructor as soon as you can to discuss your situation. Together you can come up with a plan to get you the help you need.*

1.2 Integers

Learning Objectives

By the end of this section, you will be able to:

> Simplify expressions with absolute value
> Add and subtract integers
> Multiply and divide integers
> Simplify expressions with integers
> Evaluate variable expressions with integers
> Translate phrases to expressions with integers
> Use integers in applications

Be Prepared!

A more thorough introduction to the topics covered in this section can be found in the *Elementary Algebra* chapter, Foundations.

Simplify Expressions with Absolute Value

A **negative numbers** is a number less than 0. The negative numbers are to the left of zero on the number line. See **Figure 1.2**.

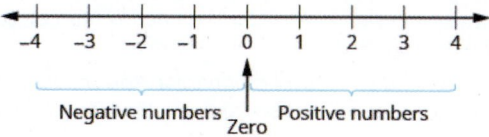

Figure 1.2 The number line shows the location of positive and negative numbers.

You may have noticed that, on the number line, the negative numbers are a mirror image of the positive numbers, with zero in the middle. Because the numbers 2 and -2 are the same distance from zero, each one is called the **opposite** of the other. The opposite of 2 is $-2,$ and the opposite of -2 is $2.$

Opposite

The **opposite** of a number is the number that is the same distance from zero on the number line but on the opposite side of zero.

Figure 1.3 illustrates the definition.

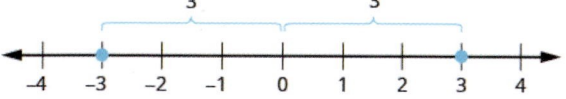

Figure 1.3 The opposite of 3 is -3.

Opposite Notation

$-a$ means the opposite of the number a

The notation $-a$ is read as "the opposite of $a.$"

We saw that numbers such as 3 and -3 are opposites because they are the same distance from 0 on the number line. They are both three units from 0. The distance between 0 and any number on the number line is called the **absolute value** of that number.

Absolute Value

The **absolute value** of a number is its distance from 0 on the number line.

The absolute value of a number n is written as $|n|$ and $|n| \geq 0$ for all numbers.

Absolute values are always greater than or equal to zero.

For example,

$$-5 \text{ is 5 units away from 0, so } |-5| = 5.$$
$$5 \text{ is 5 units away from 0, so } |5| = 5.$$

Figure 1.4 illustrates this idea.

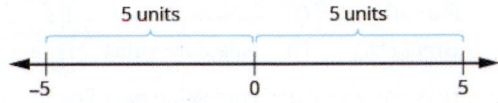

Figure 1.4 The numbers 5 and -5 are 5 units away from 0.

The absolute value of a number is never negative because distance cannot be negative. The only number with absolute value equal to zero is the number zero itself because the distance from 0 to 0 on the number line is zero units.

In the next example, we'll order expressions with absolute values.

EXAMPLE 1.12

Fill in $<$, $>$, or $=$ for each of the following pairs of numbers:

ⓐ $|-5|\_\_ - |-5|$ ⓑ $8\_\_ - |-8|$ ⓒ $-9\_\_ - |-9|$ ⓓ $-(-16)\_\_|-16|$.

✓ **Solution**

ⓐ

$$|-5| \_\_ -|-5|$$
Simplify. $$5 \_\_ -5$$
Order. $$5 > -5$$
$$|-5| > -|-5|$$

ⓑ

$$8 \_\_ -|-8|$$
Simplify. $$8 \_\_ -8$$
Order. $$8 > -8$$
$$8 > -|-8|$$

ⓒ

$$-9 \_\_ -|-9|$$
Simplify. $$-9 \_\_ -9$$
Order. $$-9 = -9$$
$$-9 = -|-9|$$

ⓓ

$$-(-16) \_\_ |-16|$$
Simplify. $$16 \_\_ 16$$
Order. $$16 = 16$$
$$-(-16) = |-16|$$

> **TRY IT :: 1.23**   Fill in $<$, $>$, or $=$ for each of the following pairs of numbers:

ⓐ $-9\_\_ - |-9|$ ⓑ $2\_\_ - |-2|$ ⓒ $-8\_\_|-8|$ ⓓ $-(-9)\_\_|-9|$.

> **TRY IT :: 1.24**   Fill in $<$, $>$, or $=$ for each of the following pairs of numbers:

ⓐ $7\_\_ -|-7|$ ⓑ $-(-10)\_\_|-10|$ ⓒ $|-4|\_\_ -|-4|$ ⓓ $-1\_\_|-1|$.

We now add absolute value bars to our list of grouping symbols. When we use the order of operations, first we simplify inside the absolute value bars as much as possible, then we take the absolute value of the resulting number.

Grouping Symbols

Parentheses	( )	Braces	{ }
Brackets	[ ]	Absolute value	| |

In the next example, we simplify the expressions inside absolute value bars first just like we do with parentheses.

EXAMPLE 1.13

Simplify: $24 - |19 - 3(6 - 2)|$.

⊘ **Solution**

$$24 - |19 - 3(6 - 2)|$$

Work inside parentheses first:

subtract 2 from 6. $24 - |19 - 3(4)|$

Multiply 3(4). $24 - |19 - 12|$

Subtract inside the absolute value bars. $24 - |7|$

Take the absolute value. $24 - 7$

Subtract. 17

> **TRY IT :: 1.25**   Simplify: $19 - |11 - 4(3 - 1)|$.

> **TRY IT :: 1.26**   Simplify: $9 - |8 - 4(7 - 5)|$.

Add and Subtract Integers

So far, we have only used the counting numbers and the whole numbers.

Counting numbers	1, 2, 3 …
Whole numbers	0, 1, 2, 3….

Our work with opposites gives us a way to define the **integers**. The whole numbers and their opposites are called the integers. The integers are the numbers … $-3, -2, -1, 0, 1, 2, 3$…

Integers

The whole numbers and their opposites are called the **integers**.

The integers are the numbers

$$\ldots -3, -2, -1, 0, 1, 2, 3\ldots,$$

Most students are comfortable with the addition and subtraction facts for positive numbers. But doing addition or subtraction with both positive and negative numbers may be more challenging.

We will use two color counters to model addition and subtraction of negatives so that you can visualize the procedures instead of memorizing the rules.

We let one color (blue) represent positive. The other color (red) will represent the negatives.

Positive   blue   ◯

Negative   red   ◯

If we have one positive counter and one negative counter, the value of the pair is zero. They form a neutral pair. The value of this neutral pair is zero.

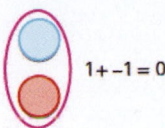

$$1 + {-1} = 0$$

We will use the counters to show how to add:

$$5 + 3 \qquad -5 + (-3) \qquad -5 + 3 \qquad 5 + (-3)$$

The first example, $5 + 3$, adds 5 positives and 3 positives—both positives.

The second example, $-5 + (-3)$, adds 5 negatives and 3 negatives—both negatives.

When the signs are the same, the counters are all the same color, and so we add them. In each case we get 8—either 8 positives or 8 negatives.

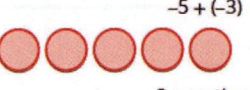

8 positives
$5 + 3 = 8$

8 negatives
$-5 + (-3) = -8$

So what happens when the signs are different? Let's add $-5 + 3$ and $5 + (-3)$.

When we use counters to model addition of positive and negative integers, it is easy to see whether there are more positive or more negative counters. So we know whether the sum will be positive or negative.

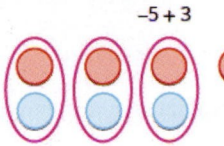
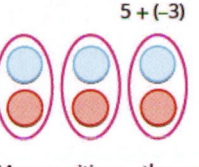

More negatives – the sum is negative.
$-5 + 3 = -2$

More positives – the sum is positive.
$5 + (-3) = 2$

EXAMPLE 1.14

Add: ⓐ $-1 + (-4)$ ⓑ $-1 + 5$ ⓒ $1 + (-5)$.

✓ **Solution**

ⓐ

$$-1 + (-4)$$

1 negative plus 4 negatives is 5 negatives -5

ⓑ

$$-1 + 5$$

There are more positives, so the sum is positive.   4

$$1 + (-5)$$

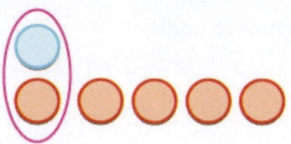

There are more negatives, so the sum is negative.   −4

> **TRY IT : : 1.27**   Add: ⓐ $-2 + (-4)$ ⓑ $-2 + 4$ ⓒ $2 + (-4)$.

> **TRY IT : : 1.28**   Add: ⓐ $-2 + (-5)$ ⓑ $-2 + 5$ ⓒ $2 + (-5)$.

We will continue to use counters to model the subtraction. Perhaps when you were younger, you read "$5 - 3$" as "5 take away 3." When you use counters, you can think of subtraction the same way!

We will use the counters to show to subtract:

$$5 - 3 \qquad -5 - (-3) \qquad -5 - 3 \qquad 5 - (-3)$$

The first example, $5 - 3$, we subtract 3 positives from 5 positives and end up with 2 positives.

In the second example, $-5 - (-3)$, we subtract 3 negatives from 5 negatives and end up with 2 negatives.

Each example used counters of only one color, and the "take away" model of subtraction was easy to apply.

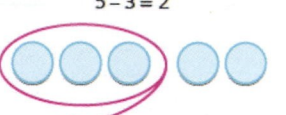 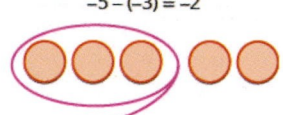

What happens when we have to subtract one positive and one negative number? We'll need to use both blue and red counters as well as some neutral pairs. If we don't have the number of counters needed to take away, we add neutral pairs. Adding a neutral pair does not change the value. It is like changing quarters to nickels—the value is the same, but it looks different.

Let's look at $-5 - 3$ and $5 - (-3)$.

	$-5 - 3$	$5 - (-3)$
Model the first number.		
We now add the needed neutral pairs.	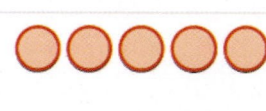 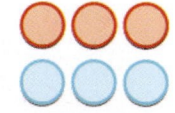	
We remove the number of counters modeled by the second number.	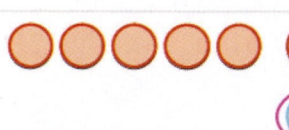 	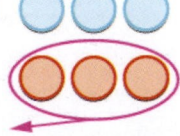

Count what is left.

$-5 - 3 = -8$	$5 - (-3) = -8$
-8	8

EXAMPLE 1.15

Subtract: ⓐ $3 - 1$ ⓑ $-3 - (-1)$ ⓒ $-3 - 1$ ⓓ $3 - (-1)$.

✓ **Solution**

ⓐ

Take 1 positive from 3 positives and get 2 positives.	2

ⓑ

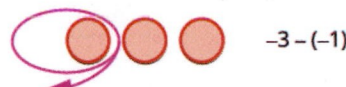

Take 1 positive from 3 negatives and get 2 negatives.	-2

ⓒ

Take 1 positive from the one added neutral pair.	-4

ⓓ

Take 1 negative from the one added neutral pair.	4

> **TRY IT : : 1.29** Subtract: ⓐ $6 - 4$ ⓑ $-6 - (-4)$ ⓒ $-6 - 4$ ⓓ $6 - (-4)$.

> **TRY IT : : 1.30** Subtract: ⓐ $7 - 4$ ⓑ $-7 - (-4)$ ⓒ $-7 - 4$ ⓓ $7 - (-4)$.

Have you noticed that *subtraction of signed numbers can be done by adding the opposite*? In the last example, $-3 - 1$ is the same as $-3 + (-1)$ and $3 - (-1)$ is the same as $3 + 1$. You will often see this idea, the Subtraction Property, written as follows:

Subtraction Property

$$a - b = a + (-b)$$

Subtracting a number is the same as adding its opposite.

EXAMPLE 1.16

Simplify: ⓐ $13 - 8$ and $13 + (-8)$ ⓑ $-17 - 9$ and $-17 + (-9)$ ⓒ $9 - (-15)$ and $9 + 15$ ⓓ $-7 - (-4)$ and $-7 + 4$.

⊘ Solution

ⓐ

	$13 - 8$	and	$13 + (-8)$
Subtract.	5		5

ⓑ

	$-17 - 9$	and	$-17 + (-9)$
Subtract.	-26		-26

ⓒ

	$9 - (-15)$	and	$9 + 15$
Subtract.	24		24

ⓓ

	$-7 - (-4)$	and	$-7 + 4$
Subtract.	-3		-3

> **TRY IT : :** 1.31
>
> Simplify: ⓐ $21 - 13$ and $21 + (-13)$ ⓑ $-11 - 7$ and $-11 + (-7)$ ⓒ $6 - (-13)$ and $6 + 13$ ⓓ $-5 - (-1)$ and $-5 + 1$.

> **TRY IT : :** 1.32
>
> Simplify: ⓐ $15 - 7$ and $15 + (-7)$ ⓑ $-14 - 8$ and $-14 + (-8)$ ⓒ $4 - (-19)$ and $4 + 19$ ⓓ $-4 - (-7)$ and $-4 + 7$.

What happens when there are more than three integers? We just use the order of operations as usual.

EXAMPLE 1.17

Simplify: $7 - (-4 - 3) - 9$.

⊘ Solution

	$7 - (-4 - 3) - 9$
Simplify inside the parentheses first.	$7 - (-7) - 9$
Subtract left to right.	$14 - 9$
Subtract.	5

> **TRY IT : :** 1.33   Simplify: $8 - (-3 - 1) - 9$.

> **TRY IT : :** 1.34   Simplify: $12 - (-9 - 6) - 14$.

Multiply and Divide Integers

Since multiplication is mathematical shorthand for repeated addition, our model can easily be applied to show multiplication of integers. Let's look at this concrete model to see what patterns we notice. We will use the same examples that we used for addition and subtraction. Here, we are using the model just to help us discover the pattern.

We remember that $a \cdot b$ means **add** a, b **times.**

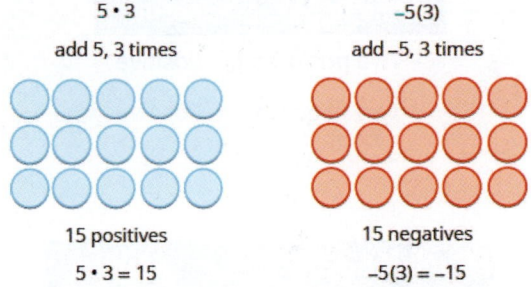

The next two examples are more interesting. What does it mean to multiply 5 by -3? It means subtract 5, 3 times.

Looking at subtraction as "taking away", it means to take away 5, 3 times. But there is nothing to take away, so we start by adding neutral pairs on the workspace.

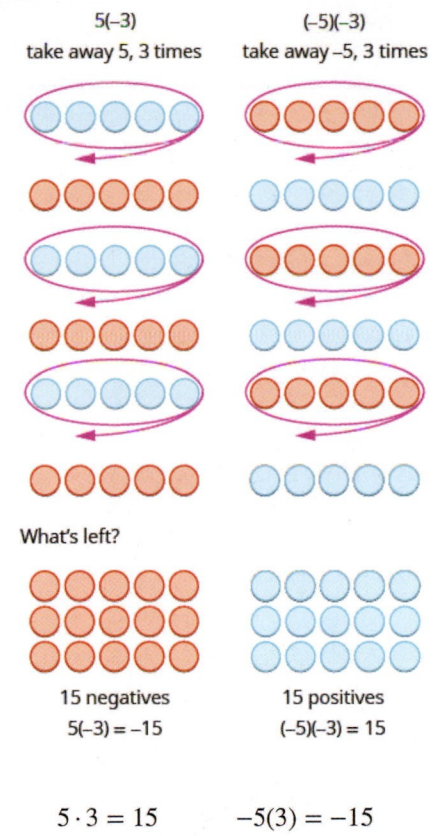

In summary:

$$5 \cdot 3 = 15 \qquad -5(3) = -15$$
$$5(-3) = -15 \qquad (-5)(-3) = 15$$

Notice that for multiplication of two signed numbers, when the

signs are the **same**, the product is **positive**.

signs are **different**, the product is **negative**.

What about division? Division is the inverse operation of multiplication. So, $15 \div 3 = 5$ because $15 \cdot 3 = 15$. In words, this expression says that 15 can be divided into 3 groups of 5 each because adding five three times gives 15. If you look at some examples of multiplying integers, you might figure out the rules for dividing integers.

$$5 \cdot 3 = 15 \qquad \text{so} \quad 15 \div 3 = 5 \qquad -5(3) = -15 \quad \text{so} \quad -15 \div 3 = -5$$
$$(-5)(-3) = 15 \quad \text{so} \quad 15 \div (-3) = -5 \qquad 5(-3) = -15 \quad \text{so} \quad -15 \div (-3) = 5$$

Division follows the same rules as multiplication with regard to signs.

Multiplication and Division of Signed Numbers

For multiplication and division of two signed numbers:

Same signs	Result
• Two positives	Positive
• Two negatives	Positive

If the signs are the same, the result is positive.

Different signs	Result
• Positive and negative	Negative
• Negative and positive	Negative

If the signs are different, the result is negative.

EXAMPLE 1.18

Multiply or divide: ⓐ $-100 \div (-4)$ ⓑ $7 \cdot 6$ ⓒ $4(-8)$ ⓓ $-27 \div 3$.

⊘ **Solution**

ⓐ

$$-100 \div (-4)$$

Divide, with signs that are
the same the quotient is positive.   25

ⓑ

$$7 \cdot 6$$

Multiply, with same signs.   42

ⓒ

$$4(-8)$$

Multiply, with different signs. -32

ⓓ

$$-27 \div 3$$

Divide, with different signs,
the quotient is negative. -9

> **TRY IT : : 1.35**   Multiply or divide: ⓐ $-115 \div (-5)$ ⓑ $5 \cdot 12$ ⓒ $9(-7)$ ⓓ $-63 \div 7$.

> **TRY IT : : 1.36**   Multiply or divide: ⓐ $-117 \div (-3)$ ⓑ $3 \cdot 13$ ⓒ $7(-4)$ ⓓ $-42 \div 6$.

When we multiply a number by 1, the result is the same number. Each time we multiply a number by -1, we get its opposite!

Multiplication by −1

$$-1a = -a$$

Multiplying a number by −1 gives its opposite.

Simplify Expressions with Integers

What happens when there are more than two numbers in an expression? The order of operations still applies when negatives are included. Remember Please Excuse My Dear Aunt Sally?

Let's try some examples. We'll simplify expressions that use all four operations with integers—addition, subtraction, multiplication, and division. Remember to follow the order of operations.

EXAMPLE 1.19

Simplify: ⓐ $(-2)^4$ ⓑ -2^4.

⊘ **Solution**

Notice the difference in parts (a) and (b). In part (a), the exponent means to raise what is in the parentheses, the −2 to the 4ᵗʰ power. In part (b), the exponent means to raise just the 2 to the 4ᵗʰ power and then take the opposite.

ⓐ

	$(-2)^4$
Write in expanded form.	$(-2)(-2)(-2)(-2)$
Multiply.	$4(-2)(-2)$
Multiply.	$-8(-2)$
Multiply.	16

ⓑ

	-2^4
Write in expanded form.	$-(2 \cdot 2 \cdot 2 \cdot 2)$
We are asked to find the opposite of 2^4.	
Multiply.	$-(4 \cdot 2 \cdot 2)$
Multiply.	$-(8 \cdot 2)$
Multiply.	-16

> **TRY IT :: 1.37**   Simplify: ⓐ $(-3)^4$ ⓑ -3^4.

> **TRY IT :: 1.38**   Simplify: ⓐ $(-7)^2$ ⓑ -7^2.

The last example showed us the difference between $(-2)^4$ and -2^4. This distinction is important to prevent future errors. The next example reminds us to multiply and divide in order left to right.

EXAMPLE 1.20

Simplify: ⓐ $8(-9) \div (-2)^3$ ⓑ $-30 \div 2 + (-3)(-7)$.

⊘ **Solution**

ⓐ

$$8(-9) \div (-2)^3$$

Exponents first.	$8(-9) \div (-8)$
Multiply.	$-72 \div (-8)$
Divide.	9

ⓑ

$$-30 \div 2 + (-3)(-7)$$

Multiply and divide left to right, so divide first.	$-15 + (-3)(-7)$
Multiply.	$-15 + 21$
Add.	6

> **TRY IT :: 1.39**   Simplify: ⓐ $12(-9) \div (-3)^3$ ⓑ $-27 \div 3 + (-5)(-6)$.

> **TRY IT :: 1.40**   Simplify: ⓐ $18(-4) \div (-2)^3$ ⓑ $-32 \div 4 + (-2)(-7)$.

Evaluate Variable Expressions with Integers

Remember that to evaluate an expression means to substitute a number for the variable in the expression. Now we can use negative numbers as well as positive numbers.

EXAMPLE 1.21

Evaluate $4x^2 - 2xy + 3y^2$ when $x = 2$, $y = -1$.

⊘ **Solution**

	$4x^2 - 2xy + 3y^2$
Substitute $x = 2$, $y = -1$. Use parentheses to show multiplication.	$4(2)^2 - 2(2)(-1) + 3(-1)^2$
Simplify exponents.	$4 \cdot 4 - (-4) + 3 \cdot 1$
Multiply.	$16 - (-4) + 3$
Subtract.	$20 + 3$
Add.	23

> **TRY IT :: 1.41**   Evaluate: $3x^2 - 2xy + 6y^2$ when $x = 1$, $y = -2$.

> **TRY IT :: 1.42**   Evaluate: $4x^2 - xy + 5y^2$ when $x = -2$, $y = 3$.

Translate Phrases to Expressions with Integers

Our earlier work translating English to algebra also applies to phrases that include both positive and negative numbers.

EXAMPLE 1.22

Translate and simplify: the sum of 8 and -12, increased by 3.

⊘ Solution

	the **sum** of 8 and −12 increased by 3
Translate.	$[8 + (-12)] + 3$
Simplify. Be careful not to confuse the brackets with an absolute value sign.	$(-4) + 3$
Add.	-1

> **TRY IT : : 1.43**   Translate and simplify the sum of 9 and −16, increased by 4.

> **TRY IT : : 1.44**   Translate and simplify the sum of −8 and −12, increased by 7.

Use Integers in Applications

We'll outline a plan to solve applications. It's hard to find something if we don't know what we're looking for or what to call it! So when we solve an application, we first need to determine what the problem is asking us to find. Then we'll write a phrase that gives the information to find it. We'll translate the phrase into an expression and then simplify the expression to get the answer. Finally, we summarize the answer in a sentence to make sure it makes sense.

EXAMPLE 1.23	HOW TO SOLVE APPLICATION PROBLEMS USING INTEGERS

The temperature in Kendallville, Indiana one morning was 11 degrees. By mid-afternoon, the temperature had dropped to −9 degrees. What was the difference in the morning and afternoon temperatures?

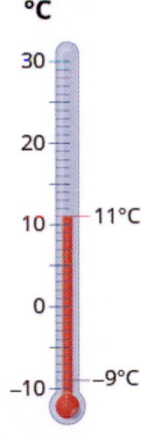

⊘ Solution

Step 1. **Read** the problem. Make sure all the words and ideas are understood.	
Step 2. **Identify** what we are asked to find.	The difference of the morning and afternoon temperatures
Step 3. Write a phrase that gives the information to find it.	the difference of 11 and –9
Step 4. **Translate** the phrase to an expression.	11 – (–9)
Step 5. Simplify the expression.	20
Step 6. Answer the question with a complete sentence.	The difference in temperatures was 20 degrees.

> **TRY IT : :** 1.45

The temperature in Anchorage, Alaska one morning was 15 degrees. By mid-afternoon the temperature had dropped to 30 degrees below zero. What was the difference in the morning and afternoon temperatures?

> **TRY IT : :** 1.46

The temperature in Denver was −6 degrees at lunchtime. By sunset the temperature had dropped to −15 degrees. What was the difference in the lunchtime and sunset temperatures?

HOW TO : : USE INTEGERS IN APPLICATIONS.

Step 1.   **Read** the problem. Make sure all the words and ideas are understood.
Step 2.   **Identify** what we are asked to find.
Step 3.   **Write a phrase** that gives the information to find it.
Step 4.   **Translate** the phrase to an expression.
Step 5.   **Simplify** the expression.
Step 6.   **Answer** the question with a complete sentence.

▶ | **MEDIA : :**

Access this online resource for additional instruction and practice with integers.

- **Subtracting Integers with Counters (https://openstax.org/l/25subintegers)**

1.2 EXERCISES

Practice Makes Perfect

Simplify Expressions with Absolute Value

In the following exercises, fill in $<$, $>$, *or* $=$ *for each of the following pairs of numbers.*

59.
ⓐ $|-7|$___ $-|-7|$
ⓑ 6___ $-|-6|$
ⓒ $|-11|$___ -11
ⓓ $-(-13)$___ $-|-13|$

60.
ⓐ $-|-9|$___ $|-9|$
ⓑ -8___ $|-8|$
ⓒ $|-1|$___ -1
ⓓ $-(-14)$___ $-|-14|$

61.
ⓐ $-|2|$___ $-|-2|$
ⓑ -12___ $-|-12|$
ⓒ $|-3|$___ -3
ⓓ $|-19|$___ $-(-19)$

62.
ⓐ $-|-4|$___ $-|4|$
ⓑ 5___ $-|-5|$
ⓒ $-|-10|$___ -10
ⓓ $-|-0|$___ $-(-0)$

In the following exercises, simplify.

63. $|15 - 7| - |14 - 6|$

64. $|17 - 8| - |13 - 4|$

65. $18 - |2(8 - 3)|$

66. $15 - |3(8 - 5)|$

67. $18 - |12 - 4(4 - 1) + 3|$

68. $27 - |19 + 4(3 - 1) - 7|$

69. $10 - 3|9 - 3(3 - 1)|$

70. $13 - 2|11 - 2(5 - 2)|$

Add and Subtract Integers

In the following exercises, simplify each expression.

71.
ⓐ $-7 + (-4)$
ⓑ $-7 + 4$
ⓒ $7 + (-4)$.

72.
ⓐ $-5 + (-9)$
ⓑ $-5 + 9$
ⓒ $5 + (-9)$

73. $48 + (-16)$

74. $34 + (-19)$

75. $-14 + (-12) + 4$

76. $-17 + (-18) + 6$

77. $19 + 2(-3 + 8)$

78. $24 + 3(-5 + 9)$

79.
ⓐ $13 - 7$
ⓑ $-13 - (-7)$
ⓒ $-13 - 7$
ⓓ $13 - (-7)$

80.
ⓐ $15 - 8$
ⓑ $-15 - (-8)$
ⓒ $-15 - 8$
ⓓ $15 - (-8)$

81. $-17 - 42$

82. $-58 - (-67)$

83. $-14 - (-27) + 9$

84. $64 + (-17) - 9$

85. ⓐ $44 - 28$ ⓑ $44 + (-28)$

86. ⓐ $35 - 16$ ⓑ $35 + (-16)$

87. ⓐ $27 - (-18)$ ⓑ $27 + 18$

88. ⓐ $46 - (-37)$ ⓑ $46 + 37$

89. $(2 - 7) - (3 - 8)$

90. $(1 - 8) - (2 - 9)$

91. $-(6 - 8) - (2 - 4)$

92. $-(4 - 5) - (7 - 8)$

93. $25 - [10 - (3 - 12)]$

94. $32 - [5 - (15 - 20)]$

Multiply and Divide Integers

In the following exercises, multiply or divide.

95.
ⓐ $-4 \cdot 8$
ⓑ $13(-5)$
ⓒ $-24 \div 6$
ⓓ $-52 \div (-4)$

96.
ⓐ $-3 \cdot 9$
ⓑ $9(-7)$
ⓒ $35 \div (-7)$
ⓓ $-84 \div (-6)$

97.
ⓐ $-28 \div 7$
ⓑ $-180 \div 15$
ⓒ $3(-13)$
ⓓ $-1(-14)$

98.
ⓐ $-36 \div 4$
ⓑ $-192 \div 12$
ⓒ $9(-7)$
ⓓ $-1(-19)$

Simplify and Evaluate Expressions with Integers

In the following exercises, simplify each expression.

99. ⓐ $(-2)^6$ ⓑ -2^6

100. ⓐ $(-3)^5$ ⓑ -3^5

101. $5(-6) + 7(-2) - 3$

102. $8(-4) + 5(-4) - 6$

103. $-3(-5)(6)$

104. $-4(-6)(3)$

105. $(8 - 11)(9 - 12)$

106. $(6 - 11)(8 - 13)$

107. $26 - 3(2 - 7)$

108. $23 - 2(4 - 6)$

109. $65 \div (-5) + (-28) \div (-7)$

110. $52 \div (-4) + (-32) \div (-8)$

111. $9 - 2[3 - 8(-2)]$

112. $11 - 3[7 - 4(-2)]$

113. $8 - |2 - 4(4 - 1) + 3|$

114. $7 - |5 - 3(4 - 1) - 6|$

115. $9 - 3|2(2 - 6) - (3 - 7)|$

116. $5 - 2|2(1 - 4) - (2 - 5)|$

117. $(-3)^2 - 24 \div (8 - 2)$

118. $(-4)^2 - 32 \div (12 - 4)$

In the following exercises, evaluate each expression.

119. $y + (-14)$ when

ⓐ $y = -33$ ⓑ $y = 30$

120. $x + (-21)$ when

ⓐ $x = -27$ ⓑ $x = 44$

121. $(x + y)^2$ when

$x = -3, y = 14$

122. $(y + z)^2$ when

$y = -3, z = 15$

123. $9a - 2b - 8$ when

$a = -6$ and $b = -3$

124. $7m - 4n - 2$ when

$m = -4$ and $n = -9$

125. $3x^2 - 4xy + 2y^2$ when

$x = -2, y = -3$

126. $4x^2 - xy + 3y^2$ when

$x = -3, y = -2$

Translate English Phrases to Algebraic Expressions

In the following exercises, translate to an algebraic expression and simplify if possible.

127. the sum of 3 and -15, increased by 7

128. the sum of -8 and -9, increased by 23

129.

ⓐ the difference of 10 and -18

ⓑ subtract 11 from -25

130.

ⓐ the difference of -5 and -30

ⓑ subtract -6 from -13

131. the quotient of -6 and the sum of a and b

132. the product of -13 and the difference of c and d

Use Integers in Applications

In the following exercises, solve.

133. Temperature On January 15, the high temperature in Anaheim, California, was 84°. That same day, the high temperature in Embarrass, Minnesota, was $-12°$. What was the difference between the temperature in Anaheim and the temperature in Embarrass?

134. Temperature On January 21, the high temperature in Palm Springs, California, was 89°, and the high temperature in Whitefield, New Hampshire, was $-31°$. What was the difference between the temperature in Palm Springs and the temperature in Whitefield?

135. Football On the first down, the Chargers had the ball on their 25-yard line. On the next three downs, they lost 6 yards, gained 10 yards, and lost 8 yards. What was the yard line at the end of the fourth down?

136. Football On the first down, the Steelers had the ball on their 30-yard line. On the next three downs, they gained 9 yards, lost 14 yards, and lost 2 yards. What was the yard line at the end of the fourth down?

137. Checking Account Mayra has $124 in her checking account. She writes a check for $152. What is the new balance in her checking account?

138. Checking Account Reymonte has a balance of $-$49 in his checking account. He deposits $281 to the account. What is the new balance?

Writing Exercises

139. Explain why the sum of -8 and 2 is negative, but the sum of 8 and -2 is positive.

140. Give an example from your life experience of adding two negative numbers.

141. In your own words, state the rules for multiplying and dividing integers.

142. Why is $-4^3 = (-4)^3$?

Self Check

ⓐ *After completing the exercises, use this checklist to evaluate your mastery of the objectives of this section.*

I can...	Confidently	With some help	No-I don't get it!
simplify expressions with absolute value.			
add and subtract integers.			
multiply and divide integers.			
simplify and evaluate expressions with integers.			
translate English phrases to algebraic expressions.			
use integers in applications.			

ⓑ *After reviewing this checklist, what will you do to become confident for all objectives?*

Fractions

Learning Objectives

By the end of this section, you will be able to:

> Simplify fractions
> Multiply and divide fractions
> Add and subtract fractions
> Use the order of operations to simplify fractions
> Evaluate variable expressions with fractions

Be Prepared!

A more thorough introduction to the topics covered in this section can be found in the *Elementary Algebra* chapter, Foundations.

Simplify Fractions

A **fraction** is a way to represent parts of a whole. The fraction $\frac{2}{3}$ represents two of three equal parts. See **Figure 1.5**. In the fraction $\frac{2}{3}$, the 2 is called the **numerator** and the 3 is called the **denominator**. The line is called the fraction bar.

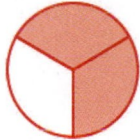

Figure 1.5 In the circle, $\frac{2}{3}$ of the circle is shaded—2 of the 3 equal parts.

Fraction

A **fraction** is written $\frac{a}{b}$, where $b \neq 0$ and

a is the **numerator** and *b* is the **denominator**.

A fraction represents parts of a whole. The denominator b is the number of equal parts the whole has been divided into, and the numerator a indicates how many parts are included.

Fractions that have the same value are **equivalent fractions**. The Equivalent Fractions Property allows us to find equivalent fractions and also simplify fractions.

Equivalent Fractions Property

If a, b, and c are numbers where $b \neq 0$, $c \neq 0$,

then $\frac{a}{b} = \frac{a \cdot c}{b \cdot c}$ and $\frac{a \cdot c}{b \cdot c} = \frac{a}{b}$.

A fraction is considered simplified if there are no common factors, other than 1, in its numerator and denominator.

For example,

$\frac{2}{3}$ is simplified because there are no common factors of 2 and 3.

$\frac{10}{15}$ is not simplified because 5 is a common factor of 10 and 15.

We simplify, or reduce, a fraction by removing the common factors of the numerator and denominator. A fraction is not simplified until all common factors have been removed. If an expression has fractions, it is not completely simplified until the fractions are simplified.

Sometimes it may not be easy to find common factors of the numerator and denominator. When this happens, a good idea is to factor the numerator and the denominator into prime numbers. Then divide out the common factors using the Equivalent Fractions Property.

EXAMPLE 1.24   HOW TO SIMPLIFY A FRACTION

Simplify: $-\dfrac{315}{770}$.

✓ **Solution**

Step 1. Rewrite the numerator and denominator to show the common factors. If needed, use a factor tree.	Rewrite 315 and 770 as the product of the primes.	$-\dfrac{315}{770}$ $-\dfrac{3\cdot3\cdot5\cdot7}{2\cdot5\cdot7\cdot11}$
Step 2. Simplify using the equivalent fractions property by dividing out common factors.	Mark the common factors 5 and 7. Divide out the common factors	$-\dfrac{3\cdot3\cdot\cancel{5}\cdot\cancel{7}}{2\cdot\cancel{5}\cdot\cancel{7}\cdot11}$ $-\dfrac{3\cdot3}{2\cdot11}$
Step 3. Multiply the remaining factors, if necessary.		$-\dfrac{9}{22}$

> **TRY IT ::** 1.47   Simplify: $-\dfrac{69}{120}$.

> **TRY IT ::** 1.48   Simplify: $-\dfrac{120}{192}$.

We now summarize the steps you should follow to simplify fractions.

HOW TO :: SIMPLIFY A FRACTION.

Step 1.   Rewrite the numerator and denominator to show the common factors. If needed, factor the numerator and denominator into prime numbers first.

Step 2.   Simplify using the Equivalent Fractions Property by dividing out common factors.

Step 3.   Multiply any remaining factors.

Multiply and Divide Fractions

Many people find multiplying and dividing fractions easier than adding and subtracting fractions.

To multiply fractions, we multiply the numerators and multiply the denominators.

Fraction Multiplication

If a, b, c, and d are numbers where $b \neq 0$, and $d \neq 0$, then

$$\frac{a}{b}\cdot\frac{c}{d}=\frac{ac}{bd}$$

To multiply fractions, multiply the numerators and multiply the denominators.

When multiplying fractions, the properties of positive and negative numbers still apply, of course. It is a good idea to determine the sign of the product as the first step. In **Example 1.25**, we will multiply negative and a positive, so the product will be negative.

When multiplying a fraction by an integer, it may be helpful to write the integer as a fraction. Any integer, a, can be written as $\dfrac{a}{1}$. So, for example, $3 = \dfrac{3}{1}$.

EXAMPLE 1.25

Multiply: $-\frac{12}{5}(-20x)$.

✓ Solution

The first step is to find the sign of the product. Since the signs are the same, the product is positive.

	$-\frac{12}{5}(-20x)$
Determine the sign of the product. The signs are the same, so the product is positive.	$\frac{12}{5}(20x)$
Write $20x$ as a fraction.	$\frac{12}{5}\left(\frac{20x}{1}\right)$
Multiply.	$\frac{12 \cdot 20x}{5 \cdot 1}$
Rewrite 20 to show the common factor 5 and divide it out.	$\frac{12 \cdot 4 \cdot \cancel{5} \cdot x}{\cancel{5} \cdot 1}$
Simplify.	$48x$

> **TRY IT : : 1.49** Multiply: $\frac{11}{3}(-9a)$.

> **TRY IT : : 1.50** Multiply: $\frac{13}{7}(-14b)$.

Now that we know how to multiply fractions, we are almost ready to divide. Before we can do that, we need some vocabulary. The **reciprocal** of a fraction is found by inverting the fraction, placing the numerator in the denominator and the denominator in the numerator. The reciprocal of $\frac{2}{3}$ is $\frac{3}{2}$. Since 4 is written in fraction form as $\frac{4}{1}$, the reciprocal of 4 is $\frac{1}{4}$.

To divide fractions, we multiply the first fraction by the reciprocal of the second.

Fraction Division

If a, b, c, and d are numbers where $b \neq 0$, $c \neq 0$, and $d \neq 0$, then

$$\frac{a}{b} \div \frac{c}{d} = \frac{a}{b} \cdot \frac{d}{c}$$

To divide fractions, we multiply the first fraction by the **reciprocal** of the second.

We need to say $b \neq 0$, $c \neq 0$, and $d \neq 0$, to be sure we don't divide by zero!

EXAMPLE 1.26

Find the quotient: $-\frac{7}{8} \div \left(-\frac{14}{27}\right)$.

⊘ **Solution**

$$-\frac{7}{8} \div \left(-\frac{14}{27}\right)$$

To divide, multiply the first fraction by the reciprocal of the second.	$-\frac{7}{18}\left(\frac{27}{14}\right)$
Determine the sign of the product, and then multiply.	$\frac{7 \cdot 27}{18 \cdot 14}$
Rewrite showing common factors.	$\frac{\cancel{7} \cdot \cancel{9} \cdot 3}{\cancel{9} \cdot 2 \cdot \cancel{7} \cdot 2}$
Remove common factors.	$\frac{3}{2 \cdot 2}$
Simplify.	$\frac{3}{4}$

> **TRY IT :: 1.51**   Divide: $-\frac{7}{27} \div \left(-\frac{35}{36}\right)$.

> **TRY IT :: 1.52**   Divide: $-\frac{5}{14} \div \left(-\frac{15}{28}\right)$.

The numerators or denominators of some fractions contain fractions themselves. A fraction in which the numerator or the denominator is a fraction is called a **complex fraction**.

Complex Fraction

A **complex fraction** is a fraction in which the numerator or the denominator contains a fraction.

Some examples of complex fractions are:

$$\frac{\frac{6}{7}}{3} \qquad \frac{\frac{3}{4}}{\frac{5}{8}} \qquad \frac{\frac{x}{2}}{\frac{5}{6}}$$

To simplify a complex fraction, remember that the fraction bar means division. For example, the complex fraction $\frac{\frac{3}{4}}{\frac{5}{8}}$

means $\frac{3}{4} \div \frac{5}{8}$.

EXAMPLE 1.27

Simplify: $\frac{\frac{x}{2}}{\frac{xy}{6}}$.

⊘ Solution

$$\frac{\frac{x}{2}}{\frac{xy}{6}}$$

Rewrite as division.

$$\frac{x}{2} \div \frac{xy}{6}$$

Multiply the first fraction by the reciprocal of the second.

$$\frac{x}{2} \cdot \frac{6}{xy}$$

Multiply.

$$\frac{x \cdot 6}{2 \cdot xy}$$

Look for common factors.

$$\frac{\not{x} \cdot 3 \cdot \not{2}}{\not{2} \cdot \not{x} \cdot y}$$

Divide common factors and simplify.

$$\frac{3}{y}$$

> **TRY IT :: 1.53**
> Simplify: $\dfrac{\frac{a}{8}}{\frac{ab}{6}}$.

> **TRY IT :: 1.54**
> Simplify: $\dfrac{\frac{p}{2}}{\frac{pq}{8}}$.

Add and Subtract Fractions

When we multiplied fractions, we just multiplied the numerators and multiplied the denominators right straight across. To add or subtract fractions, they must have a common denominator.

Fraction Addition and Subtraction

If a, b, and c are numbers where $c \neq 0$, then

$$\frac{a}{c} + \frac{b}{c} = \frac{a+b}{c} \quad \text{and} \quad \frac{a}{c} - \frac{b}{c} = \frac{a-b}{c}$$

To add or subtract fractions, add or subtract the numerators and place the result over the common denominator.

The **least common denominator** (LCD) of two fractions is the smallest number that can be used as a common denominator of the fractions. The LCD of the two fractions is the least common multiple (LCM) of their denominators.

Least Common Denominator

The **least common denominator** (LCD) of two fractions is the least common multiple (LCM) of their denominators.

After we find the least common denominator of two fractions, we convert the fractions to equivalent fractions with the LCD. Putting these steps together allows us to add and subtract fractions because their denominators will be the same!

EXAMPLE 1.28   HOW TO ADD OR SUBTRACT FRACTIONS

Add: $\dfrac{7}{12} + \dfrac{5}{18}$.

⊘ **Solution**

Step 1. Do they have a common denominator? • No—rewrite each fraction with the LCD (least common denominator).	No. Find the LCD of 12, 18.	$\begin{array}{l} 12 = 2 \cdot 2 \cdot 3 \\ 18 = 2 \cdot \quad 3 \cdot 3 \\ \hline LCD = 2 \cdot 2 \cdot 3 \cdot 3 \\ LCD = 36 \end{array}$ LCD is 36.
	We multiply the numerator and denominator of each fraction by the factor needed to get the denominator to be 36. Do not simplify the equivalent fractions! If you do, you'll get back to the original fractions and lose the common denominator!	$\dfrac{7}{12} + \dfrac{5}{18}$ $\dfrac{7 \cdot 3}{12 \cdot 3} + \dfrac{5 \cdot 2}{18 \cdot 2}$ $\dfrac{21}{36} + \dfrac{10}{36}$
Step 2. Add or subtract the fractions.	Add.	$\dfrac{31}{36}$
Step 3. Simplify, if possible.	Since 31 is prime, its only factors are 1 and 31. Since 31 does not go into 36, the answer is simplified.	

> **TRY IT ::** 1.55   Add: $\dfrac{7}{12} + \dfrac{11}{15}$.

> **TRY IT ::** 1.56   Add: $\dfrac{13}{15} + \dfrac{17}{20}$.

HOW TO :: ADD OR SUBTRACT FRACTIONS.

Step 1.   Do they have a common denominator?
   ◦ Yes—go to step 2.
   ◦ No—rewrite each fraction with the LCD (least common denominator).
      ▪ Find the LCD.
      ▪ Change each fraction into an equivalent fraction with the LCD as its denominator.
Step 2.   Add or subtract the fractions.
Step 3.   Simplify, if possible.

We now have all four operations for fractions. **Table 1.18** summarizes fraction operations.

Fraction Multiplication	Fraction Division
$\dfrac{a}{b} \cdot \dfrac{c}{d} = \dfrac{ac}{bd}$	$\dfrac{a}{b} \div \dfrac{c}{d} = \dfrac{a}{b} \cdot \dfrac{d}{c}$
Multiply the numerators and multiply the denominators	Multiply the first fraction by the reciprocal of the second.
Fraction Addition	**Fraction Subtraction**
$\dfrac{a}{c} + \dfrac{b}{c} = \dfrac{a+b}{c}$	$\dfrac{a}{c} - \dfrac{b}{c} = \dfrac{a-b}{c}$
Add the numerators and place the sum over the common denominator.	Subtract the numerators and place the difference over the common denominator.
To multiply or divide fractions, an LCD is NOT needed. To add or subtract fractions, an LCD is needed.	

Table 1.18

When starting an exercise, always identify the operation and then recall the methods needed for that operation.

EXAMPLE 1.29

Simplify: ⓐ $\dfrac{5x}{6} - \dfrac{3}{10}$ ⓑ $\dfrac{5x}{6} \cdot \dfrac{3}{10}$.

✓ **Solution**

First ask, "What is the operation?" Identifying the operation will determine whether or not we need a common denominator. Remember, we need a common denominator to add or subtract, but not to multiply or divide.

ⓐ

What is the operation? The operation is subtraction.

Do the fractions have a common denominator? No. $\dfrac{5x}{6} - \dfrac{3}{10}$

Find the LCD of 6 and 10   The LCD is 30.
$$6 = 2 \cdot 3$$
$$10 = 2 \cdot 5$$
$$\text{LCD} = 2 \cdot 3 \cdot 5$$
$$\text{LCD} = 30$$

Rewrite each fraction as an equivalent fraction with the LCD. $\dfrac{5x \cdot 5}{6 \cdot 5} - \dfrac{3 \cdot 3}{10 \cdot 3}$

$$\dfrac{25x}{30} - \dfrac{9}{30}$$

Subtract the numerators and place the $\dfrac{25x - 9}{30}$
difference over the common denominators.

Simplify, if possible. There are no common factors.
The fraction is simplified.

ⓑ

What is the operation? Multiplication.

$$\frac{25x}{6} \cdot \frac{3}{10}$$

To multiply fractions, multiply the numerators
and multiply the denominators.

$$\frac{25x \cdot 3}{6 \cdot 10}$$

Rewrite, showing common factors.

Remove common factors.

$$\frac{\cancel{5}x \cdot \cancel{3}}{2 \cdot \cancel{3} \cdot 2 \cdot \cancel{5}}$$

Simplify.

$$\frac{x}{4}$$

Notice, we needed an LCD to add $\frac{25x}{6} - \frac{3}{10}$, but not to multiply $\frac{25x}{6} \cdot \frac{3}{10}$.

> **TRY IT :: 1.57**   Simplify: ⓐ $\frac{3a}{4} - \frac{8}{9}$ ⓑ $\frac{3a}{4} \cdot \frac{8}{9}$.

> **TRY IT :: 1.58**   Simplify: ⓐ $\frac{4k}{5} - \frac{1}{6}$ ⓑ $\frac{4k}{5} \cdot \frac{1}{6}$.

Use the Order of Operations to Simplify Fractions

The fraction bar in a fraction acts as grouping symbol. The order of operations then tells us to simplify the numerator and then the denominator. Then we divide.

HOW TO :: SIMPLIFY AN EXPRESSION WITH A FRACTION BAR.

Step 1.   Simplify the expression in the numerator. Simplify the expression in the denominator.

Step 2.   Simplify the fraction.

Where does the negative sign go in a fraction? Usually the negative sign is in front of the fraction, but you will sometimes see a fraction with a negative numerator, or sometimes with a negative denominator. Remember that fractions represent division. When the numerator and denominator have different signs, the quotient is negative.

$$\frac{-1}{3} = -\frac{1}{3} \qquad \frac{\text{negative}}{\text{positive}} = \text{negative}$$

$$\frac{1}{-3} = -\frac{1}{3} \qquad \frac{\text{positive}}{\text{negative}} = \text{negative}$$

Placement of Negative Sign in a Fraction

For any positive numbers a and b,

$$\frac{-a}{b} = \frac{a}{-b} = -\frac{a}{b}$$

EXAMPLE 1.30

Simplify: $\frac{4(-3) + 6(-2)}{-3(2) - 2}$.

⊘ **Solution**

The fraction bar acts like a grouping symbol. So completely simplify the numerator and the denominator separately.

$$\frac{4(-3) + 6(-2)}{-3(2) - 2}$$

Multiply.
$$\frac{-12 + (-12)}{-6 - 2}$$

Simplify.
$$\frac{-24}{-8}$$

Divide.
$$3$$

> **TRY IT : : 1.59** Simplify: $\dfrac{8(-2) + 4(-3)}{-5(2) + 3}$.

> **TRY IT : : 1.60** Simplify: $\dfrac{7(-1) + 9(-3)}{-5(3) - 2}$.

Now we'll look at complex fractions where the numerator or denominator contains an expression that can be simplified. So we first must completely simplify the numerator and denominator separately using the order of operations. Then we divide the numerator by the denominator as the fraction bar means division.

EXAMPLE 1.31   HOW TO SIMPLIFY COMPLEX FRACTIONS

Simplify: $\dfrac{\left(\frac{1}{2}\right)^2}{4 + 3^2}$.

✓ **Solution**

Step 1. Simplify the numerator.	$\dfrac{\left(\frac{1}{2}\right)^2}{4 + 3^2}$ $\dfrac{\frac{1}{4}}{4 + 3^2}$
Step 2. Simplify the denominator.	$\dfrac{\frac{1}{4}}{4 + 9}$ $\dfrac{\frac{1}{4}}{13}$
Step 3. Divide the numerator by the denominator. Simplify if possible	$\dfrac{1}{4} \div \dfrac{13}{1}$ $\dfrac{1}{4} \cdot \dfrac{1}{13}$ $\dfrac{1}{52}$

> **TRY IT : : 1.61** Simplify: $\dfrac{\left(\frac{1}{3}\right)^2}{2^3 + 2}$.

> **TRY IT : : 1.62** Simplify: $\dfrac{1 + 4^2}{\left(\frac{1}{4}\right)^2}$.

HOW TO :: SIMPLIFY COMPLEX FRACTIONS.

Step 1.   Simplify the numerator.

Step 2.   Simplify the denominator.

Step 3.   Divide the numerator by the denominator. Simplify if possible.

EXAMPLE 1.32

Simplify: $\dfrac{\frac{1}{2} + \frac{2}{3}}{\frac{3}{4} - \frac{1}{6}}$.

⊘ **Solution**

It may help to put parentheses around the numerator and the denominator.

$$\frac{\left(\frac{1}{2} + \frac{2}{3}\right)}{\left(\frac{3}{4} - \frac{1}{6}\right)}$$

Simplify the numerator (LCD = 6) and simplify the denominator (LCD = 12).

$$\frac{\left(\frac{3}{6} + \frac{4}{6}\right)}{\left(\frac{9}{12} - \frac{2}{12}\right)}$$

Simplify.

$$\frac{\left(\frac{7}{6}\right)}{\left(\frac{7}{12}\right)}$$

Divide the numerator by the denominator.

$$\frac{7}{6} \div \frac{7}{12}$$

Simplify.

$$\frac{7}{6} \cdot \frac{12}{7}$$

Divide out common factors.

$$\frac{\cancel{7} \cdot \cancel{6} \cdot 2}{\cancel{6} \cdot \cancel{7} \cdot 1}$$

Simplify.

$$2$$

> **TRY IT :: 1.63**   Simplify: $\dfrac{\frac{1}{3} + \frac{1}{2}}{\frac{3}{4} - \frac{1}{3}}$.

> **TRY IT :: 1.64**   Simplify: $\dfrac{\frac{2}{3} - \frac{1}{2}}{\frac{1}{4} + \frac{1}{3}}$.

Evaluate Variable Expressions with Fractions

We have evaluated expressions before, but now we can evaluate expressions with fractions. Remember, to evaluate an expression, we substitute the value of the variable into the expression and then simplify.

EXAMPLE 1.33

Evaluate $2x^2 y$ when $x = \frac{1}{4}$ and $y = -\frac{2}{3}$.

⊘ **Solution**

Substitute the values into the expression.

	$2x^2y$
Substitute $\frac{1}{4}$ for x and $-\frac{2}{3}$ for y.	$2\left(\frac{1}{4}\right)^{2}\left(-\frac{2}{3}\right)$
Simplify exponents first.	$2\left(\frac{1}{16}\right)\left(-\frac{2}{3}\right)$
Multiply; divide out the common factors. Notice we write 16 as $2 \cdot 2 \cdot 4$ to make it easy to remove common factors.	$-\dfrac{\cancel{2} \cdot 1 \cdot \cancel{2}}{\cancel{2} \cdot \cancel{2} \cdot 4 \cdot 3}$
Simplify.	$-\dfrac{1}{12}$

> **TRY IT :: 1.65**   Evaluate $3ab^2$ when $a = -\frac{2}{3}$ and $b = -\frac{1}{2}$.

> **TRY IT :: 1.66**   Evaluate $4c^3d$ when $c = -\frac{1}{2}$ and $d = -\frac{4}{3}$.

▶ **MEDIA ::**

Access this online resource for additional instruction and practice with fractions.

- **Adding Fractions with Unlike Denominators (https://openstax.org/l/25addfractions)**

1.3 EXERCISES

Practice Makes Perfect

Simplify Fractions

In the following exercises, simplify.

143. $-\dfrac{108}{63}$

144. $-\dfrac{104}{48}$

145. $\dfrac{120}{252}$

146. $\dfrac{182}{294}$

147. $\dfrac{14x^2}{21y}$

148. $\dfrac{24a}{32b^2}$

149. $-\dfrac{210a^2}{110b^2}$

150. $-\dfrac{30x^2}{105y^2}$

Multiply and Divide Fractions

In the following exercises, perform the indicated operation.

151. $-\dfrac{3}{4}\left(-\dfrac{4}{9}\right)$

152. $-\dfrac{3}{8}\cdot\dfrac{4}{15}$

153. $\left(-\dfrac{14}{15}\right)\left(\dfrac{9}{20}\right)$

154. $\left(-\dfrac{9}{10}\right)\left(\dfrac{25}{33}\right)$

155. $\left(-\dfrac{63}{84}\right)\left(-\dfrac{44}{90}\right)$

156. $\left(-\dfrac{33}{60}\right)\left(-\dfrac{40}{88}\right)$

157. $\dfrac{3}{7}\cdot 21n$

158. $\dfrac{5}{6}\cdot 30m$

159. $\dfrac{3}{4}\div\dfrac{x}{11}$

160. $\dfrac{2}{5}\div\dfrac{y}{9}$

161. $\dfrac{5}{18}\div\left(-\dfrac{15}{24}\right)$

162. $\dfrac{7}{18}\div\left(-\dfrac{14}{27}\right)$

163. $\dfrac{8u}{15}\div\dfrac{12v}{25}$

164. $\dfrac{12r}{25}\div\dfrac{18s}{35}$

165. $\dfrac{3}{4}\div(-12)$

166. $-15\div\left(-\dfrac{5}{3}\right)$

In the following exercises, simplify.

167. $\dfrac{-\frac{8}{21}}{\frac{12}{35}}$

168. $\dfrac{-\frac{9}{16}}{\frac{33}{40}}$

169. $\dfrac{-\frac{4}{5}}{2}$

170. $\dfrac{\frac{5}{3}}{10}$

171. $\dfrac{\frac{m}{3}}{\frac{n}{2}}$

172. $\dfrac{-\frac{3}{8}}{-\frac{y}{12}}$

Add and Subtract Fractions

In the following exercises, add or subtract.

173. $\dfrac{7}{12}+\dfrac{5}{8}$

174. $\dfrac{5}{12}+\dfrac{3}{8}$

175. $\dfrac{7}{12}-\dfrac{9}{16}$

176. $\dfrac{7}{16}-\dfrac{5}{12}$

177. $-\dfrac{13}{30}+\dfrac{25}{42}$

178. $-\dfrac{23}{30}+\dfrac{5}{48}$

179. $-\frac{39}{56} - \frac{22}{35}$

180. $-\frac{33}{49} - \frac{18}{35}$

181. $-\frac{2}{3} - \left(-\frac{3}{4}\right)$

182. $-\frac{3}{4} - \left(-\frac{4}{5}\right)$

183. $\frac{x}{3} + \frac{1}{4}$

184. $\frac{x}{5} - \frac{1}{4}$

185.
ⓐ $\frac{2}{3} + \frac{1}{6}$

ⓑ $\frac{2}{3} \div \frac{1}{6}$

186.
ⓐ $-\frac{2}{5} - \frac{1}{8}$

ⓑ $-\frac{2}{5} \cdot \frac{1}{8}$

187.
ⓐ $\frac{5n}{6} \div \frac{8}{15}$

ⓑ $\frac{5n}{6} - \frac{8}{15}$

188.
ⓐ $\frac{3a}{8} \div \frac{7}{12}$

ⓑ $\frac{3a}{8} - \frac{7}{12}$

189.
ⓐ $-\frac{4x}{9} - \frac{5}{6}$

ⓑ $-\frac{4k}{9} \cdot \frac{5}{6}$

190.
ⓐ $-\frac{3y}{8} - \frac{4}{3}$

ⓑ $-\frac{3y}{8} \cdot \frac{4}{3}$

191.
ⓐ $-\frac{5a}{3} + \left(-\frac{10}{6}\right)$

ⓑ $-\frac{5a}{3} \div \left(-\frac{10}{6}\right)$

192.
ⓐ $\frac{2b}{5} + \frac{8}{15}$

ⓑ $\frac{2b}{5} \div \frac{8}{15}$

Use the Order of Operations to Simplify Fractions

In the following exercises, simplify.

193. $\frac{5 \cdot 6 - 3 \cdot 4}{4 \cdot 5 - 2 \cdot 3}$

194. $\frac{8 \cdot 9 - 7 \cdot 6}{5 \cdot 6 - 9 \cdot 2}$

195. $\frac{5^2 - 3^2}{3 - 5}$

196. $\frac{6^2 - 4^2}{4 - 6}$

197. $\frac{7 \cdot 4 - 2(8 - 5)}{9 \cdot 3 - 3 \cdot 5}$

198. $\frac{9 \cdot 7 - 3(12 - 8)}{8 \cdot 7 - 6 \cdot 6}$

199. $\frac{9(8 - 2) - 3(15 - 7)}{6(7 - 1) - 3(17 - 9)}$

200. $\frac{8(9 - 2) - 4(14 - 9)}{7(8 - 3) - 3(16 - 9)}$

201. $\frac{2^3 + 4^2}{\left(\frac{2}{3}\right)^2}$

202. $\frac{3^3 - 3^2}{\left(\frac{3}{4}\right)^2}$

203. $\frac{\left(\frac{3}{5}\right)^2}{\left(\frac{3}{7}\right)^2}$

204. $\frac{\left(\frac{3}{4}\right)^2}{\left(\frac{5}{8}\right)^2}$

205. $\frac{2}{\frac{1}{3} + \frac{1}{5}}$

206. $\frac{5}{\frac{1}{4} + \frac{1}{3}}$

207. $\frac{\frac{7}{8} - \frac{2}{3}}{\frac{1}{2} + \frac{3}{8}}$

208. $\frac{\frac{3}{4} - \frac{3}{5}}{\frac{1}{4} + \frac{2}{5}}$

Mixed Practice

In the following exercises, simplify.

209. $-\frac{3}{8} \div \left(-\frac{3}{10}\right)$

210. $-\frac{3}{12} \div \left(-\frac{5}{9}\right)$

211. $-\frac{3}{8} + \frac{5}{12}$

212. $-\dfrac{1}{8}+\dfrac{7}{12}$

213. $-\dfrac{7}{15}-\dfrac{y}{4}$

214. $-\dfrac{3}{8}-\dfrac{x}{11}$

215. $\dfrac{11}{12a}\cdot\dfrac{9a}{16}$

216. $\dfrac{10y}{13}\cdot\dfrac{8}{15y}$

217. $\dfrac{1}{2}+\dfrac{2}{3}\cdot\dfrac{5}{12}$

218. $\dfrac{1}{3}+\dfrac{2}{5}\cdot\dfrac{3}{4}$

219. $1-\dfrac{3}{5}\div\dfrac{1}{10}$

220. $1-\dfrac{5}{6}\div\dfrac{1}{12}$

221. $\dfrac{3}{8}-\dfrac{1}{6}+\dfrac{3}{4}$

222. $\dfrac{2}{5}+\dfrac{5}{8}-\dfrac{3}{4}$

223. $12\left(\dfrac{9}{20}-\dfrac{4}{15}\right)$

224. $8\left(\dfrac{15}{16}-\dfrac{5}{6}\right)$

225. $\dfrac{\frac{5}{8}+\frac{1}{6}}{\frac{19}{24}}$

226. $\dfrac{\frac{1}{6}+\frac{3}{10}}{\frac{14}{30}}$

227. $\left(\dfrac{5}{9}+\dfrac{1}{6}\right)\div\left(\dfrac{2}{3}-\dfrac{1}{2}\right)$

228. $\left(\dfrac{3}{4}+\dfrac{1}{6}\right)\div\left(\dfrac{5}{8}-\dfrac{1}{3}\right)$

Evaluate Variable Expressions with Fractions

In the following exercises, evaluate.

229. $\dfrac{7}{10}-w$ when

ⓐ $w=\dfrac{1}{2}$ ⓑ $w=-\dfrac{1}{2}$

230. $\dfrac{5}{12}-w$ when

ⓐ $w=\dfrac{1}{4}$ ⓑ $w=-\dfrac{1}{4}$

231. $2x^2y^3$ when

$x=-\dfrac{2}{3}$ and $y=-\dfrac{1}{2}$

232. $8u^2v^3$ when

$u=-\dfrac{3}{4}$ and $v=-\dfrac{1}{2}$

233. $\dfrac{a+b}{a-b}$ when

$a=-3,\ b=8$

234. $\dfrac{r-s}{r+s}$ when

$r=10,\ s=-5$

Writing Exercises

235. Why do you need a common denominator to add or subtract fractions? Explain.

236. How do you find the LCD of 2 fractions?

237. Explain how you find the reciprocal of a fraction.

238. Explain how you find the reciprocal of a negative number.

Self Check

ⓐ *After completing the exercises, use this checklist to evaluate your mastery of the objectives of this section.*

I can...	Confidently	With some help	No-I don't get it!
simplify fractions.			
multiply and divide fractions.			
add and subtract fractions.			
use the order of operations to simplify fractions.			
evaluate variable expressions with fractions.			

ⓑ *What does this checklist tell you about your mastery of this section? What steps will you take to improve?*

1.4 Decimals

Learning Objectives

By the end of this section, you will be able to:

› Round decimals
› Add and subtract decimals
› Multiply and divide decimals
› Convert decimals, fractions, and percents
› Simplify expressions with square roots
› Identify integers, rational numbers, irrational numbers, and real numbers
› Locate fractions and decimals on the number line

Be Prepared!

A more thorough introduction to the topics covered in this section can be found in the *Elementary Algebra* chapter, Foundations.

Round Decimals

Decimals are another way of writing fractions whose denominators are powers of ten.

$$0.1 \ = \ \frac{1}{10} \qquad \text{is "one tenth"}$$

$$0.01 \ = \ \frac{1}{100} \qquad \text{is "one hundredth"}$$

$$0.001 \ = \ \frac{1}{1000} \qquad \text{is "one thousandth"}$$

$$0.0001 \ = \ \frac{1}{10,000} \qquad \text{is "one ten-thousandth"}$$

Just as in whole numbers, each digit of a decimal corresponds to the place value based on the powers of ten. **Figure 1.6** shows the names of the place values to the left and right of the decimal point.

Place Value										
Hundred thousands	Ten thousands	Thousands	Hundreds	Tens	Ones	Tenths	Hundredths	Thousandths	Ten-thousandths	Hundred-thousandths

Figure 1.6

When we work with decimals, it is often necessary to round the number to the nearest required place value. We summarize the steps for rounding a decimal here.

HOW TO : : ROUND DECIMALS.

Step 1.   Locate the given place value and mark it with an arrow.

Step 2.   Underline the digit to the right of the place value.

Step 3.   Is the underlined digit greater than or equal to 5?

       ∘   Yes: add 1 to the digit in the given place value.

       ∘   No: do <u>not</u> change the digit in the given place value

Step 4.   Rewrite the number, deleting all digits to the right of the rounding digit.

EXAMPLE 1.34

Round 18.379 to the nearest ⓐ hundredth ⓑ tenth ⓒ whole number.

✓ **Solution**

Round 18.379.

ⓐ to the nearest hundredth

Locate the hundredths place with an arrow.	hundredths place ↓ 18.379
Underline the digit to the right of the given place value.	hundredths place ↓ 18.37<u>9</u>
Because 9 is greater than or equal to 5, add 1 to the 7.	18.379 ↗ delete add 1
Rewrite the number, deleting all digits to the right of the rounding digit.	18.38
Notice that the deleted digits were NOT replaced with zeros.	So, 18.379 rounded to the nearest hundredth is 18.38.

ⓑ to the nearest tenth

Locate the tenths place with an arrow.	tenths place ↓ 18.379
Underline the digit to the right of the given place value.	tenths place ↓ 18.3<u>7</u>9
Because 7 is greater than or equal to 5, add 1 to the 3.	18.379 ↗ delete add 1
Rewrite the number, deleting all digits to the right of the rounding digit.	18.4

Notice that the deleted digits were NOT replaced with zeros.

So, 18.379 rounded to the nearest tenth is 18.4.

ⓒ to the nearest whole number

Locate the ones place with an arrow.	ones place ↓ 18.379
Underline the digit to the right of the given place value.	ones place ↓ 18.379
Since 3 is not greater than or equal to 5, do not add 1 to the 8.	18.379 delete do not add 1
Rewrite the number, deleting all digits to the right of the rounding digit.	18
	So, 18.379 rounded to the nearest whole number is 18.

> **TRY IT :: 1.67**   Round 6.582 to the nearest ⓐ hundredth ⓑ tenth ⓒ whole number.

> **TRY IT :: 1.68**   Round 15.2175 to the nearest ⓐ thousandth ⓑ hundredth ⓒ tenth.

Add and Subtract Decimals

To add or subtract decimals, we line up the decimal points. By lining up the decimal points this way, we can add or subtract the corresponding place values. We then add or subtract the numbers as if they were whole numbers and then place the decimal point in the sum.

HOW TO :: ADD OR SUBTRACT DECIMALS.

Step 1.   Determine the sign of the sum or difference.

Step 2.   Write the numbers so the decimal points line up vertically.

Step 3.   Use zeros as placeholders, as needed.

Step 4.   Add or subtract the numbers as if they were whole numbers. Then place the decimal point in the answer under the decimal points in the given numbers.

Step 5.   Write the sum or difference with the appropriate sign.

EXAMPLE 1.35

Add or subtract: ⓐ $-23.5 - 41.38$ ⓑ $14.65 - 20.$

⊘ **Solution**

ⓐ

$$-23.5 - 41.38$$

The difference will be negative. To subtract, we add the numerals. Write the numbers so the decimal points line up vertically.

$$\begin{array}{r} 23.5 \\ +41.38 \\ \hline \end{array}$$

Put 0 as a placeholder after the 5 in 23.5. Remember, $\frac{5}{10} = \frac{50}{100}$ so $0.5 = 0.50$.

$$\begin{array}{r} 23.50 \\ +41.38 \\ \hline \end{array}$$

Add the numbers as if they were whole numbers. Then place the decimal point in the sum.

$$\begin{array}{r} 23.50 \\ +41.38 \\ \hline 64.88 \end{array}$$

Write the result with the correct sign.

$$-23.5 - 41.38 = -64.88$$

ⓑ

$$14.65 - 20$$

The difference will be negative. To subtract, we subtract 14.65 from 20.

Write the numbers so the decimal points line up vertically.

$$\begin{array}{r} 20 \\ -14.65 \\ \hline \end{array}$$

Remember, 20 is a whole number, so place the decimal point after the 0.

Put in zeros to the right as placeholders.

$$\begin{array}{r} 20.00 \\ -14.65 \\ \hline \end{array}$$

Subtract and place the decimal point in the answer.

$$\begin{array}{r} 9\quad\ 9 \\ 1\ \cancel{10}\ \ \cancel{10}\ 10 \\ 2\quad 0\ .\ \ 0\quad 0 \\ -1\quad 4\ .\ \ 6\quad 5 \\ \hline 5\ .\ \ 3\quad 5 \end{array}$$

Write the result with the correct sign.

$$14.65 - 20 = -5.35$$

> **TRY IT :: 1.69**   Add or subtract: ⓐ $-4.8 - 11.69$ ⓑ $9.58 - 10$.

> **TRY IT :: 1.70**   Add or subtract: ⓐ $-5.123 - 18.47$ ⓑ $37.42 - 50$.

Multiply and Divide Decimals

When we multiply signed decimals, first we determine the sign of the product and then multiply as if the numbers were both positive. We multiply the numbers temporarily ignoring the decimal point and then count the number of decimal points in the factors and that sum tells us the number of decimal places in the product. Finally, we write the product with the appropriate sign.

HOW TO : : MULTIPLY DECIMALS.

Step 1.   Determine the sign of the product.

Step 2.   Write in vertical format, lining up the numbers on the right. Multiply the numbers as if they were whole numbers, temporarily ignoring the decimal points.

Step 3.   Place the decimal point. The number of decimal places in the product is the sum of the number of decimal places in the factors.

Step 4.   Write the product with the appropriate sign.

EXAMPLE 1.36

Multiply: $(-3.9)(4.075)$.

 Solution

$$(-3.9)(4.075)$$

The signs are different. The product will be negative.	The product will be negative.
Write in vertical format, lining up the numbers on the right.	$\begin{array}{r} 4.075 \\ \times\ 3.9 \\ \hline \end{array}$
Multiply.	$\begin{array}{r} 4.075 \\ \times\ 3.9 \\ \hline 36675 \\ 12225\ \\ \hline 158925 \end{array}$
Add the number of decimal places in the factors $(1 + 3)$. Place the decimal point 4 places from the right. $(-3.9) \quad (4.075)$ 1 place 3 places	$\begin{array}{r} 4.075 \\ \times\ 3.9 \\ \hline 36675 \\ 12225\ \\ \hline 15.8925 \end{array}$ 4 places
The signs are the different, so the product is negative.	$(-3.9)(4.075) = -15.8925$

 TRY IT : : 1.71   Multiply: $-4.5(6.107)$.

 TRY IT : : 1.72   Multiply: $-10.79(8.12)$.

Often, especially in the sciences, you will multiply decimals by powers of 10 (10, 100, 1000, etc). If you multiply a few products on paper, you may notice a pattern relating the number of zeros in the power of 10 to number of decimal places we move the decimal point to the right to get the product.

HOW TO :: MULTIPLY A DECIMAL BY A POWER OF TEN.

Step 1.   Move the decimal point to the right the same number of places as the
number of zeros in the power of 10.

Step 2.   Add zeros at the end of the number as needed.

EXAMPLE 1.37

Multiply: 5.63 by ⓐ 10 ⓑ 100 ⓒ 1000.

✓ **Solution**

By looking at the number of zeros in the multiple of ten, we see the number of places we need to move the decimal to the right.

ⓐ

	5.63 (10)
There is 1 zero in 10, so move the decimal point 1 place to the right.	5.63
	5.63

ⓑ

	5.63(100)
There are 2 zeroes in 100, so move the decimal point 2 places to the right.	5.63
	563

ⓒ

	5.63(1,000)
There are 3 zeroes in 1,000, so move the decimal point 3 place to the right.	5.63
A zero must be added to the end.	5,630

> **TRY IT :: 1.73**   Multiply 2.58 by ⓐ 10 ⓑ 100 ⓒ 1000.

> **TRY IT :: 1.74**   Multiply 14.2 by ⓐ 10 ⓑ 100 ⓒ 1000.

Just as with multiplication, division of signed decimals is very much like dividing whole numbers. We just have to figure out where the decimal point must be placed and the sign of the quotient. When dividing signed decimals, first determine the sign of the quotient and then divide as if the numbers were both positive. Finally, write the quotient with the appropriate sign.

We review the notation and vocabulary for division:

$$\underset{\text{dividend}}{a} \div \underset{\text{divisor}}{b} = \underset{\text{quotient}}{c} \qquad \underset{\text{divisor}}{b} \overline{\smash{)}\underset{\text{dividend}}{a}}^{\,\overset{\text{quotient}}{c}}$$

We'll write the steps to take when dividing decimals for easy reference.

HOW TO :: DIVIDE DECIMALS.

Step 1. Determine the sign of the quotient.

Step 2. Make the divisor a whole number by "moving" the decimal point all the way to the right. "Move" the decimal point in the dividend the same number of places—adding zeros as needed.

Step 3. Divide. Place the decimal point in the quotient above the decimal point in the dividend.

Step 4. Write the quotient with the appropriate sign.

EXAMPLE 1.38

Divide: $-25.65 \div (-0.06)$.

⊘ **Solution**

Remember, you can "move" the decimals in the divisor and dividend because of the Equivalent Fractions Property.

	$-25.65 \div (-0.06)$
The signs are the same.	The quotient is positive.
Make the divisor a whole number by "moving" the decimal point all the way to the right.	
"Move" the decimal point in the dividend the same number of places.	$0.06\overline{)25.65}$
Divide. Place the decimal point in the quotient above the decimal point in the dividend.	$\begin{array}{r} 427.5 \\ 006.\overline{)2565.0} \\ -24 \\ \hline 16 \\ -12 \\ \hline 45 \\ -42 \\ \hline 30 \\ 30 \\ \hline \end{array}$
Write the quotient with the appropriate sign.	$-25.65 \div (-0.06) = 427.5$

> **TRY IT :: 1.75**   Divide: $-23.492 \div (-0.04)$.

> **TRY IT :: 1.76**   Divide: $-4.11 \div (-0.12)$.

Convert Decimals, Fractions, and Percents

In our work, it is often necessary to change the form of a number. We may have to change fractions to decimals or decimals to percent.

We convert decimals into fractions by identifying the place value of the last (farthest right) digit. In the decimal 0.03. the 3 is in the hundredths place, so 100 is the denominator of the fraction equivalent to 0.03.

$$0.03 = \frac{3}{100}$$

The steps to take to convert a decimal to a fraction are summarized in the procedure box.

HOW TO :: CONVERT A DECIMAL TO A PROPER FRACTION AND A FRACTION TO A DECIMAL.

Step 1.   To convert a decimal to a proper fraction, determine the place value of the final digit.

Step 2.   Write the fraction.
- numerator—the "numbers" to the right of the decimal point
- denominator—the place value corresponding to the final digit

Step 3.   To convert a fraction to a decimal, divide the numerator of the fraction by the denominator of the fraction.

EXAMPLE 1.39

Write: ⓐ 0.374 as a fraction ⓑ $-\frac{5}{8}$ as a decimal.

✓ **Solution**

ⓐ

0.374

Determine the place value of the final digit.	0.3	7	4
	tenths	hundredths	thousandths

Write the fraction for 0.374: The numerator is 374. The denominator is 1,000.	$\frac{374}{1000}$
Simplify the fraction.	$\frac{2 \cdot 187}{2 \cdot 500}$
Divide out the common factors.	$\frac{187}{500}$

so, $0.374 = \frac{187}{500}$

ⓑ Since a fraction bar means division, we begin by writing the fraction $\frac{5}{8}$ as $8\overline{)5}$. Now divide.

$$
\begin{array}{r}
0.625 \\
8\overline{)5.000} \\
\underline{48} \\
20 \\
\underline{16} \\
40 \\
\underline{40} \\
\end{array}
$$

so, $-\frac{5}{8} = -0.625$

> **TRY IT :: 1.77**   Write: ⓐ 0.234 as a fraction ⓑ $-\frac{7}{8}$ as a decimal.

> **TRY IT :: 1.78**   Write: ⓐ 0.024 as a fraction ⓑ $-\frac{3}{8}$ as a decimal.

A **percent** is a ratio whose denominator is 100. Percent means per hundred. We use the percent symbol, %, to show percent. Since a percent is a ratio, it can easily be expressed as a fraction. Percent means per 100, so the denominator of the fraction is 100. We then change the fraction to a decimal by dividing the numerator by the denominator. After doing this many times, you may see the pattern.

To convert a percent number to a decimal number, we move the decimal point two places to the left.

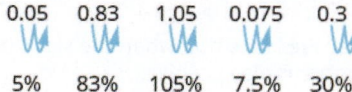

To convert a decimal to a percent, remember that percent means per hundred. If we change the decimal to a fraction whose denominator is 100, it is easy to change that fraction to a percent. After many conversions, you may recognize the pattern.

To convert a decimal to a percent, we move the decimal point two places to the right and then add the percent sign.

0.05	0.83	1.05	0.075	0.3
5%	83%	105%	7.5%	30%

HOW TO : : CONVERT A PERCENT TO A DECIMAL AND A DECIMAL TO A PERCENT.

Step 1.   To convert a percent to a decimal, move the decimal point two places to the left after removing the percent sign.

Step 2.   To convert a decimal to a percent, move the decimal point two places to the right and then add the percent sign.

EXAMPLE 1.40

Convert each:

ⓐ percent to a decimal: 62%, 135%, and 13.7%.

ⓑ decimal to a percent: 0.51, 1.25, and 0.093.

✓ **Solution**

ⓐ

	62%	135%	35.7%
Move the decimal point two places to the left.	0.62	1.35	0.357

ⓑ

	0.51	1.25	0.093
Move the decimal point two places to the right.	51%	125%	9.3%

 TRY IT : : 1.79   Convert each:

ⓐ percent to a decimal: 9%, 87%, and 3.9%.

ⓑ decimal to a percent: 0.17, 1.75, and 0.0825.

TRY IT : : 1.80   Convert each:

ⓐ percent to a decimal: 3%, 91%, and 8.3%.

ⓑ decimal to a percent: 0.41, 2.25, and 0.0925.

Simplify Expressions with Square Roots

Remember that when a number n is multiplied by itself, we write n^2 and read it "n squared." The result is called the **square of a number** n. For example, 8^2 is read "8 squared" and 64 is called the *square* of 8. Similarly, 121 is the square of 11 because 11^2 is 121. It will be helpful to learn to recognize the perfect square numbers.

> **Square of a number**
>
> If $n^2 = m$, then m is the **square** of n.

What about the squares of negative numbers? We know that when the signs of two numbers are the same, their product is positive. So the square of any negative number is also positive.

$$(-3)^2 = 9 \qquad (-8)^2 = 64 \qquad (-11)^2 = 121 \qquad (-15)^2 = 225$$

Because $10^2 = 100$, we say 100 is the square of 10. We also say that 10 is a *square root* of 100. A number whose square is m is called a **square root of a number** m.

> **Square Root of a Number**
>
> If $n^2 = m$, then n is a **square root** of m.

Notice $(-10)^2 = 100$ also, so -10 is also a square root of 100. Therefore, both 10 and -10 are square roots of 100. So, every positive number has two square roots—one positive and one negative. The radical sign, $\sqrt{m}$, denotes the positive square root. The positive square root is called the **principal square root**. When we use the radical sign that always means we want the principal square root.

> **Square Root Notation**
>
> $\sqrt{m}$ is read "the square root of m."
>
> radical sign $\longrightarrow \sqrt{m} \longleftarrow$ radicand
>
> If $m = n^2$, then $\sqrt{m} = n$, for $n \geq 0$.

The square root of m, $\sqrt{m}$, is the positive number whose square is m.

We know that every positive number has two square roots and the radical sign indicates the positive one. We write $\sqrt{100} = 10$. If we want to find the negative square root of a number, we place a negative in front of the radical sign. For example, $-\sqrt{100} = -10$. We read $-\sqrt{100}$ as "the opposite of the principal square root of 10."

EXAMPLE 1.41

Simplify: ⓐ $\sqrt{25}$ ⓑ $\sqrt{121}$ ⓒ $-\sqrt{144}$.

✓ **Solution**

ⓐ

$$\sqrt{25}$$

Since $5^2 = 25$ 5

ⓑ

$$\sqrt{121}$$

Since $11^2 = 121$ 11

ⓒ

The negative is in front of
the radical sign.

$$-\sqrt{144}$$
$$-12$$

> **TRY IT :: 1.81**   Simplify: ⓐ $\sqrt{36}$ ⓑ $\sqrt{169}$ ⓒ $-\sqrt{225}$.

> **TRY IT :: 1.82**   Simplify: ⓐ $\sqrt{16}$ ⓑ $\sqrt{196}$ ⓒ $-\sqrt{100}$.

Identify Integers, Rational Numbers, Irrational Numbers, and Real Numbers

We have already described numbers as *counting numbers*, *whole numbers*, and *integers*. What is the difference between these types of numbers? Difference could be confused with subtraction. How about asking how we distinguish between these types of numbers?

Counting numbers	1, 2, 3, 4, …..
Whole numbers	0, 1, 2, 3, 4, ….
Integers	…. − 3, −2, −1, 0, 1, 2, 3, ….

What type of numbers would we get if we started with all the integers and then included all the fractions? The numbers we would have form the set of rational numbers. A **rational number** is a number that can be written as a ratio of two integers.

In general, any decimal that ends after a number of digits (such as 7.3 or -1.2684) is a rational number. We can use the place value of the last digit as the denominator when writing the decimal as a fraction. The decimal for $\frac{1}{3}$ is the number $0.\overline{3}$. The bar over the 3 indicates that the number 3 repeats infinitely. Continuously has an important meaning in calculus. The number(s) under the bar is called the repeating block and it repeats continuously.

Since all integers can be written as a fraction whose denominator is 1, the integers (and so also the counting and whole numbers. are rational numbers.

Every rational number can be written both as a ratio of integers $\frac{p}{q}$, where p and q are integers and $q \neq 0$, and as a decimal that stops or repeats.

Rational Number

A **rational number** is a number of the form $\frac{p}{q}$, where p and q are integers and $q \neq 0$.

Its decimal form stops or repeats.

Are there any decimals that do not stop or repeat? Yes! The number π (the Greek letter *pi*, pronounced "pie"), which is very important in describing circles, has a decimal form that does not stop or repeat. We use three dots (…) to indicate the decimal does not stop or repeat.

$$\pi = 3.141592654...$$

The square root of a number that is not a perfect square is a decimal that does not stop or repeat.

A numbers whose decimal form does not stop or repeat cannot be written as a fraction of integers. We call this an **irrational number**.

Irrational Number

An **irrational number** is a number that cannot be written as the ratio of two integers.

Its decimal form does not stop and does not repeat.

Let's summarize a method we can use to determine whether a number is rational or irrational.

Rational or Irrational

If the decimal form of a number

- *repeats or stops*, the number is a **rational number**.
- *does not repeat and does not stop*, the number is an **irrational number**.

We have seen that all counting numbers are whole numbers, all whole numbers are integers, and all integers are rational numbers. The irrational numbers are numbers whose decimal form does not stop and does not repeat. When we put together the rational numbers and the irrational numbers, we get the set of **real numbers**.

Real Number

A **real number** is a number that is either rational or irrational.

Later in this course we will introduce numbers beyond the real numbers. **Figure 1.7** illustrates how the number sets we've used so far fit together.

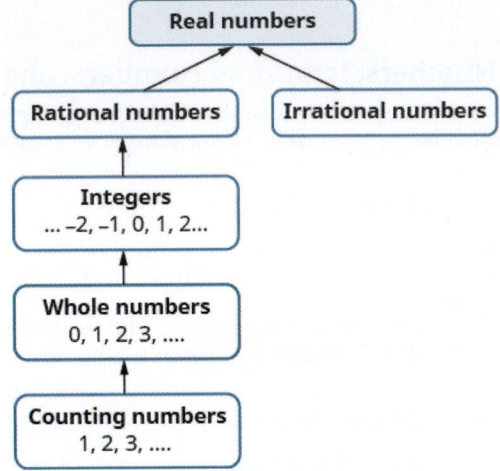

Figure 1.7 This chart shows the number sets that make up the set of real numbers.

Does the term "real numbers" seem strange to you? Are there any numbers that are not "real," and, if so, what could they be? Can we simplify $\sqrt{-25}$? Is there a number whose square is -25?

$$(\quad)^2 = -25?$$

None of the numbers that we have dealt with so far has a square that is -25. Why? Any positive number squared is positive. Any negative number squared is positive. So we say there is no real number equal to $\sqrt{-25}$. The square root of a negative number is not a real number.

EXAMPLE 1.42

Given the numbers $-7, \frac{14}{5}, 8, \sqrt{5}, 5.9, -\sqrt{64},$ list the ⓐ whole numbers ⓑ integers ⓒ rational numbers ⓓ irrational numbers ⓔ real numbers.

⊘ Solution

ⓐ Remember, the whole numbers are $0, 1, 2, 3, \ldots,$ so 8 is the only whole number given.

ⓑ The integers are the whole numbers and their opposites (which includes 0). So the whole number 8 is an integer, and -7 is the opposite of a whole number so it is an integer, too. Also, notice that 64 is the square of 8 so $-\sqrt{64} = -8$. So the integers are $-7, 8,$ and $-\sqrt{64}$.

ⓒ Since all integers are rational, then $-7, 8,$ and $-\sqrt{64}$ are rational. Rational numbers also include fractions and decimals that repeat or stop, so $\frac{14}{5}$ and 5.9 are rational. So the list of rational numbers is $-7, \frac{14}{5}, 8, 5.9,$ and $-\sqrt{64}$.

ⓓ Remember that 5 is not a perfect square, so $\sqrt{5}$ is irrational.

ⓔ All the numbers listed are real numbers.

> **TRY IT : : 1.83**

Given the numbers $-3, -\sqrt{2}, 0.\overline{3}, \frac{9}{5}, 4, \sqrt{49}$, list the ⓐ whole numbers ⓑ integers ⓒ rational numbers ⓓ irrational numbers ⓔ real numbers.

> **TRY IT : : 1.84**

Given numbers $-\sqrt{25}, -\frac{3}{8}, -1, 6, \sqrt{121}, 2.041975...$, list the ⓐ whole numbers ⓑ integers ⓒ rational numbers ⓓ irrational numbers ⓔ real numbers.

Locate Fractions and Decimals on the Number Line

We now want to include fractions and decimals on the number line. Let's start with fractions and locate $\frac{1}{5}, -\frac{4}{5}, 3, \frac{7}{4}, -\frac{9}{2}, -5$ and $\frac{8}{3}$ on the number line.

We'll start with the whole numbers 3 and -5 because they are the easiest to plot. See **Figure 1.8**.

The proper fractions listed are $\frac{1}{5}$ and $-\frac{4}{5}$. We know the proper fraction $\frac{1}{5}$ has value less than one and so would be located between 0 and 1. The denominator is 5, so we divide the unit from 0 to 1 into 5 equal parts $\frac{1}{5}, \frac{2}{5}, \frac{3}{5}, \frac{4}{5}$. We plot $\frac{1}{5}$.

Similarly, $-\frac{4}{5}$ is between 0 and -1. After dividing the unit into 5 equal parts we plot $-\frac{4}{5}$.

Finally, look at the improper fractions $\frac{7}{4}, \frac{9}{2}, \frac{8}{3}$. Locating these points may be easier if you change each of them to a mixed number.

$$\frac{7}{4} = 1\frac{3}{4} \qquad -\frac{9}{2} = -4\frac{1}{2} \qquad \frac{8}{3} = 2\frac{2}{3}$$

Figure 1.8 shows the number line with all the points plotted.

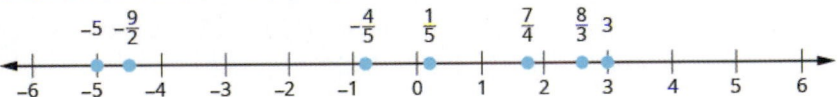

Figure 1.8

Locate and label the following on a number line: $4, \frac{3}{4}, -\frac{1}{4}, -3, \frac{6}{5}, -\frac{5}{2}$, and $\frac{7}{3}$.

✓ **Solution**

Locate and plot the integers, $4, -3$.

Locate the proper fraction $\frac{3}{4}$ first. The fraction $\frac{3}{4}$ is between 0 and 1. Divide the distance between 0 and 1 into four equal parts, then we plot $\frac{3}{4}$. Similarly plot $-\frac{1}{4}$.

Now locate the improper fractions $\frac{6}{5}, -\frac{5}{2}$, and $\frac{7}{3}$. It is easier to plot them if we convert them to mixed numbers and then plot them as described above: $\frac{6}{5} = 1\frac{1}{5}, -\frac{5}{2} = -2\frac{1}{2}, \frac{7}{3} = 2\frac{1}{3}$.

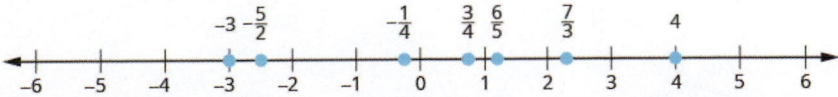

> **TRY IT : : 1.85**   Locate and label the following on a number line: $-1, \frac{1}{3}, \frac{6}{5}, -\frac{7}{4}, \frac{9}{2}, 5, -\frac{8}{3}$.

> **TRY IT : : 1.86**   Locate and label the following on a number line: $-2, \frac{2}{3}, \frac{7}{5}, -\frac{7}{4}, \frac{7}{2}, 3, -\frac{7}{3}$.

Since decimals are forms of fractions, locating decimals on the number line is similar to locating fractions on the number line.

EXAMPLE 1.44

Locate on the number line: ⓐ 0.4 ⓑ -0.74.

✓ Solution

ⓐ The decimal number 0.4 is equivalent to $\frac{4}{10}$, a proper fraction, so 0.4 is located between 0 and 1. On a number line, divide the interval between 0 and 1 into 10 equal parts. Now label the parts 0.1, 0.2, 0.3, 0.4, 0.5, 0.6, 0.7, 0.8, 0.9, 1.0. We write 0 as 0.0 and 1 as 1.0, so that the numbers are consistently in tenths. Finally, mark 0.4 on the number line.

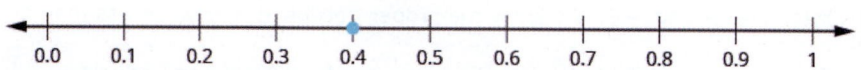

ⓑ The decimal -0.74 is equivalent to $-\frac{74}{100}$, so it is located between 0 and -1. On a number line, mark off and label the hundredths in the interval between 0 and -1.

> **TRY IT : : 1.87**   Locate on the number line: ⓐ 0.6 ⓑ -0.25.

> **TRY IT : : 1.88**   Locate on the number line: ⓐ 0.9 ⓑ -0.75.

▶ **MEDIA : :**

Access this online resource for additional instruction and practice with decimals.

- **Arithmetic Basics: Dividing Decimals (https://openstax.org/l/25dividedecim)**

1.4 EXERCISES

Practice Makes Perfect

Round Decimals

In the following exercises, round each number to the nearest ⓐ hundredth ⓑ tenth ⓒ whole number.

239. 5.781

240. 1.638

241. 0.299

242. 0.697

243. 63.479

244. 84.281

Add and Subtract Decimals

In the following exercises, add or subtract.

245. $-16.53 - 24.38$

246. $-19.47 - 32.58$

247. $-38.69 + 31.47$

248. $-29.83 + 19.76$

249. $72.5 - 100$

250. $86.2 - 100$

251. $91.75 - (-10.462)$

252. $94.69 - (-12.678)$

253. $55.01 - 3.7$

254. $59.08 - 4.6$

255. $2.51 - 7.4$

256. $3.84 - 6.1$

Multiply and Divide Decimals

In the following exercises, multiply.

257. $94.69 - (-12.678)$

258. $(-8.5)(1.69)$

259. $(-5.18)(-65.23)$

260. $(-9.16)(-68.34)$

261. $(0.06)(21.75)$

262. $(0.08)(52.45)$

263. $(9.24)(10)$

264. $(6.531)(10)$

265. $(0.025)(100)$

266. $(0.037)(100)$

267. $(55.2)(1000)$

268. $(99.4)(1000)$

In the following exercises, divide. Round money monetary answers to the nearest cent.

269. $117.25 \div 48$

270. $109.24 \div 36$

271. $1.44 \div (-0.3)$

272. $-1.15 \div (-0.05)$

273. $5.2 \div 2.5$

274. $14 \div 0.35$

Convert Decimals, Fractions and Percents

In the following exercises, write each decimal as a fraction.

275. 0.04

276. 1.464

277. 0.095

278. -0.375

In the following exercises, convert each fraction to a decimal.

279. $\dfrac{17}{20}$

280. $\dfrac{17}{4}$

281. $-\dfrac{310}{25}$

282. $-\dfrac{18}{11}$

In the following exercises, convert each percent to a decimal.

283. 71%

284. 150%

285. 39.3%

286. 7.8%

In the following exercises, convert each decimal to a percent.

287. 1.56

288. 3

289. 0.0625

290. 2.254

Simplify Expressions with Square Roots

In the following exercises, simplify.

291. $\sqrt{64}$

292. $\sqrt{169}$

293. $\sqrt{144}$

294. $-\sqrt{4}$

295. $-\sqrt{100}$

296. $-\sqrt{121}$

Identify Integers, Rational Numbers, Irrational Numbers, and Real Numbers

In the following exercises, list the ⓐ whole numbers, ⓑ integers, ⓒ rational numbers, ⓓ irrational numbers, ⓔ real numbers for each set of numbers.

297.
$-8,\ 0,\ 1.95286...,\ \frac{12}{5},\ \sqrt{36},\ 9$

298.
$-9,\ -3\frac{4}{9},\ -\sqrt{9},\ 0.4\overline{09},\ \frac{11}{6},\ 7$

299.
$-\sqrt{100},\ -7,\ -\frac{8}{3},\ -1,\ 0.77,\ 3\frac{1}{4}$

300.
$-6,\ -\frac{5}{2},\ 0,\ 0.\overline{714285},\ 2\frac{1}{5},\ \sqrt{14}$

Locate Fractions and Decimals on the Number Line

In the following exercises, locate the numbers on a number line.

301. $\frac{3}{10},\ \frac{7}{2},\ \frac{11}{6},\ 4$

302. $\frac{7}{10},\ \frac{5}{2},\ \frac{13}{8},\ 3$

303. $\frac{3}{4},\ -\frac{3}{4},\ 1\frac{2}{3},\ -1\frac{2}{3},\ \frac{5}{2},\ -\frac{5}{2}$

304. $\frac{2}{5},\ -\frac{2}{5},\ 1\frac{3}{4},\ -1\frac{3}{4},\ \frac{8}{3},\ -\frac{8}{3}$

305. ⓐ 0.8 ⓑ −1.25

306. ⓐ −0.9 ⓑ −2.75

307. ⓐ −1.6 ⓑ 3.25

308. ⓐ 3.1 ⓑ −3.65

Writing Exercises

309. How does knowing about U.S. money help you learn about decimals?

310. When the Szetos sold their home, the selling price was 500% of what they had paid for the house 30 years ago. Explain what 500% means in this context.

311. In your own words, explain the difference between a rational number and an irrational number.

312. Explain how the sets of numbers (counting, whole, integer, rational, irrationals, reals) are related to each other.

Self Check

ⓐ *Use this checklist to evaluate your mastery of the objectives of this section.*

I can...	Confidently	With some help	No-I don't get it!
round decimals.			
add and subtract decimals.			
multiply and divide decimals.			
convert decimals, fractions and percents.			
simplify expressions with square roots.			
identify integers, rational numbers, irrational numbers and real numbers.			
locate fractions and decimals on the number line.			

ⓑ *On a scale of 1-10, how would you rate your mastery of this section in light of your responses on the checklist? How can you improve this?*

1.5 | Properties of Real Numbers

Learning Objectives

By the end of this section, you will be able to:

> Use the commutative and associative properties
> Use the properties of identity, inverse, and zero
> Simplify expressions using the Distributive Property

Be Prepared!

A more thorough introduction to the topics covered in this section can be found in the *Elementary Algebra* chapter, Foundations.

Use the Commutative and Associative Properties

The order we add two numbers doesn't affect the result. If we add $8 + 9$ or $9 + 8$, the results are the same—they both equal 17. So, $8 + 9 = 9 + 8$. The order in which we add does not matter!

Similarly, when multiplying two numbers, the order does not affect the result. If we multiply $9 \cdot 8$ or $8 \cdot 9$ the results are the same—they both equal 72. So, $9 \cdot 8 = 8 \cdot 9$. The order in which we multiply does not matter!

These examples illustrate the Commutative Property.

Commutative Property

of Addition	If a and b are real numbers, then	$a + b = b + a$.
of Multiplication	If a and b are real numbers, then	$a \cdot b = b \cdot a$.

When adding or multiplying, changing the *order* gives the same result.

The Commutative Property has to do with order. We subtract $9 - 8$ and $8 - 9$, and see that $9 - 8 \neq 8 - 9$. Since changing the order of the subtraction does not give the same result, we know that *subtraction is not commutative*. *Division is not commutative either*. Since $12 \div 3 \neq 3 \div 12$, changing the order of the division did not give the same result. The commutative properties apply only to addition and multiplication!

> Addition and multiplication *are* commutative.

> Subtraction and division are *not* commutative.

When adding three numbers, changing the grouping of the numbers gives the same result. For example, $(7 + 8) + 2 = 7 + (8 + 2)$, since each side of the equation equals 17.

This is true for multiplication, too. For example, $\left(5 \cdot \frac{1}{3}\right) \cdot 3 = 5 \cdot \left(\frac{1}{3} \cdot 3\right)$, since each side of the equation equals 5.

These examples illustrate the Associative Property.

Associative Property

of Addition	If a, b, and c are real numbers, then	$(a + b) + c = a + (b + c)$.
of Multiplication	If a, b, and c are real numbers, then	$(a \cdot b) \cdot c = a \cdot (b \cdot c)$.

When adding or multiplying, changing the *grouping* gives the same result.

The Associative Property has to do with grouping. If we change how the numbers are grouped, the result will be the same. Notice it is the same three numbers in the same order—the only difference is the grouping.

We saw that subtraction and division were not commutative. They are not associative either.

$$(10 - 3) - 2 \neq 10 - (3 - 2) \qquad (24 \div 4) \div 2 \neq 24 \div (4 \div 2)$$
$$7 - 2 \neq 10 - 1 \qquad\qquad 6 \div 2 \neq 24 \div 2$$

$$5 \neq 9 \qquad\qquad\qquad 3 \neq 12$$

When simplifying an expression, it is always a good idea to plan what the steps will be. In order to combine like terms in the next example, we will use the Commutative Property of addition to write the like terms together.

EXAMPLE 1.45

Simplify: $18p + 6q + 15p + 5q$.

✓ Solution

$$18p + 6q + 15p + 5q$$

Use the Commutative Property of addition to reorder so that like terms are together.

$$18p + 15p + 6q + 5q$$

Add like terms.

$$33p + 11q$$

> **TRY IT : : 1.89** Simplify: $23r + 14s + 9r + 15s$.

> **TRY IT : : 1.90** Simplify: $37m + 21n + 4m - 15n$.

When we have to simplify algebraic expressions, we can often make the work easier by applying the Commutative Property or Associative Property first.

EXAMPLE 1.46

Simplify: $\left(\frac{5}{13} + \frac{3}{4}\right) + \frac{1}{4}$.

✓ Solution

$$\left(\frac{5}{13} + \frac{3}{4}\right) + \frac{1}{4}$$

Notice that the last 2 terms have a common denominator, so change the grouping.

$$\frac{5}{13} + \left(\frac{3}{4} + \frac{1}{4}\right)$$

Add in parentheses first.

$$\frac{5}{13} + \left(\frac{4}{4}\right)$$

Simplify the fraction.

$$\frac{5}{13} + 1$$

Add.

$$1\frac{5}{13}$$

Convert to an improper fraction.

$$\frac{18}{13}$$

> **TRY IT : : 1.91** Simplify: $\left(\frac{7}{15} + \frac{5}{8}\right) + \frac{3}{8}$.

> **TRY IT : : 1.92** Simplify: $\left(\frac{2}{9} + \frac{7}{12}\right) + \frac{5}{12}$.

Use the Properties of Identity, Inverse, and Zero

What happens when we add 0 to any number? Adding 0 doesn't change the value. For this reason, we call 0 the **additive identity**. The **Identity Property of Addition** that states that for any real number a, $a + 0 = a$ and $0 + a = a$.

What happens when we multiply any number by one? Multiplying by 1 doesn't change the value. So we call 1 the **multiplicative identity**. The **Identity Property of Multiplication** that states that for any real number a, $a \cdot 1 = a$ and $1 \cdot a = a$.

We summarize the Identity Properties here.

Identity Property

of Addition	For any real number a : $a + 0 = a$	$0 + a = a$
0 is the **additive identity**		
of Multiplication	For any real number a : $a \cdot 1 = a$	$1 \cdot a = a$
1 is the **multiplicative identity**		

What number added to 5 gives the additive identity, 0? We know

$$5 + (-5) = 0$$

The missing number was the opposite of the number!

We call $-a$ the **additive inverse** of a. *The opposite of a number is its additive inverse.* A number and its opposite add to zero, which is the additive identity. This leads to the **Inverse Property of Addition** that states for any real number a, $a + (-a) = 0$.

What number multiplied by $\frac{2}{3}$ gives the multiplicative identity, 1? In other words, $\frac{2}{3}$ times what results in 1? We know

$$\frac{2}{3} \cdot \frac{3}{2} = 1$$

The missing number was the reciprocal of the number!

We call $\frac{1}{a}$ the **multiplicative inverse** of a. *The reciprocal of a number is its multiplicative inverse.* This leads to the **Inverse Property of Multiplication** that states that for any real number a, $a \neq 0$, $a \cdot \frac{1}{a} = 1$.

We'll formally state the inverse properties here.

Inverse Property

of Addition	For any real number a, $\qquad a + (-a) = 0$
	$-a$ is the **additive inverse** of a
	A number and its *opposite* add to zero.
of Multiplication	For any real number a, $\qquad a \neq 0$, $a \cdot \frac{1}{a} = 1$.
	$\frac{1}{a}$ is the **multiplicative inverse** of a.
	A number and its *reciprocal* multiply to one.

The Identity Property of addition says that when we add 0 to any number, the result is that same number. What happens when we multiply a number by 0? Multiplying by 0 makes the product equal zero.

What about division involving zero? What is $0 \div 3$? Think about a real example: If there are no cookies in the cookie jar and 3 people are to share them, how many cookies does each person get? There are no cookies to share, so each person gets 0 cookies. So, $0 \div 3 = 0$.

We can check division with the related multiplication fact. So we know $0 \div 3 = 0$ because $0 \cdot 3 = 0$.

Now think about dividing *by* zero. What is the result of dividing 4 by 0? Think about the related multiplication fact:

$$4 \div 0 = \boxed{?} \quad \text{means} \quad \boxed{?} \cdot 0 = 4$$

Is there a number that multiplied by 0 gives 4? Since any real number multiplied by 0 gives 0, there is no real number that can be multiplied by 0 to obtain 4. We conclude that there is no answer to $4 \div 0$ and so we say that division by 0 is **undefined**.

We summarize the properties of zero here.

Properties of Zero

Multiplication by Zero: For any real number a,

$$a \cdot 0 = 0 \quad 0 \cdot a = 0 \qquad \text{The product of any number and 0 is 0.}$$

Division by Zero: For any real number a, $a \neq 0$

$$\frac{0}{a} = 0 \qquad \text{Zero divided by any real number, except itself, is zero.}$$

$$\frac{a}{0} \text{ is undefined} \qquad \text{Division by zero is undefined.}$$

We will now practice using the properties of identities, inverses, and zero to simplify expressions.

EXAMPLE 1.47

Simplify: $-84n + (-73n) + 84n$.

⊘ **Solution**

$$-84n + (-73n) + 84n$$

Notice that the first and third terms are opposites; use the Commutative Property of addition to re-order the terms.

$$-84n + 84n + (-73n)$$

Add left to right.

$$0 + (-73n)$$

Add.

$$-73n$$

> **TRY IT : : 1.93**   Simplify: $-27a + (-48a) + 27a$.

> **TRY IT : : 1.94**   Simplify: $39x + (-92x) + (-39x)$.

Now we will see how recognizing reciprocals is helpful. Before multiplying left to right, look for reciprocals—their product is 1.

EXAMPLE 1.48

Simplify: $\frac{7}{15} \cdot \frac{8}{23} \cdot \frac{15}{7}$.

⊘ **Solution**

$$\frac{7}{15} \cdot \frac{8}{23} \cdot \frac{15}{7}$$

Notice the first and third terms are reciprocals, so use the Commutative Property of multiplication to re-order the factors.

$$\frac{7}{15} \cdot \frac{15}{7} \cdot \frac{8}{23}$$

Multiply left to right.

$$1 \cdot \frac{8}{23}$$

Multiply.

$$\frac{8}{23}$$

> **TRY IT : : 1.95**   Simplify: $\frac{9}{16} \cdot \frac{5}{49} \cdot \frac{16}{9}$.

> **TRY IT : : 1.96**   Simplify: $\frac{6}{17} \cdot \frac{11}{25} \cdot \frac{17}{6}$.

The next example makes us aware of the distinction between dividing 0 by some number or some number being divided by 0.

EXAMPLE 1.49

Simplify: ⓐ $\frac{0}{n+5}$, where $n \neq -5$ ⓑ $\frac{10-3p}{0}$, where $10-3p \neq 0$.

✓ Solution

ⓐ

$$\frac{0}{n+5}$$

Zero divided by any real number except itself is 0.   0

ⓑ

$$\frac{10-3p}{0}$$

Division by 0 is undefined.   undefined

> **TRY IT :: 1.97**   Simplify: ⓐ $\frac{0}{m+7}$, where $m \neq -7$ ⓑ $\frac{18-6c}{0}$, where $18-6c \neq 0$.

> **TRY IT :: 1.98**   Simplify: ⓐ $\frac{0}{d-4}$, where $d \neq 4$ ⓑ $\frac{15-4q}{0}$, where $15-4q \neq 0$.

Simplify Expressions Using the Distributive Property

Suppose that three friends are going to the movies. They each need $9.25—that's 9 dollars and 1 quarter—to pay for their tickets. How much money do they need all together?

You can think about the dollars separately from the quarters. They need 3 times $9 so $27 and 3 times 1 quarter, so 75 cents. In total, they need $27.75. If you think about doing the math in this way, you are using the Distributive Property.

> **Distributive Property**
>
> If a, b, and c are real numbers, then $a(b+c) = ab + ac$
> $(b+c)a = ba + ca$
> $a(b-c) = ab - ac$
> $(b-c)a = ba - ca$

In algebra, we use the Distributive Property to remove parentheses as we simplify expressions.

EXAMPLE 1.50

Simplify: $3(x+4)$.

✓ Solution

$$3(x+4)$$

Distribute. $3 \cdot x + 3 \cdot 4$
Multiply. $3x + 12$

> **TRY IT :: 1.99**   Simplify: $4(x+2)$.

> **TRY IT :: 1.100**   Simplify: $6(x+7)$.

Some students find it helpful to draw in arrows to remind them how to use the Distributive Property. Then the first step in **Example 1.50** would look like this:

$3(x+4)$

EXAMPLE 1.51

Simplify: $8\left(\frac{3}{8}x + \frac{1}{4}\right)$.

✓ **Solution**

$$8\left(\frac{3}{8}x + \frac{1}{4}\right)$$

Distribute.	$8 \cdot \frac{3}{8}x + 8 \cdot \frac{1}{4}$
Multiply.	$3x + 2$

>	**TRY IT :: 1.101**	Simplify: $6\left(\frac{5}{6}y + \frac{1}{2}\right)$.

>	**TRY IT :: 1.102**	Simplify: $12\left(\frac{1}{3}n + \frac{3}{4}\right)$.

Using the Distributive Property as shown in the next example will be very useful when we solve money applications in later chapters.

EXAMPLE 1.52

Simplify: $100(0.3 + 0.25q)$.

✓ **Solution**

$$100(0.3 + 0.25q)$$

Distribute.	$100(0.3) + 100(0.25q)$
Multiply.	$30 + 25q$

>	**TRY IT :: 1.103**	Simplify: $100(0.7 + 0.15p)$.

>	**TRY IT :: 1.104**	Simplify: $100(0.04 + 0.35d)$.

When we distribute a negative number, we need to be extra careful to get the signs correct!

EXAMPLE 1.53

Simplify: $-11(4 - 3a)$.

✓ **Solution**

$$-11(4 - 3a)$$

Distribute.	$-11 \cdot 4 - (-11) \cdot 3a$
Multiply.	$-44 - (-33a)$
Simplify.	$-44 + 33a$

Notice that you could also write the result as $33a - 44$. Do you know why?

> **TRY IT : :** 1.105   Simplify: $-5(2 - 3a)$.

> **TRY IT : :** 1.106   Simplify: $-7(8 - 15y)$.

In the next example, we will show how to use the Distributive Property to find the opposite of an expression.

EXAMPLE 1.54

Simplify: $-(y + 5)$.

⊘ **Solution**

$$-(y + 5)$$

Multiplying by -1 results in the opposite.	$-1(y + 5)$
Distribute.	$-1 \cdot y + (-1) \cdot 5$
Simplify.	$-y + (-5)$
Simplify.	$-y - 5$

> **TRY IT : :** 1.107   Simplify: $-(z - 11)$.

> **TRY IT : :** 1.108   Simplify: $-(x - 4)$.

There will be times when we'll need to use the Distributive Property as part of the order of operations. Start by looking at the parentheses. If the expression inside the parentheses cannot be simplified, the next step would be multiply using the Distributive Property, which removes the parentheses. The next two examples will illustrate this.

EXAMPLE 1.55

Simplify: $8 - 2(x + 3)$

⊘ **Solution**

We follow the order of operations. Multiplication comes before subtraction, so we will distribute the 2 first and then subtract.

$$8 - 2(x + 3)$$

Distribute.	$8 - 2 \cdot x - 2 \cdot 3$
Multiply.	$8 - 2x - 6$
Combine like terms.	$-2x + 2$

> **TRY IT : :** 1.109   Simplify: $9 - 3(x + 2)$.

> **TRY IT : :** 1.110   Simplify: $7x - 5(x + 4)$.

EXAMPLE 1.56

Simplify: $4(x - 8) - (x + 3)$.

⊘ **Solution**

$$4(x - 8) - (x + 3)$$

Distribute.	$4x - 32 - x - 3$
Combine like terms.	$3x - 35$

> **TRY IT : : 1.111**   Simplify: $6(x-9)-(x+12)$.

> **TRY IT : : 1.112**   Simplify: $8(x-1)-(x+5)$.

All the properties of real numbers we have used in this chapter are summarized here.

Commutative Property
When adding or multiplying, changing the *order* gives the same result

    of addition If a, b are real numbers, then $\qquad\qquad\qquad\qquad\qquad\qquad a+b \;=\; b+a$

    of multiplication If a, b are real numbers, then $\qquad\qquad\qquad\qquad\qquad a\cdot b \;=\; b\cdot a$

Associative Property
When adding or multiplying, changing the *grouping* gives the same result.

    of addition If a, b, and c are real numbers, then $\qquad\qquad\qquad (a+b)+c \;=\; a+(b+c)$

    of multiplication If a, b, and c are real numbers, then $\qquad\qquad (a\cdot b)\cdot c \;=\; a\cdot(b\cdot c)$

Distributive Property

    If a, b, and c are real numbers, then $\qquad\qquad\qquad\qquad\qquad a(b+c) \;=\; ab+ac$

$$\qquad\qquad (b+c)a \;=\; ba+ca$$
$$\qquad\qquad a(b-c) \;=\; ab-ac$$
$$\qquad\qquad (b-c)a \;=\; ba-ca$$

Identity Property

    of addition For any real number a: $\qquad\qquad\qquad\qquad\qquad\qquad a+0=a$

        0 is the **additive identity** $\qquad\qquad\qquad\qquad\qquad\qquad\qquad 0+a=a$

    of multiplication For any real number a: $\qquad\qquad\qquad\qquad\qquad a\cdot 1=a$

        1 is the **multiplicative identity** $\qquad\qquad\qquad\qquad\qquad\qquad 1\cdot a=a$

Inverse Property

    of addition For any real number a, $a + (-a) = 0$

        $-a$ is the **additive inverse** of a

        A number and its *opposite* add to zero.

    of multiplication For any real number a, $a \neq 0$ $a \cdot \dfrac{1}{a} = 1$

        $\dfrac{1}{a}$ is the **multiplicative inverse** of a

        A number and its *reciprocal* multiply to one.

Properties of Zero

    For any real number a, $a \cdot 0 = 0$

                                            $0 \cdot a = 0$

    For any real number a, $a \neq 0$, $\dfrac{0}{a} = 0$

    For any real number a, $\dfrac{a}{0}$ is undefined

1.5 EXERCISES

Practice Makes Perfect

Use the Commutative and Associative Properties

In the following exercises, simplify.

313. $43m + (-12n) + (-16m) + (-9n)$

314. $-22p + 17q + (-35p) + (-27q)$

315. $\frac{3}{8}g + \frac{1}{12}h + \frac{7}{8}g + \frac{5}{12}h$

316. $\frac{5}{6}a + \frac{3}{10}b + \frac{1}{6}a + \frac{9}{10}b$

317. $6.8p + 9.14q + (-4.37p) + (-0.88q)$

318. $9.6m + 7.22n + (-2.19m) + (-0.65n)$

319. $-24 \cdot 7 \cdot \frac{3}{8}$

320. $-36 \cdot 11 \cdot \frac{4}{9}$

321. $\left(\frac{5}{6} + \frac{8}{15}\right) + \frac{7}{15}$

322. $\left(\frac{11}{12} + \frac{4}{9}\right) + \frac{5}{9}$

323. $17(0.25)(4)$

324. $36(0.2)(5)$

325. $[2.48(12)](0.5)$

326. $[9.731(4)](0.75)$

327. $12\left(\frac{5}{6}p\right)$

328. $20\left(\frac{3}{5}q\right)$

Use the Properties of Identity, Inverse and Zero

In the following exercises, simplify.

329. $19a + 44 - 19a$

330. $27c + 16 - 27c$

331. $\frac{1}{2} + \frac{7}{8} + \left(-\frac{1}{2}\right)$

332. $\frac{2}{5} + \frac{5}{12} + \left(-\frac{2}{5}\right)$

333. $10(0.1d)$

334. $100(0.01p)$

335. $\frac{3}{20} \cdot \frac{49}{11} \cdot \frac{20}{3}$

336. $\frac{13}{18} \cdot \frac{25}{7} \cdot \frac{18}{13}$

337. $\frac{0}{u - 4.99}$, where $u \neq 4.99$

338. $0 \div \left(y - \frac{1}{6}\right)$, where $x \neq \frac{1}{6}$

339. $\frac{32 - 5a}{0}$, where $32 - 5a \neq 0$

340. $\frac{28 - 9b}{0}$, where $28 - 9b \neq 0$

341. $\left(\frac{3}{4} + \frac{9}{10}m\right) \div 0$, where $\frac{3}{4} + \frac{9}{10}m \neq 0$

342. $\left(\frac{5}{16}n - \frac{3}{7}\right) \div 0$, where $\frac{5}{16}n - \frac{3}{7} \neq 0$

Simplify Expressions Using the Distributive Property

In the following exercises, simplify using the Distributive Property.

343. $8(4y + 9)$

344. $9(3w + 7)$

345. $6(c - 13)$

346. $7(y - 13)$

347. $\frac{1}{4}(3q + 12)$

348. $\frac{1}{5}(4m + 20)$

349. $9\left(\frac{5}{9}y - \frac{1}{3}\right)$

350. $10\left(\frac{3}{10}x - \frac{2}{5}\right)$

351. $12\left(\frac{1}{4} + \frac{2}{3}r\right)$

352. $12\left(\frac{1}{6} + \frac{3}{4}s\right)$

353. $15 \cdot \frac{3}{5}(4d + 10)$

354. $18 \cdot \frac{5}{6}(15h + 24)$

355. $r(s - 18)$

356. $u(v - 10)$

357. $(y + 4)p$

358. $(a + 7)x$

359. $-7(4p + 1)$

360. $-9(9a + 4)$

361. $-3(x - 6)$

362. $-4(q - 7)$

363. $-(3x - 7)$

364. $-(5p - 4)$

365. $16 - 3(y + 8)$

366. $18 - 4(x + 2)$

367. $4 - 11(3c - 2)$

368. $9 - 6(7n - 5)$

369. $22 - (a + 3)$

370. $8 - (r - 7)$

371. $(5m - 3) - (m + 7)$

372. $(4y - 1) - (y - 2)$

373. $9(8x - 3) - (-2)$

374. $4(6x - 1) - (-8)$

375. $5(2n + 9) + 12(n - 3)$

376. $9(5u + 8) + 2(u - 6)$

377. $14(c - 1) - 8(c - 6)$

378. $11(n - 7) - 5(n - 1)$

379. $6(7y + 8) - (30y - 15)$

380. $7(3n + 9) - (4n - 13)$

Writing Exercises

381. In your own words, state the Associative Property of addition.

382. What is the difference between the additive inverse and the multiplicative inverse of a number?

383. Simplify $8\left(x - \frac{1}{4}\right)$ using the Distributive Property and explain each step.

384. Explain how you can multiply $4(\$5.97)$ without paper or calculator by thinking of $\$5.97$ as $6 - 0.03$ and then using the Distributive Property.

Self Check

ⓐ *After completing the exercises, use this checklist to evaluate your mastery of the objectives of this section.*

I can...	Confidently	With some help	No-I don't get it!
use the commutative and associative properties.			
use the properties of identity, inverse and zero.			
simplify expressions using the distributive property.			

ⓑ *After reviewing this checklist, what will you do to become confident for all objectives?*

CHAPTER 1 REVIEW

KEY TERMS

absolute value The absolute value of a number is its distance from 0 on the number line.

additive identity The number 0 is the additive identity because adding 0 to any number does not change its value.

additive inverse The opposite of a number is its additive inverse.

coefficient The coefficient of a term is the constant that multiplies the variable in a term.

complex fraction A fraction in which the numerator or the denominator is a fraction is called a complex fraction.

composite number A composite number is a counting number that is not prime. It has factors other than 1 and the number itself.

constant A constant is a number whose value always stays the same.

denominator In a fraction, written $\frac{a}{b}$, where $b \neq 0$, the denominator b is the number of equal parts the whole has been divided into.

divisible by a number If a number m is a multiple of n, then m is divisible by n.

equation An equation is two expressions connected by an equal sign.

equivalent fractions Equivalent fractions are fractions that have the same value.

evaluate an expression To evaluate an expression means to find the value of the expression when the variables are replaced by a given number.

expression An expression is a number, a variable, or a combination of numbers and variables using operation symbols.

factors If $a \cdot b = m$, then a and b are factors of m.

fraction A fraction is written $\frac{a}{b}$, where $b \neq 0$, and a is the numerator and b is the denominator. A fraction represents parts of a whole.

integers The whole numbers and their opposites are called the integers.

irrational number An irrational number is a number that cannot be written as the ratio of two integers. Its decimal form does not stop and does not repeat.

least common denominator The least common denominator (LCD) of two fractions is the least common multiple (LCM) of their denominators.

least common multiple The least common multiple (LCM) of two numbers is the smallest number that is a multiple of both numbers.

like terms Terms that are either constants or have the same variables raised to the same powers are called like terms.

multiple of a number A number is a multiple of n if it is the product of a counting number and n.

multiplicative identity The number 1 is the multiplicative identity because multiplying 1 by any number does not change its value.

multiplicative inverse The reciprocal of a number is its multiplicative inverse.

negative numbers Numbers less than 0 are negative numbers.

numerator In a fraction, written $\frac{a}{b}$, where $b \neq 0$, the numerator a indicates how many parts are included.

opposite The opposite of a number is the number that is the same distance from zero on the number line but on the opposite side of zero.

order of operations The order of operations are established guidelines for simplifying an expression.

percent A percent is a ratio whose denominator is 100.

prime factorization The prime factorization of a number is the product of prime numbers that equals the number.

prime number A prime number is a counting number greater than 1 whose only factors are 1 and the number itself.

principal square root The positive square root is called the principal square root.

rational number A rational number is a number of the form $\frac{p}{q}$, where p and q are integers and $q \neq 0$. Its decimal form stops or repeats.

real number A real number is a number that is either rational or irrational.

reciprocal The reciprocal of a fraction is found by inverting the fraction, placing the numerator in the denominator and the denominator in the numerator.

simplify an expression To simplify an expression means to do all the math possible.

square of a number If $n^2 = m,$ then m is the square of n.

square root of a number If $n^2 = m,$ then n is a square root of m.

term A term is a constant, or the product of a constant and one or more variables.

variable A variable is a letter that represents a number whose value may change.

KEY CONCEPTS

1.1 Use the Language of Algebra

- **Divisibility Tests**
  A number is divisible by:

  > 2 if the last digit is 0, 2, 4, 6, or 8.
  > 3 if the sum of the digits is divisible by 3.
  > 5 if the last digit is 5 or 0.
  > 6 if it is divisible by both 2 and 3.
  > 10 if it ends with 0.

- **How to find the prime factorization of a composite number.**

  Step 1. Find two factors whose product is the given number, and use these numbers to create two branches.

  Step 2. If a factor is prime, that branch is complete. Circle the prime, like a bud on the tree.

  Step 3. If a factor is not prime, write it as the product of two factors and continue the process.

  Step 4. Write the composite number as the product of all the circled primes.

- **How To Find the least common multiple using the prime factors method.**

  Step 1. Write each number as a product of primes.

  Step 2. List the primes of each number. Match primes vertically when possible.

  Step 3. Bring down the columns.

  Step 4. Multiply the factors.

- **Equality Symbol**
  $a = b$ is read "a is equal to b."
  The symbol "=" is called the equal sign.

- **Inequality**

$a < b$ is read "a is less than b" a is to the left of b on the number line	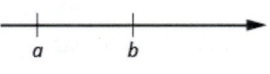
$a > b$ is read "a is greater than b" a is to the right of b on the number line	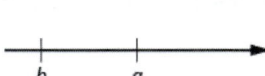

- **Inequality Symbols**

Inequality Symbols	Words
$a \neq b$	a is *not equal to b.*
$a < b$	a is *less than b.*
$a \leq b$	a is *less than or equal to b.*
$a > b$	a is *greater than b.*
$a \geq b$	a is *greater than or equal to b.*

Table 1.33

- **Grouping Symbols**

  Parentheses   ( )

  Brackets   [ ]

  Braces   { }

- **Exponential Notation**

  a^n means multiply a by itself, n times.

  The expression a^n is read a to the nth power.

- **Simplify an Expression**

  To simplify an expression, do all operations in the expression.

- **How to use the order of operations.**

  Step 1. Parentheses and Other Grouping Symbols

  - Simplify all expressions inside the parentheses or other grouping symbols, working on the innermost parentheses first.

  Step 2. Exponents

  - Simplify all expressions with exponents.

  Step 3. Multiplication and Division

  - Perform all multiplication and division in order from left to right. These operations have equal priority.

  Step 4. Addition and Subtraction

  - Perform all addition and subtraction in order from left to right. These operations have equal priority.

- **How to combine like terms.**

  Step 1. Identify like terms.

  Step 2. Rearrange the expression so like terms are together.

  Step 3. Add or subtract the coefficients and keep the same variable for each group of like terms.

Operation	Phrase	Expression
Addition	*a* plus *b* the sum of *a* and *b* *a* increased by *b* *b* more than *a* the total of *a* and *b* *b* added to *a*	$a + b$
Subtraction	*a* minus *b* the difference of *a* and *b* *a* decreased by *b* *b* less than *a* *b* subtracted from *a*	$a - b$
Multiplication	*a* times *b* the product of *a* and *b* twice *a*	$a \cdot b,\ ab,\ a(b),\ (a)(b)$ $2a$
Division	*a* divided by *b* the quotient of *a* and *b* the ratio of *a* and *b* *b* divided into *a*	$a \div b,\ a/b,\ \dfrac{a}{b},\ b\overline{)a}$

Table 1.34

1.2 Integers

- **Opposite Notation**

  $-a$ means the opposite of the number a

  The notation $-a$ is read as "the opposite of a."

- **Absolute Value**
  The absolute value of a number is its distance from 0 on the number line.
  The absolute value of a number n is written as $|n|$ and $|n| \geq 0$ for all numbers.

  Absolute values are always greater than or equal to zero.

- **Grouping Symbols**

Parentheses	( )	Braces	{ }
Brackets	[ ]	Absolute value	| |

- **Subtraction Property**
  $a - b = a + (-b)$

  Subtracting a number is the same as adding its opposite.

- **Multiplication and Division of Signed Numbers**
  For multiplication and division of two signed numbers:

Same signs	Result
• Two positives	Positive
• Two negatives	Positive

  If the signs are the same, the result is positive.

Different signs	Result
• Positive and negative	Negative
• Negative and positive	Negative

If the signs are different, the result is negative.

- **Multiplication by** -1

  $$-1a = -a$$

  Multiplying a number by -1 gives its opposite.

- **How to Use Integers in Applications.**

  Step 1. **Read** the problem. Make sure all the words and ideas are understood

  Step 2. **Identify** what we are asked to find.

  Step 3. **Write a phrase** that gives the information to find it.

  Step 4. **Translate** the phrase to an expression.

  Step 5. **Simplify** the expression.

  Step 6. **Answer** the question with a complete sentence.

1.3 Fractions

- **Equivalent Fractions Property**

  If a, b, and c are numbers where $b \neq 0$, $c \neq 0$, then

  $$\frac{a}{b} = \frac{a \cdot c}{b \cdot c} \text{ and } \frac{a \cdot c}{b \cdot c} = \frac{a}{b}.$$

- **How to simplify a fraction.**

  Step 1. Rewrite the numerator and denominator to show the common factors.
  If needed, factor the numerator and denominator into prime numbers first.

  Step 2. Simplify using the Equivalent Fractions Property by dividing out common factors.

  Step 3. Multiply any remaining factors.

- **Fraction Multiplication**

  If a, b, c, and d are numbers where $b \neq 0$, and $d \neq 0$, then

  $$\frac{a}{b} \cdot \frac{c}{d} = \frac{ac}{bd}.$$

  To multiply fractions, multiply the numerators and multiply the denominators.

- **Fraction Division**

  If a, b, c, and d are numbers where $b \neq 0$, $c \neq 0$, and $d \neq 0$, then

  $$\frac{a}{b} \div \frac{c}{d} = \frac{a}{b} \cdot \frac{d}{c}.$$

  To divide fractions, we multiply the first fraction by the reciprocal of the second.

- **Fraction Addition and Subtraction**

  If a, b, and c are numbers where $c \neq 0$, then

  $$\frac{a}{c} + \frac{b}{c} = \frac{a+b}{c} \text{ and } \frac{a}{c} - \frac{b}{c} = \frac{a-b}{c}.$$

  To add or subtract fractions, add or subtract the numerators and place the result over the common denominator.

- **How to add or subtract fractions.**

  Step 1. Do they have a common denominator?

  - Yes—go to step 2.
  - No—rewrite each fraction with the LCD (least common denominator).
    - Find the LCD.
    - Change each fraction into an equivalent fraction with the LCD as its denominator.

  Step 2. Add or subtract the fractions.

Step 3. Simplify, if possible.

- **How to simplify an expression with a fraction bar.**

  Step 1. Simplify the expression in the numerator. Simplify the expression in the denominator.

  Step 2. Simplify the fraction.

- **Placement of Negative Sign in a Fraction**

  For any positive numbers a and b,
  $$\frac{-a}{b} = \frac{a}{-b} = -\frac{a}{b}.$$

- **How to simplify complex fractions.**

  Step 1. Simplify the numerator.

  Step 2. Simplify the denominator.

  Step 3. Divide the numerator by the denominator. Simplify if possible.

1.4 Decimals

- **How to round decimals.**

  Step 1. Locate the given place value and mark it with an arrow.

  Step 2. Underline the digit to the right of the place value.

  Step 3. Is the underlined digit greater than or equal to 5?

    - Yes: add 1 to the digit in the given place value.

    - No: do <u>not</u> change the digit in the given place value

  Step 4. Rewrite the number, deleting all digits to the right of the rounding digit.

- **How to add or subtract decimals.**

  Step 1. Determine the sign of the sum or difference.

  Step 2. Write the numbers so the decimal points line up vertically.

  Step 3. Use zeros as placeholders, as needed.

  Step 4. Add or subtract the numbers as if they were whole numbers. Then place the decimal point in the answer under the decimal points in the given numbers.

  Step 5. Write the sum or difference with the appropriate sign

- **How to multiply decimals.**

  Step 1. Determine the sign of the product.

  Step 2. Write in vertical format, lining up the numbers on the right. Multiply the numbers as if they were whole numbers, temporarily ignoring the decimal points.

  Step 3. Place the decimal point. The number of decimal places in the product is the sum of the number of decimal places in the factors.

  Step 4. Write the product with the appropriate sign.

- **How to multiply a decimal by a power of ten.**

  Step 1. Move the decimal point to the right the same number of places as the number of zeros in the power of 10.

  Step 2. Add zeros at the end of the number as needed.

- **How to divide decimals.**

  Step 1. Determine the sign of the quotient.

  Step 2. Make the divisor a whole number by "moving" the decimal point all the way to the right. "Move" the decimal point in the dividend the same number of places—adding zeros as needed.

  Step 3. Divide. Place the decimal point in the quotient above the decimal point in the dividend.

  Step 4. Write the quotient with the appropriate sign.

- **How to convert a decimal to a proper fraction and a fraction to a decimal.**

  Step 1. To convert a decimal to a proper fraction, determine the place value of the final digit.

  Step 2. Write the fraction.

    - numerator—the "numbers" to the right of the decimal point

    - denominator—the place value corresponding to the final digit

Step 3. To convert a fraction to a decimal, divide the numerator of the fraction by the denominator of the fraction.

- **How to convert a percent to a decimal and a decimal to a percent.**

    Step 1. To convert a percent to a decimal, move the decimal point two places to the left after removing the percent sign.

    Step 2. To convert a decimal to a percent, move the decimal point two places to the right and then add the percent sign.

- **Square Root Notation**

    $\sqrt{m}$ is read "the square root of m."

    If $m = n^2$, then $\sqrt{m} = n$, for $n \geq 0$.

    The square root of m, $\sqrt{m}$, is the positive number whose square is m.

- **Rational or Irrational**

    If the decimal form of a number

    - *repeats or stops*, the number is a rational number.

    - *does not repeat and does not stop*, the number is an irrational number.

- **Real Numbers**

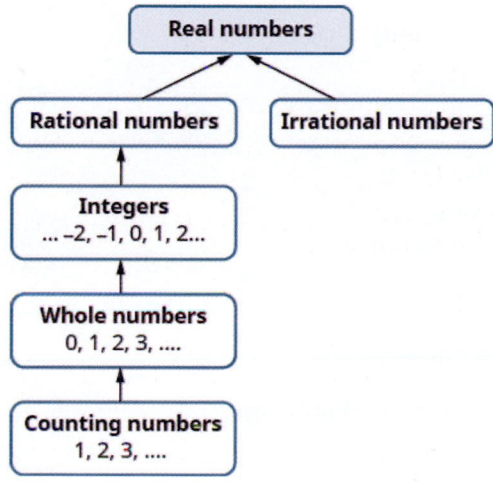

Figure 1.9

1.5 Properties of Real Numbers

Commutative Property
When adding or multiplying, changing the *order* gives the same result

of addition If a, b are real numbers, then $\qquad\qquad a + b = b + a$

of multiplication If a, b are real numbers, then $\qquad\qquad a \cdot b = b \cdot a$

Associative Property
When adding or multiplying, changing the *grouping* gives the same result.

of addition If a, b, and c are real numbers, then $\qquad\qquad (a + b) + c = a + (b + c)$

of multiplication If a, b, and c are real numbers, then $\qquad\qquad (a \cdot b) \cdot c = a \cdot (b \cdot c)$

Distributive Property

If a, b, and c are real numbers, then

$$a(b + c) = ab + ac$$
$$(b + c)a = ba + ca$$
$$a(b - c) = ab - ac$$
$$(b - c)a = ba - ca$$

Identity Property

of addition For any real number a:

$$a + 0 = a$$

0 is the **additive identity**

$$0 + a = a$$

of multiplication For any real number a:

$$a \cdot 1 = a$$

1 is the **multiplicative identity**

$$1 \cdot a = a$$

Inverse Property

of addition For any real number a,

$$a + (-a) = 0$$

$-a$ is the **additive inverse** of a
A number and its *opposite* add to zero.

of multiplication For any real number a, $a \neq 0$

$$a \cdot \frac{1}{a} = 1$$

$\frac{1}{a}$ is the **multiplicative inverse** of a

A number and its *reciprocal* multiply to one.

Properties of Zero

For any real number a,

$$a \cdot 0 = 0$$
$$0 \cdot a = 0$$

For any real number a, $a \neq 0$,

$$\frac{0}{a} = 0$$

For any real number a,

$\frac{a}{0}$ is undefined

REVIEW EXERCISES

1.1 Use the Language of Algebra

Identify Multiples and Factors

385. Use the divisibility tests to determine whether 180 is divisible by 2, by 3, by 5, by 6, and by 10.

386. Find the prime factorization of 252.

387. Find the least common multiple of 24 and 40.

In the following exercises, simplify each expression.

388. $24 \div 3 + 4(5 - 2)$

389. $7 + 3[6 - 4(5 - 4)] - 3^2$

Evaluate an Expression

In the following exercises, evaluate the following expressions.

390. When $x = 4$, ⓐ x^3 ⓑ 5^x ⓒ $2x^2 - 5x + 3$

391. $2x^2 - 4xy - 3y^2$ when $x = 3$, $y = 1$

Simplify Expressions by Combining Like Terms

In the following exercises, simplify the following expressions by combining like terms.

392. $12y + 7 + 2y - 5$

393. $14x^2 - 9x + 11 - 8x^2 + 8x - 6$

Translate an English Phrase to an Algebraic Expression

In the following exercises, translate the phrases into algebraic expressions.

394.

ⓐ the sum of $4ab^2$ and $7a^3b^2$

ⓑ the product of $6y^2$ and $3y$

ⓒ twelve more than $5x$

ⓓ $5y$ less than $8y^2$

395.

ⓐ eleven times the difference of y and two

ⓑ the difference of eleven times y and two

396. Dushko has nickels and pennies in his pocket. The number of pennies is four less than five the number of nickels. Let n represent the number of nickels. Write an expression for the number of pennies.

1.2 Integers

Simplify Expressions with Absolute Value

In the following exercise, fill in $<$, $>$, or $=$ for each of the following pairs of numbers.

397.

ⓐ $-|7|$___ $-|-7|$

ⓑ -8___ $-|-8|$

ⓒ $|-13|$___ -13

ⓓ $|-12|$___ $-(-12)$

In the following exercises, simplify.

398. $9 - |3(4 - 8)|$

399. $12 - 3|1 - 4(4 - 2)|$

Add and Subtract Integers

In the following exercises, simplify each expression.

400. $-12 + (-8) + 7$

401.

ⓐ $15 - 7$

ⓑ $-15 - (-7)$

ⓒ $-15 - 7$

ⓓ $15 - (-7)$

402. $-11 - (-12) + 5$

403. ⓐ $23 - (-17)$ ⓑ $23 + 17$

404. $-(7 - 11) - (3 - 5)$

Multiply and Divide Integers

In the following exercise, multiply or divide.

405. ⓐ $-27 \div 9$ ⓑ $120 \div (-8)$

ⓒ $4(-14)$ ⓓ $-1(-17)$

Simplify and Evaluate Expressions with Integers

In the following exercises, simplify each expression.

406. ⓐ $(-7)^3$ ⓑ -7^3

407. $(7 - 11)(6 - 13)$

408. $63 \div (-9) + (-36) \div (-4)$

409. $6 - 3|4(1 - 2) - (7 - 5)|$

410. $(-2)^4 - 24 \div (13 - 5)$

For the following exercises, evaluate each expression.

411. $(y + z)^2$ when
$y = -4, z = 7$

412. $3x^2 - 2xy + 4y^2$ when
$x = -2, y = -3$

Translate English Phrases to Algebraic Expressions

In the following exercises, translate to an algebraic expression and simplify if possible.

413. the sum of -4 and -9,
increased by 23

414. ⓐ the difference of 17 and
-8 ⓑ subtract 17 from -25

Use Integers in Applications

In the following exercise, solve.

415. Temperature On July 10, the
high temperature in Phoenix,
Arizona, was 109°, and the high
temperature in Juneau, Alaska, was
63°. What was the difference
between the temperature in Palm
Springs and the temperature in
Whitefield?

1.3 Fractions

Simplify Fractions

In the following exercises, simplify.

416. $\dfrac{204}{228}$

417. $-\dfrac{270x^3}{198y^2}$

Multiply and Divide Fractions

In the following exercises, perform the indicated operation.

418. $\left(-\dfrac{14}{15}\right)\left(\dfrac{10}{21}\right)$

419. $\dfrac{6x}{25} \div \dfrac{9y}{20}$

420. $\dfrac{-\frac{4}{9}}{\frac{8}{21}}$

Add and Subtract Fractions

In the following exercises, perform the indicated operation.

421. $\dfrac{5}{18} + \dfrac{7}{12}$

422. $\dfrac{11}{36} - \dfrac{15}{48}$

423. ⓐ $\dfrac{5}{8} + \dfrac{3}{4}$ ⓑ $\dfrac{5}{8} \div \dfrac{3}{4}$

424. ⓐ $-\dfrac{3y}{10} - \dfrac{5}{6}$ ⓑ $-\dfrac{3y}{10} \cdot \dfrac{5}{6}$

Use the Order of Operations to Simplify Fractions

In the following exercises, simplify.

425. $\dfrac{4 \cdot 3 - 2 \cdot 5}{-6 \cdot 3 + 2 \cdot 3}$

426. $\dfrac{4(7-3) - 2(4-9)}{-3(4+2) + 7(3-6)}$

427. $\dfrac{4^3 - 4^2}{\left(\frac{4}{5}\right)^2}$

Evaluate Variable Expressions with Fractions

In the following exercises, evaluate.

428. $4x^2 y^2$ when

$x = \dfrac{2}{3}$ and $y = -\dfrac{3}{4}$

429. $\dfrac{a+b}{a-b}$ when

$a = -4, \quad b = 6$

1.4 Decimals

Round Decimals

430. Round 6.738 to the nearest

ⓐ hundredth ⓑ tenth ⓒ whole number.

Add and Subtract Decimals

In the following exercises, perform the indicated operation.

431. $-23.67 + 29.84$

432. $54.3 - 100$

433. $79.38 - (-17.598)$

Multiply and Divide Decimals

In the following exercises, perform the indicated operation.

434. $(-2.8)(3.97)$

435. $(-8.43)(-57.91)$

436. $(53.48)(10)$

437. $(0.563)(100)$

438. $\$118.35 \div 2.6$

439. $1.84 \div (-0.8)$

Convert Decimals, Fractions and Percents

In the following exercises, write each decimal as a fraction.

440. $\dfrac{13}{20}$

441. $-\dfrac{240}{25}$

In the following exercises, convert each fraction to a decimal.

442. $-\dfrac{5}{8}$

443. $\dfrac{14}{11}$

In the following exercises, convert each decimal to a percent.

444. 2.43

445. 0.0475

Simplify Expressions with Square Roots

In the following exercises, simplify.

446. $\sqrt{289}$

447. $\sqrt{-121}$

Identify Integers, Rational Numbers, Irrational Numbers, and Real Numbers

In the following exercise, list the ⓐ whole numbers ⓑ integers ⓒ rational numbers ⓓ irrational numbers ⓔ real numbers for each set of numbers

448.

$$-8,\ 0,\ 1.95286...,\ \frac{12}{5},\ \sqrt{36},\ 9$$

Locate Fractions and Decimals on the Number Line

In the following exercises, locate the numbers on a number line.

449. $\frac{3}{4},\ -\frac{3}{4},\ 1\frac{1}{3},\ -1\frac{2}{3},\ \frac{7}{2},\ -\frac{5}{2}$ **450.** ⓐ 3.2 ⓑ -1.35

1.5 Properties of Real Numbers

Use the Commutative and Associative Properties

In the following exercises, simplify.

451. $\frac{5}{8}x + \frac{5}{12}y + \frac{1}{8}x + \frac{7}{12}y$ **452.** $-32 \cdot 9 \cdot \frac{5}{8}$ **453.** $\left(\frac{11}{15} + \frac{3}{8}\right) + \frac{5}{8}$

Use the Properties of Identity, Inverse and Zero

In the following exercises, simplify.

454. $\frac{4}{7} + \frac{8}{15} + \left(-\frac{4}{7}\right)$ **455.** $\frac{13}{15} \cdot \frac{9}{17} \cdot \frac{15}{13}$ **456.** $\frac{0}{x-3},\ x \neq 3$

457. $\frac{5x-7}{0},\ 5x-7 \neq 0$

Simplify Expressions Using the Distributive Property

In the following exercises, simplify using the Distributive Property.

458. $8(a-4)$ **459.** $12\left(\frac{2}{3}b + \frac{5}{6}\right)$ **460.** $18 \cdot \frac{5}{6}(2x-5)$

461. $(x-5)p$ **462.** $-4(y-3)$ **463.** $12 - 6(x+3)$

464. $6(3x-4) - (-5)$ **465.** $5(2y+3) - (4y-1)$

PRACTICE TEST

466. Find the prime factorization of 756.

467. Combine like terms: $5n + 8 + 2n - 1$

468. Evaluate when $x = -2$ and $y = 3$: $\dfrac{|3x - 4y|}{6}$

469. Translate to an algebraic expression and simplify:

ⓐ eleven less than negative eight

ⓑ the difference of -8 and -3, increased by 5

470. Dushko has nickels and pennies in his pocket. The number of pennies is seven less than four times the number of nickels. Let n represent the number of nickels. Write an expression for the number of pennies.

471. Round 28.1458 to the nearest

ⓐ hundredth ⓑ thousandth

472. Convert

ⓐ $\dfrac{5}{11}$ to a decimal ⓑ 1.15 to a percent

473. Locate $\dfrac{3}{5}$, 2.8, and $-\dfrac{5}{2}$ on a number line.

In the following exercises, simplify each expression.

474. $8 + 3[6 - 3(5 - 2)] - 4^2$

475. $-(4 - 9) - (9 - 5)$

476. $56 \div (-8) + (-27) \div (-3)$

477. $16 - 2|3(1 - 4) - (8 - 5)|$

478. $-5 + 2(-3)^2 - 9$

479. $\dfrac{180}{204}$

480. $-\dfrac{7}{18} + \dfrac{5}{12}$

481. $\dfrac{4}{5} \div \left(-\dfrac{12}{25}\right)$

482. $\dfrac{9 - 3 \cdot 9}{15 - 9}$

483. $\dfrac{4(-3 + 2(3 - 6))}{3(11 - 3(2 + 3))}$

484. $\dfrac{5}{13} \cdot 47 \cdot \dfrac{13}{5}$

485. $\dfrac{-\dfrac{5}{9}}{\dfrac{10}{21}}$

486. $-4.8 + (-6.7)$

487. $34.6 - 100$

488. $-12.04 \cdot (4.2)$

489. $-8 \div 0.05$

490. $\sqrt{-121}$

491. $\left(\dfrac{8}{13} + \dfrac{5}{7}\right) + \dfrac{2}{7}$

492. $5x + (-8y) - 6x + 3y$

493. ⓐ $\dfrac{0}{9}$ ⓑ $\dfrac{11}{0}$

494. $-3(8x - 5)$

495. $6(3y - 1) - (5y - 3)$

2 SOLVING LINEAR EQUATIONS

Figure 2.1 This drone is flying high in the sky while its pilot remains safely on the ground. (credit: "Unsplash" / Pixabay)

Chapter Outline

Introduction

Imagine being a pilot, but not just any pilot—a drone pilot. Drones, or unmanned aerial vehicles, are devices that can be flown remotely. They contain sensors that can relay information to a command center where the pilot is located. Larger drones can also carry cargo. In the near future, several companies hope to use drones to deliver materials and piloting a drone will become an important career. Law enforcement and the military are using drones rather than send personnel into dangerous situations. Building and piloting a drone requires the ability to program a set of actions, including taking off, turning, and landing. This, in turn, requires the use of linear equations. In this chapter, you will explore linear equations, develop a strategy for solving them, and relate them to real-world situations.

2.1 Use a General Strategy to Solve Linear Equations

Learning Objectives

By the end of this section, you will be able to:

> Solve linear equations using a general strategy
> Classify equations
> Solve equations with fraction or decimal coefficients

Be Prepared!

Before you get started, take this readiness quiz.

1. Simplify: $\frac{3}{2}(12x + 20)$.

   If you missed this problem, review **Example 1.51**.

2. Simplify: $5 - 2(n + 1)$.

   If you missed this problem, review **Example 1.55**.

3. Find the LCD of $\frac{5}{6}$ and $\frac{1}{4}$.

If you missed this problem, review **Example 1.28**.

Solve Linear Equations Using a General Strategy

Solving an equation is like discovering the answer to a puzzle. The purpose in solving an equation is to find the value or values of the variable that makes it a true statement. Any value of the variable that makes the equation true is called a **solution** to the equation. It is the answer to the puzzle!

Solution of an Equation

A **solution** of an equation is a value of a variable that makes a true statement when substituted into the equation.

To determine whether a number is a solution to an equation, we substitute the value for the variable in the equation. If the resulting equation is a true statement, then the number is a solution of the equation.

HOW TO :: DETERMINE WHETHER A NUMBER IS A SOLUTION TO AN EQUATION.

Step 1.   Substitute the number for the variable in the equation.

Step 2.   Simplify the expressions on both sides of the equation.

Step 3.   Determine whether the resulting equation is true.

○   If it is true, the number is a solution.

○   If it is not true, the number is not a solution.

EXAMPLE 2.1

Determine whether the values are solutions to the equation: $5y + 3 = 10y - 4$.

ⓐ $y = \frac{3}{5}$ ⓑ $y = \frac{7}{5}$

✓ **Solution**

Since a solution to an equation is a value of the variable that makes the equation true, begin by substituting the value of the solution for the variable.

ⓐ

$$5y + 3 = 10y - 4$$

Substitute $\frac{3}{5}$ for y.	$5\left(\frac{3}{5}\right) + 3 \overset{?}{=} 10\left(\frac{3}{5}\right) - 4$
Multiply.	$3 + 3 \overset{?}{=} 6 - 4$
Simplify.	$6 \neq 2$

Since $y = \frac{3}{5}$ does not result in a true equation, $y = \frac{3}{5}$ is not a solution to the equation $5y + 3 = 10y - 4$.

ⓑ

$$5y + 3 = 10y - 4$$

Substitute $\frac{7}{5}$ for y.	$5\left(\frac{7}{5}\right) + 3 \overset{?}{=} 10\left(\frac{7}{5}\right) - 4$
Multiply.	$7 + 3 \overset{?}{=} 14 - 4$
Simplify.	$10 = 10 \checkmark$

Since $y = \frac{7}{5}$ results in a true equation, $y = \frac{7}{5}$ is a solution to the equation $5y + 3 = 10y - 4$.

> **TRY IT :: 2.1**   Determine whether the values are solutions to the equation: $9y + 2 = 6y + 3$.
>
> ⓐ $y = \frac{4}{3}$ ⓑ $y = \frac{1}{3}$

> **TRY IT :: 2.2**   Determine whether the values are solutions to the equation: $4x - 2 = 2x + 1$.
>
> ⓐ $x = \frac{3}{2}$ ⓑ $x = -\frac{1}{2}$

There are many types of equations that we will learn to solve. In this section we will focus on a **linear equation**.

Linear Equation

A **linear equation** is an equation in one variable that can be written, where a and b are real numbers and $a \neq 0$, as:

$$ax + b = 0$$

To solve a linear equation it is a good idea to have an overall strategy that can be used to solve any linear equation. In the next example, we will give the steps of a general strategy for solving any linear equation. Simplifying each side of the equation as much as possible first makes the rest of the steps easier.

EXAMPLE 2.2   HOW TO SOLVE A LINEAR EQUATION USING A GENERAL STRATEGY

Solve: $7(n - 3) - 8 = -15$.

⊘ **Solution**

Step 1. Simplify each side of the equation as much as possible.	Use the Distributive Property. Notice that each side of the equation is now simplified as much as possible.	$7(n - 3) - 8 = -15$ $7n - 21 - 8 = -15$ $7n - 29 = -15$
Step 2. Collect all variable terms on one side of the equation.	Nothing to do—all n's are on the left side.	
Step 3. Collect constant terms on the other side of the equation.	To get constants only on the right, add 29 to each side. Simplify.	$7n - 29 + 29 = -15 + 29$ $7n = 14$
Step 4. Make the coefficient of the variable term equal to 1.	Divide each side by 7. Simplify.	$\frac{7n}{7} = \frac{14}{7}$ $n = 2$

Step 5. Check the solution.		Check:
	Let $n = 2$	$7(n-3) - 8 = -15$
	Subtract.	$7(2-3) - 8 \overset{?}{=} -15$
		$7(-1) - 8 \overset{?}{=} -15$
		$-7 - 8 \overset{?}{=} -15$
		$-15 = -15 \checkmark$

> **TRY IT :: 2.3**   Solve: $2(m - 4) + 3 = -1$.

> **TRY IT :: 2.4**   Solve: $5(a - 3) + 5 = -10$.

These steps are summarized in the General Strategy for Solving Linear Equations below.

HOW TO :: SOLVE LINEAR EQUATIONS USING A GENERAL STRATEGY.

Step 1.   Simplify each side of the equation as much as possible.
           Use the Distributive Property to remove any parentheses.
           Combine like terms.

Step 2.   Collect all the variable terms on one side of the equation.
           Use the Addition or Subtraction Property of Equality.

Step 3.   Collect all the constant terms on the other side of the equation.
           Use the Addition or Subtraction Property of Equality.

Step 4.   Make the coefficient of the variable term equal to 1.
           Use the Multiplication or Division Property of Equality.
           State the solution to the equation.

Step 5.   Check the solution.
           Substitute the solution into the original equation to make sure the result is a true statement.

EXAMPLE 2.3

Solve: $\frac{2}{3}(3m - 6) = 5 - m$.

⊘ **Solution**

	$\frac{2}{3}(3m - 6) = 5 - m$
Distribute.	$2m - 4 = 5 - m$
Add m to both sides to get the variables only on the left.	$2m + m - 4 = 5 - m + m$
Simplify.	$3m - 4 = 5$
Add 4 to both sides to get constants only on the right.	$3m - 4 + 4 = 5 + 4$
Simplify.	$3m = 9$
Divide both sides by three.	$\frac{3m}{3} = \frac{9}{3}$
Simplify.	$m = 3$
Check:	$\frac{2}{3}(3m - 6) = 5 - m$

Let $m = 3$.	$\frac{2}{3}(3 \cdot 3 - 6) \overset{?}{=} 5 - 3$
	$\frac{2}{3}(9 - 6) \overset{?}{=} 2$
	$\frac{2}{3}(3) \overset{?}{=} 2$
	$2 = 2 \checkmark$

> **TRY IT :: 2.5**   Solve: $\frac{1}{3}(6u + 3) = 7 - u$.

> **TRY IT :: 2.6**   Solve: $\frac{2}{3}(9x - 12) = 8 + 2x$.

We can solve equations by getting all the variable terms to either side of the equal sign. By collecting the variable terms on the side where the coefficient of the variable is larger, we avoid working with some negatives. This will be a good strategy when we solve inequalities later in this chapter. It also helps us prevent errors with negatives.

EXAMPLE 2.4

Solve: $4(x - 1) - 2 = 5(2x + 3) + 6$.

⊘ **Solution**

	$4(x - 1) - 2 = 5(2x + 3) + 6$
Distribute.	$4x - 4 - 2 = 10x + 15 + 6$
Combine like terms.	$4x - 6 = 10x + 21$
Subtract $4x$ from each side to get the variables only on the right since $10 > 4$.	$4x - 4x - 6 = 10x - 4x + 21$
Simplify.	$-6 = 6x + 21$
Subtract 21 from each side to get the constants on left.	$-6 - 21 = 6x + 21 - 21$
Simplify.	$-27 = 6x$
Divide both sides by 6.	$\frac{-27}{6} = \frac{6x}{6}$
Simplify.	$-\frac{9}{2} = x$
Check:	$4(x - 1) - 2 = 5(2x + 3) + 6$
Let $x = -\frac{9}{2}$.	$4\left(-\frac{9}{2} - 1\right) - 2 \overset{?}{=} 5\left(2\left(-\frac{9}{2}\right) + 3\right) + 6$
	$4\left(-\frac{11}{2}\right) - 2 \overset{?}{=} 5(-9 + 3) + 6$
	$-22 - 2 \overset{?}{=} 5(-6) + 6$
	$-24 \overset{?}{=} -30 + 6$
	$-24 = -24 \checkmark$

> **TRY IT :: 2.7**   Solve: $6(p - 3) - 7 = 5(4p + 3) - 12$.

> **TRY IT :: 2.8**   Solve: $8(q + 1) - 5 = 3(2q - 4) - 1$.

EXAMPLE 2.5

Solve: $10[3 - 8(2s - 5)] = 15(40 - 5s)$.

⊘ **Solution**

	$10[3 - 8(2s - 5)] = 15(40 - 5s)$
Simplify from the innermost parentheses first.	$10[3 - 16s + 40] = 15(40 - 5s)$
Combine like terms in the brackets.	$10[43 - 16s] = 15(40 - 5s)$
Distribute.	$430 - 160s = 600 - 75s$
Add $160s$ to both sides to get the variables to the right.	$430 - 160s + 160s = 600 - 75s + 160s$
Simplify.	$430 = 600 + 85s$
Subtract 600 from both sides to get the constants to the left.	$430 - 600 = 600 + 85s - 600$
Simplify.	$-170 = 85s$
Divide both sides by 85.	$\dfrac{-170}{85} = \dfrac{85s}{85}$
Simplify.	$-2 = s$
Check:	$10[3 - 8(2s - 5)] = 15(40 - 5s)$
Let $s = -2$.	$10[3 - 8(2(-2) - 5)] \overset{?}{=} 15(40 - 5(-2))$
	$10[3 - 8(-4 - 5)] \overset{?}{=} 15(40 + 10)$
	$10[3 - 8(-9)] \overset{?}{=} 15(50)$
	$10(3 + 72) \overset{?}{=} 750$
	$10(75) \overset{?}{=} 750$
	$750 = 750 \checkmark$

> **TRY IT :: 2.9**   Solve: $6[4 - 2(7y - 1)] = 8(13 - 8y)$.

> **TRY IT :: 2.10**   Solve: $12[1 - 5(4z - 1)] = 3(24 + 11z)$.

Classify Equations

Whether or not an equation is true depends on the value of the variable. The equation $7x + 8 = -13$ is true when we replace the variable, x, with the value -3, but not true when we replace x with any other value. An equation like this is called a **conditional equation**. All the equations we have solved so far are conditional equations.

Conditional Equation

An equation that is true for one or more values of the variable and false for all other values of the variable is a

conditional equation.

Now let's consider the equation $7y + 14 = 7(y + 2)$. Do you recognize that the left side and the right side are equivalent? Let's see what happens when we solve for y.

Solve:

$$7y + 14 = 7(y + 2)$$

Distribute.	$7y + 14 = 7y + 14$
Subtract $7y$ to each side to get the y's to one side.	$7y - 7y + 14 = 7y - 7y + 14$
Simplify—the y's are eliminated.	$14 = 14$
	But $14 = 14$ is true.

This means that the equation $7y + 14 = 7(y + 2)$ is true for any value of y. We say the solution to the equation is all of the real numbers. An equation that is true for any value of the variable is called an **identity**.

Identity

An equation that is true for any value of the variable is called an **identity**.
The solution of an identity is all real numbers.

What happens when we solve the equation $-8z = -8z + 9$?

Solve:

$$-8z = -8z + 9$$

Add $8z$ to both sides to leave the constant alone on the right.	$-8z + 8z = -8z + 8z + 9$
Simplify—the z's are eliminated.	$0 \neq 9$
	But $0 \neq 9$.

Solving the equation $-8z = -8z + 9$ led to the false statement $0 = 9$. The equation $-8z = -8z + 9$ will not be true for any value of z. It has no solution. An equation that has no solution, or that is false for all values of the variable, is called a **contradiction**.

Contradiction

An equation that is false for all values of the variable is called a **contradiction**.
A contradiction has no solution.

The next few examples will ask us to classify an equation as conditional, an identity, or as a contradiction.

EXAMPLE 2.6

Classify the equation as a conditional equation, an identity, or a contradiction and then state the solution: $6(2n - 1) + 3 = 2n - 8 + 5(2n + 1)$.

⊘ Solution

	$6(2n - 1) + 3 = 2n - 8 + 5(2n + 1)$
Distribute.	$12n - 6 + 3 = 2n - 8 + 10n + 5$
Combine like terms.	$12n - 3 = 12n - 3$
Subtract $12n$ from each side to get the n's to one side.	$12n - 12n - 3 = 12n - 12n - 3$
Simplify.	$-3 = -3$
This is a true statement.	The equation is an identity.
	The solution is all real numbers.

> **TRY IT :: 2.11**
>
> Classify the equation as a conditional equation, an identity, or a contradiction and then state the solution: $4 + 9(3x - 7) = -42x - 13 + 23(3x - 2)$.

> **TRY IT :: 2.12**
>
> Classify the equation as a conditional equation, an identity, or a contradiction and then state the solution: $8(1 - 3x) + 15(2x + 7) = 2(x + 50) + 4(x + 3) + 1$.

EXAMPLE 2.7

Classify the equation as a conditional equation, an identity, or a contradiction and then state the solution: $8 + 3(a - 4) = 0$.

⊘ Solution

	$8 + 3(a - 4) = 0$
Distribute.	$8 + 3a - 12 = 0$
Combine like terms.	$3a - 4 = 0$
Add 4 to both sides.	$3a - 4 + 4 = 0 + 4$
Simplify.	$3a = 4$
Divide.	$\frac{3a}{3} = \frac{4}{3}$
Simplify.	$a = \frac{4}{3}$
The equation is true when $a = \frac{4}{3}$.	This is a conditional equation.
	The solution is $a = \frac{4}{3}$.

> **TRY IT :: 2.13**
>
> Classify the equation as a conditional equation, an identity, or a contradiction and then state the solution: $11(q + 3) - 5 = 19$.

> **TRY IT : : 2.14**

Classify the equation as a conditional equation, an identity, or a contradiction and then state the solution: $6+14(k-8) = 95$.

EXAMPLE 2.8

Classify the equation as a conditional equation, an identity, or a contradiction and then state the solution: $5m + 3(9 + 3m) = 2(7m - 11)$.

⊘ **Solution**

	$5m + 3(9 + 3m) = 2(7m - 11)$
Distribute.	$5m + 27 + 9m = 14m - 22$
Combine like terms.	$14m + 27 = 14m - 22$
Subtract $14m$ from both sides.	$14m + 27 - 14m = 14m - 22 - 14m$
Simplify.	$27 \neq -22$
But $27 \neq -22$.	The equation is a contradiction.
	It has no solution.

> **TRY IT : : 2.15**

Classify the equation as a conditional equation, an identity, or a contradiction and then state the solution: $12c + 5(5 + 3c) = 3(9c - 4)$.

> **TRY IT : : 2.16**

Classify the equation as a conditional equation, an identity, or a contradiction and then state the solution: $4(7d + 18) = 13(3d - 2) - 11d$.

We summarize the methods for classifying equations in the table.

Type of equation	What happens when you solve it?	Solution
Conditional Equation	True for one or more values of the variables and false for all other values	One or more values
Identity	True for any value of the variable	All real numbers
Contradiction	False for all values of the variable	No solution

Table 2.9

Solve Equations with Fraction or Decimal Coefficients

We could use the General Strategy to solve the next example. This method would work fine, but many students do not feel very confident when they see all those fractions. So, we are going to show an alternate method to solve equations with fractions. This alternate method eliminates the fractions.

We will apply the Multiplication Property of Equality and multiply both sides of an equation by the least common denominator (LCD) of all the fractions in the equation. The result of this operation will be a new equation, equivalent to the first, but without fractions. This process is called *clearing* the equation of fractions.

To clear an equation of decimals, we think of all the decimals in their fraction form and then find the LCD of those denominators.

EXAMPLE 2.9   HOW TO SOLVE EQUATIONS WITH FRACTION OR DECIMAL COEFFICIENTS

Solve: $\frac{1}{12}x + \frac{5}{6} = \frac{3}{4}$.

✓ **Solution**

Step 1. Find the least common denominator of *all* the fractions and decimals in the equation.	What is the LCD of $\frac{1}{12}$, $\frac{5}{6}$, and $\frac{3}{4}$?	$\frac{1}{12}x + \frac{5}{6} = \frac{3}{4}$ LCD = 12
Step 2. Multiply both sides of the equation by that LCD. This clears the fractions and decimals.	Multiply both sides of the equation by the LCD, 12. Use the Distributive Property. Simplify—and notice, no more fractions!	$12\left(\frac{1}{12}x + \frac{5}{6}\right) = 12\left(\frac{3}{4}\right)$ $12 \cdot \frac{1}{12}x + 12 \cdot \frac{5}{6} = 12 \cdot \frac{3}{4}$ $x + 10 = 9$
Step 3. Solve using the General Strategy for Solving Linear Equations.	To isolate the variable term, subtract 10. Simplify.	$x + 10 - 10 = 9 - 10$ $x = -1$ **Check:** $\frac{1}{12}x + \frac{5}{6} = \frac{3}{4}$ $\frac{1}{12}(-1) + \frac{5}{6} \overset{?}{=} \frac{3}{4}$ $-\frac{1}{12} + \frac{5}{6} \overset{?}{=} \frac{3}{4}$ $-\frac{1}{12} + \frac{10}{12} \overset{?}{=} \frac{9}{12}$ $\frac{9}{12} = \frac{9}{12}$ ✓

> **TRY IT :: 2.17**   Solve: $\frac{1}{4}x + \frac{1}{2} = \frac{5}{8}$.

> **TRY IT :: 2.18**   Solve: $\frac{1}{8}x + \frac{1}{2} = \frac{1}{4}$.

Notice in the previous example, once we cleared the equation of fractions, the equation was like those we solved earlier in this chapter. We changed the problem to one we already knew how to solve. We then used the General Strategy for Solving Linear Equations.

   HOW TO :: SOLVE EQUATIONS WITH FRACTION OR DECIMAL COEFFICIENTS.

Step 1.   Find the least common denominator (LCD) of *all* the fractions and decimals (in fraction form) in the equation.

Step 2.   Multiply both sides of the equation by that LCD. This clears the fractions and decimals.

Step 3.   Solve using the General Strategy for Solving Linear Equations.

EXAMPLE 2.10

Solve: $5 = \frac{1}{2}y + \frac{2}{3}y - \frac{3}{4}y$.

⊘ Solution

We want to clear the fractions by multiplying both sides of the equation by the LCD of all the fractions in the equation.

Find the LCD of all fractions in the equation.	$5 = \frac{1}{2}y + \frac{2}{3}y - \frac{3}{4}y$
The LCD is 12.	
Multiply both sides of the equation by 12.	$12(5) = 12 \cdot \left(\frac{1}{2}y + \frac{2}{3}y - \frac{3}{4}y\right)$
Distribute.	$12(5) = 12 \cdot \frac{1}{2}y + 12 \cdot \frac{2}{3}y - 12 \cdot \frac{3}{4}y$
Simplify—notice, no more fractions.	$60 = 6y + 8y - 9y$
Combine like terms.	$60 = 5y$
Divide by five.	$\frac{60}{5} = \frac{5y}{5}$
Simplify.	$12 = y$
Check:	$5 = \frac{1}{2}y + \frac{2}{3}y - \frac{3}{4}y$
Let $y = 12$.	$5 \overset{?}{=} \frac{1}{2}(12) + \frac{2}{3}(12) - \frac{3}{4}(12)$
	$5 \overset{?}{=} 6 + 8 - 9$
	$5 = 5\checkmark$

> **TRY IT :: 2.19**   Solve: $7 = \frac{1}{2}x + \frac{3}{4}x - \frac{2}{3}x$.

> **TRY IT :: 2.20**   Solve: $-1 = \frac{1}{2}u + \frac{1}{4}u - \frac{2}{3}u$.

In the next example, we'll distribute before we clear the fractions.

EXAMPLE 2.11

Solve: $\frac{1}{2}(y - 5) = \frac{1}{4}(y - 1)$.

⊘ Solution

	$\frac{1}{2}(y - 5) = \frac{1}{4}(y - 1)$
Distribute.	$\frac{1}{2} \cdot y - \frac{1}{2} \cdot 5 = \frac{1}{4} \cdot y - \frac{1}{4} \cdot 1$
Simplify.	$\frac{1}{2}y - \frac{5}{2} = \frac{1}{4}y - \frac{1}{4}$
Multiply by the LCD, four.	$4\left(\frac{1}{2}y - \frac{5}{2}\right) = 4\left(\frac{1}{4}y - \frac{1}{4}\right)$
Distribute.	$4 \cdot \frac{1}{2}y - 4 \cdot \frac{5}{2} = 4 \cdot \frac{1}{4}y - 4 \cdot \frac{1}{4}$
Simplify.	$2y - 10 = y - 1$
Collect the variables to the left.	$2y - y - 10 = y - y - 1$

Simplify.	$y - 10 = -1$
Collect the constants to the right.	$y - 10 + 10 = -1 + 10$
Simplify.	$y = 9$

An alternate way to solve this equation is to clear the fractions without distributing first. If you multiply the factors correctly, this method will be easier.

	$\frac{1}{2}(y - 5) = \frac{1}{4}(y - 1)$
Multiply by the LCD, 4.	$4 \cdot \frac{1}{2}(y - 5) = 4 \cdot \frac{1}{4}(y - 1)$
Multiply four times the fractions.	$2(y - 5) = 1(y - 1)$
Distribute.	$2y - 10 = y - 1$
Collect the variables to the left.	$2y - y - 10 = y - y - 1$
Simplify.	$y - 10 = -1$
Collect the constants to the right.	$y - 10 + 10 = -1 + 10$
Simplify.	$y = 9$
Check:	$\frac{1}{2}(y - 5) = \frac{1}{4}(y - 1)$
Let $y = 9$.	$\frac{1}{2}(9 - 5) \stackrel{?}{=} \frac{1}{4}(9 - 1)$
Finish the check on your own.	

> **TRY IT :: 2.21**   Solve: $\frac{1}{5}(n + 3) = \frac{1}{4}(n + 2)$.

> **TRY IT :: 2.22**   Solve: $\frac{1}{2}(m - 3) = \frac{1}{4}(m - 7)$.

When you multiply both sides of an equation by the LCD of the fractions, make sure you multiply each term by the LCD—even if it does not contain a fraction.

EXAMPLE 2.12

Solve: $\frac{4q + 3}{2} + 6 = \frac{3q + 5}{4}$

⊘ **Solution**

	$\frac{4q + 3}{2} + 6 = \frac{3q + 5}{4}$
Multiply both sides by the LCD, 4.	$4\left(\frac{4q + 3}{2} + 6\right) = 4\left(\frac{3q + 5}{4}\right)$
Distribute.	$4\left(\cdot \frac{4q + 3}{2}\right) + 4 \cdot 6 = 4 \cdot \left(\frac{3q + 5}{4}\right)$
Simplify.	$2(4q + 3) + 24 = 3q + 5$
	$8q + 6 + 24 = 3q + 5$

	$8q + 30 = 3q + 5$
Collect the variables to the left.	$8q - 3q + 30 = 3q - 3q + 5$
Simplify.	$5q + 30 = 5$
Collect the constants to the right.	$5q + 30 - 30 = 5 - 30$
Simplify.	$5q = -25$
Divide both sides by five.	$\dfrac{5q}{5} = \dfrac{-25}{5}$
Simplify.	$q = -5$
Check:	$\dfrac{4q + 3}{2} + 6 = \dfrac{3q + 5}{4}$
Let $q = -5$.	$\dfrac{4(-5) + 3}{2} + 6 \overset{?}{=} \dfrac{3(-5) + 5}{4}$
Finish the check on your own.	

> **TRY IT :: 2.23** Solve: $\dfrac{3r + 5}{6} + 1 = \dfrac{4r + 3}{3}$.

> **TRY IT :: 2.24** Solve: $\dfrac{2s + 3}{2} + 1 = \dfrac{3s + 2}{4}$.

Some equations have decimals in them. This kind of equation may occur when we solve problems dealing with money or percentages. But decimals can also be expressed as fractions. For example, $0.7 = \dfrac{7}{10}$ and $0.29 = \dfrac{29}{100}$. So, with an equation with decimals, we can use the same method we used to clear fractions—multiply both sides of the equation by the least common denominator.

The next example uses an equation that is typical of the ones we will see in the money applications in a later section. Notice that we will clear all decimals by multiplying by the LCD of their fraction form.

EXAMPLE 2.13

Solve: $0.25x + 0.05(x + 3) = 2.85$.

⊘ Solution

Look at the decimals and think of the equivalent fractions:

$$0.25 = \frac{25}{100}, \qquad 0.05 = \frac{5}{100}, \qquad 2.85 = 2\frac{85}{100}.$$

Notice, the LCD is 100. By multiplying by the LCD we will clear the decimals from the equation.

	$0.25x + 0.05(x + 3) = 2.85$
Distribute first.	$0.25x + 0.05x + 0.15 = 2.85$
Combine like terms.	$0.30x + 0.15 = 2.85$
To clear decimals, multiply by 100.	$100(0.30x + 0.15) = 100(2.85)$
Distribute.	$30x + 15 = 285$
Subtract 15 from both sides.	$30x + 15 - 15 = 285 - 15$
Simplify.	$30x = 270$
Divide by 30.	$\dfrac{30x}{30} = \dfrac{270}{30}$
Simplify.	$x = 9$

Check it yourself by substituting $x = 9$ into the original equation.

> **TRY IT : :** 2.25   Solve: $0.25n + 0.05(n + 5) = 2.95$.

> **TRY IT : :** 2.26   Solve: $0.10d + 0.05(d - 5) = 2.15$.

2.1 EXERCISES

Practice Makes Perfect

Solve Equations Using the General Strategy

In the following exercises, determine whether the given values are solutions to the equation.

1. $6y + 10 = 12y$

ⓐ $y = \frac{5}{3}$

ⓑ $y = -\frac{1}{2}$

2. $4x + 9 = 8x$

ⓐ $x = -\frac{7}{8}$

ⓑ $x = \frac{9}{4}$

3. $8u - 1 = 6u$

ⓐ $u = -\frac{1}{2}$

ⓑ $u = \frac{1}{2}$

4. $9v - 2 = 3v$

ⓐ $v = -\frac{1}{3}$

ⓑ $v = \frac{1}{3}$

In the following exercises, solve each linear equation.

5. $15(y - 9) = -60$

6. $-16(3n + 4) = 32$

7. $-(w - 12) = 30$

8. $-(t - 19) = 28$

9. $51 + 5(4 - q) = 56$

10. $-6 + 6(5 - k) = 15$

11. $3(10 - 2x) + 54 = 0$

12. $-2(11 - 7x) + 54 = 4$

13. $\frac{2}{3}(9c - 3) = 22$

14. $\frac{3}{5}(10x - 5) = 27$

15. $\frac{1}{5}(15c + 10) = c + 7$

16. $\frac{1}{4}(20d + 12) = d + 7$

17. $3(4n - 1) - 2 = 8n + 3$

18. $9(2m - 3) - 8 = 4m + 7$

19. $12 + 2(5 - 3y) = -9(y - 1) - 2$

20. $-15 + 4(2 - 5y) = -7(y - 4) + 4$

21. $5 + 6(3s - 5) = -3 + 2(8s - 1)$

22. $-12 + 8(x - 5) = -4 + 3(5x - 2)$

23. $4(p - 4) - (p + 7) = 5(p - 3)$

24. $3(a - 2) - (a + 6) = 4(a - 1)$

25. $4[5 - 8(4c - 3)] = 12(1 - 13c) - 8$

26. $5[9 - 2(6d - 1)] = 11(4 - 10d) - 139$

27. $3[-9 + 8(4h - 3)] = 2(5 - 12h) - 19$

28. $3[-14 + 2(15k - 6)] = 8(3 - 5k) - 24$

29.
$5[2(m + 4) + 8(m - 7)] = 2[3(5 + m) - (21 - 3m)]$

30.
$10[5(n + 1) + 4(n - 1)] = 11[7(5 + n) - (25 - 3n)]$

Classify Equations

In the following exercises, classify each equation as a conditional equation, an identity, or a contradiction and then state the solution.

31. $23z + 19 = 3(5z - 9) + 8z + 46$

32. $15y + 32 = 2(10y - 7) - 5y + 46$

33. $18(5j - 1) + 29 = 47$

34. $24(3d - 4) + 100 = 52$

35. $22(3m - 4) = 8(2m + 9)$

36. $30(2n - 1) = 5(10n + 8)$

37. $7v + 42 = 11(3v + 8) - 2(13v - 1)$

38. $18u - 51 = 9(4u + 5) - 6(3u - 10)$

39. $45(3y - 2) = 9(15y - 6)$

40. $60(2x - 1) = 15(8x + 5)$

41. $9(14d + 9) + 4d = 13(10d + 6) + 3$

42. $11(8c + 5) - 8c = 2(40c + 25) + 5$

Solve Equations with Fraction or Decimal Coefficients

In the following exercises, solve each equation with fraction coefficients.

43. $\frac{1}{4}x - \frac{1}{2} = -\frac{3}{4}$

44. $\frac{3}{4}x - \frac{1}{2} = \frac{1}{4}$

45. $\frac{5}{6}y - \frac{2}{3} = -\frac{3}{2}$

46. $\frac{5}{6}y - \frac{1}{3} = -\frac{7}{6}$

47. $\frac{1}{2}a + \frac{3}{8} = \frac{3}{4}$

48. $\frac{5}{8}b + \frac{1}{2} = -\frac{3}{4}$

49. $2 = \frac{1}{3}x - \frac{1}{2}x + \frac{2}{3}x$

50. $2 = \frac{3}{5}x - \frac{1}{3}x + \frac{2}{5}x$

51. $\frac{1}{3}w + \frac{5}{4} = w - \frac{1}{4}$

52. $\frac{1}{2}a - \frac{1}{4} = \frac{1}{6}a + \frac{1}{12}$

53. $\frac{1}{3}b + \frac{1}{5} = \frac{2}{5}b - \frac{3}{5}$

54. $\frac{1}{3}x + \frac{2}{5} = \frac{1}{5}x - \frac{2}{5}$

55. $\frac{1}{4}(p - 7) = \frac{1}{3}(p + 5)$

56. $\frac{1}{5}(q + 3) = \frac{1}{2}(q - 3)$

57. $\frac{1}{2}(x + 4) = \frac{3}{4}$

58. $\frac{1}{3}(x + 5) = \frac{5}{6}$

59. $\frac{4n + 8}{4} = \frac{n}{3}$

60. $\frac{3p + 6}{3} = \frac{p}{2}$

61. $\frac{3x + 4}{2} + 1 = \frac{5x + 10}{8}$

62. $\frac{10y - 2}{3} + 3 = \frac{10y + 1}{9}$

63. $\frac{7u - 1}{4} - 1 = \frac{4u + 8}{5}$

64. $\frac{3v - 6}{2} + 5 = \frac{11v - 4}{5}$

In the following exercises, solve each equation with decimal coefficients.

65. $0.4x + 0.6 = 0.5x - 1.2$

66. $0.7x + 0.4 = 0.6x + 2.4$

67. $0.9x - 1.25 = 0.75x + 1.75$

68. $1.2x - 0.91 = 0.8x + 2.29$

69. $0.05n + 0.10(n + 8) = 2.15$

70. $0.05n + 0.10(n + 7) = 3.55$

71. $0.10d + 0.25(d + 5) = 4.05$

72. $0.10d + 0.25(d + 7) = 5.25$

Everyday Math

73. Fencing Micah has 74 feet of fencing to make a dog run in his yard. He wants the length to be 2.5 feet more than the width. Find the length, L, by solving the equation $2L + 2(L - 2.5) = 74$.

74. Stamps Paula bought \$22.82 worth of 49-cent stamps and 21-cent stamps. The number of 21-cent stamps was eight less than the number of 49-cent stamps. Solve the equation $0.49s + 0.21(s - 8) = 22.82$ for s, to find the number of 49-cent stamps Paula bought.

Writing Exercises

75. Using your own words, list the steps in the general strategy for solving linear equations.

76. Explain why you should simplify both sides of an equation as much as possible before collecting the variable terms to one side and the constant terms to the other side.

77. What is the first step you take when solving the equation $3 - 7(y - 4) = 38$? Why is this your first step?

78. If an equation has several fractions, how does multiplying both sides by the LCD make it easier to solve?

79. If an equation has fractions only on one side, why do you have to multiply both sides of the equation by the LCD?

80. For the equation $0.35x + 2.1 = 3.85$, how do you clear the decimal?

Self Check

ⓐ *After completing the exercises, use this checklist to evaluate your mastery of the objectives of this section.*

I can...	Confidently	With some help	No-I don't get it!
solve linear equations using a general strategy.			
classify equations.			
solve equations with fraction or decimal coefficients.			

ⓑ *If most of your checks were:*

...confidently. Congratulations! You have achieved the objectives in this section. Reflect on the study skills you used so that you can continue to use them. What did you do to become confident of your ability to do these things? Be specific.

...with some help. This must be addressed quickly because topics you do not master become potholes in your road to success. In math every topic builds upon previous work. It is important to make sure you have a strong foundation before you move on. Who can you ask for help? Your fellow classmates and instructor are good resources. Is there a place on campus where math tutors are available? Can your study skills be improved?

...no - I don't get it! This is a warning sign and you must not ignore it. You should get help right away or you will quickly be overwhelmed. See your instructor as soon as you can to discuss your situation. Together you can come up with a plan to get you the help you need.

2.2 Use a Problem Solving Strategy

Learning Objectives

By the end of this section, you will be able to:
> Use a problem solving strategy for word problems
> Solve number word problems
> Solve percent applications
> Solve simple interest applications

Be Prepared!

Before you get started, take this readiness quiz.

1. Translate "six less than twice *x*" into an algebraic expression.
   If you missed this problem, review **Example 1.8**.

2. Convert 4.5% to a decimal.
   If you missed this problem, review **Example 1.40**.

3. Convert 0.6 to a percent.
   If you missed this problem, review **Example 1.40**.

Have you ever had any negative experiences in the past with word problems? When we feel we have no control, and continue repeating negative thoughts, we set up barriers to success. Realize that your negative experiences with word problems are in your past. To move forward you need to calm your fears and change your negative feelings.

Start with a fresh slate and begin to think positive thoughts. Repeating some of the following statements may be helpful to turn your thoughts positive. Thinking positive thoughts is a first step towards success.

> I think I can! I think I can!

> While word problems were hard in the past, I think I can try them now.

> I am better prepared now—I think I will begin to understand word problems.

> I am able to solve equations because I practiced many problems and I got help when I needed it—I can try that with word problems.

> It may take time, but I can begin to solve word problems.

You are now well prepared and you are ready to succeed. If you take control and believe you can be successful, you will be able to master word problems.

Use a Problem Solving Strategy for Word Problems

Now that we can solve equations, we are ready to apply our new skills to word problems. We will develop a strategy we can use to solve any word problem successfully.

EXAMPLE 2.14

Normal yearly snowfall at the local ski resort is 12 inches more than twice the amount it received last season. The normal yearly snowfall is 62 inches. What was the snowfall last season at the ski resort?

 Solution

Step 1. Read the problem.

Step 2. Identify what you are looking for.	What was the snowfall last season?
Step 3. Name what we are looking for and choose a variable to represent it.	Let $s =$ the snowfall last season.
Step 4. Translate. Restate the problem in one sentence with all the important information.	The normal snowfall was twice more than the amount last year
Translate into an equation.	62 = $2s + 12$

Step 5. Solve the equation.	$62 = 2s + 12$
Subtract 12 from each side.	$62 - 12 = 2s + 12 - 12$
Simplify.	$50 = 2s$
Divide each side by two.	$\dfrac{50}{2} = \dfrac{2s}{2}$
Simplify.	$25 = s$
Step 6. Check: First, is our answer reasonable? Yes, having 25 inches of snow seems OK. The problem says the normal snowfall is twelve inches more than twice the number of last season. Twice 25 is 50 and 12 more than that is 62.	
Step 7. Answer the question.	The snowfall last season was 25 inches.

 TRY IT : : 2.27

Guillermo bought textbooks and notebooks at the bookstore. The number of textbooks was three more than twice the number of notebooks. He bought seven textbooks. How many notebooks did he buy?

 TRY IT : : 2.28

Gerry worked Sudoku puzzles and crossword puzzles this week. The number of Sudoku puzzles he completed is eight more than twice the number of crossword puzzles. He completed 22 Sudoku puzzles. How many crossword puzzles did he do?

We summarize an effective strategy for problem solving.

 HOW TO : : USE A PROBLEM SOLVING STRATEGY FOR WORD PROBLEMS.

Step 1. **Read** the problem. Make sure all the words and ideas are understood.

Step 2. **Identify** what you are looking for.

Step 3. **Name** what you are looking for. Choose a variable to represent that quantity.

Step 4. **Translate** into an equation. It may be helpful to restate the problem in one sentence with all the important information. Then, translate the English sentence into an algebra equation.

Step 5. **Solve** the equation using proper algebra techniques.

Step 6. **Check** the answer in the problem to make sure it makes sense.

Step 7. **Answer** the question with a complete sentence.

Solve Number Word Problems

We will now apply the problem solving strategy to "number word problems." Number word problems give some clues about one or more numbers and we use these clues to write an equation. Number word problems provide good practice for using the Problem Solving Strategy.

EXAMPLE 2.15

The sum of seven times a number and eight is thirty-six. Find the number.

✓ **Solution**

Step 1. Read the problem.

Step 2. Identify what you are looking for.	the number
Step 3. Name what you are looking for and choose a variable to represent it.	Let n = the number.

Step 4. Translate:
Restate the problem as one sentence.
Translate into an equation.

The sum of seven times a number and 8   is   36
$7n + 8$ $= \quad 36$

Step 5. Solve the equation.
Subtract eight from each side and simplify.
Divide each side by seven and simplify.

$$7n + 8 = 36$$
$$7n = 28$$
$$n = 4$$

Step 6. Check.
Is the sum of seven times four plus eight equal to 36?

$$7 \cdot 4 + 8 \overset{?}{=} 36$$
$$28 + 8 \overset{?}{=} 36$$
$$36 = 36 \checkmark$$

Step 7. Answer the question.	The number is 4.

Did you notice that we left out some of the steps as we solved this equation? If you're not yet ready to leave out these steps, write down as many as you need.

> **TRY IT : : 2.29**   The sum of four times a number and two is fourteen. Find the number.

> **TRY IT : : 2.30**   The sum of three times a number and seven is twenty-five. Find the number.

Some number word problems ask us to find two or more numbers. It may be tempting to name them all with different variables, but so far, we have only solved equations with one variable. In order to avoid using more than one variable, we will define the numbers in terms of the same variable. Be sure to read the problem carefully to discover how all the numbers relate to each other.

EXAMPLE 2.16

The sum of two numbers is negative fifteen. One number is nine less than the other. Find the numbers.

✓ **Solution**

Step 1. Read the problem.

Step 2. Identify what you are looking for.	two numbers
Step 3. Name what you are looking for by choosing a variable to represent the first number. "One number is nine less than the other."	Let n = 1st number. $n - 9 = 2$nd number

Step 4. Translate.
Write as one sentence.
Translate into an equation.

The sum of two numbers is negative fifteen.
1st number + 2nd number   is   negative fifteen
$n \quad + \quad n - 9 \quad = \quad -15$

Step 5. Solve the equation.

Combine like terms.

Add nine to each side and simplify.

Simplify.

$$n + n - 9 = -15$$
$$2n - 9 = -15$$
$$2n = -6$$
$$n = -3 \quad \text{1}^{\text{st}} \text{ number}$$

$$n - 9 \quad \text{2}^{\text{nd}} \text{ number}$$
$$-3 - 9$$
$$-12$$

Step 6. Check.

Is -12 nine less than -3?

$$-3 - 9 \stackrel{?}{=} -12$$
$$-12 = -12 \checkmark$$

Is their sum -15?

$$-3 + (-12) \stackrel{?}{=} -15$$
$$-15 = -15 \checkmark$$

Step 7. Answer the question.   The numbers are -3 and -12.

> **TRY IT :: 2.31**

The sum of two numbers is negative twenty-three. One number is seven less than the other. Find the numbers.

> **TRY IT :: 2.32**

The sum of two numbers is negative eighteen. One number is forty more than the other. Find the numbers.

Some number problems involve **consecutive integers**. Consecutive integers are integers that immediately follow each other. Examples of **consecutive integers** are:

$$1, \quad 2, \quad 3, \quad 4$$
$$-10, \quad -9, \quad -8, \quad -7$$
$$150, \quad 151, \quad 152, \quad 153$$

Notice that each number is one more than the number preceding it. Therefore, if we define the first integer as n, the next consecutive integer is $n + 1$. The one after that is one more than $n + 1$, so it is $n + 1 + 1$, which is $n + 2$.

$$n \qquad\qquad \text{1}^{\text{st}} \text{ integer}$$
$$n + 1 \qquad\qquad \text{2}^{\text{nd}} \text{ consecutive integer}$$
$$n + 2 \qquad\qquad \text{3}^{\text{rd}} \text{ consecutive integer} \quad \text{etc.}$$

We will use this notation to represent consecutive integers in the next example.

EXAMPLE 2.17

Find three consecutive integers whose sum is -54.

⊘ Solution

Step 1. Read the problem.

Step 2. Identify what you are looking for.	three consecutive integers

Step 3. Name each of the three numbers

Let $n = 1^{st}$ integer.

$n + 1 = 2^{nd}$ consecutive integer

$n + 2 = 3^{rd}$ consecutive integer

Step 4. Translate.
Restate as one sentence.
Translate into an equation.

The sum of the three integers is -54.

$n + n + 1 + n + 2 = -54$

Step 5. Solve the equation.
Combine like terms.
Subtract three from each side.
Divide each side by three.

$$n + n + 1 + n + 2 = -54$$
$$3n + 3 = -54$$
$$3n = -57$$
$$n = -19 \quad 1^{st} \text{ integer}$$

$n + 1 \quad 2^{nd}$ integer
$-19 + 1$
-18

$n + 2 \quad 3^{rd}$ integer
$-19 + 2$
-17

Step 6. Check.

$$-19 + (-18) + (-17) = -54$$
$$-54 = -54 ✓$$

Step 7. Answer the question.

The three consecutive integers are $-17, -18,$ and -19.

> **TRY IT : : 2.33**   Find three consecutive integers whose sum is -96.

> **TRY IT : : 2.34**   Find three consecutive integers whose sum is -36.

Now that we have worked with consecutive integers, we will expand our work to include **consecutive even integers** and **consecutive odd integers**. Consecutive even integers are even integers that immediately follow one another. Examples of consecutive even integers are:

$$24, 26, 28$$
$$-12, -10, -8$$

Notice each integer is two more than the number preceding it. If we call the first one n, then the next one is $n + 2$. The one after that would be $n + 2 + 2$ or $n + 4$.

n	1st even integer
$n + 2$	2nd consecutive even integer
$n + 4$	3rd consecutive even integer etc.

Consecutive odd integers are odd integers that immediately follow one another. Consider the consecutive odd integers 63, 65, and 67.

$$63, 65, 67$$
$$n, n + 2, n + 4$$

n	1st odd integer
$n + 2$	2nd consecutive odd integer
$n + 4$	3rd consecutive odd integer etc.

Does it seem strange to have to add two (an even number) to get the next odd number? Do we get an odd number or an even number when we add 2 to 3? to 11? to 47?

Whether the problem asks for consecutive even numbers or odd numbers, you do not have to do anything different. The pattern is still the same—to get to the next odd or the next even integer, add two.

EXAMPLE 2.18

Find three consecutive even integers whose sum is 120.

⊘ **Solution**

Step 1. Read the problem.	
Step 2. Identify what you are looking for.	three consecutive even integers
Step 3. Name.	Let $n = 1^{st}$ even integer.
	$n + 2 = 2^{nd}$ consecutive even integer
	$n + 4 = 3^{rd}$ consecutive even integer
Step 4. Translate.	
Restate as one sentence.	The sum of the three even integers is 120.
Translate into an equation.	$n + n + 2 + n + 4 = 120$
Step 5. Solve the equation.	$n + n + 2 + n + 4 = 120$
Combine like terms.	$3n + 6 = 120$
Subtract 6 from each side.	$3n = 114$
Divide each side by 3.	$n = 38 \quad 1^{st}$ integer
	$n + 2 \quad 2^{nd}$ integer
	$38 + 2$
	40
	$n + 4 \quad 3^{rd}$ integer
	$38 + 4$
	42
Step 6. Check.	

$$38 + 40 + 42 \stackrel{?}{=} 120$$
$$120 = 120 \checkmark$$

Step 7. Answer the question.	The three consecutive integers are 38, 40, and 42.

> **TRY IT : : 2.35**   Find three consecutive even integers whose sum is 102.

> **TRY IT : : 2.36**   Find three consecutive even integers whose sum is −24.

When a number problem is in a real life context, we still use the same strategies that we used for the previous examples.

EXAMPLE 2.19

A married couple together earns $110,000 a year. The wife earns $16,000 less than twice what her husband earns. What does the husband earn?

⊘ **Solution**

Step 1. Read the problem.

Step 2. Identify what you are looking for.   How much does the husband earn?

Step 3. Name.

Choose a variable to represent   Let h = the amount the husband earns.
the amount the husband earns.
The wife earns $16,000 less than twice that.

Step 4. Translate.

Restate the problem in one sentence $2h - 16,000$ = the amount the wife earns
with all the important information.   Together the husband and wife earn $110,000.
Translate into an equation.

$$h + 2h - 16,000 \;=\; 110,000$$

Step 5. Solve the equation. $$h + 2h - 16,000 \;=\; 110,000$$
Combine like terms. $$3h - 16,000 \;=\; 110,000$$
Add 16,000 to both sides and simplify. $$3h \;=\; 126,000$$
Divide each side by three. $$h \;=\; 42,000$$

$42,000 amount husband earns

$$2h - 16,000 \text{ amount wife earns}$$
$$2(42,000) - 16,000$$
$$84,000 - 16,000$$
$$68,000$$

Step 6. Check:

If the wife earns $68,000 and the husband
earns $42,000, is that $110,000? Yes!

Step 7. Answer the question.   The husband earns $42,000 a year.

 TRY IT : : 2.37

According to the National Automobile Dealers Association, the average cost of a car in 2014 was $28,400. This was $1,600 less than six times the cost in 1975. What was the average cost of a car in 1975?

 TRY IT : : 2.38

US Census data shows that the median price of new home in the U.S. in November 2014 was $280,900. This was $10,700 more than 14 times the price in November 1964. What was the median price of a new home in November 1964?

Solve Percent Applications

There are several methods to solve percent equations. In algebra, it is easiest if we just translate English sentences into algebraic equations and then solve the equations. Be sure to change the given percent to a decimal before you use it in the equation.

EXAMPLE 2.20

Translate and solve:

ⓐ What number is 45% of 84?

ⓑ 8.5% of what amount is $4.76?

ⓒ 168 is what percent of 112?

Solution

ⓐ

	What number is 45% of 84?
Translate into algebra. Let n = the number.	$n = 0.45 \cdot 84$
Multiply.	$n = 37.8$
	37.8 is 45% of 84.

ⓑ

	8.5% of what amount is $4.76?
Translate. Let n = the amount.	$0.085 \cdot n = 4.76$
Multiply.	$0.085n = 4.76$
Divide both sides by 0.085 and simplify.	$n = 56$
	8.5% of $56 is $4.76

ⓒ

	168 is what percent of 112?
We are asked to find percent, so we must have our result in percent form.	
Translate into algebra. Let p = the percent.	$168 = p \cdot 112$
Multiply.	$168 = 112\,p$
Divide both sides by 112 and simplify.	$1.5 = p$
Convert to percent.	$150\% = p$
	168 is 150% of 112.

> **TRY IT : : 2.39**
>
> Translate and solve: ⓐ What number is 45% of 80? ⓑ 7.5% of what amount is $1.95? ⓒ 110 is what percent of 88?

> **TRY IT : : 2.40**
>
> Translate and solve: ⓐ What number is 55% of 60? ⓑ 8.5% of what amount is $3.06? ⓐ 126 is what percent of 72?

Now that we have a problem solving strategy to refer to, and have practiced solving basic percent equations, we are ready to solve percent applications. Be sure to ask yourself if your final answer makes sense—since many of the applications we will solve involve everyday situations, you can rely on your own experience.

EXAMPLE 2.21

The label on Audrey's yogurt said that one serving provided 12 grams of protein, which is 24% of the recommended daily amount. What is the total recommended daily amount of protein?

⊘ Solution

What are you asked to find?	What total amount of protein is recommended?
Choose a variable to represent it.	Let $a =$ total amount of protein.
Write a sentence that gives the information to find it.	12 g is 24% of the total amount
Translate into an equation.	$12 = 0.24 \cdot a$
Solve.	$50 = a$
Check: Does this make sense? Yes, 24% is about $\frac{1}{4}$ of the total and 12 is about $\frac{1}{4}$ of 50.	
Write a complete sentence to answer the question.	The amount of protein that is recommended is 50 g.

> **TRY IT :: 2.41**
>
> One serving of wheat square cereal has 7 grams of fiber, which is 28% of the recommended daily amount. What is the total recommended daily amount of fiber?

> **TRY IT :: 2.42**
>
> One serving of rice cereal has 190 mg of sodium, which is 8% of the recommended daily amount. What is the total recommended daily amount of sodium?

Remember to put the answer in the form requested. In the next example we are looking for the percent.

EXAMPLE 2.22

Veronica is planning to make muffins from a mix. The package says each muffin will be 240 calories and 60 calories will be from fat. What percent of the total calories is from fat?

⊘ Solution

What are you asked to find?	What percent of the total calories is fat?
Choose a variable to represent it.	Let $p =$ percent of fat.
Write a sentence that gives the information to find it.	What percent of 240 is 60?
Translate the sentence into an equation.	$p \cdot 240 = 60$
Multiply.	$240\,p = 60$
Divide both sides by 240.	$p = 0.25$
Put in percent form.	$p = 25\%$
Check: does this make sense? Yes, 25% is one-fourth; 60 is one-fourth of 240. So, 25% makes sense.	
Write a complete sentence to answer the question.	Of the total calories in each muffin, 25% is fat.

> **TRY IT :: 2.43**

Mitzi received some gourmet brownies as a gift. The wrapper said each 28% brownie was 480 calories, and had 240 calories of fat. What percent of the total calories in each brownie comes from fat? Round the answer to the nearest whole percent.

> **TRY IT :: 2.44**

The mix Ricardo plans to use to make brownies says that each brownie will be 190 calories, and 76 calories are from fat. What percent of the total calories are from fat? Round the answer to the nearest whole percent.

It is often important in many fields—business, sciences, pop culture—to talk about how much an amount has increased or decreased over a certain period of time. This increase or decrease is generally expressed as a percent and called the **percent change**.

To find the percent change, first we find the amount of change, by finding the difference of the new amount and the original amount. Then we find what percent the amount of change is of the original amount.

HOW TO :: FIND PERCENT CHANGE.

Step 1. Find the amount of change.
$$\text{change} = \text{new amount} - \text{original amount}$$

Step 2. Find what percent the amount of change is of the original amount.
change is what percent of the original amount?

EXAMPLE 2.23

Recently, the California governor proposed raising community college fees from $36 a unit to $46 a unit. Find the percent change. (Round to the nearest tenth of a percent.)

⊘ **Solution**

Find the amount of change.	$46 - 36 = 10$
Find the percent.	Change is what percent of the original amount?
Let $p =$ the percent.	10 is what percent of 36?
Translate to an equation.	$10 \;=\; p \;\cdot\; 36$
Simplify.	$10 = 36\,p$
Divide both sides by 36.	$0.278 \approx p$
Change to percent form; round to the nearest tenth	$27.8\% \approx p$
Write a complete sentence to answer the question.	The new fees are approximately a 27.8% increase over the old fees.

Remember to round the division to the nearest thousandth in order to round the percent to the nearest tenth.

> **TRY IT :: 2.45**

Find the percent change. (Round to the nearest tenth of a percent.) In 2011, the IRS increased the deductible mileage cost to 55.5 cents from 51 cents.

TRY IT :: 2.46

> Find the percent change. (Round to the nearest tenth of a percent.) In 1995, the standard bus fare in Chicago was $1.50. In 2008, the standard bus fare was 2.25.

Applications of discount and mark-up are very common in retail settings.

When you buy an item on sale, the original price has been discounted by some dollar amount. The **discount rate**, usually given as a percent, is used to determine the amount of the discount. To determine the **amount of discount**, we multiply the discount rate by the original price.

The price a retailer pays for an item is called the **original cost**. The retailer then adds a **mark-up** to the original cost to get the **list price**, the price he sells the item for. The mark-up is usually calculated as a percent of the original cost. To determine the amount of mark-up, multiply the mark-up rate by the original cost.

Discount

$$\text{amount of discount} = \text{discount rate} \cdot \text{original price}$$
$$\text{sale price} = \text{original amount} - \text{discount price}$$

The sale price should always be less than the original price.

Mark-up

$$\text{amount of mark-up} = \text{mark-up rate} \cdot \text{original price}$$
$$\text{list price} = \text{original cost} - \text{mark-up}$$

The list price should always be more than the original cost.

EXAMPLE 2.24

Liam's art gallery bought a painting at an original cost of $750. Liam marked the price up 40%. Find the amount of mark-up and ⓑ the list price of the painting.

✓ **Solution**

Identify what you are asked to find, and choose a variable to represent it.	What is the amount of mark-up? Let $m =$ the amount of mark-up.
Write a sentence that gives the information to find it.	The mark-up is 40% of the $750 original cost
Translate into an equation.	$m = 0.40 \times 750$
Solve the equation.	$m = 300$
Write a complete sentence.	The mark-up on the painting was $300.

Identify what you are asked to find, and choose a variable to represent it.	What is the list price? Let $p =$ the list price.
Write a sentence that gives the information to find it.	The list price is original cost plus the mark-up
Translate into an equation.	$p = 750 + 300$

Solve the equation.	$p = 1,050$
Check.	Is the list price more than the original cost? Is $1,050 more than $750? Yes.
Write a complete sentence.	The list price of the painting was $1,050.

> **TRY IT ::: 2.47**
>
> Find ⓐ the amount of mark-up and ⓑ the list price: Jim's music store bought a guitar at original cost $1,200. Jim marked the price up 50%.

> **TRY IT ::: 2.48**
>
> Find ⓐ the amount of mark-up and ⓑ the list price: The Auto Resale Store bought Pablo's Toyota for $8,500. They marked the price up 35%.

Solve Simple Interest Applications

Interest is a part of our daily lives. From the interest earned on our savings to the interest we pay on a car loan or credit card debt, we all have some experience with interest in our lives.

The amount of money you initially deposit into a bank is called the **principal**, P, and the bank pays you **interest**, I. When you take out a loan, you pay interest on the amount you borrow, also called the principal.

In either case, the interest is computed as a certain percent of the principal, called the **rate of interest**, r. The rate of interest is usually expressed as a percent per year, and is calculated by using the decimal equivalent of the percent. The variable t, (for time) represents the number of years the money is saved or borrowed.

Interest is calculated as simple interest or compound interest. Here we will use simple interest.

Simple Interest

If an amount of money, P, called the principal, is invested or borrowed for a period of t years at an annual interest rate r, the amount of interest, I, earned or paid is

$$I = Prt \qquad \text{where} \qquad \begin{aligned} I &= \text{interest} \\ P &= \text{principal} \\ r &= \text{rate} \\ t &= \text{time} \end{aligned}$$

Interest earned or paid according to this formula is called **simple interest**.

The formula we use to calculate interest is $I = Prt$. To use the formula we substitute in the values for variables that are given, and then solve for the unknown variable. It may be helpful to organize the information in a chart.

EXAMPLE 2.25

Areli invested a principal of $950 in her bank account that earned simple interest at an interest rate of 3%. How much interest did she earn in five years?

⊘ Solution

$$\begin{aligned} I &= ? \\ P &= \$950 \\ r &= 3\% \\ t &= 5 \text{ years} \end{aligned}$$

Identify what you are asked to find, and choose a variable to represent it.	What is the simple interest? Let I = interest.
Write the formula.	$I = Prt$
Substitute in the given information.	$I = (950)(0.03)(5)$
Simplify.	$I = 142.5$
Check.	
Is $142.50 a reasonable amount of interest on $950?	
Yes.	
Write a complete sentence.	The interest is $142.50.

> **TRY IT ::** 2.49

   Nathaly deposited $12,500 in her bank account where it will earn 4% simple interest. How much interest will Nathaly earn in five years?

> **TRY IT ::** 2.50

   Susana invested a principal of $36,000 in her bank account that earned simple interest at an interest rate of 6.5%. How much interest did she earn in three years?

There may be times when we know the amount of interest earned on a given principal over a certain length of time, but we do not know the rate.

EXAMPLE 2.26

Hang borrowed $7,500 from her parents to pay her tuition. In five years, she paid them $1,500 interest in addition to the $7,500 she borrowed. What was the rate of simple interest?

⊘ **Solution**

$$I = \$1500$$
$$P = \$7500$$
$$r = ?$$
$$t = 5 \text{ years}$$

Identify what you are asked to find, and choose a variable to represent it.	What is the rate of simple interest? Let r = rate of interest.
Write the formula.	$I = Prt$
Substitute in the given information.	$1,500 = (7,500)r(5)$
Multiply.	$1,500 = 37,500r$
Divide.	$0.04 = r$
Change to percent form.	$4\% = r$
Check.	
$I = Prt$	
$1,500 \overset{?}{=} (7,500)(0.04)(5)$	
$1,500 = 1,500 \checkmark$	
Write a complete sentence.	The rate of interest was 4%.

> **TRY IT ::** 2.51

   Jim lent his sister $5,000 to help her buy a house. In three years, she paid him the $5,000, plus $900 interest. What was the rate of simple interest?

> **TRY IT : : 2.52**

> Loren lent his brother $3,000 to help him buy a car. In four years, his brother paid him back the $3,000 plus $660 in interest. What was the rate of simple interest?

In the next example, we are asked to find the principal—the amount borrowed.

EXAMPLE 2.27

Sean's new car loan statement said he would pay $4,866,25 in interest from a simple interest rate of 8.5% over five years. How much did he borrow to buy his new car?

 Solution

$I = 4{,}866.25$
$P = ?$
$r = 8.5\%$
$t = 5 \text{ years}$

Identify what you are asked to find, and choose a variable to represent it.	What is the amount borrowed (the principal)?
	Let $P = $ principal borrowed.
Write the formula.	$I = Prt$
Substitute in the given information.	$4{,}866.25 = P(0.085)(5)$
Multiply.	$4{,}866.25 = 0.425P$
Divide.	$11{,}450 = P$
Check.	

$$I = Prt$$
$$4{,}866.25 \overset{?}{=} (11{,}450)(0.085)(5)$$
$$4{,}866.25 = 4{,}866.25 \checkmark$$

Write a complete sentence.	The principal was $11,450.

> **TRY IT : : 2.53**

> Eduardo noticed that his new car loan papers stated that with a 7.5% simple interest rate, he would pay $6,596.25 in interest over five years. How much did he borrow to pay for his car?

> **TRY IT : : 2.54**

> In five years, Gloria's bank account earned $2,400 interest at 5% simple interest. How much had she deposited in the account?

▶ **MEDIA : :**

Access this online resource for additional instruction and practice with using a problem solving strategy.

- **Begining Arithmetic Problems (https://openstax.org/l/37begalgwordpro)**

 ## 2.2 EXERCISES

Practice Makes Perfect

Use a Problem Solving Strategy for Word Problems

81. List five positive thoughts you can say to yourself that will help you approach word problems with a positive attitude. You may want to copy them on a sheet of paper and put it in the front of your notebook, where you can read them often.

82. List five negative thoughts that you have said to yourself in the past that will hinder your progress on word problems. You may want to write each one on a small piece of paper and rip it up to symbolically destroy the negative thoughts.

In the following exercises, solve using the problem solving strategy for word problems. Remember to write a complete sentence to answer each question.

83. There are 16 girls in a school club. The number of girls is four more than twice the number of boys. Find the number of boys.

84. There are 18 Cub Scouts in Troop 645. The number of scouts is three more than five times the number of adult leaders. Find the number of adult leaders.

85. Huong is organizing paperback and hardback books for her club's used book sale. The number of paperbacks is 12 less than three times the number of hardbacks. Huong had 162 paperbacks. How many hardback books were there?

86. Jeff is lining up children's and adult bicycles at the bike shop where he works. The number of children's bicycles is nine less than three times the number of adult bicycles. There are 42 adult bicycles. How many children's bicycles are there?

Solve Number Word Problems

In the following exercises, solve each number word problem.

87. The difference of a number and 12 is three. Find the number.

88. The difference of a number and eight is four. Find the number.

89. The sum of three times a number and eight is 23. Find the number.

90. The sum of twice a number and six is 14. Find the number.

91. The difference of twice a number and seven is 17. Find the number.

92. The difference of four times a number and seven is 21. Find the number.

93. Three times the sum of a number and nine is 12. Find the number.

94. Six times the sum of a number and eight is 30. Find the number.

95. One number is six more than the other. Their sum is 42. Find the numbers.

96. One number is five more than the other. Their sum is 33. Find the numbers.

97. The sum of two numbers is 20. One number is four less than the other. Find the numbers.

98. The sum of two numbers is 27. One number is seven less than the other. Find the numbers.

99. One number is 14 less than another. If their sum is increased by seven, the result is 85. Find the numbers.

100. One number is 11 less than another. If their sum is increased by eight, the result is 71. Find the numbers.

101. The sum of two numbers is 14. One number is two less than three times the other. Find the numbers.

102. The sum of two numbers is zero. One number is nine less than twice the other. Find the numbers.

103. The sum of two consecutive integers is 77. Find the integers.

104. The sum of two consecutive integers is 89. Find the integers.

105. The sum of three consecutive integers is 78. Find the integers.

106. The sum of three consecutive integers is 60. Find the integers.

107. Find three consecutive integers whose sum is -36.

108. Find three consecutive integers whose sum is -3.

109. Find three consecutive even integers whose sum is 258.

110. Find three consecutive even integers whose sum is 222.

111. Find three consecutive odd integers whose sum is -213.

112. Find three consecutive odd integers whose sum is -267.

113. Philip pays $1,620 in rent every month. This amount is $120 more than twice what his brother Paul pays for rent. How much does Paul pay for rent?

114. Marc just bought an SUV for $54,000. This is $7,400 less than twice what his wife paid for her car last year. How much did his wife pay for her car?

115. Laurie has $46,000 invested in stocks and bonds. The amount invested in stocks is $8,000 less than three times the amount invested in bonds. How much does Laurie have invested in bonds?

116. Erica earned a total of $50,450 last year from her two jobs. The amount she earned from her job at the store was $1,250 more than three times the amount she earned from her job at the college. How much did she earn from her job at the college?

Solve Percent Applications

In the following exercises, translate and solve.

117. ⓐ What number is 45% of 120? ⓑ 81 is 75% of what number? ⓐ What percent of 260 is 78?

118. ⓐ What number is 65% of 100? ⓑ 93 is 75% of what number? ⓐ What percent of 215 is 86?

119. ⓐ 250% of 65 is what number? ⓑ 8.2% of what amount is $2.87? ⓐ 30 is what percent of 20?

120. ⓐ 150% of 90 is what number? ⓑ 6.4% of what amount is $2.88? ⓐ 50 is what percent of 40?

In the following exercises, solve.

121. Geneva treated her parents to dinner at their favorite restaurant. The bill was $74.25. Geneva wants to leave 16% of the total bill as a tip. How much should the tip be?

122. When Hiro and his co-workers had lunch at a restaurant near their work, the bill was $90.50. They want to leave 18% of the total bill as a tip. How much should the tip be?

123. One serving of oatmeal has 8 grams of fiber, which is 33% of the recommended daily amount. What is the total recommended daily amount of fiber?

124. One serving of trail mix has 67 grams of carbohydrates, which is 22% of the recommended daily amount. What is the total recommended daily amount of carbohydrates?

125. A bacon cheeseburger at a popular fast food restaurant contains 2070 milligrams (mg) of sodium, which is 86% of the recommended daily amount. What is the total recommended daily amount of sodium?

126. A grilled chicken salad at a popular fast food restaurant contains 650 milligrams (mg) of sodium, which is 27% of the recommended daily amount. What is the total recommended daily amount of sodium?

127. The nutrition fact sheet at a fast food restaurant says the fish sandwich has 380 calories, and 171 calories are from fat. What percent of the total calories is from fat?

128. The nutrition fact sheet at a fast food restaurant says a small portion of chicken nuggets has 190 calories, and 114 calories are from fat. What percent of the total calories is from fat?

129. Emma gets paid $3,000 per month. She pays $750 a month for rent. What percent of her monthly pay goes to rent?

130. Dimple gets paid $3,200 per month. She pays $960 a month for rent. What percent of her monthly pay goes to rent?

In the following exercises, solve.

131. Tamanika received a raise in her hourly pay, from $15.50 to $17.36. Find the percent change.

132. Ayodele received a raise in her hourly pay, from $24.50 to $25.48. Find the percent change.

133. Annual student fees at the University of California rose from about $4,000 in 2000 to about $12,000 in 2010. Find the percent change.

134. The price of a share of one stock rose from $12.50 to $50. Find the percent change.

135. A grocery store reduced the price of a loaf of bread from $2.80 to $2.73. Find the percent change.

136. The price of a share of one stock fell from $8.75 to $8.54. Find the percent change.

137. Hernando's salary was $49,500 last year. This year his salary was cut to $44,055. Find the percent change.

138. In ten years, the population of Detroit fell from 950,000 to about 712,500. Find the percent change.

In the following exercises, find ⓐ the amount of discount and ⓑ the sale price.

139. Janelle bought a beach chair on sale at 60% off. The original price was $44.95.

140. Errol bought a skateboard helmet on sale at 40% off. The original price was $49.95.

In the following exercises, find ⓐ the amount of discount and ⓑ the discount rate (Round to the nearest tenth of a percent if needed.)

141. Larry and Donna bought a sofa at the sale price of $1,344. The original price of the sofa was $1,920.

142. Hiroshi bought a lawnmower at the sale price of $240. The original price of the lawnmower is $300.

In the following exercises, find ⓐ the amount of the mark-up and ⓑ the list price.

143. Daria bought a bracelet at original cost $16 to sell in her handicraft store. She marked the price up 45%. What was the list price of the bracelet?

144. Regina bought a handmade quilt at original cost $120 to sell in her quilt store. She marked the price up 55%. What was the list price of the quilt?

145. Tom paid $0.60 a pound for tomatoes to sell at his produce store. He added a 33% mark-up. What price did he charge his customers for the tomatoes?

146. Flora paid her supplier $0.74 a stem for roses to sell at her flower shop. She added an 85% mark-up. What price did she charge her customers for the roses?

Solve Simple Interest Applications

In the following exercises, solve.

147. Casey deposited $1,450 in a bank account that earned simple interest at an interest rate of 4%. How much interest was earned in two years?

148. Terrence deposited $5,720 in a bank account that earned simple interest at an interest rate of 6%. How much interest was earned in four years?

149. Robin deposited $31,000 in a bank account that earned simple interest at an interest rate of 5.2%. How much interest was earned in three years?

150. Carleen deposited $16,400 in a bank account that earned simple interest at an interest rate of 3.9% How much interest was earned in eight years?

151. Hilaria borrowed $8,000 from her grandfather to pay for college. Five years later, she paid him back the $8,000, plus $1,200 interest. What was the rate of simple interest?

152. Kenneth lent his niece $1,200 to buy a computer. Two years later, she paid him back the $1,200, plus $96 interest. What was the rate of simple interest?

153. Lebron lent his daughter $20,000 to help her buy a condominium. When she sold the condominium four years later, she paid him the $20,000, plus $3,000 interest. What was the rate of simple interest?

154. Pablo borrowed $50,000 to start a business. Three years later, he repaid the $50,000, plus $9,375 interest. What was the rate of simple interest?

155. In 10 years, a bank account that paid 5.25% simple interest earned $18,375 interest. What was the principal of the account?

156. In 25 years, a bond that paid 4.75% simple interest earned $2,375 interest. What was the principal of the bond?

157. Joshua's computer loan statement said he would pay $1,244.34 in simple interest for a three-year loan at 12.4%. How much did Joshua borrow to buy the computer?

158. Margaret's car loan statement said she would pay $7,683.20 in simple interest for a five-year loan at 9.8%. How much did Margaret borrow to buy the car?

Everyday Math

159. Tipping At the campus coffee cart, a medium coffee costs $1.65. MaryAnne brings $2.00 with her when she buys a cup of coffee and leaves the change as a tip. What percent tip does she leave?

160. Tipping Four friends went out to lunch and the bill came to $53.75 They decided to add enough tip to make a total of $64, so that they could easily split the bill evenly among themselves. What percent tip did they leave?

Writing Exercises

161. What has been your past experience solving word problems? Where do you see yourself moving forward?

162. Without solving the problem "44 is 80% of what number" think about what the solution might be. Should it be a number that is greater than 44 or less than 44? Explain your reasoning.

163. After returning from vacation, Alex said he should have packed 50% fewer shorts and 200% more shirts. Explain what Alex meant.

164. Because of road construction in one city, commuters were advised to plan that their Monday morning commute would take 150% of their usual commuting time. Explain what this means.

Self Check

ⓐ After completing the exercises, use this checklist to evaluate your mastery of the objective of this section.

I can...	Confidently	With some help	No-I don't get it!
use a problem-solving strategy for word problems.			
solve number problems.			
solve percent applications.			
solve simple interest applications.			

ⓑ After reviewing this checklist, what will you do to become confident for all objectives?

2.3 Solve a Formula for a Specific Variable

Learning Objectives

By the end of this section, you will be able to:

> Solve a formula for a specific variable
> Use formulas to solve geometry applications

Be Prepared!

Before you get started, take this readiness quiz.

1. Evaluate $2(x + 3)$ when $x = 5$.
   If you missed this problem, review **Example 1.6**.

2. The length of a rectangle is three less than the width. Let w represent the width. Write an expression for the length of the rectangle.
   If you missed this problem, review **Example 1.10**.

3. Evaluate $\frac{1}{2}bh$ when $b = 14$ and $h = 9$.
   If you missed this problem, review **Example 1.33**.

Solve a Formula for a Specific Variable

We have all probably worked with some geometric formulas in our study of mathematics. Formulas are used in so many fields, it is important to recognize formulas and be able to manipulate them easily.

It is often helpful to solve a formula for a specific variable. If you need to put a formula in a spreadsheet, it is not unusual to have to solve it for a specific variable first. We isolate that variable on one side of the equals sign with a coefficient of one and all other variables and constants are on the other side of the equal sign.

Geometric formulas often need to be solved for another variable, too. The formula $V = \frac{1}{3}\pi r^2 h$ is used to find the volume of a right circular cone when given the radius of the base and height. In the next example, we will solve this formula for the height.

EXAMPLE 2.28

Solve the formula $V = \frac{1}{3}\pi r^2 h$ for h.

✓ **Solution**

Write the formula.	$V = \frac{1}{3}\pi r^2 h$
Remove the fraction on the right.	$3 \cdot V = 3 \cdot \frac{1}{3}\pi r^2 h$
Simplify.	$3V = \pi r^2 h$
Divide both sides by πr^2.	$\frac{3V}{\pi r^2} = h$

We could now use this formula to find the height of a right circular cone when we know the volume and the radius of the base, by using the formula $h = \frac{3V}{\pi r^2}$.

 **TRY IT :: 2.55**   Use the formula $A = \frac{1}{2}bh$ to solve for b.

 **TRY IT :: 2.56**   Use the formula $A = \frac{1}{2}bh$ to solve for h.

In the sciences, we often need to change temperature from Fahrenheit to Celsius or vice versa. If you travel in a foreign country, you may want to change the Celsius temperature to the more familiar Fahrenheit temperature.

EXAMPLE 2.29

Solve the formula $C = \frac{5}{9}(F - 32)$ for F.

⊘ **Solution**

Write the formula.	$C = \frac{5}{9}(F - 32)$
Remove the fraction on the right.	$\frac{9}{5}C = \frac{9}{5} \cdot \frac{5}{9}(F - 32)$
Simplify.	$\frac{9}{5}C = (F - 32)$
Add 32 to both sides.	$\frac{9}{5}C + 32 = F$

We can now use the formula $F = \frac{9}{5}C + 32$ to find the Fahrenheit temperature when we know the Celsius temperature.

> **TRY IT :: 2.57**   Solve the formula $F = \frac{9}{5}C + 32$ for C.

> **TRY IT :: 2.58**   Solve the formula $A = \frac{1}{2}h(b + B)$ for b.

The next example uses the formula for the surface area of a right cylinder.

EXAMPLE 2.30

Solve the formula $S = 2\pi r^2 + 2\pi rh$ for h.

⊘ **Solution**

Write the formula.	$S = 2\pi r^2 + 2\pi rh$
Isolate the h term by subtracting $2\pi r^2$ from each side.	$S - 2\pi r^2 = 2\pi r^2 - 2\pi r^2 + 2\pi rh$
Simplify.	$S - 2\pi r^2 = 2\pi rh$
Solve for h by dividing both sides by $2\pi r$.	$\frac{S - 2\pi r^2}{2\pi r} = \frac{2\pi rh}{2\pi r}$
Simplify.	$\frac{S - 2\pi r^2}{2\pi r} = h$

> **TRY IT :: 2.59**   Solve the formula $A = P + Prt$ for t.

> **TRY IT :: 2.60**   Solve the formula $A = P + Prt$ for r.

Sometimes we might be given an equation that is solved for y and need to solve it for x, or vice versa. In the following example, we're given an equation with both x and y on the same side and we'll solve it for y.

EXAMPLE 2.31

Solve the formula $8x + 7y = 15$ for y.

 Solution

We will isolate y on one side of the equation.	$8x + 7y = 15$
Subtract $6x$ from both sides to isolate the term with y.	$8x - 8x + 7y = 15 - 8x$
Simplify.	$7y = 15 - 8x$
Divide both sides by 7 to make the coefficient of y one.	$\dfrac{7y}{7} = \dfrac{15 - 8x}{7}$
Simplify.	$y = \dfrac{15 - 8x}{7}$

> **TRY IT :: 2.61** Solve the formula $4x + 7y = 9$ for y.

> **TRY IT :: 2.62** Solve the formula $5x + 8y = 1$ for y.

Use Formulas to Solve Geometry Applications

In this objective we will use some common geometry formulas. We will adapt our problem solving strategy so that we can solve geometry applications. The geometry formula will name the variables and give us the equation to solve.

In addition, since these applications will all involve shapes of some sort, most people find it helpful to draw a figure and label it with the given information. We will include this in the first step of the problem solving strategy for geometry applications.

HOW TO :: SOLVE GEOMETRY APPLICATIONS.

Step 1. **Read** the problem and make sure all the words and ideas are understood.

Step 2. **Identify** what you are looking for.

Step 3. **Name** what we are looking for by choosing a variable to represent it. Draw the figure and label it with the given information.

Step 4. **Translate** into an equation by writing the appropriate formula or model for the situation. Substitute in the given information.

Step 5. **Solve** the equation using good algebra techniques.

Step 6. **Check** the answer in the problem and make sure it makes sense.

Step 7. **Answer** the question with a complete sentence.

When we solve geometry applications, we often have to use some of the properties of the figures. We will review those properties as needed.

The next example involves the area of a triangle. The area of a triangle is one-half the base times the height. We can write this as $A = \frac{1}{2}bh$, where b = length of the base and h = height.

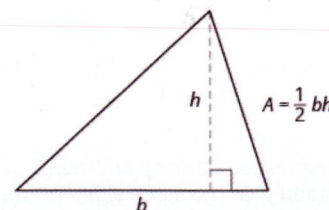

EXAMPLE 2.32

The area of a triangular painting is 126 square inches. The base is 18 inches. What is the height?

✓ **Solution**

Step 1. Read the problem.	
Step 2. Identify what you are looking for.	height of a triangle
Step 3. Name.	
Choose a variable to represent it.	Let $h =$ the height.
Draw the figure and label it with the given information.	Area = 126 sq. in.

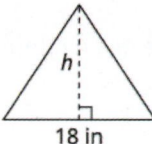

18 in

Step 4. Translate.	
Write the appropriate formula.	$A = \frac{1}{2}bh$
Substitute in the given information.	$126 = \frac{1}{2} \cdot 18 \cdot h$
Step 5. Solve the equation.	$126 = 9h$
Divide both sides by 9.	$14 = h$
Step 6. Check.	

$$A = \frac{1}{2}bh$$
$$126 \overset{?}{=} \frac{1}{2} \cdot 18 \cdot 14$$
$$126 = 126 \checkmark$$

Step 7. Answer the question.	The height of the triangle is 14 inches.

> **TRY IT : : 2.63**
>
> The area of a triangular church window is 90 square meters. The base of the window is 15 meters. What is the window's height?

> **TRY IT : : 2.64**
>
> A triangular tent door has area 15 square feet. The height is five feet. What is the length of the base?

In the next example, we will work with a right triangle. To solve for the measure of each angle, we need to use two triangle properties. In any triangle, the sum of the measures of the angles is $180°$. We can write this as a formula: $m \angle A + m \angle B + m \angle C = 180$. Also, since the triangle is a right triangle, we remember that a right triangle has one $90°$ angle.

Here, we will have to define one angle in terms of another. We will wait to draw the figure until we write expressions for all the angles we are looking for.

EXAMPLE 2.33

The measure of one angle of a right triangle is 40 degrees more than the measure of the smallest angle. Find the measures of all three angles.

⊘ **Solution**

Step 1. Read the problem.	
Step 2. Identify what you are looking for.	the measures of all three angles
Step 3. Name. Choose a variable to represent it.	Let a = 1st angle.
	$a + 40$ = 2nd angle
	90 = 3rd angle (the right angle)

Draw the figure and label it with the given information.

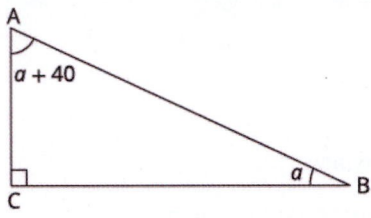

Step 4. Translate.

Write the appropriate formula.	$m\angle A + m\angle B + m\angle C = 180$
Substitute into the formula.	$a + (a + 40) + 90 = 180$

Step 5. Solve the equation.

$$2a + 130 = 180$$
$$2a = 50$$
$$a = 25 \text{ first angle}$$

$$a + 40 \text{ second angle}$$
$$25 + 40$$
$$65$$

$$90 \quad \text{third angle}$$

Step 6. Check.

$$25 + 65 + 90 \overset{?}{=} 180$$
$$180 = 180 \checkmark$$

Step 7. Answer the question.	The three angles measure $25°$, $65°$, and $90°$.

> **TRY IT :: 2.65**
>
> The measure of one angle of a right triangle is 50 more than the measure of the smallest angle. Find the measures of all three angles.

> **TRY IT :: 2.66**
>
> The measure of one angle of a right triangle is 30 more than the measure of the smallest angle. Find the measures of all three angles.

The next example uses another important geometry formula. The **Pythagorean Theorem** tells how the lengths of the three sides of a right triangle relate to each other. Writing the formula in every exercise and saying it aloud as you write it may help you memorize the Pythagorean Theorem.

The Pythagorean Theorem

In any right triangle, where a and b are the lengths of the legs, and c is the length of the hypotenuse, the sum of the squares of the lengths of the two legs equals the square of the length of the hypotenuse.

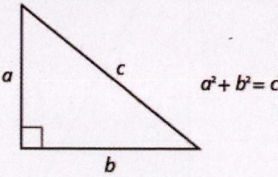

We will use the Pythagorean Theorem in the next example.

EXAMPLE 2.34

Use the Pythagorean Theorem to find the length of the other leg in

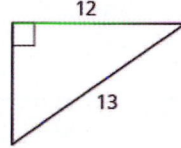

⊘ **Solution**

Step 1. Read the problem.

Step 2. Identify what you are looking for.   the length of the leg of the triangle

Step 3. Name.

Choose a variable to represent it.   Let a = the leg of the triangle.

Label side a.

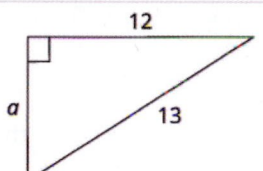

Step 4. Translate.

Write the appropriate formula.
Substitute.

$$a^2 + b^2 = c^2$$
$$a^2 + 12^2 = 13^2$$

Step 5. Solve the equation.
Isolate the variable term.
Use the definition of square root.
Simplify.

$$a^2 + 144 = 169$$
$$a^2 = 25$$
$$a = \sqrt{25}$$
$$a = 5$$

Step 6. Check.
$$5^2 + 12^2 \overset{?}{=} 13^2$$
$$25 + 144 \overset{?}{=} 169$$
$$169 = 169 \checkmark$$

Step 7. Answer the question.   The length of the leg is 5.

> **TRY IT :: 2.67**   Use the Pythagorean Theorem to find the length of the leg in the figure.

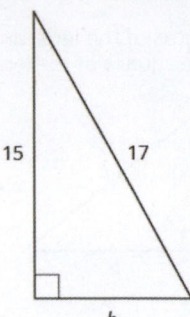

> **TRY IT :: 2.68**   Use the Pythagorean Theorem to find the length of the leg in the figure.

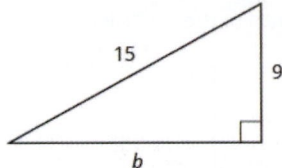

The next example is about the perimeter of a rectangle. Since the perimeter is just the distance around the rectangle, we find the sum of the lengths of its four sides—the sum of two lengths and two widths. We can write is as $P = 2L + 2W$ where L is the length and W is the width. To solve the example, we will need to define the length in terms of the width.

EXAMPLE 2.35

The length of a rectangle is six centimeters more than twice the width. The perimeter is 96 centimeters. Find the length and width.

 **Solution**

Step 1. Read the problem.	
Step 2. Identify what we are looking for.	the length and the width
Step 3. Name. Choose a variable to represent the width. The length is six more than twice the width.	Let $w =$ width. $2w + 6 =$ length

$$w \boxed{}$$
$$2w + 6$$

$$P = 96 \text{ cm}$$

Step 4. Translate.	
Write the appropriate formula.	$P = 2L + 2W$
Substitute in the given information.	$96 = 2(2w + 6) + 2w$

Step 5. Solve the equation.

$$96 = 4w + 12 + 2w$$
$$96 = 6w + 12$$
$$84 = 6w$$
$$14 = w \text{ (width)}$$

$$2w + 6 \text{ (length)}$$
$$2(14) + 6$$
$$34 \quad \text{The length is 34 cm.}$$

Step 6. Check.

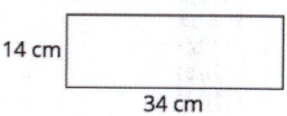

14 cm

34 cm

$$P = 2L + 2W$$
$$96 \overset{?}{=} 2 \cdot 34 + 2 \cdot 14$$
$$96 = 96 \checkmark$$

Step 7. Answer the question.   The length is 34 cm and the width is 14 cm.

> **TRY IT : : 2.69**

The length of a rectangle is seven more than twice the width. The perimeter is 110 inches. Find the length and width.

> **TRY IT : : 2.70**

The width of a rectangle is eight yards less than twice the length. The perimeter is 86 yards. Find the length and width.

The next example is about the perimeter of a triangle. Since the perimeter is just the distance around the triangle, we find the sum of the lengths of its three sides. We can write this as $P = a + b + c$, where a, b, and c are the lengths of the sides.

EXAMPLE 2.36

One side of a triangle is three inches more than the first side. The third side is two inches more than twice the first. The perimeter is 29 inches. Find the length of the three sides of the triangle.

 Solution

Step 1. Read the problem.

Step 2. Identify what we are looking for.   the lengths of the three sides of a triangle

Step 3. Name. Choose a variable to represent the length of the first side.

$$\text{Let } x = \text{length of } 1^{\text{st}} \text{ side.}$$
$$x + 3 = \text{length of } 2^{\text{nd}} \text{ side}$$
$$2x + 2 = \text{length of } 3^{\text{rd}} \text{ side}$$

Perimeter is 29 in.

Step 4. Translate.
Write the appropriate formula.
Substitute in the given information.

$$P = a + b + c$$
$$29 = x + (x + 3) + (2x + 2)$$

Step 5. Solve the equation.

$$29 = 4x + 5$$
$$24 = 4x$$
$$6 = x \quad \text{length of first side}$$

$$x + 3 \quad \text{length of second side}$$
$$6 + 3$$
$$9$$

$$2x + 2 \quad \text{length of second side}$$
$$2 \cdot 6 + 2$$
$$14$$

Step 6. Check.

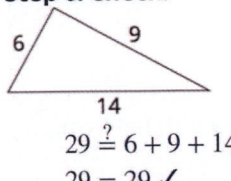

$$29 \overset{?}{=} 6 + 9 + 14$$
$$29 = 29 \checkmark$$

Step 7. Answer the question.

The lengths of the sides of the triangle are 6, 9, and 14 inches.

> **TRY IT : :** 2.71

One side of a triangle is seven inches more than the first side. The third side is four inches less than three times the first. The perimeter is 28 inches. Find the length of the three sides of the triangle.

> **TRY IT : :** 2.72

One side of a triangle is three feet less than the first side. The third side is five feet less than twice the first. The perimeter is 20 feet. Find the length of the three sides of the triangle.

EXAMPLE 2.37

The perimeter of a rectangular soccer field is 360 feet. The length is 40 feet more than the width. Find the length and width.

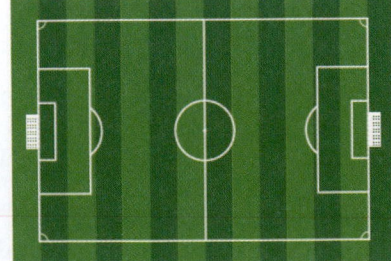

✓ **Solution**

Step 1. Read the problem.

Step 2. Identify what we are looking for.   the length and width of the soccer field

Step 3. Name. Choose a variable to represent it. The length is 40 feet more than the width. Draw the figure and label it with the given information.

Let w = width.
$w + 40 =$ length
Perimeter = 360 feet

$W + 40$

Step 4. Translate.
Write the appropriate formula and substitute.

$P = 2L + 2W$
$360 = 2(w + 40) + 2w$

Step 5. Solve the equation.

$360 = 2w + 80 + 2w$

$360 = 4w + 80$

$280 = 4w$

$70 = w$ the width of the field

$w + 40$ the length of the field

$70 + 40$

110

Step 6. Check.

$P = 2L + 2W$

$360 \overset{?}{=} 2(110) + 2(70)$

$360 = 360 ✓$

Step 7. Answer the question.

The length of the soccer field is 110 feet and the width is 70 feet.

> **TRY IT : : 2.73**

> The perimeter of a rectangular swimming pool is 200 feet. The length is 40 feet more than the width. Find the length and width.

> **TRY IT : : 2.74**

> The length of a rectangular garden is 30 yards more than the width. The perimeter is 300 yards. Find the length and width.

Applications of these geometric properties can be found in many everyday situations as shown in the next example.

EXAMPLE 2.38

Kelvin is building a gazebo and wants to brace each corner by placing a 10" piece of wood diagonally as shown.

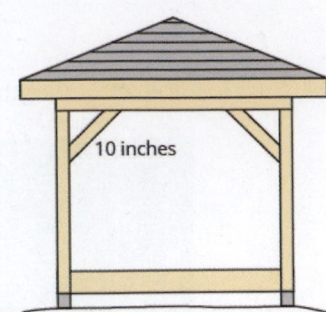

How far from the corner should he fasten the wood if wants the distances from the corner to be equal? Approximate to the nearest tenth of an inch.

✓ Solution

Step 1. Read the problem.

Step 2. Identify what we are looking for.	the distance from the corner that the bracket should be attached

Step 3. Name. Choose a variable to represent it. Draw the figure and label it with the given information.	Let $x =$ the distance from the corner. 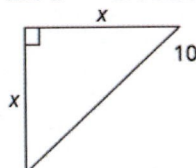

Step 4. Translate. Write the appropriate formula and substitute.	$a^2 + b^2 = c^2$ $x^2 + x^2 = 10^2$

Step 5. Solve the equation. Isolate the variable. Use the definition of square root. Simplify. Approximate to the nearest tenth.	$2x^2 = 100$ $x^2 = 50$ $x = \sqrt{50}$ $x \approx 7.1$

Step 6. Check.

$$a^2 + b^2 = c^2$$
$$(7.1)^2 + (7.1)^2 \approx 10^2 \quad \text{Yes.}$$

Step 7. Answer the question.	Kelvin should fasten each piece of wood approximately 7.1" from the corner.

> **TRY IT : : 2.75**

John puts the base of a 13-foot ladder five feet from the wall of his house as shown in the figure. How far up the wall does the ladder reach?

> **TRY IT : :** 2.76

Randy wants to attach a 17-foot string of lights to the top of the 15 foot mast of his sailboat, as shown in the figure. How far from the base of the mast should he attach the end of the light string?

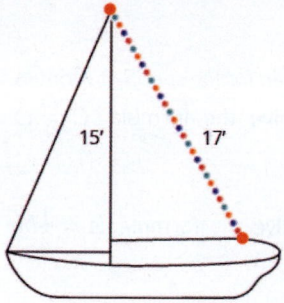

▶ **MEDIA : :**

Access this online resource for additional instruction and practice with solving for a variable in literal equations.

- **Solving Literal Equations (https://openstax.org/l/37literalequat)**

2.3 EXERCISES
Practice Makes Perfect

Solve a Formula for a Specific Variable

In the following exercises, solve the given formula for the specified variable.

165. Solve the formula $C = \pi d$ for d.

166. Solve the formula $C = \pi d$ for π.

167. Solve the formula $V = LWH$ for L.

168. Solve the formula $V = LWH$ for H.

169. Solve the formula $A = \frac{1}{2}bh$ for b.

170. Solve the formula $A = \frac{1}{2}bh$ for h.

171. Solve the formula $A = \frac{1}{2}d_1 d_2$ for d_1.

172. Solve the formula $A = \frac{1}{2}d_1 d_2$ for d_2.

173. Solve the formula $A = \frac{1}{2}h(b_1 + b_2)$ for b_1.

174. Solve the formula $A = \frac{1}{2}h(b_1 + b_2)$ for b_2.

175. Solve the formula $h = 54t + \frac{1}{2}at^2$ for a.

176. Solve the formula $h = 48t + \frac{1}{2}at^2$ for a.

177. Solve $180 = a + b + c$ for a.

178. Solve $180 = a + b + c$ for c.

179. Solve the formula $A = \frac{1}{2}pl + B$ for p.

180. Solve the formula $A = \frac{1}{2}pl + B$ for l.

181. Solve the formula $P = 2L + 2W$ for L.

182. Solve the formula $P = 2L + 2W$ for W.

In the following exercises, solve for the formula for y.

183. Solve the formula $8x + y = 15$ for y.

184. Solve the formula $9x + y = 13$ for y.

185. Solve the formula $-4x + y = -6$ for y.

186. Solve the formula $-5x + y = -1$ for y.

187. Solve the formula $x - y = -4$ for y.

188. Solve the formula $x - y = -3$ for y.

189. Solve the formula $4x + 3y = 7$ for y.

190. Solve the formula $3x + 2y = 11$ for y.

191. Solve the formula $2x + 3y = 12$ for y.

192. Solve the formula $5x + 2y = 10$ for y.

193. Solve the formula $3x - 2y = 18$ for y.

194. Solve the formula $4x - 3y = 12$ for y.

Use Formulas to Solve Geometry Applications

In the following exercises, solve using a geometry formula.

195. A triangular flag has area 0.75 square feet and height 1.5 foot. What is its base?

196. A triangular window has area 24 square feet and height six feet. What is its base?

197. What is the base of a triangle with area 207 square inches and height 18 inches?

198. What is the height of a triangle with area 893 square inches and base 38 inches?

199. The two smaller angles of a right triangle have equal measures. Find the measures of all three angles.

200. The measure of the smallest angle of a right triangle is $20°$ less than the measure of the next larger angle. Find the measures of all three angles.

201. The angles in a triangle are such that one angle is twice the smallest angle, while the third angle is three times as large as the smallest angle. Find the measures of all three angles.

202. The angles in a triangle are such that one angle is 20 more than the smallest angle, while the third angle is three times as large as the smallest angle. Find the measures of all three angles.

In the following exercises, use the Pythagorean Theorem to find the length of the hypotenuse.

203.

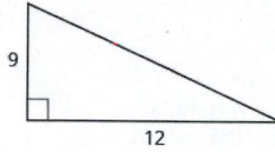

9

12

204.

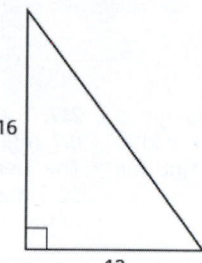

16

12

205.

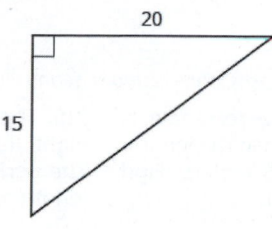

20

15

206.

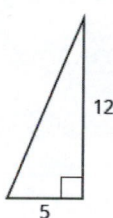

12

5

In the following exercises, use the Pythagorean Theorem to find the length of the leg. Round to the nearest tenth if necessary.

207.

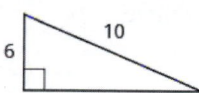

10

6

208.

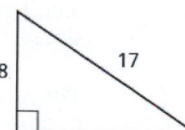

8

17

209.

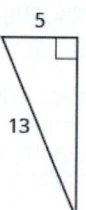

5

13

210.

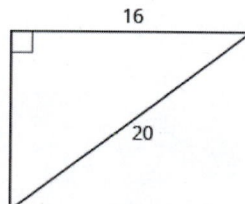

16

20

211.

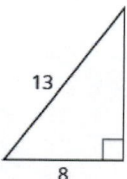

13

8

212.

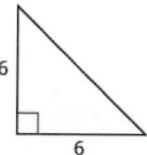

6

6

213.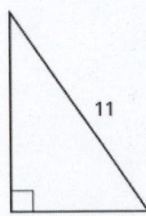

214.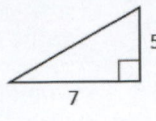

In the following exercises, solve using a geometry formula.

215. The width of a rectangle is seven meters less than the length. The perimeter is 58 meters. Find the length and width.

216. The length of a rectangle is eight feet more than the width. The perimeter is 60 feet. Find the length and width.

217. The width of the rectangle is 0.7 meters less than the length. The perimeter of a rectangle is 52.6 meters. Find the dimensions of the rectangle.

218. The length of the rectangle is 1.1 meters less than the width. The perimeter of a rectangle is 49.4 meters. Find the dimensions of the rectangle.

219. The perimeter of a rectangle of 150 feet. The length of the rectangle is twice the width. Find the length and width of the rectangle.

220. The length of the rectangle is three times the width. The perimeter of a rectangle is 72 feet. Find the length and width of the rectangle.

221. The length of the rectangle is three meters less than twice the width. The perimeter of a rectangle is 36 meters. Find the dimensions of the rectangle.

222. The length of a rectangle is five inches more than twice the width. The perimeter is 34 inches. Find the length and width.

223. The perimeter of a triangle is 39 feet. One side of the triangle is one foot longer than the second side. The third side is two feet longer than the second side. Find the length of each side.

224. The perimeter of a triangle is 35 feet. One side of the triangle is five feet longer than the second side. The third side is three feet longer than the second side. Find the length of each side.

225. One side of a triangle is twice the smallest side. The third side is five feet more than the shortest side. The perimeter is 17 feet. Find the lengths of all three sides.

226. One side of a triangle is three times the smallest side. The third side is three feet more than the shortest side. The perimeter is 13 feet. Find the lengths of all three sides.

227. The perimeter of a rectangular field is 560 yards. The length is 40 yards more than the width. Find the length and width of the field.

228. The perimeter of a rectangular atrium is 160 feet. The length is 16 feet more than the width. Find the length and width of the atrium.

229. A rectangular parking lot has perimeter 250 feet. The length is five feet more than twice the width. Find the length and width of the parking lot.

230. A rectangular rug has perimeter 240 inches. The length is 12 inches more than twice the width. Find the length and width of the rug.

In the following exercises, solve. Approximate answers to the nearest tenth, if necessary.

231. A 13-foot string of lights will be attached to the top of a 12-foot pole for a holiday display as shown. How far from the base of the pole should the end of the string of lights be anchored?

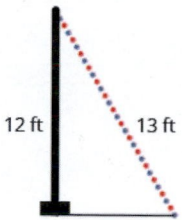

232. Pam wants to put a banner across her garage door diagonally, as shown, to congratulate her son for his college graduation. The garage door is 12 feet high and 16 feet wide. How long should the banner be to fit the garage door?

233. Chi is planning to put a diagonal path of paving stones through her flower garden as shown. The flower garden is a square with side 10 feet. What will the length of the path be?

234. Brian borrowed a 20-foot extension ladder to use when he paints his house. If he sets the base of the ladder six feet from the house as shown, how far up will the top of the ladder reach?

Everyday Math

235. Converting temperature While on a tour in Greece, Tatyana saw that the temperature was 40° Celsius. Solve for F in the formula $C = \frac{5}{9}(F - 32)$ to find the Fahrenheit temperature.

236. Converting temperature Yon was visiting the United States and he saw that the temperature in Seattle one day was 50° Fahrenheit. Solve for C in the formula $F = \frac{9}{5}C + 32$ to find the Celsius temperature.

237. Christa wants to put a fence around her triangular flowerbed. The sides of the flowerbed are six feet, eight feet and 10 feet. How many feet of fencing will she need to enclose her flowerbed?

238. Jose just removed the children's play set from his back yard to make room for a rectangular garden. He wants to put a fence around the garden to keep the dog out. He has a 50-foot roll of fence in his garage that he plans to use. To fit in the backyard, the width of the garden must be 10 feet. How long can he make the other side?

Writing Exercises

239. If you need to put tile on your kitchen floor, do you need to know the perimeter or the area of the kitchen? Explain your reasoning.

240. If you need to put a fence around your backyard, do you need to know the perimeter or the area of the backyard? Explain your reasoning.

241. Look at the two figures below.

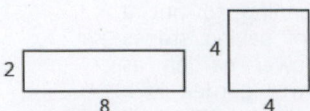

ⓐ Which figure looks like it has the larger area? Which looks like it has the larger perimeter?

ⓑ Now calculate the area and perimeter of each figure. Which has the larger area? Which has the larger perimeter?

ⓒ Were the results of part (b) the same as your answers in part (a)? Is that surprising to you?

242. Write a geometry word problem that relates to your life experience, then solve it and explain all your steps.

Self Check

ⓐ *After completing the exercises, use this checklist to evaluate your mastery of the objectives of this section.*

I can...	Confidently	With some help	No-I don't get it!
solve a formula for a specific variable.			
use formulas to solve geometry applications.			

ⓑ *What does this checklist tell you about your mastery of this section? What steps will you take to improve?*

2.4 Solve Mixture and Uniform Motion Applications

Learning Objectives

By the end of this section, you will be able to:

› Solve coin word problems
› Solve ticket and stamp word problems
› Solve mixture word problems
› Solve uniform motion applications

Be Prepared!

Before you get started, take this readiness quiz.

1. Simplify: $0.25x + 0.10(x + 4)$.

   If you missed this problem, review Example 1.52.

2. The number of adult tickets is three more than twice the number of children tickets. Let c represent the number of children tickets. Write an expression for the number of adult tickets.
   If you missed this problem, review Example 1.11.

3. Convert 4.2% to a decimal.
   If you missed this problem, review Example 1.40.

Solve Coin Word Problems

Using algebra to find the number of nickels and pennies in a piggy bank may seem silly. You may wonder why we just don't open the bank and count them. But this type of problem introduces us to some techniques that will be useful as we move forward in our study of mathematics.

If we have a pile of dimes, how would we determine its value? If we count the number of dimes, we'll know how many we have—the *number* of dimes. But this does not tell us the *value* of all the dimes. Say we counted 23 dimes, how much are they worth? Each dime is worth $0.10—that is the *value* of one dime. To find the total value of the pile of 23 dimes, multiply 23 by $0.10 to get $2.30.

The number of dimes times the value of each dime equals the total value of the dimes.

$$number \cdot value \;=\; total\ value$$
$$23 \cdot \$0.10 \;=\; \$2.30$$

This method leads to the following model.

Total Value of Coins

For the same type of coin, the total value of a number of coins is found by using the model

$$number \cdot value = total\ value$$

- *number* is the number of coins
- *value* is the value of each coin
- *total value* is the total value of all the coins

If we had several types of coins, we could continue this process for each type of coin, and then we would know the total value of each type of coin. To get the total value of *all* the coins, add the total value of each type of coin.

EXAMPLE 2.39

Jesse has $3.02 worth of pennies and nickels in his piggy bank. The number of nickels is three more than eight times the number of pennies. How many nickels and how many pennies does Jesse have?

⊘ Solution

Step 1. Read the problem. Determine the types of coins involved. Create a table. Write in the value of each type of coin.	pennies and nickels Pennies are worth $0.10. Nickels are worth $0.05.
Step 2. Identify what we are looking for.	the number of pennies and nickels
Step 3. Name. Represent the number of each type of coin using variables. The number of nickels is defined in terms of the number of pennies, so start with pennies. The number of nickels is three more than eight times the number of pennies.	Let $p =$ number of pennies. $8p + 3 =$ number of nickels

In the chart, multiply the number and the value to get the total value of each type of coin.

Type	Number	• Value ($)	= Total Value ($)
pennies	p	0.01	$0.01p$
nickels	$8p + 3$	0.05	$0.05(8p + 3)$
			$3.02

Step 4. Translate. Write the equation by adding the total value of all the types of coins.	$0.01p + 0.05(8p + 3) = 3.02$
Step 5. Solve the equation.	$0.01p + 0.40p + 0.15 = 3.02$ $0.41p + 0.15 = 3.02$ $0.41p = 2.87$ $p = 7$ pennies
How many nickels?	$8p + 3$ $8(7) + 3$ 59 nickels

Step 6. Check the answer in the problem and make sure it makes sense.
Jesse has 7 pennies and 59 nickels.
Is the total value $3.02?

$$7(0.01) + 59(0.05) \overset{?}{=} 3.02$$
$$3.02 = 3.02 ✓$$

> **TRY IT : : 2.77**

  Jesse has $6.55 worth of quarters and nickels in his pocket. The number of nickels is five more than two times the number of quarters. How many nickels and how many quarters does Jesse have?

> **TRY IT : : 2.78**

  Elane has $7.00 total in dimes and nickels in her coin jar. The number of dimes that Elane has is seven less than three times the number of nickels. How many of each coin does Elane have?

The steps for solving a coin word problem are summarized below.

HOW TO : : SOLVE COIN WORD PROBLEMS.

Step 1. **Read** the problem. Make sure all the words and ideas are understood.
- Determine the types of coins involved.
- Create a table to organize the information.
  - Label the columns "type," "number," "value," and "total value."
  - List the types of coins.
  - Write in the value of each type of coin.
  - Write in the total value of all the coins.

Type	Number •	Value ($) =	Total Value ($)

Step 2. **Identify** what you are looking for.

Step 3. **Name** what you are looking for. Choose a variable to represent that quantity.
- Use variable expressions to represent the number of each type of coin and write them in the table.
- Multiply the number times the value to get the total value of each type of coin.

Step 4. **Translate** into an equation.
- It may be helpful to restate the problem in one sentence with all the important information. Then, translate the sentence into an equation.
- Write the equation by adding the total values of all the types of coins.

Step 5. **Solve** the equation using good algebra techniques.

Step 6. **Check** the answer in the problem and make sure it makes sense.

Step 7. **Answer** the question with a complete sentence.

Solve Ticket and Stamp Word Problems

Problems involving tickets or stamps are very much like coin problems. Each type of ticket and stamp has a value, just like each type of coin does. So to solve these problems, we will follow the same steps we used to solve coin problems.

EXAMPLE 2.40

Danny paid $15.75 for stamps. The number of 49-cent stamps was five less than three times the number of 35-cent stamps. How many 49-cent stamps and how many 35-cent stamps did Danny buy?

⊘ **Solution**

Step 1. Determine the types of stamps involved.	49-cent stamps and 35-cent stamps
Step 2. Identify we are looking for.	the number of 49-cent stamps and the number of 35-cent stamps
Step 3. Write variable expressions to represent the number of each type of stamp.	Let x = number of 35-cent stamps.
"The number of 49-cent stamps was five less than three times the number of 35-cent stamps."	$3x - 5 =$ number of 49-cent stamps

Type	Number $\cdot$	Value($) $=$	Total Value ($)
49 cent stamps	$3x - 5$	0.49	$0.49(3x - 5)$
35 cent stamps	x	0.35	$0.35x$
			15.75

Step 4. Write the equation from the total values.

$$0.49(3x - 5) + 0.35x = 15.75$$

Step 5. Solve the equation.

$$1.47x - 2.45 + 0.35x = 15.75$$
$$1.82x - 2.45 = 15.75$$
$$1.82x = 18.2$$

How many 49-cent stamps?

$$x = 10 \quad \text{35-cent stamps}$$
$$3x - 5$$
$$3(10) - 5$$
$$25 \quad \text{49-cent stamps}$$

Step 6. Check.

$$10(0.35) + 25(0.49) \overset{?}{=} 15.75$$
$$3.50 + 12.25 \overset{?}{=} 15.75$$
$$15.75 = 15.75 \checkmark$$

Step 7. Answer the question with a complete sentence.

Danny bought ten 35-cent stamps and twenty-five 49-cent stamps.

> **TRY IT :: 2.79**
>
> Eric paid $19.88 for stamps. The number of 49-cent stamps was eight more than twice the number of 35-cent stamps. How many 49-cent stamps and how many 35-cent stamps did Eric buy?

> **TRY IT :: 2.80**
>
> Kailee paid $14.74 for stamps. The number of 49-cent stamps was four less than three times the number of 20-cent stamps. How many 49-cent stamps and how many 20-cent stamps did Kailee buy?

In most of our examples so far, we have been told that one quantity is four more than twice the other, or something similar. In our next example, we have to relate the quantities in a different way.

Suppose Aniket sold a total of 100 tickets. Each ticket was either an adult ticket or a child ticket. If he sold 20 child tickets, how many adult tickets did he sell?

Did you say "80"? How did you figure that out? Did you subtract 20 from 100?

If he sold 45 child tickets, how many adult tickets did he sell?

Did you say "55"? How did you find it? By subtracting 45 from 100?

Now, suppose Aniket sold x child tickets. Then how many adult tickets did he sell? To find out, we would follow the same logic we used above. In each case, we subtracted the number of child tickets from 100 to get the number of adult tickets. We now do the same with x.

We have summarized this in the table.

Child tickets	Adult tickets
20	80
45	55
75	25
x	$100 - x$

We will apply this technique in the next example.

EXAMPLE 2.41

A whale-watching ship had 40 paying passengers on board. The total revenue collected from tickets was $1,196. Full-fare passengers paid $32 each and reduced-fare passengers paid $26 each. How many full-fare passengers and how many reduced-fare passengers were on the ship?

✓ **Solution**

Step 1. Determine the types of tickets involved.	full-fare tickets and reduced-fare tickets
Step 2. Identify what we are looking for.	the number of full-fare tickets and reduced-fare tickets
Step 3. Name. Represent the number of each type of ticket using variables.	Let f = the number of full-fare tickets. $40 - f$ = the number of reduced-fare tickets

We know the total number of tickets sold was 40. This means the number of reduced-fare tickets is 40 less the number of full-fare tickets. Multiply the number times the value to get the total value of each type of ticket.

Type	Number	· Value ($)	= Total Value ($)
Full-fare	f	32	$32f$
Reduced-fare	$40 - f$	26	$26(40 - f)$
			1,196

Step 4. Translate. Write the equation by adding the total values of each type of ticket.	$32f + 26(40 - f) = 1,196$
Step 5. Solve the equation.	$32f + 1,040 - 26f = 1,196$ $6f = 156$ $f = 26$ full-fare tickets
How many reduced-fare?	$40 - f$ $40 - 26$ 14 reduced-fare tickets

Step 6. Check the answer.
There were 26 full-fare tickets at $32 each and 14 reduced-fare tickets at $26 each. Is the total value $116?

$$26 \cdot 32 = 832$$
$$14 \cdot 26 = \underline{364}$$
$$1,196 ✓$$

Step 7. Answer the question.	They sold 26 full-fare and 14 reduced-fare tickets.

> **TRY IT :: 2.81**
>
> During her shift at the museum ticket booth, Leah sold 115 tickets for a total of $1,163. Adult tickets cost $12 and student tickets cost $5. How many adult tickets and how many student tickets did Leah sell?

> **TRY IT :: 2.82**
>
> Galen sold 810 tickets for his church's carnival for a total revenue of $2,820. Children's tickets cost $3 each and adult tickets cost $5 each. How many children's tickets and how many adult tickets did he sell?

Solve Mixture Word Problems

Now we'll solve some more general applications of the mixture model. In mixture problems, we are often mixing two quantities, such as raisins and nuts, to create a mixture, such as trail mix. In our tables we will have a row for each item to be mixed as well as one for the final mixture.

EXAMPLE 2.42

Henning is mixing raisins and nuts to make 25 pounds of trail mix. Raisins cost $4.50 a pound and nuts cost $8 a pound. If Henning wants his cost for the trail mix to be $6.60 a pound, how many pounds of raisins and how many pounds of nuts should he use?

⊘ **Solution**

Step 1. Determine what is being mixed.	The 25 pounds of trail mix will come from mixing raisins and nuts.
Step 2. Identify what we are looking for.	the number of pounds of raisins and nuts
Step 3. Represent the number of each type of ticket using variables.	Let $x =$ number of pounds of raisins. $25 - x =$ number of pounds of nuts

As before, we fill in a chart to organize our information.
We enter the price per pound for each item.
We multiply the number times the value to get the total value.

Type	Number of pounds	·	Price per Pound ($)	=	Total Value ($)
Raisins	x		4.50		$4.5x$
Nuts	$25 - x$		8		$8(25 - x)$
Trail mix	25		6.60		$25(6.60)$

Notice that the last column in the table gives the information for the total amount of the mixture.

Step 4. Translate into an equation.	The value of the raisins plus the value of the nuts will be the value of the trail mix.
Step 5. Solve the equation.	$$4.5x + 8(25 - x) = 25(6.60)$$

$$4.5x + 200 - 8x = 165$$
$$-3.5x = -35$$
$$x = 10 \quad \text{pounds of raisins}$$

Find the number of pounds of nuts.	$25 - x$

$$25 - 10$$
$$15 \quad \text{pounds of nuts}$$

Step 6. Check.

$$4.5(10) + 8(15) \stackrel{?}{=} 25(6.60)$$
$$45 + 120 \stackrel{?}{=} 165$$
$$165 = 165 ✓$$

Step 7. Answer the question.	Henning mixed ten pounds of raisins with 15 pounds of nuts.

> **TRY IT : : 2.83**
>
> Orlando is mixing nuts and cereal squares to make a party mix. Nuts sell for $7 a pound and cereal squares sell for $4 a pound. Orlando wants to make 30 pounds of party mix at a cost of $6.50 a pound, how many pounds of nuts and how many pounds of cereal squares should he use?

TRY IT : : 2.84

Becca wants to mix fruit juice and soda to make a punch. She can buy fruit juice for $3 a gallon and soda for $4 a gallon. If she wants to make 28 gallons of punch at a cost of $3.25 a gallon, how many gallons of fruit juice and how many gallons of soda should she buy?

Solve Uniform Motion Applications

When you are driving down the interstate using your cruise control, the speed of your car stays the same—it is uniform. We call a problem in which the speed of an object is constant a uniform motion application. We will use the distance, rate, and time formula, $D = rt$, to compare two scenarios, such as two vehicles travelling at different rates or in opposite directions.

Our problem solving strategies will still apply here, but we will add to the first step. The first step will include drawing a diagram that shows what is happening in the example. Drawing the diagram helps us understand what is happening so that we will write an appropriate equation. Then we will make a table to organize the information, like we did for the coin, ticket, and stamp applications.

The steps are listed here for easy reference:

 HOW TO : : SOLVE A UNIFORM MOTION APPLICATION.

Step 1. **Read** the problem. Make sure all the words and ideas are understood.
   ◦ Draw a diagram to illustrate what is happening.
   ◦ Create a table to organize the information.
      ▪ Label the columns rate, time, distance.
      ▪ List the two scenarios.
      ▪ Write in the information you know.

	Rate •	Time =	Distance

Step 2. **Identify** what you are looking for.

Step 3. **Name** what you are looking for. Choose a variable to represent that quantity.
   ◦ Complete the chart.
   ◦ Use variable expressions to represent that quantity in each row.
   ◦ Multiply the rate times the time to get the distance.

Step 4. **Translate** into an equation.
   ◦ Restate the problem in one sentence with all the important information.
   ◦ Then, translate the sentence into an equation.

Step 5. **Solve** the equation using good algebra techniques.

Step 6. **Check** the answer in the problem and make sure it makes sense.

Step 7. **Answer** the question with a complete sentence.

EXAMPLE 2.43

Wayne and Dennis like to ride the bike path from Riverside Park to the beach. Dennis's speed is seven miles per hour faster than Wayne's speed, so it takes Wayne two hours to ride to the beach while it takes Dennis 1.5 hours for the ride. Find the speed of both bikers.

 **Solution**

Step 1. Read the problem. Make sure all the words and ideas are understood.
   • Draw a diagram to illustrate what it happening. Shown below is a sketch of what is happening in the example.

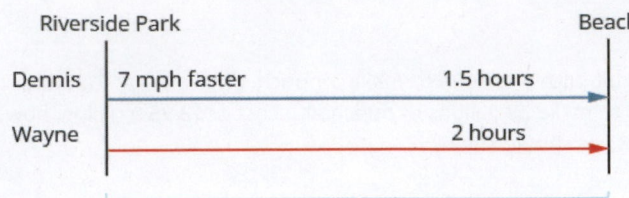

- Create a table to organize the information.
  - Label the columns "Rate," "Time," and "Distance."
  - List the two scenarios.
  - Write in the information you know.

	Rate (mph) •	Time (hrs) =	Distance (miles)
Dennis		1.5	
Wayne		2	

Step 2. Identify what you are looking for.

You are asked to find the speed of both bikers.

Notice that the distance formula uses the word "rate," but it is more common to use "speed" when we talk about vehicles in everyday English.

Step 3. Name what we are looking for. Choose a variable to represent that quantity.

- Complete the chart
- Use variable expressions to represent that quantity in each row.
  We are looking for the speed of the bikers. Let's let r represent Wayne's speed. Since Dennis' speed is 7 mph faster, we represent that as $r + 7$

$r + 7$ = Dennis' speed

r = Wayne's speed

Fill in the speeds into the chart.

	Rate (mph) •	Time (hrs) =	Distance (miles)
Dennis	$r + 7$	1.5	
Wayne	r	2	

- Multiply the rate times the time to get the distance.

	Rate (mph) •	Time (hrs) =	Distance (miles)
Dennis	$r + 7$	1.5	$1.5(r + 7)$
Wayne	r	2	$2r$

Step 4. Translate into an equation.

- Restate the problem in one sentence with all the important information.
- Then, translate the sentence into an equation.
  The equation to model this situation will come from the relation between the distances. Look at the diagram we drew above. How is the distance travelled by Dennis related to the distance travelled by Wayne?
  Since both bikers leave from Riverside and travel to the beach, they travel the same distance. So we write:

distance traveled by Dennis   =   distance traveled by Wayne

Translate to an equation. $1.5(r + 7)$ = $2r$

Step 5. Solve the equation using algebra techniques.

Now solve this equation.

$$1.5(r + 7) = 2r$$
$$1.5r + 10.5 = 2r$$
$$10.5 = 0.5r$$
$$21 = r$$

So Wayne's speed is 21 mph.

Find Dennis' speed.

$$r + 7$$
$$21 + 7$$
$$28$$

Dennis' speed 28 mph.

Step 6. Check the answer in the problem and make sure it makes sense.

Dennis 28 mph(1.5 hours) = 42 miles

Wayne 21 mph(2 hours) = 42 miles ✓

Step 7. Answer the question with a complete sentence.

Wayne rode at 21 mph and Dennis rode at 28 mph.

 TRY IT : : 2.85

An express train and a local train leave Pittsburgh to travel to Washington, D.C. The express train can make the trip in four hours and the local train takes five hours for the trip. The speed of the express train is 12 miles per hour faster than the speed of the local train. Find the speed of both trains.

 TRY IT : : 2.86

Jeromy can drive from his house in Cleveland to his college in Chicago in 4.5 hours. It takes his mother six hours to make the same drive. Jeromy drives 20 miles per hour faster than his mother. Find Jeromy's speed and his mother's speed.

In **Example 2.43**, we had two bikers traveling the same distance. In the next example, two people drive toward each other until they meet.

EXAMPLE 2.44

Carina is driving from her home in Anaheim to Berkeley on the same day her brother is driving from Berkeley to Anaheim, so they decide to meet for lunch along the way in Buttonwillow. The distance from Anaheim to Berkeley is 410 miles. It takes Carina three hours to get to Buttonwillow, while her brother drives four hours to get there. Carina's average speed is 15 miles per hour faster than her brother's average speed. Find Carina's and her brother's average speeds.

⊘ **Solution**

Step 1. Read the problem. Make sure all the words and ideas are understood.
- Draw a diagram to illustrate what it happening. Below shows a sketch of what is happening in the example.

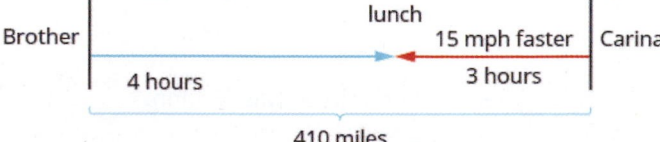

- Create a table to organize the information.
  - Label the columns rate, time, distance.
  - List the two scenarios.
  - Write in the information you know.

	Rate (mph) •	Time (hrs) =	Distance (miles)
Carina		3	
Brother		4	
			410

Step 2. Identify what we are looking for.

We are asked to find the average speeds of Carina and her brother.

Step 3. Name what we are looking for. Choose a variable to represent that quantity.

- Complete the chart.
- Use variable expressions to represent that quantity in each row.
  We are looking for their average speeds. Let's let r represent the average speed of Carina. Since the brother's speed is 15 mph faster, we represent that as $r + 15$.

  Fill in the speeds into the chart.
- Multiply the rate times the time to get the distance.

	Rate (mph) •	Time (hrs) =	Distance (miles)
Carina	r	3	$3r$
Brother	$r + 15$	4	$4(r+15)$
			410

Step 4. Translate into an equation.

- Restate the problem in one sentence with all the important information.
- Then, translate the sentence into an equation.
  Again, we need to identify a relationship between the distances in order to write an equation. Look at the diagram we created above and notice the relationship between the distance Carina traveled and the distance her brother traveled.
  The distance Carina traveled plus the distance her brother travel must add up to 410 miles. So we write:

$$\underbrace{\text{distance traveled by Carina}} + \underbrace{\text{distance traveled by her brother}} = 410$$

Translate to an equation. $3r$ $+$ $4(r + 15)$ $= 410$

Step 5. Solve the equation using algebra techniques.

Now solve this equation.
$$3r + 4(r + 15) = 410$$
$$3r + 4r + 60 = 410$$
$$7r + 60 = 410$$
$$7r = 350$$
$$r = 50$$

So Carina's speed was 50 mph.

Her brother's speed is $r + 10$.
$$r + 15$$
$$50 + 15$$
$$65$$

Her brother's speed was 65 mph.

Step 6. Check the answer in the problem and make sure it makes sense.

Carina drove 50 mph(3 hours) =   150 miles

Her brother drove 65 mph(4 hours) = 260 miles

410 miles ✓

Step 7. Answer the question with a complete sentence.

Carina drove 50 mph and her brother 65 mph.

TRY IT : : 2.87

Christopher and his parents live 115 miles apart. They met at a restaurant between their homes to celebrate his mother's birthday. Christopher drove one and a half hours while his parents drove one hour to get to the restaurant. Christopher's average speed was ten miles per hour faster than his parents' average speed. What were the average speeds of Christopher and of his parents as they drove to the restaurant?

 TRY IT : : 2.88

Ashley goes to college in Minneapolis, 234 miles from her home in Sioux Falls. She wants her parents to bring her more winter clothes, so they decide to meet at a restaurant on the road between Minneapolis and Sioux Falls. Ashley and her parents both drove two hours to the restaurant. Ashley's average speed was seven miles per hour faster than her parents' average speed. Find Ashley's and her parents' average speed.

As you read the next example, think about the relationship of the distances traveled. Which of the previous two examples is more similar to this situation?

EXAMPLE 2.45

Two truck drivers leave a rest area on the interstate at the same time. One truck travels east and the other one travels west. The truck traveling west travels at 70 mph and the truck traveling east has an average speed of 60 mph. How long will they travel before they are 325 miles apart?

⊘ **Solution**

Step 1. Read the problem. Make all the words and ideas are understood.
- Draw a diagram to illustrate what it happening.

- Create a table to organize the information.
  - Label the columns rate, time, distance.
  - List the two scenarios.
  - Write in the information you know.

	Rate (mph) •	Time (hrs) =	Distance (miles)
West	70		
East	60		
			325

Step 2. Identify what we are looking for.

We are asked to find the amount of time the trucks will travel until they are 325 miles apart.

Step 3. Name what we are looking for. Choose a variable to represent that quantity.

- Complete the chart.
- Use variable expressions to represent that quantity in each row.
  We are looking for the time travelled. Both trucks will travel the same amount of time.
  Let's call the time t. Since their speeds are different, they will travel different distances.

- Multiply the rate times the time to get the distance.

	Rate (mph) •	Time (hrs) =	Distance (miles)
West	70	t	$70t$
East	60	t	$60t$
			325

Step 4. Translate into an equation.

- Restate the problem in one sentence with all the important information.
- Then, translate the sentence into an equation.
  We need to find a relation between the distances in order to write an equation. Looking at the diagram, what is the relationship between the distances each of the trucks will travel?
  The distance travelled by the truck going west plus the distance travelled by the truck going east must add up to 325 miles. So we write:

distance traveled by westbound truck	+ distance traveled by eastbound truck	= 325
Translate to an equation. $70t$	+ $60t$	= 325

Step 5. Solve the equation using algebra techniques.

$$\text{Now solve this equation} \quad 70t + 60t = 325$$
$$130t = 325$$
$$t = 2.5$$

So it will take the trucks 2.5 hours to be 325 miles apart.

Step 6. Check the answer in the problem and make sure it makes sense.

$$\text{Truck going West} \quad 70 \text{ mph}(2.5 \text{ hours}) = 175 \text{ miles}$$
$$\text{Truck going East} \quad 60 \text{ mph}(2.5 \text{hours}) = \underline{150 \text{ miles}}$$
$$325 \text{ miles } \checkmark$$

Step 7. Answer the question with a complete sentence.
It will take the trucks 2.5 hours to be 325 miles apart.

> **TRY IT : :** 2.89

  Pierre and Monique leave their home in Portland at the same time. Pierre drives north on the turnpike at a speed of 75 miles per hour while Monique drives south at a speed of 68 miles per hour. How long will it take them to be 429 miles apart?

> **TRY IT : :** 2.90

  Thanh and Nhat leave their office in Sacramento at the same time. Thanh drives north on I-5 at a speed of 72 miles per hour. Nhat drives south on I-5 at a speed of 76 miles per hour. How long will it take them to be 330 miles apart?

It is important to make sure that the units match when we use the distance rate and time formula. For instance, if the rate is in miles per hour, then the time must be in hours.

EXAMPLE 2.46

When Naoko walks to school, it takes her 30 minutes. If she rides her bike, it takes her 15 minutes. Her speed is three miles per hour faster when she rides her bike than when she walks. What is her speed walking and her speed riding her bike?

⊘ **Solution**

First, we draw a diagram that represents the situation to help us see what is happening.

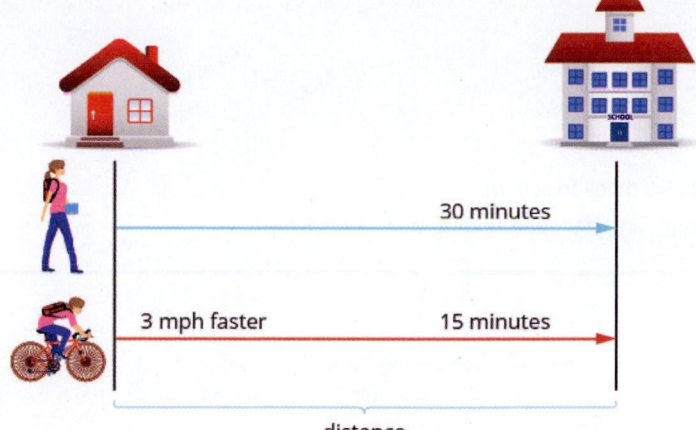

distance

We are asked to find her speed walking and riding her bike. Let's call her walking speed r. Since her biking speed is three miles per hour faster, we will call that speed $r + 3$. We write the speeds in the chart.

The speed is in miles per hour, so we need to express the times in hours, too, in order for the units to be the same. Remember, 1 hour is 60 minutes. So:

$$30 \text{ minutes is } \frac{30}{60} \text{ or } \frac{1}{2} \text{ hour}$$

$$15 \text{ minutes is } \frac{15}{60} \text{ or } \frac{1}{4} \text{ hour}$$

We write the times in the chart.

Next, we multiply rate times time to fill in the distance column.

	Rate (mph)	•	Time (hrs)	=	Distance (miles)
Walk	r		$\frac{1}{2}$		$\frac{1}{2}r$
Bike	$r + 3$		$\frac{1}{4}$		$\frac{1}{4}(r + 3)$

The equation will come from the fact that the distance from Naoko's home to her school is the same whether she is walking or riding her bike.
So we say:

distance walked = distance covered by bike

Translate to an equation.	$\frac{1}{2}r$	$=$	$\frac{1}{4}(r + 3)$
Solve this equation.		$\frac{1}{2}r = \frac{1}{4}(r + 3)$	
Clear the fractions by multiplying by the LCD of all the fractions in the equation.		$8 \cdot \frac{1}{2}r = 8 \cdot \frac{1}{4}(r + 3)$	
Simplify.		$4r = 2(r + 3)$	
		$4r = 2r + 6$	
		$2r = 6$	
		$r = 3 \text{ mph } \text{ walking speed}$	
		$r + 3 \qquad \text{biking speed}$	
		$3 + 3$	
		6	

Let's check if this works.

Walk 3 mph (0.5 hour)   =   1.5 miles
Bike 6 mph (0.25 hour)   =   1.5 miles

Yes, either way Naoko travels 1.5 miles to school.

Naoko's walking speed is 3 mph and her speed riding her bike is 6 mph.

 TRY IT : : 2.91

Suzy takes 50 minutes to hike uphill from the parking lot to the lookout tower. It takes her 30 minutes to hike back down to the parking lot. Her speed going downhill is 1.2 miles per hour faster than her speed going uphill. Find Suzy's uphill and downhill speeds.

 TRY IT : : 2.92

Llewyn takes 45 minutes to drive his boat upstream from the dock to his favorite fishing spot. It takes him 30 minutes to drive the boat back downstream to the dock. The boat's speed going downstream is four miles per hour faster than its speed going upstream. Find the boat's upstream and downstream speeds.

In the distance, rate and time formula, time represents the actual amount of elapsed time (in hours, minutes, etc.). If a problem gives us starting and ending times as clock times, we must find the elapsed time in order to use the formula.

EXAMPLE 2.47

Cruz is training to compete in a triathlon. He left his house at 6:00 and ran until 7:30. Then he rode his bike until 9:45. He covered a total distance of 51 miles. His speed when biking was 1.6 times his speed when running. Find Cruz's biking and running speeds.

 Solution

A diagram will help us model this trip.

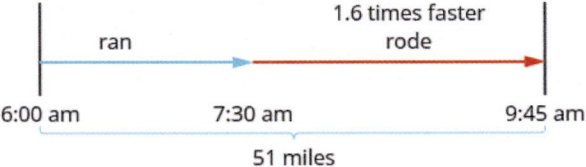

Next, we create a table to organize the information. We know the total distance is 51 miles. We are looking for the rate of speed for each part of the trip. The rate while biking is 1.6 times the rate of running. If we let r = the rate running, then the rate biking is $1.6r$.

The times here are given as clock times. Cruz started from home at 6:00 a.m. and started biking at 7:30 a.m. So he spent 1.5 hours running. Then he biked from 7:30 a.m until 9:45 a.m. So he spent 2.25 hours biking.

Now, we multiply the rates by the times.

	Rate (mph)	•	Time (hrs)	=	Distance (miles)
run	r		1.5		$1.5r$
bike	$1.6r$		2.25		$2.25(1.6r)$
					51

By looking at the diagram, we can see that the sum of the distance running and the distance biking is 255 miles.

distance running + distance biking = 51

Translate to an equation.	$1.5r$ $+$ $2.25(1.6r)$ $= 51$
Solve this equation.	$1.5r + 2.25(1.6r) = 51$
	$1.5r + 3.6r = 51$
	$5.1r = 51$
	$r = 10$ mph running
	$1.6r$ biking speed
	$1.6(10)$
	16 mph

Check.

Run   10 mph (1.5 hours) = 15 mi
Bike 16 mph (2.25 hours) = 36 mi
                                    51 mi

> **TRY IT : : 2.93**

Hamilton loves to travel to Las Vegas, 255 miles from his home in Orange County. On his last trip, he left his house at 2:00 p.m. The first part of his trip was on congested city freeways. At 4:00 pm, the traffic cleared and he was able to drive through the desert at a speed 1.75 times faster than when he drove in the congested area. He arrived in Las Vegas at 6:30 p.m. How fast was he driving during each part of his trip?

> **TRY IT : : 2.94**

Phuong left home on his bicycle at 10:00. He rode on the flat street until 11:15, then rode uphill until 11:45. He rode a total of 31 miles. His speed riding uphill was 0.6 times his speed on the flat street. Find his speed biking uphill and on the flat street.

2.4 EXERCISES

Practice Makes Perfect

Solve Coin Word Problems

In the following exercises, solve each coin word problem.

243. Michaela has $2.05 in dimes and nickels in her change purse. She has seven more dimes than nickels. How many coins of each type does she have?

244. Liliana has $2.10 in nickels and quarters in her backpack. She has 12 more nickels than quarters. How many coins of each type does she have?

245. In a cash drawer there is $125 in $5 and $10 bills. The number of $10 bills is twice the number of $5 bills. How many of each type of bill is in the drawer?

246. Sumanta has $175 in $5 and $10 bills in his drawer. The number of $5 bills is three times the number of $10 bills. How many of each are in the drawer?

247. Chi has $11.30 in dimes and quarters. The number of dimes is three more than three times the number of quarters. How many of each are there?

248. Alison has $9.70 in dimes and quarters. The number of quarters is eight more than four times the number of dimes. How many of each coin does she have?

249. Mukul has $3.75 in quarters, dimes and nickels in his pocket. He has five more dimes than quarters and nine more nickels than quarters. How many of each coin are in his pocket?

250. Vina has $4.70 in quarters, dimes and nickels in her purse. She has eight more dimes than quarters and six more nickels than quarters. How many of each coin are in her purse?

Solve Ticket and Stamp Word Problems

In the following exercises, solve each ticket or stamp word problem.

251. The first day of a water polo tournament the total value of tickets sold was $17,610. One-day passes sold for $20 and tournament passes sold for $30. The number of tournament passes sold was 37 more than the number of day passes sold. How many day passes and how many tournament passes were sold?

252. At the movie theater, the total value of tickets sold was $2,612.50. Adult tickets sold for $10 each and senior/child tickets sold for $7.50 each. The number of senior/child tickets sold was 25 less than twice the number of adult tickets sold. How many senior/child tickets and how many adult tickets were sold?

253. Julie went to the post office and bought both $0.41 stamps and $0.26 postcards. She spent $51.40. The number of stamps was 20 more than twice the number of postcards. How many of each did she buy?

254. Jason went to the post office and bought both $0.41 stamps and $0.26 postcards and spent $10.28 The number of stamps was four more than twice the number of postcards. How many of each did he buy?

255. Hilda has $210 worth of $10 and $12 stock shares. The number of $10 shares is five more than twice the number of $12 shares. How many of each type of share does she have?

256. Mario invested $475 in $45 and $25 stock shares. The number of $25 shares was five less than three times the number of $45 shares. How many of each type of share did he buy?

257. The ice rink sold 95 tickets for the afternoon skating session, for a total of $828. General admission tickets cost $10 each and youth tickets cost $8 each. How many general admission tickets and how many youth tickets were sold?

258. For the 7:30 show time, 140 movie tickets were sold. Receipts from the $13 adult tickets and the $10 senior tickets totaled $1,664. How many adult tickets and how many senior tickets were sold?

259. The box office sold 360 tickets to a concert at the college. The total receipts were $4,170. General admission tickets cost $15 and student tickets cost $10. How many of each kind of ticket was sold?

260. Last Saturday, the museum box office sold 281 tickets for a total of $3,954. Adult tickets cost $15 and student tickets cost $12. How many of each kind of ticket was sold?

Solve Mixture Word Problems

In the following exercises, solve each mixture word problem.

261. Macario is making 12 pounds of nut mixture with macadamia nuts and almonds. Macadamia nuts cost $9 per pound and almonds cost $5.25 per pound. How many pounds of macadamia nuts and how many pounds of almonds should Macario use for the mixture to cost $6.50 per pound to make?

262. Carmen wants to tile the floor of his house. He will need 1,000 square feet of tile. He will do most of the floor with a tile that costs $1.50 per square foot, but also wants to use an accent tile that costs $9.00 per square foot. How many square feet of each tile should he plan to use if he wants the overall cost to be $3 per square foot?

263. Riley is planning to plant a lawn in his yard. He will need nine pounds of grass seed. He wants to mix Bermuda seed that costs $4.80 per pound with Fescue seed that costs $3.50 per pound. How much of each seed should he buy so that the overall cost will be $4.02 per pound?

264. Vartan was paid $25,000 for a cell phone app that he wrote and wants to invest it to save for his son's education. He wants to put some of the money into a bond that pays 4% annual interest and the rest into stocks that pay 9% annual interest. If he wants to earn 7.4% annual interest on the total amount, how much money should he invest in each account?

265. Vern sold his 1964 Ford Mustang for $55,000 and wants to invest the money to earn him 5.8% interest per year. He will put some of the money into Fund A that earns 3% per year and the rest in Fund B that earns 10% per year. How much should he invest into each fund if he wants to earn 5.8% interest per year on the total amount?

266. Dominic pays 7% interest on his $15,000 college loan and 12% interest on his $11,000 car loan. What average interest rate does he pay on the total $26,000 he owes? (Round your answer to the nearest tenth of a percent.)

267. Liam borrowed a total of $35,000 to pay for college. He pays his parents 3% interest on the $8,000 he borrowed from them and pays the bank 6.8% on the rest. What average interest rate does he pay on the total $35,000? (Round your answer to the nearest tenth of a percent.)

Solve Uniform Motion Applications

In the following exercises, solve.

268. Lilah is moving from Portland to Seattle. It takes her three hours to go by train. Mason leaves the train station in Portland and drives to the train station in Seattle with all Lilah's boxes in his car. It takes him 2.4 hours to get to Seattle, driving at 15 miles per hour faster than the speed of the train. Find Mason's speed and the speed of the train.

269. Kathy and Cheryl are walking in a fundraiser. Kathy completes the course in 4.8 hours and Cheryl completes the course in eight hours. Kathy walks two miles per hour faster than Cheryl. Find Kathy's speed and Cheryl's speed.

270. Two busses go from Sacramento to San Diego. The express bus makes the trip in 6.8 hours and the local bus takes 10.2 hours for the trip. The speed of the express bus is 25 mph faster than the speed of the local bus. Find the speed of both busses.

271. A commercial jet and a private airplane fly from Denver to Phoenix. It takes the commercial jet 5.8% hours for the flight, and it takes the private airplane 1.8 hours. The speed of the commercial jet is 210 miles per hour faster than the speed of the private airplane. Find the speed of both airplanes.

272. Saul drove his truck three hours from Dallas towards Kansas City and stopped at a truck stop to get dinner. At the truck stop he met Erwin, who had driven four hours from Kansas City towards Dallas. The distance between Dallas and Kansas City is 542 miles, and Erwin's speed was eight miles per hour slower than Saul's speed. Find the speed of the two truckers.

273. Charlie and Violet met for lunch at a restaurant between Memphis and New Orleans. Charlie had left Memphis and drove 4.8 hours towards New Orleans. Violet had left New Orleans and drove two hours towards Memphis, at a speed 10 miles per hour faster than Charlie's speed. The distance between Memphis and New Orleans is 394 miles. Find the speed of the two drivers.

274. Sisters Helen and Anne live 332 miles apart. For Thanksgiving, they met at their other sister's house partway between their homes. Helen drove 3.2 hours and Anne drove 2.8 hours. Helen's average speed was four miles per hour faster than Anne's. Find Helen's average speed and Anne's average speed.

275. Ethan and Leo start riding their bikes at the opposite ends of a 65-mile bike path. After Ethan has ridden 1.5 hours and Leo has ridden two hours, they meet on the path. Ethan's speed is six miles per hour faster than Leo's speed. Find the speed of the two bikers.

276. Elvira and Aletheia live 3.1 miles apart on the same street. They are in a study group that meets at a coffee shop between their houses. It took Elvira half an hour and Aletheia two-thirds of an hour to walk to the coffee shop. Aletheia's speed is 0.6 miles per hour slower than Elvira's speed. Find both women's walking speeds.

277. DaMarcus and Fabian live 23 miles apart and play soccer at a park between their homes. DaMarcus rode his bike for three-quarters of an hour and Fabian rode his bike for half an hour to get to the park. Fabian's speed was six miles per hour faster than DaMarcus' speed. Find the speed of both soccer players.

278. Cindy and Richard leave their dorm in Charleston at the same time. Cindy rides her bicycle north at a speed of 18 miles per hour. Richard rides his bicycle south at a speed of 14 miles per hour. How long will it take them to be 96 miles apart?

279. Matt and Chris leave their uncle's house in Phoenix at the same time. Matt drives west on I-60 at a speed of 76 miles per hour. Chris drives east on I-60 at a speed of 82 miles per hour. How many hours will it take them to be 632 miles apart?

280. Two busses leave Billings at the same time. The Seattle bus heads west on I-90 at a speed of 73 miles per hour while the Chicago bus heads east at a speed of 79 miles an hour. How many hours will it take them to be 532 miles apart?

281. Two boats leave the same dock in Cairo at the same time. One heads north on the Mississippi River while the other heads south. The northbound boat travels four miles per hour. The southbound boat goes eight miles per hour. How long will it take them to be 54 miles apart?

282. Lorena walks the path around the park in 30 minutes. If she jogs, it takes her 20 minutes. Her jogging speed is 1.5 miles per hour faster than her walking speed. Find Lorena's walking speed and jogging speed.

283. Julian rides his bike uphill for 45 minutes, then turns around and rides back downhill. It takes him 15 minutes to get back to where he started. His uphill speed is 3.2 miles per hour slower than his downhill speed. Find Julian's uphill and downhill speed.

284. Cassius drives his boat upstream for 45 minutes. It takes him 30 minutes to return downstream. His speed going upstream is three miles per hour slower than his speed going downstream. Find his upstream and downstream speeds.

285. It takes Darline 20 minutes to drive to work in light traffic. To come home, when there is heavy traffic, it takes her 36 minutes. Her speed in light traffic is 24 miles per hour faster than her speed in heavy traffic. Find her speed in light traffic and in heavy traffic.

286. At 1:30, Marlon left his house to go to the beach, a distance of 7.6 miles. He rode his skateboard until 2:15, and then walked the rest of the way. He arrived at the beach at 3:00. Marlon's speed on his skateboard is 2.5 times his walking speed. Find his speed when skateboarding and when walking.

287. Aaron left at 9:15 to drive to his mountain cabin 108 miles away. He drove on the freeway until 10:45 and then drove on a mountain road. He arrived at 11:05. His speed on the freeway was three times his speed on the mountain road. Find Aaron's speed on the freeway and on the mountain road.

288. Marisol left Los Angeles at 2:30 to drive to Santa Barbara, a distance of 95 miles. The traffic was heavy until 3:20. She drove the rest of the way in very light traffic and arrived at 4:20. Her speed in heavy traffic was 40 miles per hour slower than her speed in light traffic. Find her speed in heavy traffic and in light traffic.

289. Lizette is training for a marathon. At 7:00 she left her house and ran until 8:15 then she walked until 11:15. She covered a total distance of 19 miles. Her running speed was five miles per hour faster than her walking speed. Find her running and walking speeds.

Everyday Math

290. John left his house in Irvine at 8:35 a.m. to drive to a meeting in Los Angeles, 45 miles away. He arrived at the meeting at 9:50 a.m.. At 6:30 p.m. he left the meeting and drove home. He arrived home at 7:18 p.m.

ⓐ What was his average speed on the drive from Irvine to Los Angeles?

ⓑ What was his average speed on the drive from Los Angeles to Irvine?

ⓒ What was the total time he spent driving to and from this meeting?

291. Sarah wants to arrive at her friend's wedding at 3:00. The distance from Sarah's house to the wedding is 95 miles. Based on usual traffic patterns, Sarah predicts she can drive the first 15 miles at 60 miles per hour, the next 10 miles at 30 miles per hour, and the remainder of the drive at 70 miles per hour.

ⓐ How long will it take Sarah to drive the first 15 miles?

ⓑ How long will it take Sarah to drive the next 10 miles?

ⓒ How long will it take Sarah to drive the rest of the trip?

ⓓ What time should Sarah leave her house?

Writing Exercises

292. Suppose you have six quarters, nine dimes, and four pennies. Explain how you find the total value of all the coins.

293. Do you find it helpful to use a table when solving coin problems? Why or why not?

294. In the table used to solve coin problems, one column is labeled "number" and another column is labeled "value." What is the difference between the "number" and the "value"?

295. When solving a uniform motion problem, how does drawing a diagram of the situation help you?

Self Check

ⓐ After completing the exercises, use this checklist to evaluate your mastery of the objectives of this section.

I can...	Confidently	With some help	No-I don't get it!
solve coin word problems.			
solve ticket and stamp word problems.			
solve mixture word problems.			
solve uniform motion applications.			

ⓑ On a scale of 1-10, how would you rate your mastery of this section in light of your responses on the checklist? How can you improve this?

2.5 | Solve Linear Inequalities

Learning Objectives

By the end of this section, you will be able to:

> Graph inequalities on the number line
> Solve linear inequalities
> Translate words to an inequality and solve
> Solve applications with linear inequalities

Be Prepared!

Before you get started, take this readiness quiz.

1. Translate from algebra to English: $15 > x$.
   If you missed this problem, review **Example 1.3**.

2. Translate to an algebraic expression: 15 is less than x.
   If you missed this problem, review **Example 1.8**.

Graph Inequalities on the Number Line

What number would make the inequality $x > 3$ true? Are you thinking, "x could be four"? That's correct, but x could be 6, too, or 37, or even 3.001. Any number greater than three is a solution to the inequality $x > 3$.

We show all the solutions to the inequality $x > 3$ on the number line by shading in all the numbers to the right of three, to show that all numbers greater than three are solutions. Because the number three itself is not a solution, we put an open parenthesis at three.

We can also represent inequalities using **interval notation**. There is no upper end to the solution to this inequality. In interval notation, we express $x > 3$ as $(3, \infty)$. The symbol ∞ is read as "**infinity**." It is not an actual number.

Figure 2.2 shows both the number line and the interval notation.

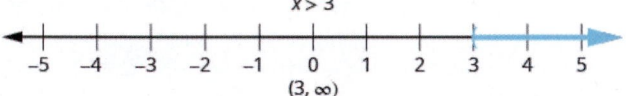

Figure 2.2 The inequality $x > 3$ is graphed on this number line and written in interval notation.

We use the left parenthesis symbol, (, to show that the endpoint of the inequality is not included. The left bracket symbol, [, shows that the endpoint is included.

The inequality $x \leq 1$ means all numbers less than or equal to one. Here we need to show that one is a solution, too. We do that by putting a bracket at $x = 1$. We then shade in all the numbers to the left of one, to show that all numbers less than one are solutions. See **Figure 2.3**.

There is no lower end to those numbers. We write $x \leq 1$ in interval notation as $(-\infty, 1]$. The symbol $-\infty$ is read as "negative infinity." **Figure 2.3** shows both the number line and interval notation.

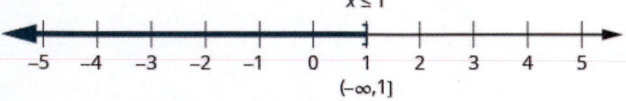

Figure 2.3 The inequality $x \leq 1$ is graphed on this number line and written in interval notation.

Inequalities, Number Lines, and Interval Notation

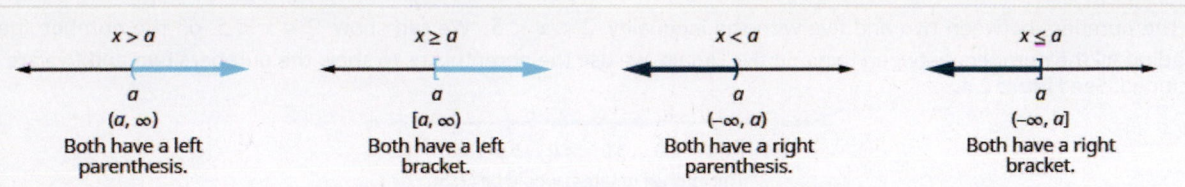

The notation for inequalities on a number line and in interval notation use the same symbols to express the endpoints of intervals.

EXAMPLE 2.48

Graph each inequality on the number line and write in interval notation.

ⓐ $x \geq -3$ ⓑ $x < 2.5$ ⓒ $x \leq -\frac{3}{5}$

✓ **Solution**

ⓐ

$x \geq -3$

Shade to the right of -3, and put a bracket at -3.	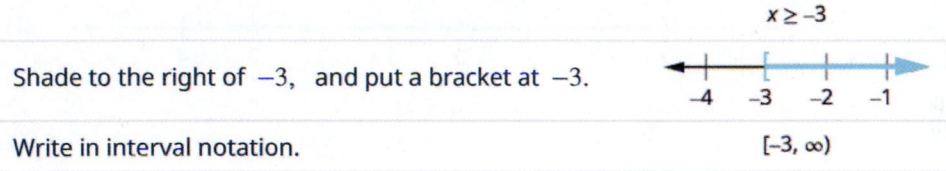
Write in interval notation.	$[-3, \infty)$

ⓑ

$x < 2.5$

Shade to the left of 2.5 and put a parenthesis at 2.5.	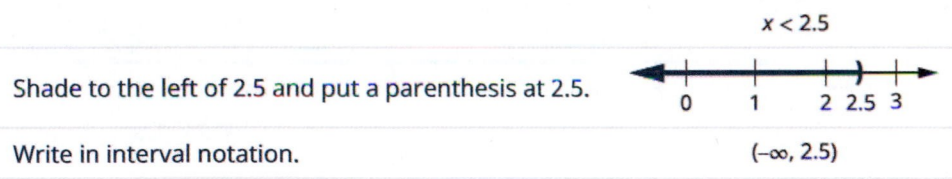
Write in interval notation.	$(-\infty, 2.5)$

ⓒ

$x \leq -\frac{3}{5}$

Shade to the left of $-\frac{3}{5}$, and put a bracket at $-\frac{3}{5}$.	
Write in interval notation.	$\left(-\infty, -\frac{3}{5}\right]$

> **TRY IT :: 2.95**
>
> Graph each inequality on the number line and write in interval notation: ⓐ $x > 2$ ⓑ $x \leq -1.5$ ⓒ $x \geq \frac{3}{4}$.

> **TRY IT :: 2.96**
>
> Graph each inequality on the number line and write in interval notation: ⓐ $x \leq -4$ ⓑ $x \geq 0.5$ ⓒ $x < -\frac{2}{3}$.

What numbers are greater than two but less than five? Are you thinking say, $2.5, 3, 3\frac{2}{3}, 4, 4, 99$? We can represent

all the numbers between two and five with the inequality $2 < x < 5$. We can show $2 < x < 5$ on the number line by shading all the numbers between two and five. Again, we use the parentheses to show the numbers two and five are not included. See Figure 2.4.

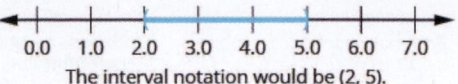

The interval notation would be (2, 5).

Figure 2.4

EXAMPLE 2.49

Graph each inequality on the number line and write in interval notation.

ⓐ $-3 < x < 4$ ⓑ $-6 \le x < -1$ ⓒ $0 \le x \le 2.5$

✓ **Solution**

ⓐ

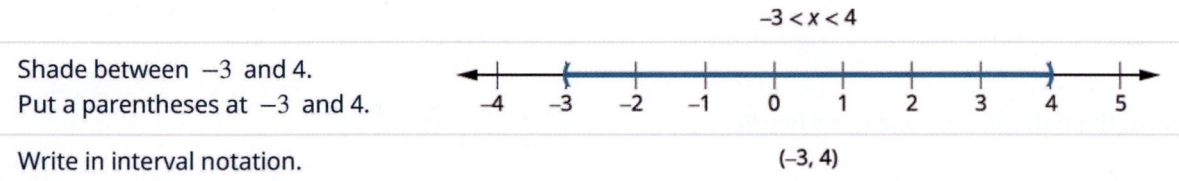

	$-3 < x < 4$
Shade between -3 and 4. Put a parentheses at -3 and 4.	
Write in interval notation.	$(-3, 4)$

ⓑ

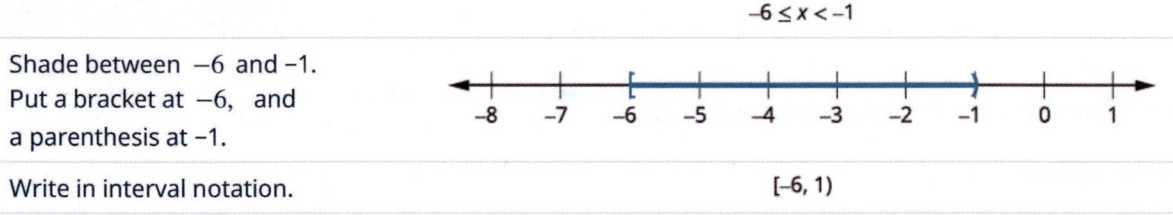

	$-6 \le x < -1$
Shade between -6 and -1. Put a bracket at -6, and a parenthesis at -1.	
Write in interval notation.	$[-6, 1)$

ⓒ

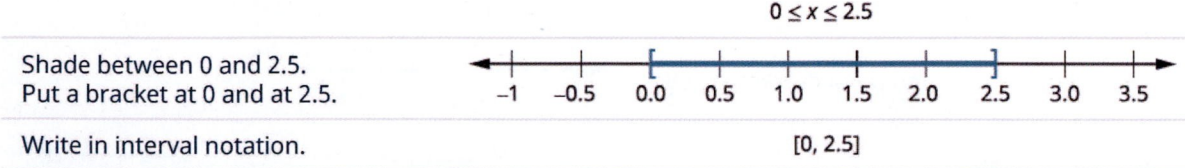

	$0 \le x \le 2.5$
Shade between 0 and 2.5. Put a bracket at 0 and at 2.5.	
Write in interval notation.	$[0, 2.5]$

> **TRY IT : : 2.97**   Graph each inequality on the number line and write in interval notation:

ⓐ $-2 < x < 1$ ⓑ $-5 \le x < -4$ ⓒ $1 \le x \le 4.25$

> **TRY IT : : 2.98**   Graph each inequality on the number line and write in interval notation:

ⓐ $-6 < x < 2$ ⓑ $-3 \le x < -1$ ⓒ $2.5 \le x \le 6$

Solve Linear Inequalities

A linear inequality is much like a linear equation—but the equal sign is replaced with an inequality sign. A **linear**

inequality is an inequality in one variable that can be written in one of the forms, $ax + b < c$, $ax + b \leq c$, $ax + b > c$, or $ax + b \geq c$.

Linear Inequality

A linear inequality is an inequality in one variable that can be written in one of the following forms where a, b, and c are real numbers and $a \neq 0$:

$$ax + b < c, \qquad ax + b \leq c, \qquad ax + b > c, \qquad ax + b \geq c.$$

When we solved linear equations, we were able to use the properties of equality to add, subtract, multiply, or divide both sides and still keep the equality. Similar properties hold true for inequalities.

We can add or subtract the same quantity from both sides of an inequality and still keep the inequality. For example:

$$-4 < 2 \qquad\qquad -4 < 2$$
$$-4 - 5 < 2 - 5 \qquad -4 + 7 < 2 + 7$$
$$-9 < -3 \text{ True} \qquad 3 < 9 \text{ True}$$

Notice that the inequality sign stayed the same.

This leads us to the Addition and Subtraction Properties of Inequality.

Addition and Subtraction Property of Inequality

For any numbers a, b, and c, if $a < b$, then

$$a + c < b + c \qquad a - c < b - c$$
$$a + c > b + c \qquad a - c > b - c$$

We can add or subtract the same quantity from both sides of an inequality and still keep the inequality.

What happens to an inequality when we divide or multiply both sides by a constant?

Let's first multiply and divide both sides by a positive number.

$$10 < 15 \qquad\qquad 10 < 15$$
$$10(5) < 15(5) \qquad \frac{10}{5} < \frac{15}{5}$$
$$50 < 75 \text{ True} \qquad 2 < 3 \text{ True}$$

The inequality signs stayed the same.

Does the inequality stay the same when we divide or multiply by a negative number?

$$10 < 15 \qquad\qquad 10 < 15$$
$$10(-5) \text{ ? } 15(-5) \qquad \frac{10}{-5} \text{ ? } \frac{15}{-5}$$
$$-50 \text{ ? } -75 \qquad -2 \text{ ? } -3$$
$$-50 > -75 \qquad -2 > -3$$

Notice that when we filled in the inequality signs, the inequality signs reversed their direction.

When we divide or multiply an inequality by a positive number, the inequality sign stays the same. When we divide or multiply an inequality by a negative number, the inequality sign reverses.

This gives us the Multiplication and Division Property of Inequality.

Multiplication and Division Property of Inequality

For any numbers a, b, and c,

multiply or divide by a positive

$$\text{if } a < b \text{ and } c > 0, \text{ then } ac < bc \text{ and } \frac{a}{c} < \frac{b}{c}.$$

$$\text{if } a > b \text{ and } c > 0, \text{ then } ac > bc \text{ and } \frac{a}{c} > \frac{b}{c}.$$

multiply or divide by a negative

$$\text{if } a < b \text{ and } c < 0, \text{ then } ac > bc \text{ and } \frac{a}{c} > \frac{b}{c}.$$

$$\text{if } a > b \text{ and } c < 0, \text{ then } ac < bc \text{ and } \frac{a}{c} < \frac{b}{c}.$$

When we divide or multiply an inequality by a:

- positive number, the inequality stays the same.
- negative number, the inequality reverses.

Sometimes when solving an inequality, as in the next example, the variable ends upon the right. We can rewrite the inequality in reverse to get the variable to the left.

$$x > a \text{ has the same meaning as } a < x$$

Think about it as "If Xander is taller than Andy, then Andy is shorter than Xander."

EXAMPLE 2.50

Solve each inequality. Graph the solution on the number line, and write the solution in interval notation.

ⓐ $x - \frac{3}{8} \leq \frac{3}{4}$ ⓑ $9y < 54$ ⓒ $-15 < \frac{3}{5}z$

✓ Solution

ⓐ

	$x - \frac{3}{8} \leq \frac{3}{4}$
Add $\frac{3}{8}$ to both sides of the inequality.	$x - \frac{3}{8} + \frac{3}{8} \leq \frac{3}{4} + \frac{3}{8}$
Simplify.	$x \leq \frac{9}{8}$
Graph the solution on the number line.	
Write the solution in interval notation.	$\left(-\infty, \frac{9}{8}\right]$

ⓑ

	$9y < 54$
Divide both sides of the inequality by 9; since 9 is positive, the inequality stays the same.	$\frac{9y}{9} < \frac{54}{9}$
Simplify.	$y < 6$
Graph the solution on the number line.	
Write the solution in interval notation.	$(-\infty, 6)$

ⓒ

$$-15 < \frac{3}{5}z$$

Multiply both sides of the inequality by $\frac{5}{3}$. Since $\frac{5}{3}$ is positive, the inequality stays the same.	$\frac{5}{3}(-15) < \frac{5}{3}\left(\frac{3}{5}z\right)$
Simplify.	$-25 < z$
Rewrite with the variable on the left.	$z > -25$
Graph the solution on the number line.	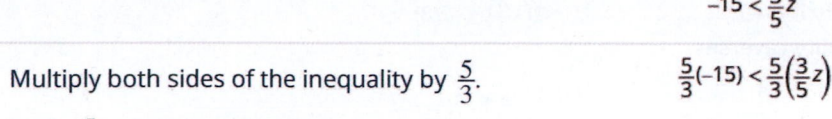
Write the solution in interval notation.	$(-25, \infty)$

> **TRY IT :: 2.99**
>
> Solve each inequality, graph the solution on the number line, and write the solution in interval notation:
>
> ⓐ $p - \frac{3}{4} \geq \frac{1}{6}$ ⓑ $9c > 72$ ⓒ $24 \leq \frac{3}{8}m$

> **TRY IT :: 2.100**
>
> Solve each inequality, graph the solution on the number line, and write the solution in interval notation:
>
> ⓐ $r - \frac{1}{3} \leq \frac{7}{12}$ ⓑ $12d \leq 60$ ⓒ $-24 < \frac{4}{3}n$

Be careful when you multiply or divide by a negative number—remember to reverse the inequality sign.

EXAMPLE 2.51

Solve each inequality, graph the solution on the number line, and write the solution in interval notation.

ⓐ $-13m \geq 65$ ⓑ $\frac{n}{-2} \geq 8$

✓ **Solution**

ⓐ

$$-13m \geq 65$$

Divide both sides of the inequality by -13. Since -13 is a negative, the inequality reverses.	$\frac{-13m}{-13} \leq \frac{65}{-13}$
Simplify.	$m \leq -5$
Graph the solution on the number line.	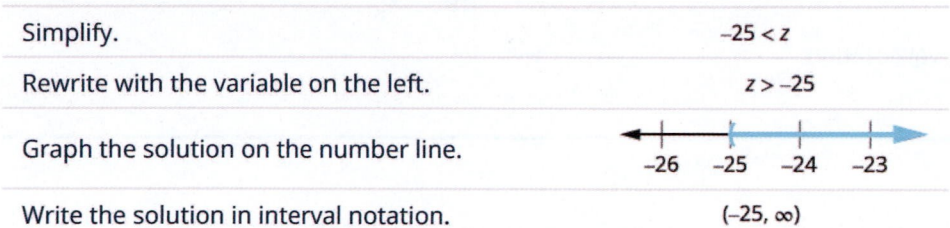
Write the solution in interval notation.	$(-\infty, -5]$

ⓑ

$$\frac{n}{-2} \geq 8$$

Multiply both sides of the inequality by -2. Since -2 is a negative, the inequality reverses.	$-2\left(\frac{n}{-2}\right) \leq -2(8)$
Simplify.	$n \leq -16$
Graph the solution on the number line.	
Write the solution in interval notation.	$(-\infty, -16]$

> **TRY IT :: 2.101**
>
> Solve each inequality, graph the solution on the number line, and write the solution in interval notation:
>
> ⓐ $-8q < 32$ ⓑ $\frac{k}{-12} \leq 15$.

> **TRY IT :: 2.102**
>
> Solve each inequality, graph the solution on the number line, and write the solution in interval notation:
>
> ⓐ $-7r \leq -70$ ⓑ $\frac{u}{-4} \geq -16$.

Most inequalities will take more than one step to solve. We follow the same steps we used in the general strategy for solving linear equations, but make sure to pay close attention when we multiply or divide to isolate the variable.

EXAMPLE 2.52

Solve the inequality $6y \leq 11y + 17$, graph the solution on the number line, and write the solution in interval notation.

⊘ **Solution**

$$6y \leq 11y + 17$$

Subtract $11y$ from both sides to collect the variables on the left.	$6y - 11y \leq 11y - 11y + 17$
Simplify.	$-5y \leq 17$
Divide both sides of the inequality by -5, and reverse the inequality.	$\frac{-5y}{-5} \geq \frac{17}{-5}$
Simplify.	$y \geq \frac{17}{-5}$
Graph the solution on the number line.	
Write the solution in interval notation.	$\left[-\frac{17}{5}, \infty\right)$

> **TRY IT :: 2.103**
>
> Solve the inequality, graph the solution on the number line, and write the solution in interval notation:
> $3q \geq 7q - 23$.

TRY IT : : 2.104

Solve the inequality, graph the solution on the number line, and write the solution in interval notation: $6x < 10x + 19$.

When solving inequalities, it is usually easiest to collect the variables on the side where the coefficient of the variable is largest. This eliminates negative coefficients and so we don't have to multiply or divide by a negative—which means we don't have to remember to reverse the inequality sign.

 **EXAMPLE 2.53**

Solve the inequality $8p + 3(p - 12) > 7p - 28$, graph the solution on the number line, and write the solution in interval notation.

⊘ **Solution**

$$8p + 3(p - 12) > 7p - 28$$

Simplify each side as much as possible.	
Distribute.	$8p + 3p - 36 > 7p - 28$
Combine like terms.	$11p - 36 > 7p - 28$
Subtract $7p$ from both sides to collect the variables on the left, since $11 > 7$.	$11p - 36 - 7p > 7p - 28 - 7p$
Simplify.	$4p - 36 > -28$
Add 36 to both sides to collect the constants on the right.	$4p - 36 + 36 > -28 + 36$
Simplify.	$4p > 8$
Divide both sides of the inequality by 4; the inequality stays the same.	$\dfrac{4p}{4} > \dfrac{8}{4}$
Simplify.	$p > 2$
Graph the solution on the number line.	
Write the solution in interval notation.	$(2, \infty)$

 TRY IT : : 2.105

Solve the inequality $9y + 2(y + 6) > 5y - 24$, graph the solution on the number line, and write the solution in interval notation.

 TRY IT : : 2.106

Solve the inequality $6u + 8(u - 1) > 10u + 32$, graph the solution on the number line, and write the solution in interval notation.

Just like some equations are identities and some are contradictions, inequalities may be identities or contradictions, too. We recognize these forms when we are left with only constants as we solve the inequality. If the result is a true statement, we have an identity. If the result is a false statement, we have a contradiction.

EXAMPLE 2.54

Solve the inequality $8x - 2(5 - x) < 4(x + 9) + 6x,$ graph the solution on the number line, and write the solution in interval notation.

⊘ **Solution**

Simplify each side as much as possible.	$8x - 2(5 - x) < 4(x + 9) + 6x$
Distribute.	$8x - 10 + 2x < 4x + 36 + 6x$
Combine like terms.	$10x - 10 < 10x + 36$
Subtract 10x from both sides to collect the variables on the left.	$10x - 10 - 10x < 10x + 36 - 10x$
Simplify.	$-10 < 36$
The x's are gone, and we have a true statement.	The inequality is an identity. The solution is all real numbers.
Graph the solution on the number line.	
Write the solution in interval notation.	$(-\infty, \infty)$

> **TRY IT : :** 2.107

Solve the inequality $4b - 3(3 - b) > 5(b - 6) + 2b$, graph the solution on the number line, and write the solution in interval notation.

> **TRY IT : :** 2.108

Solve the inequality $9h - 7(2 - h) < 8(h + 11) + 8h$, graph the solution on the number line, and write the solution in interval notation.

We can clear fractions in inequalities much as we did in equations. Again, be careful with the signs when multiplying or dividing by a negative.

EXAMPLE 2.55

Solve the inequality $\frac{1}{3}a - \frac{1}{8}a > \frac{5}{24}a + \frac{3}{4},$ graph the solution on the number line, and write the solution in interval notation.

⊘ **Solution**

$$\frac{1}{3}a - \frac{1}{8}a > \frac{5}{24}a + \frac{3}{4}$$

Multiply both sides by the LCD, 24, to clear the fractions.	$24\left(\frac{1}{3}a - \frac{1}{8}a\right) > 24\left(\frac{5}{24}a + \frac{3}{4}\right)$
Simplify.	$8a - 3a > 5a + 18$
Combine like terms.	$5a > 5a + 18$
Subtract $5a$ from both sides to collect the variables on the left.	$5a - 5a > 5a - 5a + 18$

Simplify.	$0 > 18$
The statement is false.	The inequality is a contradiction. There is no solution.
Graph the solution on the number line.	−1 0 1 2
Write the solution in interval notation.	There is no solution.

> **TRY IT :: 2.109**

Solve the inequality $\frac{1}{4}x - \frac{1}{12}x > \frac{1}{6}x + \frac{7}{8}$, graph the solution on the number line, and write the solution in interval notation.

> **TRY IT :: 2.110**

Solve the inequality $\frac{2}{5}z - \frac{1}{3}z < \frac{1}{15}z - \frac{3}{5}$, graph the solution on the number line, and write the solution in interval notation.

Translate to an Inequality and Solve

To translate English sentences into inequalities, we need to recognize the phrases that indicate the inequality. Some words are easy, like "more than" and "less than." But others are not as obvious. Table 2.60 shows some common phrases that indicate inequalities.

>	≥	<	≤
is greater than	is greater than or equal to	is less than	is less than or equal to
is more than	is at least	is smaller than	is at most
is larger than	is no less than	has fewer than	is no more than
exceeds	is the minimum	is lower than	is the maximum

Table 2.60

EXAMPLE 2.56

Translate and solve. Then graph the solution on the number line, and write the solution in interval notation.

Twenty-seven less than x is at least 48.

⊘ **Solution**

Twenty-seven less than x is at least 48.

Translate.	$x - 27 \geq 48$
Solve—add 27 to both sides.	$x - 27 + 27 \geq 48 + 27$
Simplify.	$x \geq 75$
Graph on the number line.	
Write in interval notation.	$[75, \infty)$

⟩ **TRY IT : : 2.111**

Translate and solve. Then graph the solution on the number line, and write the solution in interval notation.

Nineteen less than p is no less than 47.

⟩ **TRY IT : : 2.112**

Translate and solve. Then graph the solution on the number line, and write the solution in interval notation.

Four more than a is at most 15.

Solve Applications with Linear Inequalities

Many real-life situations require us to solve inequalities. The method we will use to solve applications with linear inequalities is very much like the one we used when we solved applications with equations.

We will read the problem and make sure all the words are understood. Next, we will identify what we are looking for and assign a variable to represent it. We will restate the problem in one sentence to make it easy to translate into an inequality. Then, we will solve the inequality.

Sometimes an application requires the solution to be a whole number, but the algebraic solution to the inequality is not a whole number. In that case, we must round the algebraic solution to a whole number. The context of the application will determine whether we round up or down.

EXAMPLE 2.57

Dawn won a mini-grant of $4,000 to buy tablet computers for her classroom. The tablets she would like to buy cost $254.12 each, including tax and delivery. What is the maximum number of tablets Dawn can buy?

⊘ Solution

Step 1. Read the problem.

Step 2. Identify what you are looking for.   the maximum number of tablets Dawn can buy

Step 3. Name what you are looking for.

Choose a variable to represent that   Let n = the number of tablets.
quantity.

Step 4. Translate. Write a sentence that gives the
information to find it. $254.12 times the number of tablets is
                                                              no more than $4,000.

Translate into an inequality. $254.12n \leq 4000$

Step 5. Solve the inequality.

But n must be a whole number of $n \leq 15.74$
tablets, so round to 15. $n \leq 15$

Step 6. Check the answer in the problem
and make sure it makes sense.

Rounding down the price to $250, 15
tablets would cost $3,750, while 16
tablets would be $4,000. So a
maximum of 15 tablets at $254.12
seems reasonable.

Step 7. Answer the question with a complete sentence.   Dawn can buy a maximum of 15 tablets.

> **TRY IT ∷ 2.113**
>
> Angie has $20 to spend on juice boxes for her son's preschool picnic. Each pack of juice boxes costs $2.63. What is the maximum number of packs she can buy?

> **TRY IT ∷ 2.114**
>
> Daniel wants to surprise his girlfriend with a birthday party at her favorite restaurant. It will cost $42.75 per person for dinner, including tip and tax. His budget for the party is $500. What is the maximum number of people Daniel can have at the party?

EXAMPLE 2.58

Taleisha's phone plan costs her $28.80 a month plus $0.20 per text message. How many text messages can she send/receive and keep her monthly phone bill no more than $50?

✓ **Solution**

Step 1. Read the problem.	
Step 2. Identify what you are looking for.	the number of text messages Taleisha can make
Step 3. Name what you are looking for.	
Choose a variable to represent that quantity.	Let t = the number of text messages.
Step 4. Translate Write a sentence that gives the information to find it.	$28.80 plus $0.20 times the number of text messages is less than or equal to $50.
Translate into an inequality.	$28.80 + 0.20t \leq 50$
Step 5. Solve the inequality.	$0.2t \leq 21.2$
	$t \leq 106$ text messages
Step 6. Check the answer in the problem and make sure it makes sense.	
Yes, $28.80 + 0.20(106) = 50$.	
Step 7. Write a sentence that answers the question.	Taleisha can send/receive no more than 106 text messages to keep her bill no more than $50.

> **TRY IT : : 2.115**
>
> Sergio and Lizeth have a very tight vacation budget. They plan to rent a car from a company that charges $75 a week plus $0.25 a mile. How many miles can they travel during the week and still keep within their $200 budget?

> **TRY IT : : 2.116**
>
> Rameen's heating bill is $5.42 per month plus $1.08 per therm. How many therms can Rameen use if he wants his heating bill to be a maximum of $87.50.

Profit is the money that remains when the costs have been subtracted from the revenue. In the next example, we will find the number of jobs a small businesswoman needs to do every month in order to make a certain amount of profit.

EXAMPLE 2.59

Felicity has a calligraphy business. She charges $2.50 per wedding invitation. Her monthly expenses are $650. How many invitations must she write to earn a profit of at least $2,800 per month?

⊘ Solution

Step 1. Read the problem.

Step 2. Identify what you are looking for.   the number of invitations Felicity needs to write

Step 3. Name what you are looking for.   Let $j =$ the number of invitations.
 Choose a variable to represent it.

Step 4. Translate. Write a sentence that
gives the information to find it.

     $2.50 times the number of invitations
minus \$650 is at least \$2,800.$

$$2.50j - 650 \geq 2,800$$

 Translate into an inequality.

Step 5. Solve the inequality.

$$2.5j \geq 3,450$$
$$j \geq 1,380 \text{ invitations}$$

Step 6. Check the answer in the problem
and make sure it makes sense.

 If Felicity wrote 1400 invitations, her
 profit would be $2.50(1400) - 650,$ or
 \$2,850. This is more than \$2800.

Step 7. Write a sentence that answers the question.   Felicity must write at least 1,380 invitations.

 TRY IT : : 2.117

 Caleb has a pet sitting business. He charges \$32 per hour. His monthly expenses are \$2,272. How many hours
 must he work in order to earn a profit of at least \$800 per month?

 TRY IT : : 2.118

 Elliot has a landscape maintenance business. His monthly expenses are \$1,100. If he charges \$60 per job, how
 many jobs must he do to earn a profit of at least \$4,000 a month?

There are many situations in which several quantities contribute to the total expense. We must make sure to account for
all the individual expenses when we solve problems like this.

EXAMPLE 2.60

Malik is planning a six-day summer vacation trip. He has \$840 in savings, and he earns \$45 per hour for tutoring. The trip
will cost him \$525 for airfare, \$780 for food and sightseeing, and \$95 per night for the hotel. How many hours must he
tutor to have enough money to pay for the trip?

⊘ **Solution**

Step 1. Read the problem.

Step 2. Identify what you are looking for.   the number of hours Malik must tutor

Step 3. Name what you are looking for.

   Choose a variable to represent that   Let h = the number of hours.
   quantity.

Step 4. Translate. Write a sentence that   The expenses must be less than or equal to
gives the information to find it.   the income. The cost of airfare plus the
                                                     cost of food and sightseeing and the hotel
                                                     bill must be less than the savings plus the
                                                     amount earned tutoring.

   Translate into an inequality. $525 + 780 + 95(6) \le 840 + 45h$

Step 5. Solve the inequality.
$$1{,}875 \le 840 + 45h$$
$$1{,}035 \le 45h$$
$$23 \le h$$
$$h \ge 23$$

Step 6. Check the answer in the problem
and make sure it makes sense.
We substitute 23 into the inequality.

   $1{,}875 \le 840 + 45h$
   $1{,}875 \le 840 + 45(23)$
   $1{,}875 \le 1875$

Step 7. Write a sentence that answers the question.   Malik must tutor at least 23 hours.

> **TRY IT : : 2.119**
>
> Brenda's best friend is having a destination wedding and the event will last three days. Brenda has $500 in savings and can earn $15 an hour babysitting. She expects to pay $350 airfare, $375 for food and entertainment and $60 a night for her share of a hotel room. How many hours must she babysit to have enough money to pay for the trip?

> **TRY IT : : 2.120**
>
> Josue wants to go on a 10-night road trip with friends next spring. It will cost him $180 for gas, $450 for food, and $49 per night to share a motel room. He has $520 in savings and can earn $30 per driveway shoveling snow. How many driveways must he shovel to have enough money to pay for the trip?

2.5 EXERCISES

Practice Makes Perfect

Graph Inequalities on the Number Line

In the following exercises, graph each inequality on the number line and write in interval notation.

296.
ⓐ $x < -2$
ⓑ $x \geq -3.5$
ⓒ $x \leq \frac{2}{3}$

297.
ⓐ $x > 3$
ⓑ $x \leq -0.5$
ⓒ $x \geq \frac{1}{3}$

298.
ⓐ $x \geq -4$
ⓑ $x < 2.5$
ⓒ $x > -\frac{3}{2}$

299.
ⓐ $x \leq 5$
ⓑ $x \geq -1.5$
ⓒ $x < -\frac{7}{3}$

300.
ⓐ $-5 < x < 2$
ⓑ $-3 \leq x < -1$
ⓒ $0 \leq x \leq 1.5$

301.
ⓐ $-2 < x < 0$
ⓑ $-5 \leq x < -3$
ⓒ $0 \leq x \leq 3.5$

302.
ⓐ $-1 < x < 3$
ⓑ $-3 < x \leq -2$
ⓒ $-1.25 \leq x \leq 0$

303.
ⓐ $-4 < x < 2$
ⓑ $-5 < x \leq -2$
ⓒ $-3.75 \leq x \leq 0$

Solve Linear Inequalities

In the following exercises, solve each inequality, graph the solution on the number line, and write the solution in interval notation.

304.
ⓐ $a + \frac{3}{4} \geq \frac{7}{10}$
ⓑ $8x > 72$
ⓒ $20 > \frac{2}{5}h$

305.
ⓐ $b + \frac{7}{8} \geq \frac{1}{6}$
ⓑ $6y < 48$
ⓒ $40 < \frac{5}{8}k$

306.
ⓐ $f - \frac{13}{20} < -\frac{5}{12}$
ⓑ $9t \geq -27$
ⓒ $\frac{7}{6}j \geq 42$

307.
ⓐ $g - \frac{11}{12} < -\frac{5}{18}$
ⓑ $7s < -28$
ⓒ $\frac{9}{4}g \leq 36$

308.
ⓐ $-5u \geq 65$
ⓑ $\frac{a}{-3} \leq 9$

309.
ⓐ $-8v \leq 96$
ⓑ $\frac{b}{-10} \geq 30$

310.
ⓐ $-9c < 126$
ⓑ $-25 < \frac{p}{-5}$

311.
ⓐ $-7d > 105$
ⓑ $-18 > \frac{q}{-6}$

In the following exercises, solve each inequality, graph the solution on the number line, and write the solution in interval notation.

312. $4v \geq 9v - 40$

313. $5u \leq 8u - 21$

314. $13q < 7q - 29$

315. $9p > 14p - 18$

316. $12x + 3(x + 7) > 10x - 24$

317. $9y + 5(y + 3) < 4y - 35$

318. $6h - 4(h - 1) \leq 7h - 11$ **319.** $4k - (k - 2) \geq 7k - 26$ **320.**
$$8m - 2(14 - m) \geq 7(m - 4) + 3m$$

321.
$6n - 12(3 - n) \leq 9(n - 4) + 9n$ **322.** $\frac{3}{4}b - \frac{1}{3}b < \frac{5}{12}b - \frac{1}{2}$ **323.**
$9u + 5(2u - 5) \geq 12(u - 1) + 7u$

324. $\frac{2}{3}g - \frac{1}{2}(g - 14) \leq \frac{1}{6}(g + 42)$ **325.**
$$\frac{4}{5}h - \frac{2}{3}(h - 9) \geq \frac{1}{15}(2h + 90)$$ **326.** $\frac{5}{6}a - \frac{1}{4}a > \frac{7}{12}a + \frac{2}{3}$

327.
$12v + 3(4v - 1) \leq 19(v - 2) + 5v$

In the following exercises, solve each inequality, graph the solution on the number line, and write the solution in interval notation.

328. $15k \leq -40$ **329.** $35k \geq -77$ **330.**
$$23p - 2(6 - 5p) > 3(11p - 4)$$

331.
$18q - 4(10 - 3q) < 5(6q - 8)$ **332.** $-\frac{9}{4}x \geq -\frac{5}{12}$ **333.** $-\frac{21}{8}y \leq -\frac{15}{28}$

334. $c + 34 < -99$ **335.** $d + 29 > -61$ **336.** $\frac{m}{18} \geq -4$

337. $\frac{n}{13} \leq -6$

Translate to an Inequality and Solve

In the following exercises, translate and solve. Then graph the solution on the number line and write the solution in interval notation.

338. Three more than h is no less than 25.   **339.** Six more than k exceeds 25.   **340.** Ten less than w is at least 39.

341. Twelve less than x is no less than 21.   **342.** Negative five times r is no more than 95.   **343.** Negative two times s is lower than 56.

344. Nineteen less than b is at most -22.   **345.** Fifteen less than a is at least -7.

Solve Applications with Linear Inequalities

In the following exercises, solve.

346. Alan is loading a pallet with boxes that each weighs 45 pounds. The pallet can safely support no more than 900 pounds. How many boxes can he safely load onto the pallet?   **347.** The elevator in Yehire's apartment building has a sign that says the maximum weight is 2100 pounds. If the average weight of one person is 150 pounds, how many people can safely ride the elevator?

348. Andre is looking at apartments with three of his friends. They want the monthly rent to be no more than $2,360. If the roommates split the rent evenly among the four of them, what is the maximum rent each will pay?   **349.** Arleen got a $20 gift card for the coffee shop. Her favorite iced drink costs $3.79. What is the maximum number of drinks she can buy with the gift card?

350. Teegan likes to play golf. He has budgeted $60 next month for the driving range. It costs him $10.55 for a bucket of balls each time he goes. What is the maximum number of times he can go to the driving range next month?

351. Ryan charges his neighbors $17.50 to wash their car. How many cars must he wash next summer if his goal is to earn at least $1,500?

352. Keshad gets paid $2,400 per month plus 6% of his sales. His brother earns $3,300 per month. For what amount of total sales will Keshad's monthly pay be higher than his brother's monthly pay?

353. Kimuyen needs to earn $4,150 per month in order to pay all her expenses. Her job pays her $3,475 per month plus 4% of her total sales. What is the minimum Kimuyen's total sales must be in order for her to pay all her expenses?

354. Andre has been offered an entry-level job. The company offered him $48,000 per year plus 3.5% of his total sales. Andre knows that the average pay for this job is $62,000. What would Andre's total sales need to be for his pay to be at least as high as the average pay for this job?

355. Nataly is considering two job offers. The first job would pay her $83,000 per year. The second would pay her $66,500 plus 15% of her total sales. What would her total sales need to be for her salary on the second offer be higher than the first?

356. Jake's water bill is $24.80 per month plus $2.20 per ccf (hundred cubic feet) of water. What is the maximum number of ccf Jake can use if he wants his bill to be no more than $60?

357. Kiyoshi's phone plan costs $17.50 per month plus $0.15 per text message. What is the maximum number of text messages Kiyoshi can use so the phone bill is no more than $56.60?

358. Marlon's TV plan costs $49.99 per month plus $5.49 per first-run movie. How many first-run movies can he watch if he wants to keep his monthly bill to be a maximum of $100?

359. Kellen wants to rent a banquet room in a restaurant for her cousin's baby shower. The restaurant charges $350 for the banquet room plus $32.50 per person for lunch. How many people can Kellen have at the shower if she wants the maximum cost to be $1,500?

360. Moshde runs a hairstyling business from her house. She charges $45 for a haircut and style. Her monthly expenses are $960. She wants to be able to put at least $1,200 per month into her savings account order to open her own salon. How many "cut & styles" must she do to save at least $1,200 per month?

361. Noe installs and configures software on home computers. He charges $125 per job. His monthly expenses are $1,600. How many jobs must he work in order to make a profit of at least $2,400?

362. Katherine is a personal chef. She charges $115 per four-person meal. Her monthly expenses are $3,150. How many four-person meals must she sell in order to make a profit of at least $1,900?

363. Melissa makes necklaces and sells them online. She charges $88 per necklace. Her monthly expenses are $3,745. How many necklaces must she sell if she wants to make a profit of at least $1,650?

364. Five student government officers want to go to the state convention. It will cost them $110 for registration, $375 for transportation and food, and $42 per person for the hotel. There is $450 budgeted for the convention in the student government savings account. They can earn the rest of the money they need by having a car wash. If they charge $5 per car, how many cars must they wash in order to have enough money to pay for the trip?

365. Cesar is planning a four-day trip to visit his friend at a college in another state. It will cost him $198 for airfare, $56 for local transportation, and $45 per day for food. He has $189 in savings and can earn $35 for each lawn he mows. How many lawns must he mow to have enough money to pay for the trip?

366. Alonzo works as a car detailer. He charges $175 per car. He is planning to move out of his parents' house and rent his first apartment. He will need to pay $120 for application fees, $950 for security deposit, and first and last months' rent at $1,140 per month. He has $1,810 in savings. How many cars must he detail to have enough money to rent the apartment?

367. Eun-Kyung works as a tutor and earns $60 per hour. She has $792 in savings. She is planning an anniversary party for her parents. She would like to invite 40 guests. The party will cost her $1,520 for food and drinks and $150 for the photographer. She will also have a favor for each of the guests, and each favor will cost $7.50. How many hours must she tutor to have enough money for the party?

Everyday Math

368. Maximum load on a stage In 2014, a high school stage collapsed in Fullerton, California, when 250 students got on stage for the finale of a musical production. Two dozen students were injured. The stage could support a maximum of 12,750 pounds. If the average weight of a student is assumed to be 140 pounds, what is the maximum number of students who could safely be on the stage?

369. Maximum weight on a boat In 2004, a water taxi sank in Baltimore harbor and five people drowned. The water taxi had a maximum capacity of 3,500 pounds (25 people with average weight 140 pounds). The average weight of the 25 people on the water taxi when it sank was 168 pounds per person. What should the maximum number of people of this weight have been?

370. Wedding budget Adele and Walter found the perfect venue for their wedding reception. The cost is $9850 for up to 100 guests, plus $38 for each additional guest. How many guests can attend if Adele and Walter want the total cost to be no more than $12,500?

371. Shower budget Penny is planning a baby shower for her daughter-in-law. The restaurant charges $950 for up to 25 guests, plus $31.95 for each additional guest. How many guests can attend if Penny wants the total cost to be no more than $1,500?

Writing Exercises

372. Explain why it is necessary to reverse the inequality when solving $-5x > 10$.

373. Explain why it is necessary to reverse the inequality when solving $\frac{n}{-3} < 12$.

374. Find your last month's phone bill and the hourly salary you are paid at your job. Calculate the number of hours of work it would take you to earn at least enough money to pay your phone bill by writing an appropriate inequality and then solving it. Do you feel this is an appropriate number of hours? Is this the appropriate phone plan for you?

375. Find out how many units you have left, after this term, to achieve your college goal and estimate the number of units you can take each term in college. Calculate the number of terms it will take you to achieve your college goal by writing an appropriate inequality and then solving it. Is this an acceptable number of terms until you meet your goal? What are some ways you could accelerate this process?

Self Check

ⓐ *After completing the exercises, use this checklist to evaluate your mastery of the objectives of this section.*

I can...	Confidently	With some help	No-I don't get it!
graph inequalities on the number line.			
solve linear inequalities.			
translate words to an inequality and solve.			
solve applications with linear inequalities.			

ⓑ *After looking at the checklist, do you think you are well-prepared for the next section? Why or why not?*

2.6 | Solve Compound Inequalities

Learning Objectives

By the end of this section, you will be able to:

> Solve compound inequalities with "and"
> Solve compound inequalities with "or"
> Solve applications with compound inequalities

Be Prepared!

Before you get started, take this readiness quiz.

1. Simplify: $\frac{2}{5}(x + 10)$.

   If you missed this problem, review **Example 1.51**.

2. Simplify: $-(x - 4)$.

   If you missed this problem, review **Example 1.54**.

Solve Compound Inequalities with "and"

Now that we know how to solve linear inequalities, the next step is to look at compound inequalities. A **compound inequality** is made up of two inequalities connected by the word "and" or the word "or." For example, the following are compound inequalities.

$$x + 3 > -4 \quad \text{and} \quad 4x - 5 \le 3$$
$$2(y + 1) < 0 \quad \text{or} \quad y - 5 \ge -2$$

Compound Inequality

A **compound inequality** is made up of two inequalities connected by the word "and" or the word "or."

To solve a compound inequality means to find all values of the variable that make the compound inequality a true statement. We solve compound inequalities using the same techniques we used to solve linear inequalities. We solve each inequality separately and then consider the two solutions.

To solve a compound inequality with the word "and," we look for all numbers that make *both* inequalities true. To solve a compound inequality with the word "or," we look for all numbers that make *either* inequality true.

Let's start with the compound inequalities with "and." Our solution will be the numbers that are solutions to *both* inequalities known as the intersection of the two inequalities. Consider the intersection of two streets—the part where the streets overlap—belongs to both streets.

To find the solution of the compound inequality, we look at the graphs of each inequality and then find the numbers that belong to both graphs—where the graphs overlap.

For the compound inequality $x > -3$ and $x \le 2$, we graph each inequality. We then look for where the graphs "overlap". The numbers that are shaded on both graphs, will be shaded on the graph of the solution of the compound inequality. See **Figure 2.5**.

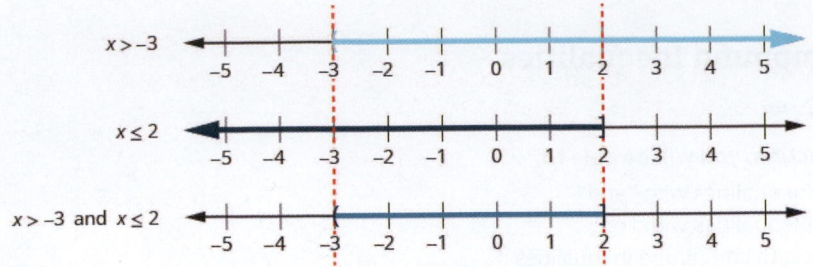

Figure 2.5

We can see that the numbers between -3 and 2 are shaded on both of the first two graphs. They will then be shaded on the solution graph.

The number -3 is not shaded on the first graph and so since it is not shaded on both graphs, it is not included on the solution graph.

The number two is shaded on both the first and second graphs. Therefore, it is be shaded on the solution graph.

This is how we will show our solution in the next examples.

EXAMPLE 2.61

Solve $6x - 3 < 9$ and $2x + 7 \geq 3$. Graph the solution and write the solution in interval notation.

✓ **Solution**

	$6x - 3 < 9$	and	$2x + 9 \geq 3$
Step 1. Solve each inequality.	$6x - 3 < 9$		$2x + 9 \geq 3$
	$6x < 12$		$2x \geq -6$
	$x < 2$	and	$x \geq -3$

Step 2. Graph each solution. Then graph the numbers that make both inequalities true. The final graph will show all the numbers that make both inequalities true—the numbers shaded on *both* of the first two graphs.

Step 3. Write the solution in interval notation.

$$[-3, 2)$$

All the numbers that make both inequalities true are the solution to the compound inequality.

> **TRY IT :: 2.121**
>
> Solve the compound inequality. Graph the solution and write the solution in interval notation: $4x - 7 < 9$ and $5x + 8 \geq 3$.

> **TRY IT :: 2.122**

Solve the compound inequality. Graph the solution and write the solution in interval notation: $3x - 4 < 5$ and $4x + 9 \geq 1$.

HOW TO :: SOLVE A COMPOUND INEQUALITY WITH "AND."

Step 1. Solve each inequality.

Step 2. Graph each solution. Then graph the numbers that make *both* inequalities true. This graph shows the solution to the compound inequality.

Step 3. Write the solution in interval notation.

EXAMPLE 2.62

Solve $3(2x + 5) \leq 18$ and $2(x - 7) < -6$. Graph the solution and write the solution in interval notation.

✓ **Solution**

	$3(2x + 5) \leq 18$	and	$2(x - 7) < -6$
Solve each inequality.	$6x + 15 \leq 18$		$2x - 14 < -6$
	$6x \leq 3$		$2x < 8$
	$x \leq \frac{1}{2}$	and	$x < 4$

Graph each solution.

$x \leq \frac{1}{2}$

$x < 4$

Graph the numbers that make both inequalities true.

$x < \frac{1}{2}$ and $x < 4$

Write the solution in interval notation.

$$\left(-\infty, \frac{1}{2}\right]$$

> **TRY IT :: 2.123**

Solve the compound inequality. Graph the solution and write the solution in interval notation: $2(3x + 1) \leq 20$ and $4(x - 1) < 2$.

> **TRY IT :: 2.124**

Solve the compound inequality. Graph the solution and write the solution in interval notation: $5(3x - 1) \leq 10$ and $4(x + 3) < 8$.

EXAMPLE 2.63

Solve $\frac{1}{3}x - 4 \geq -2$ and $-2(x - 3) \geq 4$. Graph the solution and write the solution in interval notation.

✓ **Solution**

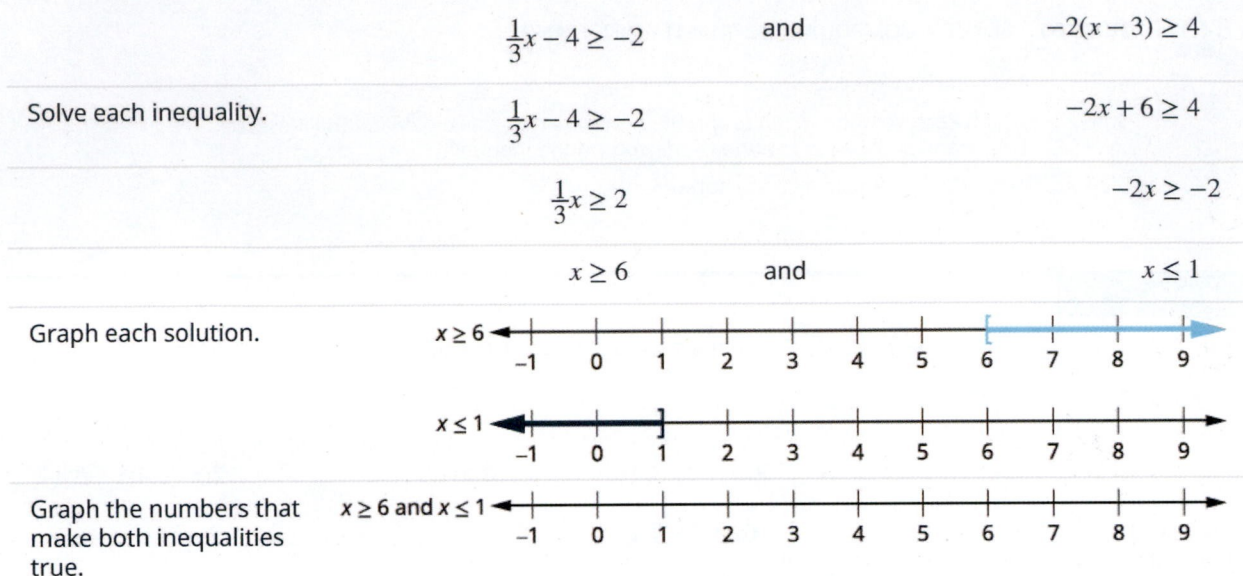

	$\frac{1}{3}x - 4 \geq -2$	and	$-2(x - 3) \geq 4$
Solve each inequality.	$\frac{1}{3}x - 4 \geq -2$		$-2x + 6 \geq 4$
	$\frac{1}{3}x \geq 2$		$-2x \geq -2$
	$x \geq 6$	and	$x \leq 1$

Graph each solution.

Graph the numbers that make both inequalities true.

There are no numbers that make both inequalities true.

This is a contradiction so there is no solution.

> **TRY IT :: 2.125**
>
> Solve the compound inequality. Graph the solution and write the solution in interval notation: $\frac{1}{4}x - 3 \geq -1$ and
>
> $-3(x - 2) \geq 2$.

> **TRY IT :: 2.126**
>
> Solve the compound inequality. Graph the solution and write the solution in interval notation: $\frac{1}{5}x - 5 \geq -3$ and
>
> $-4(x - 1) \geq -2$.

Sometimes we have a compound inequality that can be written more concisely. For example, $a < x$ and $x < b$ can be written simply as $a < x < b$ and then we call it a double inequality. The two forms are equivalent.

Double Inequality

A double inequality is a compound inequality such as $a < x < b$. It is equivalent to $a < x$ and $x < b$.

	$a < x < b$	is equivalent to	$a < x$	and	$x < b$
Other forms:	$a \leq x \leq b$	is equivalent to	$a \leq x$	and	$x \leq b$
	$a > x > b$	is equivalent to	$a > x$	and	$x > b$
	$a \geq x \geq b$	is equivalent to	$a \geq x$	and	$x \geq b$

To solve a double inequality we perform the same operation on all three "parts" of the double inequality with the goal of isolating the variable in the center.

EXAMPLE 2.64

Solve $-4 \le 3x - 7 < 8$. Graph the solution and write the solution in interval notation.

✓ **Solution**

	$-4 \le 3x - 7 < 8$
Add 7 to all three parts.	$-4 + 7 \le 3x - 7 + 7 < 8 + 7$
Simplify.	$3 \le 3x < 15$
Divide each part by three.	$\dfrac{3}{3} \le \dfrac{3x}{3} < \dfrac{15}{3}$
Simplify.	$1 \le x < 5$
Graph the solution.	
Write the solution in interval notation.	$[1, 5)$

When written as a double inequality, $1 \le x < 5$, it is easy to see that the solutions are the numbers caught between one and five, including one, but not five. We can then graph the solution immediately as we did above.

Another way to graph the solution of $1 \le x < 5$ is to graph both the solution of $x \ge 1$ and the solution of $x < 5$. We would then find the numbers that make both inequalities true as we did in previous examples.

> **TRY IT :: 2.127**
>
> Solve the compound inequality. Graph the solution and write the solution in interval notation: $-5 \le 4x - 1 < 7$.

> **TRY IT :: 2.128**
>
> Solve the compound inequality. Graph the solution and write the solution in interval notation: $-3 < 2x - 5 \le 1$.

Solve Compound Inequalities with "or"

To solve a compound inequality with "or", we start out just as we did with the compound inequalities with "and"—we solve the two inequalities. Then we find all the numbers that make *either* inequality true.

Just as the United States is the union of all of the 50 states, the solution will be the union of all the numbers that make either inequality true. To find the solution of the compound inequality, we look at the graphs of each inequality, find the numbers that belong to either graph and put all those numbers together.

To write the solution in interval notation, we will often use the union symbol, ∪ to show the union of the solutions shown in the graphs.

   HOW TO :: SOLVE A COMPOUND INEQUALITY WITH "OR."

    Step 1.   Solve each inequality.

    Step 2.   Graph each solution. Then graph the numbers that make either inequality true.

    Step 3.   Write the solution in interval notation.

EXAMPLE 2.65

Solve $5 - 3x \le -1$ or $8 + 2x \le 5$. Graph the solution and write the solution in interval notation.

✓ Solution

	$5 - 3x \leq -1$	or	$8 + 2x \leq 5$
Solve each inequality.	$5 - 3x \leq -1$		$8 + 2x \leq 5$
	$-3x \leq -6$		$2x \leq -3$
	$x \geq 2$	or	$x \leq -\dfrac{3}{2}$

Graph each solution.

$x \geq 2$

$x \leq -\dfrac{3}{2}$

Graph numbers that make either inequality true.

$x \geq 2$ or $x \leq -\dfrac{3}{2}$

$$\left(-\infty, \ -\dfrac{3}{2}\right] \cup [2, \ \infty)$$

> **TRY IT : : 2.129**

Solve the compound inequality. Graph the solution and write the solution in interval notation: $1 - 2x \leq -3$ or $7 + 3x \leq 4$.

> **TRY IT : : 2.130**

Solve the compound inequality. Graph the solution and write the solution in interval notation: $2 - 5x \leq -3$ or $5 + 2x \leq 3$.

EXAMPLE 2.66

Solve $\dfrac{2}{3}x - 4 \leq 3$ or $\dfrac{1}{4}(x + 8) \geq -1$. Graph the solution and write the solution in interval notation.

✓ Solution

	$\dfrac{2}{3}x - 4 \leq 3$	or	$\dfrac{1}{4}(x + 8) \geq -1$
Solve each inequality.	$3\left(\dfrac{2}{3}x - 4\right) \leq 3(3)$		$4 \cdot \dfrac{1}{4}(x + 8) \geq 4 \cdot (-1)$
	$2x - 12 \leq 9$		$x + 8 \geq -4$
	$2x \leq 21$		$x \geq -12$
	$x \leq \dfrac{21}{2}$		

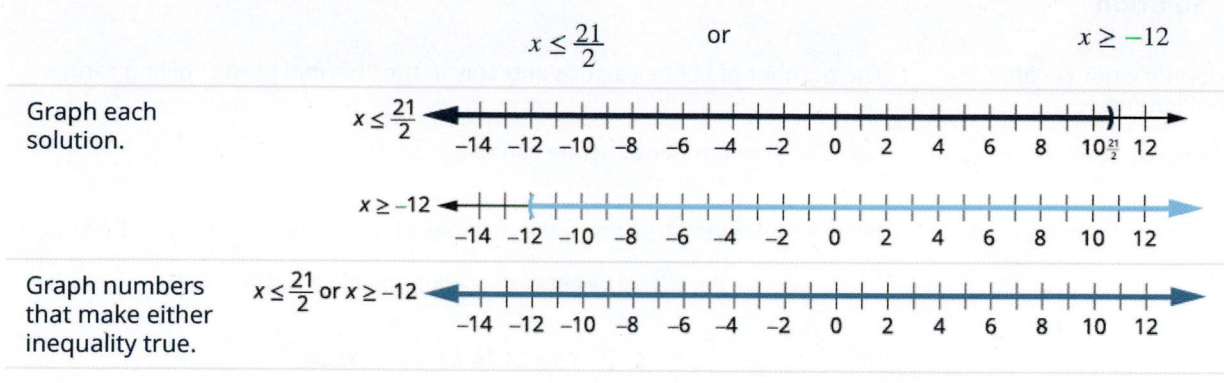

$$x \le \frac{21}{2} \qquad \text{or} \qquad x \ge -12$$

The solution covers all real numbers.

$$(-\infty, \infty)$$

> **TRY IT : : 2.131**

Solve the compound inequality. Graph the solution and write the solution in interval notation: $\frac{3}{5}x - 7 \le -1$ or $\frac{1}{3}(x + 6) \ge -2$.

> **TRY IT : : 2.132**

Solve the compound inequality. Graph the solution and write the solution in interval notation: $\frac{3}{4}x - 3 \le 3$ or $\frac{2}{5}(x + 10) \ge 0$.

Solve Applications with Compound Inequalities

Situations in the real world also involve compound inequalities. We will use the same problem solving strategy that we used to solve linear equation and inequality applications.

Recall the problem solving strategies are to first read the problem and make sure all the words are understood. Then, identify what we are looking for and assign a variable to represent it. Next, restate the problem in one sentence to make it easy to translate into a compound inequality. Last, we will solve the compound inequality.

EXAMPLE 2.67

Due to the drought in California, many communities have tiered water rates. There are different rates for Conservation Usage, Normal Usage and Excessive Usage. The usage is measured in the number of hundred cubic feet (hcf) the property owner uses.

During the summer, a property owner will pay $24.72 plus $1.54 per hcf for Normal Usage. The bill for Normal Usage would be between or equal to $57.06 and $171.02. How many hcf can the owner use if he wants his usage to stay in the normal range?

⊘ **Solution**

Identify what we are looking for.	The number of hcf he can use and stay in the "normal usage" billing range.
Name what we are looking for.	Let $x =$ the number of hcf he can use.
Translate to an inequality.	Bill is $24.72 plus $1.54 times the number of hcf he uses or $24.72 + 1.54x$.
	His bill will be between or equal to $57.06 and $171.02
	$57.06 \leq 24.74 + 1.54x \leq 171.02$
Solve the inequality.	$57.06 \leq 24.74 + 1.54x \leq 171.02$
	$57.06 - 24.72 \leq 24.74 - 24.72 + 1.54x \leq 171.02 - 24.72$
	$32.34 \leq 1.54x \leq 146.3$
	$\dfrac{32.34}{1.54} \leq \dfrac{1.54x}{1.54} \leq \dfrac{146.3}{1.54}$
	$21 \leq x \leq 95$
Answer the question.	The property owner can use 21–95 hcf and still fall within the "normal usage" billing range.

> **TRY IT : :** 2.133

Due to the drought in California, many communities now have tiered water rates. There are different rates for Conservation Usage, Normal Usage and Excessive Usage. The usage is measured in the number of hundred cubic feet (hcf) the property owner uses.

During the summer, a property owner will pay $24.72 plus $1.32 per hcf for Conservation Usage. The bill for Conservation Usage would be between or equal to $31.32 and $52.12. How many hcf can the owner use if she wants her usage to stay in the conservation range?

> **TRY IT : :** 2.134

Due to the drought in California, many communities have tiered water rates. There are different rates for Conservation Usage, Normal Usage and Excessive Usage. The usage is measured in the number of hundred cubic feet (hcf) the property owner uses.

During the winter, a property owner will pay $24.72 plus $1.54 per hcf for Normal Usage. The bill for Normal Usage would be between or equal to $49.36 and $86.32. How many hcf will he be allowed to use if he wants his usage to stay in the normal range?

▶ **MEDIA : :**

Access this online resource for additional instruction and practice with solving compound inequalities.

- **Compound inequalities (https://openstax.org/l/37compinequalit)**

2.6 EXERCISES

Practice Makes Perfect

Solve Compound Inequalities with "and"

In the following exercises, solve each inequality, graph the solution, and write the solution in interval notation.

376. $x < 3$ and $x \geq 1$

377. $x \leq 4$ and $x > -2$

378. $x \geq -4$ and $x \leq -1$

379. $x > -6$ and $x < -3$

380. $5x - 2 < 8$ and $6x + 9 \geq 3$

381. $4x - 1 < 7$ and $2x + 8 \geq 4$

382. $4x + 6 \leq 2$ and $2x + 1 \geq -5$

383. $4x - 2 \leq 4$ and $7x - 1 > -8$

384. $2x - 11 < 5$ and $3x - 8 > -5$

385. $7x - 8 < 6$ and $5x + 7 > -3$

386. $4(2x - 1) \leq 12$ and $2(x + 1) < 4$

387. $5(3x - 2) \leq 5$ and $3(x + 3) < 3$

388. $3(2x - 3) > 3$ and $4(x + 5) \geq 4$

389. $-3(x + 4) < 0$ and $-1(3x - 1) \leq 7$

390. $\frac{1}{2}(3x - 4) \leq 1$ and $\frac{1}{3}(x + 6) \leq 4$

391. $\frac{3}{4}(x - 8) \leq 3$ and $\frac{1}{5}(x - 5) \leq 3$

392. $5x - 2 \leq 3x + 4$ and $3x - 4 \geq 2x + 1$

393. $\frac{3}{4}x - 5 \geq -2$ and $-3(x + 1) \geq 6$

394. $\frac{2}{3}x - 6 \geq -4$ and $-4(x + 2) \geq 0$

395. $\frac{1}{2}(x - 6) + 2 < -5$ and $4 - \frac{2}{3}x < 6$

396. $-5 \leq 4x - 1 < 7$

397. $-3 < 2x - 5 \leq 1$

398. $5 < 4x + 1 < 9$

399. $-1 < 3x + 2 < 8$

400. $-8 < 5x + 2 \leq -3$

401. $-6 \leq 4x - 2 < -2$

Solve Compound Inequalities with "or"

In the following exercises, solve each inequality, graph the solution on the number line, and write the solution in interval notation.

402. $x \leq -2$ or $x > 3$

403. $x \leq -4$ or $x > -3$

404. $x < 2$ or $x \geq 5$

405. $x < 0$ or $x \geq 4$

406. $2 + 3x \leq 4$ or $5 - 2x \leq -1$

407. $4 - 3x \leq -2$ or $2x - 1 \leq -5$

408. $2(3x - 1) < 4$ or $3x - 5 > 1$

409. $3(2x - 3) < -5$ or $4x - 1 > 3$

410. $\frac{3}{4}x - 2 > 4$ or $4(2 - x) > 0$

411. $\frac{2}{3}x - 3 > 5$ or $3(5 - x) > 6$

412. $3x - 2 > 4$ or $5x - 3 \leq 7$

413. $2(x + 3) \geq 0$ or $3(x + 4) \leq 6$

414. $\frac{1}{2}x - 3 \leq 4$ or

$\frac{1}{3}(x - 6) \geq -2$

415. $\frac{3}{4}x + 2 \leq -1$ or

$\frac{1}{2}(x + 8) \geq -3$

Mixed practice

In the following exercises, solve each inequality, graph the solution on the number line, and write the solution in interval notation.

416. $3x + 7 \leq 1$ and

$2x + 3 \geq -5$

417. $6(2x - 1) > 6$ and

$5(x + 2) \geq 0$

418. $4 - 7x \geq -3$ or

$5(x - 3) + 8 > 3$

419. $\frac{1}{2}x - 5 \leq 3$ or

$\frac{1}{4}(x - 8) \geq -3$

420. $-5 \leq 2x - 1 < 7$

421. $\frac{1}{5}(x - 5) + 6 < 4$ and

$3 - \frac{2}{3}x < 5$

422. $4x - 2 > 6$ or

$3x - 1 \leq -2$

423. $6x - 3 \leq 1$ and

$5x - 1 > -6$

424. $-2(3x - 4) \leq 2$ and

$-4(x - 1) < 2$

425. $-5 \leq 3x - 2 \leq 4$

Solve Applications with Compound Inequalities

In the following exercises, solve.

426. Penelope is playing a number game with her sister June. Penelope is thinking of a number and wants June to guess it. Five more than three times her number is between 2 and 32. Write a compound inequality that shows the range of numbers that Penelope might be thinking of.

427. Gregory is thinking of a number and he wants his sister Lauren to guess the number. His first clue is that six less than twice his number is between four and forty-two. Write a compound inequality that shows the range of numbers that Gregory might be thinking of.

428. Andrew is creating a rectangular dog run in his back yard. The length of the dog run is 18 feet. The perimeter of the dog run must be at least 42 feet and no more than 72 feet. Use a compound inequality to find the range of values for the width of the dog run.

429. Elouise is creating a rectangular garden in her back yard. The length of the garden is 12 feet. The perimeter of the garden must be at least 36 feet and no more than 48 feet. Use a compound inequality to find the range of values for the width of the garden.

Everyday Math

430. Blood Pressure A person's blood pressure is measured with two numbers. The systolic blood pressure measures the pressure of the blood on the arteries as the heart beats. The diastolic blood pressure measures the pressure while the heart is resting.

ⓐ Let x be your systolic blood pressure. Research and then write the compound inequality that shows you what a normal systolic blood pressure should be for someone your age.

ⓑ Let y be your diastolic blood pressure. Research and then write the compound inequality that shows you what a normal diastolic blood pressure should be for someone your age.

431. Body Mass Index (BMI) is a measure of body fat is determined using your height and weight.

ⓐ Let x be your BMI. Research and then write the compound inequality to show the BMI range for you to be considered normal weight.

ⓑ Research a BMI calculator and determine your BMI. Is it a solution to the inequality in part (a)?

Writing Exercises

432. In your own words, explain the difference between the properties of equality and the properties of inequality.

433. Explain the steps for solving the compound inequality $2 - 7x \geq -5$ or $4(x - 3) + 7 > 3$.

Self Check

ⓐ *After completing the exercises, use this checklist to evaluate your mastery of the objectives of this section.*

I can...	Confidently	With some help	No-I don't get it!
solve compound inequalities with "and".			
solve compound inequalities with "or".			
solve applications with compound inequalities.			

ⓑ *What does this checklist tell you about your mastery of this section? What steps will you take to improve?*

Solve Absolute Value Inequalities

Learning Objectives

By the end of this section, you will be able to:

› Solve absolute value equations
› Solve absolute value inequalities with "less than"
› Solve absolute value inequalities with "greater than"
› Solve applications with absolute value

Be Prepared!

Before you get started, take this readiness quiz.

1. Evaluate: $-|7|$.
   If you missed this problem, review **Example 1.12**.

2. Fill in $<$, $>$, or $=$ for each of the following pairs of numbers.

   ⓐ $|-8|$___ $-|-8|$ ⓑ 12___ $-|-12|$ ⓒ $|-6|$___ -6 ⓓ $-(-15)$___ $-|-15|$
   If you missed this problem, review **Example 1.12**.

3. Simplify: $14 - 2|8 - 3(4 - 1)|$.
   If you missed this problem, review **Example 1.13**.

Solve Absolute Value Equations

As we prepare to solve absolute value equations, we review our definition of **absolute value**.

Absolute Value

The absolute value of a number is its distance from zero on the number line.

The absolute value of a number n is written as $|n|$ and $|n| \geq 0$ for all numbers.

Absolute values are always greater than or equal to zero.

We learned that both a number and its opposite are the same distance from zero on the number line. Since they have the same distance from zero, they have the same absolute value. For example:

-5 is 5 units away from 0, so $|-5| = 5$.

5 is 5 units away from 0, so $|5| = 5$.

Figure 2.6 illustrates this idea.

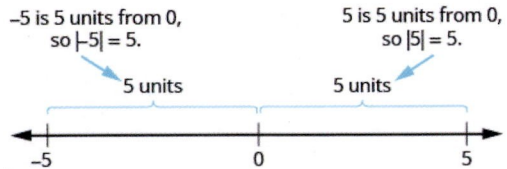

Figure 2.6 The numbers 5 and -5 are both five units away from zero.

For the equation $|x| = 5$, we are looking for all numbers that make this a true statement. We are looking for the numbers whose distance from zero is 5. We just saw that both 5 and -5 are five units from zero on the number line. They are the solutions to the equation.

$$\text{If} \qquad |x| = 5$$
$$\text{then} \qquad x = -5 \text{ or } x = 5$$

The solution can be simplified to a single statement by writing $x = \pm 5$. This is read, "x is equal to positive or negative 5".

We can generalize this to the following property for absolute value equations.

Absolute Value Equations

For any algebraic expression, u, and any positive real number, a,

$$\text{if} \qquad |u| = a$$
$$\text{then} \qquad u = -a \text{ or } u = a$$

Remember that an absolute value cannot be a negative number.

EXAMPLE 2.68

Solve: ⓐ $|x| = 8$ ⓑ $|y| = -6$ ⓒ $|z| = 0$

⊘ **Solution**

ⓐ

$$|x| = 8$$
Write the equivalent equations. $\qquad x = -8 \text{ or } x = 8$
$$x = \pm 8$$

ⓑ

$$|y| = -6$$
No solution

Since an absolute value is always positive, there are no solutions to this equation.

ⓒ

$$|z| = 0$$
Write the equivalent equations. $\qquad z = -0 \text{ or } z = 0$
Since $-0 = 0$, $\qquad z = 0$

Both equations tell us that $z = 0$ and so there is only one solution.

> **TRY IT : : 2.135**   Solve: ⓐ $|x| = 2$ ⓑ $|y| = -4$ ⓒ $|z| = 0$

> **TRY IT : : 2.136**   Solve: ⓐ $|x| = 11$ ⓑ $|y| = -5$ ⓒ $|z| = 0$

To solve an absolute value equation, we first isolate the absolute value expression using the same procedures we used to solve linear equations. Once we isolate the absolute value expression we rewrite it as the two equivalent equations.

EXAMPLE 2.69   HOW TO SOLVE ABSOLUTE VALUE EQUATIONS

Solve $|5x - 4| - 3 = 8$.

⊘ **Solution**

Step 1. Isolate the absolute value expression.	Add 3 to both sides.	$\begin{aligned}|5x - 4| - 3 &= 8 \\ |5x - 4| &= 11\end{aligned}$
Step 2. Write the equivalent equations.		$5x - 4 = -11 \text{ or } 5x - 4 = 11$
Step 3. Solve each equation.	Add 4 to each side. Divide each side by 5.	$\begin{aligned}5x &= -7 & 5x &= 15 \\ x &= -\frac{7}{5} \text{ or } & x &= 3\end{aligned}$

| Step 4. Check each solution. | Substitute 3 and $-\dfrac{7}{5}$ into the original equation. | $\begin{aligned}|5x-4|-3&=8\\ &\ x=3\\ |5\cdot 3-4|-3&\overset{?}{=}8\\ |15-4|-3&\overset{?}{=}8\\ |11|-3&\overset{?}{=}8\\ 11-3&\overset{?}{=}8\\ 8&=8\ \checkmark\\ |5x-4|-3&=8\\ &\ x=-\dfrac{7}{5}\\ \left|5\left(-\dfrac{7}{5}\right)-4\right|-3&\overset{?}{=}8\\ |-7-4|-3&\overset{?}{=}8\\ |-11|-3&\overset{?}{=}8\\ 11-3&\overset{?}{=}8\\ 8&=8\ \checkmark\end{aligned}$ |

> **TRY IT :: 2.137**   Solve: $|3x - 5| - 1 = 6$.

> **TRY IT :: 2.138**   Solve: $|4x - 3| - 5 = 2$.

The steps for solving an absolute value equation are summarized here.

HOW TO :: SOLVE ABSOLUTE VALUE EQUATIONS.

Step 1.   Isolate the absolute value expression.

Step 2.   Write the equivalent equations.

Step 3.   Solve each equation.

Step 4.   Check each solution.

EXAMPLE 2.70

Solve $2|x - 7| + 5 = 9$.

⊘ **Solution**

$$2|x - 7| + 5 = 9$$

Isolate the absolute value expression.	$2|x - 7| = 4$
	$|x - 7| = 2$
Write the equivalent equations.	$x - 7 = -2 \ \text{ or } \ x - 7 = 2$
Solve each equation.	$x = 5 \quad \text{or} \quad x = 9$

Check:

$$2|x - 7| + 5 = 9 \qquad\qquad 2|x - 7| + 5 = 9$$

$$2|5 - 7| + 5 \stackrel{?}{=} 9 \qquad\qquad 2|9 - 7| + 5 \stackrel{?}{=} 9$$

$$2|-2| + 5 \stackrel{?}{=} 9 \qquad\qquad 2|2| + 5 \stackrel{?}{=} 9$$

$$2 \cdot 2 + 5 \stackrel{?}{=} 9 \qquad\qquad 2 \cdot 2 + 5 \stackrel{?}{=} 9$$

$$4 + 5 \stackrel{?}{=} 9 \qquad\qquad 4 + 5 \stackrel{?}{=} 9$$

$$9 = 9 \checkmark \qquad\qquad 9 = 9 \checkmark$$

> **TRY IT ::** 2.139   Solve: $3|x - 4| - 4 = 8$.

> **TRY IT ::** 2.140   Solve: $2|x - 5| + 3 = 9$.

Remember, an absolute value is always positive!

EXAMPLE 2.71

Solve: $\left|\frac{2}{3}x - 4\right| + 11 = 3$.

⊘ **Solution**

$$\left|\frac{2}{3}x - 4\right| + 11 = 3$$

Isolate the absolute value term. $\qquad \left|\frac{2}{3}x - 4\right| = -8$

An absolute value cannot be negative. $\qquad$ No solution

> **TRY IT ::** 2.141   Solve: $\left|\frac{3}{4}x - 5\right| + 9 = 4$.

> **TRY IT ::** 2.142   Solve: $\left|\frac{5}{6}x + 3\right| + 8 = 6$.

Some of our absolute value equations could be of the form $|u| = |v|$ where u and v are algebraic expressions. For example, $|x - 3| = |2x + 1|$.

How would we solve them? If two algebraic expressions are equal in absolute value, then they are either equal to each other or negatives of each other. The property for absolute value equations says that for any algebraic expression, u, and a positive real number, a, if $|u| = a$, then $u = -a$ or $u = a$.

This tell us that

$$\text{if} \qquad |u| = |v|$$
$$\text{then} \qquad u = -v \quad \text{or} \quad u = v$$

This leads us to the following property for equations with two absolute values.

Equations with Two Absolute Values

For any algebraic expressions, u and v,

$$\text{if} \qquad |u| = |v|$$
$$\text{then} \qquad u = -v \text{ or } u = v$$

When we take the opposite of a quantity, we must be careful with the signs and to add parentheses where needed.

EXAMPLE 2.72

Solve: $|5x - 1| = |2x + 3|$.

✓ **Solution**

$$|5x - 1| = |2x + 3|$$

Write the equivalent equations. $5x - 1 = -(2x + 3)$ or $5x - 1 = 2x + 3$

Solve each equation. $5x - 1 = -2x - 3$ or $3x - 1 = 3$

$$7x - 1 = -3$$ $$3x = 4$$

$$7x = -2$$ $$x = \frac{4}{3}$$

$$x = -\frac{2}{7} \qquad \text{or} \qquad x = \frac{4}{3}$$

Check.
We leave the check to you.

> **TRY IT :: 2.143**   Solve: $|7x - 3| = |3x + 7|$.

> **TRY IT :: 2.144**   Solve: $|6x - 5| = |3x + 4|$.

Solve Absolute Value Inequalities with "Less Than"

Let's look now at what happens when we have an absolute value inequality. Everything we've learned about solving inequalities still holds, but we must consider how the absolute value impacts our work.

Again we will look at our definition of absolute value. The absolute value of a number is its distance from zero on the number line. For the equation $|x| = 5$, we saw that both 5 and -5 are five units from zero on the number line. They are the solutions to the equation.

$$|x| = 5$$
$$x = -5 \qquad \text{or} \qquad x = 5$$

What about the inequality $|x| \le 5$? Where are the numbers whose distance is less than or equal to 5? We know -5 and 5 are both five units from zero. All the numbers between -5 and 5 are less than five units from zero. See **Figure 2.7**.

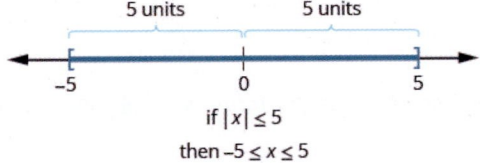

if $|x| \le 5$
then $-5 \le x \le 5$

Figure 2.7

In a more general way, we can see that if $|u| \le a$, then $-a \le u \le a$. See **Figure 2.8**.

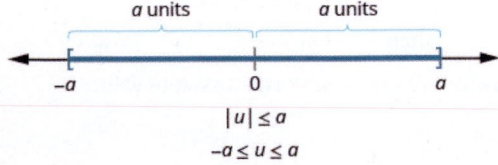

$|u| \le a$
$-a \le u \le a$

Figure 2.8

This result is summarized here.

Absolute Value Inequalities with $<$ or $\le$

For any algebraic expression, u, and any positive real number, a,

| | if | $|u| < a,$ | then $-a < u < a$ |
| | if | $|u| \leq a,$ | then $-a \leq u \leq a$ |

After solving an inequality, it is often helpful to check some points to see if the solution makes sense. The graph of the solution divides the number line into three sections. Choose a value in each section and substitute it in the original inequality to see if it makes the inequality true or not. While this is not a complete check, it often helps verify the solution.

EXAMPLE 2.73

Solve $|x| < 7$. Graph the solution and write the solution in interval notation.

⊘ **Solution**

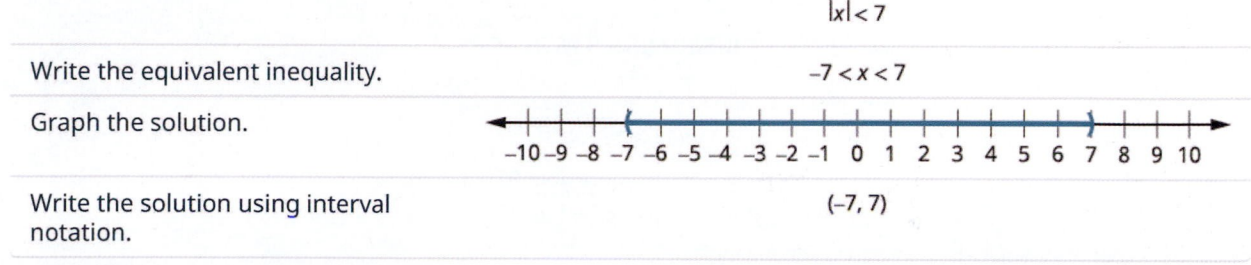

	$	x	< 7$
Write the equivalent inequality.	$-7 < x < 7$
Graph the solution.	
Write the solution using interval notation.	$(-7, 7)$

Check:

To verify, check a value in each section of the number line showing the solution. Choose numbers such as -8, 1, and 9.

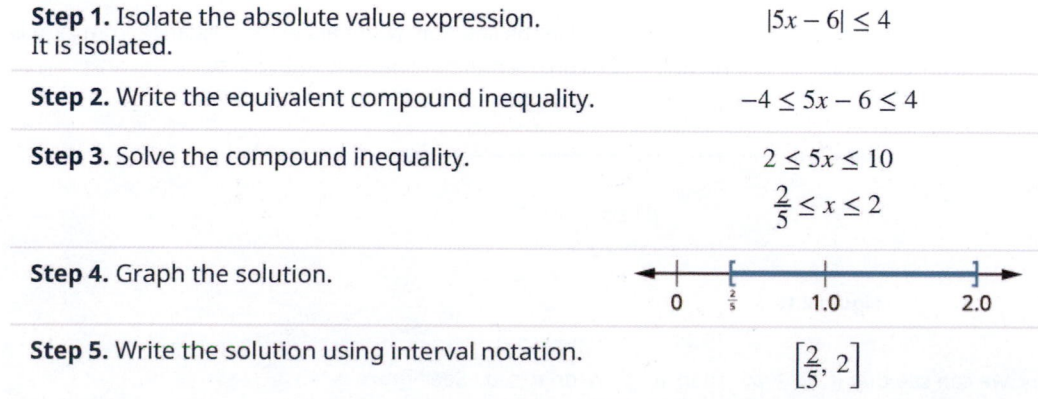

> **TRY IT :: 2.145**   Graph the solution and write the solution in interval notation: $|x| < 9$.

> **TRY IT :: 2.146**   Graph the solution and write the solution in interval notation: $|x| < 1$.

EXAMPLE 2.74

Solve $|5x - 6| \leq 4$. Graph the solution and write the solution in interval notation.

⊘ **Solution**

Step 1. Isolate the absolute value expression. It is isolated.	$	5x - 6	\leq 4$
Step 2. Write the equivalent compound inequality.	$-4 \leq 5x - 6 \leq 4$
Step 3. Solve the compound inequality.	$2 \leq 5x \leq 10$ $\frac{2}{5} \leq x \leq 2$
Step 4. Graph the solution.	
Step 5. Write the solution using interval notation.	$\left[\frac{2}{5}, 2\right]$

Check:
The check is left to you.

> **TRY IT :: 2.147**   Solve $|2x - 1| \leq 5$. Graph the solution and write the solution in interval notation:

> **TRY IT :: 2.148**   Solve $|4x - 5| \leq 3$. Graph the solution and write the solution in interval notation:

HOW TO :: SOLVE ABSOLUTE VALUE INEQUALITIES WITH < OR ≤.

Step 1.   Isolate the absolute value expression.

Step 2.   Write the equivalent compound inequality.

$$|u| < a \qquad \text{is equivalent to} \qquad -a < u < a$$
$$|u| \leq a \qquad \text{is equivalent to} \qquad -a \leq u \leq a$$

Step 3.   Solve the compound inequality.

Step 4.   Graph the solution

Step 5.   Write the solution using interval notation.

Solve Absolute Value Inequalities with "Greater Than"

What happens for absolute value inequalities that have "greater than"? Again we will look at our definition of absolute value. The absolute value of a number is its distance from zero on the number line.

We started with the inequality $|x| \leq 5$. We saw that the numbers whose distance is less than or equal to five from zero on the number line were -5 and 5 and all the numbers between -5 and 5. See **Figure 2.9**.

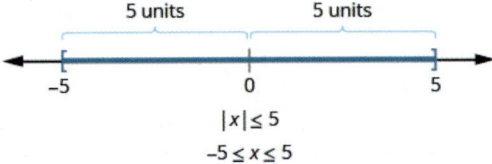

Figure 2.9

Now we want to look at the inequality $|x| \geq 5$. Where are the numbers whose distance from zero is greater than or equal to five?

Again both -5 and 5 are five units from zero and so are included in the solution. Numbers whose distance from zero is greater than five units would be less than -5 and greater than 5 on the number line. See **Figure 2.10**.

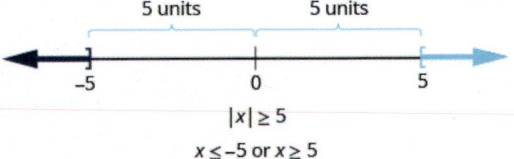

Figure 2.10

In a more general way, we can see that if $|u| \geq a$, then $u \leq -a$ or $u \leq a$. See **Figure 2.11**.

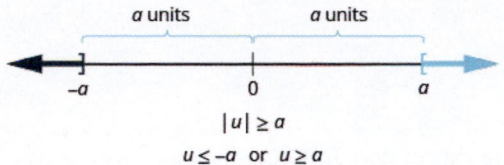

Figure 2.11

This result is summarized here.

Absolute Value Inequalities with > or ≥

For any algebraic expression, u, and any positive real number, a,

$$\text{if} \qquad |u| > a, \qquad \text{then } u < -a \text{ or } u > a$$
$$\text{if} \qquad |u| \geq a, \qquad \text{then } u \leq -a \text{ or } u \geq a$$

EXAMPLE 2.75

Solve $|x| > 4$. Graph the solution and write the solution in interval notation.

✓ **Solution**

$$|x| > 4$$

Write the equivalent inequality.	$x < -4$ or $x > 4$
Graph the solution.	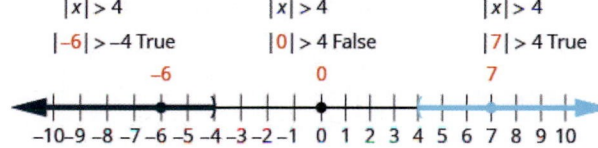
Write the solution using interval notation.	$(-\infty, -4) \cup (4, \infty)$
Check:	

To verify, check a value in each section of the number line showing the solution. Choose numbers such as -6, 0, and 7.

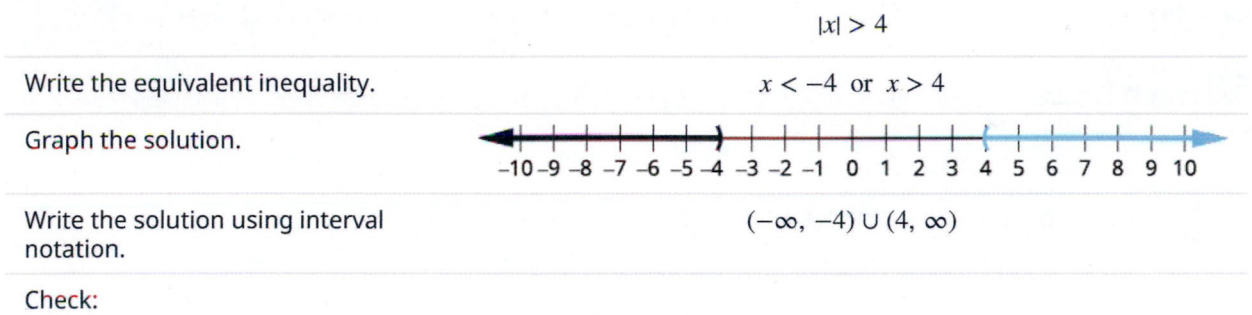

> **TRY IT :: 2.149**   Solve $|x| > 2$. Graph the solution and write the solution in interval notation.

> **TRY IT :: 2.150**   Solve $|x| > 1$. Graph the solution and write the solution in interval notation.

EXAMPLE 2.76

Solve $|2x - 3| \geq 5$. Graph the solution and write the solution in interval notation.

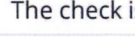 **Solution**

$$|2x - 3| \geq 5$$

Step 1. Isolate the absolute value expression. It is isolated.	
Step 2. Write the equivalent compound inequality.	$2x - 3 \leq -5 \ \text{or} \ 2x - 3 \geq 5$
Step 3. Solve the compound inequality.	$2x \leq -2 \ \text{or} \ 2x \geq 8$ $x \leq -1 \ \text{or} \ x \geq 4$
Step 4. Graph the solution.	
Step 5. Write the solution using interval notation.	$(-\infty, -1] \cup [4, \infty)$

Check:
The check is left to you.

> **TRY IT : : 2.151**   Solve $|4x - 3| \geq 5$. Graph the solution and write the solution in interval notation.

> **TRY IT : : 2.152**   Solve $|3x - 4| \geq 2$. Graph the solution and write the solution in interval notation.

 HOW TO : : SOLVE ABSOLUTE VALUE INEQUALITIES WITH > OR ≥.

Step 1.   Isolate the absolute value expression.

Step 2.   Write the equivalent compound inequality.

$$|u| > a \qquad \text{is equivalent to} \qquad u < -a \ \text{or} \ u > a$$
$$|u| \geq a \qquad \text{is equivalent to} \qquad u \leq -a \ \text{or} \ u \geq a$$

Step 3.   Solve the compound inequality.

Step 4.   Graph the solution

Step 5.   Write the solution using interval notation.

Solve Applications with Absolute Value

Absolute value inequalities are often used in the manufacturing process. An item must be made with near perfect specifications. Usually there is a certain *tolerance* of the difference from the specifications that is allowed. If the difference from the specifications exceeds the tolerance, the item is rejected.

$$|\text{actual-ideal}| \leq \text{tolerance}$$

EXAMPLE 2.77

The ideal diameter of a rod needed for a machine is 60 mm. The actual diameter can vary from the ideal diameter by 0.075 mm. What range of diameters will be acceptable to the customer without causing the rod to be rejected?

⊘ Solution

Use an absolute value inequality to express this situation.

Let $x =$ the actual measurement.

$|\text{actual-ideal}| \leq \text{tolerance}$

$$|x - 60| \leq 0.075$$

Rewrite as a compound inequality.

$$-0.075 \leq x - 60 \leq 0.075$$

Solve the inequality.

$$59.925 \leq x \leq 60.075$$

Answer the question.

The diameter of the rod can be between 59.925 mm and 60.075 mm.

> **TRY IT : : 2.153**

The ideal diameter of a rod needed for a machine is 80 mm. The actual diameter can vary from the ideal diameter by 0.009 mm. What range of diameters will be acceptable to the customer without causing the rod to be rejected?

> **TRY IT : : 2.154**

The ideal diameter of a rod needed for a machine is 75 mm. The actual diameter can vary from the ideal diameter by 0.05 mm. What range of diameters will be acceptable to the customer without causing the rod to be rejected?

▶ **MEDIA : :**

Access this online resource for additional instruction and practice with solving linear absolute value equations and inequalities.

- **Solving Linear Absolute Value Equations and Inequalities (https://openstax.org/l/37solvlinabsol)**

2.7 EXERCISES

Practice Makes Perfect

Solve Absolute Value Equations

In the following exercises, solve.

434. ⓐ $|x| = 6$ ⓑ $|y| = -3$
ⓒ $|z| = 0$

435. ⓐ $|x| = 4$ ⓑ $|y| = -5$
ⓒ $|z| = 0$

436. ⓐ $|x| = 7$ ⓑ $|y| = -11$
ⓒ $|z| = 0$

437. ⓐ $|x| = 3$ ⓑ $|y| = -1$
ⓒ $|z| = 0$

438. $|2x - 3| - 4 = 1$

439. $|4x - 1| - 3 = 0$

440. $|3x - 4| + 5 = 7$

441. $|4x + 7| + 2 = 5$

442. $4|x - 1| + 2 = 10$

443. $3|x - 4| + 2 = 11$

444. $3|4x - 5| - 4 = 11$

445. $3|x + 2| - 5 = 4$

446. $-2|x - 3| + 8 = -4$

447. $-3|x - 4| + 4 = -5$

448. $\left|\frac{3}{4}x - 3\right| + 7 = 2$

449. $\left|\frac{3}{5}x - 2\right| + 5 = 2$

450. $\left|\frac{1}{2}x + 5\right| + 4 = 1$

451. $\left|\frac{1}{4}x + 3\right| + 3 = 1$

452. $|3x - 2| = |2x - 3|$

453. $|4x + 3| = |2x + 1|$

454. $|6x - 5| = |2x + 3|$

455. $|6 - x| = |3 - 2x|$

Solve Absolute Value Inequalities with "less than"

In the following exercises, solve each inequality. Graph the solution and write the solution in interval notation.

456. $|x| < 5$

457. $|x| < 1$

458. $|x| \leq 8$

459. $|x| \leq 3$

460. $|3x - 3| \leq 6$

461. $|2x - 5| \leq 3$

462. $|2x + 3| + 5 < 4$

463. $|3x - 7| + 3 < 1$

464. $|4x - 3| < 1$

465. $|6x - 5| < 7$

466. $|x - 4| \leq -1$

467. $|5x + 1| \leq -2$

Solve Absolute Value Inequalities with "greater than"

In the following exercises, solve each inequality. Graph the solution and write the solution in interval notation.

468. $|x| > 3$

469. $|x| > 6$

470. $|x| \geq 2$

471. $|x| \geq 5$

472. $|3x - 8| > -1$

473. $|x - 5| > -2$

474. $|3x - 2| > 4$

475. $|2x - 1| > 5$

476. $|x + 3| \geq 5$

477. $|x - 7| \geq 1$

478. $3|x| + 4 \geq 1$

479. $5|x| + 6 \geq 1$

In the following exercises, solve. For each inequality, also graph the solution and write the solution in interval notation.

480. $2|x + 6| + 4 = 8$

481. $|6x - 5| = |2x + 3|$

482. $|3x - 4| \geq 2$

483. $|2x - 5| + 2 = 3$

484. $|4x - 3| < 5$

485. $|3x + 1| - 3 = 7$

486. $|7x + 2| + 8 < 4$

487. $5|2x - 1| - 3 = 7$

488. $|8 - x| = |4 - 3x|$

489. $|x - 7| > -3$

Solve Applications with Absolute Value

In the following exercises, solve.

490. A chicken farm ideally produces 200,000 eggs per day. But this total can vary by as much as 25,000 eggs. What is the maximum and minimum expected production at the farm?

491. An organic juice bottler ideally produces 215,000 bottle per day. But this total can vary by as much as 7,500 bottles. What is the maximum and minimum expected production at the bottling company?

492. In order to insure compliance with the law, Miguel routinely overshoots the weight of his tortillas by 0.5 gram. He just received a report that told him that he could be losing as much as $100,000 per year using this practice. He now plans to buy new equipment that guarantees the thickness of the tortilla within 0.005 inches. If the ideal thickness of the tortilla is 0.04 inches, what thickness of tortillas will be guaranteed?

493. At Lilly's Bakery, the ideal weight of a loaf of bread is 24 ounces. By law, the actual weight can vary from the ideal by 1.5 ounces. What range of weight will be acceptable to the inspector without causing the bakery being fined?

Writing Exercises

494. Write a graphical description of the absolute value of a number.

495. In your own words, explain how to solve the absolute value inequality, $|3x - 2| \geq 4$.

Self Check

ⓐ *After completing the exercises, use this checklist to evaluate your mastery of the objectives of this section.*

I can...	Confidently	With some help	No-I don't get it!
solve absolute value equations.			
solve absolute value inequalities with "less than".			
solve absolute value inequalities with "greater than".			
solve applications with absolute value.			

ⓑ *What does this checklist tell you about your mastery of this section? What steps will you take to improve?*

CHAPTER 2 REVIEW

KEY TERMS

compound inequality A compound inequality is made up of two inequalities connected by the word "and" or the word "or."

conditional equation An equation that is true for one or more values of the variable and false for all other values of the variable is a conditional equation.

contradiction An equation that is false for all values of the variable is called a contradiction. A contradiction has no solution.

identity An equation that is true for any value of the variable is called an Identity. The solution of an identity is all real numbers.

linear equation A linear equation is an equation in one variable that can be written, where a and b are real numbers and $a \neq 0$, as $ax + b = 0$.

solution of an equation A solution of an equation is a value of a variable that makes a true statement when substituted into the equation.

KEY CONCEPTS

2.1 Use a General Strategy to Solve Linear Equations

- **How to determine whether a number is a solution to an equation**

  Step 1. Substitute the number in for the variable in the equation.

  Step 2. Simplify the expressions on both sides of the equation.

  Step 3. Determine whether the resulting equation is true.
  If it is true, the number is a solution.
  If it is not true, the number is not a solution.

- **How to Solve Linear Equations Using a General Strategy**

  Step 1. Simplify each side of the equation as much as possible.
  Use the Distributive Property to remove any parentheses.
  Combine like terms.

  Step 2. Collect all the variable terms on one side of the equation.
  Use the Addition or Subtraction Property of Equality.

  Step 3. Collect all the constant terms on the other side of the equation.
  Use the Addition or Subtraction Property of Equality.

  Step 4. Make the coefficient of the variable term equal to 1.
  Use the Multiplication or Division Property of Equality.
  State the solution to the equation.

  Step 5. Check the solution.
  Substitute the solution into the original equation to make sure the result is a true statement.

- **How to Solve Equations with Fraction or Decimal Coefficients**

  Step 1. Find the least common denominator (LCD) of *all* the fractions and decimals (in fraction form) in the equation.

  Step 2. Multiply both sides of the equation by that LCD. This clears the fractions and decimals.

  Step 3. Solve using the General Strategy for Solving Linear Equations.

2.2 Use a Problem Solving Strategy

- **How To Use a Problem Solving Strategy for Word Problems**

  Step 1. **Read** the problem. Make sure all the words and ideas are understood.

  Step 2. **Identify** what you are looking for.

  Step 3. **Name** what you are looking for. Choose a variable to represent that quantity.

  Step 4. **Translate** into an equation. It may be helpful to restate the problem in one sentence with all the important information. Then, translate the English sentence into an algebra equation.

Step 5. **Solve** the equation using proper algebra techniques.

Step 6. **Check** the answer in the problem to make sure it makes sense.

Step 7. **Answer** the question with a complete sentence.

- **How To Find Percent Change**

  Step 1. Find the amount of change
  $$\text{change} = \text{new amount} - \text{original amount}$$

  Step 2. Find what percent the amount of change is of the original amount.
  change is what percent of the original amount?

- **Discount**
  $$\text{amount of discount} = \text{discount rate} \cdot \text{original price}$$
  $$\text{sale price} = \text{original amount} - \text{discount}$$

- **Mark-up**
  $$\text{amount of mark-up} = \text{mark-up rate} \cdot \text{original cost}$$
  $$\text{list price} = \text{original cost} + \text{mark up}$$

- **Simple Interest**
  If an amount of money, P, called the principal, is invested or borrowed for a period of t years at an annual interest rate r, the amount of interest, I, earned or paid is:

  $$I = Prt \quad \text{where} \quad \begin{aligned} I &= \text{interest} \\ P &= \text{principal} \\ r &= \text{rate} \\ t &= \text{time} \end{aligned}$$

2.3 Solve a Formula for a Specific Variable

- **How To Solve Geometry Applications**

  Step 1. **Read** the problem and make sure all the words and ideas are understood.

  Step 2. **Identify** what you are looking for.

  Step 3. **Name** what you are looking for by choosing a variable to represent it. Draw the figure and label it with the given information.

  Step 4. **Translate** into an equation by writing the appropriate formula or model for the situation. Substitute in the given information.

  Step 5. **Solve** the equation using good algebra techniques.

  Step 6. **Check** the answer in the problem and make sure it makes sense.

  Step 7. **Answer** the question with a complete sentence.

- **The Pythagorean Theorem**

  ◦ In any right triangle, where a and b are the lengths of the legs, and c is the length of the hypotenuse, the sum of the squares of the lengths of the two legs equals the square of the length of the hypotenuse.

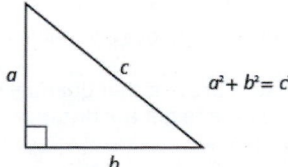

$$a^2 + b^2 = c^2$$

2.4 Solve Mixture and Uniform Motion Applications

- **Total Value of Coins**
  For the same type of coin, the total value of a number of coins is found by using the model
  $$\text{number} \cdot \text{value} = \text{total value}$$

  ◦ *number* is the number of coins

- ◦ *value* is the value of each coin
- ◦ *total value* is the total value of all the coins

- **How to solve coin word problems.**

  Step 1. **Read** the problem. Make sure all the words and ideas are understood.
  Determine the types of coins involved.
  Create a table to organize the information.
  > Label the columns "type," "number," "value," "total value."
  > List the types of coins.
  > Write in the value of each type of coin.
  > Write in the total value of all the coins.

Type	Number	•	Value ($)	=	Total Value ($)

  Step 2. **Identify** what you are looking for.

  Step 3. **Name** what you are looking for. Choose a variable to represent that quantity.
  Use variable expressions to represent the number of each type of coin and write them in the table.
  Multiply the number times the value to get the total value of each type of coin.

  Step 4. **Translate** into an equation.
  It may be helpful to restate the problem in one sentence with all the important information. Then, translate the sentence into an equation.
  Write the equation by adding the total values of all the types of coins.

  Step 5. **Solve** the equation using good algebra techniques.

  Step 6. **Check** the answer in the problem and make sure it makes sense.

  Step 7. **Answer** the question with a complete sentence.

- **How To Solve a Uniform Motion Application**

  Step 1. **Read** the problem. Make sure all the words and ideas are understood.
  Draw a diagram to illustrate what it happening.
  Create a table to organize the information.
  > Label the columns rate, time, distance.
  > List the two scenarios.
  > Write in the information you know.

	Rate	•	Time	=	Distance

  Step 2. **Identify** what you are looking for.

  Step 3. **Name** what you are looking for. Choose a variable to represent that quantity.
  Complete the chart.
  Use variable expressions to represent that quantity in each row.
  Multiply the rate times the time to get the distance.

  Step 4. **Translate** into an equation.
  Restate the problem in one sentence with all the important information.
  Then, translate the sentence into an equation.

  Step 5. **Solve** the equation using good algebra techniques.

  Step 6. **Check** the answer in the problem and make sure it makes sense.

  Step 7. **Answer** the question with a complete sentence.

2.5 Solve Linear Inequalities

- **Inequalities, Number Lines, and Interval Notation**

  $x > a$ $\qquad$ $x \geq a$ $\qquad$ $x < a$ $\qquad$ $x \leq a$

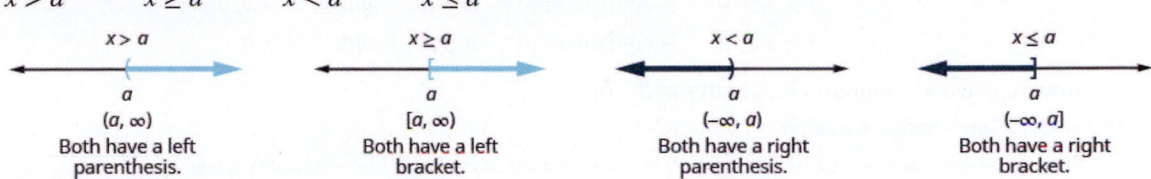

$x > a$	$x \geq a$	$x < a$	$x \leq a$
(a, ∞)	$[a, \infty)$	$(-\infty, a)$	$(-\infty, a]$
Both have a left parenthesis.	Both have a left bracket.	Both have a right parenthesis.	Both have a right bracket.

- **Linear Inequality**
  - A **linear inequality** is an inequality in one variable that can be written in one of the following forms where a, b, and c are real numbers and $a \neq 0$:

  $$ax + b < c, \qquad ax + b \leq c, \qquad ax + b > c, \qquad ax + b \geq c.$$

- **Addition and Subtraction Property of Inequality**
  - For any numbers a, b, and c, if $a < b$, then

  $$a + c < b + c \qquad a - c < b - c$$
  $$a + c > b + c \qquad a - c > b - c$$

  - We can add or subtract the same quantity from both sides of an inequality and still keep the inequality.

- **Multiplication and Division Property of Inequality**
  - For any numbers a, b, and c,
    multiply or divide by a **positive**

    if $a < b$ and $c > 0$, then $ac < bc$ and $\frac{a}{c} < \frac{b}{c}$.

    if $a > b$ and $c > 0$, then $ac > bc$ and $\frac{a}{c} > \frac{b}{c}$.

    multiply or divide by a **negative**

    if $a < b$ and $c < 0$, then $ac > bc$ and $\frac{a}{c} > \frac{b}{c}$.

    if $a > b$ and $c < 0$, then $ac < bc$ and $\frac{a}{c} < \frac{b}{c}$.

- **Phrases that indicate inequalities**

>	≥	<	≤
is greater than	is greater than or equal to	is less than	is less than or equal to
is more than	is at least	is smaller than	is at most
is larger than	is no less than	has fewer than	is no more than
exceeds	is the minimum	is lower than	is the maximum

2.6 Solve Compound Inequalities

- **How to solve a compound inequality with "and"**

  Step 1. Solve each inequality.

  Step 2. Graph each solution. Then graph the numbers that make *both* inequalities true. This graph shows the solution to the compound inequality.

  Step 3. Write the solution in interval notation.

- **Double Inequality**

◦ A **double inequality** is a compound inequality such as $a < x < b$. It is equivalent to $a < x$ and $x < b$.

	$a < x < b$	is equivalent to	$a < x$	and	$x < b$
Other forms:	$a \leq x \leq b$	is equivalent to	$a \leq x$	and	$x \leq b$
	$a > x > b$	is equivalent to	$a > x$	and	$x > b$
	$a \geq x \geq b$	is equivalent to	$a \geq x$	and	$x \geq b$

- **How to solve a compound inequality with "or"**

  Step 1. Solve each inequality.

  Step 2. Graph each solution. Then graph the numbers that make either inequality true.

  Step 3. Write the solution in interval notation.

2.7 Solve Absolute Value Inequalities

- **Absolute Value**
  The absolute value of a number is its distance from 0 on the number line.
  The absolute value of a number n is written as $|n|$ and $|n| \geq 0$ for all numbers.

  Absolute values are always greater than or equal to zero.

- **Absolute Value Equations**
  For any algebraic expression, u, and any positive real number, a,
  if $|u| = a$
  then $u = -a$ or $u = a$
  Remember that an absolute value cannot be a negative number.

- **How to Solve Absolute Value Equations**

  Step 1. Isolate the absolute value expression.

  Step 2. Write the equivalent equations.

  Step 3. Solve each equation.

  Step 4. Check each solution.

- **Equations with Two Absolute Values**
  For any algebraic expressions, u and v,
  if $|u| = |v|$
  then $u = -v$ or $u = v$

- **Absolute Value Inequalities with** $<$ or $\leq$
  For any algebraic expression, u, and any positive real number, a,
  if $|u| < a$,   then $-a < u < a$
  if $|u| \leq a$,   then $-a \leq u \leq a$

- **How To Solve Absolute Value Inequalities with** $<$ or $\leq$

  Step 1. Isolate the absolute value expression.

  Step 2. Write the equivalent compound inequality.
	$|u| < a$	is equivalent to	$-a < u < a$
	$|u| \leq a$	is equivalent to	$-a \leq u \leq a$

  Step 3. Solve the compound inequality.

  Step 4. Graph the solution

  Step 5. Write the solution using interval notation

- **Absolute Value Inequalities with** $>$ or $\geq$
  For any algebraic expression, u, and any positive real number, a,
  if $|u| > a$,   then $u < -a$ or $u > a$
  if $|u| \geq a$,   then $u \leq -a$ or $u \geq a$

- **How To Solve Absolute Value Inequalities with** $>$ or $\geq$

  Step 1. Isolate the absolute value expression.

Step 2. Write the equivalent compound inequality.

$|u| > a$ is equivalent to $u < -a$ or $u > a$

$|u| \geq a$ is equivalent to $u \leq -a$ or $u \geq a$

Step 3. Solve the compound inequality.

Step 4. Graph the solution

Step 5. Write the solution using interval notation

REVIEW EXERCISES

2.1 Use a General Strategy to Solve Linear Equations

Solve Equations Using the General Strategy for Solving Linear Equations

In the following exercises, determine whether each number is a solution to the equation.

496. $10x - 1 = 5x,\ x = \frac{1}{5}$

497. $-12n + 5 = 8n,\ n = -\frac{5}{4}$

In the following exercises, solve each linear equation.

498. $6(x + 6) = 24$

499. $-(s + 4) = 18$

500. $23 - 3(y - 7) = 8$

501. $\frac{1}{3}(6m + 21) = m - 7$

502. $4(3.5y + 0.25) = 365$

503. $0.25(q - 8) = 0.1(q + 7)$

504. $8(r - 2) = 6(r + 10)$

505. $5 + 7(2 - 5x) = 2(9x + 1) - (13x - 57)$

506. $(9n + 5) - (3n - 7) = 20 - (4n - 2)$

507. $2[-16 + 5(8k - 6)] = 8(3 - 4k) - 32$

Classify Equations

In the following exercises, classify each equation as a conditional equation, an identity, or a contradiction and then state the solution.

508. $17y - 3(4 - 2y) = 11(y - 1) + 12y - 1$

509. $9u + 32 = 15(u - 4) - 3(2u + 21)$

510. $-8(7m + 4) = -6(8m + 9)$

Solve Equations with Fraction or Decimal Coefficients

In the following exercises, solve each equation.

511. $\frac{2}{5}n - \frac{1}{10} = \frac{7}{10}$

512. $\frac{3}{4}a - \frac{1}{3} = \frac{1}{2}a + \frac{5}{6}$

513. $\frac{1}{2}(k + 3) = \frac{1}{3}(k + 16)$

514. $\frac{5y - 1}{3} + 4 = \frac{-8y + 4}{6}$

515. $0.8x - 0.3 = 0.7x + 0.2$

516. $0.10d + 0.05(d - 4) = 2.05$

2.2 Use a Problem-Solving Strategy

Use a Problem Solving Strategy for Word Problems

In the following exercises, solve using the problem solving strategy for word problems.

517. Three-fourths of the people at a concert are children. If there are 87 children, what is the total number of people at the concert?

518. There are nine saxophone players in the band. The number of saxophone players is one less than twice the number of tuba players. Find the number of tuba players.

Solve Number Word Problems

In the following exercises, solve each number word problem.

519. The sum of a number and three is forty-one. Find the number.

520. One number is nine less than another. Their sum is negative twenty-seven. Find the numbers.

521. One number is two more than four times another. Their sum is negative thirteen. Find the numbers.

522. The sum of two consecutive integers is -135. Find the numbers.

523. Find three consecutive even integers whose sum is 234.

524. Find three consecutive odd integers whose sum is 51.

525. Koji has $5,502 in his savings account. This is $30 less than six times the amount in his checking account. How much money does Koji have in his checking account?

Solve Percent Applications

In the following exercises, translate and solve.

526. What number is 67% of 250?

527. 12.5% of what number is 20?

528. What percent of 125 is 150?

In the following exercises, solve.

529. The bill for Dino's lunch was $19.45. He wanted to leave 20% of the total bill as a tip. How much should the tip be?

530. Dolores bought a crib on sale for $350. The sale price was 40% of the original price. What was the original price of the crib?

531. Jaden earns $2,680 per month. He pays $938 a month for rent. What percent of his monthly pay goes to rent?

532. Angel received a raise in his annual salary from $55,400 to $56,785. Find the percent change.

533. Rowena's monthly gasoline bill dropped from $83.75 last month to $56.95 this month. Find the percent change.

534. Emmett bought a pair of shoes on sale at 40% off from an original price of $138. Find ⓐ the amount of discount and ⓑ the sale price.

535. Lacey bought a pair of boots on sale for $95. The original price of the boots was $200. Find ⓐ the amount of discount and ⓑ the discount rate. (Round to the nearest tenth of a percent, if needed.)

536. Nga and Lauren bought a chest at a flea market for $50. They re-finished it and then added a 350% mark-up. Find ⓐ the amount of the mark-up and ⓑ the list price.

Solve Simple Interest Applications

In the following exercises, solve.

537. Winston deposited $3,294 in a bank account with interest rate 2.6% How much interest was earned in five years?

538. Moira borrowed $4,500 from her grandfather to pay for her first year of college. Three years later, she repaid the $4,500 plus $243 interest. What was the rate of interest?

539. Jaime's refrigerator loan statement said he would pay $1,026 in interest for a four-year loan at 13.5%. How much did Jaime borrow to buy the refrigerator?

2.3 Solve a formula for a Specific Variable

Solve a Formula for a Specific Variable

In the following exercises, solve the formula for the specified variable.

540. Solve the formula $V = LWH$ for L.

541. Solve the formula $A = \frac{1}{2}d_1 d_2$ for d_2.

542. Solve the formula $h = 48t + \frac{1}{2}at^2$ for t.

543. Solve the formula $4x - 3y = 12$ for y.

Use Formulas to Solve Geometry Applications

In the following exercises, solve using a geometry formula.

544. What is the height of a triangle with area 67.5 square meters and base 9 meters?

545. The measure of the smallest angle in a right triangle is $45°$ less than the measure of the next larger angle. Find the measures of all three angles.

546. The perimeter of a triangle is 97 feet. One side of the triangle is eleven feet more than the smallest side. The third side is six feet more than twice the smallest side. Find the lengths of all sides.

547. Find the length of the hypotenuse.

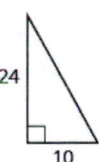

548. Find the length of the missing side. Round to the nearest tenth, if necessary.

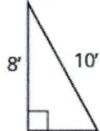

549. Sergio needs to attach a wire to hold the antenna to the roof of his house, as shown in the figure. The antenna is eight feet tall and Sergio has 10 feet of wire. How far from the base of the antenna can he attach the wire? Approximate to the nearest tenth, if necessary.

550. Seong is building shelving in his garage. The shelves are 36 inches wide and 15 inches tall. He wants to put a diagonal brace across the back to stabilize the shelves, as shown. How long should the brace be?

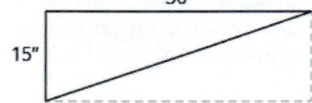

551. The length of a rectangle is 12 cm more than the width. The perimeter is 74 cm. Find the length and the width.

552. The width of a rectangle is three more than twice the length. The perimeter is 96 inches. Find the length and the width.

553. The perimeter of a triangle is 35 feet. One side of the triangle is five feet longer than the second side. The third side is three feet longer than the second side. Find the length of each side.

2.4 Solve Mixture and Uniform Motion Applications

Solve Coin Word Problems

In the following exercises, solve.

554. Paulette has $140 in $5 and $10 bills. The number of $10 bills is one less than twice the number of $5 bills. How many of each does she have?

555. Lenny has $3.69 in pennies, dimes, and quarters. The number of pennies is three more than the number of dimes. The number of quarters is twice the number of dimes. How many of each coin does he have?

Solve Ticket and Stamp Word Problems

In the following exercises, solve each ticket or stamp word problem.

556. Tickets for a basketball game cost $2 for students and $5 for adults. The number of students was three less than 10 times the number of adults. The total amount of money from ticket sales was $619. How many of each ticket were sold?

557. 125 tickets were sold for the jazz band concert for a total of $1,022. Student tickets cost $6 each and general admission tickets cost $10 each. How many of each kind of ticket were sold?

558. Yumi spent $34.15 buying stamps. The number of $0.56 stamps she bought was 10 less than four times the number of $0.41 stamps. How many of each did she buy?

Solve Mixture Word Problems

In the following exercises, solve.

559. Marquese is making 10 pounds of trail mix from raisins and nuts. Raisins cost $3.45 per pound and nuts cost $7.95 per pound. How many pounds of raisins and how many pounds of nuts should Marquese use for the trail mix to cost him $6.96 per pound?

560. Amber wants to put tiles on the backsplash of her kitchen counters. She will need 36 square feet of tile. She will use basic tiles that cost $8 per square foot and decorator tiles that cost $20 per square foot. How many square feet of each tile should she use so that the overall cost of the backsplash will be $10 per square foot?

561. Enrique borrowed $23,500 to buy a car. He pays his uncle 2% interest on the $4,500 he borrowed from him, and he pays the bank 11.5% interest on the rest. What average interest rate does he pay on the total $23,500? (Round your answer to the nearest tenth of a percent.)

Solve Uniform Motion Applications

In the following exercises, solve.

562. When Gabe drives from Sacramento to Redding it takes him 2.2 hours. It takes Elsa two hours to drive the same distance. Elsa's speed is seven miles per hour faster than Gabe's speed. Find Gabe's speed and Elsa's speed.

563. Louellen and Tracy met at a restaurant on the road between Chicago and Nashville. Louellen had left Chicago and drove 3.2 hours towards Nashville. Tracy had left Nashville and drove 4 hours towards Chicago, at a speed one mile per hour faster than Louellen's speed. The distance between Chicago and Nashville is 472 miles. Find Louellen's speed and Tracy's speed.

564. Two busses leave Amarillo at the same time. The Albuquerque bus heads west on the I-40 at a speed of 72 miles per hour, and the Oklahoma City bus heads east on the I-40 at a speed of 78 miles per hour. How many hours will it take them to be 375 miles apart?

565. Kyle rowed his boat upstream for 50 minutes. It took him 30 minutes to row back downstream. His speed going upstream is two miles per hour slower than his speed going downstream. Find Kyle's upstream and downstream speeds.

566. At 6:30, Devon left her house and rode her bike on the flat road until 7:30. Then she started riding uphill and rode until 8:00. She rode a total of 15 miles. Her speed on the flat road was three miles per hour faster than her speed going uphill. Find Devon's speed on the flat road and riding uphill.

567. Anthony drove from New York City to Baltimore, which is a distance of 192 miles. He left at 3:45 and had heavy traffic until 5:30. Traffic was light for the rest of the drive, and he arrived at 7:30. His speed in light traffic was four miles per hour more than twice his speed in heavy traffic. Find Anthony's driving speed in heavy traffic and light traffic.

2.5 Solve Linear Inequalities

Graph Inequalities on the Number Line

In the following exercises, graph the inequality on the number line and write in interval notation.

568. $x < -1$

569. $x \geq -2.5$

570. $x \leq \frac{5}{4}$

571. $x > 2$

572. $-2 < x < 0$

573. $-5 \leq x < -3$

574. $0 \leq x \leq 3.5$

Solve Linear Inequalities

In the following exercises, solve each inequality, graph the solution on the number line, and write the solution in interval notation.

575. $n - 12 \leq 23$

576. $a + \frac{2}{3} \geq \frac{7}{12}$

577. $9x > 54$

578. $\frac{q}{-2} \geq -24$

579. $6p > 15p - 30$

580. $9h - 7(h - 1) \leq 4h - 23$

581.
$5n - 15(4 - n) < 10(n - 6) + 10n$

582. $\frac{3}{8}a - \frac{1}{12}a > \frac{5}{12}a + \frac{3}{4}$

Translate Words to an Inequality and Solve

In the following exercises, translate and solve. Then write the solution in interval notation and graph on the number line.

583. Five more than z is at most 19.

584. Three less than c is at least 360.

585. Nine times n exceeds 42.

586. Negative two times a is no more than eight.

Solve Applications with Linear Inequalities

In the following exercises, solve.

587. Julianne has a weekly food budget of $231 for her family. If she plans to budget the same amount for each of the seven days of the week, what is the maximum amount she can spend on food each day?

588. Rogelio paints watercolors. He got a $100 gift card to the art supply store and wants to use it to buy 12" × 16" canvases. Each canvas costs $10.99. What is the maximum number of canvases he can buy with his gift card?

589. Briana has been offered a sales job in another city. The offer was for $42,500 plus 8% of her total sales. In order to make it worth the move, Briana needs to have an annual salary of at least $66,500. What would her total sales need to be for her to move?

590. Renee's car costs her $195 per month plus $0.09 per mile. How many miles can Renee drive so that her monthly car expenses are no more than $250?

591. Costa is an accountant. During tax season, he charges $125 to do a simple tax return. His expenses for buying software, renting an office, and advertising are $6,000. How many tax returns must he do if he wants to make a profit of at least $8,000?

592. Jenna is planning a five-day resort vacation with three of her friends. It will cost her $279 for airfare, $300 for food and entertainment, and $65 per day for her share of the hotel. She has $550 saved towards her vacation and can earn $25 per hour as an assistant in her uncle's photography studio. How many hours must she work in order to have enough money for her vacation?

2.6 Solve Compound Inequalities

Solve Compound Inequalities with "and"

In each of the following exercises, solve each inequality, graph the solution, and write the solution in interval notation.

593. $x \le 5$ and $x > -3$

594. $4x - 2 \le 4$ and $7x - 1 > -8$

595. $5(3x - 2) \le 5$ and $4(x + 2) < 3$

596. $\frac{3}{4}(x - 8) \le 3$ and $\frac{1}{5}(x - 5) \le 3$

597. $\frac{3}{4}x - 5 \ge -2$ and $-3(x + 1) \ge 6$

598. $-5 \le 4x - 1 < 7$

Solve Compound Inequalities with "or"

In the following exercises, solve each inequality, graph the solution on the number line, and write the solution in interval notation.

599. $5 - 2x \le -1$ or $6 + 3x \le 4$

600. $3(2x - 3) < -5$ or $4x - 1 > 3$

601. $\frac{3}{4}x - 2 > 4$ or $4(2 - x) > 0$

602. $2(x + 3) \ge 0$ or $3(x + 4) \le 6$

603. $\frac{1}{2}x - 3 \le 4$ or $\frac{1}{3}(x - 6) \ge -2$

Solve Applications with Compound Inequalities

In the following exercises, solve.

604. Liam is playing a number game with his sister Audry. Liam is thinking of a number and wants Audry to guess it. Five more than three times her number is between 2 and 32. Write a compound inequality that shows the range of numbers that Liam might be thinking of.

605. Elouise is creating a rectangular garden in her back yard. The length of the garden is 12 feet. The perimeter of the garden must be at least 36 feet and no more than 48 feet. Use a compound inequality to find the range of values for the width of the garden.

2.7 Solve Absolute Value Inequalities

Solve Absolute Value Equations

In the following exercises, solve.

606. $|x| = 8$

607. $|y| = -14$

608. $|z| = 0$

609. $|3x - 4| + 5 = 7$

610. $4|x - 1| + 2 = 10$

611. $-2|x - 3| + 8 = -4$

612. $\left|\frac{1}{2}x + 5\right| + 4 = 1$

613. $|6x - 5| = |2x + 3|$

Solve Absolute Value Inequalities with "less than"

In the following exercises, solve each inequality. Graph the solution and write the solution in interval notation.

614. $|x| \leq 8$

615. $|2x - 5| \leq 3$

616. $|6x - 5| < 7$

617. $|5x + 1| \leq -2$

Solve Absolute Value Inequalities with "greater than"

In the following exercises, solve. Graph the solution and write the solution in interval notation.

618. $|x| > 6$

619. $|x| \geq 2$

620. $|x - 5| > -2$

621. $|x - 7| \geq 1$

622. $3|x| + 4 \geq 1$

Solve Applications with Absolute Value

In the following exercises, solve.

623. A craft beer brewer needs 215,000 bottle per day. But this total can vary by as much as 5,000 bottles. What is the maximum and minimum expected usage at the bottling company?

624. At Fancy Grocery, the ideal weight of a loaf of bread is 16 ounces. By law, the actual weight can vary from the ideal by 1.5 ounces. What range of weight will be acceptable to the inspector without causing the bakery being fined?

PRACTICE TEST

In the following exercises, solve each equation.

625. $-5(2x + 1) = 45$

626.
$\frac{1}{4}(12m + 28) = 6 + 2(3m + 1)$

627.
$8(3a + 5) - 7(4a - 3) = 20 - 3a$

628. $0.1d + 0.25(d + 8) = 4.1$

629.
$14n - 3(4n + 5) = -9 + 2(n - 8)$

630.
$3(3u + 2) + 4[6 - 8(u - 1)] = 3(u - 2)$

631. $\frac{3}{4}x - \frac{2}{3} = \frac{1}{2}x + \frac{5}{6}$

632. $|3x - 4| = 8$

633. $|2x - 1| = |4x + 3|$

634. Solve the formula
$x + 2y = 5$ for y.

In the following exercises, graph the inequality on the number line and write in interval notation.

635. $x \geq -3.5$

636. $x < \frac{11}{4}$

637. $-2 \leq x < 5$

In the following exercises, solve each inequality, graph the solution on the number line, and write the solution in interval notation.

638. $8k \geq 5k - 120$

639. $3c - 10(c - 2) < 5c + 16$

640. $\frac{3}{4}x - 5 \geq -2$ and

$-3(x + 1) \geq 6$

641. $3(2x - 3) < -5$ or

$4x - 1 > 3$

642. $\frac{1}{2}x - 3 \leq 4$ or

$\frac{1}{3}(x - 6) \geq -2$

643. $|4x - 3| \geq 5$

In the following exercises, translate to an equation or inequality and solve.

644. Four less than twice x is 16.

645. Find the length of the missing side.

646. One number is four more than twice another. Their sum is -47. Find the numbers.

647. The sum of two consecutive odd integers is -112. Find the numbers.

648. Marcus bought a television on sale for $626.50 The original price of the television was $895. Find ⓐ the amount of discount and

ⓑ the discount rate.

649. Bonita has $2.95 in dimes and quarters in her pocket. If she has five more dimes than quarters, how many of each coin does she have?

650. Kim is making eight gallons of punch from fruit juice and soda. The fruit juice costs $6.04 per gallon and the soda costs $4.28 per gallon. How much fruit juice and how much soda should she use so that the punch costs $5.71 per gallon?

651. The measure of one angle of a triangle is twice the measure of the smallest angle. The measure of the third angle is three times the measure of the smallest angle. Find the measures of all three angles.

652. The length of a rectangle is five feet more than four times the width. The perimeter is 60 feet. Find the dimensions of the rectangle.

653. Two planes leave Dallas at the same time. One heads east at a speed of 428 miles per hour. The other plane heads west at a speed of 382 miles per hour. How many hours will it take them to be 2,025 miles apart?

654. Leon drove from his house in Cincinnati to his sister's house in Cleveland, a distance of 252 miles. It took him $4\frac{1}{2}$ hours. For the first half hour, he had heavy traffic, and the rest of the time his speed was five miles per hour less than twice his speed in heavy traffic. What was his speed in heavy traffic?

655. Sara has a budget of $1,000 for costumes for the 18 members of her musical theater group. What is the maximum she can spend for each costume?

Introduction

Imagine visiting a faraway city or even outer space from the comfort of your living room. It could be possible using virtual reality. This technology creates realistic images that make you feel as if you are truly immersed in the scene and even enable you to interact with them. It is being developed for fun applications, such as video games, but also for architects to plan buildings, car companies to design prototypes, the military to train, and medical students to learn.

Developing virtual reality devices requires modeling the environment using graphs and mathematical relationships. In this chapter, you will graph different relationships and learn ways to describe and analyze graphs.

3.1 Graph Linear Equations in Two Variables

Learning Objectives

By the end of this section, you will be able to:

> Plot points in a rectangular coordinate system
> Graph a linear equation by plotting points
> Graph vertical and horizontal lines
> Find the x- and y-intercepts
> Graph a line using the intercepts

Be Prepared!

Before you get started, take this readiness quiz.

1. Evaluate $5x - 4$ when $x = -1$.
   If you missed this problem, review **Example 1.6**.

2. Evaluate $3x - 2y$ when $x = 4$, $y = -3$.
   If you missed this problem, review **Example 1.21**.

3. Solve for y: $8 - 3y = 20$.

If you missed this problem, review Example 2.2.

Plot Points on a Rectangular Coordinate System

Just like maps use a grid system to identify locations, a grid system is used in algebra to show a relationship between two variables in a rectangular coordinate system. The rectangular coordinate system is also called the *xy*-plane or the "coordinate plane."

The rectangular coordinate system is formed by two intersecting number lines, one horizontal and one vertical. The horizontal number line is called the *x*-axis. The vertical number line is called the *y*-axis. These axes divide a plane into four regions, called quadrants. The quadrants are identified by Roman numerals, beginning on the upper right and proceeding counterclockwise. See Figure 3.2.

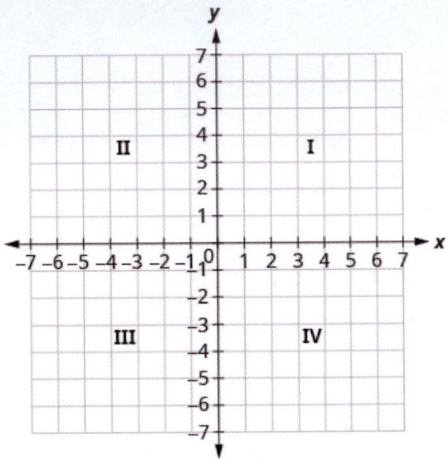

Figure 3.2

In the rectangular coordinate system, every point is represented by an **ordered pair**. The first number in the ordered pair is the *x*-coordinate of the point, and the second number is the *y*-coordinate of the point. The phrase "ordered pair" means that the order is important.

Ordered Pair

An **ordered pair**, (x, y) gives the coordinates of a point in a rectangular coordinate system. The first number is the *x*-coordinate. The second number is the *y*-coordinate.

$$(x, y)$$

x-coordinate   *y*-coordinate

What is the ordered pair of the point where the axes cross? At that point both coordinates are zero, so its ordered pair is $(0, 0)$. The point $(0, 0)$ has a special name. It is called the **origin**.

The Origin

The point $(0, 0)$ is called the **origin**. It is the point where the *x*-axis and *y*-axis intersect.

We use the coordinates to locate a point on the *xy*-plane. Let's plot the point $(1, 3)$ as an example. First, locate 1 on the *x*-axis and lightly sketch a vertical line through $x = 1$. Then, locate 3 on the *y*-axis and sketch a horizontal line through $y = 3$. Now, find the point where these two lines meet—that is the point with coordinates $(1, 3)$. See Figure 3.3.

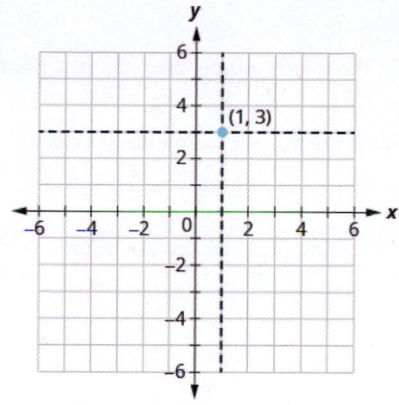

Figure 3.3

Notice that the vertical line through $x = 1$ and the horizontal line through $y = 3$ are not part of the graph. We just used them to help us locate the point $(1, 3)$.

When one of the coordinate is zero, the point lies on one of the axes. In **Figure 3.4** the point $(0, 4)$ is on the y-axis and the point $(-2, 0)$ is on the x-axis.

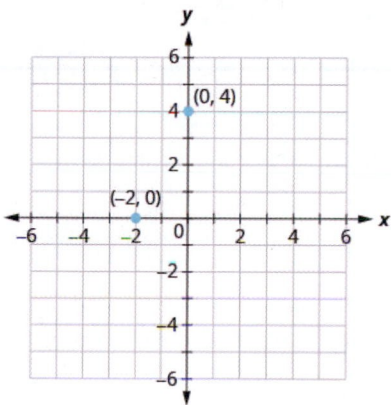

Figure 3.4

Points on the Axes

Points with a y-coordinate equal to 0 are on the x-axis, and have coordinates $(a, 0)$.

Points with an x-coordinate equal to 0 are on the y-axis, and have coordinates $(0, b)$.

EXAMPLE 3.1

Plot each point in the rectangular coordinate system and identify the quadrant in which the point is located:

ⓐ $(-5, 4)$ ⓑ $(-3, -4)$ ⓒ $(2, -3)$ ⓓ $(0, -1)$ ⓔ $\left(3, \frac{5}{2}\right)$.

⊘ Solution

The first number of the coordinate pair is the x-coordinate, and the second number is the y-coordinate. To plot each point, sketch a vertical line through the x-coordinate and a horizontal line through the y-coordinate. Their intersection is the point.

ⓐ Since $x = -5$, the point is to the left of the y-axis. Also, since $y = 4$, the point is above the x-axis. The point $(-5, 4)$ is in Quadrant II.

ⓑ Since $x = -3$, the point is to the left of the y-axis. Also, since $y = -4$, the point is below the x-axis. The point

$(-3, -4)$ is in Quadrant III.

ⓒ Since $x = 2,$ the point is to the right of the y-axis. Since $y = -3,$ the point is below the x-axis. The point $(2, -3)$ is in Quadrant IV.

ⓓ Since $x = 0,$ the point whose coordinates are $(0, -1)$ is on the y-axis.

ⓔ Since $x = 3,$ the point is to the right of the y-axis. Since $y = \frac{5}{2},$ the point is above the x-axis. (It may be helpful to write $\frac{5}{2}$ as a mixed number or decimal.) The point $\left(3, \frac{5}{2}\right)$ is in Quadrant I.

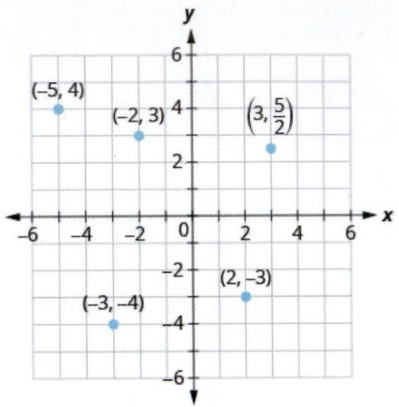

> **TRY IT : : 3.1**

Plot each point in a rectangular coordinate system and identify the quadrant in which the point is located:

ⓐ $(-2, 1)$ ⓑ $(-3, -1)$ ⓒ $(4, -4)$ ⓓ $(-4, 4)$ ⓔ $\left(-4, \frac{3}{2}\right)$

> **TRY IT : : 3.2**

Plot each point in a rectangular coordinate system and identify the quadrant in which the point is located:

ⓐ $(-4, 1)$ ⓑ $(-2, 3)$ ⓒ $(2, -5)$ ⓓ $(-2, 5)$ ⓔ $\left(-3, \frac{5}{2}\right)$

The signs of the x-coordinate and y-coordinate affect the location of the points. You may have noticed some patterns as you graphed the points in the previous example. We can summarize sign patterns of the quadrants in this way:

Quadrants

Quadrant I	Quadrant II	Quadrant III	Quadrant IV
(x, y)	(x, y)	(x, y)	(x, y)
$(+, +)$	$(-, +)$	$(-, -)$	$(+, -)$

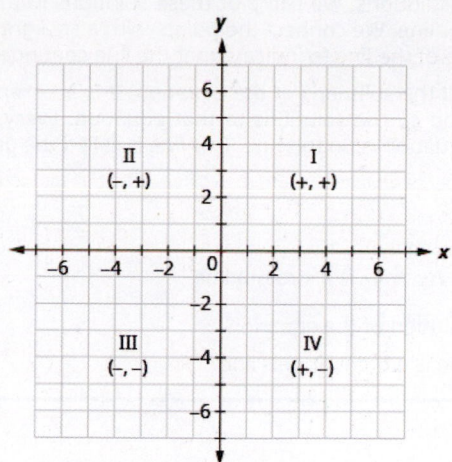

Up to now, all the equations you have solved were equations with just one variable. In almost every case, when you solved the equation you got exactly one solution. But equations can have more than one variable. Equations with two variables may be of the form $Ax + By = C$. An equation of this form is called a **linear equation** in two variables.

Linear Equation

An equation of the form $Ax + By = C$, where A and B are not both zero, is called a **linear equation** in two variables.

Here is an example of a linear equation in two variables, x and y.

$$Ax + By = C$$

$$x + 4y = 8$$

$$A = 1, B = 4, C = 8$$

The equation $y = -3x + 5$ is also a linear equation. But it does not appear to be in the form $Ax + By = C$. We can use the Addition Property of Equality and rewrite it in $Ax + By = C$ form.

	$y = -3x + 5$
Add to both sides.	$y + 3x = 3x + 5 + 3x$
Simplify.	$y + 3x = 5$
Use the Commutative Property to put it in $Ax + By = C$ form.	$3x + y = 5$

By rewriting $y = -3x + 5$ as $3x + y = 5$, we can easily see that it is a linear equation in two variables because it is of the form $Ax + By = C$. When an equation is in the form $Ax + By = C$, we say it is in **standard form of a linear equation**.

Standard Form of Linear Equation

A linear equation is in **standard form** when it is written $Ax + By = C$.

Most people prefer to have A, B, and C be integers and $A \geq 0$ when writing a linear equation in standard form, although it is not strictly necessary.

Linear equations have infinitely many solutions. For every number that is substituted for x there is a corresponding y value. This pair of values is a **solution** to the linear equation and is represented by the ordered pair (x, y). When we substitute these values of x and y into the equation, the result is a true statement, because the value on the left side is equal to the value on the right side.

Solution of a Linear Equation in Two Variables

An ordered pair (x, y) is a **solution** of the linear equation $Ax + By = C$, if the equation is a true statement when the x- and y-values of the ordered pair are substituted into the equation.

Linear equations have infinitely many solutions. We can plot these solutions in the rectangular coordinate system. The points will line up perfectly in a straight line. We connect the points with a straight line to get the graph of the equation. We put arrows on the ends of each side of the line to indicate that the line continues in both directions.

A graph is a visual representation of all the solutions of the equation. It is an example of the saying, "A picture is worth a thousand words." The line shows you *all* the solutions to that equation. Every point on the line is a solution of the equation. And, every solution of this equation is on this line. This line is called the graph of the equation. Points *not* on the line are not solutions!

Graph of a Linear Equation

The graph of a linear equation $Ax + By = C$ is a straight line.

- Every point on the line is a solution of the equation.
- Every solution of this equation is a point on this line.

EXAMPLE 3.2

The graph of $y = 2x - 3$ is shown.

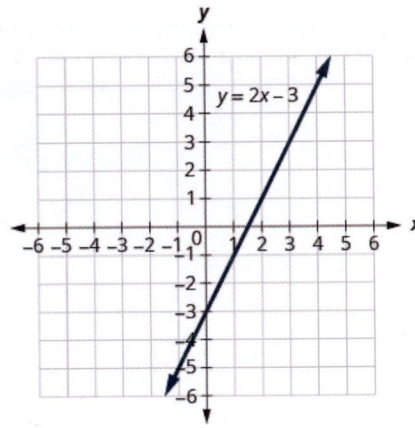

For each ordered pair, decide:

ⓐ Is the ordered pair a solution to the equation?

ⓑ Is the point on the line?

A: $(0, -3)$ B: $(3, 3)$ C: $(2, -3)$ D: $(-1, -5)$

✓ Solution

Substitute the *x*- and *y*-values into the equation to check if the ordered pair is a solution to the equation.

ⓐ

A: $(0, -3)$	B: $(3, 3)$	C: $(2, -3)$	D: $(-1, -5)$
$y = 2x - 3$	$y = 2x - 3$	$y = 2x - 3$	$y = 2x - 3$
$-3 \overset{?}{=} 2(0) - 3$	$3 \overset{?}{=} 2(3) - 3$	$-3 \overset{?}{=} 2(2) - 3$	$-5 \overset{?}{=} 2(-1) - 3$
$-3 = -3$ ✓	$3 = 3$ ✓	$-3 \neq 1$	$-5 = -5$ ✓
$(0, -3)$ is a solution.	$(3, 3)$ is a solution.	$(2, -3)$ is not a solution.	$(-1, -5)$ is a solution.

ⓑ Plot the points $(0, -3)$, $(3, 3)$, $(2, -3)$, and $(-1, -5)$.

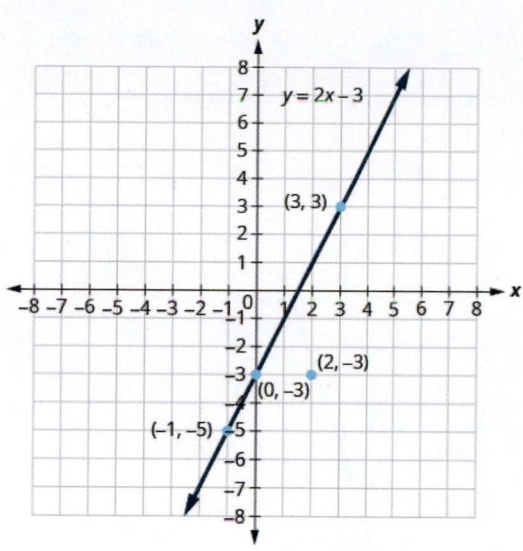

The points $(0, 3)$, $(3, -3)$, and $(-1, -5)$ are on the line $y = 2x - 3$, and the point $(2, -3)$ is not on the line. The points that are solutions to $y = 2x - 3$ are on the line, but the point that is not a solution is not on the line.

> **TRY IT :: 3.3**   Use graph of $y = 3x - 1$. For each ordered pair, decide:

ⓐ Is the ordered pair a solution to the equation?

ⓑ Is the point on the line?

A $(0, -1)$ B $(2, 5)$

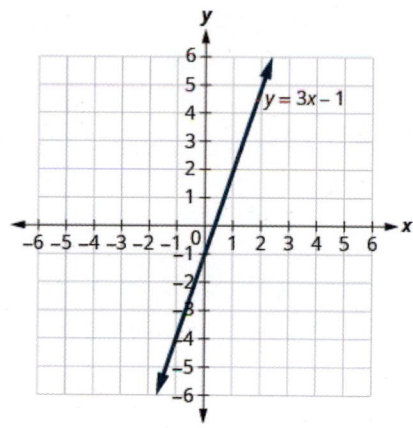

> **TRY IT : : 3.4**   Use graph of $y = 3x - 1$. For each ordered pair, decide:

ⓐ Is the ordered pair a solution to the equation?

ⓑ Is the point on the line?

A $(3, -1)$ B $(-1, -4)$

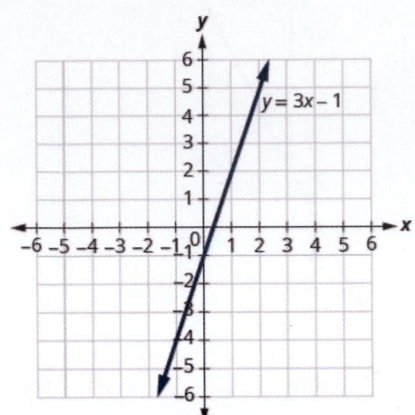

Graph a Linear Equation by Plotting Points

There are several methods that can be used to graph a linear equation. The first method we will use is called plotting points, or the Point-Plotting Method. We find three points whose coordinates are solutions to the equation and then plot them in a rectangular coordinate system. By connecting these points in a line, we have the graph of the linear equation.

EXAMPLE 3.3   HOW TO GRAPH A LINEAR EQUATION BY PLOTTING POINTS

Graph the equation $y = 2x + 1$ by plotting points.

⊘ **Solution**

Step 1. Find three points whose coordinates are solutions to the equation.	You can choose any values for *x* or *y*. In this case, since *y* is isolated on the left side of the equation, it is easier to choose values for *x*.	$y = 2x + 1$ $x = 0$ $y = 2x + 1$ $y = 2 \cdot 0 + 1$ $y = 0 + 1$ $y = 1$ $x = 1$ $y = 2x + 1$ $y = 2 \cdot 1 + 1$ $y = 2 + 1$ $y = 3$ $x = -2$ $y = 2x + 1$ $y = 2(-2) + 1$ $y = -4 + 1$ $y = -3$
Organize the solutions in a table.	Put the three solutions in a table.	<table><tr><td colspan=3>$y = 2x + 1$</td></tr><tr><td>x</td><td>y</td><td>(x, y)</td></tr><tr><td>0</td><td>1</td><td>(0, 1)</td></tr><tr><td>1</td><td>3</td><td>(1, 3)</td></tr><tr><td>-2</td><td>-3</td><td>(-2, -3)</td></tr></table>
Step 2. Plot the points in a rectangular coordinate system. Check that the points line up. If they do not, carefully check your work!	Plot: (0, 1), (1, 3), (-2, -3). Do the points line up? Yes, the points line up.	

Step 3. Draw the line through the three points. Extend the line to fill the grid and put arrows on both ends of the line.	This line is the graph of $y = 2x + 1$.	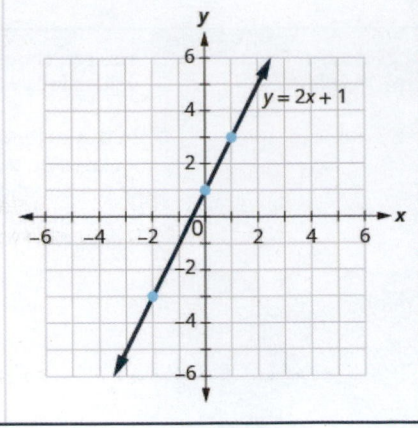

> **TRY IT :: 3.5**   Graph the equation by plotting points: $y = 2x - 3$.

> **TRY IT :: 3.6**   Graph the equation by plotting points: $y = -2x + 4$.

The steps to take when graphing a linear equation by plotting points are summarized here.

 HOW TO :: GRAPH A LINEAR EQUATION BY PLOTTING POINTS.

Step 1.   Find three points whose coordinates are solutions to the equation. Organize them in a table.

Step 2.   Plot the points in a rectangular coordinate system. Check that the points line up. If they do not, carefully check your work.

Step 3.   Draw the line through the three points. Extend the line to fill the grid and put arrows on both ends of the line.

It is true that it only takes two points to determine a line, but it is a good habit to use three points. If you only plot two points and one of them is incorrect, you can still draw a line but it will not represent the solutions to the equation. It will be the wrong line.

If you use three points, and one is incorrect, the points will not line up. This tells you something is wrong and you need to check your work. Look at the difference between these illustrations.

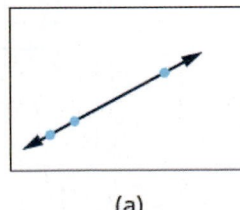   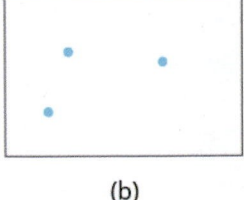

(a)   (b)

When an equation includes a fraction as the coefficient of x, we can still substitute any numbers for x. But the arithmetic is easier if we make "good" choices for the values of x. This way we will avoid fractional answers, which are hard to graph precisely.

EXAMPLE 3.4

Graph the equation: $y = \frac{1}{2}x + 3$.

⊘ **Solution**

Find three points that are solutions to the equation. Since this equation has the fraction $\frac{1}{2}$ as a coefficient of x, we will choose values of x carefully. We will use zero as one choice and multiples of 2 for the other choices. Why are multiples of two a good choice for values of x? By choosing multiples of 2 the multiplication by $\frac{1}{2}$ simplifies to a whole number

$$x = 0 \qquad\qquad x = 2 \qquad\qquad x = 4$$

$$y = \frac{1}{2}x + 3 \qquad y = \frac{1}{2}x + 3 \qquad y = \frac{1}{2}x + 3$$

$$y = \frac{1}{2}(0) + 3 \qquad y = \frac{1}{2}(2) + 3 \qquad y = \frac{1}{2}(4) + 3$$

$$y = 0 + 3 \qquad\quad y = 1 + 3 \qquad\quad y = 2 + 3$$

$$y = 3 \qquad\qquad y = 4 \qquad\qquad y = 5$$

The points are shown in Table 3.1.

$y = \frac{1}{2}x + 3$		
x	y	(x, y)
0	3	$(0, 3)$
2	4	$(2, 4)$
4	5	$(4, 5)$

Table 3.1

Plot the points, check that they line up, and draw the line.

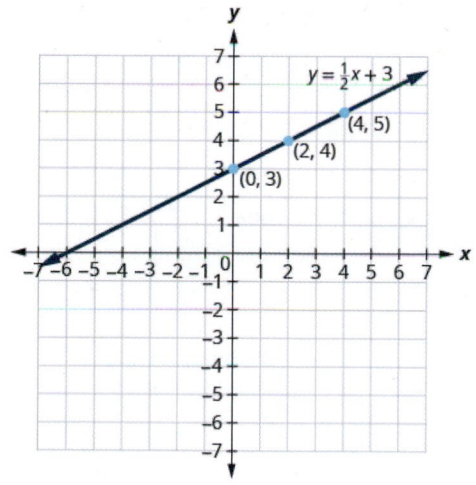

> **TRY IT ::** 3.7   Graph the equation: $y = \frac{1}{3}x - 1$.

> **TRY IT ::** 3.8   Graph the equation: $y = \frac{1}{4}x + 2$.

Graph Vertical and Horizontal Lines

Some linear equations have only one variable. They may have just x and no y, or just y without an x. This changes how we make a table of values to get the points to plot.

Let's consider the equation $x = -3$. This equation has only one variable, x. The equation says that x is *always* equal to

−3, so its value does not depend on y. No matter what is the value of y, the value of x is always −3.

So to make a table of values, write −3 in for all the x-values. Then choose any values for y. Since x does not depend on y, you can choose any numbers you like. But to fit the points on our coordinate graph, we'll use 1, 2, and 3 for the y-coordinates. See Table 3.2.

$x = -3$		
x	y	(x, y)
−3	1	(−3, 1)
−3	2	(−3, 2)
−3	3	(−3, 3)

Table 3.2

Plot the points from the table and connect them with a straight line. Notice that we have graphed a **vertical line**.

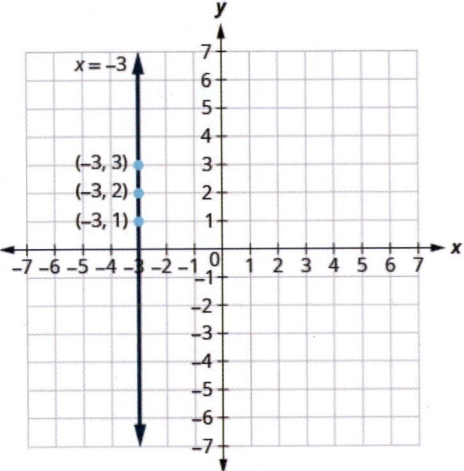

What if the equation has y but no x? Let's graph the equation $y = 4$. This time the y-value is a constant, so in this equation, y does not depend on x. Fill in 4 for all the y's in Table 3.3 and then choose any values for x. We'll use 0, 2, and 4 for the x-coordinates.

$y = 4$		
x	y	(x, y)
0	4	(0, 4)
2	4	(2, 4)
4	4	(4, 4)

Table 3.3

In this figure, we have graphed a **horizontal line** passing through the y-axis at 4.

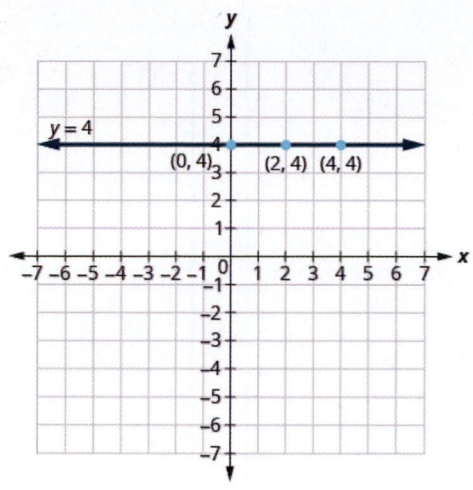

Vertical and Horizontal Lines

A **vertical line** is the graph of an equation of the form $x = a$.

The line passes through the x-axis at $(a, 0)$.

A **horizontal line** is the graph of an equation of the form $y = b$.

The line passes through the y-axis at $(0, b)$.

EXAMPLE 3.5

Graph: ⓐ $x = 2$ ⓑ $y = -1$.

✓ Solution

ⓐ The equation has only one variable, x, and x is always equal to 2. We create a table where x is always 2 and then put in any values for y. The graph is a vertical line passing through the x-axis at 2.

$x = 2$		
x	y	(x, y)
2	1	$(2, 1)$
2	2	$(2, 2)$
2	3	$(2, 3)$

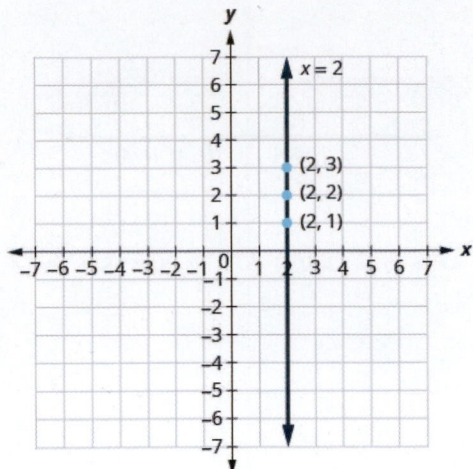

ⓑ Similarly, the equation $y = -1$ has only one variable, y. The value of y is constant. All the ordered pairs in the next table have the same y-coordinate. The graph is a horizontal line passing through the y-axis at -1.

$y = -1$		
x	y	(x, y)
0	-1	$(0, -1)$
3	-1	$(3, -1)$
-3	-1	$(-3, -1)$

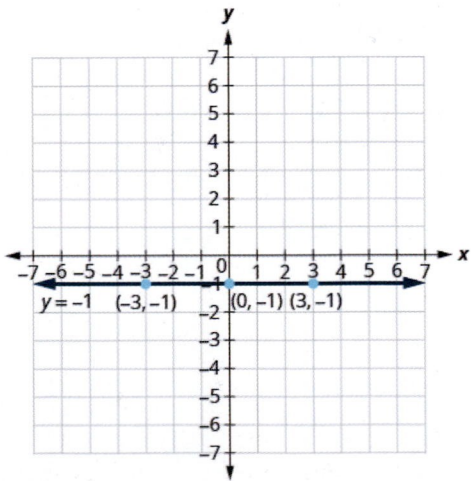

> **TRY IT :: 3.9**   Graph the equations: ⓐ $x = 5$ ⓑ $y = -4$.

> **TRY IT :: 3.10**   Graph the equations: ⓐ $x = -2$ ⓑ $y = 3$.

What is the difference between the equations $y = 4x$ and $y = 4$?

The equation $y = 4x$ has both x and y. The value of y depends on the value of x, so the y-coordinate changes according to the value of *x*. The equation $y = 4$ has only one variable. The value of *y* is constant, it does not depend on the value of *x*, so the *y*-coordinate is always 4.

y = 4x		
x	*y*	(x, y)
0	0	(0, 0)
1	4	(1, 4)
2	8	(2, 8)

y = 4		
x	*y*	(x, y)
0	4	(0, 4)
1	4	(1, 4)
2	4	(2, 4)

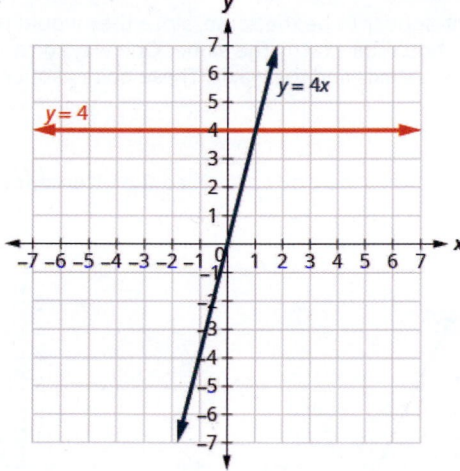

Notice, in the graph, the equation $y = 4x$ gives a slanted line, while $y = 4$ gives a horizontal line.

EXAMPLE 3.6

Graph $y = -3x$ and $y = -3$ in the same rectangular coordinate system.

⊘ Solution

We notice that the first equation has the variable *x*, while the second does not. We make a table of points for each equation and then graph the lines. The two graphs are shown.

y = -3x		
x	*y*	(x, y)
0	0	(0, 0)
1	-3	(1, -3)
2	-6	(2, -6)

y = -3		
x	*y*	(x, y)
0	-3	(0, -3)
1	-3	(1, -3)
2	-3	(2, -3)

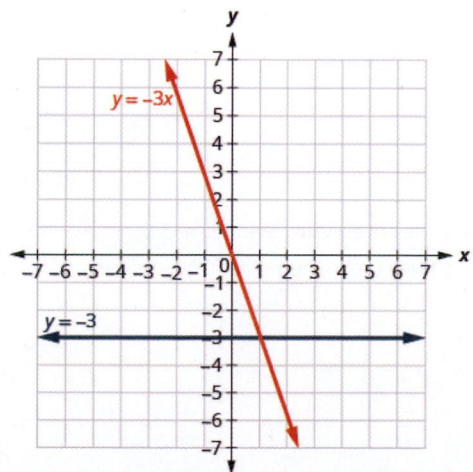

> **TRY IT ::** 3.11   Graph the equations in the same rectangular coordinate system: $y = -4x$ and $y = -4$.

> **TRY IT ::** 3.12   Graph the equations in the same rectangular coordinate system: $y = 3$ and $y = 3x$.

Find *x*- and *y*-intercepts

Every linear equation can be represented by a unique line that shows all the solutions of the equation. We have seen that when graphing a line by plotting points, you can use any three solutions to graph. This means that two people graphing the line might use different sets of three points.

At first glance, their two lines might not appear to be the same, since they would have different points labeled. But if all the work was done correctly, the lines should be exactly the same. One way to recognize that they are indeed the same line is to look at where the line crosses the *x*-axis and the *y*-axis. These points are called the **intercepts of a line**.

> ### Intercepts of a Line
>
> The points where a line crosses the *x*-axis and the *y*-axis are called the **intercepts of the line**.

Let's look at the graphs of the lines.

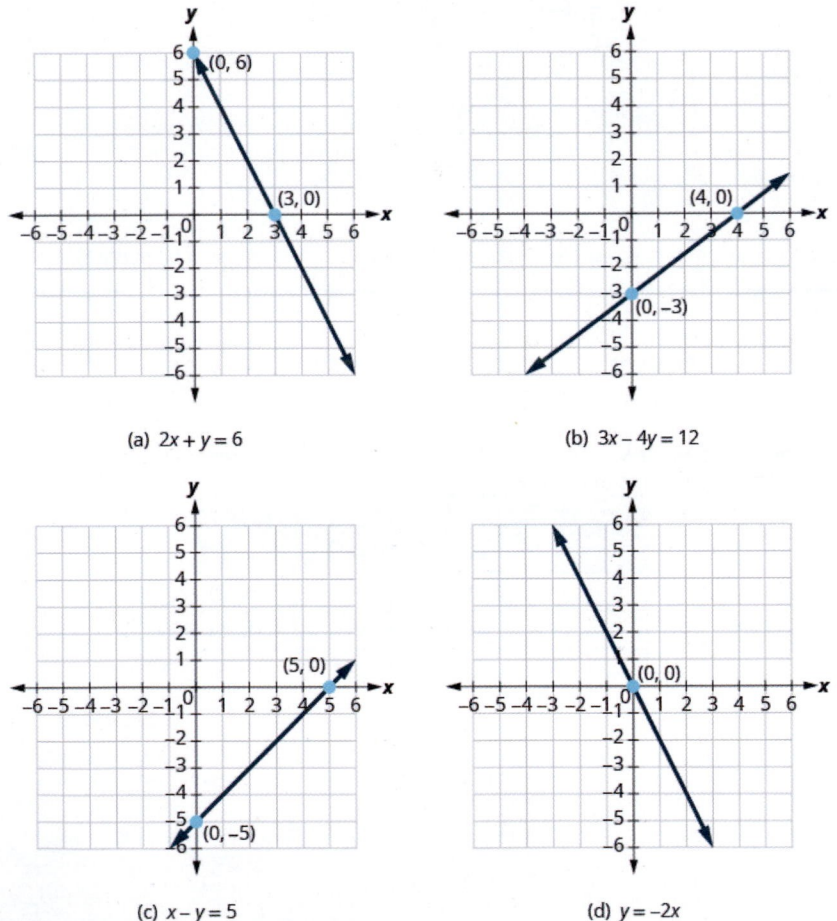

(a) $2x + y = 6$

(b) $3x - 4y = 12$

(c) $x - y = 5$

(d) $y = -2x$

First, notice where each of these lines crosses the *x*-axis. See **Table 3.6**.

Now, let's look at the points where these lines cross the *y*-axis.

Figure	The line crosses the x-axis at:	Ordered pair for this point	The line crosses the y-axis at:	Ordered pair for this point
Figure (a)	3	(3, 0)	6	(0, 6)
Figure (b)	4	(4, 0)	−3	(0, −3)
Figure (c)	5	(5, 0)	−5	(0, 5)
Figure (d)	0	(0, 0)	0	(0, 0)
General Figure	a	$(a, 0)$	b	$(0, b)$

Table 3.6

Do you see a pattern?

For each line, the y-coordinate of the point where the line crosses the x-axis is zero. The point where the line crosses the x-axis has the form $(a, 0)$ and is called the *x-intercept* of the line. The x-intercept occurs when y is zero.

In each line, the x-coordinate of the point where the line crosses the y-axis is zero. The point where the line crosses the y-axis has the form $(0, b)$ and is called the *y-intercept* of the line. The y-intercept occurs when x is zero.

x-intercept and y-intercept of a Line

The x-intercept is the point $(a, 0)$ where the line crosses the x-axis.

The y-intercept is the point $(0, b)$ where the line crosses the y-axis.

x	y
a	0
0	b

- The x-intercept occurs when y is zero.
- The y-intercept occurs when x is zero.

EXAMPLE 3.7

Find the x- and y-intercepts on each graph shown.

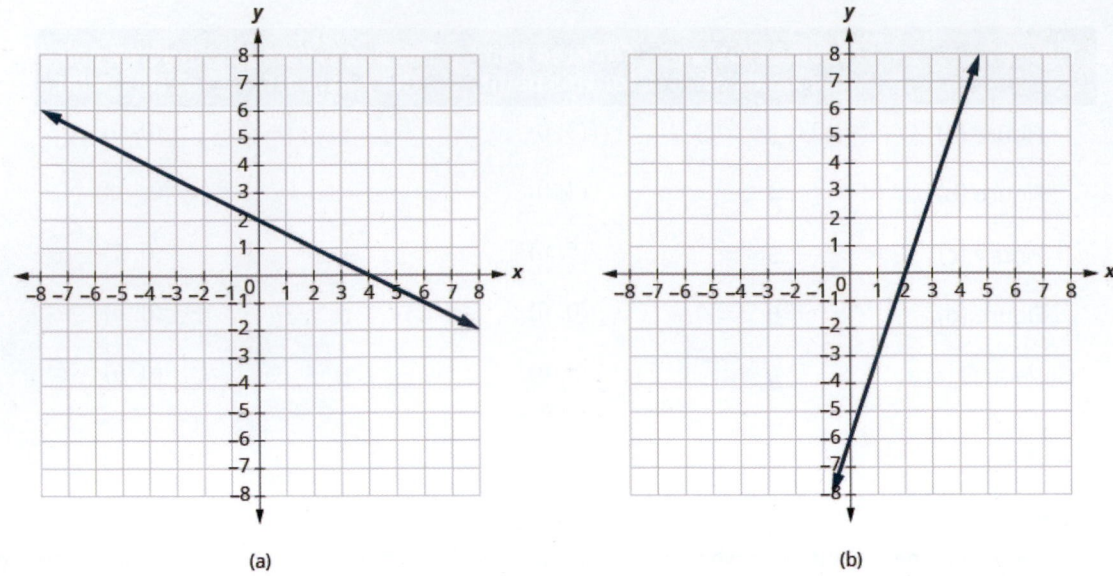

(a)   (b)

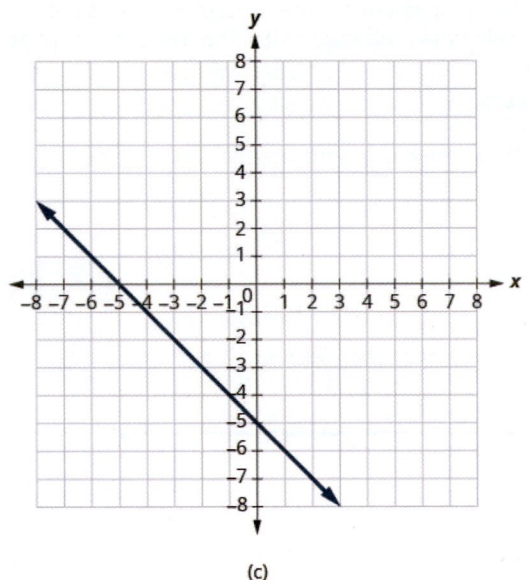

(c)

⊘ **Solution**

ⓐ The graph crosses the x-axis at the point $(4, 0)$. The x-intercept is $(4, 0)$.
The graph crosses the y-axis at the point $(0, 2)$. The y-intercept is $(0, 2)$.

ⓑ The graph crosses the x-axis at the point $(2, 0)$. The x-intercept is $(2, 0)$.
The graph crosses the y-axis at the point $(0, -6)$. The y-intercept is $(0, -6)$.

ⓒ The graph crosses the x-axis at the point $(-5, 0)$. The x-intercept is $(-5, 0)$.
The graph crosses the y-axis at the point $(0, -5)$. The y-intercept is $(0, -5)$.

> **TRY IT** : : 3.13   Find the *x*- and *y*-intercepts on the graph.

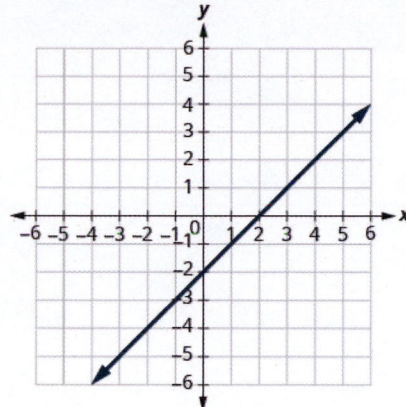

> **TRY IT** : : 3.14   Find the *x*- and *y*-intercepts on the graph.

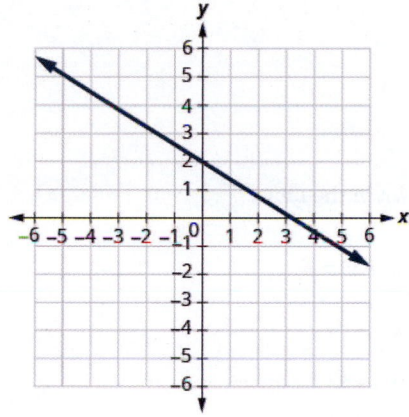

Recognizing that the *x*-intercept occurs when *y* is zero and that the *y*-intercept occurs when *x* is zero, gives us a method to find the intercepts of a line from its equation. To find the *x*-intercept, let $y = 0$ and solve for *x*. To find the *y*-intercept, let $x = 0$ and solve for *y*.

Find the *x*- and *y*-intercepts from the Equation of a Line

Use the equation of the line. To find:

- the *x*-intercept of the line, let $y = 0$ and solve for *x*.

- the *y*-intercept of the line, let $x = 0$ and solve for *y*.

EXAMPLE 3.8

Find the intercepts of $2x + y = 8$.

⊘ Solution

We will let $y = 0$ to find the *x*-intercept, and let $x = 0$ to find the *y*-intercept. We will fill in a table, which reminds us of what we need to find.

2x + y = 8		
x	**y**	
	0	*x*-intercept
0		*y*-intercept

To find the *x*-intercept, let $y = 0$.

	$2x + y = 8$
Let $y = 0$.	$2x + 0 = 8$
Simplify.	$2x = 8$
	$x = 4$
The *x*-intercept is:	$(4, 0)$
To find the *y*-intercept, let $x = 0$.	
	$2x + y = 8$
Let $x = 0$.	$2 \cdot 0 + y = 8$
Simplify.	$0 + y = 8$
	$y = 8$
The *y*-intercept is:	$(0, 8)$

The intercepts are the points $(4, 0)$ and $(0, 8)$ as shown in the table.

$2x + y = 8$	
x	*y*
4	0
0	8

> **TRY IT : :** 3.15   Find the intercepts: $3x + y = 12$.

> **TRY IT : :** 3.16   Find the intercepts: $x + 4y = 8$.

Graph a Line Using the Intercepts

To graph a linear equation by plotting points, you need to find three points whose coordinates are solutions to the equation. You can use the *x*- and *y*- intercepts as two of your three points. Find the intercepts, and then find a third point to ensure accuracy. Make sure the points line up—then draw the line. This method is often the quickest way to graph a line.

EXAMPLE 3.9   HOW TO GRAPH A LINE USING THE INTERCEPTS

Graph $-x + 2y = 6$ using the intercepts.

✓ **Solution**

Step 1. Find the x- and y-intercepts of the line. Let y = 0 and solve for x.	Find the x-intercept.	Let $y = 0$ $-x + 2y = 6$ $-x + 2(0) = 6$ $-x = 6$ $x = -6$ The x-intercept is (–6, 0).
Let x = 0 and solve for y.	Find the y-intercept.	Let $x = 0$. $-x + 2y = 6$ $-0 + 2y = 6$ $2y = 6$ $y = 3$ The y-intercept is (0, 3).
Step 2. Find another solution to the equation.	We'll use x = 2.	Let $x = 2$. $-x + 2y = 6$ $-2 + 2y = 6$ $2y = 8$ $y = 4$ A third point is (2, 4).

Step 3. Plot the three points. Check that the points line up.

x	y	(x, y)
–6	0	(–6, 0)
0	3	(0, 3)
2	4	(2, 4)

Step 4. Draw the line.

See the graph.

> **TRY IT : : 3.17**   Graph using the intercepts: $x - 2y = 4$.

TRY IT :: 3.18   Graph using the intercepts: $-x + 3y = 6$.

The steps to graph a linear equation using the intercepts are summarized here.

 HOW TO :: GRAPH A LINEAR EQUATION USING THE INTERCEPTS.

Step 1.   Find the x- and y-intercepts of the line.

   ◦ Let $y = 0$ and solve for x.

   ◦ Let $x = 0$ and solve for y.

Step 2.   Find a third solution to the equation.

Step 3.   Plot the three points and check that they line up.

Step 4.   Draw the line.

EXAMPLE 3.10

Graph $4x - 3y = 12$ using the intercepts.

⊘ **Solution**

Find the intercepts and a third point.

x-intercept, let $y = 0$	y-intercept, let $x = 0$	third point, let $y = 4$
$4x - 3y = 12$	$4x - 3y = 12$	$4x - 3y = 12$
$4x - 3(0) = 12$	$4(0) - 3y = 12$	$4x - 3(4) = 12$
$4x = 12$	$-3y = 12$	$4x - 12 = 12$
$x = 3$	$y = -4$	$4x = 24$
		$x = 6$

We list the points in the table and show the graph.

$4x - 3y = 12$		
x	y	(x, y)
3	0	$(3, 0)$
0	−4	$(0, -4)$
6	4	$(6, 4)$

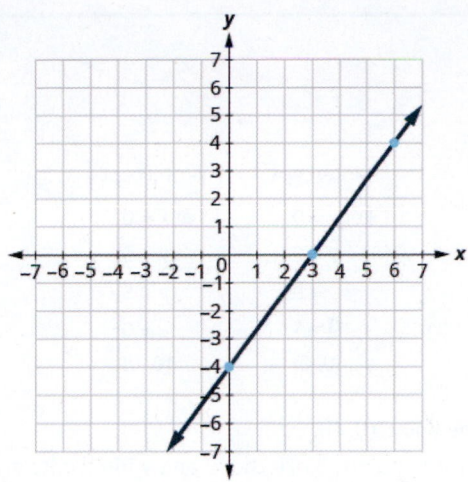

> **TRY IT : :** 3.19   Graph using the intercepts: $5x - 2y = 10$.

> **TRY IT : :** 3.20   Graph using the intercepts: $3x - 4y = 12$.

When the line passes through the origin, the *x*-intercept and the *y*-intercept are the same point.

EXAMPLE 3.11

Graph $y = 5x$ using the intercepts.

✓ **Solution**

x-intercept	*y*-intercept
Let $y = 0$.	Let $x = 0$.
$y = 5x$	$y = 5x$
$0 = 5x$	$y = 5 \cdot 0$
$0 = x$	$y = 0$
$(0, 0)$	$(0, 0)$

This line has only one intercept. It is the point $(0, 0)$.

To ensure accuracy, we need to plot three points. Since the *x*- and *y*-intercepts are the same point, we need *two* more points to graph the line.

Let $x = 1$.	Let $x = -1$.
$y = 5x$	$y = 5x$
$y = 5 \cdot 1$	$y = 5(-1)$
$y = 5$	$y = -5$

The resulting three points are summarized in the table.

\multicolumn{3}{c}{$y = 5x$}		
x	**y**	**(x, y)**
0	0	$(0, 0)$
1	5	$(1, 5)$
-1	-5	$(-1, -5)$

Plot the three points, check that they line up, and draw the line.

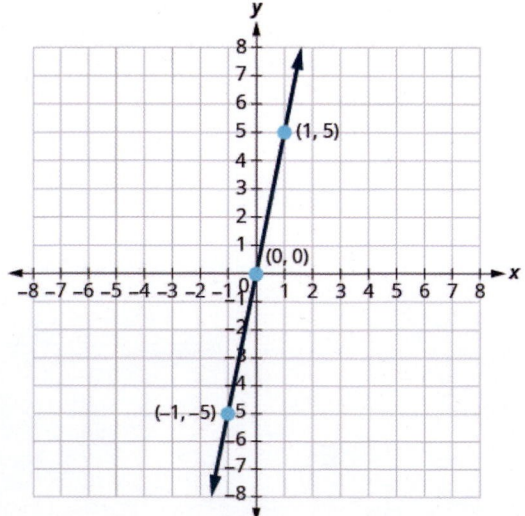

> **TRY IT : :** 3.21   Graph using the intercepts: $y = 4x$.

> **TRY IT : :** 3.22   Graph the intercepts: $y = -x$.

3.1 EXERCISES

Practice Makes Perfect

Plot Points in a Rectangular Coordinate System

In the following exercises, plot each point in a rectangular coordinate system and identify the quadrant in which the point is located.

1. ⓐ $(-4, 2)$ ⓑ $(-1, -2)$ ⓒ $(3, -5)$ ⓓ $(-3, 0)$
ⓔ $\left(\frac{5}{3}, 2\right)$

2. ⓐ $(-2, -3)$ ⓑ $(3, -3)$ ⓒ $(-4, 1)$ ⓓ $(4, -1)$
ⓔ $\left(\frac{3}{2}, 1\right)$

3. ⓐ $(3, -1)$ ⓑ $(-3, 1)$ ⓒ $(-2, 0)$ ⓓ $(-4, -3)$
ⓔ $\left(1, \frac{14}{5}\right)$

4. ⓐ $(-1, 1)$ ⓑ $(-2, -1)$ ⓒ $(2, 0)$ ⓓ $(1, -4)$
ⓔ $\left(3, \frac{7}{2}\right)$

In the following exercises, for each ordered pair, decide

ⓐ is the ordered pair a solution to the equation? ⓑ is the point on the line?

5. $y = x + 2$;
A: $(0, 2)$; B: $(1, 2)$; C: $(-1, 1)$; D: $(-3, -1)$.

6. $y = x - 4$;
A: $(0, -4)$; B: $(3, -1)$; C: $(2, 2)$; D: $(1, -5)$.

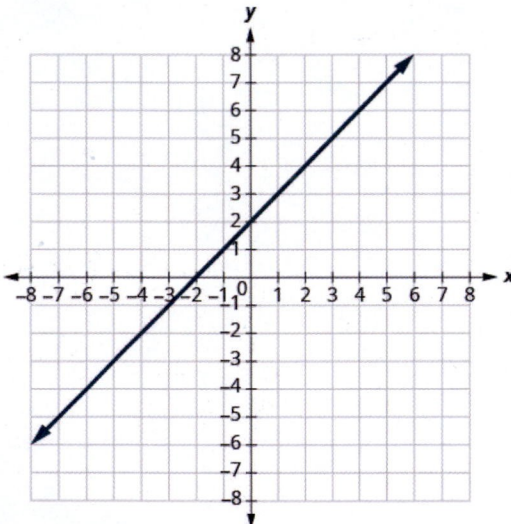

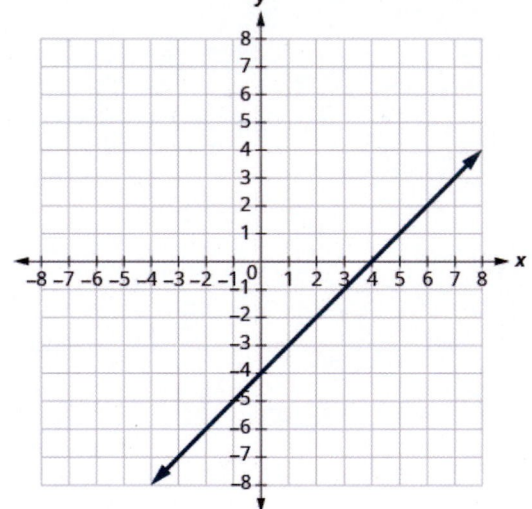

7. $y = \frac{1}{2}x - 3$;

A: $(0, -3)$; B: $(2, -2)$; C: $(-2, -4)$; D: $(4, 1)$

8. $y = \frac{1}{3}x + 2$;

A: $(0, 2)$; B: $(3, 3)$; C: $(-3, 2)$; D: $(-6, 0)$.

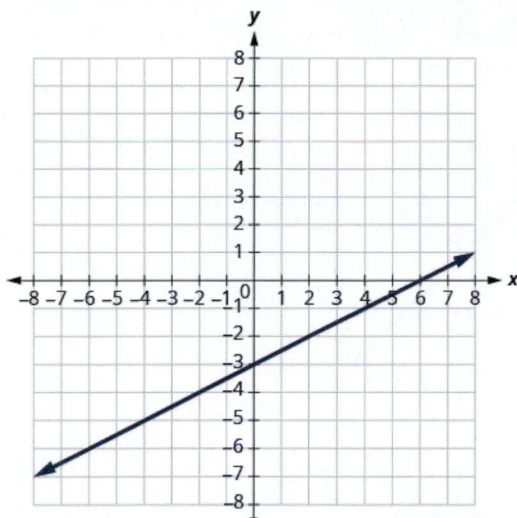

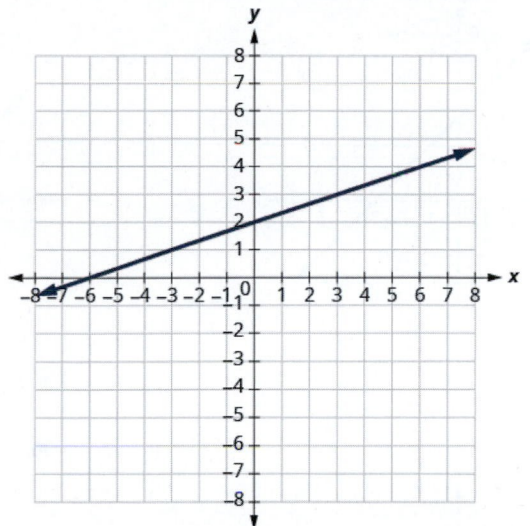

Graph a Linear Equation by Plotting Points

In the following exercises, graph by plotting points.

9. $y = x + 2$

10. $y = x - 3$

11. $y = 3x - 1$

12. $y = -2x + 2$

13. $y = -x - 3$

14. $y = -x - 2$

15. $y = 2x$

16. $y = -2x$

17. $y = \frac{1}{2}x + 2$

18. $y = \frac{1}{3}x - 1$

19. $y = \frac{4}{3}x - 5$

20. $y = \frac{3}{2}x - 3$

21. $y = -\frac{2}{5}x + 1$

22. $y = -\frac{4}{5}x - 1$

23. $y = -\frac{3}{2}x + 2$

24. $y = -\frac{5}{3}x + 4$

Graph Vertical and Horizontal lines

In the following exercises, graph each equation.

25. ⓐ $x = 4$ ⓑ $y = 3$

26. ⓐ $x = 3$ ⓑ $y = 1$

27. ⓐ $x = -2$ ⓑ $y = -5$

28. ⓐ $x = -5$ ⓑ $y = -2$

In the following exercises, graph each pair of equations in the same rectangular coordinate system.

29. $y = 2x$ and $y = 2$

30. $y = 5x$ and $y = 5$

31. $y = -\frac{1}{2}x$ and $y = -\frac{1}{2}$

32. $y = -\frac{1}{3}x$ and $y = -\frac{1}{3}$

Find *x*- and *y*-Intercepts

In the following exercises, find the x- and y-intercepts on each graph.

33.

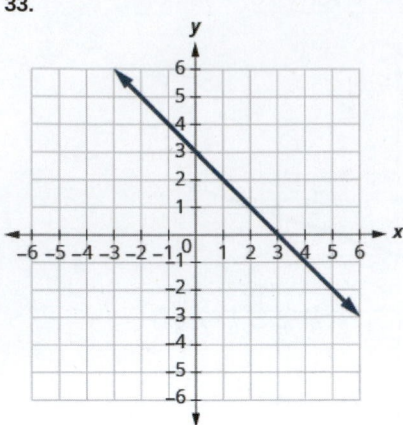

34.

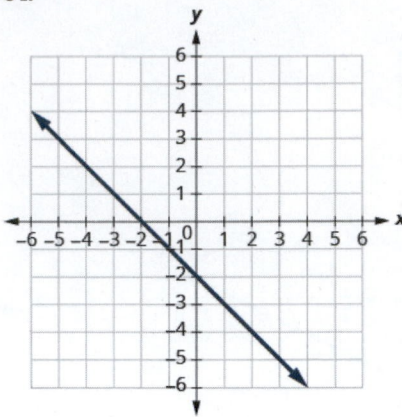

35.

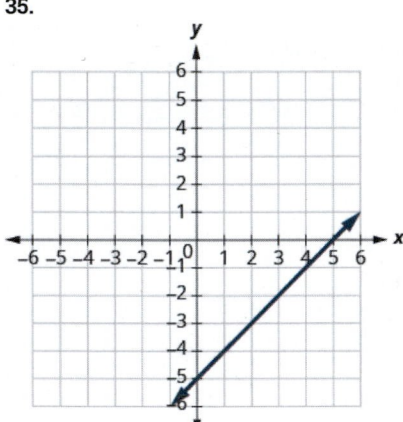

36.

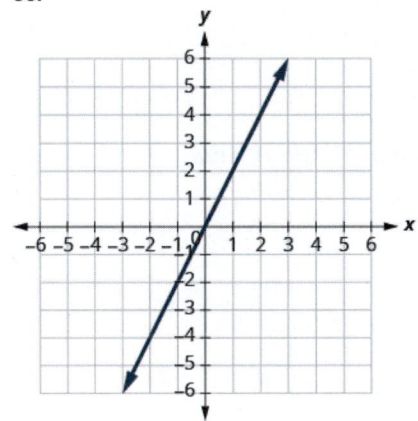

In the following exercises, find the intercepts for each equation.

37. $x - y = 5$

38. $x - y = -4$

39. $3x + y = 6$

40. $x - 2y = 8$

41. $4x - y = 8$

42. $5x - y = 5$

43. $2x + 5y = 10$

44. $3x - 2y = 12$

Graph a Line Using the Intercepts

In the following exercises, graph using the intercepts.

45. $-x + 4y = 8$

46. $x + 2y = 4$

47. $x + y = -3$

48. $x - y = -4$

49. $4x + y = 4$

50. $3x + y = 3$

51. $3x - y = -6$

52. $2x - y = -8$

53. $2x + 4y = 12$

54. $3x - 2y = 6$

55. $2x - 5y = -20$

56. $3x - 4y = -12$

57. $y = -2x$

58. $y = 5x$

59. $y = x$

60. $y = -x$

Mixed Practice

In the following exercises, graph each equation.

61. $y = \frac{3}{2}x$

62. $y = -\frac{2}{3}x$

63. $y = -\frac{1}{2}x + 3$

64. $y = \frac{1}{4}x - 2$

65. $4x + y = 2$

66. $5x + 2y = 10$

67. $y = -1$

68. $x = 3$

Writing Exercises

69. Explain how you would choose three x-values to make a table to graph the line $y = \frac{1}{5}x - 2$.

70. What is the difference between the equations of a vertical and a horizontal line?

71. Do you prefer to use the method of plotting points or the method using the intercepts to graph the equation $4x + y = -4$? Why?

72. Do you prefer to use the method of plotting points or the method using the intercepts to graph the equation $y = \frac{2}{3}x - 2$? Why?

Self Check

ⓐ *After completing the exercises, use this checklist to evaluate your mastery of the objectives of this section.*

I can...	Confidently	With some help	No-I don't get it!
plot points on a rectangular coordinate system.			
graph a linear equation by plotting points.			
graph vertical and horizontal lines.			
find x- and y-intercepts.			
graph a line using the intercepts.			

ⓑ *If most of your checks were:*

Confidently. Congratulations! You have achieved the objectives in this section. Reflect on the study skills you used so that you can continue to use them. What did you do to become confident of your ability to do these things? Be specific.

With some help. This must be addressed quickly because topics you do not master become potholes in your road to success. In math every topic builds upon previous work. It is important to make sure you have a strong foundation before you move on. Who can you ask for help? Your fellow classmates and instructor are good resources. Is there a place on campus where math tutors are available? Can your study skills be improved?

No, I don't get it. This is a warning sign and you must address it. You should get help right away or you will quickly be overwhelmed. See your instructor as soon as you can to discuss your situation. Together you can come up with a plan to get you the help you need.

3.2 | Slope of a Line

Learning Objectives

By the end of this section, you will be able to:

› Find the slope of a line
› Graph a line given a point and the slope
› Graph a line using its slope and intercept
› Choose the most convenient method to graph a line
› Graph and interpret applications of slope–intercept
› Use slopes to identify parallel and perpendicular lines

Be Prepared!

Before you get started, take this readiness quiz.

1. Simplify: $\frac{(1-4)}{(8-2)}$.

   If you missed this problem, review **Example 1.30**.

2. Divide: $\frac{0}{4}, \frac{4}{0}$.

   If you missed this problem, review **Example 1.49**.

3. Simplify: $\frac{15}{-3}, \frac{-15}{3}, \frac{-15}{-3}$.

   If you missed this problem, review **Example 1.30**.

Find the Slope of a Line

When you graph linear equations, you may notice that some lines tilt up as they go from left to right and some lines tilt down. Some lines are very steep and some lines are flatter.

In mathematics, the measure of the steepness of a line is called the *slope* of the line.

The concept of slope has many applications in the real world. In construction the pitch of a roof, the slant of the plumbing pipes, and the steepness of the stairs are all applications of slope. and as you ski or jog down a hill, you definitely experience slope.

We can assign a numerical value to the slope of a line by finding the ratio of the rise and run. The *rise* is the amount the vertical distance changes while the *run* measures the horizontal change, as shown in this illustration. Slope is a rate of change. See **Figure 3.5**.

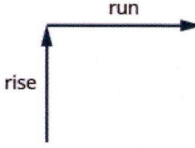

Figure 3.5

Slope of a Line

The slope of a line is $m = \frac{\text{rise}}{\text{run}}$.

The rise measures the vertical change and the run measures the horizontal change.

To find the slope of a line, we locate two points on the line whose coordinates are integers. Then we sketch a right triangle where the two points are vertices and one side is horizontal and one side is vertical.

To find the slope of the line, we measure the distance along the vertical and horizontal sides of the triangle. The vertical distance is called the *rise* and the horizontal distance is called the *run*,

HOW TO :: FIND THE SLOPE OF A LINE FROM ITS GRAPH USING $m = \frac{\text{rise}}{\text{run}}$.

Step 1.   Locate two points on the line whose coordinates are integers.

Step 2.   Starting with one point, sketch a right triangle, going from the first point to the second point.

Step 3.   Count the rise and the run on the legs of the triangle.

Step 4.   Take the ratio of rise to run to find the slope: $m = \frac{\text{rise}}{\text{run}}$.

EXAMPLE 3.12

Find the slope of the line shown.

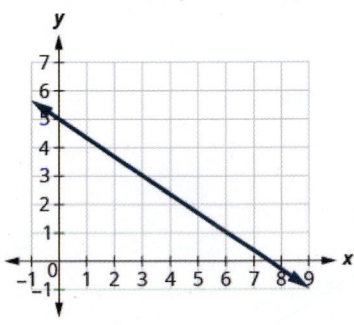

⊘ **Solution**

Locate two points on the graph whose coordinates are integers.	$(0, 5)$ and $(3, 3)$

Starting at $(0, 5)$, sketch a right triangle to $(3, 3)$ as shown in this graph.

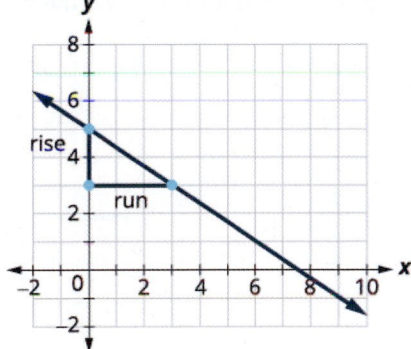

Count the rise— since it goes down, it is negative.	The rise is -2.
Count the run.	The run is 3.
Use the slope formula.	$m = \frac{\text{rise}}{\text{run}}$
Substitute the values of the rise and run.	$m = \frac{-2}{3}$
Simplify.	$m = -\frac{2}{3}$
	The slope of the line is $-\frac{2}{3}$.

So *y* decreases by 2 units as *x* increases by 3 units.

> | **TRY IT : :** 3.23   Find the slope of the line shown.

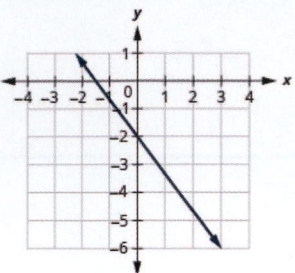

> | **TRY IT : :** 3.24   Find the slope of the line shown.

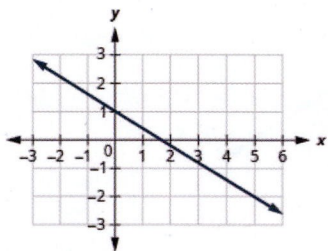

How do we find the slope of horizontal and vertical lines? To find the slope of the horizontal line, $y = 4$, we could graph the line, find two points on it, and count the rise and the run. Let's see what happens when we do this, as shown in the graph below.

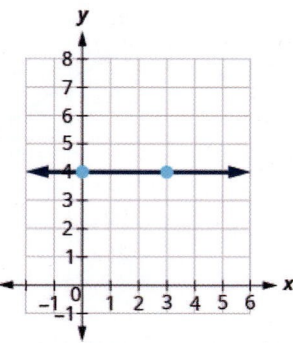

What is the rise?   The rise is 0.
What is the run?   The run is 3.
What is the slope? $m = \dfrac{\text{rise}}{\text{run}}$

$$m = \frac{0}{3}$$

$$m = 0$$

The slope of the horizontal line $y = 4$ is 0.

Let's also consider a vertical line, the line $x = 3$, as shown in the graph.

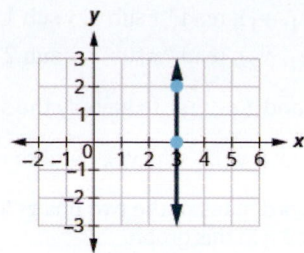

What is the rise?   The rise is 2.

What is the run?   The run is 0.

What is the slope? $m = \dfrac{\text{rise}}{\text{run}}$

$$m = \dfrac{2}{0}$$

The slope is undefined since division by zero is undefined. So we say that the slope of the vertical line $x = 3$ is undefined.

All horizontal lines have slope 0. When the y-coordinates are the same, the rise is 0.

The slope of any vertical line is undefined. When the x-coordinates of a line are all the same, the run is 0.

Slope of a Horizontal and Vertical Line

The slope of a horizontal line, $y = b$, is 0.

The slope of a vertical line, $x = a$, is undefined.

EXAMPLE 3.13

Find the slope of each line: ⓐ $x = 8$ ⓑ $y = -5$.

✓ Solution

ⓐ $x = 8$

This is a vertical line. Its slope is undefined.

ⓑ $y = -5$

This is a horizontal line. It has slope 0.

> **TRY IT : : 3.25**   Find the slope of the line: $x = -4$.

> **TRY IT : : 3.26**   Find the slope of the line: $y = 7$.

Quick Guide to the Slopes of Lines

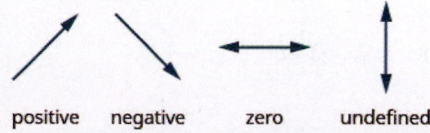

positive   negative   zero   undefined

Sometimes we'll need to find the slope of a line between two points when we don't have a graph to count out the rise and the run. We could plot the points on grid paper, then count out the rise and the run, but as we'll see, there is a way to find the slope without graphing. Before we get to it, we need to introduce some algebraic notation.

We have seen that an ordered pair (x, y) gives the coordinates of a point. But when we work with slopes, we use two points. How can the same symbol (x, y) be used to represent two different points? Mathematicians use subscripts to distinguish the points.

(x_1, y_1) read "x sub 1, y sub 1"

(x_2, y_2) read "x sub 2, y sub 2"

We will use (x_1, y_1) to identify the first point and (x_2, y_2) to identify the second point.

If we had more than two points, we could use (x_3, y_3), (x_4, y_4), and so on.

Let's see how the rise and run relate to the coordinates of the two points by taking another look at the slope of the line between the points $(2, 3)$ and $(7, 6)$, as shown in this graph.

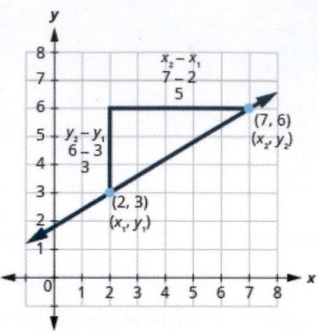

Since we have two points, we will use subscript notation.

$$\binom{x_1,\ y_1}{2,\ 3}\binom{x_2,\ y_2}{7,\ 6}$$

$$m = \frac{\text{rise}}{\text{run}}$$

On the graph, we counted the rise of 3 and the run of 5.

$$m = \frac{3}{5}$$

Notice that the rise of 3 can be found by subtracting the y-coordinates, 6 and 3, and the run of 5 can be found by

subtracting the x-coordinates 7 and 2.

We rewrite the rise and run by putting in the coordinates.

$$m = \frac{6 - 3}{7 - 2}$$

But 6 is y_2, the y-coordinate of the second point and 3 is y_1, the y-coordinate

of the first point. So we can rewrite the slope using subscript notation.

$$m = \frac{y_2 - y_1}{7 - 2}$$

Also 7 is the x-coordinate of the second point and 2 is the x-coordinate

of the first point. So again we rewrite the slope using subscript notation.

$$m = \frac{y_2 - y_1}{x_2 - x_1}$$

We've shown that $m = \frac{y_2 - y_1}{x_2 - x_1}$ is really another version of $m = \frac{\text{rise}}{\text{run}}$. We can use this formula to find the slope of a line

when we have two points on the line.

Slope of a line between two points

The slope of the line between two points (x_1, y_1) and (x_2, y_2) is:

$$m = \frac{y_2 - y_1}{x_2 - x_1}.$$

The slope is:

y of the second point minus y of the first point

over

x of the second point minus x of the first point.

EXAMPLE 3.14

Use the slope formula to find the slope of the line through the points $(-2, -3)$ and $(-7, 4)$.

⊘ Solution

We'll call $(-2, -3)$ point #1 and $(-7, 4)$ point #2.

$$\begin{pmatrix} x_1, & y_1 \\ -2, & -3 \end{pmatrix}\begin{pmatrix} x_2, & y_2 \\ -7, & 4 \end{pmatrix}$$

Use the slope formula.

$$m = \frac{y_2 - y_1}{x_2 - x_1}$$

Substitute the values.
y of the second point minus y of the first point

x of the second point minus x of the first point

$$m = \frac{4 - (-3)}{-7 - (-2)}$$

$$m = \frac{7}{-5}$$

Simplify.

$$m = -\frac{7}{5}$$

Let's verify this slope on the graph shown.

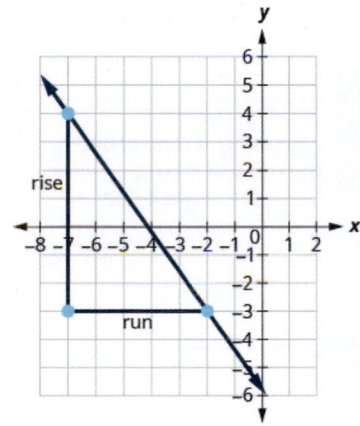

$$m = \frac{\text{rise}}{\text{run}}$$

$$m = \frac{7}{-5}$$

$$m = -\frac{7}{5}$$

> **TRY IT ::** 3.27

Use the slope formula to find the slope of the line through the pair of points: $(-3, 4)$ and $(2, -1)$.

> **TRY IT ::** 3.28

Use the slope formula to find the slope of the line through the pair of points: $(-2, 6)$ and $(-3, -4)$.

Graph a Line Given a Point and the Slope

Up to now, in this chapter, we have graphed lines by plotting points, by using intercepts, and by recognizing horizontal and vertical lines.

We can also graph a line when we know one point and the slope of the line. We will start by plotting the point and then use the definition of slope to draw the graph of the line.

EXAMPLE 3.15   HOW TO GRAPH A LINE GIVEN A POINT AND THE SLOPE

Graph the line passing through the point $(1, -1)$ whose slope is $m = \frac{3}{4}$.

⊘ **Solution**

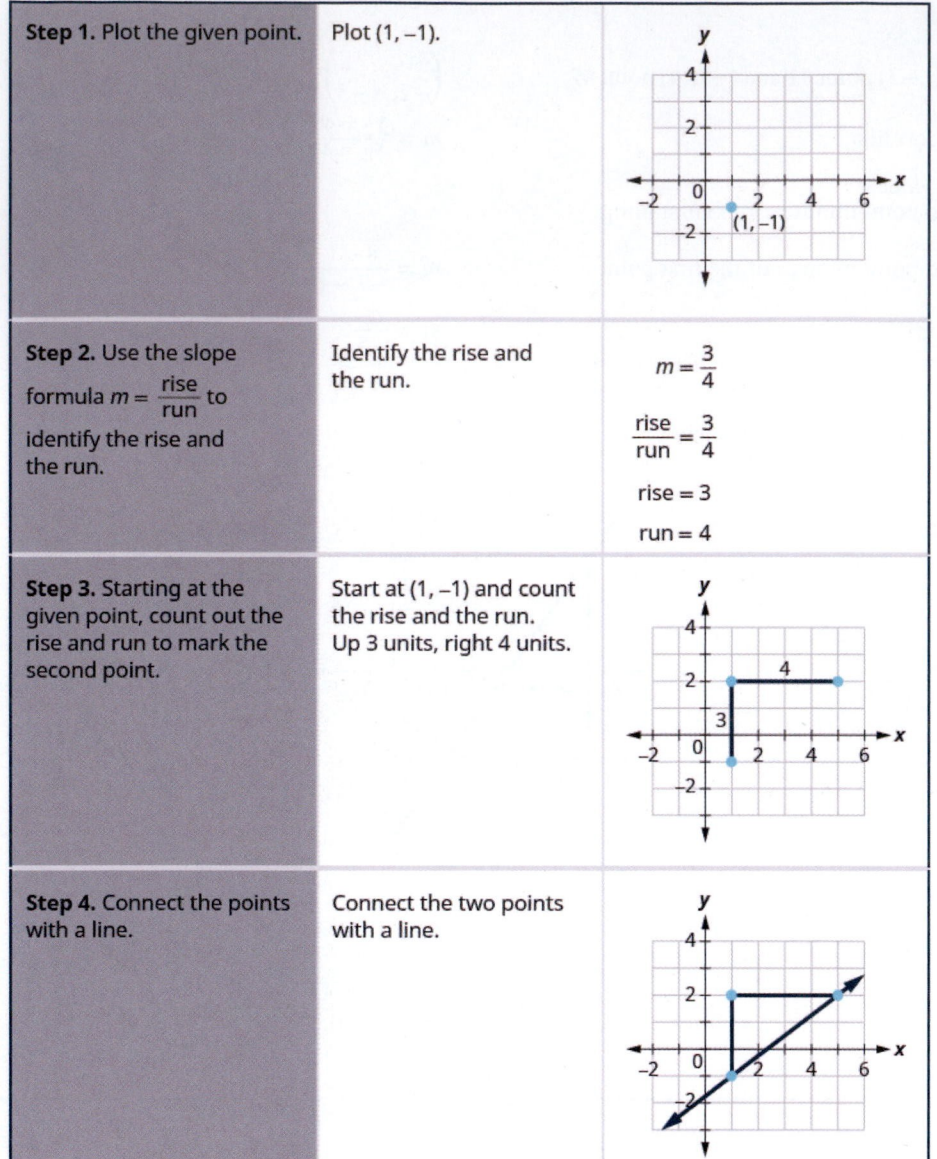

Step 1. Plot the given point.	Plot (1, –1).	
Step 2. Use the slope formula $m = \dfrac{\text{rise}}{\text{run}}$ to identify the rise and the run.	Identify the rise and the run.	$m = \dfrac{3}{4}$ $\dfrac{\text{rise}}{\text{run}} = \dfrac{3}{4}$ rise = 3 run = 4
Step 3. Starting at the given point, count out the rise and run to mark the second point.	Start at (1, –1) and count the rise and the run. Up 3 units, right 4 units.	
Step 4. Connect the points with a line.	Connect the two points with a line.	

You can check your work by finding a third point. Since the slope is $m = \dfrac{3}{4}$, it can also be written as $m = \dfrac{-3}{-4}$ (negative divided by negative is positive!). Go back to $(1, -1)$ and count out the rise, -3, and the run, -4.

> **TRY IT ::** 3.29   Graph the line passing through the point $(2, -2)$ with the slope $m = \dfrac{4}{3}$.

> **TRY IT ::** 3.30   Graph the line passing through the point $(-2, 3)$ with the slope $m = \dfrac{1}{4}$.

HOW TO :: GRAPH A LINE GIVEN A POINT AND THE SLOPE.

Step 1. Plot the given point.

Step 2. Use the slope formula $m = \frac{\text{rise}}{\text{run}}$ to identify the rise and the run.

Step 3. Starting at the given point, count out the rise and run to mark the second point.

Step 4. Connect the points with a line.

Graph a Line Using its Slope and Intercept

We have graphed linear equations by plotting points, using intercepts, recognizing horizontal and vertical lines, and using one point and the slope of the line. Once we see how an equation in slope–intercept form and its graph are related, we'll have one more method we can use to graph lines.

See **Figure 3.6**. Let's look at the graph of the equation $y = \frac{1}{2}x + 3$ and find its slope and y-intercept.

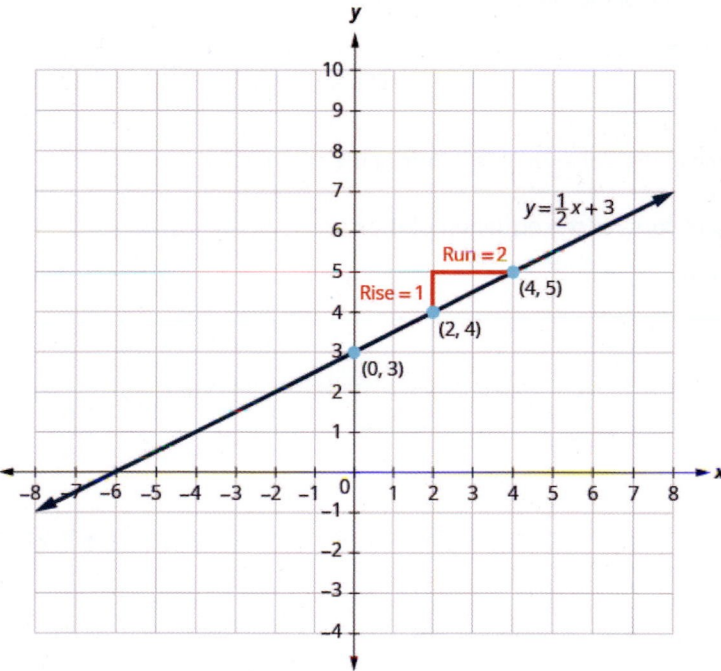

Figure 3.6

The red lines in the graph show us the rise is 1 and the run is 2. Substituting into the slope formula:

$$m = \frac{\text{rise}}{\text{run}}$$

$$m = \frac{1}{2}$$

The y-intercept is $(0, 3)$.

Look at the equation of this line.

$$y = \frac{1}{2}x + 3$$

Look at the slope and y-intercept.

slope $m = \frac{1}{2}$ and y-intercept $(0, 3)$.

When a linear equation is solved for y, the coefficient of the x term is the slope and the constant term is the y-coordinate of the y-intercept. We say that the equation $y = \frac{1}{2}x + 3$ is in slope-intercept form. Sometimes the slope-intercept form is called the "y-form."

$$m = \frac{1}{2}; \text{ } y\text{-intercept is } (0, 3)$$

$$y = \frac{1}{2}x + 3$$

$$y = mx + b$$

> ### Slope Intercept Form of an Equation of a Line
>
> The slope-intercept form of an equation of a line with slope m and y-intercept, $(0, b)$ is $y = mx + b$.

Let's practice finding the values of the slope and y-intercept from the equation of a line.

EXAMPLE 3.16

Identify the slope and y-intercept of the line from the equation:

ⓐ $y = -\frac{4}{7}x - 2$ ⓑ $x + 3y = 9$

⊘ Solution

ⓐ We compare our equation to the slope-intercept form of the equation.

Write the slope–intercept form of the equation of the line.	$y = mx + b$
Write the equation of the line.	$y = -\frac{4}{7}x - 2$
Identify the slope.	$m = -\frac{4}{7}$
Identify the y-intercept.	y-intercept is $(0, -2)$

ⓑ When an equation of a line is not given in slope–intercept form, our first step will be to solve the equation for y.

Solve for y.	$x + 3y = 9$
Subtract x from each side.	$3y = -x + 9$
Divide both sides by 3.	$\frac{3y}{3} = \frac{-x + 9}{3}$
Simplify.	$y = -\frac{1}{3}x + 3$
Write the slope–intercept form of the equation of the line.	$y = mx + b$
Write the equation of the line.	$y = -\frac{1}{3}x + 3$
Identify the slope.	$m = -\frac{1}{3}$
Identify the y-intercept.	y-intercept is $(0, 3)$

> **TRY IT ::** 3.31

Identify the slope and y-intercept from the equation of the line.

ⓐ $y = \frac{2}{5}x - 1$ ⓑ $x + 4y = 8$

> **TRY IT ::** 3.32

Identify the slope and y-intercept from the equation of the line.

ⓐ $y = -\frac{4}{3}x + 1$ ⓑ $3x + 2y = 12$

We have graphed a line using the slope and a point. Now that we know how to find the slope and y-intercept of a line from its equation, we can use the y-intercept as the point, and then count out the slope from there.

EXAMPLE 3.17

Graph the line of the equation $y = -x + 4$ using its slope and y-intercept.

⊘ **Solution**

	$y = mx + b$
The equation is in slope–intercept form.	$y = -x + 4$
Identify the slope and y-intercept.	$m = -1$ y-intercept is $(0, 4)$
Plot the y-intercept.	See the graph.
Identify the rise over the run.	$m = \frac{-1}{1}$
Count out the rise and run to mark the second point.	rise -1, run 1

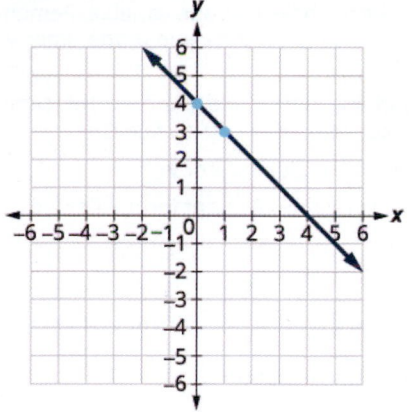

Draw the line as shown in the graph.

> **TRY IT ::** 3.33

Graph the line of the equation $y = -x - 3$ using its slope and y-intercept.

> **TRY IT ::** 3.34

Graph the line of the equation $y = -x - 1$ using its slope and y-intercept.

Now that we have graphed lines by using the slope and y-intercept, let's summarize all the methods we have used to graph lines.

Methods to Graph Lines			
Point Plotting	**Slope–Intercept**	**Intercepts**	**Recognize Vertical and Horizontal Lines**
	$y = mx + b$		
Find three points. Plot the points, make sure they line up, then draw the line.	Find the slope and y-intercept. Start at the y-intercept, then count the slope to get a second point.	Find the intercepts and a third point. Plot the points, make sure they line up, then draw the line.	The equation has only one variable. $x = a$ vertical $y = b$ horizontal

Choose the Most Convenient Method to Graph a Line

Now that we have seen several methods we can use to graph lines, how do we know which method to use for a given equation?

While we could plot points, use the slope–intercept form, or find the intercepts for *any* equation, if we recognize the most convenient way to graph a certain type of equation, our work will be easier.

Generally, plotting points is not the most efficient way to graph a line. Let's look for some patterns to help determine the most convenient method to graph a line.

Here are five equations we graphed in this chapter, and the method we used to graph each of them.

Equation	**Method**
#1 $x = 2$	Vertical line
#2 $y = -1$	Horizontal line
#3 $-x + 2y = 6$	Intercepts
#4 $4x - 3y = 12$	Intercepts
#5 $y = -x + 4$	Slope–intercept

Equations #1 and #2 each have just one variable. Remember, in equations of this form the value of that one variable is constant; it does not depend on the value of the other variable. Equations of this form have graphs that are vertical or horizontal lines.

In equations #3 and #4, both x and y are on the same side of the equation. These two equations are of the form $Ax + By = C$. We substituted $y = 0$ to find the x- intercept and $x = 0$ to find the y-intercept, and then found a third point by choosing another value for x or y.

Equation #5 is written in slope–intercept form. After identifying the slope and y-intercept from the equation we used them to graph the line.

This leads to the following strategy.

Strategy for Choosing the Most Convenient Method to Graph a Line

Consider the form of the equation.

- If it only has one variable, it is a vertical or horizontal line.
  - $x = a$ is a vertical line passing through the x-axis at a.
  - $y = b$ is a horizontal line passing through the y-axis at b.
- If y is isolated on one side of the equation, in the form $y = mx + b$, graph by using the slope and y-intercept.
  - Identify the slope and y-intercept and then graph.
- If the equation is of the form $Ax + By = C$, find the intercepts.
  - Find the x- and y-intercepts, a third point, and then graph.

EXAMPLE 3.18

Determine the most convenient method to graph each line:

ⓐ $y = 5$ ⓑ $4x - 5y = 20$ ⓒ $x = -3$ ⓓ $y = -\frac{5}{9}x + 8$

✓ Solution

ⓐ $y = 5$

This equation has only one variable, y. Its graph is a horizontal line crossing the y-axis at 5.

ⓑ $4x - 5y = 20$

This equation is of the form $Ax + By = C$. The easiest way to graph it will be to find the intercepts and one more point.

ⓒ $x = -3$

There is only one variable, x. The graph is a vertical line crossing the x-axis at -3.

ⓓ $y = -\frac{5}{9}x + 8$

Since this equation is in $y = mx + b$ form, it will be easiest to graph this line by using the slope and y-intercepts.

> **TRY IT ::3.35**   Determine the most convenient method to graph each line:

ⓐ $3x + 2y = 12$ ⓑ $y = 4$ ⓒ $y = \frac{1}{5}x - 4$ ⓓ $x = -7$.

> **TRY IT ::3.36**   Determine the most convenient method to graph each line:

ⓐ $x = 6$ ⓑ $y = -\frac{3}{4}x + 1$ ⓒ $y = -8$ ⓓ $4x - 3y = -1$.

Graph and Interpret Applications of Slope–Intercept

Many real-world applications are modeled by linear equations. We will take a look at a few applications here so you can see how equations written in slope–intercept form relate to real world situations.

Usually, when a linear equation models uses real-world data, different letters are used for the variables, instead of using only x and y. The variable names remind us of what quantities are being measured.

Also, we often will need to extend the axes in our rectangular coordinate system to bigger positive and negative numbers to accommodate the data in the application.

EXAMPLE 3.19

The equation $F = \frac{9}{5}C + 32$ is used to convert temperatures, C, on the Celsius scale to temperatures, F, on the Fahrenheit scale.

ⓐ Find the Fahrenheit temperature for a Celsius temperature of 0.

ⓑ Find the Fahrenheit temperature for a Celsius temperature of 20.

ⓒ Interpret the slope and F-intercept of the equation.

ⓓ Graph the equation.

✓ Solution

ⓐ

Find the Fahrenheit temperature for a Celsius temperature of 0.	$F = \frac{9}{5}C + 32$
Find F when $C = 0$.	$F = \frac{9}{5}(0) + 32$
Simplify.	$F = 32$

ⓑ

Find the Fahrenheit temperature for a Celsius temperature of 20. $F = \frac{9}{5}C + 32$

Find F when $C = 20$. $F = \frac{9}{5}(20) + 32$

Simplify. $F = 36 + 32$

Simplify. $F = 68$

ⓒ

Interpret the slope and F-intercept of the equation.
Even though this equation uses F and C, it is still in slope–intercept form.

$$y = mx + b$$

$$F = mC + b$$

$$F = \frac{9}{5}C + 32$$

The slope, $\frac{9}{5}$, means that the temperature Fahrenheit (F) increases 9 degrees when the temperature Celsius (C)

increases 5 degrees.
The F-intercept means that when the temperature is $0°$ on the Celsius scale, it is $32°$ on the Fahrenheit scale.

ⓓ Graph the equation.
We'll need to use a larger scale than our usual. Start at the F-intercept $(0, 32)$, and then count out the rise of 9 and the run of 5 to get a second point as shown in the graph.

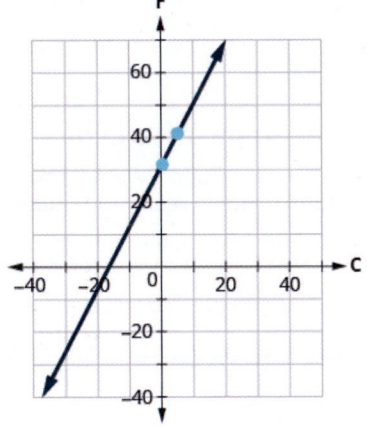

> **TRY IT :: 3.37**

The equation $h = 2s + 50$ is used to estimate a woman's height in inches, h, based on her shoe size, s.

ⓐ Estimate the height of a child who wears women's shoe size 0.

ⓑ Estimate the height of a woman with shoe size 8.

ⓒ Interpret the slope and h-intercept of the equation.

ⓓ Graph the equation.

TRY IT : : 3.38

The equation $T = \frac{1}{4}n + 40$ is used to estimate the temperature in degrees Fahrenheit, T, based on the number of cricket chirps, n, in one minute.

ⓐ Estimate the temperature when there are no chirps.

ⓑ Estimate the temperature when the number of chirps in one minute is 100.

ⓒ Interpret the slope and T-intercept of the equation.

ⓓ Graph the equation.

The cost of running some types business have two components—a *fixed cost* and a *variable cost*. The fixed cost is always the same regardless of how many units are produced. This is the cost of rent, insurance, equipment, advertising, and other items that must be paid regularly. The variable cost depends on the number of units produced. It is for the material and labor needed to produce each item.

EXAMPLE 3.20

Sam drives a delivery van. The equation $C = 0.5m + 60$ models the relation between his weekly cost, C, in dollars and the number of miles, m, that he drives.

ⓐ Find Sam's cost for a week when he drives 0 miles.

ⓑ Find the cost for a week when he drives 250 miles.

ⓒ Interpret the slope and C-intercept of the equation.

ⓓ Graph the equation.

 Solution

ⓐ

Find Sam's cost for a week when he drives 0 miles.	$C = 0.5m + 60$
Find C when $m = 0$.	$C = 0.5(0) + 60$
Simplify.	$C = 60$
	Sam's costs are \$60 when he drives 0 miles.

ⓑ

Find the cost for a week when he drives 250 miles.	$C = 0.5m + 60$
Find C when $m = 250$.	$C = 0.5(250) + 60$
Simplify.	$C = 185$
	Sam's costs are \$185 when he drives 250 miles.

ⓒ Interpret the slope and C-intercept of the equation.

$$y = mx + b$$

$$C = 0.5p + 60$$

The slope, 0.5, means that the weekly cost, C, increases by \$0.50 when the number of miles driven, n, increases by 1. The C-intercept means that when the number of miles driven is 0, the weekly cost is \$60.

ⓓ Graph the equation.

We'll need to use a larger scale than our usual. Start at the C-intercept $(0, 60)$.

To count out the slope $m = 0.5$, we rewrite it as an equivalent fraction that will make our graphing easier.

Rewrite as a fraction.
$$m = 0.5$$
$$m = \frac{0.5}{1}$$

Multiply numerator and
denominator by 100.
$$m = \frac{0.5(100)}{1(100)}$$

Simplify.
$$m = \frac{50}{100}$$

So to graph the next point go up 50 from the intercept of 60 and then to the right 100. The second point will be (100, 110).

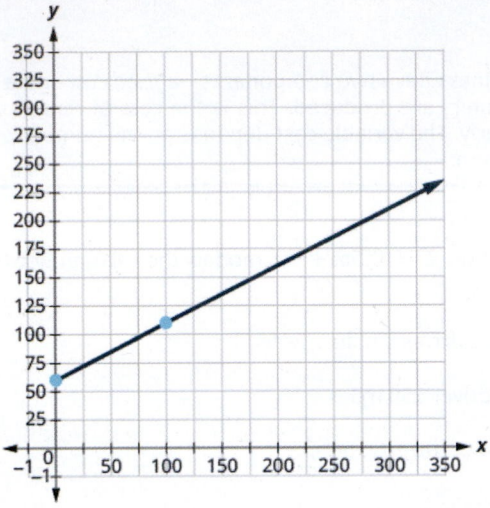

> **TRY IT :: 3.39**

Stella has a home business selling gourmet pizzas. The equation $C = 4p + 25$ models the relation between her weekly cost, C, in dollars and the number of pizzas, p, that she sells.

ⓐ Find Stella's cost for a week when she sells no pizzas.

ⓑ Find the cost for a week when she sells 15 pizzas.

ⓒ Interpret the slope and C-intercept of the equation.

ⓓ Graph the equation.

> **TRY IT :: 3.40**

Loreen has a calligraphy business. The equation $C = 1.8n + 35$ models the relation between her weekly cost, C, in dollars and the number of wedding invitations, n, that she writes.

ⓐ Find Loreen's cost for a week when she writes no invitations.

ⓑ Find the cost for a week when she writes 75 invitations.

ⓒ Interpret the slope and C-intercept of the equation.

ⓓ Graph the equation.

Use Slopes to Identify Parallel and Perpendicular Lines

Two lines that have the same slope are called **parallel lines**. Parallel lines have the same steepness and never intersect.

We say this more formally in terms of the rectangular coordinate system. Two lines that have the same slope and different y-intercepts are called parallel lines. See Figure 3.7.

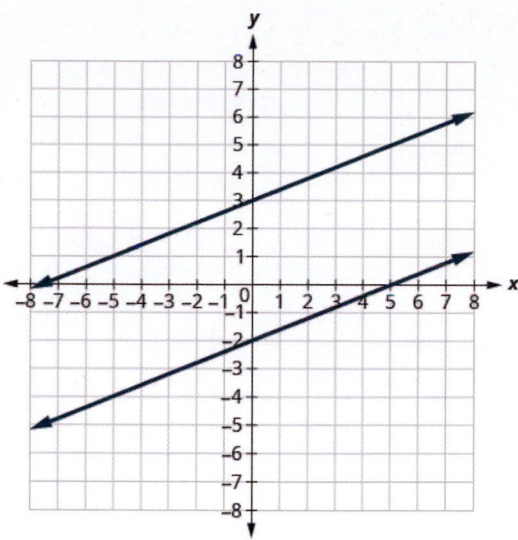

Figure 3.7

Verify that both lines have the same slope, $m = \frac{2}{5}$, and different y-intercepts.

What about vertical lines? The slope of a vertical line is undefined, so vertical lines don't fit in the definition above. We say that vertical lines that have different x-intercepts are parallel, like the lines shown in this graph.

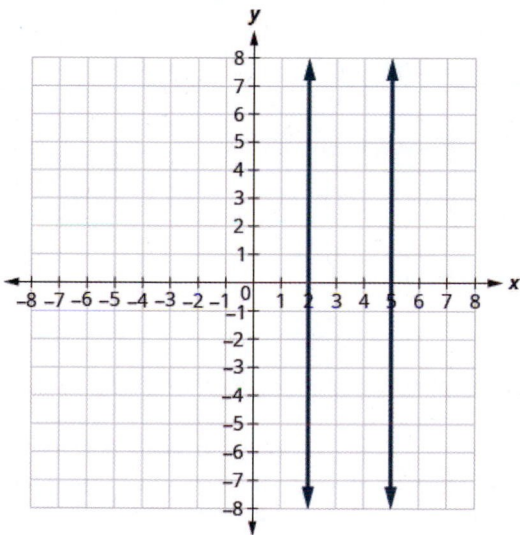

Figure 3.8

Parallel Lines

Parallel lines are lines in the same plane that do not intersect.

- Parallel lines have the same slope and different y-intercepts.
- If m_1 and m_2 are the slopes of two parallel lines then $m_1 = m_2$.
- Parallel vertical lines have different x-intercepts.

Since parallel lines have the same slope and different y-intercepts, we can now just look at the slope–intercept form of the equations of lines and decide if the lines are parallel.

EXAMPLE 3.21

Use slopes and y-intercepts to determine if the lines are parallel:

ⓐ $3x - 2y = 6$ and $y = \frac{3}{2}x + 1$ ⓑ $y = 2x - 3$ and $-6x + 3y = -9$.

✓ Solution

ⓐ

$$3x - 2y = 6 \qquad \text{and} \qquad y = \frac{3}{2}x + 1$$

Solve the first equation for y.
$$-2y = -3x + 6$$
$$\frac{-2y}{-2} = \frac{-3x + 6}{-2}$$

The equation is now in slope–intercept form.
$$y = \frac{3}{2}x - 3$$

The equation of the second line is already
in slope–intercept form.
$$y = \frac{3}{2}x + 1$$

Identify the slope and y-intercept of both lines.

$$y = \frac{3}{2}x - 3 \qquad\qquad y = \frac{3}{2}x + 1$$
$$y = mx + b \qquad\qquad y = mx + b$$
$$m = \frac{3}{2} \qquad\qquad y = \frac{3}{2}$$
$$y\text{-intercept is } (0, -3) \qquad y\text{-intercept is } (0, 1)$$

The lines have the same slope and different y-intercepts and so they are parallel.
You may want to graph the lines to confirm whether they are parallel.

ⓑ

$$y = 2x - 3 \qquad \text{and} \qquad -6x + 3y = -9$$

The first equation is already in slope–intercept form.
$$y = 2x - 3$$

$$-6x + 3y = -9$$
$$3y = 6x - 9$$

Solve the second equation for y.
$$\frac{3y}{3} = \frac{6x - 9}{3}$$
$$y = 2x - 3$$

The second equation is now in slope–intercept form.
$$y = 2x - 3$$

Identify the slope and y-intercept of both lines.

$$y = 2x - 3 \qquad\qquad y = 2x - 3$$
$$y = mx + b \qquad\qquad y = mx + b$$
$$m = 2 \qquad\qquad m = 2$$
$$y\text{-intercept is } (0, -3) \qquad y\text{-intercept is } (0, -3)$$

The lines have the same slope, but they also have the same y-intercepts. Their equations represent the same line and we say the lines are coincident. They are not parallel; they are the same line.

> **TRY IT :: 3.41**   Use slopes and y-intercepts to determine if the lines are parallel:

ⓐ $2x + 5y = 5$ and $y = -\frac{2}{5}x - 4$ ⓑ $y = -\frac{1}{2}x - 1$ and $x + 2y = -2$.

> **TRY IT :: 3.42**   Use slopes and y-intercepts to determine if the lines are parallel:

ⓐ $4x - 3y = 6$ and $y = \frac{4}{3}x - 1$ ⓑ $y = \frac{3}{4}x - 3$ and $3x - 4y = 12$.

EXAMPLE 3.22

Use slopes and y-intercepts to determine if the lines are parallel:

ⓐ $y = -4$ and $y = 3$ ⓑ $x = -2$ and $x = -5$.

✓ Solution

ⓐ $y = -4$ and $y = 3$

We recognize right away from the equations that these are horizontal lines, and so we know their slopes are both 0. Since the horizontal lines cross the y-axis at $y = -4$ and at $y = 3$, we know the y-intercepts are $(0, -4)$ and $(0, 3)$.

The lines have the same slope and different y-intercepts and so they are parallel.

ⓑ $x = -2$ and $x = -5$

We recognize right away from the equations that these are vertical lines, and so we know their slopes are undefined. Since the vertical lines cross the x-axis at $x = -2$ and $x = -5$, we know the y-intercepts are $(-2, 0)$ and $(-5, 0)$.

The lines are vertical and have different x-intercepts and so they are parallel.

> **TRY IT ∷ 3.43**   Use slopes and y-intercepts to determine if the lines are parallel:

ⓐ $y = 8$ and $y = -6$ ⓑ $x = 1$ and $x = -5$.

> **TRY IT ∷ 3.44**   Use slopes and y-intercepts to determine if the lines are parallel:

ⓐ $y = 1$ and $y = -5$ ⓑ $x = 8$ and $x = -6$.

Let's look at the lines whose equations are $y = \frac{1}{4}x - 1$ and $y = -4x + 2$, shown in **Figure 3.9**.

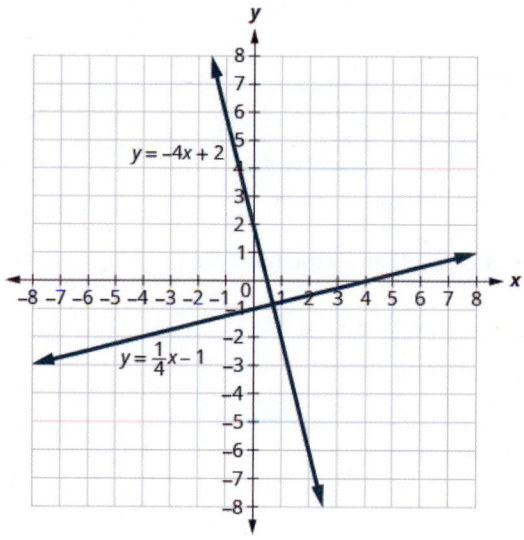

Figure 3.9

These lines lie in the same plane and intersect in right angles. We call these lines perpendicular.

If we look at the slope of the first line, $m_1 = \frac{1}{4}$, and the slope of the second line, $m_2 = -4$, we can see that they are *negative reciprocals* of each other. If we multiply them, their product is -1.

$$m_1 \cdot m_2$$
$$\frac{1}{4}(-4)$$
$$-1$$

This is always true for **perpendicular lines** and leads us to this definition.

Perpendicular Lines

Perpendicular lines are lines in the same plane that form a right angle.

- If m_1 and m_2 are the slopes of two perpendicular lines, then:

  ◦ their slopes are negative reciprocals of each other, $m_1 = -\dfrac{1}{m_2}$.

  ◦ the product of their slopes is -1, $m_1 \cdot m_2 = -1$.

- A vertical line and a horizontal line are always perpendicular to each other.

We were able to look at the slope–intercept form of linear equations and determine whether or not the lines were parallel. We can do the same thing for perpendicular lines.

We find the slope–intercept form of the equation, and then see if the slopes are opposite reciprocals. If the product of the slopes is -1, the lines are perpendicular.

EXAMPLE 3.23

Use slopes to determine if the lines are perpendicular:

ⓐ $y = -5x - 4$ and $x - 5y = 5$ ⓑ $7x + 2y = 3$ and $2x + 7y = 5$

✓ Solution

ⓐ

The first equation is in slope–intercept form. $y = -5x - 4$

Solve the second equation for y.

$$
\begin{aligned}
x - 5y &= 5 \\
-5y &= -x + 5 \\
\frac{-5y}{-5} &= \frac{-x + 5}{-5} \\
y &= \tfrac{1}{5}x - 1
\end{aligned}
$$

Identify the slope of each line.

$$
\begin{aligned}
y &= -5x - 4 \\
y &= mx + b \\
m_1 &= -5
\end{aligned}
\qquad
\begin{aligned}
y &= \tfrac{1}{5}x - 1 \\
y &= mx + b \\
m_2 &= \tfrac{1}{5}
\end{aligned}
$$

The slopes are negative reciprocals of each other, so the lines are perpendicular. We check by multiplying the slopes, Since $-5\left(\tfrac{1}{5}\right) = -1$, it checks.

ⓑ

Solve the equations for y.

$$
\begin{aligned}
7x + 2y &= 3 \\
2y &= -7x + 3 \\
\frac{2y}{2} &= \frac{-7x + 3}{2} \\
y &= -\tfrac{7}{2}x + \tfrac{3}{2} \\
y &= mx + b
\end{aligned}
\qquad
\begin{aligned}
2x + 7y &= 5 \\
7y &= -2x + 5 \\
\frac{7y}{7} &= \frac{-2x + 5}{7} \\
y &= -\tfrac{2}{7}x + \tfrac{5}{7} \\
y &= mx + b
\end{aligned}
$$

Identify the slope of each line.

$$
m_1 = -\tfrac{7}{2} \qquad\qquad m_1 = -\tfrac{2}{7}
$$

The slopes are reciprocals of each other, but they have the same sign. Since they are not negative reciprocals, the lines are not perpendicular.

> **TRY IT ::** 3.45   Use slopes to determine if the lines are perpendicular:

ⓐ $y = -3x + 2$ and $x - 3y = 4$ ⓑ $5x + 4y = 1$ and $4x + 5y = 3$.

> **TRY IT : :** 3.46   Use slopes to determine if the lines are perpendicular:

ⓐ $y = 2x - 5$ and $x + 2y = -6$ ⓑ $2x - 9y = 3$ and $9x - 2y = 1$.

3.2 EXERCISES

Practice Makes Perfect

Find the Slope of a Line

In the following exercises, find the slope of each line shown.

73.

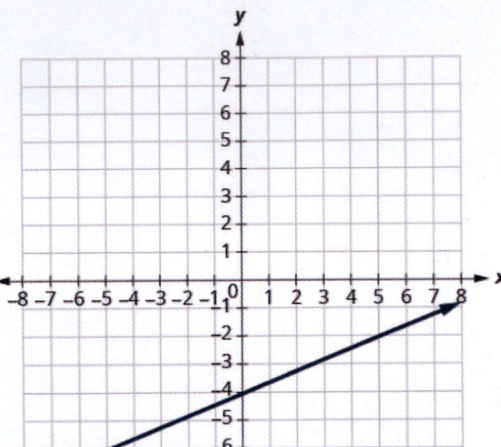

74.

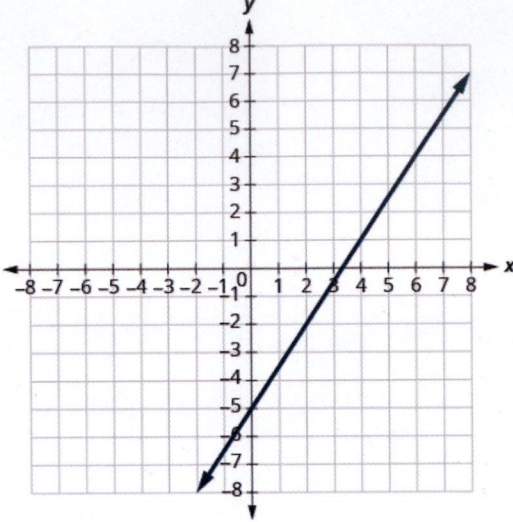

75.

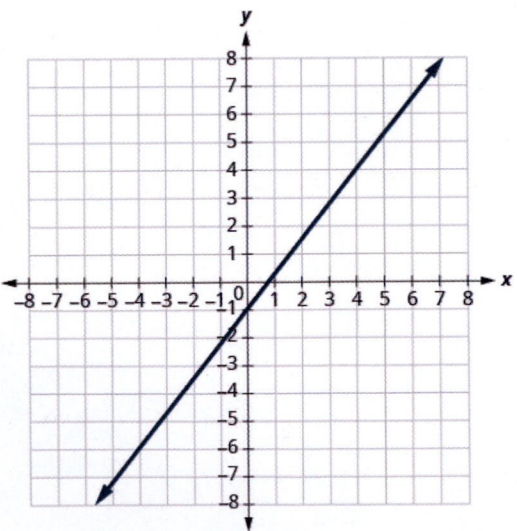

76.

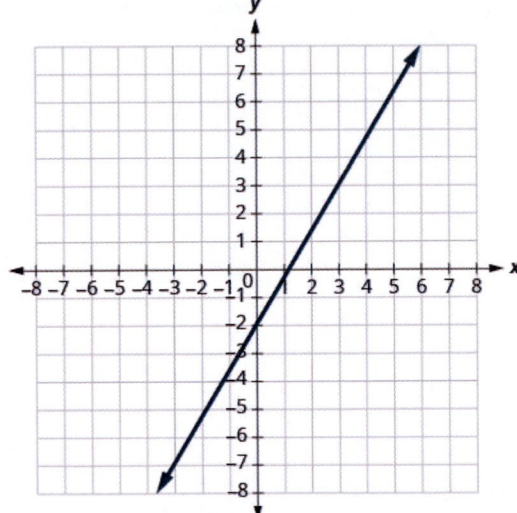

77.

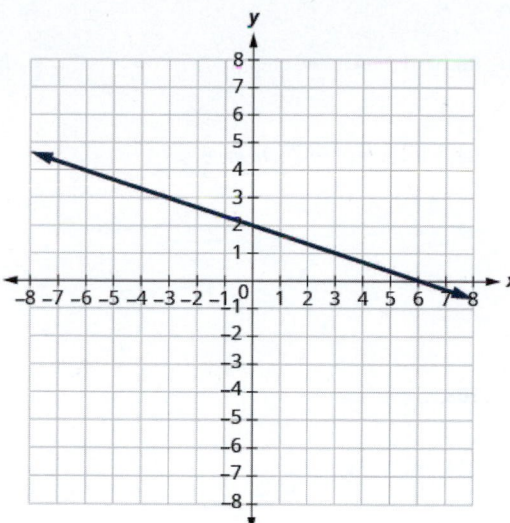

78.

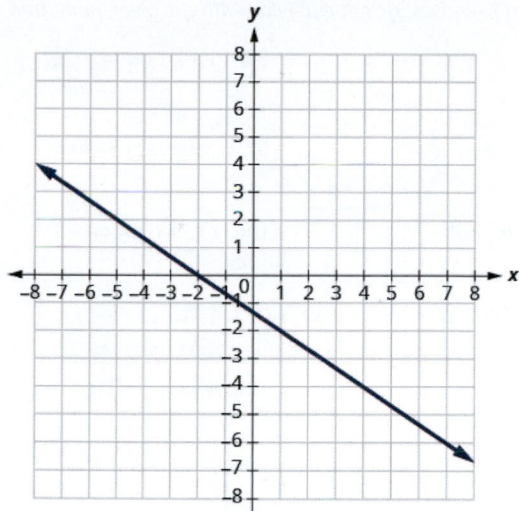

79.

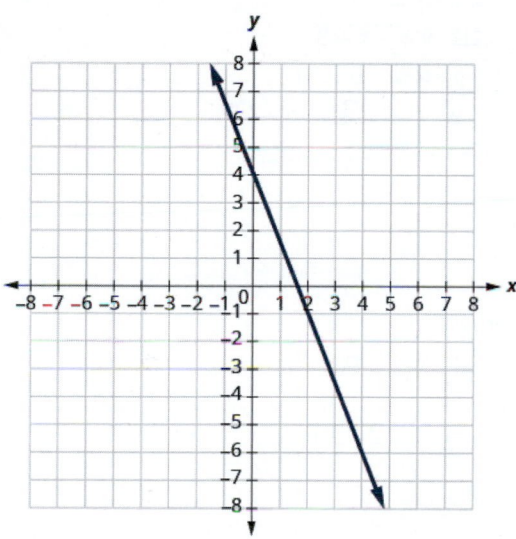

80.

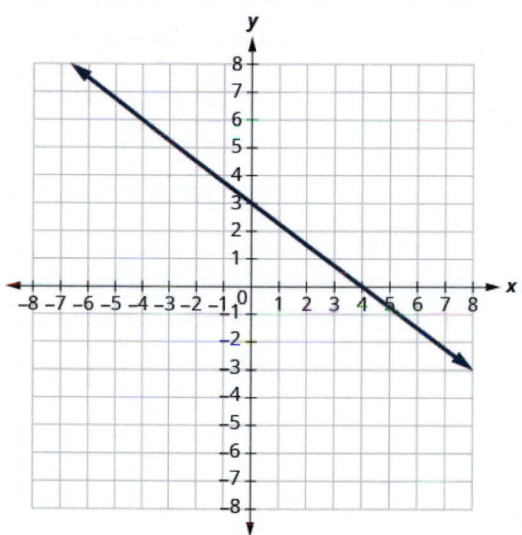

In the following exercises, find the slope of each line.

81. $y = 3$

82. $y = -2$

83. $x = -5$

84. $x = 4$

In the following exercises, use the slope formula to find the slope of the line between each pair of points.

85. $(2, 5), (4, 0)$

86. $(3, 6), (8, 0)$

87. $(-3, 3), (4, -5)$

88. $(-2, 4), (3, -1)$

89. $(-1, -2), (2, 5)$

90. $(-2, -1), (6, 5)$

91. $(4, -5), (1, -2)$

92. $(3, -6), (2, -2)$

Graph a Line Given a Point and the Slope

In the following exercises, graph each line with the given point and slope.

93. $(2, 5)$; $m = -\frac{1}{3}$

94. $(1, 4)$; $m = -\frac{1}{2}$

95. $(-1, -4)$; $m = \frac{4}{3}$

96. $(-3, -5)$; $m = \frac{3}{2}$

97. y-intercept 3; $m = -\frac{2}{5}$

98. x-intercept -2; $m = \frac{3}{4}$

99. $(-4, 2)$; $m = 4$

100. $(1, 5)$; $m = -3$

Graph a Line Using Its Slope and Intercept

In the following exercises, identify the slope and y-intercept of each line.

101. $y = -7x + 3$

102. $y = 4x - 10$

103. $3x + y = 5$

104. $4x + y = 8$

105. $6x + 4y = 12$

106. $8x + 3y = 12$

107. $5x - 2y = 6$

108. $7x - 3y = 9$

In the following exercises, graph the line of each equation using its slope and y-intercept.

109. $y = 3x - 1$

110. $y = 2x - 3$

111. $y = -x + 3$

112. $y = -x - 4$

113. $y = -\frac{2}{5}x - 3$

114. $y = -\frac{3}{5}x + 2$

115. $3x - 2y = 4$

116. $3x - 4y = 8$

Choose the Most Convenient Method to Graph a Line

In the following exercises, determine the most convenient method to graph each line.

117. $x = 2$

118. $y = 5$

119. $y = -3x + 4$

120. $x - y = 5$

121. $x - y = 1$

122. $y = \frac{2}{3}x - 1$

123. $3x - 2y = -12$

124. $2x - 5y = -10$

Graph and Interpret Applications of Slope–Intercept

125. The equation $P = 31 + 1.75w$ models the relation between the amount of Tuyet's monthly water bill payment, P, in dollars, and the number of units of water, w, used.

ⓐ Find Tuyet's payment for a month when 0 units of water are used.

ⓑ Find Tuyet's payment for a month when 12 units of water are used.

ⓒ Interpret the slope and P-intercept of the equation.

ⓓ Graph the equation.

126. The equation $P = 28 + 2.54w$ models the relation between the amount of R and y's monthly water bill payment, P, in dollars, and the number of units of water, w, used.

ⓐ Find the payment for a month when R and y used 0 units of water.

ⓑ Find the payment for a month when R and y used 15 units of water.

ⓒ Interpret the slope and P-intercept of the equation.

ⓓ Graph the equation.

127. Bruce drives his car for his job. The equation $R = 0.575m + 42$ models the relation between the amount in dollars, R, that he is reimbursed and the number of miles, m, he drives in one day.

ⓐ Find the amount Bruce is reimbursed on a day when he drives 0 miles.

ⓑ Find the amount Bruce is reimbursed on a day when he drives 220 miles.

ⓒ Interpret the slope and R-intercept of the equation.

ⓓ Graph the equation.

128. Janelle is planning to rent a car while on vacation. The equation $C = 0.32m + 15$ models the relation between the cost in dollars, C, per day and the number of miles, m, she drives in one day.

ⓐ Find the cost if Janelle drives the car 0 miles one day.

ⓑ Find the cost on a day when Janelle drives the car 400 miles.

ⓒ Interpret the slope and C-intercept of the equation.

ⓓ Graph the equation.

129. Cherie works in retail and her weekly salary includes commission for the amount she sells. The equation $S = 400 + 0.15c$ models the relation between her weekly salary, S, in dollars and the amount of her sales, c, in dollars.

ⓐ Find Cherie's salary for a week when her sales were $0.

ⓑ Find Cherie's salary for a week when her sales were $3,600.

ⓒ Interpret the slope and S-intercept of the equation.

ⓓ Graph the equation.

130. Patel's weekly salary includes a base pay plus commission on his sales. The equation $S = 750 + 0.09c$ models the relation between his weekly salary, S, in dollars and the amount of his sales, c, in dollars.

ⓐ Find Patel's salary for a week when his sales were 0.

ⓑ Find Patel's salary for a week when his sales were 18,540.

ⓒ Interpret the slope and S-intercept of the equation.

ⓓ Graph the equation.

131. Costa is planning a lunch banquet. The equation $C = 450 + 28g$ models the relation between the cost in dollars, C, of the banquet and the number of guests, g.

ⓐ Find the cost if the number of guests is 40.

ⓑ Find the cost if the number of guests is 80.

ⓒ Interpret the slope and C-intercept of the equation.

ⓓ Graph the equation.

132. Margie is planning a dinner banquet. The equation $C = 750 + 42g$ models the relation between the cost in dollars, C of the banquet and the number of guests, g.

ⓐ Find the cost if the number of guests is 50.

ⓑ Find the cost if the number of guests is 100.

ⓒ Interpret the slope and C-intercept of the equation.

ⓓ Graph the equation.

Use Slopes to Identify Parallel and Perpendicular Lines

In the following exercises, use slopes and y-intercepts to determine if the lines are parallel, perpendicular, or neither.

133.
$y = \frac{3}{4}x - 3; \quad 3x - 4y = -2$

134.
$3x - 4y = -2; \quad y = \frac{3}{4}x - 3$

135. $2x - 4y = 6; \quad x - 2y = 3$

136.
$8x + 6y = 6; \quad 12x + 9y = 12$

137. $x = 5; \quad x = -6$

138. $x = -3; \quad x = -2$

139. $4x - 2y = 5; \quad 3x + 6y = 8$

140.
$8x - 2y = 7; \quad 3x + 12y = 9$

141.
$3x - 6y = 12; \quad 6x - 3y = 3$

142.
$9x - 5y = 4; \quad 5x + 9y = -1$

143.
$7x - 4y = 8; \quad 4x + 7y = 14$

144. $5x - 2y = 11; \quad 5x - y = 7$

145. $3x - 2y = 8; \quad 2x + 3y = 6$

146. $2x + 3y = 5; \quad 3x - 2y = 7$

147. $3x - 2y = 1; \quad 2x - 3y = 2$

148. $2x + 4y = 3; \quad 6x + 3y = 2$

149. $y = 2; \quad y = 6$

150. $y = -1; \quad y = 2$

Writing Exercises

151. How does the graph of a line with slope $m = \frac{1}{2}$ differ from the graph of a line with slope $m = 2$?

152. Why is the slope of a vertical line "undefined"?

153. Explain how you can graph a line given a point and its slope.

154. Explain in your own words how to decide which method to use to graph a line.

Self Check

ⓐ *After completing the exercises, use this checklist to evaluate your mastery of the objectives of this section.*

I can...	Confidently	With some help	No-I don't get it!
find the slope of a line.			
graph a line given a point and the slope.			
graph a line using its slope and intercept.			
choose the most convenient method to graph a line.			
graph and interpret applications of slope-intercept.			
use slopes to identify parallel and perpendicular lines.			

ⓑ *After reviewing this checklist, what will you do to become confident for all objectives?*

3.3 Find the Equation of a Line

Learning Objectives

By the end of this section, you will be able to:

> Find an equation of the line given the slope and y-intercept
> Find an equation of the line given the slope and a point
> Find an equation of the line given two points
> Find an equation of a line parallel to a given line
> Find an equation of a line perpendicular to a given line

Be Prepared!

Before you get started, take this readiness quiz.

1. Solve: $\frac{2}{5}(x + 15)$.

   If you missed this problem, review **Example 1.50**.

2. Simplify: $-3(x - (-2))$.

   If you missed this problem, review **Example 1.53**.

3. Solve for y: $y - 3 = -2(x + 1)$.

   If you missed this problem, review **Example 2.31**.

How do online companies know that "you may also like" a particular item based on something you just ordered? How can economists know how a rise in the minimum wage will affect the unemployment rate? How do medical researchers create drugs to target cancer cells? How can traffic engineers predict the effect on your commuting time of an increase or decrease in gas prices? It's all mathematics.

The physical sciences, social sciences, and the business world are full of situations that can be modeled with linear equations relating two variables. To create a mathematical model of a linear relation between two variables, we must be able to find the equation of the line. In this section, we will look at several ways to write the equation of a line. The specific method we use will be determined by what information we are given.

Find an Equation of the Line Given the Slope and *y*-Intercept

We can easily determine the slope and intercept of a line if the equation is written in slope-intercept form, $y = mx + b$. Now we will do the reverse—we will start with the slope and y-intercept and use them to find the equation of the line.

EXAMPLE 3.24

Find the equation of a line with slope -9 and y-intercept $(0, -4)$.

 **Solution**

Since we are given the slope and y-intercept of the line, we can substitute the needed values into the slope-intercept form, $y = mx + b$.

Name the slope.	$m = -9$
Name the y-intercept.	y-intercept $(0, -4)$
Substitute the values into $y = mx + b$.	$y = mx + b$
	$y = -9x + (-4)$
	$y = -9x - 4$

> **TRY IT :: 3.47**   Find the equation of a line with slope $\frac{2}{5}$ and y-intercept $(0, 4)$.

> TRY IT :: 3.48   Find the equation of a line with slope -1 and y-intercept $(0, -3)$.

Sometimes, the slope and intercept need to be determined from the graph.

EXAMPLE 3.25

Find the equation of the line shown in the graph.

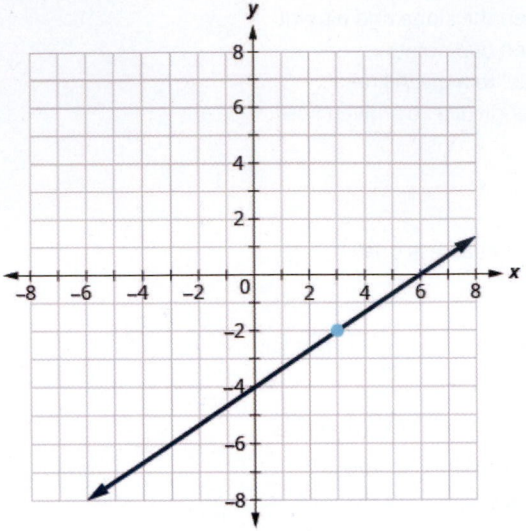

⊘ **Solution**

We need to find the slope and y-intercept of the line from the graph so we can substitute the needed values into the slope-intercept form, $y = mx + b$.

To find the slope, we choose two points on the graph.

The y-intercept is $(0, -4)$ and the graph passes through $(3, -2)$.

Find the slope, by counting the rise and run.	$m = \dfrac{\text{rise}}{\text{run}}$
	$m = \dfrac{2}{3}$
Find the y-intercept.	y-intercept $(0, -4)$
Substitute the values into $y = mx + b$.	$y = mx + b$
	$y = \dfrac{2}{3}x - 4$

> **TRY IT : : 3.49**   Find the equation of the line shown in the graph.

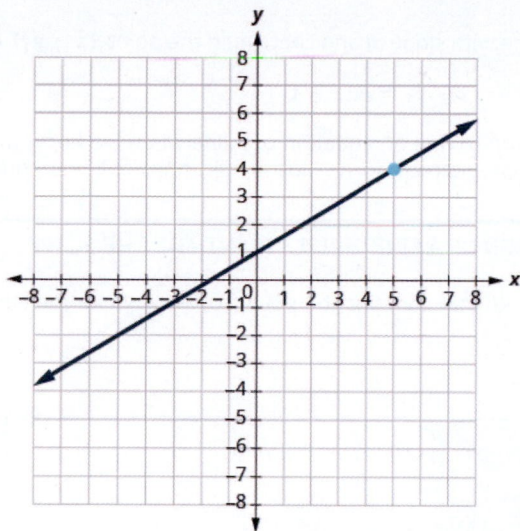

> **TRY IT : : 3.50**   Find the equation of the line shown in the graph.

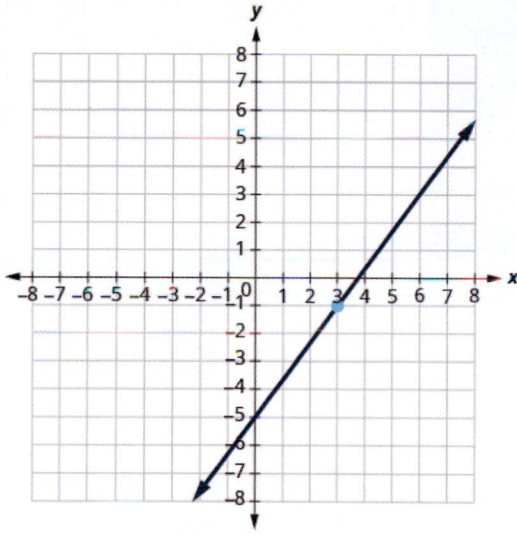

Find an Equation of the Line Given the Slope and a Point

Finding an equation of a line using the slope-intercept form of the equation works well when you are given the slope and y-intercept or when you read them off a graph. But what happens when you have another point instead of the y-intercept?

We are going to use the slope formula to derive another form of an equation of the line.

Suppose we have a line that has slope m and that contains some specific point (x_1, y_1) and some other point, which we will just call (x, y). We can write the slope of this line and then change it to a different form.

$$m = \frac{y - y_1}{x - x_1}$$

Multiply both sides of the equation by $x - x_1$. $$m(x - x_1) = \left(\frac{y - y_1}{x - x_1}\right)(x - x_1)$$

Simplify. $$m(x - x_1) = y - y_1$$

Rewrite the equation with the y terms on the left. $$y - y_1 = m(x - x_1)$$

This format is called the **point-slope form** of an equation of a line.

Point-slope Form of an Equation of a Line

The **point-slope form** of an equation of a line with slope m and containing the point (x_1, y_1) is:

$$y - y_1 = m(x - x_1)$$

We can use the point-slope form of an equation to find an equation of a line when we know the slope and at least one point. Then, we will rewrite the equation in slope-intercept form. Most applications of linear equations use the the slope-intercept form.

EXAMPLE 3.26   HOW TO FIND AN EQUATION OF A LINE GIVEN A POINT AND THE SLOPE

Find an equation of a line with slope $m = -\frac{1}{3}$ that contains the point $(6, -4)$. Write the equation in slope-intercept form.

⊘ **Solution**

Step 1. Identify the slope.	The slope is given.	$m = -\dfrac{1}{3}$
Step 2. Identify the point.	The point is given.	$\begin{pmatrix} x, & y_1 \\ 6, & -4 \end{pmatrix}$
Step 3. Substitute the values into the point–slope form, $y - y_1 = m(x - x_1)$.	Simplify.	$y - y_1 = m(x - x_1)$ $y - (-4) = -\dfrac{1}{3}(x - 6)$ $y + 4 = -\dfrac{1}{3}x + 2$
Step 4. Write the equation in slope–intercept form.		$y = -\dfrac{1}{3}x - 2$

> **TRY IT :: 3.51**   Find the equation of a line with slope $m = -\frac{2}{5}$ and containing the point $(10, -5)$.

> **TRY IT :: 3.52**   Find the equation of a line with slope $m = -\frac{3}{4}$, and containing the point $(4, -7)$.

We list the steps for easy reference.

HOW TO :: TO FIND AN EQUATION OF A LINE GIVEN THE SLOPE AND A POINT.

Step 1.   Identify the slope.

Step 2.   Identify the point.

Step 3.   Substitute the values into the point-slope form, $y - y_1 = m(x - x_1)$.

Step 4.   Write the equation in slope-intercept form.

EXAMPLE 3.27

Find an equation of a horizontal line that contains the point $(-2, -6)$. Write the equation in slope-intercept form.

⊘ **Solution**

Every horizontal line has slope 0. We can substitute the slope and points into the point-slope form, $y - y_1 = m(x - x_1)$.

Identify the slope.	$m = 0$
Identify the point.	$\begin{pmatrix} x_1 & y_1 \\ -2, & -6 \end{pmatrix}$
Substitute the values into $y - y_1 = m(x - x_1)$.	$y - y_1 = m(x - x_1)$
	$y - (-6) = 0(x - (-2))$
Simplify.	$y + 6 = 0$
	$y = -6$
Write in slope-intercept form.	It is in y-form, but could be written $y = 0x - 6$.

Did we end up with the form of a horizontal line, $y = a$?

> **TRY IT ::** 3.53 Find the equation of a horizontal line containing the point $(-3, 8)$.

> **TRY IT ::** 3.54 Find the equation of a horizontal line containing the point $(-1, 4)$.

Find an Equation of the Line Given Two Points

When real-world data is collected, a linear model can be created from two data points. In the next example we'll see how to find an equation of a line when just two points are given.

So far, we have two options for finding an equation of a line: slope-intercept or point-slope. When we start with two points, it makes more sense to use the point-slope form.

But then we need the slope. Can we find the slope with just two points? Yes. Then, once we have the slope, we can use it and one of the given points to find the equation.

EXAMPLE 3.28 HOW TO FIND THE EQUATION OF A LINE GIVEN TWO POINTS

Find an equation of a line that contains the points $(-3, -1)$ and $(2, -2)$ Write the equation in slope-intercept form.

⊘ **Solution**

Step 1. Find the slope using the given points.	Find the slope of the line through $(-3, -1)$ and $(2, -2)$.	$m = \dfrac{y_2 - y_1}{x_2 - x_1}$ $m = \dfrac{-2 - (-1)}{2 - (-3)}$ $m = \dfrac{-1}{5}$ $m = -\dfrac{1}{5}$
Step 2. Choose one point.	Choose either point.	$\begin{pmatrix} x_1 & y_1 \\ 2, & -2 \end{pmatrix}$
Step 3. Substitute the values into the point–slope form, $y - y_1 = m(x - x_1)$.	Simplify.	$y - y_1 = m(x - x_1)$ $y - (-2) = -\dfrac{1}{5}(x - 2)$ $y + 2 = -\dfrac{1}{5}x + \dfrac{2}{5}$

| **Step 4.** Write the equation in slope–intercept form. | | $y = -\dfrac{1}{5}x - \dfrac{8}{5}$ |

> **TRY IT ::** 3.55   Find the equation of a line containing the points $(-2, -4)$ and $(1, -3)$.

> **TRY IT ::** 3.56   Find the equation of a line containing the points $(-4, -3)$ and $(1, -5)$.

The steps are summarized here.

 HOW TO :: TO FIND AN EQUATION OF A LINE GIVEN TWO POINTS.

Step 1.   Find the slope using the given points. $m = \dfrac{y_2 - y_1}{x_2 - x_1}$

Step 2.   Choose one point.

Step 3.   Substitute the values into the point-slope form: $y - y_1 = m(x - x_1)$.

Step 4.   Write the equation in slope-intercept form.

EXAMPLE 3.29

Find an equation of a line that contains the points $(-3, 5)$ and $(-3, 4)$. Write the equation in slope-intercept form.

⊘ **Solution**

Again, the first step will be to find the slope.

Find the slope of the line through $(-3, 5)$ and $(-3, 4)$.

$$m = \frac{y_2 - y_1}{x_2 - x_1}$$
$$m = \frac{4 - 5}{-3 - (-3)}$$
$$m = \frac{-1}{0}$$

The slope is undefined.

This tells us it is a vertical line. Both of our points have an x-coordinate of -2. So our equation of the line is $x = -2$. Since there is no y, we cannot write it in slope-intercept form.

You may want to sketch a graph using the two given points. Does your graph agree with our conclusion that this is a vertical line?

> **TRY IT ::** 3.57   Find the equation of a line containing the points $(5, 1)$ and $(5, -4)$.

> **TRY IT ::** 3.58   Find the equaion of a line containing the points $(-4, 4)$ and $(-4, 3)$.

We have seen that we can use either the slope-intercept form or the point-slope form to find an equation of a line. Which form we use will depend on the information we are given.

To Write an Equation of a Line		
If given:	**Use:**	**Form:**
Slope and y-intercept	slope-intercept	$y = mx + b$
Slope and a point	point-slope	$y - y_1 = m(x - x_1)$
Two points	point-slope	$y - y_1 = m(x - x_1)$

Find an Equation of a Line Parallel to a Given Line

Suppose we need to find an equation of a line that passes through a specific point and is parallel to a given line. We can use the fact that parallel lines have the same slope. So we will have a point and the slope—just what we need to use the point-slope equation.

First, let's look at this graphically.

This graph shows $y = 2x - 3$. We want to graph a line parallel to this line and passing through the point $(-2, 1)$.

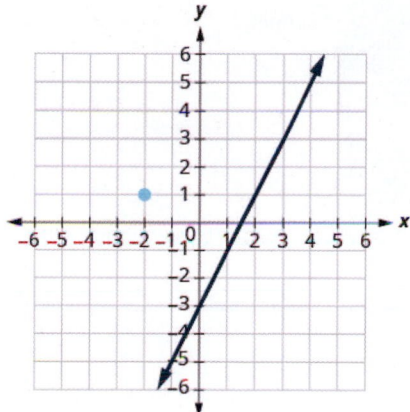

We know that parallel lines have the same slope. So the second line will have the same slope as $y = 2x - 3$. That slope is $m_\parallel = 2$. We'll use the notation $m_\parallel$ to represent the slope of a line parallel to a line with slope m. (Notice that the subscript $\parallel$ looks like two parallel lines.)

The second line will pass through $(-2, 1)$ and have $m = 2$.

To graph the line, we start at $(-2, 1)$ and count out the rise and run.

With $m = 2$ (or $m = \frac{2}{1}$), we count out the rise 2 and the run 1. We draw the line, as shown in the graph.

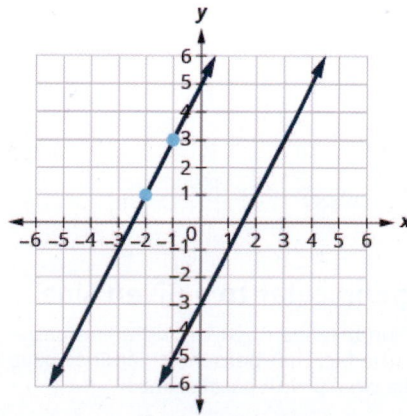

Do the lines appear parallel? Does the second line pass through $(-2, 1)$?

We were asked to graph the line, now let's see how to do this algebraically.

We can use either the slope-intercept form or the point-slope form to find an equation of a line. Here we know one point and can find the slope. So we will use the point-slope form.

EXAMPLE 3.30   HOW TO FIND THE EQUATION OF A LINE PARALLEL TO A GIVEN LINE AND A POINT

Find an equation of a line parallel to $y = 2x - 3$ that contains the point $(-2, 1)$. Write the equation in slope-intercept form.

✓ **Solution**

Step 1. Find the slope of the given line.	The line is in slope–intercept form, $y = 2x - 3$.	$m = 2$
Step 2. Find the slope of the parallel line.	Parallel lines have the same slope.	$m_{\shortparallel} = 2$
Step 3. Identify the point.	The given point is $(-2, 1)$.	$\begin{pmatrix} x_{\shortparallel}, & y_{\shortparallel} \\ -2, & 1 \end{pmatrix}$
Step 4. Substitute the values into the point–slope form, $y - y_{\shortparallel} = m(x - x_{\shortparallel})$.	Simplify.	$y - y_{\shortparallel} = m(x - x_{\shortparallel})$ $y - 1 = 2(x - (-2))$ $y - 1 = 2(x + 2)$ $y - 1 = 2x + 4$
Step 5. Write the equation in slope–intercept form.		$y = 2x + 5$

Look at graph with the parallel lines shown previously. Does this equation make sense? What is the y-intercept of the line? What is the slope?

> **TRY IT : :** 3.59

    Find an equation of a line parallel to the line $y = 3x + 1$ that contains the point $(4, 2)$. Write the equation in slope-intercept form.

> **TRY IT : :** 3.60   Find an equation of a line parallel to the line $y = \frac{1}{2}x - 3$ that contains the point $(6, 4)$.

    Write the equation in slope-intercept form.

    HOW TO : : FIND AN EQUATION OF A LINE PARALLEL TO A GIVEN LINE.

    Step 1.   Find the slope of the given line.

    Step 2.   Find the slope of the parallel line.

    Step 3.   Identify the point.

    Step 4.   Substitute the values into the point-slope form: $y - y_1 = m(x - x_1)$.

    Step 5.   Write the equation in slope-intercept form.

Find an Equation of a Line Perpendicular to a Given Line

Now, let's consider perpendicular lines. Suppose we need to find a line passing through a specific point and which is perpendicular to a given line. We can use the fact that perpendicular lines have slopes that are negative reciprocals. We will again use the point-slope equation, like we did with parallel lines.

This graph shows $y = 2x - 3$. Now, we want to graph a line perpendicular to this line and passing through $(-2, 1)$.

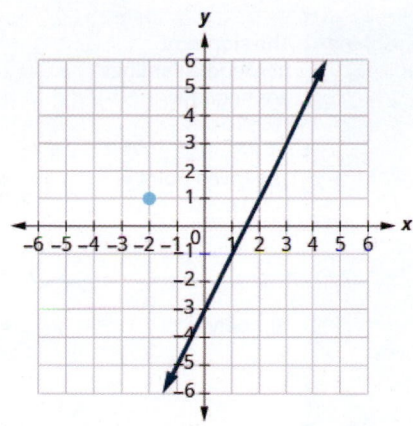

We know that perpendicular lines have slopes that are negative reciprocals.

We'll use the notation $m_{\perp}$ to represent the slope of a line perpendicular to a line with slope m. (Notice that the subscript $\perp$ looks like the right angles made by two perpendicular lines.)

$$y = 2x - 3 \qquad \text{perpendicular line}$$

$$m = 2 \qquad m_{\perp} = -\frac{1}{2}$$

We now know the perpendicular line will pass through $(-2, 1)$ with $m_{\perp} = -\frac{1}{2}$.

To graph the line, we will start at $(-2, 1)$ and count out the rise -1 and the run 2. Then we draw the line.

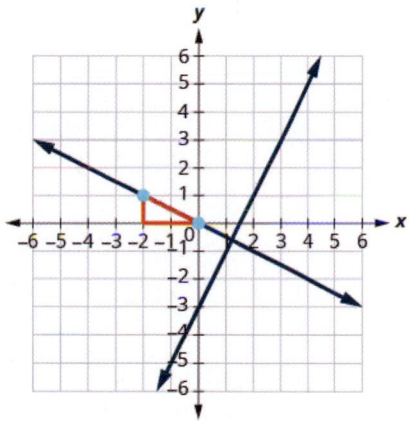

Do the lines appear perpendicular? Does the second line pass through $(-2, 1)$?

We were asked to graph the line, now, let's see how to do this algebraically.

We can use either the slope-intercept form or the point-slope form to find an equation of a line. In this example we know one point, and can find the slope, so we will use the point-slope form.

EXAMPLE 3.31   HOW TO FIND THE EQUATION OF A LINE PERPENDICULAR TO A GIVEN LINE AND A POINT

Find an equation of a line perpendicular to $y = 2x - 3$ that contains the point $(-2, 1)$. Write the equation in slope-intercept form.

✓ **Solution**

| Step 1. Find the slope of the given line. | The line is in slope–intercept form, $y = 2x - 3$. | $m = 2$ |

Step 2. Find the slope of the perpendicular line.	The slopes of perpendicular lines are negative reciprocals.	$m_\perp = -\dfrac{1}{2}$
Step 3. Identify the point.	The given point is (–2, 1).	$\left(\overset{x,\ \ y,}{-2,\ 1}\right)$
Step 4. Substitute the values into the point–slope form, $y - y_1 = m(x - x_1)$.	Simplify.	$y - y_1 = m(x - x_1)$ $y - 1 = -\dfrac{1}{2}(x - (-2))$ $y - 1 = -\dfrac{1}{2}(x + 2)$ $y - 1 = -\dfrac{1}{2}x - 1$
Step 5. Write the equation in slope–intercept form.		$y = -\dfrac{1}{2}x$

> **TRY IT ::** 3.61
>
> Find an equation of a line perpendicular to the line $y = 3x + 1$ that contains the point (4, 2). Write the equation in slope-intercept form.

> **TRY IT ::** 3.62
>
> Find an equation of a line perpendicular to the line $y = \dfrac{1}{2}x - 3$ that contains the point (6, 4). Write the equation in slope-intercept form.

HOW TO :: FIND AN EQUATION OF A LINE PERPENDICULAR TO A GIVEN LINE.

Step 1. Find the slope of the given line.

Step 2. Find the slope of the perpendicular line.

Step 3. Identify the point.

Step 4. Substitute the values into the point-slope form, $y - y_1 = m(x - x_1)$.

Step 5. Write the equation in slope-intercept form.

EXAMPLE 3.32

Find an equation of a line perpendicular to $x = 5$ that contains the point $(3, -2)$. Write the equation in slope-intercept form.

⊘ **Solution**

Again, since we know one point, the point-slope option seems more promising than the slope-intercept option. We need the slope to use this form, and we know the new line will be perpendicular to $x = 5$. This line is vertical, so its perpendicular will be horizontal. This tells us the $m_\perp = 0$.

Identify the point. $(3, -2)$

Identify the slope of the perpendicular line.

$$m_\perp = 0$$

Substitute the values into $y - y_1 = m(x - x_1)$.

$$y - y_1 = m(x - x_1)$$
$$y - (-2) = 0(x - 3)$$

Simplify.

$$y + 2 = 0$$
$$y = -2$$

Sketch the graph of both lines. On your graph, do the lines appear to be perpendicular?

> **TRY IT : : 3.63**

Find an equation of a line that is perpendicular to the line $x = 4$ that contains the point $(4, -5)$. . Write the equation in slope-intercept form.

> **TRY IT : : 3.64**

Find an equation of a line that is perpendicular to the line $x = 2$ that contains the point $(2, -1)$. Write the equation in slope-intercept form.

In **Example 3.32**, we used the point-slope form to find the equation. We could have looked at this in a different way. We want to find a line that is perpendicular to $x = 5$ that contains the point $(3, -2)$. This graph shows us the line $x = 5$ and the point $(3, -2)$.

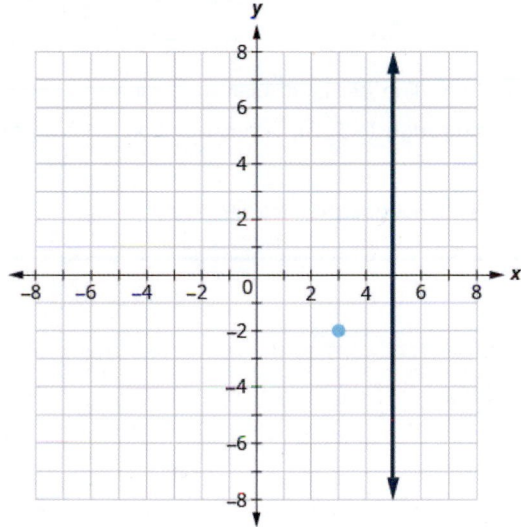

We know every line perpendicular to a vertical line is horizontal, so we will sketch the horizontal line through $(3, -2)$.

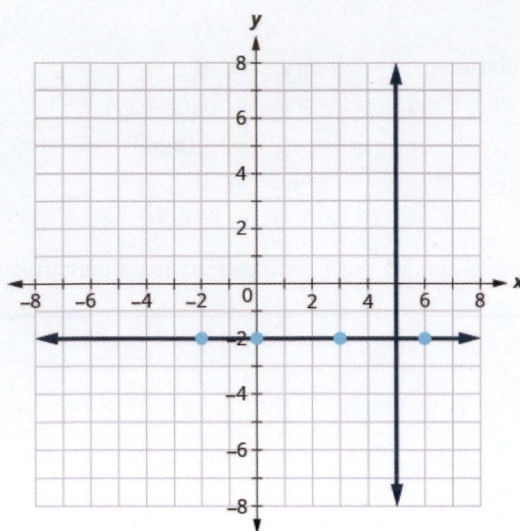

Do the lines appear perpendicular?

If we look at a few points on this horizontal line, we notice they all have y-coordinates of -2. So, the equation of the line perpendicular to the vertical line $x = 5$ is $y = -2$.

EXAMPLE 3.33

Find an equation of a line that is perpendicular to $y = -3$ that contains the point $(-3, 5)$. Write the equation in slope-intercept form.

⊘ Solution

The line $y = -3$ is a horizontal line. Any line perpendicular to it must be vertical, in the form $x = a$. Since the perpendicular line is vertical and passes through $(-3, 5)$, every point on it has an x-coordinate of -3. The equation of the perpendicular line is $x = -3$

You may want to sketch the lines. Do they appear perpendicular?

> **TRY IT : : 3.65**
>
> Find an equation of a line that is perpendicular to the line $y = 1$ that contains the point $(-5, 1)$. Write the equation in slope-intercept form.

> **TRY IT : : 3.66**
>
> Find an equation of a line that is perpendicular to the line $y = -5$ that contains the point $(-4, -5)$. Write the equation in slope-intercept form.

> ▶ **MEDIA : :**
>
> Access these online resources for additional instruction and practice with finding the equation of a line.
>
> - **Write an Equation of Line Given its slope and Y-Intercept (https://openstax.org/l/37slopeycept)**
> - **Using Point Slope Form to Write the Equation of a Line, Find the equation given slope and point (https://openstax.org/l/37slopepoint)**
> - **Find the equation given two points (https://openstax.org/l/37twoptspline)**
> - **Find the equation of perpendicular and parallel lines (https://openstax.org/l/37perpenpara)**

3.3 EXERCISES

Practice Makes Perfect

Find an Equation of the Line Given the Slope and *y*-Intercept

In the following exercises, find the equation of a line with given slope and y-intercept. Write the equation in slope-intercept form.

155. slope 3 and
y-intercept $(0, 5)$

156. slope 8 and
y-intercept $(0, -6)$

157. slope -3 and
y-intercept $(0, -1)$

158. slope -1 and
y-intercept $(0, 3)$

159. slope $\frac{1}{5}$ and
y-intercept $(0, -5)$

160. slope $-\frac{3}{4}$ and
y-intercept $(0, -2)$

161. slope 0 and
y-intercept $(0, -1)$

162. slope -4 and
y-intercept $(0, 0)$

In the following exercises, find the equation of the line shown in each graph. Write the equation in slope-intercept form.

163.

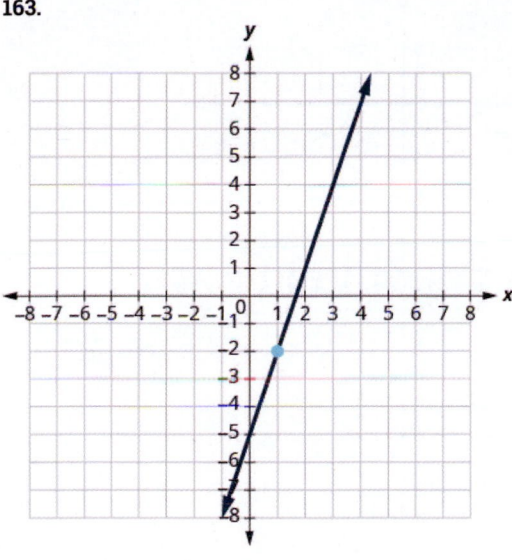

164.

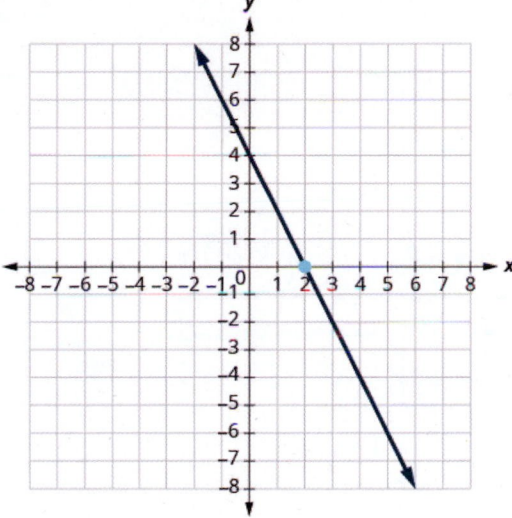

165.

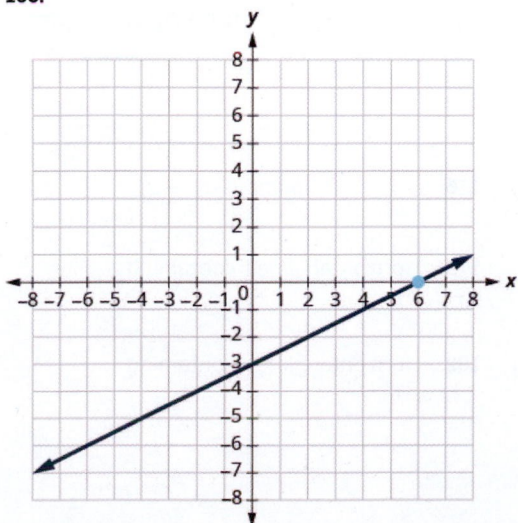

166.

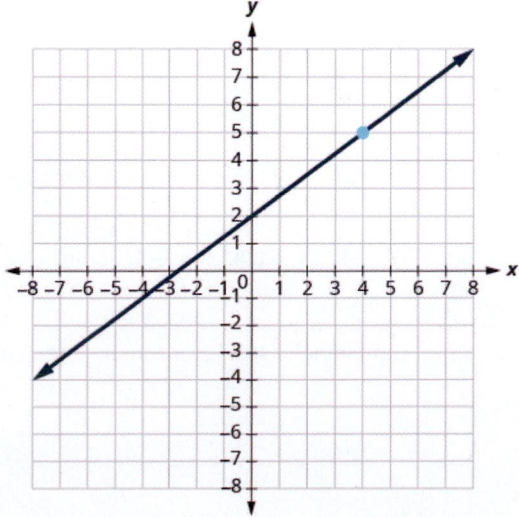

167.

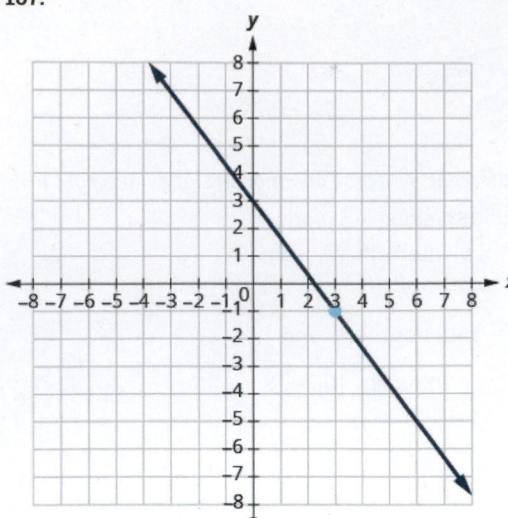

168.

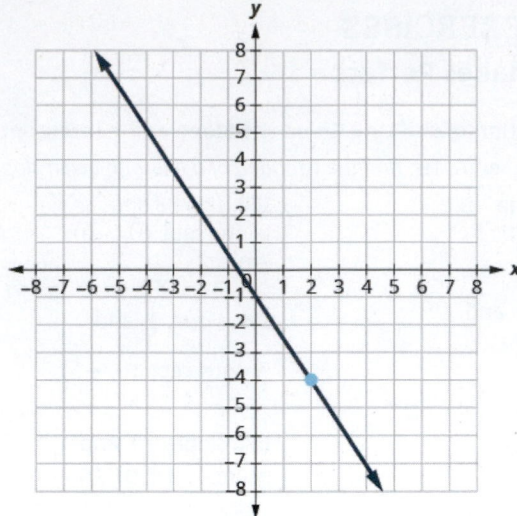

169.

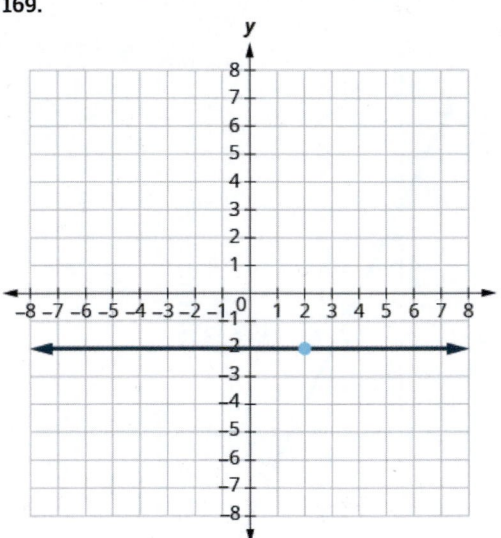

170.

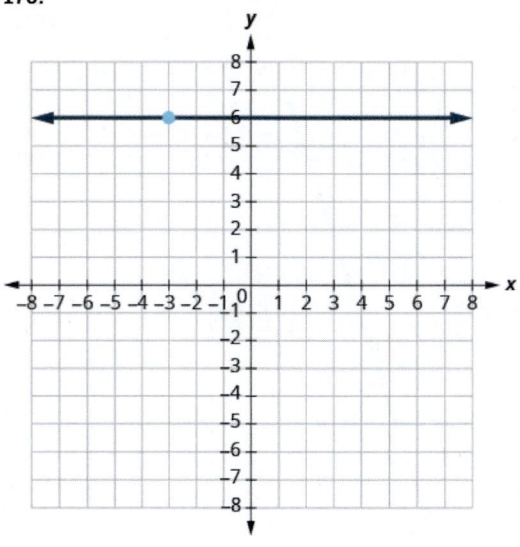

Find an Equation of the Line Given the Slope and a Point

In the following exercises, find the equation of a line with given slope and containing the given point. Write the equation in slope-intercept form.

171. $m = \frac{5}{8}$, point $(8, 3)$

172. $m = \frac{5}{6}$, point $(6, 7)$

173. $m = -\frac{3}{5}$, point $(10, -5)$

174. $m = -\frac{3}{4}$, point $(8, -5)$

175. $m = -\frac{3}{2}$, point $(-4, -3)$

176. $m = -\frac{5}{2}$, point $(-8, -2)$

177. $m = -7$, point $(-1, -3)$

178. $m = -4$, point $(-2, -3)$

179. Horizontal line containing $(-2, 5)$

180. Horizontal line containing $(-2, -3)$

181. Horizontal line containing $(-1, -7)$

182. Horizontal line containing $(4, -8)$

Find an Equation of the Line Given Two Points

In the following exercises, find the equation of a line containing the given points. Write the equation in slope-intercept form.

183. $(2, 6)$ and $(5, 3)$

184. $(4, 3)$ and $(8, 1)$

185. $(-3, -4)$ and $(5 - 2)$.

186. $(-5, -3)$ and $(4, -6)$.

187. $(-1, 3)$ and $(-6, -7)$.

188. $(-2, 8)$ and $(-4, -6)$.

189. $(0, 4)$ and $(2, -3)$.

190. $(0, -2)$ and $(-5, -3)$.

191. $(7, 2)$ and $(7, -2)$.

192. $(-2, 1)$ and $(-2, -4)$.

193. $(3, -4)$ and $(5, -4)$.

194. $(-6, -3)$ and $(-1, -3)$

Find an Equation of a Line Parallel to a Given Line

In the following exercises, find an equation of a line parallel to the given line and contains the given point. Write the equation in slope-intercept form.

195. line $y = 4x + 2$,
point $(1, 2)$

196. line $y = -3x - 1$,
point $(2, -3)$.

197. line $2x - y = 6$,
point $(3, 0)$.

198. line $2x + 3y = 6$,
point $(0, 5)$.

199. line $x = -4$,
point $(-3, -5)$.

200. line $x - 2 = 0$,
point $(1, -2)$.

201. line $y = 5$,
point $(2, -2)$

202. line $y + 2 = 0$,
point $(3, -3)$

Find an Equation of a Line Perpendicular to a Given Line

In the following exercises, find an equation of a line perpendicular to the given line and contains the given point. Write the equation in slope-intercept form.

203. line $y = -2x + 3$,
point $(2, 2)$

204. line $y = -x + 5$,
point $(3, 3)$

205. line $y = \frac{3}{4}x - 2$,
point $(-3, 4)$

206. line $y = \frac{2}{3}x - 4$,
point $(2, -4)$

207. line $2x - 3y = 8$,
point $(4, -1)$

208. line $4x - 3y = 5$,
point $(-3, 2)$

209. line $2x + 5y = 6$,
point $(0, 0)$

210. line $4x + 5y = -3$,
point $(0, 0)$

211. line $x = 3$,
point $(3, 4)$

212. line $x = -5$,
point $(1, -2)$

213. line $x = 7$,
point $(-3, -4)$

214. line $x = -1$,
point $(-4, 0)$

215. line $y - 3 = 0$,
point $(-2, -4)$

216. line $y - 6 = 0$,
point $(-5, -3)$

217. line *y*-axis,
point $(3, 4)$

218. line *y*-axis,
point $(2, 1)$

Mixed Practice

In the following exercises, find the equation of each line. Write the equation in slope-intercept form.

219. Containing the points $(4, 3)$ and $(8, 1)$

220. Containing the points $(-2, 0)$ and $(-3, -2)$

221. $m = \frac{1}{6}$, containing point $(6, 1)$

222. $m = \frac{5}{6}$, containing point $(6, 7)$

223. Parallel to the line $4x + 3y = 6$, containing point $(0, -3)$

224. Parallel to the line $2x + 3y = 6$, containing point $(0, 5)$

225. $m = -\frac{3}{4}$, containing point $(8, -5)$

226. $m = -\frac{3}{5}$, containing point $(10, -5)$

227. Perpendicular to the line $y - 1 = 0$, point $(-2, 6)$

228. Perpendicular to the line y-axis, point $(-6, 2)$

229. Parallel to the line $x = -3$, containing point $(-2, -1)$

230. Parallel to the line $x = -4$, containing point $(-3, -5)$

231. Containing the points $(-3, -4)$ and $(2, -5)$

232. Containing the points $(-5, -3)$ and $(4, -6)$

233. Perpendicular to the line $x - 2y = 5$, point $(-2, 2)$

234. Perpendicular to the line $4x + 3y = 1$, point $(0, 0)$

Writing Exercises

235. Why are all horizontal lines parallel?

236. Explain in your own words why the slopes of two perpendicular lines must have opposite signs.

Self Check

ⓐ *After completing the exercises, use this checklist to evaluate your mastery of the objectives of this section.*

I can...	Confidently	With some help	No-I don't get it!
find the equation of the line given the slope and y-intercept.			
find an equation of the line given the slope and a point.			
find an equation of the line given two points.			
find an equation of a line parallel to a given line.			
find an equation of a line perpendicular to a given line.			

ⓑ *What does this checklist tell you about your mastery of this section? What steps will you take to improve?*

3.4 | Graph Linear Inequalities in Two Variables

Learning Objectives

By the end of this section, you will be able to:

› Verify solutions to an inequality in two variables.
› Recognize the relation between the solutions of an inequality and its graph.
› Graph linear inequalities in two variables
› Solve applications using linear inequalities in two variables

Be Prepared!

Before you get started, take this readiness quiz.

1. Graph $x > 2$ on a number line.
    If you missed this problem, review **Example 2.48**.

2. Solve: $4x + 3 > 23$.
    If you missed this problem, review **Example 2.52**.

3. Translate: $8 < x > 3$.
    If you missed this problem, review **Example 2.56**.

Verify Solutions to an Inequality in Two Variables

Previously we learned to solve inequalities with only one variable. We will now learn about inequalities containing two variables. In particular we will look at **linear inequalities** in two variables which are very similar to linear equations in two variables.

Linear inequalities in two variables have many applications. If you ran a business, for example, you would want your revenue to be greater than your costs—so that your business made a profit.

Linear Inequality

A **linear inequality** is an inequality that can be written in one of the following forms:

$$Ax + By > C \quad Ax + By \geq C \quad Ax + By < C \quad Ax + By \leq C$$

Where A and B are not both zero.

Recall that an inequality with one variable had many solutions. For example, the solution to the inequality $x > 3$ is any number greater than 3. We showed this on the number line by shading in the number line to the right of 3, and putting an open parenthesis at 3. See **Figure 3.10**.

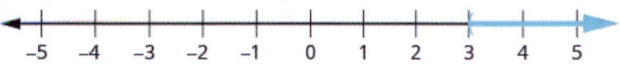

Figure 3.10

Similarly, linear inequalities in two variables have many solutions. Any ordered pair (x, y) that makes an inequality true when we substitute in the values is a **solution to a linear inequality**.

Solution to a Linear Inequality

An ordered pair (x, y) is a **solution to a linear inequality** if the inequality is true when we substitute the values of x and y.

EXAMPLE 3.34

Determine whether each ordered pair is a solution to the inequality $y > x + 4$:

ⓐ $(0, 0)$ ⓑ $(1, 6)$ ⓒ $(2, 6)$ ⓓ $(-5, -15)$ ⓔ $(-8, 12)$

⊘ **Solution**

ⓐ

(0, 0)	$y > x + 4$
Substitute 0 for *x* and 0 for *y*.	$0 \overset{?}{>} 0 + 4$
Simplify.	$0 \not> 4$
	So, $(0, 0)$ is not a solution to $y > x + 4$.

ⓑ

(1, 6)	$y > x + 4$
Substitute 1 for *x* and 6 for *y*.	$6 \overset{?}{>} 1 + 4$
Simplify.	$6 > 5$
	So, $(1, 6)$ is a solution to $y > x + 4$.

ⓒ

(2, 6)	$y > x + 4$
Substitute 2 for *x* and 6 for *y*.	$6 \overset{?}{>} 2 + 4$
Simplify.	$6 \not> 6$
	So, $(2, 6)$ is not a solution to $y > x + 4$.

ⓓ

(−5, −15)	$y > x + 4$
Substitute –5 for *x* and –15 for *y*.	$-15 \overset{?}{>} -5 + 4$
Simplify.	$-15 \not> -1$
	So, $(-5, -15)$ is not a solution to $y > x + 4$.

ⓔ

$(-8, 12)$	$y > x + 4$
Substitute –8 for x and 12 for y.	$12 \overset{?}{>} -8 + 4$
Simplify.	$12 > -4$
	So, $(-8, 12)$ is a solution to $y > x + 4$.

> **TRY IT :: 3.67**   Determine whether each ordered pair is a solution to the inequality $y > x - 3$:
>
> ⓐ $(0, 0)$ ⓑ $(4, 9)$ ⓒ $(-2, 1)$ ⓓ $(-5, -3)$ ⓔ $(5, 1)$

> **TRY IT :: 3.68**   Determine whether each ordered pair is a solution to the inequality $y < x + 1$:
>
> ⓐ $(0, 0)$ ⓑ $(8, 6)$ ⓒ $(-2, -1)$ ⓓ $(3, 4)$ ⓔ $(-1, -4)$

Recognize the Relation Between the Solutions of an Inequality and its Graph

Now, we will look at how the solutions of an inequality relate to its graph.

Let's think about the number line in shown previously again. The point $x = 3$ separated that number line into two parts. On one side of 3 are all the numbers less than 3. On the other side of 3 all the numbers are greater than 3. See **Figure 3.11**.

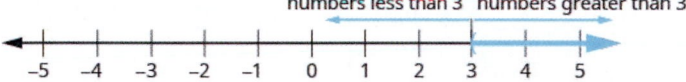

Figure 3.11 The solution to $x > 3$ is the shaded part of the number line to the right of $x = 3$.

Similarly, the line $y = x + 4$ separates the plane into two regions. On one side of the line are points with $y < x + 4$. On the other side of the line are the points with $y > x + 4$. We call the line $y = x + 4$ a **boundary line**.

Boundary Line

The line with equation $Ax + By = C$ is the **boundary line** that separates the region where $Ax + By > C$ from the region where $Ax + By < C$.

For an inequality in one variable, the endpoint is shown with a parenthesis or a bracket depending on whether or not a is included in the solution:

Similarly, for an inequality in two variables, the boundary line is shown with a solid or dashed line to show whether or not it the line is included in the solution.

$Ax + By < C$	$Ax + By \leq C$
$Ax + By > C$	$Ax + By \geq C$
Boundary line is $Ax + By = C$	Boundary line is $Ax + By = C$
Boundary line is not included in solution.	Boundary line is included in solution.
Boundary line is dashed.	**Boundary line is solid.**

Now, let's take a look at what we found in **Example 3.34**. We'll start by graphing the line $y = x + 4$, and then we'll plot the five points we tested, as shown in the graph. See **Figure 3.12**.

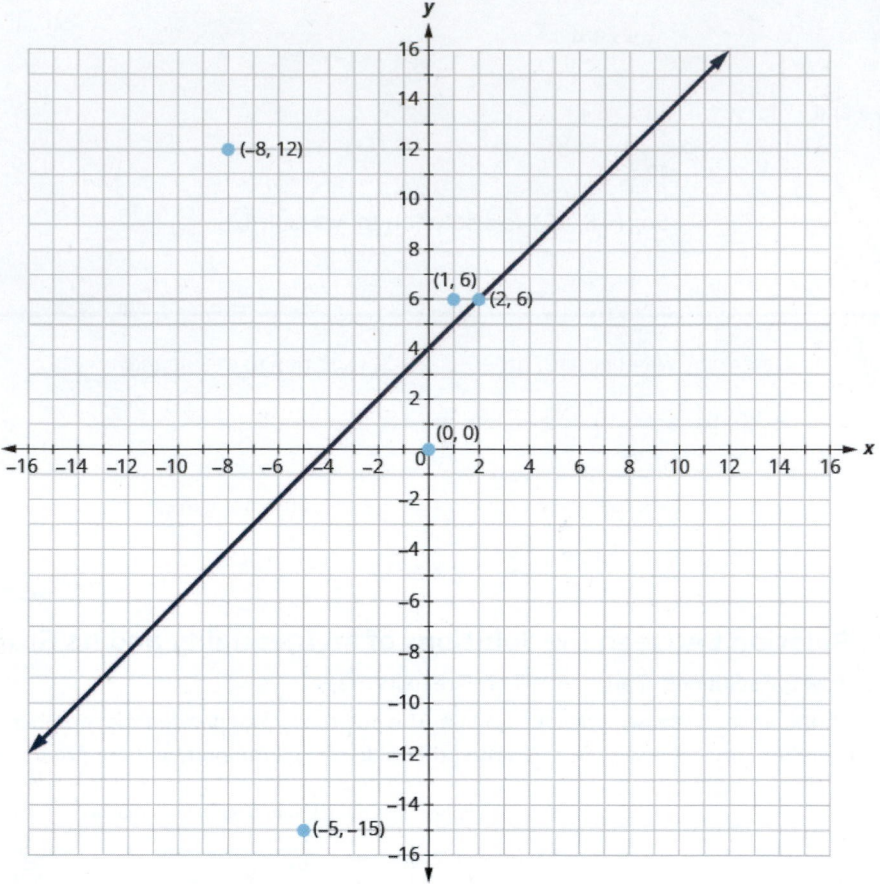

Figure 3.12

In **Example 3.34** we found that some of the points were solutions to the inequality $y > x + 4$ and some were not.

Which of the points we plotted are solutions to the inequality $y > x + 4$?

The points $(1, 6)$ and $(-8, 12)$ are solutions to the inequality $y > x + 4$. Notice that they are both on the same side of the boundary line $y = x + 4$.

The two points $(0, 0)$ and $(-5, -15)$ are on the other side of the boundary line $y = x + 4$, and they are not solutions to the inequality $y > x + 4$. For those two points, $y < x + 4$.

What about the point $(2, 6)$? Because $6 = 2 + 4$, the point is a solution to the equation $y = x + 4$, but not a solution to the inequality $y > x + 4$. So the point $(2, 6)$ is on the boundary line.

Let's take another point above the boundary line and test whether or not it is a solution to the inequality $y > x + 4$. The point $(0, 10)$ clearly looks to above the boundary line, doesn't it? Is it a solution to the inequality?

$$
\begin{aligned}
y &> x + 4 \\
10 &\overset{?}{>} 0 + 4 \\
10 &> 4
\end{aligned}
$$

So, $(0, 10)$ is a solution to $y > x + 4$.

Any point you choose above the boundary line is a solution to the inequality $y > x + 4$. All points above the boundary line are solutions.

Similarly, all points below the boundary line, the side with $(0, 0)$ and $(-5, -15)$, are not solutions to $y > x + 4$, as shown in **Figure 3.13**.

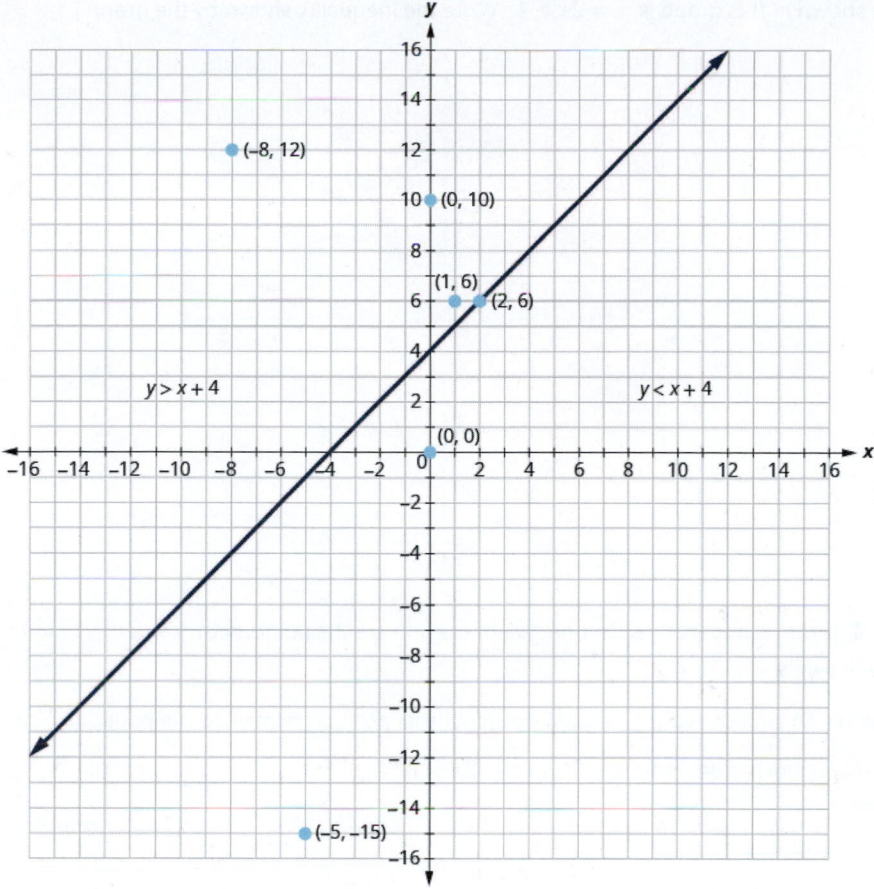

Figure 3.13

The graph of the inequality $y > x + 4$ is shown in below.

The line $y = x + 4$ divides the plane into two regions. The shaded side shows the solutions to the inequality $y > x + 4$.

The points on the boundary line, those where $y = x + 4$, are not solutions to the inequality $y > x + 4$, so the line itself is not part of the solution. We show that by making the line dashed, not solid.

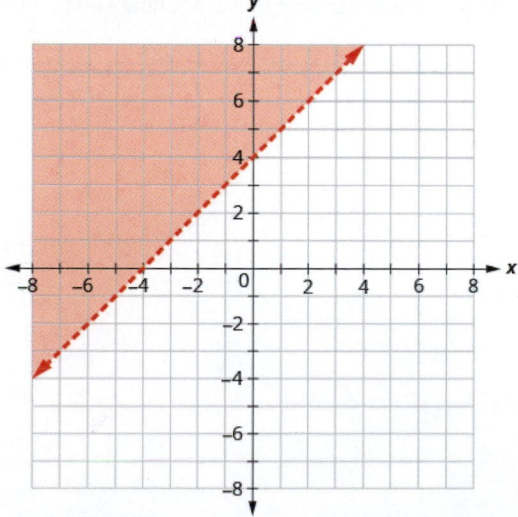

The boundary line shown in this graph is $y = 2x - 1$. Write the inequality shown by the graph.

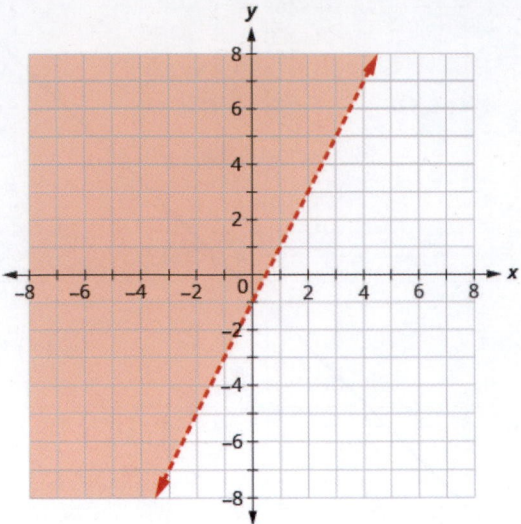

⊘ Solution

The line $y = 2x - 1$ is the boundary line. On one side of the line are the points with $y > 2x - 1$ and on the other side of the line are the points with $y < 2x - 1$.

Let's test the point $(0, 0)$ and see which inequality describes its position relative to the boundary line.

At $(0, 0)$, which inequality is true: $y > 2x - 1$ or $y < 2x - 1$?

$$y > 2x - 1 \qquad\qquad y < 2x - 1$$
$$0 \overset{?}{>} 2 \cdot 0 - 1 \qquad\qquad 0 \overset{?}{<} 2 \cdot 0 - 1$$
$$0 > -1 \text{ True} \qquad\qquad 0 < -1 \text{ False}$$

Since, $y > 2x - 1$ is true, the side of the line with $(0, 0)$, is the solution. The shaded region shows the solution of the inequality $y > 2x - 1$.

Since the boundary line is graphed with a solid line, the inequality includes the equal sign.
The graph shows the inequality $y \geq 2x - 1$.

We could use any point as a test point, provided it is not on the line. Why did we choose $(0, 0)$? Because it's the easiest to evaluate. You may want to pick a point on the other side of the boundary line and check that $y < 2x - 1$.

> **TRY IT : :** 3.69   Write the inequality shown by the graph with the boundary line $y = -2x + 3$.

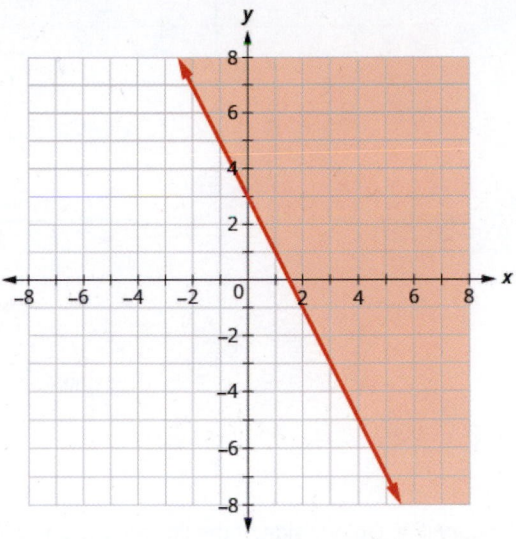

> **TRY IT : :** 3.70   Write the inequality shown by the graph with the boundary line $y = \frac{1}{2}x - 4$.

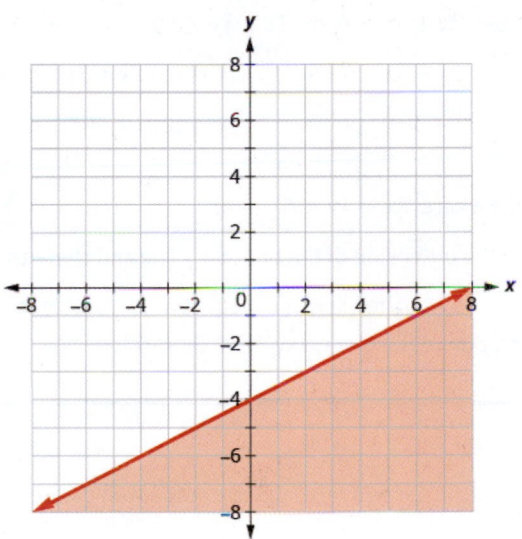

EXAMPLE 3.36

The boundary line shown in this graph is $2x + 3y = 6$. Write the inequality shown by the graph.

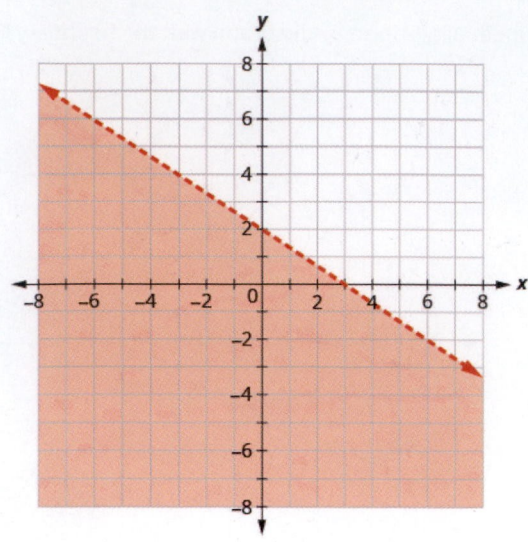

⊘ Solution

The line $2x + 3y = 6$ is the boundary line. On one side of the line are the points with $2x + 3y > 6$ and on the other side of the line are the points with $2x + 3y < 6$.

Let's test the point $(0, 0)$ and see which inequality describes its side of the boundary line.

At $(0, 0)$, which inequality is true: $2x + 3y > 6$ or $2x + 3y < 6$?

$$2x + 3y \ > \ 6 \qquad\qquad 2x + 3y \ < \ 6$$
$$2(0) + 3(0) \ \overset{?}{>} \ 6 \qquad\qquad 2(0) + 3(0) \ \overset{?}{<} \ 6$$
$$0 \ > \ 6 \ \text{False} \qquad\qquad 0 \ < \ 6 \ \text{True}$$

So the side with $(0, 0)$ is the side where $2x + 3y < 6$.

(You may want to pick a point on the other side of the boundary line and check that $2x + 3y > 6$.)

Since the boundary line is graphed as a dashed line, the inequality does not include an equal sign.

The shaded region shows the solution to the inequality $2x + 3y < 6$.

> **TRY IT ::** 3.71

Write the inequality shown by the shaded region in the graph with the boundary line $x - 4y = 8$.

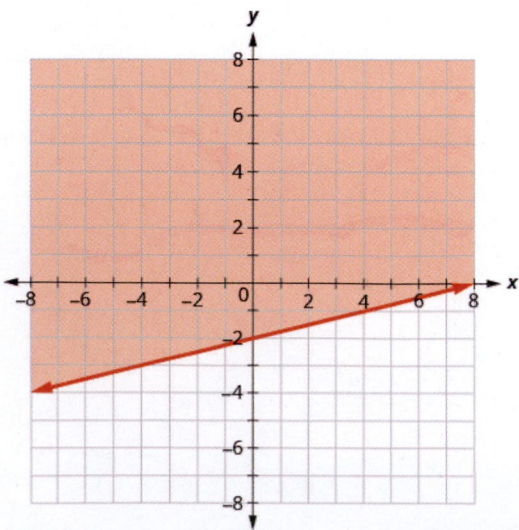

> **TRY IT : : 3.72**

Write the inequality shown by the shaded region in the graph with the boundary line $3x - y = 6$.

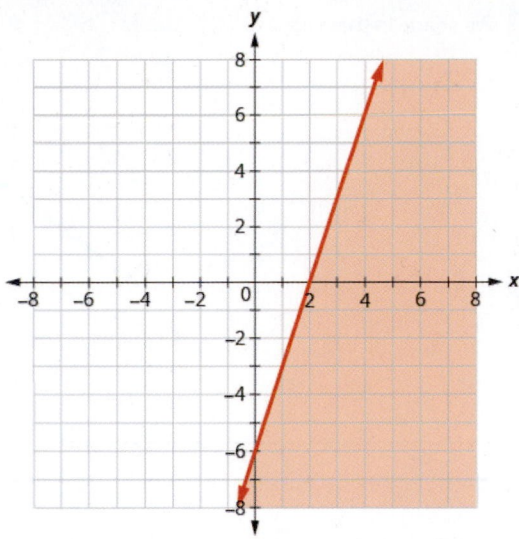

Graph Linear Inequalities in Two Variables

Now that we know what the graph of a linear inequality looks like and how it relates to a boundary equation we can use this knowledge to graph a given linear inequality.

EXAMPLE 3.37   HOW TO GRAPH A LINEAR EQUATION IN TWO VARIABLES

Graph the linear inequality $y \geq \frac{3}{4}x - 2$.

⊘ **Solution**

Step 1. Identify and graph the boundary line. • If the inequality is ≤ or ≥, the boundary line is solid. • If the inequality is < or >, the boundary line is dashed.	Replace the inequality sign with an equal sign to find the boundary line. Graph the boundary line $y = \frac{3}{4}x - 2$. The inequality sign is ≥, so we draw a solid line.	
Step 2. Test a point that is not on the boundary line. Is it a solution of the inequality?	We'll test (0, 0). Is it a solution of the inequality?	At (0, 0), is $y \geq \frac{3}{4}x - 2$? $0 \overset{?}{\geq} \frac{3}{4}(0) - 2$ $0 \geq -2$ So, (0, 0) is a solution.

| Step 3. Shade in one side of the boundary line.
• If the test point is a solution, shade in the side that includes the point.
• If the test point is not a solution, shade in the opposite side. | The test point (0, 0) is a solution to $y \geq \frac{3}{4}x - 2$. So we shade in that side. | 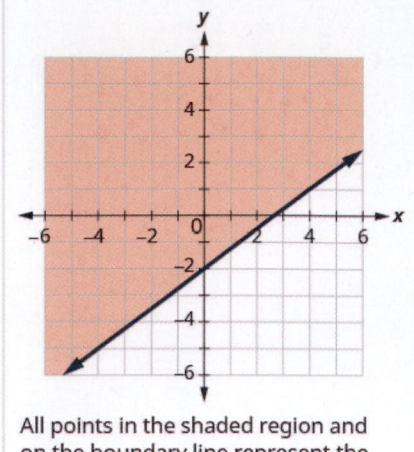 |
| | | All points in the shaded region and on the boundary line represent the solutions to $y \geq \frac{3}{4}x - 2$. |

 TRY IT :: 3.73
Graph the linear inequality $y > \frac{5}{2}x - 4$.

 TRY IT :: 3.74
Graph the linear inequality $y < \frac{2}{3}x - 5$.

The steps we take to graph a linear inequality are summarized here.

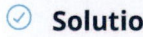 **HOW TO ::** GRAPH A LINEAR INEQUALITY IN TWO VARIABLES.

Step 1.   Identify and graph the boundary line.

   ◦ If the inequality is $\leq$ or $\geq$, the boundary line is solid.

   ◦ If the inequality is $<$ or $>$, the boundary line is dashed.

Step 2.   Test a point that is not on the boundary line. Is it a solution of the inequality?

Step 3.   Shade in one side of the boundary line.

   ◦ If the test point is a solution, shade in the side that includes the point.

   ◦ If the test point is not a solution, shade in the opposite side.

EXAMPLE 3.38

Graph the linear inequality $x - 2y < 5$.

⊘ **Solution**

First, we graph the boundary line $x - 2y = 5$. The inequality is $<$ so we draw a dashed line.

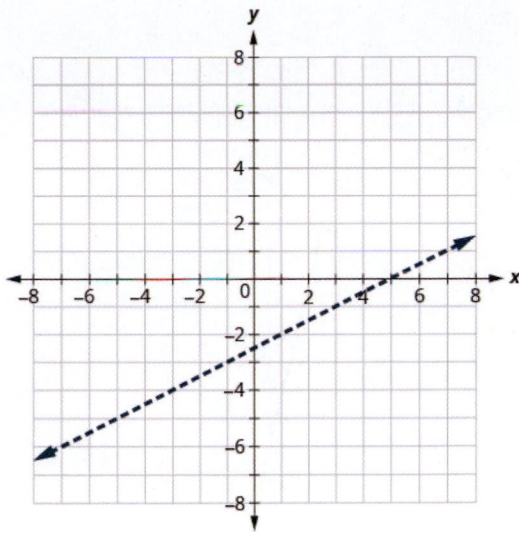

Then, we test a point. We'll use $(0, 0)$ again because it is easy to evaluate and it is not on the boundary line.

Is $(0, 0)$ a solution of $x - 2y < 5$?

$$0 - 2(0) \overset{?}{<} 5$$

$$0 - 0 \overset{?}{<} 5$$

$$0 < 5$$

The point $(0, 0)$ is a solution of $x - 2y < 5$, so we shade in that side of the boundary line.

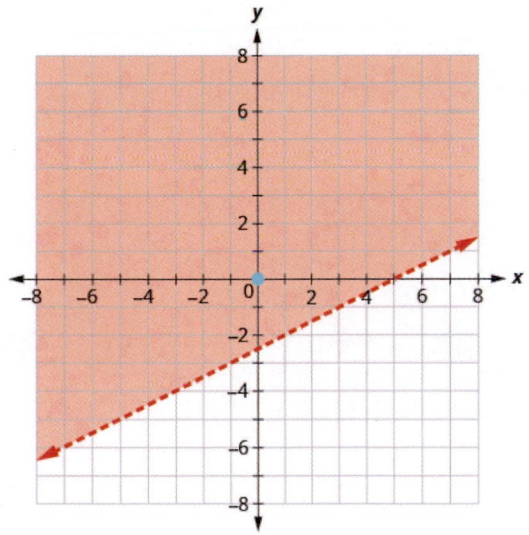

All points in the shaded region, but not those on the boundary line, represent the solutions to $x - 2y < 5$.

> **TRY IT :: 3.75**   Graph the linear inequality: $2x - 3y < 6$.

> **TRY IT :: 3.76**   Graph the linear inequality: $2x - y > 3$.

What if the boundary line goes through the origin? Then, we won't be able to use $(0, 0)$ as a test point. No problem—we'll just choose some other point that is not on the boundary line.

EXAMPLE 3.39

Graph the linear inequality: $y \leq -4x$.

⊘ **Solution**

First, we graph the boundary line $y = -4x$. It is in slope–intercept form, with $m = -4$ and $b = 0$. The inequality is $\leq$ so we draw a solid line.

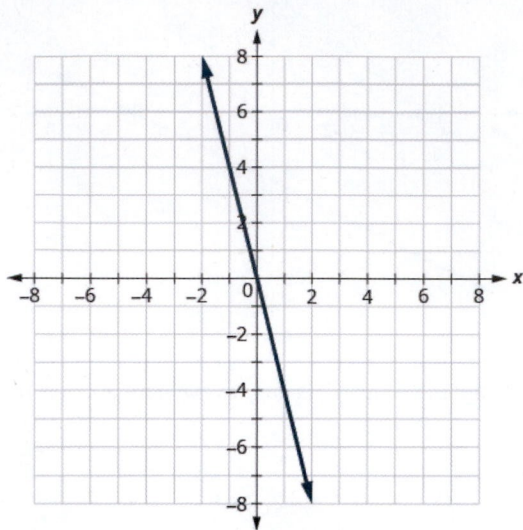

Now we need a test point. We can see that the point $(1, 0)$ is not on the boundary line.

Is $(1, 0)$ a solution of $y \leq -4x$?

$$0 \overset{?}{\leq} -4(1)$$

$$0 \nleq -4$$

The point $(1, 0)$ is not a solution to $y \leq -4x$, so we shade in the opposite side of the boundary line.

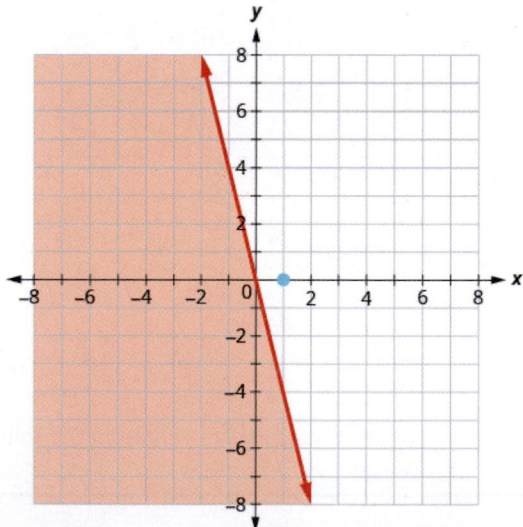

All points in the shaded region and on the boundary line represent the solutions to $y \leq -4x$.

> **TRY IT ::** 3.77   Graph the linear inequality: $y > -3x$.

> **TRY IT ::** 3.78   Graph the linear inequality: $y \geq -2x$.

Some linear inequalities have only one variable. They may have an x but no y, or a y but no x. In these cases, the boundary line will be either a vertical or a horizontal line.

Recall that:

$$x = a \qquad \text{vertical line}$$
$$y = b \qquad \text{horizontal line}$$

EXAMPLE 3.40

Graph the linear inequality: $y > 3$.

⊘ Solution

First, we graph the boundary line $y = 3$. It is a horizontal line. The inequality is $>$ so we draw a dashed line.

We test the point $(0, 0)$.

$$y > 3$$
$$0 \cancel{>} 3$$

So, $(0, 0)$ is not a solution to $y > 3$.

So we shade the side that does not include $(0, 0)$ as shown in this graph.

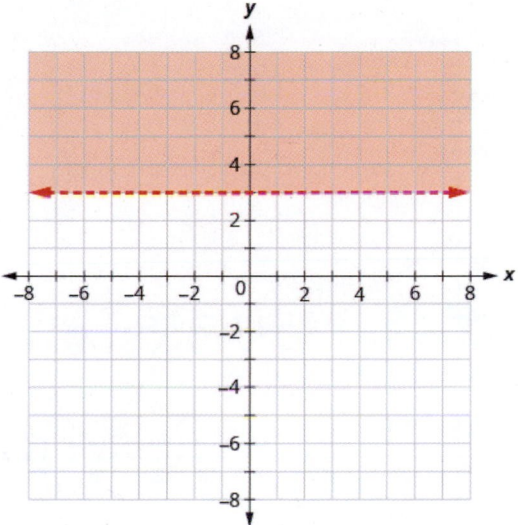

All points in the shaded region, but not those on the boundary line, represent the solutions to $y > 3$.

> | **TRY IT : :** 3.79   Graph the linear inequality: $y < 5$.

> | **TRY IT : :** 3.80   Graph the linear inequality: $y \leq -1$.

Solve Applications using Linear Inequalities in Two Variables

Many fields use linear inequalities to model a problem. While our examples may be about simple situations, they give us an opportunity to build our skills and to get a feel for how thay might be used.

EXAMPLE 3.41

Hilaria works two part time jobs in order to earn enough money to meet her obligations of at least $240 a week. Her job in food service pays $10 an hour and her tutoring job on campus pays $15 an hour. How many hours does Hilaria need to work at each job to earn at least $240?

ⓐ Let x be the number of hours she works at the job in food service and let y be the number of hours she works tutoring. Write an inequality that would model this situation.

ⓑ Graph the inequality.

ⓒ Find three ordered pairs (x, y) that would be solutions to the inequality. Then, explain what that means for Hilaria.

⊘ Solution

ⓐ We let x be the number of hours she works at the job in food service and let y be the number of hours she works tutoring.

She earns $10 per hour at the job in food service and $15 an hour tutoring. At each job, the number of hours multiplied by the hourly wage will gives the amount earned at that job.

Amount earned at the food service job plus the amount earned tutoring is at least $240

$$10x \qquad + \qquad 15y \qquad \geq 240$$

ⓑ To graph the inequality, we put it in slope–intercept form.

$$10x + 15y \geq 240$$
$$15y \geq -10x + 240$$
$$y \geq -\frac{2}{3}x + 16$$

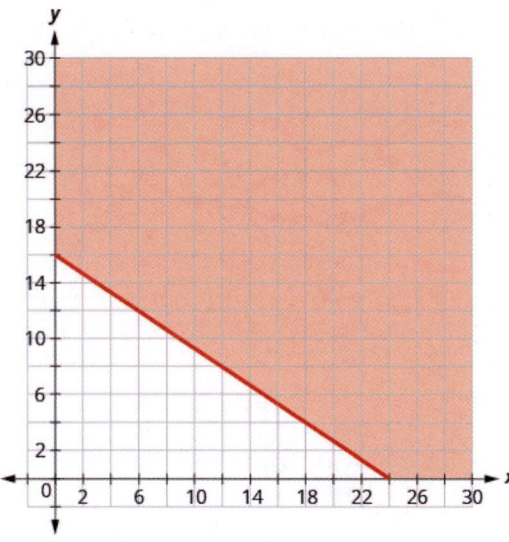

ⓒ From the graph, we see that the ordered pairs $(15, 10)$, $(0, 16)$, $(24, 0)$ represent three of infinitely many solutions. Check the values in the inequality.

$(15, 10)$	$(0, 16)$	$(24, 0)$
$10x + 15y \geq 240$	$10x + 15y \geq 240$	$10x + 15y \geq 240$
$10(15) + 15(10) \overset{?}{\geq} 240$	$10(0) + 15(16) \overset{?}{\geq} 240$	$10(24) + 15(0) \overset{?}{\geq} 240$
$300 \geq 240$ True	$240 \geq 240$ True	$240 \geq 240$ True

For Hilaria, it means that to earn at least $240, she can work 15 hours tutoring and 10 hours at her fast-food job, earn all her money tutoring for 16 hours, or earn all her money while working 24 hours at the job in food service.

> **TRY IT : : 3.81**

Hugh works two part time jobs. One at a grocery store that pays $10 an hour and the other is babysitting for $13 hour. Between the two jobs, Hugh wants to earn at least $260 a week. How many hours does Hugh need to work at each job to earn at least $260?

ⓐ Let x be the number of hours he works at the grocery store and let y be the number of hours he works babysitting. Write an inequality that would model this situation.

ⓑ Graph the inequality.

ⓒ Find three ordered pairs (x, y) that would be solutions to the inequality. Then, explain what that means for Hugh.

> **TRY IT : : 3.82**

Veronica works two part time jobs in order to earn enough money to meet her obligations of at least $280 a week. Her job at the day spa pays $10 an hour and her administrative assistant job on campus pays $17.50 an hour. How many hours does Veronica need to work at each job to earn at least $280?

ⓐ Let x be the number of hours she works at the day spa and let y be the number of hours she works as administrative assistant. Write an inequality that would model this situation.

ⓑ Graph the inequality.

ⓒ Find three ordered pairs (x, y) that would be solutions to the inequality. Then, explain what that means for Veronica

▶ **MEDIA : :**

Access this online resource for additional instruction and practice with graphing linear inequalities in two variables.

- **Graphing Linear Inequalities in Two Variables (https://openstax.org/l/37linineqgraphs)**

3.4 EXERCISES

Practice Makes Perfect

Verify Solutions to an Inequality in Two Variables

In the following exercises, determine whether each ordered pair is a solution to the given inequality.

237. Determine whether each ordered pair is a solution to the inequality $y > x - 1$:

ⓐ $(0, 1)$

ⓑ $(-4, -1)$

ⓒ $(4, 2)$

ⓓ $(3, 0)$

ⓔ $(-2, -3)$

238. Determine whether each ordered pair is a solution to the inequality $y > x - 3$:

ⓐ $(0, 0)$

ⓑ $(2, 1)$

ⓒ $(-1, -5)$

ⓓ $(-6, -3)$

ⓔ $(1, 0)$

239. Determine whether each ordered pair is a solution to the inequality $y < 3x + 2$:

ⓐ $(0, 3)$

ⓑ $(-3, -2)$

ⓒ $(-2, 0)$

ⓓ $(0, 0)$

ⓔ $(-1, 4)$

240. Determine whether each ordered pair is a solution to the inequality $y < -2x + 5$:

ⓐ $(-3, 0)$

ⓑ $(1, 6)$

ⓒ $(-6, -2)$

ⓓ $(0, 1)$

ⓔ $(5, -4)$

241. Determine whether each ordered pair is a solution to the inequality $3x - 4y > 4$:

ⓐ $(5, 1)$

ⓑ $(-2, 6)$

ⓒ $(3, 2)$

ⓓ $(10, -5)$

ⓔ $(0, 0)$

242. Determine whether each ordered pair is a solution to the inequality $2x + 3y > 2$:

ⓐ $(1, 1)$

ⓑ $(4, -3)$

ⓒ $(0, 0)$

ⓓ $(-8, 12)$

ⓔ $(3, 0)$

Recognize the Relation Between the Solutions of an Inequality and its Graph

In the following exercises, write the inequality shown by the shaded region.

243. Write the inequality shown by the graph with the boundary line $y = 3x - 4$.

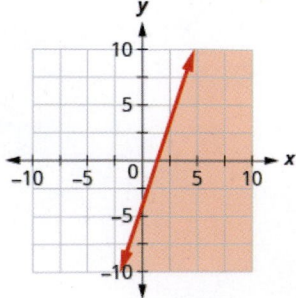

244. Write the inequality shown by the graph with the boundary line $y = 2x - 4$.

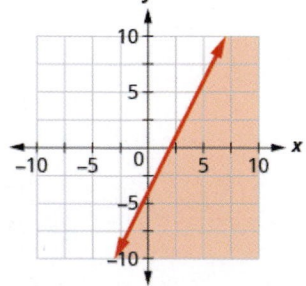

245. Write the inequality shown by the graph with the boundary line $y = -\frac{1}{2}x + 1$.

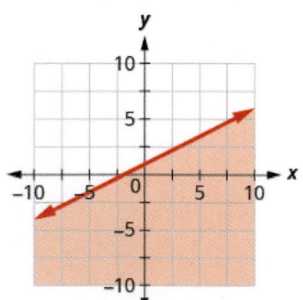

246. Write the inequality shown by the graph with the boundary line $y = -\frac{1}{3}x - 2$.

247. Write the inequality shown by the shaded region in the graph with the boundary line $x + y = 5$.

248. Write the inequality shown by the shaded region in the graph with the boundary line $x + y = 3$.

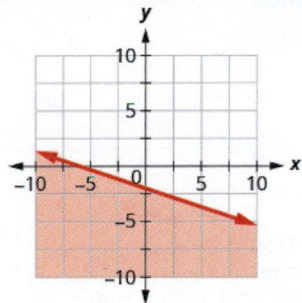

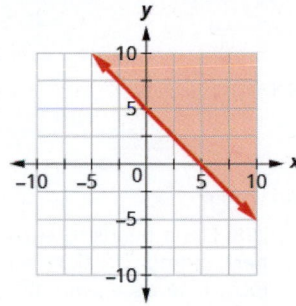

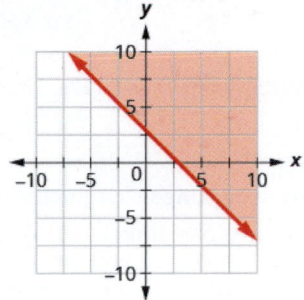

249. Write the inequality shown by the shaded region in the graph with the boundary line $3x - y = 6$.

250. Write the inequality shown by the shaded region in the graph with the boundary line $2x - y = 4$.

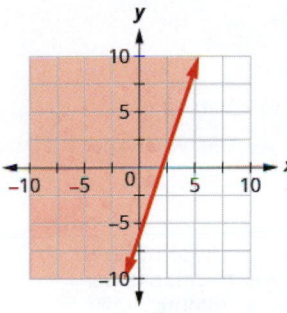

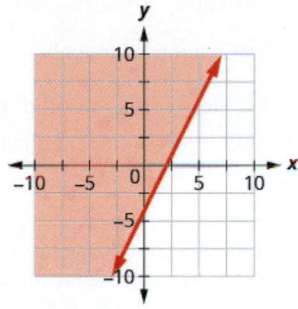

Graph Linear Inequalities in Two Variables

In the following exercises, graph each linear inequality.

251. Graph the linear inequality: $y > \frac{2}{3}x - 1$.

252. Graph the linear inequality: $y < \frac{3}{5}x + 2$.

253. Graph the linear inequality: $y \leq -\frac{1}{2}x + 4$.

254. Graph the linear inequality: $y \geq -\frac{1}{3}x - 2$.

255. Graph the linear inequality: $x - y \leq 3$.

256. Graph the linear inequality: $x - y \geq -2$.

257. Graph the linear inequality: $4x + y > -4$.

258. Graph the linear inequality: $x + 5y < -5$.

259. Graph the linear inequality: $3x + 2y \geq -6$.

260. Graph the linear inequality: $4x + 2y \geq -8$.

261. Graph the linear inequality: $y > 4x$.

262. Graph the linear inequality: $y \leq -3x$.

263. Graph the linear inequality: $y < -10$.

264. Graph the linear inequality: $y \geq 2$.

265. Graph the linear inequality: $x \leq 5$.

266. Graph the linear inequality: $x \geq 0$.

267. Graph the linear inequality: $x - y < 4$.

268. Graph the linear inequality: $x - y < -3$.

269. Graph the linear inequality: $y \geq \frac{3}{2}x$.

270. Graph the linear inequality: $y \leq \frac{5}{4}x$.

271. Graph the linear inequality: $y > -2x + 1$.

272. Graph the linear inequality: $y < -3x - 4$.

273. Graph the linear inequality: $2x + y \geq -4$.

274. Graph the linear inequality: $x + 2y \leq -2$.

275. Graph the linear inequality: $2x - 5y > 10$.

276. Graph the linear inequality: $4x - 3y > 12$.

Solve Applications using Linear Inequalities in Two Variables

277. Harrison works two part time jobs. One at a gas station that pays $11 an hour and the other is IT troubleshooting for $16.50 an hour. Between the two jobs, Harrison wants to earn at least $330 a week. How many hours does Harrison need to work at each job to earn at least $330?

ⓐ Let x be the number of hours he works at the gas station and let y be the number of (hours he works troubleshooting. Write an inequality that would model this situation.

ⓑ Graph the inequality.

ⓒ Find three ordered pairs (x, y) that would be solutions to the inequality. Then, explain what that means for Harrison.

278. Elena needs to earn at least $450 a week during her summer break to pay for college. She works two jobs. One as a swimming instructor that pays $9 an hour and the other as an intern in a genetics lab for $22.50 per hour. How many hours does Elena need to work at each job to earn at least $450 per week?

ⓐ Let x be the number of hours she works teaching swimming and let y be the number of hours she works as an intern. Write an inequality that would model this situation.

ⓑ Graph the inequality.

ⓒ Find three ordered pairs (x, y) that would be solutions to the inequality. Then, explain what that means for Elena.

279. The doctor tells Laura she needs to exercise enough to burn 500 calories each day. She prefers to either run or bike and burns 15 calories per minute while running and 10 calories a minute while biking.

ⓐ If x is the number of minutes that Laura runs and y is the number minutes she bikes, find the inequality that models the situation.

ⓑ Graph the inequality.

ⓒ List three solutions to the inequality. What options do the solutions provide Laura?

280. Armando's workouts consist of kickboxing and swimming. While kickboxing, he burns 10 calories per minute and he burns 7 calories a minute while swimming. He wants to burn 600 calories each day.

ⓐ If x is the number of minutes that Armando will kickbox and y is the number minutes he will swim, find the inequality that will help Armando create a workout for today.

ⓑ Graph the inequality.

ⓒ List three solutions to the inequality. What options do the solutions provide Armando?

Writing Exercises

281. Lester thinks that the solution of any inequality with a $>$ sign is the region above the line and the solution of any inequality with a $<$ sign is the region below the line. Is Lester correct? Explain why or why not.

282. Explain why, in some graphs of linear inequalities, the boundary line is solid but in other graphs it is dashed.

Self Check

ⓐ *After completing the exercises, use this checklist to evaluate your mastery of the objectives of this section.*

I can...	Confidently	With some help	No-I don't get it!
verify solutions to an inequality in two variables.			
recognize the relation between the solutions of an inequality and its graph.			
graph linear inequalities.			

ⓑ *On a scale of 1–10, how would you rate your mastery of this section in light of your responses on the checklist? How can you improve this?*

Relations and Functions

Learning Objectives

By the end of this section, you will be able to:

> Find the domain and range of a relation
> Determine if a relation is a function
> Find the value of a function

Be Prepared!

Before you get started, take this readiness quiz.

1. Evaluate $3x - 5$ when $x = -2$.
   If you missed this problem, review **Example 1.6**.

2. Evaluate $2x^2 - x - 3$ when $x = a$.
   If you missed this problem, review **Example 1.6**.

3. Simplify: $7x - 1 - 4x + 5$.
   If you missed this problem, review **Example 1.7**.

Find the Domain and Range of a Relation

As we go about our daily lives, we have many data items or quantities that are paired to our names. Our social security number, student ID number, email address, phone number and our birthday are matched to our name. There is a relationship between our name and each of those items.

When your professor gets her class roster, the names of all the students in the class are listed in one column and then the student ID number is likely to be in the next column. If we think of the correspondence as a set of ordered pairs, where the first element is a student name and the second element is that student's ID number, we call this a **relation**.

(Student name, Student ID #)

The set of all the names of the students in the class is called the **domain** of the relation and the set of all student ID numbers paired with these students is the range of the relation.

There are many similar situations where one variable is paired or matched with another. The set of ordered pairs that records this matching is a relation.

Relation

A **relation** is any set of ordered pairs, (x, y). All the x-values in the ordered pairs together make up the **domain**. All the y-values in the ordered pairs together make up the **range**.

EXAMPLE 3.42

For the relation $\{(1, 1), (2, 4), (3, 9), (4, 16), (5, 25)\}$:

ⓐ Find the domain of the relation.

ⓑ Find the range of the relation.

⊘ **Solution**

$$\{(1, 1), (2, 4), (3, 9), (4, 16), (5, 25)\}$$

ⓐ The domain is the set of all x-values of the relation. $\{1, 2, 3, 4, 5\}$

ⓑ The range is the set of all y-values of the relation. $\{1, 4, 9, 16, 25\}$

> **TRY IT : : 3.83**   For the relation $\{(1, 1), (2, 8), (3, 27), (4, 64), (5, 125)\}$:

ⓐ Find the domain of the relation.

ⓑ Find the range of the relation.

> **TRY IT : : 3.84**   For the relation $\{(1, 3), (2, 6), (3, 9), (4, 12), (5, 15)\}$:

ⓐ Find the domain of the relation.

ⓑ Find the range of the relation.

Mapping

A **mapping** is sometimes used to show a relation. The arrows show the pairing of the elements of the domain with the elements of the range.

EXAMPLE 3.43

Use the **mapping** of the relation shown to ⓐ list the ordered pairs of the relation, ⓑ find the domain of the relation, and ⓒ find the range of the relation.

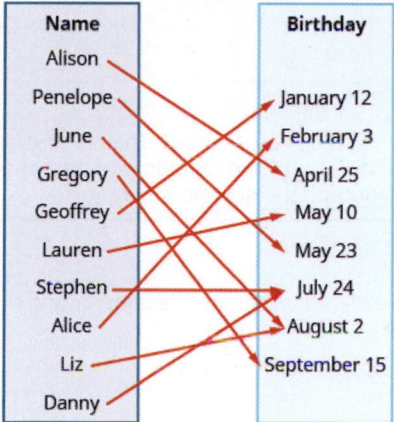

✓ Solution

ⓐ The arrow shows the matching of the person to their birthday. We create ordered pairs with the person's name as the x-value and their birthday as the y-value.

{(Alison, April 25), (Penelope, May 23), (June, August 2), (Gregory, September 15), (Geoffrey, January 12), (Lauren, May 10), (Stephen, July 24), (Alice, February 3), (Liz, August 2), (Danny, July 24)}

ⓑ The domain is the set of all x-values of the relation.

{Alison, Penelope, June, Gregory, Geoffrey, Lauren, Stephen, Alice, Liz, Danny}

ⓒ The range is the set of all y-values of the relation.

{January 12, February 3, April 25, May 10, May 23, July 24, August 2, September 15}

> **TRY IT : : 3.85**

Use the mapping of the relation shown to ⓐ list the ordered pairs of the relation ⓑ find the domain of the relation ⓒ find the range of the relation.

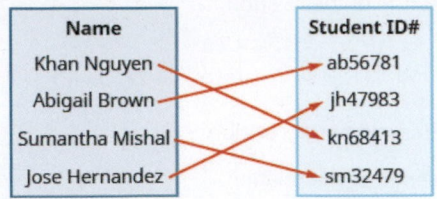

> **TRY IT : : 3.86**

Use the mapping of the relation shown to ⓐ list the ordered pairs of the relation ⓑ find the domain of the relation ⓒ find the range of the relation.

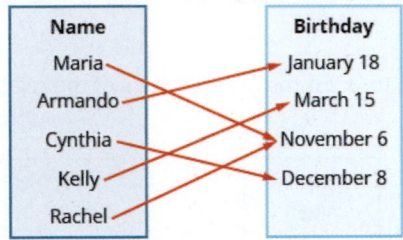

A graph is yet another way that a relation can be represented. The set of ordered pairs of all the points plotted is the relation. The set of all *x*-coordinates is the domain of the relation and the set of all *y*-coordinates is the range. Generally we write the numbers in ascending order for both the domain and range.

EXAMPLE 3.44

Use the graph of the relation to ⓐ list the ordered pairs of the relation ⓑ find the domain of the relation ⓒ find the range of the relation.

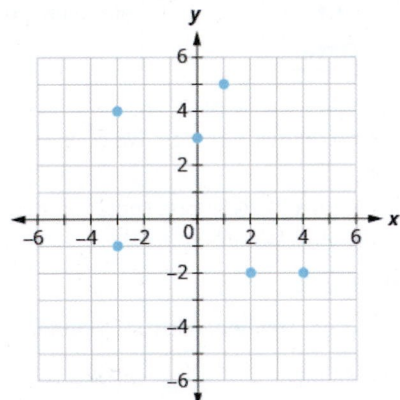

✓ **Solution**

ⓐ The ordered pairs of the relation are: $\{(1, 5), (-3, -1), (4, -2), (0, 3), (2, -2), (-3, 4)\}.$

ⓑ The domain is the set of all *x*-values of the relation: $\{-3, 0, 1, 2, 4\}.$

Notice that while -3 repeats, it is only listed once.

ⓒ The range is the set of all *y*-values of the relation: $\{-2, -1, 3, 4, 5\}.$

Notice that while -2 repeats, it is only listed once.

> **TRY IT : : 3.87**

Use the graph of the relation to ⓐ list the ordered pairs of the relation ⓑ find the domain of the relation ⓒ find the range of the relation.

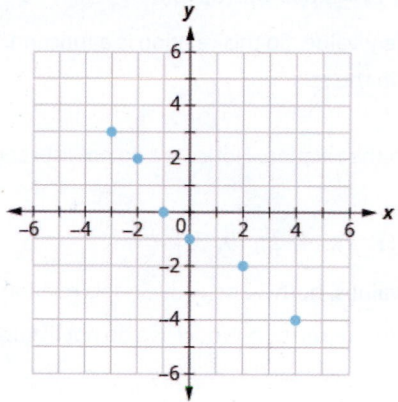

> **TRY IT : : 3.88**

Use the graph of the relation to ⓐ list the ordered pairs of the relation ⓑ find the domain of the relation ⓒ find the range of the relation.

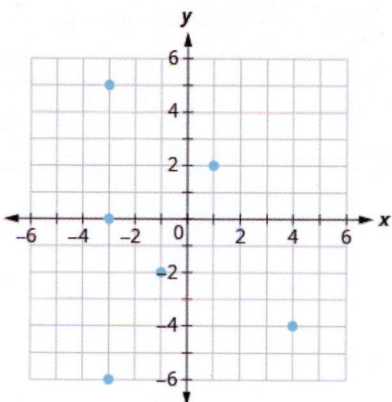

Determine if a Relation is a Function

A special type of relation, called a **function**, occurs extensively in mathematics. A function is a relation that assigns to each element in its domain exactly one element in the range. For each ordered pair in the relation, each *x*-value is matched with only one *y*-value.

Function

A **function** is a relation that assigns to each element in its domain exactly one element in the range.

The birthday example from **Example 3.43** helps us understand this definition. Every person has a birthday but no one has two birthdays. It is okay for two people to share a birthday. It is okay that Danny and Stephen share July 24th as their birthday and that June and Liz share August 2nd. Since each person has exactly one birthday, the relation in **Example 3.43** is a function.

The relation shown by the graph in **Example 3.44** includes the ordered pairs $(-3, -1)$ and $(-3, 4)$. Is that okay in a function? No, as this is like one person having two different birthdays.

EXAMPLE 3.45

Use the set of ordered pairs to (i) determine whether the relation is a function (ii) find the domain of the relation (iii) find the range of the relation.

ⓐ $\{(-3, 27), (-2, 8), (-1, 1), (0, 0), (1, 1), (2, 8), (3, 27)\}$

ⓑ {(9, −3), (4, −2), (1, −1), (0, 0), (1, 1), (4, 2), (9, 3)}

⊘ Solution

ⓐ {(−3, 27), (−2, 8), (−1, 1), (0, 0), (1, 1), (2, 8), (3, 27)}

(i) Each x-value is matched with only one y-value. So this relation is a function.

(ii) The domain is the set of all x-values in the relation.
The domain is: {−3, −2, −1, 0, 1, 2, 3}.

(iii) The range is the set of all y-values in the relation. Notice we do not list range values twice.
The range is: {27, 8, 1, 0}.

ⓑ {(9, −3), (4, −2), (1, −1), (0, 0), (1, 1), (4, 2), (9, 3)}

(i) The x-value 9 is matched with two y-values, both 3 and −3. So this relation is not a function.

(ii) The domain is the set of all x-values in the relation. Notice we do not list domain values twice.
The domain is: {0, 1, 2, 4, 9}.

(iii) The range is the set of all y-values in the relation.
The range is: {−3, −2, −1, 0, 1, 2, 3}.

> **TRY IT : : 3.89**

Use the set of ordered pairs to (i) determine whether the relation is a function (ii) find the domain of the relation (iii) find the range of the function.

ⓐ {(−3, −6), (−2, −4), (−1, −2), (0, 0), (1, 2), (2, 4), (3, 6)}

ⓑ {(8, −4), (4, −2), (2, −1), (0, 0), (2, 1), (4, 2), (8, 4)}

> **TRY IT : : 3.90**

Use the set of ordered pairs to (i) determine whether the relation is a function (ii) find the domain of the relation (iii) find the range of the relation.

ⓐ {(27, −3), (8, −2), (1, −1), (0, 0), (1, 1), (8, 2), (27, 3)}

ⓑ {(7, −3), (−5, −4), (8, −0), (0, 0), (−6, 4), (−2, 2), (−1, 3)}

EXAMPLE 3.46

Use the mapping to ⓐ determine whether the relation is a function ⓑ find the domain of the relation ⓒ find the range of the relation.

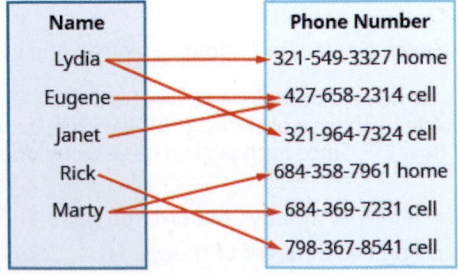

⊘ Solution

ⓐ Both Lydia and Marty have two phone numbers. So each x-value is not matched with only one y-value. So this relation is not a function.

ⓑ The domain is the set of all x-values in the relation. The domain is: {Lydia, Eugene, Janet, Rick, Marty}

ⓒ The range is the set of all *y*-values in the relation. The range is:

{321-549-3327,   427-658-2314,   321-964-7324,   684-358-7961,   684-369-7231,   798-367-8541}

> **TRY IT ::** 3.91

Use the mapping to ⓐ determine whether the relation is a function ⓑ find the domain of the relation ⓒ find the range of the relation.

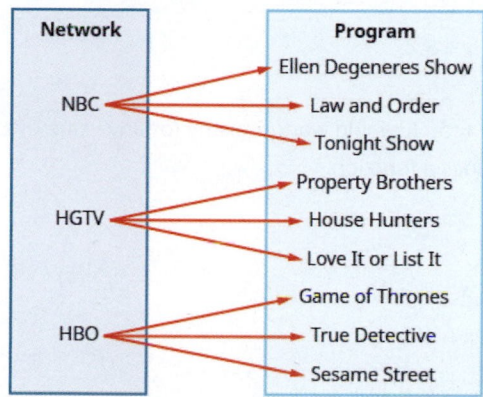

> **TRY IT ::** 3.92

Use the mapping to ⓐ determine whether the relation is a function ⓑ find the domain of the relation ⓒ find the range of the relation.

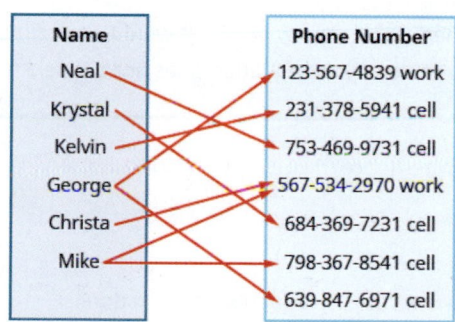

In algebra, more often than not, functions will be represented by an equation. It is easiest to see if the equation is a function when it is solved for *y*. If each value of *x* results in only one value of *y*, then the equation defines a function.

EXAMPLE 3.47

Determine whether each equation is a function.

ⓐ $2x + y = 7$ ⓑ $y = x^2 + 1$ ⓒ $x + y^2 = 3$

✓ **Solution**

ⓐ $2x + y = 7$

For each value of *x*, we multiply it by -2 and then add 7 to get the *y*-value

$$y = -2x + 7$$

For example, if $x = 3$: $y = -2 \cdot 3 + 7$

$$y = 1$$

We have that when $x = 3$, then $y = 1$. It would work similarly for any value of x. Since each value of x, corresponds to only one value of y the equation defines a function.

ⓑ $y = x^2 + 1$

For each value of x, we square it and then add 1 to get the y-value.

	$y = x^2 + 1$
For example, if $x = 2$:	$y = 2^2 + 1$
	$y = 5$

We have that when $x = 2$, then $y = 5$. It would work similarly for any value of x. Since each value of x, corresponds to only one value of y the equation defines a function.

ⓒ

	$x + y^2 = 3$
Isolate the y term.	$y^2 = -x + 3$
Let's substitute $x = 2$.	$y^2 = -2 + 3$
	$y^2 = 1$
This give us two values for y.	$y = 1 \; y = -1$

We have shown that when $x = 2$, then $y = 1$ and $y = -1$. It would work similarly for any value of x. Since each value of x does not corresponds to only one value of y the equation does not define a function.

> | **TRY IT : : 3.93**   Determine whether each equation is a function.

ⓐ $4x + y = -3$ ⓑ $x + y^2 = 1$ ⓒ $y - x^2 = 2$

> | **TRY IT : : 3.94**   Determine whether each equation is a function.

ⓐ $x + y^2 = 4$ ⓑ $y = x^2 - 7$ ⓒ $y = 5x - 4$

Find the Value of a Function

It is very convenient to name a function and most often we name it f, g, h, F, G, or H. In any function, for each x-value from the domain we get a corresponding y-value in the range. For the function f, we write this range value y as $f(x)$. This is called function notation and is read f of x or the value of f at x. In this case the parentheses does not indicate multiplication.

> **Function Notation**
>
> For the function $y = f(x)$
>
> f is the name of the function
> x is the domain value
> $f(x)$ is the range value y corresponding to the value x
>
> We read $f(x)$ as f of x or the value of f at x.

We call x the independent variable as it can be any value in the domain. We call y the dependent variable as its value depends on x.

Independent and Dependent Variables

For the function $y = f(x)$,

x is the independent variable as it can be any value in the domain

y the dependent variable as its value depends on x

Much as when you first encountered the variable x, function notation may be rather unsettling. It seems strange because it is new. You will feel more comfortable with the notation as you use it.

Let's look at the equation $y = 4x - 5$. To find the value of y when $x = 2$, we know to substitute $x = 2$ into the equation and then simplify.

$$y = 4x - 5$$

Let $x = 2$. $y = 4 \cdot 2 - 5$

$$y = 3$$

The value of the function at $x = 2$ is 3.

We do the same thing using function notation, the equation $y = 4x - 5$ can be written as $f(x) = 4x - 5$. To find the value when $x = 2$, we write:

$$f(x) = 4x - 5$$

Let $x = 2$. $f(2) = 4 \cdot 2 - 5$

$$f(2) = 3$$

The value of the function at $x = 2$ is 3.

This process of finding the value of $f(x)$ for a given value of x is called *evaluating the function*.

EXAMPLE 3.48

For the function $f(x) = 2x^2 + 3x - 1$, evaluate the function.

ⓐ $f(3)$ ⓑ $f(-2)$ ⓒ $f(a)$

✓ **Solution**

ⓐ

$$f(x) = 2x^2 + 3x - 1$$

To evaluate $f(3)$, substitute 3 for x. $f(3) = 2(3)^2 + 3 \cdot 3 - 1$

Simplify. $f(3) = 2 \cdot 9 + 3 \cdot 3 - 1$

$$f(3) = 18 + 9 - 1$$

$$f(3) = 26$$

ⓑ

$$f(x) = 2x^2 + 3x - 1$$

To evaluate $f(-2)$, substitute -2 for x.	$f(-2) = 2(-2)^2 + 3(-2) - 1$
Simplify.	$f(-2) = 2 \cdot 4 + (-6) - 1$
	$f(-2) = 8 + (-6) - 1$
	$f(-2) = 1$

ⓒ

$$f(x) = 2x^2 + 3x - 1$$

To evaluate $f(a)$, substitute a for x.	$f(a) = 2(a)^2 + 3 \cdot a - 1$
Simplify.	$f(a) = 2a^2 + 3a - 1$

> **TRY IT :: 3.95**   For the function $f(x) = 3x^2 - 2x + 1$, evaluate the function.
>
> ⓐ $f(3)$ ⓑ $f(-1)$ ⓒ $f(t)$

> **TRY IT :: 3.96**   For the function $f(x) = 2x^2 + 4x - 3$, evaluate the function.
>
> ⓐ $f(2)$ ⓑ $f(-3)$ ⓒ $f(h)$

In the last example, we found $f(x)$ for a constant value of x. In the next example, we are asked to find $g(x)$ with values of x that are variables. We still follow the same procedure and substitute the variables in for the x.

EXAMPLE 3.49

For the function $g(x) = 3x - 5$, evaluate the function.

ⓐ $g(h^2)$ ⓑ $g(x + 2)$ ⓒ $g(x) + g(2)$

✓ **Solution**

ⓐ

$$g(x) = 3x - 5$$

To evaluate $g(h^2)$, substitute h^2 for x.	$g(h^2) = 3h^2 - 5$
	$g(h^2) = 3h^2 - 5$

ⓑ

$$g(x) = 3x - 5$$

To evaluate $g(x+2)$, substitute $x+2$ for x.	$g(x+2) = 3(x+2) - 5$
Simplify.	$g(x+2) = 3x + 6 - 5$
	$g(x+2) = 3x + 1$

ⓒ

$$g(x) = 3x - 5$$

To evaluate $g(x) + g(2)$, first find $g(2)$.	$g(2) = 3 \cdot 2 - 5$
	$g(2) = 1$
Now find $g(x) + g(2)$	$g(x) + g(2) = \underbrace{3x - 5}_{g(x)} + \underbrace{1}_{g(2)}$
Simplify.	$g(x) + g(2) = 3x - 5 + 1$
	$g(x) + g(2) = 3x - 4$

Notice the difference between part ⓑ and ⓒ. We get $g(x+2) = 3x+1$ and $g(x) + g(2) = 3x - 4$. So we see that $g(x+2) \neq g(x) + g(2)$.

> **TRY IT :: 3.97** For the function $g(x) = 4x - 7$, evaluate the function.
>
>   ⓐ $g(m^2)$ ⓑ $g(x-3)$ ⓒ $g(x) - g(3)$

> **TRY IT :: 3.98** For the function $h(x) = 2x + 1$, evaluate the function.
>
>   ⓐ $h(k^2)$ ⓑ $h(x+1)$ ⓒ $h(x) + h(1)$

Many everyday situations can be modeled using functions.

EXAMPLE 3.50

The number of unread emails in Sylvia's account is 75. This number grows by 10 unread emails a day. The function $N(t) = 75 + 10t$ represents the relation between the number of emails, N, and the time, t, measured in days.

ⓐ Determine the independent and dependent variable.

ⓑ Find $N(5)$. Explain what this result means.

✓ **Solution**

ⓐ The number of unread emails is a function of the number of days. The number of unread emails, N, depends on the number of days, t. Therefore, the variable N, is the dependent variable and the variable t is the independent variable.

ⓑ Find $N(5)$. Explain what this result means.

$$N(t) = 75 + 10t$$

Substitute in $t = 5$.	$N(5) = 75 + 10 \cdot 5$
Simplify.	$N(5) = 75 + 50$
	$N(5) = 125$

Since 5 is the number of days, $N(5)$, is the number of unread emails after 5 days. After 5 days, there are 125 unread emails in the account.

> **TRY IT : : 3.99**

The number of unread emails in Bryan's account is 100. This number grows by 15 unread emails a day. The function $N(t) = 100 + 15t$ represents the relation between the number of emails, N, and the time, t, measured in days.

ⓐ Determine the independent and dependent variable.

ⓑ Find $N(7)$. Explain what this result means.

> **TRY IT : : 3.100**

The number of unread emails in Anthony's account is 110. This number grows by 25 unread emails a day. The function $N(t) = 110 + 25t$ represents the relation between the number of emails, N, and the time, t, measured in days.

ⓐ Determine the independent and dependent variable.

ⓑ Find $N(14)$. Explain what this result means.

▶ **MEDIA : :**
Access this online resource for additional instruction and practice with relations and functions.

- **Introduction to Functions (https://openstax.org/l/37introfunction)**

3.5 EXERCISES

Practice Makes Perfect

Find the Domain and Range of a Relation

In the following exercises, for each relation ⓐ find the domain of the relation ⓑ find the range of the relation.

283. {(1, 4), (2, 8), (3, 12), (4, 16), (5, 20)}

284. {(1, −2), (2, −4), (3, −6), (4, −8), (5, −10)}

285. {(1, 7), (5, 3), (7, 9), (−2, −3), (−2, 8)}

286. {(11, 3), (−2, −7), (4, −8), (4, 17), (−6, 9)}

In the following exercises, use the mapping of the relation to ⓐ list the ordered pairs of the relation, ⓑ find the domain of the relation, and ⓒ find the range of the relation.

287.

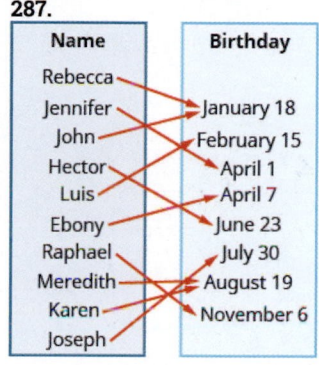

288.

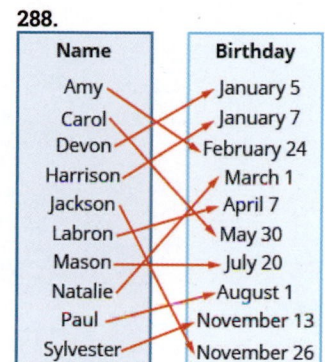

289. For a woman of height 5′4″ the mapping below shows the corresponding Body Mass Index (BMI). The body mass index is a measurement of body fat based on height and weight. A BMI of 18.5 − 24.9 is considered healthy.

Weight (lbs)	BMI
+100	18.9
110	22.3
120	17.2
130	24.0
140	25.7
150	20.6
160	27.5

290. For a man of height 5′11″ the mapping below shows the corresponding Body Mass Index (BMI). The body mass index is a measurement of body fat based on height and weight. A BMI of 18.5 − 24.9 is considered healthy.

Weight (lbs)	BMI
130	22.3
140	19.5
150	20.9
160	27.9
170	25.1
180	26.5
190	23.7
200	18.1

In the following exercises, use the graph of the relation to ⓐ *list the ordered pairs of the relation* ⓑ *find the domain of the relation* ⓒ *find the range of the relation.*

291.

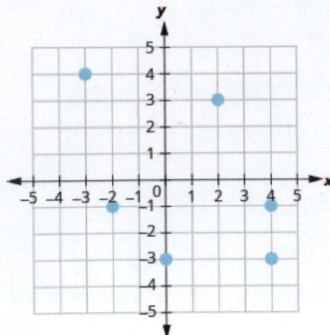

292.

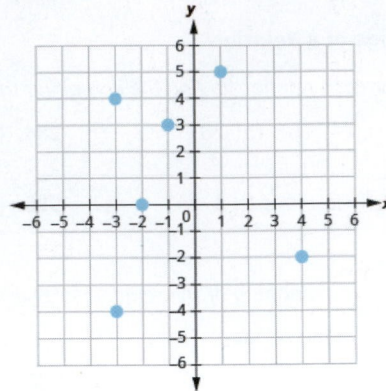

293.

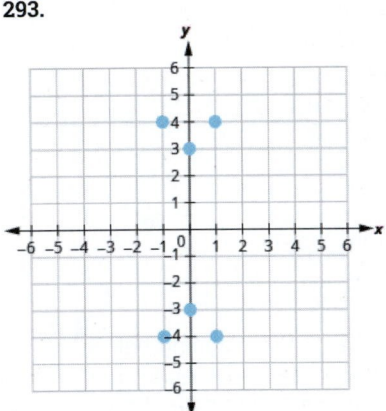

294.

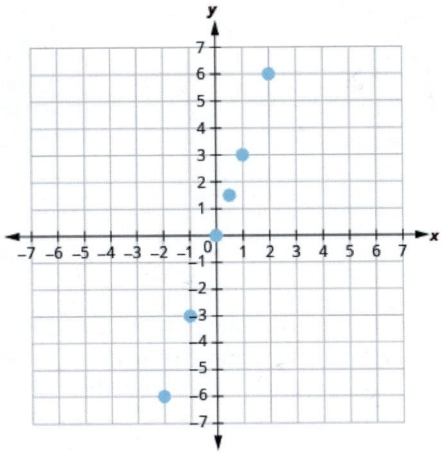

Determine if a Relation is a Function

In the following exercises, use the set of ordered pairs to ⓐ *determine whether the relation is a function,* ⓑ *find the domain of the relation, and* ⓒ *find the range of the relation.*

295. $\{(-3, 9), (-2, 4), (-1, 1), (0, 0), (1, 1), (2, 4), (3, 9)\}$

296. $\{(9, -3), (4, -2), (1, -1), (0, 0), (1, 1), (4, 2), (9, 3)\}$

297. $\{(-3, 27), (-2, 8), (-1, 1), (0, 0), (1, 1), (2, 8), (3, 27)\}$

298. $\{(9, -3), (4, -2), (1, -1), (0, 0), (1, 1), (4, 2), (9, 3)\}$

In the following exercises, use the mapping to ⓐ *determine whether the relation is a function,* ⓑ *find the domain of the function, and* ⓒ *find the range of the function.*

299.

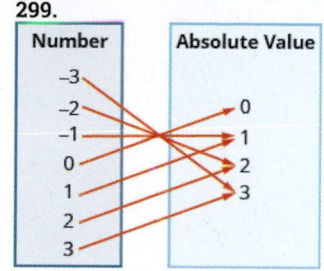

300.

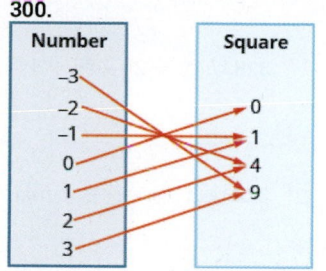

301.

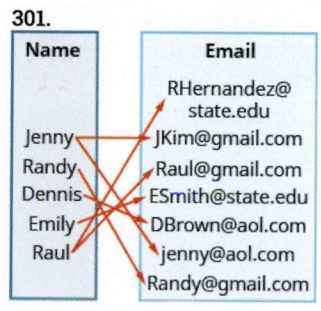

302.

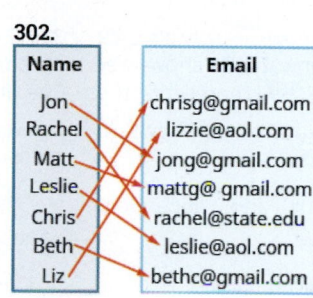

In the following exercises, determine whether each equation is a function.

303.

ⓐ $2x + y = -3$

ⓑ $y = x^2$

ⓒ $x + y^2 = -5$

304.

ⓐ $y = 3x - 5$

ⓑ $y = x^3$

ⓒ $2x + y^2 = 4$

305.

ⓐ $y - 3x^3 = 2$

ⓑ $x + y^2 = 3$

ⓒ $3x - 2y = 6$

306.

ⓐ $2x - 4y = 8$

ⓑ $-4 = x^2 - y$

ⓒ $y^2 = -x + 5$

Find the Value of a Function

In the following exercises, evaluate the function: ⓐ *f(2)* ⓑ *f(−1)* ⓒ *f(a).*

307. $f(x) = 5x - 3$

308. $f(x) = 3x + 4$

309. $f(x) = -4x + 2$

310. $f(x) = -6x - 3$

311. $f(x) = x^2 - x + 3$

312. $f(x) = x^2 + x - 2$

313. $f(x) = 2x^2 - x + 3$

314. $f(x) = 3x^2 + x - 2$

In the following exercises, evaluate the function: ⓐ *g(h²)* ⓑ *g(x + 2)* ⓒ *g(x) + g(2).*

315. $g(x) = 2x + 1$

316. $g(x) = 5x - 8$

317. $g(x) = -3x - 2$

318. $g(x) = -8x + 2$

319. $g(x) = 3 - x$

320. $g(x) = 7 - 5x$

In the following exercises, evaluate the function.

321. $f(x) = 3x^2 - 5x$; $f(2)$

322. $g(x) = 4x^2 - 3x$; $g(3)$

323. $F(x) = 2x^2 - 3x + 1$; $F(-1)$

324. $G(x) = 3x^2 - 5x + 2$; $G(-2)$

325. $h(t) = 2|t - 5| + 4$; $f(-4)$

326. $h(y) = 3|y - 1| - 3$; $h(-4)$

327. $f(x) = \dfrac{x+2}{x-1}$; $f(2)$

328. $g(x) = \dfrac{x-2}{x+2}$; $g(4)$

In the following exercises, solve.

329. The number of unwatched shows in Sylvia's DVR is 85. This number grows by 20 unwatched shows per week. The function $N(t) = 85 + 20t$ represents the relation between the number of unwatched shows, *N*, and the time, *t*, measured in weeks.

ⓐ Determine the independent and dependent variable.

ⓑ Find $N(4)$. Explain what this result means

330. Every day a new puzzle is downloaded into Ken's account. Right now he has 43 puzzles in his account. The function $N(t) = 43 + t$ represents the relation between the number of puzzles, *N*, and the time, *t*, measured in days.

ⓐ Determine the independent and dependent variable.

ⓑ Find $N(30)$. Explain what this result means.

331. The daily cost to the printing company to print a book is modeled by the function $C(x) = 3.25x + 1500$ where *C* is the total daily cost and *x* is the number of books printed.

ⓐ Determine the independent and dependent variable.

ⓑ Find $N(0)$. Explain what this result means.

ⓒ Find $N(1000)$. Explain what this result means.

332. The daily cost to the manufacturing company is modeled by the function $C(x) = 7.25x + 2500$ where $C(x)$ is the total daily cost and *x* is the number of items manufactured.

ⓐ Determine the independent and dependent variable.

ⓑ Find $C(0)$. Explain what this result means.

ⓒ Find $C(1000)$. Explain what this result means.

Writing Exercises

333. In your own words, explain the difference between a relation and a function.

334. In your own words, explain what is meant by domain and range.

335. Is every relation a function? Is every function a relation?

336. How do you find the value of a function?

Self Check

ⓐ *After completing the exercises, use this checklist to evaluate your mastery of the objectives of this section.*

I can...	Confidently	With some help	No-I don't get it!
find the domain and range of a relation.			
determine if a relation is a function.			
find the value of a function.			

ⓑ *After looking at the checklist, do you think you are well-prepared for the next section? Why or why not?*

3.6 | Graphs of Functions

Learning Objectives

By the end of this section, you will be able to:

> Use the vertical line test
> Identify graphs of basic functions
> Read information from a graph of a function

Be Prepared!

Before you get started, take this readiness quiz.

1. Evaluate: ⓐ 2^3 ⓑ 3^2.
   If you missed this problem, review **Example 1.5**.

2. Evaluate: ⓐ $|7|$ ⓑ $|-3|$.
   If you missed this problem, review **Example 1.14**.

3. Evaluate: ⓐ $\sqrt{4}$ ⓑ $\sqrt{16}$.
   If you missed this problem, review **Example 1.41**.

Use the Vertical Line Test

In the last section we learned how to determine if a relation is a function. The relations we looked at were expressed as a set of ordered pairs, a mapping or an equation. We will now look at how to tell if a graph is that of a function.

An ordered pair (x, y) is a solution of a linear equation, if the equation is a true statement when the x- and y-values of the ordered pair are substituted into the equation.

The graph of a linear equation is a straight line where every point on the line is a solution of the equation and every solution of this equation is a point on this line.

In **Figure 3.14**, we can see that, in graph of the equation $y = 2x - 3$, for every x-value there is only one y-value, as shown in the accompanying table.

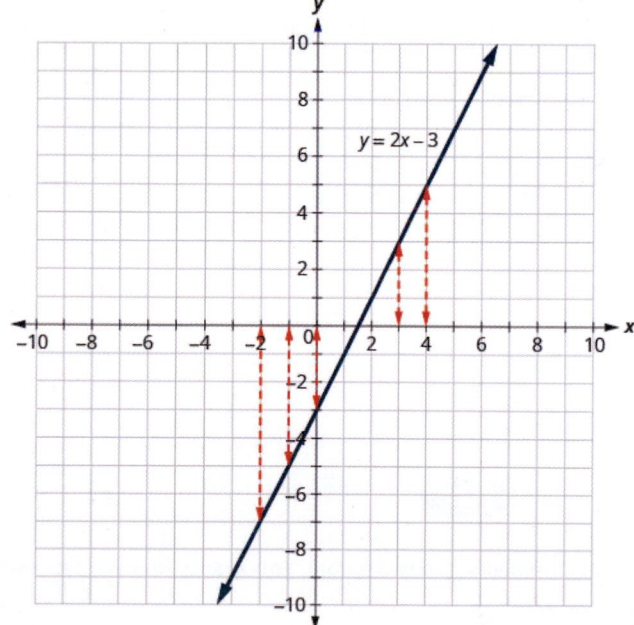

$y = 2x - 3$		
x	y	(x, y)
−2	−7	(−2, −7)
−1	−5	(−1, −5)
0	−3	(0, −3)
3	3	(3, 3)
4	5	(4, 5)

Figure 3.14

A relation is a function if every element of the domain has exactly one value in the range. So the relation defined by the equation $y = 2x - 3$ is a function.

If we look at the graph, each vertical dashed line only intersects the line at one point. This makes sense as in a function, for every *x*-value there is only one *y*-value.

If the vertical line hit the graph twice, the *x*-value would be mapped to two *y*-values, and so the graph would not represent a function.

This leads us to the vertical line test. A set of points in a rectangular coordinate system is the graph of a function if every vertical line intersects the graph in at most one point. If any vertical line intersects the graph in more than one point, the graph does not represent a function.

Vertical Line Test

A set of points in a rectangular coordinate system is the graph of a function if every vertical line intersects the graph in at most one point.

If any vertical line intersects the graph in more than one point, the graph does not represent a function.

EXAMPLE 3.51

Determine whether each graph is the graph of a function.

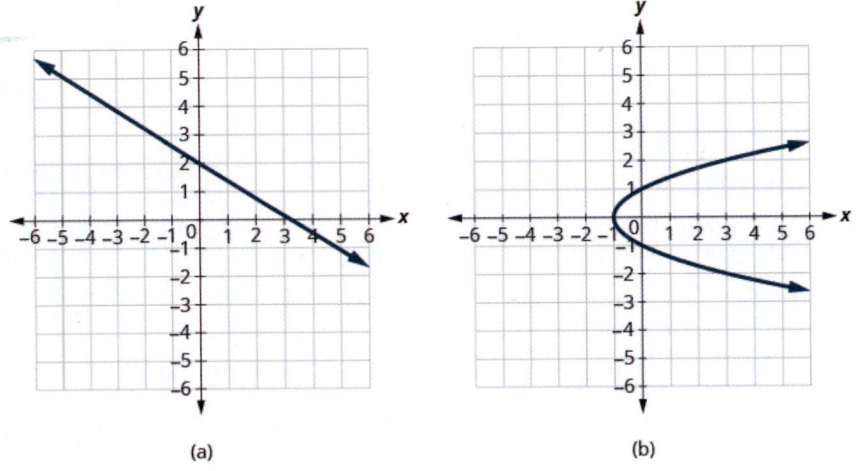

(a)   (b)

✓ **Solution**

ⓐ Since any vertical line intersects the graph in at most one point, the graph is the graph of a function.

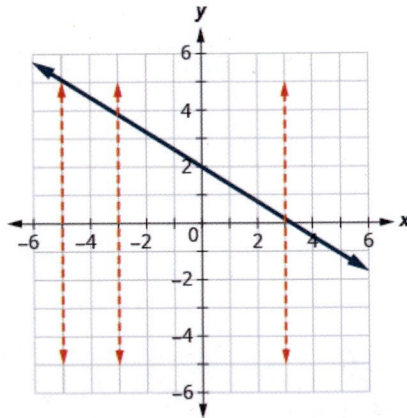

ⓑ One of the vertical lines shown on the graph, intersects it in two points. This graph does not represent a function.

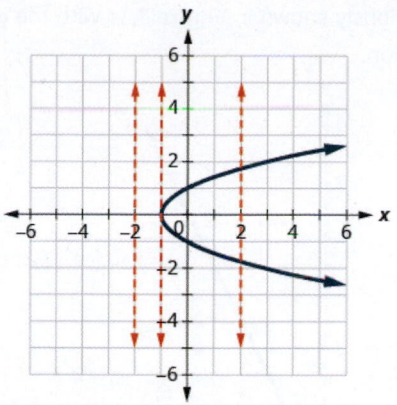

> **TRY IT : : 3.101**   Determine whether each graph is the graph of a function.

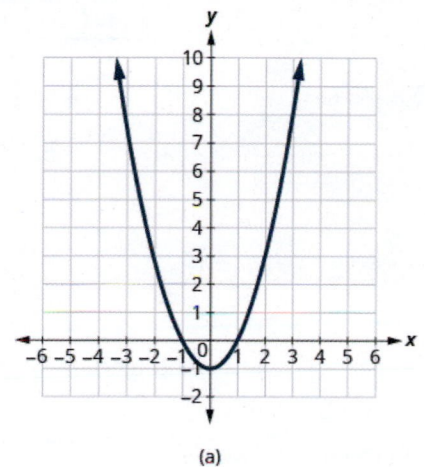

(a)

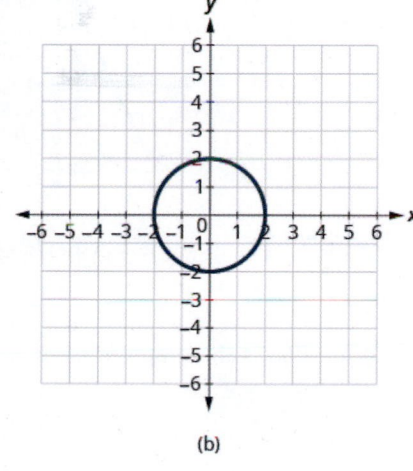

(b)

> **TRY IT : : 3.102**   Determine whether each graph is the graph of a function.

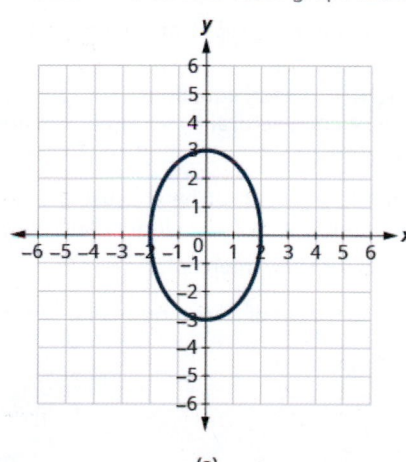

(a)

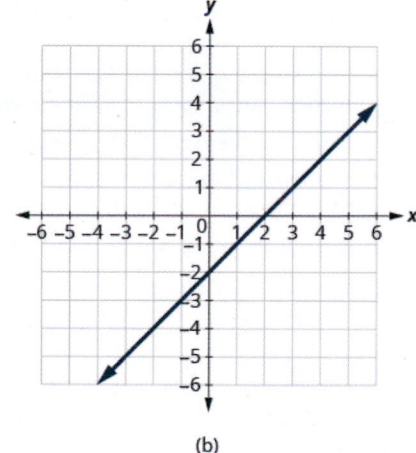

(b)

Identify Graphs of Basic Functions

We used the equation $y = 2x - 3$ and its graph as we developed the vertical line test. We said that the relation defined by the equation $y = 2x - 3$ is a function.

We can write this as in function notation as $f(x) = 2x - 3$. It still means the same thing. The graph of the function is the graph of all ordered pairs (x, y) where $y = f(x)$. So we can write the ordered pairs as $(x, f(x))$. It looks different but the graph will be the same.

Compare the graph of $y = 2x - 3$ previously shown in **Figure 3.14** with the graph of $f(x) = 2x - 3$ shown in **Figure 3.15**. Nothing has changed but the notation.

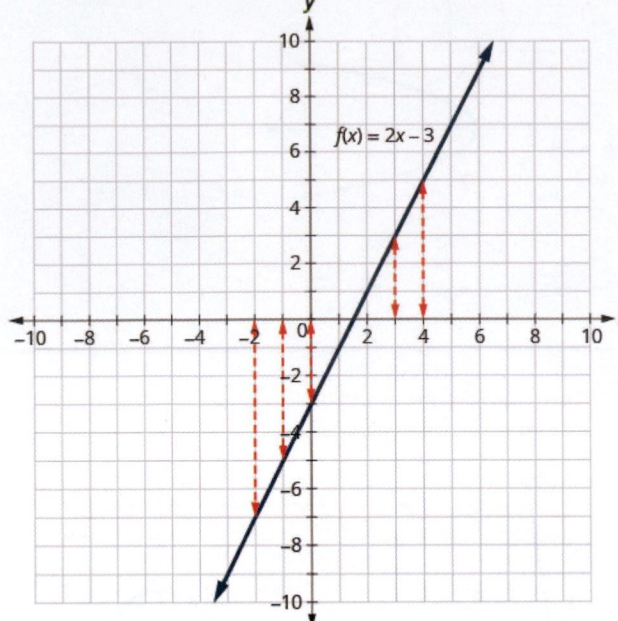

$f(x) = 2x - 3$		
x	$f(x)$	$(x, f(x))$
−2	−7	(−2, −7)
−1	−5	(−1, −5)
0	−3	(0, −3)
3	3	(3, 3)
4	5	(4, 5)

Figure 3.15

Graph of a Function

The graph of a function is the graph of all its ordered pairs, (x, y) or using function notation, $(x, f(x))$ where $y = f(x)$.

> f name of function
> x x-coordinate of the ordered pair
> $f(x)$ y-coordinate of the ordered pair

As we move forward in our study, it is helpful to be familiar with the graphs of several basic functions and be able to identify them.

Through our earlier work, we are familiar with the graphs of linear equations. The process we used to decide if $y = 2x - 3$ is a function would apply to all linear equations. All non-vertical linear equations are functions. Vertical lines are not functions as the x-value has infinitely many y-values.

We wrote linear equations in several forms, but it will be most helpful for us here to use the slope-intercept form of the linear equation. The slope-intercept form of a linear equation is $y = mx + b$. In function notation, this linear function becomes $f(x) = mx + b$ where m is the slope of the line and b is the y-intercept.

The domain is the set of all real numbers, and the range is also the set of all real numbers.

Linear Function

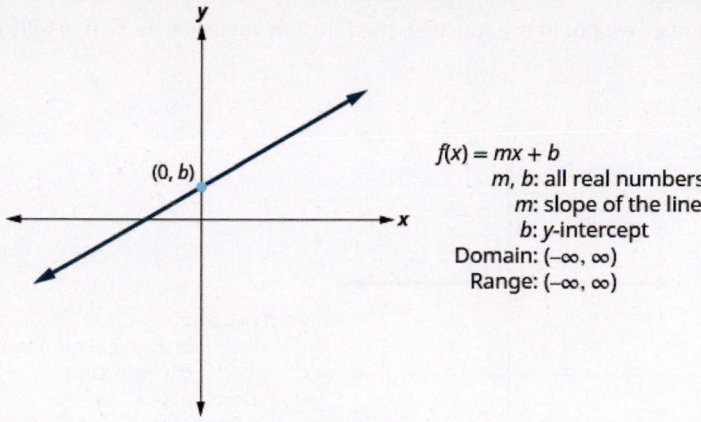

$$f(x) = mx + b$$
m, b: all real numbers
m: slope of the line
b: y-intercept
Domain: $(-\infty, \infty)$
Range: $(-\infty, \infty)$

We will use the graphing techniques we used earlier, to graph the basic functions.

EXAMPLE 3.52

Graph: $f(x) = -2x - 4$.

⊘ **Solution**

$$f(x) = -2x - 4$$

We recognize this as a linear function.

Find the slope and y-intercept.

$$m = -2$$
$$b = -4$$

Graph using the slope intercept.

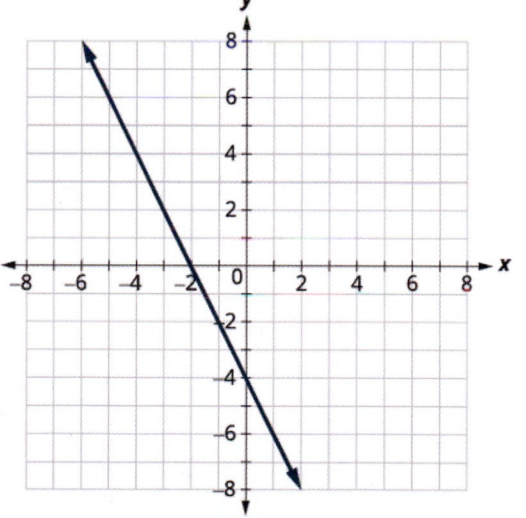

> | **TRY IT ::** 3.103   Graph: $f(x) = -3x - 1$

> | **TRY IT ::** 3.104   Graph: $f(x) = -4x - 5$

The next function whose graph we will look at is called the constant function and its equation is of the form $f(x) = b$, where b is any real number. If we replace the $f(x)$ with y, we get $y = b$. We recognize this as the horizontal line whose y-intercept is b. The graph of the function $f(x) = b$, is also the horizontal line whose y-intercept is b.

Notice that for any real number we put in the function, the function value will be b. This tells us the range has only one value, b.

Constant Function

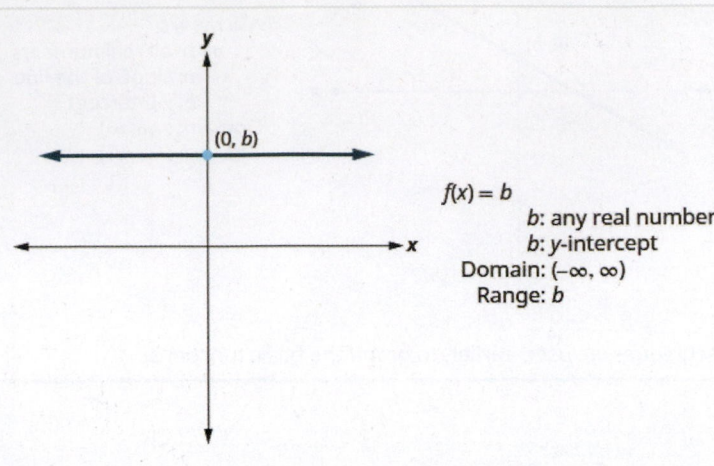

$f(x) = b$
b: any real number
b: y-intercept
Domain: $(-\infty, \infty)$
Range: b

EXAMPLE 3.53

Graph: $f(x) = 4$.

⊘ **Solution**

$$f(x) = 4$$

We recognize this as a constant function.

The graph will be a horizontal line through $(0, 4)$.

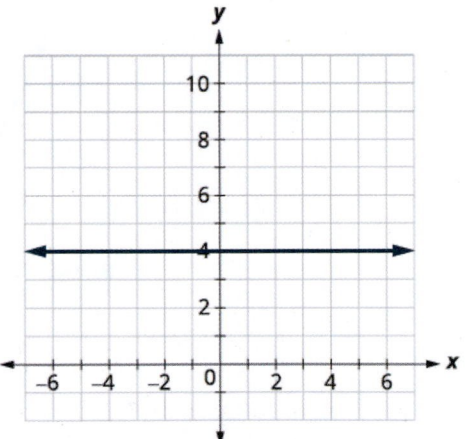

> **TRY IT : :** 3.105   Graph: $f(x) = -2$.

> **TRY IT : :** 3.106   Graph: $f(x) = 3$.

The identity function, $f(x) = x$ is a special case of the linear function. If we write it in linear function form, $f(x) = 1x + 0$, we see the slope is 1 and the y-intercept is 0.

Identity Function

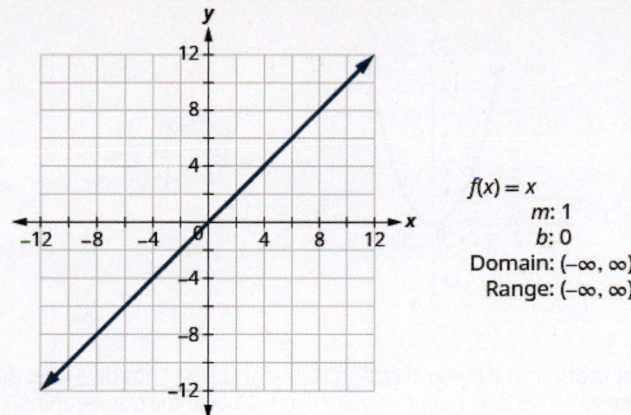

$f(x) = x$
$m: 1$
$b: 0$
Domain: $(-\infty, \infty)$
Range: $(-\infty, \infty)$

The next function we will look at is not a linear function. So the graph will not be a line. The only method we have to graph this function is point plotting. Because this is an unfamiliar function, we make sure to choose several positive and negative values as well as 0 for our x-values.

EXAMPLE 3.54

Graph: $f(x) = x^2$.

✓ Solution

We choose x-values. We substitute them in and then create a chart as shown.

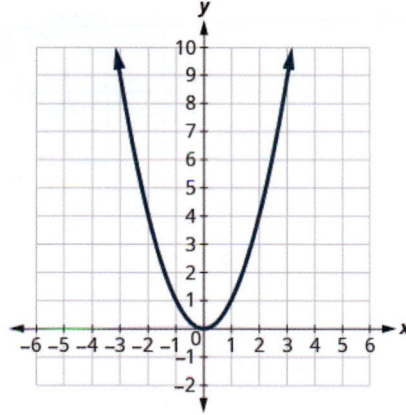

x	$f(x) = x^2$	$(x, f(x))$
−3	9	(−3, 9)
−2	4	(−2, 4)
−1	1	(−1, 1)
0	0	(0, 0)
1	1	(1, 1)
2	4	(2, 4)
3	9	(3, 9)

> **TRY IT :: 3.107**   Graph: $f(x) = x^2$.

> **TRY IT :: 3.108** $f(x) = -x^2$

Looking at the result in **Example 3.54**, we can summarize the features of the square function. We call this graph a parabola. As we consider the domain, notice any real number can be used as an x-value. The domain is all real numbers.

The range is not all real numbers. Notice the graph consists of values of y never go below zero. This makes sense as the square of any number cannot be negative. So, the range of the square function is all non-negative real numbers.

Square Function

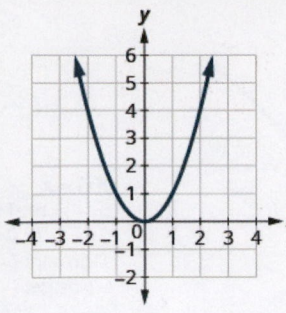

$f(x) = x^2$
Domain: $(-\infty, \infty)$
Range: $[0, \infty)$

The next function we will look at is also not a linear function so the graph will not be a line. Again we will use point plotting, and make sure to choose several positive and negative values as well as 0 for our x-values.

EXAMPLE 3.55

Graph: $f(x) = x^3$.

⊘ **Solution**

We choose x-values. We substitute them in and then create a chart.

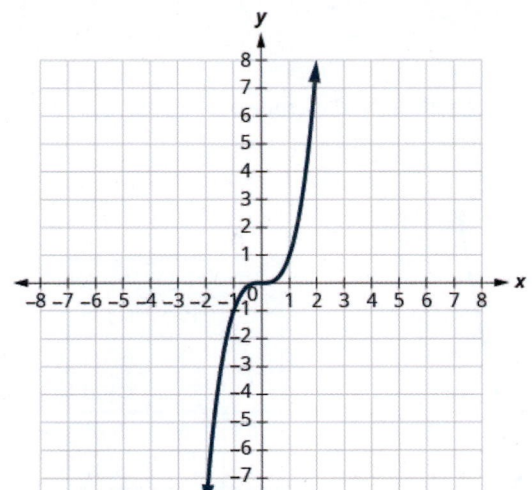

x	$f(x) = x^3$	$(x, f(x))$
-2	-8	$(-2, -8)$
-1	-1	$(-1, -1)$
0	0	$(0, 0)$
1	1	$(1, 1)$
2	8	$(2, 8)$

> **TRY IT ::** 3.109   Graph: $f(x) = x^3$.

> **TRY IT ::** 3.110   Graph: $f(x) = -x^3$.

Looking at the result in **Example 3.55**, we can summarize the features of the cube function. As we consider the domain, notice any real number can be used as an x-value. The domain is all real numbers.

The range is all real numbers. This makes sense as the cube of any non-zero number can be positive or negative. So, the range of the cube function is all real numbers.

Cube Function

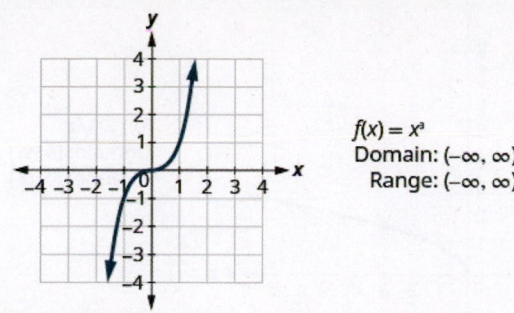

$f(x) = x^3$
Domain: $(-\infty, \infty)$
Range: $(-\infty, \infty)$

The next function we will look at does not square or cube the input values, but rather takes the square root of those values.

Let's graph the function $f(x) = \sqrt{x}$ and then summarize the features of the function. Remember, we can only take the square root of non-negative real numbers, so our domain will be the non-negative real numbers.

EXAMPLE 3.56

$f(x) = \sqrt{x}$

⊘ **Solution**

We choose x-values. Since we will be taking the square root, we choose numbers that are perfect squares, to make our work easier. We substitute them in and then create a chart.

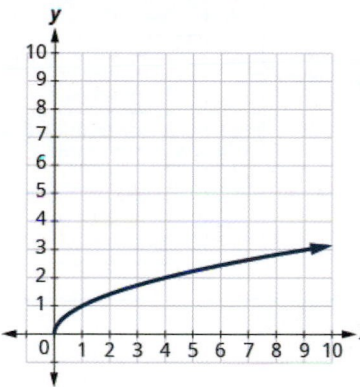

x	$f(x) = \sqrt{x}$	$(x, f(x))$
0	0	(0, 0)
1	1	(1, 1)
4	2	(4, 2)
9	3	(9, 3)

> **TRY IT : : 3.111**   Graph: $f(x) = \sqrt{x}$.

> **TRY IT : : 3.112**   Graph: $f(x) = -\sqrt{x}$.

Square Root Function

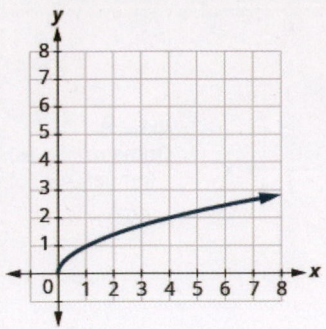

$f(x) = \sqrt{x}$
Domain: $[0, \infty)$
Range: $[0, \infty)$

Our last basic function is the absolute value function, $f(x) = |x|$. Keep in mind that the absolute value of a number is its distance from zero. Since we never measure distance as a negative number, we will never get a negative number in the range.

EXAMPLE 3.57

Graph: $f(x) = |x|$.

⊘ **Solution**

We choose x-values. We substitute them in and then create a chart.

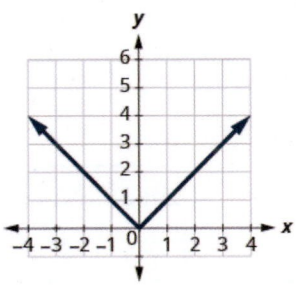

x	$f(x) = |x|$	$(x, f(x))$
−3	3	(−3, 3)
−2	2	(−2, 2)
−1	1	(−1, 1)
0	0	(0, 0)
1	1	(1, 1)
2	2	(2, 2)
3	3	(3, 3)

> **TRY IT : : 3.113**   Graph: $f(x) = |x|$.

> **TRY IT : : 3.114**   Graph: $f(x) = -|x|$.

Absolute Value Function

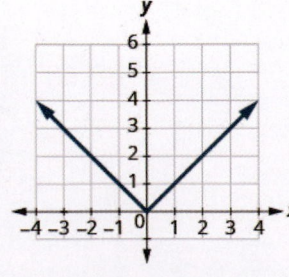

$f(x) = |x|$
Domain: $(-\infty, \infty)$
Range: $[0, \infty)$

Read Information from a Graph of a Function

In the sciences and business, data is often collected and then graphed. The graph is analyzed, information is obtained from the graph and then often predictions are made from the data.

We will start by reading the domain and range of a function from its graph.

Remember the domain is the set of all the *x*-values in the ordered pairs in the function. To find the domain we look at the graph and find all the values of *x* that have a corresponding value on the graph. Follow the value *x* up or down vertically. If you hit the graph of the function then *x* is in the domain.

Remember the range is the set of all the *y*-values in the ordered pairs in the function. To find the range we look at the graph and find all the values of *y* that have a corresponding value on the graph. Follow the value *y* left or right horizontally. If you hit the graph of the function then *y* is in the range.

EXAMPLE 3.58

Use the graph of the function to find its domain and range. Write the domain and range in interval notation.

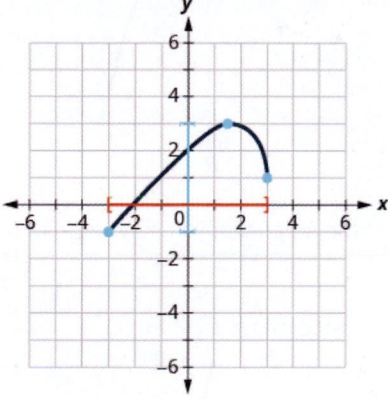

✓ Solution

To find the domain we look at the graph and find all the values of *x* that correspond to a point on the graph. The domain is highlighted in red on the graph. The domain is $[-3, 3]$.

To find the range we look at the graph and find all the values of *y* that correspond to a point on the graph. The range is highlighted in blue on the graph. The range is $[-1, 3]$.

> **TRY IT ::** 3.115

Use the graph of the function to find its domain and range. Write the domain and range in interval notation.

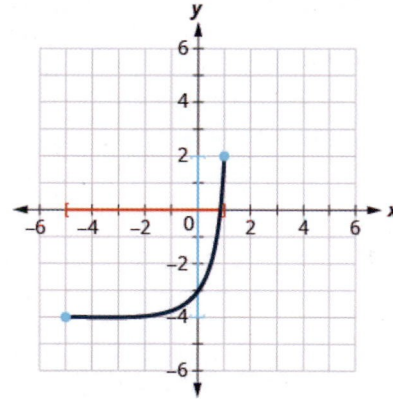

> **TRY IT ::** 3.116

Use the graph of the function to find its domain and range. Write the domain and range in interval notation.

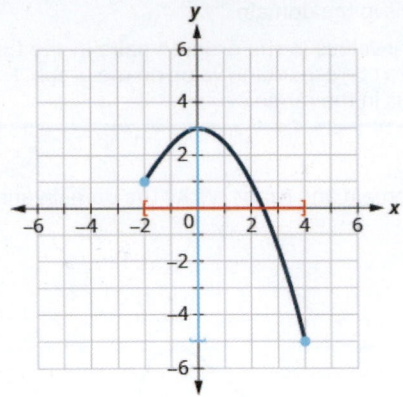

We are now going to read information from the graph that you may see in future math classes.

EXAMPLE 3.59

Use the graph of the function to find the indicated values.

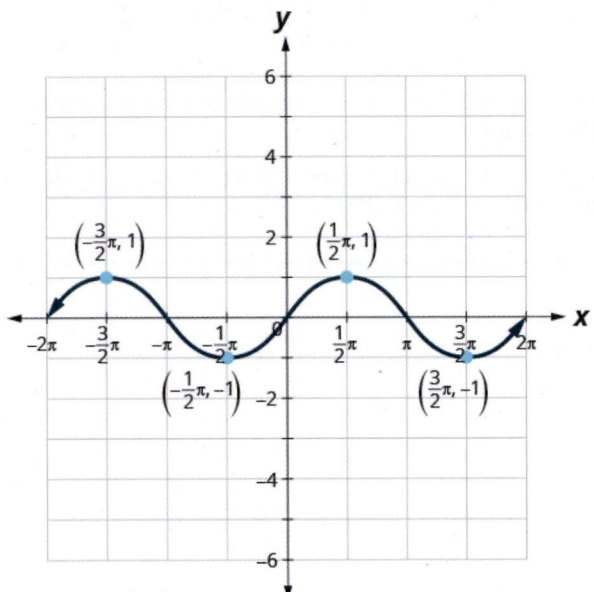

ⓐ Find: $f(0)$.

ⓑ Find: $f\left(\frac{3}{2}\pi\right)$.

ⓒ Find: $f\left(-\frac{1}{2}\pi\right)$.

ⓓ Find the values for x when $f(x) = 0$.

ⓔ Find the x-intercepts.

ⓕ Find the y-intercepts.

ⓖ Find the domain. Write it in interval notation.

ⓗ Find the range. Write it in interval notation.

✓ **Solution**

ⓐ When $x = 0$, the function crosses the y-axis at 0. So, $f(0) = 0$.

ⓑ When $x = \frac{3}{2}\pi$, the y-value of the function is -1. So, $f\left(\frac{3}{2}\pi\right) = -1$.

ⓒ When $x = -\frac{1}{2}\pi$, the y-value of the function is -1. So, $f\left(-\frac{1}{2}\pi\right) = -1$.

ⓓ The function is 0 at the points, $(-2\pi, 0)$, $(-\pi, 0)$, $(0, 0)$, $(\pi, 0)$, $(2\pi, 0)$. The x-values when $f(x) = 0$ are $-2\pi, -\pi, 0, \pi, 2\pi$.

ⓔ The x-intercepts occur when $y = 0$. So the x-intercepts occur when $f(x) = 0$. The x-intercepts are $(-2\pi, 0)$, $(-\pi, 0)$, $(0, 0)$, $(\pi, 0)$, $(2\pi, 0)$.

ⓕ The y-intercepts occur when $x = 0$. So the y-intercepts occur at $f(0)$. The y-intercept is $(0, 0)$.

ⓖ This function has a value when x is from -2π to 2π. Therefore, the domain in interval notation is $[-2\pi, 2\pi]$.

ⓗ This function values, or y-values go from -1 to 1. Therefore, the range, in interval notation, is $[-1, 1]$.

> **TRY IT :: 3.117**   Use the graph of the function to find the indicated values.

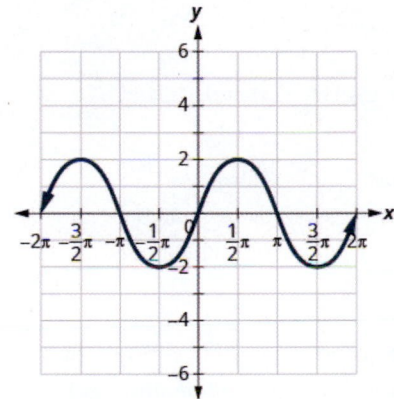

ⓐ Find: $f(0)$.

ⓑ Find: $f\left(\frac{1}{2}\pi\right)$.

ⓒ Find: $f\left(-\frac{3}{2}\pi\right)$.

ⓓ Find the values for x when $f(x) = 0$.

ⓔ Find the x-intercepts.

ⓕ Find the y-intercepts.

ⓖ Find the domain. Write it in interval notation.

ⓗ Find the range. Write it in interval notation.

> TRY IT :: 3.118   Use the graph of the function to find the indicated values.

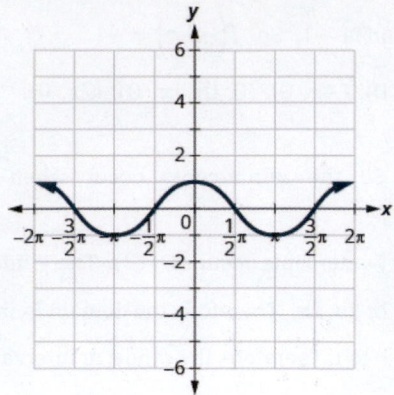

ⓐ Find: $f(0)$.

ⓑ Find: $f(\pi)$.

ⓒ Find: $f(-\pi)$.

ⓓ Find the values for x when $f(x) = 0$.

ⓔ Find the x-intercepts.

ⓕ Find the y-intercepts.

ⓖ Find the domain. Write it in interval notation.

ⓗ Find the range. Write it in interval notation.

▶ **MEDIA ::**

Access this online resource for additional instruction and practice with graphs of functions.

• **Find Domain and Range (https://openstax.org/l/37domainrange)**

3.6 EXERCISES

Practice Makes Perfect

Use the Vertical Line Test

In the following exercises, determine whether each graph is the graph of a function.

337. (a)

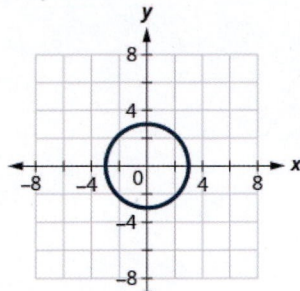

(b)

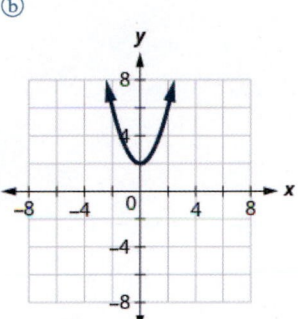

338. (a)

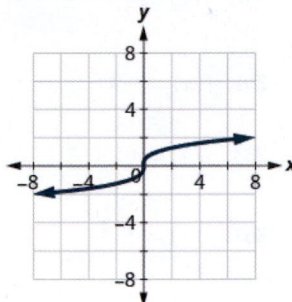

(b)

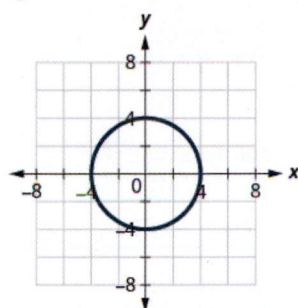

339. (a)

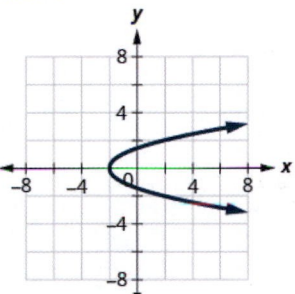

(b)

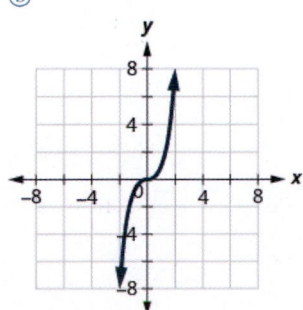

340. (a)

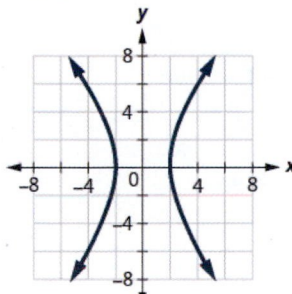

(b)

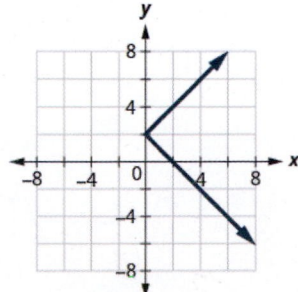

Identify Graphs of Basic Functions

In the following exercises, ⓐ graph each function ⓑ state its domain and range. Write the domain and range in interval notation.

341. $f(x) = 3x + 4$

342. $f(x) = 2x + 5$

343. $f(x) = -x - 2$

344. $f(x) = -4x - 3$

345. $f(x) = -2x + 2$

346. $f(x) = -3x + 3$

347. $f(x) = \frac{1}{2}x + 1$

348. $f(x) = \frac{2}{3}x - 2$

349. $f(x) = 5$

350. $f(x) = 2$

351. $f(x) = -3$

352. $f(x) = -1$

353. $f(x) = 2x$

354. $f(x) = 3x$

355. $f(x) = -2x$

356. $f(x) = -3x$

357. $f(x) = 3x^2$

358. $f(x) = 2x^2$

359. $f(x) = -3x^2$

360. $f(x) = -2x^2$

361. $f(x) = \frac{1}{2}x^2$

362. $f(x) = \frac{1}{3}x^2$

363. $f(x) = x^2 - 1$

364. $f(x) = x^2 + 1$

365. $f(x) = -2x^3$

366. $f(x) = 2x^3$

367. $f(x) = x^3 + 2$

368. $f(x) = x^3 - 2$

369. $f(x) = 2\sqrt{x}$

370. $f(x) = -2\sqrt{x}$

371. $f(x) = \sqrt{x - 1}$

372. $f(x) = \sqrt{x + 1}$

373. $f(x) = 3|x|$

374. $f(x) = -2|x|$

375. $f(x) = |x| + 1$

376. $f(x) = |x| - 1$

Read Information from a Graph of a Function

In the following exercises, use the graph of the function to find its domain and range. Write the domain and range in interval notation.

377.

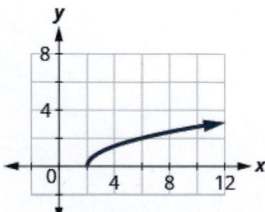

378.

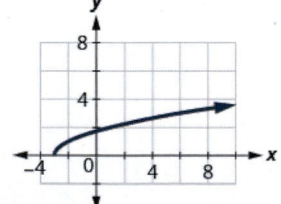

379.

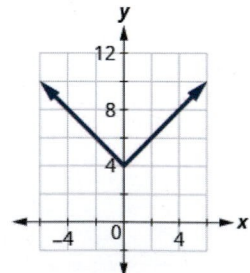

380.

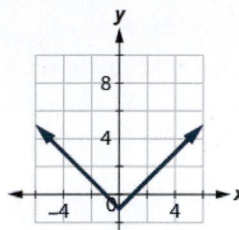

381.

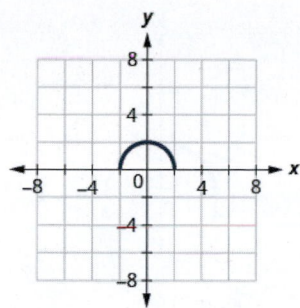

382.

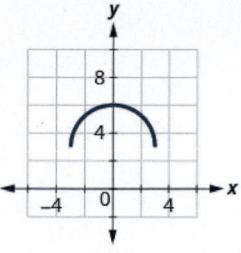

In the following exercises, use the graph of the function to find the indicated values.

383.

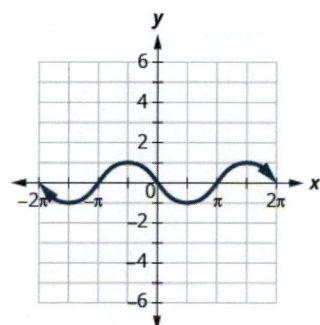

ⓐ Find: $f(0)$.

ⓑ Find: $f\left(\frac{1}{2}\pi\right)$.

ⓒ Find: $f\left(-\frac{3}{2}\pi\right)$.

ⓓ Find the values for x when $f(x) = 0$.

ⓔ Find the x-intercepts.

ⓕ Find the y-intercepts.

ⓖ Find the domain. Write it in interval notation.

ⓗ Find the range. Write it in interval notation.

384.

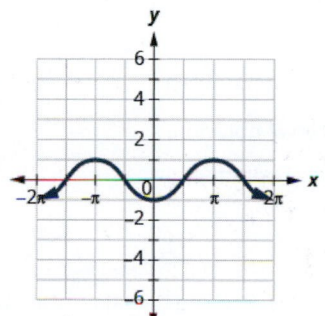

ⓐ Find: $f(0)$.

ⓑ Find: $f(\pi)$.

ⓒ Find: $f(-\pi)$.

ⓓ Find the values for x when $f(x) = 0$.

ⓔ Find the x-intercepts.

ⓕ Find the y-intercepts.

ⓖ Find the domain. Write it in interval notation.

ⓗ Find the range. Write it in interval notation

385.

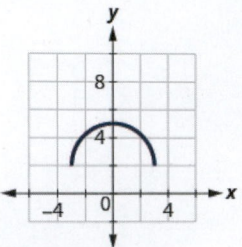

ⓐ Find: $f(0)$.

ⓑ Find: $f(-3)$.

ⓒ Find: $f(3)$.

ⓓ Find the values for x when $f(x) = 0$.

ⓔ Find the x-intercepts.

ⓕ Find the y-intercepts.

ⓖ Find the domain. Write it in interval notation.

ⓗ Find the range. Write it in interval notation.

386.

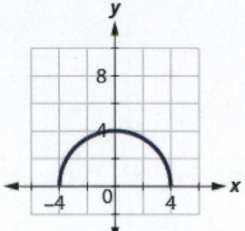

ⓐ Find: $f(0)$.

ⓑ Find the values for x when $f(x) = 0$.

ⓒ Find the x-intercepts.

ⓓ Find the y-intercepts.

ⓔ Find the domain. Write it in interval notation.

ⓕ Find the range. Write it in interval notation

Writing Exercises

387. Explain in your own words how to find the domain from a graph.

388. Explain in your own words how to find the range from a graph.

389. Explain in your own words how to use the vertical line test.

390. Draw a sketch of the square and cube functions. What are the similarities and differences in the graphs?

Self Check

ⓐ After completing the exercises, use this checklist to evaluate your mastery of the objectives of this section.

I can...	Confidently	With some help	No-I don't get it!
use the vertical line test.			
identify graphs of basic functions.			
read information from a graph.			

ⓑ After reviewing this checklist, what will you do to become confident for all objectives?

CHAPTER 3 REVIEW

KEY TERMS

boundary line The line with equation $Ax + By = C$ is the boundary line that separates the region where $Ax + By > C$ from the region where $Ax + By < C$.

domain of a relation The domain of a relation is all the x-values in the ordered pairs of the relation.

function A function is a relation that assigns to each element in its domain exactly one element in the range.

horizontal line A horizontal line is the graph of an equation of the form $y = b$. The line passes through the y-axis at $(0, b)$.

intercepts of a line The points where a line crosses the x-axis and the y-axis are called the intercepts of the line.

linear equation An equation of the form $Ax + By = C$, where A and B are not both zero, is called a linear equation in two variables.

linear inequality A linear inequality is an inequality that can be written in one of the following forms: $Ax + By > C$, $Ax + By \geq C$, $Ax + By < C$, or $Ax + By \leq C$, where A and B are not both zero.

mapping A mapping is sometimes used to show a relation. The arrows show the pairing of the elements of the domain with the elements of the range.

ordered pair An ordered pair, (x, y) gives the coordinates of a point in a rectangular coordinate system. The first number is the x-coordinate. The second number is the y-coordinate.

origin The point $(0, 0)$ is called the origin. It is the point where the x-axis and y-axis intersect.

parallel lines Parallel lines are lines in the same plane that do not intersect.

perpendicular lines Perpendicular lines are lines in the same plane that form a right angle.

point-slope form The point-slope form of an equation of a line with slope m and containing the point (x_1, y_1) is $y - y_1 = m(x - x_1)$.

range of a relation The range of a relation is all the y-values in the ordered pairs of the relation.

relation A relation is any set of ordered pairs, (x, y). All the x-values in the ordered pairs together make up the domain. All the y-values in the ordered pairs together make up the range.

solution of a linear equation in two variables An ordered pair (x, y) is a solution of the linear equation $Ax + By = C$, if the equation is a true statement when the x- and y-values of the ordered pair are substituted into the equation.

solution to a linear inequality An ordered pair (x, y) is a solution to a linear inequality if the inequality is true when we substitute the values of x and y.

standard form of a linear equation A linear equation is in standard form when it is written $Ax + By = C$.

vertical line A vertical line is the graph of an equation of the form $x = a$. The line passes through the x-axis at $(a, 0)$.

KEY CONCEPTS

3.1 Graph Linear Equations in Two Variables

- **Points on the Axes**
  - Points with a y-coordinate equal to 0 are on the x-axis, and have coordinates $(a, 0)$.
  - Points with an x-coordinate equal to 0 are on the y-axis, and have coordinates $(0, b)$.

- **Quadrant**

Quadrant I	Quadrant II	Quadrant III	Quadrant IV
(x, y)	(x, y)	(x, y)	(x, y)
$(+, +)$	$(-, +)$	$(-, -)$	$(+, -)$

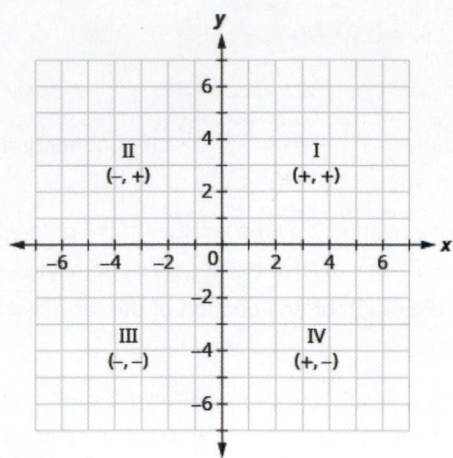

- **Graph of a Linear Equation:** The graph of a linear equation $Ax + By = C$ is a straight line.

  Every point on the line is a solution of the equation.
  Every solution of this equation is a point on this line.

- **How to graph a linear equation by plotting points.**

  Step 1. Find three points whose coordinates are solutions to the equation. Organize them in a table.

  Step 2. Plot the points in a rectangular coordinate system. Check that the points line up. If they do not, carefully check your work.

  Step 3. Draw the line through the three points. Extend the line to fill the grid and put arrows on both ends of the line.

- ***x*-intercept and *y*-intercept of a Line**

  - The *x*-intercept is the point $(a, 0)$ where the line crosses the *x*-axis.

  - The *y*-intercept is the point $(0, b)$ where the line crosses the *y*-axis.

• The *x*-intercept occurs when *y* is zero.	
• The *y*-intercept occurs when *x* is zero.	

x	y
a	0
0	b

- **Find the *x*- and *y*-intercepts from the Equation of a Line**

  - Use the equation of the line. To find:
    the *x*-intercept of the line, let $y = 0$ and solve for *x*.

    the *y*-intercept of the line, let $x = 0$ and solve for *y*.

- **How to graph a linear equation using the intercepts.**

  Step 1. Find the *x*- and *y*-intercepts of the line.
  Let $y = 0$ and solve for *x*.

  Let $x = 0$ and solve for *y*.

  Step 2. Find a third solution to the equation.

  Step 3. Plot the three points and check that they line up.

  Step 4. Draw the line

3.2 Slope of a Line

- **Slope of a Line**

  - The slope of a line is $m = \dfrac{\text{rise}}{\text{run}}$.

  - The rise measures the vertical change and the run measures the horizontal change.

- **How to find the slope of a line from its graph using** $m = \frac{\text{rise}}{\text{run}}$.

  Step 1. Locate two points on the line whose coordinates are integers.

  Step 2. Starting with one point, sketch a right triangle, going from the first point to the second point.

  Step 3. Count the rise and the run on the legs of the triangle.

  Step 4. Take the ratio of rise to run to find the slope: $m = \frac{\text{rise}}{\text{run}}$.

- **Slope of a line between two points.**

  - The slope of the line between two points (x_1, y_1) and (x_2, y_2) is:

  $$m = \frac{y_2 - y_1}{x_2 - x_1}.$$

- **How to graph a line given a point and the slope.**

  Step 1. Plot the given point.

  Step 2. Use the slope formula $m = \frac{\text{rise}}{\text{run}}$ to identify the rise and the run.

  Step 3. Starting at the given point, count out the rise and run to mark the second point.

  Step 4. Connect the points with a line.

- **Slope Intercept Form of an Equation of a Line**

  - The slope–intercept form of an equation of a line with slope m and y-intercept, $(0, b)$ is $y = mx + b$

Methods to Graph Lines			
Point Plotting	**Slope–Intercept**	**Intercepts**	**Recognize Vertical and Horizontal Lines**
x | y	$y = mx + b$	x | y 0 | | 0	
Find three points. Plot the points, make sure they line up, then draw the line.	Find the slope and y-intercept. Start at the y-intercept, then count the slope to get a second point.	Find the intercepts and a third point. Plot the points, make sure they line up, then draw the line.	The equation has only one variable. $x = a$ vertical $y = b$ horizontal

- **Parallel Lines**

  - Parallel lines are lines in the same plane that do not intersect. Parallel lines have the same slope and different y-intercepts. If m_1 and m_2 are the slopes of two parallel lines then $m_1 = m_2$.

    Parallel vertical lines have different x-intercepts.

- **Perpendicular Lines**

  - Perpendicular lines are lines in the same plane that form a right angle.

  - If m_1 and m_2 are the slopes of two perpendicular lines, then:

    their slopes are negative reciprocals of each other, $m_1 = -\frac{1}{m_2}$.

    the product of their slopes is -1, $m_1 \cdot m_2 = -1$.

  - A vertical line and a horizontal line are always perpendicular to each other.

3.3 Find the Equation of a Line

- **How to find an equation of a line given the slope and a point.**

  Step 1. Identify the slope.

  Step 2. Identify the point.

  Step 3. Substitute the values into the point-slope form, $y - y_1 = m(x - x_1)$.

Step 4. Write the equation in slope-intercept form.

- **How to find an equation of a line given two points.**

  Step 1. Find the slope using the given points. $m = \dfrac{y_2 - y_1}{x_2 - x_1}$

  Step 2. Choose one point.

  Step 3. Substitute the values into the point-slope form: $y - y_1 = m(x - x_1)$.

  Step 4. Write the equation in slope-intercept form.

To Write an Equation of a Line		
If given:	**Use:**	**Form:**
Slope and y-intercept	**slope-intercept**	$y = mx + b$
Slope and a point	**point-slope**	$y - y_1 = m(x - x_1)$
Two points	**point-slope**	$y - y_1 = m(x - x_1)$

- **How to find an equation of a line parallel to a given line.**

  Step 1. Find the slope of the given line.

  Step 2. Find the slope of the parallel line.

  Step 3. Identify the point.

  Step 4. Substitute the values into the point-slope form: $y - y_1 = m(x - x_1)$.

  Step 5. Write the equation in slope-intercept form

- **How to find an equation of a line perpendicular to a given line.**

  Step 1. Find the slope of the given line.

  Step 2. Find the slope of the perpendicular line.

  Step 3. Identify the point.

  Step 4. Substitute the values into the point-slope form, $y - y_1 = m(x - x_1)$

  Step 5. Write the equation in slope-intercept form.

3.4 Graph Linear Inequalities in Two Variables

- **How to graph a linear inequality in two variables.**

  Step 1. Identify and graph the boundary line.
  If the inequality is $\leq$ or $\geq$, the boundary line is solid.
  If the inequality is $<$ or $>$, the boundary line is dashed.

  Step 2. Test a point that is not on the boundary line. Is it a solution of the inequality?

  Step 3. Shade in one side of the boundary line.
  If the test point is a solution, shade in the side that includes the point.
  If the test point is not a solution, shade in the opposite side.

3.5 Relations and Functions

- **Function Notation:** For the function $y = f(x)$

  ○ f is the name of the function

  ○ x is the domain value

  ○ $f(x)$ is the range value y corresponding to the value x

We read $f(x)$ as f of x or the value of f at x.

- **Independent and Dependent Variables:** For the function $y = f(x)$,
    - x is the independent variable as it can be any value in the domain
    - y is the dependent variable as its value depends on x

3.6 Graphs of Functions

- **Vertical Line Test**
    - A set of points in a rectangular coordinate system is the graph of a function if every vertical line intersects the graph in at most one point.
    - If any vertical line intersects the graph in more than one point, the graph does not represent a function.
- **Graph of a Function**
    - The graph of a function is the graph of all its ordered pairs, (x, y) or using function notation, $(x, f(x))$ where $y = f(x)$.

$$f \quad \text{name of function}$$
$$x \quad x\text{-coordinate of the ordered pair}$$
$$f(x) \quad y\text{-coordinate of the ordered pair}$$

- **Linear Function**

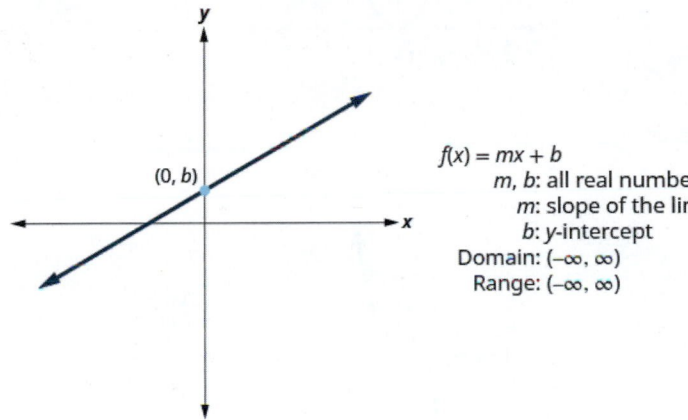

$f(x) = mx + b$
m, b: all real numbers
m: slope of the line
b: y-intercept
Domain: $(-\infty, \infty)$
Range: $(-\infty, \infty)$

- **Constant Function**

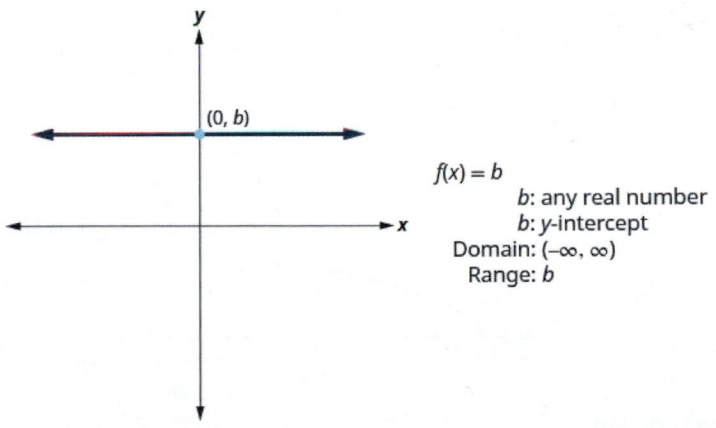

$f(x) = b$
b: any real number
b: y-intercept
Domain: $(-\infty, \infty)$
Range: b

- **Identity Function**

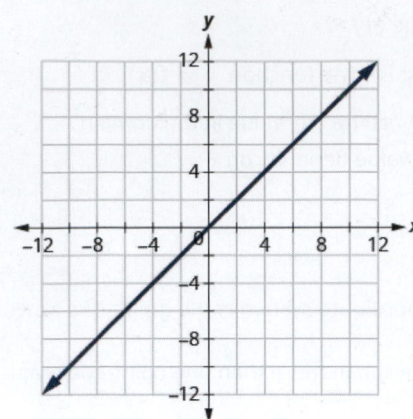

$f(x) = x$
m: 1
b: 0
Domain: $(-\infty, \infty)$
Range: $(-\infty, \infty)$

- **Square Function**

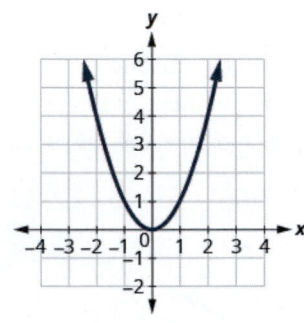

$f(x) = x^2$
Domain: $(-\infty, \infty)$
Range: $[0, \infty)$

- **Cube Function**

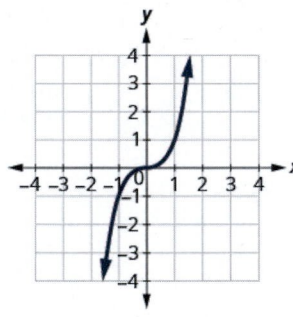

$f(x) = x^3$
Domain: $(-\infty, \infty)$
Range: $(-\infty, \infty)$

- **Square Root Function**

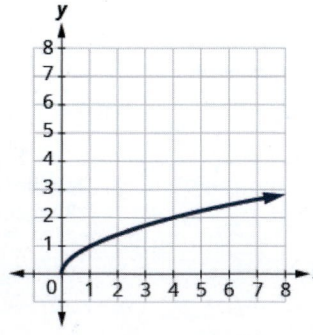

$f(x) = \sqrt{x}$
Domain: $[0, \infty)$
Range: $[0, \infty)$

- **Absolute Value Function**

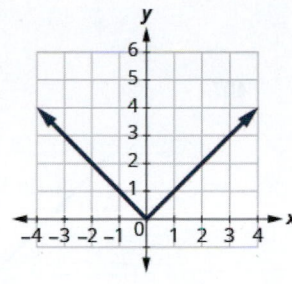

$f(x) = |x|$
Domain: $(-\infty, \infty)$
Range: $[0, \infty)$

REVIEW EXERCISES

3.1 Graph Linear Equations in Two Variables

Plot Points in a Rectangular Coordinate System

In the following exercises, plot each point in a rectangular coordinate system.

391.
ⓐ $(-1, -5)$
ⓑ $(-3, 4)$
ⓒ $(2, -3)$
ⓓ $\left(1, \frac{5}{2}\right)$

392.
ⓐ $(-2, 0)$
ⓑ $(0, -4)$
ⓒ $(0, 5)$
ⓓ $(3, 0)$

In the following exercises, determine which ordered pairs are solutions to the given equations.

393. $5x + y = 10$;
ⓐ $(5, 1)$
ⓑ $(2, 0)$
ⓒ $(4, -10)$

394. $y = 6x - 2$;
ⓐ $(1, 4)$
ⓑ $\left(\frac{1}{3}, 0\right)$
ⓒ $(6, -2)$

Graph a Linear Equation by Plotting Points

In the following exercises, graph by plotting points.

395. $y = 4x - 3$

396. $y = -3x$

397. $y = \frac{1}{2}x + 3$

398. $y = -\frac{4}{5}x - 1$

399. $x - y = 6$

400. $2x + y = 7$

401. $3x - 2y = 6$

Graph Vertical and Horizontal lines

In the following exercises, graph each equation.

402. $y = -2$

403. $x = 3$

In the following exercises, graph each pair of equations in the same rectangular coordinate system.

404. $y = -2x$ and $y = -2$

405. $y = \frac{4}{3}x$ and $y = \frac{4}{3}$

Find x- and y-Intercepts

In the following exercises, find the x- and y-intercepts.

406.

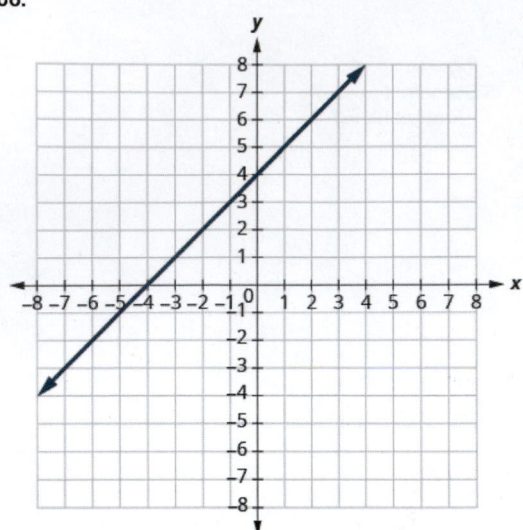

407.

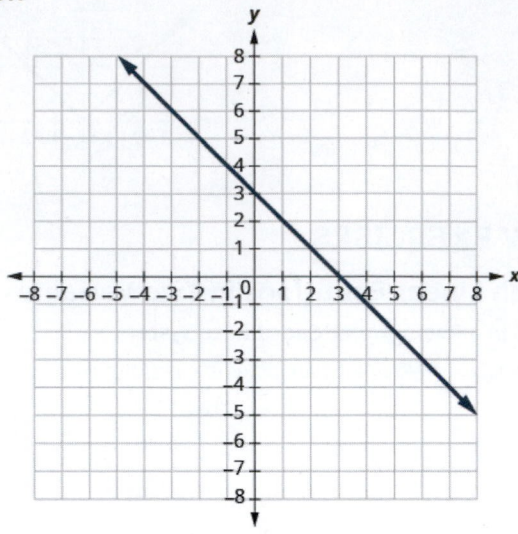

In the following exercises, find the intercepts of each equation.

408. $x - y = -1$

409. $x + 2y = 6$

410. $2x + 3y = 12$

411. $y = \frac{3}{4}x - 12$

412. $y = 3x$

Graph a Line Using the Intercepts

In the following exercises, graph using the intercepts.

413. $-x + 3y = 3$

414. $x - y = 4$

415. $2x - y = 5$

416. $2x - 4y = 8$

417. $y = 4x$

3.2 Slope of a Line

Find the Slope of a Line

In the following exercises, find the slope of each line shown.

418.

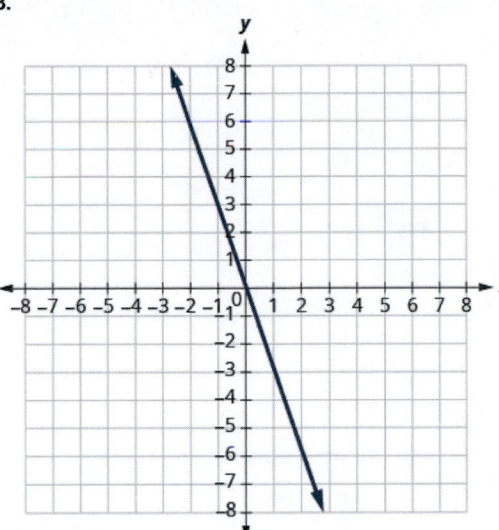

419.

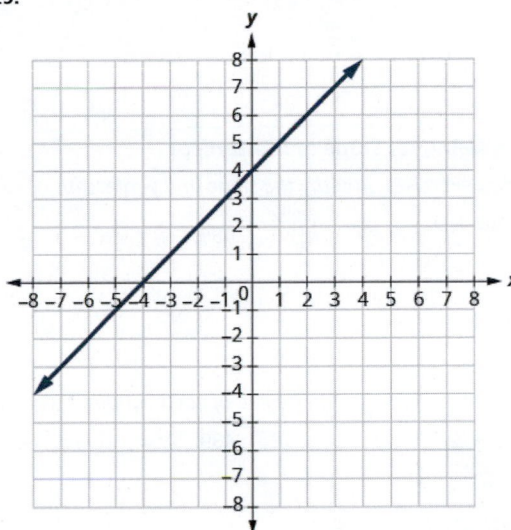

420.

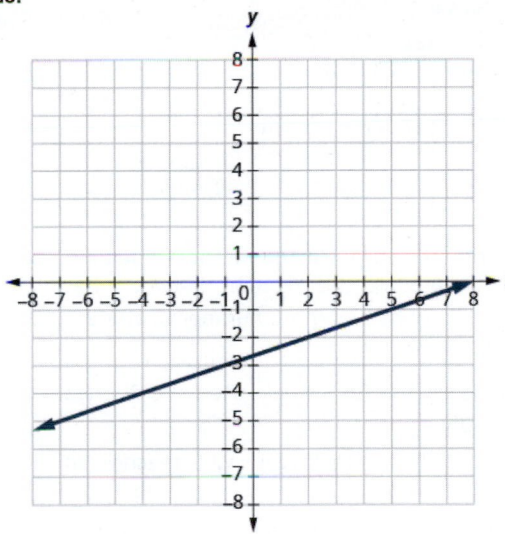

421.

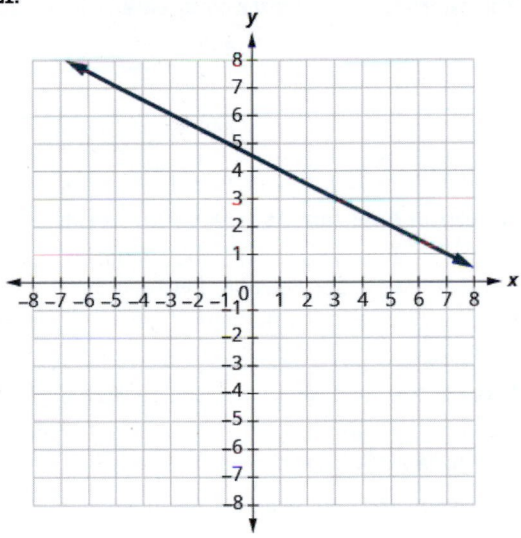

In the following exercises, find the slope of each line.

422. $y = 2$

423. $x = 5$

424. $x = -3$

425. $y = -1$

Use the Slope Formula to find the Slope of a Line between Two Points

In the following exercises, use the slope formula to find the slope of the line between each pair of points.

426. $(-1, -1), (0, 5)$

427. $(3.5), (4, -1)$

428. $(-5, -2), (3, 2)$

429. $(2, 1), (4, 6)$

Graph a Line Given a Point and the Slope

In the following exercises, graph each line with the given point and slope.

430. $(2, -2)$; $m = \frac{5}{2}$

431. $(-3, 4)$; $m = -\frac{1}{3}$

432. x-intercept -4; $m = 3$

433. y-intercept 1; $m = -\frac{3}{4}$

Graph a Line Using Its Slope and Intercept

In the following exercises, identify the slope and y-intercept of each line.

434. $y = -4x + 9$

435. $y = \frac{5}{3}x - 6$

436. $5x + y = 10$

437. $4x - 5y = 8$

In the following exercises, graph the line of each equation using its slope and y-intercept.

438. $y = 2x + 3$

439. $y = -x - 1$

440. $y = -\frac{2}{5}x + 3$

441. $4x - 3y = 12$

In the following exercises, determine the most convenient method to graph each line.

442. $x = 5$

443. $y = -3$

444. $2x + y = 5$

445. $x - y = 2$

446. $y = \frac{2}{2}x + 2$

447. $y = \frac{3}{4}x - 1$

Graph and Interpret Applications of Slope-Intercept

448. Katherine is a private chef. The equation $C = 6.5m + 42$ models the relation between her weekly cost, C, in dollars and the number of meals, m, that she serves.

ⓐ Find Katherine's cost for a week when she serves no meals.

ⓑ Find the cost for a week when she serves 14 meals.

ⓒ Interpret the slope and C-intercept of the equation.

ⓓ Graph the equation.

449. Marjorie teaches piano. The equation $P = 35h - 250$ models the relation between her weekly profit, P, in dollars and the number of student lessons, s, that she teaches.

ⓐ Find Marjorie's profit for a week when she teaches no student lessons.

ⓑ Find the profit for a week when she teaches 20 student lessons.

ⓒ Interpret the slope and P-intercept of the equation.

ⓓ Graph the equation.

Use Slopes to Identify Parallel and Perpendicular Lines

In the following exercises, use slopes and y-intercepts to determine if the lines are parallel, perpendicular, or neither.

450. $4x - 3y = -1$; $y = \frac{4}{3}x - 3$

451. $y = 5x - 1$; $10x + 2y = 0$

452. $3x - 2y = 5$; $2x + 3y = 6$

453. $2x - y = 8$; $x - 2y = 4$

3.3 Find the Equation of a Line

Find an Equation of the Line Given the Slope and *y*-Intercept

In the following exercises, find the equation of a line with given slope and y-intercept. Write the equation in slope–intercept form.

454. slope $\frac{1}{3}$ and *y*-intercept $(0, -6)$

455. slope -5 and *y*-intercept $(0, -3)$

456. slope 0 and *y*-intercept $(0, 4)$

457. slope -2 and *y*-intercept $(0, 0)$

In the following exercises, find the equation of the line shown in each graph. Write the equation in slope–intercept form.

458.

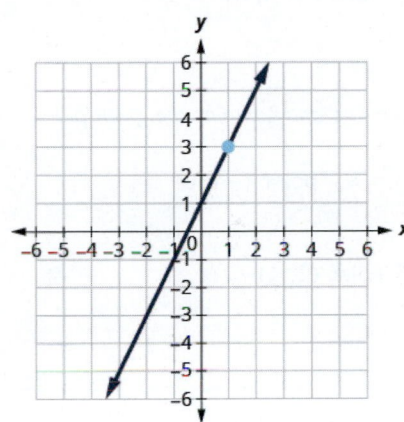

459.

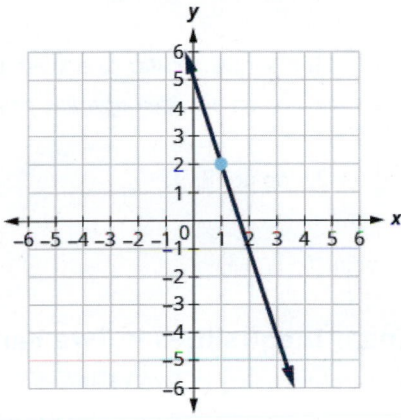

460.

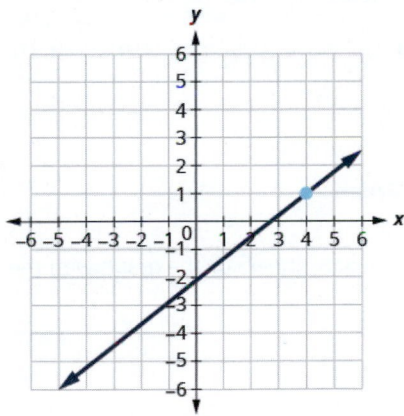

461.

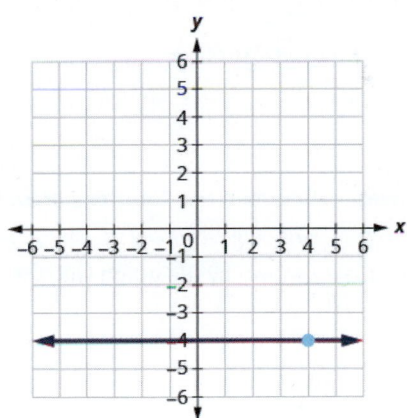

Find an Equation of the Line Given the Slope and a Point

In the following exercises, find the equation of a line with given slope and containing the given point. Write the equation in slope–intercept form.

462. $m = -\frac{1}{4}$, point $(-8, 3)$

463. $m = \frac{3}{5}$, point $(10, 6)$

464. Horizontal line containing $(-2, 7)$

465. $m = -2$, point $(-1, -3)$

Find an Equation of the Line Given Two Points

In the following exercises, find the equation of a line containing the given points. Write the equation in slope-intercept form.

466. $(2, 10)$ and $(-2, -2)$

467. $(7, 1)$ and $(5, 0)$

468. $(3, 8)$ and $(3, -4)$

469. $(5, 2)$ and $(-1, 2)$

Find an Equation of a Line Parallel to a Given Line

In the following exercises, find an equation of a line parallel to the given line and contains the given point. Write the equation in slope-intercept form.

470. line $y = -3x + 6$, point $(1, -5)$

471. line $2x + 5y = -10$, point $(10, 4)$

472. line $x = 4$, point $(-2, -1)$

473. line $y = -5$, point $(-4, 3)$

Find an Equation of a Line Perpendicular to a Given Line

In the following exercises, find an equation of a line perpendicular to the given line and contains the given point. Write the equation in slope-intercept form.

474. line $y = -\frac{4}{5}x + 2$, point $(8, 9)$

475. line $2x - 3y = 9$, point $(-4, 0)$

476. line $y = 3$, point $(-1, -3)$

477. line $x = -5$ point $(2, 1)$

3.4 Graph Linear Inequalities in Two Variables

Verify Solutions to an Inequality in Two Variables

In the following exercises, determine whether each ordered pair is a solution to the given inequality.

478. Determine whether each ordered pair is a solution to the inequality $y < x - 3$:

ⓐ $(0, 1)$ ⓑ $(-2, -4)$ ⓒ $(5, 2)$ ⓓ $(3, -1)$
ⓔ $(-1, -5)$

479. Determine whether each ordered pair is a solution to the inequality $x + y > 4$:

ⓐ $(6, 1)$ ⓑ $(-3, 6)$ ⓒ $(3, 2)$ ⓓ $(-5, 10)$ ⓔ $(0, 0)$

Recognize the Relation Between the Solutions of an Inequality and its Graph

In the following exercises, write the inequality shown by the shaded region.

480. Write the inequality shown by the graph with the boundary line $y = -x + 2$.

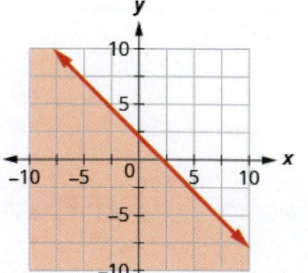

481. Write the inequality shown by the graph with the boundary line $y = \frac{2}{3}x - 3$.

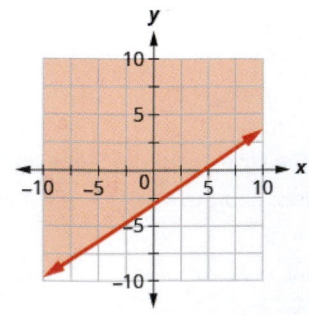

482. Write the inequality shown by the shaded region in the graph with the boundary line $x + y = -4$.

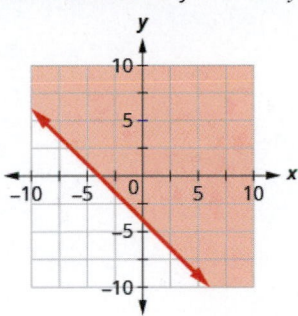

483. Write the inequality shown by the shaded region in the graph with the boundary line $x - 2y = 6$.

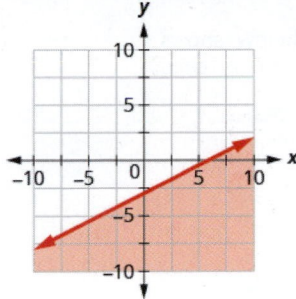

Graph Linear Inequalities in Two Variables

In the following exercises, graph each linear inequality.

484. Graph the linear inequality $y > \frac{2}{5}x - 4$.

485. Graph the linear inequality $y \leq -\frac{1}{4}x + 3$.

486. Graph the linear inequality $x - y \leq 5$.

487. Graph the linear inequality $3x + 2y > 10$.

488. Graph the linear inequality $y \leq -3x$.

489. Graph the linear inequality $y < 6$.

Solve Applications using Linear Inequalities in Two Variables

490. Shanthie needs to earn at least $500 a week during her summer break to pay for college. She works two jobs. One as a swimming instructor that pays $10 an hour and the other as an intern in a law office for $25 hour. How many hours does Shanthie need to work at each job to earn at least $500 per week?

ⓐ Let x be the number of hours she works teaching swimming and let y be the number of hours she works as an intern. Write an inequality that would model this situation.

ⓑ Graph the inequality.

ⓒ Find three ordered pairs (x, y) that would be solutions to the inequality. Then, explain what that means for Shanthie.

491. Atsushi he needs to exercise enough to burn 600 calories each day. He prefers to either run or bike and burns 20 calories per minute while running and 15 calories a minute while biking.

ⓐ If x is the number of minutes that Atsushi runs and y is the number minutes he bikes, find the inequality that models the situation.

ⓑ Graph the inequality.

ⓒ List three solutions to the inequality. What options do the solutions provide Atsushi?

3.5 Relations and Functions

Find the Domain and Range of a Relation

In the following exercises, for each relation, ⓐ find the domain of the relation ⓑ find the range of the relation.

492. $\{(5, -2), (5, -4), (7, -6), (8, -8), (9, -10)\}$

493. $\{(-3, 7), (-2, 3), (-1, 9), (0, -3), (-1, 8)\}$

In the following exercise, use the mapping of the relation to ⓐ list the ordered pairs of the relation ⓑ find the domain of the relation ⓒ find the range of the relation.

494. The mapping below shows the average weight of a child according to age.

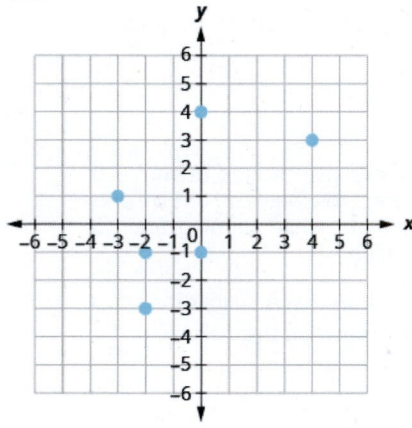

In the following exercise, use the graph of the relation to ⓐ list the ordered pairs of the relation ⓑ find the domain of the relation ⓒ find the range of the relation.

495.

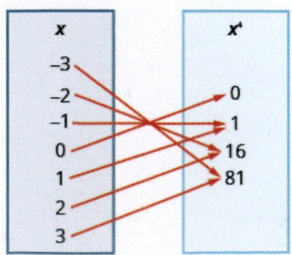

Determine if a Relation is a Function

In the following exercises, use the set of ordered pairs to ⓐ determine whether the relation is a function ⓑ find the domain of the relation ⓒ find the range of the relation.

496. $\{(9, -5), (4, -3), (1, -1),$
$(0, 0), (1, 1), (4, 3), (9, 5)\}$

497. $\{(-3, 27), (-2, 8), (-1, 1),$
$(0, 0), (1, 1), (2, 8), (3, 27)\}$

In the following exercises, use the mapping to ⓐ determine whether the relation is a function ⓑ find the domain of the function ⓒ find the range of the function.

498.

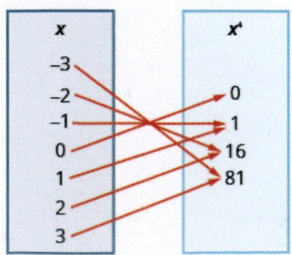

499.

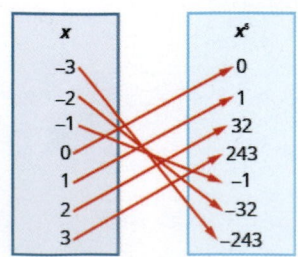

In the following exercises, determine whether each equation is a function.

500. $2x + y = -3$

501. $y = x^2$

502. $y = 3x - 5$

503. $y = x^3$

504. $2x + y^2 = 4$

Find the Value of a Function

In the following exercises, evaluate the function:

ⓐ $f(-2)$ ⓑ $f(3)$ ⓒ $f(a)$.

505. $f(x) = 3x - 4$

506. $f(x) = -2x + 5$

507. $f(x) = x^2 - 5x + 6$

508. $f(x) = 3x^2 - 2x + 1$

In the following exercises, evaluate the function.

509. $g(x) = 3x^2 - 5x;\ g(2)$

510. $F(x) = 2x^2 - 3x + 1;$ $F(-1)$

511. $h(t) = 4|t - 1| + 2;\ h(-3)$

512. $f(x) = \frac{x + 2}{x - 1};\ f(3)$

3.6 Graphs of Functions

Use the Vertical line Test

In the following exercises, determine whether each graph is the graph of a function.

513.

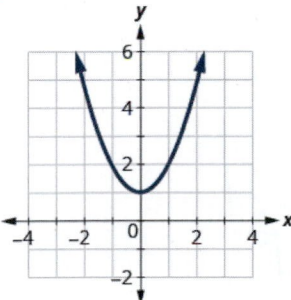

514.

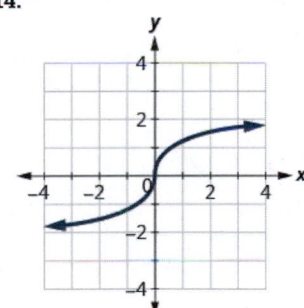

515.

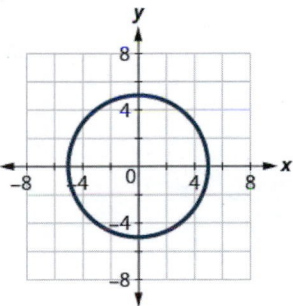

516.

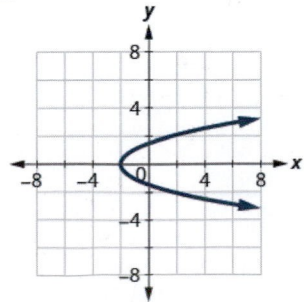

517.

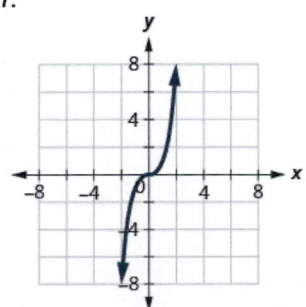

518.

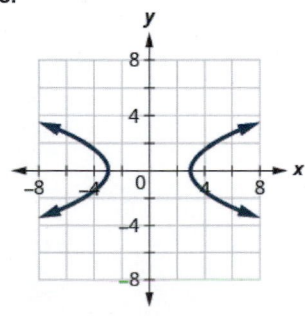

519.

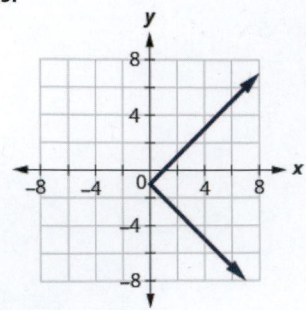

Identify Graphs of Basic Functions

In the following exercises, ⓐ graph each function ⓑ state its domain and range. Write the domain and range in interval notation.

520. $f(x) = 5x + 1$

521. $f(x) = -4x - 2$

522. $f(x) = \frac{2}{3}x - 1$

523. $f(x) = -6$

524. $f(x) = 2x$

525. $f(x) = 3x^2$

526. $f(x) = -\frac{1}{2}x^2$

527. $f(x) = x^2 + 2$

528. $f(x) = x^3 - 2$

529. $f(x) = \sqrt{x + 2}$

530. $f(x) = -|x|$

531. $f(x) = |x| + 1$

Read Information from a Graph of a Function

In the following exercises, use the graph of the function to find its domain and range. Write the domain and range in interval notation

532.

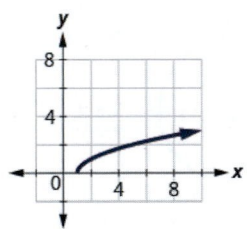

533.

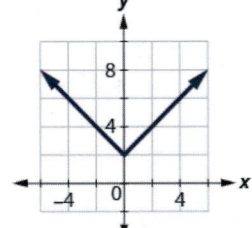

534.

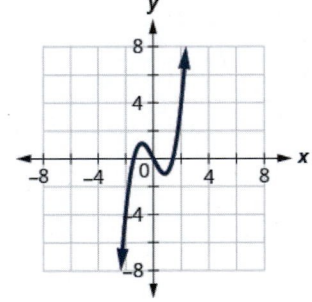

In the following exercises, use the graph of the function to find the indicated values.

535.

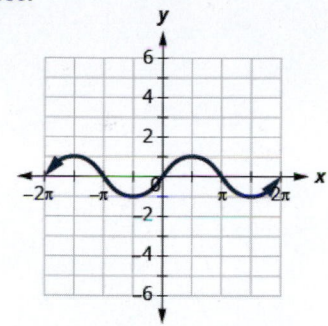

536.

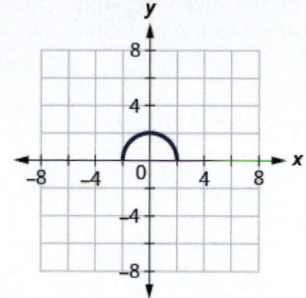

ⓐ Find $f(0)$.

ⓑ Find $f\left(\frac{1}{2}\pi\right)$.

ⓒ Find $f\left(-\frac{3}{2}\pi\right)$.

ⓓ Find the values for x when $f(x) = 0$.

ⓔ Find the x-intercepts.

ⓕ Find the y-intercepts.

ⓖ Find the domain. Write it in interval notation.

ⓗ Find the range. Write it in interval notation.

ⓐ Find $f(0)$.

ⓑ Find the values for x when $f(x) = 0$.

ⓒ Find the x-intercepts.

ⓓ Find the y-intercepts.

ⓔ Find the domain. Write it in interval notation.

ⓕ Find the range. Write it in interval notation.

PRACTICE TEST

537. Plot each point in a rectangular coordinate system.

ⓐ $(2, 5)$

ⓑ $(-1, -3)$

ⓒ $(0, 2)$

ⓓ $\left(-4, \frac{3}{2}\right)$

ⓔ $(5, 0)$

538. Which of the given ordered pairs are solutions to the equation $3x - y = 6$?

ⓐ $(3, 3)$ ⓑ $(2, 0)$ ⓒ $(4, -6)$

Find the slope of each line shown.

539. ⓐ

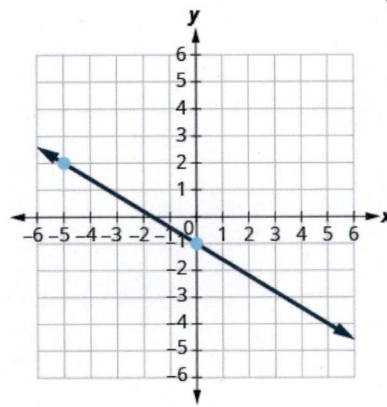

ⓑ

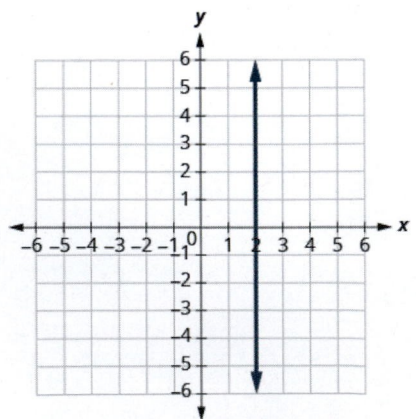

540. Find the slope of the line between the points $(5, 2)$ and $(-1, -4)$.

541. Graph the line with slope $\frac{1}{2}$ containing the point $(-3, -4)$.

542. Find the intercepts of $4x + 2y = -8$ and graph.

Graph the line for each of the following equations.

543. $y = \frac{5}{3}x - 1$

544. $y = -x$

545. $y = 2$

Find the equation of each line. Write the equation in slope-intercept form.

546. slope $-\frac{3}{4}$ and y-intercept $(0, -2)$

547. $m = 2$, point $(-3, -1)$

548. containing $(10, 1)$ and $(6, -1)$

549. perpendicular to the line $y = \frac{5}{4}x + 2$, containing the point $(-10, 3)$

550. Write the inequality shown by the graph with the boundary line $y = -x - 3$.

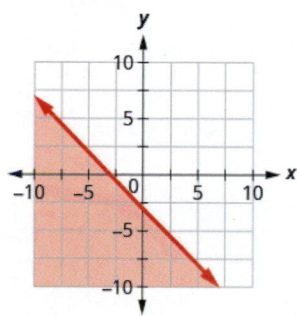

Graph each linear inequality.

551. $y > \frac{3}{2}x + 5$

552. $x - y \geq -4$

553. $y \leq -5x$

554. Hiro works two part time jobs in order to earn enough money to meet her obligations of at least $450 a week. Her job at the mall pays $10 an hour and her administrative assistant job on campus pays $15 an hour. How many hours does Hiro need to work at each job to earn at least $450?

ⓐ Let x be the number of hours she works at the mall and let y be the number of hours she works as administrative assistant. Write an inequality that would model this situation.

ⓑ Graph the inequality .

ⓒ Find three ordered pairs (x, y) that would be solutions to the inequality. Then explain what that means for Hiro.

555. Use the set of ordered pairs to ⓐ determine whether the relation is a function, ⓑ find the domain of the relation, and ⓒ find the range of the relation.

$\{(-3, 27), (-2, 8), (-1, 1), (0, 0),$
$(1, 1), (2, 8), (3, 27)\}$

556. Evaluate the function: ⓐ $f(-1)$ ⓑ $f(2)$ ⓒ $f(c)$.

$f(x) = 4x^2 - 2x - 3$

557. For $h(y) = 3|y - 1| - 3$, evaluate $h(-4)$.

558. Determine whether the graph is the graph of a function. Explain your answer.

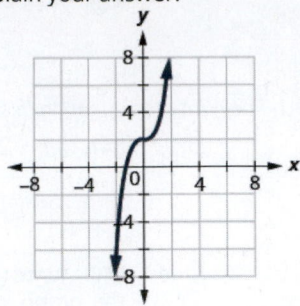

In the following exercises, ⓐ graph each function ⓑ state its domain and range. Write the domain and range in interval notation.

559. $f(x) = x^2 + 1$

560. $f(x) = \sqrt{x + 1}$

561.

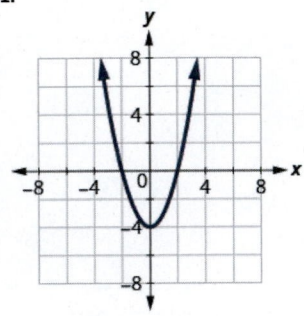

ⓑ Find the y-intercepts.

ⓒ Find $f(-1)$.

ⓓ Find $f(1)$.

ⓔ Find the domain. Write it in interval notation.

ⓕ Find the range. Write it in interval notation.

Figure 4.1 In the future, car drivers may become passengers because cars will be able to drive themselves. (credit: jingoba/Pixabay)

Chapter Outline

Introduction

Climb into your car. Put on your seatbelt. Choose your destination and then...relax. That's right. You don't have to do anything else because you are in an autonomous car, or one that navigates its way to your destination! No cars are fully autonomous at the moment and so you theoretically still need to have your hands on the wheel. Self-driving cars may help ease traffic congestion, prevent accidents, and lower pollution. The technology is thanks to computer programmers who are developing software to control the navigation of the car. These programmers rely on their understanding of mathematics, including relationships between equations. In this chapter, you will learn how to solve systems of linear equations in different ways and use them to analyze real-world situations.

4.1 Solve Systems of Linear Equations with Two Variables

Learning Objectives

By the end of this section, you will be able to:

> Determine whether an ordered pair is a solution of a system of equations
> Solve a system of linear equations by graphing
> Solve a system of equations by substitution
> Solve a system of equations by elimination
> Choose the most convenient method to solve a system of linear equations

Be Prepared!

Before you get started, take this readiness quiz.

1. For the equation $y = \frac{2}{3}x - 4$,

   ⓐ Is $(6, 0)$ a solution? ⓑ Is $(-3, -2)$ a solution?

If you missed this problem, review **Example 3.2**.

2. Find the slope and y-intercept of the line $3x - y = 12$.

   If you missed this problem, review **Example 3.16**.

3. Find the x- and y-intercepts of the line $2x - 3y = 12$.

   If you missed this problem, review **Example 3.8**.

Determine Whether an Ordered Pair is a Solution of a System of Equations

In **Solving Linear Equations**, we learned how to solve linear equations with one variable. Now we will work with two or more linear equations grouped together, which is known as a **system of linear equations**.

System of Linear Equations

When two or more linear equations are grouped together, they form a **system of linear equations**.

In this section, we will focus our work on systems of two linear equations in two unknowns. We will solve larger systems of equations later in this chapter.

An example of a system of two linear equations is shown below. We use a brace to show the two equations are grouped together to form a system of equations.

$$\begin{cases} 2x + y = 7 \\ x - 2y = 6 \end{cases}$$

A linear equation in two variables, such as $2x + y = 7$, has an infinite number of solutions. Its graph is a line. Remember, every point on the line is a solution to the equation and every solution to the equation is a point on the line.

To solve a system of two linear equations, we want to find the values of the variables that are solutions to *both* equations. In other words, we are looking for the ordered pairs (x, y) that make both equations true. These are called the **solutions of a system of equations**.

Solutions of a System of Equations

The **solutions of a system of equations** are the values of the variables that make *all* the equations true. A solution of a system of two linear equations is represented by an ordered pair (x, y).

To determine if an ordered pair is a solution to a system of two equations, we substitute the values of the variables into each equation. If the ordered pair makes both equations true, it is a solution to the system.

EXAMPLE 4.1

Determine whether the ordered pair is a solution to the system $\begin{cases} x - y = -1 \\ 2x - y = -5 \end{cases}$.

ⓐ $(-2, -1)$ ⓑ $(-4, -3)$

✓ Solution

ⓐ

$$\begin{cases} x - y = -1 \\ 2x - y = -5 \end{cases}$$

We substitute $x = -2$ and $y = -1$ into both equations.

$$x - y = -1 \qquad\qquad 2x - y = -5$$
$$-2 - (-1) \overset{?}{=} -1 \qquad 2 \cdot -2 - (-1) \overset{?}{=} -5$$
$$-1 = -1\ \checkmark \qquad\qquad\quad 5 \neq -5$$

$(-2, -1)$ does not make both $(-2, -1)$ is not a solution.
equations true.

ⓑ

We substitute $x = -4$ and $y = -3$ into both equations.

$$x - y = -1 \qquad\qquad 2x - y = -5$$
$$-4 - (-3) \overset{?}{=} -1 \qquad 2 \cdot (-4) - (-3) \overset{?}{=} -5$$
$$-1 = -1 \checkmark \qquad\qquad -5 = -5 \checkmark$$

$(-4, -3)$ does make both equations true. $(-4, -3)$ is a solution.

> **TRY IT : : 4.1**
>
> Determine whether the ordered pair is a solution to the system $\begin{cases} 3x + y = 0 \\ x + 2y = -5 \end{cases}$.
>
> ⓐ $(1, -3)$ ⓑ $(0, 0)$

> **TRY IT : : 4.2**
>
> Determine whether the ordered pair is a solution to the system $\begin{cases} x - 3y = -8 \\ -3x - y = 4 \end{cases}$.
>
> ⓐ $(2, -2)$ ⓑ $(-2, 2)$

Solve a System of Linear Equations by Graphing

In this section, we will use three methods to solve a system of linear equations. The first method we'll use is graphing.

The graph of a linear equation is a line. Each point on the line is a solution to the equation. For a system of two equations, we will graph two lines. Then we can see all the points that are solutions to each equation. And, by finding what the lines have in common, we'll find the solution to the system.

Most linear equations in one variable have one solution, but we saw that some equations, called contradictions, have no solutions and for other equations, called identities, all numbers are solutions.

Similarly, when we solve a system of two linear equations represented by a graph of two lines in the same plane, there are three possible cases, as shown.

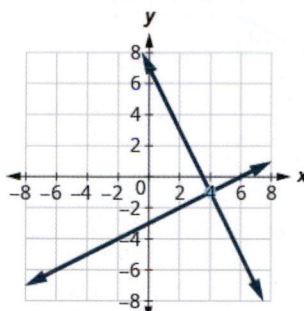

The lines intersect.
Intersecting lines have one point in common.
There is one solution to this system.

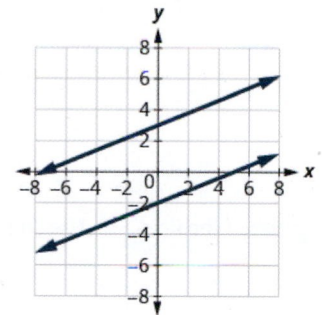

The lines are parallel.
Parallel lines have no points in common.
There is no solution to this system.

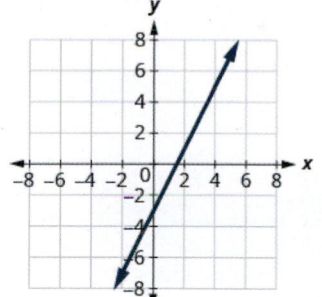

Both equations give the same line.
Because we have just one line, there are infinitely many solutions.

Figure 4.2

Each time we demonstrate a new method, we will use it on the same system of linear equations. At the end of the section you'll decide which method was the most convenient way to solve this system.

EXAMPLE 4.2 HOW TO SOLVE A SYSTEM OF EQUATIONS BY GRAPHING

Solve the system by graphing $\begin{cases} 2x + y = 7 \\ x - 2y = 6 \end{cases}$.

⊘ Solution

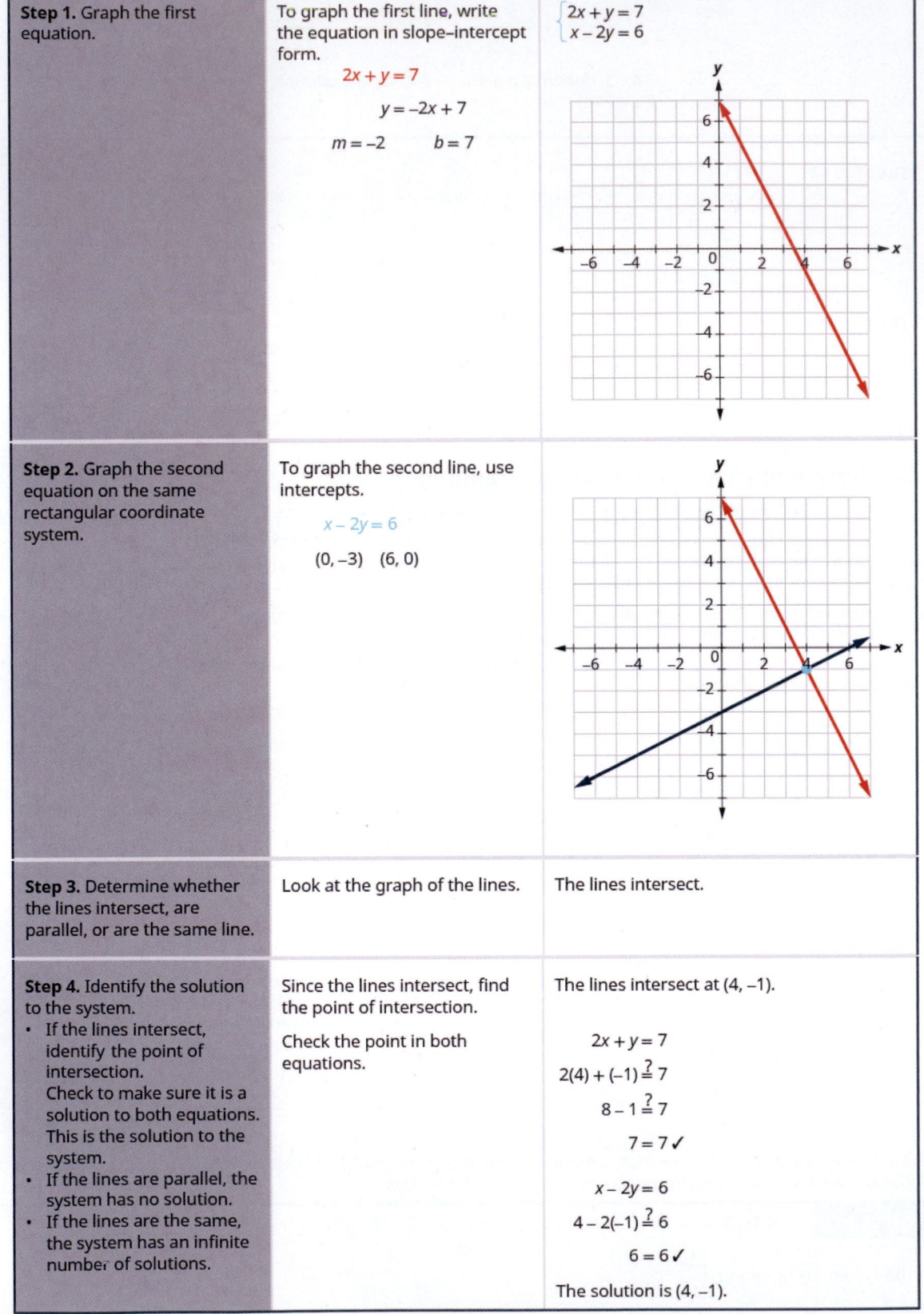

Step 1. Graph the first equation.	To graph the first line, write the equation in slope–intercept form. $2x + y = 7$ $y = -2x + 7$ $m = -2 \qquad b = 7$	$\begin{cases} 2x + y = 7 \\ x - 2y = 6 \end{cases}$
Step 2. Graph the second equation on the same rectangular coordinate system.	To graph the second line, use intercepts. $x - 2y = 6$ $(0, -3) \quad (6, 0)$	
Step 3. Determine whether the lines intersect, are parallel, or are the same line.	Look at the graph of the lines.	The lines intersect.
Step 4. Identify the solution to the system. • If the lines intersect, identify the point of intersection. Check to make sure it is a solution to both equations. This is the solution to the system. • If the lines are parallel, the system has no solution. • If the lines are the same, the system has an infinite number of solutions.	Since the lines intersect, find the point of intersection. Check the point in both equations.	The lines intersect at $(4, -1)$. $2x + y = 7$ $2(4) + (-1) \overset{?}{=} 7$ $8 - 1 \overset{?}{=} 7$ $7 = 7 \checkmark$ $x - 2y = 6$ $4 - 2(-1) \overset{?}{=} 6$ $6 = 6 \checkmark$ The solution is $(4, -1)$.

> **TRY IT :: 4.3**

Solve the system by graphing: $\begin{cases} x - 3y = -3 \\ x + y = 5 \end{cases}$.

> **TRY IT :: 4.4**

Solve the system by graphing: $\begin{cases} -x + y = 1 \\ 3x + 2y = 12 \end{cases}$.

The steps to use to solve a system of linear equations by graphing are shown here.

HOW TO :: SOLVE A SYSTEM OF LINEAR EQUATIONS BY GRAPHING.

Step 1.   Graph the first equation.

Step 2.   Graph the second equation on the same rectangular coordinate system.

Step 3.   Determine whether the lines intersect, are parallel, or are the same line.

Step 4.   Identify the solution to the system.

- If the lines intersect, identify the point of intersection. This is the solution to the system.
- If the lines are parallel, the system has no solution.
- If the lines are the same, the system has an infinite number of solutions.

Step 5.   Check the solution in both equations.

In the next example, we'll first re-write the equations into slope–intercept form as this will make it easy for us to quickly graph the lines.

EXAMPLE 4.3

Solve the system by graphing: $\begin{cases} 3x + y = -1 \\ 2x + y = 0 \end{cases}$.

⊘ **Solution**

We'll solve both of these equations for y so that we can easily graph them using their slopes and y-intercepts.

$$\begin{cases} 3x + y = -1 \\ 2x + y = 0 \end{cases}$$

Solve the first equation for y.	$3x + y = -1$ $y = -3x - 1$
Find the slope and y-intercept.	$m = -3$ $b = -1$
Solve the second equation for y.	$2x + y = 0$ $y = -2x$
Find the slope and y-intercept.	$m = -2$ $b = 0$

Graph the lines.

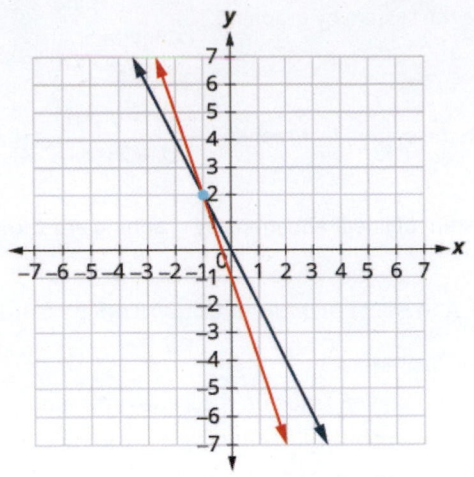

Determine the point of intersection.   The lines intersect at $(-1, 2)$.

Check the solution in both equations.

$$3x + y = -1 \qquad\qquad 2x + y = 0$$
$$3(-1) + 2 \overset{?}{=} -1 \qquad 2(-1) + 2 \overset{?}{=} 0$$
$$-1 = -1 \checkmark \qquad\qquad 0 = 0 \checkmark$$

The solution is $(-1, 2)$.

> **TRY IT :: 4.5**

Solve the system by graphing: $\begin{cases} -x + y = 1 \\ 2x + y = 10 \end{cases}$.

> **TRY IT :: 4.6**

Solve the system by graphing: $\begin{cases} 2x + y = 6 \\ x + y = 1 \end{cases}$.

In all the systems of linear equations so far, the lines intersected and the solution was one point. In the next two examples, we'll look at a system of equations that has no solution and at a system of equations that has an infinite number of solutions.

EXAMPLE 4.4

Solve the system by graphing: $\begin{cases} y = \frac{1}{2}x - 3 \\ x - 2y = 4 \end{cases}$.

⊘ **Solution**

$$\begin{cases} y = \frac{1}{2}x - 3 \\ x - 2y = 4 \end{cases}$$

To graph the first equation, we will use its slope and y-intercept.

$$y = \frac{1}{2}x - 3$$
$$m = \frac{1}{2}$$
$$b = -3$$

To graph the second equation, we will use the intercepts.

$$x - 2y = 4$$

x	y
0	-2
4	0

Graph the lines.

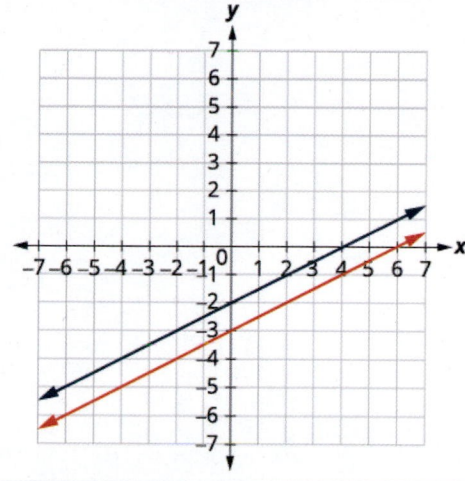

Determine the points of intersection.	The lines are parallel. Since no point is on both lines, there is no ordered pair that makes both equations true. There is no solution to this system.

> **TRY IT : : 4.7**

Solve the system by graphing: $\begin{cases} y = -\frac{1}{4}x + 2 \\ x + 4y = -8 \end{cases}$.

> **TRY IT : : 4.8**

Solve the system by graphing: $\begin{cases} y = 3x - 1 \\ 6x - 2y = 6 \end{cases}$.

Sometimes the equations in a system represent the same line. Since every point on the line makes both equations true, there are infinitely many ordered pairs that make both equations true. There are infinitely many solutions to the system.

EXAMPLE 4.5

Solve the system by graphing: $\begin{cases} y = 2x - 3 \\ -6x + 3y = -9 \end{cases}$.

⊘ **Solution**

$$\begin{cases} y = 2x - 3 \\ -6x + 3y = -9 \end{cases}$$

Find the slope and y-intercept of the first equation.	$y = 2x - 3$ $m = 2$ $b = -3$

Find the intercepts of the second equation.	$-6x + 3y = -9$

x	y
0	−3
$\frac{3}{2}$	0

Graph the lines.

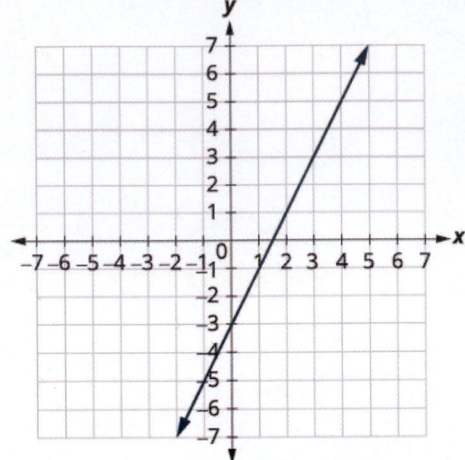

The lines are the same!
Since every point on the line makes both equations true, there are infinitely many ordered pairs that make both equations true. There are infinitely many solutions to this system.

If you write the second equation in slope-intercept form, you may recognize that the equations have the same slope and same y-intercept.

> | **TRY IT : : 4.9**

Solve the system by graphing: $\begin{cases} y = -3x - 6 \\ 6x + 2y = -12 \end{cases}$.

> | **TRY IT : : 4.10**

Solve the system by graphing: $\begin{cases} y = \frac{1}{2}x - 4 \\ 2x - 4y = 16 \end{cases}$.

When we graphed the second line in the last example, we drew it right over the first line. We say the two lines are **coincident**. Coincident lines have the same slope and same y-intercept.

Coincident Lines

Coincident lines have the same slope and same y-intercept.

The systems of equations in **Example 4.2** and **Example 4.3** each had two intersecting lines. Each system had one solution.

In **Example 4.5**, the equations gave coincident lines, and so the system had infinitely many solutions.

The systems in those three examples had at least one solution. A system of equations that has at least one solution is called a *consistent* system.

A system with parallel lines, like **Example 4.4**, has no solution. We call a system of equations like this *inconsistent*. It has no solution.

Consistent and Inconsistent Systems

A **consistent system of equations** is a system of equations with at least one solution.

An **inconsistent system of equations** is a system of equations with no solution.

We also categorize the equations in a system of equations by calling the equations *independent* or *dependent*. If two equations are independent, they each have their own set of solutions. Intersecting lines and parallel lines are independent.

If two equations are dependent, all the solutions of one equation are also solutions of the other equation. When we graph two dependent equations, we get coincident lines.

Let's sum this up by looking at the graphs of the three types of systems. See below and Table 4.4.

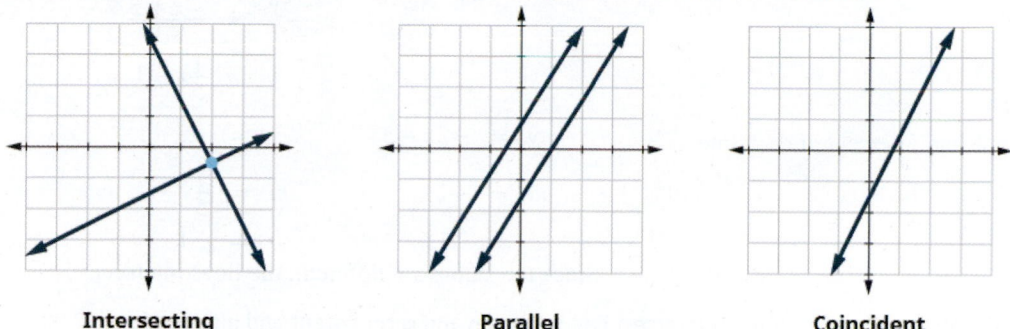

| | **Intersecting** | **Parallel** | **Coincident** |

Lines	Intersecting	Parallel	Coincident
Number of solutions	1 point	No solution	Infinitely many
Consistent/inconsistent	Consistent	Inconsistent	Consistent
Dependent/ independent	Independent	Independent	Dependent

Table 4.4

EXAMPLE 4.6

Without graphing, determine the number of solutions and then classify the system of equations.

ⓐ $\begin{cases} y = 3x - 1 \\ 6x - 2y = 12 \end{cases}$ ⓑ $\begin{cases} 2x + y = -3 \\ x - 5y = 5 \end{cases}$

✓ **Solution**

ⓐ We will compare the slopes and intercepts of the two lines.

$$\begin{cases} y = 3x - 1 \\ 6x - 2y = 12 \end{cases}$$

The first equation is already in slope-intercept form.

$$y = 3x - 1$$

Write the second equation in slope-intercept form.

$$6x - 2y = 12$$
$$-2y = -6x + 12$$
$$\frac{-2y}{-2} = \frac{-6x + 12}{-2}$$
$$y = 3x - 6$$

Find the slope and intercept of each line.

$$
\begin{array}{ll}
y = 3x - 1 & y = 3x - 6 \\
m = 3 & m = 3 \\
b = -1 & b = -6
\end{array}
$$

Since the slopes are the same and y-intercepts are different, the lines are parallel.

A system of equations whose graphs are parallel lines has no solution and is inconsistent and independent.

ⓑ We will compare the slope and intercepts of the two lines.

$$\begin{cases} 2x + y = -3 \\ x - 5y = 5 \end{cases}$$

Write both equations in slope–intercept form.

$$
\begin{array}{ll}
2x + y = -3 & x - 5y = 5 \\
y = -2x - 3 & -5y = -x + 5 \\
 & \dfrac{-5y}{-5} = \dfrac{-x+5}{-5} \\
 & y = \dfrac{1}{5}x - 1
\end{array}
$$

Find the slope and intercept of each line.

$$
\begin{array}{ll}
y = -2x - 3 & y = \dfrac{1}{5}x - 1 \\
m = -2 & m = \dfrac{1}{5} \\
b = -3 & b = -1
\end{array}
$$

Since the slopes are different, the lines intersect.

A system of equations whose graphs are intersect has 1 solution and is consistent and independent.

> **TRY IT :: 4.11**
>
> Without graphing, determine the number of solutions and then classify the system of equations.
>
> ⓐ $\begin{cases} y = -2x - 4 \\ 4x + 2y = 9 \end{cases}$ ⓑ $\begin{cases} 3x + 2y = 2 \\ 2x + y = 1 \end{cases}$

> **TRY IT :: 4.12**
>
> Without graphing, determine the number of solutions and then classify the system of equations.
>
> ⓐ $\begin{cases} y = \dfrac{1}{3}x - 5 \\ x - 3y = 6 \end{cases}$ ⓑ $\begin{cases} x + 4y = 12 \\ -x + y = 3 \end{cases}$

Solving systems of linear equations by graphing is a good way to visualize the types of solutions that may result. However, there are many cases where solving a system by graphing is inconvenient or imprecise. If the graphs extend beyond the small grid with x and y both between -10 and 10, graphing the lines may be cumbersome. And if the solutions to the system are not integers, it can be hard to read their values precisely from a graph.

Solve a System of Equations by Substitution

We will now solve systems of linear equations by the substitution method.

We will use the same system we used first for graphing.

$$\begin{cases} 2x + y = 7 \\ x - 2y = 6 \end{cases}$$

We will first solve one of the equations for either x or y. We can choose either equation and solve for either variable—but we'll try to make a choice that will keep the work easy.

Then we substitute that expression into the other equation. The result is an equation with just one variable—and we know how to solve those!

After we find the value of one variable, we will substitute that value into one of the original equations and solve for the other variable. Finally, we check our solution and make sure it makes both equations true.

EXAMPLE 4.7   HOW TO SOLVE A SYSTEM OF EQUATIONS BY SUBSTITUTION

Solve the system by substitution: $\begin{cases} 2x + y = 7 \\ x - 2y = 6 \end{cases}$.

✓ Solution

Step 1. Solve one of the equations for either variable.		$\begin{cases} 2x + y = 7 \\ x - 2y = 6 \end{cases}$
	We'll solve the first equation for y.	$2x + y = 7$ $y = 7 - 2x$
Step 2. Substitute the expression from Step 1 into the other equation.	We replace y in the second equation with the expression $7 - 2x$.	$x - 2y = 6$ $x - 2(7 - 2x) = 6$
Step 3. Solve the resulting equation.	Now we have an equation with just 1 variable. We know how to solve this!	$x - 2(7 - 2x) = 6$ $x - 14 + 4x = 6$ $5x = 20$ $x = 4$
Step 4. Substitute the solution from Step 3 into one of the original equations to find the other variable.	We'll use the first equation and replace x with 4.	$2x + y = 7$ $2(4) + y = 7$ $8 + y = 7$ $y = -1$
Step 5. Write the solution as an ordered pair.	The ordered pair is (x, y).	$(4, -1)$
Step 6. Check that the ordered pair is a solution to **both** original equations.	Substitute $x = 4$, $y = -1$ into both equations and make sure they are both true.	$2x + y = 7 \qquad\qquad x - 2y = 6$ $2(4) + (-1) \overset{?}{=} 7 \qquad 4 - 2(-1) \overset{?}{=} 6$ $7 = 7 \checkmark \qquad\qquad 6 = 6 \checkmark$ Both equations are true. $(4, -1)$ is the solution to the system.

> **TRY IT :: 4.13**   Solve the system by substitution: $\begin{cases} -2x + y = -11 \\ x + 3y = 9 \end{cases}$.

> **TRY IT :: 4.14**   Solve the system by substitution: $\begin{cases} 2x + y = -1 \\ 4x + 3y = 3 \end{cases}$.

HOW TO :: SOLVE A SYSTEM OF EQUATIONS BY SUBSTITUTION.

Step 1.   Solve one of the equations for either variable.

Step 2.   Substitute the expression from Step 1 into the other equation.

Step 3.   Solve the resulting equation.

Step 4.   Substitute the solution in Step 3 into either of the original equations to find the other variable.

Step 5.   Write the solution as an ordered pair.

Step 6.   Check that the ordered pair is a solution to **both** original equations.

Be very careful with the signs in the next example.

EXAMPLE 4.8

Solve the system by substitution: $\begin{cases} 4x + 2y = 4 \\ 6x - y = 8 \end{cases}$.

⊘ Solution

We need to solve one equation for one variable. We will solve the first equation for y.

$$\begin{cases} 4x + 2y = 4 \\ 6x - y = 8 \end{cases}$$

Solve the first equation for y. Substitute $-2x + 2$ for y in the second equation.	$4x + 2y = 4$ $2y = -4x + 4$ $y = -2x + 2$ $6x - y = 8$
Replace the y with $-2x + 2$.	$6x - (-2x + 2) = 8$
Solve the equation for x.	$6x + 2x - 2 = 8$ $8x - 2 = 8$ $8x = 10$ $x = \dfrac{5}{4}$
Substitute $x = \dfrac{5}{4}$ into $4x + 2y = 4$ to find y.	$4x + 2y = 4$ $4\left(\dfrac{5}{4}\right) + 2y = 4$ $5 + 2y = 4$ $2y = -1$ $y = -\dfrac{1}{2}$

The ordered pair is $\left(\dfrac{5}{4}, -\dfrac{1}{2}\right)$.

Check the ordered pair in both equations.

$$4x + 2y = 4 \qquad\qquad 6x - y = 8$$
$$4\left(\tfrac{5}{4}\right) + 2\left(-\tfrac{1}{2}\right) \overset{?}{=} 4 \qquad 6\left(\tfrac{5}{4}\right) - \left(-\tfrac{1}{2}\right) \overset{?}{=} 8$$
$$5 - 1 \overset{?}{=} 4 \qquad\qquad \tfrac{15}{4} - \left(-\tfrac{1}{2}\right) \overset{?}{=} 8$$
$$4 = 4 \checkmark \qquad\qquad \tfrac{16}{2} \overset{?}{=} 8$$
$$8 = 8 \checkmark$$

The solution is $\left(\dfrac{5}{4}, -\dfrac{1}{2}\right)$.

> **TRY IT :: 4.15**
>
> Solve the system by substitution: $\begin{cases} x - 4y = -4 \\ -3x + 4y = 0 \end{cases}$.

> **TRY IT :: 4.16**
>
> Solve the system by substitution: $\begin{cases} 4x - y = 0 \\ 2x - 3y = 5 \end{cases}$.

Solve a System of Equations by Elimination

We have solved systems of linear equations by graphing and by substitution. Graphing works well when the variable coefficients are small and the solution has integer values. Substitution works well when we can easily solve one equation for one of the variables and not have too many fractions in the resulting expression.

The third method of solving systems of linear equations is called the Elimination Method. When we solved a system by substitution, we started with two equations and two variables and reduced it to one equation with one variable. This is what we'll do with the elimination method, too, but we'll have a different way to get there.

The Elimination Method is based on the Addition Property of Equality. The Addition Property of Equality says that when you add the same quantity to both sides of an equation, you still have equality. We will extend the Addition Property of Equality to say that when you add equal quantities to both sides of an equation, the results are equal.

For any expressions a, b, c, and d.

$$\begin{aligned} \text{if} \quad a &= b \\ \text{and} \quad c &= d \\ \text{then} \quad a + c &= b + d. \end{aligned}$$

To solve a system of equations by elimination, we start with both equations in standard form. Then we decide which variable will be easiest to eliminate. How do we decide? We want to have the coefficients of one variable be opposites, so that we can add the equations together and eliminate that variable.

Notice how that works when we add these two equations together:

$$\begin{array}{r} 3x + y = 5 \\ 2x - y = 0 \\ \hline 5x \quad\;\; = 5 \end{array}$$

The y's add to zero and we have one equation with one variable.

Let's try another one:

$$\begin{cases} x + 4y = 2 \\ 2x + 5y = -2 \end{cases}$$

This time we don't see a variable that can be immediately eliminated if we add the equations.

But if we multiply the first equation by -2, we will make the coefficients of x opposites. We must multiply every term on both sides of the equation by -2.

$$\begin{cases} -2(x + 4y) = -2(2) \\ 2x + 5y = -2 \end{cases}$$

Then rewrite the system of equations.

$$\begin{cases} -2x - 8y = -4 \\ 2x + 5y = -2 \end{cases}$$

Now we see that the coefficients of the x terms are opposites, so x will be eliminated when we add these two equations.

$$\begin{array}{r} -2x - 8y = -4 \\ 2x + 5y = -2 \\ \hline -3y = -6 \end{array}$$

Once we get an equation with just one variable, we solve it. Then we substitute that value into one of the original equations to solve for the remaining variable. And, as always, we check our answer to make sure it is a solution to both of the original equations.

Now we'll see how to use elimination to solve the same system of equations we solved by graphing and by substitution.

EXAMPLE 4.9 HOW TO SOLVE A SYSTEM OF EQUATIONS BY ELIMINATION

Solve the system by elimination: $\begin{cases} 2x + y = 7 \\ x - 2y = 6 \end{cases}$.

⊘ **Solution**

| Step 1. Write both equations in standard form.
 • If any coefficients are fractions, clear them. | Both equations are in standard form, $Ax + By = C$. There are no fractions. | $\begin{cases} 2x + y = 7 \\ x - 2y = 6 \end{cases}$ |

Step 2. Make the coefficients of one variable opposites. • Decide which variable you will eliminate. • Multiply one or both equations so that the coefficients of that variable are opposites.	We can eliminate the y's by multiplying the first equation by 2. Multiply both sides of $2x + y = 7$ by 2.	$\begin{cases} 2x + y = 7 \\ x - 2y = 6 \end{cases}$ $\begin{cases} 2(2x + y) = 2(7) \\ x - 2y = 6 \end{cases}$
Step 3. Add the equations resulting from Step 2 to eliminate one variable.	We add the x's, y's, and constants.	$\begin{aligned} 4x + 2y &= 14 \\ x - 2y &= 6 \\ \hline 5x &= 20 \end{aligned}$
Step 4. Solve for the remaining variable.	Solve for x.	$x = 4$
Step 5. Substitute the solution from Step 4 into one of the original equations. Then solve for the other variable.	Substitute $x = 4$ into the second equation, $x - 2y = 6$. Then solve for y.	$\begin{aligned} x - 2y &= 6 \\ 4 - 2y &= 6 \\ -2y &= 2 \\ y &= -1 \end{aligned}$
Step 6. Write the solution as an ordered pair.	Write it as (x, y).	$(4, -1)$
Step 7. Check that the ordered pair is a solution to **both** original equations.	Substitute $x = 4$, $y = -1$ into $2x + y = 7$ and $x - 2y = 6$. Do they make both equations true? Yes!	$\begin{aligned} 2x + y &= 7 & x - 2y &= 6 \\ 2(4) + (-1) &\overset{?}{=} 7 & 4 - 2(-1) &\overset{?}{=} 6 \\ 7 &= 7 \checkmark & 6 &= 6 \checkmark \end{aligned}$ The solution is $(4, -1)$.

> **TRY IT :: 4.17**
>
> Solve the system by elimination: $\begin{cases} 3x + y = 5 \\ 2x - 3y = 7 \end{cases}$.

> **TRY IT :: 4.18**
>
> Solve the system by elimination: $\begin{cases} 4x + y = -5 \\ -2x - 2y = -2 \end{cases}$.

The steps are listed here for easy reference.

HOW TO :: SOLVE A SYSTEM OF EQUATIONS BY ELIMINATION.

Step 1.   Write both equations in standard form. If any coefficients are fractions, clear them.

Step 2.   Make the coefficients of one variable opposites.

   ◦ Decide which variable you will eliminate.

   ◦ Multiply one or both equations so that the coefficients of that variable are opposites.

Step 3.   Add the equations resulting from Step 2 to eliminate one variable.

Step 4.   Solve for the remaining variable.

Step 5.   Substitute the solution from Step 4 into one of the original equations. Then solve for the other variable.

Step 6.   Write the solution as an ordered pair.

Step 7.   Check that the ordered pair is a solution to **both** original equations.

Now we'll do an example where we need to multiply both equations by constants in order to make the coefficients of one variable opposites.

Solve the system by elimination: $\begin{cases} 4x - 3y = 9 \\ 7x + 2y = -6 \end{cases}$

✓ Solution

In this example, we cannot multiply just one equation by any constant to get opposite coefficients. So we will strategically multiply both equations by different constants to get the opposites.

	$\begin{cases} 4x - 3y = 9 \\ 7x + 2y = -6 \end{cases}$
Both equations are in standard form. To get opposite coefficients of y, we will multiply the first equation by 2 and the second equation by 3.	$\begin{cases} 2(4x - 3y) = 2(9) \\ 3(7x + 2y) = 3(-6) \end{cases}$
Simplify.	$\begin{cases} 8x - 6y = 18 \\ 21x + 6y = -18 \end{cases}$
Add the two equations to eliminate y.	$\begin{array}{r} 8x - 6y = 18 \\ 21x + 6y = -18 \\ \hline 29x \quad\quad = 0 \end{array}$
Solve for x.	$x = \boxed{0}$ $7x + 2y = -6$
Substitute $x = 0$ into one of the original equations.	$7 \cdot 0 + 2y = -6$
Solve for y.	$\begin{array}{r} 2y = -6 \\ y = -3 \end{array}$
Write the solution as an ordered pair.	The ordered pair is $(0, -3)$.
Check that the ordered pair is a solution to **both** original equations.	

$$\begin{array}{ll} 4x - 3y = 9 & 7x + 2y = -6 \\ 4(0) - 3(-3) \overset{?}{=} 9 & 7(0) + 2(-3) \overset{?}{=} -6 \\ \quad\quad 9 = 9 \checkmark & \quad\quad -6 = -6 \checkmark \end{array}$$

The solution is $(0, -3)$.

TRY IT : : 4.19

Solve the system by elimination: $\begin{cases} 3x - 4y = -9 \\ 5x + 3y = 14 \end{cases}$.

TRY IT : : 4.20

Solve each system by elimination: $\begin{cases} 7x + 8y = 4 \\ 3x - 5y = 27 \end{cases}$.

When the system of equations contains fractions, we will first clear the fractions by multiplying each equation by the LCD of all the fractions in the equation.

Solve the system by elimination: $\begin{cases} x + \frac{1}{2}y = 6 \\ \frac{3}{2}x + \frac{2}{3}y = \frac{17}{2} \end{cases}$.

✓ Solution

In this example, both equations have fractions. Our first step will be to multiply each equation by the LCD of all the fractions in the equation to clear the fractions.

	$\begin{cases} x + \frac{1}{2}y = 6 \\ \frac{3}{2}x + \frac{2}{3}y = \frac{17}{2} \end{cases}$
To clear the fractions, multiply each equation by its LCD.	$\begin{cases} 2\left(x + \frac{1}{2}y\right) = 2(6) \\ 6\left(\frac{3}{2}x + \frac{2}{3}y\right) = 6\left(\frac{17}{2}\right) \end{cases}$
Simplify.	$\begin{cases} 2x + y = 12 \\ 9x + 4y = 51 \end{cases}$

Now we are ready to eliminate one of the variables. Notice that both equations are in standard form.

We can eliminate y by multiplying the top equation by -4.

$\begin{cases} -4(2x + y) = -4(12) \\ 9x + 4y = 51 \end{cases}$

Simplify and add.

$\begin{array}{r} -8x - 4y = -48 \\ 9x + 4y = 51 \\ \hline x = 3 \end{array}$

Substitute $x = 3$ into one of the original equations.

$x + \frac{1}{2}y = 6$

Solve for y.

$3 + \frac{1}{2}y = 6$

$\frac{1}{2}y = 3$

$y = 6$

Write the solution as an ordered pair.

The ordered pair is $(3, 6)$.

Check that the ordered pair is a solution to both original equations.

$x + \frac{1}{2}y = 6$	$\frac{3}{2}x + \frac{2}{3}y = \frac{17}{2}$
$3 + \frac{1}{2}(6) \overset{?}{=} 6$	$\frac{3}{2}(3) + \frac{2}{3}(6) \overset{?}{=} \frac{17}{2}$
$3 + 3 \overset{?}{=} 6$	$\frac{9}{2} + 4 \overset{?}{=} \frac{17}{2}$
$6 = 6 \checkmark$	$\frac{9}{2} + \frac{8}{2} \overset{?}{=} \frac{17}{2}$
	$\frac{17}{2} = \frac{17}{2} \checkmark$

The solution is $(3, 6)$.

> **TRY IT : : 4.21**

Solve each system by elimination: $\begin{cases} \frac{1}{3}x - \frac{1}{2}y = 1 \\ \frac{3}{4}x - y = \frac{5}{2} \end{cases}$.

> **TRY IT ::** 4.22

Solve each system by elimination: $\begin{cases} x + \frac{3}{5}y = -\frac{1}{5} \\ -\frac{1}{2}x - \frac{2}{3}y = \frac{5}{6} \end{cases}$.

When we solved the system by graphing, we saw that not all systems of linear equations have a single ordered pair as a solution. When the two equations were really the same line, there were infinitely many solutions. We called that a consistent system. When the two equations described parallel lines, there was no solution. We called that an inconsistent system.

The same is true using substitution or elimination. If the equation at the end of substitution or elimination is a true statement, we have a consistent but dependent system and the system of equations has infinitely many solutions. If the equation at the end of substitution or elimination is a false statement, we have an inconsistent system and the system of equations has no solution.

EXAMPLE 4.12

Solve the system by elimination: $\begin{cases} 3x + 4y = 12 \\ y = 3 - \frac{3}{4}x \end{cases}$.

⊘ **Solution**

$$\begin{cases} 3x + 4y = 12 \\ y = 3 - \frac{3}{4}x \end{cases}$$

Write the second equation in standard form.	$\begin{cases} 3x + 4y = 12 \\ \frac{3}{4}x + y = 3 \end{cases}$
Clear the fractions by multiplying the second equation by 4.	$\begin{cases} 3x + 4y = 12 \\ 4\left(\frac{3}{4}x + y\right) = 4(3) \end{cases}$
Simplify.	$\begin{cases} 3x + 4y = 12 \\ 3x + 4y = 12 \end{cases}$
To eliminate a variable, we multiply the second equation by -1. Simplify and add.	$\begin{cases} 3x + 4y = 12 \\ \underline{-3x - 4y = -12} \\ \quad\quad 0 = 0 \end{cases}$

This is a true statement. The equations are consistent but dependent. Their graphs would be the same line. The system has infinitely many solutions.

After we cleared the fractions in the second equation, did you notice that the two equations were the same? That means we have coincident lines.

> **TRY IT ::** 4.23

Solve the system by elimination: $\begin{cases} 5x - 3y = 15 \\ y = -5 + \frac{5}{3}x \end{cases}$.

> **TRY IT ::** 4.24

Solve the system by elimination: $\begin{cases} x + 2y = 6 \\ y = -\frac{1}{2}x + 3 \end{cases}$.

Choose the Most Convenient Method to Solve a System of Linear Equations

When you solve a system of linear equations in in an application, you will not be told which method to use. You will need to make that decision yourself. So you'll want to choose the method that is easiest to do and minimizes your chance of making mistakes.

Choose the Most Convenient Method to Solve a System of Linear Equations

Graphing	Substitution	Elimination
Use when you need a picture of the situation.	Use when one equation is already solved or can be easily solved for one variable.	Use when the equations are in standard form.

EXAMPLE 4.13

For each system of linear equations, decide whether it would be more convenient to solve it by substitution or elimination. Explain your answer.

(a) $\begin{cases} 3x + 8y = 40 \\ 7x - 4y = -32 \end{cases}$ (b) $\begin{cases} 5x + 6y = 12 \\ y = \frac{2}{3}x - 1 \end{cases}$

✓ **Solution**

(a)

$$\begin{cases} 3x + 8y = 40 \\ 7x - 4y = -32 \end{cases}$$

Since both equations are in standard form, using elimination will be most convenient.

(b)

$$\begin{cases} 5x + 6y = 12 \\ y = \frac{2}{3}x - 1 \end{cases}$$

Since one equation is already solved for y, using substitution will be most convenient.

> **TRY IT : : 4.25**

For each system of linear equations decide whether it would be more convenient to solve it by substitution or elimination. Explain your answer.

(a) $\begin{cases} 4x - 5y = -32 \\ 3x + 2y = -1 \end{cases}$ (b) $\begin{cases} x = 2y - 1 \\ 3x - 5y = -7 \end{cases}$

> **TRY IT : : 4.26**

For each system of linear equations decide whether it would be more convenient to solve it by substitution or elimination. Explain your answer.

(a) $\begin{cases} y = 2x - 1 \\ 3x - 4y = -6 \end{cases}$ (b) $\begin{cases} 6x - 2y = 12 \\ 3x + 7y = -13 \end{cases}$

4.1 EXERCISES

Practice Makes Perfect

Determine Whether an Ordered Pair is a Solution of a System of Equations

In the following exercises, determine if the following points are solutions to the given system of equations.

1. $\begin{cases} 2x - 6y = 0 \\ 3x - 4y = 5 \end{cases}$

ⓐ $(3, 1)$

ⓑ $(-3, 4)$

2. $\begin{cases} -3x + y = 8 \\ -x + 2y = -9 \end{cases}$

ⓐ $(-5, -7)$

ⓑ $(-5, 7)$

3. $\begin{cases} x + y = 2 \\ y = \frac{3}{4}x \end{cases}$

ⓐ $\left(\frac{8}{7}, \frac{6}{7}\right)$

ⓑ $\left(1, \frac{3}{4}\right)$

4. $\begin{cases} 2x + 3y = 6 \\ y = \frac{2}{3}x + 2 \end{cases}$

ⓐ $(-6, 2)$

ⓑ $(-3, 4)$

Solve a System of Linear Equations by Graphing

In the following exercises, solve the following systems of equations by graphing.

5. $\begin{cases} 3x + y = -3 \\ 2x + 3y = 5 \end{cases}$

6. $\begin{cases} -x + y = 2 \\ 2x + y = -4 \end{cases}$

7. $\begin{cases} y = x + 2 \\ y = -2x + 2 \end{cases}$

8. $\begin{cases} y = x - 2 \\ y = -3x + 2 \end{cases}$

9. $\begin{cases} y = \frac{3}{2}x + 1 \\ y = -\frac{1}{2}x + 5 \end{cases}$

10. $\begin{cases} y = \frac{2}{3}x - 2 \\ y = -\frac{1}{3}x - 5 \end{cases}$

11. $\begin{cases} x + y = -4 \\ -x + 2y = -2 \end{cases}$

12. $\begin{cases} -x + 3y = 3 \\ x + 3y = 3 \end{cases}$

13. $\begin{cases} -2x + 3y = 3 \\ x + 3y = 12 \end{cases}$

14. $\begin{cases} 2x - y = 4 \\ 2x + 3y = 12 \end{cases}$

15. $\begin{cases} x + 3y = -6 \\ y = -\frac{4}{3}x + 4 \end{cases}$

16. $\begin{cases} -x + 2y = -6 \\ y = -\frac{1}{2}x - 1 \end{cases}$

17. $\begin{cases} -2x + 4y = 4 \\ y = \frac{1}{2}x \end{cases}$

18. $\begin{cases} 3x + 5y = 10 \\ y = -\frac{3}{5}x + 1 \end{cases}$

19. $\begin{cases} 4x - 3y = 8 \\ 8x - 6y = 14 \end{cases}$

20. $\begin{cases} x + 3y = 4 \\ -2x - 6y = 3 \end{cases}$

21. $\begin{cases} x = -3y + 4 \\ 2x + 6y = 8 \end{cases}$

22. $\begin{cases} 4x = 3y + 7 \\ 8x - 6y = 14 \end{cases}$

23. $\begin{cases} 2x + y = 6 \\ -8x - 4y = -24 \end{cases}$

24. $\begin{cases} 5x + 2y = 7 \\ -10x - 4y = -14 \end{cases}$

Without graphing, determine the number of solutions and then classify the system of equations.

25. $\begin{cases} y = \frac{2}{3}x + 1 \\ -2x + 3y = 5 \end{cases}$

26. $\begin{cases} y = \frac{3}{2}x + 1 \\ 2x - 3y = 7 \end{cases}$

27. $\begin{cases} 5x + 3y = 4 \\ 2x - 3y = 5 \end{cases}$

28. $\begin{cases} y = -\frac{1}{2}x + 5 \\ x + 2y = 10 \end{cases}$

29. $\begin{cases} 5x - 2y = 10 \\ y = \frac{5}{2}x - 5 \end{cases}$

Solve a System of Equations by Substitution

In the following exercises, solve the systems of equations by substitution.

30. $\begin{cases} 2x + y = -4 \\ 3x - 2y = -6 \end{cases}$

31. $\begin{cases} 2x + y = -2 \\ 3x - y = 7 \end{cases}$

32. $\begin{cases} x - 2y = -5 \\ 2x - 3y = -4 \end{cases}$

33. $\begin{cases} x - 3y = -9 \\ 2x + 5y = 4 \end{cases}$

34. $\begin{cases} 5x - 2y = -6 \\ y = 3x + 3 \end{cases}$

35. $\begin{cases} -2x + 2y = 6 \\ y = -3x + 1 \end{cases}$

36. $\begin{cases} 2x + 5y = 1 \\ y = \frac{1}{3}x - 2 \end{cases}$

37. $\begin{cases} 3x + 4y = 1 \\ y = -\frac{2}{5}x + 2 \end{cases}$

38. $\begin{cases} 2x + y = 5 \\ x - 2y = -15 \end{cases}$

39. $\begin{cases} 4x + y = 10 \\ x - 2y = -20 \end{cases}$

40. $\begin{cases} y = -2x - 1 \\ y = -\frac{1}{3}x + 4 \end{cases}$

41. $\begin{cases} y = x - 6 \\ y = -\frac{3}{2}x + 4 \end{cases}$

42. $\begin{cases} x = 2y \\ 4x - 8y = 0 \end{cases}$

43. $\begin{cases} 2x - 16y = 8 \\ -x - 8y = -4 \end{cases}$

44. $\begin{cases} y = \frac{7}{8}x + 4 \\ -7x + 8y = 6 \end{cases}$

45. $\begin{cases} y = -\frac{2}{3}x + 5 \\ 2x + 3y = 11 \end{cases}$

Solve a System of Equations by Elimination

In the following exercises, solve the systems of equations by elimination.

46. $\begin{cases} 5x + 2y = 2 \\ -3x - y = 0 \end{cases}$

47. $\begin{cases} 6x - 5y = -1 \\ 2x + y = 13 \end{cases}$

48. $\begin{cases} 2x - 5y = 7 \\ 3x - y = 17 \end{cases}$

49. $\begin{cases} 5x - 3y = -1 \\ 2x - y = 2 \end{cases}$

50. $\begin{cases} 3x - 5y = -9 \\ 5x + 2y = 16 \end{cases}$

51. $\begin{cases} 4x - 3y = 3 \\ 2x + 5y = -31 \end{cases}$

52. $\begin{cases} 3x + 8y = -3 \\ 2x + 5y = -3 \end{cases}$

53. $\begin{cases} 11x + 9y = -5 \\ 7x + 5y = -1 \end{cases}$

54. $\begin{cases} 3x + 8y = 67 \\ 5x + 3y = 60 \end{cases}$

55. $\begin{cases} 2x + 9y = -4 \\ 3x + 13y = -7 \end{cases}$

56. $\begin{cases} \frac{1}{3}x - y = -3 \\ x + \frac{5}{2}y = 2 \end{cases}$

57. $\begin{cases} x + \frac{1}{2}y = \frac{3}{2} \\ \frac{1}{5}x - \frac{1}{5}y = 3 \end{cases}$

58. $\begin{cases} x + \frac{1}{3}y = -1 \\ \frac{1}{3}x + \frac{1}{2}y = 1 \end{cases}$

59. $\begin{cases} \frac{1}{3}x - y = -3 \\ \frac{2}{3}x + \frac{5}{2}y = 3 \end{cases}$

60. $\begin{cases} 2x + y = 3 \\ 6x + 3y = 9 \end{cases}$

61. $\begin{cases} x - 4y = -1 \\ -3x + 12y = 3 \end{cases}$

62. $\begin{cases} -3x - y = 8 \\ 6x + 2y = -16 \end{cases}$

63. $\begin{cases} 4x + 3y = 2 \\ 20x + 15y = 10 \end{cases}$

Choose the Most Convenient Method to Solve a System of Linear Equations

In the following exercises, decide whether it would be more convenient to solve the system of equations by substitution or elimination.

64.

ⓐ $\begin{cases} 8x - 15y = -32 \\ 6x + 3y = -5 \end{cases}$

ⓑ $\begin{cases} x = 4y - 3 \\ 4x - 2y = -6 \end{cases}$

65.

ⓐ $\begin{cases} y = 7x - 5 \\ 3x - 2y = 16 \end{cases}$

ⓑ $\begin{cases} 12x - 5y = -42 \\ 3x + 7y = -15 \end{cases}$

66.

ⓐ $\begin{cases} y = 4x + 9 \\ 5x - 2y = -21 \end{cases}$

ⓑ $\begin{cases} 9x - 4y = 24 \\ 3x + 5y = -14 \end{cases}$

67.

ⓐ $\begin{cases} 14x - 15y = -30 \\ 7x + 2y = 10 \end{cases}$

ⓑ $\begin{cases} x = 9y - 11 \\ 2x - 7y = -27 \end{cases}$

Writing Exercises

68. In a system of linear equations, the two equations have the same intercepts. Describe the possible solutions to the system.

69. Solve the system of equations by substitution and explain all your steps in words: $\begin{cases} 3x + y = 12 \\ x = y - 8 \end{cases}$.

70. Solve the system of equations by elimination and explain all your steps in words: $\begin{cases} 5x + 4y = 10 \\ 2x = 3y + 27 \end{cases}$.

71. Solve the system of equations $\begin{cases} x + y = 10 \\ x - y = 6 \end{cases}$

ⓐ by graphing ⓑ by substitution

ⓒ Which method do you prefer? Why?

Self Check

After completing the exercises, use this checklist to evaluate your mastery of the objectives of this section.

I can...	Confidently	With some help	No-I don't get it!
determine whether an ordered pair is a solution of a system of equations.			
solve a system of linear equations by graphing.			
solve a system of equations by substitution.			
solve a system of equations by elimination.			
choose the most convenient method to solve a system of linear equations.			

If most of your checks were:

...confidently. *Congratulations! You have achieved the objectives in this section. Reflect on the study skills you used so that you can continue to use them. What did you do to become confident of your ability to do these things? Be specific.*

...with some help. *This must be addressed quickly because topics you do not master become potholes in your road to success. In math every topic builds upon previous work. It is important to make sure you have a strong foundation before you move on. Who can you ask for help? Your fellow classmates and instructor are good resources. Is there a place on campus where math tutors are available? Can your study skills be improved?*

...no - I don't get it! *This is a warning sign and you must not ignore it. You should get help right away or you will quickly be overwhelmed. See your instructor as soon as you can to discuss your situation. Together you can come up with a plan to get you the help you need.*

4.2   Solve Applications with Systems of Equations

Learning Objectives

By the end of this section, you will be able to:

› Solve direct translation applications
› Solve geometry applications
› Solve uniform motion applications

Be Prepared!

Before you get started, take this readiness quiz.

1. The sum of twice a number and nine is 31. Find the number.
   If you missed this problem, review **Example 2.15**.

2. Twins Jon and Ron together earned \$96,000 last year. Ron earned \$8000 more than three times what Jon earned. How much did each of the twins earn?
   If you missed this problem, review **Example 2.19**.

3. An express train and a local train leave Pittsburgh to travel to Washington, D.C. The express train can make the trip in four hours and the local train takes five hours for the trip. The speed of the express train is 12 miles per hour faster than the speed of the local train. Find the speed of both trains.
   If you missed this problem, review **Example 2.43**.

Solve Direct Translation Applications

Systems of linear equations are very useful for solving applications. Some people find setting up word problems with two variables easier than setting them up with just one variable. To solve an application, we'll first translate the words into a system of linear equations. Then we will decide the most convenient method to use, and then solve the system.

HOW TO : : SOLVE APPLICATIONS WITH SYSTEMS OF EQUATIONS.

Step 1.   **Read** the problem. Make sure all the words and ideas are understood.

Step 2.   **Identify** what we are looking for.

Step 3.   **Name** what we are looking for. Choose variables to represent those quantities.

Step 4.   **Translate** into a system of equations.

Step 5.   **Solve** the system of equations using good algebra techniques.

Step 6.   **Check** the answer in the problem and make sure it makes sense.

Step 7.   **Answer** the question with a complete sentence.

We solved number problems with one variable earlier. Let's see how differently it works using two variables.

EXAMPLE 4.14

The sum of two numbers is zero. One number is nine less than the other. Find the numbers.

 Solution

Step 1. Read the problem.	
Step 2. Identify what we are looking for.	We are looking for two numbers.
Step 3. Name what we are looking for.	Let $n =$ the first number. $m =$ the second number
Step 4. Translate into a system of equations.	The sum of two numbers is zero.

$$n + m = 0$$

One number is nine less than the other.

$$n = m - 9$$

The system is:	$\begin{cases} n + m = 0 \\ n = m - 9 \end{cases}$

Step 5. Solve the system of equations. We will use substitution since the second equation is solved for n.

Substitute $m - 9$ for n in the first equation.

$$n = \boxed{m - 9}$$
$$n + m = 0$$

Solve for m.

$$m - 9 + m = 0$$

$$2m - 9 = 0$$

$$2m = 9$$

Substitute $m = \dfrac{9}{2}$ into the second equation and then solve for n.

$$m = \boxed{\dfrac{9}{2}}$$

$$n = m - 9$$

$$m = \dfrac{9}{2} - 9$$

$$m = \dfrac{9}{2} - \dfrac{18}{2}$$

$$n = -\dfrac{9}{2}$$

Step 6. Check the answer in the problem.

Do these numbers make sense in the problem? We will leave this to you!

Step 7. Answer the question.

The numbers are $\dfrac{9}{2}$ and $-\dfrac{9}{2}$.

> **TRY IT : : 4.27**   The sum of two numbers is 10. One number is 4 less than the other. Find the numbers.

> **TRY IT : : 4.28**   The sum of two numbers is -6. One number is 10 less than the other. Find the numbers.

EXAMPLE 4.15

Heather has been offered two options for her salary as a trainer at the gym. Option A would pay her $25,000 plus $15 for each training session. Option B would pay her $10,000 + 40 for each training session. How many training sessions would make the salary options equal?

✓ Solution

Step 1. Read the problem.

Step 2. Identify what we are looking for.	We are looking for the number of training sessions that would make the pay equal.
Step 3. Name what we are looking for.	Let $s =$ Heather's salary. $n =$ the number of training sessions
Step 4. Translate into a system of equations.	Option A would pay her $25,000 plus $15 for each training session.
	$$s = 25,000 + 15n$$
	Option B would pay her $10,000 + $40 for each training session.
	$$s = 10,000 + 40n$$
The system is shown.	$$\begin{cases} s = 25,000 + 15n \\ s = 10,000 + 40n \end{cases}$$
Step 5. Solve the system of equations. We will use substitution.	$$s = 25,000 + 15n$$ $$s = 10,000 + 40n$$
Substitute 25,000 + 15n for s in the second equation.	$$25,000 + 15n = 10,000 + 40n$$
Solve for n.	$$25,000 = 10,000 + 25n$$ $$15,000 = 25n$$ $$600 = n$$
Step 6. Check the answer.	Are 600 training sessions a year reasonable? Are the two options equal when $n = 600$?
Step 7. Answer the question.	The salary options would be equal for 600 training sessions.

> **TRY IT : : 4.29**
>
> Geraldine has been offered positions by two insurance companies. The first company pays a salary of $12,000 plus a commission of $100 for each policy sold. The second pays a salary of $20,000 plus a commission of $50 for each policy sold. How many policies would need to be sold to make the total pay the same?

> **TRY IT : : 4.30**
>
> Kenneth currently sells suits for company A at a salary of $22,000 plus a $10 commission for each suit sold. Company B offers him a position with a salary of $28,000 plus a $4 commission for each suit sold. How many suits would Kenneth need to sell for the options to be equal?

As you solve each application, remember to analyze which method of solving the system of equations would be most convenient.

EXAMPLE 4.16

Translate to a system of equations and then solve:

When Jenna spent 10 minutes on the elliptical trainer and then did circuit training for 20 minutes, her fitness app says she burned 278 calories. When she spent 20 minutes on the elliptical trainer and 30 minutes circuit training she burned 473 calories. How many calories does she burn for each minute on the elliptical trainer? How many calories for each minute of circuit training?

✓ Solution

Step 1. Read the problem.	
Step 2. Identify what we are looking for.	We are looking for the number of calories burned each minute on the elliptical trainer and each minute of circuit training.
Step 3. Name what we are looking for.	Let $e =$ number of calories burned per minute on the elliptical trainer. $c =$ number of calories burned per minute while circuit training
Step 4. Translate into a system of equations.	10 minutes on the elliptical and circuit training for 20 minutes, burned 278 calories
	$$10e + 20c = 278$$
	20 minutes on the elliptical and 30 minutes of circuit training burned 473 calories
	$$20e + 30c = 473$$
The system is:	$$\begin{cases} 10e + 20c = 278 \\ 20e + 30c = 473 \end{cases}$$
Step 5. Solve the system of equations.	
Multiply the first equation by –2 to get opposite coefficients of e.	$$\begin{cases} -2(10e + 20c) = -2(278) \\ 20e + 30c = 473 \end{cases}$$
Simplify and add the equations. Solve for c.	$$\begin{array}{r} -20e - 40c = -556 \\ 20e + 30c = 473 \\ \hline -10c = -83 \\ c = 8.3 \end{array}$$
Substitute $c = 8.3$ into one of the original equations to solve for e.	$$\begin{array}{r} 10e + 20c = 278 \\ 10e + 20(8.3) = 278 \\ 10e + 166 = 278 \\ 10e = 112 \\ e = 11.2 \end{array}$$
Step 6. Check the answer in the problem.	Check the math on your own.
	$$\begin{cases} 10(11.2) + 20(8.3) \overset{?}{=} 278 \\ 20(11.2) + 30(8.3) \overset{?}{=} 473 \end{cases}$$
Step 7. Answer the question.	Jenna burns 8.3 calories per minute circuit training and 11.2 calories per minute while on the elliptical trainer.

TRY IT :: 4.31

Translate to a system of equations and then solve:

Mark went to the gym and did 40 minutes of Bikram hot yoga and 10 minutes of jumping jacks. He burned 510 calories. The next time he went to the gym, he did 30 minutes of Bikram hot yoga and 20 minutes of jumping jacks burning 470 calories. How many calories were burned for each minute of yoga? How many calories were burned for each minute of jumping jacks?

> **TRY IT :: 4.32**

Translate to a system of equations and then solve:

Erin spent 30 minutes on the rowing machine and 20 minutes lifting weights at the gym and burned 430 calories. During her next visit to the gym she spent 50 minutes on the rowing machine and 10 minutes lifting weights and burned 600 calories. How many calories did she burn for each minutes on the rowing machine? How many calories did she burn for each minute of weight lifting?

Solve Geometry Applications

We will now solve geometry applications using systems of linear equations. We will need to add complementary angles and supplementary angles to our list some properties of angles.

The measures of two **complementary angles** add to 90 degrees. The measures of two **supplementary angles** add to 180 degrees.

Complementary and Supplementary Angles

Two angles are complementary if the sum of the measures of their angles is 90 degrees.

Two angles are supplementary if the sum of the measures of their angles is 180 degrees.

If two angles are complementary, we say that *one angle is the complement of the other.*

If two angles are supplementary, we say that *one angle is the supplement of the other.*

EXAMPLE 4.17

Translate to a system of equations and then solve.

The difference of two complementary angles is 26 degrees. Find the measures of the angles.

 **Solution**

Step 1. Read the problem.

Step 2. Identify what we are looking for.

We are looking for the measure of each angle.

Step 3. Name what we are looking for.

Let $x =$ the measure of the first angle.

$\quad y =$ the measure of the second angle

Step 4. Translate into a system of equations.

The angles are complementary.

$$x + y = 90$$

The difference of the two angles is 26 degrees.

$$x - y = 26$$

The system is shown.

$$\begin{cases} x + y = 90 \\ x - y = 26 \end{cases}$$

Step 5. Solve the system of equations by elimination.

$$\begin{cases} x + y = 90 \\ \underline{x - y = 26} \end{cases}$$
$$\begin{aligned} 2x \quad &= 116 \\ x &= 58 \end{aligned}$$

Substitute $x = 58$ into the first equation.

$$x + y = 90$$
$$58 + y = 90$$
$$y = 32$$

Step 6. Check the answer in the problem.

$$58 + 32 = 90 ✓$$
$$58 - 32 = 26 ✓$$

Step 7. Answer the question.

The angle measures are 58 and 32 degrees.

> **TRY IT :: 4.33**   Translate to a system of equations and then solve:

The difference of two complementary angles is 20 degrees. Find the measures of the angles.

> **TRY IT :: 4.34**   Translate to a system of equations and then solve:

The difference of two complementary angles is 80 degrees. Find the measures of the angles.

In the next example, we remember that the measures of supplementary angles add to 180.

EXAMPLE 4.18

Translate to a system of equations and then solve:

Two angles are supplementary. The measure of the larger angle is twelve degrees less than five times the measure of the smaller angle. Find the measures of both angles.

 **Solution**

Step 1. Read the problem.

Step 2. Identify what we are looking for.   We are looking for measure of each angle.

| **Step 3. Name** what we are looking for. | Let $x=$ the measure of the first angle. |
| | $y=$ the measure of the second angle |

Step 4. Translate into a system of equations.	The angles are supplementary.
	$x+y=180$
	The larger angle is twelve less than five times the smaller angle.
	$y=5x-12$

| The system is shown:
Step 5. Solve the system of equations substitution. | $\begin{cases} x+y=180 \\ y=5x-12 \end{cases}$ |
| | $x+y=180$ |

| Substitute $5x-12$ for y in the first equation.
Solve for x. | $x+5x-12=180$
$6x-12=180$ |

| Substitute 32 for x in the second equation, then solve for y. | $6x=192$
$x=32$
$y=5x-12$
$y=5\cdot32-12$
$y=160-12$
$y=148$ |

| **Step 6. Check** the answer in the problem. | $32+148=180$ ✓
$5\cdot32-12=148$ ✓ |

| **Step 7. Answer** the question. | The angle measures are 148 and 32 degrees. |

> **TRY IT :: 4.35**
>
> Translate to a system of equations and then solve:
>
> Two angles are supplementary. The measure of the larger angle is 12 degrees more than three times the smaller angle. Find the measures of the angles.

> **TRY IT :: 4.36**
>
> Translate to a system of equations and then solve:
>
> Two angles are supplementary. The measure of the larger angle is 18 less than twice the measure of the smaller angle. Find the measures of the angles.

Recall that the angles of a triangle add up to 180 degrees. A right triangle has one angle that is 90 degrees. What does that tell us about the other two angles? In the next example we will be finding the measures of the other two angles.

EXAMPLE 4.19

The measure of one of the small angles of a right triangle is ten more than three times the measure of the other small angle. Find the measures of both angles.

⊘ **Solution**

We will draw and label a figure.

Step 1. Read the problem.	
Step 2. Identify what you are looking for.	We are looking for the measures of the angles.
Step 3. Name what we are looking for.	Let $a =$ the measure of the first angle. $b =$ the measure of the second angle
Step 4. Translate into a system of equations.	The measure of one of the small angles of a right triangle is ten more than three times the measure of the other small angle.
	$$a = 3b + 10$$
	The sum of the measures of the angles of a triangle is 180.
	$$a + b + 90 = 180$$
The system is shown.	$$\begin{cases} a = 3b + 10 \\ a + b + 90 = 180 \end{cases}$$
Step 5. Solve the system of equations. We will use substitution since the first equation is solved for a.	$$a = \boxed{3b + 10}$$ $$a + b + 90 = 180$$
Substitute $3b + 10$ for a in the second equation.	$$(3b + 10) + b + 90 = 180$$
Solve for b.	$$4b + 100 = 180$$ $$4b = 80$$ $$b = 20$$ $$a = 3b + 10$$
Substitute $b = 20$ into the first equation and then solve for a.	$$a = 3 \cdot 20 + 10$$ $$a = 70$$
Step 6. Check the answer in the problem.	We will leave this to you!
Step 7. Answer the question.	The measures of the small angles are 20 and 70 degrees.

> **TRY IT : : 4.37**
>
> The measure of one of the small angles of a right triangle is 2 more than 3 times the measure of the other small angle. Find the measure of both angles.

> **TRY IT : : 4.38**
>
> The measure of one of the small angles of a right triangle is 18 less than twice the measure of the other small angle. Find the measure of both angles.

Often it is helpful when solving geometry applications to draw a picture to visualize the situation.

EXAMPLE 4.20

Translate to a system of equations and then solve:

Randall has 125 feet of fencing to enclose the part of his backyard adjacent to his house. He will only need to fence around three sides, because the fourth side will be the wall of the house. He wants the length of the fenced yard (parallel to the house wall) to be 5 feet more than four times as long as the width. Find the length and the width.

⊘ Solution

Step 1. Read the problem.

Step 2. Identify what you are looking for.	We are looking for the length and width.

Step 3. Name what we are looking for.	Let $L =$ the length of the fenced yard. $W =$ the width of the fenced yard

Step 4. Translate into a system of equations.	One lenth and two widths equal 125.

$$L + 2W = 125$$

The length will be 5 feet more than four times the width.

$$L = 4W + 5$$

The system is shown. **Step 5. Solve** The system of equations by substitution.	$\begin{cases} L + 2W = 125 \\ \quad L = 4W + 5 \end{cases}$ $L + 2W = 125$

Substitute $L = 4W + 5$ into the first equation, then solve for W.	$4W + 5 + 2W = 125$

$$6W + 5 = 125$$
$$6W = 120$$
$$W = 20$$

Substitute 20 for W in the second equation, then solve for L.	$L = 4W + 5$ $L = 4 \cdot 20 + 5$ $L = 80 + 5$ $L = 85$

Step 6. Check the answer in the problem.	$20 + 85 + 20 = 125$ $85 = 4 \cdot 20 + 5$ $85 = 85$ ✓

Step 7. Answer the equation.	The length is 85 feet and the width is 20 feet.

> **TRY IT :: 4.39**

Translate to a system of equations and then solve:

Mario wants to put a fence around the pool in his backyard. Since one side is adjacent to the house, he will only need to fence three sides. There are two long sides and the one shorter side is parallel to the house. He needs 155 feet of fencing to enclose the pool. The length of the long side is 10 feet less than twice the width. Find the length and width of the pool area to be enclosed.

TRY IT : : 4.40

Translate to a system of equations and then solve:

Alexis wants to build a rectangular dog run in her yard adjacent to her neighbor's fence. She will use 136 feet of fencing to completely enclose the rectangular dog run. The length of the dog run along the neighbor's fence will be 16 feet less than twice the width. Find the length and width of the dog run.

Solve uniform motion applications

We used a table to organize the information in uniform motion problems when we introduced them earlier. We'll continue using the table here. The basic equation was $D = rt$ where D is the distance traveled, r is the rate, and t is the time.

Our first example of a uniform motion application will be for a situation similar to some we have already seen, but now we can use two variables and two equations.

EXAMPLE 4.21

Translate to a system of equations and then solve:

Joni left St. Louis on the interstate, driving west towards Denver at a speed of 65 miles per hour. Half an hour later, Kelly left St. Louis on the same route as Joni, driving 78 miles per hour. How long will it take Kelly to catch up to Joni?

 Solution

A diagram is useful in helping us visualize the situation.

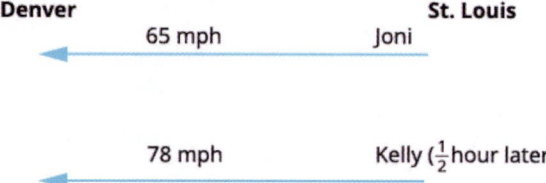

Identify and name what we are looking for. A chart will help us organize the data. We know the rates of both Joni and Kelly, and so we enter them in the chart. We are looking for the length of time Kelly, k, and Joni, j, will each drive.

	Rate · Time = Distance		
Joni	65	j	$65j$
Kelly	78	k	$78k$

Since $D = r \cdot t$ we can fill in the Distance column.

Translate into a system of equations.

To make the system of equations, we must recognize that Kelly and Joni will drive the same distance. So,

$$65j = 78k$$

Also, since Kelly left later, her time will be $\frac{1}{2}$ hour less than Joni's time. So,

$$k = j - \frac{1}{2}$$

Now we have the system.

$$\begin{cases} k = j - \frac{1}{2} \\ 65j = 78k \end{cases}$$

Solve the system of equations by substitution.

Substitute $k = j - \frac{1}{2}$ into the second equation,

then solve for j.

$$
\begin{aligned}
65j &= 78k \\
65j &= 78\left(j - \frac{1}{2}\right) \\
65j &= 78j - 39 \\
-13j &= -39 \\
j &= 3
\end{aligned}
$$

To find Kelly's time, substitute $j = 3$ into the first equation, then solve for k.

$$k = j - \frac{1}{2}$$

$$k = 3 - \frac{1}{2}$$

$$k = \frac{5}{2} \text{ or } k = 2\frac{1}{2}$$

Check the answer in the problem.

Joni   3 hours (65 mph) = 195 miles

Kelly $2\frac{1}{2}$ hours $\left(78 \text{ mph}\right)$ = 195 miles

Yes, they will have traveled the same distance when they meet.

Answer the question.

Kelly will catch up to Joni in $2\frac{1}{2}$ hours. By then, Joni will have traveled 3 hours.

> **TRY IT : : 4.41**

Translate to a system of equations and then solve:

Mitchell left Detroit on the interstate driving south towards Orlando at a speed of 60 miles per hour. Clark left Detroit 1 hour later traveling at a speed of 75 miles per hour, following the same route as Mitchell. How long will it take Clark to catch Mitchell?

> **TRY IT : : 4.42**

Translate to a system of equations and then solve:

Charlie left his mother's house traveling at an average speed of 36 miles per hour. His sister Sally left 15 minutes $\left(\frac{1}{4} \text{ hour}\right)$ later traveling the same route at an average speed of 42 miles per hour. How long before Sally catches up to Charlie?

Many real-world applications of uniform motion arise because of the effects of currents—of water or air—on the actual speed of a vehicle. Cross-country airplane flights in the United States generally take longer going west than going east because of the prevailing wind currents.

Let's take a look at a boat travelling on a river. Depending on which way the boat is going, the current of the water is either slowing it down or speeding it up.

The images below show how a river current affects the speed at which a boat is actually travelling. We'll call the speed of the boat in still water b and the speed of the river current c.

The boat is going downstream, in the same direction as the river current. The current helps push the boat, so the boat's actual speed is faster than its speed in still water. The actual speed at which the boat is moving is $b + c$.

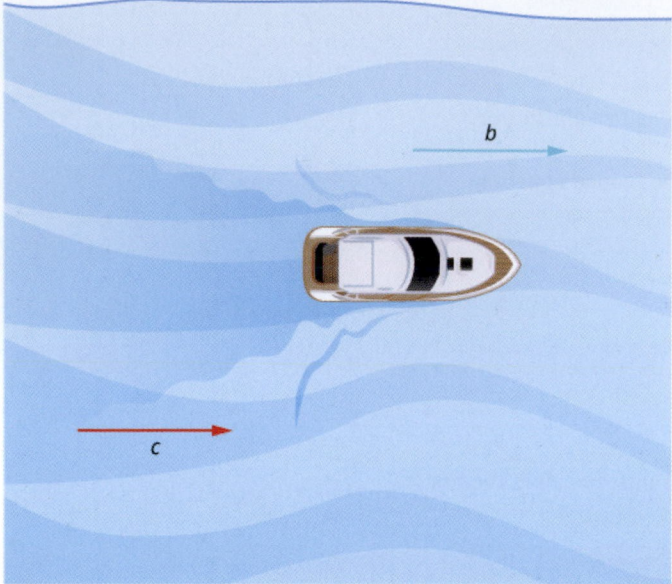

Now, the boat is going upstream, opposite to the river current. The current is going against the boat, so the boat's actual speed is slower than its speed in still water. The actual speed of the boat is $b - c$.

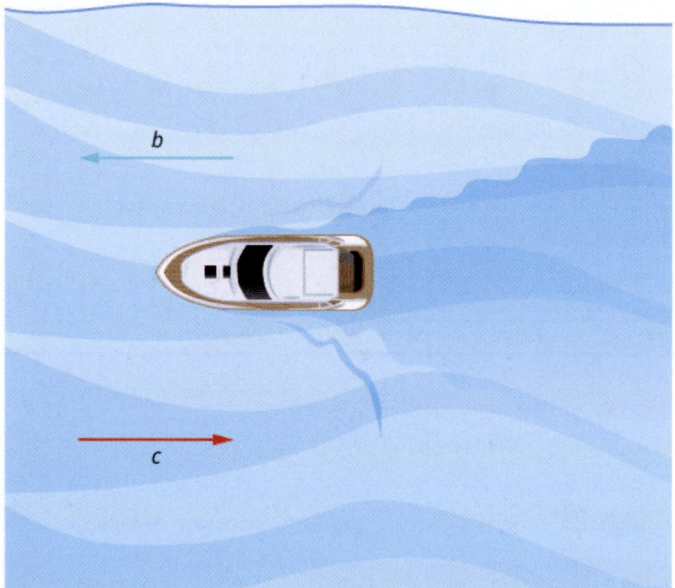

We'll put some numbers to this situation in the next example.

EXAMPLE 4.22

Translate to a system of equations and then solve.

A river cruise ship sailed 60 miles downstream for 4 hours and then took 5 hours sailing upstream to return to the dock. Find the speed of the ship in still water and the speed of the river current.

⊘ Solution

Read the problem.	This is a uniform motion problem and a picture will help us visualize the situation.

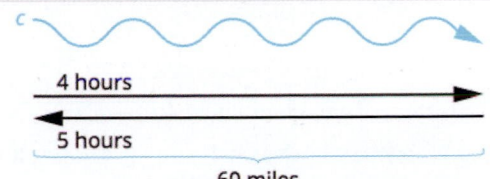

Identify what we are looking for.	We are looking for the speed of the ship in still water and the speed of the current.
Name what we are looking for.	Let $s =$ the rate of the ship in still water. $c =$ the rate of the current

A chart will help us organize the information. The ship goes downstream and then upstream. Going downstream, the current helps the ship and so the ship's actual rate is $s + c$. Going upstream, the current slows the ship and so the actual rate is $s - c$.

	Rate · Time = Distance		
downstream	$s + c$	4	60
upstream	$s - c$	5	60

Downstream it takes 4 hours.
Upstream it takes 5 hours.
Each way the distance is 60 miles.

Translate into a system of equations. Since rate times time is distance, we can write the system of equations.	$\begin{cases} 4(s + c) = 60 \\ 5(s - c) = 60 \end{cases}$
Solve the system of equations. Distribute to put both equations in standard form, then solve by elimination.	$\begin{cases} 4s + 4c = 60 \\ 5s - 5c = 60 \end{cases}$
Multiply the top equation by 5 and the bottom equation by 4. Add the equations, then solve for s.	$\begin{aligned} 20s + 20c &= 300 \\ \underline{20s - 20c} &= \underline{240} \\ 40s &= 540 \\ s &= 13.5 \end{aligned}$ $4(s + c) = 60$
Substitute $s = 13.5$ into of the original equations.	$\begin{aligned} 4(13.5 + c) &= 60 \\ 54 + 4c &= 60 \\ 4c &= 6 \\ c &= 1.5 \end{aligned}$

Check the answer in the problem.
The downstream rate would be
  $13.5 + 1.5 = 15$ mph.
In 4 hours the ship would travel
  $15 \cdot 4 = 60$ miles.
The upstream rate would be
  $13.5 - 1.5 = 12$ mph.
In 5 hours the ship would travel
  $12 \cdot 5 = 60$ miles.

Answer the question.	The rate of the ship is 13.5 mph and the rate of the current is 1.5 mph.

> **TRY IT : :** 4.43

Translate to a system of equations and then solve:

A Mississippi river boat cruise sailed 120 miles upstream for 12 hours and then took 10 hours to return to the dock. Find the speed of the river boat in still water and the speed of the river current.

> **TRY IT : :** 4.44

Translate to a system of equations and then solve:

Jason paddled his canoe 24 miles upstream for 4 hours. It took him 3 hours to paddle back. Find the speed of the canoe in still water and the speed of the river current.

Wind currents affect airplane speeds in the same way as water currents affect boat speeds. We'll see this in the next example. A wind current in the same direction as the plane is flying is called a *tailwind*. A wind current blowing against the direction of the plane is called a *headwind*.

EXAMPLE 4.23

Translate to a system of equations and then solve:

A private jet can fly 1,095 miles in three hours with a tailwind but only 987 miles in three hours into a headwind. Find the speed of the jet in still air and the speed of the wind.

⊘ **Solution**

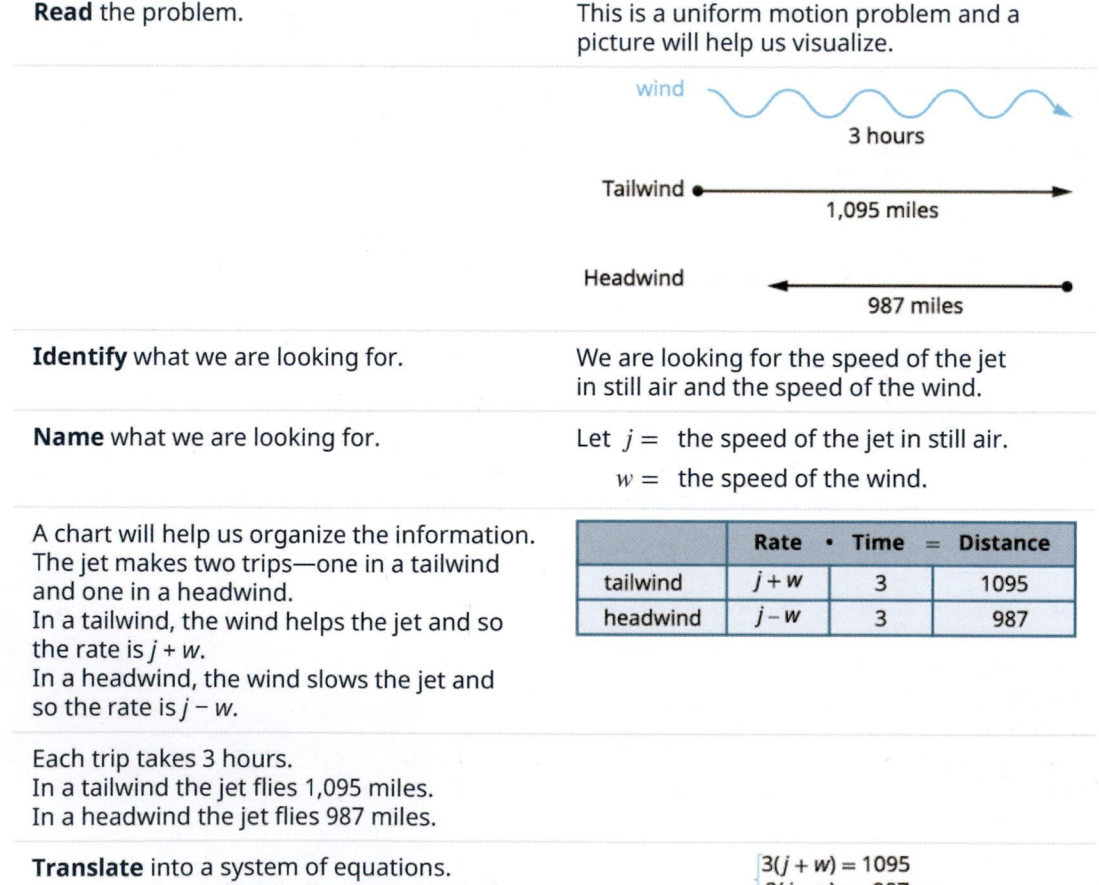

Read the problem.	This is a uniform motion problem and a picture will help us visualize.
Identify what we are looking for.	We are looking for the speed of the jet in still air and the speed of the wind.
Name what we are looking for.	Let $j =$ the speed of the jet in still air. $w =$ the speed of the wind.

A chart will help us organize the information. The jet makes two trips—one in a tailwind and one in a headwind.
In a tailwind, the wind helps the jet and so the rate is $j + w$.
In a headwind, the wind slows the jet and so the rate is $j - w$.

	Rate • **Time** = **Distance**		
tailwind	$j + w$	3	1095
headwind	$j - w$	3	987

Each trip takes 3 hours.
In a tailwind the jet flies 1,095 miles.
In a headwind the jet flies 987 miles.

Translate into a system of equations. Since rate times time is distance, we get the system of equations.	$\begin{cases} 3(j + w) = 1095 \\ 3(j - w) = 987 \end{cases}$

Solve the system of equations. Distribute, then solve by elimination. Add, and solve for j.	$\left\{\begin{array}{l}3j + 3w = 1095 \\ 3j - 3w = 987\end{array}\right.$ $6j = 2082$ $j = 347$ $4(j + w) = 1095$
Substitute $j = 347$ into one of the original equations, then solve for w.	$3(347 + w) = 1095$ $1041 + 3w = 1095$ $3w = 54$ $w = 18$
Check the answer in the problem. With the tailwind, the actual rate of the jet would be $\quad 347 + 18 = 365\ $ mph. In 3 hours the jet would travel $\quad 365 \cdot 3 = 1{,}095\ $ miles Going into the headwind, the jet's actual rate would be $\quad 347 - 18 = 329\ $ mph. In 3 hours the jet would travel $\quad 329 \cdot 3 = 987\ $ miles.	
Answer the question.	The rate of the jet is 347 mph and the rate of the wind is 18 mph.

> **TRY IT : : 4.45**

Translate to a system of equations and then solve:

A small jet can fly 1,325 miles in 5 hours with a tailwind but only 1,035 miles in 5 hours into a headwind. Find the speed of the jet in still air and the speed of the wind.

> **TRY IT : : 4.46**

Translate to a system of equations and then solve:

A commercial jet can fly 1,728 miles in 4 hours with a tailwind but only 1,536 miles in 4 hours into a headwind. Find the speed of the jet in still air and the speed of the wind.

▶ **MEDIA : :**

Access this online resource for additional instruction and practice with systems of equations.

- **Systems of Equations (https://openstax.org/l/37sysequations)**

4.2 EXERCISES

Practice Makes Perfect

Direct Translation Applications

In the following exercises, translate to a system of equations and solve.

72. The sum of two number is 15. One number is 3 less than the other. Find the numbers.

73. The sum of two number is 30. One number is 4 less than the other. Find the numbers.

74. The sum of two number is −16. One number is 20 less than the other. Find the numbers.

75. The sum of two number is −26. One number is 12 less than the other. Find the numbers.

76. The sum of two numbers is 65. Their difference is 25. Find the numbers.

77. The sum of two numbers is 37. Their difference is 9. Find the numbers.

78. The sum of two numbers is −27. Their difference is −59. Find the numbers.

79. The sum of two numbers is −45. Their difference is −89. Find the numbers.

80. Maxim has been offered positions by two car companies. The first company pays a salary of $10,000 plus a commission of $1000 for each car sold. The second pays a salary of $20,000 plus a commission of $500 for each car sold. How many cars would need to be sold to make the total pay the same?

81. Jackie has been offered positions by two cable companies. The first company pays a salary of $14,000 plus a commission of $100 for each cable package sold. The second pays a salary of $20,000 plus a commission of $25 for each cable package sold. How many cable packages would need to be sold to make the total pay the same?

82. Amara currently sells televisions for company A at a salary of $17,000 plus a $100 commission for each television she sells. Company B offers her a position with a salary of $29,000 plus a $20 commission for each television she sells. How televisions would Amara need to sell for the options to be equal?

83. Mitchell currently sells stoves for company A at a salary of $12,000 plus a $150 commission for each stove he sells. Company B offers him a position with a salary of $24,000 plus a $50 commission for each stove he sells. How many stoves would Mitchell need to sell for the options to be equal?

84. Two containers of gasoline hold a total of fifty gallons. The big container can hold ten gallons less than twice the small container. How many gallons does each container hold?

85. June needs 48 gallons of punch for a party and has two different coolers to carry it in. The bigger cooler is five times as large as the smaller cooler. How many gallons can each cooler hold?

86. Shelly spent 10 minutes jogging and 20 minutes cycling and burned 300 calories. The next day, Shelly swapped times, doing 20 minutes of jogging and 10 minutes of cycling and burned the same number of calories. How many calories were burned for each minute of jogging and how many for each minute of cycling?

87. Drew burned 1800 calories Friday playing one hour of basketball and canoeing for two hours. Saturday he spent two hours playing basketball and three hours canoeing and burned 3200 calories. How many calories did he burn per hour when playing basketball? How many calories did he burn per hour when canoeing?

88. Troy and Lisa were shopping for school supplies. Each purchased different quantities of the same notebook and thumb drive. Troy bought four notebooks and five thumb drives for $116. Lisa bought two notebooks and three thumb dives for $68. Find the cost of each notebook and each thumb drive.

89. Nancy bought seven pounds of oranges and three pounds of bananas for $17. Her husband later bought three pounds of oranges and six pounds of bananas for $12. What was the cost per pound of the oranges and the bananas?

90. Andrea is buying some new shirts and sweaters. She is able to buy 3 shirts and 2 sweaters for $114 or she is able to buy 2 shirts and 4 sweaters for $164. How much does a shirt cost? How much does a sweater cost?

91. Peter is buying office supplies. He is able to buy 3 packages of paper and 4 staplers for $40 or he is able to buy 5 packages of paper and 6 staplers for $62. How much does a package of paper cost? How much does a stapler cost?

92. The total amount of sodium in 2 hot dogs and 3 cups of cottage cheese is 4720 mg. The total amount of sodium in 5 hot dogs and 2 cups of cottage cheese is 6300 mg. How much sodium is in a hot dog? How much sodium is in a cup of cottage cheese?

93. The total number of calories in 2 hot dogs and 3 cups of cottage cheese is 960 calories. The total number of calories in 5 hot dogs and 2 cups of cottage cheese is 1190 calories. How many calories are in a hot dog? How many calories are in a cup of cottage cheese?

94. Molly is making strawberry infused water. For each ounce of strawberry juice, she uses three times as many ounces of water as juice. How many ounces of strawberry juice and how many ounces of water does she need to make 64 ounces of strawberry infused water?

95. Owen is making lemonade from concentrate. The number of quarts of water he needs is 4 times the number of quarts of concentrate. How many quarts of water and how many quarts of concentrate does Owen need to make 100 quarts of lemonade?

Solve Geometry Applications

In the following exercises, translate to a system of equations and solve.

96. The difference of two complementary angles is 55 degrees. Find the measures of the angles.

97. The difference of two complementary angles is 17 degrees. Find the measures of the angles.

98. Two angles are complementary. The measure of the larger angle is twelve less than twice the measure of the smaller angle. Find the measures of both angles.

99. Two angles are complementary. The measure of the larger angle is ten more than four times the measure of the smaller angle. Find the measures of both angles.

100. The difference of two supplementary angles is 8 degrees. Find the measures of the angles.

101. The difference of two supplementary angles is 88 degrees. Find the measures of the angles.

102. Two angles are supplementary. The measure of the larger angle is four more than three times the measure of the smaller angle. Find the measures of both angles.

103. Two angles are supplementary. The measure of the larger angle is five less than four times the measure of the smaller angle. Find the measures of both angles.

104. The measure of one of the small angles of a right triangle is 14 more than 3 times the measure of the other small angle. Find the measure of both angles.

105. The measure of one of the small angles of a right triangle is 26 more than 3 times the measure of the other small angle. Find the measure of both angles.

106. The measure of one of the small angles of a right triangle is 15 less than twice the measure of the other small angle. Find the measure of both angles.

107. The measure of one of the small angles of a right triangle is 45 less than twice the measure of the other small angle. Find the measure of both angles.

108. Wayne is hanging a string of lights 45 feet long around the three sides of his patio, which is adjacent to his house. The length of his patio, the side along the house, is five feet longer than twice its width. Find the length and width of the patio.

109. Darrin is hanging 200 feet of Christmas garland on the three sides of fencing that enclose his front yard. The length is five feet less than three times the width. Find the length and width of the fencing.

110. A frame around a family portrait has a perimeter of 90 inches. The length is fifteen less than twice the width. Find the length and width of the frame.

111. The perimeter of a toddler play area is 100 feet. The length is ten more than three times the width. Find the length and width of the play area.

Solve Uniform Motion Applications

In the following exercises, translate to a system of equations and solve.

112. Sarah left Minneapolis heading east on the interstate at a speed of 60 mph. Her sister followed her on the same route, leaving two hours later and driving at a rate of 70 mph. How long will it take for Sarah's sister to catch up to Sarah?

113. College roommates John and David were driving home to the same town for the holidays. John drove 55 mph, and David, who left an hour later, drove 60 mph. How long will it take for David to catch up to John?

114. At the end of spring break, Lucy left the beach and drove back towards home, driving at a rate of 40 mph. Lucy's friend left the beach for home 30 minutes (half an hour) later, and drove 50 mph. How long did it take Lucy's friend to catch up to Lucy?

115. Felecia left her home to visit her daughter driving 45 mph. Her husband waited for the dog sitter to arrive and left home twenty minutes (1/3 hour) later. He drove 55 mph to catch up to Felecia. How long before he reaches her?

116. The Jones family took a 12-mile canoe ride down the Indian River in two hours. After lunch, the return trip back up the river took three hours. Find the rate of the canoe in still water and the rate of the current.

117. A motor boat travels 60 miles down a river in three hours but takes five hours to return upstream. Find the rate of the boat in still water and the rate of the current.

118. A motor boat traveled 18 miles down a river in two hours but going back upstream, it took 4.5 hours due to the current. Find the rate of the motor boat in still water and the rate of the current. (Round to the nearest hundredth.)

119. A river cruise boat sailed 80 miles down the Mississippi River for four hours. It took five hours to return. Find the rate of the cruise boat in still water and the rate of the current.

120. A small jet can fly 1072 miles in 4 hours with a tailwind but only 848 miles in 4 hours into a headwind. Find the speed of the jet in still air and the speed of the wind.

121. A small jet can fly 1435 miles in 5 hours with a tailwind but only 1,215 miles in 5 hours into a headwind. Find the speed of the jet in still air and the speed of the wind.

122. A commercial jet can fly 868 miles in 2 hours with a tailwind but only 792 miles in 2 hours into a headwind. Find the speed of the jet in still air and the speed of the wind.

123. A commercial jet can fly 1,320 miles in 3 hours with a tailwind but only 1170 miles in 3 hours into a headwind. Find the speed of the jet in still air and the speed of the wind.

Writing Exercises

124. Write an application problem similar to **Example 4.14**. Then translate to a system of equations and solve it.

125. Write a uniform motion problem similar to **Example 4.15** that relates to where you live with your friends or family members. Then translate to a system of equations and solve it.

Self Check

ⓐ *After completing the exercises, use this checklist to evaluate your mastery of the objectives of this section.*

I can...	Confidently	With some help	No-I don't get it!
solve direct translation applications.			
solve geometry applications.			
solve uniform motion applications.			

ⓑ *After reviewing this checklist, what will you do to become confident for all objectives?*

Solve Mixture Applications with Systems of Equations

Learning Objectives

By the end of this section, you will be able to:

> Solve mixture applications
> Solve interest applications
> Solve applications of cost and revenue functions

Be Prepared!

Before you get started, take this readiness quiz.

1. Multiply: $4.025(1,562)$.

   If you missed this problem, review **Example 1.36**.

2. Write 8.2% as a decimal.
   If you missed this problem, review **Example 1.40**.

3. Earl's dinner bill came to $32.50 and he wanted to leave an 18% tip. How much should the tip be?
   If you missed this problem, review **Example 2.20**.

Solve Mixture Applications

Mixture application involve combining two or more quantities. When we solved mixture applications with coins and tickets earlier, we started by creating a table so we could organize the information. For a coin example with nickels and dimes, the table looked like this:

	Number •	Value =	Total Value
nickels			
dimes			

Using one variable meant that we had to relate the number of nickels and the number of dimes. We had to decide if we were going to let n be the number of nickels and then write the number of dimes in terms of n, or if we would let d be the number of dimes and write the number of nickels in terms of d.

Now that we know how to solve systems of equations with two variables, we'll just let n be the number of nickels and d be the number of dimes. We'll write one equation based on the total value column, like we did before, and the other equation will come from the number column.

For the first example, we'll do a ticket problem where the ticket prices are in whole dollars, so we won't need to use decimals just yet.

EXAMPLE 4.24

Translate to a system of equations and solve:

A science center sold 1,363 tickets on a busy weekend. The receipts totaled $12,146. How many $12 adult tickets and how many $7 child tickets were sold?

⊘ **Solution**

Step 1. Read the problem.	We will create a table to organize the information.
Step 2. Identify what we are looking for.	We are looking for the number of adult tickets and the number of child tickets sold.
Step 3. Name what we are looking for.	Let $a =$ the number of adult tickets. $c =$ the number of child tickets
A table will help us organize the data. We have two types of tickets, adult and child.	Write in a and c for the number of tickets.

| Write the total number of tickets sold at the bottom of the Number column. | Altogether 1,363 were sold. |

| Write the value of each type of ticket in the Value column. | The value of each adult ticket is $12. The value of each child tickets is $7. |

| The number times the value gives the total value, so the total value of adult tickets is $a \cdot 12 = 12a$, and the total value of child tickets is $c \cdot 7 = 7c$. | Fill in the Total Value column. |

Altogether the total value of the tickets was $12,146.

Type	Number	$\cdot$	Value	=	Total Value
adult	a		12		$11a$
child	c		7		$8c$
	1,363				12,146

Step 4. Translate into a system of equations.

| The Number column and the Total value column give us the system of equations. | $\begin{cases} a + c = 1{,}363 \\ 12a + 7c = 12{,}146 \end{cases}$ |

| We will use the elimination method to solve this system. Multiply the first equation by -7. | $\begin{cases} -7(a + c) = -7(1{,}363) \\ 12a + 7c = 12{,}146 \end{cases}$ |

| Simplify and add, then solve for a. | $\begin{array}{rl} -7a - 7c = -9{,}541 \\ \underline{12a + 7c = 12{,}146} \\ 5a \quad\;\; = 2{,}605 \\ a \quad\;\; = \boxed{521} \end{array}$ |

| Substitute $a = 521$ into the first equation, then solve for c. | $\begin{array}{l} a + c = 1{,}363 \\ 521 + c = 1{,}363 \\ c = 842 \end{array}$ |

Step 6. Check the answer in the problem.
521 adult at $12 per ticket makes $ 6,252
   842 child at $7 per ticket makes $58,994
        The total receipts are $12,146✓

| **Step 7. Answer** the question. | The science center sold 521 adult tickets and 842 child tickets. |

> **TRY IT :: 4.47**

Translate to a system of equations and solve:

The ticket office at the zoo sold 553 tickets one day. The receipts totaled $3,936. How many $9 adult tickets and how many $6 child tickets were sold?

> **TRY IT :: 4.48**

Translate to a system of equations and solve:

The box office at a movie theater sold 147 tickets for the evening show, and receipts totaled $1,302. How many $11 adult and how many $8 child tickets were sold?

In the next example, we'll solve a coin problem. Now that we know how to work with systems of two variables, naming the variables in the 'number' column will be easy.

EXAMPLE 4.25

Translate to a system of equations and solve:

Juan has a pocketful of nickels and dimes. The total value of the coins is $8.10. The number of dimes is 9 less than twice the number of nickels. How many nickels and how many dimes does Juan have?

⊘ **Solution**

Step 1. Read the problem.
We will create a table to organize the information.

Step 2. Identify what we are looking for.	We are looking for the number of nickels and the number of dimes.

Step 3. Name what we are looking for.	Let $n =$ the number of nickels.
	$d =$ the number of dimes

A table will help us organize the data. We have two types of coins, nickels and dimes.	Write n and d for the number of each type of coin.

Fill in the Value column with the value of each type of coin.	The value of each nickel is $0.05. The value of each dime is $0.10.

The number times the value gives the total value, so, the total value of the nickels is $n(0.05) = 0.05n$ and the total value of dimes is $d(0.10) = 0.10d$.

Altogether the total value of the coins is $8.10.

Type	Number	•	Value	=	Total Value
nickels	n		0.05		$0.05n$
dimes	d		0.10		$0.10d$
					8.10

Step 4. Translate into a system of equations.

The Total Value column gives one equation.	$0.05n + 0.10d = 8.10$

We also know the number of dimes is 9 less than twice the number of nickels.

Translate to get the second equation.	$d = 2n - 9$

Now we have the system to solve.	$\begin{cases} 0.05n + 0.10d = 8.10 \\ d = 2n - 9 \end{cases}$

Step 5. Solve the system of equations
We will use the substitution method.

Substitute $d = 2n - 9$ into the first equation.	$0.05n + 0.10d = 8.10$
	$0.05n + 0.10(2n - 9) = 8.10$

Simplify and solve for n.	$0.05n + 0.2n - 0.90 = 8.10$
	$0.25n - 0.90 = 8.10$
	$0.25n = 9.00$
	$n = 36$

To find the number of dimes, substitute $n = 36$ into the second equation.	$d = 2n - 9$
	$d = 2 \cdot 36 - 9$
	$d = 63$

Step 6. Check the answer in the problem
$$63 \text{ dimes at } \$0.10 = \$6.30$$
$$36 \text{ nickels at } \$0.05 = \$1.80$$
$$\text{Total } = \$8.10 \checkmark$$

Step 7. Answer the question.	Juan has 36 nickels and 63 dimes.

> ⟩ **TRY IT : :** 4.49

Translate to a system of equations and solve:

Matilda has a handful of quarters and dimes, with a total value of $8.55. The number of quarters is 3 more than twice the number of dimes. How many dimes and how many quarters does she have?

> ⟩ **TRY IT : :** 4.50

Translate to a system of equations and solve:

Priam has a collection of nickels and quarters, with a total value of $7.30. The number of nickels is six less than three times the number of quarters. How many nickels and how many quarters does he have?

Some mixture applications involve combining foods or drinks. Example situations might include combining raisins and nuts to make a trail mix or using two types of coffee beans to make a blend.

EXAMPLE 4.26

Translate to a system of equations and solve:

Carson wants to make 20 pounds of trail mix using nuts and chocolate chips. His budget requires that the trail mix costs him $7.60. per pound. Nuts cost $9.00 per pound and chocolate chips cost $2.00 per pound. How many pounds of nuts and how many pounds of chocolate chips should he use?

⊘ **Solution**

Step 1. Read the problem.
We will create a table to organize the information.

Step 2. Identify what we are looking for.	We are looking for the number of pounds of nuts and the number of pounds of chocolate chips.
Step 3. Name what we are looking for.	Let $n =$ the number of pound of nuts. $c =$ the number of pounds of chips

Carson will mix nuts and chocolate chips to get trail mix.
Write in n and c for the number of pounds of nuts and chocolate chips.

Type	Number of pounds	•	Value	=	Total Value
Nuts	n		9.00		$9n$
Chocolate chips	c		2.00		$2c$
Trail mix	20		7.60		$7.60(20) = 152$

There will be 20 pounds of trail mix.
Put the price per pound of each item in the Value column.
Fill in the last column using
    Number • Value = Total Value

Step 4. Translate into a system of equations.
We get the equations from the Number and Total Value columns.

$$\begin{cases} n + c = 20 \\ 9n + 2c = 152 \end{cases}$$

Step 5. Solve the system of equations
We will use elimination to solve the system.
Multiply the first equation by -2 to eliminate c.

$$\begin{cases} -2(n + c) = -2(20) \\ 9n + 2c = 152 \end{cases}$$

Simplify and add.
Solve for n.

$$\begin{cases} -2n - 2c = -40 \\ 9n + 2c = 152 \end{cases}$$
$$\overline{\ 7n\ = 112}$$
$$n = 16$$

To find the number of pounds of chocolate chips, substitute $n = 16$ into the first equation, then solve for c.

$$n + c = 20$$
$$16 + c = 20$$
$$c = 4$$

Step 6. Check the answer in the problem.

$$16 + 4 \ = \ 20 \ ✓$$
$$9 \cdot 16 + 2 \cdot 4 \ = \ 152 \ ✓$$

Step 7. Answer the question.

Carson should mix 16 pounds of nuts with 4 pounds of chocolate chips to create the trail mix.

 TRY IT : : 4.51

Translate to a system of equations and solve:

Greta wants to make 5 pounds of a nut mix using peanuts and cashews. Her budget requires the mixture to cost her $6 per pound. Peanuts are $4 per pound and cashews are $9 per pound. How many pounds of peanuts and how many pounds of cashews should she use?

 TRY IT : : 4.52

Translate to a system of equations and solve:

Sammy has most of the ingredients he needs to make a large batch of chili. The only items he lacks are beans and ground beef. He needs a total of 20 pounds combined of beans and ground beef and has a budget of $3 per pound. The price of beans is $1 per pound and the price of ground beef is $5 per pound. How many pounds of beans and how many pounds of ground beef should he purchase?

Another application of mixture problems relates to concentrated cleaning supplies, other chemicals, and mixed drinks. The concentration is given as a percent. For example, a 20% concentrated household cleanser means that 20% of the total amount is cleanser, and the rest is water. To make 35 ounces of a 20% concentration, you mix 7 ounces (20% of 35) of the cleanser with 28 ounces of water.

For these kinds of mixture problems, we'll use "percent" instead of "value" for one of the columns in our table.

EXAMPLE 4.27

Translate to a system of equations and solve:

Sasheena is lab assistant at her community college. She needs to make 200 milliliters of a 40% solution of sulfuric acid for a lab experiment. The lab has only 25% and 50% solutions in the storeroom. How much should she mix of the 25% and the 50% solutions to make the 40% solution?

 Solution

Step 1. Read the problem. A figure may help us visualize the situation, then we will create a table to organize the information.

Sasheena must mix some of the 25% solution and some of the 50% solution together to get 200 ml of the 40% solution.

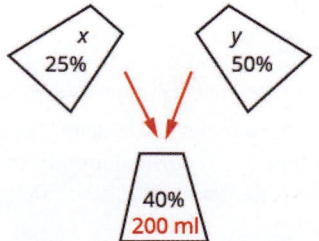

Step 2. Identify what we are looking for.	We are looking for how much of each solution she needs.

Step 3. Name what we are looking for.	Let $x =$ number of ml of 25% solution. $y =$ number of ml of 50% solution

A table will help us organize the data. She will mix x ml of 25% with y ml of 50% to get 200 ml of 40% solution. We write the percents as decimals in the chart.
We multiply the number of units times the concentration to get the total amount of sulfuric acid in each solution.

Type	Number of units	• Concentration %	= Amount
25%	x	0.25	0.25x
50%	y	0.50	0.50y
40%	200	0.40	0.40(200)

Step 4. Translate into a system of equations. We get the equations from the Number column and the Amount column. Now we have the system.	$\begin{cases} x + \quad y = 200 \\ 0.25x + 0.50y = 0.40(200) \end{cases}$

Step 5. Solve the system of equations We will solve the system by elimination. Multiply the first equation by -0.5 to eliminate y.	$\begin{cases} -0.5(x + y) = -0.5(200) \\ 0.25x + 0.50y = 80 \end{cases}$

Simplify and add to solve for x.	$\begin{aligned} -0.5\ x - 0.5y &= -100 \\ 0.25x + 0.5y &= \quad 80 \\ \hline -0.25x \qquad\quad &= -20 \\ x \qquad\quad &= \quad 80 \end{aligned}$

To solve for y, substitute $x = 80$ into the first equation.	$\begin{aligned} x + y &= 200 \\ 80 + y &= 200 \\ y &= 120 \end{aligned}$

Step 6. Check the answer in the problem.
$$80 + 120 = 200 ✓$$
$$0.25(80) + 0.50(120) = 200 ✓$$
$$\text{Yes!}$$

Step 7. Answer the question.	Sasheena should mix 80 ml of the 25% solution with 120 ml of the 50% solution to get the 200 ml of the 40% solution.

> **TRY IT :: 4.53**

Translate to a system of equations and solve:

LeBron needs 150 milliliters of a 30% solution of sulfuric acid for a lab experiment but only has access to a 25% and a 50% solution. How much of the 25% and how much of the 50% solution should he mix to make the 30% solution?

> **TRY IT :: 4.54**

Translate to a system of equations and solve:

Anatole needs to make 250 milliliters of a 25% solution of hydrochloric acid for a lab experiment. The lab only has a 10% solution and a 40% solution in the storeroom. How much of the 10% and how much of the 40% solutions should he mix to make the 25% solution?

Solve Interest Applications

The formula to model simple interest applications is $I = Prt$. Interest, I, is the product of the principal, P, the rate, r, and the time, t. In our work here, we will calculate the interest earned in one year, so t will be 1.

We modify the column titles in the mixture table to show the formula for interest, as you'll see in the next example.

EXAMPLE 4.28

Translate to a system of equations and solve:

Adnan has \$40,000 to invest and hopes to earn 7.1% interest per year. He will put some of the money into a stock fund that earns 8% per year and the rest into bonds that earns 3% per year. How much money should he put into each fund?

⊘ Solution

Step 1. Read the problem.	A chart will help us organize the information.
Step 2. Identify what we are looking for.	We are looking for the amount to invest in each fund.
Step 3. Name what we are looking for.	Let $s =$ the amount invested in stocks. $\quad\ b =$ the amount invested in stocks

Write the interest rate as a decimal for each fund.
Multiply: Principal · Rate · Time

Account	Principal ·	Rate ·	Time =	Interest
Stock fund	s	0.08	1	$0.08s$
Bonds	b	0.03	1	$0.03b$
Total	40,000	0.071		0.071(40,000)

Step 4. Translate into a system of equations.
We get our system of equations from the Principal column and the Interest column.

$$\begin{cases} s + b = 40{,}000 \\ 0.08s + 0.03b = 0.071(40{,}000) \end{cases}$$

Step 5. Solve the system of equations by elimination.
Multiply the top equation by −0.03.

$$\begin{cases} -0.03(s+b) = -0.03(40{,}000) \\ 0.08s + 0.03b = 2{,}840 \end{cases}$$

Simplify and add to solve for s.

$$\begin{aligned} -0.03s - 0.03b &= -1{,}200 \\ 0.08s + 0.03b &= \ \ 2{,}840 \\ \hline 0.05s &= \ \ 1{,}640 \\ s &= 32{,}800 \end{aligned}$$

To find b, substitute $s = 32{,}800$ into the first equation.

$$\begin{aligned} s + b &= 40{,}000 \\ 32{,}800 + b &= 40{,}000 \\ b &= 7{,}200 \end{aligned}$$

Step 6. Check the answer in the problem. — We leave the check to you.

Step 7. Answer the question. — Adnan should invest \$32,000 in stock and \$7,200 in bonds.

Did you notice that the Principal column represents the total amount of money invested while the Interest column represents only the interest earned? Likewise, the first equation in our system, $s + b = 40{,}000$, represents the total amount of money invested and the second equation, $0.08s + 0.03b = 0.071(40{,}000)$, represents the interest earned.

> **TRY IT :: 4.55**
>
> Translate to a system of equations and solve:
>
> Leon had \$50,000 to invest and hopes to earn 6.2% interest per year. He will put some of the money into a stock fund that earns 7% per year and the rest in to a savings account that earns 2% per year. How much money should he put into each fund?

TRY IT :: 4.56

Translate to a system of equations and solve:

Julius invested $7000 into two stock investments. One stock paid 11% interest and the other stock paid 13% interest. He earned 12.5% interest on the total investment. How much money did he put in each stock?

The next example requires that we find the principal given the amount of interest earned.

EXAMPLE 4.29

Translate to a system of equations and solve:

Rosie owes $21,540 on her two student loans. The interest rate on her bank loan is 10.5% and the interest rate on the federal loan is 5.9%. The total amount of interest she paid last year was $1,669.68. What was the principal for each loan?

 Solution

Step 1. Read the problem.	A chart will help us organize the information.
Step 2. Identify what we are looking for.	We are looking for the principal of each loan.
Step 3. Name what we are looking for.	Let $b =$ the principal for the bank loan. $f =$ the principal on the federal loan

The total loans are $21,540.

Record the interest rates as decimals in the chart.
Multiply using the formula $I = Prt$ to get the Interest.

Account	Principal	• Rate	• Time	= Interest
Bank	b	0.105	1	0.105b
Federal	f	0.059	1	0.059f
Total	21,540			1669.68

Step 4. Translate into a system of equations.
The system of equations comes from the Principal column and the Interest column.

$$\begin{cases} b + f = 21,540 \\ 0.105b + 0.059f = 1669.68 \end{cases}$$

Step 5. Solve the system of equations
We will use substitution to solve.
Solve the first equation for b.

$$b + f = 21,540$$
$$b = -f + 21,540$$

$$0.105b + 0.059f = 1669.68$$

Substitute $b = -f + 21.540$ into the second equation.

$$0.105(-f + 21,540) + 0.059f = 1669.68$$
$$-0.105f + 2261.70 + 0.059f = 1669.68$$

Simplify and solve for f.

$$-0.046f + 2261.70 = 1669.68$$
$$-0.046f = -592.02$$
$$f = 12,870$$

To find b, substitute $f = 12,870$ into the first equation.

$$b + f = 21,540$$
$$b + 12,870 = 21,540$$
$$b = 8,670$$

Step 6. Check the answer in the problem. | We leave the check to you.

Step 7. Answer the question. | The principal of the federal loan is $12,870 and the principal for the bank loan is $8,670.

TRY IT :: 4.57

Translate to a system of equations and solve:

Laura owes $18,000 on her student loans. The interest rate on the bank loan is 2.5% and the interest rate on the federal loan is 6.9%. The total amount of interest she paid last year was $1,066. What was the principal for each loan?

 TRY IT :: 4.58

Translate to a system of equations and solve:

Jill's Sandwich Shoppe owes $65,200 on two business loans, one at 4.5% interest and the other at 7.2% interest. The total amount of interest owed last year was $3,582. What was the principal for each loan?

Solve applications of cost and revenue functions

Suppose a company makes and sells x units of a product. The cost to the company is the total costs to produce x units. This is the cost to manufacture for each unit times x, the number of units manufactured, plus the fixed costs.

The **revenue** is the money the company brings in as a result of selling x units. This is the selling price of each unit times the number of units sold.

When the costs equal the revenue we say the business has reached the **break-even point**.

Cost and Revenue Functions

The **cost function** is the cost to manufacture each unit times x, the number of units manufactured, plus the fixed costs.

$$C(x) = (\text{cost per unit}) \cdot x + \text{fixed costs}$$

The **revenue** function is the selling price of each unit times x, the number of units sold.

$$R(x) = (\text{selling price per unit}) \cdot x$$

The **break-even point** is when the revenue equals the costs.

$$C(x) = R(x)$$

EXAMPLE 4.30

The manufacturer of a weight training bench spends $105 to build each bench and sells them for $245. The manufacturer also has fixed costs each month of $7,000.

ⓐ Find the cost function C when x benches are manufactured.

ⓑ Find the revenue function R when x benches are sold.

ⓒ Show the break-even point by graphing both the Revenue and Cost functions on the same grid.

ⓓ Find the break-even point. Interpret what the break-even point means.

⊘ Solution

ⓐ The manufacturer has $7,000 of fixed costs no matter how many weight training benches it produces. In addition to the fixed costs, the manufacturer also spends $105 to produce each bench. Suppose x benches are sold.

Write the general Cost function formula. $C(x) = (\text{cost per unit}) \cdot x + \text{fixed costs}$
Substitute in the cost values. $C(x) = 105x + 7000$

ⓑ The manufacturer sells each weight training bench for $245. We get the total revenue by multiplying the revenue per unit times the number of units sold.

Write the general Revenue function. $R(x) = (\text{selling price per unit}) \cdot x$
Substitute in the revenue per unit. $R(x) = 245x$

ⓒ Essentially we have a system of linear equations. We will show the graph of the system as this helps make the idea of a break-even point more visual.

$$\begin{cases} C(x) = 105x + 7000 \\ R(x) = 245x \end{cases} \quad \text{or} \quad \begin{cases} y = 105x + 7000 \\ y = 245x \end{cases}$$

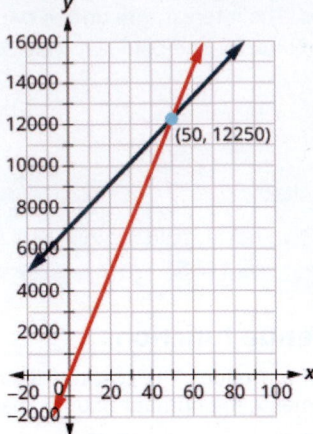

ⓓ To find the actual value, we remember the break-even point occurs when costs equal revenue.

Write the break-even formula.

$$\begin{aligned} C(x) &= R(x) \\ 105x + 7000 &= 245x \\ 7000 &= 140x \\ 50 &= x \end{aligned}$$

Solve.

When 50 benches are sold, the costs equal the revenue.

$$C(x) = 105x + 7000 \qquad R(x) = 245x$$

$$C(50) = 105(50) + 7000 \qquad R(50) = 245 \cdot 50$$

$$C(50) = 12{,}250 \qquad R(50) = 12{,}250$$

When 50 benches are sold, the revenue and costs are both $12,250. Notice this corresponds to the ordered pair (50, 12,250).

> **TRY IT :: 4.59**
>
> The manufacturer of a weight training bench spends $15 to build each bench and sells them for $32. The manufacturer also has fixed costs each month of $25,500.
>
> ⓐ Find the cost function C when x benches are manufactured.
>
> ⓑ Find the revenue function R when x benches are sold.
>
> ⓒ Show the break-even point by graphing both the Revenue and Cost functions on the same grid.
>
> ⓓ Find the break-even point. Interpret what the break-even point means.

> **TRY IT :: 4.60**
>
> The manufacturer of a weight training bench spends $120 to build each bench and sells them for $170. The manufacturer also has fixed costs each month of $150,000.
>
> ⓐ Find the cost function C when x benches are manufactured.
>
> ⓑ Find the revenue function R when x benches are sold.
>
> ⓒ Show the break-even point by graphing both the Revenue and Cost functions on the same grid.
>
> ⓓ Find the break-even point. Interpret what the break-even point means.

> ▶ **MEDIA ::**
>
> Access this online resource for additional instruction and practice with interest and mixtures.
>
> • **Interest and Mixtures (https://openstax.org/l/37intmixure)**

4.3 EXERCISES

Practice Makes Perfect

Solve Mixture Applications

In the following exercises, translate to a system of equations and solve.

126. Tickets to a Broadway show cost $35 for adults and $15 for children. The total receipts for 1650 tickets at one performance were $47,150. How many adult and how many child tickets were sold?

127. Tickets for the Cirque du Soleil show are $70 for adults and $50 for children. One evening performance had a total of 300 tickets sold and the receipts totaled $17,200. How many adult and how many child tickets were sold?

128. Tickets for an Amtrak train cost $10 for children and $22 for adults. Josie paid $1200 for a total of 72 tickets. How many children tickets and how many adult tickets did Josie buy?

129. Tickets for a Minnesota Twins baseball game are $69 for Main Level seats and $39 for Terrace Level seats. A group of sixteen friends went to the game and spent a total of $804 for the tickets. How many of Main Level and how many Terrace Level tickets did they buy?

130. Tickets for a dance recital cost $15 for adults and $7 dollars for children. The dance company sold 253 tickets and the total receipts were $2771. How many adult tickets and how many child tickets were sold?

131. Tickets for the community fair cost $12 for adults and $5 dollars for children. On the first day of the fair, 312 tickets were sold for a total of $2204. How many adult tickets and how many child tickets were sold?

132. Brandon has a cup of quarters and dimes with a total value of $3.80. The number of quarters is four less than twice the number of quarters. How many quarters and how many dimes does Brandon have?

133. Sherri saves nickels and dimes in a coin purse for her daughter. The total value of the coins in the purse is $0.95. The number of nickels is two less than five times the number of dimes. How many nickels and how many dimes are in the coin purse?

134. Peter has been saving his loose change for several days. When he counted his quarters and nickels, he found they had a total value $13.10. The number of quarters was fifteen more than three times the number of dimes. How many quarters and how many dimes did Peter have?

135. Lucinda had a pocketful of dimes and quarters with a value of $6.20. The number of dimes is eighteen more than three times the number of quarters. How many dimes and how many quarters does Lucinda have?

136. A cashier has 30 bills, all of which are $10 or $20 bills. The total value of the money is $460. How many of each type of bill does the cashier have?

137. A cashier has 54 bills, all of which are $10 or $20 bills. The total value of the money is $910. How many of each type of bill does the cashier have?

138. Marissa wants to blend candy selling for $1.80 per pound with candy costing $1.20 per pound to get a mixture that costs her $1.40 per pound to make. She wants to make 90 pounds of the candy blend. How many pounds of each type of candy should she use?

139. How many pounds of nuts selling for $6 per pound and raisins selling for $3 per pound should Kurt combine to obtain 120 pounds of trail mix that cost him $5 per pound?

140. Hannah has to make twenty-five gallons of punch for a potluck. The punch is made of soda and fruit drink. The cost of the soda is $1.79 per gallon and the cost of the fruit drink is $2.49 per gallon. Hannah's budget requires that the punch cost $2.21 per gallon. How many gallons of soda and how many gallons of fruit drink does she need?

141. Joseph would like to make twelve pounds of a coffee blend at a cost of $6 per pound. He blends Ground Chicory at $5 a pound with Jamaican Blue Mountain at $9 per pound. How much of each type of coffee should he use?

142. Julia and her husband own a coffee shop. They experimented with mixing a City Roast Columbian coffee that cost $7.80 per pound with French Roast Columbian coffee that cost $8.10 per pound to make a twenty-pound blend. Their blend should cost them $7.92 per pound. How much of each type of coffee should they buy?

143. Twelve-year old Melody wants to sell bags of mixed candy at her lemonade stand. She will mix M&M's that cost $4.89 per bag and Reese's Pieces that cost $3.79 per bag to get a total of twenty-five bags of mixed candy. Melody wants the bags of mixed candy to cost her $4.23 a bag to make. How many bags of M&M's and how many bags of Reese's Pieces should she use?

144. Jotham needs 70 liters of a 50% solution of an alcohol solution. He has a 30% and an 80% solution available. How many liters of the 30% and how many liters of the 80% solutions should he mix to make the 50% solution?

145. Joy is preparing 15 liters of a 25% saline solution. She only has 40% and 10% solution in her lab. How many liters of the 40% and how many liters of the 10% should she mix to make the 25% solution?

146. A scientist needs 65 liters of a 15% alcohol solution. She has available a 25% and a 12% solution. How many liters of the 25% and how many liters of the 12% solutions should she mix to make the 15% solution?

147. A scientist needs 120 milliliters of a 20% acid solution for an experiment. The lab has available a 25% and a 10% solution. How many liters of the 25% and how many liters of the 10% solutions should the scientist mix to make the 20% solution?

148. A 40% antifreeze solution is to be mixed with a 70% antifreeze solution to get 240 liters of a 50% solution. How many liters of the 40% and how many liters of the 70% solutions will be used?

149. A 90% antifreeze solution is to be mixed with a 75% antifreeze solution to get 360 liters of an 85% solution. How many liters of the 90% and how many liters of the 75% solutions will be used?

Solve Interest Applications

In the following exercises, translate to a system of equations and solve.

150. Hattie had $3000 to invest and wants to earn 10.6% interest per year. She will put some of the money into an account that earns 12% per year and the rest into an account that earns 10% per year. How much money should she put into each account?

151. Carol invested $2560 into two accounts. One account paid 8% interest and the other paid 6% interest. She earned 7.25% interest on the total investment. How much money did she put in each account?

152. Sam invested $48,000, some at 6% interest and the rest at 10%. How much did he invest at each rate if he received $4000 in interest in one year?

153. Arnold invested $64,000, some at 5.5% interest and the rest at 9%. How much did he invest at each rate if he received $4500 in interest in one year?

154. After four years in college, Josie owes $65, 800 in student loans. The interest rate on the federal loans is 4.5% and the rate on the private bank loans is 2%. The total interest she owes for one year was $2878.50. What is the amount of each loan?

155. Mark wants to invest $10,000 to pay for his daughter's wedding next year. He will invest some of the money in a short term CD that pays 12% interest and the rest in a money market savings account that pays 5% interest. How much should he invest at each rate if he wants to earn $1095 in interest in one year?

156. A trust fund worth $25,000 is invested in two different portfolios. This year, one portfolio is expected to earn 5.25% interest and the other is expected to earn 4%. Plans are for the total interest on the fund to be $1150 in one year. How much money should be invested at each rate?

157. A business has two loans totaling $85,000. One loan has a rate of 6% and the other has a rate of 4.5% This year, the business expects to pay $4,650 in interest on the two loans. How much is each loan?

Solve Applications of Cost and Revenue Functions

158. The manufacturer of an energy drink spends $1.20 to make each drink and sells them for $2. The manufacturer also has fixed costs each month of $8,000.

ⓐ Find the cost function C when x energy drinks are manufactured.

ⓑ Find the revenue function R when x drinks are sold.

ⓒ Show the break-even point by graphing both the Revenue and Cost functions on the same grid.

ⓓ Find the break-even point. Interpret what the break-even point means.

159. The manufacturer of a water bottle spends $5 to build each bottle and sells them for $10. The manufacturer also has fixed costs each month of $6500. ⓐ Find the cost function C when x bottles are manufactured. ⓑ Find the revenue function R when x bottles are sold. ⓒ Show the break-even point by graphing both the Revenue and Cost functions on the same grid. ⓓ Find the break-even point. Interpret what the break-even point means.

Writing Exercises

160. Take a handful of two types of coins, and write a problem similar to **Example 4.25** relating the total number of coins and their total value. Set up a system of equations to describe your situation and then solve it.

161. In **Example 4.28**, we used elimination to solve the system of equations

$$\begin{cases} s + b = 40{,}000 \\ 0.08s + 0.03b = 0.071(40{,}000) \end{cases}$$

Could you have used substitution or elimination to solve this system? Why?

Self Check

ⓐ *After completing the exercises, use this checklist to evaluate your mastery of the objectives of this section.*

I can...	Confidently	With some help	No-I don't get it!
solve mixture applications.			
solve interest applications.			

ⓑ *What does this checklist tell you about your mastery of this section? What steps will you take to improve?*

4.4 **Solve Systems of Equations with Three Variables**

Learning Objectives

By the end of this section, you will be able to:

> Determine whether an ordered triple is a solution of a system of three linear equations with three variables
> Solve a system of linear equations with three variables
> Solve applications using systems of linear equations with three variables

---

Be Prepared!

Before you get started, take this readiness quiz.

1. Evaluate $5x - 2y + 3z$ when $x = -2$, $y = -4$, and $z = 3$.
   If you missed this problem, review **Example 1.21**.

2. Classify the equations as a conditional equation, an identity, or a contradiction and then state the solution.
$$\begin{cases} -2x + y = -11 \\ x + 3y = 9 \end{cases}.$$
   If you missed this problem, review **Example 2.6**.

3. Classify the equations as a conditional equation, an identity, or a contradiction and then state the solution.
$$\begin{cases} 7x + 8y = 4 \\ 3x - 5y = 27 \end{cases}.$$
   If you missed this problem, review **Example 2.8**.

Determine Whether an Ordered Triple is a Solution of a System of Three Linear Equations with Three Variables

In this section, we will extend our work of solving a system of linear equations. So far we have worked with systems of equations with two equations and two variables. Now we will work with systems of three equations with three variables. But first let's review what we already know about solving equations and systems involving up to two variables.

We learned earlier that the graph of a linear equation, $ax + by = c$, is a line. Each point on the line, an ordered pair (x, y), is a solution to the equation. For a system of two equations with two variables, we graph two lines. Then we can see that all the points that are solutions to each equation form a line. And, by finding what the lines have in common, we'll find the solution to the system.

Most linear equations in one variable have one solution, but we saw that some equations, called contradictions, have no solutions and for other equations, called identities, all numbers are solutions

We know when we solve a system of two linear equations represented by a graph of two lines in the same plane, there are three possible cases, as shown.

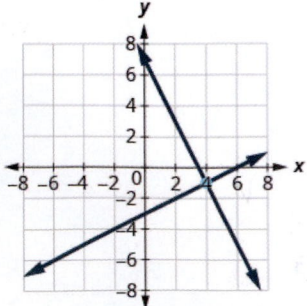

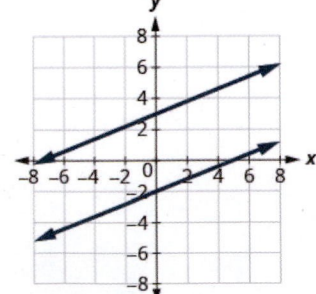

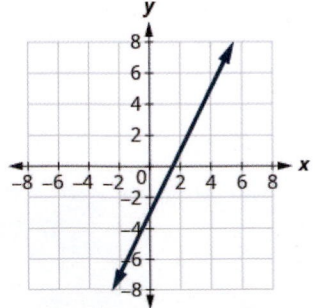

One solution

The lines intersect.
Intersecting lines have one point in common.
There is one solution to this system.

Consistent Independent

No solution

The lines are parallel.
Parallel lines have no points in common.
There is no solution to this system.

Inconsistent

Infinitely many solutions

Both equations give the same line.
Because we have just one line, there are infinitely many solutions.

Consistent Dependent

Similarly, for a linear equation with three variables $ax + by + cz = d,$ every solution to the equation is an ordered triple, (x, y, z) that makes the equation true.

Linear Equation in Three Variables

A linear equation with three variables, where *a, b, c,* and *d* are real numbers and *a, b,* and *c* are not all 0, is of the form

$$ax + by + cz = d$$

Every solution to the equation is an ordered triple, (x, y, z) that makes the equation true.

All the points that are solutions to one equation form a plane in three-dimensional space. And, by finding what the planes have in common, we'll find the solution to the system.

When we solve a system of three linear equations represented by a graph of three planes in space, there are three possible cases.

<u>**One solution**</u>

Consistent system and Independent equations

The 3 planes intersect.

The three intersecting planes have one point in common.

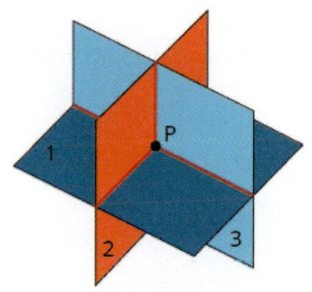

<u>**No solution**</u>

Inconsistent system

The planes are parallel.

Parallel planes have no points in common.

Two planes are coincident and parallel to the third plane.

The planes have no points in common.

Two planes are parallel and each intersect the third plane.

The planes have no points in common.

Each plane intersects the other two, but all three share no points.

The planes have no points in common.

<u>Infinitely many solutions</u>
Consistent system and dependent equations

Three planes intersect in one line.

There is just one line, so there are infinitely many solutions.

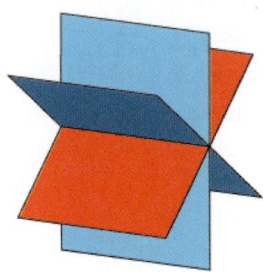

Two planes are coincident and intersect the third plane in a line.

There is just one line, so there are infinitely many solutions.

Three planes are coincident.

There is just one plane, so there are infinitely many solutions.

To solve a system of three linear equations, we want to find the values of the variables that are solutions to all three equations. In other words, we are looking for the ordered triple (x, y, z) that makes all three equations true. These are called the **solutions of the system of three linear equations with three variables**.

Solutions of a System of Linear Equations with Three Variables

Solutions of a system of equations are the values of the variables that make all the equations true. A solution is represented by an **ordered triple** (x, y, z).

To determine if an ordered triple is a solution to a system of three equations, we substitute the values of the variables into each equation. If the ordered triple makes all three equations true, it is a solution to the system.

EXAMPLE 4.31

Determine whether the ordered triple is a solution to the system: $\begin{cases} x - y + z = 2 \\ 2x - y - z = -6 \\ 2x + 2y + z = -3 \end{cases}$.

ⓐ $(-2, -1, 3)$ ⓑ $(-4, -3, 4)$

✓ **Solution**

ⓐ

$$\begin{cases} x - y + z = 2 \\ 2x - y - z = -6 \\ 2x + 2y + z = -3 \end{cases}$$

We substitute $x = -2$ and $y = -1$ and $z = 3$ into all three equations.

$x - y + z = 2$	$2x - y - z = -6$	$2x + 2y + z = -3$
$-2 - (-1) + 3 \overset{?}{=} 2$	$2 \cdot (-2) - (-1) - 3 \overset{?}{=} -6$	$2 \cdot (-2) + 2(-1) + 3 \overset{?}{=} -3$
$2 = 2 \checkmark$	$-6 = -6 \checkmark$	$-3 = -3 \checkmark$

$(-2, -1, 3)$ does make all three equations true. $(-2, -1, 3)$ is a solution.

ⓑ

$$\begin{cases} x - y + z = 2 \\ 2x - y - z = -6 \\ 2x + 2y + z = -3 \end{cases}$$

We substitute $x = -4$ and $y = -3$ and $z = 4$ into all three equations.

$x - y + z = 2$	$2x - y - z = -6$	$2x + 2y + z = -3$
$-4 - (-3) + 4 \overset{?}{=} 2$	$2 \cdot (-4) - (-3) - 4 \overset{?}{=} -6$	$2 \cdot (-4) + 2(-3) + 4 \overset{?}{=} -3$
$3 \neq 2$	$-9 \neq -6$	$-10 \neq -3$

$(-4, -3, 4)$ does not make all three equations true. $(-4, -3, 4)$ is not a solution.

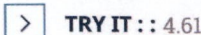 **TRY IT : : 4.61**

Determine whether the ordered triple is a solution to the system: $\begin{cases} 3x + y + z = 2 \\ x + 2y + z = -3. \\ 3x + y + 2z = 4 \end{cases}$

ⓐ $(1, -3, 2)$ ⓑ $(4, -1, -5)$

TRY IT : : 4.62

Determine whether the ordered triple is a solution to the system: $\begin{cases} x - 3y + z = -5 \\ -3x - y - z = 1. \\ 2x - 2y + 3z = 1 \end{cases}$

ⓐ $(2, -2, 3)$ ⓑ $(-2, 2, 3)$

Solve a System of Linear Equations with Three Variables

To solve a system of linear equations with three variables, we basically use the same techniques we used with systems that had two variables. We start with two pairs of equations and in each pair we eliminate the same variable. This will then give us a system of equations with only two variables and then we know how to solve that system!

Next, we use the values of the two variables we just found to go back to the original equation and find the third variable. We write our answer as an ordered triple and then check our results.

EXAMPLE 4.32 HOW TO SOLVE A SYSTEM OF EQUATIONS WITH THREE VARIABLES BY ELIMINATION

Solve the system by elimination: $\begin{cases} x - 2y + z = 3 \\ 2x + y + z = 4 \\ 3x + 4y + 3z = -1 \end{cases}$.

✓ **Solution**

Step 1. Write the equations in standard form. • If any coefficients are fractions, clear them.	The equations are in standard form. There are no fractions.	$\begin{cases} x - 2y + z = 3 & (1) \\ 2x + y + z = 4 & (2) \\ 3x + 4y + 3z = -1 & (3) \end{cases}$
Step 2. Eliminate the same variable from two equations. • Decide which variable you will eliminate. • Work with a pair of equations to eliminate the chosen variable. • Multiply one or both equations so that the coefficients of that variable are opposites. • Add the equations to eliminate one variable.	We can eliminate the y's from equations (1) and (2) by multiplying equation (2) by 2. Multiply. We add the x's, y's, and constants. The new equation, (4), has only x and z.	$x - 2y + z = 3 \quad (1)$ $2(2x + y + z) = (4)2 \quad (2)$ $x - 2y + z = 3 \quad (1)$ $4x + 2y + 2z = 8 \quad (2)$ $\begin{array}{l} x - 2y + z = 3 \quad (1) \\ \underline{4x + 2y + 2z = 8} \quad (2) \\ 5x + 3z = 11 \quad (4) \end{array}$
Step 3. Repeat Step 2 using two other equations and eliminate the same variable as in Step 2.	We can again eliminate the y's using the equations (1) and (3) by multiplying equation (1) by 2. Add the new equations and the result will be equation (5).	$2(x - 2y + z) = (3)2 \quad (1)$ $3x + 4y + 3z = -1 \quad (3)$ $2x - 4y + 2z = 6 \quad (1)$ $\underline{3x + 4y + 3z = -1} \quad (3)$ $5x + 5z = 5 \quad (5)$

Step 4. The two new equations form a system of two equations with two variables. Solve this system.	Use equations (4) and (5) to eliminate either x or z.	$\begin{cases} 5x + 3z = 11 & (4) \\ 5x + 5z = 5 & (5) \end{cases}$
	Multiply (5) by –1.	$\begin{array}{ll} 5x + 3z = 11 & (4) \\ -1(5x + 5z) = 5(-1) & (5) \end{array}$
		$\begin{array}{ll} 5x + 3z = 11 & (4) \\ \underline{-5x - 5z = -5} & (5) \\ -2z = 6 & \\ z = -3 & \end{array}$
	Use either (4) or (5) to solve for x using substitution.	$\begin{array}{ll} 5x + 3z = 11 & (4) \\ 5x + 3(-3) = 11 & \\ 5x - 9 = 11 & \\ 5x = 20 & \\ x = 4 & \end{array}$
Step 5. Use the values of the two variables found in Step 4 to find the third variable.	Use an original equation to find y.	$\begin{array}{ll} 2x + y + z = 4 & (2) \\ 2 \cdot 4 + y + z = 4 & \\ 8 + y - 3 = 4 & \\ y = -1 & \end{array}$
Step 6. Write the solution as an ordered triple.	Write it as (x, y, z).	$(4, -1, -3)$
Step 7. Check that the ordered triple is a solution to **all three** original equations.	Substitute $x = 4$, $y = -1$ and $z = -3$ into all three equations. Do they make the equations true?	We'll leave the checks to you. The solution is $(4, -1, -3)$.

> **TRY IT : :** 4.63

Solve the system by elimination: $\begin{cases} 3x + y - z = 2 \\ 2x - 3y - 2z = 1. \\ 4x - y - 3z = 0 \end{cases}$

> **TRY IT : :** 4.64

Solve the system by elimination: $\begin{cases} 4x + y + z = -1 \\ -2x - 2y + z = 2. \\ 2x + 3y - z = 1 \end{cases}$

The steps are summarized here.

HOW TO :: SOLVE A SYSTEM OF LINEAR EQUATIONS WITH THREE VARIABLES.

Step 1.   Write the equations in standard form
  ◦ If any coefficients are fractions, clear them.

Step 2.   Eliminate the same variable from two equations.
  ◦ Decide which variable you will eliminate.
  ◦ Work with a pair of equations to eliminate the chosen variable.
  ◦ Multiply one or both equations so that the coefficients of that variable are opposites.
  ◦ Add the equations resulting from Step 2 to eliminate one variable

Step 3.   Repeat Step 2 using two other equations and eliminate the same variable as in Step 2.

Step 4.   The two new equations form a system of two equations with two variables. Solve this system.

Step 5.   Use the values of the two variables found in Step 4 to find the third variable.

Step 6.   Write the solution as an ordered triple.

Step 7.   Check that the ordered triple is a solution to **all three** original equations.

EXAMPLE 4.33

Solve: $\begin{cases} 3x - 4z = 0 \\ 3y + 2z = -3 \\ 2x + 3y = -5 \end{cases}$.

✓ **Solution**

$$\begin{cases} 3x - 4z = 0 \quad (1) \\ 3y + 2z = -3 \ (2) \\ 2x + 3y = -5 \ (3) \end{cases}$$

We can eliminate z from equations (1) and (2) by multiplying equation (2) by 2 and then adding the resulting equations.

$$\begin{array}{lll} 3x \quad\ - 4z = 0 \quad (1) & 3x \quad\ - 4z = 0 & 3x \quad\quad - 4z = 0 \\ 3y + 2z = -3 \quad (2) & 2(3y + 2z) = (-3)2 & \underline{\quad 6y + 4z = -6 \quad} \\ & & 3x + 6y \quad\quad = -6 \quad (4) \end{array}$$

Notice that equations (3) and (4) both have the variables x and y. We will solve this new system for x and y.

$$\begin{array}{lll} 2x + 3y = -5 \quad (3) & -2(2x + 3y) = -2(-5) & -4x - 6y = 10 \\ 3x + 6y = -6 \quad (4) & 3x + 6y = -6 & \underline{\quad 3x + 6y = -6 \quad} \\ & & -x \quad\quad = 4 \\ & & \quad\quad x = -4 \end{array}$$

To solve for *y*, we substitute $x = -4$ into equation (3).

$$\begin{array}{l} 2x + 3y = -5 \quad (3) \\ 2(-4) + 3y = -5 \\ -8 + 3y = -5 \\ 3y = 3 \\ y = 1 \end{array}$$

We now have $x = -4$ and $y = 1$. We need to solve for *z*. We can substitute $x = -4$ into equation (1) to find *z*.

$$\begin{array}{l} 3x - 4z = 0 \quad (1) \\ 3(-4) - 4z = 0 \\ -12 - 4z = 0 \\ -4z = 12 \\ z = -3 \end{array}$$

We write the solution as an ordered triple. $\qquad$ $(-4, 1, -3)$

We check that the solution makes all three equations true.

$$3x - 4z = 0 \,(1) \qquad\qquad 3y + 2z = -3 \,(2) \qquad\qquad 2x + 3y = -5 \,(3)$$
$$3(-4) - 4(-3) \overset{?}{=} 0 \qquad 3(1) + 2(-3) \overset{?}{=} -3 \qquad 2(-4) + 3(1) \overset{?}{=} -5$$
$$0 = 0 \,\checkmark \qquad\qquad -3 = -3 \,\checkmark \qquad\qquad -5 = -5 \,\checkmark$$

The solution is $(-4, 1, -3)$.

> **TRY IT :: 4.65**

Solve: $\begin{cases} 3x - 4z = -1 \\ 2y + 3z = 2 \\ 2x + 3y = 6 \end{cases}$.

> **TRY IT :: 4.66**

Solve: $\begin{cases} 4x - 3z = -5 \\ 3y + 2z = 7 \\ 3x + 4y = 6 \end{cases}$.

When we solve a system and end up with no variables and a false statement, we know there are no solutions and that the system is inconsistent. The next example shows a system of equations that is inconsistent.

EXAMPLE 4.34

Solve the system of equations: $\begin{cases} x + 2y - 3z = -1 \\ x - 3y + z = 1 \\ 2x - y - 2z = 2 \end{cases}$.

⊘ Solution

$$\begin{cases} x + 2y - 3z = -1 \ (1) \\ x - 3y + z = 1 \quad (2) \\ 2x - y - 2z = 2 \quad (3) \end{cases}$$

Use equation (1) and (2) to eliminate z.

$$\begin{array}{ll} x + 2y - 3z = -1 & (1) \\ x - 3y + z = 1 & (2) \end{array} \xrightarrow{\text{multiply by 3}} \begin{array}{l} x + 2y - 3z = -1 \\ 3x - 9y + 3z = 3 \\ \hline 4x - 7y = 2 \quad (4) \end{array}$$

Use (2) and (3) to eliminate z again.

$$\begin{array}{ll} x - 3y + z = 1 & (2) \\ 2x - y - 2z = 2 & (3) \end{array} \xrightarrow{\text{multiply by 2}} \begin{array}{l} 2x - 6y + 2z = 2 \\ 2x - y - 2z = 2 \\ \hline 4x - 7y = 4 \quad (5) \end{array}$$

Use (4) and (5) to eliminate a variable.

$$\begin{array}{ll} 4x - 7y = 2 & (4) \\ 4x - 7y = 4 & (5) \end{array} \xrightarrow{\text{multiply by } -1} \begin{array}{l} 4x - 7y = 2 \\ -4x + 7y = -4 \\ \hline 0 = -2 \text{ false} \end{array}$$

There is no solution.

We are left with a false statement and this tells us the system is inconsistent and has no solution.

> **TRY IT :: 4.67**

Solve the system of equations: $\begin{cases} x + 2y + 6z = 5 \\ -x + y - 2z = 3 \\ x - 4y - 2z = 1 \end{cases}$.

> **TRY IT :: 4.68**

Solve the system of equations: $\begin{cases} 2x - 2y + 3z = 6 \\ 4x - 3y + 2z = 0 \\ -2x + 3y - 7z = 1 \end{cases}$.

When we solve a system and end up with no variables but a true statement, we know there are infinitely many solutions. The system is consistent with dependent equations. Our solution will show how two of the variables depend on the third.

EXAMPLE 4.35

Solve the system of equations: $\begin{cases} x + 2y - z = 1 \\ 2x + 7y + 4z = 11. \\ x + 3y + z = 4 \end{cases}$

⊘ **Solution**

$$\begin{cases} x + 2y - z = 1 & (1) \\ 2x + 7y + 4z = 11 & (2) \\ x + 3y + z = 4 & (3) \end{cases}$$

Use equation (1) and (3) to eliminate x.

$$\begin{array}{ll} x + 2y - z = 1 & (1) \\ x + 3y + z = 4 & (3) \end{array} \xrightarrow{\text{multiply by } -1} \begin{array}{ll} -x - 2y + z = -1 & (1) \\ \underline{x + 3y + z = 4} & (3) \\ y + 2z = 3 & (4) \end{array}$$

Use equation (1) and (2) to eliminate x again.

$$\begin{array}{ll} x + 2y - z = 1 & (1) \\ 2x + 7y + 4z = 11 & (2) \end{array} \xrightarrow{\text{multiply by } -2} \begin{array}{ll} -2x - 4y + 2z = -2 & (1) \\ \underline{2x + 7y + 4z = 11} & (2) \\ 3y + 6z = 9 & (5) \end{array}$$

Use equation (4) and (5) to eliminate y.

$$\begin{array}{ll} y + 2z = 3 & (4) \\ 3y + 6z = 9 & (5) \end{array} \xrightarrow{\text{multiply by } -3} \begin{array}{ll} -3y - 6z = -9 & (4) \\ \underline{3y + 6z = 9} & (5) \\ 0 = 0 \end{array}$$

There are infinitely many solutions.

Solve equation (4) for y.	Represent the solution showing how x and y are dependent on z. $$y + 2z = 3$$ $$y = -2z + 3$$
Use equation (1) to solve for x.	$$x + 2y - z = 1$$
Substitute $y = -2z + 3$.	$$\begin{aligned} x + 2(-2z + 3) - z &= 1 \\ x - 4z + 6 - z &= 1 \\ x - 5z + 6 &= 1 \\ x &= 5z - 5 \end{aligned}$$

The true statement $0 = 0$ tells us that this is a dependent system that has infinitely many solutions. The solutions are of the form (x, y, z) where $x = 5z - 5$; $y = -2z + 3$ and z is any real number.

> **TRY IT :: 4.69**

Solve the system by equations: $\begin{cases} x+y-z=0 \\ 2x+4y-2z=6. \\ 3x+6y-3z=9 \end{cases}$

> **TRY IT :: 4.70**

Solve the system by equations: $\begin{cases} x-y-z=1 \\ -x+2y-3z=-4. \\ 3x-2y-7z=0 \end{cases}$

Solve Applications using Systems of Linear Equations with Three Variables

Applications that are modeled by a systems of equations can be solved using the same techniques we used to solve the systems. Many of the application are just extensions to three variables of the types we have solved earlier.

EXAMPLE 4.36

The community college theater department sold three kinds of tickets to its latest play production. The adult tickets sold for \$15, the student tickets for \$10 and the child tickets for \$8. The theater department was thrilled to have sold 250 tickets and brought in \$2,825 in one night. The number of student tickets sold is twice the number of adult tickets sold. How many of each type did the department sell?

Solution

We will use a chart to organize the information.

Type	Number	·	Value	=	Total Value
adult	x		15		$15x$
student	y		10		$10y$
child	z		8		$8z$
	250				2825

Number of students is twice number of adults.

Rewrite the equation in standard form.

$$\begin{aligned} y &= 2x \\ 2x - y &= 0 \end{aligned}$$

Write the system of equations.

$$\begin{cases} x+\ y+\ z=250 & (1) \\ 15x+10y+8z=2825 & (2) \\ -2x+\ y\ \ \ \ =0 & (3) \end{cases}$$

Use equations (1) and (2) to eliminate z.

$$\begin{aligned} x+\ y+\ z&=250 \quad (1) \\ 15x+10y+8z&=2825 \quad (2) \end{aligned}$$

Multiply by –8

$$\begin{aligned} -8x-\ 8y-8z&=250 \quad (1) \\ 15x+10y+8z&=2825 \quad (2) \\ \hline 7x+\ 2y\ \ \ \ \ &=825 \quad (4) \end{aligned}$$

Use (3) and (4) to eliminate y.

$$\begin{aligned} -2x+\ y\ \ \ &=0 \quad (3) \\ 7x+\ 2y\ \ \ &=825 \quad (4) \end{aligned}$$

Multiply by –2

$$\begin{aligned} 4x-\ 2y\ \ \ &=0 \quad (3) \\ 7x+\ 2y\ \ \ &=825 \quad (4) \\ \hline 11x\ \ \ \ \ \ &=825 \end{aligned}$$

Solve for x.

$$x\ \ \ \ \ = 75 \text{ adult tickets}$$

Use equation (3) to find y.

$$-2x+y=0$$

Substitute $x = 75$.	$-2(75) + y = 0$
	$-150 + y = 0$
	$y = 150$ student tickets

| Use equation (1) to find z. | $x + y + z = 250$ |

Substitute in the values $x = 75$, $y = 150$.	$75 + 150 + z = 250$
	$225 + z = 250$
	$z = 25$ child tickets

| Write the solution. | The theater department sold 75 adult tickets, 150 student tickets, and 25 child tickets. |

> **TRY IT : :** 4.71

The community college fine arts department sold three kinds of tickets to its latest dance presentation. The adult tickets sold for $20, the student tickets for $12 and the child tickets for $10.The fine arts department was thrilled to have sold 350 tickets and brought in $4,650 in one night. The number of child tickets sold is the same as the number of adult tickets sold. How many of each type did the department sell?

> **TRY IT : :** 4.72

The community college soccer team sold three kinds of tickets to its latest game. The adult tickets sold for $10, the student tickets for $8 and the child tickets for $5. The soccer team was thrilled to have sold 600 tickets and brought in $4,900 for one game. The number of adult tickets is twice the number of child tickets. How many of each type did the soccer team sell?

> ▶ **MEDIA : :**

Access this online resource for additional instruction and practice with solving a linear system in three variables with no or infinite solutions.

- **Solving a Linear System in Three Variables with No or Infinite Solutions (https://openstax.org/l/37linsys3var)**
- **3 Variable Application (https://openstax.org/l/37variableapp)**

4.4 EXERCISES

Practice Makes Perfect

Determine Whether an Ordered Triple is a Solution of a System of Three Linear Equations with Three Variables

In the following exercises, determine whether the ordered triple is a solution to the system.

162. $\begin{cases} 2x - 6y + z = 3 \\ 3x - 4y - 3z = 2 \\ 2x + 3y - 2z = 3 \end{cases}$

ⓐ $(3, 1, 3)$

ⓑ $(4, 3, 7)$

163. $\begin{cases} -3x + y + z = -4 \\ -x + 2y - 2z = 1 \\ 2x - y - z = -1 \end{cases}$

ⓐ $(-5, -7, 4)$

ⓑ $(5, 7, 4)$

164. $\begin{cases} y - 10z = -8 \\ 2x - y = 2 \\ x - 5z = 3 \end{cases}$

ⓐ $(7, 12, 2)$

ⓑ $(2, 2, 1)$

165. $\begin{cases} x + 3y - z = 15 \\ y = \frac{2}{3}x - 2 \\ x - 3y + z = -2 \end{cases}$

ⓐ $\left(-6, 5, \frac{1}{2}\right)$

ⓑ $\left(5, \frac{4}{3}, -3\right)$

Solve a System of Linear Equations with Three Variables

In the following exercises, solve the system of equations.

166. $\begin{cases} 5x + 2y + z = 5 \\ -3x - y + 2z = 6 \\ 2x + 3y - 3z = 5 \end{cases}$

167. $\begin{cases} 6x - 5y + 2z = 3 \\ 2x + y - 4z = 5 \\ 3x - 3y + z = -1 \end{cases}$

168. $\begin{cases} 2x - 5y + 3z = 8 \\ 3x - y + 4z = 7 \\ x + 3y + 2z = -3 \end{cases}$

169. $\begin{cases} 5x - 3y + 2z = -5 \\ 2x - y - z = 4 \\ 3x - 2y + 2z = -7 \end{cases}$

170. $\begin{cases} 3x - 5y + 4z = 5 \\ 5x + 2y + z = 0 \\ 2x + 3y - 2z = 3 \end{cases}$

171. $\begin{cases} 4x - 3y + z = 7 \\ 2x - 5y - 4z = 3 \\ 3x - 2y - 2z = -7 \end{cases}$

172. $\begin{cases} 3x + 8y + 2z = -5 \\ 2x + 5y - 3z = 0 \\ x + 2y - 2z = -1 \end{cases}$

173. $\begin{cases} 11x + 9y + 2z = -9 \\ 7x + 5y + 3z = -7 \\ 4x + 3y + z = -3 \end{cases}$

174. $\begin{cases} \frac{1}{3}x - y - z = 1 \\ x + \frac{5}{2}y + z = -2 \\ 2x + 2y + \frac{1}{2}z = -4 \end{cases}$

175. $\begin{cases} x + \frac{1}{2}y + \frac{1}{2}z = 0 \\ \frac{1}{5}x - \frac{1}{5}y + z = 0 \\ \frac{1}{3}x - \frac{1}{3}y + 2z = -1 \end{cases}$

176. $\begin{cases} x + \frac{1}{3}y - 2z = -1 \\ \frac{1}{3}x + y + \frac{1}{2}z = 0 \\ \frac{1}{2}x + \frac{1}{3}y - \frac{1}{2}z = -1 \end{cases}$

177. $\begin{cases} \frac{1}{3}x - y + \frac{1}{2}z = 4 \\ \frac{2}{3}x + \frac{5}{2}y - 4z = 0 \\ x - \frac{1}{2}y + \frac{3}{2}z = 2 \end{cases}$

178. $\begin{cases} x + 2z = 0 \\ 4y + 3z = -2 \\ 2x - 5y = 3 \end{cases}$

179. $\begin{cases} 2x + 5y = 4 \\ 3y - z = 3 \\ 4x + 3z = -3 \end{cases}$

180. $\begin{cases} 2y + 3z = -1 \\ 5x + 3y = -6 \\ 7x + z = 1 \end{cases}$

181. $\begin{cases} 3x - z = -3 \\ 5y + 2z = -6 \\ 4x + 3y = -8 \end{cases}$

182. $\begin{cases} 4x - 3y + 2z = 0 \\ -2x + 3y - 7z = 1 \\ 2x - 2y + 3z = 6 \end{cases}$

183. $\begin{cases} x - 2y + 2z = 1 \\ -2x + y - z = 2 \\ x - y + z = 5 \end{cases}$

184. $\begin{cases} 2x + 3y + z = 12 \\ x + y + z = 9 \\ 3x + 4y + 2z = 20 \end{cases}$

185. $\begin{cases} x + 4y + z = -8 \\ 4x - y + 3z = 9 \\ 2x + 7y + z = 0 \end{cases}$

186. $\begin{cases} x + 2y + z = 4 \\ x + y - 2z = 3 \\ -2x - 3y + z = -7 \end{cases}$

187. $\begin{cases} x + y - 2z = 3 \\ -2x - 3y + z = -7 \\ x + 2y + z = 4 \end{cases}$

188. $\begin{cases} x + y - 3z = -1 \\ y - z = 0 \\ -x + 2y = 1 \end{cases}$

189. $\begin{cases} x - 2y + 3z = 1 \\ x + y - 3z = 7 \\ 3x - 4y + 5z = 7 \end{cases}$

Solve Applications using Systems of Linear Equations with Three Variables

In the following exercises, solve the given problem.

190. The sum of the measures of the angles of a triangle is 180. The sum of the measures of the second and third angles is twice the measure if the first angle. The third angle is twelve more than the second. Find the measures of the three angles.

191. The sum of the measures of the angles of a triangle is 180. The sum of the measures of the second and third angles is three the measure if the first angle. The third angle is fifteen more than the second. Find the measures of the three angles.

192. After watching a major musical production at the theater, the patrons can purchase souvenirs. If a family purchases 4 t-shirts, the video and 1 stuffed animal their total is $135.

A couple buys 2 t-shirts, the video and 3 stuffed animal for their nieces and spends $115. Another couple buys 2 t-shirts, the video and 1 stuffed animal and their total is $85. What is the cost of each item?

193. The church youth group is selling snacks to raise money to attend their convention. Amy sold 2 pounds of candy, 3 boxes of cookies and 1 can of popcorn for a total sales of $65. Brian sold 4 pounds of candy, 6 boxes of cookies and 3 cans of popcorn for a total sales of $140. Paulina sold 8 pounds of candy, 8 boxes of cookies and 5 can of popcorn for a total sales of $250. What is the cost of each item?

Writing Exercises

194. In your own words explain the steps to solve a system of linear equations with three variables by elimination.

195. How can you tell when a system of three linear equations with three variables has no solution? Infinitely many solutions?

Self Check

ⓐ *After completing the exercises, use this checklist to evaluate your mastery of the objectives of this section.*

I can...	Confidently	With some help	No-I don't get it!
determine whether an ordered triple is a solution of a system of three linear equations with three variables.			
solve a system of linear equations with three variables.			
solve applications using systems of linear equations with three variables.			

ⓑ *On a scale of 1-10, how would you rate your mastery of this section in light of your responses on the checklist? How can you improve this?*

4.5 Solve Systems of Equations Using Matrices

Learning Objectives

By the end of this section, you will be able to:

> Write the augmented matrix for a system of equations
> Use row operations on a matrix
> Solve systems of equations using matrices

Be Prepared!

Before you get started, take this readiness quiz.

1. Solve: $3(x + 2) + 4 = 4(2x - 1) + 9$.
   If you missed this problem, review **Example 2.2**.

2. Solve: $0.25p + 0.25(x + 4) = 5.20$.
   If you missed this problem, review **Example 2.13**.

3. Evaluate when $x = -2$ and $y = 3$: $2x^2 - xy + 3y^2$.
   If you missed this problem, review **Example 1.21**.

Write the Augmented Matrix for a System of Equations

Solving a system of equations can be a tedious operation where a simple mistake can wreak havoc on finding the solution. An alternative method which uses the basic procedures of elimination but with notation that is simpler is available. The method involves using a **matrix**. A matrix is a rectangular array of numbers arranged in rows and columns.

Matrix

A **matrix** is a rectangular array of numbers arranged in rows and columns.

A matrix with m rows and n columns has order $m \times n$. The matrix on the left below has 2 rows and 3 columns and so it has order 2×3. We say it is a 2 by 3 matrix.

$$2 \text{ rows} \left\{ \overset{3 \text{ columns}}{\begin{bmatrix} -3 & -2 & 2 \\ -1 & 4 & 5 \end{bmatrix}} \right. $$
$$2 \times 3 \text{ matrix}$$

$$3 \text{ rows} \left\{ \overset{4 \text{ columns}}{\begin{bmatrix} 2 & -5 & 3 & -1 \\ -4 & 0 & 2 & -2 \\ 0 & -2 & -4 & 1 \end{bmatrix}} \right. $$
$$3 \times 4 \text{ matrix}$$

Each number in the matrix is called an element or entry in the matrix.

We will use a matrix to represent a system of linear equations. We write each equation in standard form and the coefficients of the variables and the constant of each equation becomes a row in the matrix. Each column then would be the coefficients of one of the variables in the system or the constants. A vertical line replaces the equal signs. We call the resulting matrix the augmented matrix for the system of equations.

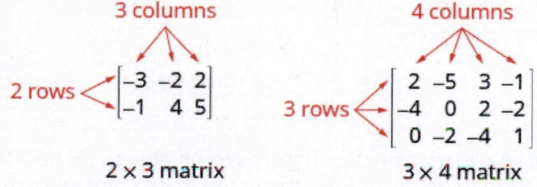

$$\begin{cases} 3x - y = -3 \\ 2x + 3y = 6 \end{cases} \longrightarrow \left[\begin{array}{cc|c} 3 & 1 & -3 \\ 2 & 3 & 6 \end{array} \right]$$

coefficients of x coefficients of y constants

Notice the first column is made up of all the coefficients of x, the second column is the all the coefficients of y, and the third column is all the constants.

EXAMPLE 4.37

Write each system of linear equations as an augmented matrix:

(a) $\begin{cases} 5x - 3y = -1 \\ y = 2x - 2 \end{cases}$ (b) $\begin{cases} 6x - 5y + 2z = 3 \\ 2x + y - 4z = 5 \\ 3x - 3y + z = -1 \end{cases}$

⊘ Solution

(a) The second equation is not in standard form. We rewrite the second equation in standard form.

$$\begin{aligned} y &= 2x - 2 \\ -2x + y &= -2 \end{aligned}$$

We replace the second equation with its standard form. In the augmented matrix, the first equation gives us the first row and the second equation gives us the second row. The vertical line replaces the equal signs.

$$\begin{matrix} x & y \end{matrix}$$

$$\begin{cases} 5x - 3y = -1 \\ 2x - y = 2 \end{cases} \qquad \left[\begin{array}{cc|c} 5 & -3 & -1 \\ 2 & -1 & 2 \end{array}\right]$$

(b) All three equations are in standard form. In the augmented matrix the first equation gives us the first row, the second equation gives us the second row, and the third equation gives us the third row. The vertical line replaces the equal signs.

$$\begin{matrix} x & y & z \end{matrix}$$

$$\begin{cases} 6x - 5y + 2z = 3 \\ 2x + y - 4z = 5 \\ 3x - 3y + z = -1 \end{cases} \qquad \left[\begin{array}{ccc|c} 6 & -5 & 2 & 3 \\ 2 & 1 & -4 & 5 \\ 3 & -3 & 1 & -1 \end{array}\right]$$

> **TRY IT : : 4.73**   Write each system of linear equations as an augmented matrix:

(a) $\begin{cases} 3x + 8y = -3 \\ 2x = -5y - 3 \end{cases}$ (b) $\begin{cases} 2x - 5y + 3z = 8 \\ 3x - y + 4z = 7 \\ x + 3y + 2z = -3 \end{cases}$

> **TRY IT : : 4.74**   Write each system of linear equations as an augmented matrix:

(a) $\begin{cases} 11x = -9y - 5 \\ 7x + 5y = -1 \end{cases}$ (b) $\begin{cases} 5x - 3y + 2z = -5 \\ 2x - y - z = 4 \\ 3x - 2y + 2z = -7 \end{cases}$

It is important as we solve systems of equations using matrices to be able to go back and forth between the system and the matrix. The next example asks us to take the information in the matrix and write the system of equations.

EXAMPLE 4.38

Write the system of equations that corresponds to the augmented matrix:

$$\left[\begin{array}{ccc|c} 4 & -3 & 3 & -1 \\ 1 & 2 & -1 & 2 \\ -2 & -1 & 3 & -4 \end{array}\right].$$

⊘ Solution

We remember that each row corresponds to an equation and that each entry is a coefficient of a variable or the constant. The vertical line replaces the equal sign. Since this matrix is a 4×3, we know it will translate into a system of three equations with three variables.

$$\begin{matrix} x & y & z \end{matrix}$$

$$\left[\begin{array}{ccc|c} 4 & -3 & 3 & -1 \\ 1 & 2 & -1 & 2 \\ -2 & -1 & 3 & -4 \end{array}\right] \longrightarrow \begin{cases} 4x - 3y + 3z = -1 \\ x + 2y - z = 2 \\ -2x - y + 3z = -4 \end{cases}$$

> **TRY IT : : 4.75**

Write the system of equations that corresponds to the augmented matrix: $\begin{bmatrix} 1 & -1 & 2 & 3 \\ 2 & 1 & -2 & 1 \\ 4 & -1 & 2 & 0 \end{bmatrix}$.

> **TRY IT : : 4.76**

Write the system of equations that corresponds to the augmented matrix: $\begin{bmatrix} 1 & 1 & 1 & 4 \\ 2 & 3 & -1 & 8 \\ 1 & 1 & -1 & 3 \end{bmatrix}$.

Use Row Operations on a Matrix

Once a system of equations is in its augmented matrix form, we will perform operations on the rows that will lead us to the solution.

To solve by elimination, it doesn't matter which order we place the equations in the system. Similarly, in the matrix we can interchange the rows.

When we solve by elimination, we often multiply one of the equations by a constant. Since each row represents an equation, and we can multiply each side of an equation by a constant, similarly we can multiply each entry in a row by any real number except 0.

In elimination, we often add a multiple of one row to another row. In the matrix we can replace a row with its sum with a multiple of another row.

These actions are called row operations and will help us use the matrix to solve a system of equations.

Row Operations

In a matrix, the following operations can be performed on any row and the resulting matrix will be equivalent to the original matrix.

1. Interchange any two rows.
2. Multiply a row by any real number except 0.
3. Add a nonzero multiple of one row to another row.

Performing these operations is easy to do but all the arithmetic can result in a mistake. If we use a system to record the row operation in each step, it is much easier to go back and check our work.

We use capital letters with subscripts to represent each row. We then show the operation to the left of the new matrix. To show interchanging a row:

$$\begin{bmatrix} 5 & -3 & | & -1 \\ 2 & -1 & | & 2 \end{bmatrix} \xrightarrow{\quad\quad} \begin{matrix} R_2 \\ R_1 \end{matrix} \begin{bmatrix} 2 & -1 & | & 2 \\ 5 & -3 & | & -1 \end{bmatrix}$$

To multiply row 2 by -3 :

$$\begin{bmatrix} 5 & -3 & | & -1 \\ 2 & -1 & | & 2 \end{bmatrix} \xrightarrow[-3R_2]{\quad\quad} \begin{bmatrix} 5 & -3 & | & -1 \\ -6 & 3 & | & -6 \end{bmatrix}$$

To multiply row 2 by -3 and add it to row 1:

$$\begin{bmatrix} 5 & -3 & | & -1 \\ 2 & -1 & | & 2 \end{bmatrix} \xrightarrow{-3R_2 + R_1} \begin{bmatrix} -1 & 0 & | & -7 \\ 2 & -1 & | & 2 \end{bmatrix}$$

EXAMPLE 4.39

Perform the indicated operations on the augmented matrix:

ⓐ Interchange rows 2 and 3.

ⓑ Multiply row 2 by 5.

ⓒ Multiply row 3 by -2 and add to row 1.

$$\begin{bmatrix} 6 & -5 & 2 & | & 3 \\ 2 & 1 & -4 & | & 5 \\ 3 & -3 & 1 & | & -1 \end{bmatrix}$$

✓ Solution

(a) We interchange rows 2 and 3.

$$\begin{bmatrix} 6 & -5 & 2 & | & 3 \\ 2 & 1 & -4 & | & 5 \\ 3 & -3 & 1 & | & -1 \end{bmatrix} \xrightarrow[R_2]{R_3} \begin{bmatrix} 6 & -5 & 2 & | & 3 \\ 3 & -3 & 1 & | & -1 \\ 2 & 1 & -4 & | & 5 \end{bmatrix}$$

(b) We multiply row 2 by 5.

$$\begin{bmatrix} 6 & -5 & 2 & | & 3 \\ 2 & 1 & -4 & | & 5 \\ 3 & -3 & 1 & | & -1 \end{bmatrix} \xrightarrow{5R_2} \begin{bmatrix} 6 & -5 & 2 & | & 3 \\ 10 & 5 & -20 & | & 25 \\ 3 & -3 & 1 & | & -1 \end{bmatrix}$$

(c) We multiply row 3 by -2 and add to row 1.

$$\begin{bmatrix} 6 & -5 & 2 & | & 3 \\ 2 & 1 & -4 & | & 5 \\ 3 & -3 & 1 & | & -1 \end{bmatrix} \xrightarrow{-2R_3 + R_1} \begin{bmatrix} (-2 \cdot 3) + 6 & (-2 \cdot (-3)) + (-5) & (-2 \cdot 1) + 2 & | & (-2 \cdot (-1)) + 3 \\ 2 & 1 & -4 & | & 5 \\ 3 & -3 & 1 & | & -1 \end{bmatrix}$$

$$\xrightarrow{-2R_3 + R_1} \begin{bmatrix} 0 & 1 & 0 & | & 5 \\ 2 & 1 & -4 & | & 5 \\ 3 & -3 & 1 & | & -1 \end{bmatrix}$$

> **TRY IT : : 4.77**   Perform the indicated operations on the augmented matrix:

(a) Interchange rows 1 and 3.

(b) Multiply row 3 by 3.

(c) Multiply row 3 by 2 and add to row 2.

$$\begin{bmatrix} 5 & -2 & -2 & | & -2 \\ 4 & -1 & -4 & | & 4 \\ -2 & 3 & 0 & | & -1 \end{bmatrix}$$

> **TRY IT : : 4.78**   Perform the indicated operations on the augmented matrix:

(a) Interchange rows 1 and 2,

(b) Multiply row 1 by 2,

(c) Multiply row 2 by 3 and add to row 1.

$$\begin{bmatrix} 2 & -3 & -2 & | & -4 \\ 4 & 1 & -3 & | & 2 \\ 5 & 0 & 4 & | & -1 \end{bmatrix}$$

Now that we have practiced the row operations, we will look at an augmented matrix and figure out what operation we will use to reach a goal. This is exactly what we did when we did elimination. We decided what number to multiply a row by in order that a variable would be eliminated when we added the rows together.

Given this system, what would you do to eliminate x?

$$\begin{cases} x - y = 2 \\ 4x - 8y = 0 \end{cases} \xrightarrow[\text{equation by } -4]{\text{multiply the first}} \begin{cases} -4x + 4y = -8 \\ 4x - 8y = 0 \end{cases} \xrightarrow{\text{then add}} \begin{array}{r} -4x + 4y = -8 \\ 4x - 8y = 0 \\ \hline -4y = -8 \end{array}$$

This next example essentially does the same thing, but to the matrix.

EXAMPLE 4.40

Perform the needed row operation that will get the first entry in row 2 to be zero in the augmented matrix: $\begin{bmatrix} 1 & -1 & | & 2 \\ 4 & -8 & | & 0 \end{bmatrix}$.

✓ **Solution**

To make the 4 a 0, we could multiply row 1 by -4 and then add it to row 2.

$$\begin{bmatrix} 1 & -1 & | & 2 \\ 4 & -8 & | & 0 \end{bmatrix} \xrightarrow{-4R_1 + R_2} \begin{bmatrix} 1 & -1 & | & 2 \\ 0 & -4 & | & -8 \end{bmatrix}$$

> **TRY IT :: 4.79**

> Perform the needed row operation that will get the first entry in row 2 to be zero in the augmented matrix: $\begin{bmatrix} 1 & -1 & | & 2 \\ 3 & -6 & | & 2 \end{bmatrix}$.

> **TRY IT :: 4.80**

> Perform the needed row operation that will get the first entry in row 2 to be zero in the augmented matrix: $\begin{bmatrix} 1 & -1 & | & 3 \\ -2 & -3 & | & 2 \end{bmatrix}$.

Solve Systems of Equations Using Matrices

To solve a system of equations using matrices, we transform the augmented matrix into a matrix in **row-echelon form** using row operations. For a consistent and independent system of equations, its augmented matrix is in row-echelon form when to the left of the vertical line, each entry on the diagonal is a 1 and all entries below the diagonal are zeros.

Row-Echelon Form

For a consistent and independent system of equations, its augmented matrix is in **row-echelon form** when to the left of the vertical line, each entry on the diagonal is a 1 and all entries below the diagonal are zeros.

$$\begin{bmatrix} 1 & a & | & b \\ 0 & 1 & | & c \end{bmatrix} \qquad \begin{bmatrix} 1 & a & b & | & d \\ 0 & 1 & c & | & e \\ 0 & 0 & 1 & | & f \end{bmatrix} \qquad a, b, c, d, e, f \text{ are real numbers}$$

Once we get the augmented matrix into row-echelon form, we can write the equivalent system of equations and read the value of at least one variable. We then substitute this value in another equation to continue to solve for the other variables. This process is illustrated in the next example.

EXAMPLE 4.41   HOW TO SOLVE A SYSTEM OF EQUATIONS USING A MATRIX

Solve the system of equations using a matrix: $\begin{cases} 3x + 4y = 5 \\ x + 2y = 1 \end{cases}$.

✓ **Solution**

Step 1. Write the augmented matrix for the system of equations.		$\begin{cases} 3x + 4y = 5 \\ x + 2y = 1 \end{cases}$ $\begin{bmatrix} 3 & 4 & | & 5 \\ 1 & 2 & | & 1 \end{bmatrix}$
Step 2. Using row operations get the entry in row 1, column 1 to be 1.	Interchange the rows, so 1 will be in row 1, column 1.	$\begin{matrix} R_2 \\ R_1 \end{matrix} \begin{bmatrix} 1 & 2 & | & 1 \\ 3 & 4 & | & 5 \end{bmatrix}$

Step 3. Using row operations, get zeros in column 1 below the 1.	Multiply row 1 by –3 and add it to row 2.	$-3R_1 + R_2$ $\begin{bmatrix} 1 & 2 & \vert & 1 \\ 0 & -2 & \vert & 2 \end{bmatrix}$
Step 4. Using row operations, get the entry in row 2, column 2 to be 1.	Multiply row 2 by $-\frac{1}{2}$.	$-\frac{1}{2}R_2$ $\begin{bmatrix} 1 & 2 & \vert & 1 \\ 0 & 1 & \vert & -1 \end{bmatrix}$
Step 5. Continue the process until the matrix is in row-echelon form.	The matrix is now in row-echelon form.	$\begin{bmatrix} 1 & 2 & \vert & 1 \\ 0 & 1 & \vert & -1 \end{bmatrix}$
Step 6. Write the corresponding system of equations.	$\begin{array}{cc} x & y \end{array}$ $\begin{bmatrix} 1 & 2 & \vert & 1 \\ 0 & 1 & \vert & -1 \end{bmatrix}$	$\begin{cases} x + 2y = 1 \\ y = -1 \end{cases}$
Step 7. Use substitution to find the remaining variables.	Substitute $y = -1$ into $x + 2y = 1$.	$y = -1$ $x + 2y = 1$ $x + 2(-1) = 1$ $x - 2 = 1$ $x = 3$
Step 8. Write the solution as an ordered pair or triple.		$(3, -1)$
Step 9. Check that the solution makes the original equations true.		We leave the check to you.

> **TRY IT : : 4.81**
>
> Solve the system of equations using a matrix: $\begin{cases} 2x + y = 7 \\ x - 2y = 6 \end{cases}$.

> **TRY IT : : 4.82**
>
> Solve the system of equations using a matrix: $\begin{cases} 2x + y = -4 \\ x - y = -2 \end{cases}$.

The steps are summarized here.

HOW TO : : SOLVE A SYSTEM OF EQUATIONS USING MATRICES.

Step 1.   Write the augmented matrix for the system of equations.

Step 2.   Using row operations get the entry in row 1, column 1 to be 1.

Step 3.   Using row operations, get zeros in column 1 below the 1.

Step 4.   Using row operations, get the entry in row 2, column 2 to be 1.

Step 5.   Continue the process until the matrix is in row-echelon form.

Step 6.   Write the corresponding system of equations.

Step 7.   Use substitution to find the remaining variables.

Step 8.   Write the solution as an ordered pair or triple.

Step 9.   Check that the solution makes the original equations true.

Here is a visual to show the order for getting the 1's and 0's in the proper position for row-echelon form.

2 × 3 matrix

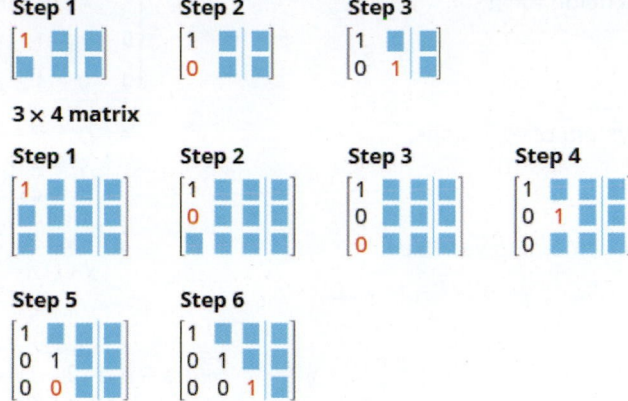

3 × 4 matrix

We use the same procedure when the system of equations has three equations.

EXAMPLE 4.42

Solve the system of equations using a matrix: $\begin{cases} 3x + 8y + 2z = -5 \\ 2x + 5y - 3z = 0 \\ x + 2y - 2z = -1 \end{cases}$.

⊘ Solution

$$\begin{cases} 3x + 8y + 2z = -5 \\ 2x + 5y - 3z = 0 \\ x + 2y - 2z = -1 \end{cases}$$

Write the augmented matrix for the equations.	$\begin{bmatrix} 3 & 8 & 2 & \vert & -5 \\ 2 & 5 & -3 & \vert & 0 \\ 1 & 2 & -2 & \vert & -1 \end{bmatrix}$
Interchange row 1 and 3 to get the entry in row 1, column 1 to be 1.	$\begin{matrix} R_3 \\ \\ R_1 \end{matrix} \begin{bmatrix} 1 & 2 & -2 & \vert & -1 \\ 2 & 5 & -3 & \vert & 0 \\ 3 & 8 & 2 & \vert & -5 \end{bmatrix}$
Using row operations, get zeros in column 1 below the 1.	$\begin{matrix} \\ -2R_1 + R_2 \\ \\ \end{matrix} \begin{bmatrix} 1 & 2 & -2 & \vert & -1 \\ 0 & 1 & 1 & \vert & 2 \\ 3 & 8 & 2 & \vert & -5 \end{bmatrix}$
	$\begin{matrix} \\ \\ -3R_1 + R_3 \end{matrix} \begin{bmatrix} 1 & 2 & -2 & \vert & -1 \\ 0 & 1 & 1 & \vert & 2 \\ 0 & 2 & 8 & \vert & -2 \end{bmatrix}$
The entry in row 2, column 2 is now 1.	
Continue the process until the matrix is in row-echelon form.	$\begin{matrix} \\ \\ -2R_2 + R_3 \end{matrix} \begin{bmatrix} 1 & 2 & -2 & \vert & -1 \\ 0 & 1 & 1 & \vert & 2 \\ 0 & 0 & 6 & \vert & -6 \end{bmatrix}$
	$\begin{matrix} \\ \\ \frac{1}{6}R_3 \end{matrix} \begin{bmatrix} 1 & 2 & -2 & \vert & -1 \\ 0 & 1 & 1 & \vert & 2 \\ 0 & 0 & 1 & \vert & -1 \end{bmatrix}$

The matrix is now in row-echelon form.	$\begin{bmatrix} 1 & 2 & -2 & \vert & -1 \\ 0 & 1 & 1 & \vert & 2 \\ 0 & 0 & 1 & \vert & -1 \end{bmatrix}$
Write the corresponding system of equations.	$\begin{cases} x + 2y - 2z = -1 \\ \quad\quad y + z = 2 \\ \quad\quad\quad\quad z = -1 \end{cases}$
Use substitution to find the remaining variables.	$\begin{aligned} y + z &= 2 \\ y + (-1) &= 2 \\ y &= 3 \end{aligned}$
	$\begin{aligned} x + 2y - 2z &= -1 \\ x + 2(3) - 2(-1) &= -1 \\ x + 6 + 2 &= -1 \\ x &= -9 \end{aligned}$
Write the solution as an ordered pair or triple.	$(-9, 3, -1)$
Check that the solution makes the original equations true.	We leave the check for you.

> **TRY IT :: 4.83**
>
> Solve the system of equations using a matrix: $\begin{cases} 2x - 5y + 3z = 8 \\ 3x - y + 4z = 7 \\ x + 3y + 2z = -3 \end{cases}$.

> **TRY IT :: 4.84**
>
> Solve the system of equations using a matrix: $\begin{cases} -3x + y + z = -4 \\ -x + 2y - 2z = 1 \\ 2x - y - z = -1 \end{cases}$.

So far our work with matrices has only been with systems that are consistent and independent, which means they have exactly one solution. Let's now look at what happens when we use a matrix for a dependent or inconsistent system.

EXAMPLE 4.43

Solve the system of equations using a matrix: $\begin{cases} x + y + 3z = 0 \\ x + 3y + 5z = 0 \\ 2x + 4z = 1 \end{cases}$.

✓ **Solution**

	$\begin{cases} x + y + 3z = 0 \\ x + 3y + 5z = 0 \\ 2x + \quad\;\; 4z = 1 \end{cases}$
Write the augmented matrix for the equations.	$\begin{bmatrix} 1 & 1 & 3 & \vert & 0 \\ 1 & 3 & 5 & \vert & 0 \\ 2 & 0 & 4 & \vert & 1 \end{bmatrix}$
The entry in row 1, column 1 is 1.	
Using row operations, get zeros in column 1 below the 1.	$-1R_1 + R_2 \quad \begin{bmatrix} 1 & 1 & 3 & \vert & 0 \\ 0 & 2 & 2 & \vert & 0 \\ 2 & 0 & 4 & \vert & 1 \end{bmatrix}$

$$-2R_1 + R_3 \begin{bmatrix} 1 & 1 & 3 & | & 0 \\ 0 & 2 & 2 & | & 0 \\ 0 & -2 & -2 & | & 1 \end{bmatrix}$$

Continue the process until the matrix is in row-echelon form.	$\frac{1}{2}R_2 \begin{bmatrix} 1 & 1 & 3 & \vert & 0 \\ 0 & 1 & 1 & \vert & 0 \\ 0 & -2 & -2 & \vert & 1 \end{bmatrix}$
Multiply row 2 by 2 and add it to row 3.	$2R_2 + R_3 \begin{bmatrix} 1 & 1 & 3 & \vert & 0 \\ 0 & 1 & 1 & \vert & 0 \\ 0 & 0 & 0 & \vert & 1 \end{bmatrix}$
At this point, we have all zeros on the left of row 3.	
Write the corresponding system of equations.	$\begin{cases} x + y + 3z = 0 \\ \quad\; y + \; z = 0 \\ \qquad\qquad 0 \neq 1 \end{cases}$

Since $0 \neq 1$ we have a false statement. Just as when we solved a system using other methods, this tells us we have an inconsistent system. There is no solution.

> **TRY IT :: 4.85**

Solve the system of equations using a matrix: $\begin{cases} x - 2y + 2z = 1 \\ -2x + y - z = 2. \\ x - y + z = 5 \end{cases}$

> **TRY IT :: 4.86**

Solve the system of equations using a matrix: $\begin{cases} 3x + 4y - 3z = -2 \\ 2x + 3y - z = -12. \\ x + y - 2z = 6 \end{cases}$

The last system was inconsistent and so had no solutions. The next example is dependent and has infinitely many solutions.

EXAMPLE 4.44

Solve the system of equations using a matrix: $\begin{cases} x - 2y + 3z = 1 \\ x + y - 3z = 7 \\ 3x - 4y + 5z = 7 \end{cases}$.

⊘ Solution

$$\begin{cases} x - 2y + 3z = 1 \\ x + \; y - 3z = 7 \\ 3x - 4y + 5z = 7 \end{cases}$$

Write the augmented matrix for the equations.	$\begin{bmatrix} 1 & -2 & 3 & \vert & 1 \\ 1 & 1 & -3 & \vert & 7 \\ 3 & -4 & 5 & \vert & 7 \end{bmatrix}$
The entry in row 1, column 1 is 1.	

Using row operations, get zeros in column 1 below the 1.

$$-1R_1 + R_2 \quad \begin{bmatrix} 1 & -2 & 3 & | & 1 \\ 0 & 3 & -6 & | & 6 \\ 3 & -4 & 5 & | & 7 \end{bmatrix}$$

$$-3R_1 + R_3 \quad \begin{bmatrix} 1 & -2 & 3 & | & 1 \\ 0 & 3 & -6 & | & 6 \\ 0 & 2 & -4 & | & 4 \end{bmatrix}$$

Continue the process until the matrix is in row-echelon form.

$$\tfrac{1}{3}R_2 \quad \begin{bmatrix} 1 & -2 & 3 & | & 1 \\ 0 & 1 & -2 & | & 2 \\ 0 & 2 & -4 & | & 4 \end{bmatrix}$$

Multiply row 2 by -2 and add it to row 3.

$$-2R_2 + R_3 \quad \begin{bmatrix} 1 & -2 & 3 & | & 1 \\ 0 & 1 & -2 & | & 2 \\ 0 & 0 & 0 & | & 0 \end{bmatrix}$$

At this point, we have all zeros in the bottom row.

Write the corresponding system of equations.

$$\begin{cases} x - 2y + 3z = 1 \\ y - 2z = 2 \\ 0 = 0 \end{cases}$$

Since $0 = 0$ we have a true statement. Just as when we solved by substitution, this tells us we have a dependent system. There are infinitely many solutions.

Solve for y in terms of z in the second equation.

$$y - 2z = 2$$
$$y = 2z + 2$$

Solve the first equation for x in terms of z.

$$x - 2y + 3z = 1$$

Substitute $y = 2z + 2$.

$$x - 2(2z + 2) + 3z = 1$$

Simplify.

$$x - 4z - 4 + 3z = 1$$

Simplify.

$$x - z - 4 = 1$$

Simplify.

$$x = z + 5$$

The system has infinitely many solutions (x, y, z), where $x = z + 5$; $y = 2z + 2$; z is any real number.

> **TRY IT : : 4.87**

Solve the system of equations using a matrix: $\begin{cases} x + y - z = 0 \\ 2x + 4y - 2z = 6. \\ 3x + 6y - 3z = 9 \end{cases}$

> **TRY IT : : 4.88**

Solve the system of equations using a matrix: $\begin{cases} x - y - z = 1 \\ -x + 2y - 3z = -4. \\ 3x - 2y - 7z = 0 \end{cases}$

▶ **MEDIA : :**

Access this online resource for additional instruction and practice with Gaussian Elimination.

- **Gaussian Elimination (https://openstax.org/l/37GaussElim)**

4.5 EXERCISES

Practice Makes Perfect

Write the Augmented Matrix for a System of Equations

In the following exercises, write each system of linear equations as an augmented matrix.

196.
ⓐ $\begin{cases} 3x - y = -1 \\ 2y = 2x + 5 \end{cases}$

ⓑ $\begin{cases} 4x + 3y = -2 \\ x - 2y - 3z = 7 \\ 2x - y + 2z = -6 \end{cases}$

197.
ⓐ $\begin{cases} 2x + 4y = -5 \\ 3x - 2y = 2 \end{cases}$

ⓑ $\begin{cases} 3x - 2y - z = -2 \\ -2x + y = 5 \\ 5x + 4y + z = -1 \end{cases}$

198.
ⓐ $\begin{cases} 3x - y = -4 \\ 2x = y + 2 \end{cases}$

ⓑ $\begin{cases} x - 3y - 4z = -2 \\ 4x + 2y + 2z = 5 \\ 2x - 5y + 7z = -8 \end{cases}$

199.
ⓐ $\begin{cases} 2x - 5y = -3 \\ 4x = 3y - 1 \end{cases}$

ⓑ $\begin{cases} 4x + 3y - 2z = -3 \\ -2x + y - 3z = 4 \\ -x - 4y + 5z = -2 \end{cases}$

Write the system of equations that that corresponds to the augmented matrix.

200. $\begin{bmatrix} 2 & -1 & | & 4 \\ 1 & -3 & | & 2 \end{bmatrix}$

201. $\begin{bmatrix} 2 & -4 & | & -2 \\ 3 & -3 & | & -1 \end{bmatrix}$

202. $\begin{bmatrix} 1 & 0 & -3 & | & -1 \\ 1 & -2 & 0 & | & -2 \\ 0 & -1 & 2 & | & 3 \end{bmatrix}$

203. $\begin{bmatrix} 2 & -2 & 0 & | & -1 \\ 0 & 2 & -1 & | & 2 \\ 3 & 0 & -1 & | & -2 \end{bmatrix}$

Use Row Operations on a Matrix

In the following exercises, perform the indicated operations on the augmented matrices.

204. $\begin{bmatrix} 6 & -4 & | & 3 \\ 3 & -2 & | & 1 \end{bmatrix}$

ⓐ Interchange rows 1 and 2

ⓑ Multiply row 2 by 3

ⓒ Multiply row 2 by −2 and add to row 1.

205. $\begin{bmatrix} 4 & -6 & | & -3 \\ 3 & 2 & | & 1 \end{bmatrix}$

ⓐ Interchange rows 1 and 2

ⓑ Multiply row 1 by 4

ⓒ Multiply row 2 by 3 and add to row 1.

206. $\begin{bmatrix} 1 & -3 & -2 & | & 4 \\ 2 & 2 & -1 & | & -3 \\ 4 & -2 & -3 & | & -1 \end{bmatrix}$

ⓐ Interchange rows 2 and 3

ⓑ Multiply row 1 by 4

ⓒ Multiply row 2 by −2 and add to row 3.

207. $\begin{bmatrix} 6 & -5 & 2 & | & 3 \\ 2 & 1 & -4 & | & 5 \\ 3 & -3 & 1 & | & -1 \end{bmatrix}$

ⓐ Interchange rows 2 and 3

ⓑ Multiply row 2 by 5

ⓒ Multiply row 3 by −2 and add to row 1.

208. Perform the needed row operation that will get the first entry in row 2 to be zero in the augmented matrix:
$\begin{bmatrix} 1 & 2 & | & 5 \\ -3 & -4 & | & -1 \end{bmatrix}$.

209. Perform the needed row operations that will get the first entry in both row 2 and row 3 to be zero in the augmented matrix:
$\begin{bmatrix} 1 & -2 & 3 & | & -4 \\ 3 & -1 & -2 & | & 5 \\ 2 & -3 & -4 & | & -1 \end{bmatrix}$.

Solve Systems of Equations Using Matrices

In the following exercises, solve each system of equations using a matrix.

210. $\begin{cases} 2x + y = 2 \\ x - y = -2 \end{cases}$

211. $\begin{cases} 3x + y = 2 \\ x - y = 2 \end{cases}$

212. $\begin{cases} -x + 2y = -2 \\ x + y = -4 \end{cases}$

213. $\begin{cases} -2x + 3y = 3 \\ x + 3y = 12 \end{cases}$

In the following exercises, solve each system of equations using a matrix.

214. $\begin{cases} 2x - 3y + z = 19 \\ -3x + y - 2z = -15 \\ x + y + z = 0 \end{cases}$

215. $\begin{cases} 2x - y + 3z = -3 \\ -x + 2y - z = 10 \\ x + y + z = 5 \end{cases}$

216. $\begin{cases} 2x - 6y + z = 3 \\ 3x + 2y - 3z = 2 \\ 2x + 3y - 2z = 3 \end{cases}$

217. $\begin{cases} 4x - 3y + z = 7 \\ 2x - 5y - 4z = 3 \\ 3x - 2y - 2z = -7 \end{cases}$

218. $\begin{cases} x + 2z = 0 \\ 4y + 3z = -2 \\ 2x - 5y = 3 \end{cases}$

219. $\begin{cases} 2x + 5y = 4 \\ 3y - z = 3 \\ 4x + 3z = -3 \end{cases}$

220. $\begin{cases} 2y + 3z = -1 \\ 5x + 3y = -6 \\ 7x + z = 1 \end{cases}$

221. $\begin{cases} 3x - z = -3 \\ 5y + 2z = -6 \\ 4x + 3y = -8 \end{cases}$

222. $\begin{cases} 2x + 3y + z = 12 \\ x + y + z = 9 \\ 3x + 4y + 2z = 20 \end{cases}$

223. $\begin{cases} x + 2y + 6z = 5 \\ -x + y - 2z = 3 \\ x - 4y - 2z = 1 \end{cases}$

224. $\begin{cases} x + 2y - 3z = -1 \\ x - 3y + z = 1 \\ 2x - y - 2z = 2 \end{cases}$

225. $\begin{cases} 4x - 3y + 2z = 0 \\ -2x + 3y - 7z = 1 \\ 2x - 2y + 3z = 6 \end{cases}$

226. $\begin{cases} x - y + 2z = -4 \\ 2x + y + 3z = 2 \\ -3x + 3y - 6z = 12 \end{cases}$

227. $\begin{cases} -x - 3y + 2z = 14 \\ -x + 2y - 3z = -4 \\ 3x + y - 2z = 6 \end{cases}$

228. $\begin{cases} x + y - 3z = -1 \\ y - z = 0 \\ -x + 2y = 1 \end{cases}$

229. $\begin{cases} x + 2y + z = 4 \\ x + y - 2z = 3 \\ -2x - 3y + z = -7 \end{cases}$

Writing Exercises

230. Solve the system of equations $\begin{cases} x + y = 10 \\ x - y = 6 \end{cases}$ ⓐ by graphing and ⓑ by substitution. ⓒ Which method do you prefer? Why?

231. Solve the system of equations $\begin{cases} 3x + y = 12 \\ x = y - 8 \end{cases}$ by substitution and explain all your steps in words.

Self Check

ⓐ After completing the exercises, use this checklist to evaluate your mastery of the objectives of this section.

I can...	Confidently	With some help	No-I don't get it!
write the augmented matrix for a system of equations.			
use row operations on a matrix.			
solve systems of equations using matrices.			
write the augmented matrix for a system of equations.			
use row operations on a matrix.			

ⓑ After looking at the checklist, do you think you are well-prepared for the next section? Why or why not?

4.6 Solve Systems of Equations Using Determinants

Learning Objectives

By the end of this section, you will be able to:

› Evaluate the determinant of a 2×2 matrix
› Evaluate the determinant of a 3×3 matrix
› Use Cramer's Rule to solve systems of equations
› Solve applications using determinants

Be Prepared!

Before you get started, take this readiness quiz.

1. Simplify: $5(-2) - (-4)(1)$.
   If you missed this problem, review **Example 1.20**.

2. Simplify: $-3(8 - 10) + (-2)(6 - 3) - 4(-3 - (-4))$.
   If you missed this problem, review **Example 1.19**.

3. Simplify: $\dfrac{-12}{-8}$.
   If you missed this problem, review **Example 1.18**.

In this section we will learn of another method to solve systems of linear equations called Cramer's rule. Before we can begin to use the rule, we need to learn some new definitions and notation.

Evaluate the Determinant of a 2×2 Matrix

If a matrix has the same number of rows and columns, we call it a **square matrix**. Each square matrix has a real number associated with it called its **determinant**. To find the determinant of the square matrix $\begin{bmatrix} a & b \\ c & d \end{bmatrix}$, we first write it as $\begin{vmatrix} a & b \\ c & d \end{vmatrix}$. To get the real number value of the determinate we subtract the products of the diagonals, as shown.

$$\begin{vmatrix} a & b \\ c & d \end{vmatrix} = ad - bc$$

Determinant

The determinant of any square matrix $\begin{bmatrix} a & b \\ c & d \end{bmatrix}$, where a, b, c, and d are real numbers, is

$$\begin{vmatrix} a & b \\ c & d \end{vmatrix} = ad - bc$$

EXAMPLE 4.45

Evaluate the determinate of ⓐ $\begin{bmatrix} 4 & -2 \\ 3 & -1 \end{bmatrix}$ ⓑ $\begin{bmatrix} -3 & -4 \\ -2 & 0 \end{bmatrix}$.

⊘ **Solution**

ⓐ

$$\begin{vmatrix} 4 & -2 \\ 3 & -1 \end{vmatrix}$$

Write the determinant.	$\begin{vmatrix} 4 & -2 \\ 3 & -1 \end{vmatrix}$
Subtract the products of the diagonals.	$4(-1) - 3(-2)$
Simplify.	$-4 + 6$
Simplify.	2

ⓑ

$$\begin{bmatrix} -3 & -4 \\ -2 & 0 \end{bmatrix}$$

Write the determinant.	$\begin{vmatrix} -3 & -4 \\ -2 & 0 \end{vmatrix}$
Subtract the products of the diagonals.	$-3(0) - (-2)(-4)$
Simplify.	$0 - 8$
Simplify.	-8

> **TRY IT :: 4.89** Evaluate the determinate of ⓐ $\begin{bmatrix} 5 & -3 \\ 2 & -4 \end{bmatrix}$ ⓑ $\begin{bmatrix} -4 & -6 \\ 0 & 7 \end{bmatrix}$.

> **TRY IT :: 4.90** Evaluate the determinate of ⓐ $\begin{bmatrix} -1 & 3 \\ -2 & 4 \end{bmatrix}$ ⓑ $\begin{bmatrix} -7 & -3 \\ -5 & 0 \end{bmatrix}$.

Evaluate the Determinant of a 3×3 Matrix

To evaluate the determinant of a 3×3 matrix, we have to be able to evaluate the **minor of an entry** in the determinant. The minor of an entry is the 2×2 determinant found by eliminating the row and column in the 3×3 determinant that contains the entry.

Minor of an entry in 3×3 a Determinant

The **minor of an entry** in a 3×3 determinant is the 2×2 determinant found by eliminating the row and column in the 3×3 determinant that contains the entry.

To find the minor of entry a_1, we eliminate the row and column which contain it. So we eliminate the first row and first column. Then we write the 2×2 determinant that remains.

$$\begin{vmatrix} a_1 & b_1 & c_1 \\ a_2 & b_2 & c_2 \\ a_3 & b_3 & c_3 \end{vmatrix} \quad \text{minor of } a_1 \quad \begin{vmatrix} b_2 & c_2 \\ b_3 & c_3 \end{vmatrix}$$

To find the minor of entry b_2, we eliminate the row and column that contain it. So we eliminate the 2nd row and 2nd column. Then we write the 2×2 determinant that remains.

$$\begin{vmatrix} a_1 & b_1 & c_1 \\ a_2 & b_2 & c_2 \\ a_3 & b_3 & c_3 \end{vmatrix} \quad \text{minor of } b_2 \quad \begin{vmatrix} a_1 & c_1 \\ a_3 & c_3 \end{vmatrix}$$

EXAMPLE 4.46

For the determinant $\begin{vmatrix} 4 & -2 & 3 \\ 1 & 0 & -3 \\ -2 & -4 & 2 \end{vmatrix}$, find and then evaluate the minor of ⓐ a_1 ⓑ b_3 ⓒ c_2.

✓ Solution

ⓐ

	$\begin{vmatrix} 4 & -2 & 3 \\ 1 & 0 & -3 \\ -2 & -4 & 2 \end{vmatrix}$
Eliminate the row and column that contains a_1.	$\begin{vmatrix} 4 & -2 & 3 \\ 1 & 0 & -3 \\ -2 & -4 & 2 \end{vmatrix}$
Write the 2×2 determinant that remains.	minor of a_1 $\begin{vmatrix} 0 & -3 \\ -4 & 2 \end{vmatrix}$
Evaluate.	$0(2) - (-3)(-4)$
Simplify.	-12

ⓑ

	$\begin{vmatrix} 4 & -2 & 3 \\ 1 & 0 & -3 \\ -2 & -4 & 2 \end{vmatrix}$
Eliminate the row and column that contains b_3.	$\begin{vmatrix} 4 & -2 & 3 \\ 1 & 0 & -3 \\ -2 & -4 & 2 \end{vmatrix}$
Write the 2×2 determinant that remains.	minor of b_3 $\begin{vmatrix} 4 & 3 \\ 1 & -3 \end{vmatrix}$
Evaluate.	$4(-3) - (1)(3)$
Simplify.	-15

ⓒ

	$\begin{vmatrix} 4 & -2 & 3 \\ 1 & 0 & -3 \\ -2 & -4 & 2 \end{vmatrix}$
Eliminate the row and column that contains c_2.	$\begin{vmatrix} 4 & -2 & 3 \\ 1 & 0 & -3 \\ -2 & -4 & 2 \end{vmatrix}$

Write the 2×2 determinant that remains.	minor of c_2	$\begin{vmatrix} 4 & -2 \\ -2 & 4 \end{vmatrix}$
Evaluate.		$4(4) - (-2)(-2)$
Simplify.		12

> **TRY IT : : 4.91**
>
> For the determinant $\begin{vmatrix} 1 & -1 & 4 \\ 0 & 2 & -1 \\ -2 & -3 & 3 \end{vmatrix}$, find and then evaluate the minor of ⓐ a_1 ⓑ b_2 ⓒ c_3.

> **TRY IT : : 4.92**
>
> For the determinant $\begin{vmatrix} -2 & -1 & 0 \\ 3 & 0 & -1 \\ -1 & -2 & 3 \end{vmatrix}$, find and then evaluate the minor of ⓐ a_2 ⓑ b_3 ⓒ c_2.

We are now ready to evaluate a 3×3 determinant. To do this we expand by minors, which allows us to evaluate the 3×3 determinant using 2×2 determinants—which we already know how to evaluate!

To evaluate a 3×3 determinant by expanding by minors along the first row, we use the following pattern:

$$\begin{vmatrix} a_1 & b_1 & c_1 \\ a_2 & b_2 & c_2 \\ a_3 & b_3 & c_3 \end{vmatrix} = a_1 \underbrace{\begin{vmatrix} b_2 & c_2 \\ b_3 & c_3 \end{vmatrix}}_{\text{minor of } a_1} - b_1 \underbrace{\begin{vmatrix} a_2 & c_2 \\ a_3 & c_3 \end{vmatrix}}_{\text{minor of } b_1} + c_1 \underbrace{\begin{vmatrix} a_2 & b_2 \\ a_3 & b_3 \end{vmatrix}}_{\text{minor of } c_1}$$

Remember, to find the minor of an entry we eliminate the row and column that contains the entry.

Expanding by Minors along the First Row to Evaluate a 3×3 Determinant

---

To evaluate a 3×3 determinant by **expanding by minors along the first row**, the following pattern:

$$\begin{vmatrix} a_1 & b_1 & c_1 \\ a_2 & b_2 & c_2 \\ a_3 & b_3 & c_3 \end{vmatrix} = a_1 \underbrace{\begin{vmatrix} b_2 & c_2 \\ b_3 & c_3 \end{vmatrix}}_{\text{minor of } a_1} - b_1 \underbrace{\begin{vmatrix} a_2 & c_2 \\ a_3 & c_3 \end{vmatrix}}_{\text{minor of } b_1} + c_1 \underbrace{\begin{vmatrix} a_2 & b_2 \\ a_3 & b_3 \end{vmatrix}}_{\text{minor of } c_1}$$

EXAMPLE 4.47

Evaluate the determinant $\begin{vmatrix} 2 & -3 & -1 \\ 3 & 2 & 0 \\ -1 & -1 & -2 \end{vmatrix}$ by expanding by minors along the first row.

⊘ Solution

$$\begin{vmatrix} 2 & -3 & -1 \\ 3 & 2 & 0 \\ -1 & -1 & -2 \end{vmatrix}$$

Expand by minors along the first row	$2\begin{vmatrix} 2 & 0 \\ -1 & -2 \end{vmatrix} - (-3)\begin{vmatrix} 3 & 0 \\ -1 & -2 \end{vmatrix} + (-1)\begin{vmatrix} 3 & 2 \\ -1 & -1 \end{vmatrix}$ minor of 2 · · · · minor of −3 · · · · minor of −1
Evaluate each determinant.	$2(-4-0) + 3(-6-0) - 1(-3-(-2))$
Simplify.	$2(-4) + 3(-6) - 1(-1)$
Simplify.	$-8 - 18 + 1$
Simplify.	-25

> **TRY IT : :** 4.93
>
> Evaluate the determinant $\begin{vmatrix} 3 & -2 & 4 \\ 0 & -1 & -2 \\ 2 & 3 & -1 \end{vmatrix}$, by expanding by minors along the first row.

> **TRY IT : :** 4.94
>
> Evaluate the determinant $\begin{vmatrix} 3 & -2 & -2 \\ 2 & -1 & 4 \\ -1 & 0 & -3 \end{vmatrix}$, by expanding by minors along the first row.

To evaluate a 3×3 determinant we can expand by minors using any row or column. Choosing a row or column other than the first row sometimes makes the work easier.

When we expand by any row or column, we must be careful about the sign of the terms in the expansion. To determine the sign of the terms, we use the following sign pattern chart.

$$\begin{vmatrix} + & - & + \\ - & + & - \\ + & - & + \end{vmatrix}$$

Sign Pattern

When expanding by minors using a row or column, the sign of the terms in the expansion follow the following pattern.

$$\begin{vmatrix} + & - & + \\ - & + & - \\ + & - & + \end{vmatrix}$$

Notice that the sign pattern in the first row matches the signs between the terms in the expansion by the first row.

$$\begin{vmatrix} + & - & + \\ - & + & - \\ + & - & + \end{vmatrix}$$

$$\begin{vmatrix} a_1 & b_1 & c_1 \\ a_2 & b_2 & c_2 \\ a_3 & b_3 & c_3 \end{vmatrix} = a_1\begin{vmatrix} b_2 & c_2 \\ b_3 & c_3 \end{vmatrix} - b_1\begin{vmatrix} a_2 & c_2 \\ a_3 & c_3 \end{vmatrix} + c_1\begin{vmatrix} a_2 & b_2 \\ a_3 & b_3 \end{vmatrix}$$

minor of a_1 · · · minor of b_1 · · · minor of c_1

Since we can expand by any row or column, how do we decide which row or column to use? Usually we try to pick a row or column that will make our calculation easier. If the determinant contains a 0, using the row or column that contains the 0 will make the calculations easier.

EXAMPLE 4.48

Evaluate the determinant $\begin{vmatrix} 4 & -1 & -3 \\ 3 & 0 & 2 \\ 5 & -4 & -3 \end{vmatrix}$ by expanding by minors.

⊘ Solution

To expand by minors, we look for a row or column that will make our calculations easier. Since 0 is in the second row and second column, expanding by either of those is a good choice. Since the second row has fewer negatives than the second column, we will expand by the second row.

$$\begin{vmatrix} 4 & -1 & -3 \\ 3 & 0 & 2 \\ 5 & -4 & -3 \end{vmatrix}$$

Expand using the second row.

Be careful of the signs.

$$\begin{vmatrix} + & - & + \\ - & + & - \\ + & - & + \end{vmatrix} \qquad -3\underbrace{\begin{vmatrix} -1 & -3 \\ -4 & -3 \end{vmatrix}}_{\text{minor of 3}} + (0)\underbrace{\begin{vmatrix} 4 & -3 \\ 5 & -3 \end{vmatrix}}_{\text{minor of 0}} - 2\underbrace{\begin{vmatrix} 4 & -1 \\ 5 & -4 \end{vmatrix}}_{\text{minor of 2}}$$

Evaluate each determinant.

$$-3(3 - 12) + 0(-12 - (-15)) - 2(-16 - (-5))$$

Simplify.

$$-3(-9) + 0 - 2(-11)$$

Simplify.

$$27 + 0 + 22$$

Add.

$$49$$

> **TRY IT : : 4.95**
>
> Evaluate the determinant $\begin{vmatrix} 2 & -1 & -3 \\ 0 & 3 & -4 \\ 3 & -4 & -3 \end{vmatrix}$ by expanding by minors.

> **TRY IT : : 4.96**
>
> Evaluate the determinant $\begin{vmatrix} -2 & -1 & -3 \\ -1 & 2 & 2 \\ 4 & -4 & 0 \end{vmatrix}$ by expanding by minors.

Use Cramer's Rule to Solve Systems of Equations

Cramer's Rule is a method of solving systems of equations using determinants. It can be derived by solving the general form of the systems of equations by elimination. Here we will demonstrate the rule for both systems of two equations with two variables and for systems of three equations with three variables.

Let's start with the systems of two equations with two variables.

Cramer's Rule for Solving a System of Two Equations

For the system of equations $\begin{cases} a_1 x + b_1 y = k_1 \\ a_2 x + b_2 y = k_2 \end{cases}$, the solution (x, y) can be determined by

$$x = \frac{D_x}{D} \text{ and } y = \frac{D_y}{D}$$

where $D = \begin{vmatrix} a_1 & b_1 \\ a_2 & b_2 \end{vmatrix}$ use the coefficients of the variables.

$D_x = \begin{vmatrix} k_1 & b_1 \\ k_2 & b_2 \end{vmatrix}$ replace the x coefficients with the constants.

$D_y = \begin{vmatrix} a_1 & k_1 \\ a_2 & k_2 \end{vmatrix}$ replace the y coefficients with the constants.

Notice that to form the determinant D, we use take the coefficients of the variables.

$$\begin{cases} a_1x + b_1y = k_1 \\ a_2x + b_2y = k_2 \end{cases} \qquad D = \begin{vmatrix} a_1 & b_1 \\ a_2 & b_2 \end{vmatrix}$$

Coefficients   Coefficient   Coefficient
                    of x of y

Notice that to form the determinant D_x and D_y, we substitute the constants for the coefficients of the variable we are finding.

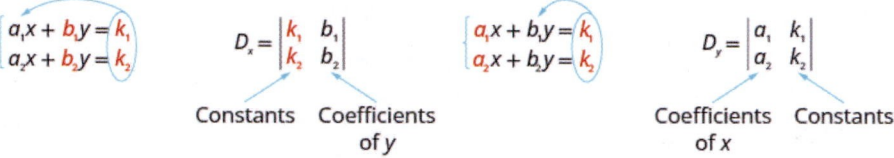

$$\begin{cases} a_1x + b_1y = k_1 \\ a_2x + b_2y = k_2 \end{cases} \qquad D_x = \begin{vmatrix} k_1 & b_1 \\ k_2 & b_2 \end{vmatrix} \qquad \begin{cases} a_1x + b_1y = k_1 \\ a_2x + b_2y = k_2 \end{cases} \qquad D_y = \begin{vmatrix} a_1 & k_1 \\ a_2 & k_2 \end{vmatrix}$$

Constants   Coefficients   Coefficients   Constants
            of y of x

EXAMPLE 4.49   HOW TO SOLVE A SYSTEM OF EQUATIONS USING CRAMER'S RULE

Solve using Cramer's Rule: $\begin{cases} 2x + y = -4 \\ 3x - 2y = -6 \end{cases}$.

⊘ Solution

Step 1. Evaluate the determinant D, using the coefficients of the variables.		$\begin{cases} 2x + y = -4 \\ 3x - 2y = -6 \end{cases}$ $D = \begin{vmatrix} 2 & 1 \\ 3 & -2 \end{vmatrix}$ $D = -4 - 3$ $D = -7$
Step 2. Evaluate the determinant D_x. Use the constants in place of the x coefficients.	We replace the coefficients of x, 2 and 3, with the constants, -4 and -6.	$D_x = \begin{vmatrix} -4 & 1 \\ -6 & -2 \end{vmatrix}$ $D_x = 8 - (-6)$ $D_x = 14$
Step 3. Evaluate the determinant D_y. Use the constants in place of the y coefficients.	We replace the coefficients of y, 1 and 2, with the constants, -4 and -6.	$D_y = \begin{vmatrix} 2 & -4 \\ 3 & -6 \end{vmatrix}$ $D_y = -12 - (-12)$ $D_y = 0$
Step 4. Find x and y.	Substitute in the values of D, D_x and D_y.	$x = \frac{D_x}{D}$ and $y = \frac{D_y}{D}$ $x = \frac{14}{-7}$ and $y = \frac{0}{-7}$ $x = -2$ and $y = 0$

Step 5. Write the solution as an ordered pair.	The ordered pair is (x, y).	$(-2, 0)$
Step 6. Check that the ordered pair is a solution to **both** original equations.	Substitute $x = -2, y = 0$ into both equations and make sure they are both true.	$(-2, 0)$ is the solution to the system.

> **TRY IT :: 4.97**
>
> Solve using Cramer's rule: $\begin{cases} 3x + y = -3 \\ 2x + 3y = 6 \end{cases}$.

> **TRY IT :: 4.98**
>
> Solve using Cramer's rule: $\begin{cases} -x + y = 2 \\ 2x + y = -4 \end{cases}$.

HOW TO :: SOLVE A SYSTEM OF TWO EQUATIONS USING CRAMER'S RULE.

Step 1.   Evaluate the determinant D, using the coefficients of the variables.

Step 2.   Evaluate the determinant D_x. Use the constants in place of the x coefficients.

Step 3.   Evaluate the determinant D_y. Use the constants in place of the y coefficients.

Step 4.
Find x and y. $x = \dfrac{D_x}{D}, \quad y = \dfrac{D_y}{D}$

Step 5.   Write the solution as an ordered pair.

Step 6.   Check that the ordered pair is a solution to both original equations.

To solve a system of three equations with three variables with Cramer's Rule, we basically do what we did for a system of two equations. However, we now have to solve for three variables to get the solution. The determinants are also going to be 3×3 which will make our work more interesting!

Cramer's Rule for Solving a System of Three Equations

For the system of equations $\begin{cases} a_1 x + b_1 y + c_1 z = k_1 \\ a_2 x + b_2 y + c_2 z = k_2, \\ a_3 x + b_3 y + c_3 z = k_3 \end{cases}$ the solution (x, y, z) can be determined by

$$x = \frac{D_x}{D}, y = \frac{D_y}{D} \text{ and } z = \frac{D_z}{D}$$

where $D = \begin{vmatrix} a_1 & b_1 & c_1 \\ a_2 & b_2 & c_2 \\ a_3 & b_3 & c_3 \end{vmatrix}$ use the coefficients of the variables.

$D_x = \begin{vmatrix} k_1 & b_1 & c_1 \\ k_2 & b_2 & c_2 \\ k_3 & b_3 & c_3 \end{vmatrix}$ replace the x coefficients with the constants.

$D_y = \begin{vmatrix} a_1 & k_1 & c_1 \\ a_2 & k_2 & c_2 \\ a_3 & k_3 & c_3 \end{vmatrix}$ replace the y coefficients with the constants.

$D_z = \begin{vmatrix} a_1 & b_1 & k_1 \\ a_2 & b_2 & k_2 \\ a_3 & b_3 & k_3 \end{vmatrix}$ replace the z coefficients with the constants.

EXAMPLE 4.50

Solve the system of equations using Cramer's Rule: $\begin{cases} 3x - 5y + 4z = 5 \\ 5x + 2y + z = 0 \\ 2x + 3y - 2z = 3 \end{cases}$.

✓ Solution

Evaluate the determinant D.	$D = \begin{vmatrix} 3 & -5 & 4 \\ 5 & 2 & 1 \\ 2 & 3 & -2 \end{vmatrix}$

Expand by minors using column 1.

Be careful of the signs. $\begin{vmatrix} + & - & + \\ - & + & - \\ + & - & + \end{vmatrix}$	$D = 3\begin{vmatrix} 2 & 1 \\ 3 & -2 \end{vmatrix} - 5\begin{vmatrix} -5 & 4 \\ 3 & -2 \end{vmatrix} + 2\begin{vmatrix} -5 & 4 \\ 2 & 1 \end{vmatrix}$
Evaluate the determinants.	$D = 3(-4-3) - 5(10-12) + 2(-5-8)$
Simplify.	$D = 3(-7) - 5(-2) + 2(-13)$
Simplify.	$D = -21 + 10 - 26$
Simplify.	$D = -37$
Evaluate the determinant D_x. Use the constants to replace the coefficients of x.	$D_x = \begin{vmatrix} 5 & -5 & 4 \\ 0 & 2 & 1 \\ 3 & 3 & -2 \end{vmatrix}$
Expand by minors using column 1.	$D_x = 5\begin{vmatrix} 2 & 1 \\ 3 & -2 \end{vmatrix} - 0\begin{vmatrix} -5 & 4 \\ 3 & -2 \end{vmatrix} + 3\begin{vmatrix} -5 & 4 \\ 2 & 1 \end{vmatrix}$
Evaluate the determinants.	$D_x = 5(-4-3) - 0(10-12) + 3(-5-8)$
Simplify.	$D_x = 5(-7) - 0 + 3(-13)$
Simplify.	$D_x = -74$
Evaluate the determinant D_y. Use the constants to replace the coefficients of y.	$D_y = \begin{vmatrix} 3 & 5 & 4 \\ 5 & 0 & 1 \\ 2 & 3 & -2 \end{vmatrix}$

Expand by minors using column 2.

Be careful of the signs. $\begin{vmatrix} + & - & + \\ - & + & - \\ + & - & + \end{vmatrix}$	$D_y = -5\begin{vmatrix} 5 & 1 \\ 2 & -2 \end{vmatrix} + 0\begin{vmatrix} 5 & 4 \\ 3 & -2 \end{vmatrix} - 3\begin{vmatrix} 3 & 4 \\ 5 & 1 \end{vmatrix}$
Evaluate the determinants.	$D_y = -5(-10-2) + 0(-10-12) - 3(3-20)$
Simplify.	$D_y = -5(-12) + 0 - 3(-17)$
Simplify.	$D_y = 60 + 0 + 51$
Simplify.	$D_y = 111$

Evaluate the determinant D_z. Use the constants to replace the coefficients of z.	$D_z = \begin{vmatrix} 3 & -5 & 5 \\ 5 & 2 & 0 \\ 2 & 3 & 3 \end{vmatrix}$
Expand by minors using column 3. Be careful of the signs. $\begin{vmatrix} + & - & + \\ - & + & - \\ + & - & + \end{vmatrix}$	$D_z = 5\begin{vmatrix} 5 & 2 \\ 2 & 3 \end{vmatrix} - 0\begin{vmatrix} 3 & -5 \\ 2 & 3 \end{vmatrix} + 3\begin{vmatrix} 3 & -5 \\ 5 & 2 \end{vmatrix}$
Evaluate the determinants.	$D_z = 5(15 - 4) - 0(9 - (-10)) + 3(6 - (-25))$
Simplify.	$D_z = 5(11) - 0 + 3(31)$
Simplify.	$D_z = 55 - 0 + 93$
Simplify.	$D_z = 148$
Find x, y, and z.	$x = \dfrac{D_x}{D}, y = \dfrac{D_y}{D}$ and $z = \dfrac{D_z}{D}$
Substitute in the values.	$x = \dfrac{-74}{-37}, y = \dfrac{111}{-37}$ and $z = \dfrac{148}{-37}$
Simplify.	$x = 2, y = -3$ and $z = -4$
Write the solution as an ordered triple.	$(2, -3, -4)$
Check that the ordered triple is a solution to **all three** original equations.	We leave the check to you.
	The solution is $(2, -3, -4)$.

> **TRY IT :: 4.99**

Solve the system of equations using Cramer's Rule: $\begin{cases} 3x + 8y + 2z = -5 \\ 2x + 5y - 3z = 0 \\ x + 2y - 2z = -1 \end{cases}$.

> **TRY IT :: 4.100**

Solve the system of equations using Cramer's Rule: $\begin{cases} 3x + y - 6z = -3 \\ 2x + 6y + 3z = 0 \\ 3x + 2y - 3z = -6 \end{cases}$.

Cramer's rule does not work when the value of the D determinant is 0, as this would mean we would be dividing by 0. But when $D = 0$, the system is either inconsistent or dependent.

When the value of $D = 0$ and D_x, D_y and D_z are all zero, the system is consistent and dependent and there are infinitely many solutions.

When the value of $D = 0$ and D_x, D_y and D_z are not all zero, the system is inconsistent and there is no solution.

Dependent and Inconsistent Systems of Equations

For any system of equations, where the **value of the determinant** $D = 0$,

Value of determinants	Type of system	Solution
$D = 0$ and D_x, D_y and D_z are all zero	consistent and dependent	infinitely many solutions
$D = 0$ and D_x, D_y and D_z are not all zero	inconsistent	no solution

In the next example, we will use the values of the determinants to find the solution of the system.

EXAMPLE 4.51

Solve the system of equations using Cramer's rule : $\begin{cases} x+3y = 4 \\ -2x-6y = 3 \end{cases}$.

⊘ Solution

$\begin{cases} x+3y = 4 \\ -2x-6y = 3 \end{cases}$

Evaluate the determinant $D,$ using the coefficients of the variables.

$D = \begin{vmatrix} 1 & 3 \\ -2 & -6 \end{vmatrix}$

$D = -6-(-6)$

$D = 0$

We cannot use Cramer's Rule to solve this system. But by looking at the value of the determinants D_x and $D_y,$ we can determine whether the system is dependent or inconsistent.

Evaluate the determinant D_x.

$D_x = \begin{vmatrix} 4 & 3 \\ 3 & -6 \end{vmatrix}$

$D_x = -24-9$

$D_x = 15$

Since all the determinants are not zero, the system is inconsistent. There is no solution.

> **TRY IT : : 4.101**
>
> Solve the system of equations using Cramer's rule: $\begin{cases} 4x-3y = 8 \\ 8x-6y = 14 \end{cases}$.

> **TRY IT : : 4.102**
>
> Solve the system of equations using Cramer's rule: $\begin{cases} x = -3y+4 \\ 2x+6y = 8 \end{cases}$.

Solve Applications using Determinants

An interesting application of determinants allows us to test if points are collinear. Three points $(x_1,\ y_1),\quad (x_2,\ y_2)$ and $(x_3,\ y_3)$ are collinear if and only if the determinant below is zero.

$$\begin{vmatrix} x_1 & y_1 & 1 \\ x_2 & y_2 & 1 \\ x_3 & y_3 & 1 \end{vmatrix} = 0$$

Test for Collinear Points

Three points $(x_1,\ y_1),\quad (x_2,\ y_2)$ and $(x_3,\ y_3)$ are collinear if and only if

$$\begin{vmatrix} x_1 & y_1 & 1 \\ x_2 & y_2 & 1 \\ x_3 & y_3 & 1 \end{vmatrix} = 0$$

We will use this property in the next example.

EXAMPLE 4.52

Determine whether the points $(5,\ -5),\quad (4,\ -3),\quad$ and $(3,\ -1)$ are collinear.

Solution

$$\begin{vmatrix} x_1 & y_1 & 1 \\ x_2 & y_2 & 1 \\ x_3 & y_3 & 1 \end{vmatrix}$$

Substitute the values into the determinant. $(5, -5), \quad (4, -3), \quad \text{and} \ (3, -1)$	$\begin{vmatrix} 5 & -5 & 1 \\ 4 & -3 & 1 \\ 3 & -1 & 1 \end{vmatrix}$
Evaluate the determinant by expanding by minors using column 3.	$1\begin{vmatrix} 4 & -3 \\ 3 & -1 \end{vmatrix} - 1\begin{vmatrix} 5 & -5 \\ 3 & -1 \end{vmatrix} + 1\begin{vmatrix} 5 & -5 \\ 4 & -3 \end{vmatrix}$
Evaluate the determinants.	$1(-4-(-9)) - 1(-5-(-15)) + 1(-15-(-20))$
Simplify.	$1(5) - 1(10) + 1(5)$
Simplify.	0
	The value of the determinate is 0, so the points are collinear.

> **TRY IT ::** 4.103   Determine whether the points $(3, -2), \quad (5, -3), \quad \text{and} \ (1, -1)$ are collinear.

> **TRY IT ::** 4.104   Determine whether the points $(-4, -1), \quad (-6, 2), \quad \text{and} \ (-2, -4)$ are collinear.

> **MEDIA ::**

Access these online resources for additional instruction and practice with solving systems of linear inequalities by graphing.

- **Solving Systems of Linear Inequalities by Graphing (https://www.openstax.org/l/37syslinineqgph)**
- **Systems of Linear Inequalities (https://www.openstax.org/l/37syslineqs)**

4.6 EXERCISES
Practice Makes Perfect

Evaluate the Determinant of a 2 × 2 Matrix

In the following exercises, evaluate the determinate of each square matrix.

232. $\begin{bmatrix} 6 & -2 \\ 3 & -1 \end{bmatrix}$

233. $\begin{bmatrix} -4 & 8 \\ -3 & 5 \end{bmatrix}$

234. $\begin{bmatrix} -3 & 5 \\ 0 & -4 \end{bmatrix}$

235. $\begin{bmatrix} -2 & 0 \\ 7 & -5 \end{bmatrix}$

Evaluate the Determinant of a 3 × 3 Matrix

In the following exercises, find and then evaluate the indicated minors.

236. $\begin{vmatrix} 3 & -1 & 4 \\ -1 & 0 & -2 \\ -4 & 1 & 5 \end{vmatrix}$
Find the minor ⓐ a_1 ⓑ b_2 ⓒ c_3

237. $\begin{vmatrix} -1 & -3 & 2 \\ 4 & -2 & -1 \\ -2 & 0 & -3 \end{vmatrix}$
Find the minor ⓐ a_1 ⓑ b_1 ⓒ c_2

238. $\begin{vmatrix} 2 & -3 & -4 \\ -1 & 2 & -3 \\ 0 & -1 & -2 \end{vmatrix}$
Find the minor ⓐ a_2 ⓑ b_2 ⓒ c_2

239. $\begin{vmatrix} -2 & -2 & 3 \\ 1 & -3 & 0 \\ -2 & 3 & -2 \end{vmatrix}$
Find the minor ⓐ a_3 ⓑ b_3 ⓒ c_3

In the following exercises, evaluate each determinant by expanding by minors along the first row.

240. $\begin{vmatrix} -2 & 3 & -1 \\ -1 & 2 & -2 \\ 3 & 1 & -3 \end{vmatrix}$

241. $\begin{vmatrix} 4 & -1 & -2 \\ -3 & -2 & 1 \\ -2 & -5 & 7 \end{vmatrix}$

242. $\begin{vmatrix} -2 & -3 & -4 \\ 5 & -6 & 7 \\ -1 & 2 & 0 \end{vmatrix}$

243. $\begin{vmatrix} 1 & 3 & -2 \\ 5 & -6 & 4 \\ 0 & -2 & -1 \end{vmatrix}$

In the following exercises, evaluate each determinant by expanding by minors.

244. $\begin{vmatrix} -5 & -1 & -4 \\ 4 & 0 & -3 \\ 2 & -2 & 6 \end{vmatrix}$

245. $\begin{vmatrix} 4 & -1 & 3 \\ 3 & -2 & 2 \\ -1 & 0 & 4 \end{vmatrix}$

246. $\begin{vmatrix} 3 & 5 & 4 \\ -1 & 3 & 0 \\ -2 & 6 & 1 \end{vmatrix}$

247. $\begin{vmatrix} 2 & -4 & -3 \\ 5 & -1 & -4 \\ 3 & 2 & 0 \end{vmatrix}$

Use Cramer's Rule to Solve Systems of Equations

In the following exercises, solve each system of equations using Cramer's Rule.

248. $\begin{cases} -2x + 3y = 3 \\ x + 3y = 12 \end{cases}$

249. $\begin{cases} x - 2y = -5 \\ 2x - 3y = -4 \end{cases}$

250. $\begin{cases} x - 3y = -9 \\ 2x + 5y = 4 \end{cases}$

251. $\begin{cases} 2x + y = -4 \\ 3x - 2y = -6 \end{cases}$

252. $\begin{cases} x - 2y = -5 \\ 2x - 3y = -4 \end{cases}$

253. $\begin{cases} x - 3y = -9 \\ 2x + 5y = 4 \end{cases}$

254. $\begin{cases} 5x - 3y = -1 \\ 2x - y = 2 \end{cases}$

255. $\begin{cases} 3x + 8y = -3 \\ 2x + 5y = -3 \end{cases}$

256. $\begin{cases} 6x - 5y + 2z = 3 \\ 2x + y - 4z = 5 \\ 3x - 3y + z = -1 \end{cases}$

257. $\begin{cases} 4x - 3y + z = 7 \\ 2x - 5y - 4z = 3 \\ 3x - 2y - 2z = -7 \end{cases}$

258. $\begin{cases} 2x - 5y + 3z = 8 \\ 3x - y + 4z = 7 \\ x + 3y + 2z = -3 \end{cases}$

259. $\begin{cases} 11x + 9y + 2z = -9 \\ 7x + 5y + 3z = -7 \\ 4x + 3y + z = -3 \end{cases}$

260. $\begin{cases} x + 2z = 0 \\ 4y + 3z = -2 \\ 2x - 5y = 3 \end{cases}$

261. $\begin{cases} 2x + 5y = 4 \\ 3y - z = 3 \\ 4x + 3z = -3 \end{cases}$

262. $\begin{cases} 2y + 3z = -1 \\ 5x + 3y = -6 \\ 7x + z = 1 \end{cases}$

263. $\begin{cases} 3x - z = -3 \\ 5y + 2z = -6 \\ 4x + 3y = -8 \end{cases}$

264. $\begin{cases} 2x + y = 3 \\ 6x + 3y = 9 \end{cases}$

265. $\begin{cases} x - 4y = -1 \\ -3x + 12y = 3 \end{cases}$

266. $\begin{cases} -3x - y = 4 \\ 6x + 2y = -16 \end{cases}$

267. $\begin{cases} 4x + 3y = 2 \\ 20x + 15y = 5 \end{cases}$

268. $\begin{cases} x + y - 3z = -1 \\ y - z = 0 \\ -x + 2y = 1 \end{cases}$

269. $\begin{cases} 2x + 3y + z = 12 \\ x + y + z = 9 \\ 3x + 4y + 2z = 20 \end{cases}$

270. $\begin{cases} 3x + 4y - 3z = -2 \\ 2x + 3y - z = -12 \\ x + y - 2z = 6 \end{cases}$

271. $\begin{cases} x - 2y + 3z = 1 \\ x + y - 3z = 7 \\ 3x - 4y + 5z = 7 \end{cases}$

Solve Applications Using Determinants

In the following exercises, determine whether the given points are collinear.

272. $(0, 1)$, $(2, 0)$, and $(-2, 2)$.

273. $(0, -5)$, $(-2, -2)$, and $(2, -8)$.

274. $(4, -3)$, $(6, -4)$, and $(2, -2)$.

275. $(-2, 1)$, $(-4, 4)$, and $(0, -2)$.

Writing Exercises

276. Explain the difference between a square matrix and its determinant. Give an example of each.

277. Explain what is meant by the minor of an entry in a square matrix.

278. Explain how to decide which row or column you will use to expand a 3×3 determinant.

279. Explain the steps for solving a system of equations using Cramer's rule.

Self Check

ⓐ *After completing the exercises, use this checklist to evaluate your mastery of the objectives of this section.*

I can...	Confidently	With some help	No-I don't get it!
evaluate the determinant of a 2 × 2 matrix.			
evaluate the determinant of a 3 × 3 matrix.			
use Cramer's rule to solve systems of equations.			
solve applications using determinants.			

ⓑ *After reviewing this checklist, what will you do to become confident for all objectives?*

4.7 **Graphing Systems of Linear Inequalities**

Learning Objectives

By the end of this section, you will be able to:

› Determine whether an ordered pair is a solution of a system of linear inequalities
› Solve a system of linear inequalities by graphing
› Solve applications of systems of inequalities

Be Prepared!

Before you get started, take this readiness quiz.

1. Solve the inequality $2a < 5a + 12$.
   If you missed this problem, review **Example 2.52**.

2. Determine whether the ordered pair $\left(3, \frac{1}{2}\right)$ is a solution to the system $y > 2x + 3$.
   If you missed this problem, review **Example 3.34**.

Determine whether an ordered pair is a solution of a system of linear inequalities

The definition of a **system of linear inequalities** is very similar to the definition of a system of linear equations.

System of Linear Inequalities

Two or more linear inequalities grouped together form a system of linear inequalities.

A system of linear inequalities looks like a system of linear equations, but it has inequalities instead of equations. A system of two linear inequalities is shown here.

$$\begin{cases} x + 4y \geq 10 \\ 3x - 2y < 12 \end{cases}$$

To solve a system of linear inequalities, we will find values of the variables that are solutions to both inequalities. We solve the system by using the graphs of each inequality and show the solution as a graph. We will find the region on the plane that contains all ordered pairs (x, y) that make both inequalities true.

Solutions of a System of Linear Inequalities

Solutions of a system of linear inequalities are the values of the variables that make all the inequalities true.

The solution of a system of linear inequalities is shown as a shaded region in the x, y coordinate system that includes all the points whose ordered pairs make the inequalities true.

To determine if an ordered pair is a solution to a system of two inequalities, we substitute the values of the variables into each inequality. If the ordered pair makes both inequalities true, it is a solution to the system.

EXAMPLE 4.53

Determine whether the ordered pair is a solution to the system $\begin{cases} x + 4y \geq 10 \\ 3x - 2y < 12 \end{cases}$.

ⓐ $(-2, 4)$ ⓑ $(3, 1)$

⊘ Solution

ⓐ Is the ordered pair $(-2, 4)$ a solution?

We substitute $x = -2$ and $y = 4$ into both inequalities.

$x + 4y \geq 10$ $3x - 2y < 12$

$-2 + 4(4) \overset{?}{\geq} 10$ $3(-2) - 2(4) \overset{?}{<} 12$

$14 \geq 10$ true $-14 < 12$ true

The ordered pair $(-2, 4)$ made both inequalities true. Therefore $(-2, 4)$ is a solution to this system.

ⓑ Is the ordered pair $(3, 1)$ a solution?

We substitute $x = 3$ and $y = 1$ into both inequalities.

$$x + 4y \geq 10 \qquad\qquad 3x - 2y < 12$$

$$3 + 4(1) \overset{?}{\geq} 10 \qquad\qquad 3(3) - 2(1) \overset{?}{<} 12$$

$$7 \geq 10 \text{ false} \qquad\qquad 7 < 12 \text{ true}$$

The ordered pair $(3, 1)$ made one inequality true, but the other one false. Therefore $(3, 1)$ is not a solution to this system.

 TRY IT :: 4.105

Determine whether the ordered pair is a solution to the system: $\begin{cases} x - 5y > 10 \\ 2x + 3y > -2 \end{cases}$.

ⓐ $(3, -1)$ ⓑ $(6, -3)$

 TRY IT :: 4.106

Determine whether the ordered pair is a solution to the system: $\begin{cases} y > 4x - 2 \\ 4x - y < 20 \end{cases}$.

ⓐ $(-2, 1)$ ⓑ $(4, -1)$

Solve a System of Linear Inequalities by Graphing

The solution to a single linear inequality is the region on one side of the boundary line that contains all the points that make the inequality true. The solution to a system of two linear inequalities is a region that contains the solutions to both inequalities. To find this region, we will graph each inequality separately and then locate the region where they are both true. The solution is always shown as a graph.

EXAMPLE 4.54 HOW TO SOLVE A SYSTEM OF LINEAR INEQUALITIES BY GRAPHING

Solve the system by graphing: $\begin{cases} y \geq 2x - 1 \\ y < x + 1 \end{cases}$.

 Solution

Step 1. Graph the first inequality. • Graph the boundary line.	We will graph $y \geq 2x - 1$. We graph the line $y = 2x - 1$. It is a solid line because the inequality sign is $\geq$.	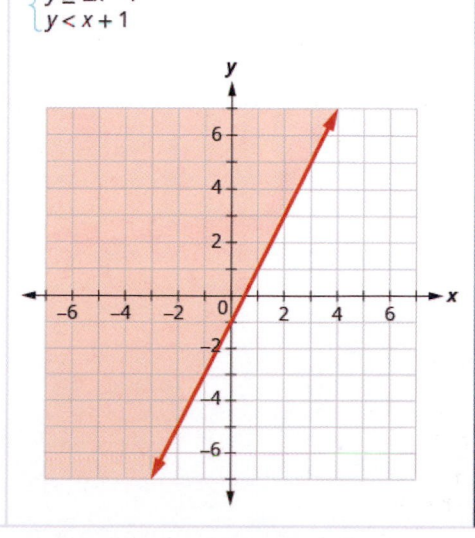
• Shade in the side of the boundary line where the inequality is true.	We choose $(0, 0)$ as a test point. It is a solution to $y \geq 2x - 1$, so we shade in above the boundary line.	

Step 2. On the same grid, graph the second inequality. • Graph the boundary line. • Shade in the side of that boundary line where the inequality is true.	We will graph $y < x + 1$ on the same grid. We graph the line $y = x + 1$. It is a dashed line because the inequality sign is $<$. Again, we use (0, 0) as a test point. It is a solution so we shade in that side of the line $y = x + 1$.	
Step 3. The solution is the region where the shading overlaps.	The point where the boundary lines intersect is not a solution because it is not a solution to $y < x + 1$.	The solution is all points in the area shaded twice—which appears as the darkest shaded region.
Step 4. Check by choosing a test point.	We'll use (–1, –1) as a test point.	Is (–1, –1) a solution to $y \geq 2x - 1$? $-1 \overset{?}{\geq} 2(-1) - 1$ $-1 \geq -3$ true Is (–1, –1) a solution to $y < x + 1$? $-1 \overset{?}{<} -1 + 1$ $-1 < 0$ true The region containing (–1, –1) is the solution to this system.

> **TRY IT : :** 4.107

Solve the system by graphing: $\begin{cases} y < 3x + 2 \\ y > -x - 1 \end{cases}$.

> **TRY IT : :** 4.108

Solve the system by graphing: $\begin{cases} y < -\frac{1}{2}x + 3 \\ y < 3x - 4 \end{cases}$.

HOW TO :: SOLVE A SYSTEM OF LINEAR INEQUALITIES BY GRAPHING.

Step 1.   Graph the first inequality.

- Graph the boundary line.
- Shade in the side of the boundary line where the inequality is true.

Step 2.   On the same grid, graph the second inequality.

- Graph the boundary line.
- Shade in the side of that boundary line where the inequality is true.

Step 3.   The solution is the region where the shading overlaps.

Step 4.   Check by choosing a test point.

EXAMPLE 4.55

Solve the system by graphing: $\begin{cases} x - y > 3 \\ y < -\frac{1}{5}x + 4 \end{cases}$.

 Solution

$$\begin{cases} x - y > 3 \\ y < -\frac{1}{5}x + 4 \end{cases}$$

Graph $x - y > 3$, by graphing $x - y = 3$ and testing a point.

The intercepts are $x = 3$ and $y = -3$ and the boundary line will be dashed.

Test $(0, 0)$ which makes the inequality false so shade (red) the side that does not contain $(0, 0)$.

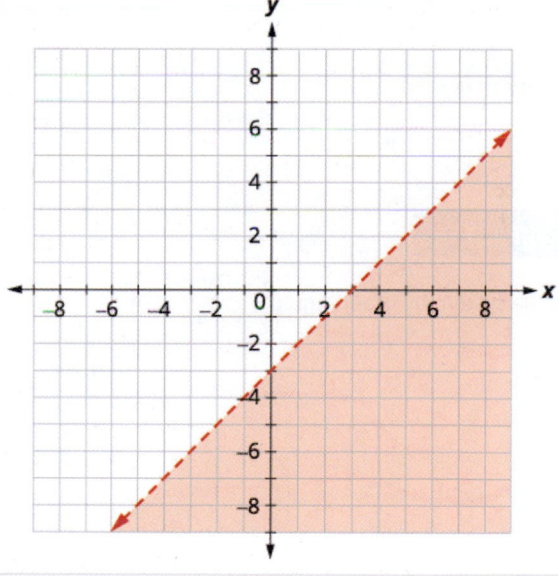

Graph $y < -\frac{1}{5}x + 4$ by graphing $y = -\frac{1}{5}x + 4$

using the slope $m = -\frac{1}{5}$ and y-intercept $b = 4$.

The boundary line will be dashed

Test $(0, 0)$ which makes the inequality true, so shade (blue) the side that contains $(0, 0)$.

Choose a test point in the solution and verify that it is a solution to both inequalties.

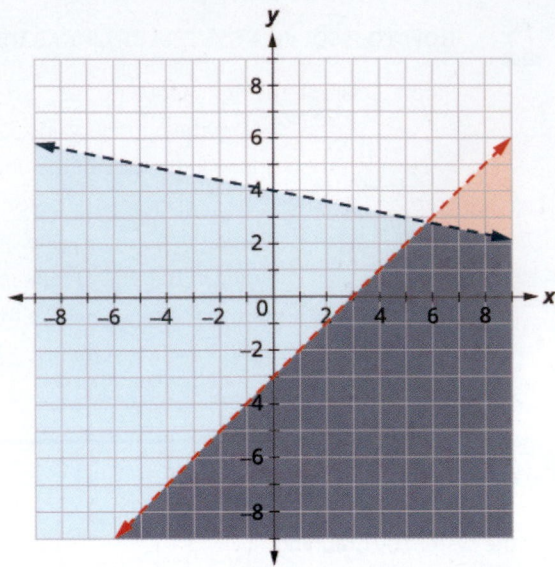

The point of intersection of the two lines is not included as both boundary lines were dashed. The solution is the area shaded twice—which appears as the darkest shaded region.

> **TRY IT : : 4.109**

Solve the system by graphing: $\begin{cases} x + y \le 2 \\ y \ge \frac{2}{3}x - 1 \end{cases}$.

> **TRY IT : : 4.110**

Solve the system by graphing: $\begin{cases} 3x - 2y \le 6 \\ y > -\frac{1}{4}x + 5 \end{cases}$.

EXAMPLE 4.56

Solve the system by graphing: $\begin{cases} x - 2y < 5 \\ y > -4 \end{cases}$.

✓ Solution

$$\begin{cases} x - 2y < 5 \\ y > -4 \end{cases}$$

Graph $x - 2y < 5$, by graphing $x - 2y = 5$
and testing a point. The intercepts are
$x = 5$ and $y = -2.5$ and the
boundary line will be dashed.

Test $(0, 0)$ which makes the inequality true, so shade
(red) the side that contains $(0, 0)$.

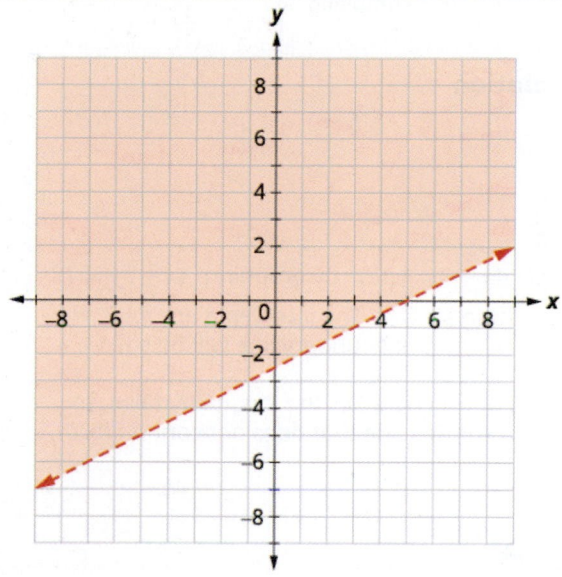

Graph $y > -4$, by graphing $y = -4$ and
recognizing that it is a horizontal line
through $y = -4$. The boundary line will
be dashed.

Test $(0, 0)$ which makes the inequality
true so shade (blue) the side that contains $(0, 0)$.

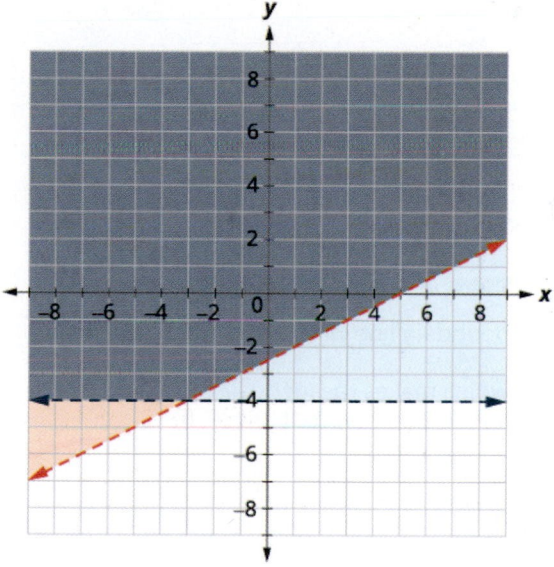

The point $(0, 0)$ is in the solution and we have already found it to be a solution of each inequality. The point of intersection of the two lines is not included as both boundary lines were dashed.

The solution is the area shaded twice—which appears as the darkest shaded region.

> **TRY IT : : 4.111**

Solve the system by graphing: $\begin{cases} y \geq 3x - 2 \\ y < -1 \end{cases}$.

> **TRY IT : : 4.112**

Solve the system by graphing: $\begin{cases} x > -4 \\ x - 2y \geq -4 \end{cases}$.

Systems of linear inequalities where the boundary lines are parallel might have no solution. We'll see this in the next example.

EXAMPLE 4.57

Solve the system by graphing: $\begin{cases} 4x + 3y \geq 12 \\ y < -\frac{4}{3}x + 1 \end{cases}$.

✓ **Solution**

$$\begin{cases} 4x + 3y \geq 12 \\ y < -\frac{4}{3}x + 1 \end{cases}$$

Graph $4x + 3y \geq 12$, by graphing $4x + 3y = 12$ and testing a point. The intercepts are $x = 3$ and $y = 4$ and the boundary line will be solid.

Test $(0, 0)$ which makes the inequality false, so shade (red) the side that does not contain $(0, 0)$.

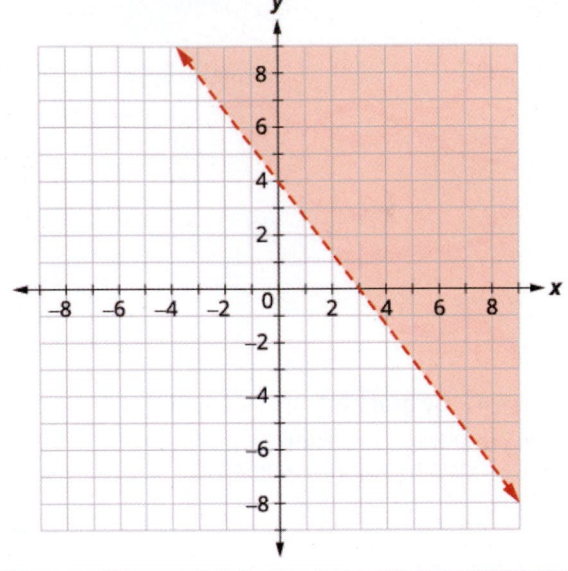

Graph $y < -\frac{4}{3}x + 1$ by graphing $y = -\frac{4}{3}x + 1$ using the slope $m = -\frac{4}{3}$ and y-intercept $b = 1$. The boundary line will be dashed.

Test $(0, 0)$ which makes the inequality true, so shade (blue) the side that contains $(0, 0)$.

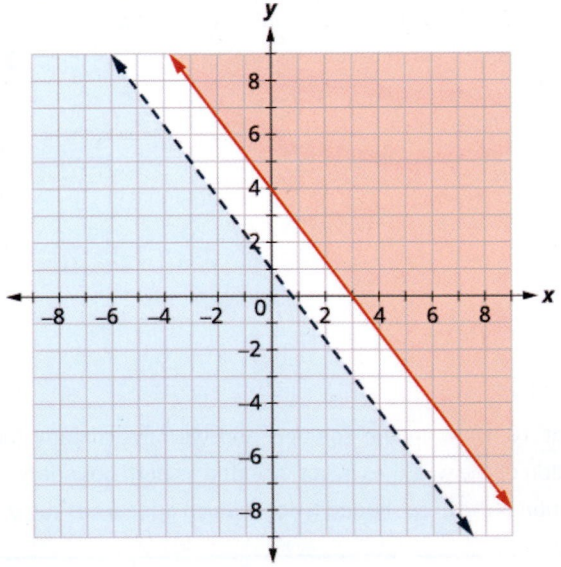

There is no point in both shaded regions, so the system has no solution.

> **TRY IT ::** 4.113
>
> Solve the system by graphing: $\begin{cases} 3x - 2y \geq 12 \\ y \geq \frac{3}{2}x + 1 \end{cases}$.

> **TRY IT ::** 4.114
>
> Solve the system by graphing: $\begin{cases} x + 3y > 8 \\ y < -\frac{1}{3}x - 2 \end{cases}$.

Some systems of linear inequalities where the boundary lines are parallel will have a solution. We'll see this in the next example.

EXAMPLE 4.58

Solve the system by graphing: $\begin{cases} y > \frac{1}{2}x - 4 \\ x - 2y < -4 \end{cases}$.

✓ **Solution**

$$\begin{cases} y > \frac{1}{2}x - 4 \\ x - 2y < -4 \end{cases}$$

Graph $y > \frac{1}{2}x - 4$ by graphing $y = \frac{1}{2}x - 4$

using the slope $m = \frac{1}{2}$ and the intercept

$b = -4$. The boundary line will be dashed.

Test $(0, 0)$ which makes the inequality true, so shade (red) the side that contains $(0, 0)$.

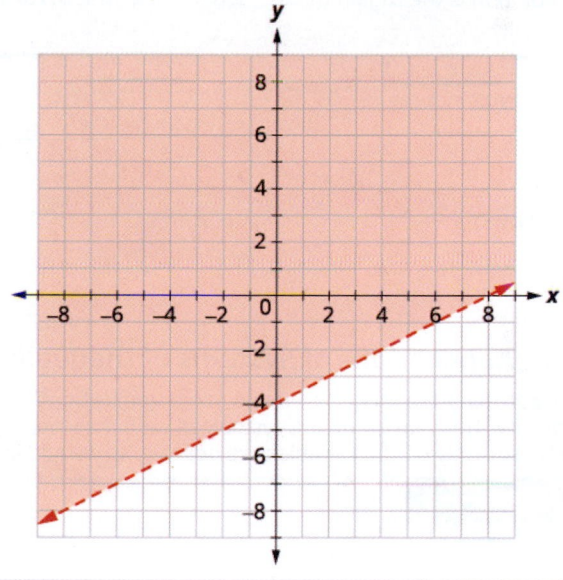

Graph $x - 2y < -4$ by graphing $x - 2y = -4$
and testing a point. The intercepts are
$x = -4$ and $y = 2$ and the boundary line will be dashed.

Choose a test point in the solution and verify
that it is a solution to both inequalties.

Test (0, 0) which makes the inequality false, so
shade (blue) the side that does not contain (0, 0).

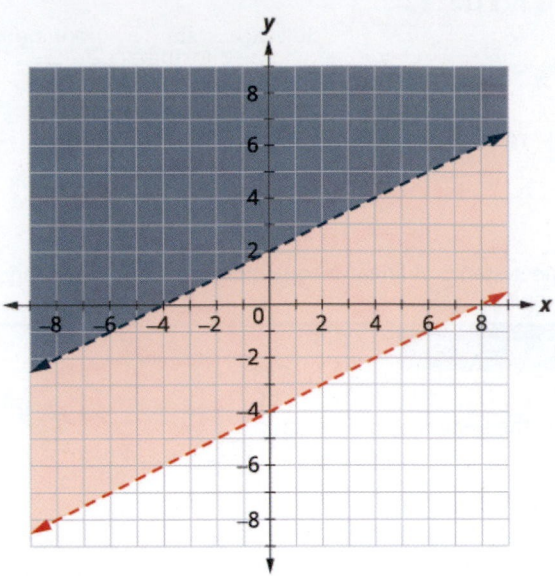

No point on the boundary lines is included in the solution as both lines are dashed.

The solution is the region that is shaded twice which is also the solution to $x - 2y < -4$.

> **TRY IT : :** 4.115

Solve the system by graphing: $\begin{cases} y \geq 3x + 1 \\ -3x + y \geq -4 \end{cases}$.

> **TRY IT : :** 4.116

Solve the system by graphing: $\begin{cases} y \leq -\frac{1}{4}x + 2 \\ x + 4y \leq 4 \end{cases}$.

Solve Applications of Systems of Inequalities

The first thing we'll need to do to solve applications of systems of inequalities is to translate each condition into an inequality. Then we graph the system, as we did above, to see the region that contains the solutions. Many situations will be realistic only if both variables are positive, so we add inequalities to the system as additional requirements.

EXAMPLE 4.59

Christy sells her photographs at a booth at a street fair. At the start of the day, she wants to have at least 25 photos to display at her booth. Each small photo she displays costs her $4 and each large photo costs her $10. She doesn't want to spend more than $200 on photos to display.

ⓐ Write a system of inequalities to model this situation.

ⓑ Graph the system.

ⓒ Could she display 10 small and 20 large photos?

ⓓ Could she display 20 large and 10 small photos?

✓ **Solution**

ⓐ

Let $x = $ the number of small photos.

$\quad\quad y = $ the number of large photos

To find the system of equations translate the information.

She wants to have at least 25 photos.

The number of small plus the number of large should be at least 25.

$$x + y \geq 25$$

$4 for each small and $10 for each large must be no more than $200

$$4x + 10y \leq 200$$

The number of small photos must be greater than or equal to 0.

$$x \geq 0$$

The number of large photos must be greater than or equal to 0.

$$y \geq 0$$

We have our system of equations.

$$\begin{cases} x + y \geq 25 \\ 4x + 10y \leq 200 \\ x \geq 0 \\ y \geq 0 \end{cases}$$

ⓑ

Since $x \geq 0$ and $y \geq 0$ (both are greater than or equal to) all solutions will be in the first quadrant. As a result, our graph shows only quadrant one.

To graph $x + y \geq 25$, graph $x + y = 25$ as a solid line.

Choose (0, 0) as a test point. Since it does not make the inequality true, shade (red) the side that does not include the point (0, 0).

To graph $4x + 10y \leq 200$, graph $4x + 10y = 200$ as a solid line.

Choose (0, 0) as a test point. Since it does make the inequality true, shade (blue) the side that include the point (0, 0).

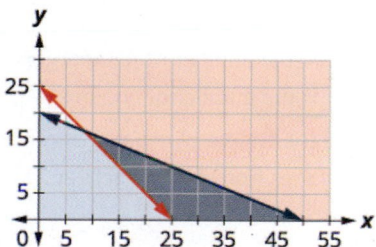

The solution of the system is the region of the graph that is shaded the darkest. The boundary line sections that border the darkly-shaded section are included in the solution as are the points on the x-axis from (25, 0) to (55, 0).

ⓒ To determine if 10 small and 20 large photos would work, we look at the graph to see if the point (10, 20) is in the solution region. We could also test the point to see if it is a solution of both equations.

It is not, Christy would not display 10 small and 20 large photos.

ⓓ To determine if 20 small and 10 large photos would work, we look at the graph to see if the point (20, 10) is in the solution region. We could also test the point to see if it is a solution of both equations.

It is, so Christy could choose to display 20 small and 10 large photos.

Notice that we could also test the possible solutions by substituting the values into each inequality.

> **TRY IT :: 4.117**

A trailer can carry a maximum weight of 160 pounds and a maximum volume of 15 cubic feet. A microwave oven weighs 30 pounds and has 2 cubic feet of volume, while a printer weighs 20 pounds and has 3 cubic feet of space.

ⓐ Write a system of inequalities to model this situation.
ⓑ Graph the system.
ⓒ Could 4 microwaves and 2 printers be carried on this trailer?
ⓓ Could 7 microwaves and 3 printers be carried on this trailer?

> **TRY IT : :** 4.118

Mary needs to purchase supplies of answer sheets and pencils for a standardized test to be given to the juniors at her high school. The number of the answer sheets needed is at least 5 more than the number of pencils. The pencils cost $2 and the answer sheets cost $1. Mary's budget for these supplies allows for a maximum cost of $400.

ⓐ Write a system of inequalities to model this situation.

ⓑ Graph the system.

ⓒ Could Mary purchase 100 pencils and 100 answer sheets?

ⓓ Could Mary purchase 150 pencils and 150 answer sheets?

When we use variables other than *x* and *y* to define an unknown quantity, we must change the names of the axes of the graph as well.

EXAMPLE 4.60

Omar needs to eat at least 800 calories before going to his team practice. All he wants is hamburgers and cookies, and he doesn't want to spend more than $5. At the hamburger restaurant near his college, each hamburger has 240 calories and costs $1.40. Each cookie has 160 calories and costs $0.50.

ⓐ Write a system of inequalities to model this situation.

ⓑ Graph the system.

ⓒ Could he eat 3 hamburgers and 1 cookie?

ⓓ Could he eat 2 hamburgers and 4 cookies?

✓ **Solution**

ⓐ

Let $h =$ the number of hamburgers.

$c =$ the number of cookies

To find the system of equations translate the information.

The calories from hamburgers at 240 calories each, plus the calories from cookies at 160 calories each must be more that 800.

$$240h + 160c \geq 800$$

The amount spent on hamburgers at $1.40 each, plus the amount spent on cookies at $0.50 each must be no more than $5.00.

$$1.40h + 0.50c \leq 5$$

The number of hamburgers must be greater than or equal to 0.

$$h \geq 0$$

The number of cookies must be greater than or equal to 0.

$$c \geq 0$$

We have our system of equations.
$$\begin{cases} 240h + 160c \geq 800 \\ 1.40h + 0.50c \leq 5 \\ h \geq 0 \\ c \geq 0 \end{cases}$$

ⓑ

Since $h > = 0$ and $c > = 0$ (both are greater than or equal to) all solutions will be in the first quadrant. As a result, our graph shows only quadrant one.

To graph $240h + 160c \geq 800$, graph $240h + 160c = 800$ as a solid line.

Choose (0, 0) as a test point. Since it does not make the inequality true, shade (red) the side that does not include the point (0, 0).

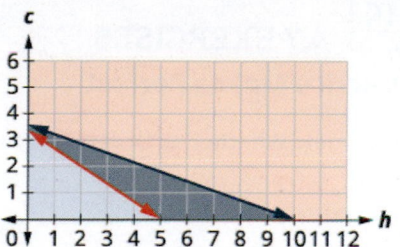

Graph $1.40h + 0.50c \leq 5$. The boundary line is $1.40h + 0.50c = 5$. We test (0, 0) and it makes the inequality true. We shade the side of the line that includes (0, 0).

The solution of the system is the region of the graph that is shaded the darkest. The boundary line sections that border the darkly shaded section are included in the solution as are the points on the *x*-axis from (5, 0) to (10, 0).

ⓒ To determine if 3 hamburgers and 2 cookies would meet Omar's criteria, we see if the point (3, 2) is in the solution region. It is, so Omar might choose to eat 3 hamburgers and 2 cookies.

ⓓ To determine if 2 hamburgers and 4 cookies would meet Omar's criteria, we see if the point (2, 4) is in the solution region. It is, Omar might choose to eat 2 hamburgers and 4 cookies.

We could also test the possible solutions by substituting the values into each inequality.

> **TRY IT : :** 4.119

Tension needs to eat at least an extra 1,000 calories a day to prepare for running a marathon. He has only $25 to spend on the extra food he needs and will spend it on $0.75 donuts which have 360 calories each and $2 energy drinks which have 110 calories.

ⓐ Write a system of inequalities that models this situation.
ⓑ Graph the system.
ⓒ Can he buy 8 donuts and 4 energy drinks and satisfy his caloric needs?
ⓓ Can he buy 1 donut and 3 energy drinks and satisfy his caloric needs?

> **TRY IT : :** 4.120

Philip's doctor tells him he should add at least 1,000 more calories per day to his usual diet. Philip wants to buy protein bars that cost $1.80 each and have 140 calories and juice that costs $1.25 per bottle and have 125 calories. He doesn't want to spend more than $12.

ⓐ Write a system of inequalities that models this situation.
ⓑ Graph the system.
ⓒ Can he buy 3 protein bars and 5 bottles of juice?
ⓓ Can he buy 5 protein bars and 3 bottles of juice?

▶ **MEDIA : :**

Access these online resources for additional instruction and practice with solving systems of linear inequalities by graphing.

- **Solving Systems of Linear Inequalities by Graphing (https://openstax.org/l/37sysineqgraph)**
- **Systems of Linear Inequalities (https://openstax.org/l/37sysineqgraph2)**

4.7 EXERCISES

Practice Makes Perfect

Determine Whether an Ordered Pair is a Solution of a System of Linear Inequalities

In the following exercises, determine whether each ordered pair is a solution to the system.

280. $\begin{cases} 3x + y > 5 \\ 2x - y \le 10 \end{cases}$

ⓐ $(3, -3)$

ⓑ $(7, 1)$

281. $\begin{cases} 4x - y < 10 \\ -2x + 2y > -8 \end{cases}$

ⓐ $(5, -2)$

ⓑ $(-1, 3)$

282. $\begin{cases} y > \frac{2}{3}x - 5 \\ x + \frac{1}{2}y \le 4 \end{cases}$

ⓐ $(6, -4)$

ⓑ $(3, 0)$

283. $\begin{cases} y < \frac{3}{2}x + 3 \\ \frac{3}{4}x - 2y < 5 \end{cases}$

ⓐ $(-4, -1)$

ⓑ $(8, 3)$

284. $\begin{cases} 7x + 2y > 14 \\ 5x - y \le 8 \end{cases}$

ⓐ $(2, 3)$

ⓑ $(7, -1)$

285. $\begin{cases} 6x - 5y < 20 \\ -2x + 7y > -8 \end{cases}$

ⓐ $(1, -3)$

ⓑ $(-4, 4)$

Solve a System of Linear Inequalities by Graphing

In the following exercises, solve each system by graphing.

286. $\begin{cases} y \le 3x + 2 \\ y > x - 1 \end{cases}$

287. $\begin{cases} y < -2x + 2 \\ y \ge -x - 1 \end{cases}$

288. $\begin{cases} y < 2x - 1 \\ y \le -\frac{1}{2}x + 4 \end{cases}$

289. $\begin{cases} y \ge -\frac{2}{3}x + 2 \\ y > 2x - 3 \end{cases}$

290. $\begin{cases} x - y > 1 \\ y < -\frac{1}{4}x + 3 \end{cases}$

291. $\begin{cases} x + 2y < 4 \\ y < x - 2 \end{cases}$

292. $\begin{cases} 3x - y \ge 6 \\ y \ge -\frac{1}{2}x \end{cases}$

293. $\begin{cases} 2x + 4y \ge 8 \\ y \le \frac{3}{4}x \end{cases}$

294. $\begin{cases} 2x - 5y < 10 \\ 3x + 4y \ge 12 \end{cases}$

295. $\begin{cases} 3x - 2y \le 6 \\ -4x - 2y > 8 \end{cases}$

296. $\begin{cases} 2x + 2y > -4 \\ -x + 3y \ge 9 \end{cases}$

297. $\begin{cases} 2x + y > -6 \\ -x + 2y \ge -4 \end{cases}$

298. $\begin{cases} x - 2y < 3 \\ y \le 1 \end{cases}$

299. $\begin{cases} x - 3y > 4 \\ y \le -1 \end{cases}$

300. $\begin{cases} y \ge -\frac{1}{2}x - 3 \\ x \le 2 \end{cases}$

301. $\begin{cases} y \le -\frac{2}{3}x + 5 \\ x \ge 3 \end{cases}$

302. $\begin{cases} y \ge \frac{3}{4}x - 2 \\ y < 2 \end{cases}$

303. $\begin{cases} y \le -\frac{1}{2}x + 3 \\ y < 1 \end{cases}$

304. $\begin{cases} 3x - 4y < 8 \\ x < 1 \end{cases}$

305. $\begin{cases} -3x + 5y > 10 \\ x > -1 \end{cases}$

306. $\begin{cases} x \ge 3 \\ y \le 2 \end{cases}$

307. $\begin{cases} x \le -1 \\ y \ge 3 \end{cases}$

308. $\begin{cases} 2x + 4y > 4 \\ y \le -\frac{1}{2}x - 2 \end{cases}$

309. $\begin{cases} x - 3y \ge 6 \\ y > \frac{1}{3}x + 1 \end{cases}$

310. $\begin{cases} -2x + 6y < 0 \\ 6y > 2x + 4 \end{cases}$

311. $\begin{cases} -3x + 6y > 12 \\ 4y \le 2x - 4 \end{cases}$

312. $\begin{cases} y \ge -3x + 2 \\ 3x + y > 5 \end{cases}$

313. $\begin{cases} y \ge \frac{1}{2}x - 1 \\ -2x + 4y \ge 4 \end{cases}$

314. $\begin{cases} y \le -\frac{1}{4}x - 2 \\ x + 4y < 6 \end{cases}$

315. $\begin{cases} y \ge 3x - 1 \\ -3x + y > -4 \end{cases}$

316. $\begin{cases} 3y > x + 2 \\ -2x + 6y > 8 \end{cases}$

317. $\begin{cases} y < \frac{3}{4}x - 2 \\ -3x + 4y < 7 \end{cases}$

Solve Applications of Systems of Inequalities

In the following exercises, translate to a system of inequalities and solve.

318. Caitlyn sells her drawings at the county fair. She wants to sell at least 60 drawings and has portraits and landscapes. She sells the portraits for $15 and the landscapes for $10. She needs to sell at least $800 worth of drawings in order to earn a profit.

ⓐ Write a system of inequalities to model this situation.

ⓑ Graph the system.

ⓒ Will she make a profit if she sells 20 portraits and 35 landscapes?

ⓓ Will she make a profit if she sells 50 portraits and 20 landscapes?

319. Jake does not want to spend more than $50 on bags of fertilizer and peat moss for his garden. Fertilizer costs $2 a bag and peat moss costs $5 a bag. Jake's van can hold at most 20 bags.

ⓐ Write a system of inequalities to model this situation.

ⓑ Graph the system.

ⓒ Can he buy 15 bags of fertilizer and 4 bags of peat moss?

ⓓ Can he buy 10 bags of fertilizer and 10 bags of peat moss?

320. Reiko needs to mail her Christmas cards and packages and wants to keep her mailing costs to no more than $500. The number of cards is at least 4 more than twice the number of packages. The cost of mailing a card (with pictures enclosed) is $3 and for a package the cost is $7.

ⓐ Write a system of inequalities to model this situation.

ⓑ Graph the system.

ⓒ Can she mail 60 cards and 26 packages?

ⓓ Can she mail 90 cards and 40 packages?

321. Juan is studying for his final exams in chemistry and algebra. he knows he only has 24 hours to study, and it will take him at least three times as long to study for algebra than chemistry.

ⓐ Write a system of inequalities to model this situation.

ⓑ Graph the system.

ⓒ Can he spend 4 hours on chemistry and 20 hours on algebra?

ⓓ Can he spend 6 hours on chemistry and 18 hours on algebra?

322. Jocelyn is pregnant and so she needs to eat at least 500 more calories a day than usual. When buying groceries one day with a budget of $15 for the extra food, she buys bananas that have 90 calories each and chocolate granola bars that have 150 calories each. The bananas cost $0.35 each and the granola bars cost $2.50 each.

ⓐ Write a system of inequalities to model this situation.

ⓑ Graph the system.

ⓒ Could she buy 5 bananas and 6 granola bars?

ⓓ Could she buy 3 bananas and 4 granola bars?

323. Mark is attempting to build muscle mass and so he needs to eat more than an additional 80 grams of protein a day. A bottle of protein water costs $3.20 and a protein bar costs $1.75. The protein water supplies 27 grams of protein and the bar supplies 16 gram. If he has $10 dollars to spend

ⓐ Write a system of inequalities to model this situation.

ⓑ Graph the system.

ⓒ Could he buy 3 bottles of protein water and 1 protein bar?

ⓓ Could he buy no bottles of protein water and 5 protein bars?

324. Jocelyn desires to increase both her protein consumption and caloric intake. She desires to have at least 35 more grams of protein each day and no more than an additional 200 calories daily. An ounce of cheddar cheese has 7 grams of protein and 110 calories. An ounce of parmesan cheese has 11 grams of protein and 22 calories.

ⓐ Write a system of inequalities to model this situation.

ⓑ Graph the system.

ⓒ Could she eat 1 ounce of cheddar cheese and 3 ounces of parmesan cheese?

ⓓ Could she eat 2 ounces of cheddar cheese and 1 ounce of parmesan cheese?

325. Mark is increasing his exercise routine by running and walking at least 4 miles each day. His goal is to burn a minimum of 1500 calories from this exercise. Walking burns 270 calories/mile and running burns 650 calories.

ⓐ Write a system of inequalities to model this situation.

ⓑ Graph the system.

ⓒ Could he meet his goal by walking 3 miles and running 1 mile?

ⓓ Could he his goal by walking 2 miles and running 2 mile

Writing Exercises

326. Graph the inequality $x - y \geq 3$. How do you know which side of the line $x - y = 3$ should be shaded?

327. Graph the system $\begin{cases} x + 2y \leq 6 \\ y \geq -\frac{1}{2}x - 4 \end{cases}$. What does the solution mean?

Self Check

ⓐ *After completing the exercises, use this checklist to evaluate your mastery of the objectives of this section.*

I can...	Confidently	With some help	No-I don't get it!
determine whether an ordered pair is a solution of a system of linear inequalities.			
solve a system of linear inequalities by graphing.			
solve applications of systems of inequalities.			

ⓑ *What does this checklist tell you about your mastery of this section? What steps will you take to improve?*

CHAPTER 4 REVIEW

KEY TERMS

break-even point The point at which the revenue equals the costs is the break-even point; $C(x) = R(x)$.

coincident lines Coincident lines have the same slope and same y-intercept.

complementary angles Two angles are complementary if the sum of the measures of their angles is 90 degrees.

consistent and inconsistent systems Consistent system of equations is a system of equations with at least one solution; inconsistent system of equations is a system of equations with no solution.

cost function The cost function is the cost to manufacture each unit times x, the number of units manufactured, plus the fixed costs; $C(x) =$ (cost per unit)x + fixed costs.

determinant Each square matrix has a real number associated with it called its determinant.

matrix A matrix is a rectangular array of numbers arranged in rows and columns.

minor of an entry in a 3×3 determinant The minor of an entry in a 3×3 determinant is the 2×2 determinant found by eliminating the row and column in the 3×3 determinant that contains the entry.

revenue The revenue is the selling price of each unit times x, the number of units sold; $R(x) =$ (selling price per unit)x.

row-echelon form A matrix is in row-echelon form when to the left of the vertical line, each entry on the diagonal is a 1 and all entries below the diagonal are zeros.

solutions of a system of equations Solutions of a system of equations are the values of the variables that make *all* the equations true; solution is represented by an ordered pair (x, y).

solutions of a system of linear equations with three variables The solutions of a system of equations are the values of the variables that make all the equations true; a solution is represented by an ordered triple (x, y, z).

square matrix A square matrix is a matrix with the same number of rows and columns.

supplementary angles Two angles are supplementary if the sum of the measures of their angles is 180 degrees.

system of linear equations When two or more linear equations are grouped together, they form a system of linear equations.

system of linear inequalities Two or more linear inequalities grouped together form a system of linear inequalities.

KEY CONCEPTS

4.1 Solve Systems of Linear Equations with Two Variables

- **How to solve a system of linear equations by graphing.**

  Step 1. Graph the first equation.

  Step 2. Graph the second equation on the same rectangular coordinate system.

  Step 3. Determine whether the lines intersect, are parallel, or are the same line.

  Step 4. Identify the solution to the system.
  If the lines intersect, identify the point of intersection. This is the solution to the system.
  If the lines are parallel, the system has no solution.
  If the lines are the same, the system has an infinite number of solutions.

  Step 5. Check the solution in both equations.

- **How to solve a system of equations by substitution.**

  Step 1. Solve one of the equations for either variable.

  Step 2. Substitute the expression from Step 1 into the other equation.

  Step 3. Solve the resulting equation.

  Step 4. Substitute the solution in Step 3 into either of the original equations to find the other variable.

  Step 5. Write the solution as an ordered pair.

  Step 6. Check that the ordered pair is a solution to **both** original equations.

- **How to solve a system of equations by elimination.**

  Step 1. Write both equations in standard form. If any coefficients are fractions, clear them.

Step 2. Make the coefficients of one variable opposites.
Decide which variable you will eliminate.
Multiply one or both equations so that the coefficients of that variable are opposites.

Step 3. Add the equations resulting from Step 2 to eliminate one variable.

Step 4. Solve for the remaining variable.

Step 5. Substitute the solution from Step 4 into one of the original equations. Then solve for the other variable.

Step 6. Write the solution as an ordered pair.

Step 7. Check that the ordered pair is a solution to **both** original equations.

Choose the Most Convenient Method to Solve a System of Linear Equations

Graphing	Substitution	Elimination
Use when you need a picture of the situation.	Use when one equation is already solved or can be easily solved for one variable.	Use when the equations are in standard form.

4.2 Solve Applications with Systems of Equations

- **How To Solve Applications with Systems of Equations**

Step 1. **Read** the problem. Make sure all the words and ideas are understood.

Step 2. **Identify** what we are looking for.

Step 3. **Name** what we are looking for. Choose variables to represent those quantities.

Step 4. **Translate** into a system of equations.

Step 5. **Solve** the system of equations using good algebra techniques.

Step 6. **Check** the answer in the problem and make sure it makes sense.

Step 7. **Answer** the question with a complete sentence.

4.3 Solve Mixture Applications with Systems of Equations

- **Cost function:** The cost function is the cost to manufacture each unit times x, the number of units manufactured, plus the fixed costs.

$$C(x) = (\text{cost per unit}) \cdot x + \text{fixed costs}$$

- **Revenue:** The revenue function is the selling price of each unit times x, the number of units sold.

$$R(x) = (\text{selling price per unit}) \cdot x$$

- **Break-even point:** The break-even point is when the revenue equals the costs.

$$C(x) = R(x)$$

4.4 Solve Systems of Equations with Three Variables

- **Linear Equation in Three Variables:** A linear equation with three variables, where a, b, c, and d are real numbers and a, b, and c are not all 0, is of the form

$$ax + by + cz = d$$

Every solution to the equation is an ordered triple, (x, y, z) that makes the equation true.

- **How to solve a system of linear equations with three variables.**

Step 1. Write the equations in standard form
If any coefficients are fractions, clear them.

Step 2. Eliminate the same variable from two equations.
Decide which variable you will eliminate.
Work with a pair of equations to eliminate the chosen variable.
Multiply one or both equations so that the coefficients of that variable are opposites.
Add the equations resulting from Step 2 to eliminate one variable

Step 3. Repeat Step 2 using two other equations and eliminate the same variable as in Step 2.

Step 4. The two new equations form a system of two equations with two variables. Solve this system.

Step 5. Use the values of the two variables found in Step 4 to find the third variable.

Step 6. Write the solution as an ordered triple.

Step 7. Check that the ordered triple is a solution to **all three** original equations.

4.5 Solve Systems of Equations Using Matrices

- **Matrix:** A matrix is a rectangular array of numbers arranged in rows and columns. A matrix with m rows and n columns has *order* $m \times n$. The matrix on the left below has 2 rows and 3 columns and so it has order 2×3. We say it is a 2 by 3 matrix.

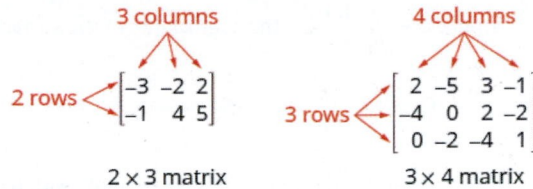

Each number in the matrix is called an *element* or *entry* in the matrix.

- **Row Operations:** In a matrix, the following operations can be performed on any row and the resulting matrix will be equivalent to the original matrix.

    ◦ Interchange any two rows

    ◦ Multiply a row by any real number except 0

    ◦ Add a nonzero multiple of one row to another row

- **Row-Echelon Form:** For a consistent and independent system of equations, its augmented matrix is in row-echelon form when to the left of the vertical line, each entry on the diagonal is a 1 and all entries below the diagonal are zeros.

$$\begin{bmatrix} 1 & a & | & b \\ 0 & 1 & | & c \end{bmatrix} \quad \begin{bmatrix} 1 & a & b & | & d \\ 0 & 1 & c & | & e \\ 0 & 0 & 1 & | & f \end{bmatrix} \quad a, b, c, d, e, f \text{ are real numbers}$$

- **How to solve a system of equations using matrices.**

    Step 1. Write the augmented matrix for the system of equations.

    Step 2. Using row operations get the entry in row 1, column 1 to be 1.

    Step 3. Using row operations, get zeros in column 1 below the 1.

    Step 4. Using row operations, get the entry in row 2, column 2 to be 1.

    Step 5. Continue the process until the matrix is in row-echelon form.

    Step 6. Write the corresponding system of equations.

    Step 7. Use substitution to find the remaining variables.

    Step 8. Write the solution as an ordered pair or triple.

    Step 9. Check that the solution makes the original equations true.

4.6 Solve Systems of Equations Using Determinants

- **Determinant:** The determinant of any square matrix $\begin{bmatrix} a & b \\ c & d \end{bmatrix}$, where a, b, c, and d are real numbers, is

$$\begin{vmatrix} a & b \\ c & d \end{vmatrix} = ad - bc$$

- **Expanding by Minors along the First Row to Evaluate a 3 × 3 Determinant:** To evaluate a 3×3 determinant by expanding by minors along the first row, the following pattern:

$$\begin{vmatrix} a_1 & b_1 & c_1 \\ a_2 & b_2 & c_2 \\ a_3 & b_3 & c_3 \end{vmatrix} = a_1 \underbrace{\begin{vmatrix} b_2 & c_2 \\ b_3 & c_3 \end{vmatrix}}_{\text{minor of } a_1} - b_1 \underbrace{\begin{vmatrix} a_2 & c_2 \\ a_3 & c_3 \end{vmatrix}}_{\text{minor of } b_1} + c_1 \underbrace{\begin{vmatrix} a_2 & b_2 \\ a_3 & b_3 \end{vmatrix}}_{\text{minor of } c_1}$$

- **Sign Pattern:** When expanding by minors using a row or column, the sign of the terms in the expansion follow the following pattern.

$$\begin{vmatrix} + & - & + \\ - & + & - \\ + & - & + \end{vmatrix}$$

- **Cramer's Rule:** For the system of equations $\begin{cases} a_1 x + b_1 y = k_1 \\ a_2 x + b_2 y = k_2 \end{cases}$, the solution (x, y) can be determined by

$$x = \frac{D_x}{D} \text{ and } y = \frac{D_y}{D}$$

where $D = \begin{vmatrix} a_1 & b_1 \\ a_2 & b_2 \end{vmatrix}$ use the coefficients of the variables.

$D_x = \begin{vmatrix} k_1 & b_1 \\ k_2 & b_2 \end{vmatrix}$ replace the x coefficients with the constants.

$D_y = \begin{vmatrix} a_1 & k_1 \\ a_2 & k_2 \end{vmatrix}$ replace the y coefficients with the constants.

Notice that to form the determinant D, we use take the coefficients of the variables.

- **How to solve a system of two equations using Cramer's rule.**

Step 1. Evaluate the determinant D, using the coefficients of the variables.

Step 2. Evaluate the determinant D_x. Use the constants in place of the x coefficients.

Step 3. Evaluate the determinant D_y. Use the constants in place of the y coefficients.

Step 4. Find x and y. $x = \dfrac{D_x}{D}, \quad y = \dfrac{D_y}{D}$.

Step 5. Write the solution as an ordered pair.

Step 6. Check that the ordered pair is a solution to **both** original equations.

Step 7. **Dependent and Inconsistent Systems of Equations:** For any system of equations, where the **value of the determinant** $D = 0$,

Value of determinants	Type of system	Solution
$D = 0$ and D_x, D_y and D_z are all zero	consistent and dependent	infinitely many solutions
$D = 0$ and D_x, D_y and D_z are not all zero	inconsistent	no solution

Step 8. **Test for Collinear Points:** Three points (x_1, y_1), (x_2, y_2), and (x_3, y_3) are collinear if and only if

$$\begin{vmatrix} x_1 & y_1 & 1 \\ x_2 & y_2 & 1 \\ x_3 & y_3 & 1 \end{vmatrix} = 0$$

4.7 Graphing Systems of Linear Inequalities

- **Solutions of a System of Linear Inequalities:** Solutions of a system of linear inequalities are the values of the variables that make all the inequalities true. The solution of a system of linear inequalities is shown as a shaded region in the x, y coordinate system that includes all the points whose ordered pairs make the inequalities true.

- **How to solve a system of linear inequalities by graphing.**

Step 1. Graph the first inequality.
Graph the boundary line.
Shade in the side of the boundary line where the inequality is true.

Step 2. On the same grid, graph the second inequality.
Graph the boundary line.
Shade in the side of that boundary line where the inequality is true.

Step 3. The solution is the region where the shading overlaps.

Step 4. Check by choosing a test point.

REVIEW EXERCISES

4.1 Solve Systems of Linear Equations with Two Variables

Determine Whether an Ordered Pair is a Solution of a System of Equations.

In the following exercises, determine if the following points are solutions to the given system of equations.

328. $\begin{cases} x + 3y = -9 \\ 2x - 4y = 12 \end{cases}$

ⓐ $(-3, -2)$

ⓑ $(0, -3)$

329. $\begin{cases} x + y = 8 \\ y = x - 4 \end{cases}$

ⓐ $(6, 2)$

ⓑ $(9, -1)$

Solve a System of Linear Equations by Graphing

In the following exercises, solve the following systems of equations by graphing.

330. $\begin{cases} 3x + y = 6 \\ x + 3y = -6 \end{cases}$

331. $\begin{cases} x + 4y = -1 \\ x = 3 \end{cases}$

332. $\begin{cases} 2x - y = 5 \\ 4x - 2y = 10 \end{cases}$

333. $\begin{cases} -x + 2y = 4 \\ y = \frac{1}{2}x - 3 \end{cases}$

In the following exercises, without graphing determine the number of solutions and then classify the system of equations.

334. $\begin{cases} y = \frac{2}{5}x + 2 \\ -2x + 5y = 10 \end{cases}$

335. $\begin{cases} 3x + 2y = 6 \\ y = -3x + 4 \end{cases}$

336. $\begin{cases} 5x - 4y = 0 \\ y = \frac{5}{4}x - 5 \end{cases}$

Solve a System of Equations by Substitution

In the following exercises, solve the systems of equations by substitution.

337. $\begin{cases} 3x - 2y = 2 \\ y = \frac{1}{2}x + 3 \end{cases}$

338. $\begin{cases} x - y = 0 \\ 2x + 5y = -14 \end{cases}$

339. $\begin{cases} y = -2x + 7 \\ y = \frac{2}{3}x - 1 \end{cases}$

340. $\begin{cases} y = -5x \\ 5x + y = 6 \end{cases}$

341. $\begin{cases} y = -\frac{1}{3}x + 2 \\ x + 3y = 6 \end{cases}$

Solve a System of Equations by Elimination

In the following exercises, solve the systems of equations by elimination

342. $\begin{cases} x + y = 12 \\ x - y = -10 \end{cases}$

343. $\begin{cases} 3x - 8y = 20 \\ x + 3y = 1 \end{cases}$

344. $\begin{cases} 9x + 4y = 2 \\ 5x + 3y = 5 \end{cases}$

345. $\begin{cases} \frac{1}{3}x - \frac{1}{2}y = 1 \\ \frac{3}{4}x - y = \frac{5}{2} \end{cases}$

346. $\begin{cases} -x + 3y = 8 \\ 2x - 6y = -20 \end{cases}$

Choose the Most Convenient Method to Solve a System of Linear Equations

In the following exercises, decide whether it would be more convenient to solve the system of equations by substitution or elimination.

347. $\begin{cases} 6x - 5y = 27 \\ 3x + 10y = -24 \end{cases}$

348. $\begin{cases} y = 3x - 9 \\ 4x - 5y = 23 \end{cases}$

4.2 Solve Applications with Systems of Equations

Solve Direct Translation Applications

In the following exercises, translate to a system of equations and solve.

349. Mollie wants to plant 200 bulbs in her garden, all irises and tulips. She wants to plant three times as many tulips as irises. How many irises and how many tulips should she plant?

350. Ashanti has been offered positions by two phone companies. The first company pays a salary of $22,000 plus a commission of $100 for each contract sold. The second pays a salary of $28,000 plus a commission of $25 for each contract sold. How many contract would need to be sold to make the total pay the same?

351. Leroy spent 20 minutes jogging and 40 minutes cycling and burned 600 calories. The next day, Leroy swapped times, doing 40 minutes of jogging and 20 minutes of cycling and burned the same number of calories. How many calories were burned for each minute of jogging and how many for each minute of cycling?

352. Troy and Lisa were shopping for school supplies. Each purchased different quantities of the same notebook and calculator. Troy bought four notebooks and five calculators for $116. Lisa bought two notebooks and three calculators for $68. Find the cost of each notebook and each thumb drive.

Solve Geometry Applications

In the following exercises, translate to a system of equations and solve.

353. The difference of two supplementary angles is 58 degrees. Find the measures of the angles.

354. Two angles are complementary. The measure of the larger angle is five more than four times the measure of the smaller angle. Find the measures of both angles.

355. The measure of one of the small angles of a right triangle is 15 less than twice the measure of the other small angle. Find the measure of both angles.

356. Becca is hanging a 28 foot floral garland on the two sides and top of a pergola to prepare for a wedding. The height is four feet less than the width. Find the height and width of the pergola.

357. The perimeter of a city rectangular park is 1428 feet. The length is 78 feet more than twice the width. Find the length and width of the park.

Solve Uniform Motion Applications

In the following exercises, translate to a system of equations and solve.

358. Sheila and Lenore were driving to their grandmother's house. Lenore left one hour after Sheila. Sheila drove at a rate of 45 mph, and Lenore drove at a rate of 60 mph. How long will it take for Lenore to catch up to Sheila?

359. Bob left home, riding his bike at a rate of 10 miles per hour to go to the lake. Cheryl, his wife, left 45 minutes ($\frac{3}{4}$ hour) later, driving her car at a rate of 25 miles per hour. How long will it take Cheryl to catch up to Bob?

360. Marcus can drive his boat 36 miles down the river in three hours but takes four hours to return upstream. Find the rate of the boat in still water and the rate of the current.

361. A passenger jet can fly 804 miles in 2 hours with a tailwind but only 776 miles in 2 hours into a headwind. Find the speed of the jet in still air and the speed of the wind.

4.3 Solve Mixture Applications with Systems of Equations

Solve Mixture Applications with Systems of Equations

For the following exercises, translate to a system of equations and solve.

362. Lynn paid a total of $2,780 for 261 tickets to the theater. Student tickets cost $10 and adult tickets cost $15. How many student tickets and how many adult tickets did Lynn buy?

363. Priam has dimes and pennies in a cup holder in his car. The total value of the coins is $4.21. The number of dimes is three less than four times the number of pennies. How many dimes and how many pennies are in the cup?

364. Yumi wants to make 12 cups of party mix using candies and nuts. Her budget requires the party mix to cost her $1.29 per cup. The candies are $2.49 per cup and the nuts are $0.69 per cup. How many cups of candies and how many cups of nuts should she use?

365. A scientist needs 70 liters of a 40% solution of alcohol. He has a 30% and a 60% solution available. How many liters of the 30% and how many liters of the 60% solutions should he mix to make the 40% solution?

Solve Interest Applications

For the following exercises, translate to a system of equations and solve.

366. Jack has $12,000 to invest and wants to earn 7.5% interest per year. He will put some of the money into a savings account that earns 4% per year and the rest into CD account that earns 9% per year. How much money should he put into each account?

367. When she graduates college, Linda will owe $43,000 in student loans. The interest rate on the federal loans is 4.5% and the rate on the private bank loans is 2%. The total interest she owes for one year was $1,585. What is the amount of each loan?

4.4 Solve Systems of Equations with Three Variables

Solve Systems of Equations with Three Variables

In the following exercises, determine whether the ordered triple is a solution to the system.

368. $\begin{cases} 3x - 4y - 3z = 2 \\ 2x - 6y + z = 3 \\ 2x + 3y - 2z = 3 \end{cases}$

ⓐ $(2, 3, -1)$

ⓑ $(3, 1, 3)$

369. $\begin{cases} y = \frac{2}{3}x - 2 \\ x + 3y - z = 15 \\ x - 3y + z = -2 \end{cases}$

ⓐ $\left(-6, 5, \frac{1}{2}\right)$

ⓑ $\left(5, \frac{4}{3}, -3\right)$

Solve a System of Linear Equations with Three Variables

In the following exercises, solve the system of equations.

370. $\begin{cases} 3x - 5y + 4z = 5 \\ 5x + 2y + z = 0 \\ 2x + 3y - 2z = 3 \end{cases}$

371. $\begin{cases} x + \frac{5}{2}y + z = -2 \\ 2x + 2y + \frac{1}{2}z = -4 \\ \frac{1}{3}x - y - z = 1 \end{cases}$

372. $\begin{cases} 5x + 3y = -6 \\ 2y + 3z = -1 \\ 7x + z = 1 \end{cases}$

373. $\begin{cases} 2x + 3y + z = 12 \\ x + y + z = 9 \\ 3x + 4y + 2z = 20 \end{cases}$

374. $\begin{cases} -x - 3y + 2z = 14 \\ -x + 2y - 3z = -4 \\ 3x + y - 2z = 6 \end{cases}$

Solve Applications using Systems of Linear Equations with Three Variables

375. After attending a major league baseball game, the patrons often purchase souvenirs. If a family purchases 4 t-shirts, a cap and 1 stuffed animal their total is $135. A couple buys 2 t-shirts, a cap and 3 stuffed animals for their nieces and spends $115. Another couple buys 2 t-shirts, a cap and 1 stuffed animal and their total is $85. What is the cost of each item?

4.5 Solve Systems of Equations Using Matrices

Write the Augmented Matrix for a System of Equations.

Write each system of linear equations as an augmented matrix.

376. $\begin{cases} 3x - y = -1 \\ -2x + 2y = 5 \end{cases}$

377. $\begin{cases} 4x + 3y = -2 \\ x - 2y - 3z = 7 \\ 2x - y + 2z = -6 \end{cases}$

Write the system of equations that that corresponds to the augmented matrix.

378. $\begin{bmatrix} 2 & -4 & | & -2 \\ 3 & -3 & | & -1 \end{bmatrix}$

379. $\begin{bmatrix} 1 & 0 & -3 & | & -1 \\ 1 & -2 & 0 & | & -2 \\ 0 & -1 & 2 & | & 3 \end{bmatrix}$

In the following exercises, perform the indicated operations on the augmented matrices.

380. $\begin{bmatrix} 4 & -6 & | & -3 \\ 3 & 2 & | & 1 \end{bmatrix}$

ⓐ Interchange rows 2 and 1.

ⓑ Multiply row 1 by 4.

ⓒ Multiply row 2 by 3 and add to row 1.

381. $\begin{bmatrix} 1 & -3 & -2 & | & 4 \\ 2 & 2 & -1 & | & -3 \\ 4 & -2 & -3 & | & -1 \end{bmatrix}$

ⓐ Interchange rows 2 and 3.

ⓑ Multiply row 1 by 2.

ⓒ Multiply row 3 by −2 and add to row 2.

Solve Systems of Equations Using Matrices

In the following exercises, solve each system of equations using a matrix.

382. $\begin{cases} 4x + y = 6 \\ x - y = 4 \end{cases}$

383. $\begin{cases} 2x - y + 3z = -3 \\ -x + 2y - z = 10 \\ x + y + z = 5 \end{cases}$

384. $\begin{cases} 2y + 3z = -1 \\ 5x + 3y = -6 \\ 7x + z = 1 \end{cases}$

385. $\begin{cases} x + 2y - 3z = -1 \\ x - 3y + z = 1 \\ 2x - y - 2z = 2 \end{cases}$

386. $\begin{cases} x + y - 3z = -1 \\ y - z = 0 \\ -x + 2y = 1 \end{cases}$

4.6 Solve Systems of Equations Using Determinants

Evaluate the Determinant of a 2 × 2 Matrix

In the following exercise, evaluate the determinate of the square matrix.

387. $\begin{bmatrix} 8 & -4 \\ 5 & -3 \end{bmatrix}$

Evaluate the Determinant of a 3 × 3 Matrix

In the following exercise, find and then evaluate the indicated minors.

388. $\begin{vmatrix} -1 & -3 & 2 \\ 4 & -2 & -1 \\ -2 & 0 & -3 \end{vmatrix}$; Find the

minor ⓐ a_1 ⓑ b_1 ⓒ c_2

In the following exercise, evaluate each determinant by expanding by minors along the first row.

389. $\begin{vmatrix} -2 & -3 & -4 \\ 5 & -6 & 7 \\ -1 & 2 & 0 \end{vmatrix}$

In the following exercise, evaluate each determinant by expanding by minors.

390. $\begin{vmatrix} 3 & 5 & 4 \\ -1 & 3 & 0 \\ -2 & 6 & 1 \end{vmatrix}$

Use Cramer's Rule to Solve Systems of Equations

In the following exercises, solve each system of equations using Cramer's rule

391. $\begin{cases} x - 3y = -9 \\ 2x + 5y = 4 \end{cases}$

392. $\begin{cases} 4x - 3y + z = 7 \\ 2x - 5y - 4z = 3 \\ 3x - 2y - 2z = -7 \end{cases}$

393. $\begin{cases} 2x + 5y = 4 \\ 3y - z = 3 \\ 4x + 3z = -3 \end{cases}$

394. $\begin{cases} x + y - 3z = -1 \\ y - z = 0 \\ -x + 2y = 1 \end{cases}$

395. $\begin{cases} 3x + 4y - 3z = -2 \\ 2x + 3y - z = -12 \\ x + y - 2z = 6 \end{cases}$

Solve Applications Using Determinants

In the following exercises, determine whether the given points are collinear.

396. $(0, 2)$, $(-1, -1)$, and $(-2, 4)$

4.7 Graphing Systems of Linear Inequalities

Determine Whether an Ordered Pair is a Solution of a System of Linear Inequalities

In the following exercises, determine whether each ordered pair is a solution to the system.

397. $\begin{cases} 4x + y > 6 \\ 3x - y \le 12 \end{cases}$

ⓐ $(2, -1)$

ⓑ $(3, -2)$

398. $\begin{cases} y > \frac{1}{3}x + 2 \\ x - \frac{1}{4}y \le 10 \end{cases}$

ⓐ $(6, 5)$

ⓑ $(15, 8)$

Solve a System of Linear Inequalities by Graphing

In the following exercises, solve each system by graphing.

399. $\begin{cases} y < 3x + 1 \\ y \ge -x - 2 \end{cases}$

400. $\begin{cases} x - y > -1 \\ y < \frac{1}{3}x - 2 \end{cases}$

401. $\begin{cases} 2x - 3y < 6 \\ 3x + 4y \ge 12 \end{cases}$

402. $\begin{cases} y \le -\frac{3}{4}x + 1 \\ x \ge -5 \end{cases}$

403. $\begin{cases} x + 3y < 5 \\ y \ge -\frac{1}{3}x + 6 \end{cases}$

404. $\begin{cases} y \ge 2x - 5 \\ -6x + 3y > -4 \end{cases}$

Solve Applications of Systems of Inequalities

In the following exercises, translate to a system of inequalities and solve.

405. Roxana makes bracelets and necklaces and sells them at the farmers' market. She sells the bracelets for $12 each and the necklaces for $18 each. At the market next weekend she will have room to display no more than 40 pieces, and she needs to sell at least $500 worth in order to earn a profit.

ⓐ Write a system of inequalities to model this situation.

ⓑ Graph the system.

ⓒ Should she display 26 bracelets and 14 necklaces?

ⓓ Should she display 39 bracelets and 1 necklace?

406. Annie has a budget of $600 to purchase paperback books and hardcover books for her classroom. She wants the number of hardcover to be at least 5 more than three times the number of paperback books. Paperback books cost $4 each and hardcover books cost $15 each.

ⓐ Write a system of inequalities to model this situation.

ⓑ Graph the system.

ⓒ Can she buy 8 paperback books and 40 hardcover books?

ⓓ Can she buy 10 paperback books and 37 hardcover books?

PRACTICE TEST

In the following exercises, solve the following systems by graphing.

407. $\begin{cases} x - y = 5 \\ x + 2y = -4 \end{cases}$

408. $\begin{cases} x - y > -2 \\ y \le 3x + 1 \end{cases}$

In the following exercises, solve each system of equations. Use either substitution or elimination.

409. $\begin{cases} x + 4y = 6 \\ -2x + y = -3 \end{cases}$

410. $\begin{cases} -3x + 4y = 25 \\ x - 5y = -23 \end{cases}$

411. $\begin{cases} x + y - z = -1 \\ 2x - y + 2z = 8 \\ -3x + 2y + z = -9 \end{cases}$

Solve the system of equations using a matrix.

412. $\begin{cases} 2x + y = 7 \\ x - 2y = 6 \end{cases}$

413. $\begin{cases} -3x + y + z = -4 \\ -x + 2y - 2z = 1 \\ 2x - y - z = -1 \end{cases}$

Solve using Cramer's rule.

414. $\begin{cases} 3x + y = -3 \\ 2x + 3y = 6 \end{cases}$

415. Evaluate the determinant by expanding by minors:
$$\begin{vmatrix} 3 & -2 & -2 \\ 2 & -1 & 4 \\ -1 & 0 & -3 \end{vmatrix}.$$

In the following exercises, translate to a system of equations and solve.

416. Greg is paddling his canoe upstream, against the current, to a fishing spot 10 miles away. If he paddles upstream for 2.5 hours and his return trip takes 1.25 hours, find the speed of the current and his paddling speed in still water.

417. A pharmacist needs 20 liters of a 2% saline solution. He has a 1% and a 5% solution available. How many liters of the 1% and how many liters of the 5% solutions should she mix to make the 2% solution?

418. Arnold invested $64,000, some at 5.5% interest and the rest at 9%. How much did he invest at each rate if he received $4,500 in interest in one year?

419. The church youth group is selling snacks to raise money to attend their convention. Amy sold 2 pounds of candy, 3 boxes of cookies and 1 can of popcorn for a total sales of $65. Brian sold 4 pounds of candy, 6 boxes of cookies and 3 cans of popcorn for a total sales of $140. Paulina sold 8 pounds of candy, 8 boxes of cookies and 5 can of popcorn for a total sales of $250. What is the cost of each item?

420. The manufacturer of a granola bar spends $1.20 to make each bar and sells them for $2. The manufacturer also has fixed costs each month of $8,000.

ⓐ Find the cost function *C* when *x* granola bars are manufactured

ⓑ Find the revenue function *R* when *x* granola bars are sold.

ⓒ Show the break-even point by graphing both the Revenue and Cost functions on the same grid.

ⓓ Find the break-even point. Interpret what the break-even point means.

421. Translate to a system of inequalities and solve.

Andi wants to spend no more than $50 on Halloween treats. She wants to buy candy bars that cost $1 each and lollipops that cost $0.50 each, and she wants the number of lollipops to be at least three times the number of candy bars.

ⓐ Write a system of inequalities to model this situation.

ⓑ Graph the system.

ⓒ Can she buy 20 candy bars and 40 lollipops?

Figure 5.1 There are many different kinds of coins in circulation, but a new type of coin exists only in the virtual world. It is the bitcoin.

Chapter Outline

 Introduction

You may have coins and paper money in your wallet, but you may soon want to acquire a type of currency called bitcoins. They exist only in a digital wallet on your computer. You can use bitcoins to pay for goods at some companies, or save them as an investment. Although the future of bitcoins is uncertain, investment brokers are beginning to investigate ways to make business predictions using this digital currency. Understanding how bitcoins are created and obtained requires an understanding of a type of function known as a polynomial function. In this chapter you will investigate polynomials and polynomial functions and learn how to perform mathematical operations on them.

5.1 Add and Subtract Polynomials

Learning Objectives

By the end of this section, you will be able to:

> Determine the degree of polynomials
> Add and subtract polynomials
> Evaluate a polynomial function for a given value
> Add and subtract polynomial functions

Be Prepared!

Before you get started, take this readiness quiz.

1. Simplify: $3x^2 + 3x + 1 + 8x^2 + 5x + 5$.
   If you missed this problem, review Example 1.7.

2. Subtract: $(5n + 8) - (2n - 1)$.
   If you missed this problem, review Example 1.5.

3. Evaluate: $4xy^2$ when $x = -2$ and $y = 5$.
   If you missed this problem, review Example 1.21.

Determine the Degree of Polynomials

We have learned that a *term* is a constant or the product of a constant and one or more variables. A **monomial** is an algebraic expression with one term. When it is of the form ax^m, where a is a constant and m is a whole number, it is called a monomial in one variable. Some examples of monomial in one variable are. Monomials can also have more than one variable such as and $-4a^2b^3c^2$.

> ### Monomial
>
> A **monomial** is an algebraic expression with one term.
>
> A monomial in one variable is a term of the form ax^m, where a is a constant and m is a whole number.

A monomial, or two or more monomials combined by addition or subtraction, is a **polynomial**. Some polynomials have special names, based on the number of terms. A monomial is a polynomial with exactly one term. A binomial has exactly two terms, and a **trinomial** has exactly three terms. There are no special names for polynomials with more than three terms.

> ### Polynomials
>
> **polynomial**—A monomial, or two or more algebraic terms combined by addition or subtraction is a polynomial.
>
> **monomial**—A polynomial with exactly one term is called a monomial.
>
> **binomial**—A polynomial with exactly two terms is called a binomial.
>
> **trinomial**—A polynomial with exactly three terms is called a trinomial.

Here are some examples of polynomials.

Polynomial	$y+1$	$4a^2 - 7ab + 2b^2$	$4x^4 + x^3 + 8x^2 - 9x + 1$	
Monomial	14	$8y^2$	$-9x^3y^5$	$-13a^3b^2c$
Binomial	$a+7b$	$4x^2 - y^2$	$y^2 - 16$	$3p^3q - 9p^2q$
Trinomial	$x^2 - 7x + 12$	$9m^2 + 2mn - 8n^2$	$6k^4 - k^3 + 8k$	$z^4 + 3z^2 - 1$

Notice that every monomial, binomial, and trinomial is also a polynomial. They are just special members of the "family" of polynomials and so they have special names. We use the words *monomial*, *binomial*, and *trinomial* when referring to these special polynomials and just call all the rest *polynomials*.

The **degree of a polynomial** and the degree of its terms are determined by the exponents of the variable.

A monomial that has no variable, just a constant, is a special case. The **degree of a constant** is 0.

> ### Degree of a Polynomial
>
> The **degree of a term** is the sum of the exponents of its variables.
>
> The **degree of a constant** is 0.
>
> The **degree of a polynomial** is the highest degree of all its terms.

Let's see how this works by looking at several polynomials. We'll take it step by step, starting with monomials, and then progressing to polynomials with more terms.

Let's start by looking at a monomial. The monomial $8ab^2$ has two variables a and b. To find the degree we need to find the sum of the exponents. The variable a doesn't have an exponent written, but remember that means the exponent is 1. The exponent of b is 2. The sum of the exponents, $1+2$, is 3 so the degree is 3.

$$8\ a\ b^2$$

exponents   1 2

degree of monomial   3 $(1+2)$

Here are some additional examples.

Monomials	14	$8ab^2$	$-9x^3y^5$	$-13a$
Degree	0	3	8	1

Binomial	$h + 7$	$7b^2 - 3b$	$x^2y^2 - 25$	$4n^3 - 8n^2$
Degree of each term	1 0	2 1	4 0	3 2
Degree of polynomial	1	2	4	3

Trinomial	$x^2 - 12x + 27$	$9a^2 + 6ab + b^2$	$6m^4 - m^3n^2 + 8mn^5$	$z^4 + 3z^2 - 1$
Degree of each term	2 1 0	2 2 2	4 5 6	4 2 0
Degree of polynomial	2	2	6	4

Polynomial	$y - 1$	$3y^2 - 2y - 5$	$4x^4 + x^3 + 8x^2 - 9x + 1$
Degree of each term	1 0	2 1 0	4 3 2 1 0
Degree of polynomial	1	2	4

Working with polynomials is easier when you list the terms in descending order of degrees. When a polynomial is written this way, it is said to be in **standard form of a polynomial**. Get in the habit of writing the term with the highest degree first.

EXAMPLE 5.1

Determine whether each polynomial is a monomial, binomial, trinomial, or other polynomial. Then, find the degree of each polynomial.

ⓐ $7y^2 - 5y + 3$ ⓑ $-2a^4b^2$ ⓒ $3x^5 - 4x^3 - 6x^2 + x - 8$ ⓓ $2y - 8xy^3$ ⓔ 15

⊘ Solution

	Polynomial	Number of terms	Type	Degree of terms	Degree of polynomial
ⓐ	$7y^2 - 5y + 3$	3	Trinomial	2, 1, 0	2
ⓑ	$-2a^4b^2$	1	Monomial	4, 2	6
ⓒ	$3x^5 - 4x^3 - 6x^2 + x - 8$	5	Polynomial	5, 3, 2, 1, 0	5
ⓓ	$2y - 8xy^3$	2	Binomial	1, 4	4
ⓔ	15	1	Monomial	0	0

> **TRY IT :: 5.1**
>
> Determine whether each polynomial is a monomial, binomial, trinomial, or other polynomial. Then, find the degree of each polynomial.
>
> ⓐ -5 ⓑ $8y^3 - 7y^2 - y - 3$ ⓒ $-3x^2y - 5xy + 9xy^3$ ⓓ $81m^2 - 4n^2$ ⓔ $-3x^6y^3z$

> **TRY IT : :** 5.2

Determine whether each polynomial is a monomial, binomial, trinomial, or other polynomial. Then, find the degree of each polynomial.

   ⓐ $64k^3 - 8$ ⓑ $9m^3 + 4m^2 - 2$ ⓒ $\frac{5}{6}$ ⓓ $8a^4 - 7a^3 b - 6a^2 b^2 - 4ab^3 + 7b^4$ ⓔ $-p^4 q^3$

Add and Subtract Polynomials

We have learned how to simplify expressions by combining like terms. Remember, like terms must have the same variables with the same exponent. Since monomials are terms, adding and subtracting monomials is the same as combining like terms. If the monomials are like terms, we just combine them by adding or subtracting the coefficients.

EXAMPLE 5.2

Add or subtract: ⓐ $25y^2 + 15y^2$ ⓑ $16pq^3 - \left(-7pq^3\right)$.

✓ **Solution**

ⓐ

$$25y^2 + 15y^2$$

Combine like terms. $40y^2$

ⓑ

$$16pq^3 - \left(-7pq^3\right)$$

Combine like terms. $23pq^3$

> **TRY IT : :** 5.3   Add or subtract: ⓐ $12q^2 + 9q^2$ ⓑ $8mn^3 - \left(-5mn^3\right)$.

> **TRY IT : :** 5.4   Add or subtract: ⓐ $-15c^2 + 8c^2$ ⓑ $-15y^2 z^3 - \left(-5y^2 z^3\right)$.

Remember that like terms must have the same variables with the same exponents.

EXAMPLE 5.3

Simplify: ⓐ $a^2 + 7b^2 - 6a^2$ ⓑ $u^2 v + 5u^2 - 3v^2$.

✓ **Solution**

ⓐ

$$a^2 + 7b^2 - 6a^2$$

Combine like terms. $-5a^2 + 7b^2$

ⓑ

$$u^2 v + 5u^2 - 3v^2$$

There are no like terms to combine. $u^2 v + 5u^2 - 3v^2$
In this case, the polynomial is unchanged.

> **TRY IT : :** 5.5   Add: ⓐ $8y^2 + 3z^2 - 3y^2$ ⓑ $m^2 n^2 - 8m^2 + 4n^2$.

> **TRY IT : :** 5.6   Add: ⓐ $3m^2 + n^2 - 7m^2$ ⓑ $pq^2 - 6p - 5q^2$.

We can think of adding and subtracting polynomials as just adding and subtracting a series of monomials. Look for the like terms—those with the same variables and the same exponent. The Commutative Property allows us to rearrange the terms to put like terms together.

EXAMPLE 5.4

Find the sum: $\left(7y^2 - 2y + 9\right) + \left(4y^2 - 8y - 7\right)$.

✓ **Solution**

Identify like terms.

$$\left(\underline{7y^2} - \underline{2y} + 9\right) + \left(\underline{4y^2} - \underline{8y} - 7\right)$$

Rewrite without the parentheses, rearranging to get the like terms together.

$$\underline{7y^2 + 4y^2} - \underline{2y - 8y} + 9 - 7$$

Combine like terms.

$$11y^2 - 10y + 2$$

> **TRY IT :: 5.7**   Find the sum: $\left(7x^2 - 4x + 5\right) + \left(x^2 - 7x + 3\right)$.

> **TRY IT :: 5.8**   Find the sum: $\left(14y^2 + 6y - 4\right) + \left(3y^2 + 8y + 5\right)$.

Be careful with the signs as you distribute while subtracting the polynomials in the next example.

EXAMPLE 5.5

Find the difference: $\left(9w^2 - 7w + 5\right) - \left(2w^2 - 4\right)$.

✓ **Solution**

$$\left(9w^2 - 7w + 5\right) - \left(2w^2 - 4\right)$$

Distribute and identify like terms.

$$\underline{9w^2} - \underline{7w} + 5 - \underline{2w^2} + 4$$

Rearrange the terms.

$$\underline{9w^2 - 2w^2} - \underline{7w} + 5 + 4$$

Combine like terms.

$$7w^2 - 7w + 9$$

> **TRY IT :: 5.9**   Find the difference: $\left(8x^2 + 3x - 19\right) - \left(7x^2 - 14\right)$.

> **TRY IT :: 5.10**   Find the difference: $\left(9b^2 - 5b - 4\right) - \left(3b^2 - 5b - 7\right)$.

To subtract a from b, we write it as $b - a$, placing the b first.

EXAMPLE 5.6

Subtract $\left(p^2 + 10pq - 2q^2\right)$ from $\left(p^2 + q^2\right)$.

✓ **Solution**

$$\left(p^2 + q^2\right) - \left(p^2 + 10pq - 2q^2\right)$$

Distribute.

$$p^2 + q^2 - p^2 - 10pq + 2q^2$$

Rearrange the terms, to put like terms together.

$$p^2 - p^2 - 10pq + q^2 + 2q^2$$

Combine like terms.

$$-10pq + 3q^2$$

> **TRY IT ::** 5.11   Subtract $(a^2 + 5ab - 6b^2)$ from $(a^2 + b^2)$.

> **TRY IT ::** 5.12   Subtract $(m^2 - 7mn - 3n^2)$ from $(m^2 + n^2)$.

EXAMPLE 5.7

Find the sum: $(u^2 - 6uv + 5v^2) + (3u^2 + 2uv)$.

✓ **Solution**

$$(u^2 - 6uv + 5v^2) + (3u^2 + 2uv)$$

Distribute.

$$u^2 - 6uv + 5v^2 + 3u^2 + 2uv$$

Rearrange the terms to put like terms together. $u^2 + 3u^2 - 6uv + 2uv + 5v^2$

Combine like terms.

$$4u^2 - 4uv + 5v^2$$

> **TRY IT ::** 5.13   Find the sum: $(3x^2 - 4xy + 5y^2) + (2x^2 - xy)$.

> **TRY IT ::** 5.14   Find the sum: $(2x^2 - 3xy - 2y^2) + (5x^2 - 3xy)$.

When we add and subtract more than two polynomials, the process is the same.

EXAMPLE 5.8

Simplify: $(a^3 - a^2b) - (ab^2 + b^3) + (a^2b + ab^2)$.

✓ **Solution**

$$(a^3 - a^2b) - (ab^2 + b^3) + (a^2b + ab^2)$$

Distribute.

$$a^3 - a^2b - ab^2 - b^3 + a^2b + ab^2$$

Rewrite without the parentheses,
rearranging to get the like terms together. $a^3 - a^2b + a^2b - ab^2 + ab^2 - b^3$

Combine like terms.

$$a^3 - b^3$$

> **TRY IT ::** 5.15   Simplify: $(x^3 - x^2y) - (xy^2 + y^3) + (x^2y + xy^2)$.

> **TRY IT ::** 5.16   Simplify: $(p^3 - p^2q) + (pq^2 + q^3) - (p^2q + pq^2)$.

Evaluate a Polynomial Function for a Given Value

A **polynomial function** is a function defined by a polynomial. For example, $f(x) = x^2 + 5x + 6$ and $g(x) = 3x - 4$ are polynomial functions, because $x^2 + 5x + 6$ and $3x - 4$ are polynomials.

Polynomial Function

A **polynomial function** is a function whose range values are defined by a polynomial.

In **Graphs and Functions**, where we first introduced functions, we learned that evaluating a function means to find the value of $f(x)$ for a given value of x. To evaluate a polynomial function, we will substitute the given value for the variable and then simplify using the order of operations.

EXAMPLE 5.9

For the function $f(x) = 5x^2 - 8x + 4$ find: ⓐ $f(4)$ ⓑ $f(-2)$ ⓒ $f(0)$.

✓ **Solution**

	$f(x) = 5x^2 - 8x + 4$
To find $f(4)$, substitute 4 for x.	$f(4) = 5(4)^2 - 8(4) + 4$
Simplify the exponents.	$f(4) = 5 \cdot 16 - 8(4) + 4$
Multiply.	$f(4) = 80 - 32 + 4$
Simplify.	$f(4) = 52$

ⓑ

	$f(x) = 5x^2 - 8x + 4$
To find $f(-2)$, substitute –2 for x.	$f(-2) = 5(-2)^2 - 8(-2) + 4$
Simplify the exponents.	$f(-2) = 5 \cdot 4 - 8(-2) + 4$
Multiply.	$f(-2) = 20 + 16 + 4$
Simplify.	$f(-2) = 40$

ⓒ

	$f(x) = 5x^2 - 8x + 4$
To find $f(0)$, substitute 0 for x.	$f(0) = 5(0)^2 - 8(0) + 4$
Simplify the exponents.	$f(0) = 5 \cdot 0 - 8(0) + 4$
Multiply.	$f(0) = 0 + 0 + 4$
Simplify.	$f(0) = 4$

> **TRY IT :: 5.17** For the function $f(x) = 3x^2 + 2x - 15$, find ⓐ $f(3)$ ⓑ $f(-5)$ ⓒ $f(0)$.

> **TRY IT :: 5.18** For the function $g(x) = 5x^2 - x - 4$, find ⓐ $g(-2)$ ⓑ $g(-1)$ ⓒ $g(0)$.

The polynomial functions similar to the one in the next example are used in many fields to determine the height of an object at some time after it is projected into the air. The polynomial in the next function is used specifically for dropping something from 250 ft.

EXAMPLE 5.10

The polynomial function $h(t) = -16t^2 + 250$ gives the height of a ball t seconds after it is dropped from a 250-foot tall building. Find the height after $t = 2$ seconds.

✓ **Solution**

$$h(t) = -16t^2 + 250$$

To find $h(2)$, substitute $t = 2$. $h(2) = -16(2)^2 + 250$

Simplify. $h(2) = -16 \cdot 4 + 250$

Simplify. $h(2) = -64 + 250$

Simplify. $h(2) = 186$

After 2 seconds the height of the ball is 186 feet.

> **TRY IT : : 5.19**
>
> The polynomial function $h(t) = -16t^2 + 150$ gives the height of a stone t seconds after it is dropped from a 150-foot tall cliff. Find the height after $t = 0$ seconds (the initial height of the object).

> **TRY IT : : 5.20**
>
> The polynomial function $h(t) = -16t^2 + 175$ gives the height of a ball t seconds after it is dropped from a 175-foot tall bridge. Find the height after $t = 3$ seconds.

Add and Subtract Polynomial Functions

Just as polynomials can be added and subtracted, polynomial functions can also be added and subtracted.

Addition and Subtraction of Polynomial Functions

For functions $f(x)$ and $g(x)$,

$$(f + g)(x) = f(x) + g(x)$$
$$(f - g)(x) = f(x) - g(x)$$

EXAMPLE 5.11

For functions $f(x) = 3x^2 - 5x + 7$ and $g(x) = x^2 - 4x - 3$, find:

ⓐ $(f + g)(x)$ ⓑ $(f + g)(3)$ ⓒ $(f - g)(x)$ ⓓ $(f - g)(-2)$.

✓ **Solution**

ⓐ

$$(f + g)(x) = f(x) + g(x)$$

Substitute $f(x) = 3x^2 - 5x + 7$ and
$g(x) = x^2 - 4x - 3$. $(f + g)(x) = (3x^2 - 5x + 7) + (x^2 - 4x - 3)$

Rewrite without the parentheses. $(f + g)(x) = 3x^2 - 5x + 7 + x^2 - 4x - 3$

Put like terms together. $(f + g)(x) = 3x^2 + x^2 - 5x - 4x + 7 - 3$

Combine like terms. $(f + g)(x) = 4x^2 - 9x + 4$

ⓑ In part (a) we found $(f + g)(x)$ and now are asked to find $(f + g)(3)$.

$$(f + g)(x) = 4x^2 - 9x + 4$$

To find $(f + g)(3)$, substitute $x = 3$. $(f + g)(3) = 4(3)^2 - 9 \cdot 3 + 4$

$$(f + g)(3) = 4 \cdot 9 - 9 \cdot 3 + 4$$

$$(f + g)(3) = 36 - 27 + 4$$

Notice that we could have found $(f + g)(3)$ by first finding the values of $f(3)$ and $g(3)$ separately and then adding the results.

Find $f(3)$.	$f(x) = 3x^2 - 5x + 7$
	$f(3) = 3(3)^2 - 5(3) + 7$
	$f(3) = 19$
Find $g(3)$.	$g(x) = x^2 - 4x - 3$
	$g(3) = 3^2 - 4(3) - 3$
	$g(3) = -6$
Find $(f + g)(3)$.	$(f + g)(x) = f(x) + g(x)$
	$(f + g)(3) = f(3) + g(3)$
Substitute $f(3) = 19$ and $g(3) = -6$.	$(f + g)(3) = 19 + (-6)$
	$(f + g)(3) = 13$

ⓒ

	$(f - g)(x) = f(x) - g(x)$
Substitute $f(x) = 3x^2 - 5x + 7$ and $g(x) = x^2 - 4x - 3$.	$(f - g)(x) = (3x^2 - 5x + 7) - (x^2 - 4x - 3)$
Rewrite without the parentheses.	$(f - g)(x) = 3x^2 - 5x + 7 - x^2 + 4x + 3$
Put like terms together.	$(f - g)(x) = 3x^2 - x^2 - 5x + 4x + 7 + 3$
Combine like terms.	$(f - g)(x) = 2x^2 - x + 10$

ⓓ

	$(f - g)(x) = 2x^2 - x + 10$
To find $(f - g)(-2)$, substitute $x = -2$.	$(f - g)(-2) = 2(-2)^2 - (-2) + 10$
	$(f - g)(-2) = 2 \cdot 4 - (-2) + 10$
	$(f - g)(-2) = 20$

> **TRY IT : :** 5.21

For functions $f(x) = 2x^2 - 4x + 3$ and $g(x) = x^2 - 2x - 6$, find: ⓐ $(f + g)(x)$ ⓑ $(f + g)(3)$ ⓒ $(f - g)(x)$ ⓓ $(f - g)(-2)$.

> **TRY IT : :** 5.22

For functions $f(x) = 5x^2 - 4x - 1$ and $g(x) = x^2 + 3x + 8$, find ⓐ $(f + g)(x)$ ⓑ $(f + g)(3)$ ⓒ $(f - g)(x)$ ⓓ $(f - g)(-2)$.

▶ **MEDIA : :**

Access this online resource for additional instruction and practice with adding and subtracting polynomials.

- **Adding and Subtracting Polynomials (https://openstax.org/l/37AddSubtrPoly)**

5.1 EXERCISES

Practice Makes Perfect

Determine the Degree of Polynomials

In the following exercises, determine if the polynomial is a monomial, binomial, trinomial, or other polynomial.

1.
ⓐ $47x^5 - 17x^2 y^3 + y^2$
ⓑ $5c^3 + 11c^2 - c - 8$
ⓒ $\frac{5}{9}ab + \frac{1}{3}b$
ⓓ 4
ⓔ $4pq + 17$

2.
ⓐ $x^2 - y^2$
ⓑ $-13c^4$
ⓒ $a^2 + 2ab - 7b^2$
ⓓ $4x^2 y^2 - 3xy + 8$
ⓔ 19

3.
ⓐ $8y - 5x$
ⓑ $y^2 - 5yz - 6z^2$
ⓒ $y^3 - 8y^2 + 2y - 16$
ⓓ $81ab^4 - 24a^2 b^2 + 3b$
ⓔ -18

4.
ⓐ $11y^2$
ⓑ -73
ⓒ $6x^2 - 3xy + 4x - 2y + y^2$
ⓓ $4y^2 + 17z^2$
ⓔ $5c^3 + 11c^2 - c - 8$

5.
ⓐ $5a^2 + 12ab - 7b^2$
ⓑ $18xy^2 z$
ⓒ $5x + 2$
ⓓ $y^3 - 8y^2 + 2y - 16$
ⓔ -24

6.
ⓐ $9y^3 - 10y^2 + 2y - 6$
ⓑ $-12p^3 q$
ⓒ $a^2 + 9ab + 18b^2$
ⓓ $20x^2 y^2 - 10a^2 b^2 + 30$
ⓔ 17

7.
ⓐ $14s - 29t$
ⓑ $z^2 - 5z - 6$
ⓒ $y^3 - 8y^2 z + 2yz^2 - 16z^3$
ⓓ $23ab^2 - 14$
ⓔ -3

8.
ⓐ $15xy$
ⓑ 15
ⓒ $6x^2 - 3xy + 4x - 2y + y^2$
ⓓ $10p - 9q$
ⓔ $m^4 + 4m^3 + 6m^2 + 4m + 1$

Add and Subtract Polynomials

In the following exercises, add or subtract the monomials.

9.
ⓐ $7x^2 + 5x^2$
ⓑ $4a - 9a$

10.
ⓐ $4y^3 + 6y^3$
ⓑ $-y - 5y$

11.
ⓐ $-12w + 18w$
ⓑ $7x^2 y - (-12x^2 y)$

12.
ⓐ $-3m + 9m$
ⓑ $15yz^2 - (-8yz^2)$

13. $7x^2 + 5x^2 + 4a - 9a$

14. $4y^3 + 6y^3 - y - 5y$

15.
$-12w + 18w + 7x^2 y - (-12x^2 y)$

16. $-3m + 9m + 15yz^2 - (-8yz^2)$

17.
ⓐ $-5b - 17b$
ⓑ $3xy - (-8xy) + 5xy$

18.
ⓐ $-10x - 35x$
ⓑ $17mn^2 - (-9mn^2) + 3mn^2$

19.
ⓐ $12a + 5b - 22a$
ⓑ $pq^2 - 4p - 3q^2$

20.
ⓐ $14x - 3y - 13x$
ⓑ $a^2 b - 4a - 5ab^2$

21.
ⓐ $2a^2 + b^2 - 6a^2$
ⓑ $x^2y - 3x + 7xy^2$

22.
ⓐ $5u^2 + 4v^2 - 6u^2$
ⓑ $12a + 8b$

23.
ⓐ $xy^2 - 5x - 5y^2$
ⓑ $19y + 5z$

24.
$12a + 5b - 22a + pq^2 - 4p - 3q^2$

25.
$14x - 3y - 13x + a^2b - 4a - 5ab^2$

26.
$2a^2 + b^2 - 6a^2 + x^2y - 3x + 7xy^2$

27. $5u^2 + 4v^2 - 6u^2 + 12a + 8b$ **28.** $xy^2 - 5x - 5y^2 + 19y + 5z$ **29.** Add: $4a, -3b, -8a$

30. Add: $4x, 3y, -3x$ **31.** Subtract $5x^6$ from $-12x^6$ **32.** Subtract $2p^4$ from $-7p^4$

In the following exercises, add the polynomials.

33.
$(5y^2 + 12y + 4) + (6y^2 - 8y + 7)$

34.
$(4y^2 + 10y + 3) + (8y^2 - 6y + 5)$

35.
$(x^2 + 6x + 8) + (-4x^2 + 11x - 9)$

36.
$(y^2 + 9y + 4) + (-2y^2 - 5y - 1)$

37. $(8x^2 - 5x + 2) + (3x^2 + 3)$

38. $(7x^2 - 9x + 2) + (6x^2 - 4)$

39. $(5a^2 + 8) + (a^2 - 4a - 9)$ **40.** $(p^2 - 6p - 18) + (2p^2 + 11)$

In the following exercises, subtract the polynomials.

41.
$(4m^2 - 6m - 3) - (2m^2 + m - 7)$

42.
$(3b^2 - 4b + 1) - (5b^2 - b - 2)$

43.
$(a^2 + 8a + 5) - (a^2 - 3a + 2)$

44.
$(b^2 - 7b + 5) - (b^2 - 2b + 9)$

45. $(12s^2 - 15s) - (s - 9)$

46. $(10r^2 - 20r) - (r - 8)$

In the following exercises, subtract the polynomials.

47. Subtract $(9x^2 + 2)$ from $(12x^2 - x + 6)$

48. Subtract $(5y^2 - y + 12)$ from $(10y^2 - 8y - 20)$

49. Subtract $(7w^2 - 4w + 2)$ from $(8w^2 - w + 6)$

50. Subtract $(5x^2 - x + 12)$ from $(9x^2 - 6x - 20)$

In the following exercises, find the difference of the polynomials.

51. Find the difference of $(w^2 + w - 42)$ and $(w^2 - 10w + 24)$

52. Find the difference of $(z^2 - 3z - 18)$ and $(z^2 + 5z - 20)$

In the following exercises, add the polynomials.

53. $(7x^2 - 2xy + 6y^2) + (3x^2 - 5xy)$

54. $(-5x^2 - 4xy - 3y^2) + (2x^2 - 7xy)$

55. $(7m^2 + mn - 8n^2) + (3m^2 + 2mn)$

56. $(2r^2 - 3rs - 2s^2) + (5r^2 - 3rs)$

In the following exercises, add or subtract the polynomials.

57. $\left(a^2 - b^2\right) - \left(a^2 + 3ab - 4b^2\right)$

58. $\left(m^2 + 2n^2\right) - \left(m^2 - 8mn - n^2\right)$

59. $\left(p^3 - 3p^2 q\right) + \left(2pq^2 + 4q^3\right) - \left(3p^2 q + pq^2\right)$

60. $\left(a^3 - 2a^2 b\right) + \left(ab^2 + b^3\right) - \left(3a^2 b + 4ab^2\right)$

61. $\left(x^3 - x^2 y\right) - \left(4xy^2 - y^3\right) + \left(3x^2 y - xy^2\right)$

62. $\left(x^3 - 2x^2 y\right) - \left(xy^2 - 3y^3\right) - \left(x^2 y - 4xy^2\right)$

Evaluate a Polynomial Function for a Given Value

In the following exercises, find the function values for each polynomial function.

63. For the function $f(x) = 8x^2 - 3x + 2$, find:

ⓐ $f(5)$ ⓑ $f(-2)$ ⓒ $f(0)$

64. For the function $f(x) = 5x^2 - x - 7$, find:

ⓐ $f(-4)$ ⓑ $f(1)$ ⓒ $f(0)$

65. For the function $g(x) = 4 - 36x$, find:

ⓐ $g(3)$ ⓑ $g(0)$ ⓒ $g(-1)$

66. For the function $g(x) = 16 - 36x^2$, find:

ⓐ $g(-1)$ ⓑ $g(0)$ ⓒ $g(2)$

In the following exercises, find the height for each polynomial function.

67. A painter drops a brush from a platform 75 feet high. The polynomial function $h(t) = -16t^2 + 75$ gives the height of the brush t seconds after it was dropped. Find the height after $t = 2$ seconds.

68. A girl is throwing a ball off the cliff into the ocean. The polynomial $h(t) = -16t^2 + 200$ gives the height of a ball t seconds after it is dropped from a 250-foot tall cliff. Find the height after $t = 3$ seconds.

69. A manufacturer of stereo sound speakers has found that the revenue received from selling the speakers at a cost of p dollars each is given by the polynomial function $R(p) = -4p^2 + 420p$. Find the revenue received when $p = 60$ dollars.

70. A manufacturer of the latest basketball shoes has found that the revenue received from selling the shoes at a cost of p dollars each is given by the polynomial $R(p) = -4p^2 + 420p$. Find the revenue received when $p = 90$ dollars.

71. The polynomial $C(x) = 6x^2 + 90x$ gives the cost, in dollars, of producing a rectangular container whose top and bottom are squares with side x feet and height 6 feet. Find the cost of producing a box with $x = 4$ feet.

72. The polynomial $C(x) = 6x^2 + 90x$ gives the cost, in dollars, of producing a rectangular container whose top and bottom are squares with side x feet and height 4 feet. Find the cost of producing a box with $x = 6$ feet.

Add and Subtract Polynomial Functions

For each function, find ⓐ $(f + g)(x)$ ⓑ $(f + g)(2)$ ⓒ $(f - g)(x)$ ⓓ $(f - g)(-3)$.

73. $f(x) = 2x^2 - 4x + 1$ and $g(x) = 5x^2 + 8x + 3$

74. $f(x) = 4x^2 - 7x + 3$ and $g(x) = 4x^2 + 2x - 1$

75. $f(x) = 3x^3 - x^2 - 2x + 3$ and $g(x) = 3x^3 - 7x$

76. $f(x) = 5x^3 - x^2 + 3x + 4$ and $g(x) = 8x^3 - 1$

Writing Exercises

77. Using your own words, explain the difference between a monomial, a binomial, and a trinomial.

78. Using your own words, explain the difference between a polynomial with five terms and a polynomial with a degree of 5.

79. Ariana thinks the sum $6y^2 + 5y^4$ is $11y^6$. What is wrong with her reasoning?

80. Is every trinomial a second degree polynomial? If not, give an example.

Self Check

ⓐ *After completing the exercises, use this checklist to evaluate your mastery of the objectives of this section.*

I can...	Confidently	With some help	No-I don't get it!
identify polynomials, monomials, binomials, and trinomials.			
determine the degree of polynomials.			
add and subtract monomials.			
add and subtract polynomials.			
evaluate a polynomial for a given value.			

ⓑ *If most of your checks were:*

...confidently. *Congratulations! You have achieved the objectives in this section. Reflect on the study skills you used so that you can continue to use them. What did you do to become confident of your ability to do these things? Be specific.*

...with some help. *This must be addressed quickly because topics you do not master become potholes in your road to success. In math every topic builds upon previous work. It is important to make sure you have a strong foundation before you move on. Who can you ask for help? Your fellow classmates and instructor are good resources. Is there a place on campus where math tutors are available? Can your study skills be improved?*

...no - I don't get it! *This is a warning sign and you must not ignore it. You should get help right away or you will quickly be overwhelmed. See your instructor as soon as you can to discuss your situation. Together you can come up with a plan to get you the help you need.*

5.2 Properties of Exponents and Scientific Notation

Learning Objectives

By the end of this section, you will be able to:

› Simplify expressions using the properties for exponents
› Use the definition of a negative exponent
› Use scientific notation

Be Prepared!

Before you get started, take this readiness quiz.

1. Simplify: $(-2)(-2)(-2)$.

   If you missed this problem, review **Example 1.19**.

2. Simplify: $\dfrac{8x}{24y}$.

   If you missed this problem, review **Example 1.24**.

3. Name the decimal $(-2.6)(4.21)$.

   If you missed this problem, review **Example 1.36**.

Simplify Expressions Using the Properties for Exponents

Remember that an exponent indicates repeated multiplication of the same quantity. For example, in the expression a^m, the *exponent m* tells us how many times we use the *base a* as a factor.

$$a^m = \underbrace{a \cdot a \cdot a \cdot \ldots \cdot a}_{m \text{ factors}} \qquad (-9)^5 = \underbrace{(-9)(-9)(-9)(-9)(-9)}_{5 \text{ factors}}$$

Let's review the vocabulary for expressions with exponents.

Exponential Notation

$$a^m \leftarrow \text{exponent}$$
$$\uparrow$$
$$\text{base}$$

a^m means multiply a, m times

$$a^m = \underbrace{a \cdot a \cdot a \cdot \ldots \cdot a}_{m \text{ factors}}$$

This is read a to the mth power.

In the expression a^m, the *exponent m* tells us how many times we use the *base a* as a factor.

When we combine like terms by adding and subtracting, we need to have the same base with the same exponent. But when you multiply and divide, the exponents may be different, and sometimes the bases may be different, too.

First, we will look at an example that leads to the **Product Property**.

$$x^2 \cdot x^3$$

What does this mean?	$\underbrace{x \cdot x}_{2 \text{ factors}} \cdot \underbrace{x \cdot x \cdot x}_{3 \text{ factors}}$
	$\underbrace{\qquad\qquad\qquad}_{5 \text{ factors}}$

$$x^5$$

Notice that 5 is the sum of the exponents, 2 and 3. We see $x^2 \cdot x^3$ is x^{2+3} or x^5.

The base stayed the same and we added the exponents. This leads to the Product Property for Exponents.

Product Property for Exponents

If a is a real number and m and n are integers, then

$$a^m \cdot a^n = a^{m+n}$$

To multiply with like bases, add the exponents.

EXAMPLE 5.12

Simplify each expression: ⓐ $y^5 \cdot y^6$ ⓑ $2^x \cdot 2^{3x}$ ⓒ $2a^7 \cdot 3a$.

✓ **Solution**

ⓐ

	$y^5 \cdot y^6$
Use the Product Property, $a^m \cdot a^n = a^{m+n}$.	y^{5+6}
Simplify.	y^{11}

ⓑ

	$2^x \cdot 2^{3x}$
Use the Product Property, $a^m \cdot a^n = a^{m+n}$.	2^{x+3x}
Simplify.	2^{4x}

ⓒ

	$2a^7 \cdot 3a$
Rewrite, $a = a^1$.	$2a^7 \cdot 3a^1$
Use the Commutative Property and use the Product Property, $a^m \cdot a^n = a^{m+n}$.	$2 \cdot 3 \cdot a^{7+1}$
Simplify.	$6a^8$

ⓓ

	$d^4 \cdot d^5 \cdot d^2$
Add the exponents, since bases are the same.	d^{4+5+2}
Simplify.	d^{11}

> **TRY IT :: 5.23**   Simplify each expression:
>
> ⓐ $b^9 \cdot b^8$ ⓑ $4^{2x} \cdot 4^x$ ⓒ $3p^5 \cdot 4p$ ⓓ $x^6 \cdot x^4 \cdot x^8$.

> **TRY IT :: 5.24**   Simplify each expression:
>
> ⓐ $x^{12} \cdot x^4$ ⓑ $10 \cdot 10^x$ ⓒ $2z \cdot 6z^7$ ⓓ $b^5 \cdot b^9 \cdot b^5$.

Now we will look at an exponent property for division. As before, we'll try to discover a property by looking at some examples.

Consider	$\dfrac{x^5}{x^2}$	and	$\dfrac{x^2}{x^3}$
What do they mean?	$\dfrac{x \cdot x \cdot x \cdot x \cdot x}{x \cdot x}$		$\dfrac{x \cdot x}{x \cdot x \cdot x}$
Use the Equivalent Fractions Property.	$\dfrac{\cancel{x} \cdot \cancel{x} \cdot x \cdot x \cdot x}{\cancel{x} \cdot \cancel{x}}$		$\dfrac{\cancel{x} \cdot \cancel{x} \cdot 1}{\cancel{x} \cdot \cancel{x} \cdot x}$
Simplify.	x^3		$\dfrac{1}{x}$

Notice, in each case the bases were the same and we subtracted exponents. We see $\dfrac{x^5}{x^2}$ is x^{5-2} or x^3. We see $\dfrac{x^2}{x^3}$

is or $\dfrac{1}{x}$. When the larger exponent was in the numerator, we were left with factors in the numerator. When the larger exponent was in the denominator, we were left with factors in the denominator--notice the numerator of 1. When all the factors in the numerator have been removed, remember this is really dividing the factors to one, and so we need a 1 in the numerator. $\dfrac{\cancel{x}}{\cancel{x}} = 1$. This leads to the **Quotient Property** for Exponents.

Quotient Property for Exponents

If a is a real number, $a \neq 0$, and m and n are integers, then

$$\frac{a^m}{a^n} = a^{m-n}, \quad m > n \quad \text{and} \quad \frac{a^m}{a^n} = \frac{1}{a^{n-m}}, \quad n > m$$

EXAMPLE 5.13

Simplify each expression: ⓐ $\dfrac{x^9}{x^7}$ ⓑ $\dfrac{3^{10}}{3^2}$ ⓒ $\dfrac{b^8}{b^{12}}$ ⓓ $\dfrac{7^3}{7^5}$.

✓ Solution

To simplify an expression with a quotient, we need to first compare the exponents in the numerator and denominator.

ⓐ

Since $9 > 7$, there are more factors of x in the numerator.	$\dfrac{x^9}{x^7}$
Use Quotient Property, $\dfrac{a^m}{a^n} = a^{m-n}$.	x^{9-7}
Simplify.	x^2

ⓑ

Since $10 > 2$, there are more factors of 3 in the numerator.	$\dfrac{3^{10}}{3^2}$
Use Quotient Property, $\dfrac{a^m}{a^n} = a^{m-n}$.	3^{10-2}
Simplify.	3^8

Notice that when the larger exponent is in the numerator, we are left with factors in the numerator.

ⓒ

Since $12 > 8$, there are more factors of b in the denominator.	$\dfrac{b^8}{b^{12}}$
Use Quotient Property, $\dfrac{a^m}{a^n} = \dfrac{1}{a^{n-m}}$.	$\dfrac{1}{b^{12-8}}$
Simplify.	$\dfrac{1}{b^4}$

ⓓ

Since $5 > 3$, there are more factors of 3 in the denominator.	$\dfrac{7^3}{7^5}$
Use Quotient Property, $\dfrac{a^m}{a^n} = \dfrac{1}{a^{n-m}}$.	$\dfrac{1}{7^{5-3}}$
Simplify.	$\dfrac{1}{7^2}$
Simplify.	$\dfrac{1}{49}$

Notice that when the larger exponent is in the denominator, we are left with factors in the denominator.

> **TRY IT : : 5.25**

Simplify each expression: ⓐ $\dfrac{x^{15}}{x^{10}}$ ⓑ $\dfrac{6^{14}}{6^5}$ ⓒ $\dfrac{x^{18}}{x^{22}}$ ⓓ $\dfrac{12^{15}}{12^{30}}$.

> **TRY IT : : 5.26**

Simplify each expression: ⓐ $\dfrac{y^{43}}{y^{37}}$ ⓑ $\dfrac{10^{15}}{10^7}$ ⓒ $\dfrac{m^7}{m^{15}}$ ⓓ $\dfrac{9^8}{9^{19}}$.

A special case of the Quotient Property is when the exponents of the numerator and denominator are equal, such as an expression like $\dfrac{a^m}{a^m}$. We know, $\dfrac{x}{x} = 1$, for any $x\,(x \neq 0)$ since any number divided by itself is 1.

The Quotient Property for Exponents shows us how to simplify $\dfrac{a^m}{a^m}$ when $m > n$ and when $n < m$ by subtracting exponents. What if $m = n$? We will simplify $\dfrac{a^m}{a^m}$ in two ways to lead us to the definition of the **Zero Exponent Property**. In general, for $a \neq 0$:

$$\frac{a^m}{a^m} \qquad \Bigg| \qquad \frac{a^m}{a^m}$$

$$a^{m-m} \qquad \Bigg| \qquad \overbrace{\frac{\cancel{a} \cdot \cancel{a} \cdot \cancel{a} \cdot \ldots \cdot \cancel{a}}{\cancel{a} \cdot \cancel{a} \cdot \cancel{a} \cdot \ldots \cdot \cancel{a}}}^{m\ factors}$$

$$a^0 \qquad \Bigg| \qquad 1$$

We see $\frac{a^m}{a^m}$ simplifies to a^0 and to 1. So $a^0 = 1$. Any non-zero base raised to the power of zero equals 1.

Zero Exponent Property

If a is a non-zero number, then $a^0 = 1$.

If a is a non-zero number, then a to the power of zero equals 1.

Any non-zero number raised to the zero power is 1.

In this text, we assume any variable that we raise to the zero power is not zero.

EXAMPLE 5.14

Simplify each expression: ⓐ 9^0 ⓑ n^0.

✓ Solution

The definition says any non-zero number raised to the zero power is 1.

ⓐ

$$9^0$$

Use the definition of the zero exponent. 1

ⓑ

$$n^0$$

Use the definition of the zero exponent. 1

To simplify the expression n raised to the zero power we just use the definition of the zero exponent. The result is 1.

> **TRY IT : : 5.27** Simplify each expression: ⓐ 11^0 ⓑ q^0.

> **TRY IT : : 5.28** Simplify each expression: ⓐ 23^0 ⓑ r^0.

Use the Definition of a Negative Exponent

We saw that the Quotient Property for Exponents has two forms depending on whether the exponent is larger in the numerator or the denominator. What if we just subtract exponents regardless of which is larger?

Let's consider $\frac{x^2}{x^5}$. We subtract the exponent in the denominator from the exponent in the numerator. We see $\frac{x^2}{x^5}$ is x^{2-5} or x^{-3}.

We can also simplify $\frac{x^2}{x^5}$ by dividing out common factors:

$$\frac{x^2}{x^5}$$

$$\frac{x \cdot x}{x \cdot x \cdot x \cdot x \cdot x}$$

$$\frac{1}{x^3}$$

This implies that $x^{-3} = \frac{1}{x^3}$ and it leads us to the definition of a *negative exponent*. If n is an integer and $a \neq 0$, then $a^{-n} = \frac{1}{a^n}$.

Let's now look at what happens to a fraction whose numerator is one and whose denominator is an integer raised to a negative exponent.

$$\frac{1}{a^{-n}}$$

Use the definition of a negative exponent, $a^{-n} = \frac{1}{a^n}$.

$$\frac{1}{\frac{1}{a^n}}$$

Simplify the complex fraction.

$$1 \cdot \frac{a^n}{1}$$

Multiply.

$$a^n$$

This implies $\frac{1}{a^{-n}} = a^n$ and is another form of the definition of **Properties of Negative Exponents**.

Properties of Negative Exponents

If n is an integer and $a \neq 0$, then $a^{-n} = \frac{1}{a^n}$ or $\frac{1}{a^{-n}} = a^n$.

The negative exponent tells us we can rewrite the expression by taking the reciprocal of the base and then changing the sign of the exponent.

Any expression that has negative exponents is not considered to be in simplest form. We will use the definition of a negative exponent and other properties of exponents to write the expression with only positive exponents.

For example, if after simplifying an expression we end up with the expression x^{-3}, we will take one more step and write $\frac{1}{x^3}$. The answer is considered to be in simplest form when it has only positive exponents.

EXAMPLE 5.15

Simplify each expression: ⓐ x^{-5} ⓑ 10^{-3} ⓒ $\frac{1}{y^{-4}}$ ⓓ $\frac{1}{3^{-2}}$.

✓ Solution

ⓐ

$$x^{-5}$$

Use the definition of a negative exponent, $a^{-n} = \frac{1}{a^n}$.

$$\frac{1}{x^5}$$

ⓑ

Use the definition of a negative exponent, $a^{-n} = \dfrac{1}{a^n}$.

$$10^{-3}$$
$$\dfrac{1}{10^3}$$

Simplify.

$$\dfrac{1}{1000}$$

ⓒ

$$\dfrac{1}{y^{-4}}$$

Use the property of a negative exponent, $\dfrac{1}{a^{-n}} = a^n$.

$$y^4$$

ⓓ

$$\dfrac{1}{3^{-2}}$$

Use the property of a negative exponent, $\dfrac{1}{a^{-n}} = a^n$.

$$3^2$$

Simplify.

$$9$$

> **TRY IT :: 5.29** Simplify each expression: ⓐ z^{-3} ⓑ 10^{-7} ⓒ $\dfrac{1}{p^{-8}}$ ⓓ $\dfrac{1}{4^{-3}}$.

> **TRY IT :: 5.30** Simplify each expression: ⓐ n^{-2} ⓑ 10^{-4} ⓒ $\dfrac{1}{q^{-7}}$ ⓓ $\dfrac{1}{2^{-4}}$.

Suppose now we have a fraction raised to a negative exponent. Let's use our definition of negative exponents to lead us to a new property.

$$\left(\dfrac{3}{4}\right)^{-2}$$

Use the definition of a negative exponent, $a^{-n} = \dfrac{1}{a^n}$.

$$\dfrac{1}{\left(\dfrac{3}{4}\right)^2}$$

Simplify the denominator.

$$\dfrac{1}{\dfrac{9}{16}}$$

Simplify the complex fraction.

$$\dfrac{16}{9}$$

But we know that $\dfrac{16}{9}$ is $\left(\dfrac{4}{3}\right)^2$.

This tells us that

$$\left(\dfrac{3}{4}\right)^{-2} = \left(\dfrac{4}{3}\right)^2$$

To get from the original fraction raised to a negative exponent to the final result, we took the reciprocal of the base—the fraction—and changed the sign of the exponent.

This leads us to the **Quotient to a Negative Power Property**.

Quotient to a Negative Power Property

If a and b are real numbers, $a \neq 0$, $b \neq 0$ and n is an integer, then

$$\text{and } \left(\dfrac{a}{b}\right)^{-n} = \left(\dfrac{b}{a}\right)^n$$

EXAMPLE 5.16

Simplify each expression: ⓐ $\left(\frac{5}{7}\right)^{-2}$ ⓑ $\left(-\frac{x}{y}\right)^{-3}$.

✓ **Solution**

ⓐ

$$\left(\frac{5}{7}\right)^{-2}$$

Use the Quotient to a Negative Exponent Property, $\left(\frac{a}{b}\right)^{-n} = \left(\frac{b}{a}\right)^{n}$.

Take the reciprocal of the fraction and change the sign of the exponent.

$$\left(\frac{7}{5}\right)^{2}$$

Simplify.

$$\frac{49}{25}$$

ⓑ

$$\left(-\frac{x}{y}\right)^{-3}$$

Use the Quotient to a Negative Exponent Property, $\left(\frac{a}{b}\right)^{-n} = \left(\frac{b}{a}\right)^{n}$.

Take the reciprocal of the fraction and change the sign of the exponent.

$$\left(-\frac{y}{x}\right)^{3}$$

Simplify.

$$-\frac{y^3}{x^3}$$

> **TRY IT ::** 5.31

Simplify each expression: ⓐ $\left(\frac{2}{3}\right)^{-4}$ ⓑ $\left(-\frac{m}{n}\right)^{-2}$.

> **TRY IT ::** 5.32

Simplify each expression: ⓐ $\left(\frac{3}{5}\right)^{-3}$ ⓑ $\left(-\frac{a}{b}\right)^{-4}$.

Now that we have negative exponents, we will use the Product Property with expressions that have negative exponents.

EXAMPLE 5.17

Simplify each expression: ⓐ $z^{-5} \cdot z^{-3}$ ⓑ $\left(m^4 n^{-3}\right)\left(m^{-5} n^{-2}\right)$ ⓒ $\left(2x^{-6} y^8\right)\left(-5x^5 y^{-3}\right)$.

✓ **Solution**

ⓐ

$$z^{-5} \cdot z^{-3}$$

Add the exponents, since the bases are the same.

$$z^{-5-3}$$

Simplify.

$$z^{-8}$$

Use the definition of a negative exponent.

$$\frac{1}{z^8}$$

ⓑ

$$\left(m^4 n^{-3}\right)\left(m^{-5} n^{-2}\right)$$

Use the Commutative Property to get like bases together.

$$m^4 m^{-5} \cdot n^{-2} n^{-3}$$

Add the exponents for each base.

$$m^{-1} \cdot n^{-5}$$

Take reciprocals and change the signs of the exponents.

$$\frac{1}{m^1} \cdot \frac{1}{n^5}$$

Simplify.

$$\frac{1}{mn^5}$$

ⓒ

$$\left(2x^{-6} y^8\right)\left(-5x^5 y^{-3}\right)$$

Rewrite with the like bases together.

$$2(-5) \cdot \left(x^{-6} x^5\right) \cdot \left(y^8 y^{-3}\right)$$

Multiply the coefficients and add the exponents of each variable.

$$-10 \cdot x^{-1} \cdot y^5$$

Use the definition of a negative exponent, $a^{-n} = \frac{1}{a^n}$.

$$-10 \cdot \frac{1}{x} \cdot y^5$$

Simplify.

$$\frac{-10y^5}{x}$$

> **TRY IT : : 5.33**   Simplify each expression:

ⓐ $z^{-4} \cdot z^{-5}$ ⓑ $\left(p^6 q^{-2}\right)\left(p^{-9} q^{-1}\right)$ ⓒ $\left(3u^{-5} v^7\right)\left(-4u^4 v^{-2}\right)$.

> **TRY IT : : 5.34**   Simplify each expression:

ⓐ $c^{-8} \cdot c^{-7}$ ⓑ $\left(r^5 s^{-3}\right)\left(r^{-7} s^{-5}\right)$ ⓒ $\left(-6c^{-6} d^4\right)\left(-5c^{-2} d^{-1}\right)$.

Now let's look at an exponential expression that contains a power raised to a power. See if you can discover a general property.

$$(x^2)^3$$

What does this mean?

$$x^2 \cdot x^2 \cdot x^2$$

How many factors altogether?

$$\underbrace{x \cdot x}_{2\ factors} \cdot \underbrace{x \cdot x}_{2\ factors} \cdot \underbrace{x \cdot x}_{2\ factors}$$

$$\underbrace{}_{6\ factors}$$

So we have

$$x^6$$

Notice the 6 is the *product* of the exponents, 2 and 3. We see that $(x^2)^3$ is $x^{2 \cdot 3}$ or x^6.

We multiplied the exponents. This leads to the **Power Property for Exponents.**

Power Property for Exponents

If a is a real number and m and n are integers, then

$$(a^m)^n = a^{m \cdot n}$$

To raise a power to a power, multiply the exponents.

EXAMPLE 5.18

Simplify each expression: ⓐ $(y^5)^9$ ⓑ $(4^4)^7$ ⓒ $(y^3)^6 (y^5)^4$.

✓ **Solution**

ⓐ

	$(y^5)^9$
Use the Power Property, $(a^m)^n = a^{m \cdot n}$.	$y^{5 \cdot 9}$
Simplify.	y^{45}

ⓑ

	$(4^4)^7$
Use the Power Property.	$4^{4 \cdot 7}$
Simplify.	4^{28}

ⓒ

	$(y^3)^6 (y^5)^4$
Use the Power Property.	$y^{18} \cdot y^{20}$
Add the exponents.	y^{38}

> **TRY IT ::5.35**   Simplify each expression: ⓐ $(b^7)^5$ ⓑ $(5^4)^3$ ⓒ $(a^4)^5 (a^7)^4$.

> **TRY IT ::5.36**   Simplify each expression: ⓐ $(z^6)^9$ ⓑ $(3^7)^7$ ⓒ $(q^4)^5 (q^3)^3$.

We will now look at an expression containing a product that is raised to a power. Can you find this pattern?

$$(2x)^3$$

What does this mean? $2x \cdot 2x \cdot 2x$

We group the like factors together. $2 \cdot 2 \cdot 2 \cdot x \cdot x \cdot x$

How many factors of 2 and of x $2^3 \cdot x^3$

Notice that each factor was raised to the power and $(2x)^3$ is $2^3 \cdot x^3$.

The exponent applies to each of the factors! This leads to the **Product to a Power Property for Exponents**.

Product to a Power Property for Exponents

If a and b are real numbers and m is a whole number, then

$$(ab)^m = a^m b^m$$

To raise a product to a power, raise each factor to that power.

EXAMPLE 5.19

Simplify each expression: ⓐ $(-3mn)^3$ ⓑ $\left(-4a^2 b\right)^0$ ⓒ $\left(6k^3\right)^{-2}$ ⓓ $\left(5x^{-3}\right)^2$.

✓ Solution

ⓐ

$$(-3mn)^3$$

Use Power of a Product Property, $(ab)^m = a^m b^m$.	$(-3)^3 m^3 n^3$
Simplify.	$-27m^3n^3$

ⓑ

$$\left(-4a^2 b\right)^0$$

Use Power of a Product Property, $(ab)^m = a^m b^m$. $(-4)^0 \left(a^2\right)^0 (b)^0$

Simplify. $1 \cdot 1 \cdot 1$

Multiply. 1

ⓒ

$$\left(6k^3\right)^{-2}$$

Use the Product to a Power Property, $(ab)^m = a^m b^m$. $(6)^{-2}\left(k^3\right)^{-2}$

Use the Power Property, $(a^m)^n = a^{m \cdot n}$. $6^{-2} k^{-6}$

Use the Definition of a negative exponent, $a^{-n} = \dfrac{1}{a^n}$. $\dfrac{1}{6^2} \cdot \dfrac{1}{k^6}$

Simplify. $\dfrac{1}{36k^6}$

ⓓ

$$\left(5x^{-3}\right)^2$$

Use the Product to a Power Property, $(ab)^m = a^m b^m$. $5^2\left(x^{-3}\right)^2$

Simplify. $25 \cdot x^{-6}$

Rewrite x^{-6} using, $a^{-n} = \dfrac{1}{a^n}$. $25 \cdot \dfrac{1}{x^6}$

Simplify. $\dfrac{25}{x^6}$

> **TRY IT :: 5.37**
>
> Simplify each expression: ⓐ $(2wx)^5$ ⓑ $\left(-11pq^3\right)^0$ ⓒ $\left(2b^3\right)^{-4}$ ⓓ $\left(8a^{-4}\right)^2$.

> **TRY IT :: 5.38**
>
> Simplify each expression: ⓐ $(-3y)^3$ ⓑ $(-8m^2 n^3)^0$ ⓒ $\left(-4x^4\right)^{-2}$ ⓓ $\left(2c^{-4}\right)^3$.

Now we will look at an example that will lead us to the Quotient to a Power Property.

This means
$$\left(\frac{x}{y}\right)^3$$
$$\frac{x}{y} \cdot \frac{x}{y} \cdot \frac{x}{y}$$

Multiply the fractions.
$$\frac{x \cdot x \cdot x}{y \cdot y \cdot y}$$

Write with exponents.
$$\frac{x^3}{y^3}$$

Notice that the exponent applies to both the numerator and the denominator.

We see that $\left(\frac{x}{y}\right)^3$ is $\frac{x^3}{y^3}$.

This leads to the **Quotient to a Power Property for Exponents**.

Quotient to a Power Property for Exponents

If a and b are real numbers, $b \neq 0$, and m is an integer, then

$$\left(\frac{a}{b}\right)^m = \frac{a^m}{b^m}$$

To raise a fraction to a power, raise the numerator and denominator to that power.

EXAMPLE 5.20

Simplify each expression:

ⓐ $\left(\frac{b}{3}\right)^4$ ⓑ $\left(\frac{k}{j}\right)^{-3}$ ⓒ $\left(\frac{2xy^2}{z}\right)^3$ ⓓ $\left(\frac{4p^{-3}}{q^2}\right)^2$.

✓ **Solution**

ⓐ

	$\left(\frac{b}{3}\right)^4$
Use Quotient to a Power Property, $\left(\frac{a}{b}\right)^m = \frac{a^m}{b^m}$.	$\frac{b^4}{3^4}$
Simplify.	$\frac{b^4}{81}$

ⓑ

	$\left(\frac{k}{j}\right)^{-3}$
Raise the numerator and denominator to the power.	$\frac{k^{-3}}{j^{-3}}$
Use the definition of negative exponent.	$\frac{1}{k^3} \cdot \frac{j^3}{1}$
Multiply.	$\frac{j^3}{k^3}$

ⓒ

$$\left(\frac{2xy^2}{z}\right)^3$$

Use Quotient to a Power Property, $\left(\frac{a}{b}\right)^m = \frac{a^m}{b^m}$. $\qquad \dfrac{\left(2xy^2\right)^3}{z^3}$

Use the Product to a Power Property, $(ab)^m = a^m b^m$. $\qquad \dfrac{8x^3y^6}{z^3}$

ⓓ

$$\left(\frac{4p^{-3}}{q^2}\right)^2$$

Use Quotient to a Power Property, $\left(\frac{a}{b}\right)^m = \frac{a^m}{b^m}$. $\qquad \dfrac{\left(4p^{-3}\right)^2}{\left(q^2\right)^2}$

Use the Product to a Power Property, $(ab)^m = a^m b^m$. $\qquad \dfrac{4^2\left(p^{-3}\right)^2}{\left(q^2\right)^2}$

Simplify using the Power Property, $(a^m)^n = a^{m \cdot n}$. $\qquad \dfrac{16p^{-6}}{q^4}$

Use the definition of negative exponent. $\qquad \dfrac{16}{q^4} \cdot \dfrac{1}{p^6}$

Simplify. $\qquad \dfrac{16}{p^6 q^4}$

> **TRY IT ::: 5.39**   Simplify each expression:

ⓐ $\left(\frac{p}{10}\right)^4$ ⓑ $\left(\frac{m}{n}\right)^{-7}$ ⓒ $\left(\frac{3ab^3}{c^2}\right)^4$ ⓓ $\left(\frac{3x^{-2}}{y^3}\right)^3$.

> **TRY IT ::: 5.40**   Simplify each expression:

ⓐ $\left(\frac{-2}{q}\right)^3$ ⓑ $\left(\frac{w}{x}\right)^{-4}$ ⓒ $\left(\frac{xy^3}{3z^2}\right)^2$ ⓓ $\left(\frac{2m^{-2}}{n^{-2}}\right)^3$.

We now have several properties for exponents. Let's summarize them and then we'll do some more examples that use more than one of the properties.

Summary of Exponent Properties

If a and b are real numbers, and m and n are integers, then

Property	Description
Product Property	$a^m \cdot a^n = a^{m+n}$
Power Property	$(a^m)^n = a^{m \cdot n}$
Product to a Power	$(ab)^n = a^m b^m$
Quotient Property	$\dfrac{a^m}{a^n} = a^{m-n}, a \neq 0$
Zero Exponent Property	$a^0 = 1, a \neq 0$
Quotient to a Power Property	$\left(\dfrac{a}{b}\right)^m = \dfrac{a^m}{b^m}, b \neq 0$
Properties of Negative Exponents	$a^{-n} = \dfrac{1}{a^n}$ and $\dfrac{1}{a^{-n}} = a^n$
Quotient to a Negative Exponent	$\left(\dfrac{a}{b}\right)^{-n} = \left(\dfrac{b}{a}\right)^n$

EXAMPLE 5.21

Simplify each expression by applying several properties:

ⓐ $(3x^2 y)^4 (2xy^2)^3$ ⓑ $\dfrac{(x^3)^4 (x^{-2})^5}{(x^6)^5}$ ⓒ $\left(\dfrac{2xy^2}{x^3 y^{-2}}\right)^2 \left(\dfrac{12xy^3}{x^3 y^{-1}}\right)^{-1}$.

✓ **Solution**

ⓐ

$$(3x^2 y)^4 (2xy^2)^3$$

Use the Product to a Power Property, $(ab)^m = a^m b^m$. $(3^4 x^8 y^4)(2^3 x^3 y^6)$

Simplify. $(81x^8 y^4)(8x^3 y^6)$

Use the Commutative Property. $81 \cdot 8 \cdot x^8 \cdot x^3 \cdot y^4 \cdot y^6$

Multiply the constants and add the exponents. $648x^{11} y^{10}$

ⓑ

$$\dfrac{(x^3)^4 (x^{-2})^5}{(x^6)^5}$$

Use the Power Property, $(a^m)^n = a^{m \cdot n}$. $\dfrac{(x^{12})(x^{-10})}{(x^{30})}$

Add the exponents in the numerator. $\dfrac{x^2}{x^{30}}$

Use the Quotient Property, $\dfrac{a^m}{a^n} = \dfrac{1}{a^{n-m}}$. $\dfrac{1}{x^{28}}$

ⓒ

	$\left(\dfrac{2xy^2}{x^3y^{-2}}\right)^2 \left(\dfrac{12xy^3}{x^3y^{-1}}\right)^{-1}$
Simplify inside the parentheses first.	$\left(\dfrac{2y^4}{x^2}\right)^2 \left(\dfrac{12y^4}{x^2}\right)^{-1}$
Use the Quotient to a Power Property, $\left(\dfrac{a}{b}\right)^m = \dfrac{a^m}{b^m}$.	$\dfrac{\left(2y^4\right)^2 \left(12y^4\right)^{-1}}{\left(x^2\right)^2 \left(x^2\right)^{-1}}$
Use the Product to a Power Property, $(ab)^m = a^m b^m$.	$\dfrac{4y^8}{x^4} \cdot \dfrac{12^{-1}y^{-4}}{x^{-2}}$
Simplify.	$\dfrac{4y^4}{12x^2}$
Simplify.	$\dfrac{y^4}{3x^2}$

> **TRY IT :: 5.41**

Simplify each expression:

ⓐ $(c^4 d^2)^5 (3cd^5)^4$ ⓑ $\dfrac{(a^{-2})^3 (a^2)^4}{(a^4)^5}$ ⓒ $\left(\dfrac{3xy^2}{x^2 y^{-3}}\right)^2 \left(\dfrac{9xy^{-3}}{x^3 y^2}\right)^{-1}$.

> **TRY IT :: 5.42**

Simplify each expression:

ⓐ $(a^3 b^2)^6 (4ab^3)^4$ ⓑ $\dfrac{(p^{-3})^4 (p^5)^3}{(p^7)^6}$ ⓒ $\left(\dfrac{4x^3 y^2}{x^2 y^{-1}}\right)^2 \left(\dfrac{8xy^{-3}}{x^2 y}\right)^{-1}$.

Use Scientific Notation

Working with very large or very small numbers can be awkward. Since our number system is base ten we can use powers of ten to rewrite very large or very small numbers to make them easier to work with. Consider the numbers 4,000 and 0.004.

Using place value, we can rewrite the numbers 4,000 and 0.004. We know that 4,000 means $4 \times 1,000$ and 0.004 means $4 \times \dfrac{1}{1,000}$.

If we write the 1,000 as a power of ten in exponential form, we can rewrite these numbers in this way:

4,000	$4 \times 1,000$	4×10^3	
0.004	$4 \times \dfrac{1}{1,000}$	$4 \times \dfrac{1}{10^3}$	4×10^{-3}

When a number is written as a product of two numbers, where the first factor is a number greater than or equal to one but less than ten, and the second factor is a power of 10 written in exponential form, it is said to be in **scientific notation**.

Scientific Notation

A number is expressed in **scientific notation** when it is of the form

$$a \times 10^n \text{ where } 1 \le a < 10 \text{ and } n \text{ is an integer.}$$

It is customary in scientific notation to use as the $\times$ multiplication sign, even though we avoid using this sign elsewhere in algebra.

If we look at what happened to the decimal point, we can see a method to easily convert from decimal notation to scientific notation.

$$4000. = 4 \times 10^3 \qquad\qquad 0.004 = 4 \times 10^{-3}$$

Moved the decimal point   Moved the decimal point
3 places to the left.   3 places to the right.

In both cases, the decimal was moved 3 places to get the first factor between 1 and 10.

The power of 10 is positive when the number is larger than 1: $4,000 = 4 \times 10^3$

The power of 10 is negative when the number is between 0 and 1: $0.004 = 4 \times 10^{-3}$

 HOW TO :: TO CONVERT A DECIMAL TO SCIENTIFIC NOTATION.

Step 1.   Move the decimal point so that the first factor is greater than or equal to 1 but less than 10.

Step 2.   Count the number of decimal places, n, that the decimal point was moved.

Step 3.   Write the number as a product with a power of 10. If the original number is.

  ◦ greater than 1, the power of 10 will be 10^n.

  ◦ between 0 and 1, the power of 10 will be 10^{-n}.

Step 4.   Check.

EXAMPLE 5.22

Write in scientific notation: ⓐ 37,000 ⓑ 0.0052.

✓ Solution

ⓐ

The original number, 37,000, is greater than 1 so we will have a positive power of 10.	37,000
Move the decimal point to get 3.7, a number between 1 and 10.	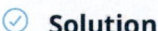 37,000
Count the number of decimal places the point was moved.	4 places
Write as a product with a power of 10.	3.7×10^4

Check:
$$3.7 \times 10^4$$
$$3.7 \times 10,000$$
$$37,000$$

$$37,000 = 3.7 \times 10^4$$

ⓑ

The original number, 0.0052, is between 0 and 1 so we will have a negative power of 10.	0.0052
Move the decimal point to get 5.2, a number between 1 and 10.	0.0052
Count the number of decimal places the point was moved.	3 places
Write as a product with a power of 10.	5.2×10^{-3}

Check:　5.2×10^{-3}

$5.2 \times \dfrac{1}{10^3}$

$5.2 \times \dfrac{1}{1000}$

5.2×0.001

0.0052

$0.0052 = 5.2 \times 10^{-3}$

> **TRY IT :: 5.43**　　Write in scientific notation: ⓐ 96,000 ⓑ 0.0078.

> **TRY IT :: 5.44**　　Write in scientific notation: ⓐ 48,300 ⓑ 0.0129.

How can we convert from scientific notation to decimal form? Let's look at two numbers written in scientific notation and see.

$$9.12 \times 10^4 \qquad\qquad 9.12 \times 10^{-4}$$
$$9.12 \times 10,000 \qquad 9.12 \times 0.0001$$
$$91,200 \qquad\qquad 0.000912$$

If we look at the location of the decimal point, we can see an easy method to convert a number from scientific notation to decimal form.

$$9.12 \times 10^4 = 91,200 \qquad\qquad 9.12 \times 10^{-4} = 0.000912$$
$$9.12\_\_\times 10^4 = 91,200 \qquad \_\_\_9.12 \times 10^{-4} = 0.000912$$

Move the decimal point 4 places to the right.　　Move the decimal point 4 places to the left.

In both cases the decimal point moved 4 places. When the exponent was positive, the decimal moved to the right. When the exponent was negative, the decimal point moved to the left.

HOW TO :: CONVERT SCIENTIFIC NOTATION TO DECIMAL FORM.

Step 1.　Determine the exponent, n, on the factor 10.

Step 2.　Move the decimal n places, adding zeros if needed.

　　◦ If the exponent is positive, move the decimal point n places to the right.

　　◦ If the exponent is negative, move the decimal point $|n|$ places to the left.

Step 3.　Check.

EXAMPLE 5.23

Convert to decimal form: ⓐ 6.2×10^3 ⓑ -8.9×10^{-2}.

✓ **Solution**

ⓐ

	6.2×10^3
Determine the exponent, n, on the factor 10.	The exponent is 3.
Since the exponent is positive, move the decimal point 3 places to the right.	6.200
Add zeros as needed for placeholders.	6,200
	$6.2 \times 10^3 = 6,200$

ⓑ

	-8.9×10^{-2}
Determine the exponent, n, on the factor 10.	The exponent is -2.
Since the exponent is negative, move the decimal point 2 places to the left.	$-\,8.9$
Add zeros as needed for placeholders.	-0.089
	$-8.9 \times 10^{-2} = -0.089$

> **TRY IT : : 5.45**   Convert to decimal form: ⓐ 1.3×10^3 ⓑ -1.2×10^{-4}.

> **TRY IT : : 5.46**   Convert to decimal form: ⓐ -9.5×10^4 ⓑ 7.5×10^{-2}.

When scientists perform calculations with very large or very small numbers, they use scientific notation. Scientific notation provides a way for the calculations to be done without writing a lot of zeros. We will see how the Properties of Exponents are used to multiply and divide numbers in scientific notation.

EXAMPLE 5.24

Multiply or divide as indicated. Write answers in decimal form: ⓐ $\left(-4 \times 10^5\right)\left(2 \times 10^{-7}\right)$ ⓑ $\dfrac{9 \times 10^3}{3 \times 10^{-2}}$.

✓ **Solution**

ⓐ

	$\left(-4 \times 10^5\right)\left(2 \times 10^{-7}\right)$
Use the Commutative Property to rearrange the factors.	$-4 \cdot 2 \cdot 10^5 \cdot 10^{-7}$
Multiply.	-8×10^{-2}
Change to decimal form by moving the decimal two places left.	-0.08

ⓑ

$$\frac{9 \times 10^3}{9 \times 10^{-2}}$$

Separate the factors, rewriting as the product of two fractions.

$$\frac{9}{3} \times \frac{10^3}{10^{-2}}$$

Divide.

$$3 \times 10^5$$

Change to decimal form by moving the decimal five places right.

$$300,000$$

> **TRY IT : : 5.47**   Multiply or divide as indicated. Write answers in decimal form:
>
> ⓐ $\left(-3 \times 10^5\right)\left(2 \times 10^{-8}\right)$ ⓑ $\frac{8 \times 10^2}{4 \times 10^{-2}}$.

> **TRY IT : : 5.48**   Multiply or divide as indicated. Write answers in decimal form:
>
> ⓐ; $\left(-3 \times 10^{-2}\right)\left(3 \times 10^{-1}\right)$ ⓑ $\frac{8 \times 10^4}{2 \times 10^{-1}}$.

▶ **MEDIA : :**

Access these online resources for additional instruction and practice with using multiplication properties of exponents.

- **Properties of Exponents (https://openstax.org/l/37Propofexpo)**
- **Negative exponents (https://openstax.org/l/37Negativeexpo)**
- **Scientific Notation (https://openstax.org/l/37SciNotation)**

5.2 EXERCISES

Practice Makes Perfect

Simplify Expressions Using the Properties for Exponents

In the following exercises, simplify each expression using the properties for exponents.

81. ⓐ $d^3 \cdot d^6$ ⓑ $4^{5x} \cdot 4^{9x}$ ⓒ $2y \cdot 4y^3$ ⓓ $w \cdot w^2 \cdot w^3$ **82.** ⓐ $x^4 \cdot x^2$ ⓑ $8^{9x} \cdot 8^3$ ⓒ $3z^{25} \cdot 5z^8$ ⓓ $y \cdot y^3 \cdot y^5$

83. ⓐ $n^{19} \cdot n^{12}$ ⓑ $3^x \cdot 3^6$ ⓒ $7w^5 \cdot 8w$ ⓓ $a^4 \cdot a^3 \cdot a^9$ **84.** ⓐ $q^{27} \cdot q^{15}$ ⓑ $5^x \cdot 5^{4x}$ ⓒ $9u^{41} \cdot 7u^{53}$
　　　　　　　　　　　　　　　　　　　　　　　　　　　　　　　　　　ⓓ $c^5 \cdot c^{11} \cdot c^2$

85. $m^x \cdot m^3$ **86.** $n^y \cdot n^2$

87. $y^a \cdot y^b$ **88.** $x^p \cdot x^q$

89. ⓐ $\dfrac{x^{18}}{x^3}$ ⓑ $\dfrac{5^{12}}{5^3}$ ⓒ $\dfrac{q^{18}}{q^{36}}$ ⓓ $\dfrac{10^2}{10^3}$ **90.** ⓐ $\dfrac{y^{20}}{y^{10}}$ ⓑ $\dfrac{7^{16}}{7^2}$ ⓒ $\dfrac{t^{10}}{t^{40}}$ ⓓ $\dfrac{8^3}{8^5}$

91. ⓐ $\dfrac{p^{21}}{p^7}$ ⓑ $\dfrac{4^{16}}{4^4}$ ⓒ $\dfrac{b}{b^9}$ ⓓ $\dfrac{4}{4^6}$ **92.** ⓐ $\dfrac{u^{24}}{u^3}$ ⓑ $\dfrac{9^{15}}{9^5}$ ⓒ $\dfrac{x}{x^7}$ ⓓ $\dfrac{10}{10^3}$

93. ⓐ 20^0 ⓑ b^0 **94.** ⓐ 13^0 ⓑ k^0

95. ⓐ -27^0 ⓑ $-\left(27^0\right)$ **96.** ⓐ -15^0 ⓑ $-\left(15^0\right)$

Use the Definition of a Negative Exponent

In the following exercises, simplify each expression.

97. ⓐ a^{-2} ⓑ 10^{-3} ⓒ $\dfrac{1}{c^{-5}}$ ⓓ $\dfrac{1}{3^{-2}}$ **98.** ⓐ b^{-4} ⓑ 10^{-2} ⓒ $\dfrac{1}{c^{-5}}$ ⓓ $\dfrac{1}{5^{-2}}$

99. ⓐ r^{-3} ⓑ 10^{-5} ⓒ $\dfrac{1}{q^{-10}}$ ⓓ $\dfrac{1}{10^{-3}}$ **100.** ⓐ s^{-8} ⓑ 10^{-2} ⓒ $\dfrac{1}{t^{-9}}$ ⓓ $\dfrac{1}{10^{-4}}$

101. ⓐ $\left(\dfrac{5}{8}\right)^{-2}$ ⓑ $\left(-\dfrac{b}{a}\right)^{-2}$ **102.** ⓐ $\left(\dfrac{3}{10}\right)^{-2}$ ⓑ $\left(-\dfrac{2}{z}\right)^{-3}$

103. ⓐ $\left(\dfrac{4}{9}\right)^{-3}$ ⓑ $\left(-\dfrac{u}{v}\right)^{-5}$ **104.** ⓐ $\left(\dfrac{7}{2}\right)^{-3}$ ⓑ $\left(-\dfrac{3}{x}\right)^{-3}$

105. ⓐ $(-5)^{-2}$ ⓑ -5^{-2} ⓒ $\left(-\dfrac{1}{5}\right)^{-2}$ ⓓ $-\left(\dfrac{1}{5}\right)^{-2}$ **106.** ⓐ -5^{-3} ⓑ $\left(-\dfrac{1}{5}\right)^{-3}$ ⓒ $-\left(\dfrac{1}{5}\right)^{-3}$ ⓓ $(-5)^{-3}$

107. ⓐ $3 \cdot 5^{-1}$ ⓑ $(3 \cdot 5)^{-1}$ **108.** ⓐ $3 \cdot 4^{-2}$ ⓑ $(3 \cdot 4)^{-2}$

In the following exercises, simplify each expression using the Product Property.

109. ⓐ $b^4 b^{-8}$ ⓑ $\left(w^4 x^{-5}\right)\left(w^{-2} x^{-4}\right)$
ⓒ $\left(-6c^{-3} d^9\right)\left(2c^4 d^{-5}\right)$

110. ⓐ $s^3 \cdot s^{-7}$ ⓑ $\left(m^3 n^{-3}\right)\left(m^{-5} n^{-1}\right)$
ⓒ $\left(-2j^{-5} k^8\right)\left(7j^2 k^{-3}\right)$

111. ⓐ $a^3 \cdot a^{-3}$ ⓑ $\left(uv^{-2}\right)\left(u^{-5} v^{-3}\right)$
ⓒ $\left(-4r^{-2} s^{-8}\right)\left(9r^4 s^3\right)$

112. ⓐ $y^5 \cdot y^{-5}$ ⓑ $\left(pq^{-4}\right)\left(p^{-6} q^{-3}\right)$
ⓒ $\left(-5m^4 n^6\right)\left(8m^{-5} n^{-3}\right)$

113. $p^5 \cdot p^{-2} \cdot p^{-4}$

114. $x^4 \cdot x^{-2} \cdot x^{-3}$

In the following exercises, simplify each expression using the Power Property.

115. ⓐ $\left(m^4\right)^2$ ⓑ $(10^3)^6$ ⓒ $\left(x^3\right)^{-4}$

116. ⓐ $(b^2)^7$ ⓑ $(3^8)^2$ ⓒ $\left(k^2\right)^{-5}$

117. ⓐ $\left(y^3\right)^x$ ⓑ $(5^x)^y$ ⓒ $\left(q^6\right)^{-8}$

118. ⓐ $\left(x^2\right)^y$ ⓑ $(7^a)^b$ ⓒ $\left(a^9\right)^{-10}$

In the following exercises, simplify each expression using the Product to a Power Property.

119. ⓐ $(-3xy)^2$ ⓑ $(6a)^0$ ⓒ $\left(5x^2\right)^{-2}$ ⓓ $\left(-4y^{-3}\right)^2$

120. ⓐ $(-4ab)^2$ ⓑ $(5x)^0$ ⓒ $\left(4y^3\right)^{-3}$ ⓓ $\left(-7y^{-3}\right)^2$

121. ⓐ $(-5ab)^3$ ⓑ $(-4pq)^0$ ⓒ $\left(-6x^3\right)^{-2}$ ⓓ $\left(3y^{-4}\right)^2$

122. ⓐ $(-3xyz)^4$ ⓑ $(-7mn)^0$ ⓒ $\left(-3x^3\right)^{-2}$
ⓓ $\left(2y^{-5}\right)^2$

In the following exercises, simplify each expression using the Quotient to a Power Property.

123. ⓐ $\left(\frac{p}{2}\right)^5$ ⓑ $\left(\frac{x}{y}\right)^{-6}$ ⓒ $\left(\frac{2xy^2}{z}\right)^3$ ⓓ $\left(\frac{4p^{-3}}{q^2}\right)^2$

124. ⓐ $\left(\frac{x}{3}\right)^4$ ⓑ $\left(\frac{a}{b}\right)^{-5}$ ⓒ $\left(\frac{2xy^2}{z}\right)^3$ ⓓ $\left(\frac{4p^{-3}}{q^2}\right)^2$

125. ⓐ $\left(\frac{a}{3b}\right)^4$ ⓑ $\left(\frac{5}{4m}\right)^{-2}$ ⓒ $\left(\frac{2xy^2}{z}\right)^3$ ⓓ $\left(\frac{4p^{-3}}{q^2}\right)^2$

126. ⓐ $\left(\frac{x}{2y}\right)^3$ ⓑ $\left(\frac{10}{3q}\right)^{-4}$ ⓒ $\left(\frac{2xy^2}{z}\right)^3$ ⓓ $\left(\frac{4p^{-3}}{q^2}\right)^2$

In the following exercises, simplify each expression by applying several properties.

127. ⓐ $(5t^2)^3 (3t)^2$ ⓑ $\dfrac{\left(t^2\right)^5 \left(t^{-4}\right)^2}{\left(t^3\right)^7}$
ⓒ $\left(\dfrac{2xy^2}{x^3 y^{-2}}\right)^2 \left(\dfrac{12xy^3}{x^3 y^{-1}}\right)^{-1}$

128. ⓐ $(10k^4)^3 (5k^6)^2$ ⓑ $\dfrac{\left(q^3\right)^6 \left(q^{-2}\right)^3}{\left(q^4\right)^8}$

129. ⓐ $(m^2 n)^2 (2mn^5)^4$ ⓑ $\dfrac{\left(-2p^{-2}\right)^4 \left(3p^4\right)^2}{\left(-6p^3\right)^2}$

130. ⓐ $(3pq^4)^2 (6p^6 q)^2$ ⓑ $\dfrac{\left(-2k^{-3}\right)^2 \left(6k^2\right)^4}{\left(9k^4\right)^2}$

Mixed Practice

In the following exercises, simplify each expression.

131. ⓐ $7n^{-1}$ ⓑ $(7n)^{-1}$ ⓒ $(-7n)^{-1}$

132. ⓐ $6r^{-1}$ ⓑ $(6r)^{-1}$ ⓒ $(-6r)^{-1}$

133. ⓐ $(3p)^{-2}$ ⓑ $3p^{-2}$ ⓒ $-3p^{-2}$

134. ⓐ $(2q)^{-4}$ ⓑ $2q^{-4}$ ⓒ $-2q^{-4}$

135. $\left(x^2\right)^4 \cdot \left(x^3\right)^2$

136. $\left(y^4\right)^3 \cdot \left(y^5\right)^2$

137. $\left(a^2\right)^6 \cdot \left(a^3\right)^8$

138. $\left(b^7\right)^5 \cdot \left(b^2\right)^6$

139. $(2m^6)^3$

140. $(3y^2)^4$

141. $(10x^2 y)^3$

142. $(2mn^4)^5$

143. $(-2a^3 b^2)^4$

144. $(-10u^2 v^4)^3$

145. $\left(\frac{2}{3}x^2 y\right)^3$

146. $\left(\frac{7}{9}pq^4\right)^2$

147. $(8a^3)^2 (2a)^4$

148. $(5r^2)^3 (3r)^2$

149. $(10p^4)^3 (5p^6)^2$

150. $(4x^3)^3 (2x^5)^4$

151. $\left(\frac{1}{2}x^2 y^3\right)^4 \left(4x^5 y^3\right)^2$

152. $\left(\frac{1}{3}m^3 n^2\right)^4 \left(9m^8 n^3\right)^2$

153. $(3m^2 n)^2 (2mn^5)^4$

154. $(2pq^4)^3 (5p^6 q)^2$

155. ⓐ $(3x)^2(5x)$ ⓑ $(2y)^3(6y)$

156. ⓐ $\left(\frac{1}{2}y^2\right)^3 \left(\frac{2}{3}y\right)^2$ ⓑ $\left(\frac{1}{2}j^2\right)^5 \left(\frac{2}{5}j^3\right)^2$

157. ⓐ $(2r^{-2})^3 (4^{-1} r)^2$ ⓑ $(3x^{-3})^3 (3^{-1} x^5)^4$

158. $\left(\dfrac{k^{-2} k^8}{k^3}\right)^2$

159. $\left(\dfrac{j^{-2} j^5}{j^4}\right)^3$

160. $\dfrac{\left(-4m^{-3}\right)^2 \left(5m^4\right)^3}{\left(-10m^6\right)^3}$

161. $\dfrac{\left(-10n^{-2}\right)^3 \left(4n^5\right)^2}{\left(2n^8\right)^2}$

Use Scientific Notation

In the following exercises, write each number in scientific notation.

162. ⓐ 57,000 ⓑ 0.026

163. ⓐ 340,000 ⓑ 0.041

164. ⓐ 8,750,000 ⓑ 0.00000871

165. ⓐ 1,290,000 ⓑ 0.00000103

In the following exercises, convert each number to decimal form.

166. ⓐ 5.2×10^2 ⓑ 2.5×10^{-2}

167. ⓐ -8.3×10^2 ⓑ 3.8×10^{-2}

168. ⓐ 7.5×10^6 ⓑ -4.13×10^{-5}

169. ⓐ 1.6×10^{10} ⓑ 8.43×10^{-6}

In the following exercises, multiply or divide as indicated. Write your answer in decimal form.

170. ⓐ $\left(3 \times 10^{-5}\right)\left(3 \times 10^9\right)$ ⓑ $\dfrac{7 \times 10^{-3}}{1 \times 10^{-7}}$

171. ⓐ $\left(2 \times 10^2\right)\left(1 \times 10^{-4}\right)$ ⓑ $\dfrac{5 \times 10^{-2}}{1 \times 10^{-10}}$

172. ⓐ $\left(7.1 \times 10^{-2}\right)\left(2.4 \times 10^{-4}\right)$ ⓑ $\dfrac{6 \times 10^4}{3 \times 10^{-2}}$

173. ⓐ $\left(3.5 \times 10^{-4}\right)\left(1.6 \times 10^{-2}\right)$ ⓑ $\dfrac{8 \times 10^6}{4 \times 10^{-1}}$

Writing Exercises

174. Use the Product Property for Exponents to explain why $x \cdot x = x^2$.

175. Jennifer thinks the quotient $\dfrac{a^{24}}{a^6}$ simplifies to a^4. What is wrong with her reasoning?

176. Explain why $-5^3 = (-5)^3$ but $-5^4 \neq (-5)^4$.

177. When you convert a number from decimal notation to scientific notation, how do you know if the exponent will be positive or negative?

Self Check

ⓐ *After completing the exercises, use this checklist to evaluate your mastery of the objectives of this section.*

I can...	Confidently	With some help	No-I don't get it!
simplify expressions using the properties for exponents.			
use the definition of a Negative Exponents.			
use scientific notation.			

ⓑ *After reviewing this checklist, what will you do to become confident for all goals?*

5.3 | Multiply Polynomials

Learning Objectives

By the end of this section, you will be able to:

› Multiply monomials
› Multiply a polynomial by a monomial
› Multiply a binomial by a binomial
› Multiply a polynomial by a polynomial
› Multiply special products
› Multiply polynomial functions

Be Prepared!

Before you get started, take this readiness quiz.

1. Distribute: $2(x + 3)$.
   If you missed this problem, review **Example 1.50**.

2. Simplify: ⓐ 9^2 ⓑ $(-9)^2$ ⓒ -9^2.
   If you missed this problem, review **Example 1.19**.

3. Evaluate: $2x^2 - 5x + 3$ for $x = -2$.
   If you missed this problem, review **Example 1.21**.

Multiply Monomials

We are ready to perform operations on polynomials. Since monomials are algebraic expressions, we can use the properties of exponents to multiply monomials.

EXAMPLE 5.25

Multiply: ⓐ $(3x^2)(-4x^3)$ ⓑ $\left(\frac{5}{6}x^3 y\right)(12xy^2)$.

⊘ **Solution**

ⓐ

$$(3x^2)(-4x^3)$$

Use the Commutative Property to rearrange the terms. $3 \cdot (-4) \cdot x^2 \cdot x^3$

Multiply. $-12x^5$

ⓑ

$$\left(\frac{5}{6}x^3 y\right)(12xy^2)$$

Use the Commutative Property to rearrange the terms. $\frac{5}{6} \cdot 12 \cdot x^3 \cdot x \cdot y \cdot y^2$

Multiply. $10x^4 y^3$

> **TRY IT :: 5.49**   Multiply: ⓐ $(5y^7)(-7y^4)$ ⓑ $\left(\frac{2}{5}a^4 b^3\right)(15ab^3)$.

> **TRY IT :: 5.50**   Multiply: ⓐ $(-6b^4)(-9b^5)$ ⓑ $\left(\frac{2}{3}r^5 s\right)(12r^6 s^7)$.

Multiply a Polynomial by a Monomial

Multiplying a polynomial by a monomial is really just applying the Distributive Property.

EXAMPLE 5.26

Multiply: ⓐ $-2y(4y^2 + 3y - 5)$ ⓑ $3x^3 y(x^2 - 8xy + y^2)$.

✓ **Solution**

ⓐ

$$-2y\,(4y^2 + 3y - 5)$$

Distribute.	$-2y \cdot 4y^2 + (-2y) \cdot 3y - (-2y) \cdot 5$
Multiply.	$-8y^3 - 6y^2 + 10y$

ⓑ

$$3x^3 y(x^2 - 8xy + y^2)$$

| Distribute. | $3x^3 y \cdot x^2 + (3x^3 y) \cdot (-8xy) + (3x^3 y) \cdot y^2$ |
| Multiply. | $3x^5 y - 24x^4 y^2 + 3x^3 y^3$ |

> **TRY IT : : 5.51** Multiply: ⓐ $-3y(5y^2 + 8y - 7)$ ⓑ $4x^2 y^2 (3x^2 - 5xy + 3y^2)$.

> **TRY IT : : 5.52** Multiply: ⓐ $4x^2(2x^2 - 3x + 5)$ ⓑ $-6a^3 b(3a^2 - 2ab + 6b^2)$.

Multiply a Binomial by a Binomial

Just like there are different ways to represent multiplication of numbers, there are several methods that can be used to multiply a binomial times a binomial. We will start by using the Distributive Property.

EXAMPLE 5.27

Multiply: ⓐ $(y + 5)(y + 8)$ ⓑ $(4y + 3)(2y - 5)$.

✓ **Solution**

ⓐ

$$(y + 5)(y + 8)$$

Distribute $(y + 8)$.	$y(y + 8) + 5(y + 8)$
Distribute again.	$y^2 + 8y + 5y + 40$
Combine like terms.	$y^2 + 13y + 40$

ⓑ

	$(4y + 3)(2y - 5)$
Distribute.	$4y(2y - 5) + 3(2y - 5)$
Distribute again.	$8y^2 - 20y + 6y - 15$
Combine like terms.	$8y^2 - 14y - 15$

> **TRY IT :: 5.53**   Multiply: ⓐ $(x + 8)(x + 9)$ ⓑ $(3c + 4)(5c - 2)$.

> **TRY IT :: 5.54**   Multiply: ⓐ $(5x + 9)(4x + 3)$ ⓑ $(5y + 2)(6y - 3)$.

If you multiply binomials often enough you may notice a pattern. Notice that the first term in the result is the product of the *first* terms in each binomial. The second and third terms are the product of multiplying the two *outer* terms and then the two *inner* terms. And the last term results from multiplying the two *last* terms,

We abbreviate "First, Outer, Inner, Last" as FOIL. The letters stand for 'First, Outer, Inner, Last'. We use this as another method of multiplying binomials. The word FOIL is easy to remember and ensures we find all four products.

Let's multiply $(x + 3)(x + 7)$ using both methods.

Distributive Property	FOIL
$(x + 3)(x + 7)$	$(x + 3)(x + 7)$
$x(x + 7) + 3(x + 7)$	
$x^2 + 7x + 3x + 21$ F O I L	$x^2 + 7x + 3x + 21$ F O I L
$x^2 + 10x + 21$	$x^2 + 10x + 21$

We summarize the steps of the FOIL method below. The FOIL method only applies to multiplying binomials, not other polynomials!

   HOW TO :: USE THE FOIL METHOD TO MULTIPLY TWO BINOMIALS.

Step 1. Multiply the *First* terms.

Step 2. Multiply the *Outer* terms.

Step 3. Multiply the *Inner* terms.

Step 4. Multiply the *Last* terms.

Step 5. Combine like terms, when possible.

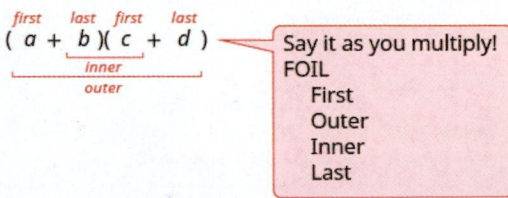

When you multiply by the FOIL method, drawing the lines will help your brain focus on the pattern and make it easier to apply.

Now we will do an example where we use the FOIL pattern to multiply two binomials.

EXAMPLE 5.28

Multiply: ⓐ $(y - 7)(y + 4)$ ⓑ $(4x + 3)(2x - 5)$.

⊘ Solution

ⓐ

$$(y - 7)(y + 4)$$

Step 1. Multiply the *First* terms. $(y - 7)(y + 4)$ $\underset{F\ \ \ O\ \ \ I\ \ \ L}{y^2 + \_ + \_ + \_}$

Step 2. Multiply the *Outer* terms. $(y - 7)(y + 4)$ $\underset{F\ \ \ O\ \ \ I\ \ \ L}{y^2 + 4y + \_ + \_}$

Step 3. Multiply the *Inner* terms. $(y - 7)(y + 4)$ $\underset{F\ \ \ O\ \ \ I\ \ \ L}{y^2 + 4y - 7y + \_}$

Step 4. Multiply the *Last* terms. $(y - 7)(y + 4)$ $\underset{F\ \ \ O\ \ \ I\ \ \ L}{y^2 + 4y - 7y - 28}$

Step 5. Combine like terms. $y^2 - 3y - 28$

ⓑ

$$(4x - 3)(2x - 5)$$

$$(4x + 3)(2x - 5)$$

Step 1. Multiply the *First* terms, $4x \cdot 2x$. $\underset{F\ \ \ O\ \ \ I\ \ \ L}{8x^2 + \_ + \_ + \_}$

Step 2. Multiply the *Outer* terms, $4x \cdot (-5)$. $\underset{F\ \ \ O\ \ \ I\ \ \ L}{8x^2 - 20x + \_ + \_}$

Step 3. Multiply the *Inner* terms, $3 \cdot 2x$. $\underset{F\ \ \ O\ \ \ I\ \ \ L}{8x^2 - 2x + 6x + \_}$

Step 4. Multiply the *Last* terms, $3 \cdot (-5)$. $\underset{F\ \ \ O\ \ \ I\ \ \ L}{8x^2 - 20x + 6x - 15}$

Step 5. Combine like terms. $8x^2 - 14x - 15$

> **TRY IT : : 5.55**   Multiply: ⓐ $(x - 7)(x + 5)$ ⓑ $(3x + 7)(5x - 2)$.

> **TRY IT : : 5.56**   Multiply: ⓐ $(b - 3)(b + 6)$ ⓑ $(4y + 5)(4y - 10)$.

The final products in the last example were trinomials because we could combine the two middle terms. This is not always the case.

EXAMPLE 5.29

Multiply: ⓐ $\left(n^2 + 4\right)(n - 1)$ ⓑ $(3pq + 5)(6pq - 11)$.

⊘ Solution

ⓐ

$$(n^2 + 4)(n - 1)$$

$$(n^2 + 4)(n - 1)$$

Step 1. Multiply the *First* terms.	$n^3 + \underline{\;\;} + \underline{\;\;} + \underline{\;\;}$ $\;\;\;F\quad\; O\quad\; I\quad\; L$
Step 2. Multiply the *Outer* terms.	$n^3 - n^2 + \underline{\;\;} + \underline{\;\;}$ $\;\;\;F\quad O\quad\;\; I\quad\; L$
Step 3. Multiply the *Inner* terms.	$n^3 - n^2 + 4n + \underline{\;\;}$ $\;\;\;F\quad O\quad\;\; I\quad\;\; L$
Step 4. Multiply the *Last* terms.	$n^3 - n^2 + 4n - 4$ $\;\;\;F\quad O\quad\;\; I\quad\;\; L$
Step 5. Combine like terms—there are none.	$n^3 - n^2 + 4n - 4$

ⓑ

$$(3pq + 5)(6pq - 11)$$

$$(3pq + 5)(6pq - 11)$$

Step 1. Multiply the *First* terms.	$18p^2q^2 + \underline{\;\;} + \underline{\;\;} + \underline{\;\;}$ $\quad\;\;\; F\quad\;\; O\quad\; I\quad\; L$
Step 2. Multiply the *Outer* terms.	$18p^2q^2 - 33pq + \underline{\;\;} + \underline{\;\;}$ $\quad\;\;\; F\quad\quad O\quad\;\; I\quad\; L$
Step 3. Multiply the *Inner* terms.	$18p^2q^2 - 33pq + 30pq + \underline{\;\;}$ $\quad\;\;\; F\quad\quad O\quad\quad I\quad\;\; L$
Step 4. Multiply the *Last* terms.	$18p^2q^2 - 33pq + 30pq - 55$ $\quad\;\;\; F\quad\quad O\quad\quad I\quad\;\; L$
Step 5. Combine like terms.	$18p^2q^2 - 3pq - 55$

> **TRY IT : :** 5.57   Multiply: ⓐ $\left(x^2 + 6\right)(x - 8)$ ⓑ $(2ab + 5)(4ab - 4)$.

> **TRY IT : :** 5.58   Multiply: ⓐ $\left(y^2 + 7\right)(y - 9)$ ⓑ $(2xy + 3)(4xy - 5)$.

The FOIL method is usually the quickest method for multiplying two binomials, but it *only* works for binomials. You can use the Distributive Property to find the product of any two polynomials. Another method that works for all polynomials is the Vertical Method. It is very much like the method you use to multiply whole numbers. Look carefully at this example of multiplying two-digit numbers.

$$\begin{array}{r} 23 \\ \times 46 \\ \hline 138 \\ 92 \\ \hline 1058 \end{array}$$

partial product	Start by multiplying 23 by 6 to get 138.
partial product	Next, multiply 23 by 4, lining up the partial product in the correct columns.
product	Last you add the partial products.

Now we'll apply this same method to multiply two binomials.

EXAMPLE 5.30

Multiply using the Vertical Method: $(3y - 1)(2y - 6)$.

✓ **Solution**

It does not matter which binomial goes on the top.

$$3y - 1$$
$$\times\ 2y - 6$$

Multiply $3y - 1$ by -6.	$-18y + 6$	partial product
Multiply $3y - 1$ by $2y$.	$6y^2 - 2y$	partial product
Add like terms.	$6y^2 - 20y + 6$	product

Notice the partial products are the same as the terms in the FOIL method.

$$(3y - 1)(2y - 6)$$
$$6y^2 - 2y - 18y + 6$$
$$6y^2 - 20y + 6$$

$$3y - 1$$
$$\times\ 2y - 6$$
$$-18y + 6$$
$$6y^2 - 2y$$
$$6y^2 - 20x + 6$$

> **TRY IT :: 5.59**   Multiply using the Vertical Method: $(5m - 7)(3m - 6)$.

> **TRY IT :: 5.60**   Multiply using the Vertical Method: $(6b - 5)(7b - 3)$.

We have now used three methods for multiplying binomials. Be sure to practice each method, and try to decide which one you prefer. The methods are listed here all together, to help you remember them.

Multiplying Two Binomials

To multiply binomials, use the:

- Distributive Property
- FOIL Method
- Vertical Method

Multiply a Polynomial by a Polynomial

We have multiplied monomials by monomials, monomials by polynomials, and binomials by binomials. Now we're ready to multiply a polynomial by a polynomial. Remember, FOIL will not work in this case, but we can use either the Distributive Property or the Vertical Method.

EXAMPLE 5.31

Multiply $(b + 3)(2b^2 - 5b + 8)$ using ⓐ the Distributive Property and ⓑ the Vertical Method.

✓ **Solution**

ⓐ

$$(b + 3)(2b^2 - 5b + 8)$$

Distribute.	$b(2b^2 - 5b + 8) + 3(2b^2 - 5b + 8)$
Multiply.	$2b^3 - 5b^2 + 8b + 6b^2 - 15b + 24$
Combine like terms.	$2b^3 + b^2 - 7b + 24$

ⓑ It is easier to put the polynomial with fewer terms on the bottom because we get fewer partial products this way.

Multiply $(2b^2 - 5b + 8)$ by 3.	$\begin{array}{r} 2b^2 - 5b + 8 \\ \times \quad\quad b + 3 \\ \hline 6b^2 - 15b + 24 \end{array}$
Multiply $(2b^2 - 5b + 8)$ by b.	
Add like terms.	$\begin{array}{r} 2b^3 - 5b^2 + 8b \\ \hline 2b^3 + b^2 - 7b + 24 \end{array}$

 TRY IT ::: 5.61   Multiply $(y - 3)(y^2 - 5y + 2)$ using ⓐ the Distributive Property and ⓑ the Vertical Method.

 TRY IT ::: 5.62

Multiply $(x + 4)(2x^2 - 3x + 5)$ using ⓐ the Distributive Property and ⓑ The Vertical Method.

We have now seen two methods you can use to multiply a polynomial by a polynomial. After you practice each method, you'll probably find you prefer one way over the other. We list both methods are listed here, for easy reference.

> **Multiplying a Polynomial by a Polynomial**
>
> To multiply a trinomial by a binomial, use the:
> - Distributive Property
> - Vertical Method

Multiply Special Products

Mathematicians like to look for patterns that will make their work easier. A good example of this is squaring binomials. While you can always get the product by writing the binomial twice and multiplying them, there is less work to do if you learn to use a pattern. Let's start by looking at three examples and look for a pattern.

Look at these results. Do you see any patterns?

$$(x + 9)^2 \qquad (y - 7)^2 \qquad (2x + 3)^2$$

$$(x + 9)(x + 9) \qquad (y - 7)(y - 7) \qquad (2x + 3)(2x + 3)$$

$$x^2 + 9x + 9x + 81 \qquad y^2 - 7y - 7y + 49 \qquad 4x^2 + 6x + 6x + 9$$

$$x^2 + 18x + 81 \qquad y^2 - 14y + 49 \qquad 4x^2 + 12x + 9$$

What about the number of terms? In each example we squared a binomial and the result was a trinomial.

$$(a + b)^2 = \underline{\quad} + \underline{\quad} + \underline{\quad}$$

Now look at the *first term* in each result. Where did it come from?

The first term is the product of the first terms of each binomial. Since the binomials are identical, it is just the square of the first term!

$$(a + b)^2 = a^2 + \underline{\quad} + \underline{\quad}$$

To get the first term of the product, square the first term.

Where did the *last term* come from? Look at the examples and find the pattern.

The last term is the product of the last terms, which is the square of the last term.

$$(a + b)^2 = \underline{\quad} + \underline{\quad} + b^2$$

To get the last term of the product, square the last term.

Finally, look at the *middle term*. Notice it came from adding the "outer" and the "inner" terms—which are both the same! So the middle term is double the product of the two terms of the binomial.

$$(a + b)^2 = \underline{} + 2ab + \underline{}$$
$$(a - b)^2 = \underline{} - 2ab + \underline{}$$

To get the middle term of the product, multiply the terms and double their product.

Putting it all together:

Binomial Squares Pattern

If *a* and *b* are real numbers,

$(a + b)^2 = a^2 + 2ab + b^2$	$(a + b)^2$	$=$	a^2	$+$	$2ab$	$+$	b^2
	(binomial)²		(first term)²		2(product of terms)		(last term)²
$(a - b)^2 = a^2 - 2ab + b^2$	$(a - b)^2$	$=$	a^2	$-$	$2ab$	$+$	b^2
	(binomial)²		(first term)²		2(product of terms)		(last term)²

To square a binomial, square the first term, square the last term , double their product.

EXAMPLE 5.32

Multiply: ⓐ $(x + 5)^2$ ⓑ $(2x - 3y)^2$.

✓ **Solution**

ⓐ

$$\left(\overset{a\ +\ b}{x + 5}\right)^2$$

Square the first term.	$\overset{a^2\ +\ 2ab\ +\ b^2}{x^2 + \underline{} + \underline{}}$
Square the last term.	$\overset{a^2\ +\ 2ab\ +\ b^2}{x^2 + \underline{} + 5^2}$
Double their product.	$\overset{a^2\ +\ 2 \cdot a \cdot b\ +\ b^2}{x^2 + 2 \cdot x \cdot 5 + 5^2}$
Simplify.	$x^2 + 10x + 25$

ⓑ

$$\left(\overset{a\ -\ b}{2x - 3y}\right)^2$$

Use the pattern.	$\overset{a^2\ -\ 2 \cdot a \cdot b\ +\ b^2}{(2x)^2 - 2 \cdot 2x \cdot 3y + (3y)^2}$
Simplify.	$4x^2 - 12xy + 9y^2$

> **TRY IT : : 5.63**   Multiply: ⓐ $(x + 9)^2$ ⓑ $(2c - d)^2$.

> **TRY IT : : 5.64**   Multiply: ⓐ $(y + 11)^2$ ⓑ $(4x - 5y)^2$.

We just saw a pattern for squaring binomials that we can use to make multiplying some binomials easier. Similarly, there is a pattern for another product of binomials. But before we get to it, we need to introduce some vocabulary.

A pair of binomials that each have the same first term and the same last term, but one is a sum and one is a difference is called a **conjugate pair** and is of the form $(a - b)$, $(a + b)$.

Conjugate Pair

A conjugate pair is two binomials of the form

$$(a - b), \ (a + b).$$

The pair of binomials each have the same first term and the same last term, but one binomial is a sum and the other is a difference.

There is a nice pattern for finding the product of conjugates. You could, of course, simply FOIL to get the product, but using the pattern makes your work easier. Let's look for the pattern by using FOIL to multiply some conjugate pairs.

$$(x + 9)(x - 9) \qquad (y - 8)(y + 8) \qquad (2x - 5)(2x + 5)$$

$$x^2 - 9x + 9x - 81 \qquad y^2 + 8y - 8y - 64 \qquad 4x^2 + 10x - 10x - 25$$

$$x^2 - 81 \qquad y^2 - 64 \qquad 4x^2 - 25$$

What do you observe about the products?

The product of the two binomials is also a binomial! Most of the products resulting from FOIL have been trinomials.

Each *first term* is the product of the first terms of the binomials, and since they are identical it is the square of the first term.

$$(a + b)(a - b) = a^2 - \underline{\quad}$$

> To get the first term, square the first term.

The *last term* came from multiplying the last terms, the square of the last term.

$$(a + b)(a - b) = a^2 - b^2$$

> To get the last term, square the last term.

Why is there no middle term? Notice the two middle terms you get from FOIL combine to 0 in every case, the result of one addition and one subtraction.

The product of conjugates is always of the form $a^2 - b^2$. This is called a **difference of squares**.

This leads to the pattern:

Product of Conjugates Pattern

If a and b are real numbers,

$$(a - b)(a + b) = a^2 - b^2 \qquad (a - b)(a + b) = a^2 \overset{\text{difference}}{-} b^2$$

$$\underset{\text{conjugates}}{\underbrace{\qquad}} \quad \underset{\text{squares}}{\qquad}$$

The product is called a difference of squares.

To multiply conjugates, square the first term, square the last term, write it as a difference of squares.

EXAMPLE 5.33

Multiply using the product of conjugates pattern: ⓐ $(2x + 5)(2x - 5)$ ⓑ $(5m - 9n)(5m + 9n)$.

✓ **Solution**

ⓐ

Are the binomials conjugates?	$(2x + 5)(2x - 5)$
It is the product of conjugates.	$\left(\overset{a\ +\ b}{2x + 5}\right)\left(\overset{a\ -\ b}{2x - 5}\right)$
Square the first term, $2x$.	$\overset{a^2\ -\ b^2}{(2x)^2 - \_\_\_}$
Square the last term, 5.	$\overset{a^2\ -\ b^2}{(2x)^2 - 5^2}$
Simplify. The product is a difference of squares.	$\overset{a^2\ -\ b^2}{4x^2 - 25}$

ⓑ

	$(5m - 9n)(5m + 9n)$
This fits the pattern.	$\left(\overset{a\ -\ b}{5m - 9n}\right)\left(\overset{a\ +\ b}{5m + 9n}\right)$
Use the pattern.	$\overset{a^2\ -\ b^2}{(5m)^2 - (9n)^2}$
Simplify.	$25m^2 - 81n^2$

> **TRY IT :: 5.65** Multiply: ⓐ $(6x + 5)(6x - 5)$ ⓑ $(4p - 7q)(4p + 7q)$.

> **TRY IT :: 5.66** Multiply: ⓐ $(2x + 7)(2x - 7)$ ⓑ $(3x - y)(3x + y)$.

We just developed special product patterns for Binomial Squares and for the Product of Conjugates. The products look similar, so it is important to recognize when it is appropriate to use each of these patterns and to notice how they differ. Look at the two patterns together and note their similarities and differences.

Comparing the Special Product Patterns

Binomial Squares	Product of Conjugates
$(a + b)^2 = a^2 + 2ab + b^2$	$(a - b)(a + b) = a^2 - b^2$
$(a - b)^2 = a^2 - 2ab + b^2$	
• Squaring a binomial	• Multiplying conjugates
• Product is a **trinomial**	• Product is a **binomial.**
• Inner and outer terms with FOIL are **the same.**	• Inner and outer terms with FOIL are **opposites.**
• Middle term is **double the product** of the terms	• There is **no** middle term.

EXAMPLE 5.34

Choose the appropriate pattern and use it to find the product:

ⓐ $(2x - 3)(2x + 3)$ ⓑ $(5x - 8)^2$ ⓒ $(6m + 7)^2$ ⓓ $(5x - 6)(6x + 5)$.

✓ Solution

ⓐ $(2x - 3)(2x + 3)$

These are conjugates. They have the same first numbers, and the same last numbers, and one binomial is a sum and the other is a difference. It fits the Product of Conjugates pattern.

$$\overset{\color{red}(a-b)}{(2x-3)}\overset{\color{red}(a+b)}{(2x+3)}$$

Use the pattern.	$\overset{\color{red}a^2\ -\ b^2}{(2x)^2 - 3^2}$
Simplify.	$4x^2 - 9$

ⓑ $(8x - 5)^2$

We are asked to square a binomial. It fits the binomial squares pattern.

$$\overset{\color{red}(a-b)^2}{(8x-5)^2}$$

Use the pattern.	$\overset{\color{red}a^2\ -\ 2ab\ +\ b^2}{(8x)^2 - 2\cdot 8x\cdot 5 + 5^2}$
Simplify.	$64x^2 - 80x + 25$

ⓒ $(6m + 7)^2$

Again, we will square a binomial so we use the binomial squares pattern.

$$\overset{\color{red}(a+b)^2}{(6m+7)^2}$$

Use the pattern.	$\overset{\color{red}a^2\ +\ 2ab\ +\ b^2}{(6m)^2 + 2\cdot 6m\cdot 7 + 7^2}$
Simplify.	$36m^2 + 84m + 49$

ⓓ $(5x - 6)(6x + 5)$

This product does not fit the patterns, so we will use FOIL.

$$(5x - 6)(6x + 5)$$

Use FOIL.	$30x^2 + 25x - 36x - 30$
Simplify.	$30x^2 - 11x - 30$

> **TRY IT : : 5.67**   Choose the appropriate pattern and use it to find the product:
>
> ⓐ $(9b - 2)(2b + 9)$ ⓑ $(9p - 4)^2$ ⓒ $(7y + 1)^2$ ⓓ $(4r - 3)(4r + 3)$.

> **TRY IT : : 5.68**   Choose the appropriate pattern and use it to find the product:
>
> ⓐ $(6x + 7)^2$ ⓑ $(3x - 4)(3x + 4)$ ⓒ $(2x - 5)(5x - 2)$ ⓓ $(6n - 1)^2$.

Multiply Polynomial Functions

Just as polynomials can be multiplied, polynomial functions can also be multiplied.

Multiplication of Polynomial Functions

For functions $f(x)$ and $g(x)$,

$$(f \cdot g)(x) = f(x) \cdot g(x)$$

EXAMPLE 5.35

For functions $f(x) = x + 2$ and $g(x) = x^2 - 3x - 4$, find: ⓐ $(f \cdot g)(x)$ ⓑ $(f \cdot g)(2)$.

✓ **Solution**

ⓐ

$$(f \cdot g)(x) = f(x) \cdot g(x)$$

Substitute for $f(x)$ and $g(x)$.　　$(f \cdot g)(x) = (x + 2)(x^2 - 3x - 4)$

Multiply the polynomials.　　$(f \cdot g)(x) = x(x^2 - 3x - 4) + 2(x^2 - 3x - 4)$

Distribute.　　$(f \cdot g)(x) = x^3 - 3x^2 - 4x + 2x^2 - 6x - 8$

Combine like terms.　　$(f \cdot g)(x) = x^3 - x^2 - 10x - 8$

ⓑ In part ⓐ we found $(f \cdot g)(x)$ and now are asked to find $(f \cdot g)(2)$.

$$(f \cdot g)(x) = x^3 - x^2 - 10x - 8$$

To find $(f \cdot g)(2)$, substitute $x = 2$.　　$(f \cdot g)(2) = 2^3 - 2^2 - 10 \cdot 2 - 8$

$$(f \cdot g)(2) = 8 - 4 - 20 - 8$$

$$(f \cdot g)(2) = -24$$

> **TRY IT : : 5.69**　　For functions $f(x) = x - 5$ and $g(x) = x^2 - 2x + 3$, find ⓐ $(f \cdot g)(x)$ ⓑ $(f \cdot g)(2)$.

> **TRY IT : : 5.70**　　For functions $f(x) = x - 7$ and $g(x) = x^2 + 8x + 4$, find ⓐ $(f \cdot g)(x)$ ⓑ $(f \cdot g)(2)$.

▶ **MEDIA : :**

Access this online resource for additional instruction and practice with multiplying polynomials.

- **Introduction to special products of binomials (https://openstax.org/l/37Introspecprod)**

5.3 EXERCISES

Practice Makes Perfect

Multiply Monomials

In the following exercises, multiply the monomials.

178.
ⓐ $\left(6y^7\right)\left(-3y^4\right)$

ⓑ $\left(\frac{4}{7}rs^2\right)\left(14rs^3\right)$

179.
ⓐ $\left(-10x^5\right)\left(-3x^3\right)$

ⓑ $\left(\frac{5}{8}x^3y\right)\left(24x^5y\right)$

180.
ⓐ $(-8u^6)(-9u)$

ⓑ $\left(\frac{2}{3}x^2y\right)\left(\frac{3}{4}xy^2\right)$

181.
ⓐ $(-6c^4)(-12c)$

ⓑ $\left(\frac{3}{5}m^3n^2\right)\left(\frac{5}{9}m^2n^3\right)$

Multiply a Polynomial by a Monomial

In the following exercises, multiply.

182.
ⓐ $-8x(x^2 + 2x - 15)$

ⓑ $5pq^3(p^2 - 2pq + 6q^2)$

183.
ⓐ $-5t(t^2 + 3t - 18)$;

ⓑ $9r^3s(r^2 - 3rs + 5s^2)$

184.
ⓐ $-8y(y^2 + 2y - 15)$

ⓑ $-4y^2z^2(3y^2 + 12yz - z^2)$

185.
ⓐ $-5m(m^2 + 3m - 18)$

ⓑ $-3x^2y^2(7x^2 + 10xy - y^2)$

Multiply a Binomial by a Binomial

In the following exercises, multiply the binomials using ⓐ the Distributive Property; ⓑ the FOIL method; ⓒ the Vertical Method.

186. $(w + 5)(w + 7)$

187. $(y + 9)(y + 3)$

188. $(4p + 11)(5p - 4)$

189. $(7q + 4)(3q - 8)$

In the following exercises, multiply the binomials. Use any method.

190. $(x + 8)(x + 3)$

191. $(y - 6)(y - 2)$

192. $(2t - 9)(10t + 1)$

193. $(6p + 5)(p + 1)$

194. $(q - 5)(q + 8)$

195. $(m + 11)(m - 4)$

196. $(7m + 1)(m - 3)$

197. $(3r - 8)(11r + 1)$

198. $(x^2 + 3)(x + 2)$

199. $(y^2 - 4)(y + 3)$

200. $(5ab - 1)(2ab + 3)$

201. $(2xy + 3)(3xy + 2)$

202. $(x^2 + 8)(x^2 - 5)$

203. $(y^2 - 7)(y^2 - 4)$

204. $(6pq - 3)(4pq - 5)$

205. $(3rs - 7)(3rs - 4)$

Multiply a Polynomial by a Polynomial

In the following exercises, multiply using ⓐ the Distributive Property; ⓑ the Vertical Method.

206. $(x + 5)(x^2 + 4x + 3)$

207. $(u + 4)(u^2 + 3u + 2)$

208. $(y + 8)(4y^2 + y - 7)$

209. $(a + 10)(3a^2 + a - 5)$

210. $(y^2 - 3y + 8)(4y^2 + y - 7)$

211.
$(2a^2 - 5a + 10)(3a^2 + a - 5)$

Multiply Special Products

In the following exercises, multiply. Use either method.

212. $(w - 7)(w^2 - 9w + 10)$

213. $(p - 4)(p^2 - 6p + 9)$

214. $(3q + 1)(q^2 - 4q - 5)$

215. $(6r + 1)(r^2 - 7r - 9)$

In the following exercises, square each binomial using the Binomial Squares Pattern.

216. $(w + 4)^2$

217. $(q + 12)^2$

218. $(3x - y)^2$

219. $(2y - 3z)^2$

220. $\left(y + \frac{1}{4}\right)^2$

221. $\left(x + \frac{2}{3}\right)^2$

222. $\left(\frac{1}{5}x - \frac{1}{7}y\right)^2$

223. $\left(\frac{1}{8}x - \frac{1}{9}y\right)^2$

224. $(3x^2 + 2)^2$

225. $(5u^2 + 9)^2$

226. $(4y^3 - 2)^2$

227. $(8p^3 - 3)^2$

In the following exercises, multiply each pair of conjugates using the Product of Conjugates Pattern.

228. $(5k + 6)(5k - 6)$

229. $(8j + 4)(8j - 4)$

230. $(11k + 4)(11k - 4)$

231. $(9c + 5)(9c - 5)$

232. $(9c - 2d)(9c + 2d)$

233. $(7w + 10x)(7w - 10x)$

234. $\left(m + \frac{2}{3}n\right)\left(m - \frac{2}{3}n\right)$

235. $\left(p + \frac{4}{5}q\right)\left(p - \frac{4}{5}q\right)$

236. $(ab - 4)(ab + 4)$

237. $(xy - 9)(xy + 9)$

238.
$(12p^3 - 11q^2)(12p^3 + 11q^2)$

239. $(15m^2 - 8n^4)(15m^2 + 8n^4)$

In the following exercises, find each product.

240. $(p - 3)(p + 3)$

241. $(t - 9)^2$

242. $(m + n)^2$

243. $(2x + y)(x - 2y)$

244. $(2r + 12)^2$

245. $(3p + 8)(3p - 8)$

246. $(7a + b)(a - 7b)$

247. $(k - 6)^2$

248. $(a^5 - 7b)^2$

249. $(x^2 + 8y)(8x - y^2)$

250. $(r^6 + s^6)(r^6 - s^6)$

251. $(y^4 + 2z)^2$

252. $(x^5 + y^5)(x^5 - y^5)$

253. $(m^3 - 8n)^2$

254. $(9p + 8q)^2$

255. $(r^2 - s^3)(r^3 + s^2)$

Mixed Practice

256. $(10y - 6) + (4y - 7)$

257. $(15p - 4) + (3p - 5)$

258. $(x^2 - 4x - 34) - (x^2 + 7x - 6)$

259. $(j^2 - 8j - 27) - (j^2 + 2j - 12)$

260. $(\frac{1}{5}f^8)(20f^3)$

261. $(\frac{1}{4}d^5)(36d^2)$

262. $(4a^3 b)(9a^2 b^6)$

263. $(6m^4 n^3)(7mn^5)$

264. $-5m(m^2 + 3m - 18)$

265. $5q^3(q^2 - 2q + 6)$

266. $(s - 7)(s + 9)$

267. $(y^2 - 2y)(y + 1)$

268. $(5x - y)(x - 4)$

269. $(6k - 1)(k^2 + 2k - 4)$

270. $(3x - 11y)(3x - 11y)$

271. $(11 - b)(11 + b)$

272. $(rs - \frac{2}{7})(rs + \frac{2}{7})$

273. $(2x^2 - 3y^4)(2x^2 + 3y^4)$

274. $(m - 15)^2$

275. $(3d + 1)^2$

276. $(4a + 10)^2$

277. $\left(3z + \frac{1}{5}\right)^2$

Multiply Polynomial Functions

278. For functions $f(x) = x + 2$ and $g(x) = 3x^2 - 2x + 4$, find ⓐ $(f \cdot g)(x)$ ⓑ $(f \cdot g)(-1)$

279. For functions $f(x) = x - 1$ and $g(x) = 4x^2 + 3x - 5$, find ⓐ $(f \cdot g)(x)$ ⓑ $(f \cdot g)(-2)$

280. For functions $f(x) = 2x - 7$ and $g(x) = 2x + 7$, find ⓐ $(f \cdot g)(x)$ ⓑ $(f \cdot g)(-3)$

281. For functions $f(x) = 7x - 8$ and $g(x) = 7x + 8$, find ⓐ $(f \cdot g)(x)$ ⓑ $(f \cdot g)(-2)$

282. For functions $f(x) = x^2 - 5x + 2$ and $g(x) = x^2 - 3x - 1$, find ⓐ $(f \cdot g)(x)$ ⓑ $(f \cdot g)(-1)$

283. For functions $f(x) = x^2 + 4x - 3$ and $g(x) = x^2 + 2x + 4$, find ⓐ $(f \cdot g)(x)$ ⓑ $(f \cdot g)(1)$

Writing Exercises

284. Which method do you prefer to use when multiplying two binomials: the Distributive Property or the FOIL method? Why? Which method do you prefer to use when multiplying a polynomial by a polynomial: the Distributive Property or the Vertical Method? Why?

285. Multiply the following:

$(x + 2)(x - 2)$

$(y + 7)(y - 7)$

$(w + 5)(w - 5)$

Explain the pattern that you see in your answers.

286. Multiply the following:

$(p + 3)(p + 3)$

$(q + 6)(q + 6)$

$(r + 1)(r + 1)$

Explain the pattern that you see in your answers.

287. Why does $(a + b)^2$ result in a trinomial, but $(a - b)(a + b)$ result in a binomial?

Self Check

ⓐ *After completing the exercises, use this checklist to evaluate your mastery of the objectives of this section.*

I can...	Confidently	With some help	No-I don't get it!
multiply monomials.			
multiply a polynomial by a monomial.			
multiply a binomial by a binomial.			
multiply a polynomial by a polynomial.			
multiply special products.			

ⓑ *What does this checklist tell you about your mastery of this section? What steps will you take to improve?*

5.4 Dividing Polynomials

Learning Objectives

By the end of this section, you will be able to:

> Dividing monomials
> Dividing a polynomial by a monomial
> Dividing polynomials using long division
> Dividing polynomials using synthetic division
> Dividing polynomial functions
> Use the remainder and factor theorems

Be Prepared!

Before you get started, take this readiness quiz.

1. Add: $\frac{3}{d} + \frac{x}{d}$.
    If you missed this problem, review **Example 1.28**.

2. Simplify: $\frac{30xy^3}{5xy}$.
    If you missed this problem, review **Example 1.25**.

3. Combine like terms: $8a^2 + 12a + 1 + 3a^2 - 5a + 4$.
    If you missed this problem, review **Example 1.7**.

Dividing Monomials

We are now familiar with all the properties of exponents and used them to multiply polynomials. Next, we'll use these properties to divide monomials and polynomials.

EXAMPLE 5.36

Find the quotient: $54a^2 b^3 \div \left(-6ab^5\right)$.

 Solution

When we divide monomials with more than one variable, we write one fraction for each variable.

$$54a^2 b^3 \div \left(-6ab^5\right)$$

Rewrite as a fraction.	$\dfrac{54a^2 b^3}{-6ab^5}$
Use fraction multiplication.	$\dfrac{54}{-6} \cdot \dfrac{a^2}{a} \cdot \dfrac{b^3}{b^5}$
Simplify and use the Quotient Property.	$-9 \cdot a \cdot \dfrac{1}{b^2}$
Multiply.	$-\dfrac{9a}{b^2}$

 TRY IT :: 5.71   Find the quotient: $-72a^7 b^3 \div \left(8a^{12} b^4\right)$.

 TRY IT :: 5.72   Find the quotient: $-63c^8 d^3 \div \left(7c^{12} d^2\right)$.

Once you become familiar with the process and have practiced it step by step several times, you may be able to simplify a fraction in one step.

EXAMPLE 5.37

Find the quotient: $\dfrac{14x^7 y^{12}}{21x^{11} y^6}$.

✓ Solution

Be very careful to simplify $\dfrac{14}{21}$ by dividing out a common factor, and to simplify the variables by subtracting their exponents.

$$\frac{14x^7 y^{12}}{21x^{11} y^6}$$

Simplify and use the Quotient Property. $\dfrac{2y^6}{3x^4}$

> **TRY IT :: 5.73**   Find the quotient: $\dfrac{28x^5 y^{14}}{49x^9 y^{12}}$.

> **TRY IT :: 5.74**   Find the quotient: $\dfrac{30m^5 n^{11}}{48m^{10} n^{14}}$.

Divide a Polynomial by a Monomial

Now that we know how to divide a monomial by a monomial, the next procedure is to divide a polynomial of two or more terms by a monomial.

The method we'll use to divide a polynomial by a monomial is based on the properties of fraction addition. So we'll start with an example to review fraction addition. The sum $\dfrac{y}{5} + \dfrac{2}{5}$ simplifies to $\dfrac{y+2}{5}$.

Now we will do this in reverse to split a single fraction into separate fractions. For example, $\dfrac{y+2}{5}$ can be written $\dfrac{y}{5} + \dfrac{2}{5}$.

This is the "reverse" of fraction addition and it states that if a, b, and c are numbers where $c \neq 0$, then $\dfrac{a+b}{c} = \dfrac{a}{c} + \dfrac{b}{c}$. We will use this to divide polynomials by monomials.

Division of a Polynomial by a Monomial

To divide a polynomial by a monomial, divide each term of the polynomial by the monomial.

EXAMPLE 5.38

Find the quotient: $\left(18x^3 y - 36xy^2\right) \div (-3xy)$.

✓ Solution

$$\left(18x^3 y - 36xy^2\right) \div (-3xy)$$

Rewrite as a fraction. $\dfrac{18x^3 y - 36xy^2}{-3xy}$

Divide each term by the divisor. Be careful with the signs! $\dfrac{18x^3 y}{-3xy} - \dfrac{36xy^2}{-3xy}$

Simplify. $-6x^2 + 12y$

> **TRY IT :: 5.75**   Find the quotient: $\left(32a^2 b - 16ab^2\right) \div (-8ab)$.

> **TRY IT : : 5.76** Find the quotient: $\left(-48a^8 b^4 - 36a^6 b^5\right) \div \left(-6a^3 b^3\right)$.

Divide Polynomials Using Long Division

Divide a polynomial by a binomial, we follow a procedure very similar to long division of numbers. So let's look carefully the steps we take when we divide a 3-digit number, 875, by a 2-digit number, 25.

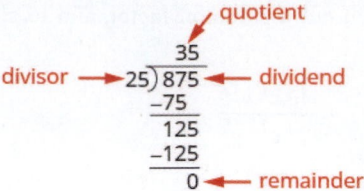

We check division by multiplying the quotient by the divisor.

If we did the division correctly, the product should equal the dividend.

$$35 \cdot 25$$
$$875 \checkmark$$

Now we will divide a trinomial by a binomial. As you read through the example, notice how similar the steps are to the numerical example above.

EXAMPLE 5.39

Find the quotient: $\left(x^2 + 9x + 20\right) \div (x + 5)$.

✓ **Solution**

$(x^2 + 9x + 20) \div (x + 5)$

Write it as a long division problem. Be sure the dividend is in standard form.	$x + 5\overline{)x^2 + 9x + 20}$
Divide x^2 by x. It may help to ask yourself, "What do I need to multiply x by to get x^2?"	$x + 5\overline{)x^2 + 9x + 20}$ (quotient x)
Put the answer, x, in the quotient over the x term. Multiply x times $x + 5$. Line up the like terms under the dividend.	$x + 5\overline{)x^2 + 9x + 20}$, $x^2 + 5x$
Subtract $x^2 + 5x$ from $x^2 + 9x$. You may find it easier to change the signs and then add. Then bring down the last term, 20.	$\frac{-x^2 + (-5x)}{4x + 20}$
Divide $4x$ by x. It may help to ask yourself, "What do I need to multiply x by to get $4x$?" Put the answer, 4, in the quotient over the constant term.	quotient $x + 4$
Multiply 4 times $x + 5$.	$4x + 20$
Subtract $4x + 20$ from $4x + 20$.	$\frac{-4x + (-20)}{0}$

Check:

Multiply the quotient by the divisor. $(x+4)(x+5)$

You should get the dividend. $x^2 + 9x + 20$ ✓

> **TRY IT :: 5.77** Find the quotient: $\left(y^2 + 10y + 21\right) \div (y+3)$.

> **TRY IT :: 5.78** Find the quotient: $\left(m^2 + 9m + 20\right) \div (m+4)$.

When we divided 875 by 25, we had no remainder. But sometimes division of numbers does leave a remainder. The same is true when we divide polynomials. In the next example, we'll have a division that leaves a remainder. We write the remainder as a fraction with the divisor as the denominator.

Look back at the dividends in previous examples. The terms were written in descending order of degrees, and there were no missing degrees. The dividend in this example will be $x^4 - x^2 + 5x - 6$. It is missing an x^3 term. We will add in $0x^3$ as a placeholder.

EXAMPLE 5.40

Find the quotient: $\left(x^4 - x^2 + 5x - 6\right) \div (x+2)$.

✓ **Solution**

Notice that there is no x^3 term in the dividend. We will add $0x^3$ as a placeholder.

$(x^4 - x^2 + 5x - 6) \div (x+2)$

Write it as a long division problem. Be sure the dividend is in standard form with placeholders for missing terms.	$x+2\overline{)x^4 + 0x^3 - x^2 + 5x - 6}$
Divide x^4 by x. Put the answer, x^3, in the quotient over the x^3 term. Multiply x^3 times $x+2$. Line up the like terms. Subtract and then bring down the next term.	$\begin{array}{r} x^3 \\ x+2\overline{)x^4 + 0x^3 - x^2 + 5x - 6} \\ \underline{-(x^4 + 2x^3)} \\ -2x^3 - x^2 \end{array}$ It may be helpful to change the signs and add.
Divide $-2x^3$ by x. Put the answer, $-2x^2$, in the quotient over the x^2 term. Multiply $-2x^2$ times $x+1$. Line up the like terms Subtract and bring down the next term.	$\begin{array}{r} x^3 - 2x^2 \\ x+2\overline{)x^4 + 0x^3 - x^2 + 5x - 6} \\ \underline{-(x^4 + 2x^3)} \\ -2x^3 - x^2 \\ \underline{-(-2x^3 - 4x^2)} \\ 3x^2 + 5x \end{array}$ It may be helpful to change the signs and add.
Divide $3x^2$ by x. Put the answer, $3x$, in the quotient over the x term. Multiply $3x$ times $x+1$. Line up the like terms. Subtract and bring down the next term.	$\begin{array}{r} x^3 - 2x^2 + 3x \\ x+2\overline{)x^4 + 0x^3 - x^2 + 5x - 6} \\ \underline{-(x^4 + 2x^3)} \\ -2x^3 - x^2 \\ \underline{-(-2x^3 - 4x^2)} \\ 3x^2 + 5x \\ \underline{-(3x^2 + 6x)} \\ -x - 6 \end{array}$ It may be helpful to change the signs and add.

Divide $-x$ by x.
Put the answer, -1, in the quotient over the
constant term.
Multiply -1 times $x + 1$. Line up the like terms.
Change the signs, add.

Write the remainder as a fraction with the divisor
as the denominator.

$$x^3 - 2x^2 + 3x - 1 - \frac{4}{x+2}$$
$$x + 2 \overline{)\, x^4 + 0x^3 - \ x^2 + 5x - 6}$$
$$\underline{-(x^4 + 2x^3)}$$
$$-2x^3 - \ x^2$$
$$\underline{-(-2x^3 - 4x^2)}$$
$$3x^2 + 5x$$
$$\underline{-(3x^2 + 6x)}$$
$$-x - 6$$
$$\underline{-(-x - 2)}$$
$$-4$$

It may be helpful to change the signs and add.

To check, multiply
$$(x + 2)\left(x^3 - 2x^2 + 3x - 1 - \frac{4}{x+2}\right).$$
The result should be $x^4 - x^2 + 5x - 6$.

> **TRY IT : : 5.79**   Find the quotient: $\left(x^4 - 7x^2 + 7x + 6\right) \div (x + 3)$.

> **TRY IT : : 5.80**   Find the quotient: $\left(x^4 - 11x^2 - 7x - 6\right) \div (x + 3)$.

In the next example, we will divide by $2a - 3$. As we divide, we will have to consider the constants as well as the variables.

EXAMPLE 5.41

Find the quotient: $\left(8a^3 + 27\right) \div (2a + 3)$.

⊘ Solution

This time we will show the division all in one step. We need to add two placeholders in order to divide.

$$(8a^3 + 27) \div (2a + 3)$$

$$4a^2 - 6a + 9$$
$$2a + 3 \overline{)\, 8a^3 + 0a^2 + 0a + 27}$$
$$\underline{-(8a^3 + 12a^2)} \quad \leftarrow 4a^2(2a + 3)$$
$$-12a^2 + 0a$$
$$\underline{-(-12a^2 - 18a)} \quad \leftarrow 6a(2a + 3)$$
$$18a + 27$$
$$\underline{-(18a + 27)} \quad \leftarrow 9(2a + 3)$$
$$0$$

To check, multiply $(2a + 3)\left(4a^2 - 6a + 9\right)$.

The result should be $8a^3 + 27$.

> **TRY IT : : 5.81**   Find the quotient: $\left(x^3 - 64\right) \div (x - 4)$.

> **TRY IT : : 5.82**   Find the quotient: $\left(125x^3 - 8\right) \div (5x - 2)$.

Divide Polynomials using Synthetic Division

As we have mentioned before, mathematicians like to find patterns to make their work easier. Since long division can be tedious, let's look back at the long division we did in **Example 5.39** and look for some patterns. We will use this as a basis for what is called synthetic division. The same problem in the synthetic division format is shown next.

$$x + 5 \overline{)\; 1x^2 + 9x + 20}$$

with $x + 4$ on top, $-x^2 + (-5x)$, then $4x + 20$, $-4x + (-20)$, 0 — same coefficients →

$$-5 \,\big|\; 1 \quad 9 \quad 20$$
$$ \; -5 \; -20$$
$$ 1 \quad 4 \quad \boxed{0} \leftarrow \text{remainder}$$

coefficients of quotient

Synthetic division basically just removes unnecessary repeated variables and numbers. Here all the x and x^2 are removed. as well as the $-x^2$ and $-4x$ as they are opposite the term above.

The first row of the synthetic division is the coefficients of the dividend. The -5 is the opposite of the 5 in the divisor.

The second row of the synthetic division are the numbers shown in red in the division problem.
The third row of the synthetic division are the numbers shown in blue in the division problem.
Notice the quotient and remainder are shown in the third row.

Synthetic division only works when the divisor is of the form $x - c$.

The following example will explain the process.

EXAMPLE 5.42

Use synthetic division to find the quotient and remainder when $2x^3 + 3x^2 + x + 8$ is divided by $x + 2$.

✓ **Solution**

Write the dividend with decreasing powers of x.	$2x^3 + 3x^2 + x + 8$
Write the coefficients of the terms as the first row of the synthetic division.	$\underline{}\,\big|\; 2 \quad 3 \quad 1 \quad 8$
Write the divisor as $x - c$ and place c in the synthetic division in the divisor box.	$-2\,\big|\; 2 \quad 3 \quad 1 \quad 8$
Bring down the first coefficient to the third row.	$-2\,\big|\; 2 \quad 3 \quad 1 \quad 8$ $ 2$
Multiply that coefficient by the divisor and place the result in the second row under the second coefficient.	$-2\,\big|\; 2 \quad 3 \quad 1 \quad 8$ $ \quad\; -4$ $ 2$
Add the second column, putting the result in the third row.	$-2\,\big|\; 2 \quad 3 \quad 1 \quad 8$ $ \quad\; -4$ $ 2 \;\; -1$
Multiply that result by the divisor and place the result in the second row under the third coefficient.	$-2\,\big|\; 2 \quad 3 \quad 1 \quad 8$ $ \quad\; -4 \;\; 2$ $ 2 \;\; -1$
Add the third column, putting the result in the third row.	$-2\,\big|\; 2 \quad 3 \quad 1 \quad 8$ $ \quad\; -4 \;\; 2$ $ 2 \;\; -1 \;\; 3$
Multiply that result by the divisor and place the result in the third row under the third coefficient.	$-2\,\big|\; 2 \quad 3 \quad 1 \quad 8$ $ \quad\; -4 \;\; 2 \;\; -6$ $ 2 \;\; -1 \;\; 3$

Add the final column, putting the result in the third row.

$$\begin{array}{r|rrrr} -2 & 2 & 3 & 1 & 8 \\ & & -4 & 2 & -6 \\ \hline & 2 & -1 & 3 & \boxed{2} \end{array}$$

The quotient is $2x^2 - 1x + 3$ and the remainder is 2.

The division is complete. The numbers in the third row give us the result. The 2   −1   3 are the coefficients of the quotient. The quotient is $2x^2 - 1x + 3$. The 2 in the box in the third row is the remainder.

Check:

$$\text{(quotient)(divisor)} + \text{remainder} = \text{dividend}$$

$$\left(2x^2 - 1x + 3\right)(x + 2) + 2 \overset{?}{=} 2x^3 + 3x^2 + x + 8$$

$$2x^3 - x^2 + 3x + 4x^2 - 2x + 6 + 2 \overset{?}{=} 2x^3 + 3x^2 + x + 8$$

$$2x^3 + 3x^2 + x + 8 = 2x^3 + 3x^2 + x + 8 \checkmark$$

> **TRY IT : :** 5.83
>
> Use synthetic division to find the quotient and remainder when $3x^3 + 10x^2 + 6x - 2$ is divided by $x + 2$.

> **TRY IT : :** 5.84
>
> Use synthetic division to find the quotient and remainder when $4x^3 + 5x^2 - 5x + 3$ is divided by $x + 2$.

In the next example, we will do all the steps together.

EXAMPLE 5.43

Use synthetic division to find the quotient and remainder when $x^4 - 16x^2 + 3x + 12$ is divided by $x + 4$.

⊘ **Solution**

The polynomial $x^4 - 16x^2 + 3x + 12$ has its term in order with descending degree but we notice there is no x^3 term. We will add a 0 as a placeholder for the x^3 term. In $x - c$ form, the divisor is $x - (-4)$.

$$\begin{array}{r|rrrrr} -4 & 1 & 0 & -16 & 3 & 12 \\ & & -4 & 16 & 0 & -12 \\ \hline & 1 & -4 & 0 & 3 & \boxed{0} \end{array}$$

We divided a 4th degree polynomial by a 1st degree polynomial so the quotient will be a 3rd degree polynomial.

Reading from the third row, the quotient has the coefficients 1   −4   0   3,   which is $x^3 - 4x^2 + 3$. The remainder is 0.

> **TRY IT : :** 5.85
>
> Use synthetic division to find the quotient and remainder when $x^4 - 16x^2 + 5x + 20$ is divided by $x + 4$.

> **TRY IT : :** 5.86
>
> Use synthetic division to find the quotient and remainder when $x^4 - 9x^2 + 2x + 6$ is divided by $x + 3$.

Divide Polynomial Functions

Just as polynomials can be divided, polynomial functions can also be divided.

Division of Polynomial Functions

For functions $f(x)$ and $g(x)$, where $g(x) \neq 0$,

$$\left(\frac{f}{g}\right)(x) = \frac{f(x)}{g(x)}$$

EXAMPLE 5.44

For functions $f(x) = x^2 - 5x - 14$ and $g(x) = x + 2$, find: ⓐ $\left(\frac{f}{g}\right)(x)$ ⓑ $\left(\frac{f}{g}\right)(-4)$.

✓ **Solution**

ⓐ

$$\left(\frac{f}{g}\right)(x) = \frac{f(x)}{g(x)} \qquad \begin{array}{r} x - 7 \\ x + 2 \overline{\smash{)}x^2 - 5x - 14} \\ \underline{-(x^2 + 2x)} \\ -7x - 14 \\ \underline{-(-7x - 14)} \\ 0 \end{array}$$

Substitute for $f(x)$ and $g(x)$.

$$\left(\frac{f}{g}\right)(x) = \frac{x^2 - 5x - 14}{x + 2}$$

Divide the polynomials.

$$\left(\frac{f}{g}\right)(x) = x - 7$$

ⓑ In part ⓐ we found $\left(\frac{f}{g}\right)(x)$ and now are asked to find $\left(\frac{f}{g}\right)(-4)$.

$$\left(\frac{f}{g}\right)(x) = x - 7$$

To find $\left(\frac{f}{g}\right)(-4)$, substitute $x = -4$.

$$\left(\frac{f}{g}\right)(-4) = -4 - 7$$

$$\left(\frac{f}{g}\right)(-4) = -11$$

> **TRY IT : : 5.87**　For functions $f(x) = x^2 - 5x - 24$ and $g(x) = x + 3$, find ⓐ $\left(\frac{f}{g}\right)(x)$ ⓑ $\left(\frac{f}{g}\right)(-3)$.

> **TRY IT : : 5.88**　For functions $f(x) = x^2 - 5x - 36$ and $g(x) = x + 4$, find ⓐ $\left(\frac{f}{g}\right)(x)$ ⓑ $\left(\frac{f}{g}\right)(-5)$.

Use the Remainder and Factor Theorem

Let's look at the division problems we have just worked that ended up with a remainder. They are summarized in the chart below. If we take the dividend from each division problem and use it to define a function, we get the functions shown in the chart. When the divisor is written as $x - c$, the value of the function at c, $f(c)$, is the same as the remainder from the division problem.

Dividend	Divisor $x - c$	Remainder	Function	$f(c)$
$x^4 - x^2 + 5x - 6$	$x - (-2)$	-4	$f(x) = x^4 - x^2 + 5x - 6$	-4
$3x^3 - 2x^2 - 10x + 8$	$x - 2$	4	$f(x) = 3x^3 - 2x^2 - 10x + 8$	4
$x^4 - 16x^2 + 3x + 15$	$x - (-4)$	3	$f(x) = x^4 - 16x^2 + 3x + 15$	3

Table 5.44

To see this more generally, we realize we can check a division problem by multiplying the quotient times the divisor and add the remainder. In function notation we could say, to get the dividend $f(x)$, we multiply the quotient, $q(x)$ times the divisor, $x - c$, and add the remainder, r.

$$f(x) = q(x)(x - c) + r$$

If we evaluate this at c, we get: $$f(c) = q(c)(c - c) + r$$

$$f(c) = q(c)(0) + r$$

$$f(c) = r$$

This leads us to the Remainder Theorem.

Remainder Theorem

If the polynomial function $f(x)$ is divided by $x - c$, then the remainder is $f(c)$.

EXAMPLE 5.45

Use the Remainder Theorem to find the remainder when $f(x) = x^3 + 3x + 19$ is divided by $x + 2$.

✓ **Solution**

To use the Remainder Theorem, we must use the divisor in the $x - c$ form. We can write the divisor $x + 2$ as $x - (-2)$. So, our c is -2.

To find the remainder, we evaluate $f(c)$ which is $f(-2)$.

$$f(x) = x^3 + 3x + 19$$

To evaluate $f(-2)$, substitute $x = -2$.	$f(-2) = (-2)^3 + 3(-2) + 19$
Simplify.	$f(-2) = -8 - 6 + 19$
	$f(-2) = 5$
	The remainder is 5 when $f(x) = x^3 + 3x + 19$ is divided by $x + 2$.

Check:
Use synthetic division to check.

```
-2| 1   0   3   19
          -2   4  -14
   ------------------
   1  -2   7  |5
```

The remainder is 5.

> **TRY IT : : 5.89**

Use the Remainder Theorem to find the remainder when $f(x) = x^3 + 4x + 15$ is divided by $x + 2$.

> **TRY IT : : 5.90**

Use the Remainder Theorem to find the remainder when $f(x) = x^3 - 7x + 12$ is divided by $x + 3$.

When we divided $8a^3 + 27$ by $2a + 3$ in **Example 5.41** the result was $4a^2 - 6a + 9$. To check our work, we multiply $4a^2 - 6a + 9$ by $2a + 3$ to get $8a^3 + 27$.

$$\left(4a^2 - 6a + 9\right)(2a + 3) = 8a^3 + 27$$

Written this way, we can see that $4a^2 - 6a + 9$ and $2a + 3$ are factors of $8a^3 + 27$. When we did the division, the remainder was zero.

Whenever a divisor, $x - c$, divides a polynomial function, $f(x)$, and resulting in a remainder of zero, we say $x - c$ is a factor of $f(x)$.

The reverse is also true. If $x - c$ is a factor of $f(x)$ then $x - c$ will divide the polynomial function resulting in a remainder of zero.

We will state this in the Factor Theorem.

Factor Theorem

For any polynomial function $f(x)$,

- if $x - c$ is a factor of $f(x)$, then $f(c) = 0$
- if $f(c) = 0$, then $x - c$ is a factor of $f(x)$

EXAMPLE 5.46

Use the Remainder Theorem to determine if $x - 4$ is a factor of $f(x) = x^3 - 64$.

⊘ **Solution**

The Factor Theorem tells us that $x - 4$ is a factor of $f(x) = x^3 - 64$ if $f(4) = 0$.

$$f(x) = x^3 - 64$$

To evaluate $f(4)$ substitute $x = 4$. $\qquad f(4) = 4^3 - 64$

Simplify. $\qquad f(4) = 64 - 64$

Subtract. $\qquad f(4) = 0$

Since $f(4) = 0$, $\quad x - 4$ is a factor of $f(x) = x^3 - 64$.

> **TRY IT : : 5.91**   Use the Factor Theorem to determine if $x - 5$ is a factor of $f(x) = x^3 - 125$.

> **TRY IT : : 5.92**   Use the Factor Theorem to determine if $x - 6$ is a factor of $f(x) = x^3 - 216$.

▶ **MEDIA : :**

Access these online resources for additional instruction and practice with dividing polynomials.

- **Dividing a Polynomial by a Binomial (https://openstax.org/l/37Polybybinom)**
- **Synthetic Division & Remainder Theorem (https://openstax.org/l/37SynDivision)**

5.4 EXERCISES

Practice Makes Perfect

Divide Monomials

In the following exercises, divide the monomials.

288. $15r^4 s^9 \div \left(15r^4 s^9\right)$

289. $20m^8 n^4 \div \left(30m^5 n^9\right)$

290. $\dfrac{18a^4 b^8}{-27a^9 b^5}$

291. $\dfrac{45x^5 y^9}{-60x^8 y^6}$

292. $\dfrac{\left(10m^5 n^4\right)\left(5m^3 n^6\right)}{25m^7 n^5}$

293. $\dfrac{\left(-18p^4 q^7\right)\left(-6p^3 q^8\right)}{-36p^{12} q^{10}}$

294. $\dfrac{\left(6a^4 b^3\right)\left(4ab^5\right)}{\left(12a^2 b\right)\left(a^3 b\right)}$

295. $\dfrac{\left(4u^2 v^5\right)\left(15u^3 v\right)}{\left(12u^3 v\right)\left(u^4 v\right)}$

Divide a Polynomial by a Monomial

In the following exercises, divide each polynomial by the monomial.

296. $\left(9n^4 + 6n^3\right) \div 3n$

297. $\left(8x^3 + 6x^2\right) \div 2x$

298. $\left(63m^4 - 42m^3\right) \div \left(-7m^2\right)$

299. $\left(48y^4 - 24y^3\right) \div \left(-8y^2\right)$

300. $\dfrac{66x^3 y^2 - 110x^2 y^3 - 44x^4 y^3}{11x^2 y^2}$

301. $\dfrac{72r^5 s^2 + 132r^4 s^3 - 96r^3 s^5}{12r^2 s^2}$

302. $\dfrac{10x^2 + 5x - 4}{-5x}$

303. $\dfrac{20y^2 + 12y - 1}{-4y}$

Divide Polynomials using Long Division

In the following exercises, divide each polynomial by the binomial.

304. $\left(y^2 + 7y + 12\right) \div (y + 3)$

305. $\left(a^2 - 2a - 35\right) \div (a + 5)$

306. $\left(6m^2 - 19m - 20\right) \div (m - 4)$

307. $\left(4x^2 - 17x - 15\right) \div (x - 5)$

308. $\left(q^2 + 2q + 20\right) \div (q + 6)$

309. $\left(p^2 + 11p + 16\right) \div (p + 8)$

310. $\left(3b^3 + b^2 + 4\right) \div (b + 1)$

311. $\left(2n^3 - 10n + 28\right) \div (n + 3)$

312. $\left(z^3 + 1\right) \div (z + 1)$

313. $\left(m^3 + 1000\right) \div (m + 10)$

314. $\left(64x^3 - 27\right) \div (4x - 3)$

315. $\left(125y^3 - 64\right) \div (5y - 4)$

Divide Polynomials using Synthetic Division

In the following exercises, use synthetic Division to find the quotient and remainder.

316. $x^3 - 6x^2 + 5x + 14$ is divided by $x + 1$

317. $x^3 - 3x^2 - 4x + 12$ is divided by $x + 2$

318. $2x^3 - 11x^2 + 11x + 12$ is divided by $x - 3$

319. $2x^3 - 11x^2 + 16x - 12$ is divided by $x - 4$

320. $x^4 + 13x^2 + 13x + 3$ is divided by $x + 3$

321. $x^4 + x^2 + 6x - 10$ is divided by $x + 2$

322. $2x^4 - 9x^3 + 5x^2 - 3x - 6$ is divided by $x - 4$ **323.** $3x^4 - 11x^3 + 2x^2 + 10x + 6$ is divided by $x - 3$

Divide Polynomial Functions

In the following exercises, divide.

324. For functions $f(x) = x^2 - 13x + 36$ and $g(x) = x - 4$, find ⓐ $\left(\frac{f}{g}\right)(x)$ ⓑ $\left(\frac{f}{g}\right)(-1)$

325. For functions $f(x) = x^2 - 15x + 45$ and $g(x) = x - 9$, find ⓐ $\left(\frac{f}{g}\right)(x)$ ⓑ $\left(\frac{f}{g}\right)(-5)$

326. For functions $f(x) = x^3 + x^2 - 7x + 2$ and $g(x) = x - 2$, find ⓐ $\left(\frac{f}{g}\right)(x)$ ⓑ $\left(\frac{f}{g}\right)(2)$

327. For functions $f(x) = x^3 + 2x^2 - 19x + 12$ and $g(x) = x - 3$, find ⓐ $\left(\frac{f}{g}\right)(x)$ ⓑ $\left(\frac{f}{g}\right)(0)$

328. For functions $f(x) = x^2 - 5x + 2$ and $g(x) = x^2 - 3x - 1$, find ⓐ $(f \cdot g)(x)$ ⓑ $(f \cdot g)(-1)$

329. For functions $f(x) = x^2 + 4x - 3$ and $g(x) = x^2 + 2x + 4$, find ⓐ $(f \cdot g)(x)$ ⓑ $(f \cdot g)(1)$

Use the Remainder and Factor Theorem

In the following exercises, use the Remainder Theorem to find the remainder.

330. $f(x) = x^3 - 8x + 7$ is divided by $x + 3$

331. $f(x) = x^3 - 4x - 9$ is divided by $x + 2$

332. $f(x) = 2x^3 - 6x - 24$ divided by $x - 3$

333. $f(x) = 7x^2 - 5x - 8$ divided by $x - 1$

In the following exercises, use the Factor Theorem to determine if $x - c$ is a factor of the polynomial function.

334. Determine whether $x + 3$ a factor of $x^3 + 8x^2 + 21x + 18$

335. Determine whether $x + 4$ a factor of $x^3 + x^2 - 14x + 8$

336. Determine whether $x - 2$ a factor of $x^3 - 7x^2 + 7x - 6$

337. Determine whether $x - 3$ a factor of $x^3 - 7x^2 + 11x + 3$

Writing Exercises

338. James divides $48y + 6$ by 6 this way: $\frac{48y + \cancel{6}}{\cancel{6}} = 48y$. What is wrong with his reasoning?

339. Divide $\frac{10x^2 + x - 12}{2x}$ and explain with words how you get each term of the quotient.

340. Explain when you can use synthetic division.

341. In your own words, write the steps for synthetic division for $x^2 + 5x + 6$ divided by $x - 2$.

Self Check

ⓐ *After completing the exercises, use this checklist to evaluate your mastery of the objectives of this section*

I can...	Confidently	With some help	No-I don't get it!
divide monomials.			
divide a polynomial by a monomial.			
divide polynomials using long division.			
divide polynomials synthetic division.			
divide polynomial functions.			
use the Remainder and Factor Theorem.			

ⓑ *On a scale of 1-10, how would you rate your mastery of this section in light of your responses on the checklist? How can you improve this?*

CHAPTER 5 REVIEW

KEY TERMS

binomial A binomial is a polynomial with exactly two terms.

conjugate pair A conjugate pair is two binomials of the form $(a - b)$, $(a + b)$. The pair of binomials each have the same first term and the same last term, but one binomial is a sum and the other is a difference.

degree of a constant The degree of any constant is 0.

degree of a polynomial The degree of a polynomial is the highest degree of all its terms.

degree of a term The degree of a term is the sum of the exponents of its variables.

monomial A monomial is an algebraic expression with one term. A monomial in one variable is a term of the form ax^m, where a is a constant and m is a whole number.

polynomial A monomial or two or more monomials combined by addition or subtraction is a polynomial.

polynomial function A polynomial function is a function whose range values are defined by a polynomial.

Power Property According to the Power Property, a to the m to the n equals a to the m times n.

Product Property According to the Product Property, a to the m times a to the n equals a to the m plus n.

Product to a Power According to the Product to a Power Property, a times b in parentheses to the m equals a to the m times b to the m.

Properties of Negative Exponents According to the Properties of Negative Exponents, a to the negative n equals 1 divided by a to the n and 1 divided by a to the negative n equals a to the n.

Quotient Property According to the Quotient Property, a to the m divided by a to the n equals a to the m minus n as long as a is not zero.

Quotient to a Negative Exponent Raising a quotient to a negative exponent occurs when a divided by b in parentheses to the power of negative n equals b divided by a in parentheses to the power of n.

Quotient to a Power Property According to the Quotient to a Power Property, a divided by b in parentheses to the power of m is equal to a to the m divided by b to the m as long as b is not zero.

standard form of a polynomial A polynomial is in standard form when the terms of a polynomial are written in descending order of degrees.

trinomial A trinomial is a polynomial with exactly three terms.

Zero Exponent Property According to the Zero Exponent Property, a to the zero is 1 as long as a is not zero.

KEY CONCEPTS

5.1 Add and Subtract Polynomials

- **Monomial**
    - A **monomial** is an algebraic expression with one term.
    - A monomial in one variable is a term of the form ax^m, where a is a constant and m is a whole number.

- **Polynomials**
    - **Polynomial**—A monomial, or two or more algebraic terms combined by addition or subtraction is a polynomial.
    - **monomial** —A polynomial with exactly one term is called a monomial.
    - **binomial** — A polynomial with exactly two terms is called a binomial.
    - **trinomial** —A polynomial with exactly three terms is called a trinomial.

- **Degree of a Polynomial**
    - The **degree of a term** is the sum of the exponents of its variables.
    - The **degree of a constant** is 0.
    - The **degree of a polynomial** is the highest degree of all its terms.

5.2 Properties of Exponents and Scientific Notation

- **Exponential Notation**

$$a^m \leftarrow \text{exponent}$$
$$\uparrow$$
$$\text{base}$$

a^m means multiply a, m times

$$a^m = \underbrace{a \cdot a \cdot a \cdot \ldots \cdot a}_{m \text{ factors}}$$

This is read a to the mth power.

In the expression a^m, the *exponent m* tells us how many times we use the *base a* as a factor.

- **Product Property for Exponents**
  If a is a real number and m and n are integers, then

$$a^m \cdot a^n = a^{m+n}$$

To multiply with like bases, add the exponents.

- **Quotient Property for Exponents**
  If a is a real number, $a \neq 0$, and m and n are integers, then

$$\frac{a^m}{a^n} = a^{m-n}, \quad m > n \quad \text{and} \quad \frac{a^m}{a^n} = \frac{1}{a^{n-m}}, \quad n > m$$

- **Zero Exponent**

  ○ If a is a non-zero number, then $a^0 = 1$.

  ○ If a is a non-zero number, then a to the power of zero equals 1.

  ○ Any non-zero number raised to the zero power is 1.

- **Negative Exponent**

  ○ If n is an integer and $a \neq 0$, then $a^{-n} = \frac{1}{a^n}$ or $\frac{1}{a^{-n}} = a^n$.

- **Quotient to a Negative Exponent Property**
  If a, b are real numbers, $a \neq 0, b \neq 0$ and n is an integer, then

$$\left(\frac{a}{b}\right)^{-n} = \left(\frac{b}{a}\right)^n$$

- **Power Property for Exponents**
  If a is a real number and m, n are integers, then

$$(a^m)^n = a^{m \cdot n}$$

To raise a power to a power, multiply the exponents.

- **Product to a Power Property for Exponents**
  If a and b are real numbers and m is a whole number, then

$$(ab)^m = a^m b^m$$

To raise a product to a power, raise each factor to that power.

- **Quotient to a Power Property for Exponents**
  If a and are real numbers, $b \neq 0$, and m is an integer, then

$$\left(\frac{a}{b}\right)^m = \frac{a^m}{b^m}$$

To raise a fraction to a power, raise the numerator and denominator to that power.

- **Summary of Exponent Properties**
  If a and b are real numbers, and m and n are integers, then

Property	Description
Product Property	$a^m \cdot a^n = a^{m+n}$
Power Property	$(a^m)^n = a^{m \cdot n}$
Product to a Power	$(ab)^n = a^m b^m$
Quotient Property	$\dfrac{a^m}{a^n} = a^{m-n}, \, a \neq 0$
Zero Exponent Property	$a^0 = 1, \, a \neq 0$
Quotient to a Power Property:	$\left(\dfrac{a}{b}\right)^m = \dfrac{a^m}{b^m}, \, b \neq 0$
Properties of Negative Exponents	$a^{-n} = \dfrac{1}{a^n}$ and $\dfrac{1}{a^{-n}} = a^n$
Quotient to a Negative Exponent	$\left(\dfrac{a}{b}\right)^{-n} = \left(\dfrac{b}{a}\right)^n$

- **Scientific Notation**
  A number is expressed in scientific notation when it is of the form

  $$a \times 10^n \text{ where } 1 \leq a < 10 \text{ and } n \text{ is an integer.}$$

- **How to convert a decimal to scientific notation.**

  Step 1. Move the decimal point so that the first factor is greater than or equal to 1 but less than 10.

  Step 2. Count the number of decimal places, $n,$ that the decimal point was moved.

  Step 3. Write the number as a product with a power of 10. If the original number is.

    - greater than 1, the power of 10 will be 10^n.

    - between 0 and 1, the power of 10 will be 10^{-n}.

  Step 4. Check.

- **How to convert scientific notation to decimal form.**

  Step 1. Determine the exponent, $n,$ on the factor 10.

  Step 2. Move the decimal n places, adding zeros if needed.

    - If the exponent is positive, move the decimal point n places to the right.

    - If the exponent is negative, move the decimal point $|n|$ places to the left.

  Step 3. Check.

5.3 Multiply Polynomials

- **How to use the FOIL method to multiply two binomials.**

  Step 1. Multiply the *First* terms.

  Step 2. Multiply the *Outer* terms.

  Step 3. Multiply the *Inner* terms.

  Step 4. Multiply the *Last* terms.

  Step 5. Combine like terms, when possible.

- **Multiplying Two Binomials:** To multiply binomials, use the:

- ◦ Distributive Property
- ◦ FOIL Method
- **Multiplying a Polynomial by a Polynomial:** To multiply a trinomial by a binomial, use the:
  - ◦ Distributive Property
  - ◦ Vertical Method
- **Binomial Squares Pattern**
  If a and b are real numbers,

$$(a+b)^2 = a^2 + 2ab + b^2$$

$$\underbrace{(a+b)^2}_{(\text{binomial})^2} = \underbrace{a^2}_{(\text{first term})^2} + \underbrace{2ab}_{2(\text{product of terms})} + \underbrace{b^2}_{(\text{last term})^2}$$

$$(a-b)^2 = a^2 - 2ab + b^2$$

$$\underbrace{(a-b)^2}_{(\text{binomial})^2} = \underbrace{a^2}_{(\text{first term})^2} - \underbrace{2ab}_{2(\text{product of terms})} + \underbrace{b^2}_{(\text{last term})^2}$$

- **Product of Conjugates Pattern**
  If a, b are real numbers

$$(a-b)(a+b) = a^2 - b^2$$

$$(a-b)(a+b) = a^2 \overset{difference}{-} b^2$$

conjugates   squares

The product is called a difference of squares.
To multiply conjugates, square the first term, square the last term, write it as a difference of squares.

- **Comparing the Special Product Patterns**

Binomial Squares	Product of Conjugates
$(a+b)^2 = a^2 + 2ab + b^2$	$(a-b)^2 = a^2 - 2ab + b^2$
$(a-b)(a+b) = a^2 - b^2$	
• Squaring a binomial	• Multiplying conjugates
• Product is a **trinomial**	• Product is a **binomial.**
• Inner and outer terms with FOIL are **the same.**	• Inner and outer terms with FOIL are **opposites.**
• Middle term is **double the product** of the terms	• There is **no** middle term.

- **Multiplication of Polynomial Functions:**
  - ◦ For functions $f(x)$ and $g(x)$,

$$(f \cdot g)(x) = f(x) \cdot g(x)$$

5.4 Dividing Polynomials

- **Division of a Polynomial by a Monomial**
  - ◦ To divide a polynomial by a monomial, divide each term of the polynomial by the monomial.
- **Division of Polynomial Functions**
  - ◦ For functions $f(x)$ and $g(x)$, where $g(x) \neq 0$,

$$\left(\frac{f}{g}\right)(x) = \frac{f(x)}{g(x)}$$

- **Remainder Theorem**
  - If the polynomial function $f(x)$ is divided by $x - c$, then the remainder is $f(c)$.
- **Factor Theorem:** For any polynomial function $f(x)$,
  - if $x - c$ is a factor of $f(x)$, then $f(c) = 0$
  - if $f(c) = 0$, then $x - c$ is a factor of $f(x)$

REVIEW EXERCISES

5.1 Add and Subtract Polynomials

Determine the Degree of Polynomials

In the following exercises, determine the type of polynomial.

342. $16x^2 - 40x - 25$

343. $5m + 9$

344. -15

345. $y^2 + 6y^3 + 9y^4$

Add and Subtract Polynomials

In the following exercises, add or subtract the polynomials.

346. $4p + 11p$

347. $-8y^3 - 5y^3$

348. $\left(4a^2 + 9a - 11\right) + \left(6a^2 - 5a + 10\right)$

349. $\left(8m^2 + 12m - 5\right) - \left(2m^2 - 7m - 1\right)$

350. $\left(y^2 - 3y + 12\right) + \left(5y^2 - 9\right)$

351. $\left(5u^2 + 8u\right) - (4u - 7)$

352. Find the sum of $8q^3 - 27$ and $q^2 + 6q - 2$.

353. Find the difference of $x^2 + 6x + 8$ and $x^2 - 8x + 15$.

In the following exercises, simplify.

354. $17mn^2 - (-9mn^2) + 3mn^2$

355. $18a - 7b - 21a$

356. $2pq^2 - 5p - 3q^2$

357. $(6a^2 + 7) + (2a^2 - 5a - 9)$

358. $(3p^2 - 4p - 9) + (5p^2 + 14)$

359. $(7m^2 - 2m - 5) - (4m^2 + m - 8)$

360. $(7b^2 - 4b + 3) - (8b^2 - 5b - 7)$

361. Subtract $\left(8y^2 - y + 9\right)$ from $\left(11y^2 - 9y - 5\right)$

362. Find the difference of $\left(z^2 - 4z - 12\right)$ and $\left(3z^2 + 2z - 11\right)$

363. $\left(x^3 - x^2 y\right) - \left(4xy^2 - y^3\right) + \left(3x^2 y - xy^2\right)$

364. $\left(x^3 - 2x^2 y\right) - \left(xy^2 - 3y^3\right) - \left(x^2 y - 4xy^2\right)$

Evaluate a Polynomial Function for a Given Value of the Variable

In the following exercises, find the function values for each polynomial function.

365. For the function $f(x) = 7x^2 - 3x + 5$ find:

ⓐ $f(5)$ ⓑ $f(-2)$ ⓒ $f(0)$

366. For the function $g(x) = 15 - 16x^2$, find:

ⓐ $g(-1)$ ⓑ $g(0)$ ⓒ $g(2)$

367. A pair of glasses is dropped off a bridge 640 feet above a river. The polynomial function $h(t) = -16t^2 + 640$ gives the height of the glasses t seconds after they were dropped. Find the height of the glasses when $t = 6$.

368. A manufacturer of the latest soccer shoes has found that the revenue received from selling the shoes at a cost of p dollars each is given by the polynomial $R(p) = -5p^2 + 360p$. Find the revenue received when $p = 110$ dollars.

Add and Subtract Polynomial Functions

In the following exercises, find ⓐ $(f + g)(x)$ ⓑ $(f + g)(3)$ ⓒ $(f - g)(x)$ ⓓ $(f - g)(-2)$

369. $f(x) = 2x^2 - 4x - 7$ and $g(x) = 2x^2 - x + 5$

370. $f(x) = 4x^3 - 3x^2 + x - 1$ and $g(x) = 8x^3 - 1$

5.2 Properties of Exponents and Scientific Notation

Simplify Expressions Using the Properties for Exponents

In the following exercises, simplify each expression using the properties for exponents.

371. $p^3 \cdot p^{10}$

372. $2 \cdot 2^6$

373. $a \cdot a^2 \cdot a^3$

374. $x \cdot x^8$

375. $y^a \cdot y^b$

376. $\dfrac{2^8}{2^2}$

377. $\dfrac{a^6}{a}$

378. $\dfrac{n^3}{n^{12}}$

379. $\dfrac{1}{x^5}$

380. 3^0

381. y^0

382. $(14t)^0$

383. $12a^0 - 15b^0$

Use the Definition of a Negative Exponent

In the following exercises, simplify each expression.

384. 6^{-2}

385. $(-10)^{-3}$

386. $5 \cdot 2^{-4}$

387. $(8n)^{-1}$

388. y^{-5}

389. 10^{-3}

390. $\dfrac{1}{a^{-4}}$

391. $\dfrac{1}{6^{-2}}$

392. -5^{-3}

393. $\left(-\dfrac{1}{5}\right)^{-3}$

394. $-\left(\dfrac{1}{2}\right)^{-3}$

395. $(-5)^{-3}$

396. $\left(\dfrac{5}{9}\right)^{-2}$

397. $\left(-\dfrac{3}{x}\right)^{-3}$

In the following exercises, simplify each expression using the Product Property.

398. $\left(y^4\right)^3$

399. $\left(3^2\right)^5$

400. $\left(a^{10}\right)^y$

401. $x^{-3} \cdot x^9$

402. $r^{-5} \cdot r^{-4}$

403. $\left(uv^{-3}\right)\left(u^{-4}v^{-2}\right)$

404. $\left(m^5\right)^{-1}$

405. $p^5 \cdot p^{-2} \cdot p^{-4}$

In the following exercises, simplify each expression using the Power Property.

406. $\left(k^{-2}\right)^{-3}$

407. $\dfrac{q^4}{q^{20}}$

408. $\dfrac{b^8}{b^{-2}}$

409. $\dfrac{n^{-3}}{n^{-5}}$

In the following exercises, simplify each expression using the Product to a Power Property.

410. $(-5ab)^3$

411. $(-4pq)^0$

412. $\left(-6x^3\right)^{-2}$

413. $\left(3y^{-4}\right)^2$

In the following exercises, simplify each expression using the Quotient to a Power Property.

414. $\left(\dfrac{3}{5x}\right)^{-2}$

415. $\left(\dfrac{3xy^2}{z}\right)^4$

416. $\left(\dfrac{4p^{-3}}{q^2}\right)^2$

In the following exercises, simplify each expression by applying several properties.

417. $(x^2 y)^2 (3xy^5)^3$

418. $\dfrac{\left(-3a^{-2}\right)^4 \left(2a^4\right)^2}{\left(-6a^2\right)^3}$

419. $\left(\dfrac{3xy^3}{4x^4 y^{-2}}\right)^2 \left(\dfrac{6xy^4}{8x^3 y^{-2}}\right)^{-1}$

In the following exercises, write each number in scientific notation.

420. 2.568

421. 5,300,000

422. 0.00814

In the following exercises, convert each number to decimal form.

423. 2.9×10^4

424. 3.75×10^{-1}

425. 9.413×10^{-5}

In the following exercises, multiply or divide as indicated. Write your answer in decimal form.

426. $\left(3 \times 10^7\right)\left(2 \times 10^{-4}\right)$

427. $\left(1.5 \times 10^{-3}\right)\left(4.8 \times 10^{-1}\right)$

428. $\dfrac{6 \times 10^9}{2 \times 10^{-1}}$

429. $\dfrac{9 \times 10^{-3}}{1 \times 10^{-6}}$

5.3 Multiply Polynomials

Multiply Monomials

In the following exercises, multiply the monomials.

430. $\left(-6p^4\right)(9p)$

431. $\left(\frac{1}{3}c^2\right)\left(30c^8\right)$

432. $\left(8x^2y^5\right)\left(7xy^6\right)$

433. $\left(\frac{2}{3}m^3n^6\right)\left(\frac{1}{6}m^4n^4\right)$

Multiply a Polynomial by a Monomial

In the following exercises, multiply.

434. $7(10 - x)$

435. $a^2\left(a^2 - 9a - 36\right)$

436. $-5y\left(125y^3 - 1\right)$

437. $(4n - 5)\left(2n^3\right)$

Multiply a Binomial by a Binomial

In the following exercises, multiply the binomials using:

ⓐ the Distributive Property ⓑ the FOIL method ⓒ the Vertical Method.

438. $(a + 5)(a + 2)$

439. $(y - 4)(y + 12)$

440. $(3x + 1)(2x - 7)$

441. $(6p - 11)(3p - 10)$

In the following exercises, multiply the binomials. Use any method.

442. $(n + 8)(n + 1)$

443. $(k + 6)(k - 9)$

444. $(5u - 3)(u + 8)$

445. $(2y - 9)(5y - 7)$

446. $(p + 4)(p + 7)$

447. $(x - 8)(x + 9)$

448. $(3c + 1)(9c - 4)$

449. $(10a - 1)(3a - 3)$

Multiply a Polynomial by a Polynomial

In the following exercises, multiply using ⓐ the Distributive Property ⓑ the Vertical Method.

450. $(x + 1)\left(x^2 - 3x - 21\right)$

451. $(5b - 2)\left(3b^2 + b - 9\right)$

In the following exercises, multiply. Use either method.

452. $(m + 6)\left(m^2 - 7m - 30\right)$

453. $(4y - 1)\left(6y^2 - 12y + 5\right)$

Multiply Special Products

In the following exercises, square each binomial using the Binomial Squares Pattern.

454. $(2x - y)^2$

455. $\left(x + \frac{3}{4}\right)^2$

456. $\left(8p^3 - 3\right)^2$

457. $(5p + 7q)^2$

In the following exercises, multiply each pair of conjugates using the Product of Conjugates.

458. $(3y + 5)(3y - 5)$

459. $(6x + y)(6x - y)$

460. $\left(a + \frac{2}{3}b\right)\left(a - \frac{2}{3}b\right)$

461. $(12x^3 - 7y^2)(12x^3 + 7y^2)$

462. $(13a^2 - 8b^4)(13a^2 + 8b^4)$

5.4 Divide Monomials

Divide Monomials

In the following exercises, divide the monomials.

463. $72p^{12} \div 8p^3$

464. $-26a^8 \div (2a^2)$

465. $\dfrac{45y^6}{-15y^{10}}$

466. $\dfrac{-30x^8}{-36x^9}$

467. $\dfrac{28a^9 b}{7a^4 b^3}$

468. $\dfrac{11u^6 v^3}{55u^2 v^8}$

469. $\dfrac{\left(5m^9 n^3\right)\left(8m^3 n^2\right)}{\left(10mn^4\right)\left(m^2 n^5\right)}$

470. $\dfrac{(42r^2 s^4)(54rs^2)}{(6rs^3)(9s)}$

Divide a Polynomial by a Monomial

In the following exercises, divide each polynomial by the monomial

471. $\left(54y^4 - 24y^3\right) \div \left(-6y^2\right)$

472. $\dfrac{63x^3 y^2 - 99x^2 y^3 - 45x^4 y^3}{9x^2 y^2}$

473. $\dfrac{12x^2 + 4x - 3}{-4x}$

Divide Polynomials using Long Division

In the following exercises, divide each polynomial by the binomial.

474. $\left(4x^2 - 21x - 18\right) \div (x - 6)$

475. $\left(y^2 + 2y + 18\right) \div (y + 5)$

476. $\left(n^3 - 2n^2 - 6n + 27\right) \div (n + 3)$

477. $\left(a^3 - 1\right) \div (a + 1)$

Divide Polynomials using Synthetic Division

In the following exercises, use synthetic Division to find the quotient and remainder.

478. $x^3 - 3x^2 - 4x + 12$ is divided by $x + 2$

479. $2x^3 - 11x^2 + 11x + 12$ is divided by $x - 3$

480. $x^4 + x^2 + 6x - 10$ is divided by $x + 2$

Divide Polynomial Functions

In the following exercises, divide.

481. For functions $f(x) = x^2 - 15x + 45$ and $g(x) = x - 9$, find ⓐ $\left(\dfrac{f}{g}\right)(x)$

ⓑ $\left(\dfrac{f}{g}\right)(-2)$

482. For functions $f(x) = x^3 + x^2 - 7x + 2$ and $g(x) = x - 2$, find ⓐ $\left(\dfrac{f}{g}\right)(x)$

ⓑ $\left(\dfrac{f}{g}\right)(3)$

Use the Remainder and Factor Theorem

In the following exercises, use the Remainder Theorem to find the remainder.

483. $f(x) = x^3 - 4x - 9$ is divided by $x + 2$

484. $f(x) = 2x^3 - 6x - 24$ divided by $x - 3$

In the following exercises, use the Factor Theorem to determine if $x - c$ is a factor of the polynomial function.

485. Determine whether $x - 2$ is a factor of $x^3 - 7x^2 + 7x - 6$.

486. Determine whether $x - 3$ is a factor of $x^3 - 7x^2 + 11x + 3$.

PRACTICE TEST

487. For the polynomial $8y^4 - 3y^2 + 1$

ⓐ Is it a monomial, binomial, or trinomial? ⓑ What is its degree?

488. $(5a^2 + 2a - 12)(9a^2 + 8a - 4)$

489. $(10x^2 - 3x + 5) - (4x^2 - 6)$

490. $\left(-\frac{3}{4}\right)^3$

491. $x^{-3}x^4$

492. $\dfrac{5^6}{5^8}$

493. $\left(47a^{18}b^{23}c^5\right)^0$

494. 4^{-1}

495. $(2y)^{-3}$

496. $p^{-3} \cdot p^{-8}$

497. $\dfrac{x^4}{x^{-5}}$

498. $\left(3x^{-3}\right)^2$

499. $\dfrac{24r^3 s}{6r^2 s^7}$

500. $\left(\dfrac{x^4 y^9}{x^{-3}}\right)^2$

501. $\left(8xy^3\right)\left(-6x^4 y^6\right)$

502. $4u\left(u^2 - 9u + 1\right)$

503. $(m + 3)(7m - 2)$

504. $(n - 8)\left(n^2 - 4n + 11\right)$

505. $(4x - 3)^2$

506. $(5x + 2y)(5x - 2y)$

507. $\left(15xy^3 - 35x^2 y\right) \div 5xy$

508. $\left(3x^3 - 10x^2 + 7x + 10\right) \div (3x + 2)$

509. Use the Factor Theorem to determine if $x + 3$ a factor of $x^3 + 8x^2 + 21x + 18$.

510. ⓐ Convert 112,000 to scientific notation. ⓑ Convert 5.25×10^{-4} to decimal form.

In the following exercises, simplify and write your answer in decimal form.

511. $\left(2.4 \times 10^8\right)\left(2 \times 10^{-5}\right)$

512. $\dfrac{9 \times 10^4}{3 \times 10^{-1}}$

513. For the function $f(x) = 6x^2 - 3x - 9$ find:

ⓐ $f(3)$ ⓑ $f(-2)$ ⓒ $f(0)$

514. For $f(x) = 2x^2 - 3x - 5$ and $g(x) = 3x^2 - 4x + 1$, find

ⓐ $(f + g)(x)$ ⓑ $(f + g)(1)$
ⓒ $(f - g)(x)$ ⓓ $(f - g)(-2)$

515. For functions $f(x) = 3x^2 - 23x - 36$ and $g(x) = x - 9$, find

ⓐ $\left(\dfrac{f}{g}\right)(x)$ ⓑ $\left(\dfrac{f}{g}\right)(3)$

516. A hiker drops a pebble from a bridge 240 feet above a canyon. The function $h(t) = -16t^2 + 240$ gives the height of the pebble t seconds after it was dropped. Find the height when $t = 3$.

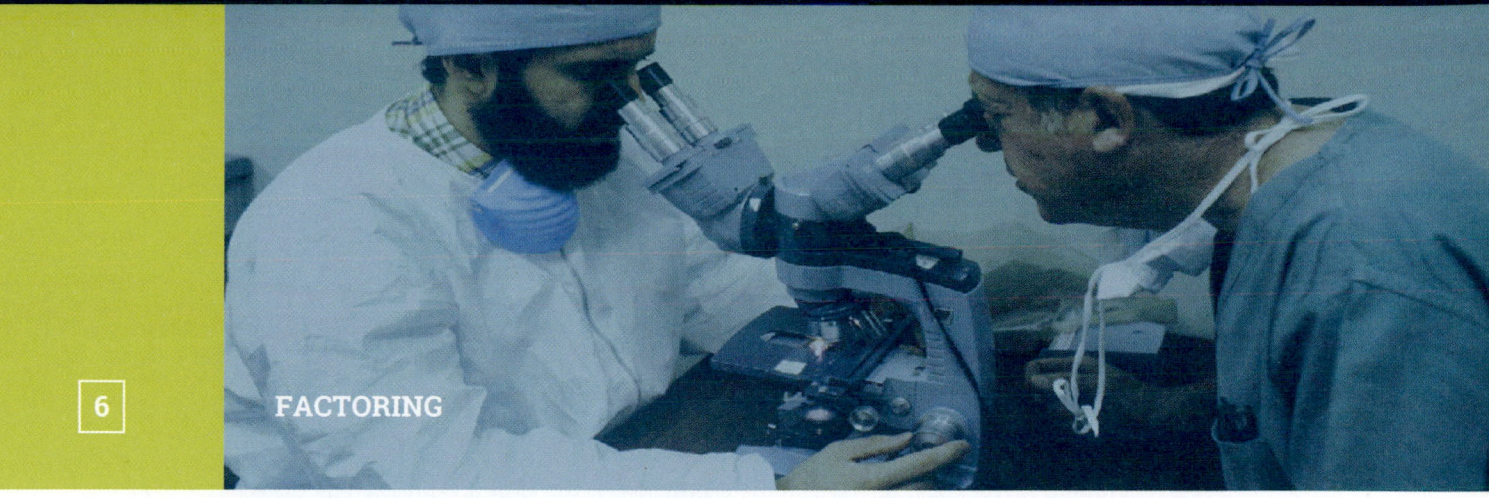

Figure 6.1 Scientists use factoring to calculate growth rates of infectious diseases such as viruses. (credit: "FotoshopTofs" / Pixabay)

Chapter Outline

 Introduction

An epidemic of a disease has broken out. Where did it start? How is it spreading? What can be done to control it? Answers to these and other questions can be found by scientists known as epidemiologists. They collect data and analyze it to study disease and consider possible control measures. Because diseases can spread at alarming rates, these scientists must use their knowledge of mathematics involving factoring. In this chapter, you will learn how to factor and apply factoring to real-life situations.

 Greatest Common Factor and Factor by Grouping

Learning Objectives

By the end of this section, you will be able to:

› Find the greatest common factor of two or more expressions
› Factor the greatest common factor from a polynomial
› Factor by grouping

Be Prepared!

Before you get started, take this readiness quiz.

1. Factor 56 into primes.
   If you missed this problem, review **Example 1.2**.

2. Find the least common multiple (LCM) of 18 and 24.
   If you missed this problem, review **Example 1.3**.

3. Multiply: $-3a(7a + 8b)$.

   If you missed this problem, review **Example 5.26**.

Find the Greatest Common Factor of Two or More Expressions

Earlier we multiplied factors together to get a product. Now, we will reverse this process; we will start with a product and then break it down into its factors. Splitting a product into factors is called **factoring**.

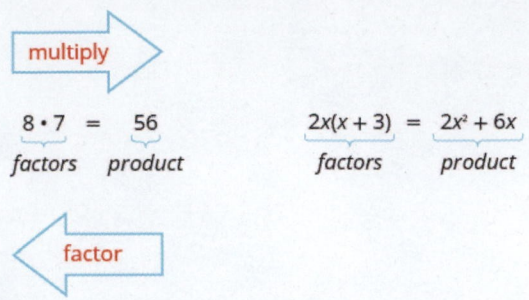

We have learned how to factor numbers to find the least common multiple (LCM) of two or more numbers. Now we will factor expressions and find the **greatest common factor** of two or more expressions. The method we use is similar to what we used to find the LCM.

Greatest Common Factor

The **greatest common factor** (GCF) of two or more expressions is the largest expression that is a factor of all the expressions.

We summarize the steps we use to find the greatest common factor.

 HOW TO :: FIND THE GREATEST COMMON FACTOR (GCF) OF TWO EXPRESSIONS.

Step 1.   Factor each coefficient into primes. Write all variables with exponents in expanded form.

Step 2.   List all factors—matching common factors in a column. In each column, circle the common factors.

Step 3.   Bring down the common factors that all expressions share.

Step 4.   Multiply the factors.

The next example will show us the steps to find the greatest common factor of three expressions.

EXAMPLE 6.1

Find the greatest common factor of $21x^3$, $9x^2$, $15x$.

⊘ **Solution**

Factor each coefficient into primes and write the variables with exponents in expanded form. Circle the common factors in each column. Bring down the common factors.	$21x^3 = 3 \cdot \quad 7 \cdot x \cdot x \cdot x$ $\;\;9x^2 = 3 \cdot 3 \cdot \quad x \cdot x$ $\;15x = 3 \cdot \quad 5 \cdot \quad x$ $\overline{\text{GCF} = 3 \cdot \qquad\quad x}$
Multiply the factors.	$\text{GCF} = 3x$
	The GCF of $21x^3$, $9x^2$ and $15x$ is $3x$.

▷ **TRY IT :: 6.1**   Find the greatest common factor: $25m^4$, $35m^3$, $20m^2$.

▷ **TRY IT :: 6.2**   Find the greatest common factor: $14x^3$, $70x^2$, $105x$.

Factor the Greatest Common Factor from a Polynomial

It is sometimes useful to represent a number as a product of factors, for example, 12 as $2 \cdot 6$ or $3 \cdot 4$. In algebra, it can also be useful to represent a polynomial in factored form. We will start with a product, such as $3x^2 + 15x$, and end with its factors, $3x(x + 5)$. To do this we apply the Distributive Property "in reverse."

We state the Distributive Property here just as you saw it in earlier chapters and "in reverse."

Distributive Property

If a, b, and c are real numbers, then

$$a(b + c) = ab + ac \quad \text{and} \quad ab + ac = a(b + c)$$

The form on the left is used to multiply. The form on the right is used to factor.

So how do you use the Distributive Property to factor a polynomial? You just find the GCF of all the terms and write the polynomial as a product!

EXAMPLE 6.2   HOW TO USE THE DISTRIBUTIVE PROPERTY TO FACTOR A POLYNOMIAL

Factor: $8m^3 - 12m^2 n + 20mn^2$.

⊘ **Solution**

Step 1. Find the GCF of all the terms of the polynomial.	Find the GCF of $8m^3$, $12m^2n$, $20mn^2$	$8m^3 = 2 \cdot 2 \cdot 2 \cdot \quad m \cdot m \cdot m$ $12m^2n = 2 \cdot 2 \cdot 3 \cdot \quad \cdot m \cdot m \cdot n$ $20mn^2 = 2 \cdot 2 \cdot \quad 5 \cdot m \cdot \quad n \cdot n$ ――――――――――― $GCF = 2 \cdot 2 \cdot \quad m$ $GCF = 4m$
Step 2. Rewrite each term as a product using the GCF.	Rewrite $8m^3$, $12m^2n$, $20mn^2$ as products of their GCF, $4m$. $8m^3 = 4m \cdot 2m^2$ $12m^2n = -4m \cdot 3m\,n$ $20mn^2 = 4m \cdot 5n^2$	$8m^3 - 12m^2n + 20mn^2$ $4m \cdot 2m^2 - 4m \cdot 3m\,n + 4m \cdot 5n^2$
Step 3. Use the "reverse" Distributive Property to factor the expression.		$4m(2m^2 - 3m\,n + 5n^2)$
Step 4. Check by multiplying the factors.		$4m(2m^2 - 3m\,n + 5n^2)$ $4m \cdot 2m^2 - 4m \cdot 3m\,n + 4m \cdot 5n^2$ $8m^3 - 12m^2n + 20mn^2$ ✓

 **TRY IT :: 6.3**   Factor: $9xy^2 + 6x^2 y^2 + 21y^3$.

 TRY IT :: 6.4   Factor: $3p^3 - 6p^2 q + 9pq^3$.

 HOW TO :: FACTOR THE GREATEST COMMON FACTOR FROM A POLYNOMIAL.

Step 1.   Find the GCF of all the terms of the polynomial.

Step 2.   Rewrite each term as a product using the GCF.

Step 3.   Use the "reverse" Distributive Property to factor the expression.

Step 4.   Check by multiplying the factors.

Factor as a Noun and a Verb

We use "factor" as both a noun and a verb:

Noun:	7 is a *factor* of 14
Verb:	*factor* 3 from $3a + 3$

EXAMPLE 6.3

Factor: $5x^3 - 25x^2$.

⊘ **Solution**

Find the GCF of $5x^3$ and $25x^2$.	$\begin{aligned} 5x^3 &= 5 \cdot x \cdot x \cdot x \\ 25x^2 &= 5 \cdot 5 \cdot x \cdot x \\ \hline \text{GCF} &= 5 \cdot \quad x \cdot x \end{aligned}$
	$\text{GCF} = 5x^2$
	$5x^3 - 25x^2$
Rewrite each term.	$5x^2 \cdot x - 5x^2 \cdot 5$
Factor the GCF.	$5x^2(x - 5)$

Check:

$$5x^2(x - 5)$$
$$5x^2 \cdot x - 5x^2 \cdot 5$$
$$5x^3 - 25x^2 \checkmark$$

>	**TRY IT : :** 6.5	Factor: $2x^3 + 12x^2$.

>	**TRY IT : :** 6.6	Factor: $6y^3 - 15y^2$.

EXAMPLE 6.4

Factor: $8x^3 y - 10x^2 y^2 + 12xy^3$.

✓ Solution

The GCF of $8x^3 y$, $-10x^2 y^2$, and $12xy^3$ is $2xy$.

$$
\begin{array}{llll}
8x^3y & = 2 \cdot 2 \cdot 2 & x \cdot x \cdot x \cdot y \\
10x^2y^2 & = 2 \cdot \quad\;\; 5 & x \cdot x \cdot & y \cdot y \\
12xy^3 & = 2 \cdot 2 \cdot 3 & x \cdot & y \cdot y \cdot y \\
\hline
GCF & = 2 \cdot & x \cdot & y
\end{array}
$$

$$GCF = 2xy$$

$$8x^3y - 10x^2y^2 + 12xy^3$$

Rewrite each term using the GCF, $2xy$.

$$2xy \cdot 4x^2 - 2xy \cdot 5xy + 2xy \cdot 6y^2$$

Factor the GCF.

$$2xy(4x^2 - 5xy + 6y^2)$$

Check:

$$2xy\left(4x^2 - 5xy + 6y^2\right)$$
$$2xy \cdot 4x^2 - 2xy \cdot 5xy + 2xy \cdot 6y^2$$
$$8x^3 y - 10x^2 y^2 + 12xy^3 \checkmark$$

> **TRY IT :: 6.7** Factor: $15x^3 y - 3x^2 y^2 + 6xy^3$.

> **TRY IT :: 6.8** Factor: $8a^3 b + 2a^2 b^2 - 6ab^3$.

When the leading coefficient is negative, we factor the negative out as part of the GCF.

EXAMPLE 6.5

Factor: $-4a^3 + 36a^2 - 8a$.

✓ Solution

The leading coefficient is negative, so the GCF will be negative.

$$-4a^3 + 36a^2 - 8a$$

Rewrite each term using the GCF, $-4a$.

$$-4a \cdot a^2 - (-4a) \cdot 9a + (-4a) \cdot 2$$

Factor the GCF.

$$-4a(a^2 - 9a + 2)$$

Check:

$$-4a\left(a^2 - 9a + 2\right)$$
$$-4a \cdot a^2 - (-4a) \cdot 9a + (-4a) \cdot 2$$
$$-4a^3 + 36a^2 - 8a \checkmark$$

> **TRY IT :: 6.9** Factor: $-4b^3 + 16b^2 - 8b$.

> **TRY IT :: 6.10** Factor: $-7a^3 + 21a^2 - 14a$.

So far our greatest common factors have been monomials. In the next example, the greatest common factor is a binomial.

EXAMPLE 6.6

Factor: $3y(y + 7) - 4(y + 7)$.

 **Solution**

The GCF is the binomial $y + 7$.

<table>
<tr><td></td><td>$3y(y + 7) - 4(y + 7)$</td></tr>
<tr><td>Factor the GCF, $(y + 7)$.</td><td>$(y + 7)(3y - 4)$</td></tr>
<tr><td>Check on your own by multiplying.</td><td></td></tr>
</table>

> **TRY IT :: 6.11**   Factor: $4m(m + 3) - 7(m + 3)$.

> **TRY IT :: 6.12**   Factor: $8n(n - 4) + 5(n - 4)$.

Factor by Grouping

Sometimes there is no common factor of all the terms of a polynomial. When there are four terms we separate the polynomial into two parts with two terms in each part. Then look for the GCF in each part. If the polynomial can be factored, you will find a common factor emerges from both parts. Not all polynomials can be factored. Just like some numbers are prime, some polynomials are prime.

EXAMPLE 6.7   HOW TO FACTOR A POLYNOMIAL BY GROUPING

Factor by grouping: $xy + 3y + 2x + 6$.

 **Solution**

Step 1. Group terms with common factors.	Is there a greatest common factor of all four terms?	$xy + 3y + 2x + 6$
	No, so let's separate the first two terms from the second two.	$xy + 3y + 2x + 6$
Step 2. Factor out the common factor in each group.	Factor the GCF from the first two terms.	$y(x + 3) + 2x + 6$
	Factor the GCF from the second two terms.	$y(x + 3) + 2(x + 3)$
Step 3. Factor the common factor from the expression.	Notice that each term has a common factor of $(x + 3)$.	$y(x + 3) + 2(x + 3)$
	Factor out the common factor.	$(x + 3)(y + 2)$
Step 4. Check.	Multiply $(x + 3)(y + 2)$. Is the product the original expression?	$(x + 3)(y + 2)$ $xy + 2x + 3y + 6$ $xy + 3y + 2x + 6$ ✓

> **TRY IT :: 6.13**   Factor by grouping: $xy + 8y + 3x + 24$.

> **TRY IT :: 6.14**   Factor by grouping: $ab + 7b + 8a + 56$.

HOW TO :: FACTOR BY GROUPING.

Step 1. Group terms with common factors.

Step 2. Factor out the common factor in each group.

Step 3. Factor the common factor from the expression.

Step 4. Check by multiplying the factors.

EXAMPLE 6.8

Factor by grouping: ⓐ $x^2 + 3x - 2x - 6$ ⓑ $6x^2 - 3x - 4x + 2$.

 **Solution**

ⓐ

There is no GCF in all four terms.	$x^2 + 3x - 2x - 6$
Separate into two parts.	$x^2 + 3x \quad -2x - 6$
Factor the GCF from both parts. Be careful with the signs when factoring the GCF from the last two terms.	$x(x + 3) - 2(x + 3)$
Factor out the common factor.	$(x + 3)(x - 2)$
Check on your own by multiplying.	

ⓑ

There is no GCF in all four terms.	$6x^2 - 3x - 4x + 2$
Separate into two parts.	$6x^2 - 3x \quad -4x + 2$
Factor the GCF from both parts.	$3x(2x - 1) - 2(2x - 1)$
Factor out the common factor.	$(2x - 1)(3x - 2)$
Check on your own by multiplying.	

 TRY IT :: 6.15   Factor by grouping: ⓐ $x^2 + 2x - 5x - 10$ ⓑ $20x^2 - 16x - 15x + 12$.

 TRY IT :: 6.16   Factor by grouping: ⓐ $y^2 + 4y - 7y - 28$ ⓑ $42m^2 - 18m - 35m + 15$.

6.1 EXERCISES

Practice Makes Perfect

Find the Greatest Common Factor of Two or More Expressions

In the following exercises, find the greatest common factor.

1. $10p^3q$, $12pq^2$

2. $8a^2b^3$, $10ab^2$

3. $12m^2n^3$, $30m^5n^3$

4. $28x^2y^4$, $42x^4y^4$

5. $10a^3$, $12a^2$, $14a$

6. $20y^3$, $28y^2$, $40y$

7. $35x^3y^2$, $10x^4y$, $5x^5y^3$

8. $27p^2q^3$, $45p^3q^4$, $9p^4q^3$

Factor the Greatest Common Factor from a Polynomial

In the following exercises, factor the greatest common factor from each polynomial.

9. $6m + 9$

10. $14p + 35$

11. $9n - 63$

12. $45b - 18$

13. $3x^2 + 6x - 9$

14. $4y^2 + 8y - 4$

15. $8p^2 + 4p + 2$

16. $10q^2 + 14q + 20$

17. $8y^3 + 16y^2$

18. $12x^3 - 10x$

19. $5x^3 - 15x^2 + 20x$

20. $8m^2 - 40m + 16$

21. $24x^3 - 12x^2 + 15x$

22. $24y^3 - 18y^2 - 30y$

23. $12xy^2 + 18x^2y^2 - 30y^3$

24. $21pq^2 + 35p^2q^2 - 28q^3$

25. $20x^3y - 4x^2y^2 + 12xy^3$

26. $24a^3b + 6a^2b^2 - 18ab^3$

27. $-2x - 4$

28. $-3b + 12$

29. $-2x^3 + 18x^2 - 8x$

30. $-5y^3 + 35y^2 - 15y$

31. $-4p^3q - 12p^2q^2 + 16pq^2$

32. $-6a^3b - 12a^2b^2 + 18ab^2$

33. $5x(x + 1) + 3(x + 1)$

34. $2x(x - 1) + 9(x - 1)$

35. $3b(b - 2) - 13(b - 2)$

36. $6m(m - 5) - 7(m - 5)$

Factor by Grouping

In the following exercises, factor by grouping.

37. $ab + 5a + 3b + 15$

38. $cd + 6c + 4d + 24$

39. $8y^2 + y + 40y + 5$

40. $6y^2 + 7y + 24y + 28$

41. $uv - 9u + 2v - 18$

42. $pq - 10p + 8q - 80$

43. $u^2 - u + 6u - 6$

44. $x^2 - x + 4x - 4$

45. $9p^2 - 3p - 20$

46. $16q^2 - 8q - 35$

47. $mn - 6m - 4n + 24$

48. $r^2 - 3r - r + 3$

49. $2x^2 - 14x - 5x + 35$

50. $4x^2 - 36x - 3x + 27$

Mixed Practice

In the following exercises, factor.

51. $-18xy^2 - 27x^2 y$

52. $-4x^3 y^5 - x^2 y^3 + 12xy^4$

53. $3x^3 - 7x^2 + 6x - 14$

54. $x^3 + x^2 - x - 1$

55. $x^2 + xy + 5x + 5y$

56. $5x^3 - 3x^2 + 5x - 3$

Writing Exercises

57. What does it mean to say a polynomial is in factored form?

58. How do you check result after factoring a polynomial?

59. The greatest common factor of 36 and 60 is 12. Explain what this means.

60. What is the GCF of y^4, y^5, and y^{10}? Write a general rule that tells you how to find the GCF of y^a, y^b, and y^c.

Self Check

ⓐ After completing the exercises, use this checklist to evaluate your mastery of the objectives of this section.

I can...	Confidently	With some help	No-I don't get it!
find the greatest common factor of two or more expressions.			
factor the greatest common factor from a polynomial.			
factor by grouping.			

ⓑ If most of your checks were:

...confidently. Congratulations! You have achieved your goals in this section! Reflect on the study skills you used so that you can continue to use them. What did you do to become confident of your ability to do these things? Be specific!

...with some help. This must be addressed quickly as topics you do not master become potholes in your road to success. Math is sequential - every topic builds upon previous work. It is important to make sure you have a strong foundation before you move on. Who can you ask for help? Your fellow classmates and instructor are good resources. Is there a place on campus where math tutors are available? Can your study skills be improved?

...no - I don't get it! This is critical and you must not ignore it. You need to get help immediately or you will quickly be overwhelmed. See your instructor as soon as possible to discuss your situation. Together you can come up with a plan to get you the help you need.

Factor Trinomials

Learning Objectives

By the end of this section, you will be able to:

> Factor trinomials of the form $x^2 + bx + c$

> Factor trinomials of the form $ax^2 + bx + c$ using trial and error

> Factor trinomials of the form $ax^2 + bx + c$ using the 'ac' method

> Factor using substitution

Be Prepared!

Before you get started, take this readiness quiz.

1. Find all the factors of 72.
   If you missed this problem, review **Example 1.2**.

2. Find the product: $(3y + 4)(2y + 5)$.

   If you missed this problem, review **Example 5.28**.

3. Simplify: $-9(6)$; $-9(-6)$.

   If you missed this problem, review **Example 1.18**.

Factor Trinomials of the Form $x^2 + bx + c$

You have already learned how to multiply binomials using FOIL. Now you'll need to "undo" this multiplication. To factor the trinomial means to start with the product, and end with the factors.

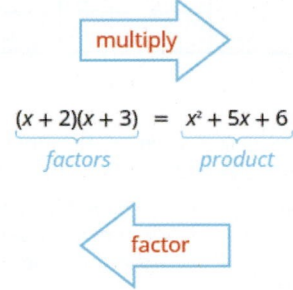

To figure out how we would factor a trinomial of the form $x^2 + bx + c$, such as $x^2 + 5x + 6$ and factor it to $(x + 2)(x + 3)$, let's start with two general binomials of the form $(x + m)$ and $(x + n)$.

$$(x + m)(x + n)$$

Foil to find the product.	$x^2 + mx + nx + mn$

Factor the GCF from the middle terms.	$x^2 + (m + n)x + mn$

Our trinomial is of the form $x^2 + bx + c$.	$\begin{array}{c} x^2 + \quad bx \quad + c \\ \hline x^2 + (m+n)x + mn \end{array}$

This tells us that to factor a trinomial of the form $x^2 + bx + c$, we need two factors $(x + m)$ and $(x + n)$ where the two numbers m and n multiply to c and add to b.

EXAMPLE 6.9 HOW TO FACTOR A TRINOMIAL OF THE FORM $x^2 + bx + c$

Factor: $x^2 + 11x + 24$.

Solution

Step 1. Write the factors as two binomials with first terms x.	Write two sets of parentheses and put x as the first term.	$x^2 + 11x + 24$ $(x \quad)(x \quad)$
Step 2. Find two numbers m and n that multiply to c, $m \cdot n = c$ add to b, $m + n = b$	Find two numbers that multiply to 24 and add to 11. <table><tr><th>Factors of 24</th><th>Sum of factors</th></tr><tr><td>1, 24</td><td>1 + 24 = 25</td></tr><tr><td>2, 12</td><td>2 + 12 = 14</td></tr><tr><td>3, 8</td><td>3 + 8 = 11*</td></tr><tr><td>4, 6</td><td>4 + 6 = 10</td></tr></table>	
Step 3. Use m and n as the last terms of the factors.	Use 3 and 8 as the last terms of the binomials.	$(x + 3)(x + 8)$
Step 4. Check by multiplying the factors.		$(x + 3)(x + 8)$ $x^2 + 8x + 3x + 24$ $x^2 + 11x + 24$ ✓

> **TRY IT ::** 6.17   Factor: $q^2 + 10q + 24$.

> **TRY IT ::** 6.18   Factor: $t^2 + 14t + 24$.

Let's summarize the steps we used to find the factors.

HOW TO :: FACTOR TRINOMIALS OF THE FORM $x^2 + bx + c$.

Step 1.
    Write the factors as two binomials with first terms x. $x^2 + bx + c$
    $(x \quad)(x \quad)$

Step 2. Find two numbers m and n that

    ◦ multiply to c, $m \cdot n = c$

    ◦ add to b, $m + n = b$

Step 3. Use m and n as the last terms of the factors. $(x + m)(x + n)$

Step 4. Check by multiplying the factors.

In the first example, all terms in the trinomial were positive. What happens when there are negative terms? Well, it depends which term is negative. Let's look first at trinomials with only the middle term negative.

How do you get a *positive product* and a *negative sum*? We use two negative numbers.

EXAMPLE 6.10

Factor: $y^2 - 11y + 28$.

⊘ **Solution**

Again, with the positive last term, 28, and the negative middle term, $-11y$, we need two negative factors. Find two numbers that multiply 28 and add to -11.

$$y^2 - 11y + 28$$

Write the factors as two binomials with first terms y. $(y\quad)(y\quad)$

Find two numbers that: multiply to 28 and add to -11.

Factors of 28	Sum of factors
$-1, -28$	$-1 + (-28) = -29$
$-2, -14$	$-2 + (-14) = -16$
$-4, -7$	$-4 + (-7) = -11*$

Use $-4, -7$ as the last terms of the binomials. $(y-4)(y-7)$

Check:

$$(y - 4)(y - 7)$$
$$y^2 - 7y - 4y + 28$$
$$y^2 - 11y + 28 ✓$$

> **TRY IT : : 6.19**   Factor: $u^2 - 9u + 18$.

> **TRY IT : : 6.20**   Factor: $y^2 - 16y + 63$.

Now, what if the last term in the trinomial is negative? Think about FOIL. The last term is the product of the last terms in the two binomials. A negative product results from multiplying two numbers with opposite signs. You have to be very careful to choose factors to make sure you get the correct sign for the middle term, too.

How do you get a *negative product* and a *positive sum*? We use one positive and one negative number.

When we factor trinomials, we must have the terms written in descending order—in order from highest degree to lowest degree.

EXAMPLE 6.11

Factor: $2x + x^2 - 48$.

⊘ Solution

$$2x + x^2 - 48$$

First we put the terms in decreasing degree order. $$x^2 + 2x - 48$$

Factors will be two binomials with first terms x. $$(x \quad)(x \quad)$$

Factors of -48	Sum of factors
$-1, 48$	$-1 + 48 = 47$
$-2, 24$	$-2 + 24 = 22$
$-3, 16$	$-3 + 16 = 13$
$-4, 12$	$-4 + 12 = 8$
$-6, 8$	$-6 + 8 = 2*$

Use $-6, 8$ as the last terms of the binomials. $$(x - 6)(x + 8)$$

Check:

$$(x - 6)(x + 8)$$
$$x^2 - 6q + 8q - 48$$
$$x^2 + 2x - 48 \checkmark$$

> **TRY IT : : 6.21**   Factor: $9m + m^2 + 18$.

> **TRY IT : : 6.22**   Factor: $-7n + 12 + n^2$.

Sometimes you'll need to factor trinomials of the form $x^2 + bxy + cy^2$ with two variables, such as $x^2 + 12xy + 36y^2$.

The first term, x^2, is the product of the first terms of the binomial factors, $x \cdot x$. The y^2 in the last term means that the second terms of the binomial factors must each contain y. To get the coefficients b and c, you use the same process summarized in **How To Factor trinomials**.

EXAMPLE 6.12

Factor: $r^2 - 8rs - 9s^2$.

⊘ Solution

We need r in the first term of each binomial and s in the second term. The last term of the trinomial is negative, so the factors must have opposite signs.

$$r^2 - 8rs - 9s^2$$

Note that the first terms are r, last terms contain s. $$(r \quad s)(r \quad s)$$

Find the numbers that multiply to -9 and add to -8.

Factors of -9	Sum of factors
1, -9	$-1 + 9 = 8$
-1, 9	$1 + (-9) = -8*$
3, -3	$3 + (-3) = 0$

Use 1, -9 as coefficients of the last terms. $(r + s)(r - 9s)$

Check:

$$(r - 9s)(r + s)$$
$$r^2 + rs - 9rs - 9s^2$$
$$r^2 - 8rs - 9s^2 \checkmark$$

 **TRY IT : : 6.23**   Factor: $a^2 - 11ab + 10b^2$.

 **TRY IT : : 6.24**   Factor: $m^2 - 13mn + 12n^2$.

Some trinomials are prime. The only way to be certain a trinomial is prime is to list all the possibilities and show that none of them work.

EXAMPLE 6.13

Factor: $u^2 - 9uv - 12v^2$.

⊘ **Solution**

We need u in the first term of each binomial and v in the second term. The last term of the trinomial is negative, so the factors must have opposite signs.

$$u^2 - 9uv - 12v^2$$

Note that the first terms are u, last terms contain v. $(u \quad v)(u \quad v)$

Find the numbers that multiply to -12 and add to -9.

Factors of -12	Sum of factors
1, -12	$1 + (-12) = -11$
-1, 12	$-1 + 12 = 11$
2, -6	$2 + (-6) = -4$
-2, 6	$-2 + 6 = 4$
3, -4	$3 + (-4) = -1$
-3, 4	$-3 + 4 = 1$

Note there are no factor pairs that give us -9 as a sum. The trinomial is prime.

 **TRY IT : : 6.25**   Factor: $x^2 - 7xy - 10y^2$.

 **TRY IT : : 6.26**   Factor: $p^2 + 15pq + 20q^2$.

Let's summarize the method we just developed to factor trinomials of the form $x^2 + bx + c$.

Strategy for Factoring Trinomials of the Form $x^2 + bx + c$

When we factor a trinomial, we look at the signs of its terms first to determine the signs of the binomial factors.

$$x^2 + bx + c$$
$$(x + m)(x + n)$$

When c is positive, m and n have the same sign.

b positive	b negative
m, n positive	m, n negative
$x^2 + 5x + 6$	$x^2 - 6x + 8$
$(x + 2)(x + 3)$	$(x - 4)(x - 2)$
same signs	same signs

When c is negative, m and n have opposite signs.

$x^2 + x - 12$	$x^2 - 2x - 15$
$(x + 4)(x - 3)$	$(x - 5)(x + 3)$
opposite signs	opposite signs

Notice that, in the case when m and n have opposite signs, the sign of the one with the larger absolute value matches the sign of b.

Factor Trinomials of the form $ax^2 + bx + c$ using Trial and Error

Our next step is to factor trinomials whose leading coefficient is not 1, trinomials of the form $ax^2 + bx + c$.

Remember to always check for a GCF first! Sometimes, after you factor the GCF, the leading coefficient of the trinomial becomes 1 and you can factor it by the methods we've used so far. Let's do an example to see how this works.

EXAMPLE 6.14

Factor completely: $4x^3 + 16x^2 - 20x$.

⊘ **Solution**

Is there a greatest common factor?

$$4x^3 + 16x^2 - 20x$$

    Yes, GCF = $4x$. Factor it.

$$4x\left(x^2 + 4x - 5\right)$$

Binomial, trinomial, or more than three terms?

    It is a trinomial. So "undo FOIL."

$$4x(x\quad)(x\quad)$$

Use a table like the one shown to find two numbers that multiply to -5 and add to 4.

$$4x(x - 1)(x + 5)$$

Factors of -5	Sum of factors
-1, 5	$-1 + 5 = 4$*
1, -5	$1 + (-5) = -4$

Check:

$$4x(x - 1)(x + 5)$$
$$4x(x^2 + 5x - x - 5)$$
$$4x(x^2 + 4x - 5)$$
$$4x^3 + 16x^2 - 20x \checkmark$$

> **TRY IT :: 6.27**   Factor completely: $5x^3 + 15x^2 - 20x$.

> **TRY IT :: 6.28**   Factor completely: $6y^3 + 18y^2 - 60y$.

What happens when the leading coefficient is not 1 and there is no GCF? There are several methods that can be used to factor these trinomials. First we will use the Trial and Error method.

Let's factor the trinomial $3x^2 + 5x + 2$.

From our earlier work, we expect this will factor into two binomials.

$$3x^2 + 5x + 2$$
$$(\quad)(\quad)$$

We know the first terms of the binomial factors will multiply to give us $3x^2$. The only factors of $3x^2$ are $1x, 3x$. We can place them in the binomials.

$$3x^2 + 5x + 2$$
1x, 3x
$$(x \quad)(3x \quad)$$

Check: Does $1x \cdot 3x = 3x^2$?

We know the last terms of the binomials will multiply to 2. Since this trinomial has all positive terms, we only need to consider positive factors. The only factors of 2 are 1, 2. But we now have two cases to consider as it will make a difference if we write 1, 2 or 2, 1.

$$3x^2 + 5x + 2 \qquad 3x^2 + 5x + 2$$
1x, 3x   1, 2   1x, 3x   1, 2
$$(x + 1)(3x + 2) \quad \text{or} \quad (x + 2)(3x + 1)$$

Which factors are correct? To decide that, we multiply the inner and outer terms.

$$3x^2 + 5x + 2 \qquad 3x^2 + 5x + 2$$
1x, 3x   1, 2   1x, 3x   1, 2
$$(x + 1)(3x + 2) \quad \text{or} \quad (x + 2)(3x + 1)$$
3x   6x
2x   1x
5x   7x

Since the middle term of the trinomial is $5x$, the factors in the first case will work. Let's use FOIL to check.

$$(x + 1)(3x + 2)$$
$$3x^2 + 2x + 3x + 2$$
$$3x^2 + 5x + 2 \checkmark$$

Our result of the factoring is:

$$3x^2 + 5x + 2$$
$$(x + 1)(3x + 2)$$

EXAMPLE 6.15   HOW TO FACTOR A TRINOMIAL USING TRIAL AND ERROR

Factor completely using trial and error: $3y^2 + 22y + 7$.

✓ **Solution**

Step 1. Write the trinomial in descending order.	The trinomial is already in descending order.	$3y^2 + 22y + 7$

Step 2. Factor any GCF.	There is no GCF.	
Step 3. Find all the factor pairs of the first term.	The only of $3y^2$ are $1y$, $3y$. Since there is only one pair, we can put them in the parentheses.	$3y^2 + 22y + 7$ _1y, 3y_ $3y^2 + 22y + 7$ _1y, 3y_ $(y\quad)(3y\quad)$
Step 4. Find all the factor pairs of the third term.	The only factors of 7 are 1, 7.	$3y^2 + 22y + 7$ _1y, 3y_ _1, 7_ $(y\quad)(3y\quad)$
Step 5. Test all the possible combinations of the factors until the correct product is found.	$3y^2 + 22y + 7$ _1y, 3y_ _1, 7_ $(y + 1)(3y + 7)$ $3y$ $7y$ $10y$ No! We need $22y$ $3y^2 + 22y + 7$ _1y, 3y_ _1, 7_ $(y + 7)(3y + 1)$ $21y$ $+y$ $22y$	**$3y^2 + 22y + 7$** | Possible factors	Product
---	---
$(y + 1)(3y + 7)$	$3y^2 + 10y + 7$
$(y + 7)(3y + 1)$	$3y^2 + 22y + 7$	
Step 6. Check by multiplying.		$(y + 7)(3y + 1)$ $3y^2 + 22y + 7 \checkmark$

> **TRY IT :: 6.29**   Factor completely using trial and error: $2a^2 + 5a + 3$.

> **TRY IT :: 6.30**   Factor completely using trial and error: $4b^2 + 5b + 1$.

HOW TO :: FACTOR TRINOMIALS OF THE FORM $ax^2 + bx + c$ USING TRIAL AND ERROR.

Step 1.   Write the trinomial in descending order of degrees as needed.

Step 2.   Factor any GCF.

Step 3.   Find all the factor pairs of the first term.

Step 4.   Find all the factor pairs of the third term.

Step 5.   Test all the possible combinations of the factors until the correct product is found.

Step 6.   Check by multiplying.

Remember, when the middle term is negative and the last term is positive, the signs in the binomials must both be negative.

EXAMPLE 6.16

Factor completely using trial and error: $6b^2 - 13b + 5$.

✓ Solution

The trinomial is already in descending order.	$6b^2 - 13b + 5$
Find the factors of the first term.	$6b^2 - 13b + 5$ $1b \cdot 6b$ $2b \cdot 3b$
Find the factors of the last term. Consider the signs. Since the last term, 5, is positive its factors must both be positive or both be negative. The coefficient of the middle term is negative, so we use the negative factors.	$6b^2 - 13b + 5$ $1b \cdot 6b$ $-1, -5$ $2b \cdot 3b$

Consider all the combinations of factors.

$6b^2 - 13b + 5$	
Possible factors	**Product**
$(b - 1)(6b - 5)$	$6b^2 - 11b + 5$
$(b - 5)(6b - 1)$	$6b^2 - 31b + 5$
$(2b - 1)(3b - 5)$	$6b^2 - 13b + 5$*
$(2b - 5)(3b - 1)$	$6b^2 - 17b + 5$

The correct factors are those whose product is the original trinomial. $(2b - 1)(3b - 5)$

Check by multiplying:

$$(2b - 1)(3b - 5)$$
$$6b^2 - 10b - 3b + 5$$
$$6b^2 - 13b + 5 \checkmark$$

> **TRY IT : :** 6.31   Factor completely using trial and error: $8x^2 - 13x + 3$.

> **TRY IT : :** 6.32   Factor completely using trial and error: $10y^2 - 37y + 7$.

When we factor an expression, we always look for a greatest common factor first. If the expression does not have a greatest common factor, there cannot be one in its factors either. This may help us eliminate some of the possible factor combinations.

EXAMPLE 6.17

Factor completely using trial and error: $18x^2 - 37xy + 15y^2$.

✓ Solution

The trinomial is already in descending order.	$18x^2 - 37xy + 15y^2$

Find the factors of the first term.

$18x^2 - 37xy + 15y^2$
1x · 18x
2x · 9x
3x · 6x

Find the factors of the last term. Consider the signs. Since 15 is positive and the coefficient of the middle term is negative, we use the negative factors.

$18x^2 - 37xy + 15y^2$
1x · 18x   −1, −5
2x · 9x   −5, −1
3x · 6x

Consider all the combinations of factors.

$18x^2 - 37xy + 15y^2$	
Possible factors	**Product**
$(x - 1y)(18x - 15y)$	Not an option
$(x - 15y)(18x - 1y)$	$18x^2 - 271xy + 15y^2$
$(x - 3y)(18x - 5y)$	$18x^2 - 59xy + 15y^2$
$(x - 5y)(18x - 3y)$	Not an option
$(2x - 1y)(9x - 15y)$	Not an option
$(2x - 15y)(9x - 1y)$	$18x^2 - 137xy + 15y^2$
$(2x - 3y)(9x - 5y)$	$18x^2 - 37xy + 15y^2$*
$(2x - 5y)(9x - 3y)$	Not an option
$(3x - 1y)(6x - 15y)$	Not an option
$(3x - 15y)(6x - 1y)$	Not an option
$(3x - 3y)(6x - 5y)$	Not an option

If the trinomial has no common factors, then neither factor can contain a common factor. That means this combination is not an option.

The correct factors are those whose product is the original trinomial.

$$(2x - 3y)(9x - 5y)$$

Check by multiplying:

$$(2x - 3y)(9x - 5y)$$
$$18x^2 - 10xy - 27xy + 15y^2$$
$$18x^2 - 37xy + 15y^2 \checkmark$$

> **TRY IT ::** 6.33   Factor completely using trial and error $18x^2 - 3xy - 10y^2$.

> **TRY IT ::** 6.34   Factor completely using trial and error: $30x^2 - 53xy - 21y^2$.

Don't forget to look for a GCF first and remember if the leading coefficient is negative, so is the GCF.

EXAMPLE 6.18

Factor completely using trial and error: $-10y^4 - 55y^3 - 60y^2$.

✓ **Solution**

	$-10y^4 - 55y^3 - 60y^2$
Notice the greatest common factor, so factor it first.	$-5y^2(2y^2 + 11y + 12)$
Factor the trinomial.	$-5y^2 \left(\begin{array}{c} 2y^2 + 11y + 12 \\ y \cdot 2y \qquad \begin{array}{c} 1 \cdot 12 \\ 2 \cdot 6 \\ 3 \cdot 4 \end{array} \end{array} \right)$

Consider all the combinations.

$2y^2 + 11y + 12$	
Possible factors	**Product**
$(y + 1)(2y + 12)$	Not an option
$(y + 12)(2y + 1)$	$2y^2 + 25y + 12$
$(y + 2)(2y + 6)$	Not an option
$(y + 6)(2y + 2)$	Not an option
$(y + 3)(2y + 4)$	Not an option
$(y + 4)(2y + 3)$	$2y^2 + 11y + 12*$

If the trinomial has no common factors, then neither factor can contain a common factor. That means this combination is not an option.

The correct factors are those whose product is the original trinomial. Remember to include the factor $-5y^2$.

$$-5y^2(y + 4)(2y + 3)$$

Check by multiplying:

$$-5y^2(y + 4)(2y + 3)$$
$$-5y^2(2y^2 + 8y + 3y + 12)$$
$$-10y^4 - 55y^3 - 60y^2 \checkmark$$

> **TRY IT : : 6.35**   Factor completely using trial and error: $15n^3 - 85n^2 + 100n$.

> **TRY IT : : 6.36**   Factor completely using trial and error: $56q^3 + 320q^2 - 96q$.

Factor Trinomials of the Form $ax^2 + bx + c$ using the "ac" Method

Another way to factor trinomials of the form $ax^2 + bx + c$ is the "ac" method. (The "ac" method is sometimes called the grouping method.) The "ac" method is actually an extension of the methods you used in the last section to factor trinomials with leading coefficient one. This method is very structured (that is step-by-step), and it always works!

EXAMPLE 6.19   HOW TO FACTOR TRINOMIALS USING THE "AC" METHOD

Factor using the 'ac' method: $6x^2 + 7x + 2$.

✓ **Solution**

Step 1. Factor any GCF.	Is there a greatest common factor? No!	$6x^2 + 7x + 2$
Step 2. Find the product ac.	$a \cdot c$ $6 \cdot 2$ 12	$ax^2 + bx + c$ $6x^2 + 7x + 2$

Step 3. Find two numbers m and n that: Multiply to ac. $m \cdot n = a \cdot c$ Add to b. $m + n = b$	Find two numbers that multiply to 12 and add to 7. Both factors must be positive. $3 \cdot 4 = 12$ $3 + 4 = 7$	
Step 4. Split the middle term using m and n. $ax^2 + bx + c$ bx $ax^2 + mx + nx + c$	Rewrite $7x$ as $3x + 4x$. It would also give the same result if we used $4x + 3x$. Notice that $6x^2 + 3x + 4x + 2$ is equal to $6x^2 + 7x + 2$. We just split the middle term to get a more useful form.	$6x^2 + 7x + 2$ $6x^2 + 3x + 4x + 2$
Step 5. Factor by grouping.		$3x(2x + 1) + 2(2x + 1)$ $(2x + 1)(3x + 2)$
Step 6. Check by multiplying the factors.		$(2x + 1)(3x + 2)$ $6x^2 + 4x + 3x + 2$ $6x^2 + 7x + 2$ ✓

> **TRY IT : :** 6.37   Factor using the 'ac' method: $6x^2 + 13x + 2$.

> **TRY IT : :** 6.38   Factor using the 'ac' method: $4y^2 + 8y + 3$.

The "ac" method is summarized here.

HOW TO : : FACTOR TRINOMIALS OF THE FORM $ax^2 + bx + c$ USING THE "AC" METHOD.

Step 1.   Factor any GCF.

Step 2.   Find the product ac.

Step 3.   Find two numbers m and n that:
  Multiply to ac $m \cdot n = a \cdot c$
  Add to b $m + n = b$
    $ax^2 + bx + c$

Step 4.   Split the middle term using m and n. $ax^2 + mx + nx + c$

Step 5.   Factor by grouping.

Step 6.   Check by multiplying the factors.

Don't forget to look for a common factor!

EXAMPLE 6.20

Factor using the 'ac' method: $10y^2 - 55y + 70$.

⊘ **Solution**

Is there a greatest common factor?	
Yes. The GCF is 5.	$10y^2 - 55y + 70$
Factor it.	$5(2y^2 - 11y + 14)$
The trinomial inside the parentheses has a leading coefficient that is not 1.	$ax^2 + bx + c$ $5(2y^2 - 11y + 14)$
Find the product ac.	$ac = 28$
Find two numbers that multiply to ac	$(-4)(-7) = 28$
and add to b.	$-4 + (-7) = -11$
Split the middle term.	$5(2y^2 - 11y + 14)$
	$5(2y^2 - 7y - 4y + 14)$
Factor the trinomial by grouping.	$5(y\,(2y - 7) - 2(y - 7))$
	$5(y - 2)(2y - 7)$

Check by multiplying all three factors.

$$5(y - 2)(2y - 7)$$
$$5\left(2y^2 - 7y - 4y + 14\right)$$
$$5\left(2y^2 - 11y + 14\right)$$
$$10y^2 - 55y + 70 \checkmark$$

> **TRY IT ∷ 6.39**   Factor using the 'ac' method: $16x^2 - 32x + 12$.

> **TRY IT ∷ 6.40**   Factor using the 'ac' method: $18w^2 - 39w + 18$.

Factor Using Substitution

Sometimes a trinomial does not appear to be in the $ax^2 + bx + c$ form. However, we can often make a thoughtful substitution that will allow us to make it fit the $ax^2 + bx + c$ form. This is called factoring by substitution. It is standard to use u for the substitution.

In the $ax^2 + bx + c$, the middle term has a variable, x, and its square, x^2, is the variable part of the first term. Look for this relationship as you try to find a substitution.

EXAMPLE 6.21

Factor by substitution: $x^4 - 4x^2 - 5$.

⊘ **Solution**

The variable part of the middle term is x^2 and its square, x^4, is the variable part of the first term. (We know $\left(x^2\right)^2 = x^4$). If we let $u = x^2$, we can put our trinomial in the $ax^2 + bx + c$ form we need to factor it.

	$x^4 - 4x^2 - 5$
Rewrite the trinomial to prepare for the substitution.	$(x^2)^2 - 4(x^2) - 5$
Let $u = x^2$ and substitute.	$u^2 - 4u - 5$
Factor the trinomial.	$(u + 1)(u - 5)$
Replace u with x^2.	$(x^2 + 1)(x^2 - 5)$

Check:

$$\left(x^2 + 1\right)\left(x^2 - 5\right)$$
$$x^4 - 5x^2 + x^2 - 5$$
$$x^4 - 4x^2 - 5 \checkmark$$

> **TRY IT ::** 6.41   Factor by substitution: $h^4 + 4h^2 - 12$.

> **TRY IT ::** 6.42   Factor by substitution: $y^4 - y^2 - 20$.

Sometimes the expression to be substituted is not a monomial.

EXAMPLE 6.22

Factor by substitution: $(x - 2)^2 + 7(x - 2) + 12$

⊘ **Solution**

The binomial in the middle term, $(x - 2)$ is squared in the first term. If we let $u = x - 2$ and substitute, our trinomial will be in $ax^2 + bx + c$ form.

	$(x - 2)^2 + 7(x - 2) + 12$
Rewrite the trinomial to prepare for the substitution.	$(x - 2)^2 + 7(x - 2) + 12$
Let $u = x - 2$ and substitute.	$u^2 + 7u + 12$
Factor the trinomial.	$(u + 3)(u + 4)$
Replace u with $x - 2$.	$((x - 2) + 3)((x - 2) + 4)$
Simplify inside the parentheses.	$(x + 1)(x + 2)$

This could also be factored by first multiplying out the $(x - 2)^2$ and the $7(x - 2)$ and then combining like terms and then factoring. Most students prefer the substitution method.

> **TRY IT ::** 6.43   Factor by substitution: $(x - 5)^2 + 6(x - 5) + 8$.

> **TRY IT ::** 6.44   Factor by substitution: $(y - 4)^2 + 8(y - 4) + 15$.

▶ | **MEDIA : :**
Access this online resource for additional instruction and practice with factoring.

 • **Factor a trinomial using the AC method (https://openstax.org/l/37ACmethod)**

 • **Factor a trinomial using the AC method (https://openstax.org/l/37ACmethod)**

6.2 EXERCISES
Practice Makes Perfect

Factor Trinomials of the Form $x^2 + bx + c$

In the following exercises, factor each trinomial of the form $x^2 + bx + c$.

61. $p^2 + 11p + 30$

62. $w^2 + 10x + 21$

63. $n^2 + 19n + 48$

64. $b^2 + 14b + 48$

65. $a^2 + 25a + 100$

66. $u^2 + 101u + 100$

67. $x^2 - 8x + 12$

68. $q^2 - 13q + 36$

69. $y^2 - 18x + 45$

70. $m^2 - 13m + 30$

71. $x^2 - 8x + 7$

72. $y^2 - 5y + 6$

73. $5p - 6 + p^2$

74. $6n - 7 + n^2$

75. $8 - 6x + x^2$

76. $7x + x^2 + 6$

77. $x^2 - 12 - 11x$

78. $-11 - 10x + x^2$

In the following exercises, factor each trinomial of the form $x^2 + bxy + cy^2$.

79. $x^2 - 2xy - 80y^2$

80. $p^2 - 8pq - 65q^2$

81. $m^2 - 64mn - 65n^2$

82. $p^2 - 2pq - 35q^2$

83. $a^2 + 5ab - 24b^2$

84. $r^2 + 3rs - 28s^2$

85. $x^2 - 3xy - 14y^2$

86. $u^2 - 8uv - 24v^2$

87. $m^2 - 5mn + 30n^2$

88. $c^2 - 7cd + 18d^2$

Factor Trinomials of the Form $ax^2 + bx + c$ Using Trial and Error

In the following exercises, factor completely using trial and error.

89. $p^3 - 8p^2 - 20p$

90. $q^3 - 5q^2 - 24q$

91. $3m^3 - 21m^2 + 30m$

92. $11n^3 - 55n^2 + 44n$

93. $5x^4 + 10x^3 - 75x^2$

94. $6y^4 + 12y^3 - 48y^2$

95. $2t^2 + 7t + 5$

96. $5y^2 + 16y + 11$

97. $11x^2 + 34x + 3$

98. $7b^2 + 50b + 7$

99. $4w^2 - 5w + 1$

100. $5x^2 - 17x + 6$

101. $4q^2 - 7q - 2$

102. $10y^2 - 53y - 11$

103. $6p^2 - 19pq + 10q^2$

104. $21m^2 - 29mn + 10n^2$

105. $4a^2 + 17ab - 15b^2$

106. $6u^2 + 5uv - 14v^2$

107. $-16x^2 - 32x - 16$

108. $-81a^2 + 153a + 18$

109. $-30q^3 - 140q^2 - 80q$

110. $-5y^3 - 30y^2 + 35y$

Factor Trinomials of the Form $ax^2 + bx + c$ using the 'ac' Method

In the following exercises, factor using the 'ac' method.

111. $5n^2 + 21n + 4$

112. $8w^2 + 25w + 3$

113. $4k^2 - 16k + 15$

114. $5s^2 - 9s + 4$

115. $6y^2 + y - 15$

116. $6p^2 + p - 22$

117. $2n^2 - 27n - 45$

118. $12z^2 - 41z - 11$

119. $60y^2 + 290y - 50$

120. $6u^2 - 46u - 16$

121. $48z^3 - 102z^2 - 45z$

122. $90n^3 + 42n^2 - 216n$

123. $16s^2 + 40s + 24$

124. $24p^2 + 160p + 96$

125. $48y^2 + 12y - 36$

126. $30x^2 + 105x - 60$

Factor Using Substitution

In the following exercises, factor using substitution.

127. $x^4 - x^2 - 12$

128. $x^4 + 2x^2 - 8$

129. $x^4 - 3x^2 - 28$

130. $x^4 - 13x^2 - 30$

131. $(x - 3)^2 - 5(x - 3) - 36$

132. $(x - 2)^2 - 3(x - 2) - 54$

133. $(3y - 2)^2 - (3y - 2) - 2$

134. $(5y - 1)^2 - 3(5y - 1) - 18$

Mixed Practice

In the following exercises, factor each expression using any method.

135. $u^2 - 12u + 36$

136. $x^2 - 14x - 32$

137. $r^2 - 20rs + 64s^2$

138. $q^2 - 29qr - 96r^2$

139. $12y^2 - 29y + 14$

140. $12x^2 + 36y - 24z$

141. $6n^2 + 5n - 4$

142. $3q^2 + 6q + 2$

143. $13z^2 + 39z - 26$

144. $5r^2 + 25r + 30$

145. $3p^2 + 21p$

146. $7x^2 - 21x$

147. $6r^2 + 30r + 36$

148. $18m^2 + 15m + 3$

149. $24n^2 + 20n + 4$

150. $4a^2 + 5a + 2$

151. $x^4 - 4x^2 - 12$

152. $x^4 - 7x^2 - 8$

153. $(x + 3)^2 - 9(x + 3) - 36$

154. $(x + 2)^2 - 25(x + 2) - 54$

Writing Exercises

155. Many trinomials of the form $x^2 + bx + c$ factor into the product of two binomials $(x + m)(x + n)$. Explain how you find the values of m and n.

156. Tommy factored $x^2 - x - 20$ as $(x + 5)(x - 4)$. Sara factored it as $(x + 4)(x - 5)$. Ernesto factored it as $(x - 5)(x - 4)$. Who is correct? Explain why the other two are wrong.

157. List, in order, all the steps you take when using the "ac" method to factor a trinomial of the form $ax^2 + bx + c$.

158. How is the "ac" method similar to the "undo FOIL" method? How is it different?

Self Check

ⓐ *After completing the exercises, use this checklist to evaluate your mastery of the objectives of this section.*

I can...	Confidently	With some help	No-I don't get it!
factor trinomials of the form $x^2 + bx + c$			
factor trinomials of the form $ax^2 + bx + c$ using trial and error.			
factor trinomials of the form $ax^2 + bx + c$ with using the "ac" method.			
factor using substitution.			

ⓑ *After reviewing this checklist, what will you do to become confident for all objectives?*

6.3 | Factor Special Products

Learning Objectives

By the end of this section, you will be able to:

> Factor perfect square trinomials
> Factor differences of squares
> Factor sums and differences of cubes

Be Prepared!

Before you get started, take this readiness quiz.

1. Simplify: $\left(3x^2\right)^3$.

   If you missed this problem, review **Example 5.18**.

2. Multiply: $(m+4)^2$.

   If you missed this problem, review **Example 5.32**.

3. Multiply: $(x-3)(x+3)$.

   If you missed this problem, review **Example 5.33**.

We have seen that some binomials and trinomials result from special products—squaring binomials and multiplying conjugates. If you learn to recognize these kinds of polynomials, you can use the special products patterns to factor them much more quickly.

Factor Perfect Square Trinomials

Some trinomials are perfect squares. They result from multiplying a binomial times itself. We squared a binomial using the Binomial Squares pattern in a previous chapter.

$$\overset{\overset{a \;+\; b}{\frown}}{(3x + 4)^2}$$

$$\overset{a^2 \;+\; 2 \cdot a \cdot b \;+\; b^2}{(3x)^2 + 2(3x \cdot 4) + 4^2}$$

$$9x^2 + 24x + 16$$

The trinomial $9x^2 + 24x + 16$ is called a *perfect square trinomial*. It is the square of the binomial $3x + 4$.

In this chapter, you will start with a perfect square trinomial and factor it into its prime factors.

You could factor this trinomial using the methods described in the last section, since it is of the form $ax^2 + bx + c$. But if you recognize that the first and last terms are squares and the trinomial fits the perfect square trinomials pattern, you will save yourself a lot of work.

Here is the pattern—the reverse of the binomial squares pattern.

Perfect Square Trinomials Pattern

If a and b are real numbers

$$a^2 + 2ab + b^2 = (a+b)^2$$
$$a^2 - 2ab + b^2 = (a-b)^2$$

To make use of this pattern, you have to recognize that a given trinomial fits it. Check first to see if the leading coefficient is a perfect square, a^2. Next check that the last term is a perfect square, b^2. Then check the middle term—is it the product, $2ab$? If everything checks, you can easily write the factors.

EXAMPLE 6.23   HOW TO FACTOR PERFECT SQUARE TRINOMIALS

Factor: $9x^2 + 12x + 4$.

⊘ Solution

Step 1. Does the trinomial fit the perfect square trinomials pattern, $a^2 + 2ab + b^2$?		
• Is the first term a perfect square? Write it as a square, a^2.	Is $9x^2$ a perfect square? Yes—write it as $(3x)^2$.	$9x^2 \;+\; 12x \;+\; 4$ $(3x)^2$
• Is the last term a perfect square? Write it as a square, b^2.	Is 4 a perfect square? Yes—write it as $(2)^2$.	$(3x)^2 \qquad\qquad (2)^2$
• Check the middle term. Is it $2ab$?	Is $12x$ twice the product of $3x$ and 2? Does it match? Yes, so we have a perfect square trinomial!	$(3x)^2 \qquad\qquad (2)^2$ $2(3x)(2)$ $12x$
Step 2. Write the square of the binomial.		$9x^2 + 12x + 4$ $a^2 \;+ 2\cdot\; a\; \cdot\; b + b^2$ $(3x)^2 + 2\cdot 3x \cdot 2 + 2^2$ $(a\; +\; b)^2$ $(3x + 2)^2$
Step 3. Check.		$(3x + 2)^2$ $(3x)^2 + 2\cdot 3x \cdot 2 + 2^2$ $9x^2 + 12x + 4 \checkmark$

> **TRY IT :: 6.45**   Factor: $4x^2 + 12x + 9$.

> **TRY IT :: 6.46**   Factor: $9y^2 + 24y + 16$.

The sign of the middle term determines which pattern we will use. When the middle term is negative, we use the pattern $a^2 - 2ab + b^2$, which factors to $(a - b)^2$.

The steps are summarized here.

HOW TO :: FACTOR PERFECT SQUARE TRINOMIALS.

Step 1. Does the trinomial fit the pattern?

  Is the first term a perfect square?
  Write it as a square.

  Is the last term a perfect square?
  Write it as a square.

  Check the middle term. Is it $2ab$?

Step 2. Write the square of the binomial.

Step 3. Check by multiplying.

$$a^2 + 2ab + b^2 \qquad\qquad a^2 - 2ab + b^2$$
$$(a)^2 \qquad\qquad\qquad (a)^2$$
$$(a)^2 \qquad (b)^2 \quad (a)^2 \qquad (b)^2$$
$$(a)^2 \searrow_{2\cdot a\cdot b}\nearrow (b)^2 \quad (a)^2 \searrow_{2\cdot a\cdot b}\nearrow (b)^2$$
$$(a + b)^2 \qquad\qquad (a - b)^2$$

We'll work one now where the middle term is negative.

EXAMPLE 6.24

Factor: $81y^2 - 72y + 16$.

✓ Solution

The first and last terms are squares. See if the middle term fits the pattern of a perfect square trinomial. The middle term is negative, so the binomial square would be $(a - b)^2$.

	$81y^2 - 72y + 16$
Are the first and last terms perfect squares?	$(9y)^2$ $(4)^2$
Check the middle term.	$(9y)^2$ $(4)^2$
	$2(9y)(4)$
	$72y$
Does it match $(a - b)^2$? Yes.	$a^2 - 2 \cdot a \cdot b + b^2$
	$(9y)^2 - 2 \cdot 9y \cdot 4 + 4^2$
Write as the square of a binomial.	$(9y - 4)^2$

Check by multiplying:

$$(9y - 4)^2$$
$$(9y)^2 - 2 \cdot 9y \cdot 4 + 4^2$$
$$81y^2 - 72y + 16 \checkmark$$

> **TRY IT :: 6.47**   Factor: $64y^2 - 80y + 25$.

> **TRY IT :: 6.48**   Factor: $16z^2 - 72z + 81$.

The next example will be a perfect square trinomial with two variables.

EXAMPLE 6.25

Factor: $36x^2 + 84xy + 49y^2$.

✓ Solution

	$36x^2 + 84xy + 49y^2$
Test each term to verify the pattern.	$a^2 + 2 \cdot a \cdot b + b^2$
	$(6x)^2 + 2 \cdot 6x \cdot 7y + (7y)^2$
Factor.	$(6x + 7y)^2$

Check by multiplying.

$$(6x + 7y)^2$$
$$(6x)^2 + 2 \cdot 6x \cdot 7y + (7y)^2$$
$$36x^2 + 84xy + 49y^2 \checkmark$$

> **TRY IT : : 6.49** Factor: $49x^2 + 84xy + 36y^2$.

> **TRY IT : : 6.50** Factor: $64m^2 + 112mn + 49n^2$.

Remember the first step in factoring is to look for a greatest common factor. Perfect square trinomials may have a GCF in all three terms and it should be factored out first. And, sometimes, once the GCF has been factored, you will recognize a perfect square trinomial.

EXAMPLE 6.26

Factor: $100x^2y - 80xy + 16y$.

✓ **Solution**

	$100x^2y - 80xy + 16y$
Is there a GCF? Yes, $4y$, so factor it out.	$4y(25x^2 - 20x + 4)$
Is this a perfect square trinomial?	
Verify the pattern.	$4y[(5x)^2 - 2 \cdot 5x \cdot 2 + 2^2]$
Factor.	$4y(5x - 2)^2$

Remember: Keep the factor 4y in the final product.

Check:

$$4y(5x - 2)^2$$
$$4y\left[(5x)^2 - 2 \cdot 5x \cdot 2 + 2^2\right]$$
$$4y\left(25x^2 - 20x + 4\right)$$
$$100x^2y - 80xy + 16y ✓$$

> **TRY IT : : 6.51** Factor: $8x^2y - 24xy + 18y$.

> **TRY IT : : 6.52** Factor: $27p^2q + 90pq + 75q$.

Factor Differences of Squares

The other special product you saw in the previous chapter was the Product of Conjugates pattern. You used this to multiply two binomials that were conjugates. Here's an example:

$$(a - b)(a + b)$$
$$(3x - 4)(3x + 4)$$

$$(a)^2 - (b)^2$$
$$(3x)^2 - (4)^2$$

$$9x^2 - 16$$

A difference of squares factors to a product of conjugates.

Difference of Squares Pattern

If a and b are real numbers,

$$a^2 - b^2 = (a-b)(a+b) \qquad a^2 \quad \overset{\textit{difference}}{-} \quad b^2 = (a-b)(a+b)$$

squares   conjugates

Remember, "difference" refers to subtraction. So, to use this pattern you must make sure you have a binomial in which two squares are being subtracted.

EXAMPLE 6.27 HOW TO FACTOR A TRINOMIAL USING THE DIFFERENCE OF SQUARES

Factor: $64y^2 - 1$.

⊘ **Solution**

Step 1. Does the binomial fit the pattern?		$64y^2 - 1$
• Is this a difference?	Yes	$64y^2 - 1$
• Are the first and last terms perfect squares?	Yes	
Step 2. Write them as squares.	Write them as x^2 and 2^2.	$\overset{a^2 \ - \ b^2}{(8y)^2 - 1^2}$
Step 3. Write the product of conjugates.		$\overset{(a \ - \ b) \ (a \ + \ b)}{(8y-1)(8y+1)}$
Step 4. Check.		$(8y-1)(8y+1)$ $64y^2 - 1 ✓$

> **TRY IT :: 6.53**   Factor: $121m^2 - 1$.

> **TRY IT :: 6.54**   Factor: $81y^2 - 1$.

  HOW TO :: FACTOR DIFFERENCES OF SQUARES.

Step 1. Does the binomial fit the pattern? $a^2 - b^2$
         Is this a difference?   ____ − ____
         Are the first and last terms perfect squares?
Step 2. Write them as squares. $(a)^2 - (b)^2$
Step 3. Write the product of conjugates. $(a-b)(a+b)$
Step 4. Check by multiplying.

It is important to remember that *sums of squares do not factor into a product of binomials*. There are no binomial factors that multiply together to get a sum of squares. After removing any GCF, the expression $a^2 + b^2$ is prime!

The next example shows variables in both terms.

EXAMPLE 6.28

Factor: $144x^2 - 49y^2$.

⊘ **Solution**

$$144x^2 - 49y^2$$

Is this a difference of squares? Yes. $\qquad (12x)^2 - (7y)^2$

Factor as the product of conjugates. $\qquad (12x - 7y)(12x + 7y)$

Check by multiplying.

$$(12x - 7y)(12x + 7y)$$
$$144x^2 - 49y^2 \checkmark$$

> │ **TRY IT :: 6.55** $\qquad$ Factor: $196m^2 - 25n^2$.

> │ **TRY IT :: 6.56** $\qquad$ Factor: $121p^2 - 9q^2$.

As always, you should look for a common factor first whenever you have an expression to factor. Sometimes a common factor may "disguise" the difference of squares and you won't recognize the perfect squares until you factor the GCF.

Also, to completely factor the binomial in the next example, we'll factor a difference of squares twice!

EXAMPLE 6.29

Factor: $48x^4 y^2 - 243y^2$.

⊘ **Solution**

$$48x^4 y^2 - 243y^2$$

Is there a GCF? Yes, $3y^2$—factor it out! $\qquad 3y^2(16x^4 - 81)$

Is the binomial a difference of squares? Yes. $\qquad 3y^2\left(\left(4x^2\right)^2 - (9)^2\right)$

Factor as a product of conjugates. $\qquad 3y^2\left(4x^2 - 9\right)\left(4x^2 + 9\right)$

Notice the first binomial is also a difference of squares! $\qquad 3y^2\left((2x)^2 - (3)^2\right)\left(4x^2 + 9\right)$

Factor it as the product of conjugates. $\qquad 3y^2(2x - 3)(2x + 3)\left(4x^2 + 9\right)$

The last factor, the sum of squares, cannot be factored.

Check by multiplying:

$$3y^2(2x - 3)(2x + 3)\left(4x^2 + 9\right)$$
$$3y^2\left(4x^2 - 9\right)\left(4x^2 + 9\right)$$
$$3y^2\left(16x^4 - 81\right)$$
$$48x^4 y^2 - 243y^2 \checkmark$$

> │ **TRY IT :: 6.57** $\qquad$ Factor: $2x^4 y^2 - 32y^2$.

> │ **TRY IT :: 6.58** $\qquad$ Factor: $7a^4 c^2 - 7b^4 c^2$.

The next example has a polynomial with 4 terms. So far, when this occurred we grouped the terms in twos and factored from there. Here we will notice that the first three terms form a perfect square trinomial.

EXAMPLE 6.30

Factor: $x^2 - 6x + 9 - y^2$.

✓ **Solution**

Notice that the first three terms form a perfect square trinomial.

	$x^2 - 6x + 9 - y^2$
Factor by grouping the first three terms.	$x^2 - 6x + 9 - y^2$
Use the perfect square trinomial pattern.	$(x - 3)^2 - y^2$
Is this a difference of squares? Yes.	
Yes—write them as squares.	$\overset{a^2}{(x-3)^2} - \overset{b^2}{y^2}$
Factor as the product of conjugates.	$\overset{(a\ -\ b)}{((x-3)-y)}\overset{(x\ +\ b)}{((x-3)+y)}$
	$(x - 3 - y)(x - 3 + y)$

You may want to rewrite the solution as $(x - y - 3)(x + y - 3)$.

⊳ **TRY IT :: 6.59**   Factor: $x^2 - 10x + 25 - y^2$.

⊳ **TRY IT :: 6.60**   Factor: $x^2 + 6x + 9 - 4y^2$.

Factor Sums and Differences of Cubes

There is another special pattern for factoring, one that we did not use when we multiplied polynomials. This is the pattern for the sum and difference of cubes. We will write these formulas first and then check them by multiplication.

$$a^3 + b^3 = (a + b)(a^2 - ab + b^2)$$
$$a^3 - b^3 = (a - b)(a^2 + ab + b^2)$$

We'll check the first pattern and leave the second to you.

	$(a + b)(a^2 - ab + b^2)$
Distribute.	$a(a^2 - ab + b^2) + b(a^2 - ab + b^2)$
Multiply.	$a^3 - a^2b + ab^2 + a^2b - ab^2 + b^3$
Combine like terms.	$a^3 + b^3$

Sum and Difference of Cubes Pattern

$$a^3 + b^3 = (a + b)(a^2 - ab + b^2)$$
$$a^3 - b^3 = (a - b)(a^2 + ab + b^2)$$

The two patterns look very similar, don't they? But notice the signs in the factors. The sign of the binomial factor matches the sign in the original binomial. And the sign of the middle term of the trinomial factor is the opposite of the sign in the original binomial. If you recognize the pattern of the signs, it may help you memorize the patterns.

$$a^3 + b^3 = (a + b)(a^2 - ab + b^2)$$

same sign

opposite signs

$$a^3 - b^3 = (a - b)(a^2 + ab + b^2)$$

same sign

opposite signs

The trinomial factor in the sum and difference of cubes pattern cannot be factored.

It be very helpful if you learn to recognize the cubes of the integers from 1 to 10, just like you have learned to recognize squares. We have listed the cubes of the integers from 1 to 10 in **Table 6.22**.

n	1	2	3	4	5	6	7	8	9	10
n^3	1	8	27	64	125	216	343	512	729	1000

Table 6.22

EXAMPLE 6.31 HOW TO FACTOR THE SUM OR DIFFERENCE OF CUBES

Factor: $x^3 + 64$.

✓ **Solution**

Step 1. Does the binomial fit the sum or difference of cubes pattern?		$x^3 + 64$
• Is it a sum or difference? • Are the first and last terms perfect cubes?	This is a sum. Yes.	$x^3 + 64$
Step 2. Write the terms as cubes.	Write them as x^3 and 4^3.	$a^3 + b^3$ $x^3 + 4^3$
Step 3. Use either the sum or difference of cubes pattern.	This is a sum of cubes.	$\overset{a\ +\ b}{(x + 4)}\overset{a^2\ -\ ab\ +\ b^2}{(x^2 - 4x + 4^2)}$
Step 4. Simplify inside the parentheses.	It is already simplified.	$(x + 4)(x^2 - 4x + 16)$
Step 5. Check by multiplying the factors.		$\begin{array}{r} x^2 - 4x + 16 \\ x + 4 \\ \hline 4x^2 - 16x + 64 \ \checkmark \\ x^3 - 4x^2 + 16x \\ \hline x^3 \qquad\qquad + 64 \end{array}$

> | **TRY IT : :** 6.61   Factor: $x^3 + 27$.

> | **TRY IT : :** 6.62   Factor: $y^3 + 8$.

HOW TO :: FACTOR THE SUM OR DIFFERENCE OF CUBES.

Step 1.   Does the binomial fit the sum or difference of cubes pattern?
          Is it a sum or difference?
          Are the first and last terms perfect cubes?

Step 2.   Write them as cubes.

Step 3.   Use either the sum or difference of cubes pattern.

Step 4.   Simplify inside the parentheses.

Step 5.   Check by multiplying the factors.

EXAMPLE 6.32

Factor: $27u^3 - 125v^3$.

 **Solution**

	$27u^3 - 125v^3$
This binomial is a difference. The first and last terms are perfect cubes.	
Write the terms as cubes.	$\overset{a^3\ -\ b^3}{(3u)^3 - (5v)^3}$
Use the difference of cubes pattern.	$\left(\overset{a\ -\ b}{3u - 5v}\right)\left(\overset{a^2\ +\ ab\ +\ b^2}{(3u)^2 + 3u \cdot 5v + (5v)^2}\right)$
Simplify.	$\left(\overset{a\ -\ b}{3u - 5v}\right)\left(\overset{a^2\ +\ ab\ +\ b^2}{9u^2 + 15uv + 25v^2}\right)$
Check by multiplying.	We'll leave the check to you.

> **TRY IT :: 6.63**   Factor: $8x^3 - 27y^3$.

> **TRY IT :: 6.64**   Factor: $1000m^3 - 125n^3$.

In the next example, we first factor out the GCF. Then we can recognize the sum of cubes.

EXAMPLE 6.33

Factor: $6x^3y + 48y^4$.

✓ Solution

$$6x^3y + 48y^4$$

Factor the common factor.	$6y(x^3 + 8y^3)$
This binomial is a sum The first and last terms are perfect cubes.	
Write the terms as cubes.	$6y\left(\overset{a^3 +\ \ b^3}{x^3 + (2y)^3}\right)$
Use the sum of cubes pattern.	$6y\left(\overset{a +\ b}{x + 2y}\right)\left(\overset{a^2 -\ \ \ ab\ \ +\ \ b^2}{x^2 - x \cdot 2y + (2y)^2}\right)$
Simplify.	$6y(x + 2y)(x^2 - 2xy + 4y^2)$

Check:

To check, you may find it easier to multiply the sum of cubes factors first, then multiply that product by $6y$. We'll leave the multiplication for you.

> **TRY IT :: 6.65**   Factor: $500p^3 + 4q^3$.

> **TRY IT :: 6.66**   Factor: $432c^3 + 686d^3$.

The first term in the next example is a binomial cubed.

EXAMPLE 6.34

Factor: $(x + 5)^3 - 64x^3$.

✓ Solution

$$(x + 5)^3 - 64x^3$$

This binomial is a difference. The first and last terms are perfect cubes.	
Write the terms as cubes.	$\overset{a^3\ \ \ \ -\ \ \ b^3}{(x + 5)^3 - (4x)^3}$
Use the difference of cubes pattern.	$\left(\overset{a\ \ \ -\ b}{(x + 5) - 4x}\right)\left(\overset{a^2\ \ \ +\ \ \ a\ \ \ \ b\ +\ b^2}{(x + 5)^2 + (x + 5) \cdot 4x + (4x)^2}\right)$
Simplify.	$(x + 5 - 4x)(x^2 + 10x + 25 + 4x^2 + 20x + 16x^2)$
	$(-3x + 5)(21x^2 + 30x + 25)$
Check by multiplying.	We'll leave the check to you.

> **TRY IT :: 6.67**   Factor: $(y + 1)^3 - 27y^3$.

> **TRY IT :: 6.68**   Factor: $(n + 3)^3 - 125n^3$.

MEDIA : :

Access this online resource for additional instruction and practice with factoring special products.

- **Factoring Binomials-Cubes #2 (https://openstax.org/l/37BinomCubes)**

6.3 EXERCISES
Practice Makes Perfect

Factor Perfect Square Trinomials

In the following exercises, factor completely using the perfect square trinomials pattern.

159. $16y^2 + 24y + 9$

160. $25v^2 + 20v + 4$

161. $36s^2 + 84s + 49$

162. $49s^2 + 154s + 121$

163. $100x^2 - 20x + 1$

164. $64z^2 - 16z + 1$

165. $25n^2 - 120n + 144$

166. $4p^2 - 52p + 169$

167. $49x^2 + 28xy + 4y^2$

168. $25r^2 + 60rs + 36s^2$

169. $100y^2 - 52y + 1$

170. $64m^2 - 34m + 1$

171. $10jk^2 + 80jk + 160j$

172. $64x^2y - 96xy + 36y$

173. $75u^4 - 30u^3v + 3u^2v^2$

174. $90p^4 + 300p^4q + 250p^2q^2$

Factor Differences of Squares

In the following exercises, factor completely using the difference of squares pattern, if possible.

175. $25v^2 - 1$

176. $169q^2 - 1$

177. $4 - 49x^2$

178. $121 - 25s^2$

179. $6p^2q^2 - 54p^2$

180. $98r^3 - 72r$

181. $24p^2 + 54$

182. $20b^2 + 140$

183. $121x^2 - 144y^2$

184. $49x^2 - 81y^2$

185. $169c^2 - 36d^2$

186. $36p^2 - 49q^2$

187. $16z^4 - 1$

188. $m^4 - n^4$

189. $162a^4b^2 - 32b^2$

190. $48m^4n^2 - 243n^2$

191. $x^2 - 16x + 64 - y^2$

192. $p^2 + 14p + 49 - q^2$

193. $a^2 + 6a + 9 - 9b^2$

194. $m^2 - 6m + 9 - 16n^2$

Factor Sums and Differences of Cubes

In the following exercises, factor completely using the sums and differences of cubes pattern, if possible.

195. $x^3 + 125$

196. $n^6 + 512$

197. $z^6 - 27$

198. $v^3 - 216$

199. $8 - 343t^3$

200. $125 - 27w^3$

201. $8y^3 - 125z^3$

202. $27x^3 - 64y^3$

203. $216a^3 + 125b^3$

204. $27y^3 + 8z^3$

205. $7k^3 + 56$

206. $6x^3 - 48y^3$

207. $2x^2 - 16x^2 y^3$ **208.** $-2x^3 y^2 - 16y^5$ **209.** $(x+3)^3 + 8x^3$

210. $(x+4)^3 - 27x^3$ **211.** $(y-5)^3 - 64y^3$ **212.** $(y-5)^3 + 125y^3$

Mixed Practice

In the following exercises, factor completely.

213. $64a^2 - 25$ **214.** $121x^2 - 144$ **215.** $27q^2 - 3$

216. $4p^2 - 100$ **217.** $16x^2 - 72x + 81$ **218.** $36y^2 + 12y + 1$

219. $8p^2 + 2$ **220.** $81x^2 + 169$ **221.** $125 - 8y^3$

222. $27u^3 + 1000$ **223.** $45n^2 + 60n + 20$ **224.** $48q^3 - 24q^2 + 3q$

225. $x^2 - 10x + 25 - y^2$ **226.** $x^2 + 12x + 36 - y^2$ **227.** $(x+1)^3 + 8x^3$

228. $(y-3)^3 - 64y^3$

Writing Exercises

229. Why was it important to practice using the binomial squares pattern in the chapter on multiplying polynomials?

230. How do you recognize the binomial squares pattern?

231. Explain why $n^2 + 25 \neq (n+5)^2$. Use algebra, words, or pictures.

232. Maribel factored $y^2 - 30y + 81$ as $(y-9)^2$. Was she right or wrong? How do you know?

Self Check

ⓐ *After completing the exercises, use this checklist to evaluate your mastery of the objectives of this section.*

I can...	Confidently	With some help	No-I don't get it!
factor perfect square trinomials.			
factor differences of squares.			
factor sums and differences of cubes.			

ⓑ *What does this checklist tell you about your mastery of this section? What steps will you take to improve?*

6.4 General Strategy for Factoring Polynomials

Learning Objectives

By the end of this section, you will be able to:

> Recognize and use the appropriate method to factor a polynomial completely

Recognize and Use the Appropriate Method to Factor a Polynomial Completely

You have now become acquainted with all the methods of factoring that you will need in this course. The following chart summarizes all the factoring methods we have covered, and outlines a strategy you should use when factoring polynomials.

General Strategy for Factoring Polynomials

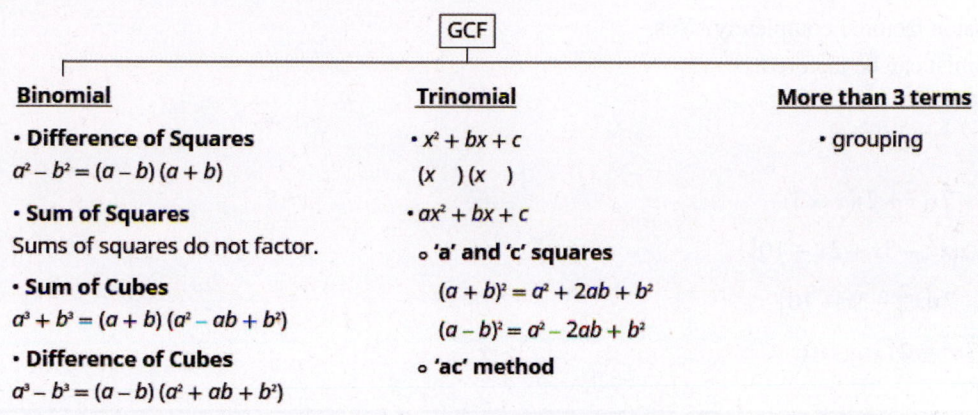

HOW TO :: USE A GENERAL STRATEGY FOR FACTORING POLYNOMIALS.

Step 1.   Is there a greatest common factor?
Factor it out.

Step 2.   Is the polynomial a binomial, trinomial, or are there more than three terms?
If it is a binomial:

- Is it a sum?
Of squares? Sums of squares do not factor.
Of cubes? Use the sum of cubes pattern.

- Is it a difference?
Of squares? Factor as the product of conjugates.
Of cubes? Use the difference of cubes pattern.

If it is a trinomial:

- Is it of the form $x^2 + bx + c$? Undo FOIL.

- Is it of the form $ax^2 + bx + c$?
If a and c are squares, check if it fits the trinomial square pattern.
Use the trial and error or "ac" method.

If it has more than three terms:

- Use the grouping method.

Step 3.   Check.
Is it factored completely?
Do the factors multiply back to the original polynomial?

Remember, a polynomial is completely factored if, other than monomials, its factors are prime!

EXAMPLE 6.35

Factor completely: $7x^3 - 21x^2 - 70x$.

✓ **Solution**

$$7x^3 - 21x^2 - 70x$$

Is there a GCF? Yes, $7x$.

Factor out the GCF.

$$7x(x^2 - 3x - 10)$$

In the parentheses, is it a binomial, trinomial, or are there more terms?

Trinomial with leading coefficient 1.

"Undo" FOIL.

$$7x(x\quad)(x\quad)$$
$$7x(x + 2)(x - 5)$$

Is the expression factored completely? Yes.

Neither binomial can be factored.

Check your answer.

Multiply.

$$7x(x + 2)(x - 5)$$
$$7x(x^2 - 5x + 2x - 10)$$
$$7x(x^2 - 3x - 10)$$
$$7x^3 - 21x^2 - 70x \checkmark$$

> **TRY IT : : 6.69**   Factor completely: $8y^3 + 16y^2 - 24y$.

> **TRY IT : : 6.70**   Factor completely: $5y^3 - 15y^2 - 270y$.

Be careful when you are asked to factor a binomial as there are several options!

EXAMPLE 6.36

Factor completely: $24y^2 - 150$.

⊘ Solution

$$24y^2 - 150$$

Is there a GCF? Yes, 6.

Factor out the GCF.

$$6(4y^2 - 25)$$

In the parentheses, is it a binomial, trinomial
or are there more than three terms? Binomial.
Is it a sum? No.

Is it a difference? Of squares or cubes? Yes, squares.

$$6\left((2y)^2 - (5)^2\right)$$

Write as a product of conjugates.

$$6(2y - 5)(2y + 5)$$

    Is the expression factored completely?
    Neither binomial can be factored.

Check:

    Multiply.

$$6(2y - 5)(2y + 5)$$
$$6(4y^2 - 25)$$
$$24y^2 - 150 \checkmark$$

> **TRY IT ∷ 6.71**   Factor completely: $16x^3 - 36x$.

> **TRY IT ∷ 6.72**   Factor completely: $27y^2 - 48$.

The next example can be factored using several methods. Recognizing the trinomial squares pattern will make your work easier.

EXAMPLE 6.37

Factor completely: $4a^2 - 12ab + 9b^2$.

⊘ Solution

$$4a^2 - 12ab + 9b^2$$

Is there a GCF? No.

Is it a binomial, trinomial, or are there more terms?

Trinomial with $a \neq 1$. But the first term is a perfect square.

Is the last term a perfect square? Yes.

$$(2a)^2 - 12ab + (3b)^2$$

Does it fit the pattern, $a^2 - 2ab + b^2$? Yes.

$$(2a)^2 \searrow \underset{-2(2a)(3b)}{-12ab+} \swarrow (3b)^2$$

Write it as a square.

$$(2a - 3b)^2$$

    Is the expression factored completely? Yes.

    The binomial cannot be factored.

Check your answer.

    Multiply.

$$(2a - 3b)^2$$
$$(2a)^2 - 2 \cdot 2a \cdot 3b + (3b)^2$$
$$4a^2 - 12ab + 9b^2 \checkmark$$

> **TRY IT :: 6.73**   Factor completely: $4x^2 + 20xy + 25y^2$.

> **TRY IT :: 6.74**   Factor completely: $9x^2 - 24xy + 16y^2$.

Remember, sums of squares do not factor, but sums of cubes do!

EXAMPLE 6.38

Factor completely $12x^3y^2 + 75xy^2$.

⊘ Solution

$$12x^3y^2 + 75xy^2$$

Is there a GCF? Yes, $3xy^2$.

Factor out the GCF.

$$3xy^2\left(4x^2 + 25\right)$$

In the parentheses, is it a binomial, trinomial, or are there more than three terms? Binomial.

Is it a sum? Of squares? Yes.   Sums of squares are prime.

    Is the expression factored completely? Yes.

Check:

    Multiply.

$$3xy^2\left(4x^2 + 25\right)$$
$$12x^3y^2 + 75xy^2 \checkmark$$

> **TRY IT :: 6.75**   Factor completely: $50x^3y + 72xy$.

TRY IT :: 6.76   Factor completely: $27xy^3 + 48xy$.

When using the sum or difference of cubes pattern, being careful with the signs.

EXAMPLE 6.39

Factor completely: $24x^3 + 81y^3$.

✓ **Solution**

Is there a GCF? Yes, 3.	$24x^3 + 81y^3$
Factor it out.	$3(8x^3 + 27y^3)$
In the parentheses, is it a binomial, trinomial, of are there more than three terms? Binomial.	
Is it a sum or difference? Sum.	
Of squares or cubes? Sum of cubes.	$3\left(\overset{a^3}{(2x)^3} + \overset{b^3}{(3y)^3}\right)$
Write it using the sum of cubes pattern.	$3\left(\overset{a\ +\ b}{2x + 3y}\right)\left(\overset{a^2\ -\quad ab\quad +\ b^2}{(2x)^2 - 2x \cdot 3y + (3y)^3}\right)$
Is the expression factored completely? Yes.	$3(2x + 3y)(4x^2 - 6xy + 9y^2)$
Check by multiplying.	

 TRY IT :: 6.77   Factor completely: $250m^3 + 432n^3$.

 TRY IT :: 6.78   Factor completely: $2p^3 + 54q^3$.

EXAMPLE 6.40

Factor completely: $3x^5y - 48xy$.

⊘ Solution

$$3x^5 y - 48xy$$

Is there a GCF? Factor out $3xy$

$$3xy(x^4 - 16)$$

Is the binomial a sum or difference? Of squares or cubes?
Write it as a difference of squares.

$$3xy\left(\left(x^2\right)^2 - (4)^2\right)$$

Factor it as a product of conjugates

$$3xy(x^2 - 4)(x^2 + 4)$$

The first binomial is again a difference of squares.

$$3xy\left((x)^2 - (2)^2\right)(x^2 + 4)$$

Factor it as a product of conjugates.

$$3xy(x - 2)(x + 2)(x^2 + 4)$$

Is the expression factored completely? Yes.
Check your answer.
Multiply.

$$3xy(x - 2)(x + 2)(x^2 + 4)$$
$$3xy(x^2 - 4)(x^2 + 4)$$
$$3xy(x^4 - 16)$$
$$3x^5 y - 48xy \checkmark$$

> **TRY IT :: 6.79**   Factor completely: $4a^5 b - 64ab$.

> **TRY IT :: 6.80**   Factor completely: $7xy^5 - 7xy$.

EXAMPLE 6.41

Factor completely: $4x^2 + 8bx - 4ax - 8ab$.

⊘ Solution

$$4x^2 + 8bx - 4ax - 8ab$$

Is there a GCF? Factor out the GCF, 4.

$$4(x^2 + 2bx - ax - 2ab)$$

There are four terms. Use grouping.

$$4[x(x + 2b) - a(x + 2b)]$$
$$4(x + 2b)(x - a)$$

Is the expression factored completely? Yes.
Check your answer.
Multiply.

$$4(x + 2b)(x - a)$$
$$4(x^2 - ax + 2bx - 2ab)$$
$$4x^2 + 8bx - 4ax - 8ab \checkmark$$

> **TRY IT :: 6.81**   Factor completely: $6x^2 - 12xc + 6bx - 12bc$.

> **TRY IT :: 6.82**   Factor completely: $16x^2 + 24xy - 4x - 6y$.

Taking out the complete GCF in the first step will always make your work easier.

EXAMPLE 6.42

Factor completely: $40x^2y + 44xy - 24y$.

✓ **Solution**

$$40x^2y + 44xy - 24y$$

Is there a GCF? Factor out the GCF, $4y$. $4y\left(10x^2 + 11x - 6\right)$

Factor the trinomial with $a \neq 1$. $4y\left(10x^2 + 11x - 6\right)$

$4y(5x - 2)(2x + 3)$

Is the expression factored completely? Yes.
Check your answer.
Multiply.

$$4y(5x - 2)(2x + 3)$$
$$4y\left(10x^2 + 11x - 6\right)$$
$$40x^2y + 44xy - 24y \checkmark$$

> **TRY IT ::** 6.83   Factor completely: $4p^2q - 16pq + 12q$.

> **TRY IT ::** 6.84   Factor completely: $6pq^2 - 9pq - 6p$.

When we have factored a polynomial with four terms, most often we separated it into two groups of two terms. Remember that we can also separate it into a trinomial and then one term.

EXAMPLE 6.43

Factor completely: $9x^2 - 12xy + 4y^2 - 49$.

⊘ Solution

$$9x^2 - 12xy + 4y^2 - 49$$

Is there a GCF? No.

With more than 3 terms, use grouping. Last 2 terms
have no GCF. Try grouping first 3 terms.

$$9x^2 - 12xy + 4y^2 - 49$$

Factor the trinomial with $a \neq 1$. But the first term is a
perfect square.

Is the last term of the trinomial a perfect square? Yes.

$$(3x)^2 - 12xy + (2y)^2 - 49$$

Does the trinomial fit the pattern, $a^2 - 2ab + b^2$? Yes.

$$(3x)^2 \searrow {-12xy + \atop -2(3x)(2y)} \swarrow (2y)^2 - 49$$

Write the trinomial as a square.

$$(3x - 2y)^2 - 49$$

Is this binomial a sum or difference? Of squares or
cubes? Write it as a difference of squares.

$$(3x - 2y)^2 - 7^2$$

Write it as a product of conjugates.

$$((3x - 2y) - 7)((3x - 2y) + 7)$$
$$(3x - 2y - 7)(3x - 2y + 7)$$

Is the expression factored completely? Yes.

Check your answer.

Multiply.

$$(3x - 2y - 7)(3x - 2y + 7)$$
$$9x^2 - 6xy - 21x - 6xy + 4y^2 + 14y + 21x - 14y - 49$$
$$9x^2 - 12xy + 4y^2 - 49 \checkmark$$

> **TRY IT : : 6.85**   Factor completely: $4x^2 - 12xy + 9y^2 - 25$.

> **TRY IT : : 6.86**   Factor completely: $16x^2 - 24xy + 9y^2 - 64$.

6.4 EXERCISES

Practice Makes Perfect

Recognize and Use the Appropriate Method to Factor a Polynomial Completely

In the following exercises, factor completely.

233. $2n^2 + 13n - 7$

234. $8x^2 - 9x - 3$

235. $a^5 + 9a^3$

236. $75m^3 + 12m$

237. $121r^2 - s^2$

238. $49b^2 - 36a^2$

239. $8m^2 - 32$

240. $36q^2 - 100$

241. $25w^2 - 60w + 36$

242. $49b^2 - 112b + 64$

243. $m^2 + 14mn + 49n^2$

244. $64x^2 + 16xy + y^2$

245. $7b^2 + 7b - 42$

246. $30n^2 + 30n + 72$

247. $3x^4 y - 81xy$

248. $4x^5 y - 32x^2 y$

249. $k^4 - 16$

250. $m^4 - 81$

251. $5x^5 y^2 - 80xy^2$

252. $48x^5 y^2 - 243xy^2$

253. $15pq - 15p + 12q - 12$

254. $12ab - 6a + 10b - 5$

255. $4x^2 + 40x + 84$

256. $5q^2 - 15q - 90$

257. $4u^5 v + 4u^2 v^3$

258. $5m^4 n + 320mn^4$

259. $4c^2 + 20cd + 81d^2$

260. $25x^2 + 35xy + 49y^2$

261. $10m^4 - 6250$

262. $3v^4 - 768$

263. $36x^2 y + 15xy - 6y$

264. $60x^2 y - 75xy + 30y$

265. $8x^3 - 27y^3$

266. $64x^3 + 125y^3$

267. $y^6 - 1$

268. $y^6 + 1$

269. $9x^2 - 6xy + y^2 - 49$

270. $16x^2 - 24xy + 9y^2 - 64$

271. $(3x + 1)^2 - 6(3x - 1) + 9$

272. $(4x - 5)^2 - 7(4x - 5) + 12$

Writing Exercises

273. Explain what it mean to factor a polynomial completely.

274. The difference of squares $y^4 - 625$ can be factored as $\left(y^2 - 25\right)\left(y^2 + 25\right)$. But it is not completely factored. What more must be done to completely factor.

275. Of all the factoring methods covered in this chapter (GCF, grouping, undo FOIL, 'ac' method, special products) which is the easiest for you? Which is the hardest? Explain your answers.

276. Create three factoring problems that would be good test questions to measure your knowledge of factoring. Show the solutions.

Self Check

ⓐ *After completing the exercises, use this checklist to evaluate your mastery of the objectives of this section.*

I can...	Confidently	With some help	No-I don't get it!
recognize and use the appropriate method to factor a polynomial completely.			

ⓑ *On a scale of 1-10, how would you rate your mastery of this section in light of your responses on the checklist? How can you improve this?*

6.5 Polynomial Equations

Learning Objectives

By the end of this section, you will be able to:

> Use the Zero Product Property
> Solve quadratic equations by factoring
> Solve equations with polynomial functions
> Solve applications modeled by polynomial equations

Be Prepared!

Before you get started, take this readiness quiz.

1. Solve: $5y - 3 = 0$.
   If you missed this problem, review **Example 2.2**.

2. Factor completely: $n^3 - 9n^2 - 22n$.
   If you missed this problem, review **Example 3.48**.

3. If $f(x) = 8x - 16$, find $f(3)$ and solve $f(x) = 0$.
   If you missed this problem, review **Example 3.59**.

We have spent considerable time learning how to factor polynomials. We will now look at polynomial equations and solve them using factoring, if possible.

A **polynomial equation** is an equation that contains a polynomial expression. The **degree of the polynomial equation** is the degree of the polynomial.

Polynomial Equation

A **polynomial equation** is an equation that contains a polynomial expression.

The **degree of the polynomial equation** is the degree of the polynomial.

We have already solved polynomial equations of degree one. Polynomial equations of degree one are linear equations are of the form $ax + b = c$.

We are now going to solve polynomial equations of degree two. A polynomial equation of degree two is called a **quadratic equation**. Listed below are some examples of quadratic equations:

$$x^2 + 5x + 6 = 0 \qquad 3y^2 + 4y = 10 \qquad 64u^2 - 81 = 0 \qquad n(n + 1) = 42$$

The last equation doesn't appear to have the variable squared, but when we simplify the expression on the left we will get $n^2 + n$.

The general form of a quadratic equation is $ax^2 + bx + c = 0$, with $a \neq 0$. (If $a = 0$, then $0 \cdot x^2 = 0$ and we are left with no quadratic term.)

Quadratic Equation

An equation of the form $ax^2 + bx + c = 0$ is called a quadratic equation.

$$a, \ b, \ \text{and } c \text{ are real numbers and } a \neq 0$$

To solve quadratic equations we need methods different from the ones we used in solving linear equations. We will look at one method here and then several others in a later chapter.

Use the Zero Product Property

We will first solve some quadratic equations by using the **Zero Product Property**. The Zero Product Property says that if the product of two quantities is zero, then at least one of the quantities is zero. The only way to get a product equal to zero is to multiply by zero itself.

Zero Product Property

If $a \cdot b = 0$, then either $a = 0$ or $b = 0$ or both.

We will now use the Zero Product Property, to solve a quadratic equation.

> **EXAMPLE 6.44** HOW TO SOLVE A QUADRATIC EQUATION USING THE ZERO PRODUCT PROPERTY

Solve: $(5n - 2)(6n - 1) = 0$.

✅ **Solution**

Step 1. Set each factor equal to zero.	The product equals zero, so at least one factor must equal zero.	$(5n - 2)(6n - 1) = 0$ $5n - 2 = 0$ or $6n - 1 = 0$
Step 2. Solve the linear equations.	Solve each equation.	$n = \dfrac{2}{5}$ $\qquad n = \dfrac{1}{6}$
Step 3. Check.	Substitute each solution separately into the original equation.	$n = \dfrac{2}{5}$ $(5n - 2)(6n - 1) = 0$ $\left(5 \cdot \dfrac{2}{5} - 2\right)\left(6 \cdot \dfrac{2}{5} - 1\right) \overset{?}{=} 0$ $(2 - 2)\left(\dfrac{12}{5} - 1\right) \overset{?}{=} 0$ $0 \cdot \dfrac{7}{5} \overset{?}{=} 0$ $0 = 0 ✓$ $n = \dfrac{1}{6}$ $(5n - 2)(6n - 1) = 0$ $\left(5 \cdot \dfrac{1}{6} - 2\right)\left(6 \cdot \dfrac{1}{6} - 1\right) \overset{?}{=} 0$ $\left(\dfrac{5}{6} - \dfrac{12}{6}\right)(1 - 1) \overset{?}{=} 0$ $\left(-\dfrac{7}{6}\right)(0) \overset{?}{=} 0$ $0 = 0 ✓$

> **TRY IT ::** 6.87   Solve: $(3m - 2)(2m + 1) = 0$.

> **TRY IT ::** 6.88   Solve: $(4p + 3)(4p - 3) = 0$.

HOW TO :: USE THE ZERO PRODUCT PROPERTY.

Step 1.   Set each factor equal to zero.

Step 2.   Solve the linear equations.

Step 3.   Check.

Solve Quadratic Equations by Factoring

The Zero Product Property works very nicely to solve quadratic equations. The quadratic equation must be factored, with zero isolated on one side. So we be sure to start with the quadratic equation in standard form, $ax^2 + bx + c = 0$. Then we factor the expression on the left.

EXAMPLE 6.45 HOW TO SOLVE A QUADRATIC EQUATION BY FACTORING

Solve: $2y^2 = 13y + 45$.

⊘ **Solution**

Step 1. Write the quadratic equation in standard form, $ax^2 + bx + c = 0$.	Write the equation in standard form.	$2y^2 = 13y + 45$ $2y^2 - 13y - 45 = 0$
Step 2. Factor the quadratic expression.	Factor $2y^2 - 13y + 45$ $(2y + 5)(y - 9)$	$(2y + 5)(y - 9) = 0$
Step 3. Use the Zero Product Property.	Set each factor equal to zero. We have two linear equations.	$2y + 5 = 0 \quad y - 9 = 0$
Step 4. Solve the linear equations.		$y = -\dfrac{5}{2} \quad y = 9$
Step 5. Check. Substitute each solution separately into the original equation.	Substitute each solution separately into the original equation.	$y = -\dfrac{5}{2}$ $2y^2 = 13y + 45$ $2\left(-\dfrac{5}{2}\right)^2 \overset{?}{=} 13\left(-\dfrac{5}{2}\right) + 45$ $2\left(\dfrac{25}{4}\right) \overset{?}{=} \left(-\dfrac{65}{2}\right) + \dfrac{90}{2}$ $\dfrac{25}{2} = \dfrac{25}{2} \checkmark$ $y = 9$ $2y^2 = 13y + 45$ $2(9)^2 \overset{?}{=} 13(9) + 45$ $2(81) \overset{?}{=} 117 + 45$ $162 = 162 \checkmark$

> **TRY IT :: 6.89**   Solve: $3c^2 = 10c - 8$.

> **TRY IT :: 6.90**   Solve: $2d^2 - 5d = 3$.

HOW TO :: SOLVE A QUADRATIC EQUATION BY FACTORING.

Step 1.   Write the quadratic equation in standard form, $ax^2 + bx + c = 0$.

Step 2.   Factor the quadratic expression.

Step 3.   Use the Zero Product Property.

Step 4.   Solve the linear equations.

Step 5.   Check. Substitute each solution separately into the original equation.

Before we factor, we must make sure the quadratic equation is in standard form.

Solving quadratic equations by factoring will make use of all the factoring techniques you have learned in this chapter! Do you recognize the special product pattern in the next example?

EXAMPLE 6.46

Solve: $169q^2 = 49$.

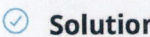 **Solution**

$$169x^2 = 49$$

Write the quadratic equation in standard form. $$169x^2 - 49 = 0$$

Factor. It is a difference of squares. $$(13x - 7)(13x + 7) = 0$$

Use the Zero Product Property to set each factor to 0.

Solve each equation.

$$13x - 7 = 0 \qquad 13x + 7 = 0$$
$$13x = 7 \qquad\quad 13x = -7$$
$$x = \frac{7}{13} \qquad\quad x = -\frac{7}{13}$$

Check:

We leave the check up to you.

> **TRY IT :: 6.91**   Solve: $25p^2 = 49$.

> **TRY IT :: 6.92**   Solve: $36x^2 = 121$.

In the next example, the left side of the equation is factored, but the right side is not zero. In order to use the Zero Product Property, one side of the equation must be zero. We'll multiply the factors and then write the equation in standard form.

EXAMPLE 6.47

Solve: $(3x - 8)(x - 1) = 3x$.

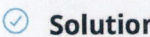 **Solution**

$$(3x - 8)(x - 1) = 3x$$

Multiply the binomials. $$3x^2 - 11x + 8 = 3x$$

Write the quadratic equation in standard form. $$3x^2 - 14x + 8 = 0$$

Factor the trinomial. $$(3x - 2)(x - 4) = 0$$

Use the Zero Product Property to set each factor to 0.

Solve each equation.

$$3x - 2 = 0 \qquad x - 4 = 0$$
$$3x = 2 \qquad\quad x = 4$$
$$x = \frac{2}{3}$$

Check your answers.   The check is left to you.

> **TRY IT :: 6.93**   Solve: $(2m + 1)(m + 3) = 12m$.

> **TRY IT :: 6.94**   Solve: $(k + 1)(k - 1) = 8$.

In the next example, when we factor the quadratic equation we will get three factors. However the first factor is a constant. We know that factor cannot equal 0.

EXAMPLE 6.48

Solve: $3x^2 = 12x + 63$.

✓ **Solution**

$$3x^2 = 12x + 63$$

Write the quadratic equation in standard form.

$$3x^2 - 12x - 63 = 0$$

Factor the greatest common factor first.

$$3(x^2 - 4x - 21) = 0$$

Factor the trinomial.

$$3(x - 7)(x + 3) = 0$$

Use the Zero Product Property to set each factor to 0.

$3 \neq 0 \qquad x - 7 = 0 \qquad x + 3 = 0$

Solve each equation.

$3 \neq 0 \qquad x = 7 \qquad x = -3$

Check your answers.

The check is left to you.

>  **TRY IT :: 6.95**   Solve: $18a^2 - 30 = -33a$.

>  **TRY IT :: 6.96**   Solve: $123b = -6 - 60b^2$.

The Zero Product Property also applies to the product of three or more factors. If the product is zero, at least one of the factors must be zero. We can solve some equations of degree greater than two by using the Zero Product Property, just like we solved quadratic equations.

EXAMPLE 6.49

Solve: $9m^3 + 100m = 60m^2$.

✓ **Solution**

$$9m^3 + 100m = 60m^2$$

Bring all the terms to one side so that the other side is zero.

$$9m^3 - 60m^2 + 100m = 0$$

Factor the greatest common factor first.

$$m(9m^2 - 60m + 100) = 0$$

Factor the trinomial.

$$m(3m - 10)(3m - 10) = 0$$

Use the Zero Product Property to set each factor to 0.

$m = 0 \qquad 3m - 10 = 0 \qquad 3m - 10 = 0$

Solve each equation.

$m = 0 \qquad m = \dfrac{10}{3} \qquad m = \dfrac{10}{3}$

Check your answers.

The check is left to you.

> **TRY IT :: 6.97**   Solve: $8x^3 = 24x^2 - 18x$.

> **TRY IT :: 6.98**   Solve: $16y^2 = 32y^3 + 2y$.

Solve Equations with Polynomial Functions

As our study of polynomial functions continues, it will often be important to know when the function will have a certain value or what points lie on the graph of the function. Our work with the Zero Product Property will be help us find these answers.

EXAMPLE 6.50

For the function $f(x) = x^2 + 2x - 2$,

ⓐ find x when $f(x) = 6$

ⓑ find two points that lie on the graph of the function.

✓ Solution

ⓐ

$$f(x) = x^2 + 2x - 2$$

Substitute 6 for $f(x)$. $\qquad 6 = x^2 + 2x - 2$

Put the quadratic in standard form. $\qquad x^2 + 2x - 8 = 0$

Factor the trinomial. $\qquad (x+4)(x-2) = 0$

Use the zero product property. $\qquad x+4 = 0 \quad$ or $\quad x-2 = 0$

Solve. $\qquad\qquad\qquad\qquad\qquad x = -4 \quad$ or $\quad x = 2$

Check:

$$f(x) = x^2 + 2x - 2 \qquad\qquad f(x) = x^2 + 2x - 2$$
$$f(-4) = (-4)^2 + 2(-4) - 2 \qquad f(2) = 2^2 + 2 \cdot 2 - 2$$
$$f(-4) = 16 - 8 - 2 \qquad\qquad f(2) = 4 + 4 - 2$$
$$f(-4) = 6 \checkmark \qquad\qquad\qquad f(2) = 6 \checkmark$$

ⓑ Since $f(-4) = 6$ and $f(2) = 6$, the points $(-4, 6)$ and $(2, 6)$ lie on the graph of the function.

> **TRY IT : : 6.99**

For the function $f(x) = x^2 - 2x - 8$,

ⓐ find x when $f(x) = 7$

ⓑ Find two points that lie on the graph of the function.

> **TRY IT : : 6.100**

For the function $f(x) = x^2 - 8x + 3$,

ⓐ find x when $f(x) = -4$

ⓑ Find two points that lie on the graph of the function.

The Zero Product Property also helps us determine where the function is zero. A value of x where the function is 0, is called a **zero of the function**.

Zero of a Function

For any function f, if $f(x) = 0$, then x is a **zero of the function**.

When $f(x) = 0$, the point $(x, 0)$ is a point on the graph. This point is an x-intercept of the graph. It is often important to know where the graph of a function crosses the axes. We will see some examples later.

EXAMPLE 6.51

For the function $f(x) = 3x^2 + 10x - 8$, find

ⓐ the zeros of the function,

ⓑ any x-intercepts of the graph of the function

ⓒ any y-intercepts of the graph of the function

✓ Solution

ⓐ To find the zeros of the function, we need to find when the function value is 0.

$$f(x) = 3x^2 + 10x - 8$$

Substitute 0 for $f(x)$. $\qquad\qquad 0 = 3x^2 + 10x - 8$

Factor the trinomial. $\qquad\qquad (x+4)(3x-2) = 0$

Use the zero product property. $\qquad x+4 \;=\; 0 \quad$ or $\quad 3x-2 \;=\; 0$

Solve. $\qquad\qquad\qquad\qquad\qquad\quad x \;=\; -4 \quad$ or $\qquad x \;=\; \dfrac{2}{3}$

ⓑ An x-intercept occurs when $y = 0$. Since $f(-4) = 0$ and $f\!\left(\frac{2}{3}\right) = 0$, the points $(-4, 0)$ and $\left(\frac{2}{3}, 0\right)$ lie on the graph. These points are x-intercepts of the function.

ⓒ A y-intercept occurs when $x = 0$. To find the y-intercepts we need to find $f(0)$.

$$f(x) = 3x^2 + 10x - 8$$

Find $f(0)$ by substituting 0 for x. $\qquad f(0) = 3 \cdot 0^2 + 10 \cdot 0 - 8$

Simplify. $\qquad\qquad\qquad\qquad\qquad f(0) = -8$

Since $f(0) = -8$, the point $(0, -8)$ lies on the graph. This point is the y-intercept of the function.

> **TRY IT :: 6.101**   For the function $f(x) = 2x^2 - 7x + 5$, find

ⓐ the zeros of the function

ⓑ any x-intercepts of the graph of the function

ⓒ any y-intercepts of the graph of the function.

> **TRY IT :: 6.102**   For the function $f(x) = 6x^2 + 13x - 15$, find

ⓐ the zeros of the function

ⓑ any x-intercepts of the graph of the function

ⓒ any y-intercepts of the graph of the function.

Solve Applications Modeled by Polynomial Equations

The problem-solving strategy we used earlier for applications that translate to linear equations will work just as well for applications that translate to polynomial equations. We will copy the problem-solving strategy here so we can use it for reference.

HOW TO :: USE A PROBLEM SOLVING STRATEGY TO SOLVE WORD PROBLEMS.

Step 1.   **Read** the problem. Make sure all the words and ideas are understood.

Step 2.   **Identify** what we are looking for.

Step 3.   **Name** what we are looking for. Choose a variable to represent that quantity.

Step 4.   **Translate** into an equation. It may be helpful to restate the problem in one sentence with all the important information. Then, translate the English sentence into an algebraic equation.

Step 5.   **Solve** the equation using appropriate algebra techniques.

Step 6.   **Check** the answer in the problem and make sure it makes sense.

Step 7.   **Answer** the question with a complete sentence.

We will start with a number problem to get practice translating words into a polynomial equation.

EXAMPLE 6.52

The product of two consecutive odd integers is 323. Find the integers.

✓ **Solution**

Step 1. Read the problem.	
Step 2. Identify what we are looking for.	We are looking for two consecutive integers.
Step 3. Name what we are looking for.	Let $n =$ the first integer.
	$n + 2 =$ next consecutive odd integer
Step 4. Translate into an equation. Restate the problem in a sentence.	The product of the two consecutive odd integers is 323.

$$n(n+2) = 323$$

$$n^2 + 2n = 323$$

Step 5. Solve the equation.	
Bring all the terms to one side.	$n^2 + 2n - 323 = 0$
Factor the trinomial.	$(n - 17)(n + 19) = 0$
Use the Zero Product Property.	$n - 17 = 0 \qquad n + 19 = 0$
Solve the equations.	$n = 17 \qquad\quad n = -19$

There are two values for n that are solutions to this problem. So there are two sets of consecutive odd integers that will work.

If the first integer is $n = 17$ If the first integer is $n = -19$

then the next odd integer is   then the next odd integer is

$$n + 2 \qquad\qquad\qquad\qquad\qquad n + 2$$

$$17 + 2 \qquad\qquad\qquad\qquad\qquad -19 + 2$$

$$19 \qquad\qquad\qquad\qquad\qquad\qquad -17$$

$$17, 19 \qquad\qquad\qquad\qquad\qquad -17, -19$$

Step 6. Check the answer.

The results are consecutive odd integers

17, 19 and $-19, -17$.

$17 \cdot 19 = 323$ ✓ $-19(-17) = 323$ ✓

Both pairs of consecutive integers are solutions.

Step 7. Answer the question   The consecutive integers are 17, 19 and $-19, -17$.

 TRY IT ∷ 6.103   The product of two consecutive odd integers is 255. Find the integers.

 TRY IT ∷ 6.104   The product of two consecutive odd integers is 483 Find the integers.

Were you surprised by the pair of negative integers that is one of the solutions to the previous example? The product of the two positive integers and the product of the two negative integers both give positive results.

In some applications, negative solutions will result from the algebra, but will not be realistic for the situation.

 **EXAMPLE 6.53**

A rectangular bedroom has an area 117 square feet. The length of the bedroom is four feet more than the width. Find the length and width of the bedroom.

✓ **Solution**

Step 1. Read the problem. In problems involving geometric figures, a sketch can help you visualize the situation.

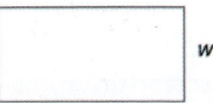

Step 2. Identify what you are looking for.	We are looking for the length and width.
Step 3. Name what you are looking for.	Let $w =$ the width of the bedroom.
The length is four feet more than the width.	$w + 4 =$ the length of the garden
Step 4. Translate into an equation.	
Restate the important information in a sentence.	The area of the bedroom is 117 square feet.
Use the formula for the area of a rectangle.	$A = l \cdot w$
Substitute in the variables.	$117 = (w + 4)w$
Step 5. Solve the equation Distribute first.	$117 = w^2 + 4w$
Get zero on one side.	$117 = w^2 + 4w$
Factor the trinomial.	$0 = w^2 + 4w - 117$
Use the Zero Product Property.	$0 = (w^2 + 13)(w - 9)$
Solve each equation.	$0 = w + 13 \qquad 0 = w - 9$
Since w is the width of the bedroom, it does not make sense for it to be negative. We eliminate that value for w.	$\cancel{-13 = w} \qquad 9 = w$
	$w = 9$ Width is 9 feet.
Find the value of the length.	$w + 4$ $9 + 4$ $\quad 13$ Length is 13 feet.

Step 6. Check the answer.
Does the answer make sense?

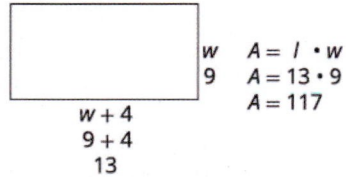

w $A = l \cdot w$
9 $A = 13 \cdot 9$
$A = 117$

$w + 4$
$9 + 4$
$\quad 13$

Yes, this makes sense.

Step 7. Answer the question.	The width of the bedroom is 9 feet and the length is 13 feet.

> **TRY IT : : 6.105**

A rectangular sign has area 30 square feet. The length of the sign is one foot more than the width. Find the length and width of the sign.

> **TRY IT : : 6.106**

A rectangular patio has area 180 square feet. The width of the patio is three feet less than the length. Find the length and width of the patio.

In the next example, we will use the Pythagorean Theorem $\left(a^2 + b^2 = c^2\right)$. This formula gives the relation between the legs and the hypotenuse of a right triangle.

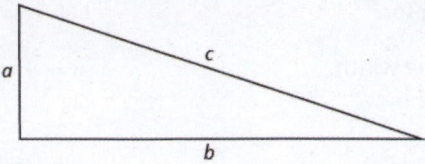

We will use this formula to in the next example.

EXAMPLE 6.54

A boat's sail is in the shape of a right triangle as shown. The hypotenuse will be 17 feet long. The length of one side will be 7 feet less than the length of the other side. Find the lengths of the sides of the sail.

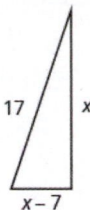

⊘ **Solution**

Step 1. Read the problem

Step 2. Identify what you are looking for.	We are looking for the lengths of the sides of the sail.
Step 3. Name what you are looking for. One side is 7 less than the other.	Let $x =$ length of a side of the sail. $x - 7 =$ length of other side
Step 4. Translate into an equation. Since this is a right triangle we can use the Pythagorean Theorem.	$a^2 + b^2 = c^2$
Substitute in the variables.	$x^2 + (x - 7)^2 = 17^2$
Step 5. Solve the equation Simplify.	$x^2 + x^2 - 14x + 49 = 289$
	$2x^2 - 14x + 49 = 289$
It is a quadratic equation, so get zero on one side.	$2x^2 - 14x - 240 = 0$
Factor the greatest common factor.	$2(x^2 - 7x - 120) = 0$
Factor the trinomial.	$2(x - 15)(x + 8) = 0$
Use the Zero Product Property.	$2 \neq 0 \quad x - 15 = 0 \quad x + 8 = 0$
Solve.	$2 \neq 0 \quad x = 15 \quad x = -8$
Since x is a side of the triangle, $x = -8$ does not make sense.	$2 \neq 0 \quad x = 15 \quad \cancel{x = -8}$
Find the length of the other side.	

If the length of one side is $x = 15$

then the length of the other side is $x - 7$

$15 - 7$

8 is the length of the other side.

Step 6. Check the answer in the problem
Do these numbers make sense?

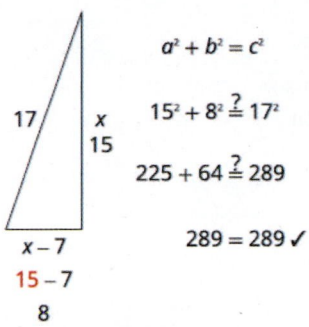

$a^2 + b^2 = c^2$

$15^2 + 8^2 \stackrel{?}{=} 17^2$

$225 + 64 \stackrel{?}{=} 289$

$289 = 289 \checkmark$

Step 7. Answer the question   The sides of the sail are 8, 15 and 17 feet.

 TRY IT : : 6.107

Justine wants to put a deck in the corner of her backyard in the shape of a right triangle. The length of one side of the deck is 7 feet more than the other side. The hypotenuse is 13. Find the lengths of the two sides of the deck.

 TRY IT : : 6.108

A meditation garden is in the shape of a right triangle, with one leg 7 feet. The length of the hypotenuse is one more than the length of the other leg. Find the lengths of the hypotenuse and the other leg.

The next example uses the function that gives the height of an object as a function of time when it is thrown from 80 feet above the ground.

EXAMPLE 6.55

Dennis is going to throw his rubber band ball upward from the top of a campus building. When he throws the rubber band ball from 80 feet above the ground, the function $h(t) = -16t^2 + 64t + 80$ models the height, h, of the ball above the ground as a function of time, t. Find:

ⓐ the zeros of this function which tell us when the ball hits the ground

ⓑ when the ball will be 80 feet above the ground

ⓒ the height of the ball at $t = 2$ seconds.

✓ Solution

ⓐ The zeros of this function are found by solving $h(t) = 0$. This will tell us when the ball will hit the ground.

$$h(t) = 0$$

Substitute in the polynomial for $h(t)$. $-16t^2 + 64t + 80 = 0$

Factor the GCF, -16. $-16(t^2 - 4t - 5) = 0$

Factor the trinomial. $-16(t - 5)(t + 1) = 0$

Use the Zero Product Property. $t - 5 = 0 \qquad\qquad t + 1 = 0$

Solve. $t = 5 \qquad\qquad\quad t = -1$

The result $t = 5$ tells us the ball will hit the ground 5 seconds after it is thrown. Since time cannot be negative, the result

$t = -1$ is discarded.

ⓑ The ball will be 80 feet above the ground when $h(t) = 80$.

$$h(t) = 80$$

Substitute in the polynomial for $h(t)$. $\qquad -16t^2 + 64t + 80 = 80$

Subtract 80 from both sides. $\qquad\qquad -16t^2 + 64t = 0$

Factor the GCF, $-16t$. $\qquad\qquad\qquad -16t(t - 4) = 0$

Use the Zero Product Property. $\qquad -16t = 0 \qquad t - 4 = 0$

Solve. $\qquad\qquad\qquad\qquad\qquad t = 0 \qquad\quad t = 4$

The ball will be at 80 feet the moment Dennis tosses the ball and then 4 seconds later, when the ball is falling.

ⓒ To find the height ball at $t = 2$ seconds we find $h(2)$.

$$h(t) = -16t^2 + 64t + 80$$

To find $h(2)$ substitute 2 for t. $\qquad h(2) = -16(2)^2 + 64 \cdot 2 + 80$

Simplify. $\qquad\qquad\qquad\qquad\qquad h(2) = 144$

After 2 seconds, the ball will be at 144 feet.

> **TRY IT : : 6.109**

Genevieve is going to throw a rock from the top a trail overlooking the ocean. When she throws the rock upward from 160 feet above the ocean, the function $h(t) = -16t^2 + 48t + 160$ models the height, h, of the rock above the ocean as a function of time, t. Find:

ⓐ the zeros of this function which tell us when the rock will hit the ocean
ⓑ when the rock will be 160 feet above the ocean.
ⓒ the height of the rock at $t = 1.5$ seconds.

> **TRY IT : : 6.110**

Calib is going to throw his lucky penny from his balcony on a cruise ship. When he throws the penny upward from 128 feet above the ground, the function $h(t) = -16t^2 + 32t + 128$ models the height, h, of the penny above the ocean as a function of time, t. Find:

ⓐ the zeros of this function which is when the penny will hit the ocean
ⓑ when the penny will be 128 feet above the ocean.
ⓒ the height the penny will be at $t = 1$ seconds which is when the penny will be at its highest point.

▶ **MEDIA : :**
Access this online resource for additional instruction and practice with quadratic equations.

- **Beginning Algebra & Solving Quadratics with the Zero Property (https://openstax.org/l/37ZeroProperty)**

6.5 EXERCISES

Practice Makes Perfect

Use the Zero Product Property

In the following exercises, solve.

277. $(3a - 10)(2a - 7) = 0$

278. $(5b + 1)(6b + 1) = 0$

279. $6m(12m - 5) = 0$

280. $2x(6x - 3) = 0$

281. $(2x - 1)^2 = 0$

282. $(3y + 5)^2 = 0$

Solve Quadratic Equations by Factoring

In the following exercises, solve.

283. $5a^2 - 26a = 24$

284. $4b^2 + 7b = -3$

285. $4m^2 = 17m - 15$

286. $n^2 = 5 - 6n$

287. $7a^2 + 14a = 7a$

288. $12b^2 - 15b = -9b$

289. $49m^2 = 144$

290. $625 = x^2$

291. $16y^2 = 81$

292. $64p^2 = 225$

293. $121n^2 = 36$

294. $100y^2 = 9$

295. $(x + 6)(x - 3) = -8$

296. $(p - 5)(p + 3) = -7$

297. $(2x + 1)(x - 3) = -4x$

298. $(y - 3)(y + 2) = 4y$

299. $(3x - 2)(x + 4) = 12x$

300. $(2y - 3)(3y - 1) = 8y$

301. $20x^2 - 60x = -45$

302. $3y^2 - 18y = -27$

303. $15x^2 - 10x = 40$

304. $14y^2 - 77y = -35$

305. $18x^2 - 9 = -21x$

306. $16y^2 + 12 = -32x$

307. $16p^3 = 24p^2 + 9p$

308. $m^3 - 2m^2 = -m$

309. $2x^3 + 72x = 24x^2$

310. $3y^3 + 48y = 24y^2$

311. $36x^3 + 24x^2 = -4x$

312. $2y^3 + 2y^2 = 12y$

Solve Equations with Polynomial Functions

In the following exercises, solve.

313. For the function, $f(x) = x^2 - 8x + 8$, ⓐ find when $f(x) = -4$ ⓑ Use this information to find two points that lie on the graph of the function.

314. For the function, $f(x) = x^2 + 11x + 20$, ⓐ find when $f(x) = -8$ ⓑ Use this information to find two points that lie on the graph of the function.

315. For the function, $f(x) = 8x^2 - 18x + 5$, ⓐ find when $f(x) = -4$ ⓑ Use this information to find two points that lie on the graph of the function.

316. For the function, $f(x) = 18x^2 + 15x - 10$, ⓐ find when $f(x) = 15$ ⓑ Use this information to find two points that lie on the graph of the function.

In the following exercises, for each function, find: ⓐ the zeros of the function ⓑ the x-intercepts of the graph of the function ⓒ the y-intercept of the graph of the function.

317. $f(x) = 9x^2 - 4$

318. $f(x) = 25x^2 - 49$

319. $f(x) = 6x^2 - 7x - 5$

320. $f(x) = 12x^2 - 11x + 2$

Solve Applications Modeled by Quadratic Equations

In the following exercises, solve.

321. The product of two consecutive odd integers is 143. Find the integers.

322. The product of two consecutive odd integers is 195. Find the integers.

323. The product of two consecutive even integers is 168. Find the integers.

324. The product of two consecutive even integers is 288. Find the integers.

325. The area of a rectangular carpet is 28 square feet. The length is three feet more than the width. Find the length and the width of the carpet.

326. A rectangular retaining wall has area 15 square feet. The height of the wall is two feet less than its length. Find the height and the length of the wall.

327. The area of a bulletin board is 55 feet. The length is four feet less than three times the width. Find the length and the width of the a bulletin board.

328. A rectangular carport has area 150 square feet. The height of the carport is five feet less than twice its length. Find the height and the length of the carport.

329. A pennant is shaped like a right triangle, with hypotenuse 10 feet. The length of one side of the pennant is two feet longer than the length of the other side. Find the length of the two sides of the pennant.

330. A stained glass window is shaped like a right triangle. The hypotenuse is 15 feet. One leg is three more than the other. Find the lengths of the legs.

331. A reflecting pool is shaped like a right triangle, with one leg along the wall of a building. The hypotenuse is 9 feet longer than the side along the building. The third side is 7 feet longer than the side along the building. Find the lengths of all three sides of the reflecting pool.

332. A goat enclosure is in the shape of a right triangle. One leg of the enclosure is built against the side of the barn. The other leg is 4 feet more than the leg against the barn. The hypotenuse is 8 feet more than the leg along the barn. Find the three sides of the goat enclosure.

333. Juli is going to launch a model rocket in her back yard. When she launches the rocket, the function $h(t) = -16t^2 + 32t$ models the height, h, of the rocket above the ground as a function of time, t. Find:

ⓐ the zeros of this function which tells us when the penny will hit the ground. ⓑ the time the rocket will be 16 feet above the ground.

334. Gianna is going to throw a ball from the top floor of her middle school. When she throws the ball from 48 feet above the ground, the function $h(t) = -16t^2 + 32t + 48$ models the height, h, of the ball above the ground as a function of time, t. Find:

ⓐ the zeros of this function which tells us when the ball will hit the ground. ⓑ the time(s) the ball will be 48 feet above the ground. ⓒ the height the ball will be at $t = 1$ seconds which is when the ball will be at its highest point.

Writing Exercises

335. Explain how you solve a quadratic equation. How many answers do you expect to get for a quadratic equation?

336. Give an example of a quadratic equation that has a GCF and none of the solutions to the equation is zero.

Self Check

ⓐ *After completing the exercises, use this checklist to evaluate your mastery of the objectives of this section.*

I can...	Confidently	With some help	No-I don't get it!
solve quadratic equations by using the Zero Product Property.			
solve quadratic equations by factoring.			
solve applications modeled by quadratic equations.			

ⓑ *Overall, after looking at the checklist, do you think you are well-prepared for the next section? Why or why not?*

CHAPTER 6 REVIEW

KEY TERMS

degree of the polynomial equation The degree of the polynomial equation is the degree of the polynomial.

factoring Splitting a product into factors is called factoring.

greatest common factor The greatest common factor (GCF) of two or more expressions is the largest expression that is a factor of all the expressions.

polynomial equation A polynomial equation is an equation that contains a polynomial expression.

quadratic equation Polynomial equations of degree two are called quadratic equations.

zero of the function A value of x where the function is 0, is called a zero of the function.

Zero Product Property The Zero Product Property says that if the product of two quantities is zero, then at least one of the quantities is zero.

KEY CONCEPTS

6.1 Greatest Common Factor and Factor by Grouping

- **How to find the greatest common factor (GCF) of two expressions.**

  Step 1. Factor each coefficient into primes. Write all variables with exponents in expanded form.

  Step 2. List all factors—matching common factors in a column. In each column, circle the common factors.

  Step 3. Bring down the common factors that all expressions share.

  Step 4. Multiply the factors.

- **Distributive Property:** If a, b, and c are real numbers, then
$$a(b + c) = ab + ac \quad \text{and} \quad ab + ac = a(b + c)$$

  The form on the left is used to multiply. The form on the right is used to factor.

- **How to factor the greatest common factor from a polynomial.**

  Step 1. Find the GCF of all the terms of the polynomial.

  Step 2. Rewrite each term as a product using the GCF.

  Step 3. Use the "reverse" Distributive Property to factor the expression.

  Step 4. Check by multiplying the factors.

- **Factor as a Noun and a Verb:** We use "factor" as both a noun and a verb.

  Noun: 7 is a *factor* of 14

  Verb: *factor* 3 from $3a + 3$

- **How to factor by grouping.**

  Step 1. Group terms with common factors.

  Step 2. Factor out the common factor in each group.

  Step 3. Factor the common factor from the expression.

  Step 4. Check by multiplying the factors.

6.2 Factor Trinomials

- **How to factor trinomials of the form $x^2 + bx + c$.**

  Step 1. Write the factors as two binomials with first terms x. $x^2 + bx + c$
  $(x \quad)(x \quad)$

  Step 2. Find two numbers m and n that
  multiply to c, $m \cdot n = c$
  add to b, $m + n = b$

  Step 3. Use m and n as the last terms of the factors. $(x + m)(x + n)$

Step 4. Check by multiplying the factors.

- **Strategy for Factoring Trinomials of the Form** $x^2 + bx + c$: When we factor a trinomial, we look at the signs of its terms first to determine the signs of the binomial factors.

$$x^2 + bx + c$$
$$(x + m)(x + n)$$

When c **is positive,** m **and** n **have the same sign.**

b positive	b negative
m, n positive	m, n negative
$x^2 + 5x + 6$	$x^2 - 6x + 8$
$(x + 2)(x + 3)$	$(x - 4)(x - 2)$
same signs	same signs

When c **is negative,** m **and** n **have opposite signs.**

$x^2 + x - 12$	$x^2 - 2x - 15$
$(x + 4)(x - 3)$	$(x - 5)(x + 3)$
opposite signs	opposite signs

Notice that, in the case when m and n have opposite signs, the sign of the one with the larger absolute value matches the sign of b.

- **How to factor trinomials of the form** $ax^2 + bx + c$ **using trial and error.**

  Step 1. Write the trinomial in descending order of degrees as needed.

  Step 2. Factor any GCF.

  Step 3. Find all the factor pairs of the first term.

  Step 4. Find all the factor pairs of the third term.

  Step 5. Test all the possible combinations of the factors until the correct product is found.

  Step 6. Check by multiplying.

- **How to factor trinomials of the form** $ax^2 + bx + c$ **using the "ac" method.**

  Step 1. Factor any GCF.

  Step 2. Find the product ac.

  Step 3. Find two numbers m and n that:
    Multiply to ac. $m \cdot n = a \cdot c$
    Add to b. $m + n = b$
    $$ax^2 + bx + c$$

  Step 4. Split the middle term using m and n. $ax^2 + mx + nx + c$

  Step 5. Factor by grouping.

  Step 6. Check by multiplying the factors.

6.3 Factor Special Products

- **Perfect Square Trinomials Pattern:** If a and b are real numbers,

$$a^2 + 2ab + b^2 = (a + b)^2$$
$$a^2 - 2ab + b^2 = (a - b)^2$$

- **How to factor perfect square trinomials.**

Step 1. Does the trinomial fit the pattern? $a^2 + 2ab + b^2$ $a^2 - 2ab + b^2$

Is the first term a perfect square? $(a)^2$ $(a)^2$
Write it as a square.

Is the last term a perfect square? $(a)^2$ $(b)^2$ $(a)^2$ $(b)^2$
Write it as a square.

Check the middle term. Is it $2ab$?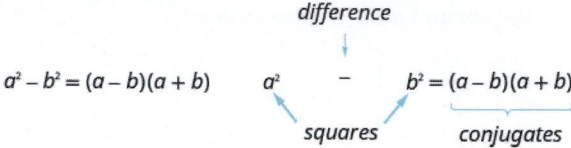

Step 2. Write the square of the binomial. $(a+b)^2$ $(a-b)^2$
Step 3. Check by multiplying.

- **Difference of Squares Pattern:** If a, b are real numbers,

$$a^2 - b^2 = (a - b)(a + b)$$

- **How to factor differences of squares.**
  Step 1. Does the binomial fit the pattern? $a^2 - b^2$

  Is this a difference? $\underline{\quad} - \underline{\quad}$
  Are the first and last terms perfect squares?

  Step 2. Write them as squares. $(a)^2 - (b)^2$
  Step 3. Write the product of conjugates. $(a - b)(a + b)$
  Step 4. Check by multiplying.

- **Sum and Difference of Cubes Pattern**
  $$a^3 + b^3 = (a + b)\left(a^2 - ab + b^2\right)$$
  $$a^3 - b^3 = (a - b)\left(a^2 + ab + b^2\right)$$

- **How to factor the sum or difference of cubes.**
  Step 1. Does the binomial fit the sum or difference of cubes pattern?
            Is it a sum or difference?
            Are the first and last terms perfect cubes?

  Step 2. Write them as cubes.

  Step 3. Use either the sum or difference of cubes pattern.

  Step 4. Simplify inside the parentheses

  Step 5. Check by multiplying the factors.

6.4 General Strategy for Factoring Polynomials

General Strategy for Factoring Polynomials

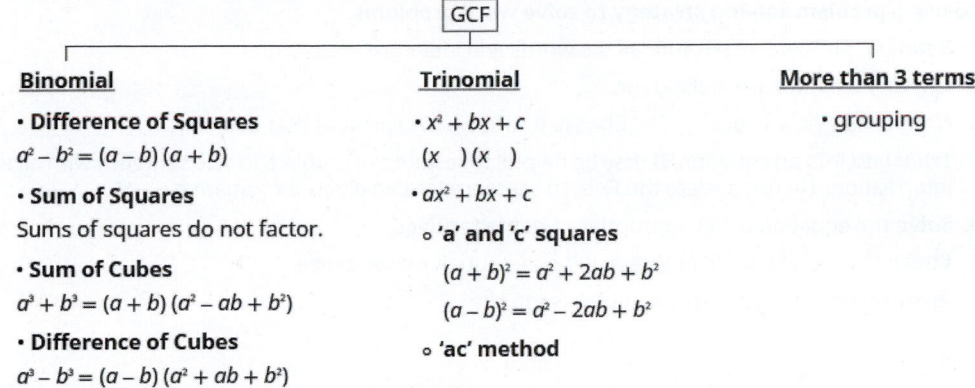

- **How to use a general strategy for factoring polynomials.**

  Step 1. Is there a greatest common factor?
  Factor it out.

  Step 2. Is the polynomial a binomial, trinomial, or are there more than three terms?
  If it is a binomial:
  Is it a sum?
  Of squares? Sums of squares do not factor.
  Of cubes? Use the sum of cubes pattern.
  Is it a difference?
  Of squares? Factor as the product of conjugates.
  Of cubes? Use the difference of cubes pattern.
  If it is a trinomial:

  Is it of the form $x^2 + bx + c$? Undo FOIL.

  Is it of the form $ax^2 + bx + c$?

  If a and c are squares, check if it fits the trinomial square pattern.
  Use the trial and error or "ac" method.
  If it has more than three terms:
  Use the grouping method.

  Step 3. Check.
  Is it factored completely?
  Do the factors multiply back to the original polynomial?

6.5 Polynomial Equations

- **Polynomial Equation:** A polynomial equation is an equation that contains a polynomial expression. The degree of the polynomial equation is the degree of the polynomial.

- **Quadratic Equation:** An equation of the form $ax^2 + bx + c = 0$ is called a quadratic equation.

$$a, b, c \text{ are real numbers and } a \neq 0$$

- **Zero Product Property:** If $a \cdot b = 0$, then either $a = 0$ or $b = 0$ or both.

- **How to use the Zero Product Property**

  Step 1. Set each factor equal to zero.

  Step 2. Solve the linear equations.

  Step 3. Check.

- **How to solve a quadratic equation by factoring.**

  Step 1. Write the quadratic equation in standard form, $ax^2 + bx + c = 0$.

  Step 2. Factor the quadratic expression.

  Step 3. Use the Zero Product Property.

Step 4. Solve the linear equations.

Step 5. Check. Substitute each solution separately into the original equation.

- **Zero of a Function:** For any function f, if $f(x) = 0$, then x is a zero of the function.

- **How to use a problem solving strategy to solve word problems.**

Step 1. **Read** the problem. Make sure all the words and ideas are understood.

Step 2. **Identify** what we are looking for.

Step 3. **Name** what we are looking for. Choose a variable to represent that quantity.

Step 4. **Translate** into an equation. It may be helpful to restate the problem in one sentence with all the important information. Then, translate the English sentence into an algebraic equation.

Step 5. **Solve** the equation using appropriate algebra techniques.

Step 6. **Check** the answer in the problem and make sure it makes sense.

Step 7. **Answer** the question with a complete sentence.

REVIEW EXERCISES

6.1 Greatest Common Factor and Factor by Grouping

Find the Greatest Common Factor of Two or More Expressions

In the following exercises, find the greatest common factor.

337. $12a^2 b^3$, $15ab^2$

338. $12m^2 n^3$, $42m^5 n^3$

339. $15y^3$, $21y^2$, $30y$

340. $45x^3 y^2$, $15x^4 y$, $10x^5 y^3$

Factor the Greatest Common Factor from a Polynomial

In the following exercises, factor the greatest common factor from each polynomial.

341. $35y + 84$

342. $6y^2 + 12y - 6$

343. $18x^3 - 15x$

344. $15m^4 + 6m^2 n$

345. $4x^3 - 12x^2 + 16x$

346. $-3x + 24$

347. $-3x^3 + 27x^2 - 12x$

348. $3x(x - 1) + 5(x - 1)$

Factor by Grouping

In the following exercises, factor by grouping.

349. $ax - ay + bx - by$

350. $x^2 y - xy^2 + 2x - 2y$

351. $x^2 + 7x - 3x - 21$

352. $4x^2 - 16x + 3x - 12$

353. $m^3 + m^2 + m + 1$

354. $5x - 5y - y + x$

6.2 Factor Trinomials

Factor Trinomials of the Form $x^2 + bx + c$

In the following exercises, factor each trinomial of the form $x^2 + bx + c$.

355. $a^2 + 14a + 33$

356. $k^2 - 16k + 60$

357. $m^2 + 3m - 54$

358. $x^2 - 3x - 10$

In the following examples, factor each trinomial of the form $x^2 + bxy + cy^2$.

359. $x^2 + 12xy + 35y^2$

360. $r^2 + 3rs - 28s^2$

361. $a^2 + 4ab - 21b^2$

362. $p^2 - 5pq - 36q^2$

363. $m^2 - 5mn + 30n^2$

Factor Trinomials of the Form $ax^2 + bx + c$ Using Trial and Error

In the following exercises, factor completely using trial and error.

364. $x^3 + 5x^2 - 24x$

365. $3y^3 - 21y^2 + 30y$

366. $5x^4 + 10x^3 - 75x^2$

367. $5y^2 + 14y + 9$

368. $8x^2 + 25x + 3$

369. $10y^2 - 53y - 11$

370. $6p^2 - 19pq + 10q^2$

371. $-81a^2 + 153a + 18$

Factor Trinomials of the Form $ax^2 + bx + c$ using the 'ac' Method

In the following exercises, factor.

372. $2x^2 + 9x + 4$

373. $18a^2 - 9a + 1$

374. $15p^2 + 2p - 8$

375. $15x^2 + 6x - 2$

376. $8a^2 + 32a + 24$

377. $3x^2 + 3x - 36$

378. $48y^2 + 12y - 36$

379. $18a^2 - 57a - 21$

380. $3n^4 - 12n^3 - 96n^2$

Factor using substitution

In the following exercises, factor using substitution.

381. $x^4 - 13x^2 - 30$

382. $(x - 3)^2 - 5(x - 3) - 36$

6.3 Factor Special Products

Factor Perfect Square Trinomials

In the following exercises, factor completely using the perfect square trinomials pattern.

383. $25x^2 + 30x + 9$

384. $36a^2 - 84ab + 49b^2$

385. $40x^2 + 360x + 810$

386. $5k^3 - 70k^2 + 245k$

387. $75u^4 - 30u^3v + 3u^2v^2$

Factor Differences of Squares

In the following exercises, factor completely using the difference of squares pattern, if possible.

388. $81r^2 - 25$

389. $169m^2 - n^2$

390. $25p^2 - 1$

391. $9 - 121y^2$

392. $20x^2 - 125$

393. $169n^3 - n$

394. $6p^2q^2 - 54p^2$

395. $24p^2 + 54$

396. $49x^2 - 81y^2$

397. $16z^4 - 1$

398. $48m^4n^2 - 243n^2$

399. $a^2 + 6a + 9 - 9b^2$

400. $x^2 - 16x + 64 - y^2$

Factor Sums and Differences of Cubes

In the following exercises, factor completely using the sums and differences of cubes pattern, if possible.

401. $a^3 - 125$

402. $b^3 - 216$

403. $2m^3 + 54$

404. $81m^3 + 3$

6.4 General Strategy for Factoring Polynomials

Recognize and Use the Appropriate Method to Factor a Polynomial Completely

In the following exercises, factor completely.

405. $24x^3 + 44x^2$

406. $24a^4 - 9a^3$

407. $16n^2 - 56mn + 49m^2$

408. $6a^2 - 25a - 9$

409. $5u^4 - 45u^2$

410. $n^4 - 81$

411. $64j^2 + 225$

412. $5x^2 + 5x - 60$

413. $b^3 - 64$

414. $m^3 + 125$

415. $2b^2 - 2bc + 5cb - 5c^2$

416. $48x^5 y^2 - 243xy^2$

417. $5q^2 - 15q - 90$

418. $4u^5 v + 4u^2 v^3$

419. $10m^4 - 6250$

420. $60x^2 y - 75xy + 30y$

421. $16x^2 - 24xy + 9y^2 - 64$

6.5 Polynomial Equations

Use the Zero Product Property

In the following exercises, solve.

422. $(a - 3)(a + 7) = 0$

423. $(5b + 1)(6b + 1) = 0$

424. $6m(12m - 5) = 0$

425. $(2x - 1)^2 = 0$

426. $3m(2m - 5)(m + 6) = 0$

Solve Quadratic Equations by Factoring

In the following exercises, solve.

427. $x^2 + 9x + 20 = 0$

428. $y^2 - y - 72 = 0$

429. $2p^2 - 11p = 40$

430. $q^3 + 3q^2 + 2q = 0$

431. $144m^2 - 25 = 0$

432. $4n^2 = 36$

433. $(x + 6)(x - 3) = -8$

434. $(3x - 2)(x + 4) = 12x$

435. $16p^3 = 24p^2 + 9p$

436. $2y^3 + 2y^2 = 12y$

Solve Equations with Polynomial Functions

In the following exercises, solve.

437. For the function, $f(x) = x^2 + 11x + 20,$ ⓐ find when $f(x) = -8$ ⓑ Use this information to find two points that lie on the graph of the function.

438. For the function, $f(x) = 9x^2 - 18x + 5,$ ⓐ find when $f(x) = -3$ ⓑ Use this information to find two points that lie on the graph of the function.

In each function, find: ⓐ *the zeros of the function* ⓑ *the x-intercepts of the graph of the function* ⓒ *the y-intercept of the graph of the function.*

439. $f(x) = 64x^2 - 49$

440. $f(x) = 6x^2 - 13x - 5$

Solve Applications Modeled by Quadratic Equations

In the following exercises, solve.

441. The product of two consecutive numbers is 399. Find the numbers.

442. The area of a rectangular shaped patio 432 square feet. The length of the patio is 6 feet more than its width. Find the length and width.

443. A ladder leans against the wall of a building. The length of the ladder is 9 feet longer than the distance of the bottom of the ladder from the building. The distance of the top of the ladder reaches up the side of the building is 7 feet longer than the distance of the bottom of the ladder from the building. Find the lengths of all three sides of the triangle formed by the ladder leaning against the building.

444. Shruti is going to throw a ball from the top of a cliff. When she throws the ball from 80 feet above the ground, the function $h(t) = -16t^2 + 64t + 80$ models the height, *h*, of the ball above the ground as a function of time, *t*. Find: ⓐ the zeros of this function which tells us when the ball will hit the ground. ⓑ the time(s) the ball will be 80 feet above the ground. ⓒ the height the ball will be at $t = 2$ seconds which is when the ball will be at its highest point.

PRACTICE TEST

In the following exercises, factor completely.

445. $80a^2 + 120a^3$

446. $5m(m-1) + 3(m-1)$

447. $x^2 + 13x + 36$

448. $p^2 + pq - 12q^2$

449. $xy - 8y + 7x - 56$

450. $40r^2 + 810$

451. $9s^2 - 12s + 4$

452. $6x^2 - 11x - 10$

453. $3x^2 - 75y^2$

454. $6u^2 + 3u - 18$

455. $x^3 + 125$

456. $32x^5 y^2 - 162xy^2$

457. $6x^4 - 19x^2 + 15$

458. $3x^3 - 36x^2 + 108x$

In the following exercises, solve

459. $5a^2 + 26a = 24$

460. The product of two consecutive integers is 156. Find the integers.

461. The area of a rectangular place mat is 168 square inches. Its length is two inches longer than the width. Find the length and width of the placemat.

462. Jing is going to throw a ball from the balcony of her condo. When she throws the ball from 80 feet above the ground, the function $h(t) = -16t^2 + 64t + 80$ models the height, h, of the ball above the ground as a function of time, t. Find: ⓐ the zeros of this function which tells us when the ball will hit the ground. ⓑ the time(s) the ball will be 128 feet above the ground. ⓒ the height the ball will be at $t = 4$ seconds.

463. For the function, $f(x) = x^2 - 7x + 5$, ⓐ find when $f(x) = -7$ ⓑ Use this information to find two points that lie on the graph of the function.

464. For the function $f(x) = 25x^2 - 81$, find: ⓐ the zeros of the function ⓑ the x-intercepts of the graph of the function ⓒ the y-intercept of the graph of the function.

Figure 7.1 American football is the most watched spectator sport in the United States. People around the country are constantly tracking statistics for football and other sports. (credit: "keijj44" / Pixabay)

Chapter Outline

Introduction

Twelve goals last season. Fifteen home runs. Nine touchdowns. Whatever the statistics, sports analysts know it. Their jobs depend on it. Compiling and analyzing sports data not only help fans appreciate their teams but also help owners and coaches decide which players to recruit, how to best use them in games, how much they should be paid, and which players to trade. Understanding this kind of data requires a knowledge of specific types of expressions and functions. In this chapter, you will work with rational expressions and perform operations on them. And you will use rational expressions and inequalities to solve real-world problems.

7.1 | Multiply and Divide Rational Expressions

Learning Objectives

By the end of this section, you will be able to:

> Determine the values for which a rational expression is undefined
> Simplify rational expressions
> Multiply rational expressions
> Divide rational expressions
> Multiply and divide rational functions

> **Be Prepared!**

> Before you get started, take this readiness quiz.

> 1. Simplify: $\dfrac{90y}{15y^2}$.

> If you missed this problem, review **Example 5.13**.

2. Multiply: $\frac{14}{15} \cdot \frac{6}{35}$.

If you missed this problem, review Example 1.25.

3. Divide: $\frac{12}{10} \div \frac{8}{25}$.

If you missed this problem, review Example 1.26.

We previously reviewed the properties of fractions and their operations. We introduced rational numbers, which are just fractions where the numerators and denominators are integers. In this chapter, we will work with fractions whose numerators and denominators are polynomials. We call this kind of expression a **rational expression**.

Rational Expression

A rational expression is an expression of the form $\frac{p}{q}$, where p and q are polynomials and $q \neq 0$.

Here are some examples of rational expressions:

$$-\frac{24}{56} \qquad \frac{5x}{12y} \qquad \frac{4x+1}{x^2-9} \qquad \frac{4x^2+3x-1}{2x-8}$$

Notice that the first rational expression listed above, $-\frac{24}{56}$, is just a fraction. Since a constant is a polynomial with degree zero, the ratio of two constants is a rational expression, provided the denominator is not zero.

We will do the same operations with rational expressions that we did with fractions. We will simplify, add, subtract, multiply, divide and use them in applications.

Determine the Values for Which a Rational Expression is Undefined

If the denominator is zero, the rational expression is undefined. The numerator of a rational expression may be 0—but not the denominator.

When we work with a numerical fraction, it is easy to avoid dividing by zero because we can see the number in the denominator. In order to avoid dividing by zero in a rational expression, we must not allow values of the variable that will make the denominator be zero.

So before we begin any operation with a rational expression, we examine it first to find the values that would make the denominator zero. That way, when we solve a rational equation for example, we will know whether the algebraic solutions we find are allowed or not.

HOW TO :: DETERMINE THE VALUES FOR WHICH A RATIONAL EXPRESSION IS UNDEFINED.

Step 1.   Set the denominator equal to zero.

Step 2.   Solve the equation.

EXAMPLE 7.1

Determine the value for which each rational expression is undefined:

ⓐ $\frac{8a^2b}{3c}$ ⓑ $\frac{4b-3}{2b+5}$ ⓒ $\frac{x+4}{x^2+5x+6}$.

⊘ **Solution**

The expression will be undefined when the denominator is zero.

ⓐ

$$\frac{8a^2 b}{3c}$$

Set the denominator equal to zero and solve
for the variable.

$$3c = 0$$

$$c = 0$$

$\frac{8a^2 b}{3c}$ is undefined for $c = 0$.

ⓑ

$$\frac{4b - 3}{2b + 5}$$

Set the denominator equal to zero and solve
for the variable.

$$2b + 5 = 0$$

$$2b = -5$$

$$b = -\frac{5}{2}$$

$\frac{4b - 3}{2b + 5}$ is undefined for $b = -\frac{5}{2}$.

ⓒ

$$\frac{x + 4}{x^2 + 5x + 6}$$

Set the denominator equal to zero and solve
for the variable.

$$x^2 + 5x + 6 = 0$$

$$(x + 2)(x + 3) = 0$$

$$x + 2 = 0 \ \text{ or } \ x + 3 = 0$$

$$x = -2 \ \text{or} \ x = -3$$

$\frac{x + 4}{x^2 + 5x + 6}$ is undefined for $x = -2$ or $x = -3$.

> | **TRY IT :: 7.1**   Determine the value for which each rational expression is undefined.

ⓐ $\frac{3y^2}{8x}$ ⓑ $\frac{8n - 5}{3n + 1}$ ⓒ $\frac{a + 10}{a^2 + 4a + 3}$

> | **TRY IT :: 7.2**   Determine the value for which each rational expression is undefined.

ⓐ $\frac{4p}{5q}$ ⓑ $\frac{y - 1}{3y + 2}$ ⓒ $\frac{m - 5}{m^2 + m - 6}$

Simplify Rational Expressions

A fraction is considered simplified if there are no common factors, other than 1, in its numerator and denominator. Similarly, a **simplified rational expression** has no common factors, other than 1, in its numerator and denominator.

Simplified Rational Expression

A rational expression is considered simplified if there are no common factors in its numerator and denominator.

For example,

$\frac{x + 2}{x + 3}$ is simplified because there are no common factors of $x + 2$ and $x + 3$.

$\frac{2x}{3x}$ is not simplified because x is a common factor of $2x$ and $3x$.

We use the Equivalent Fractions Property to simplify numerical fractions. We restate it here as we will also use it to simplify rational expressions.

Equivalent Fractions Property

If a, b, and c are numbers where $b \neq 0$, $c \neq 0$,

$$\text{then } \frac{a}{b} = \frac{a \cdot c}{b \cdot c} \quad \text{and} \quad \frac{a \cdot c}{b \cdot c} = \frac{a}{b}.$$

Notice that in the Equivalent Fractions Property, the values that would make the denominators zero are specifically disallowed. We see $b \neq 0$, $c \neq 0$ clearly stated.

To simplify rational expressions, we first write the numerator and denominator in factored form. Then we remove the common factors using the Equivalent Fractions Property.

Be very careful as you remove common factors. Factors are multiplied to make a product. You can remove a factor from a product. You cannot remove a term from a sum.

$\dfrac{2 \cdot \cancel{3} \cdot \cancel{7}}{\cancel{3} \cdot 5 \cdot \cancel{7}}$	$\dfrac{3x\cancel{(x-9)}}{5\cancel{(x-9)}}$ where $x \neq 9$	$\dfrac{x+5}{x}$
$\dfrac{2}{5}$	$\dfrac{3x}{5}$	**NO COMMON FACTORS**
We removed the common factors 3 and 7. They are factors of the product.	We removed the common factor $(x-9)$. It is a factor of the product.	While there is an x in both the numerator and denominator, the x in the numerator is a term of a sum!

Removing the x's from $\dfrac{x+5}{x}$ would be like cancelling the 2's in the fraction $\dfrac{2+5}{2}$!

EXAMPLE 7.2   HOW TO SIMPLIFY A RATIONAL EXPRESSION

Simplify: $\dfrac{x^2 + 5x + 6}{x^2 + 8x + 12}$.

✓ **Solution**

Step 1. Factor the numerator and denominator completely.	Factor $x^2 + 5x + 6$ and $x^2 + 8x + 12$.	$\dfrac{x^2 + 5x + 6}{x^2 + 8x + 12}$ $\dfrac{(x+2)(x+3)}{(x+2)(x+6)}$
Step 2. Simplify by dividing out common factors.	Remove the common factor $x + 2$ from the numerator and the denominator.	$\dfrac{\cancel{(x+2)}(x+3)}{\cancel{(x+2)}(x+6)}$ $\dfrac{(x+3)}{(x+6)}$ $x \neq -2 \ x \neq -6$

> **TRY IT :: 7.3**   Simplify: $\dfrac{x^2 - x - 2}{x^2 - 3x + 2}$.

> **TRY IT :: 7.4**   Simplify: $\dfrac{x^2 - 3x - 10}{x^2 + x - 2}$.

We now summarize the steps you should follow to simplify rational expressions.

HOW TO :: SIMPLIFY A RATIONAL EXPRESSION.

Step 1.   Factor the numerator and denominator completely.

Step 2.   Simplify by dividing out common factors.

Usually, we leave the simplified rational expression in factored form. This way, it is easy to check that we have removed *all* the common factors.

We'll use the methods we have learned to factor the polynomials in the numerators and denominators in the following examples.

Every time we write a rational expression, we should make a statement disallowing values that would make a denominator zero. However, to let us focus on the work at hand, we will omit writing it in the examples.

EXAMPLE 7.3

Simplify: $\dfrac{3a^2 - 12ab + 12b^2}{6a^2 - 24b^2}$.

✓ Solution

$$\frac{3a^2 - 12ab + 12b^2}{6a^2 - 24b^2}$$

Factor the numerator and denominator, first factoring out the GCF.

$$\frac{3\left(a^2 - 4ab + 4b^2\right)}{6\left(a^2 - 4b^2\right)}$$

$$\frac{3(a - 2b)(a - 2b)}{6(a + 2b)(a - 2b)}$$

Remove the common factors of $a - 2b$ and 3.

$$\frac{\cancel{3}(a - 2b)\cancel{(a - 2b)}}{\cancel{3} \cdot 2(a + 2b)\cancel{(a - 2b)}}$$

$$\frac{a - 2b}{2(a + 2b)}$$

> **TRY IT :: 7.5**   Simplify: $\dfrac{2x^2 - 12xy + 18y^2}{3x^2 - 27y^2}$.

> **TRY IT :: 7.6**   Simplify: $\dfrac{5x^2 - 30xy + 25y^2}{2x^2 - 50y^2}$.

Now we will see how to simplify a rational expression whose numerator and denominator have opposite factors. We previously introduced opposite notation: the opposite of a is $-a$ and $-a = -1 \cdot a$.

The numerical fraction, say $\dfrac{7}{-7}$ simplifies to -1. We also recognize that the numerator and denominator are opposites.

The fraction $\dfrac{a}{-a}$, whose numerator and denominator are opposites also simplifies to -1.

Let's look at the expression $b - a$. $\qquad\qquad b - a$

Rewrite. $\qquad\qquad\qquad\qquad\qquad\qquad -a + b$

Factor out -1. $\qquad\qquad\qquad\qquad\qquad -1(a - b)$

This tells us that $b - a$ is the opposite of $a - b$.

In general, we could write the opposite of $a - b$ as $b - a$. So the rational expression $\dfrac{a - b}{b - a}$ simplifies to -1.

Opposites in a Rational Expression

The opposite of $a - b$ is $b - a$.

$$\frac{a - b}{b - a} = -1 \quad a \neq b$$

An expression and its opposite divide to -1.

We will use this property to simplify rational expressions that contain opposites in their numerators and denominators. Be careful not to treat $a + b$ and $b + a$ as opposites. Recall that in addition, order doesn't matter so $a + b = b + a$. So if $a \neq -b$, then $\frac{a + b}{b + a} = 1$.

EXAMPLE 7.4

Simplify: $\dfrac{x^2 - 4x - 32}{64 - x^2}$.

Solution

$$\frac{x^2 - 4x - 32}{64 - x^2}$$

Factor the numerator and the denominator.	$\dfrac{(x - 8)(x + 4)}{(8 - x)(8 + x)}$
Recognize the factors that are opposites.	$(-1)\dfrac{(x - 8)(x + 4)}{(8 - x)(8 + x)}$
Simplify.	$-\dfrac{x + 4}{x + 8}$

> **TRY IT :: 7.7** Simplify: $\dfrac{x^2 - 4x - 5}{25 - x^2}$.

> **TRY IT :: 7.8** Simplify: $\dfrac{x^2 + x - 2}{1 - x^2}$.

Multiply Rational Expressions

To multiply rational expressions, we do just what we did with numerical fractions. We multiply the numerators and multiply the denominators. Then, if there are any common factors, we remove them to simplify the result.

Multiplication of Rational Expressions

If p, q, r, and s are polynomials where $q \neq 0$, $s \neq 0$, then

$$\frac{p}{q} \cdot \frac{r}{s} = \frac{pr}{qs}$$

To multiply rational expressions, multiply the numerators and multiply the denominators.

Remember, throughout this chapter, we will assume that all numerical values that would make the denominator be zero are excluded. We will not write the restrictions for each rational expression, but keep in mind that the denominator can never be zero. So in this next example, $x \neq 0$, $x \neq 3$, and $x \neq 4$.

EXAMPLE 7.5   HOW TO MULTIPLY RATIONAL EXPRESSIONS

Simplify: $\dfrac{2x}{x^2 - 7x + 12} \cdot \dfrac{x^2 - 9}{6x^2}$.

Solution

Step 1. Factor each numerator and denominator completely.	Factor $x^2 - 9$ and $x^2 - 7x + 12$.	$\dfrac{2x}{x^2 - 7x + 12} \cdot \dfrac{x^2 - 9}{6x^2}$
		$\dfrac{2x}{(x - 3)(x - 4)} \cdot \dfrac{(x - 3)(x + 3)}{6x^2}$

Step 2. Multiply the numerators and denominators.	Multiply the numerators and denominators. It is helpful to write the monomials first.	$\dfrac{2x(x-3)(x+3)}{6x^2(x-3)(x-4)}$
Step 3. Simplify by dividing out common factors.	Divide out the common factors. Leave the denominator in factored form.	$\dfrac{2\!\!\!/\,x(x\!\!\!\!-\!\!\!\!3)(x+3)}{2\!\!\!/\cdot 3\cdot x\cdot x(x\!\!\!\!-\!\!\!\!3)(x-4)}$ $\dfrac{(x+3)}{3x(x-4)}$

> **TRY IT :: 7.9**

Simplify: $\dfrac{5x}{x^2+5x+6}\cdot\dfrac{x^2-4}{10x}$.

> **TRY IT :: 7.10**

Simplify: $\dfrac{9x^2}{x^2+11x+30}\cdot\dfrac{x^2-36}{3x^2}$.

 HOW TO :: MULTIPLY RATIONAL EXPRESSIONS.

Step 1.   Factor each numerator and denominator completely.

Step 2.   Multiply the numerators and denominators.

Step 3.   Simplify by dividing out common factors.

EXAMPLE 7.6

Multiply: $\dfrac{3a^2-8a-3}{a^2-25}\cdot\dfrac{a^2+10a+25}{3a^2-14a-5}$.

⊘ Solution

$$\dfrac{3a^2-8a-3}{a^2-25}\cdot\dfrac{a^2+10a+25}{3a^2-14a-5}$$

Factor the numerators and denominators and then multiply.

$$\dfrac{(3a+1)(a-3)(a+5)(a+5)}{(a-5)(a+5)(3a+1)(a-5)}$$

Simplify by dividing out common factors.

$$\dfrac{(3a+1)(a-3)(a+5)(a+5)}{(a-5)(a+5)(3a+1)(a-5)}$$

Simplify.

$$\dfrac{(a-3)(a+5)}{(a-5)(a-5)}$$

Rewrite $(a-5)(a-5)$ using an exponent.

$$\dfrac{(a-3)(a+5)}{(a-5)^2}$$

> **TRY IT :: 7.11**

Simplify: $\dfrac{2x^2+5x-12}{x^2-16}\cdot\dfrac{x^2-8x+16}{2x^2-13x+15}$.

> **TRY IT :: 7.12**

Simplify: $\dfrac{4b^2+7b-2}{1-b^2}\cdot\dfrac{b^2-2b+1}{4b^2+15b-4}$.

Divide Rational Expressions

Just like we did for numerical fractions, to divide rational expressions, we multiply the first fraction by the reciprocal of the second.

Division of Rational Expressions

If p, q, r, and s are polynomials where $q \neq 0$, $r \neq 0$, $s \neq 0$, then

$$\frac{p}{q} \div \frac{r}{s} = \frac{p}{q} \cdot \frac{s}{r}$$

To divide rational expressions, multiply the first fraction by the reciprocal of the second.

Once we rewrite the division as multiplication of the first expression by the reciprocal of the second, we then factor everything and look for common factors.

EXAMPLE 7.7   HOW TO DIVIDE RATIONAL EXPRESSIONS

Divide: $\dfrac{p^3 + q^3}{2p^2 + 2pq + 2q^2} \div \dfrac{p^2 - q^2}{6}$.

✓ Solution

Step 1. Rewrite the division as the product of the first rational expression and the reciprocal of the second.	"Flip" the second fraction and change the division sign to multiplication.	$\dfrac{p^3 + q^3}{2p^2 + 2pq + 2q^2} \div \dfrac{p^2 - q^2}{6}$ $\dfrac{p^3 + q^3}{2p^2 + 2pq + 2q^2} \cdot \dfrac{6}{p^2 - q^2}$
Step 2. Factor the numerators and denominators completely.	Factor the numerators and denominators.	$\dfrac{(p + q)(p^2 - pq + q^2)}{2(p^2 + pq + q^2)} \cdot \dfrac{2 \cdot 3}{(p - q)(p + q)}$
Step 3. Multiply the numerators and denominators.	Multiply the numerators and multiply the denominators.	$\dfrac{(p + q)(p^2 - pq + q^2) \, 2 \cdot 3}{2(p^2 + pq + q^2)(p - q)(p + q)}$
Step 4. Simplify by dividing out common factors.	Divide out the common factors.	$\dfrac{\cancel{(p + q)}(p^2 - pq + q^2)\,\cancel{2} \cdot 3}{\cancel{2}(p^2 + pq + q^2)(p - q)\cancel{(p + q)}}$ $\dfrac{3(p^2 - pq + q^2)}{(p - q)(p^2 + pq + q^2)}$

> **TRY IT :: 7.13**
> Simplify: $\dfrac{x^3 - 8}{3x^2 - 6x + 12} \div \dfrac{x^2 - 4}{6}$.

> **TRY IT :: 7.14**
> Simplify: $\dfrac{2z^2}{z^2 - 1} \div \dfrac{z^3 - z^2 + z}{z^3 + 1}$.

HOW TO :: DIVIDE RATIONAL EXPRESSIONS.

Step 1.   Rewrite the division as the product of the first rational expression and the reciprocal of the second.

Step 2.   Factor the numerators and denominators completely.

Step 3.   Multiply the numerators and denominators together.

Step 4.   Simplify by dividing out common factors.

Recall from **Use the Language of Algebra** that a complex fraction is a fraction that contains a fraction in the numerator, the denominator or both. Also, remember a fraction bar means division. A complex fraction is another way of writing division of two fractions.

EXAMPLE 7.8

Divide: $\dfrac{\frac{6x^2 - 7x + 2}{4x - 8}}{\frac{2x^2 - 7x + 3}{x^2 - 5x + 6}}$.

✓ Solution

$$\dfrac{\frac{6x^2 - 7x + 2}{4x - 8}}{\frac{2x^2 - 7x + 3}{x^2 - 5x + 6}}$$

Rewrite with a division sign.

$$\frac{6x^2 - 7x + 2}{4x - 8} \div \frac{2x^2 - 7x + 3}{x^2 - 5x + 6}$$

Rewrite as product of first times reciprocal of second.

$$\frac{6x^2 - 7x + 2}{4x - 8} \cdot \frac{x^2 - 5x + 6}{2x^2 - 7x + 3}$$

Factor the numerators and the denominators, and then multiply.

$$\frac{(2x - 1)(3x - 2)(x - 2)(x - 3)}{4(x - 2)(2x - 1)(x - 3)}$$

Simplify by dividing out common factors.

$$\frac{(2x - 1)(3x - 2)(x - 2)(x - 3)}{4(x - 2)(2x - 1)(x - 3)}$$

Simplify.

$$\frac{3x - 2}{4}$$

> **TRY IT :: 7.15**
>
> Simplify: $\dfrac{\frac{3x^2 + 7x + 2}{4x + 24}}{\frac{3x^2 - 14x - 5}{x^2 + x - 30}}$.

> **TRY IT :: 7.16**
>
> Simplify: $\dfrac{\frac{y^2 - 36}{2y^2 + 11y - 6}}{\frac{2y^2 - 2y - 60}{8y - 4}}$.

If we have more than two rational expressions to work with, we still follow the same procedure. The first step will be to rewrite any division as multiplication by the reciprocal. Then, we factor and multiply.

EXAMPLE 7.9

Perform the indicated operations: $\dfrac{3x - 6}{4x - 4} \cdot \dfrac{x^2 + 2x - 3}{x^2 - 3x - 10} \div \dfrac{2x + 12}{8x + 16}$.

✓ **Solution**

$$\frac{3x-6}{4x-4} \cdot \frac{x^2+2x-3}{x^2-3x-10} \div \frac{2x+12}{8x+16}$$

Rewrite the division as multiplication by the reciprocal.	$\frac{3x-6}{4x-4} \cdot \frac{x^2+2x-3}{x^2-3x-10} \cdot \frac{8x+16}{2x+12}$
Factor the numerators and the denominators.	$\frac{3(x-2)}{4(x-1)} \cdot \frac{(x+3)(x-1)}{(x+2)(x-5)} \cdot \frac{8(x+2)}{2(x+6)}$
Multiply the fractions. Bringing the constants to the front will help when removing common factors.	
Simplify by dividing out common factors.	$\frac{3 \cdot 8(x-2)(x+3)\cancel{(x-1)}\cancel{(x+2)}}{4 \cdot 2\cancel{(x-1)}\cancel{(x+2)}(x-5)(x+6)}$
Simplify.	$\frac{3(x-2)(x+3)}{(x-5)(x+6)}$

> **TRY IT :: 7.17**

Perform the indicated operations: $\frac{4m+4}{3m-15} \cdot \frac{m^2-3m-10}{m^2-4m-32} \div \frac{12m-36}{6m-48}$.

> **TRY IT :: 7.18**

Perform the indicated operations: $\frac{2n^2+10n}{n-1} \div \frac{n^2+10n+24}{n^2+8n-9} \cdot \frac{n+4}{8n^2+12n}$.

Multiply and Divide Rational Functions

We started this section stating that a rational expression is an expression of the form $\frac{p}{q}$, where p and q are polynomials and $q \neq 0$. Similarly, we define a **rational function** as a function of the form $R(x) = \frac{p(x)}{q(x)}$ where $p(x)$ and $q(x)$ are polynomial functions and $q(x)$ is not zero.

> ### Rational Function
>
> A rational function is a function of the form
>
> $$R(x) = \frac{p(x)}{q(x)}$$
>
> where $p(x)$ and $q(x)$ are polynomial functions and $q(x)$ is not zero.

The domain of a rational function is all real numbers except for those values that would cause division by zero. We must eliminate any values that make $q(x) = 0$.

 HOW TO :: DETERMINE THE DOMAIN OF A RATIONAL FUNCTION.

Step 1.   Set the denominator equal to zero.

Step 2.   Solve the equation.

Step 3.   The domain is all real numbers excluding the values found in Step 2.

EXAMPLE 7.10

Find the domain of $R(x) = \dfrac{2x^2 - 14x}{4x^2 - 16x - 48}$.

✓ Solution

The domain will be all real numbers except those values that make the denominator zero. We will set the denominator equal to zero , solve that equation, and then exclude those values from the domain.

Set the denominator to zero.
$$4x^2 - 16x - 48 = 0$$

Factor, first factor out the GCF.
$$4(x^2 - 4x - 12) = 0$$
$$4(x - 6)(x + 2) = 0$$

Use the Zero Product Property.
$$4 \neq 0 \quad x - 6 = 0 \quad x + 2 = 0$$

Solve.
$$x = 6 \quad x = -2$$

The domain of $R(x)$ is all real numbers where $x \neq 6$ and $x \neq -2$.

> **TRY IT :: 7.19**

Find the domain of $R(x) = \dfrac{2x^2 - 10x}{4x^2 - 16x - 20}$.

> **TRY IT :: 7.20**

Find the domain of $R(x) = \dfrac{4x^2 - 16x}{8x^2 - 16x - 64}$.

To multiply rational functions, we multiply the resulting rational expressions on the right side of the equation using the same techniques we used to multiply rational expressions.

EXAMPLE 7.11

Find $R(x) = f(x) \cdot g(x)$ where $f(x) = \dfrac{2x - 6}{x^2 - 8x + 15}$ and $g(x) = \dfrac{x^2 - 25}{2x + 10}$.

✓ Solution

$$R(x) = f(x) \cdot g(x)$$

$$R(x) = \frac{2x - 6}{x^2 - 8x + 15} \cdot \frac{x^2 - 25}{2x + 10}$$

Factor each numerator and denominator.
$$R(x) = \frac{2(x - 3)}{(x - 3)(x - 5)} \cdot \frac{(x - 5)(x + 5)}{2(x + 5)}$$

Multiply the numerators and denominators.
$$R(x) = \frac{2(x - 3)(x - 5)(x + 5)}{2(x - 3)(x - 5)(x + 5)}$$

Remove common factors.
$$R(x) = \frac{\cancel{2}\cancel{(x-3)}\cancel{(x-5)}\cancel{(x+5)}}{\cancel{2}\cancel{(x-3)}\cancel{(x-5)}\cancel{(x+5)}}$$

Simplify.
$$R(x) = 1$$

> **TRY IT :: 7.21**

Find $R(x) = f(x) \cdot g(x)$ where $f(x) = \dfrac{3x - 21}{x^2 - 9x + 14}$ and $g(x) = \dfrac{2x^2 - 8}{3x + 6}$.

> **TRY IT :: 7.22**

Find $R(x) = f(x) \cdot g(x)$ where $f(x) = \dfrac{x^2 - x}{3x^2 + 27x - 30}$ and $g(x) = \dfrac{x^2 - 100}{x^2 - 10x}$.

To divide rational functions, we divide the resulting rational expressions on the right side of the equation using the same techniques we used to divide rational expressions.

EXAMPLE 7.12

Find $R(x) = \dfrac{f(x)}{g(x)}$ where $f(x) = \dfrac{3x^2}{x^2 - 4x}$ and $g(x) = \dfrac{9x^2 - 45x}{x^2 - 7x + 10}$.

✓ **Solution**

$$R(x) = \frac{f(x)}{g(x)}$$

Substitute in the functions $f(x),\ g(x)$.

$$R(x) = \frac{\dfrac{3x^2}{x^2 - 4x}}{\dfrac{9x^2 - 45x}{x^2 - 7x + 10}}$$

Rewrite the division as the product of $f(x)$ and the reciprocal of $g(x)$.

$$R(x) = \frac{3x^2}{x^2 - 4x} \cdot \frac{x^2 - 7x + 10}{9x^2 - 45x}$$

Factor the numerators and denominators and then multiply.

$$R(x) = \frac{3 \cdot x \cdot x \cdot (x-5)(x-2)}{x(x-4) \cdot 3 \cdot 3 \cdot x \cdot (x-5)}$$

Simplify by dividing out common factors.

$$R(x) = \frac{\cancel{3} \cdot \cancel{x} \cdot \cancel{x} \cancel{(x-5)}(x-2)}{\cancel{x}(x-4) \cdot \cancel{3} \cdot 3 \cdot \cancel{x}\cancel{(x-5)}}$$

$$R(x) = \frac{x-2}{3(x-4)}$$

> **TRY IT ::** 7.23

Find $R(x) = \dfrac{f(x)}{g(x)}$ where $f(x) = \dfrac{2x^2}{x^2 - 8x}$ and $g(x) = \dfrac{8x^2 + 24x}{x^2 + x - 6}$.

> **TRY IT ::** 7.24

Find $R(x) = \dfrac{f(x)}{g(x)}$ where $f(x) = \dfrac{15x^2}{3x^2 + 33x}$ and $g(x) = \dfrac{5x - 5}{x^2 + 9x - 22}$.

7.1 EXERCISES

Practice Makes Perfect

Determine the Values for Which a Rational Expression is Undefined

In the following exercises, determine the values for which the rational expression is undefined.

1.
ⓐ $\dfrac{2x^2}{z}$

ⓑ $\dfrac{4p-1}{6p-5}$

ⓒ $\dfrac{n-3}{n^2+2n-8}$

2.
ⓐ $\dfrac{10m}{11n}$

ⓑ $\dfrac{6y+13}{4y-9}$

ⓒ $\dfrac{b-8}{b^2-36}$

3.
ⓐ $\dfrac{4x^2y}{3y}$

ⓑ $\dfrac{3x-2}{2x+1}$

ⓒ $\dfrac{u-1}{u^2-3u-28}$

4.
ⓐ $\dfrac{5pq^2}{9q}$

ⓑ $\dfrac{7a-4}{3a+5}$

ⓒ $\dfrac{1}{x^2-4}$

Simplify Rational Expressions

In the following exercises, simplify each rational expression.

5. $-\dfrac{44}{55}$

6. $\dfrac{56}{63}$

7. $\dfrac{8m^3n}{12mn^2}$

8. $\dfrac{36v^3w^2}{27vw^3}$

9. $\dfrac{8n-96}{3n-36}$

10. $\dfrac{12p-240}{5p-100}$

11. $\dfrac{x^2+4x-5}{x^2-2x+1}$

12. $\dfrac{y^2+3y-4}{y^2-6y+5}$

13. $\dfrac{a^2-4}{a^2+6a-16}$

14. $\dfrac{y^2-2y-3}{y^2-9}$

15. $\dfrac{p^3+3p^2+4p+12}{p^2+p-6}$

16. $\dfrac{x^3-2x^2-25x+50}{x^2-25}$

17. $\dfrac{8b^2-32b}{2b^2-6b-80}$

18. $\dfrac{-5c^2-10c}{-10c^2+30c+100}$

19. $\dfrac{3m^2+30mn+75n^2}{4m^2-100n^2}$

20. $\dfrac{5r^2+30rs-35s^2}{r^2-49s^2}$

21. $\dfrac{a-5}{5-a}$

22. $\dfrac{5-d}{d-5}$

23. $\dfrac{20-5y}{y^2-16}$

24. $\dfrac{4v-32}{64-v^2}$

25. $\dfrac{w^3+216}{w^2-36}$

26. $\dfrac{v^3+125}{v^2-25}$

27. $\dfrac{z^2-9z+20}{16-z^2}$

28. $\dfrac{a^2-5z-36}{81-a^2}$

Multiply Rational Expressions

In the following exercises, multiply the rational expressions.

29. $\dfrac{12}{16} \cdot \dfrac{4}{10}$

30. $\dfrac{32}{5} \cdot \dfrac{16}{24}$

31. $\dfrac{5x^2 y^4}{12xy^3} \cdot \dfrac{6x^2}{20y^2}$

32. $\dfrac{12a^3 b}{b^2} \cdot \dfrac{2ab^2}{9b^3}$

33. $\dfrac{5p^2}{p^2 - 5p - 36} \cdot \dfrac{p^2 - 16}{10p}$

34. $\dfrac{3q^2}{q^2 + q - 6} \cdot \dfrac{q^2 - 9}{9q}$

35. $\dfrac{2y^2 - 10y}{y^2 + 10y + 25} \cdot \dfrac{y + 5}{6y}$

36. $\dfrac{z^2 + 3z}{z^2 - 3z - 4} \cdot \dfrac{z - 4}{z^2}$

37. $\dfrac{28 - 4b}{3b - 3} \cdot \dfrac{b^2 + 8b - 9}{b^2 - 49}$

38. $\dfrac{72m - 12m^2}{8m + 32} \cdot \dfrac{m^2 + 10m + 24}{m^2 - 36}$

39. $\dfrac{5c^2 + 9c + 2}{c^2 - 25} \cdot \dfrac{c^2 + 10c + 25}{3c^2 - 14c - 5}$

40. $\dfrac{2d^2 + d - 3}{d^2 - 16} \cdot \dfrac{d^2 - 8d + 16}{2d^2 - 9d - 18}$

41. $\dfrac{6m^2 - 2m - 10}{9 - m^2} \cdot \dfrac{m^2 - 6m + 9}{6m^2 + 29m - 20}$

42. $\dfrac{2n^2 - 3n - 14}{25 - n^2} \cdot \dfrac{n^2 - 10n + 25}{2n^2 - 13n + 21}$

Divide Rational Expressions

In the following exercises, divide the rational expressions.

43. $\dfrac{v - 5}{11 - v} \div \dfrac{v^2 - 25}{v - 11}$

44. $\dfrac{10 + w}{w - 8} \div \dfrac{100 - w^2}{8 - w}$

45. $\dfrac{3s^2}{s^2 - 16} \div \dfrac{s^3 - 4s^2 + 16s}{s^3 - 64}$

46. $\dfrac{r^2 - 9}{15} \div \dfrac{r^3 - 27}{5r^2 + 15r + 45}$

47. $\dfrac{p^3 + q^3}{3p^2 + 3pq + 3q^2} \div \dfrac{p^2 - q^2}{12}$

48. $\dfrac{v^3 - 8w^3}{2v^2 + 4vw + 8w^2} \div \dfrac{v^2 - 4w^2}{4}$

49. $\dfrac{x^2 + 3x - 10}{4x} \div (2x^2 + 20x + 50)$

50. $\dfrac{2y^2 - 10yz - 48z^2}{2y - 1} \div (4y^2 - 32yz)$

51. $\dfrac{\dfrac{2a^2 - a - 21}{5a + 20}}{\dfrac{a^2 + 7a + 12}{a^2 + 8a + 16}}$

52. $\dfrac{\dfrac{3b^2 + 2b - 8}{12b + 18}}{\dfrac{3b^2 + 2b - 8}{2b^2 - 7b - 15}}$

53. $\dfrac{\dfrac{12c^2 - 12}{2c^2 - 3c + 1}}{\dfrac{4c + 4}{6c^2 - 13c + 5}}$

54. $\dfrac{\dfrac{4d^2 + 7d - 2}{35d + 10}}{\dfrac{d^2 - 4}{7d^2 - 12d - 4}}$

For the following exercises, perform the indicated operations.

55. $\dfrac{10m^2 + 80m}{3m - 9} \cdot \dfrac{m^2 + 4m - 21}{m^2 - 9m + 20} \div \dfrac{5m^2 + 10m}{2m - 10}$

56. $\dfrac{4n^2 + 32n}{3n + 2} \cdot \dfrac{3n^2 - n - 2}{n^2 + n - 30} \div \dfrac{108n^2 - 24n}{n + 6}$

57. $\dfrac{12p^2 + 3p}{p + 3} \div \dfrac{p^2 + 2p - 63}{p^2 - p - 12} \cdot \dfrac{p - 7}{9p^3 - 9p^2}$

58. $\dfrac{6q + 3}{9q^2 - 9q} \div \dfrac{q^2 + 14q + 33}{q^2 + 4q - 5} \cdot \dfrac{4q^2 + 12q}{12q + 6}$

Multiply and Divide Rational Functions

In the following exercises, find the domain of each function.

59. $R(x) = \dfrac{x^3 - 2x^2 - 25x + 50}{x^2 - 25}$

60. $R(x) = \dfrac{x^3 + 3x^2 - 4x - 12}{x^2 - 4}$

61. $R(x) = \dfrac{3x^2 + 15x}{6x^2 + 6x - 36}$

62. $R(x) = \dfrac{8x^2 - 32x}{2x^2 - 6x - 80}$

For the following exercises, find $R(x) = f(x) \cdot g(x)$ where $f(x)$ and $g(x)$ are given.

63. $f(x) = \dfrac{6x^2 - 12x}{x^2 + 7x - 18}$

$g(x) = \dfrac{x^2 - 81}{3x^2 - 27x}$

64. $f(x) = \dfrac{x^2 - 2x}{x^2 + 6x - 16}$

$g(x) = \dfrac{x^2 - 64}{x^2 - 8x}$

65. $f(x) = \dfrac{4x}{x^2 - 3x - 10}$

$g(x) = \dfrac{x^2 - 25}{8x^2}$

66. $f(x) = \dfrac{2x^2 + 8x}{x^2 - 9x + 20}$

$g(x) = \dfrac{x - 5}{x^2}$

For the following exercises, find $R(x) = \dfrac{f(x)}{g(x)}$ where $f(x)$ and $g(x)$ are given.

67. $f(x) = \dfrac{27x^2}{3x - 21}$

$g(x) = \dfrac{3x^2 + 18x}{x^2 + 13x + 42}$

68. $f(x) = \dfrac{24x^2}{2x - 8}$

$g(x) = \dfrac{4x^3 + 28x^2}{x^2 + 11x + 28}$

69. $f(x) = \dfrac{16x^2}{4x + 36}$

$g(x) = \dfrac{4x^2 - 24x}{x^2 + 4x - 45}$

70. $f(x) = \dfrac{24x^2}{2x - 4}$

$g(x) = \dfrac{12x^2 + 36x}{x^2 - 11x + 18}$

Writing Exercises

71. Explain how you find the values of x for which the rational expression $\dfrac{x^2 - x - 20}{x^2 - 4}$ is undefined.

72. Explain all the steps you take to simplify the rational expression $\dfrac{p^2 + 4p - 21}{9 - p^2}$.

73. ⓐ Multiply $\dfrac{7}{4} \cdot \dfrac{9}{10}$ and explain all your steps.

ⓑ Multiply $\dfrac{n}{n-3} \cdot \dfrac{9}{n+3}$ and explain all your steps.

ⓒ Evaluate your answer to part ⓑ when $n = 7$. Did you get the same answer you got in part ⓐ? Why or why not?

74. ⓐ Divide $\dfrac{24}{5} \div 6$ and explain all your steps.

ⓑ Divide $\dfrac{x^2-1}{x} \div (x+1)$ and explain all your steps.

ⓒ Evaluate your answer to part ⓑ when $x = 5$. Did you get the same answer you got in part ⓐ? Why or why not?

Self Check

ⓐ *After completing the exercises, use this checklist to evaluate your mastery of the objectives of this section.*

I can...	Confidently	With some help	No-I don't get it!
determine the values for which a rational expression is undefined.			
simplify rational expressions.			
multiply rational expressions.			
divide rational expressions.			
multiply and divide rational functions.			

ⓑ *If most of your checks were:*

...confidently. *Congratulations! You have achieved your goals in this section! Reflect on the study skills you used so that you can continue to use them. What did you do to become confident of your ability to do these things? Be specific!*

...with some help. *This must be addressed quickly as topics you do not master become potholes in your road to success. Math is sequential - every topic builds upon previous work. It is important to make sure you have a strong foundation before you move on. Who can you ask for help? Your fellow classmates and instructor are good resources. Is there a place on campus where math tutors are available? Can your study skills be improved?*

...no - I don't get it! *This is critical and you must not ignore it. You need to get help immediately or you will quickly be overwhelmed. See your instructor as soon as possible to discuss your situation. Together you can come up with a plan to get you the help you need.*

7.2 Add and Subtract Rational Expressions

Learning Objectives

By the end of this section, you will be able to:
- Add and subtract rational expressions with a common denominator
- Add and subtract rational expressions whose denominators are opposites
- Find the least common denominator of rational expressions
- Add and subtract rational expressions with unlike denominators
- Add and subtract rational functions

Be Prepared!

Before you get started, take this readiness quiz.

1. Add: $\frac{7}{10} + \frac{8}{15}$.

   If you missed this problem, review Example 1.29.

2. Subtract: $\frac{3x}{4} - \frac{8}{9}$.

   If you missed this problem, review Example 1.28.

3. Subtract: $6(2x + 1) - 4(x - 5)$.

   If you missed this problem, review Example 1.56.

Add and Subtract Rational Expressions with a Common Denominator

What is the first step you take when you add numerical fractions? You check if they have a common denominator. If they do, you add the numerators and place the sum over the common denominator. If they do not have a common denominator, you find one before you add.

It is the same with rational expressions. To add rational expressions, they must have a common denominator. When the denominators are the same, you add the numerators and place the sum over the common denominator.

Rational Expression Addition and Subtraction

If p, q, and r are polynomials where $r \neq 0$, then

$$\frac{p}{r} + \frac{q}{r} = \frac{p+q}{r} \quad \text{and} \quad \frac{p}{r} - \frac{q}{r} = \frac{p-q}{r}$$

To add or subtract rational expressions with a common denominator, add or subtract the numerators and place the result over the common denominator.

We always simplify rational expressions. Be sure to factor, if possible, after you subtract the numerators so you can identify any common factors.

Remember, too, we do not allow values that would make the denominator zero. What value of x should be excluded in the next example?

EXAMPLE 7.13

Add: $\frac{11x + 28}{x + 4} + \frac{x^2}{x + 4}$.

 Solution

Since the denominator is $x + 4$, we must exclude the value $x = -4$.

$$\frac{11x+28}{x+4}+\frac{x^2}{x+4}, \quad x \neq -4$$

The fractions have a common denominator, so add the numerators and place the sum over the common denominator.

$$\frac{11x+28+x^2}{x+4}$$

Write the degrees in descending order.

$$\frac{x^2+11x+28}{x+4}$$

Factor the numerator.

$$\frac{(x+4)(x+7)}{x+4}$$

Simplify by removing common factors.

$$\frac{\cancel{(x+4)}(x+7)}{\cancel{x+4}}$$

Simplify.

$$x+7$$

The expression simplifies to $x+7$ but the original expression had a denominator of $x+4$ so $x \neq -4$.

> **TRY IT :: 7.25**
>
> Simplify: $\dfrac{9x+14}{x+7}+\dfrac{x^2}{x+7}$.

> **TRY IT :: 7.26**
>
> Simplify: $\dfrac{x^2+8x}{x+5}+\dfrac{15}{x+5}$.

To subtract rational expressions, they must also have a common denominator. When the denominators are the same, you subtract the numerators and place the difference over the common denominator. Be careful of the signs when you subtract a binomial or trinomial.

EXAMPLE 7.14

Subtract: $\dfrac{5x^2-7x+3}{x^2-3x+18}-\dfrac{4x^2+x-9}{x^2-3x+18}$.

✓ Solution

$$\frac{5x^2 - 7x + 3}{x^2 - 3x + 18} - \frac{4x^2 + x - 9}{x^2 - 3x + 18}$$

Subtract the numerators and place the difference over the common denominator.	$\dfrac{5x^2 - 7x + 3 - \left(4x^2 + x - 9\right)}{x^2 - 3x + 18}$
Distribute the sign in the numerator.	$\dfrac{5x^2 - 7x + 3 - 4x^2 - x + 9}{x^2 - 3x - 18}$
Combine like terms.	$\dfrac{x^2 - 8x + 12}{x^2 - 3x - 18}$
Factor the numerator and the denominator.	$\dfrac{(x - 2)(x - 6)}{(x + 3)(x - 6)}$
Simplify by removing common factors.	$\dfrac{(x - 2)\cancel{(x - 6)}}{(x + 3)\cancel{(x - 6)}}$
	$\dfrac{(x - 2)}{(x + 3)}$

> **TRY IT :: 7.27**   Subtract: $\dfrac{4x^2 - 11x + 8}{x^2 - 3x + 2} - \dfrac{3x^2 + x - 3}{x^2 - 3x + 2}$.

> **TRY IT :: 7.28**   Subtract: $\dfrac{6x^2 - x + 20}{x^2 - 81} - \dfrac{5x^2 + 11x - 7}{x^2 - 81}$.

Add and Subtract Rational Expressions Whose Denominators are Opposites

When the denominators of two rational expressions are opposites, it is easy to get a common denominator. We just have to multiply one of the fractions by $\frac{-1}{-1}$.

Let's see how this works.

$$\frac{7}{d} + \frac{5}{-d}$$

Multiply the second fraction by $\frac{-1}{-1}$.	$\dfrac{7}{d} + \dfrac{(-1)5}{(-1)(-d)}$
The denominators are the same.	$\dfrac{7}{d} + \dfrac{-5}{d}$
Simplify.	$\dfrac{2}{d}$

Be careful with the signs as you work with the opposites when the fractions are being subtracted.

EXAMPLE 7.15

Subtract: $\dfrac{m^2 - 6m}{m^2 - 1} - \dfrac{3m + 2}{1 - m^2}$.

✓ **Solution**

$$\dfrac{m^2 - 6m}{m^2 - 1} - \dfrac{3m + 2}{1 - m^2}$$

The denominators are opposites, so multiply the second fraction by $\frac{-1}{-1}$.	$\dfrac{m^2 - 6m}{m^2 - 1} - \dfrac{-1(3m + 2)}{-1(1 - m^2)}$
Simplify the second fraction.	$\dfrac{m^2 - 6m}{m^2 - 1} - \dfrac{-3m - 2}{m^2 - 1}$
The denominators are the same. Subtract the numerators.	$\dfrac{m^2 - 6m - (-3m - 2)}{m^2 - 1}$
Distribute.	$\dfrac{m^2 - 6m + 3m + 2}{m^2 - 1}$
Combine like terms.	$\dfrac{m^2 - 3m + 2}{m^2 - 1}$
Factor the numerator and denominator.	$\dfrac{(m - 1)(m - 2)}{(m - 1)(m + 1)}$
Simplify by removing common factors.	$\dfrac{\cancel{(m - 1)}(m - 2)}{\cancel{(m - 1)}(m + 1)}$
Simplify.	$\dfrac{m - 2}{m + 1}$

> **TRY IT :: 7.29**   Subtract: $\dfrac{y^2 - 5y}{y^2 - 4} - \dfrac{6y - 6}{4 - y^2}$.

> **TRY IT :: 7.30**   Subtract: $\dfrac{2n^2 + 8n - 1}{n^2 - 1} - \dfrac{n^2 - 7n - 1}{1 - n^2}$.

Find the Least Common Denominator of Rational Expressions

When we add or subtract rational expressions with unlike denominators, we will need to get common denominators. If we review the procedure we used with numerical fractions, we will know what to do with rational expressions.

Let's look at this example: $\dfrac{7}{12} + \dfrac{5}{18}$. Since the denominators are not the same, the first step was to find the least common denominator (LCD).

To find the LCD of the fractions, we factored 12 and 18 into primes, lining up any common primes in columns. Then we "brought down" one prime from each column. Finally, we multiplied the factors to find the LCD.

When we add numerical fractions, once we found the LCD, we rewrote each fraction as an equivalent fraction with the LCD by multiplying the numerator and denominator by the same number. We are now ready to add.

$$\dfrac{7}{12} + \dfrac{5}{18} \qquad \begin{array}{l} 12 = 2 \cdot 2 \cdot 3 \\ 18 = 2 \cdot 3 \cdot 3 \\ \hline \text{LCD} = 2 \cdot 2 \cdot 3 \cdot 3 \\ \text{LCD} = 36 \end{array}$$

$$\dfrac{7 \cdot 3}{12 \cdot 3} + \dfrac{5 \cdot 2}{18 \cdot 2}$$

$$\dfrac{21}{36} + \dfrac{10}{36}$$

We do the same thing for rational expressions. However, we leave the LCD in factored form.

HOW TO :: FIND THE LEAST COMMON DENOMINATOR OF RATIONAL EXPRESSIONS.

Step 1.   Factor each denominator completely.

Step 2.   List the factors of each denominator. Match factors vertically when possible.

Step 3.   Bring down the columns by including all factors, but do not include common factors twice.

Step 4.   Write the LCD as the product of the factors.

Remember, we always exclude values that would make the denominator zero. What values of x should we exclude in this next example?

EXAMPLE 7.16

ⓐ Find the LCD for the expressions $\dfrac{8}{x^2 - 2x - 3}, \dfrac{3x}{x^2 + 4x + 3}$ and ⓑ rewrite them as equivalent rational expressions with the lowest common denominator.

 Solution

ⓐ

Find the LCD for $\dfrac{8}{x^2 - 2x - 3}, \dfrac{3x}{x^2 + 4x + 3}$.	
Factor each denominator completely, lining up common factors.	$x^2 - 2x - 3 = (x + 1)(x - 3)$
Bring down the columns.	$\dfrac{x^2 + 4x + 3 = (x + 1)(x + 3)}{\text{LCD} \qquad = (x + 1)(x - 3)(x + 3)}$
Write the LCD as the product of the factors.	The LCD is $(x + 1)(x - 3)(x + 3)$.

ⓑ

	$\dfrac{8}{x^2 - 2x - 3}, \dfrac{3x}{x^2 + 4x + 3}$
Factor each denominator.	$\dfrac{8}{(x + 1)(x - 3)}, \dfrac{3x}{(x + 1)(x + 3)}$
Multiply each denominator by the 'missing' LCD factor and multiply each numerator by the same factor.	$\dfrac{8(x + 3)}{(x + 1)(x - 3)(x + 3)}, \dfrac{3x(x - 3)}{(x + 1)(x + 3)(x - 3)}$
Simplify the numerators.	$\dfrac{8x + 24}{(x + 1)(x - 3)(x + 3)}, \dfrac{3x^2 - 9x}{(x + 1)(x + 3)(x - 3)}$

> **TRY IT :: 7.31**
>
> ⓐ Find the LCD for the expressions $\dfrac{2}{x^2 - x - 12}, \dfrac{1}{x^2 - 16}$ ⓑ rewrite them as equivalent rational expressions with the lowest common denominator.

> **TRY IT :: 7.32**
>
> ⓐ Find the LCD for the expressions $\dfrac{3x}{x^2 - 3x + 10}, \dfrac{5}{x^2 + 3x + 2}$ ⓑ rewrite them as equivalent rational expressions with the lowest common denominator.

Add and Subtract Rational Expressions with Unlike Denominators

Now we have all the steps we need to add or subtract rational expressions with unlike denominators.

EXAMPLE 7.17   HOW TO ADD RATIONAL EXPRESSIONS WITH UNLIKE DENOMINATORS

Add: $\dfrac{3}{x-3} + \dfrac{2}{x-2}$.

✓ **Solution**

Step 1. Determine if the expressions have a common denominator. • Yes–go to step 2. • No–Rewrite each rational expression with the LCD. • Find the LCD. • Rewrite each rational expression as an equivalent rational expression with the LCD.	No. Find the LCD of $(x-3)$ and $(x-2)$. Change into equivalent rational expressions with the LCD, $(x-3)$ and $(x-2)$. Keep the denominators factored!	$x-3 : (x-3)\bigcirc$ $\dfrac{x-2 \quad\bigcirc \quad : \quad (x-2)}{\text{LCD} : (x-3)(x-2)}$ $\dfrac{3}{x-3} + \dfrac{2}{x-2}$ $\dfrac{3\,(x-2)}{(x-3)(x-2)} + \dfrac{2\,(x-3)}{(x-2)(x-3)}$ $\dfrac{3x-6}{(x-3)(x-2)} + \dfrac{2x-6}{(x-2)(x-3)}$
Step 2. Add or subtract the rational expressions.	Add the numerators and place the sum over the common denominator.	$\dfrac{3x-6+2x-6}{(x-3)(x-2)}$
Step 3. Simplify, if possible.	Because $5x-12$ cannot be factored, the answer is simplified.	$\dfrac{5x-12}{(x-3)(x-2)}$

> **TRY IT :: 7.33**   Add: $\dfrac{2}{x-2} + \dfrac{5}{x+3}$.

> **TRY IT :: 7.34**   Add: $\dfrac{4}{m+3} + \dfrac{3}{m+4}$.

The steps used to add rational expressions are summarized here.

HOW TO :: ADD OR SUBTRACT RATIONAL EXPRESSIONS.

Step 1.   Determine if the expressions have a common denominator.
- **Yes** – go to step 2.
- **No** – Rewrite each rational expression with the LCD.
  - Find the LCD.
  - Rewrite each rational expression as an equivalent rational expression with the LCD.

Step 2.   Add or subtract the rational expressions.

Step 3.   Simplify, if possible.

Avoid the temptation to simplify too soon. In the example above, we must leave the first rational expression as $\dfrac{3x-6}{(x-3)(x-2)}$ to be able to add it to $\dfrac{2x-6}{(x-2)(x-3)}$. Simplify *only* after you have combined the numerators.

EXAMPLE 7.18

Add: $\dfrac{8}{x^2 - 2x - 3} + \dfrac{3x}{x^2 + 4x + 3}$.

✓ Solution

	$\dfrac{8}{x^2 - 2x - 3} + \dfrac{3x}{x^2 + 4x + 3}$
Do the expressions have a common denominator?	No.
Rewrite each expression with the LCD.	

$$x^2 - 2x - 3 = (x + 1)(x - 3)$$

Find the LCD. $\quad \underline{x^2 + 4x + 3 = (x + 1) \quad (x + 3)}$

$$\text{LCD} = (x + 1)(x - 3)(x + 3)$$

Rewrite each rational expression as an equivalent rational expression with the LCD.	$\dfrac{8(x + 3)}{(x + 1)(x - 3)(x + 3)} + \dfrac{3x(x - 3)}{(x + 1)(x + 3)(x - 3)}$
Simplify the numerators.	$\dfrac{8x + 24}{(x + 1)(x - 3)(x + 3)} + \dfrac{3x^2 - 9x}{(x + 1)(x + 3)(x - 3)}$
Add the rational expressions.	$\dfrac{8x + 24 + 3x^2 - 9x}{(x + 1)(x - 3)(x + 3)}$
Simplify the numerator.	$\dfrac{3x^2 - x + 24}{(x + 1)(x - 3)(x + 3)}$
	The numerator is prime, so there are no common factors.

> **TRY IT :: 7.35**   Add: $\dfrac{1}{m^2 - m - 2} + \dfrac{5m}{m^2 + 3m + 2}$.

> **TRY IT :: 7.36**   Add: $\dfrac{2n}{n^2 - 3n - 10} + \dfrac{6}{n^2 + 5n + 6}$.

The process we use to subtract rational expressions with different denominators is the same as for addition. We just have to be very careful of the signs when subtracting the numerators.

EXAMPLE 7.19

Subtract: $\dfrac{8y}{y^2 - 16} - \dfrac{4}{y - 4}$.

✓ **Solution**

$$\frac{8y}{y^2-16}-\frac{4}{y-4}$$

Do the expressions have a common denominator?	No.
Rewrite each expression with the LCD.	

Find the LCD.
$$y^2-16=(y-4)(y+4)$$
$$y-4=y-4$$
$$\text{LCD}=(y-4)(y+4)$$

Rewrite each rational expression as an equivalent rational expression with the LCD.	$\dfrac{8y}{(y-4)(y+4)}-\dfrac{4(y+4)}{(y-4)(y+4)}$
Simplify the numerators.	$\dfrac{8y}{(y-4)(y+4)}-\dfrac{4y+16}{(y-4)(y+4)}$
Subtract the rational expressions.	$\dfrac{8y-4y-16}{(y-4)(y+4)}$
Simplify the numerator.	$\dfrac{4y-16}{(y-4)(y+4)}$
Factor the numerator to look for common factors.	$\dfrac{4(y-4)}{(y-4)(y+4)}$
Remove common factors	$\dfrac{4\cancel{(y-4)}}{\cancel{(y-4)}(y+4)}$
Simplify.	$\dfrac{4}{(y+4)}$

> **TRY IT :: 7.37**   Subtract: $\dfrac{2x}{x^2-4}-\dfrac{1}{x+2}$.

> **TRY IT :: 7.38**   Subtract: $\dfrac{3}{z+3}-\dfrac{6z}{z^2-9}$.

There are lots of negative signs in the next example. Be extra careful.

EXAMPLE 7.20

Subtract: $\dfrac{-3n-9}{n^2+n-6}-\dfrac{n+3}{2-n}$.

⊘ **Solution**

$$\frac{-3n-9}{n^2+n-6} - \frac{n+3}{2-n}$$

Factor the denominator.	$$\frac{-3n-9}{(n-2)(n+3)} - \frac{n+3}{2-n}$$
Since $n-2$ and $2-n$ are opposites, we will multiply the second rational expression by $\frac{-1}{-1}$.	$$\frac{-3n-9}{(n-2)(n+3)} - \frac{(-1)(n+3)}{(-1)(2-n)}$$
Write $(-1)(2-n)$ as $n-2$.	$$\frac{-3n-9}{(n-2)(n+3)} - \frac{(-1)(n+3)}{(n-2)}$$
Simplify. Remember, $a-(-b)=a+b$.	$$\frac{-3n-9}{(n-2)(n+3)} + \frac{(n+3)}{(n-2)}$$
Do the rational expressions have a common denominator? No.	

Find the LCD.

$$\begin{aligned} n^2+n-6 &= (n-2)(n+3) \\ n-2 &= (n-2) \\ \hline \text{LCD} &= (n-2)(n+3) \end{aligned}$$

Rewrite each rational expression as an equivalent rational expression with the LCD.	$$\frac{-3n-9}{(n-2)(n+3)} + \frac{(n+3)(n+3)}{(n-2)(n+3)}$$
Simplify the numerators.	$$\frac{-3n-9}{(n-2)(n+3)} + \frac{n^2+6n+9}{(n-2)(n+3)}$$
Add the rational expressions.	$$\frac{-3n-9+n^2+6n+9}{(n-2)(n+3)}$$
Simplify the numerator.	$$\frac{n^2+3n}{(n-2)(n+3)}$$
Factor the numerator to look for common factors.	$$\frac{n\cancel{(n+3)}}{(n-2)\cancel{(n+3)}}$$
Simplify.	$$\frac{n}{(n-2)}$$

> **TRY IT :: 7.39** Subtract : $\dfrac{3x-1}{x^2-5x-6} - \dfrac{2}{6-x}$.

> **TRY IT :: 7.40** Subtract: $\dfrac{-2y-2}{y^2+2y-8} - \dfrac{y-1}{2-y}$.

Things can get very messy when both fractions must be multiplied by a binomial to get the common denominator.

EXAMPLE 7.21

Subtract: $\dfrac{4}{a^2+6a+5} - \dfrac{3}{a^2+7a+10}$.

⊘ Solution

$$\frac{4}{a^2 + 6a + 5} - \frac{3}{a^2 + 7a + 10}$$

Factor the denominators.

$$\frac{4}{(a+1)(a+5)} - \frac{3}{(a+2)(a+5)}$$

Do the rational expressions have a
common denominator? No.

Find the LCD.

$$\begin{aligned} a^2 + 6a + 5 &= (a+1)(a+5) \\ a^2 + 7a + 10 &= \qquad\ (a+5)(a+2) \\ \text{LCD} &= (a+1)(a+5)(a+2) \end{aligned}$$

Rewrite each rational expression as an
equivalent rational expression with the LCD.

$$\frac{4(a+2)}{(a+1)(a+5)(a+2)} - \frac{3(a+1)}{(a+2)(a+5)(a+1)}$$

Simplify the numerators.

$$\frac{4a+8}{(a+1)(a+5)(a+2)} - \frac{3a+3}{(a+2)(a+5)(a+1)}$$

Subtract the rational expressions.

$$\frac{4a+8-(3a+3)}{(a+1)(a+5)(a+2)}$$

Simplify the numerator.

$$\frac{4a+8-3a+3}{(a+1)(a+5)(a+2)}$$

$$\frac{a+5}{(a+1)(a+5)(a+2)}$$

Look for common factors.

$$\frac{\cancel{(a+5)}}{(a+1)\cancel{(a+5)}(a+2)}$$

Simplify.

$$\frac{1}{(a+1)(a+2)}$$

> **TRY IT : : 7.41**   Subtract: $\dfrac{3}{b^2 - 4b - 5} - \dfrac{2}{b^2 - 6b + 5}$.

> **TRY IT : : 7.42**   Subtract: $\dfrac{4}{x^2 - 4} - \dfrac{3}{x^2 - x - 2}$.

We follow the same steps as before to find the LCD when we have more than two rational expressions. In the next example, we will start by factoring all three denominators to find their LCD.

EXAMPLE 7.22

Simplify: $\dfrac{2u}{u-1} + \dfrac{1}{u} - \dfrac{2u-1}{u^2 - u}$.

⊘ Solution

$$\frac{2u}{u-1} + \frac{1}{u} - \frac{2u-1}{u^2 - u}$$

Do the expressions have a common denominator? No.
Rewrite each expression with the LCD.

<table>
<tr><td>Find the LCD.</td><td>

$$\begin{aligned} u - 1 &= (u - 1) \\ u &= u \\ \underline{u^2 - u} &= \underline{u(u - 1)} \\ \text{LCD} &= u(u - 1) \end{aligned}$$

</td></tr>
</table>

Rewrite each rational expression as an equivalent rational expression with the LCD.	$\dfrac{2u \cdot u}{(u-1)u} + \dfrac{1 \cdot (u-1)}{u \cdot (u-1)} - \dfrac{2u-1}{u(u-1)}$
	$\dfrac{2u^2}{(u-1)u} + \dfrac{u-1}{u \cdot (u-1)} - \dfrac{2u-1}{u(u-1)}$
Write as one rational expression.	$\dfrac{2u^2 + u - 1 - 2u + 1}{u(u-1)}$
Simplify.	$\dfrac{2u^2 - u}{u(u-1)}$
Factor the numerator, and remove common factors.	$\dfrac{\cancel{u}(2u-1)}{\cancel{u}(u-1)}$
Simplify.	$\dfrac{2u-1}{u-1}$

> **TRY IT :: 7.43**   Simplify: $\dfrac{v}{v+1} + \dfrac{3}{v-1} - \dfrac{6}{v^2-1}$.

> **TRY IT :: 7.44**   Simplify: $\dfrac{3w}{w+2} + \dfrac{2}{w+7} - \dfrac{17w+4}{w^2+9w+14}$.

Add and subtract rational functions

To add or subtract rational functions, we use the same techniques we used to add or subtract polynomial functions.

EXAMPLE 7.23

Find $R(x) = f(x) - g(x)$ where $f(x) = \dfrac{x+5}{x-2}$ and $g(x) = \dfrac{5x+18}{x^2-4}$.

✓ **Solution**

	$R(x) = f(x) - g(x)$
Substitute in the functions $f(x),\ g(x)$.	$R(x) = \dfrac{x+5}{x-2} - \dfrac{5x+18}{x^2-4}$
Factor the denominators.	$R(x) = \dfrac{x+5}{x-2} - \dfrac{5x+18}{(x-2)(x+2)}$
Do the expressions have a common denominator? No. Rewrite each expression with the LCD.	
Find the LCD.	$\begin{aligned} x - 2 &= (x-2) \\ \underline{x^2 - 4} &= \underline{(x-2)(x+2)} \\ \text{LCD} &= (x-2)(x+2) \end{aligned}$
Rewrite each rational expression as an equivalent rational expression with the LCD.	$R(x) = \dfrac{(x+5)(x+2)}{(x-2)(x+2)} - \dfrac{5x+18}{(x-2)(x+2)}$

Write as one rational expression.	$R(x) = \dfrac{(x+5)(x+2) - (5x+18)}{(x-2)(x+2)}$
Simplify.	$R(x) = \dfrac{x^2 + 7x + 10 - 5x - 18}{(x-2)(x+2)}$
	$R(x) = \dfrac{x^2 + 2x - 8}{(x-2)(x+2)}$
Factor the numerator, and remove common factors.	$R(x) = \dfrac{(x+4)\cancel{(x-2)}}{\cancel{(x-2)}(x+2)}$
Simplify.	$R(x) = \dfrac{(x+4)}{(x+2)}$

> **TRY IT : : 7.45**   Find $R(x) = f(x) - g(x)$ where $f(x) = \dfrac{x+1}{x+3}$ and $g(x) = \dfrac{x+17}{x^2 - x - 12}$.

> **TRY IT : : 7.46**   Find $R(x) = f(x) + g(x)$ where $f(x) = \dfrac{x-4}{x+3}$ and $g(x) = \dfrac{4x+6}{x^2 - 9}$.

▶ **MEDIA : :**

Access this online resource for additional instruction and practice with adding and subtracting rational expressions.

- **Add and Subtract Rational Expressions- Unlike Denominators (https://openstax.org/l/ 37AddSubRatExp)**

7.2 EXERCISES

Practice Makes Perfect

Add and Subtract Rational Expressions with a Common Denominator

In the following exercises, add.

75. $\dfrac{2}{15} + \dfrac{7}{15}$

76. $\dfrac{7}{24} + \dfrac{11}{24}$

77. $\dfrac{3c}{4c-5} + \dfrac{5}{4c-5}$

78. $\dfrac{7m}{2m+n} + \dfrac{4}{2m+n}$

79. $\dfrac{2r^2}{2r-1} + \dfrac{15r-8}{2r-1}$

80. $\dfrac{3s^2}{3s-2} + \dfrac{13s-10}{3s-2}$

81. $\dfrac{2w^2}{w^2-16} + \dfrac{8w}{w^2-16}$

82. $\dfrac{7x^2}{x^2-9} + \dfrac{21x}{x^2-9}$

In the following exercises, subtract.

83. $\dfrac{9a^2}{3a-7} - \dfrac{49}{3a-7}$

84. $\dfrac{25b^2}{5b-6} - \dfrac{36}{5b-6}$

85. $\dfrac{3m^2}{6m-30} - \dfrac{21m-30}{6m-30}$

86. $\dfrac{2n^2}{4n-32} - \dfrac{18n-16}{4n-32}$

87.
$\dfrac{6p^2+3p+4}{p^2+4p-5} - \dfrac{5p^2+p+7}{p^2+4p-5}$

88.
$\dfrac{5q^2+3q-9}{q^2+6q+8} - \dfrac{4q^2+9q+7}{q^2+6q+8}$

89.
$\dfrac{5r^2+7r-33}{r^2-49} - \dfrac{4r^2+5r+30}{r^2-49}$

90. $\dfrac{7t^2-t-4}{t^2-25} - \dfrac{6t^2+12t-44}{t^2-25}$

Add and Subtract Rational Expressions whose Denominators are Opposites

In the following exercises, add or subtract.

91. $\dfrac{10v}{2v-1} + \dfrac{2v+4}{1-2v}$

92. $\dfrac{20w}{5w-2} + \dfrac{5w+6}{2-5w}$

93.
$\dfrac{10x^2+16x-7}{8x-3} + \dfrac{2x^2+3x-1}{3-8x}$

94.
$\dfrac{6y^2+2y-11}{3y-7} + \dfrac{3y^2-3y+17}{7-3y}$

95. $\dfrac{z^2+6z}{z^2-25} - \dfrac{3z+20}{25-z^2}$

96. $\dfrac{a^2+3a}{a^2-9} - \dfrac{3a-27}{9-a^2}$

97.
$\dfrac{2b^2+30b-13}{b^2-49} - \dfrac{2b^2-5b-8}{49-b^2}$

98.
$\dfrac{c^2+5c-10}{c^2-16} - \dfrac{c^2-8c-10}{16-c^2}$

Find the Least Common Denominator of Rational Expressions

In the following exercises, ⓐ find the LCD for the given rational expressions ⓑ rewrite them as equivalent rational expressions with the lowest common denominator.

99. $\dfrac{5}{x^2-2x-8}, \dfrac{2x}{x^2-x-12}$

100. $\dfrac{8}{y^2+12y+35}, \dfrac{3y}{y^2+y-42}$

101. $\dfrac{9}{z^2+2z-8}, \dfrac{4z}{z^2-4}$

102. $\dfrac{6}{a^2 + 14a + 45}, \dfrac{5a}{a^2 - 81}$ **103.** $\dfrac{4}{b^2 + 6b + 9}, \dfrac{2b}{b^2 - 2b - 15}$ **104.** $\dfrac{5}{c^2 - 4c + 4}, \dfrac{3c}{c^2 - 7c + 10}$

105.

$\dfrac{2}{3d^2 + 14d - 5}, \dfrac{5d}{3d^2 - 19d + 6}$

106.

$\dfrac{3}{5m^2 - 3m - 2}, \dfrac{6m}{5m^2 + 17m + 6}$

Add and Subtract Rational Expressions with Unlike Denominators

In the following exercises, perform the indicated operations.

107. $\dfrac{7}{10x^2 y} + \dfrac{4}{15xy^2}$ **108.** $\dfrac{1}{12a^3 b^2} + \dfrac{5}{9a^2 b^3}$ **109.** $\dfrac{3}{r + 4} + \dfrac{2}{r - 5}$

110. $\dfrac{4}{s - 7} + \dfrac{5}{s + 3}$ **111.** $\dfrac{5}{3w - 2} + \dfrac{2}{w + 1}$ **112.** $\dfrac{4}{2x + 5} + \dfrac{2}{x - 1}$

113. $\dfrac{2y}{y + 3} + \dfrac{3}{y - 1}$ **114.** $\dfrac{3z}{z - 2} + \dfrac{1}{z + 5}$ **115.** $\dfrac{5b}{a^2 b - 2a^2} + \dfrac{2b}{b^2 - 4}$

116. $\dfrac{4}{cd + 3c} + \dfrac{1}{d^2 - 9}$ **117.** $\dfrac{-3m}{3m - 3} + \dfrac{5m}{m^2 + 3m - 4}$ **118.** $\dfrac{8}{4n + 4} + \dfrac{6}{n^2 - n - 2}$

119. $\dfrac{3r}{r^2 + 7r + 6} + \dfrac{9}{r^2 + 4r + 3}$ **120.** $\dfrac{2s}{s^2 + 2s - 8} + \dfrac{4}{s^2 + 3s - 10}$ **121.** $\dfrac{t}{t - 6} - \dfrac{t - 2}{t + 6}$

122. $\dfrac{x - 3}{x + 6} - \dfrac{x}{x + 3}$ **123.** $\dfrac{5a}{a + 3} - \dfrac{a + 2}{a + 6}$ **124.** $\dfrac{3b}{b - 2} - \dfrac{b - 6}{b - 8}$

125. $\dfrac{6}{m + 6} - \dfrac{12m}{m^2 - 36}$ **126.** $\dfrac{4}{n + 4} - \dfrac{8n}{n^2 - 16}$ **127.** $\dfrac{-9p - 17}{p^2 - 4p - 21} - \dfrac{p + 1}{7 - p}$

128. $\dfrac{7q + 8}{q^2 - 2q - 24} - \dfrac{q + 2}{4 - q}$ **129.** $\dfrac{-2r - 16}{r^2 + 6r - 16} - \dfrac{5}{2 - r}$ **130.** $\dfrac{2t - 30}{t^2 + 6t - 27} - \dfrac{2}{3 - t}$

131. $\dfrac{2x + 7}{10x - 1} + 3$ **132.** $\dfrac{8y - 4}{5y + 2} - 6$ **133.** $\dfrac{3}{x^2 - 3x - 4} - \dfrac{2}{x^2 - 5x + 4}$

134.

$\dfrac{4}{x^2 - 6x + 5} - \dfrac{3}{x^2 - 7x + 10}$ **135.** $\dfrac{5}{x^2 + 8x - 9} - \dfrac{4}{x^2 + 10x + 9}$ **136.**

$\dfrac{3}{2x^2 + 5x + 2} - \dfrac{1}{2x^2 + 3x + 1}$

137. $\dfrac{5a}{a - 2} + \dfrac{9}{a} - \dfrac{2a + 18}{a^2 - 2a}$ **138.** $\dfrac{2b}{b - 5} + \dfrac{3}{2b} - \dfrac{2b - 15}{2b^2 - 10b}$ **139.** $\dfrac{c}{c + 2} + \dfrac{5}{c - 2} - \dfrac{11c}{c^2 - 4}$

140.

$\dfrac{6d}{d - 5} + \dfrac{1}{d + 4} - \dfrac{7d - 5}{d^2 - d - 20}$ **141.** $\dfrac{3d}{d + 2} + \dfrac{4}{d} - \dfrac{d + 8}{d^2 + 2d}$ **142.**

$\dfrac{2q}{q + 5} + \dfrac{3}{q - 3} - \dfrac{13q + 15}{q^2 + 2q - 15}$

Add and Subtract Rational Functions

In the following exercises, find ⓐ $R(x) = f(x) + g(x)$ ⓑ $R(x) = f(x) - g(x)$.

143. $f(x) = \dfrac{-5x - 5}{x^2 + x - 6}$ and

$g(x) = \dfrac{x + 1}{2 - x}$

144. $f(x) = \dfrac{-4x - 24}{x^2 + x - 30}$ and

$g(x) = \dfrac{x + 7}{5 - x}$

145. $f(x) = \dfrac{6x}{x^2 - 64}$ and

$g(x) = \dfrac{3}{x - 8}$

146. $f(x) = \dfrac{5}{x + 7}$ and

$g(x) = \dfrac{10x}{x^2 - 49}$

Writing Exercises

147. Donald thinks that $\dfrac{3}{x} + \dfrac{4}{x}$ is $\dfrac{7}{2x}$. Is Donald correct? Explain.

148. Explain how you find the Least Common Denominator of $x^2 + 5x + 4$ and $x^2 - 16$.

149. Felipe thinks $\dfrac{1}{x} + \dfrac{1}{y}$ is $\dfrac{2}{x + y}$.

ⓐ Choose numerical values for *x* and *y* and evaluate $\dfrac{1}{x} + \dfrac{1}{y}$.

ⓑ Evaluate $\dfrac{2}{x + y}$ for the same values of *x* and *y* you used in part ⓐ.

ⓒ Explain why Felipe is wrong.

ⓓ Find the correct expression for $\dfrac{1}{x} + \dfrac{1}{y}$.

150. Simplify the expression $\dfrac{4}{n^2 + 6n + 9} - \dfrac{1}{n^2 - 9}$ and explain all your steps.

Self Check

ⓐ *After completing the exercises, use this checklist to evaluate your mastery of the objectives of this section.*

I can...	Confidently	With some help	No-I don't get it!
add and subtract rational expressions with a common denominator.			
add and subtract rational expressions whose denominators are opposites.			
find the least common denominator of rational expressions.			
add and subtract rational expressions with unlike denominators.			
add or subtract rational functions.			

ⓑ *After reviewing this checklist, what will you do to become confident for all objectives?*

7.3   Simplify Complex Rational Expressions

Learning Objectives

By the end of this section, you will be able to:
› Simplify a complex rational expression by writing it as division
› Simplify a complex rational expression by using the LCD

Be Prepared!

Before you get started, take this readiness quiz.

1.   Simplify: $\dfrac{\frac{3}{5}}{\frac{9}{10}}$.

     If you missed this problem, review **Example 1.27**.

2.   Simplify: $\dfrac{1 - \frac{1}{3}}{4^2 + 4 \cdot 5}$.

     If you missed this problem, review **Example 1.31**.

3.   Solve: $\dfrac{1}{2x} + \dfrac{1}{4} = \dfrac{1}{8}$.

     If you missed this problem, review **Example 2.9**.

Simplify a Complex Rational Expression by Writing it as Division

Complex fractions are fractions in which the numerator or denominator contains a fraction. We previously simplified complex fractions like these:

$$\frac{\frac{3}{4}}{\frac{5}{8}} \qquad \frac{\frac{x}{2}}{\frac{xy}{6}}$$

In this section, we will simplify complex rational expressions, which are rational expressions with rational expressions in the numerator or denominator.

Complex Rational Expression

A **complex rational expression** is a rational expression in which the numerator and/or the denominator contains a rational expression.

Here are a few complex rational expressions:

$$\frac{\frac{4}{y-3}}{\frac{8}{y^2-9}} \qquad \frac{\frac{1}{x}+\frac{1}{y}}{\frac{x}{y}-\frac{y}{x}} \qquad \frac{\frac{2}{x+6}}{\frac{4}{x-6}-\frac{4}{x^2-36}}$$

Remember, we always exclude values that would make any denominator zero.

We will use two methods to simplify complex rational expressions.

We have already seen this complex rational expression earlier in this chapter.

$$\frac{\frac{6x^2-7x+2}{4x-8}}{\frac{2x^2-8x+3}{x^2-5x+6}}$$

We noted that fraction bars tell us to divide, so rewrote it as the division problem:

$$\left(\frac{6x^2-7x+2}{4x-8}\right) \div \left(\frac{2x^2-8x+3}{x^2-5x+6}\right).$$

Then, we multiplied the first rational expression by the reciprocal of the second, just like we do when we divide two fractions.

This is one method to simplify complex rational expressions. We make sure the complex rational expression is of the form where one fraction is over one fraction. We then write it as if we were dividing two fractions.

EXAMPLE 7.24

Simplify the complex rational expression by writing it as division: $\dfrac{\frac{6}{x-4}}{\frac{3}{x^2-16}}$.

✓ **Solution**

$$\dfrac{\frac{6}{x-4}}{\frac{3}{x^2-16}}$$

Rewrite the complex fraction as division.

$$\frac{6}{x-4} \div \frac{3}{x^2-16}$$

Rewrite as the product of first times the reciprocal of the second.

$$\frac{6}{x-4} \cdot \frac{x^2-16}{3}$$

Factor.

$$\frac{3 \cdot 2}{x-4} \cdot \frac{(x-4)(x+4)}{3}$$

Multiply.

$$\frac{3 \cdot 2(x-4)(x+4)}{3(x-4)}$$

Remove common factors.

$$\frac{\cancel{3} \cdot 2\cancel{(x-4)}(x+4)}{\cancel{3}\cancel{(x-4)}}$$

Simplify.

$$2(x+4)$$

Are there any value(s) of x that should not be allowed? The original complex rational expression had denominators of $x-4$ and x^2-16. This expression would be undefined if $x=4$ or $x=-4$.

> **TRY IT :: 7.47**

Simplify the complex rational expression by writing it as division: $\dfrac{\frac{2}{x^2-1}}{\frac{3}{x+1}}$.

> **TRY IT :: 7.48**

Simplify the complex rational expression by writing it as division: $\dfrac{\frac{1}{x^2-7x+12}}{\frac{2}{x-4}}$.

Fraction bars act as grouping symbols. So to follow the Order of Operations, we simplify the numerator and denominator as much as possible before we can do the division.

EXAMPLE 7.25

Simplify the complex rational expression by writing it as division: $\dfrac{\frac{1}{3}+\frac{1}{6}}{\frac{1}{2}-\frac{1}{3}}$.

✓ **Solution**

$$\dfrac{\frac{1}{3}+\frac{1}{6}}{\frac{1}{2}-\frac{1}{3}}$$

Simplify the numerator and denominator. Find the LCD and add the fractions in the numerator. Find the LCD and subtract the fractions in the denominator.	$\dfrac{\frac{1\cdot2}{3\cdot2}+\frac{1}{6}}{\frac{1\cdot3}{2\cdot3}-\frac{1\cdot2}{3\cdot2}}$
Simplify the numerator and denominator.	$\dfrac{\frac{2}{6}+\frac{1}{6}}{\frac{3}{6}-\frac{2}{6}}$
Rewrite the complex rational expression as a division problem.	$\dfrac{3}{6}\div\dfrac{1}{6}$
Multiply the first by the reciprocal of the second.	$\dfrac{3}{6}\cdot\dfrac{6}{1}$
Simplify.	3

> **TRY IT : : 7.49**

Simplify the complex rational expression by writing it as division: $\dfrac{\frac{1}{2}+\frac{2}{3}}{\frac{5}{6}+\frac{1}{12}}$.

> **TRY IT : : 7.50**

Simplify the complex rational expression by writing it as division: $\dfrac{\frac{3}{4}-\frac{1}{3}}{\frac{1}{8}+\frac{5}{6}}$.

We follow the same procedure when the complex rational expression contains variables.

EXAMPLE 7.26   HOW TO SIMPLIFY A COMPLEX RATIONAL EXPRESSION USING DIVISION

Simplify the complex rational expression by writing it as division: $\dfrac{\frac{1}{x}+\frac{1}{y}}{\frac{x}{y}-\frac{y}{x}}$.

✓ Solution

Step 1. Simplify the numerator and denominator.	We will simplify the sum in the numerator and difference in the denominator.	$\dfrac{\dfrac{1}{x} + \dfrac{1}{y}}{\dfrac{x}{y} - \dfrac{y}{x}}$
	Find a common denominator and add the fractions in the numerator.	$\dfrac{\dfrac{1 \cdot y}{x \cdot y} + \dfrac{1 \cdot x}{y \cdot x}}{\dfrac{x \cdot x}{y \cdot x} - \dfrac{y \cdot y}{x \cdot y}}$
		$\dfrac{\dfrac{y}{xy} + \dfrac{x}{xy}}{\dfrac{x^2}{xy} - \dfrac{y^2}{xy}}$
	Find a common denominator and subtract the fractions in the denominator.	$\dfrac{\dfrac{y + x}{xy}}{\dfrac{x^2 - y^2}{xy}}$
	We now have just one rational expression in the numerator and one in the denominator.	
Step 2. Rewrite the complex rational expression as a division problem.	We write the numerator divided by the denominator.	$\left(\dfrac{y + x}{xy}\right) \div \left(\dfrac{x^2 - y^2}{xy}\right)$
Step 3. Divide the expressions.	Multiply the first by the reciprocal of the second.	$\left(\dfrac{y + x}{xy}\right) \cdot \left(\dfrac{xy}{x^2 - y^2}\right)$
	Factor any expressions if possible.	$\dfrac{xy(y + x)}{xy(x - y)(x + y)}$
	Remove common factors.	$\dfrac{\cancel{xy}\cancel{(y + x)}}{\cancel{xy}(x - y)\cancel{(x + y)}}$
	Simplify.	$\dfrac{1}{x - y}$

> **TRY IT :: 7.51**
>
> Simplify the complex rational expression by writing it as division: $\dfrac{\dfrac{1}{x} + \dfrac{1}{y}}{\dfrac{1}{x} - \dfrac{1}{y}}$.

> **TRY IT :: 7.52**
>
> Simplify the complex rational expression by writing it as division: $\dfrac{\dfrac{1}{a} + \dfrac{1}{b}}{\dfrac{1}{a^2} - \dfrac{1}{b^2}}$.

We summarize the steps here.

HOW TO :: SIMPLIFY A COMPLEX RATIONAL EXPRESSION BY WRITING IT AS DIVISION.

Step 1.   Simplify the numerator and denominator.

Step 2.   Rewrite the complex rational expression as a division problem.

Step 3.   Divide the expressions.

EXAMPLE 7.27

Simplify the complex rational expression by writing it as division: $\dfrac{n - \dfrac{4n}{n+5}}{\dfrac{1}{n+5} + \dfrac{1}{n-5}}$.

✓ Solution

$$\dfrac{n - \dfrac{4n}{n+5}}{\dfrac{1}{n+5} + \dfrac{1}{n-5}}$$

Simplify the numerator and denominator. Find common denominators for the numerator and denominator.	$\dfrac{\dfrac{n(n+5)}{1(n+5)} - \dfrac{4n}{n+5}}{\dfrac{1(n-5)}{(n+5)(n-5)} + \dfrac{1(n+5)}{(n-5)(n+5)}}$
Simplify the numerators.	$\dfrac{\dfrac{n^2+5n}{n+5} - \dfrac{4n}{n+5}}{\dfrac{n-5}{(n+5)(n-5)} + \dfrac{n+5}{(n-5)(n+5)}}$
Subtract the rational expressions in the numerator and add in the denominator.	$\dfrac{\dfrac{n^2+5n-4n}{n+5}}{\dfrac{n-5+n+5}{(n+5)(n-5)}}$
Simplify. (We now have one rational expression over one rational expression.)	$\dfrac{\dfrac{n^2+n}{n+5}}{\dfrac{2n}{(n+5)(n-5)}}$
Rewrite as fraction division.	$\dfrac{n^2+n}{n+5} \div \dfrac{2n}{(n+5)(n-5)}$
Multiply the first times the reciprocal of the second.	$\dfrac{n^2+n}{n+5} \cdot \dfrac{(n+5)(n-5)}{2n}$
Factor any expressions if possible.	$\dfrac{n(n+1)(n+5)(n-5)}{(n+5)2n}$
Remove common factors.	$\dfrac{\cancel{n}(n+1)\cancel{(n+5)}(n-5)}{\cancel{(n+5)}2\cancel{n}}$
Simplify.	$\dfrac{(n+1)(n-5)}{2}$

> **TRY IT :: 7.53**
>
> Simplify the complex rational expression by writing it as division: $\dfrac{b - \dfrac{3b}{b+5}}{\dfrac{2}{b+5} + \dfrac{1}{b-5}}$.

> **TRY IT :: 7.54**
>
> Simplify the complex rational expression by writing it as division: $\dfrac{1 - \dfrac{3}{c+4}}{\dfrac{1}{c+4} + \dfrac{c}{3}}$.

Simplify a Complex Rational Expression by Using the LCD

We "cleared" the fractions by multiplying by the LCD when we solved equations with fractions. We can use that strategy here to simplify complex rational expressions. We will multiply the numerator and denominator by the LCD of all the rational expressions.

Let's look at the complex rational expression we simplified one way in **Example 7.25**. We will simplify it here by multiplying the numerator and denominator by the LCD. When we multiply by $\dfrac{\text{LCD}}{\text{LCD}}$ we are multiplying by 1, so the value stays the same.

EXAMPLE 7.28

Simplify the complex rational expression by using the LCD: $\dfrac{\frac{1}{3}+\frac{1}{6}}{\frac{1}{2}-\frac{1}{3}}$.

✓ **Solution**

$$\dfrac{\frac{1}{3}+\frac{1}{6}}{\frac{1}{2}-\frac{1}{3}}$$

The LCD of all the fractions in the whole expression is 6.	
Clear the fractions by multiplying the numerator and denominator by that LCD.	$\dfrac{6\cdot\left(\frac{1}{3}+\frac{1}{6}\right)}{6\cdot\left(\frac{1}{2}-\frac{1}{3}\right)}$
Distribute.	$\dfrac{6\cdot\frac{1}{3}+6\cdot\frac{1}{6}}{6\cdot\frac{1}{2}-6\cdot\frac{1}{3}}$
Simplify.	$\dfrac{2+1}{3-2}$
	$\dfrac{3}{1}$
	3

> **TRY IT :: 7.55**
> Simplify the complex rational expression by using the LCD: $\dfrac{\frac{1}{2}+\frac{1}{5}}{\frac{1}{10}+\frac{1}{5}}$.

> **TRY IT :: 7.56**
> Simplify the complex rational expression by using the LCD: $\dfrac{\frac{1}{4}+\frac{3}{8}}{\frac{1}{2}-\frac{5}{16}}$.

We will use the same example as in **Example 7.26**. Decide which method works better for you.

EXAMPLE 7.29   HOW TO SIMPLIFY A COMPLEX RATIONAL EXPRESSING USING THE LCD

Simplify the complex rational expression by using the LCD: $\dfrac{\frac{1}{x}+\frac{1}{y}}{\frac{x}{y}-\frac{y}{x}}$.

✓ **Solution**

Step 1. Find the LCD of all fractions in the complex rational expression.	The LCD of all the fractions is xy.	$\dfrac{\frac{1}{x}+\frac{1}{y}}{\frac{x}{y}-\frac{y}{x}}$
Step 2. Multiply the numerator and denominator by the LCD.	Multiply both the numerator and denominator by xy.	$\dfrac{xy\cdot\left(\frac{1}{x}+\frac{1}{y}\right)}{xy\cdot\left(\frac{x}{y}-\frac{y}{x}\right)}$

Step 3. Simplify the expression.	Distribute.	$\dfrac{xy\cdot\frac{1}{x}+xy\cdot\frac{1}{y}}{xy\cdot\frac{x}{y}-xy\cdot\frac{y}{x}}$
		$\dfrac{y+x}{x^2-y^2}$
	Simplify.	$\dfrac{\cancel{(y+x)}}{(x-y)\cancel{(x+y)}}$
	Remove common factors.	$\dfrac{1}{x-y}$

> **TRY IT :: 7.57**

Simplify the complex rational expression by using the LCD: $\dfrac{\frac{1}{a}+\frac{1}{b}}{\frac{a}{b}+\frac{b}{a}}$.

> **TRY IT :: 7.58**

Simplify the complex rational expression by using the LCD: $\dfrac{\frac{1}{x^2}-\frac{1}{y^2}}{\frac{1}{x}+\frac{1}{y}}$.

 HOW TO :: SIMPLIFY A COMPLEX RATIONAL EXPRESSION BY USING THE LCD.

Step 1.   Find the LCD of all fractions in the complex rational expression.

Step 2.   Multiply the numerator and denominator by the LCD.

Step 3.   Simplify the expression.

Be sure to start by factoring all the denominators so you can find the LCD.

EXAMPLE 7.30

Simplify the complex rational expression by using the LCD: $\dfrac{\frac{2}{x+6}}{\frac{4}{x-6}-\frac{4}{x^2-36}}$.

✓ **Solution**

$$\dfrac{\frac{2}{x+6}}{\frac{4}{x-6}-\frac{4}{x^2-36}}$$

Find the LCD of all fractions in the complex rational expression. The LCD is $x^2-36=(x+6)(x-6)$.

Multiply the numerator and denominator by the LCD.

$$\dfrac{(x+6)(x-6)\frac{2}{x+6}}{(x+6)(x-6)\left(\frac{4}{x-6}-\frac{4}{(x+6)(x-6)}\right)}$$

Simplify the expression.

Distribute in the denominator.	$$\dfrac{(x+6)(x-6)\dfrac{2}{x+6}}{(x+6)(x-6)\left(\dfrac{4}{x-6}\right)-(x+6)(x-6)\left(\dfrac{4}{(x+6)(x-6)}\right)}$$
Simplify.	$$\dfrac{(x+6)(x-6)\dfrac{2}{x+6}}{(x+6)(x-6)\left(\dfrac{4}{x-6}\right)-(x+6)(x-6)\left(\dfrac{4}{(x+6)(x-6)}\right)}$$
Simplify.	$$\dfrac{2(x-6)}{4(x+6)-4}$$
To simplify the denominator, distribute and combine like terms.	$$\dfrac{2(x-6)}{4x+20}$$
Factor the denominator.	$$\dfrac{2(x-6)}{4(x+5)}$$
Remove common factors.	$$\dfrac{2(x-6)}{2\cdot 2(x+5)}$$
Simplify.	$$\dfrac{x-6}{2(x+5)}$$

Notice that there are no more factors common to the numerator and denominator.

> **TRY IT :: 7.59**

Simplify the complex rational expression by using the LCD: $\dfrac{\dfrac{3}{x+2}}{\dfrac{5}{x-2}-\dfrac{3}{x^2-4}}$.

> **TRY IT :: 7.60**

Simplify the complex rational expression by using the LCD: $\dfrac{\dfrac{2}{x-7}-\dfrac{1}{x+7}}{\dfrac{6}{x+7}-\dfrac{1}{x^2-49}}$.

Be sure to factor the denominators first. Proceed carefully as the math can get messy!

EXAMPLE 7.31

Simplify the complex rational expression by using the LCD: $\dfrac{\dfrac{4}{m^2-7m+12}}{\dfrac{3}{m-3}-\dfrac{2}{m-4}}$.

⊘ **Solution**

$$\dfrac{\dfrac{4}{m^2 - 7m + 12}}{\dfrac{3}{m-3} - \dfrac{2}{m-4}}$$

Find the LCD of all fractions in the complex rational expression.	
The LCD is $(m-3)(m-4)$.	
Multiply the numerator and denominator by the LCD.	$\dfrac{(m-3)(m-4)\dfrac{4}{(m-3)(m-4)}}{(m-3)(m-4)\left(\dfrac{3}{m-3} - \dfrac{2}{m-4}\right)}$
Simplify.	$\dfrac{\cancel{(m-3)}\cancel{(m-4)}\dfrac{4}{\cancel{(m-3)}\cancel{(m-4)}}}{(m-3)(m-4)\left(\dfrac{3}{\cancel{m-3}}\right) - (m-3)(m-4)\left(\dfrac{2}{\cancel{m-4}}\right)}$
Simplify.	$\dfrac{4}{3(m-4) - 2(m-3)}$
Distribute.	$\dfrac{4}{3m - 12 - 2m + 6}$
Combine like terms.	$\dfrac{4}{m-6}$

> **TRY IT : : 7.61**
>
> Simplify the complex rational expression by using the LCD: $\dfrac{\dfrac{3}{x^2 + 7x + 10}}{\dfrac{4}{x+2} + \dfrac{1}{x+5}}$.

> **TRY IT : : 7.62**
>
> Simplify the complex rational expression by using the LCD: $\dfrac{\dfrac{4y}{y+5} + \dfrac{2}{y+6}}{\dfrac{3y}{y^2 + 11y + 30}}$.

EXAMPLE 7.32

Simplify the complex rational expression by using the LCD: $\dfrac{\dfrac{y}{y+1}}{1 + \dfrac{1}{y-1}}$.

⊘ **Solution**

$$\dfrac{\dfrac{y}{y+1}}{1 + \dfrac{1}{y-1}}$$

Find the LCD of all fractions in the complex rational expression.	
The LCD is $(y+1)(y-1)$.	

Multiply the numerator and denominator by the LCD.	$$\dfrac{(y+1)(y-1)\dfrac{y}{y+1}}{(y+1)(y-1)\left(1+\dfrac{1}{y-1}\right)}$$
Distribute in the denominator and simplify.	$$\dfrac{(y+1)(y-1)\dfrac{y}{y+1}}{(y+1)(y-1)(1)+(y+1)(y-1)\left(\dfrac{1}{y-1}\right)}$$
Simplify.	$$\dfrac{(y-1)y}{(y+1)(y-1)+(y+1)}$$
Simplify the denominator and leave the numerator factored.	$$\dfrac{y(y-1)}{y^2-1+y+1}$$
	$$\dfrac{y(y-1)}{y^2+y}$$
Factor the denominator and remove factors common with the numerator.	$$\dfrac{y(y-1)}{y(y+1)}$$
Simplify.	$$\dfrac{y-1}{y+1}$$

> **TRY IT : : 7.63**
>
> Simplify the complex rational expression by using the LCD: $\dfrac{\dfrac{x}{x+3}}{1+\dfrac{1}{x+3}}$.

> **TRY IT : : 7.64**
>
> Simplify the complex rational expression by using the LCD: $\dfrac{1+\dfrac{1}{x-1}}{\dfrac{3}{x+1}}$.

▶ **MEDIA : :**

Access this online resource for additional instruction and practice with complex fractions.

- **Complex Fractions (https://openstax.org/l/37CompFrac)**

7.3 EXERCISES

Practice Makes Perfect

Simplify a Complex Rational Expression by Writing it as Division

In the following exercises, simplify each complex rational expression by writing it as division.

151. $\dfrac{\frac{2a}{a+4}}{\frac{4a^2}{a^2-16}}$

152. $\dfrac{\frac{3b}{b-5}}{\frac{b^2}{b^2-25}}$

153. $\dfrac{\frac{5}{c^2+5c-14}}{\frac{10}{c+7}}$

154. $\dfrac{\frac{8}{d^2+9d+18}}{\frac{12}{d+6}}$

155. $\dfrac{\frac{1}{2}+\frac{5}{6}}{\frac{2}{3}+\frac{7}{9}}$

156. $\dfrac{\frac{1}{2}+\frac{3}{4}}{\frac{3}{5}+\frac{7}{10}}$

157. $\dfrac{\frac{2}{3}-\frac{1}{9}}{\frac{3}{4}+\frac{5}{6}}$

158. $\dfrac{\frac{1}{2}-\frac{1}{6}}{\frac{2}{3}+\frac{3}{4}}$

159. $\dfrac{\frac{n}{m}+\frac{1}{n}}{\frac{1}{n}-\frac{n}{m}}$

160. $\dfrac{\frac{1}{p}+\frac{p}{q}}{\frac{q}{p}-\frac{1}{q}}$

161. $\dfrac{\frac{1}{r}+\frac{1}{t}}{\frac{1}{r^2}-\frac{1}{t^2}}$

162. $\dfrac{\frac{2}{v}+\frac{2}{w}}{\frac{1}{v^2}-\frac{1}{w^2}}$

163. $\dfrac{x-\frac{2x}{x+3}}{\frac{1}{x+3}+\frac{1}{x-3}}$

164. $\dfrac{y-\frac{2y}{y-4}}{\frac{2}{y-4}+\frac{2}{y+4}}$

165. $\dfrac{2-\frac{2}{a+3}}{\frac{1}{a+3}+\frac{a}{2}}$

166. $\dfrac{4+\frac{4}{b-5}}{\frac{1}{b-5}+\frac{b}{4}}$

Simplify a Complex Rational Expression by Using the LCD

In the following exercises, simplify each complex rational expression by using the LCD.

167. $\dfrac{\frac{1}{3}+\frac{1}{8}}{\frac{1}{4}+\frac{1}{12}}$

168. $\dfrac{\frac{1}{4}+\frac{1}{9}}{\frac{1}{6}+\frac{1}{12}}$

169. $\dfrac{\frac{5}{6}+\frac{2}{9}}{\frac{7}{18}-\frac{1}{3}}$

170. $\dfrac{\frac{1}{6}+\frac{4}{15}}{\frac{3}{5}-\frac{1}{2}}$

171. $\dfrac{\frac{c}{d}+\frac{1}{d}}{\frac{1}{d}-\frac{d}{c}}$

172. $\dfrac{\frac{1}{m}+\frac{m}{n}}{\frac{n}{m}-\frac{1}{n}}$

173. $\dfrac{\frac{1}{p}+\frac{1}{q}}{\frac{1}{p^2}-\frac{1}{q^2}}$

174. $\dfrac{\frac{2}{r}+\frac{2}{t}}{\frac{1}{r^2}-\frac{1}{t^2}}$

175. $\dfrac{\frac{2}{x+5}}{\frac{3}{x-5}+\frac{1}{x^2-25}}$

176. $\dfrac{\frac{5}{y-4}}{\frac{3}{y+4}+\frac{2}{y^2-16}}$

177. $\dfrac{\frac{5}{z^2-64}+\frac{3}{z+8}}{\frac{1}{z+8}+\frac{2}{z-8}}$

178. $\dfrac{\frac{3}{s+6}+\frac{5}{s-6}}{\frac{1}{s^2-36}+\frac{4}{s+6}}$

179. $\dfrac{\frac{4}{a^2-2a-15}}{\frac{1}{a-5}+\frac{2}{a+3}}$

180. $\dfrac{\frac{5}{b^2-6b-27}}{\frac{3}{b-9}+\frac{1}{b+3}}$

181. $\dfrac{\frac{5}{c+2}-\frac{3}{c+7}}{\frac{5c}{c^2+9c+14}}$

182. $\dfrac{\dfrac{6}{d-4} - \dfrac{2}{d+7}}{\dfrac{2d}{d^2+3d-28}}$

183. $\dfrac{2 + \dfrac{1}{p-3}}{\dfrac{5}{p-3}}$

184. $\dfrac{\dfrac{n}{n-2}}{3 + \dfrac{5}{n-2}}$

185. $\dfrac{\dfrac{m}{m+5}}{4 + \dfrac{1}{m-5}}$

186. $\dfrac{7 + \dfrac{2}{q-2}}{\dfrac{1}{q+2}}$

In the following exercises, simplify each complex rational expression using either method.

187. $\dfrac{\dfrac{3}{4} - \dfrac{2}{7}}{\dfrac{1}{2} + \dfrac{5}{14}}$

188. $\dfrac{\dfrac{v}{w} + \dfrac{1}{v}}{\dfrac{1}{v} - \dfrac{v}{w}}$

189. $\dfrac{\dfrac{2}{a+4}}{\dfrac{1}{a^2-16}}$

190. $\dfrac{\dfrac{3}{b^2-3b-40}}{\dfrac{5}{b+5} - \dfrac{2}{b-8}}$

191. $\dfrac{\dfrac{3}{m} + \dfrac{3}{n}}{\dfrac{1}{m^2} - \dfrac{1}{n^2}}$

192. $\dfrac{\dfrac{2}{r-9}}{\dfrac{1}{r+9} + \dfrac{3}{r^2-81}}$

193. $\dfrac{x - \dfrac{3x}{x+2}}{\dfrac{3}{x+2} + \dfrac{3}{x-2}}$

194. $\dfrac{\dfrac{y}{y+3}}{2 + \dfrac{1}{y-3}}$

Writing Exercises

195. In this section, you learned to simplify the complex fraction $\dfrac{\dfrac{3}{x+2}}{\dfrac{x}{x^2-4}}$ two ways: rewriting it as a division problem or multiplying the numerator and denominator by the LCD. Which method do you prefer? Why?

196. Efraim wants to start simplifying the complex fraction $\dfrac{\dfrac{1}{a} + \dfrac{1}{b}}{\dfrac{1}{a} - \dfrac{1}{b}}$ by cancelling the variables from the numerator and denominator, $\dfrac{\dfrac{1}{\cancel{a}} + \dfrac{1}{\cancel{b}}}{\dfrac{1}{\cancel{a}} - \dfrac{1}{\cancel{b}}}$. Explain what is wrong with Efraim's plan.

Self Check

ⓐ *After completing the exercises, use this checklist to evaluate your mastery of the objectives of this section.*

I can...	Confidently	With some help	No-I don't get it!
simplify a complex rational expression by writing it as division.			
simplify a complex rational expression by using the LCD.			

ⓑ *After looking at the checklist, do you think you are well-prepared for the next section? Why or why not?*

Solve Rational Equations

Learning Objectives

By the end of this section, you will be able to:

> Solve rational equations
> Use rational functions
> Solve a rational equation for a specific variable

Be Prepared!

Before you get started, take this readiness quiz.

1. Solve: $\frac{1}{6}x + \frac{1}{2} = \frac{1}{3}$.
   If you missed this problem, review Example 2.9.

2. Solve: $n^2 - 5n - 36 = 0$.
   If you missed this problem, review Example 6.45.

3. Solve the formula $5x + 2y = 10$ for y.
   If you missed this problem, review Example 2.31.

After defining the terms 'expression' and 'equation' earlier, we have used them throughout this book. We have *simplified* many kinds of *expressions* and *solved* many kinds of *equations*. We have simplified many rational expressions so far in this chapter. Now we will *solve* a **rational equation**.

Rational Equation

A **rational equation** is an equation that contains a rational expression.

You must make sure to know the difference between rational expressions and rational equations. The equation contains an equal sign.

Rational Expression	Rational Equation
$\frac{1}{8}x + \frac{1}{2}$	$\frac{1}{8}x + \frac{1}{2} = \frac{1}{4}$
$\frac{y+6}{y^2-36}$	$\frac{y+6}{y^2-36} = y+1$
$\frac{1}{n-3} + \frac{1}{n+4}$	$\frac{1}{n-3} + \frac{1}{n+4} = \frac{15}{n^2+n-12}$

Solve Rational Equations

We have already solved linear equations that contained fractions. We found the LCD of all the fractions in the equation and then multiplied both sides of the equation by the LCD to "clear" the fractions.

We will use the same strategy to solve rational equations. We will multiply both sides of the equation by the LCD. Then, we will have an equation that does not contain rational expressions and thus is much easier for us to solve. But because the original equation may have a variable in a denominator, we must be careful that we don't end up with a solution that would make a denominator equal to zero.

So before we begin solving a rational equation, we examine it first to find the values that would make any denominators zero. That way, when we solve a rational equation we will know if there are any algebraic solutions we must discard.

An algebraic solution to a rational equation that would cause any of the rational expressions to be undefined is called an **extraneous solution to a rational equation**.

Extraneous Solution to a Rational Equation

An **extraneous solution to a rational equation** is an algebraic solution that would cause any of the expressions in the original equation to be undefined.

We note any possible extraneous solutions, *c*, by writing $x \neq c$ next to the equation.

EXAMPLE 7.33   HOW TO SOLVE A RATIONAL EQUATION

Solve: $\frac{1}{x} + \frac{1}{3} = \frac{5}{6}$.

✓ **Solution**

Step 1. Note any value of the variable that would make any denominator zero.	If $x = 0$, then $\frac{1}{x}$ is undefined.	
	So we'll write $x \neq 0$ next to the equation.	$\frac{1}{x} + \frac{1}{3} = \frac{5}{6},\, x \neq 0$
Step 2. Find the least common denominator of *all* denominators in the equation.	Find the LCD of $\frac{1}{x}$, $\frac{1}{3}$, and $\frac{5}{6}$.	The LCD is $6x$.
Step 3. Clear the fractions by multiplying both sides of the equation by the LCD.	Multiply both sides of the equation by the LCD, $6x$.	$6x \cdot \left(\frac{1}{x} + \frac{1}{3}\right) = 6x \cdot \left(\frac{5}{6}\right)$
	Use the Distributive Property.	$6x \cdot \frac{1}{x} + 6x \cdot \frac{1}{3} = 6x \cdot \left(\frac{5}{6}\right)$
	Simplify – and notice, no more fractions!	$6 + 2x = 5x$
Step 4. Solve the resulting equation.	Simplify.	$6 = 3x$ $2 = x$
Step 5. Check. • If any values found in Step 1 are algebraic solutions, discard them. • Check any remaining solutions in the original equation.	We did not get 0 as an algebraic solution. We substitute $x = 2$ into the original equation.	$\frac{1}{x} + \frac{1}{3} = \frac{5}{6}$ $\frac{1}{2} + \frac{1}{3} \overset{?}{=} \frac{5}{6}$ $\frac{3}{6} + \frac{2}{6} \overset{?}{=} \frac{5}{6}$ $\frac{5}{6} = \frac{5}{6}$ ✓ The solution is $x = 2$.

> **TRY IT : : 7.65**   Solve: $\frac{1}{y} + \frac{2}{3} = \frac{1}{5}$.

> **TRY IT : : 7.66**   Solve: $\frac{2}{3} + \frac{1}{5} = \frac{1}{x}$.

The steps of this method are shown.

HOW TO :: SOLVE EQUATIONS WITH RATIONAL EXPRESSIONS.

Step 1. Note any value of the variable that would make any denominator zero.

Step 2. Find the least common denominator of *all* denominators in the equation.

Step 3. Clear the fractions by multiplying both sides of the equation by the LCD.

Step 4. Solve the resulting equation.

Step 5. Check:

- If any values found in Step 1 are algebraic solutions, discard them.

- Check any remaining solutions in the original equation.

We always start by noting the values that would cause any denominators to be zero.

EXAMPLE 7.34   HOW TO SOLVE A RATIONAL EQUATION USING THE ZERO PRODUCT PROPERTY

Solve: $1 - \dfrac{5}{y} = -\dfrac{6}{y^2}$.

 **Solution**

	$1 - \dfrac{5}{y} = -\dfrac{6}{y^2}$
Note any value of the variable that would make any denominator zero.	$1 - \dfrac{5}{y} = -\dfrac{6}{y^2}, y \neq 0$
Find the least common denominator of all denominators in the equation. The LCD is y^2.	
Clear the fractions by multiplying both sides of the equation by the LCD.	$y^2\left(1 - \dfrac{5}{y}\right) = y^2\left(-\dfrac{6}{y^2}\right)$
Distribute.	$y^2 \cdot 1 - y^2\left(\dfrac{5}{y}\right) = y^2\left(-\dfrac{6}{y^2}\right)$
Multiply.	$y^2 - 5y = -6$
Solve the resulting equation. First write the quadratic equation in standard form.	$y^2 - 5y + 6 = 0$
Factor.	$(y - 2)(y - 3) = 0$
Use the Zero Product Property.	$y - 2 = 0$ or $y - 3 = 0$
Solve.	$y = 2$ or $y = 3$

Check.
We did not get 0 as an algebraic solution.

Check $y = 2$ and $y = 3$ in the original equation.

$$1 - \frac{5}{y} = -\frac{6}{y^2} \qquad 1 - \frac{5}{y} = -\frac{6}{y^2}$$

$$1 - \frac{5}{2} \overset{?}{=} -\frac{6}{2^2} \qquad 1 - \frac{5}{3} \overset{?}{=} -\frac{6}{3^2}$$

$$1 - \frac{5}{2} \overset{?}{=} -\frac{6}{4} \qquad 1 - \frac{5}{3} \overset{?}{=} -\frac{6}{9}$$

$$\frac{2}{2} - \frac{5}{2} \overset{?}{=} -\frac{6}{4} \qquad \frac{3}{3} - \frac{5}{3} \overset{?}{=} -\frac{6}{9}$$

$$-\frac{3}{2} \overset{?}{=} -\frac{6}{4} \qquad -\frac{2}{3} \overset{?}{=} -\frac{6}{9}$$

$$-\frac{3}{2} = -\frac{3}{2} \checkmark \qquad -\frac{2}{3} = -\frac{2}{3} \checkmark$$

The solution is $y = 2, \quad y = 3$.

> **TRY IT :: 7.67** Solve: $1 - \frac{2}{x} = \frac{15}{x^2}$.

> **TRY IT :: 7.68** Solve: $1 - \frac{4}{y} = \frac{12}{y^2}$.

In the next example, the last denominators is a difference of squares. Remember to factor it first to find the LCD.

EXAMPLE 7.35

Solve: $\dfrac{2}{x+2} + \dfrac{4}{x-2} = \dfrac{x-1}{x^2-4}$.

✓ **Solution**

$$\frac{2}{x+2} + \frac{4}{x-2} = \frac{x-1}{x^2-4}$$

Note any value of the variable that would make any denominator zero.

$$\frac{2}{x+2} + \frac{4}{x-2} = \frac{x-1}{(x+2)(x-2)}, \; x \neq -2, x \neq 2$$

Find the least common denominator of all denominators in the equation.
The LCD is $(x+2)(x-2)$.

Clear the fractions by multiplying both sides of the equation by the LCD.

$$(x+2)(x-2)\left(\frac{2}{x+2} + \frac{4}{x-2}\right) = (x+2)(x-2)\left(\frac{x-1}{x^2-4}\right)$$

Distribute.

$$(x+2)(x-2)\frac{2}{x+2} + (x+2)(x-2)\frac{4}{x-2} = (x+2)(x-2)\left(\frac{x-1}{x^2-4}\right)$$

Remove common factors.	$(x+2)(x-2)\dfrac{2}{x+2} + (x+2)(x-2)\dfrac{4}{x-2} = (x+2)(x-2)\left(\dfrac{x-1}{x^2-4}\right)$
Simplify.	$2(x-2) + 4(x+2) = x-1$
Distribute.	$2x - 4 + 4x + 8 = x - 1$
Solve.	$6x + 4 = x - 1$
	$5x = -5$
	$x = -1$

Check:
We did not get 2 or –2 as algebraic solutions.

Check $x = -1$ in the original equation.

$$\frac{2}{x+2} + \frac{4}{x-2} = \frac{x-1}{x^2-4}$$

$$\frac{2}{(-1)+2} + \frac{4}{(-1)-2} \stackrel{?}{=} \frac{(-1)-1}{(-1)^2-4}$$

$$\frac{2}{1} + \frac{4}{-3} \stackrel{?}{=} \frac{-2}{-3}$$

$$\frac{6}{3} - \frac{4}{3} \stackrel{?}{=} \frac{2}{3}$$

$$\frac{2}{3} = \frac{2}{3} \checkmark$$

The solution is $x = -1$.

> **TRY IT : : 7.69**   Solve: $\dfrac{2}{x+1} + \dfrac{1}{x-1} = \dfrac{1}{x^2-1}$.

> **TRY IT : : 7.70**   Solve: $\dfrac{5}{y+3} + \dfrac{2}{y-3} = \dfrac{5}{y^2-9}$.

In the next example, the first denominator is a trinomial. Remember to factor it first to find the LCD.

EXAMPLE 7.36

Solve: $\dfrac{m+11}{m^2 - 5m + 4} = \dfrac{5}{m-4} - \dfrac{3}{m-1}$.

✓ Solution

$$\frac{m+11}{m^2-5m+4} = \frac{5}{m-4} = \frac{3}{m-1}$$

Note any value of the variable that would make any denominator zero. Use the factored form of the quadratic denominator.	$\dfrac{m+11}{(m-4)(m-1)} = \dfrac{5}{m-4} - \dfrac{3}{m-1}, \; m \neq 4, \, m \neq 1$
Find the least common denominator of all denominators in the equation. The LCD is $(m-4)(m-1)$.	
Clear the fractions by multiplying both sides of the equation by the LCD.	$(m-4)(m-1)\left(\dfrac{m+11}{(m-4)(m-1)}\right) = (m-4)(m-1)\left(\dfrac{5}{m-4} - \dfrac{3}{m-1}\right)$
Distribute.	$(m-4)(m-1)\left(\dfrac{m+11}{(m-4)(m-1)}\right) = (m-4)(m-1)\dfrac{5}{m-4} - (m-4)(m-1)\dfrac{3}{m-1}$
Remove common factors.	$\cancel{(m-4)}\cancel{(m-1)}\left(\dfrac{m+11}{\cancel{(m-4)}\cancel{(m-1)}}\right) = \cancel{(m-4)}(m-1)\dfrac{5}{\cancel{m-4}} - (m-4)\cancel{(m-1)}\dfrac{3}{\cancel{m-1}}$
Simplify.	$m+11 = 5(m-1) - 3(m-4)$
Solve the resulting equation.	$m+11 = 5m-5-3m+12$
	$4 = m$
Check. The only algebraic solution was 4, but we said that 4 would make a denominator equal to zero. The algebraic solution is an extraneous solution.	
	There is no solution to this equation.

> **TRY IT :: 7.71**   Solve: $\dfrac{x+13}{x^2-7x+10} = \dfrac{6}{x-5} - \dfrac{4}{x-2}$.

> **TRY IT :: 7.72**   Solve: $\dfrac{y-6}{y^2+3y-4} = \dfrac{2}{y+4} + \dfrac{7}{y-1}$.

The equation we solved in the previous example had only one algebraic solution, but it was an extraneous solution. That left us with no solution to the equation. In the next example we get two algebraic solutions. Here one or both could be extraneous solutions.

EXAMPLE 7.37

Solve: $\dfrac{y}{y+6} = \dfrac{72}{y^2-36} + 4$.

⊘ **Solution**

$$\frac{y}{y+6} = \frac{72}{y^2-36} + 4$$

Factor all the denominators,
so we can note any value of
the variable that would make
any denominator zero.

$$\frac{y}{y+6} = \frac{72}{(y-6)(y+6)} + 4, y \neq 6, y \neq -6$$

Find the least common denominator.
The LCD is $(y-6)(y+6)$.

Clear the fractions.

$$(y-6)(y+6)\left(\frac{y}{y+6}\right) = (y-6)(y+6)\left(\frac{72}{(y-6)(y+6)} + 4\right)$$

Simplify.

$$(y-6) \cdot y = 72 + (y-6)(y+6) \cdot 4$$

Simplify.

$$y(y-6) = 72 + 4(y^2 - 36)$$

Solve the resulting equation.

$$y^2 - 6y = 72 + 4y^2 - 144$$

$$0 = 3y^2 + 6y - 72$$

$$0 = 3(y^2 + 2y - 24)$$

$$0 = 3(y+6)(y-4)$$

$$y = -6, y = 4$$

Check.

$y = -6$ is an extraneous solution.
Check $y = 4$ in the original equation.

$$\frac{y}{y+6} = \frac{72}{y^2-36} + 4$$

$$\frac{4}{4+6} \overset{?}{=} \frac{72}{4^2-36} + 4$$

$$\frac{4}{10} \overset{?}{=} \frac{72}{-20} + 4$$

$$\frac{4}{10} \overset{?}{=} -\frac{36}{10} + \frac{40}{10}$$

$$\frac{4}{10} = \frac{4}{10} \checkmark$$

The solution is $y = 4$.

> **TRY IT :: 7.73**   Solve: $\dfrac{x}{x+4} = \dfrac{32}{x^2-16} + 5$.

> **TRY IT :: 7.74**   Solve: $\dfrac{y}{y+8} = \dfrac{128}{y^2-64} + 9$.

In some cases, all the algebraic solutions are extraneous.

EXAMPLE 7.38

Solve: $\dfrac{x}{2x-2} - \dfrac{2}{3x+3} = \dfrac{5x^2 - 2x + 9}{12x^2 - 12}$.

⊘ Solution

$$\frac{x}{2x-2} - \frac{2}{3x+3} = \frac{5x^2 - 2x + 9}{12x^2 - 12}$$

We will start by factoring all denominators, to make it easier to identify extraneous solutions and the LCD.	$\dfrac{x}{2(x-1)} - \dfrac{2}{3(x+1)} = \dfrac{5x^2 - 2x + 9}{12(x-1)(x+1)}$
Note any value of the variable that would make any denominator zero.	$\dfrac{x}{2(x-1)} - \dfrac{2}{3(x+1)} = \dfrac{5x^2 - 2x + 9}{12(x-1)(x+1)}, x \neq 1, x \neq -1$
Find the least common denominator. The LCD is $12(x-1)(x+1)$.	
Clear the fractions.	$12(x-1)(x+1)\left(\dfrac{x}{2(x-1)} - \dfrac{2}{3(x+1)}\right) = 12(x-1)(x+1)\left(\dfrac{5x^2 - 2x + 9}{12(x-1)(x+1)}\right)$
Simplify.	$6(x+1) \cdot x - 4(x-1) \cdot 2 = 5x^2 - 2x + 9$
Simplify.	$6x(x+1) - 4 \cdot 2(x-1) = 5x^2 - 2x + 9$
Solve the resulting equation.	$6x^2 + 6x - 8x + 8 = 5x^2 - 2x + 9$
	$x^2 - 1 = 0$
	$(x-1)(x+1) = 0$
	$x = 1 \text{ or } x = -1$
Check.	
$x = 1$ and $x = -1$ are extraneous solutions.	
	The equation has no solution.

> **TRY IT : : 7.75**
Solve: $\dfrac{y}{5y-10} - \dfrac{5}{3y+6} = \dfrac{2y^2 - 19y + 54}{15y^2 - 60}$.

> **TRY IT : : 7.76**
Solve: $\dfrac{z}{2z+8} - \dfrac{3}{4z-8} = \dfrac{3z^2 - 16z - 16}{8z^2 + 2z - 64}$.

EXAMPLE 7.39

Solve: $\dfrac{4}{3x^2 - 10x + 3} + \dfrac{3}{3x^2 + 2x - 1} = \dfrac{2}{x^2 - 2x - 3}$.

⊘ **Solution**

$$\frac{4}{3x^2 - 10x + 3} + \frac{3}{3x^2 + 2x - 1} = \frac{2}{x^2 - 2x - 3}$$

Factor all the denominators, so we can note any value of the variable that would make any denominator zero.	$$\frac{4}{(3x-1)(x-3)} + \frac{3}{(3x-1)(x+1)} = \frac{2}{(x-3)(x+1)}$$ $$x \neq -1,\ x \neq \tfrac{1}{3},\ x \neq 3$$
Find the least common denominator. The LCD is $(3x-1)(x+1)(x-3)$.	
Clear the fractions.	
	$$(3x-1)(x+1)(x-3)\left(\frac{4}{(3x-1)(x-3)} + \frac{3}{(3x-1)(x+1)}\right) = (3x-1)(x+1)(x-3)\left(\frac{2}{(x-3)(x+1)}\right)$$
Simplify.	$$4(x+1) + 3(x-3) = 2(3x-1)$$
Distribute.	$$4x + 4 + 3x - 9 = 6x - 2$$
Simplify.	$$7x - 5 = 6x - 2$$
	$$x = 3$$
The only algebraic solution was $x = 3$, but we said that $x = 3$ would make a denominator equal to zero. The algebraic solution is an extraneous solution.	
	There is no solution to this equation.

> **TRY IT :: 7.77** Solve: $\dfrac{15}{x^2 + x - 6} - \dfrac{3}{x-2} = \dfrac{2}{x+3}$.

> **TRY IT :: 7.78** Solve: $\dfrac{5}{x^2 + 2x - 3} - \dfrac{3}{x^2 + x - 2} = \dfrac{1}{x^2 + 5x + 6}$.

Use Rational Functions

Working with functions that are defined by rational expressions often lead to rational equations. Again, we use the same techniques to solve them.

EXAMPLE 7.40

For rational function, $f(x) = \dfrac{2x - 6}{x^2 - 8x + 15}$, ⓐ find the domain of the function, ⓑ solve $f(x) = 1$, and ⓒ find the points on the graph at this function value.

⊘ **Solution**

ⓐ The domain of a rational function is all real numbers except those that make the rational expression undefined. So to find them, we will set the denominator equal to zero and solve.

$$x^2 - 8x + 15 = 0$$

Factor the trinomial.

Use the Zero Product Property.

Solve.

$$(x - 3)(x - 5) = 0$$
$$x - 3 = 0 \quad x - 5 = 0.$$
$$x = 3 \quad x = 5$$

The domain is all real numbers except $x \neq 3$, $x \neq 5$.

ⓑ

	$f(x) = 1$
Substitute in the rational expression.	$\dfrac{2x - 6}{x^2 - 8x + 15} = 1$
Factor the denominator.	$\dfrac{2x - 6}{(x - 3)(x - 5)} = 1$
Multiply both sides by the LCD, $(x - 3)(x - 5)$.	$(x - 3)(x - 5)\left(\dfrac{2x - 6}{(x - 3)(x - 5)}\right) = (x - 3)(x - 5)(1)$
Simplify.	$2x - 6 = x^2 - 8x + 15$
Solve.	$0 = x^2 - 10x + 21$
Factor.	$0 = (x - 7)(x - 3)$
Use the Zero Product Property.	$x - 7 = 0 \qquad x - 3 = 0$
Solve.	$x = 7 \qquad x = 3$

ⓒ The value of the function is 1 when $x = 7$, $x = 3$. So the points on the graph of this function when $f(x) = 1$, will be $(7, 1)$, $(3, 1)$.

> **TRY IT : : 7.79**
>
> For rational function, $f(x) = \dfrac{8 - x}{x^2 - 7x + 12}$, ⓐ find the domain of the function ⓑ solve $f(x) = 3$ ⓒ find the
>
> points on the graph at this function value.

> **TRY IT : : 7.80**
>
> For rational function, $f(x) = \dfrac{x - 1}{x^2 - 6x + 5}$, ⓐ find the domain of the function ⓑ solve $f(x) = 4$ ⓒ find the
>
> points on the graph at this function value.

Solve a Rational Equation for a Specific Variable

When we solved linear equations, we learned how to solve a formula for a specific variable. Many formulas used in business, science, economics, and other fields use rational equations to model the relation between two or more variables. We will now see how to solve a rational equation for a specific variable.

When we developed the point-slope formula from our slope formula, we cleared the fractions by multiplying by the LCD.

$$m = \frac{y - y_1}{x - x_1}$$

Multiply both sides of the equation by $x - x_1$.	$m(x - x_1) = \left(\dfrac{y - y_1}{x - x_1}\right)(x - x_1)$
Simplify.	$m(x - x_1) = y - y_1$
Rewrite the equation with the y terms on the left.	$y - y_1 = m(x - x_1)$

In the next example, we will use the same technique with the formula for slope that we used to get the point-slope form of an equation of a line through the point $(2, 3)$. We will add one more step to solve for y.

EXAMPLE 7.41

Solve: $m = \dfrac{y-2}{x-3}$ for y.

✓ **Solution**

$$m = \frac{y-2}{x-3}$$

Note any value of the variable that would make any denominator zero.	$m = \dfrac{y-2}{x-3},\, x \neq 3$
Clear the fractions by multiplying both sides of the equation by the LCD, $x-3$.	$(x-3)m = (x-3)\left(\dfrac{y-2}{x-3}\right)$
Simplify.	$xm - 3m = y - 2$
Isolate the term with y.	$xm - 3m + 2 = y$

> **TRY IT :: 7.81**   Solve: $m = \dfrac{y-5}{x-4}$ for y.

> **TRY IT :: 7.82**   Solve: $m = \dfrac{y-1}{x+5}$ for y.

Remember to multiply both sides by the LCD in the next example.

EXAMPLE 7.42

Solve: $\dfrac{1}{c} + \dfrac{1}{m} = 1$ for c.

✓ **Solution**

$$\frac{1}{c} + \frac{1}{m} = 1 \text{ for } c$$

Note any value of the variable that would make any denominator zero.	$\dfrac{1}{c} + \dfrac{1}{m} = 1,\, c \neq 0,\, m \neq 0$
Clear the fractions by multiplying both sides of the equations by the LCD, cm.	$cm\left(\dfrac{1}{c} + \dfrac{1}{m}\right) = cm(1)$
Distribute.	$cm\left(\dfrac{1}{c}\right) + cm\dfrac{1}{m} = cm(1)$
Simplify.	$m + c = cm$
Collect the terms with c to the right.	$m = cm - c$
Factor the expression on the right.	$m = c(m-1)$
To isolate c, divide both sides by $m-1$.	$\dfrac{m}{m-1} = \dfrac{c(m-1)}{m-1}$
Simplify by removing common factors.	$\dfrac{m}{m-1} = c$

Notice that even though we excluded $c = 0$, $m = 0$ from the original equation, we must also now state that $m \neq 1$.

> **TRY IT :: 7.83**   Solve: $\frac{1}{a} + \frac{1}{b} = c$ for a.

> **TRY IT :: 7.84**   Solve: $\frac{2}{x} + \frac{1}{3} = \frac{1}{y}$ for y.

▶ **MEDIA ::**

Access this online resource for additional instruction and practice with equations with rational expressions.

- **Equations with Rational Expressions (https://openstax.org/l/37EqRatExp)**

7.4 EXERCISES

Practice Makes Perfect

Solve Rational Equations

In the following exercises, solve each rational equation.

197. $\frac{1}{a} + \frac{2}{5} = \frac{1}{2}$

198. $\frac{6}{3} - \frac{2}{d} = \frac{4}{9}$

199. $\frac{4}{5} + \frac{1}{4} = \frac{2}{v}$

200. $\frac{3}{8} + \frac{2}{y} = \frac{1}{4}$

201. $1 - \frac{2}{m} = \frac{8}{m^2}$

202. $1 + \frac{4}{n} = \frac{21}{n^2}$

203. $1 + \frac{9}{p} = \frac{-20}{p^2}$

204. $1 - \frac{7}{q} = \frac{-6}{q^2}$

205. $\frac{5}{3v - 2} = \frac{7}{4v}$

206. $\frac{8}{2w + 1} = \frac{3}{w}$

207. $\frac{3}{x + 4} + \frac{7}{x - 4} = \frac{8}{x^2 - 16}$

208. $\frac{5}{y - 9} + \frac{1}{y + 9} = \frac{18}{y^2 - 81}$

209. $\frac{8}{z - 10} - \frac{7}{z + 10} = \frac{5}{z^2 - 100}$

210. $\frac{9}{a + 11} - \frac{6}{a - 11} = \frac{6}{a^2 - 121}$

211. $\frac{-10}{q - 2} - \frac{7}{q + 4} = 1$

212. $\frac{2}{s + 7} - \frac{3}{s - 3} = 1$

213. $\frac{v - 10}{v^2 - 5v + 4} = \frac{3}{v - 1} - \frac{6}{v - 4}$

214. $\frac{w + 8}{w^2 - 11w + 28} = \frac{5}{w - 7} + \frac{2}{w - 4}$

215. $\frac{x - 10}{x^2 + 8x + 12} = \frac{3}{x + 2} + \frac{4}{x + 6}$

216. $\frac{y - 5}{y^2 - 4y - 5} = \frac{1}{y + 1} + \frac{1}{y - 5}$

217. $\frac{b + 3}{3b} + \frac{b}{24} = \frac{1}{b}$

218. $\frac{c + 3}{12c} + \frac{c}{36} = \frac{1}{4c}$

219. $\frac{d}{d + 3} = \frac{18}{d^2 - 9} + 4$

220. $\frac{m}{m + 5} = \frac{50}{m^2 - 25} + 6$

221. $\frac{n}{n + 2} - 3 = \frac{8}{n^2 - 4}$

222. $\frac{p}{p + 7} - 8 = \frac{98}{p^2 - 49}$

223. $\frac{q}{3q - 9} - \frac{3}{4q + 12} = \frac{7q^2 + 6q + 63}{24q^2 - 216}$

224. $\frac{r}{3r - 15} - \frac{1}{4r + 20} = \frac{3r^2 + 17r + 40}{12r^2 - 300}$

225. $\frac{s}{2s + 6} - \frac{2}{5s + 5} = \frac{5s^2 - 3s - 7}{10s^2 + 40s + 30}$

226. $\frac{t}{6t - 12} - \frac{5}{2t + 10} = \frac{t^2 - 23t + 70}{12t^2 + 36t - 120}$

227. $\dfrac{2}{x^2 + 2x - 8} - \dfrac{1}{x^2 + 9x + 20} = \dfrac{4}{x^2 + 3x - 10}$

228. $\dfrac{5}{x^2 + 4x + 3} + \dfrac{2}{x^2 + x - 6} = \dfrac{3}{x^2 - x - 2}$

229. $\dfrac{3}{x^2 - 5x - 6} + \dfrac{3}{x^2 - 7x + 6} = \dfrac{6}{x^2 - 1}$

230. $\dfrac{2}{x^2 + 2x - 3} + \dfrac{3}{x^2 + 4x + 3} = \dfrac{6}{x^2 - 1}$

Solve Rational Equations that Involve Functions

231. For rational function, $f(x) = \dfrac{x - 2}{x^2 + 6x + 8}$,

ⓐ find the domain of the function

ⓑ solve $f(x) = 5$

ⓒ find the points on the graph at this function value.

232. For rational function, $f(x) = \dfrac{x + 1}{x^2 - 2x - 3}$,

ⓐ find the domain of the function

ⓑ solve $f(x) = 1$

ⓒ find the points on the graph at this function value.

233. For rational function, $f(x) = \dfrac{2 - x}{x^2 - 7x + 10}$,

ⓐ find the domain of the function

ⓑ solve $f(x) = 2$

ⓒ find the points on the graph at this function value.

234. For rational function, $f(x) = \dfrac{5 - x}{x^2 + 5x + 6}$,

ⓐ find the domain of the function

ⓑ solve $f(x) = 3$

ⓒ the points on the graph at this function value.

Solve a Rational Equation for a Specific Variable

In the following exercises, solve.

235. $\dfrac{C}{r} = 2\pi$ for r.

236. $\dfrac{I}{r} = P$ for r.

237. $\dfrac{v + 3}{w - 1} = \dfrac{1}{2}$ for w.

238. $\dfrac{x + 5}{2 - y} = \dfrac{4}{3}$ for y.

239. $a = \dfrac{b + 3}{c - 2}$ for c.

240. $m = \dfrac{n}{2 - n}$ for n.

241. $\dfrac{1}{p} + \dfrac{2}{q} = 4$ for p.

242. $\dfrac{3}{s} + \dfrac{1}{t} = 2$ for s.

243. $\dfrac{2}{v} + \dfrac{1}{5} = \dfrac{3}{w}$ for w.

244. $\dfrac{6}{x} + \dfrac{2}{3} = \dfrac{1}{y}$ for y.

245. $\dfrac{m + 3}{n - 2} = \dfrac{4}{5}$ for n.

246. $r = \dfrac{s}{3 - t}$ for t.

247. $\dfrac{E}{c} = m^2$ for c.

248. $\dfrac{R}{T} = W$ for T.

249. $\dfrac{3}{x} - \dfrac{5}{y} = \dfrac{1}{4}$ for y.

250. $c = \dfrac{2}{a} + \dfrac{b}{5}$ for a.

Writing Exercises

251. Your class mate is having trouble in this section. Write down the steps you would use to explain how to solve a rational equation.

252. Alek thinks the equation $\dfrac{y}{y + 6} = \dfrac{72}{y^2 - 36} + 4$ has two solutions, $y = -6$ and $y = 4$. Explain why Alek is wrong.

Self Check

ⓐ *After completing the exercises, use this checklist to evaluate your mastery of the objectives of this section.*

I can...	Confidently	With some help	No-I don't get it!
solve rational equations.			
solve rational equations involving functions.			
solve rational equations for a specific variable.			

ⓑ *On a scale of 1 − 10, how would you rate your mastery of this section in light of your responses on the checklist? How can you improve this?*

7.5 | Solve Applications with Rational Equations

Learning Objectives

By the end of this section, you will be able to:

- › Solve proportions
- › Solve similar figure applications
- › Solve uniform motion applications
- › Solve work applications
- › Solve direct variation problems
- › Solve inverse variation problems

Be Prepared!

Before you get started, take this readiness quiz.

1. Solve: $2(n - 1) - 4 = -10$.
   If you missed this problem, review **Example 2.13**.

2. An express train and a charter bus leave Chicago to travel to Champaign. The express train can make the trip in two hours and the bus takes five hours for the trip. The speed of the express train is 42 miles per hour faster than the speed of the bus. Find the speed of the bus.
   If you missed this problem, review **Example 2.43**.

3. Solve $\frac{1}{3}x + \frac{1}{4}x = \frac{5}{6}$.
   If you missed this problem, review **Example 2.9**.

Solve Proportions

When two rational expressions are equal, the equation relating them is called a **proportion**.

Proportion

A **proportion** is an equation of the form $\frac{a}{b} = \frac{c}{d}$, where $b \neq 0$, $d \neq 0$.

The proportion is read "*a* is to *b* as *c* is to *d*."

The equation $\frac{1}{2} = \frac{4}{8}$ is a proportion because the two fractions are equal. The proportion $\frac{1}{2} = \frac{4}{8}$ is read "1 is to 2 as 4 is to 8."

Since a proportion is an equation with rational expressions, we will solve proportions the same way we solved rational equations. We'll multiply both sides of the equation by the LCD to clear the fractions and then solve the resulting equation.

EXAMPLE 7.43

Solve: $\frac{n}{n + 14} = \frac{5}{7}$.

 Solution

$$\frac{n}{n + 14} = \frac{5}{7} \qquad n \neq -14$$

Multiply both sides by LCD.	$7(n + 14)\left(\frac{n}{n + 14}\right) = 7(n + 14)\left(\frac{5}{7}\right)$
Remove common factors on each side.	$7n = 5(n + 14)$
Simplify.	$7n = 5n + 70$
Solve for *n*.	$2n = 70$

$$n = 35$$

Check.	
	$\dfrac{n}{n+14} = \dfrac{5}{7}$
Substitute $n = 35$.	$\dfrac{35}{35+14} \overset{?}{=} \dfrac{5}{7}$
Simplify.	$\dfrac{35}{49} \overset{?}{=} \dfrac{5}{7}$
Show common factors.	$\dfrac{5 \cdot 7}{7 \cdot 7} \overset{?}{=} \dfrac{5}{7}$
Simplify.	$\dfrac{5}{7} = \dfrac{5}{7}\ \checkmark$

> **TRY IT :: 7.85**   Solve the proportion: $\dfrac{y}{y+55} = \dfrac{3}{8}$.

> **TRY IT :: 7.86**   Solve the proportion: $\dfrac{z}{z-84} = -\dfrac{1}{5}$.

Notice in the last example that when we cleared the fractions by multiplying by the LCD, the result is the same as if we had cross-multiplied.

$$\dfrac{n}{n+14} = \dfrac{5}{7} \qquad\qquad \dfrac{n}{n+14} = \dfrac{5}{7}$$

$$7(n+14)\left(\dfrac{n}{n+14}\right) = 7(n+14)\left(\dfrac{5}{7}\right) \qquad \dfrac{n}{n+14} \diagup \dfrac{5}{7}$$

$$7n = 5(n+14) \qquad\qquad 7n = 5(n+14)$$

For any proportion, $\dfrac{a}{b} = \dfrac{c}{d}$, we get the same result when we clear the fractions by multiplying by the LCD as when we cross-multiply.

$$\dfrac{a}{b} = \dfrac{c}{d} \qquad\qquad \dfrac{a}{b} = \dfrac{c}{d}$$

$$bd\left(\dfrac{a}{b} = \dfrac{c}{d}\right)bd \qquad\qquad \dfrac{a}{b} \diagup \dfrac{c}{d}$$

$$ad = bc \qquad\qquad ad = bc$$

To solve applications with proportions, we will follow our usual strategy for solving applications But when we set up the proportion, we must make sure to have the units correct—the units in the numerators must match each other and the units in the denominators must also match each other.

EXAMPLE 7.44

When pediatricians prescribe acetaminophen to children, they prescribe 5 milliliters (ml) of acetaminophen for every 25 pounds of the child's weight. If Zoe weighs 80 pounds, how many milliliters of acetaminophen will her doctor prescribe?

⊘ Solution

Identify what we are asked to find, and choose a variable to represent it.	How many ml of acetaminophen will the doctor prescribe?
	Let a = ml of acetaminophen.
Write a sentence that gives the information to find it.	If 5 ml is prescribed for every 25 pounds, how much will be prescribed for 80 pounds?
Translate into a proportion—be careful of the units.	
$\dfrac{\text{ml}}{\text{pounds}} = \dfrac{\text{ml}}{\text{pounds}}$	$\dfrac{5}{25} = \dfrac{a}{80}$
Multiply both sides by the LCD, 400.	$400\left(\dfrac{5}{25}\right) = 400\left(\dfrac{a}{80}\right)$
Remove common factors on each side.	$25 \cdot 16\left(\dfrac{5}{25}\right) = 80 \cdot 5\left(\dfrac{a}{80}\right)$
Simplify, but don't multiply on the left. Notice what the next step will be.	$16 \cdot 5 = 5a$
Solve for a.	$\dfrac{16 \cdot 5}{5} = \dfrac{5a}{5}$
	$16 = a$

Check.
Is the answer reasonable?

Yes, since 80 is about 3 times 25, the medicine should be about 3 times 5. So 16 ml makes sense.
Substitute $a = 16$ in the original proportion.

$$\dfrac{5}{25} = \dfrac{a}{80}$$

$$\dfrac{5}{25} \overset{?}{=} \dfrac{16}{80}$$

$$\dfrac{1}{5} = \dfrac{1}{5} \checkmark$$

Write a complete sentence.	The pediatrician would prescribe 16 ml of acetaminophen to Zoe.

> **TRY IT : : 7.87**

> Pediatricians prescribe 5 milliliters (ml) of acetaminophen for every 25 pounds of a child's weight. How many milliliters of acetaminophen will the doctor prescribe for Emilia, who weighs 60 pounds?

> **TRY IT : : 7.88**

> For every 1 kilogram (kg) of a child's weight, pediatricians prescribe 15 milligrams (mg) of a fever reducer. If Isabella weighs 12 kg, how many milligrams of the fever reducer will the pediatrician prescribe?

Solve similar figure applications

When you shrink or enlarge a photo on a phone or tablet, figure out a distance on a map, or use a pattern to build a bookcase or sew a dress, you are working with **similar figures**. If two figures have exactly the same shape, but different sizes, they are said to be similar. One is a scale model of the other. All their corresponding angles have the same measures and their corresponding sides have the same ratio.

Similar Figures

Two figures are similar if the measures of their corresponding angles are equal and their corresponding sides have the same ratio.

For example, the two triangles in **Figure 7.2** are similar. Each side of $\triangle ABC$ is four times the length of the corresponding side of $\triangle XYZ$.

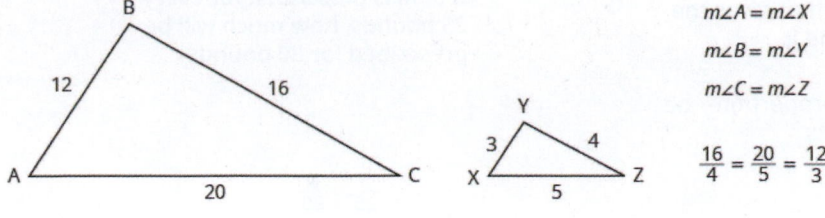

Figure 7.2

This is summed up in the Property of Similar Triangles.

Property of Similar Triangles

If $\triangle ABC$ is similar to $\triangle XYZ$, then their corresponding angle measure are equal and their corresponding sides have the same ratio.

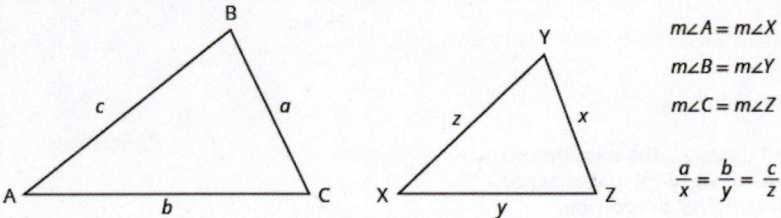

To solve applications with similar figures we will follow the Problem-Solving Strategy for Geometry Applications we used earlier.

EXAMPLE 7.45

On a map, San Francisco, Las Vegas, and Los Angeles form a triangle. The distance between the cities is measured in inches. The figure on the left below represents the triangle formed by the cities on the map. If the actual distance from Los Angeles to Las Vegas is 270 miles, find the distance from Los Angeles to San Francisco.

⊘ **Solution**

Since the triangles are similar, the corresponding sides are proportional.

Read the problem. Draw the figures and label it with the given information.	The figures are shown above.
Identify what we are looking for.	the actual distance from Los Angeles to San Francisco
Name the variables.	Let x = distance from Los Angeles to San Francisco.
Translate into an equation. Since the triangles are similar, the corresponding sides are proportional. We'll make the numerators "miles" and the denominators "inches".	$\dfrac{x \text{ miles}}{1.3 \text{ inches}} = \dfrac{270 \text{ miles}}{1 \text{ inch}}$
Solve the equation.	$1.3\left(\dfrac{x}{1.3}\right) = 1.3\left(\dfrac{270}{1}\right)$
	$x = 351$

Check.
On the map, the distance from Los Angeles to San Francisco is more than the distance from Los Angeles to Las Vegas. Since 351 is more than 270 the answer makes sense.

Check $x = 351$ in the original proportion.
Use a calculator.

$\dfrac{x \text{ miles}}{1.3 \text{ inches}} = \dfrac{270 \text{ miles}}{1 \text{ inch}}$

$\dfrac{351 \text{ miles}}{1.3 \text{ inches}} \overset{?}{=} \dfrac{270 \text{ miles}}{1 \text{ inch}}$

$\dfrac{270 \text{ miles}}{1 \text{ inch}} = \dfrac{270 \text{ miles}}{1 \text{ inch}} \checkmark$

Answer the question.	The distance from Los Angeles to San Francisco is 351 miles.

On the map, Seattle, Portland, and Boise form a triangle. The distance between the cities is measured in inches. The figure on the left below represents the triangle formed by the cities on the map. The actual distance from Seattle to Boise is 400 miles.

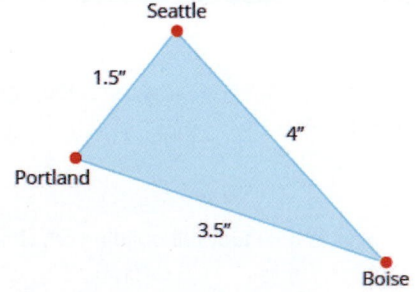

TRY IT :: 7.89   Find the actual distance from Seattle to Portland.

TRY IT :: 7.90   Find the actual distance from Portland to Boise.

We can use similar figures to find heights that we cannot directly measure.

EXAMPLE 7.46

Tyler is 6 feet tall. Late one afternoon, his shadow was 8 feet long. At the same time, the shadow of a tree was 24 feet long. Find the height of the tree.

✓ **Solution**

Read the problem and draw a figure.

We are looking for h, the height of the tree.

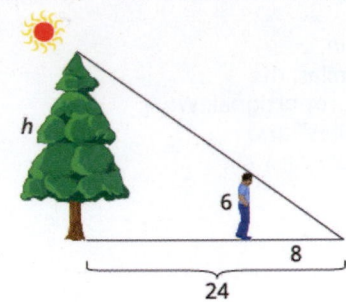

We will use similar triangles to write an equation.	
The small triangle is similar to the large triangle.	$\frac{h}{24} = \frac{6}{8}$
Solve the proportion.	$24\left(\frac{6}{8}\right) = 24\left(\frac{h}{24}\right)$
	$18 = h$
Simplify.	
Check.	

Tyler's height is less than his shadow's length so it makes sense that the tree's height is less than the length of its shadow. Check $h = 18$ in the original proportion.

$$\frac{6}{8} = \frac{h}{24}$$

$$\frac{6}{8} \overset{?}{=} \frac{18}{24}$$

$$\frac{3}{4} = \frac{3}{4} ✓$$

> **TRY IT : : 7.91**

A telephone pole casts a shadow that is 50 feet long. Nearby, an 8 foot tall traffic sign casts a shadow that is 10 feet long. How tall is the telephone pole?

> **TRY IT : : 7.92**

A pine tree casts a shadow of 80 feet next to a 30 foot tall building which casts a 40 feet shadow. How tall is the pine tree?

Solve Uniform Motion Applications

We have solved uniform motion problems using the formula $D = rt$ in previous chapters. We used a table like the one below to organize the information and lead us to the equation.

	Rate	•	Time	=	Distance

The formula $D = rt$ assumes we know r and t and use them to find D. If we know D and r and need to find t, we would solve the equation for t and get the formula $t = \frac{D}{r}$.

We have also explained how flying with or against the wind affects the speed of a plane. We will revisit that idea in the next example.

EXAMPLE 7.47

An airplane can fly 200 miles into a 30 mph headwind in the same amount of time it takes to fly 300 miles with a 30 mph tailwind. What is the speed of the airplane?

⊘ Solution

This is a uniform motion situation. A diagram will help us visualize the situation.

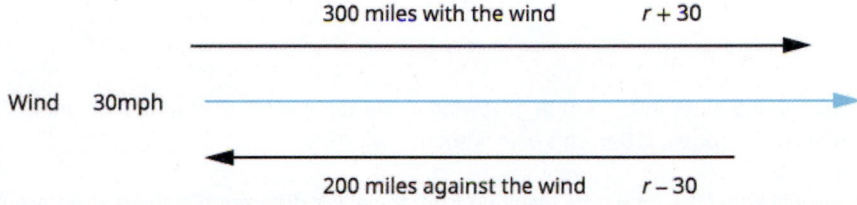

We fill in the chart to organize the information.

We are looking for the speed of the airplane.   Let r = the speed of the airplane.

When the plane flies with the wind,
the wind increases its speed and so the rate is $r + 30$.

When the plane flies against the wind,
the wind decreases its speed and the rate is $r - 30$.

Write in the rates.
Write in the distances.

	Rate	•	Time	=	Distance
Headwind	$r - 30$		$\frac{200}{r-30}$		200
Tailwind	$r + 30$		$\frac{300}{r+30}$		300

Since $D = r \cdot t$, we solve for t and get $t = \frac{D}{r}$.

We divide the distance by the rate in each row, and place the expression in the time column.

We know the times are equal and so we write our equation.

$$\frac{200}{r - 30} = \frac{300}{r + 30}$$

We multiply both sides by the LCD.

$$(r + 30)(r - 30)\left(\frac{200}{r - 30}\right) = (r + 30)(r - 30)\left(\frac{300}{r + 30}\right)$$

Simplify.

$$(r + 30)(200) = (r - 30)300$$

$$200r + 6000 = 300r - 9000$$

Solve.

$$15000 = 100r$$

Check.
Is 150 mph a reasonable speed for an airplane? Yes.
If the plane is traveling 150 mph and the wind is 30 mph,

Tailwind $\quad 150 + 30 = 180$ mph $\quad \dfrac{300}{180} = \dfrac{5}{3}$ hours

Headwind $\quad 150 - 30 = 120$ mph $\quad \dfrac{200}{120} = \dfrac{5}{3}$ hours

The times are equal, so it checks.	The plane was traveling 150 mph.

> **TRY IT : : 7.93**
>
> Link can ride his bike 20 miles into a 3 mph headwind in the same amount of time he can ride 30 miles with a 3 mph tailwind. What is Link's biking speed?

> **TRY IT : : 7.94**
>
> Danica can sail her boat 5 miles into a 7 mph headwind in the same amount of time she can sail 12 miles with a 7 mph tailwind. What is the speed of Danica's boat without a wind?

In the next example, we will know the total time resulting from travelling different distances at different speeds.

EXAMPLE 7.48

Jazmine trained for 3 hours on Saturday. She ran 8 miles and then biked 24 miles. Her biking speed is 4 mph faster than her running speed. What is her running speed?

✓ **Solution**

This is a uniform motion situation. A diagram will help us visualize the situation.

We fill in the chart to organize the information.

We are looking for Jazmine's running speed.	Let r = Jazmine's running speed.
Her biking speed is 4 miles faster than her running speed.	$r + 4$ = her biking speed

The distances are given, enter them into the chart.

Since $D = r \cdot t$, we solve for t and get $t = \dfrac{D}{r}$.

We divide the distance by the rate in each row, and place the expression in the time column.

	Rate	·	Time	=	Distance
Run	r		$\dfrac{8}{r}$		8
Bike	$r + 4$		$\dfrac{24}{r+4}$		24
			3		

Write a word sentence.	Her time plus the time biking is 3 hours.

Translate the sentence to get the equation.

$$\frac{8}{r} + \frac{24}{r+4} = 3$$

Solve.

$$r(r+4)\left(\frac{8}{r}+\frac{24}{r+4}\right) = 3 \cdot r(r+4)$$
$$8(r+4) + 24r = 3r(r+4)$$
$$8r + 32 + 24r = 3r^2 + 12r$$
$$32 + 32r = 3r^2 + 12r$$
$$0 = 3r^2 - 20r - 32$$
$$0 = (3r+4)(r-8)$$

$$(3r+4) = 0 \quad (r-8) = 0$$

$$r = \cancel{-\frac{4}{3}} \quad r = 8$$

Check.
A negative speed does not make sense in this problem,
so $r = 8$ is the solution.

Is 8 mph a reasonable running speed? Yes.
If Jazmine's running rate is 4, then her biking rate,
$r + 4$, which is $8 + 4 = 12$.

Run 8 mph $\dfrac{8 \text{ miles}}{8 \text{ mph}} = 1 \text{ hour}$

Bike 12 mph $\dfrac{24 \text{ miles}}{12 \text{ mph}} = 2 \text{ hours}$

Total 3 hours.   Jazmine's running speed is 8 mph.

> **TRY IT : :** 7.95

Dennis went cross-country skiing for 6 hours on Saturday. He skied 20 mile uphill and then 20 miles back downhill, returning to his starting point. His uphill speed was 5 mph slower than his downhill speed. What was Dennis' speed going uphill and his speed going downhill?

> **TRY IT : :** 7.96

Joon drove 4 hours to his home, driving 208 miles on the interstate and 40 miles on country roads. If he drove 15 mph faster on the interstate than on the country roads, what was his rate on the country roads?

Once again, we will use the uniform motion formula solved for the variable t.

EXAMPLE 7.49

Hamilton rode his bike downhill 12 miles on the river trail from his house to the ocean and then rode uphill to return home. His uphill speed was 8 miles per hour slower than his downhill speed. It took him 2 hours longer to get home than it took him to get to the ocean. Find Hamilton's downhill speed.

⊘ **Solution**

This is a uniform motion situation. A diagram will help us visualize the situation.

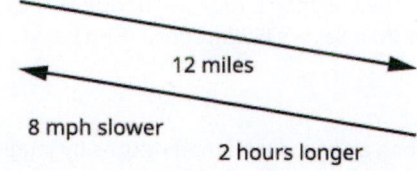

12 miles

8 mph slower

2 hours longer

We fill in the chart to organize the information.

We are looking for Hamilton's downhill speed.	Let h = Hamilton's downhill speed.
His uphill speed is 8 miles per hour slower. Enter the rates into the chart.	$h - 8$ = Hamilton's uphill speed

The distance is the same in both directions. 12 miles.

Since $D = r \cdot t$, we solve for t and get $t = \frac{D}{r}$.

We divide the distance by the rate in each row, and place the expression in the time column.

	Rate	·	Time	=	Distance
Downhill	h		$\frac{12}{h}$		12
Uphill	$h - 8$		$\frac{12}{h-8}$		12

Write a word sentence about the line.	He took 2 hours longer uphill than downhill. The uphill time is 2 more than the downhill time.

Translate the sentence to get the equation.

$$\frac{12}{h-8} = \frac{12}{h} + 2$$

Solve.

$$h(h-8)\left(\frac{12}{h-8}\right) = h(h-8)\left(\frac{12}{h} + 2\right)$$

$$12h = 12(h-8) + 2h(h-8)$$

$$12h = 12h - 96 + 2h^2 - 16h$$

$$0 = 2h^2 - 16h - 96$$

$$0 = 2(h^2 - 8h - 48)$$

$$0 = 2(h-12)(h+4)$$

$$h - 12 = 0 \quad h + 4 = 0$$

$$h = 12 \quad \cancel{h = -4}$$

Check.
Is 12 mph a reasonable speed for biking downhill? Yes.

Downhill 12 mph $\frac{12 \text{ miles}}{12 \text{ mph}} = 1$ hour

Uphill $12 - 8 = 4$ mph $\frac{12 \text{ miles}}{4 \text{ mph}} = 3$ hours.

The uphill time is 2 hours more that the downhill time.

Hamilton's downhill speed is 12 mph.

> **TRY IT ∷ 7.97**

Kayla rode her bike 75 miles home from college one weekend and then rode the bus back to college. It took her 2 hours less to ride back to college on the bus than it took her to ride home on her bike, and the average speed of the bus was 10 miles per hour faster than Kayla's biking speed. Find Kayla's biking speed.

> **TRY IT ∷ 7.98**

Victoria jogs 12 miles to the park along a flat trail and then returns by jogging on an 20 mile hilly trail. She jogs 1 mile per hour slower on the hilly trail than on the flat trail, and her return trip takes her two hours longer. Find her rate of jogging on the flat trail.

Solve Work Applications

The weekly gossip magazine has a big story about the Princess' baby and the editor wants the magazine to be printed as soon as possible. She has asked the printer to run an extra printing press to get the printing done more quickly. Press #1 takes 6 hours to do the job and Press #2 takes 12 hours to do the job. How long will it take the printer to get the magazine printed with both presses running together?

This is a typical 'work' application. There are three quantities involved here—the time it would take each of the two presses to do the job alone and the time it would take for them to do the job together.

If Press #1 can complete the job in 6 hours, in one hour it would complete $\frac{1}{6}$ of the job.

If Press #2 can complete the job in 12 hours, in one hour it would complete $\frac{1}{12}$ of the job.

We will let t be the number of hours it would take the presses to print the magazines with both presses running together. So in 1 hour working together they have completed $\frac{1}{t}$ of the job.

We can model this with the word equation and then translate to a rational equation. To find the time it would take the presses to complete the job if they worked together, we solve for t.

A chart will help us organize the information. We are looking for how many hours it would take to complete the job with both presses running together.

Let t = the number of hours needed to complete the job together.

Enter the hours per job for Press #1, Press #2, and when they work together.

If a job on Press #1 takes 6 hours, then in 1 hour $\frac{1}{6}$ of the job is completed.

	Number of hours to complete the job.	Part of job completed/hour
Press #1	6	$\frac{1}{6}$
Press #2	12	$\frac{1}{12}$
Together	t	$\frac{1}{t}$

Similarly find the part of the job completed/ hours for Press #2 and when thet both together.

Write a word sentence.

The part completed by Press #1 plus the part completed by Press #2 equals the amount completed together.

Work completed by

Press #1 + Press #2 = Together

Translate into an equation.

$$\frac{1}{6} \quad + \quad \frac{1}{12} \quad = \quad \frac{1}{t}$$

Solve.

$$\frac{1}{6} + \frac{1}{12} = \frac{1}{t}$$

Mutiply by the LCD, $12t$

$$12t\left(\frac{1}{6} + \frac{1}{12}\right) = 12t\left(\frac{1}{t}\right)$$

Simplify.

$$2t + t = 12$$
$$3t = 12$$
$$t = 4$$

When both presses are running it takes 4 hours to do the job.

Keep in mind, it should take less time for two presses to complete a job working together than for either press to do it alone.

EXAMPLE 7.50

Suppose Pete can paint a room in 10 hours. If he works at a steady pace, in 1 hour he would paint $\frac{1}{10}$ of the room. If

Alicia would take 8 hours to paint the same room, then in 1 hour she would paint $\frac{1}{8}$ of the room. How long would it take

Pete and Alicia to paint the room if they worked together (and didn't interfere with each other's progress)?

⊘ **Solution**

This is a 'work' application. A chart will help us organize the information. We are looking for the numbers of hours it will take them to paint the room together.

In one hour Pete did $\frac{1}{10}$ of the job. Alicia did $\frac{1}{8}$ of the job. And together they did $\frac{1}{t}$ of the job.

		Number of complete hours to the job.	Part of job completed/hour
Let t be the number of hours needed to paint the room together.			
Enter the hours per job for Pete, Alicia, and when they work together.	Pete	10	$\frac{1}{10}$
In 1 hour working together, they have completed $\frac{1}{t}$ of the job.	Alicia	8	$\frac{1}{8}$
Similarly, find the part of the job completed/hour by Pete and then by Alicia.	Together	t	$\frac{1}{t}$

Write a word sentence.	The work completed by Pete plus the work completed by Alicia equals the total work completed.

Work completed by:

Pete + Alicia = Together

$$\frac{1}{10} + \frac{1}{8} = \frac{1}{t}$$

$$\frac{1}{10} + \frac{1}{8} = \frac{1}{t}$$

Multiply by the LCD, 40t.	$40t\left(\frac{1}{10} + \frac{1}{8}\right) = 40t\left(\frac{1}{t}\right)$
Distribute.	$40t \cdot \frac{1}{10} + 40t \cdot \frac{1}{8} = 40t\left(\frac{1}{t}\right)$
Simplify and solve.	$4t + 5t = 40$ $9t = 40$ $t = \frac{40}{9}$
We'll write as a mixed number so that we can convert it to hours and minutes.	$t = 4\frac{4}{9}$ hours
Remember, 1 hour = 60 minutes.	$t = 4$ hours $+ \frac{4}{9}$ (60 minutes)
Multiply, and then round to the nearest minute.	$t = 4$ hours $+ 27$ minutes

It would take Pete and Alica about
4 hours and 27 minutes to paint the room.

> **TRY IT : : 7.99**

One gardener can mow a golf course in 4 hours, while another gardener can mow the same golf course in 6 hours. How long would it take if the two gardeners worked together to mow the golf course?

> **TRY IT : : 7.100**

Daria can weed the garden in 7 hours, while her mother can do it in 3. How long will it take the two of them working together?

EXAMPLE 7.51

Ra'shon can clean the house in 7 hours. When his sister helps him it takes 3 hours. How long does it take his sister when she cleans the house alone?

⊘ Solution

This is a work problem. A chart will help us organize the information.

We are looking for how many hours it would take Ra'shon's sister to complete the job by herself.

Let s be the number of hours Ra'shon's sister takes to clean the house alone.

Enter the hours per job for Ra'shon, his sister, and when they work together.

If Ra'shon takes 7 hours, then in 1 hour $\frac{1}{7}$ of the job is completed.
If Ra'shon's sister takes s hours, then in 1 hour $\frac{1}{s}$ of the job is completed.

	Number of hours to clean the house	Part of job completed/hour
Ra'shon	7	$\frac{1}{7}$
His sister	s	$\frac{1}{s}$
Together	3	$\frac{1}{3}$

Write a word sentence.

The part completed by Ra'shon plus the part by his sister equals the amount completed together.

Translate to an equation.

Work completed by
Ra'shon + His sister = Together
$$\frac{1}{7} \quad + \quad \frac{1}{s} \quad = \quad \frac{1}{3}$$

Solve.

$$\frac{1}{7} + \frac{1}{s} = \frac{1}{3}$$

Multiply by the LCD, 21s.

$$21s\left(\frac{1}{7} + \frac{1}{s}\right) = \left(\frac{1}{3}\right)21s$$

$$3s + 21 = 7s$$

$$-4s = -21$$

Simplify.

$$s = \frac{-21}{-4} = \frac{21}{4}$$

Write as a mixed number to convert it to hours and minutes.

$$s = 5\frac{1}{4}\text{ hours}$$

There are 60 minutes in 1 hour.

$$s = 5\text{ hours} + \frac{1}{4}(60\text{ minutes})$$

$$s = 5\text{ hours} + 15\text{ minutes}$$

It would take Ra'shon's sister 5 hours and 15 minutes to clean the house alone.

> **TRY IT :: 7.101**
>
> Alice can paint a room in 6 hours. If Kristina helps her it takes them 4 hours to paint the room. How long would it take Kristina to paint the room by herself?

> **TRY IT :: 7.102**
>
> Tracy can lay a slab of concrete in 3 hours, with Jordan's help they can do it in 2 hours. If Jordan works alone, how long will it take?

Solve Direct Variation Problems

When two quantities are related by a proportion, we say they are *proportional* to each other. Another way to express this relation is to talk about the *variation* of the two quantities. We will discuss direct variation and inverse variation in this section.

Lindsay gets paid $15 per hour at her job. If we let *s* be her salary and *h* be the number of hours she has worked, we could model this situation with the equation

$$s = 15h$$

Lindsay's salary is the product of a constant, 15, and the number of hours she works. We say that Lindsay's salary *varies directly* with the number of hours she works. Two variables vary directly if one is the product of a constant and the other.

Direct Variation

For any two variables *x* and *y*, *y* varies directly with *x* if

$$y = kx, \quad \text{where } k \neq 0$$

The constant *k* is called the constant of variation.

In applications using direct variation, generally we will know values of one pair of the variables and will be asked to find the equation that relates *x* and *y*. Then we can use that equation to find values of *y* for other values of *x*.

We'll list the steps here.

HOW TO :: SOLVE DIRECT VARIATION PROBLEMS.

Step 1.   Write the formula for direct variation.

Step 2.   Substitute the given values for the variables.

Step 3.   Solve for the constant of variation.

Step 4.   Write the equation that relates *x* and *y* using the constant of variation.

Now we'll solve an application of direct variation.

EXAMPLE 7.52

When Raoul runs on the treadmill at the gym, the number of calories, *c*, he burns varies directly with the number of minutes, *m*, he uses the treadmill. He burned 315 calories when he used the treadmill for 18 minutes.

ⓐ Write the equation that relates *c* and *m*. ⓑ How many calories would he burn if he ran on the treadmill for 25 minutes?

⊘ **Solution**

ⓐ

The number of calories, *c*, varies directly with the number of minutes, *m*, on the treadmill, and $c = 315$ when $m = 18$.

Write the formula for direct variation. $y = kx$

We will use c in place of y and m in place of x.	$c = km$
Substitute the given values for the variables.	$315 = k \cdot 18$
Solve for the constant of variation.	$\dfrac{315}{18} = \dfrac{k \cdot 18}{18}$ $17.5 = k$
Write the equation that relates c and m.	$c = km$
Substitute in the constant of variation.	$c = 17.5m$

ⓑ

	Find c when $m = 25$.
Write the equation that relates c and m.	$c = 17.5m$
Substitute the given value for m.	$c = 17.5(25)$
Simplify.	$c = 437.5$

Raoul would burn 437.5 calories if he used the treadmill for 25 minutes.

> **TRY IT : : 7.103**
>
> The number of calories, c, burned varies directly with the amount of time, t, spent exercising. Arnold burned 312 calories in 65 minutes exercising.
>
> ⓐ Write the equation that relates c and t. ⓑ How many calories would he burn if he exercises for 90 minutes?

> **TRY IT : : 7.104**
>
> The distance a moving body travels, d, varies directly with time, t, it moves. A train travels 100 miles in 2 hours
>
> ⓐ Write the equation that relates d and t. ⓑ How many miles would it travel in 5 hours?

Solve Inverse Variation Problems

Many applications involve two variable that *vary inversely*. As one variable increases, the other decreases. The equation that relates them is $y = \dfrac{k}{x}$.

Inverse Variation

For any two variables x and y, y varies inversely with x if

$$y = \frac{k}{x}, \ \text{ where } k \neq 0$$

The constant k is called the constant of variation.

The word 'inverse' in inverse variation refers to the multiplicative inverse. The multiplicative inverse of x is $\dfrac{1}{x}$.

We solve inverse variation problems in the same way we solved direct variation problems. Only the general form of the equation has changed. We will copy the procedure box here and just change 'direct' to 'inverse'.

HOW TO :: SOLVE INVERSE VARIATION PROBLEMS.

Step 1.   Write the formula for inverse variation.

Step 2.   Substitute the given values for the variables.

Step 3.   Solve for the constant of variation.

Step 4.   Write the equation that relates x and y using the constant of variation.

EXAMPLE 7.53

The frequency of a guitar string varies inversely with its length. A 26 in.-long string has a frequency of 440 vibrations per second.

ⓐ Write the equation of variation. ⓑ How many vibrations per second will there be if the string's length is reduced to 20 inches by putting a finger on a fret?

 Solution

ⓐ

	The frequency varies inversely with the length.
Name the variables.	Let f = frequency. L = length
Write the formula for inverse variation.	$y = \frac{k}{x}$
We will use f in place of y and L in place of x.	$f = \frac{k}{L}$
Substitute the given values for the variables.	$f = 440$ when $L = 26$ $440 = \frac{k}{26}$
Solve for the constant of variation	$26(440) = 26\left(\frac{k}{26}\right)$ $11{,}440 = k$
Write the equation that relates f and L.	$f = \frac{k}{L}$
Substitute the constant of variation	$f = \frac{11{,}440}{L}$

ⓑ

	Find f when $L = 20$.
Write the equation that relates f and L.	$f = \dfrac{11{,}440}{L}$
Substitute the given value for L.	$f = \dfrac{11{,}440}{20}$
Simplify.	$f = 572$ A 20″-guitar string has frequency 572 vibrations per second.

> **TRY IT : : 7.105**

The number of hours it takes for ice to melt varies inversely with the air temperature. Suppose a block of ice melts in 2 hours when the temperature is 65 degrees Celsius.

ⓐ Write the equation of variation. ⓑ How many hours would it take for the same block of ice to melt if the temperature was 78 degrees?

> **TRY IT : : 7.106**

Xander's new business found that the daily demand for its product was inversely proportional to the price, p.

When the price is $5, the demand is 700 units.

ⓐ Write the equation of variation. ⓑ What is the demand if the price is raised to $7?

▶ **MEDIA : :**

Access this online resource for additional instruction and practice with applications of rational expressions

- **Applications of Rational Expressions (https://openstax.org/l/37AppRatExp)**

7.5 EXERCISES

Practice Makes Perfect

Solve Proportions

In the following exercises, solve each proportion.

253. $\dfrac{x}{56} = \dfrac{7}{8}$

254. $\dfrac{56}{72} = \dfrac{y}{9}$

255. $\dfrac{98}{154} = \dfrac{-7}{p}$

256. $\dfrac{72}{156} = \dfrac{-6}{q}$

257. $\dfrac{a}{a+12} = \dfrac{4}{7}$

258. $\dfrac{b}{b-16} = \dfrac{11}{9}$

259. $\dfrac{m+90}{25} = \dfrac{m+30}{15}$

260. $\dfrac{n+10}{4} = \dfrac{40-n}{6}$

261. $\dfrac{2p+4}{8} = \dfrac{p+18}{6}$

262. $\dfrac{q-2}{2} = \dfrac{2q-7}{18}$

In the following exercises, solve.

263. Kevin wants to keep his heart rate at 160 beats per minute while training. During his workout he counts 27 beats in 10 seconds.

ⓐ How many beats per minute is this?

ⓑ Has Kevin met his target heart rate?

264. Jesse's car gets 30 miles per gallon of gas.

ⓐ If Las Vegas is 285 miles away, how many gallons of gas are needed to get there and then home?

ⓑ If gas is $3.09 per gallon, what is the total cost of the gas for the trip?

265. Pediatricians prescribe 5 milliliters (ml) of acetaminophen for every 25 pounds of a child's weight. How many milliliters of acetaminophen will the doctor prescribe for Jocelyn, who weighs 45 pounds?

266. A veterinarian prescribed Sunny, a 65-pound dog, an antibacterial medicine in case an infection emerges after her teeth were cleaned. If the dosage is 5 mg for every pound, how much medicine was Sunny given?

267. A new energy drink advertises 106 calories for 8 ounces. How many calories are in 12 ounces of the drink?

268. One 12-ounce can of soda has 150 calories. If Josiah drinks the big 32-ounce size from the local mini-mart, how many calories does he get?

269. Kyra is traveling to Canada and will change $250 US dollars into Canadian dollars. At the current exchange rate, $1 US is equal to $1.3 Canadian. How many Canadian dollars will she get for her trip?

270. Maurice is traveling to Mexico and needs to exchange $450 into Mexican pesos. If each dollar is worth 12.29 pesos, how many pesos will he get for his trip?

271. Ronald needs a morning breakfast drink that will give him at least 390 calories. Orange juice has 130 calories in one cup. How many cups does he need to drink to reach his calorie goal?

272. Sonya drinks a 32-ounce energy drink containing 80 calories per 12 ounce. How many calories did she drink?

273. Phil wants to fertilize his lawn. Each bag of fertilizer covers about 4,000 square feet of lawn. Phil's lawn is approximately 13,500 square feet. How many bags of fertilizer will he have to buy?

274. An oatmeal cookie recipe calls for $\dfrac{1}{2}$ cup of butter to make 4 dozen cookies. Hilda needs to make 10 dozen cookies for the bake sale. How many cups of butter will she need?

Solve Similar Figure Applications

In the following exercises, the triangles are similar. Find the length of the indicated side.

275.

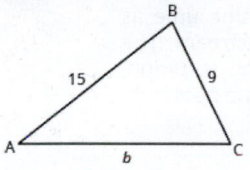

ⓐ side *x*

ⓑ side *b*

276.

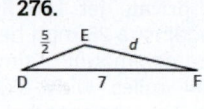

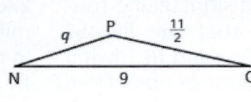

ⓐ side *d*

ⓑ side *q*

In the following exercises, use the map shown. On the map, New York City, Chicago, and Memphis form a triangle. The actual distance from New York to Chicago is 800 miles.

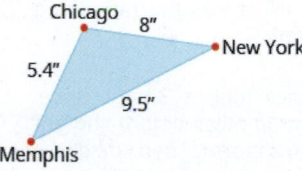

277. Find the actual distance from New York to Memphis.

278. Find the actual distance from Chicago to Memphis.

In the following exercises, use the map shown. On the map, Atlanta, Miami, and New Orleans form a triangle. The actual distance from Atlanta to New Orleans is 420 miles.

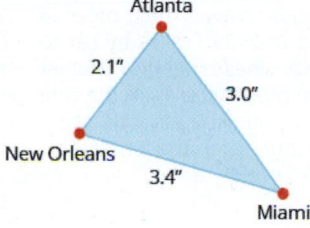

279. Find the actual distance from New Orleans to Miami.

280. Find the actual distance from Atlanta to Miami.

In the following exercises, answer each question.

281. A 2-foot-tall dog casts a 3-foot shadow at the same time a cat casts a one foot shadow. How tall is the cat ?

282. Larry and Tom were standing next to each other in the backyard when Tom challenged Larry to guess how tall he was. Larry knew his own height is 6.5 feet and when they measured their shadows, Larry's shadow was 8 feet and Tom's was 7.75 feet long. What is Tom's height?

283. The tower portion of a windmill is 212 feet tall. A six foot tall person standing next to the tower casts a seven-foot shadow. How long is the windmill's shadow?

284. The height of the Statue of Liberty is 305 feet. Nikia, who is standing next to the statue, casts a 6-foot shadow and she is 5 feet tall. How long should the shadow of the statue be?

Solve Uniform Motion Applications

In the following exercises, solve the application problem provided.

285. Mary takes a sightseeing tour on a helicopter that can fly 450 miles against a 35-mph headwind in the same amount of time it can travel 702 miles with a 35-mph tailwind. Find the speed of the helicopter.

286. A private jet can fly 1,210 miles against a 25-mph headwind in the same amount of time it can fly 1694 miles with a 25-mph tailwind. Find the speed of the jet.

287. A boat travels 140 miles downstream in the same time as it travels 92 miles upstream. The speed of the current is 6mph. What is the speed of the boat?

288. Darrin can skateboard 2 miles against a 4-mph wind in the same amount of time he skateboards 6 miles with a 4-mph wind. Find the speed Darrin skateboards with no wind.

289. Jane spent 2 hours exploring a mountain with a dirt bike. First, she rode 40 miles uphill. After she reached the peak she rode for 12 miles along the summit. While going uphill, she went 5 mph slower than when she was on the summit. What was her rate along the summit?

290. Laney wanted to lose some weight so she planned a day of exercising. She spent a total of 2 hours riding her bike and jogging. She biked for 12 miles and jogged for 6 miles. Her rate for jogging was 10 mph less than biking rate. What was her rate when jogging?

291. Byron wanted to try out different water craft. He went 62 miles downstream in a motor boat and 27 miles downstream on a jet ski. His speed on the jet ski was 10 mph faster than in the motor boat. Bill spent a total of 4 hours on the water. What was his rate of speed in the motor boat?

292. Nancy took a 3-hour drive. She went 50 miles before she got caught in a storm. Then she drove 68 miles at 9 mph less than she had driven when the weather was good. What was her speed driving in the storm?

293. Chester rode his bike uphill 24 miles and then back downhill at 2 mph faster than his uphill. If it took him 2 hours longer to ride uphill than downhill, what was his uphill rate?

294. Matthew jogged to his friend's house 12 miles away and then got a ride back home. It took him 2 hours longer to jog there than ride back. His jogging rate was 25 mph slower than the rate when he was riding. What was his jogging rate?

295. Hudson travels 1080 miles in a jet and then 240 miles by car to get to a business meeting. The jet goes 300 mph faster than the rate of the car, and the car ride takes 1 hour longer than the jet. What is the speed of the car?

296. Nathan walked on an asphalt pathway for 12 miles. He walked the 12 miles back to his car on a gravel road through the forest. On the asphalt he walked 2 miles per hour faster than on the gravel. The walk on the gravel took one hour longer than the walk on the asphalt. How fast did he walk on the gravel.

297. John can fly his airplane 2800 miles with a wind speed of 50 mph in the same time he can travel 2400 miles against the wind. If the speed of the wind is 50 mph, find the speed of his airplane.

298. Jim's speedboat can travel 20 miles upstream against a 3-mph current in the same amount of time it travels 22 miles downstream with a 3-mph current speed . Find the speed of the Jim's boat.

299. Hazel needs to get to her granddaughter's house by taking an airplane and a rental car. She travels 900 miles by plane and 250 miles by car. The plane travels 250 mph faster than the car. If she drives the rental car for 2 hours more than she rode the plane, find the speed of the car.

300. Stu trained for 3 hours yesterday. He ran 14 miles and then biked 40 miles. His biking speed is 6 mph faster than his running speed. What is his running speed?

301. When driving the 9-hour trip home, Sharon drove 390 miles on the interstate and 150 miles on country roads. Her speed on the interstate was 15 more than on country roads. What was her speed on country roads?

302. Two sisters like to compete on their bike rides. Tamara can go 4 mph faster than her sister, Samantha. If it takes Samantha 1 hours longer than Tamara to go 80 miles, how fast can Samantha ride her bike?

303. Dana enjoys taking her dog for a walk, but sometimes her dog gets away, and she has to run after him. Dana walked her dog for 7 miles but then had to run for 1 mile, spending a total time of 2.5 hours with her dog. Her running speed was 3 mph faster than her walking speed. Find her walking speed.

304. Ken and Joe leave their apartment to go to a football game 45 miles away. Ken drives his car 30 mph faster Joe can ride his bike. If it takes Joe 2 hours longer than Ken to get to the game, what is Joe's speed?

Solve Work Applications

305. Mike, an experienced bricklayer, can build a wall in 3 hours, while his son, who is learning, can do the job in 6 hours. How long does it take for them to build a wall together?

306. It takes Sam 4 hours to rake the front lawn while his brother, Dave, can rake the lawn in 2 hours. How long will it take them to rake the lawn working together?

307. Mia can clean her apartment in 6 hours while her roommate can clean the apartment in 5 hours. If they work together, how long would it take them to clean the apartment?

308. Brian can lay a slab of concrete in 6 hours, while Greg can do it in 4 hours. If Brian and Greg work together, how long will it take?

309. Josephine can correct her students test papers in 5 hours, but if her teacher's assistant helps, it would take them 3 hours. How long would it take the assistant to do it alone?

310. Washing his dad's car alone, eight year old Levi takes 2.5 hours. If his dad helps him, then it takes 1 hour. How long does it take Levi's dad to wash the car by himself?

311. At the end of the day Dodie can clean her hair salon in 15 minutes. Ann, who works with her, can clean the salon in 30 minutes. How long would it take them to clean the shop if they work together?

312. Ronald can shovel the driveway in 4 hours, but if his brother Donald helps it would take 2 hours. How long would it take Donald to shovel the driveway alone?

Solve Direct Variation Problems

In the following exercises, solve.

313. If y varies directly as x and $y = 14$, when $x = 3$. find the equation that relates x and y.

314. If a varies directly as b and $a = 16$, when $b = 4$. find the equation that relates a and b.

315. If p varies directly as q and $p = 9.6$, when $q = 3$. find the equation that relates p and q.

316. If v varies directly as w and $v = 8$, when $w = \frac{1}{2}$. find the equation that relates v and w.

317. The price, P, that Eric pays for gas varies directly with the number of gallons, g, he buys. It costs him $50 to buy 20 gallons of gas.

ⓐ Write the equation that relates P and g.

ⓑ How much would 33 gallons cost Eric?

318. Joseph is traveling on a road trip. The distance, d, he travels before stopping for lunch varies directly with the speed, v, he travels. He can travel 120 miles at a speed of 60 mph.

ⓐ Write the equation that relates d and v.

ⓑ How far would he travel before stopping for lunch at a rate of 65 mph?

319. The mass of a liquid varies directly with its volume. A liquid with mass 16 kilograms has a volume of 2 liters.

ⓐ Write the equation that relates the mass to the volume.

ⓑ What is the volume of this liquid if its mass is 128 kilograms?

320. The length that a spring stretches varies directly with a weight placed at the end of the spring. When Sarah placed a 10-pound watermelon on a hanging scale, the spring stretched 5 inches.

ⓐ Write the equation that relates the length of the spring to the weight.

ⓑ What weight of watermelon would stretch the spring 6 inches?

321. The maximum load a beam will support varies directly with the square of the diagonal of the beam's cross-section. A beam with diagonal 6 inch will support a maximum load of 108 pounds.

ⓐ Write the equation that relates the load to the diagonal of the cross-section.

ⓑ What load will a beam with a 10-inch diagonal support?

322. The area of a circle varies directly as the square of the radius. A circular pizza with a radius of 6 inches has an area of 113.04 square inches.

ⓐ Write the equation that relates the area to the radius.

ⓑ What is the area of a personal pizza with a radius 4 inches?

Solve Inverse Variation Problems

In the following exercises, solve.

323. If y varies inversely with x and $y = 5$ when $x = 4$, find the equation that relates x and y.

324. If p varies inversely with q and $p = 2$ when $q = 1$, find the equation that relates p and q.

325. If v varies inversely with w and $v = 6$ when $w = \frac{1}{2}$, find the equation that relates v and w.

326. If a varies inversely with b and $a = 12$ when $b = \frac{1}{3}$, find the equation that relates a and b.

In the following exercises, write an inverse variation equation to solve the following problems.

327. The fuel consumption (mpg) of a car varies inversely with its weight. A Toyota Corolla weighs 2800 pounds getting 33 mpg on the highway.

ⓐ Write the equation that relates the mpg to the car's weight.

ⓑ What would the fuel consumption be for a Toyota Sequoia that weighs 5500 pounds?

328. A car's value varies inversely with its age. Jackie bought a 10-year-old car for $2,400.

ⓐ Write the equation that relates the car's value to its age.

ⓑ What will be the value of Jackie's car when it is 15 years old?

329. The time required to empty a tank varies inversely as the rate of pumping. It took Ada 5 hours to pump her flooded basement using a pump that was rated at 200 gpm (gallons per minute).

ⓐ Write the equation that relates the number of hours to the pump rate.

ⓑ How long would it take Ada to pump her basement if she used a pump rated at 400 gpm?

330. On a string instrument, the length of a string varies inversely as the frequency of its vibrations. An 11-inch string on a violin has a frequency of 400 cycles per second.

ⓐ Write the equation that relates the string length to its frequency.

ⓑ What is the frequency of a 10 inch string?

331. Paul, a dentist, determined that the number of cavities that develops in his patient's mouth each year varies inversely to the number of minutes spent brushing each night. His patient, Lori, had four cavities when brushing her teeth 30 seconds (0.5 minutes) each night.

ⓐ Write the equation that relates the number of cavities to the time spent brushing.

ⓑ How many cavities would Paul expect Lori to have if she had brushed her teeth for 2 minutes each night?

332. Boyle's law states that if the temperature of a gas stays constant, then the pressure varies inversely to the volume of the gas. Braydon, a scuba diver, has a tank that holds 6 liters of air under a pressure of 220 psi.

ⓐ Write the equation that relates pressure to volume.

ⓑ If the pressure increases to 330 psi, how much air can Braydon's tank hold?

333. The cost of a ride service varies directly with the distance traveled. It costs $35 for a ride from the city center to the airport, 14 miles away.

ⓐ Write the equation that relates the cost, c, with the number of miles, m.

ⓑ What would it cost to travel 22 miles with this service?

334. The number of hours it takes Jack to drive from Boston to Bangor is inversely proportional to his average driving speed. When he drives at an average speed of 40 miles per hour, it takes him 6 hours for the trip.

ⓐ Write the equation that relates the number of hours, h, with the speed, s.

ⓑ How long would the trip take if his average speed was 75 miles per hour?

Writing Exercises

335. Marisol solves the proportion $\frac{144}{a} = \frac{9}{4}$ by 'cross multiplying,' so her first step looks like $4 \cdot 144 = 9 \cdot a$. Explain how this differs from the method of solution shown in **Example 7.44**.

336. Paula and Yuki are roommates. It takes Paula 3 hours to clean their apartment. It takes Yuki 4 hours to clean the apartment. The equation $\frac{1}{3} + \frac{1}{4} = \frac{1}{t}$ can be used to find t, the number of hours it would take both of them, working together, to clean their apartment. Explain how this equation models the situation.

337. In your own words, explain the difference between direct variation and inverse variation.

338. Make up an example from your life experience of inverse variation.

Self Check

ⓐ After completing the exercises, use this checklist to evaluate your mastery of the objectives of this section.

I can...	Confidently	With some help	No-I don't get it!
solve proportions.			
solve similar figure applications.			
solve uniform motion applications.			
solve work applications.			
solve direct variation problems.			
solve inverse variation problems.			

ⓑ *After looking at the checklist, do you think you are well-prepared for the next section? Why or why not?*

7.6 Solve Rational Inequalities

Learning Objectives

By the end of this section, you will be able to:

› Solve rational inequalities
› Solve an inequality with rational functions

Be Prepared!

Before you get started, take this readiness quiz.

1. Find the value of $x - 5$ when ⓐ $x = 6$ ⓑ $x = -3$ ⓒ $x = 5$.
   If you missed this problem, review **Example 1.6**.

2. Solve: $8 - 2x < 12$.
   If you missed this problem, review **Example 2.52**.

3. Write in interval notation: $-3 \leq x < 5$.
   If you missed this problem, review **Example 2.49**.

Solve Rational Inequalities

We learned to solve linear inequalities after learning to solve linear equations. The techniques were very much the same with one major exception. When we multiplied or divided by a negative number, the inequality sign reversed.

Having just learned to solve rational equations we are now ready to solve rational inequalities. A **rational inequality** is an inequality that contains a rational expression.

Rational Inequality

A **rational inequality** is an inequality that contains a rational expression.

Inequalities such as $\frac{3}{2x} > 1$, $\frac{2x}{x-3} < 4$, $\frac{2x-3}{x-6} \geq x$, and $\frac{1}{4} - \frac{2}{x^2} \leq \frac{3}{x}$ are rational inequalities as they each contain

a rational expression.

When we solve a rational inequality, we will use many of the techniques we used solving linear inequalities. We especially must remember that when we multiply or divide by a negative number, the inequality sign must reverse.

Another difference is that we must carefully consider what value might make the rational expression undefined and so must be excluded.

When we solve an equation and the result is $x = 3$, we know there is one solution, which is 3.

When we solve an inequality and the result is $x > 3$, we know there are many solutions. We graph the result to better help show all the solutions, and we start with 3. Three becomes a **critical point** and then we decide whether to shade to the left or right of it. The numbers to the right of 3 are larger than 3, so we shade to the right.

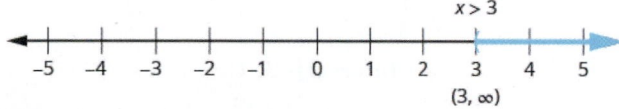

$$x > 3$$

$$(3, \infty)$$

To solve a rational inequality, we first must write the inequality with only one quotient on the left and 0 on the right.

Next we determine the critical points to use to divide the number line into intervals. A **critical point** is a number which make the rational expression zero or undefined.

We then will evaluate the factors of the numerator and denominator, and find the quotient in each interval. This will identify the interval, or intervals, that contains all the solutions of the rational inequality.

We write the solution in interval notation being careful to determine whether the endpoints are included.

EXAMPLE 7.54

Solve and write the solution in interval notation: $\frac{x-1}{x+3} \geq 0$.

⊘ Solution

Step 1. Write the inequality as one quotient on the left and zero on the right.

Our inequality is in this form.
$$\frac{x-1}{x+3} \geq 0$$

Step 2. Determine the critical points—the points where the rational expression will be zero or undefined.

The rational expression will be zero when the numerator is zero. Since $x - 1 = 0$ when $x = 1$, then 1 is a critical point.

The rational expression will be undefined when the denominator is zero. Since $x + 3 = 0$ when $x = -3$, then -3 is a critical point.

The critical points are 1 and -3.

Step 3. Use the critical points to divide the number line into intervals.

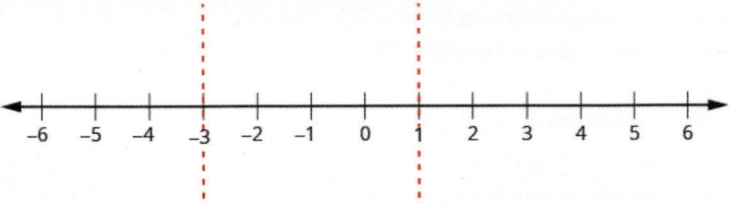

The number line is divided into three intervals:

$$(-\infty, -3) \qquad (-3, 1) \qquad (1, \infty)$$

Step 4. Test a value in each interval. Above the number line show the sign of each factor of the rational expression in each interval. Below the number line show the sign of the quotient.

To find the sign of each factor in an interval, we choose any point in that interval and use it as a test point. Any point in the interval will give the expression the same sign, so we can choose any point in the interval.

Interval $(-\infty, -3)$

The number -4 is in the interval $(-\infty, -3)$. Test $x = -4$ in the expression in the numerator and the denominator.

the numerator $\quad x - 1$ the denominator $\quad x + 3$

$-4 - 1$ $-4 + 3$

-5 3

Negative   Negative

Above the number line, mark the factor $x - 1$ negative and mark the factor $x + 3$ negative.

Since a negative divided by a negative is positive, mark the quotient positive in the interval $(-\infty, -3)$.

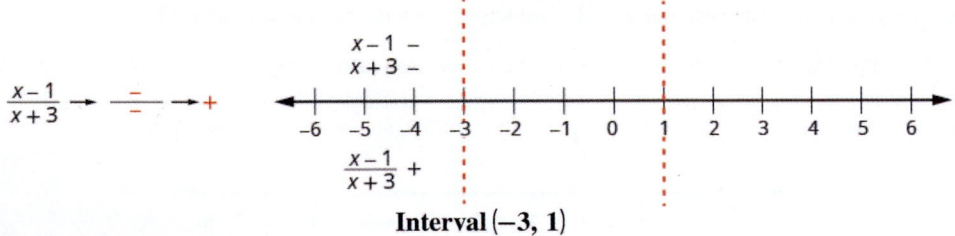

Interval $(-3, 1)$

The number 0 is in the interval $(-3, 1)$. Test $x = 0$.

the numerator $\quad x - 1$ the denominator $\quad x + 3$

$0 - 1$ $0 + 3$

-1 3

Negative   Positive

Above the number line, mark the factor $x - 1$ negative and mark $x + 3$ positive.

Since a negative divided by a positive is negative, the quotient is marked negative in the interval $(-3, 1)$.

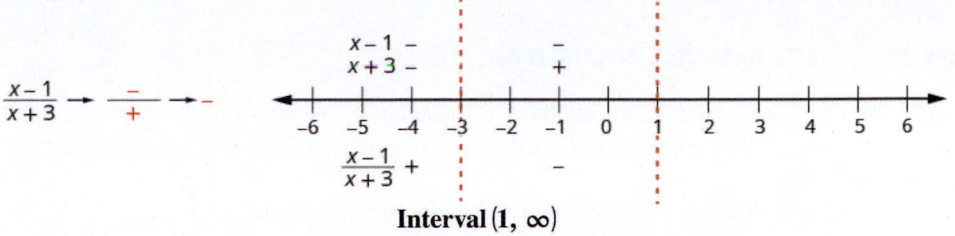

Interval $(1, \infty)$

The number 2 is in the interval $(1, \infty)$. Test $x = 2$.

the numerator	$x - 1$	the denominator	$x + 3$
	$2 - 1$		$2 + 3$
	1		5
	Positive		Positive

Above the number line, mark the factor $x - 1$ positive and mark $x + 3$ positive.

Since a positive divided by a positive is positive, mark the quotient positive in the interval $(1, \infty)$.

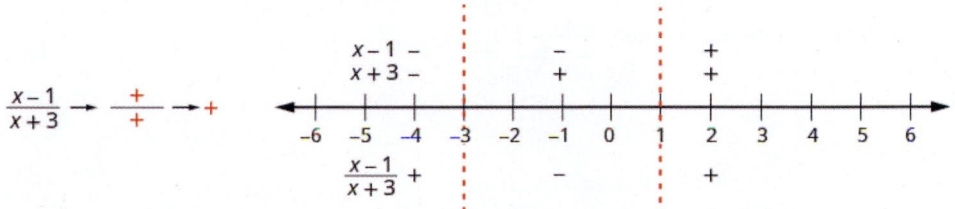

Step 5. Determine the intervals where the inequality is correct. Write the solution in interval notation.

We want the quotient to be greater than or equal to zero, so the numbers in the intervals $(-\infty, -3)$ and $(1, \infty)$ are solutions.

But what about the critical points?

The critical point $x = -3$ makes the denominator 0, so it must be excluded from the solution and we mark it with a parenthesis.

The critical point $x = 1$ makes the whole rational expression 0. The inequality requires that the rational expression be greater than or equal to. So, 1 is part of the solution and we will mark it with a bracket.

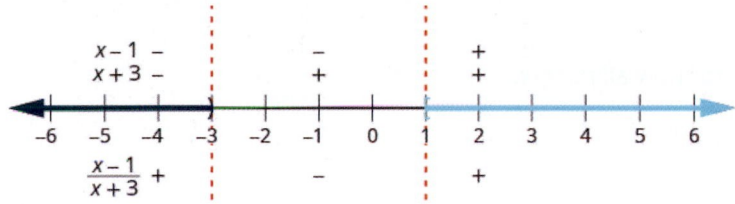

Recall that when we have a solution made up of more than one interval we use the union symbol, $\cup$, to connect the two intervals. The solution in interval notation is $(-\infty, -3) \cup [1, \infty)$.

> **TRY IT :: 7.107**   Solve and write the solution in interval notation: $\dfrac{x-2}{x+4} \geq 0$.

> **TRY IT :: 7.108**   Solve and write the solution in interval notation: $\dfrac{x+2}{x-4} \geq 0$.

We summarize the steps for easy reference.

HOW TO :: SOLVE A RATIONAL INEQUALITY.

Step 1.   Write the inequality as one quotient on the left and zero on the right.

Step 2.   Determine the critical points–the points where the rational expression will be zero or undefined.

Step 3.   Use the critical points to divide the number line into intervals.

Step 4.   Test a value in each interval. Above the number line show the sign of each factor of the numerator and denominator in each interval. Below the number line show the sign of the quotient.

Step 5.   Determine the intervals where the inequality is correct. Write the solution in interval notation.

The next example requires that we first get the rational inequality into the correct form.

EXAMPLE 7.55

Solve and write the solution in interval notation: $\dfrac{4x}{x-6} < 1$.

 **Solution**

$$\frac{4x}{x-6} < 1$$

Subtract 1 to get zero on the right.	$\dfrac{4x}{x-6} - 1 < 0$
Rewrite 1 as a fraction using the LCD.	$\dfrac{4x}{x-6} - \dfrac{x-6}{x-6} < 0$
Subtract the numerators and place the difference over the common denominator.	$\dfrac{4x-(x-6)}{x-6} < 0$
Simplify.	$\dfrac{3x+6}{x-6} < 0$
Factor the numerator to show all factors.	$\dfrac{3(x+2)}{x-6} < 0$

Find the critical points.

The quotient will be zero when the numerator is zero. The quotient is undefined when the denominator is zero.	$x+2 = 0 \qquad x-6 = 0$ $\qquad x = -2 \qquad\quad x = 6$

Use the critical points to divide the number line into intervals.

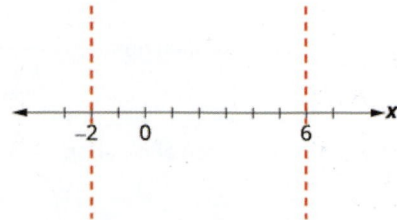

Test a value in each interval.

	$(-\infty, -2)$	$(-2, 6)$	$(6, \infty)$
$x + 2$	$x + 2$ $-3 + 2$ -1 $-$	$x + 2$ $0 + 2$ 2 $+$	$x + 2$ $7 + 2$ 9 $+$
$x - 6$	$x - 6$ $-3 - 6$ -9 $-$	$x - 6$ $0 - 6$ -6 $-$	$x - 6$ $7 - 6$ 1 $+$

Above the number line show the sign of each factor of the rational expression in each interval. Below the number line show the sign of the quotient.

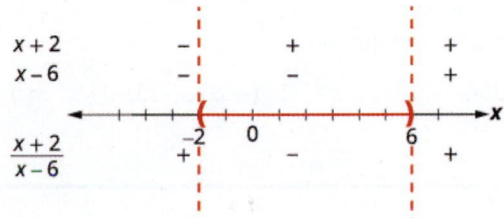

Determine the intervals where the inequality is correct. We want the quotient to be negative, so the solution includes the points between –2 and 6. Since the inequality is strictly less than, the endpoints are not included.

We write the solution in interval notation as (–2, 6).

> **TRY IT :: 7.109**   Solve and write the solution in interval notation: $\dfrac{3x}{x-3} < 1$.

> **TRY IT :: 7.110**   Solve and write the solution in interval notation: $\dfrac{3x}{x-4} < 2$.

In the next example, the numerator is always positive, so the sign of the rational expression depends on the sign of the denominator.

EXAMPLE 7.56

Solve and write the solution in interval notation: $\dfrac{5}{x^2 - 2x - 15} > 0$.

✓ **Solution**

The inequality is in the correct form.

$$\frac{5}{x^2 - 2x - 15} > 0$$

Factor the denominator.

$$\frac{5}{(x + 3)(x - 5)} > 0$$

Find the critical points.
The quotient is 0 when the numerator is 0.
Since the numerator is always 5, the quotient cannot be 0.

The quotient will be undefined when the denominator is zero.	$(x + 3)(x - 5) = 0$ $x = -3, \ x = 5$

Use the critical points to divide the number line into intervals.

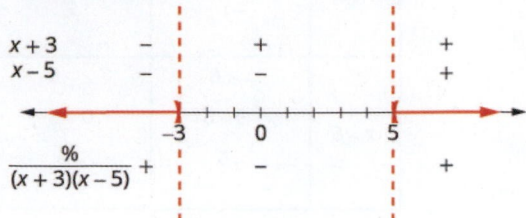

Test values in each interval.
Above the number line show the sign of each
factor of the denominator in each interval.
Below the number line, show the sign of the quotient.

Write the solution in interval notation.	$(-\infty, \ -3) \cup (5, \ \infty)$

> **TRY IT : : 7.111**   Solve and write the solution in interval notation: $\dfrac{1}{x^2 + 2x - 8} > 0$.

> **TRY IT : : 7.112**   Solve and write the solution in interval notation: $\dfrac{3}{x^2 + x - 12} > 0$.

The next example requires some work to get it into the needed form.

EXAMPLE 7.57

Solve and write the solution in interval notation: $\dfrac{1}{3} - \dfrac{2}{x^2} < \dfrac{5}{3x}$.

✓ **Solution**

$$\frac{1}{3} - \frac{2}{x^2} < \frac{5}{3x}$$

Subtract $\dfrac{5}{3x}$ to get zero on the right.	$\dfrac{1}{3} - \dfrac{2}{x^2} - \dfrac{5}{3x} < 0$
Rewrite to get each fraction with the LCD $3x^2$.	$\dfrac{1 \cdot x^2}{3 \cdot x^2} - \dfrac{2 \cdot 3}{x^2 \cdot 3} - \dfrac{5 \cdot x}{3x \cdot x} < 0$
Simplify.	$\dfrac{x^2}{3x^2} - \dfrac{6}{3x^2} - \dfrac{5x}{3x^2} < 0$
Subtract the numerators and place the difference over the common denominator.	$\dfrac{x^2 - 5x - 6}{3x^2} < 0$
Factor the numerator.	$\dfrac{(x - 6)(x + 1)}{3x^2} < 0$

Find the critical points.

$$3x^2 = 0 \qquad x - 6 = 0 \qquad x + 1 = 0$$
$$x = 0 \qquad x = 6 \qquad x = -1$$

Use the critical points to divide the number line into intervals.

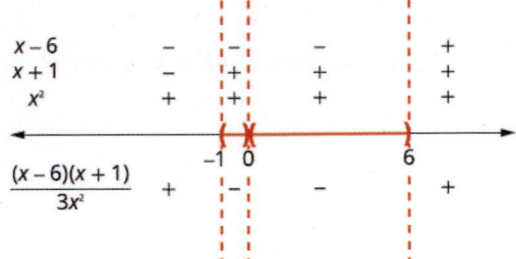

Above the number line show the sign of each factor in each interval. Below the number line, show the sign of the quotient.

Since, 0 is excluded, the solution is the two intervals, $(-1, 0)$ and $(0, 6)$.

$$(-1, 0) \cup (0, 6)$$

> **TRY IT :: 7.113**   Solve and write the solution in interval notation: $\frac{1}{2} + \frac{4}{x^2} < \frac{3}{x}$.

> **TRY IT :: 7.114**   Solve and write the solution in interval notation: $\frac{1}{3} + \frac{6}{x^2} < \frac{3}{x}$.

Solve an Inequality with Rational Functions

When working with rational functions, it is sometimes useful to know when the function is greater than or less than a particular value. This leads to a rational inequality.

EXAMPLE 7.58

Given the function $R(x) = \frac{x+3}{x-5}$, find the values of x that make the function less than or equal to 0.

⊘ **Solution**

We want the function to be less than or equal to 0.

$$R(x) \leq 0$$

Substitute the rational expression for $R(x)$.	$\dfrac{x+3}{x-5} \leq 0 \qquad x \neq 5$
Find the critical points.	$\begin{aligned} x + 3 &= 0 & x - 5 &= 0 \\ x &= -3 & x &= 5 \end{aligned}$
Use the critical points to divide the number line into intervals.	

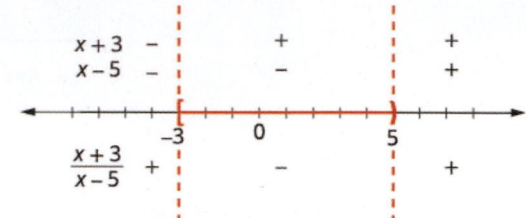

Test values in each interval. Above the number line, show the sign of each factor in each interval. Below the number line, show the sign of the quotient	
Write the solution in interval notation. Since 5 is excluded we, do not include it in the interval.	$[-3, 5)$

> **TRY IT :: 7.115**

Given the function $R(x) = \dfrac{x-2}{x+4}$, find the values of x that make the function less than or equal to 0.

> **TRY IT :: 7.116**

Given the function $R(x) = \dfrac{x+1}{x-4}$, find the values of x that make the function less than or equal to 0.

In economics, the function $C(x)$ is used to represent the cost of producing x units of a commodity. The average cost per unit can be found by dividing $C(x)$ by the number of items x. Then, the average cost per unit is $c(x) = \dfrac{C(x)}{x}$.

EXAMPLE 7.59

The function $C(x) = 10x + 3000$ represents the cost to produce x, number of items. Find ⓐ the average cost function, $c(x)$ ⓑ how many items should be produced so that the average cost is less than \$40.

✓ **Solution**

ⓐ

$$C(x) = 10x + 3000$$

The average cost function is $c(x) = \dfrac{C(x)}{x}$.

To find the average cost function, divide the cost function by x.

$$c(x) = \dfrac{C(x)}{x}$$

$$c(x) = \dfrac{10x + 3000}{x}$$

The average cost function is $c(x) = \dfrac{10x + 3000}{x}$.

We want the function $c(x)$ to be less than 40.

$$c(x) < 40$$

Substitute the rational expression for $c(x)$.

$$\frac{10x + 3000}{x} < 40 \quad x \neq 0$$

Subtract 40 to get 0 on the right.

$$\frac{10x + 3000}{x} - 40 < 0$$

Rewrite the left side as one quotient by finding the LCD and performing the subtraction.

$$\frac{10x + 3000}{x} - 40\left(\frac{x}{x}\right) < 0$$

$$\frac{10x + 3000}{x} - \frac{40x}{x} < 0$$

$$\frac{10x + 3000 - 40x}{x} < 0$$

$$\frac{-30x + 3000}{x} < 0$$

Factor the numerator to show all factors.

$$\frac{-30(x - 100)}{x} < 0$$

Find the critical points.

$$-30(x - 100) = 0 \qquad x = 0$$
$$-30 \neq 0 \quad x - 100 = 0$$
$$x = 100$$

More than 100 items must be produced to keep the average cost below $40 per item.

> **TRY IT : : 7.117**
>
> The function $C(x) = 20x + 6000$ represents the cost to produce x, number of items. Find ⓐ the average cost function, $c(x)$ ⓑ how many items should be produced so that the average cost is less than $60?

> **TRY IT : : 7.118**
>
> The function $C(x) = 5x + 900$ represents the cost to produce x, number of items. Find ⓐ the average cost function, $c(x)$ ⓑ how many items should be produced so that the average cost is less than $20?

7.6 EXERCISES

Practice Makes Perfect

Solve Rational Inequalities

In the following exercises, solve each rational inequality and write the solution in interval notation.

339. $\dfrac{x-3}{x+4} \geq 0$

340. $\dfrac{x+6}{x-5} \geq 0$

341. $\dfrac{x+1}{x-3} \leq 0$

342. $\dfrac{x-4}{x+2} \leq 0$

343. $\dfrac{x-7}{x-1} > 0$

344. $\dfrac{x+8}{x+3} > 0$

345. $\dfrac{x-6}{x+5} < 0$

346. $\dfrac{x+5}{x-2} < 0$

347. $\dfrac{3x}{x-5} < 1$

348. $\dfrac{5x}{x-2} < 1$

349. $\dfrac{6x}{x-6} > 2$

350. $\dfrac{3x}{x-4} > 2$

351. $\dfrac{2x+3}{x-6} \leq 1$

352. $\dfrac{4x-1}{x-4} \leq 1$

353. $\dfrac{3x-2}{x-4} \geq 2$

354. $\dfrac{4x-3}{x-3} \geq 2$

355. $\dfrac{1}{x^2+7x+12} > 0$

356. $\dfrac{1}{x^2-4x-12} > 0$

357. $\dfrac{3}{x^2-5x+4} < 0$

358. $\dfrac{4}{x^2+7x+12} < 0$

359. $\dfrac{2}{2x^2+x-15} \geq 0$

360. $\dfrac{6}{3x^2-2x-5} \geq 0$

361. $\dfrac{-2}{6x^2-13x+6} \leq 0$

362. $\dfrac{-1}{10x^2+11x-6} \leq 0$

363. $\dfrac{1}{2} + \dfrac{12}{x^2} > \dfrac{5}{x}$

364. $\dfrac{1}{3} + \dfrac{1}{x^2} > \dfrac{4}{3x}$

365. $\dfrac{1}{2} - \dfrac{4}{x^2} \leq \dfrac{1}{x}$

366. $\dfrac{1}{2} - \dfrac{3}{2x^2} \geq \dfrac{1}{x}$

367. $\dfrac{1}{x^2-16} < 0$

368. $\dfrac{4}{x^2-25} > 0$

369. $\dfrac{4}{x-2} \geq \dfrac{3}{x+1}$

370. $\dfrac{5}{x-1} \leq \dfrac{4}{x+2}$

Solve an Inequality with Rational Functions

In the following exercises, solve each rational function inequality and write the solution in interval notation.

371. Given the function $R(x) = \dfrac{x-5}{x-2}$, find the values of x that make the function less than or equal to 0.

372. Given the function $R(x) = \dfrac{x+1}{x+3}$, find the values of x that make the function less than or equal to 0.

373. Given the function $R(x) = \dfrac{x-6}{x+2}$, find the values of x that make the function less than or equal to 0.

374. Given the function $R(x) = \dfrac{x+1}{x-4}$, find the values of x that make the function less than or equal to 0.

Writing Exercises

375. Write the steps you would use to explain solving rational inequalities to your little brother.

376. Create a rational inequality whose solution is $(-\infty, -2] \cup [4, \infty)$.

Self Check

ⓐ *After completing the exercises, use this checklist to evaluate your mastery of the objectives of this section.*

I can...	Confidently	With some help	No-I don't get it!
solve rational inequalities.			
solve an inequality with rational functions.			

ⓑ *After reviewing this checklist, what will you do to become confident for all objectives?*

CHAPTER 7 REVIEW

KEY TERMS

complex rational expression A complex rational expression is a rational expression in which the numerator and/or denominator contains a rational expression.

critical point of a rational inequality The critical point of a rational inequality is a number which makes the rational expression zero or undefined.

extraneous solution to a rational equation An extraneous solution to a rational equation is an algebraic solution that would cause any of the expressions in the original equation to be undefined.

proportion When two rational expressions are equal, the equation relating them is called a proportion.

rational equation A rational equation is an equation that contains a rational expression.

rational expression A rational expression is an expression of the form $\frac{p}{q}$, where p and q are polynomials and $q \neq 0$.

rational function A rational function is a function of the form $R(x) = \frac{p(x)}{q(x)}$ where $p(x)$ and $q(x)$ are polynomial functions and $q(x)$ is not zero.

rational inequality A rational inequality is an inequality that contains a rational expression.

similar figures Two figures are similar if the measures of their corresponding angles are equal and their corresponding sides have the same ratio.

simplified rational expression A simplified rational expression has no common factors, other than 1, in its numerator and denominator.

KEY CONCEPTS

7.1 Multiply and Divide Rational Expressions

- **Determine the values for which a rational expression is undefined.**

  Step 1. Set the denominator equal to zero.

  Step 2. Solve the equation.

- **Equivalent Fractions Property**
  If a, b, and c are numbers where $b \neq 0$, $c \neq 0$, then $\frac{a}{b} = \frac{a \cdot c}{b \cdot c}$ and $\frac{a \cdot c}{b \cdot c} = \frac{a}{b}$.

- **How to simplify a rational expression.**

  Step 1. Factor the numerator and denominator completely.

  Step 2. Simplify by dividing out common factors.

- **Opposites in a Rational Expression**
  The opposite of $a - b$ is $b - a$.
  $$\frac{a - b}{b - a} = -1 \qquad\qquad a \neq b$$
  An expression and its opposite divide to -1.

- **Multiplication of Rational Expressions**
  If p, q, r, and s are polynomials where $q \neq 0$, $s \neq 0$, then
  $$\frac{p}{q} \cdot \frac{r}{s} = \frac{pr}{qs}$$

- **How to multiply rational expressions.**

  Step 1. Factor each numerator and denominator completely.

  Step 2. Multiply the numerators and denominators.

  Step 3. Simplify by dividing out common factors.

- **Division of Rational Expressions**
  If p, q, r, and s are polynomials where $q \neq 0$, $r \neq 0$, $s \neq 0$, then
  $$\frac{p}{q} \div \frac{r}{s} = \frac{p}{q} \cdot \frac{s}{r}$$

- **How to divide rational expressions.**

  Step 1. Rewrite the division as the product of the first rational expression and the reciprocal of the second.

  Step 2. Factor the numerators and denominators completely.

  Step 3. Multiply the numerators and denominators together.

  Step 4. Simplify by dividing out common factors.

- **How to determine the domain of a rational function.**

  Step 1. Set the denominator equal to zero.

  Step 2. Solve the equation.

  Step 3. The domain is all real numbers excluding the values found in Step 2.

7.2 Add and Subtract Rational Expressions

- **Rational Expression Addition and Subtraction**
  If p, q, and r are polynomials where $r \neq 0$, then

$$\frac{p}{r} + \frac{q}{r} = \frac{p+q}{r} \text{ and } \frac{p}{r} - \frac{q}{r} = \frac{p-q}{r}$$

- **How to find the least common denominator of rational expressions.**

  Step 1. Factor each expression completely.

  Step 2. List the factors of each expression. Match factors vertically when possible.

  Step 3. Bring down the columns.

  Step 4. Write the LCD as the product of the factors.

- **How to add or subtract rational expressions.**

  Step 1. Determine if the expressions have a common denominator.

  - Yes – go to step 2.
  - No – Rewrite each rational expression with the LCD.
    - Find the LCD.
    - Rewrite each rational expression as an equivalent rational expression with the LCD.

  Step 2. Add or subtract the rational expressions.

  Step 3. Simplify, if possible.

7.3 Simplify Complex Rational Expressions

- **How to simplify a complex rational expression by writing it as division.**

  Step 1. Simplify the numerator and denominator.

  Step 2. Rewrite the complex rational expression as a division problem.

  Step 3. Divide the expressions.

- **How to simplify a complex rational expression by using the LCD.**

  Step 1. Find the LCD of all fractions in the complex rational expression.

  Step 2. Multiply the numerator and denominator by the LCD.

  Step 3. Simplify the expression.

7.4 Solve Rational Equations

- **How to solve equations with rational expressions.**

  Step 1. Note any value of the variable that would make any denominator zero.

  Step 2. Find the least common denominator of all denominators in the equation.

  Step 3. Clear the fractions by multiplying both sides of the equation by the LCD.

  Step 4. Solve the resulting equation.

  Step 5. Check:
  - If any values found in Step 1 are algebraic solutions, discard them.
  - Check any remaining solutions in the original equation.

7.5 Solve Applications with Rational Equations

- A proportion is an equation of the form $\frac{a}{b} = \frac{c}{d}$, where $b \neq 0$, $d \neq 0$. The proportion is read "a is to b as c is to d."

- **Property of Similar Triangles**
  If $\triangle ABC$ is similar to $\triangle XYZ$, then their corresponding angle measure are equal and their corresponding sides have the same ratio.

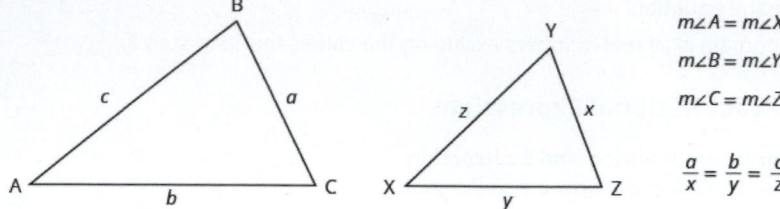

$$m\angle A = m\angle X$$
$$m\angle B = m\angle Y$$
$$m\angle C = m\angle Z$$

$$\frac{a}{x} = \frac{b}{y} = \frac{c}{z}$$

- **Direct Variation**
  - For any two variables x and y, y varies directly with x if $y = kx$, where $k \neq 0$. The constant k is called the constant of variation.
  - How to solve direct variation problems.
    Step 1. Write the formula for direct variation.
    Step 2. Substitute the given values for the variables.
    Step 3. Solve for the constant of variation.
    Step 4. Write the equation that relates x and y.

- **Inverse Variation**

  - For any two variables x and y, y varies inversely with x if $y = \frac{k}{x}$, where $k \neq 0$. The constant k is called the constant of variation.
  - How to solve inverse variation problems.
    Step 1. Write the formula for inverse variation.
    Step 2. Substitute the given values for the variables.
    Step 3. Solve for the constant of variation.
    Step 4. Write the equation that relates x and y.

7.6 Solve Rational Inequalities

- **Solve a rational inequality.**
  Step 1. Write the inequality as one quotient on the left and zero on the right.
  Step 2. Determine the critical points–the points where the rational expression will be zero or undefined.
  Step 3. Use the critical points to divide the number line into intervals.
  Step 4. Test a value in each interval. Above the number line show the sign of each factor of the rational expression in each interval. Below the number line show the sign of the quotient.
  Step 5. Determine the intervals where the inequality is correct. Write the solution in interval notation.

REVIEW EXERCISES

7.1 Simplify, Multiply, and Divide Rational Expressions

Determine the Values for Which a Rational Expression is Undefined

In the following exercises, determine the values for which the rational expression is undefined.

377. $\frac{5a + 3}{3a - 2}$

378. $\frac{b - 7}{b^2 - 25}$

379. $\frac{5x^2 y^2}{8y}$

380. $\dfrac{x-3}{x^2-x-30}$

Simplify Rational Expressions

In the following exercises, simplify.

381. $\dfrac{18}{24}$

382. $\dfrac{9m^4}{18mn^3}$

383. $\dfrac{x^2+7x+12}{x^2+8x+16}$

384. $\dfrac{7v-35}{25-v^2}$

Multiply Rational Expressions

In the following exercises, multiply.

385. $\dfrac{5}{8}\cdot\dfrac{4}{15}$

386. $\dfrac{3xy^2}{8y^3}\cdot\dfrac{16y^2}{24x}$

387. $\dfrac{72x-12x^2}{8x+32}\cdot\dfrac{x^2+10x+24}{x^2-36}$

388. $\dfrac{6y^2-2y-10}{9-y^2}\cdot\dfrac{y^2-6y+9}{6y^2+29y-20}$

Divide Rational Expressions

In the following exercises, divide.

389. $\dfrac{x^2-4x+12}{x^2+8x+12}\div\dfrac{x^2-36}{3x}$

390. $\dfrac{y^2-16}{4}\div\dfrac{y^3-64}{2y^2+8y+32}$

391. $\dfrac{11+w}{w-9}\div\dfrac{121-w^2}{9-w}$

392. $\dfrac{3y^2-12y-63}{4y+3}\div(6y^2-42y)$

393. $\dfrac{\frac{c^2-64}{3c^2+26c+16}}{\frac{c^2-4c-32}{15c+10}}$

394. $\dfrac{8a^2+16a}{a-4}\cdot\dfrac{a^2+2a-24}{a^2+7a+10}\div\dfrac{2a^2-6a}{a+5}$

Multiply and Divide Rational Functions

395. Find $R(x)=f(x)\cdot g(x)$ where $f(x)=\dfrac{9x^2+9x}{x^2-3x-4}$ and $g(x)=\dfrac{x^2-16}{3x^2+12x}$.

396. Find $R(x)=\dfrac{f(x)}{g(x)}$ where $f(x)=\dfrac{27x^2}{3x-21}$ and $g(x)=\dfrac{9x^2+54x}{x^2-x-42}$.

7.2 Add and Subtract Rational Expressions

Add and Subtract Rational Expressions with a Common Denominator

In the following exercises, perform the indicated operations.

397. $\dfrac{7}{15}+\dfrac{8}{15}$

398. $\dfrac{4a^2}{2a-1}-\dfrac{1}{2a-1}$

399. $\dfrac{y^2+10y}{y+5}+\dfrac{25}{y+5}$

400. $\dfrac{7x^2}{x^2-9}+\dfrac{21x}{x^2-9}$

401. $\dfrac{x^2}{x-7}-\dfrac{3x+28}{x-7}$

402. $\dfrac{y^2}{y+11}-\dfrac{121}{y+11}$

403. $\dfrac{4q^2 - q + 3}{q^2 + 6q + 5} - \dfrac{3q^2 - q - 6}{q^2 + 6q + 5}$

404.

$\dfrac{5t + 4t + 3}{t^2 - 25} - \dfrac{4t^2 - 8t - 32}{t^2 - 25}$

Add and Subtract Rational Expressions Whose Denominators Are Opposites

In the following exercises, add and subtract.

405. $\dfrac{18w}{6w - 1} + \dfrac{3w - 2}{1 - 6w}$

406. $\dfrac{a^2 + 3a}{a^2 - 4} - \dfrac{3a - 8}{4 - a^2}$

407.

$\dfrac{2b^2 + 3b - 15}{b^2 - 49} - \dfrac{b^2 + 16b - 1}{49 - b^2}$

408.

$\dfrac{8y^2 - 10y + 7}{2y - 5} + \dfrac{2y^2 + 7y + 2}{5 - 2y}$

Find the Least Common Denominator of Rational Expressions

In the following exercises, find the LCD.

409. $\dfrac{7}{a^2 - 3a - 10}, \dfrac{3a}{a^2 - a - 20}$

410. $\dfrac{6}{n^2 - 4}, \dfrac{2n}{n^2 - 4n + 4}$

411.

$\dfrac{5}{3p^2 + 17p - 6}, \dfrac{2m}{3p^2 - 23p - 8}$

Add and Subtract Rational Expressions with Unlike Denominators

In the following exercises, perform the indicated operations.

412. $\dfrac{7}{5a} + \dfrac{3}{2b}$

413. $\dfrac{2}{c - 2} + \dfrac{9}{c + 3}$

414. $\dfrac{3x}{x^2 - 9} + \dfrac{5}{x^2 + 6x + 9}$

415.

$\dfrac{2x}{x^2 + 10x + 24} + \dfrac{3x}{x^2 + 8x + 16}$

416. $\dfrac{5q}{p^2q - p^2} + \dfrac{4q}{q^2 - 1}$

417. $\dfrac{3y}{y + 2} - \dfrac{y + 2}{y + 8}$

418. $\dfrac{-3w - 15}{w^2 + w - 20} - \dfrac{w + 2}{4 - w}$

419. $\dfrac{7m + 3}{m + 2} - 5$

420. $\dfrac{n}{n + 3} + \dfrac{2}{n - 3} - \dfrac{n - 9}{n^2 - 9}$

421. $\dfrac{8a}{a^2 - 64} - \dfrac{4}{a + 8}$

422. $\dfrac{5}{12x^2 y} + \dfrac{7}{20xy^3}$

Add and Subtract Rational Functions

In the following exercises, find $R(x) = f(x) + g(x)$ where $f(x)$ and $g(x)$ are given.

423. $f(x) = \dfrac{2x^2 + 12x - 11}{x^2 + 3x - 10}, \quad g(x) = \dfrac{x + 1}{2 - x}$

424. $f(x) = \dfrac{-4x + 31}{x^2 + x - 30}, \quad g(x) = \dfrac{5}{x + 6}$

In the following exercises, find $R(x) = f(x) - g(x)$ where $f(x)$ and $g(x)$ are given.

425. $f(x) = \dfrac{4x}{x^2 - 121}, \quad g(x) = \dfrac{2}{x - 11}$

426. $f(x) = \dfrac{7}{x + 6}, \quad g(x) = \dfrac{14x}{x^2 - 36}$

7.3 Simplify Complex Rational Expressions

Simplify a Complex Rational Expression by Writing It as Division

In the following exercises, simplify.

427. $\dfrac{\frac{7x}{x+2}}{\frac{14x^2}{x^2-4}}$

428. $\dfrac{\frac{2}{5}+\frac{5}{6}}{\frac{1}{3}+\frac{1}{4}}$

429. $\dfrac{x-\frac{3x}{x+5}}{\frac{1}{x+5}+\frac{1}{x-5}}$

430. $\dfrac{\frac{2}{m}+\frac{m}{n}}{\frac{n}{m}-\frac{1}{n}}$

Simplify a Complex Rational Expression by Using the LCD

In the following exercises, simplify.

431. $\dfrac{\frac{1}{3}+\frac{1}{8}}{\frac{1}{4}+\frac{1}{12}}$

432. $\dfrac{\frac{3}{a^2}-\frac{1}{b}}{\frac{1}{a}+\frac{1}{b^2}}$

433. $\dfrac{\frac{2}{z^2-49}+\frac{1}{z+7}}{\frac{9}{z+7}+\frac{12}{z-7}}$

434. $\dfrac{\frac{3}{y^2-4y-32}}{\frac{2}{y-8}+\frac{1}{y+4}}$

7.4 7.4 Solve Rational Equations

Solve Rational Equations

In the following exercises, solve.

435. $\dfrac{1}{2}+\dfrac{2}{3}=\dfrac{1}{x}$

436. $1-\dfrac{2}{m}=\dfrac{8}{m^2}$

437. $\dfrac{1}{b-2}+\dfrac{1}{b+2}=\dfrac{3}{b^2-4}$

438. $\dfrac{3}{q+8}-\dfrac{2}{q-2}=1$

439. $\dfrac{v-15}{v^2-9v+18}=\dfrac{4}{v-3}+\dfrac{2}{v-6}$

440. $\dfrac{z}{12}+\dfrac{z+3}{3z}=\dfrac{1}{z}$

Solve Rational Equations that Involve Functions

441. For rational function, $f(x)=\dfrac{x+2}{x^2-6x+8}$, ⓐ find the domain of the function ⓑ solve $f(x)=1$ ⓒ find the points on the graph at this function value.

442. For rational function, $f(x)=\dfrac{2-x}{x^2+7x+10}$, ⓐ find the domain of the function ⓑ solve $f(x)=2$ ⓒ find the points on the graph at this function value.

Solve a Rational Equation for a Specific Variable

In the following exercises, solve for the indicated variable.

443. $\dfrac{V}{l}=hw$ for l.

444. $\dfrac{1}{x}-\dfrac{2}{y}=5$ for y.

445. $x=\dfrac{y+5}{z-7}$ for z.

446. $P=\dfrac{k}{V}$ for V.

7.5 Solve Applications with Rational Equations

Solve Proportions

In the following exercises, solve.

447. $\dfrac{x}{4} = \dfrac{3}{5}$

448. $\dfrac{3}{y} = \dfrac{9}{5}$

449. $\dfrac{s}{s+20} = \dfrac{3}{7}$

450. $\dfrac{t-3}{5} = \dfrac{t+2}{9}$

Solve Using Proportions

In the following exercises, solve.

451. Rachael had a 21-ounce strawberry shake that has 739 calories. How many calories are there in a 32-ounce shake?

452. Leo went to Mexico over Christmas break and changed $525 dollars into Mexican pesos. At that time, the exchange rate had $1 US is equal to 16.25 Mexican pesos. How many Mexican pesos did he get for his trip?

Solve Similar Figure Applications

In the following exercises, solve.

453. ΔABC is similar to ΔXYZ. The lengths of two sides of each triangle are given in the figure. Find the lengths of the third sides.

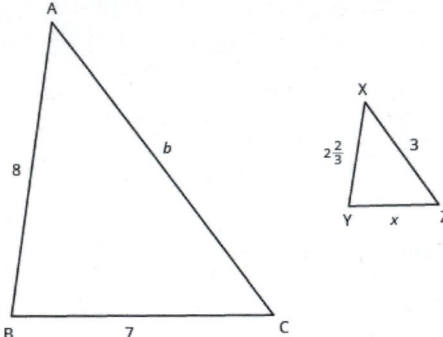

454. On a map of Europe, Paris, Rome, and Vienna form a triangle whose sides are shown in the figure below. If the actual distance from Rome to Vienna is 700 miles, find the distance from

ⓐ Paris to Rome

ⓑ Paris to Vienna

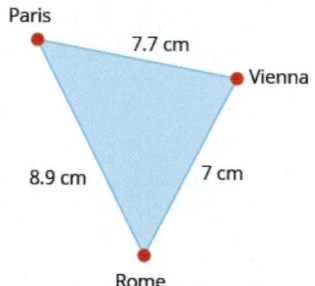

455. Francesca is 5.75 feet tall. Late one afternoon, her shadow was 8 feet long. At the same time, the shadow of a nearby tree was 32 feet long. Find the height of the tree.

456. The height of a lighthouse in Pensacola, Florida is 150 feet. Standing next to the statue, 5.5-foot-tall Natasha cast a 1.1-foot shadow. How long would the shadow of the lighthouse be?

Solve Uniform Motion Applications

In the following exercises, solve.

457. When making the 5-hour drive home from visiting her parents, Lolo ran into bad weather. She was able to drive 176 miles while the weather was good, but then driving 10 mph slower, went 81 miles when it turned bad. How fast did she drive when the weather was bad?

458. Mark is riding on a plane that can fly 490 miles with a tailwind of 20 mph in the same time that it can fly 350 miles against a tailwind of 20 mph. What is the speed of the plane?

459. Josue can ride his bicycle 8 mph faster than Arjun can ride his bike. It takes Luke 3 hours longer than Josue to ride 48 miles. How fast can John ride his bike?

460. Curtis was training for a triathlon. He ran 8 kilometers and biked 32 kilometers in a total of 3 hours. His running speed was 8 kilometers per hour less than his biking speed. What was his running speed?

Solve Work Applications

In the following exercises, solve.

461. Brandy can frame a room in 1 hour, while Jake takes 4 hours. How long could they frame a room working together?

462. Prem takes 3 hours to mow the lawn while her cousin, Barb, takes 2 hours. How long will it take them working together?

463. Jeffrey can paint a house in 6 days, but if he gets a helper he can do it in 4 days. How long would it take the helper to paint the house alone?

464. Marta and Deb work together writing a book that takes them 90 days. If Sue worked alone it would take her 120 days. How long would it take Deb to write the book alone?

Solve Direct Variation Problems

In the following exercises, solve.

465. If y varies directly as x when $y = 9$ and $x = 3$, find x when $y = 21$.

466. If y varies inversely as x when $y = 20$ and $x = 2$, find y when $x = 4$.

467. Vanessa is traveling to see her fiancé. The distance, d, varies directly with the speed, v, she drives. If she travels 258 miles driving 60 mph, how far would she travel going 70 mph?

468. If the cost of a pizza varies directly with its diameter, and if an 8" diameter pizza costs $12, how much would a 6" diameter pizza cost?

469. The distance to stop a car varies directly with the square of its speed. It takes 200 feet to stop a car going 50 mph. How many feet would it take to stop a car going 60 mph?

Solve Inverse Variation Problems

In the following exercises, solve.

470. If m varies inversely with the square of n, when $m = 4$ and $n = 6$ find m when $n = 2$.

471. The number of tickets for a music fundraiser varies inversely with the price of the tickets. If Madelyn has just enough money to purchase 12 tickets for $6, how many tickets can Madelyn afford to buy if the price increased to $8?

472. On a string instrument, the length of a string varies inversely with the frequency of its vibrations. If an 11-inch string on a violin has a frequency of 360 cycles per second, what frequency does a 12-inch string have?

7.6 Solve Rational Inequalities

Solve Rational Inequalities

In the following exercises, solve each rational inequality and write the solution in interval notation.

473. $\dfrac{x-3}{x+4} \le 0$

474. $\dfrac{5x}{x-2} > 1$

475. $\dfrac{3x-2}{x-4} \le 2$

476. $\dfrac{1}{x^2 - 4x - 12} < 0$

477. $\dfrac{1}{2} - \dfrac{4}{x^2} \ge \dfrac{1}{x}$

478. $\dfrac{4}{x-2} < \dfrac{3}{x+1}$

Solve an Inequality with Rational Functions

In the following exercises, solve each rational function inequality and write the solution in interval notation

479. Given the function, $R(x) = \frac{x-5}{x-2}$, find the values of x that make the function greater than or equal to 0.

480. Given the function, $R(x) = \frac{x+1}{x+3}$, find the values of x that make the function less than or equal to 0.

481. The function $C(x) = 150x + 100{,}000$ represents the cost to produce x, number of items. Find ⓐ the average cost function, $c(x)$ ⓑ how many items should be produced so that the average cost is less than $160.

482. Tillman is starting his own business by selling tacos at the beach. Accounting for the cost of his food truck and ingredients for the tacos, the function $C(x) = 2x + 6{,}000$ represents the cost for Tillman to produce x, tacos. Find ⓐ the average cost function, $c(x)$ for Tillman's Tacos ⓑ how many tacos should Tillman produce so that the average cost is less than $4.

PRACTICE TEST

In the following exercises, simplify.

483. $\dfrac{4a^2 b}{12ab^2}$

484. $\dfrac{6x - 18}{x^2 - 9}$

In the following exercises, perform the indicated operation and simplify.

485. $\dfrac{4x}{x+2} \cdot \dfrac{x^2 + 5x + 6}{12x^2}$

486. $\dfrac{2y^2}{y^2 - 1} \div \dfrac{y^3 - y^2 + y}{y^3 - 1}$

487.
$\dfrac{6x^2 - x + 20}{x^2 - 81} - \dfrac{5x^2 + 11x - 7}{x^2 - 81}$

488. $\dfrac{-3a}{3a - 3} + \dfrac{5a}{a^2 + 3a - 4}$

489.
$\dfrac{2n^2 + 8n - 1}{n^2 - 1} - \dfrac{n^2 - 7n - 1}{1 - n^2}$

490.
$\dfrac{10x^2 + 16x - 7}{8x - 3} + \dfrac{2x^2 + 3x - 1}{3 - 8x}$

491. $\dfrac{\frac{1}{m} - \frac{1}{n}}{\frac{1}{n} + \frac{1}{m}}$

In the following exercises, solve each equation.

492. $\dfrac{1}{x} + \dfrac{3}{4} = \dfrac{5}{8}$

493. $\dfrac{1}{z - 5} + \dfrac{1}{z + 5} = \dfrac{1}{z^2 - 25}$

494.
$\dfrac{z}{2z + 8} - \dfrac{3}{4z - 8} = \dfrac{3z^2 - 16z - 16}{8z^2 + 2z - 64}$

In the following exercises, solve each rational inequality and write the solution in interval notation.

495. $\dfrac{6x}{x - 6} \le 2$

496. $\dfrac{2x + 3}{x - 6} > 1$

497. $\dfrac{1}{2} + \dfrac{12}{x^2} \ge \dfrac{5}{x}$

In the following exercises, find $R(x)$ given $f(x) = \dfrac{x - 4}{x^2 - 3x - 10}$ and $g(x) = \dfrac{x - 5}{x^2 - 2x - 8}$.

498. $R(x) = f(x) - g(x)$

499. $R(x) = f(x) \cdot g(x)$

500. $R(x) = f(x) \div g(x)$

501. Given the function,
$R(x) = \dfrac{2}{2x^2 + x - 15}$, find the values of x that make the function less than or equal to 0.

In the following exercises, solve.

502. If y varies directly with x, and $x = 5$ when $y = 30$, find x when $y = 42$.

503. If y varies inversely with the square of x and $x = 3$ when $y = 9$, find y when $x = 4$.

504. Matheus can ride his bike for 30 miles with the wind in the same amount of time that he can go 21 miles against the wind. If the wind's speed is 6 mph, what is Matheus' speed on his bike?

505. Oliver can split a truckload of logs in 8 hours, but working with his dad they can get it done in 3 hours. How long would it take Oliver's dad working alone to split the logs?

506. The volume of a gas in a container varies inversely with the pressure on the gas. If a container of nitrogen has a volume of 29.5 liters with 2000 psi, what is the volume if the tank has a 14.7 psi rating? Round to the nearest whole number.

507. The cities of Dayton, Columbus, and Cincinnati form a triangle in southern Ohio. The diagram gives the map distances between these cities in inches.

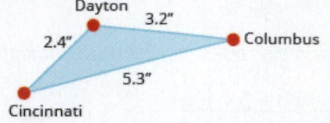

The actual distance from Dayton to Cincinnati is 48 miles. What is the actual distance between Dayton and Columbus?

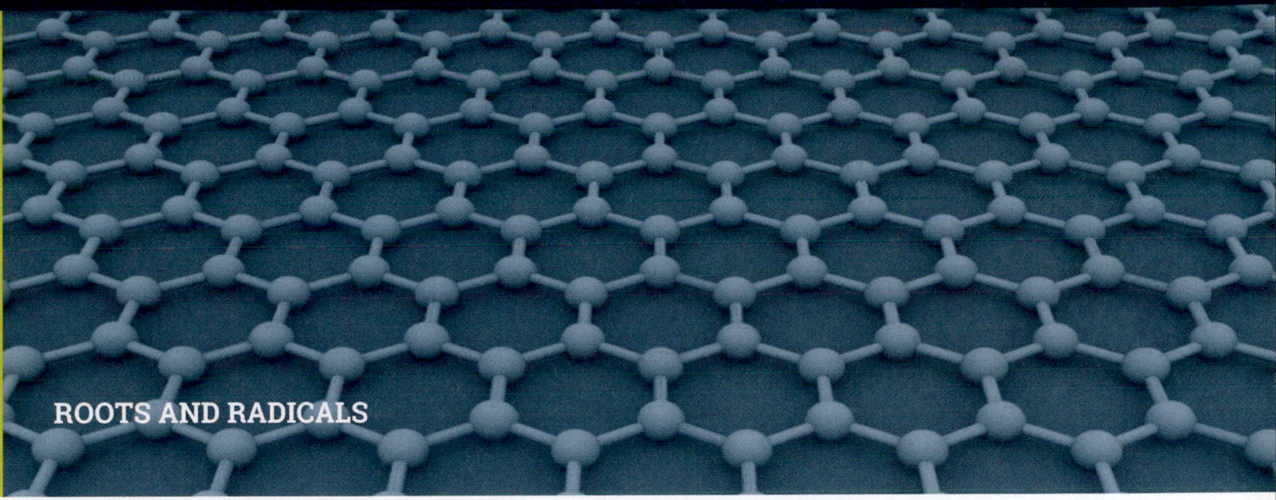

8 ROOTS AND RADICALS

Figure 8.1 Graphene is an incredibly strong and flexible material made from carbon. It can also conduct electricity. Notice the hexagonal grid pattern. (credit: "AlexanderAIUS" / Wikimedia Commons)

Chapter Outline

Introduction

Imagine charging your cell phone is less than five seconds. Consider cleaning radioactive waste from contaminated water. Think about filtering salt from ocean water to make an endless supply of drinking water. Ponder the idea of bionic devices that can repair spinal injuries. These are just of few of the many possible uses of a material called graphene. Materials scientists are developing a material made up of a single layer of carbon atoms that is stronger than any other material, completely flexible, and conducts electricity better than most metals. Research into this type of material requires a solid background in mathematics, including understanding roots and radicals. In this chapter, you will learn to simplify expressions containing roots and radicals, perform operations on radical expressions and equations, and evaluate radical functions.

8.1 Simplify Expressions with Roots

Learning Objectives

By the end of this section, you will be able to:

- Simplify expressions with roots
- Estimate and approximate roots
- Simplify variable expressions with roots

Be Prepared!

Before you get started, take this readiness quiz.

1. Simplify: ⓐ $(-9)^2$ ⓑ -9^2 ⓒ $(-9)^3$.
   If you missed this problem, review Example 2.21.

2. Round 3.846 to the nearest hundredth.
   If you missed this problem, review **Example 1.34**.

3. Simplify: ⓐ $x^3 \cdot x^3$ ⓑ $y^2 \cdot y^2 \cdot y^2$ ⓒ $z^3 \cdot z^3 \cdot z^3 \cdot z^3$.
   If you missed this problem, review **Example 5.12**.

Simplify Expressions with Roots

In **Foundations**, we briefly looked at square roots. Remember that when a real number n is multiplied by itself, we write n^2 and read it 'n squared'. This number is called the **square** of n, and n is called the **square root**. For example,

$$13^2 \text{ is read "13 squared"}$$

$$169 \text{ is called the } square \text{ of } 13, \text{ since } 13^2 = 169$$

$$13 \text{ is a } square\ root \text{ of } 169$$

Square and Square Root of a number

Square

> If $n^2 = m,$ then m is the **square** of n.

Square Root

> If $n^2 = m,$ then n is a **square root** of m.

Notice $(-13)^2 = 169$ also, so -13 is also a square root of 169. Therefore, both 13 and -13 are square roots of 169.

So, every positive number has two square roots—one positive and one negative. What if we only wanted the positive square root of a positive number? We use a *radical sign*, and write, $\sqrt{m},$ which denotes the positive square root of m. The positive square root is also called the **principal square root**.

We also use the radical sign for the square root of zero. Because $0^2 = 0,$ $\sqrt{0} = 0.$ Notice that zero has only one square root.

Square Root Notation

> $\sqrt{m}$ is read "the square root of m".
>
> If $n^2 = m,$ then $n = \sqrt{m},$ for $n \geq 0.$

radical sign $\longrightarrow \sqrt{m} \longleftarrow$ radicand

We know that every positive number has two square roots and the radical sign indicates the positive one. We write $\sqrt{169} = 13.$ If we want to find the negative square root of a number, we place a negative in front of the radical sign. For example, $-\sqrt{169} = -13.$

EXAMPLE 8.1

Simplify: ⓐ $\sqrt{144}$ ⓑ $-\sqrt{289}$.

✓ **Solution**

ⓐ

	$\sqrt{144}$
Since $12^2 = 144.$	12

ⓑ

	$-\sqrt{289}$
Since $17^2 = 289$ and the negative is in front of the radical sign.	-17

> **TRY IT ::** 8.1   Simplify: ⓐ $-\sqrt{64}$ ⓑ $\sqrt{225}$.

> **TRY IT ::** 8.2   Simplify: ⓐ $\sqrt{100}$ ⓑ $-\sqrt{121}$.

Can we simplify $\sqrt{-49}$? Is there a number whose square is -49?

$$(\ )^2 = -49$$

Any positive number squared is positive. Any negative number squared is positive. There is no real number equal to $\sqrt{-49}$. The square root of a negative number is not a real number.

EXAMPLE 8.2

Simplify: ⓐ $\sqrt{-196}$ ⓑ $-\sqrt{64}$.

✓ **Solution**

ⓐ

	$\sqrt{-196}$
There is no real number whose square is -196.	$\sqrt{-196}$ is not a real number.

ⓑ

	$-\sqrt{64}$
The negative is in front of the radical.	-8

> **TRY IT ::** 8.3   Simplify: ⓐ $\sqrt{-169}$ ⓑ $-\sqrt{81}$.

> **TRY IT ::** 8.4   Simplify: ⓐ $-\sqrt{49}$ ⓑ $\sqrt{-121}$.

So far we have only talked about squares and square roots. Let's now extend our work to include higher powers and higher roots.

Let's review some vocabulary first.

We write: We say:

n^2 n squared

n^3 n cubed

n^4 n to the fourth power

n^5 n to the fifth power

The terms 'squared' and 'cubed' come from the formulas for area of a square and volume of a cube.

It will be helpful to have a table of the powers of the integers from −5 to 5. See **Figure 8.2**.

Number	Square	Cube	Fourth power	Fifth power
n	n^2	n^3	n^4	n^5
1	1	1	1	1
2	4	8	16	32
3	9	27	81	243
4	16	64	256	1024
5	25	125	625	3125
x	x^2	x^3	x^4	x^5
x^2	x^4	x^6	x^8	x^{10}

Number	Square	Cube	Fourth power	Fifth power
n	n^2	n^3	n^4	n^5
−1	1	−1	1	−1
−2	4	−8	16	−32
−3	9	−27	81	−243
−4	16	−64	256	−1024
−5	25	−125	625	−3125

Figure 8.2

Notice the signs in the table. All powers of positive numbers are positive, of course. But when we have a negative number, the *even* powers are positive and the *odd* powers are negative. We'll copy the row with the powers of −2 to help you see this.

n	n^2	n^3	n^4	n^5
−2	4	−8	16	−32

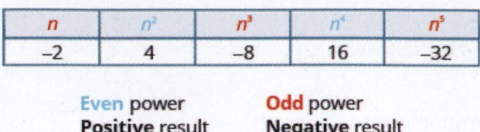

Even power   **Odd** power
Positive result   **Negative** result

We will now extend the square root definition to higher roots.

n^{th} Root of a Number

If $b^n = a,$ then b is an nth root of a.

The principal n^{th} root of a is written $\sqrt[n]{a}$.

n is called the **index** of the radical.

Just like we use the word 'cubed' for b^3, we use the term 'cube root' for $\sqrt[3]{a}$.

We can refer to **Figure 8.2** to help find higher roots.

$$4^3 = 64 \qquad\qquad \sqrt[3]{64} = 4$$
$$3^4 = 81 \qquad\qquad \sqrt[4]{81} = 3$$
$$(-2)^5 = -32 \qquad\qquad \sqrt[5]{-32} = -2$$

Could we have an even root of a negative number? We know that the square root of a negative number is not a real number. The same is true for any even root. *Even* roots of negative numbers are not real numbers. *Odd* roots of negative numbers are real numbers.

Properties of $\sqrt[n]{a}$

When n is an even number and

- $a \geq 0,$ then $\sqrt[n]{a}$ is a real number.

- $a < 0,$ then $\sqrt[n]{a}$ is not a real number.

When n is an odd number, $\sqrt[n]{a}$ is a real number for all values of a.

We will apply these properties in the next two examples.

EXAMPLE 8.3

Simplify: ⓐ $\sqrt[3]{64}$ ⓑ $\sqrt[4]{81}$ ⓒ $\sqrt[5]{32}$.

✓ Solution

ⓐ

$$\sqrt[3]{64}$$

Since $4^3 = 64$. 4

ⓑ

$$\sqrt[4]{81}$$

Since $(3)^4 = 81$. 3

ⓒ

$$\sqrt[5]{32}$$

Since $(2)^5 = 32$. 2

> **TRY IT : : 8.5**   Simplify: ⓐ $\sqrt[3]{27}$ ⓑ $\sqrt[4]{256}$ ⓒ $\sqrt[5]{243}$.

> **TRY IT : : 8.6**   Simplify: ⓐ $\sqrt[3]{1000}$ ⓑ $\sqrt[4]{16}$ ⓒ $\sqrt[5]{243}$.

In this example be alert for the negative signs as well as even and odd powers.

EXAMPLE 8.4

Simplify: ⓐ $\sqrt[3]{-125}$ ⓑ $\sqrt[4]{-16}$ ⓒ $\sqrt[5]{-243}$.

✓ **Solution**

ⓐ

Since $(-5)^3 = -125$.

$$\sqrt[3]{-125}$$
$$-5$$

ⓑ

Think, $(?)^4 = -16$. No real number raised to the fourth power is negative.

$$\sqrt[4]{-16}$$

Not a real number.

ⓒ

Since $(-3)^5 = -243$.

$$\sqrt[5]{-243}$$
$$-3$$

> **TRY IT : : 8.7**   Simplify: ⓐ $\sqrt[3]{-27}$ ⓑ $\sqrt[4]{-256}$ ⓒ $\sqrt[5]{-32}$.

> **TRY IT : : 8.8**   Simplify: ⓐ $\sqrt[3]{-216}$ ⓑ $\sqrt[4]{-81}$ ⓒ $\sqrt[5]{-1024}$.

Estimate and Approximate Roots

When we see a number with a radical sign, we often don't think about its numerical value. While we probably know that the $\sqrt{4} = 2$, what is the value of $\sqrt{21}$ or $\sqrt[3]{50}$? In some situations a quick estimate is meaningful and in others it is convenient to have a decimal approximation.

To get a numerical estimate of a square root, we look for perfect square numbers closest to the radicand. To find an estimate of $\sqrt{11}$, we see 11 is between perfect square numbers 9 and 16, *closer* to 9. Its square root then will be between 3 and 4, but closer to 3.

Number	Square Root
4	2
9	3
16	4
25	5

Number	Cube Root
8	2
27	3
64	4
125	5

$$9 < 11 < 16$$
$$3 < \sqrt{11} < 4$$

$$64 < 91 < 125$$
$$4 < \sqrt[3]{91} < 5$$

Similarly, to estimate $\sqrt[3]{91}$, we see 91 is between perfect cube numbers 64 and 125. The cube root then will be between 4 and 5.

EXAMPLE 8.5

Estimate each root between two consecutive whole numbers: ⓐ $\sqrt{105}$ ⓑ $\sqrt[3]{43}$.

✓ **Solution**

ⓐ Think of the perfect square numbers closest to 105. Make a small table of these perfect squares and their squares roots.

$\sqrt{105}$

Number	Square Root
81	9
100	10
105 → 121 ← $\sqrt{105}$	11
144	12

Locate 105 between two consecutive perfect squares.	$100 < 105 < 121$
$\sqrt{105}$ is between their square roots.	$10 < \sqrt{105} < 11$

ⓑ Similarly we locate 43 between two perfect cube numbers.

$\sqrt[3]{43}$

Number	Cube Root
8	2
27	3
43 → 64 ← $\sqrt[3]{43}$	4
125	5

Locate 43 between two consecutive perfect cubes.	$27 < 43 < 64$
$\sqrt[3]{43}$ is between their cube roots.	$3 < \sqrt[3]{43} < 4$

> **TRY IT : : 8.9**   Estimate each root between two consecutive whole numbers:
>
> ⓐ $\sqrt{38}$ ⓑ $\sqrt[3]{93}$

> **TRY IT : : 8.10**   Estimate each root between two consecutive whole numbers:
>
> ⓐ $\sqrt{84}$ ⓑ $\sqrt[3]{152}$

There are mathematical methods to approximate square roots, but nowadays most people use a calculator to find square roots. To find a square root you will use the $\sqrt{x}$ key on your calculator. To find a cube root, or any root with higher index, you will use the $\sqrt[y]{x}$ key.

When you use these keys, you get an approximate value. It is an approximation, accurate to the number of digits shown on your calculator's display. The symbol for an approximation is $\approx$ and it is read 'approximately'.

Suppose your calculator has a 10 digit display. You would see that

$$\sqrt{5} \approx 2.236067978 \text{ rounded to two decimal places is } \sqrt{5} \approx 2.24$$

$$\sqrt[4]{93} \approx 3.105422799 \text{ rounded to two decimal places is } \sqrt[4]{93} \approx 3.11$$

How do we know these values are approximations and not the exact values? Look at what happens when we square them:

$$(2.236067978)^2 = 5.000000002 \qquad (3.105422799)^4 = 92.999999991$$
$$(2.24)^2 = 5.0176 \qquad (3.11)^4 = 93.54951841$$

Their squares are close to 5, but are not exactly equal to 5. The fourth powers are close to 93, but not equal to 93.

EXAMPLE 8.6

Round to two decimal places: ⓐ $\sqrt{17}$ ⓑ $\sqrt[3]{49}$ ⓒ $\sqrt[4]{51}$.

✓ **Solution**

ⓐ

	$\sqrt{17}$
Use the calculator square root key.	$4.123105626\ldots$
Round to two decimal places.	4.12
	$\sqrt{17} \approx 4.12$

ⓑ

	$\sqrt[3]{49}$
Use the calculator $\sqrt[y]{x}$ key.	$3.659305710\ldots$
Round to two decimal places.	3.66
	$\sqrt[3]{49} \approx 3.66$

ⓒ

	$\sqrt[4]{51}$
Use the calculator $\sqrt[y]{x}$ key.	$2.6723451177\ldots$
Round to two decimal places.	2.67
	$\sqrt[4]{51} \approx 2.67$

> **TRY IT :: 8.11**   Round to two decimal places:
>
> ⓐ $\sqrt{11}$ ⓑ $\sqrt[3]{71}$ ⓒ $\sqrt[4]{127}$.

> **TRY IT :: 8.12**   Round to two decimal places:
>
> ⓐ $\sqrt{13}$ ⓑ $\sqrt[3]{84}$ ⓒ $\sqrt[4]{98}$.

Simplify Variable Expressions with Roots

The odd root of a number can be either positive or negative. For example,

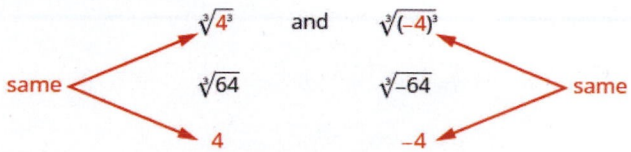

In either case, when n is odd, $\sqrt[n]{a^n} = a$.

But what about an even root? We want the principal root, so $\sqrt[4]{625} = 5$.

But notice,

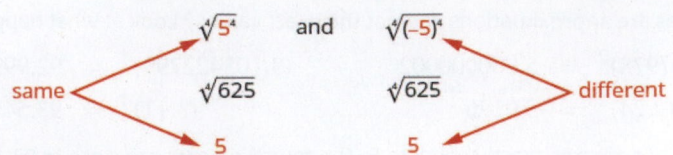

Here we see that sometimes when n is even, $\sqrt[n]{a^n} \neq a$.

How can we make sure the fourth root of -5 raised to the fourth power is 5? We can use the absolute value. $|-5| = 5$. So we say that when n is even $\sqrt[n]{a^n} = |a|$. This guarantees the principal root is positive.

Simplifying Odd and Even Roots

For any integer $n \geq 2$,

$$\text{when the index } n \text{ is odd} \qquad \sqrt[n]{a^n} = a$$
$$\text{when the index } n \text{ is even} \qquad \sqrt[n]{a^n} = |a|$$

We must use the absolute value signs when we take an even root of an expression with a variable in the radical.

EXAMPLE 8.7

Simplify: ⓐ $\sqrt{x^2}$ ⓑ $\sqrt[3]{n^3}$ ⓒ $\sqrt[4]{p^4}$ ⓓ $\sqrt[5]{y^5}$.

⊘ **Solution**

ⓐ We use the absolute value to be sure to get the positive root.

$$\sqrt{x^2}$$

Since the index n is even, $\sqrt[n]{a^n} = |a|$.
$$|x|$$

ⓑ This is an odd indexed root so there is no need for an absolute value sign.

$$\sqrt[3]{m^3}$$

Since the index n is odd, $\sqrt[n]{a^n} = a$.
$$m$$

ⓒ

$$\sqrt[4]{p^4}$$

Since the index n is even $\sqrt[n]{a^n} = |a|$.
$$|p|$$

ⓓ

$$\sqrt[5]{y^5}$$

Since the index n is odd, $\sqrt[n]{a^n} = a$.
$$y$$

> **TRY IT :: 8.13**
>
> Simplify: ⓐ $\sqrt{b^2}$ ⓑ $\sqrt[3]{w^3}$ ⓒ $\sqrt[4]{m^4}$ ⓓ $\sqrt[5]{q^5}$.

> **TRY IT :: 8.14**
>
> Simplify: ⓐ $\sqrt{y^2}$ ⓑ $\sqrt[3]{p^3}$ ⓒ $\sqrt[4]{z^4}$ ⓓ $\sqrt[5]{q^5}$.

What about square roots of higher powers of variables? The Power Property of Exponents says $(a^m)^n = a^{m \cdot n}$. So if we square a^m, the exponent will become $2m$.

$$(a^m)^2 = a^{2m}$$

Looking now at the square root,

$$\sqrt{a^{2m}}$$

Since $(a^m)^2 = a^{2m}$. $\sqrt{(a^m)^2}$

Since n is even $\sqrt[n]{a^n} = |a|$. $|a^m|$

So $\sqrt{a^{2m}} = |a^m|$.

We apply this concept in the next example.

EXAMPLE 8.8

Simplify: ⓐ $\sqrt{x^6}$ ⓑ $\sqrt{y^{16}}$.

✓ **Solution**

ⓐ

$$\sqrt{x^6}$$

Since $(x^3)^2 = x^6$. $\sqrt{(x^3)^2}$

Since the index n is even $\sqrt{a^n} = |a|$. $|x^3|$

ⓑ

$$\sqrt{y^{16}}$$

Since $(y^8)^2 = y^{16}$. $\sqrt{(y^8)^2}$

Since the index n is even $\sqrt[n]{a^n} = |a|$. y^8

In this case the absolute value sign is not needed as y^8 is positive.

> **TRY IT :: 8.15**   Simplify: ⓐ $\sqrt{y^{18}}$ ⓑ $\sqrt{z^{12}}$.

> **TRY IT :: 8.16**   Simplify: ⓐ $\sqrt{m^4}$ ⓑ $\sqrt{b^{10}}$.

The next example uses the same idea for highter roots.

EXAMPLE 8.9

Simplify: ⓐ $\sqrt[3]{y^{18}}$ ⓑ $\sqrt[4]{z^8}$.

✓ **Solution**

ⓐ

$$\sqrt[3]{y^{18}}$$

Since $(y^6)^3 = y^{18}$. $\sqrt[3]{(y^6)^3}$

Since n is odd, $\sqrt[n]{a^n} = a$. y^6

ⓑ

	$\sqrt[4]{z^8}$
Since $\left(z^2\right)^4 = z^8$.	$\sqrt[4]{\left(z^2\right)^4}$
Since z^2 is positive, we do not need an absolute value sign.	z^2

> **TRY IT : : 8.17**   Simplify: ⓐ $\sqrt[4]{u^{12}}$ ⓑ $\sqrt[3]{v^{15}}$.

> **TRY IT : : 8.18**   Simplify: ⓐ $\sqrt[5]{c^{20}}$ ⓑ $\sqrt[6]{d^{24}}$

In the next example, we now have a coefficient in front of the variable. The concept $\sqrt{a^{2m}} = |a^m|$ works in much the same way.

$$\sqrt{16r^{22}} = 4\left|r^{11}\right| \text{ because } \left(4r^{11}\right)^2 = 16r^{22}.$$

But notice $\sqrt{25u^8} = 5u^4$ and no absolute value sign is needed as u^4 is always positive.

EXAMPLE 8.10

Simplify: ⓐ $\sqrt{16n^2}$ ⓑ $-\sqrt{81c^2}$.

✓ **Solution**

ⓐ

	$\sqrt{16n^2}$
Since $(4n)^2 = 16n^2$.	$\sqrt{(4n)^2}$
Since the index n is even $\sqrt[n]{a^n} = |a|$.	$4|n|$

ⓑ

	$-\sqrt{81c^2}$
Since $(9c)^2 = 81c^2$.	$-\sqrt{(9c)^2}$
Since the index n is even $\sqrt[n]{a^n} = |a|$.	$-9|c|$

> **TRY IT : : 8.19**   Simplify: ⓐ $\sqrt{64x^2}$ ⓑ $-\sqrt{100p^2}$.

> **TRY IT : : 8.20**   Simplify: ⓐ $\sqrt{169y^2}$ ⓑ $-\sqrt{121y^2}$.

This example just takes the idea farther as it has roots of higher index.

EXAMPLE 8.11

Simplify: ⓐ $\sqrt[3]{64p^6}$ ⓑ $\sqrt[4]{16q^{12}}$.

✓ Solution

ⓐ

$$\sqrt[3]{64p^6}$$

Rewrite $64p^6$ as $\left(4p^2\right)^3$.

$$\sqrt[3]{\left(4p^2\right)^3}$$

Take the cube root.

$$4p^2$$

ⓑ

$$\sqrt[4]{16q^{12}}$$

Rewrite the radicand as a fourth power.

$$\sqrt[4]{\left(2q^3\right)^4}$$

Take the fourth root.

$$2\left|q^3\right|$$

> **TRY IT ::** 8.21   Simplify: ⓐ $\sqrt[3]{27x^{27}}$ ⓑ $\sqrt[4]{81q^{28}}$.

> **TRY IT ::** 8.22   Simplify: ⓐ $\sqrt[3]{125q^9}$ ⓑ $\sqrt[5]{243q^{25}}$.

The next examples have two variables.

EXAMPLE 8.12

Simplify: ⓐ $\sqrt{36x^2y^2}$ ⓑ $\sqrt{121a^6b^8}$ ⓒ $\sqrt[3]{64p^{63}q^9}$.

✓ Solution

ⓐ

$$\sqrt{36x^2y^2}$$

Since $(6xy)^2 = 36x^2y^2$

$$\sqrt{(6xy)^2}$$

Take the square root.

$$6|xy|$$

ⓑ

$$\sqrt{121a^6b^8}$$

Since $\left(11a^3b^4\right)^2 = 121a^6b^8$

$$\sqrt{\left(11a^3b^4\right)^2}$$

Take the square root.

$$11\left|a^3\right|b^4$$

ⓒ

$$\sqrt[3]{64p^{63}q^9}$$

Since $\left(4p^{21}q^3\right)^3 = 64p^{63}q^9$

$$\sqrt[3]{\left(4p^{21}q^3\right)^3}$$

Take the cube root.

$$4p^{21}q^3$$

> **TRY IT : :** 8.23

Simplify: ⓐ $\sqrt{100a^2 b^2}$ ⓑ $\sqrt{144p^{12}q^{20}}$ ⓒ $\sqrt[3]{8x^{30}y^{12}}$

> **TRY IT : :** 8.24

Simplify: ⓐ $\sqrt{225m^2 n^2}$ ⓑ $\sqrt{169x^{10}y^{14}}$ ⓒ $\sqrt[3]{27w^{36}z^{15}}$

▶ **MEDIA : :**

Access this online resource for additional instruction and practice with simplifying expressions with roots.

- **Simplifying Variables Exponents with Roots using Absolute Values (https://openstax.org/l/37SimVarAbVal)**

8.1 EXERCISES

Practice Makes Perfect

Simplify Expressions with Roots

In the following exercises, simplify.

1. ⓐ $\sqrt{64}$ ⓑ $-\sqrt{81}$

2. ⓐ $\sqrt{169}$ ⓑ $-\sqrt{100}$

3. ⓐ $\sqrt{196}$ ⓑ $-\sqrt{1}$

4. ⓐ $\sqrt{144}$ ⓑ $-\sqrt{121}$

5. ⓐ $\sqrt{\frac{4}{9}}$ ⓑ $-\sqrt{0.01}$

6. ⓐ $\sqrt{\frac{64}{121}}$ ⓑ $-\sqrt{0.16}$

7. ⓐ $\sqrt{-121}$ ⓑ $-\sqrt{289}$

8. ⓐ $-\sqrt{400}$ ⓑ $\sqrt{-36}$

9. ⓐ $-\sqrt{225}$ ⓑ $\sqrt{-9}$

10. ⓐ $\sqrt{-49}$ ⓑ $-\sqrt{256}$

11. ⓐ $\sqrt[3]{216}$ ⓑ $\sqrt[4]{256}$

12. ⓐ $\sqrt[3]{27}$ ⓑ $\sqrt[4]{16}$ ⓒ $\sqrt[5]{243}$

13. ⓐ $\sqrt[3]{512}$ ⓑ $\sqrt[4]{81}$ ⓒ $\sqrt[5]{1}$

14. ⓐ $\sqrt[3]{125}$ ⓑ $\sqrt[4]{1296}$ ⓒ $\sqrt[5]{1024}$

15. ⓐ $\sqrt[3]{-8}$ ⓑ $\sqrt[4]{-81}$ ⓒ $\sqrt[5]{-32}$

16.
ⓐ $\sqrt[3]{-64}$
ⓑ $\sqrt[4]{-16}$
ⓒ $\sqrt[5]{-243}$

17.
ⓐ $\sqrt[3]{-125}$
ⓑ $\sqrt[4]{-1296}$
ⓒ $\sqrt[5]{-1024}$

18.
ⓐ $\sqrt[3]{-512}$
ⓑ $\sqrt[4]{-81}$
ⓒ $\sqrt[5]{-1}$

Estimate and Approximate Roots

In the following exercises, estimate each root between two consecutive whole numbers.

19. ⓐ $\sqrt{70}$ ⓑ $\sqrt[3]{71}$

20. ⓐ $\sqrt{55}$ ⓑ $\sqrt[3]{119}$

21. ⓐ $\sqrt{200}$ ⓑ $\sqrt[3]{137}$

22. ⓐ $\sqrt{172}$ ⓑ $\sqrt[3]{200}$

In the following exercises, approximate each root and round to two decimal places.

23. ⓐ $\sqrt{19}$ ⓑ $\sqrt[3]{89}$ ⓒ $\sqrt[4]{97}$

24. ⓐ $\sqrt{21}$ ⓑ $\sqrt[3]{93}$ ⓒ $\sqrt[4]{101}$

25. ⓐ $\sqrt{53}$ ⓑ $\sqrt[3]{147}$ ⓒ $\sqrt[4]{452}$

26. ⓐ $\sqrt{47}$ ⓑ $\sqrt[3]{163}$ ⓒ $\sqrt[4]{527}$

Simplify Variable Expressions with Roots

In the following exercises, simplify using absolute values as necessary.

27. ⓐ $\sqrt[5]{u^5}$ ⓑ $\sqrt[8]{v^8}$

28. ⓐ $\sqrt[3]{a^3}$ ⓑ $\sqrt[9]{b^9}$

29. ⓐ $\sqrt[4]{y^4}$ ⓑ $\sqrt[7]{m^7}$

30. ⓐ $\sqrt[8]{k^8}$ ⓑ $\sqrt[6]{p^6}$

31. ⓐ $\sqrt{x^6}$ ⓑ $\sqrt{y^{16}}$

32. ⓐ $\sqrt{a^{14}}$ ⓑ $\sqrt{w^{24}}$

33. ⓐ $\sqrt{x^{24}}$ ⓑ $\sqrt{y^{22}}$

34. ⓐ $\sqrt{a^{12}}$ ⓑ $\sqrt{b^{26}}$

35. ⓐ $\sqrt[3]{x^9}$ ⓑ $\sqrt[4]{y^{12}}$

36. ⓐ $\sqrt[5]{a^{10}}$ ⓑ $\sqrt[3]{b^{27}}$

37. ⓐ $\sqrt[4]{m^8}$ ⓑ $\sqrt[5]{n^{20}}$

38. ⓐ $\sqrt[6]{r^{12}}$ ⓑ $\sqrt[3]{s^{30}}$

39. (a) $\sqrt{49x^2}$ (b) $-\sqrt{81x^{18}}$ **40.** (a) $\sqrt{100y^2}$ (b) $-\sqrt{100m^{32}}$ **41.** (a) $\sqrt{121m^{20}}$ (b) $-\sqrt{64a^2}$

42.
(a) $\sqrt{81x^{36}}$
(b) $-\sqrt{25x^2}$

43.
(a) $\sqrt[4]{16x^8}$
(b) $\sqrt[6]{64y^{12}}$

44.
(a) $\sqrt[3]{-8c^9}$
(b) $\sqrt[3]{125d^{15}}$

45.
(a) $\sqrt[3]{216a^6}$
(b) $\sqrt[5]{32b^{20}}$

46.
(a) $\sqrt[7]{128r^{14}}$
(b) $\sqrt[4]{81s^{24}}$

47.
(a) $\sqrt{144x^2y^2}$
(b) $\sqrt{169w^8y^{10}}$
(c) $\sqrt[3]{8a^{51}b^6}$

48.
(a) $\sqrt{196a^2b^2}$
(b) $\sqrt{81p^{24}q^6}$
(c) $\sqrt[3]{27p^{45}q^9}$

49.
(a) $\sqrt{121a^2b^2}$
(b) $\sqrt{9c^8d^{12}}$
(c) $\sqrt[3]{64x^{15}y^{66}}$

50.
(a) $\sqrt{225x^2y^2z^2}$
(b) $\sqrt{36r^6s^{20}}$
(c) $\sqrt[3]{125y^{18}z^{27}}$

Writing Exercises

51. Why is there no real number equal to $\sqrt{-64}$?

52. What is the difference between 9^2 and $\sqrt{9}$?

53. Explain what is meant by the nth root of a number.

54. Explain the difference of finding the nth root of a number when the index is even compared to when the index is odd.

Self Check

ⓐ *After completing the exercises, use this checklist to evaluate your mastery of the objectives of this section.*

I can...	Confidently	With some help	No-I don't get it!
simplify expressions with roots.			
estimate and approximate roots.			
simplify variable expressions with roots.			

ⓑ *If most of your checks were:*

...confidently. *Congratulations! You have achieved the objectives in this section. Reflect on the study skills you used so that you can continue to use them. What did you do to become confident of your ability to do these things? Be specific.*

...with some help. *This must be addressed quickly because topics you do not master become potholes in your road to success. In math every topic builds upon previous work. It is important to make sure you have a strong foundation before you move on. Who can you ask for help? Your fellow classmates and instructor are good resources. Is there a place on campus where math tutors are available? Can your study skills be improved?*

...no - I don't get it! *This is a warning sign and you must not ignore it. You should get help right away or you will quickly be overwhelmed. See your instructor as soon as you can to discuss your situation. Together you can come up with a plan to get you the help you need.*

8.2 | Simplify Radical Expressions

Learning Objectives

By the end of this section, you will be able to:
- › Use the Product Property to simplify radical expressions
- › Use the Quotient Property to simplify radical expressions

Be Prepared!

Before you get started, take this readiness quiz.

1. Simplify: $\dfrac{x^9}{x^4}$.

   If you missed this problem, review **Example 5.13**.

2. Simplify: $\dfrac{y^3}{y^{11}}$.

   If you missed this problem, review **Example 5.13**.

3. Simplify: $\left(n^2\right)^6$.

   If you missed this problem, review **Example 5.17**.

Use the Product Property to Simplify Radical Expressions

We will simplify radical expressions in a way similar to how we simplified fractions. A fraction is simplified if there are no common factors in the numerator and denominator. To simplify a fraction, we look for any common factors in the numerator and denominator.

A radical expression, $\sqrt[n]{a}$, is considered simplified if it has no factors of m^n. So, to simplify a radical expression, we look for any factors in the radicand that are powers of the index.

Simplified Radical Expression

For real numbers a and m, and $n \geq 2$,

$$\sqrt[n]{a} \text{ is considered simplified if } a \text{ has no factors of } m^n$$

For example, $\sqrt{5}$ is considered simplified because there are no perfect square factors in 5. But $\sqrt{12}$ is not simplified because 12 has a perfect square factor of 4.

Similarly, $\sqrt[3]{4}$ is simplified because there are no perfect cube factors in 4. But $\sqrt[3]{24}$ is not simplified because 24 has a perfect cube factor of 8.

To simplify radical expressions, we will also use some properties of roots. The properties we will use to simplify radical expressions are similar to the properties of exponents. We know that $(ab)^n = a^n b^n$. The corresponding of **Product Property of Roots** says that $\sqrt[n]{ab} = \sqrt[n]{a} \cdot \sqrt[n]{b}$.

Product Property of nth Roots

If $\sqrt[n]{a}$ and $\sqrt[n]{b}$ are real numbers, and $n \geq 2$ is an integer, then

$$\sqrt[n]{ab} = \sqrt[n]{a} \cdot \sqrt[n]{b} \quad \text{and} \quad \sqrt[n]{a} \cdot \sqrt[n]{b} = \sqrt[n]{ab}$$

We use the Product Property of Roots to remove all perfect square factors from a square root.

EXAMPLE 8.13 SIMPLIFY SQUARE ROOTS USING THE PRODUCT PROPERTY OF ROOTS

Simplify: $\sqrt{98}$.

✓ **Solution**

Step 1. Find the largest factor in the radicand that is a perfect power of the index. Rewrite the radicand as a product of two factors, using that factor.	We see that 49 is the largest factor of 98 that has a power of 2. In other words 49 is the largest perfect square factor of 98. $98 = 49 \cdot 2$ Always write the perfect square factor first.	$\sqrt{98}$ $\sqrt{49 \cdot 2}$
Step 2. Use the product rule to rewrite the radical as the product of two radicals.		$\sqrt{49} \cdot \sqrt{2}$
Step 3. Simplify the root of the perfect power.		$7\sqrt{2}$

> **TRY IT : : 8.25**   Simplify: $\sqrt{48}$.

> **TRY IT : : 8.26**   Simplify: $\sqrt{45}$.

Notice in the previous example that the simplified form of $\sqrt{98}$ is $7\sqrt{2}$, which is the product of an integer and a square root. We always write the integer in front of the square root.

Be careful to write your integer so that it is not confused with the index. The expression $7\sqrt{2}$ is very different from $\sqrt[7]{2}$.

HOW TO : : SIMPLIFY A RADICAL EXPRESSION USING THE PRODUCT PROPERTY.

Step 1.   Find the largest factor in the radicand that is a perfect power of the index. Rewrite the radicand as a product of two factors, using that factor.

Step 2.   Use the product rule to rewrite the radical as the product of two radicals.

Step 3.   Simplify the root of the perfect power.

We will apply this method in the next example. It may be helpful to have a table of perfect squares, cubes, and fourth powers.

EXAMPLE 8.14

Simplify: ⓐ $\sqrt{500}$ ⓑ $\sqrt[3]{16}$ ⓒ $\sqrt[4]{243}$.

✓ **Solution**

ⓐ

	$\sqrt{500}$
Rewrite the radicand as a product using the largest perfect square factor.	$\sqrt{100 \cdot 5}$
Rewrite the radical as the product of two radicals	$\sqrt{100} \cdot \sqrt{5}$
Simplify.	$10\sqrt{5}$

ⓑ

	$\sqrt[3]{16}$
Rewrite the radicand as a product using the greatest perfect cube factor. $2^3 = 8$	$\sqrt[3]{8 \cdot 2}$
Rewrite the radical as the product of two radicals.	$\sqrt[3]{8} \cdot \sqrt[3]{2}$
Simplify.	$2\sqrt[3]{2}$

ⓒ

	$\sqrt[4]{243}$
Rewrite the radicand as a product using the greatest perfect fourth power factor. $3^4 = 81$	$\sqrt[4]{81 \cdot 3}$
Rewrite the radical as the product of two radicals	$\sqrt[4]{81} \cdot \sqrt[4]{3}$
Simplify.	$3\sqrt[4]{3}$

> **TRY IT : : 8.27**   Simplify: ⓐ $\sqrt{288}$ ⓑ $\sqrt[3]{81}$ ⓒ $\sqrt[4]{64}$.

> **TRY IT : : 8.28**   Simplify: ⓐ $\sqrt{432}$ ⓑ $\sqrt[3]{625}$ ⓒ $\sqrt[4]{729}$.

The next example is much like the previous examples, but with variables. Don't forget to use the absolute value signs when taking an even root of an expression with a variable in the radical.

EXAMPLE 8.15

Simplify: ⓐ $\sqrt{x^3}$ ⓑ $\sqrt[3]{x^4}$ ⓒ $\sqrt[4]{x^7}$.

✓ **Solution**

ⓐ

	$\sqrt{x^3}$
Rewrite the radicand as a product using the largest perfect square factor.	$\sqrt{x^2 \cdot x}$
Rewrite the radical as the product of two radicals.	$\sqrt{x^2} \cdot \sqrt{x}$
Simplify.	$|x|\sqrt{x}$

ⓑ

	$\sqrt[3]{x^4}$
Rewrite the radicand as a product using the largest perfect cube factor.	$\sqrt[3]{x^3 \cdot x}$.
Rewrite the radical as the product of two radicals.	$\sqrt[3]{x^3} \cdot \sqrt[3]{x}$
Simplify.	$x\sqrt[3]{x}$

ⓒ

$$\sqrt[4]{x^7}$$

Rewrite the radicand as a product
using the greatest perfect fourth power
factor.

$$\sqrt[4]{x^4 \cdot x^3}$$

Rewrite the radical as the product of two
radicals.

$$\sqrt[4]{x^4} \cdot \sqrt[4]{x^3}$$

Simplify.

$$|x| \sqrt[4]{x^3}$$

> **TRY IT :: 8.29**   Simplify: ⓐ $\sqrt{b^5}$ ⓑ $\sqrt[4]{y^6}$ ⓒ $\sqrt[3]{z^5}$

> **TRY IT :: 8.30**   Simplify: ⓐ $\sqrt{p^9}$ ⓑ $\sqrt[5]{y^8}$ ⓒ $\sqrt[6]{q^{13}}$

We follow the same procedure when there is a coefficient in the radicand. In the next example, both the constant and the variable have perfect square factors.

EXAMPLE 8.16

Simplify: ⓐ $\sqrt{72n^7}$ ⓑ $\sqrt[3]{24x^7}$ ⓒ $\sqrt[4]{80y^{14}}$.

⊘ **Solution**

ⓐ

$$\sqrt{72n^7}$$

Rewrite the radicand as a product
using the largest perfect square factor.

$$\sqrt{36n^6 \cdot 2n}$$

Rewrite the radical as the product of two
radicals.

$$\sqrt{36n^6} \cdot \sqrt{2n}$$

Simplify.

$$6|n^3| \sqrt{2n}$$

ⓑ

$$\sqrt[3]{24x^7}$$

Rewrite the radicand as a product
using perfect cube factors.

$$\sqrt[3]{8x^6 \cdot 3x}$$

Rewrite the radical as the product of two
radicals.

$$\sqrt[3]{8x^6} \cdot \sqrt[3]{3x}$$

Rewrite the first radicand as $(2x^2)^3$.

$$\sqrt[3]{(2x^2)^3} \cdot \sqrt[3]{3x}$$

Simplify.

$$2x^2 \sqrt[3]{3x}$$

ⓒ

	$\sqrt[4]{80y^{14}}$
Rewrite the radicand as a product using perfect fourth power factors.	$\sqrt[4]{16y^{12} \cdot 5\,y^2}$
Rewrite the radical as the product of two radicals.	$\sqrt[4]{16y^{12}} \cdot \sqrt[4]{5y^2}$
Rewrite the first radicand as $\left(2y^3\right)^4$.	$\sqrt[4]{\left(2\,y^3\right)^4} \cdot \sqrt[4]{5y^2}$
Simplify.	$2\left|y^3\right| \sqrt[4]{5\,y^2}$

 TRY IT :: 8.31

Simplify: ⓐ $\sqrt{32y^5}$ ⓑ $\sqrt[3]{54p^{10}}$ ⓒ $\sqrt[4]{64q^{10}}$.

 **TRY IT :: 8.32**

Simplify: ⓐ $\sqrt{75a^9}$ ⓑ $\sqrt[3]{128m^{11}}$ ⓒ $\sqrt[4]{162n^7}$.

In the next example, we continue to use the same methods even though there are more than one variable under the radical.

EXAMPLE 8.17

Simplify: ⓐ $\sqrt{63u^3v^5}$ ⓑ $\sqrt[3]{40x^4y^5}$ ⓒ $\sqrt[4]{48x^4y^7}$.

 **Solution**

ⓐ

	$\sqrt{63u^3v^5}$
Rewrite the radicand as a product using the largest perfect square factor.	$\sqrt{9u^2v^4 \cdot 7uv}$
Rewrite the radical as the product of two radicals.	$\sqrt{9u^2v^4} \cdot \sqrt{7uv}$
Rewrite the first radicand as $\left(3uv^2\right)^2$.	$\sqrt{\left(3uv^2\right)^2} \cdot \sqrt{7uv}$
Simplify.	$3|u|v^2 \sqrt{7uv}$

ⓑ

	$\sqrt[3]{40x^4y^5}$
Rewrite the radicand as a product using the largest perfect cube factor.	$\sqrt[3]{8x^3y^3 \cdot 5xy^2}$
Rewrite the radical as the product of two radicals.	$\sqrt[3]{8x^3y^3} \cdot \sqrt[3]{5xy^2}$
Rewrite the first radicand as $(2xy)^3$.	$\sqrt[3]{(2xy)^3} \cdot \sqrt[3]{5xy^2}$
Simplify.	$2xy \sqrt[3]{5xy^2}$

 ⓒ

<table>
<tr><td>Rewrite the radicand as a product using the largest perfect fourth power factor.</td><td>$\sqrt[4]{48x^4y^7}$
$\sqrt[4]{16x^4y^4 \cdot 3y^3}$</td></tr>
</table>

$$\sqrt[4]{48x^4y^7}$$

Rewrite the radicand as a product using the largest perfect fourth power factor.

$$\sqrt[4]{16x^4y^4 \cdot 3y^3}$$

Rewrite the radical as the product of two radicals.

$$\sqrt[4]{16x^4y^4} \cdot \sqrt[4]{3y^3}$$

Rewrite the first radicand as $(2xy)^4$.

$$\sqrt[4]{(2xy)^4} \cdot \sqrt[4]{3y^3}$$

Simplify.

$$2|xy|\sqrt[4]{3y^3}$$

> **TRY IT :: 8.33**
Simplify: ⓐ $\sqrt{98a^7b^5}$ ⓑ $\sqrt[3]{56x^5y^4}$ ⓒ $\sqrt[4]{32x^5y^8}$.

> **TRY IT :: 8.34**
Simplify: ⓐ $\sqrt{180m^9n^{11}}$ ⓑ $\sqrt[3]{72x^6y^5}$ ⓒ $\sqrt[4]{80x^7y^4}$.

EXAMPLE 8.18

Simplify: ⓐ $\sqrt[3]{-27}$ ⓑ $\sqrt[4]{-16}$.

⊘ Solution

ⓐ

$$\sqrt[3]{-27}$$

Rewrite the radicand as a product using perfect cube factors.

$$\sqrt[3]{(-3)^3}$$

Take the cube root.

$$-3$$

ⓑ

$$\sqrt[4]{-16}$$

There is no real number n where $n^4 = -16$.   Not a real number.

> **TRY IT :: 8.35**
Simplify: ⓐ $\sqrt[3]{-64}$ ⓑ $\sqrt[4]{-81}$.

> **TRY IT :: 8.36**
Simplify: ⓐ $\sqrt[3]{-625}$ ⓑ $\sqrt[4]{-324}$.

We have seen how to use the order of operations to simplify some expressions with radicals. In the next example, we have the sum of an integer and a square root. We simplify the square root but cannot add the resulting expression to the integer since one term contains a radical and the other does not. The next example also includes a fraction with a radical in the numerator. Remember that in order to simplify a fraction you need a common factor in the numerator and denominator.

EXAMPLE 8.19

Simplify: ⓐ $3 + \sqrt{32}$ ⓑ $\dfrac{4 - \sqrt{48}}{2}$.

⊘ Solution

ⓐ

	$3 + \sqrt{32}$
Rewrite the radicand as a product using the largest perfect square factor.	$3 + \sqrt{16 \cdot 2}$
Rewrite the radical as the product of two radicals.	$3 + \sqrt{16} \cdot \sqrt{2}$
Simplify.	$3 + 4\sqrt{2}$

The terms cannot be added as one has a radical and the other does not. Trying to add an integer and a radical is like trying to add an integer and a variable. They are not like terms!

ⓑ

	$\dfrac{4 - \sqrt{48}}{2}$
Rewrite the radicand as a product using the largest perfect square factor.	$\dfrac{4 - \sqrt{16 \cdot 3}}{2}$
Rewrite the radical as the product of two radicals.	$\dfrac{4 - \sqrt{16} \cdot \sqrt{3}}{2}$
Simplify.	$\dfrac{4 - 4\sqrt{3}}{2}$
Factor the common factor from the numerator.	$\dfrac{4(1 - \sqrt{3})}{2}$
Remove the common factor, 2, from the numerator and denominator.	$\dfrac{\cancel{2} \cdot 2(1 - \sqrt{3})}{\cancel{2}}$
Simplify.	$2(1 - \sqrt{3})$

> **TRY IT :: 8.37**   Simplify: ⓐ $5 + \sqrt{75}$ ⓑ $\dfrac{10 - \sqrt{75}}{5}$

> **TRY IT :: 8.38**   Simplify: ⓐ $2 + \sqrt{98}$ ⓑ $\dfrac{6 - \sqrt{45}}{3}$

Use the Quotient Property to Simplify Radical Expressions

Whenever you have to simplify a radical expression, the first step you should take is to determine whether the radicand is a perfect power of the index. If not, check the numerator and denominator for any common factors, and remove them. You may find a fraction in which both the numerator and the denominator are perfect powers of the index.

EXAMPLE 8.20

Simplify: ⓐ $\sqrt{\dfrac{45}{80}}$ ⓑ $\sqrt[3]{\dfrac{16}{54}}$ ⓒ $\sqrt[4]{\dfrac{5}{80}}$.

✓ **Solution**

ⓐ

	$\sqrt{\dfrac{45}{80}}$
Simplify inside the radical first. Rewrite showing the common factors of the numerator and denominator.	$\sqrt{\dfrac{5 \cdot 9}{5 \cdot 16}}$
Simplify the fraction by removing common factors.	$\sqrt{\dfrac{9}{16}}$
Simplify. Note $\left(\dfrac{3}{4}\right)^2 = \dfrac{9}{16}$.	$\dfrac{3}{4}$

ⓑ
$$\sqrt[3]{\dfrac{16}{54}}$$

Simplify inside the radical first.
Rewrite showing the common factors of
the numerator and denominator.

$$\sqrt[3]{\dfrac{2\cdot 8}{2\cdot 27}}$$

Simplify the fraction by removing
common factors.

$$\sqrt[3]{\dfrac{8}{27}}$$

Simplify. Note $\left(\dfrac{2}{3}\right)^3 = \dfrac{8}{27}$.

$$\dfrac{2}{3}$$

ⓒ
$$\sqrt[4]{\dfrac{5}{80}}$$

Simplify inside the radical first.
Rewrite showing the common factors of
the numerator and denominator.

$$\sqrt[4]{\dfrac{5\cdot 1}{5\cdot 16}}$$

Simplify the fraction by removing
common factors.

$$\sqrt[4]{\dfrac{1}{16}}$$

Simplify. Note $\left(\dfrac{1}{2}\right)^4 = \dfrac{1}{16}$.

$$\dfrac{1}{2}$$

> **TRY IT : : 8.39**

Simplify: ⓐ $\sqrt{\dfrac{75}{48}}$ ⓑ $\sqrt[3]{\dfrac{54}{250}}$ ⓒ $\sqrt[4]{\dfrac{32}{162}}$.

> **TRY IT : : 8.40**

Simplify: ⓐ $\sqrt{\dfrac{98}{162}}$ ⓑ $\sqrt[3]{\dfrac{24}{375}}$ ⓒ $\sqrt[4]{\dfrac{4}{324}}$.

In the last example, our first step was to simplify the fraction under the radical by removing common factors. In the next example we will use the Quotient Property to simplify under the radical. We divide the like bases by subtracting their exponents,

$$\dfrac{a^m}{a^n} = a^{m-n}, \quad a \neq 0$$

EXAMPLE 8.21

Simplify: ⓐ $\sqrt{\dfrac{m^6}{m^4}}$ ⓑ $\sqrt[3]{\dfrac{a^8}{a^5}}$ ⓒ $\sqrt[4]{\dfrac{a^{10}}{a^2}}$.

⊘ **Solution**

ⓐ
$$\sqrt{\dfrac{m^6}{m^4}}$$

Simplify the fraction inside the radical first.
Divide the like bases by subtracting the
exponents.

$$\sqrt{m^2}$$

Simplify.

$$|m|$$

ⓑ

$$\sqrt[3]{\dfrac{a^8}{a^5}}$$

Use the Quotient Property of exponents to simplify the fraction under the radical first.	$\sqrt[3]{a^3}$
Simplify.	a

ⓒ

$$\sqrt[4]{\dfrac{a^{10}}{a^2}}$$

Use the Quotient Property of exponents to simplify the fraction under the radical first.	$\sqrt[4]{a^8}$
Rewrite the radicand using perfect fourth power factors.	$\sqrt[4]{\left(a^2\right)^4}$
Simplify.	a^2

> **TRY IT :: 8.41**   Simplify: ⓐ $\sqrt{\dfrac{a^8}{a^6}}$ ⓑ $\sqrt[4]{\dfrac{x^7}{x^3}}$ ⓒ $\sqrt[4]{\dfrac{y^{17}}{y^5}}$.

> **TRY IT :: 8.42**   Simplify: ⓐ $\sqrt{\dfrac{x^{14}}{x^{10}}}$ ⓑ $\sqrt[3]{\dfrac{m^{13}}{m^7}}$ ⓒ $\sqrt[5]{\dfrac{n^{12}}{n^2}}$.

Remember the Quotient to a Power Property? It said we could raise a fraction to a power by raising the numerator and denominator to the power separately.

$$\left(\dfrac{a}{b}\right)^m = \dfrac{a^m}{b^m},\ b \neq 0$$

We can use a similar property to simplify a root of a fraction. After removing all common factors from the numerator and denominator, if the fraction is not a perfect power of the index, we simplify the numerator and denominator separately.

Quotient Property of Radical Expressions

If $\sqrt[n]{a}$ and $\sqrt[n]{b}$ are real numbers, $b \neq 0$, and for any integer $n \geq 2$ then,

$$\sqrt[n]{\dfrac{a}{b}} = \dfrac{\sqrt[n]{a}}{\sqrt[n]{b}} \quad \text{and} \quad \dfrac{\sqrt[n]{a}}{\sqrt[n]{b}} = \sqrt[n]{\dfrac{a}{b}}$$

EXAMPLE 8.22   HOW TO SIMPLIFY THE QUOTIENT OF RADICAL EXPRESSIONS

Simplify: $\sqrt{\dfrac{27m^3}{196}}$.

✓ **Solution**

Step 1. Simplify the fraction in the radicand, if possible.	$\dfrac{27m^3}{196}$ cannot be simplified.	$\sqrt{\dfrac{27m^3}{196}}$

| Step 2. Use the Quotient Property to rewrite the radical as the quotient of two radicals. | We rewrite $\sqrt{\dfrac{27m^3}{196}}$ as the quotient of $\sqrt{27m^3}$ and $\sqrt{196}$. | $\dfrac{\sqrt{27m^3}}{\sqrt{196}}$ |
| Step 3. Simplify the radicals in the numerator and the denominator. | $9m^2$ and 196 are perfect squares. | $\dfrac{\sqrt{9m^2}\cdot\sqrt{3m}}{\sqrt{196}}$ $\dfrac{3m\sqrt{3m}}{14}$ |

> **TRY IT : : 8.43**

Simplify: $\sqrt{\dfrac{24p^3}{49}}$.

> **TRY IT : : 8.44**

Simplify: $\sqrt{\dfrac{48x^5}{100}}$.

   HOW TO : : SIMPLIFY A SQUARE ROOT USING THE QUOTIENT PROPERTY.

Step 1.   Simplify the fraction in the radicand, if possible.

Step 2.   Use the Quotient Property to rewrite the radical as the quotient of two radicals.

Step 3.   Simplify the radicals in the numerator and the denominator.

EXAMPLE 8.23

Simplify: ⓐ $\sqrt{\dfrac{45x^5}{y^4}}$ ⓑ $\sqrt[3]{\dfrac{24x^7}{y^3}}$ ⓒ $\sqrt[4]{\dfrac{48x^{10}}{y^8}}$.

 Solution

ⓐ

$$\sqrt{\dfrac{45x^5}{y^4}}$$

We cannot simplify the fraction in the radicand. Rewrite using the Quotient Property.

$$\dfrac{\sqrt{45x^5}}{\sqrt{y^4}}$$

Simplify the radicals in the numerator and the denominator.

$$\dfrac{\sqrt{9x^4}\cdot\sqrt{5x}}{y^2}$$

Simplify.

$$\dfrac{3x^2\sqrt{5x}}{y^2}$$

ⓑ

$$\sqrt[3]{\dfrac{24x^7}{y^3}}$$

The fraction in the radicand cannot be simplified. Use the Quotient Property to write as two radicals.

$$\dfrac{\sqrt[3]{24x^7}}{\sqrt[3]{y^3}}$$

Rewrite each radicand as a product using perfect cube factors.

$$\dfrac{\sqrt[3]{8x^6 \cdot 3x}}{\sqrt[3]{y^3}}$$

Rewrite the numerator as the product of two radicals.

$$\dfrac{\sqrt[3]{\left(2x^2\right)^3} \cdot \sqrt[3]{3x}}{\sqrt[3]{y^3}}$$

Simplify.

$$\dfrac{2x^2\sqrt[3]{3x}}{y}$$

ⓒ

$$\sqrt[4]{\dfrac{48x^{10}}{y^8}}$$

The fraction in the radicand cannot be simplified.

$$\dfrac{\sqrt[4]{48x^{10}}}{\sqrt[4]{y^8}}$$

Use the Quotient Property to write as two radicals. Rewrite each radicand as a product using perfect fourth power factors.

$$\dfrac{\sqrt[4]{16x^8 \cdot 3x^2}}{\sqrt[4]{y^8}}$$

Rewrite the numerator as the product of two radicals.

$$\dfrac{\sqrt[4]{\left(2x^2\right)^4} \cdot \sqrt[4]{3x^2}}{\sqrt[4]{\left(y^2\right)^4}}$$

Simplify.

$$\dfrac{2x^2\sqrt[4]{3x^2}}{y^2}$$

> **TRY IT :: 8.45**

Simplify: ⓐ $\sqrt{\dfrac{80m^3}{n^6}}$ ⓑ $\sqrt[3]{\dfrac{108c^{10}}{d^6}}$ ⓒ $\sqrt[4]{\dfrac{80x^{10}}{y^4}}$.

> **TRY IT :: 8.46**

Simplify: ⓐ $\sqrt{\dfrac{54u^7}{v^8}}$ ⓑ $\sqrt[3]{\dfrac{40r^3}{s^6}}$ ⓒ $\sqrt[4]{\dfrac{162m^{14}}{n^{12}}}$.

Be sure to simplify the fraction in the radicand first, if possible.

EXAMPLE 8.24

Simplify: ⓐ $\sqrt{\dfrac{18p^5q^7}{32pq^2}}$ ⓑ $\sqrt[3]{\dfrac{16x^5y^7}{54x^2y^2}}$ ⓒ $\sqrt[4]{\dfrac{5a^8b^6}{80a^3b^2}}$.

✓ Solution

ⓐ

$$\sqrt{\frac{18p^5q^7}{32pq^2}}$$

Simplify the fraction in the radicand, if possible.

$$\sqrt{\frac{9p^4q^5}{16}}$$

Rewrite using the Quotient Property.

$$\frac{\sqrt{9p^4q^5}}{\sqrt{16}}$$

Simplify the radicals in the numerator and the denominator.

$$\frac{\sqrt{9p^4q^4}\cdot\sqrt{q}}{4}$$

Simplify.

$$\frac{3p^2q^2\sqrt{q}}{4}$$

ⓑ

$$\sqrt[3]{\frac{16x^5y^7}{54x^2y^2}}$$

Simplify the fraction in the radicand, if possible.

$$\sqrt[3]{\frac{8x^3y^5}{27}}$$

Rewrite using the Quotient Property.

$$\frac{\sqrt[3]{8x^3y^5}}{\sqrt[3]{27}}$$

Simplify the radicals in the numerator and the denominator.

$$\frac{\sqrt[3]{8x^3y^3}\cdot\sqrt[3]{y^2}}{\sqrt[3]{27}}$$

Simplify.

$$\frac{2xy\sqrt[3]{y^2}}{3}$$

ⓒ

$$\sqrt[4]{\frac{5a^8b^6}{80a^3b^2}}$$

Simplify the fraction in the radicand, if possible.

$$\sqrt[4]{\frac{a^5b^4}{16}}$$

Rewrite using the Quotient Property.

$$\frac{\sqrt[4]{a^5b^4}}{\sqrt[4]{16}}$$

Simplify the radicals in the numerator and the denominator.

$$\frac{\sqrt[4]{a^4b^4}\cdot\sqrt[4]{a}}{\sqrt[4]{16}}$$

Simplify.

$$\frac{|ab|\sqrt[4]{a}}{2}$$

> **TRY IT : : 8.47**

Simplify: ⓐ $\sqrt{\dfrac{50x^5y^3}{72x^4y}}$ ⓑ $\sqrt[3]{\dfrac{16x^5y^7}{54x^2y^2}}$ ⓒ $\sqrt[4]{\dfrac{5a^8b^6}{80a^3b^2}}$.

> **TRY IT : : 8.48**

Simplify: ⓐ $\sqrt{\dfrac{48m^7n^2}{100m^5n^8}}$ ⓑ $\sqrt[3]{\dfrac{54x^7y^5}{250x^2y^2}}$ ⓒ $\sqrt[4]{\dfrac{32a^9b^7}{162a^3b^3}}$.

In the next example, there is nothing to simplify in the denominators. Since the index on the radicals is the same, we can use the Quotient Property again, to combine them into one radical. We will then look to see if we can simplify the expression.

EXAMPLE 8.25

Simplify: ⓐ $\dfrac{\sqrt{48a^7}}{\sqrt{3a}}$ ⓑ $\dfrac{\sqrt[3]{-108}}{\sqrt[3]{2}}$ ⓒ $\dfrac{\sqrt[4]{96x^7}}{\sqrt[4]{3x^2}}$.

⊘ **Solution**

ⓐ

$$\frac{\sqrt{48a^7}}{\sqrt{3a}}$$

The denominator cannot be simplified, so use the Quotient Property to write as one radical.

$$\sqrt{\frac{48a^7}{3a}}$$

Simplify the fraction under the radical.

$$\sqrt{16a^6}$$

Simplify.

$$4|a^3|$$

ⓑ

$$\frac{\sqrt[3]{-108}}{\sqrt[3]{2}}$$

The denominator cannot be simplified, so use the Quotient Property to write as one radical.

$$\sqrt[3]{\frac{-108}{2}}$$

Simplify the fraction under the radical.

$$\sqrt[3]{-54}$$

Rewrite the radicand as a product using perfect cube factors.

$$\sqrt[3]{(-3)^3 \cdot 2}$$

Rewrite the radical as the product of two radicals.

$$\sqrt[3]{(-3)^3} \cdot \sqrt[3]{2}$$

Simplify.

$$-3\sqrt[3]{2}$$

 ©

$$\frac{\sqrt[4]{96x^7}}{\sqrt[4]{3x^2}}$$

The denominator cannot be simplified, so use the Quotient Property to write as one radical.

$$\sqrt[4]{\frac{96x^7}{3x^2}}$$

Simplify the fraction under the radical.

$$\sqrt[4]{32x^5}$$

Rewrite the radicand as a product using perfect fourth power factors.

$$\sqrt[4]{16x^4} \cdot \sqrt[4]{2x}$$

Rewrite the radical as the product of two radicals.

$$\sqrt[4]{(2x)^4} \cdot \sqrt[4]{2x}$$

Simplify.

$$2|x|\sqrt[4]{2x}$$

> **TRY IT : : 8.49**

Simplify: ⓐ $\dfrac{\sqrt{98z^5}}{\sqrt{2z}}$ ⓑ $\dfrac{\sqrt[3]{-500}}{\sqrt[3]{2}}$ ⓒ $\dfrac{\sqrt[4]{486m^{11}}}{\sqrt[4]{3m^5}}$.

> **TRY IT : : 8.50**

Simplify: ⓐ $\dfrac{\sqrt{128m^9}}{\sqrt{2m}}$ ⓑ $\dfrac{\sqrt[3]{-192}}{\sqrt[3]{3}}$ ⓒ $\dfrac{\sqrt[4]{324n^7}}{\sqrt[4]{2n^3}}$.

▶ **MEDIA : :**

Access these online resources for additional instruction and practice with simplifying radical expressions.

- **Simplifying Square Root and Cube Root with Variables (https://openstax.org/l/37SimRtwithVar1)**
- **Express a Radical in Simplified Form-Square and Cube Roots with Variables and Exponents (https://openstax.org/l/37SimRtwithVar2)**
- **Simplifying Cube Roots (https://openstax.org/l/37SimRtwithVar3)**

8.2 EXERCISES

Practice Makes Perfect

Use the Product Property to Simplify Radical Expressions

In the following exercises, use the Product Property to simplify radical expressions.

55. $\sqrt{27}$

56. $\sqrt{80}$

57. $\sqrt{125}$

58. $\sqrt{96}$

59. $\sqrt{147}$

60. $\sqrt{450}$

61. $\sqrt{800}$

62. $\sqrt{675}$

63. ⓐ $\sqrt[4]{32}$ ⓑ $\sqrt[5]{64}$

64. ⓐ $\sqrt[3]{625}$ ⓑ $\sqrt[6]{128}$

65. ⓐ $\sqrt[5]{64}$ ⓑ $\sqrt[3]{256}$

66. ⓐ $\sqrt[4]{3125}$ ⓑ $\sqrt[3]{81}$

In the following exercises, simplify using absolute value signs as needed.

67.
ⓐ $\sqrt{y^{11}}$
ⓑ $\sqrt[3]{r^5}$
ⓒ $\sqrt[4]{s^{10}}$

68.
ⓐ $\sqrt{m^{13}}$
ⓑ $\sqrt[5]{u^7}$
ⓒ $\sqrt[6]{v^{11}}$

69.
ⓐ $\sqrt{n^{21}}$
ⓑ $\sqrt[3]{q^8}$
ⓒ $\sqrt[8]{n^{10}}$

70.
ⓐ $\sqrt{r^{25}}$
ⓑ $\sqrt[5]{p^8}$
ⓒ $\sqrt[4]{m^5}$

71.
ⓐ $\sqrt{125r^{13}}$
ⓑ $\sqrt[3]{108x^5}$
ⓒ $\sqrt[4]{48y^6}$

72.
ⓐ $\sqrt{80s^{15}}$
ⓑ $\sqrt[5]{96a^7}$
ⓒ $\sqrt[6]{128b^7}$

73.
ⓐ $\sqrt{242m^{23}}$
ⓑ $\sqrt[4]{405m10}$
ⓒ $\sqrt[5]{160n^8}$

74.
ⓐ $\sqrt{175n^{13}}$
ⓑ $\sqrt[5]{512p^5}$
ⓒ $\sqrt[4]{324q^7}$

75.
ⓐ $\sqrt{147m^7n^{11}}$
ⓑ $\sqrt[3]{48x^6y^7}$
ⓒ $\sqrt[4]{32x^5y^4}$

76.
ⓐ $\sqrt{96r^3s^3}$
ⓑ $\sqrt[3]{80x^7y^6}$
ⓒ $\sqrt[4]{80x^8y^9}$

77.
ⓐ $\sqrt{192q^3r^7}$
ⓑ $\sqrt[3]{54m^9n^{10}}$
ⓒ $\sqrt[4]{81a^9b^8}$

78.
ⓐ $\sqrt{150m^9n^3}$
ⓑ $\sqrt[3]{81p^7q^8}$
ⓒ $\sqrt[4]{162c^{11}d^{12}}$

79.
ⓐ $\sqrt[3]{-864}$
ⓑ $\sqrt[4]{-256}$

80.
ⓐ $\sqrt[5]{-486}$
ⓑ $\sqrt[6]{-64}$

81.
ⓐ $\sqrt[5]{-32}$
ⓑ $\sqrt[8]{-1}$

82.
ⓐ $\sqrt[3]{-8}$
ⓑ $\sqrt[4]{-16}$

83.
ⓐ $5 + \sqrt{12}$
ⓑ $\dfrac{10 - \sqrt{24}}{2}$

84.
ⓐ $8 + \sqrt{96}$
ⓑ $\dfrac{8 - \sqrt{80}}{4}$

85.

ⓐ $1 + \sqrt{45}$

ⓑ $\dfrac{3 + \sqrt{90}}{3}$

86.

ⓐ $3 + \sqrt{125}$

ⓑ $\dfrac{15 + \sqrt{75}}{5}$

Use the Quotient Property to Simplify Radical Expressions

In the following exercises, use the Quotient Property to simplify square roots.

87. ⓐ $\sqrt{\dfrac{45}{80}}$ ⓑ $\sqrt[3]{\dfrac{8}{27}}$ ⓒ $\sqrt[4]{\dfrac{1}{81}}$

88. ⓐ $\sqrt{\dfrac{72}{98}}$ ⓑ $\sqrt[3]{\dfrac{24}{81}}$ ⓒ $\sqrt[4]{\dfrac{6}{96}}$

89. ⓐ $\sqrt{\dfrac{100}{36}}$ ⓑ $\sqrt[3]{\dfrac{81}{375}}$ ⓒ $\sqrt[4]{\dfrac{1}{256}}$

90. ⓐ $\sqrt{\dfrac{121}{16}}$ ⓑ $\sqrt[3]{\dfrac{16}{250}}$ ⓒ $\sqrt[4]{\dfrac{32}{162}}$

91. ⓐ $\sqrt{\dfrac{x^{10}}{x^6}}$ ⓑ $\sqrt[3]{\dfrac{p^{11}}{p^2}}$ ⓒ $\sqrt[4]{\dfrac{q^{17}}{q^{13}}}$

92. ⓐ $\sqrt{\dfrac{p^{20}}{p^{10}}}$ ⓑ $\sqrt[5]{\dfrac{d^{12}}{d^7}}$ ⓒ $\sqrt[8]{\dfrac{m^{12}}{m^4}}$

93. ⓐ $\sqrt{\dfrac{y^4}{y^8}}$ ⓑ $\sqrt[5]{\dfrac{u^{21}}{u^{11}}}$ ⓒ $\sqrt[6]{\dfrac{v^{30}}{v^{12}}}$

94. ⓐ $\sqrt{\dfrac{q^8}{q^{14}}}$ ⓑ $\sqrt[3]{\dfrac{r^{14}}{r^5}}$ ⓒ $\sqrt[4]{\dfrac{c^{21}}{c^9}}$

95. $\sqrt{\dfrac{96x^7}{121}}$

96. $\sqrt{\dfrac{108y^4}{49}}$

97. $\sqrt{\dfrac{300m^5}{64}}$

98. $\sqrt{\dfrac{125n^7}{169}}$

99. $\sqrt{\dfrac{98r^5}{100}}$

100. $\sqrt{\dfrac{180s^{10}}{144}}$

101. $\sqrt{\dfrac{28q^6}{225}}$

102. $\sqrt{\dfrac{150r^3}{256}}$

103.

ⓐ $\sqrt{\dfrac{75r^9}{s^8}}$

ⓑ $\sqrt[3]{\dfrac{54a^8}{b^3}}$

ⓒ $\sqrt[4]{\dfrac{64c^5}{d^4}}$

104.

ⓐ $\sqrt{\dfrac{72x^5}{y^6}}$

ⓑ $\sqrt[5]{\dfrac{96r^{11}}{s^5}}$

ⓒ $\sqrt[6]{\dfrac{128u^7}{v^{12}}}$

105.

ⓐ $\sqrt{\dfrac{28p^7}{q^2}}$

ⓑ $\sqrt[3]{\dfrac{81s^8}{t^3}}$

ⓒ $\sqrt[4]{\dfrac{64p^{15}}{q^{12}}}$

106.

ⓐ $\sqrt{\dfrac{45r^3}{s^{10}}}$

ⓑ $\sqrt[3]{\dfrac{625u^{10}}{v^3}}$

ⓒ $\sqrt[4]{\dfrac{729c^{21}}{d^8}}$

107.

ⓐ $\sqrt{\dfrac{32x^5y^3}{18x^3y}}$

ⓑ $\sqrt[3]{\dfrac{5x^6y^9}{40x^5y^3}}$

ⓒ $\sqrt[4]{\dfrac{5a^8b^6}{80a^3b^2}}$

108.

ⓐ $\sqrt{\dfrac{75r^6s^8}{48rs^4}}$

ⓑ $\sqrt[3]{\dfrac{24x^8y^4}{81x^2y}}$

ⓒ $\sqrt[4]{\dfrac{32m^9n^2}{162mn^2}}$

109.

ⓐ $\sqrt{\dfrac{27p^2q}{108p^4q^3}}$

ⓑ $\sqrt[3]{\dfrac{16c^5d^7}{250c^2d^2}}$

ⓒ $\sqrt[6]{\dfrac{2m^9n^7}{128m^3n}}$

110.

ⓐ $\sqrt{\dfrac{50r^5s^2}{128r^2s^6}}$

ⓑ $\sqrt[3]{\dfrac{24m^9n^7}{375m^4n}}$

ⓒ $\sqrt[4]{\dfrac{81m^2n^8}{256m^1n^2}}$

111.

ⓐ $\dfrac{\sqrt{45p^9}}{\sqrt{5q^2}}$

ⓑ $\dfrac{\sqrt[4]{64}}{\sqrt[4]{2}}$

ⓒ $\dfrac{\sqrt[5]{128x^8}}{\sqrt[5]{2x^2}}$

112.

ⓐ $\dfrac{\sqrt{80q^5}}{\sqrt{5q}}$

ⓑ $\dfrac{\sqrt[3]{-625}}{\sqrt[3]{5}}$

ⓒ $\dfrac{\sqrt[4]{80m^7}}{\sqrt[4]{5m}}$

113.

ⓐ $\dfrac{\sqrt{50m^7}}{\sqrt{2m}}$

ⓑ $\sqrt[3]{\dfrac{1250}{2}}$

ⓒ $\sqrt[4]{\dfrac{486y^9}{2y^3}}$

114.

ⓐ $\dfrac{\sqrt{72n^{11}}}{\sqrt{2n}}$

ⓑ $\sqrt[3]{\dfrac{162}{6}}$

ⓒ $\sqrt[4]{\dfrac{160r^{10}}{5r^3}}$

Writing Exercises

115. Explain why $\sqrt{x^4} = x^2$. Then explain why $\sqrt{x^{16}} = x^8$.

116. Explain why $7 + \sqrt{9}$ is not equal to $\sqrt{7+9}$.

117. Explain how you know that $\sqrt[5]{x^{10}} = x^2$.

118. Explain why $\sqrt[4]{-64}$ is not a real number but $\sqrt[3]{-64}$ is.

Self Check

ⓐ After completing the exercises, use this checklist to evaluate your mastery of the objectives of this section.

I can...	Confidently	With some help	No-I don't get it!
use the Product Property to simplify radical expressions.			
use the Quotient Property to simplify radical expressions.			

ⓑ After reviewing this checklist, what will you do to become confident for all objectives?

8.3 | Simplify Rational Exponents

Learning Objectives

By the end of this section, you will be able to:

- Simplify expressions with $a^{\frac{1}{n}}$
- Simplify expressions with $a^{\frac{m}{n}}$
- Use the properties of exponents to simplify expressions with rational exponents

Be Prepared!

Before you get started, take this readiness quiz.

1. Add: $\frac{7}{15} + \frac{5}{12}$.
   If you missed this problem, review **Example 1.28**.

2. Simplify: $\left(4x^2 y^5\right)^3$.
   If you missed this problem, review **Example 5.18**.

3. Simplify: 5^{-3}.
   If you missed this problem, review **Example 5.14**.

Simplify Expressions with $a^{\frac{1}{n}}$

Rational exponents are another way of writing expressions with radicals. When we use rational exponents, we can apply the properties of exponents to simplify expressions.

The Power Property for Exponents says that $(a^m)^n = a^{m \cdot n}$ when m and n are whole numbers. Let's assume we are now not limited to whole numbers.

Suppose we want to find a number p such that $(8^p)^3 = 8$. We will use the Power Property of Exponents to find the value of p.

$$(8^p)^3 = 8$$

Multiply the exponents on the left.	$8^{3p} = 8$
Write the exponent 1 on the right.	$8^{3p} = 8^1$
Since the bases are the same, the exponents must be equal.	$3p = 1$
Solve for p.	$p = \frac{1}{3}$

So $\left(8^{\frac{1}{3}}\right)^3 = 8$. But we know also $\left(\sqrt[3]{8}\right)^3 = 8$. Then it must be that $8^{\frac{1}{3}} = \sqrt[3]{8}$.

This same logic can be used for any positive integer exponent n to show that $a^{\frac{1}{n}} = \sqrt[n]{a}$.

Rational Exponent $a^{\frac{1}{n}}$

If $\sqrt[n]{a}$ is a real number and $n \geq 2$, then

$$a^{\frac{1}{n}} = \sqrt[n]{a}$$

The denominator of the rational exponent is the index of the radical.

There will be times when working with expressions will be easier if you use rational exponents and times when it will be easier if you use radicals. In the first few examples, you'll practice converting expressions between these two notations.

EXAMPLE 8.26

Write as a radical expression: ⓐ $x^{\frac{1}{2}}$ ⓑ $y^{\frac{1}{3}}$ ⓒ $z^{\frac{1}{4}}$.

✓ Solution

We want to write each expression in the form $\sqrt[n]{a}$.

ⓐ

$$x^{\frac{1}{2}}$$

The denominator of the rational exponent is 2, so the index of the radical is 2. We do not show the index when it is 2.

$$\sqrt{x}$$

ⓑ

$$y^{\frac{1}{3}}$$

The denominator of the exponent is 3, so the index is 3.

$$\sqrt[3]{y}$$

ⓒ

$$z^{\frac{1}{4}}$$

The denominator of the exponent is 4, so the index is 4.

$$\sqrt[4]{z}$$

> **TRY IT :: 8.51**
>
> Write as a radical expression: ⓐ $t^{\frac{1}{2}}$ ⓑ $m^{\frac{1}{3}}$ ⓒ $r^{\frac{1}{4}}$.

> **TRY IT :: 8.52**
>
> Write as a radial expression: ⓐ $b^{\frac{1}{6}}$ ⓑ $z^{\frac{1}{5}}$ ⓒ $p^{\frac{1}{4}}$.

In the next example, we will write each radical using a rational exponent. It is important to use parentheses around the entire expression in the radicand since the entire expression is raised to the rational power.

EXAMPLE 8.27

Write with a rational exponent: ⓐ $\sqrt{5y}$ ⓑ $\sqrt[3]{4x}$ ⓒ $3\sqrt[4]{5z}$.

✓ Solution

We want to write each radical in the form $a^{\frac{1}{n}}$.

ⓐ

$$\sqrt{5y}$$

No index is shown, so it is 2.
The denominator of the exponent will be 2.
Put parentheses around the entire expression $5y$.

$$(5y)^{\frac{1}{2}}$$

ⓑ

$$\sqrt[3]{4x}$$

The index is 3, so the denominator of the
exponent is 3. Include parentheses $(4x)$.

$$(4x)^{\frac{1}{3}}$$

ⓒ

$$3\sqrt[4]{5z}$$

The index is 4, so the denominator of the
exponent is 4. Put parentheses only around
the $5z$ since 3 is not under the radical sign.

$$3(5z)^{\frac{1}{4}}$$

> **TRY IT :: 8.53**   Write with a rational exponent: ⓐ $\sqrt{10m}$ ⓑ $\sqrt[5]{3n}$ ⓒ $3\sqrt[4]{6y}$.

> **TRY IT :: 8.54**   Write with a rational exponent: ⓐ $\sqrt[7]{3k}$ ⓑ $\sqrt[4]{5j}$ ⓒ $8\sqrt[3]{2a}$.

In the next example, you may find it easier to simplify the expressions if you rewrite them as radicals first.

EXAMPLE 8.28

Simplify: ⓐ $25^{\frac{1}{2}}$ ⓑ $64^{\frac{1}{3}}$ ⓒ $256^{\frac{1}{4}}$.

✓ **Solution**

ⓐ

$$25^{\frac{1}{2}}$$

Rewrite as a square root. $\sqrt{25}$
Simplify. 5

ⓑ

$$64^{\frac{1}{3}}$$

Rewrite as a cube root. $\sqrt[3]{64}$

Recognize 64 is a perfect cube. $\sqrt[3]{4^3}$
Simplify. 4

ⓒ

$$256^{\frac{1}{4}}$$

Rewrite as a fourth root. $\sqrt[4]{256}$

Recognize 256 is a perfect fourth power. $\sqrt[4]{4^4}$
Simplify. 4

> **TRY IT :: 8.55**   Simplify: ⓐ $36^{\frac{1}{2}}$ ⓑ $8^{\frac{1}{3}}$ ⓒ $16^{\frac{1}{4}}$.

> **TRY IT ::** 8.56

Simplify: ⓐ $100^{\frac{1}{2}}$ ⓑ $27^{\frac{1}{3}}$ ⓒ $81^{\frac{1}{4}}$.

Be careful of the placement of the negative signs in the next example. We will need to use the property $a^{-n} = \frac{1}{a^n}$ in one case.

EXAMPLE 8.29

Simplify: ⓐ $(-16)^{\frac{1}{4}}$ ⓑ $-16^{\frac{1}{4}}$ ⓒ $(16)^{-\frac{1}{4}}$.

⊘ **Solution**

ⓐ

$$(-16)^{\frac{1}{4}}$$

Rewrite as a fourth root.

$$\sqrt[4]{-16}$$

$$\sqrt[4]{(-2)^4}$$

Simplify.   No real solution.

ⓑ

$$-16^{\frac{1}{4}}$$

The exponent only applies to the 16.
Rewrite as a fouth root.

$$-\sqrt[4]{16}$$

Rewrite 16 as 2^4.

$$-\sqrt[4]{2^4}$$

Simplify. -2

ⓒ

$$(16)^{-\frac{1}{4}}$$

Rewrite using the property $a^{-n} = \frac{1}{a^n}$.

$$\frac{1}{(16)^{\frac{1}{4}}}$$

Rewrite as a fourth root.

$$\frac{1}{\sqrt[4]{16}}$$

Rewrite 16 as 2^4.

$$\frac{1}{\sqrt[4]{2^4}}$$

Simplify. $\frac{1}{2}$

> **TRY IT ::** 8.57

Simplify: ⓐ $(-64)^{-\frac{1}{2}}$ ⓑ $-64^{\frac{1}{2}}$ ⓒ $(64)^{-\frac{1}{2}}$.

> **TRY IT ::** 8.58

Simplify: ⓐ $(-256)^{\frac{1}{4}}$ ⓑ $-256^{\frac{1}{4}}$ ⓒ $(256)^{-\frac{1}{4}}$.

Simplify Expressions with $a^{\frac{m}{n}}$

We can look at $a^{\frac{m}{n}}$ in two ways. Remember the Power Property tells us to multiply the exponents and so $\left(a^{\frac{1}{n}}\right)^m$ and

$(a^m)^{\frac{1}{n}}$ both equal $a^{\frac{m}{n}}$. If we write these expressions in radical form, we get

$$a^{\frac{m}{n}} = \left(a^{\frac{1}{n}}\right)^m = (\sqrt[n]{a})^m \quad \text{and} \quad a^{\frac{m}{n}} = (a^m)^{\frac{1}{n}} = \sqrt[n]{a^m}$$

This leads us to the following definition.

Rational Exponent $a^{\frac{m}{n}}$

For any positive integers m and n,

$$a^{\frac{m}{n}} = (\sqrt[n]{a})^m \quad \text{and} \quad a^{\frac{m}{n}} = \sqrt[n]{a^m}$$

Which form do we use to simplify an expression? We usually take the root first—that way we keep the numbers in the radicand smaller, before raising it to the power indicated.

EXAMPLE 8.30

Write with a rational exponent: ⓐ $\sqrt{y^3}$ ⓑ $\left(\sqrt[3]{2x}\right)^4$ ⓒ $\sqrt{\left(\frac{3a}{4b}\right)^3}$.

✓ **Solution**

We want to use $a^{\frac{m}{n}} = \sqrt[n]{a^m}$ to write each radical in the form $a^{\frac{m}{n}}$.

 ⓐ

$$\sqrt{y^3}$$

The numerator of the exponent is the exponent, 3.

The denominator of the exponent is the index of the radical, 2. $y^{\frac{3}{2}}$

 ⓑ

$$\left(\sqrt[3]{2x}\right)^4$$

The numerator of the exponent is the exponent, 4.

The denominator of the exponent is the index of the radical, 3. $(2x)^{\frac{4}{3}}$

 ⓒ

$$\sqrt{\left(\frac{3a}{4b}\right)^3}$$

The numerator of the exponent is the exponent, 3.

The denominator of the exponent is the index of the radical, 2. $\left(\frac{3a}{4b}\right)^{\frac{3}{2}}$

> **TRY IT : : 8.59**

Write with a rational exponent: ⓐ $\sqrt{x^5}$ ⓑ $\left(\sqrt[4]{3y}\right)^3$ ⓒ $\sqrt{\left(\frac{2m}{3n}\right)^5}$.

> **TRY IT : : 8.60**

Write with a rational exponent: ⓐ $\sqrt[5]{a^2}$ ⓑ $\left(\sqrt[3]{5ab}\right)^5$ ⓒ $\sqrt{\left(\frac{7xy}{z}\right)^3}$.

Remember that $a^{-n} = \frac{1}{a^n}$. The negative sign in the exponent does not change the sign of the expression.

EXAMPLE 8.31

Simplify: ⓐ $125^{\frac{2}{3}}$ ⓑ $16^{-\frac{3}{2}}$ ⓒ $32^{-\frac{2}{5}}$.

✓ **Solution**

We will rewrite the expression as a radical first using the defintion, $a^{\frac{m}{n}} = \left(\sqrt[n]{a}\right)^m$. This form lets us take the root first and so we keep the numbers in the radicand smaller than if we used the other form.

ⓐ

	$125^{\frac{2}{3}}$
The power of the radical is the numerator of the exponent, 2. The index of the radical is the denominator of the exponent, 3.	$\left(\sqrt[3]{125}\right)^2$
Simplify.	$(5)^2$
	25

ⓑ We will rewrite each expression first using $a^{-n} = \frac{1}{a^n}$ and then change to radical form.

	$16^{-\frac{3}{2}}$
Rewrite using $a^{-n} = \frac{1}{a^n}$	$\frac{1}{16^{\frac{3}{2}}}$
Change to radical form. The power of the radical is the numerator of the exponent, 3. The index is the denominator of the exponent, 2.	$\frac{1}{\left(\sqrt{16}\right)^3}$
Simplify.	$\frac{1}{4^3}$
	$\frac{1}{64}$

ⓒ

	$32^{-\frac{2}{5}}$
Rewrite using $a^{-n} = \frac{1}{a^n}$.	$\dfrac{1}{32^{\frac{2}{5}}}$
Change to radical form.	$\dfrac{1}{\left(\sqrt[5]{32}\right)^2}$
Rewrite the radicand as a power.	$\dfrac{1}{\left(\sqrt[5]{2^5}\right)^2}$
Simplify.	$\dfrac{1}{2^2}$
	$\dfrac{1}{4}$

> **TRY IT :: 8.61**

Simplify: ⓐ $27^{\frac{2}{3}}$ ⓑ $81^{-\frac{3}{2}}$ ⓒ $16^{-\frac{3}{4}}$.

> **TRY IT :: 8.62**

Simplify: ⓐ $4^{\frac{3}{2}}$ ⓑ $27^{-\frac{2}{3}}$ ⓒ $625^{-\frac{3}{4}}$.

EXAMPLE 8.32

Simplify: ⓐ $-25^{\frac{3}{2}}$ ⓑ $-25^{-\frac{3}{2}}$ ⓒ $(-25)^{\frac{3}{2}}$.

✓ **Solution**

ⓐ

	$-25^{\frac{3}{2}}$
Rewrite in radical form.	$-\left(\sqrt{25}\right)^3$
Simplify the radical.	$-(5)^3$
Simplify.	-125

ⓑ

	$-25^{-\frac{3}{2}}$
Rewrite using $a^{-n} = \frac{1}{a^n}$.	$-\left(\dfrac{1}{25^{\frac{3}{2}}}\right)$
Rewrite in radical form.	$-\left(\dfrac{1}{(\sqrt{25})^3}\right)$
Simplify the radical.	$-\left(\dfrac{1}{(5)^3}\right)$
Simplify.	$-\dfrac{1}{125}$

ⓒ

$$(-25)^{\frac{3}{2}}$$

Rewrite in radical form. $(\sqrt{-25})^3$

There is no real number whose square root is -25.

Not a real number.

> **TRY IT : :** 8.63

Simplify: ⓐ $-16^{\frac{3}{2}}$ ⓑ $-16^{-\frac{3}{2}}$ ⓒ $(-16)^{-\frac{3}{2}}$.

> **TRY IT : :** 8.64

Simplify: ⓐ $-81^{\frac{3}{2}}$ ⓑ $-81^{-\frac{3}{2}}$ ⓒ $(-81)^{-\frac{3}{2}}$.

Use the Properties of Exponents to Simplify Expressions with Rational Exponents

The same properties of exponents that we have already used also apply to rational exponents. We will list the Properties of Exponenets here to have them for reference as we simplify expressions.

Properties of Exponents

If a and b are real numbers and m and n are rational numbers, then

Product Property	$a^m \cdot a^n = a^{m+n}$
Power Property	$(a^m)^n = a^{m \cdot n}$
Product to a Power	$(ab)^m = a^m b^m$
Quotient Property	$\dfrac{a^m}{a^n} = a^{m-n},\ a \neq 0$
Zero Exponent Definition	$a^0 = 1,\ a \neq 0$
Quotient to a Power Property	$\left(\dfrac{a}{b}\right)^m = \dfrac{a^m}{b^m},\ b \neq 0$
Negative Exponent Property	$a^{-n} = \dfrac{1}{a^n},\ a \neq 0$

We will apply these properties in the next example.

EXAMPLE 8.33

Simplify: ⓐ $x^{\frac{1}{2}} \cdot x^{\frac{5}{6}}$ ⓑ $\left(z^9\right)^{\frac{2}{3}}$ ⓒ $\dfrac{x^{\frac{1}{3}}}{x^{\frac{5}{3}}}$.

✓ Solution

ⓐ The Product Property tells us that when we multiply the same base, we add the exponents.

$$x^{\frac{1}{2}} \cdot x^{\frac{5}{6}}$$

The bases are the same, so we add the exponents.

$$x^{\frac{1}{2}+\frac{5}{6}}$$

Add the fractions.

$$x^{\frac{8}{6}}$$

Simplify the exponent.

$$x^{\frac{4}{3}}$$

ⓑ The Power Property tells us that when we raise a power to a power, we multiply the exponents.

$$\left(z^9\right)^{\frac{2}{3}}$$

To raise a power to a power, we multiply the exponents.

$$z^{9 \cdot \frac{2}{3}}$$

Simplify.

$$z^6$$

ⓒ The Quotient Property tells us that when we divide with the same base, we subtract the exponents.

$$\frac{x^{\frac{1}{3}}}{x^{\frac{5}{3}}}$$

$$\frac{x^{\frac{1}{3}}}{x^{\frac{5}{3}}}$$

To divide with the same base, we subtract the exponents.

$$\frac{1}{x^{\frac{5}{3}-\frac{1}{3}}}$$

Simplify.

$$\frac{1}{x^{\frac{4}{3}}}$$

> **TRY IT ::** 8.65

Simplify: ⓐ $x^{\frac{1}{6}} \cdot x^{\frac{4}{3}}$ ⓑ $\left(x^6\right)^{\frac{4}{3}}$ ⓒ $\frac{x^{\frac{2}{3}}}{x^{\frac{5}{3}}}$.

> **TRY IT ::** 8.66

Simplify: ⓐ $y^{\frac{3}{4}} \cdot y^{\frac{5}{8}}$ ⓑ $\left(m^9\right)^{\frac{2}{9}}$ ⓒ $\frac{d^{\frac{1}{5}}}{d^{\frac{6}{5}}}$.

Sometimes we need to use more than one property. In the next example, we will use both the Product to a Power Property and then the Power Property.

EXAMPLE 8.34

Simplify: ⓐ $\left(27u^{\frac{1}{2}}\right)^{\frac{2}{3}}$ ⓑ $\left(m^{\frac{2}{3}}n^{\frac{1}{2}}\right)^{\frac{3}{2}}$.

✓ **Solution**

ⓐ

$$\left(27u^{\frac{1}{2}}\right)^{\frac{2}{3}}$$

First we use the Product to a Power Property.

$$(27)^{\frac{2}{3}}\left(u^{\frac{1}{2}}\right)^{\frac{2}{3}}$$

Rewrite 27 as a power of 3.

$$\left(3^3\right)^{\frac{2}{3}}\left(u^{\frac{1}{2}}\right)^{\frac{2}{3}}$$

To raise a power to a power, we multiply the exponents.

$$\left(3^2\right)\left(u^{\frac{1}{3}}\right)$$

Simplify.

$$9u^{\frac{1}{3}}$$

ⓑ

$$\left(m^{\frac{2}{3}}n^{\frac{1}{2}}\right)^{\frac{3}{2}}$$

First we use the Product to a Power Property.

$$\left(m^{\frac{2}{3}}\right)^{\frac{3}{2}}\left(n^{\frac{1}{2}}\right)^{\frac{3}{2}}$$

To raise a power to a power, we multiply the exponents.

$$mn^{\frac{3}{4}}$$

> **TRY IT : : 8.67**
>
> Simplify: ⓐ $\left(32x^{\frac{1}{3}}\right)^{\frac{3}{5}}$ ⓑ $\left(x^{\frac{3}{4}}y^{\frac{1}{2}}\right)^{\frac{2}{3}}$.

> **TRY IT : : 8.68**
>
> Simplify: ⓐ $\left(81n^{\frac{2}{5}}\right)^{\frac{3}{2}}$ ⓑ $\left(a^{\frac{3}{2}}b^{\frac{1}{2}}\right)^{\frac{4}{3}}$.

We will use both the Product Property and the Quotient Property in the next example.

EXAMPLE 8.35

Simplify: ⓐ $\dfrac{x^{\frac{3}{4}} \cdot x^{-\frac{1}{4}}}{x^{-\frac{6}{4}}}$ ⓑ $\left(\dfrac{16\,x^{\frac{4}{3}} y^{-\frac{5}{6}}}{x^{-\frac{2}{3}} y^{\frac{1}{6}}}\right)^{\frac{1}{2}}$.

✓ **Solution**

ⓐ

$$\dfrac{x^{\frac{3}{4}} \cdot x^{-\frac{1}{4}}}{x^{-\frac{6}{4}}}$$

Use the Product Property in the numerator, add the exponents.

$$\dfrac{x^{\frac{2}{4}}}{x^{-\frac{6}{4}}}$$

Use the Quotient Property, subtract the exponents.

$$x^{\frac{8}{4}}$$

Simplify.

$$x^2$$

ⓑ Follow the order of operations to simplify inside the parenthese first.

$$\left(\dfrac{16\,x^{\frac{4}{3}} y^{-\frac{5}{6}}}{x^{-\frac{2}{3}} y^{\frac{1}{6}}}\right)^{\frac{1}{2}}$$

Use the Quotient Property, subtract the exponents.

$$\left(\dfrac{16x^{\frac{6}{3}}}{y^{\frac{6}{6}}}\right)^{\frac{1}{2}}$$

Simplify.

$$\left(\dfrac{16x^2}{y}\right)^{\frac{1}{2}}$$

Use the Product to a Power Property, multiply the exponents.

$$\dfrac{4x}{y^{\frac{1}{2}}}$$

> **TRY IT :: 8.69**

Simplify: ⓐ $\dfrac{m^{\frac{2}{3}} \cdot m^{-\frac{1}{3}}}{m^{-\frac{5}{3}}}$ ⓑ $\left(\dfrac{25m^{\frac{1}{6}} n^{\frac{11}{6}}}{m^{\frac{2}{3}} n^{-\frac{1}{6}}}\right)^{\frac{1}{2}}$.

> **TRY IT :: 8.70**

Simplify: ⓐ $\dfrac{u^{\frac{4}{5}} \cdot u^{-\frac{2}{5}}}{u^{-\frac{13}{5}}}$ ⓑ $\left(\dfrac{27x^{\frac{4}{5}} y^{\frac{1}{6}}}{x^{\frac{1}{5}} y^{-\frac{5}{6}}}\right)^{\frac{1}{3}}$.

MEDIA : :

Access these online resources for additional instruction and practice with simplifying rational exponents.

- **Review-Rational Exponents (https://openstax.org/l/37RatExpont1)**
- **Using Laws of Exponents on Radicals: Properties of Rational Exponents (https://openstax.org/l/37RatExpont2)**

8.3 EXERCISES
Practice Makes Perfect

Simplify expressions with $a^{\frac{1}{n}}$

In the following exercises, write as a radical expression.

119. ⓐ $x^{\frac{1}{2}}$ ⓑ $y^{\frac{1}{3}}$ ⓒ $z^{\frac{1}{4}}$

120. ⓐ $r^{\frac{1}{2}}$ ⓑ $s^{\frac{1}{3}}$ ⓒ $t^{\frac{1}{4}}$

121. ⓐ $u^{\frac{1}{5}}$ ⓑ $v^{\frac{1}{9}}$ ⓒ $w^{\frac{1}{20}}$

122. ⓐ $g^{\frac{1}{7}}$ ⓑ $h^{\frac{1}{5}}$ ⓒ $j^{\frac{1}{25}}$

In the following exercises, write with a rational exponent.

123. ⓐ $\sqrt[7]{x}$ ⓑ $\sqrt[9]{y}$ ⓒ $\sqrt[5]{f}$

124. ⓐ $\sqrt[8]{r}$ ⓑ $\sqrt[10]{s}$ ⓒ $\sqrt[4]{t}$

125. ⓐ $\sqrt[3]{7c}$ ⓑ $\sqrt[7]{12d}$ ⓒ $2\sqrt[4]{6b}$

126. ⓐ $\sqrt[4]{5x}$ ⓑ $\sqrt[8]{9y}$ ⓒ $7\sqrt[5]{3z}$

127. ⓐ $\sqrt{21p}$ ⓑ $\sqrt[4]{8q}$ ⓒ $4\sqrt[6]{36r}$

128. ⓐ $\sqrt[3]{25a}$ ⓑ $\sqrt{3b}$ ⓒ $\sqrt[8]{40c}$

In the following exercises, simplify.

129.
ⓐ $81^{\frac{1}{2}}$
ⓑ $125^{\frac{1}{3}}$
ⓒ $64^{\frac{1}{2}}$

130.
ⓐ $625^{\frac{1}{4}}$
ⓑ $243^{\frac{1}{5}}$
ⓒ $32^{\frac{1}{5}}$

131.
ⓐ $16^{\frac{1}{4}}$
ⓑ $16^{\frac{1}{2}}$
ⓒ $625^{\frac{1}{4}}$

132.
ⓐ $64^{\frac{1}{3}}$
ⓑ $32^{\frac{1}{5}}$
ⓒ $81^{\frac{1}{4}}$

133.
ⓐ $(-216)^{\frac{1}{3}}$
ⓑ $-216^{\frac{1}{3}}$
ⓒ $(216)^{-\frac{1}{3}}$

134.
ⓐ $(-1000)^{\frac{1}{3}}$
ⓑ $-1000^{\frac{1}{3}}$
ⓒ $(1000)^{-\frac{1}{3}}$

135.
ⓐ $(-81)^{\frac{1}{4}}$
ⓑ $-81^{\frac{1}{4}}$
ⓒ $(81)^{-\frac{1}{4}}$

136.
ⓐ $(-49)^{\frac{1}{2}}$
ⓑ $-49^{\frac{1}{2}}$
ⓒ $(49)^{-\frac{1}{2}}$

137.
ⓐ $(-36)^{\frac{1}{2}}$
ⓑ $-36^{\frac{1}{2}}$
ⓒ $(36)^{-\frac{1}{2}}$

138.
ⓐ $(-16)^{\frac{1}{4}}$
ⓑ $-16^{\frac{1}{4}}$
ⓒ $16^{-\frac{1}{4}}$

139.
ⓐ $(-100)^{\frac{1}{2}}$
ⓑ $-100^{\frac{1}{2}}$
ⓒ $(100)^{-\frac{1}{2}}$

140.
ⓐ $(-32)^{\frac{1}{5}}$
ⓑ $(243)^{-\frac{1}{5}}$
ⓒ $-125^{\frac{1}{3}}$

Simplify Expressions with $a^{\frac{m}{n}}$

In the following exercises, write with a rational exponent.

141.

ⓐ $\sqrt{m^5}$

ⓑ $\left(\sqrt[3]{3y}\right)^7$

ⓒ $\sqrt[5]{\left(\frac{4x}{5y}\right)^3}$

142.

ⓐ $\sqrt[4]{r^7}$

ⓑ $\left(\sqrt[5]{2pq}\right)^3$

ⓒ $\sqrt[4]{\left(\frac{12m}{7n}\right)^3}$

143.

ⓐ $\sqrt[5]{u^2}$

ⓑ $\left(\sqrt[3]{6x}\right)^5$

ⓒ $\sqrt[4]{\left(\frac{18a}{5b}\right)^7}$

144.

ⓐ $\sqrt[3]{a}$

ⓑ $\left(\sqrt[4]{21v}\right)^3$

ⓒ $\sqrt[4]{\left(\frac{2xy}{5z}\right)^2}$

In the following exercises, simplify.

145.

ⓐ $64^{\frac{5}{2}}$

ⓑ $81^{\frac{-3}{2}}$

ⓒ $(-27)^{\frac{2}{3}}$

146.

ⓐ $25^{\frac{3}{2}}$

ⓑ $9^{-\frac{3}{2}}$

ⓒ $(-64)^{\frac{2}{3}}$

147.

ⓐ $32^{\frac{2}{5}}$

ⓑ $27^{-\frac{2}{3}}$

ⓒ $(-25)^{\frac{1}{2}}$

148.

ⓐ $100^{\frac{3}{2}}$

ⓑ $49^{-\frac{5}{2}}$

ⓒ $(-100)^{\frac{3}{2}}$

149.

ⓐ $-9^{\frac{3}{2}}$

ⓑ $-9^{-\frac{3}{2}}$

ⓒ $(-9)^{\frac{3}{2}}$

150.

ⓐ $-64^{\frac{3}{2}}$

ⓑ $-64^{-\frac{3}{2}}$

ⓒ $(-64)^{\frac{3}{2}}$

Use the Laws of Exponents to Simplify Expressions with Rational Exponents

In the following exercises, simplify.

151.

ⓐ $c^{\frac{1}{4}} \cdot c^{\frac{5}{8}}$

ⓑ $\left(p^{12}\right)^{\frac{3}{4}}$

ⓒ $\dfrac{r^{\frac{4}{5}}}{r^{\frac{9}{5}}}$

152.

ⓐ $6^{\frac{5}{2}} \cdot 6^{\frac{1}{2}}$

ⓑ $\left(b^{15}\right)^{\frac{3}{5}}$

ⓒ $\dfrac{w^{\frac{2}{7}}}{w^{\frac{9}{7}}}$

153.

ⓐ $y^{\frac{1}{2}} \cdot y^{\frac{3}{4}}$

ⓑ $\left(x^{12}\right)^{\frac{2}{3}}$

ⓒ $\dfrac{m^{\frac{5}{8}}}{m^{\frac{13}{8}}}$

154.

(a) $q^{\frac{2}{3}} \cdot q^{\frac{5}{6}}$

(b) $\left(h^{6}\right)^{\frac{4}{3}}$

(c) $\dfrac{n^{\frac{3}{5}}}{n^{\frac{8}{5}}}$

155.

(a) $\left(27q^{\frac{3}{2}}\right)^{\frac{4}{3}}$

(b) $\left(a^{\frac{1}{3}} b^{\frac{2}{3}}\right)^{\frac{3}{2}}$

156.

(a) $\left(64s^{\frac{3}{7}}\right)^{\frac{1}{6}}$

(b) $\left(m^{\frac{4}{3}} n^{\frac{1}{2}}\right)^{\frac{3}{4}}$

157.

(a) $\left(16\,u^{\frac{1}{3}}\right)^{\frac{3}{4}}$

(b) $\left(4\,p^{\frac{1}{3}} q^{\frac{1}{2}}\right)^{\frac{3}{2}}$

158.

(a) $\left(625\,n^{\frac{8}{3}}\right)^{\frac{3}{4}}$

(b) $\left(9\,x^{\frac{2}{5}} y^{\frac{3}{5}}\right)^{\frac{5}{2}}$

159.

(a) $\dfrac{r^{\frac{5}{2}} \cdot r^{-\frac{1}{2}}}{r^{-\frac{3}{2}}}$

(b) $\left(\dfrac{36\,s^{\frac{1}{5}} t^{-\frac{3}{2}}}{s^{-\frac{9}{5}} t^{\frac{1}{2}}}\right)^{\frac{1}{2}}$

160.

(a) $\dfrac{a^{\frac{3}{4}} \cdot a^{-\frac{1}{4}}}{a^{-\frac{10}{4}}}$

(b) $\left(\dfrac{27\,b^{\frac{2}{3}} c^{-\frac{5}{2}}}{b^{-\frac{7}{3}} c^{\frac{1}{2}}}\right)^{\frac{1}{3}}$

161.

(a) $\dfrac{c^{\frac{5}{3}} \cdot c^{-\frac{1}{3}}}{c^{-\frac{2}{3}}}$

(b) $\left(\dfrac{8\,x^{\frac{5}{3}} y^{-\frac{1}{2}}}{27\,x^{-\frac{4}{3}} y^{\frac{5}{2}}}\right)^{\frac{1}{3}}$

162.

(a) $\dfrac{m^{\frac{7}{4}} \cdot m^{-\frac{5}{4}}}{m^{-\frac{2}{4}}}$

(b) $\left(\dfrac{16\,m^{\frac{1}{5}} n^{\frac{3}{2}}}{81\,m^{\frac{9}{5}} n^{-\frac{1}{2}}}\right)^{\frac{1}{4}}$

Writing Exercises

163. Show two different algebraic methods to simplify $4^{\frac{3}{2}}$. Explain all your steps.

164. Explain why the expression $(-16)^{\frac{3}{2}}$ cannot be evaluated.

Self Check

ⓐ *After completing the exercises, use this checklist to evaluate your mastery of the objectives of this section.*

I can...	Confidently	With some help	No-I don't get it!
simplify expressions with $a^{\frac{1}{n}}$.			
simplify expressions with $a^{\frac{m}{n}}$.			
use the Laws of Exponents to simply expressions with rational exponents.			

ⓑ *What does this checklist tell you about your mastery of this section? What steps will you take to improve?*

8.4 Add, Subtract, and Multiply Radical Expressions

Learning Objectives

By the end of this section, you will be able to:

> Add and subtract radical expressions
> Multiply radical expressions
> Use polynomial multiplication to multiply radical expressions

Be Prepared!

Before you get started, take this readiness quiz.

1. Add: $3x^2 + 9x - 5 - \left(x^2 - 2x + 3\right)$.
   If you missed this problem, review **Example 5.5**.

2. Simplify: $(2 + a)(4 - a)$.
   If you missed this problem, review **Example 5.28**.

3. Simplify: $(9 - 5y)^2$.
   If you missed this problem, review **Example 5.31**.

Add and Subtract Radical Expressions

Adding radical expressions with the same index and the same radicand is just like adding like terms. We call radicals with the same index and the same radicand **like radicals** to remind us they work the same as like terms.

Like Radicals

Like radicals are radical expressions with the same index and the same radicand.

We add and subtract like radicals in the same way we add and subtract like terms. We know that $3x + 8x$ is $11x$. Similarly we add $3\sqrt{x} + 8\sqrt{x}$ and the result is $11\sqrt{x}$.

Think about adding like terms with variables as you do the next few examples. When you have like radicals, you just add or subtract the coefficients. When the radicals are not like, you cannot combine the terms.

EXAMPLE 8.36

Simplify: ⓐ $2\sqrt{2} - 7\sqrt{2}$ ⓑ $5\sqrt[3]{y} + 4\sqrt[3]{y}$ ⓒ $7\sqrt[4]{x} - 2\sqrt[4]{y}$.

 Solution

ⓐ

$$2\sqrt{2} - 7\sqrt{2}$$

Since the radicals are like, we subtract the coefficients.

$$-5\sqrt{2}$$

ⓑ

$$5\sqrt[3]{y} + 4\sqrt[3]{y}$$

Since the radicals are like, we add the coefficients.

$$9\sqrt[3]{y}$$

ⓒ

$$7\sqrt[4]{x} - 2\sqrt[4]{y}$$

The indices are the same but the radicals are different. These are not like radicals. Since the radicals are not like, we cannot subtract them.

> **TRY IT :: 8.71** Simplify: ⓐ $8\sqrt{2} - 9\sqrt{2}$ ⓑ $4\sqrt[3]{x} + 7\sqrt[3]{x}$ ⓒ $3\sqrt[4]{x} - 5\sqrt[4]{y}$.

> **TRY IT : : 8.72**   Simplify: ⓐ $5\sqrt{3} - 9\sqrt{3}$ ⓑ $5\sqrt[3]{y} + 3\sqrt[3]{y}$ ⓒ $5\sqrt[4]{m} - 2\sqrt[3]{m}$.

For radicals to be like, they must have the same index and radicand. When the radicands contain more than one variable, as long as all the variables and their exponents are identical, the radicands are the same.

EXAMPLE 8.37

Simplify: ⓐ $2\sqrt{5n} - 6\sqrt{5n} + 4\sqrt{5n}$ ⓑ $\sqrt[4]{3xy} + 5\sqrt[4]{3xy} - 4\sqrt[4]{3xy}$.

✓ Solution

ⓐ

$$2\sqrt{5n} - 6\sqrt{5n} + 4\sqrt{5n}$$

Since the radicals are like, we combine them. $0\sqrt{5n}$

Simplify. 0

ⓑ

$$\sqrt[4]{3xy} + 5\sqrt[4]{3xy} - 4\sqrt[4]{3xy}$$

Since the radicals are like, we combine them. $2\sqrt[4]{3xy}$

> **TRY IT : : 8.73**   Simplify: ⓐ $\sqrt{7x} - 7\sqrt{7x} + 4\sqrt{7x}$ ⓑ $4\sqrt[4]{5xy} + 2\sqrt[4]{5xy} - 7\sqrt[4]{5xy}$.

> **TRY IT : : 8.74**   Simplify: ⓐ $4\sqrt{3y} - 7\sqrt{3y} + 2\sqrt{3y}$ ⓑ $6\sqrt[3]{7mn} + \sqrt[3]{7mn} - 4\sqrt[3]{7mn}$.

Remember that we always simplify radicals by removing the largest factor from the radicand that is a power of the index. Once each radical is simplified, we can then decide if they are like radicals.

EXAMPLE 8.38

Simplify: ⓐ $\sqrt{20} + 3\sqrt{5}$ ⓑ $\sqrt[3]{24} - \sqrt[3]{375}$ ⓒ $\frac{1}{2}\sqrt[4]{48} - \frac{2}{3}\sqrt[4]{243}$.

✓ Solution

ⓐ

$$\sqrt{20} + 3\sqrt{5}$$

Simplify the radicals, when possible. $\sqrt{4} \cdot \sqrt{5} + 3\sqrt{5}$

$$2\sqrt{5} + 3\sqrt{5}$$

Combine the like radicals. $5\sqrt{5}$

ⓑ

$$\sqrt[3]{24} - \sqrt[3]{375}$$

Simplify the radicals. $\sqrt[3]{8} \cdot \sqrt[3]{3} - \sqrt[3]{125} \cdot \sqrt[3]{3}$

$$2\sqrt[3]{3} - 5\sqrt[3]{3}$$

Combine the like radicals. $-3\sqrt[3]{3}$

ⓒ

Simplify the radicals.

$$\frac{1}{2}\sqrt[4]{48} - \frac{2}{3}\sqrt[4]{243}$$

$$\frac{1}{2}\sqrt[4]{16} \cdot \sqrt[4]{3} - \frac{2}{3}\sqrt[4]{81} \cdot \sqrt[4]{3}$$

$$\frac{1}{2} \cdot 2 \cdot \sqrt[4]{3} - \frac{2}{3} \cdot 3 \cdot \sqrt[4]{3}$$

$$\sqrt[4]{3} - 2\sqrt[4]{3}$$

Combine the like radicals.

$$-\sqrt[4]{3}$$

> **TRY IT :: 8.75**   Simplify: ⓐ $\sqrt{18} + 6\sqrt{2}$ ⓑ $6\sqrt[3]{16} - 2\sqrt[3]{250}$ ⓒ $\frac{2}{3}\sqrt[3]{81} - \frac{1}{2}\sqrt[3]{24}$.

> **TRY IT :: 8.76**   Simplify: ⓐ $\sqrt{27} + 4\sqrt{3}$ ⓑ $4\sqrt[3]{5} - 7\sqrt[3]{40}$ ⓒ $\frac{1}{2}\sqrt[3]{128} - \frac{5}{3}\sqrt[3]{54}$.

In the next example, we will remove both constant and variable factors from the radicals. Now that we have practiced taking both the even and odd roots of variables, it is common practice at this point for us to assume all variables are greater than or equal to zero so that absolute values are not needed. We will use this assumption thoughout the rest of this chapter.

EXAMPLE 8.39

Simplify: ⓐ $9\sqrt{50m^2} - 6\sqrt{48m^2}$ ⓑ $\sqrt[3]{54n^5} - \sqrt[3]{16n^5}$.

⊘ **Solution**

ⓐ

$$9\sqrt{50m^2} - 6\sqrt{48m^2}$$

Simplify the radicals.

$$9\sqrt{25m^2} \cdot \sqrt{2} - 6\sqrt{16m^2} \cdot \sqrt{3}$$

$$9 \cdot 5m \cdot \sqrt{2} - 6 \cdot 4m \cdot \sqrt{3}$$

$$45m\sqrt{2} - 24m\sqrt{3}$$

The radicals are not like and so cannot be combined.

ⓑ

$$\sqrt[3]{54n^5} - \sqrt[3]{16n^5}$$

Simplify the radicals.

$$\sqrt[3]{27n^3} \cdot \sqrt[3]{2n^2} - \sqrt[3]{8n^3} \cdot \sqrt[3]{2n^2}$$

$$3n\sqrt[3]{2n^2} - 2n\sqrt[3]{2n^2}$$

Combine the like radicals.

$$n\sqrt[3]{2n^2}$$

> **TRY IT :: 8.77**   Simplify: ⓐ $\sqrt{32m^7} - \sqrt{50m^7}$ ⓑ $\sqrt[3]{135x^7} - \sqrt[3]{40x^7}$.

> **TRY IT :: 8.78**   Simplify: ⓐ $\sqrt{27p^3} - \sqrt{48p^3}$ ⓑ $\sqrt[3]{256y^5} - \sqrt[3]{32n^5}$.

Multiply Radical Expressions

We have used the Product Property of Roots to simplify square roots by removing the perfect square factors. We can use the Product Property of Roots 'in reverse' to multiply square roots. Remember, we assume all variables are greater than or equal to zero.

We will rewrite the Product Property of Roots so we see both ways together.

Product Property of Roots

For any real numbers, $\sqrt[n]{a}$ and $\sqrt[n]{b}$, and for any integer $n \geq 2$

$$\sqrt[n]{ab} = \sqrt[n]{a} \cdot \sqrt[n]{b} \quad \text{and} \quad \sqrt[n]{a} \cdot \sqrt[n]{b} = \sqrt[n]{ab}$$

When we multiply two radicals they must have the same index. Once we multiply the radicals, we then look for factors that are a power of the index and simplify the radical whenever possible.

Multiplying radicals with coefficients is much like multiplying variables with coefficients. To multiply $4x \cdot 3y$ we multiply the coefficients together and then the variables. The result is 12xy. Keep this in mind as you do these examples.

EXAMPLE 8.40

Simplify: ⓐ $(6\sqrt{2})(3\sqrt{10})$ ⓑ $\left(-5\sqrt[3]{4}\right)\left(-4\sqrt[3]{6}\right)$.

⊘ **Solution**

ⓐ

$$(6\sqrt{2})(3\sqrt{10})$$

Multiply using the Product Property.	$18\sqrt{20}$
Simplify the radical.	$18\sqrt{4} \cdot \sqrt{5}$
Simplify.	$18 \cdot 2 \cdot \sqrt{5}$
	$36\sqrt{5}$

ⓑ

$$\left(-5\sqrt[3]{4}\right)\left(-4\sqrt[3]{6}\right)$$

Multiply using the Product Property.	$20\sqrt[3]{24}$
Simplify the radical.	$20\sqrt[3]{8} \cdot \sqrt[3]{3}$
Simplify.	$20 \cdot 2 \cdot \sqrt[3]{3}$
	$40\sqrt[3]{3}$

> **TRY IT :: 8.79** Simplify: ⓐ $(3\sqrt{2})(2\sqrt{30})$ ⓑ $\left(2\sqrt[3]{18}\right)\left(-3\sqrt[3]{6}\right)$.

> **TRY IT :: 8.80** Simplify: ⓐ $(3\sqrt{3})(3\sqrt{6})$ ⓑ $\left(-4\sqrt[3]{9}\right)\left(3\sqrt[3]{6}\right)$.

We follow the same procedures when there are variables in the radicands.

EXAMPLE 8.41

Simplify: ⓐ $\left(10\sqrt{6p^3}\right)\left(4\sqrt{3p}\right)$ ⓑ $\left(2\sqrt[4]{20y^2}\right)\left(3\sqrt[4]{28y^3}\right)$.

⊘ **Solution**

ⓐ

$$\left(10\sqrt{6p^3}\right)\left(4\sqrt{3p}\right)$$

Multiply. $\qquad 40\sqrt{18p^4}$

Simplify the radical. $\qquad 40\sqrt{9p^4}\cdot\sqrt{2}$

Simplify. $\qquad 40\cdot 3p^2\cdot\sqrt{3}$

$$120p^2\sqrt{3}$$

ⓑ When the radicands involve large numbers, it is often advantageous to factor them in order to find the perfect powers.

$$\left(2\sqrt[4]{20y^2}\right)\left(3\sqrt[4]{28y^3}\right)$$

Multiply. $\qquad 6\sqrt[4]{4\cdot 5\cdot 4\cdot 7y^5}$

Simplify the radical. $\qquad 6\sqrt[4]{16y^4}\cdot\sqrt[4]{35y}$

Simplify. $\qquad 6\cdot 2y\sqrt[4]{35y}$

Multiply. $\qquad 12y\sqrt[4]{35y}$

> **TRY IT :: 8.81**   Simplify: ⓐ $\left(6\sqrt{6x^2}\right)\left(8\sqrt{30x^4}\right)$ ⓑ $\left(-4\sqrt[4]{12y^3}\right)\left(-\sqrt[4]{8y^3}\right)$.

> **TRY IT :: 8.82**   Simplify: ⓐ $\left(2\sqrt{6y^4}\right)\left(12\sqrt{30y}\right)$ ⓑ $\left(-4\sqrt[4]{9a^3}\right)\left(3\sqrt[4]{27a^2}\right)$.

Use Polynomial Multiplication to Multiply Radical Expressions

In the next a few examples, we will use the Distributive Property to multiply expressions with radicals. First we will distribute and then simplify the radicals when possible.

EXAMPLE 8.42

Simplify: ⓐ $\sqrt{6}\left(\sqrt{2}+\sqrt{18}\right)$ ⓑ $\sqrt[3]{9}\left(5-\sqrt[3]{18}\right)$.

⊘ **Solution**

ⓐ
$$\sqrt{6}\left(\sqrt{2}+\sqrt{18}\right)$$

Multiply. $\qquad \sqrt{12}+\sqrt{108}$

Simplify. $\qquad \sqrt{4}\cdot\sqrt{3}+\sqrt{36}\cdot\sqrt{3}$

Simplify. $\qquad 2\sqrt{3}+6\sqrt{3}$

Combine like radicals. $\qquad 8\sqrt{3}$

ⓑ
$$\sqrt[3]{9}\left(5-\sqrt[3]{18}\right)$$

Distribute. $\qquad 5\sqrt[3]{9}-\sqrt[3]{162}$

Simplify. $\qquad 5\sqrt[3]{9}-\sqrt[3]{27}\cdot\sqrt[3]{6}$

Simplify. $\qquad 5\sqrt[3]{9}-3\sqrt[3]{6}$

> **TRY IT : : 8.83**   Simplify: ⓐ $\sqrt{6}(1 + 3\sqrt{6})$ ⓑ $\sqrt[3]{4}\left(-2 - \sqrt[3]{6}\right)$.

> **TRY IT : : 8.84**   Simplify: ⓐ $\sqrt{8}(2 - 5\sqrt{8})$ ⓑ $\sqrt[3]{3}\left(-\sqrt[3]{9} - \sqrt[3]{6}\right)$.

When we worked with polynomials, we multiplied binomials by binomials. Remember, this gave us four products before we combined any like terms. To be sure to get all four products, we organized our work—usually by the FOIL method.

EXAMPLE 8.43

Simplify: ⓐ $(3 - 2\sqrt{7})(4 - 2\sqrt{7})$ ⓑ $\left(\sqrt[3]{x} - 2\right)\left(\sqrt[3]{x} + 4\right)$.

✓ Solution

ⓐ

$$(3 - 2\sqrt{7})(4 - 2\sqrt{7})$$

Multiply	$12 - 6\sqrt{7} - 8\sqrt{7} + 4\cdot 7$
Simplify.	$12 - 6\sqrt{7} - 8\sqrt{7} + 28$
Combine like terms.	$40 - 14\sqrt{7}$

ⓑ

$$\left(\sqrt[3]{x} - 2\right)\left(\sqrt[3]{x} + 4\right)$$

| Multiply. | $\sqrt[3]{x^2} + 4\sqrt[3]{x} - 2\sqrt[3]{x} - 8$ |
| Combine like terms. | $\sqrt[3]{x^2} + 2\sqrt[3]{x} - 8$ |

> **TRY IT : : 8.85**   Simplify: ⓐ $(6 - 3\sqrt{7})(3 + 4\sqrt{7})$ ⓑ $\left(\sqrt[3]{x} - 2\right)\left(\sqrt[3]{x} - 3\right)$.

> **TRY IT : : 8.86**   Simplify: ⓐ $(2 - 3\sqrt{11})(4 - \sqrt{11})$ ⓑ $\left(\sqrt[3]{x} + 1\right)\left(\sqrt[3]{x} + 3\right)$.

EXAMPLE 8.44

Simplify: $\left(3\sqrt{2} - \sqrt{5}\right)\left(\sqrt{2} + 4\sqrt{5}\right)$.

✓ Solution

$$\left(3\sqrt{2} - \sqrt{5}\right)\left(\sqrt{2} + 4\sqrt{5}\right)$$

Multiply.	$3\cdot 2 + 12\sqrt{10} - \sqrt{10} - 4\cdot 5$
Simplify.	$6 + 12\sqrt{10} - \sqrt{10} - 20$
Combine like terms.	$-14 + 11\sqrt{10}$

> **TRY IT : : 8.87**   Simplify: $\left(5\sqrt{3} - \sqrt{7}\right)\left(\sqrt{3} + 2\sqrt{7}\right)$

> **TRY IT : : 8.88**   Simplify: $\left(\sqrt{6} - 3\sqrt{8}\right)\left(2\sqrt{6} + \sqrt{8}\right)$

Recognizing some special products made our work easier when we multiplied binomials earlier. This is true when we multiply radicals, too. The special product formulas we used are shown here.

Special Products

Binomial Squares	**Product of Conjugates**
$(a+b)^2 = a^2 + 2ab + b^2$	$(a+b)(a-b) = a^2 - b^2$
$(a-b)^2 = a^2 - 2ab + b^2$	

We will use the special product formulas in the next few examples. We will start with the Product of Binomial Squares Pattern.

EXAMPLE 8.45

Simplify: ⓐ $(2 + \sqrt{3})^2$ ⓑ $(4 - 2\sqrt{5})^2$.

✓ **Solution**

Be sure to include the $2ab$ term when squaring a binomial.

ⓐ

	$(a + b)^2$
	$(2 + \sqrt{3})^2$
Multiply, using the Product of Binomial Squares Pattern.	$a^2 + 2 \; a \;\; b + b^2$
	$2^2 + 2 \cdot 2 \cdot \sqrt{3} + (\sqrt{3})^2$
Simplify.	$4 + 4\sqrt{3} + 3$
Combine like terms.	$7 + 4\sqrt{3}$

ⓑ

	$(a - b)^2$
	$(4 - 2\sqrt{5})^2$
Multiply, using the Product of Binomial Squares Pattern.	$a^2 - 2 \; a \;\; b + b^2$
	$4^2 - 2 \cdot 4 \cdot 2\sqrt{5} + (2\sqrt{5})^2$
Simplify.	$16 - 16\sqrt{5} + 4 \cdot 5$
	$16 - 16\sqrt{5} + 20$
Combine like terms.	$36 - 16\sqrt{5}$

> **TRY IT :: 8.89**   Simplify: ⓐ $(10 + \sqrt{2})^2$ ⓑ $(1 + 3\sqrt{6})^2$.

> **TRY IT :: 8.90**   Simplify: ⓐ $(6 - \sqrt{5})^2$ ⓑ $(9 - 2\sqrt{10})^2$.

In the next example, we will use the Product of Conjugates Pattern. Notice that the final product has no radical.

EXAMPLE 8.46

Simplify: $(5 - 2\sqrt{3})(5 + 2\sqrt{3})$.

✓ **Solution**

$$(a - \ b) \ (a + \ b)$$
$$(5 - 2\sqrt{3})(5 + 2\sqrt{3})$$

Multiply, using the Product of Conjugates Pattern.	$a^2 - b^2$ $5^2 - (2\sqrt{3})^2$
Simplify.	$25 - 4 \cdot 3$
	13

> | **TRY IT :: 8.91**   Simplify: $(3 - 2\sqrt{5})(3 + 2\sqrt{5})$

> | **TRY IT :: 8.92**   Simplify: $(4 + 5\sqrt{7})(4 - 5\sqrt{7})$.

▶ | **MEDIA ::**

Access these online resources for additional instruction and practice with adding, subtracting, and multiplying radical expressions.

- **Multiplying Adding Subtracting Radicals (https://openstax.org/l/37Radicals1)**
- **Multiplying Special Products: Square Binomials Containing Square Roots (https://openstax.org/l/37Radicals2)**
- **Multiplying Conjugates (https://openstax.org/l/37Radicals3)**

8.4 EXERCISES

Practice Makes Perfect

Add and Subtract Radical Expressions

In the following exercises, simplify.

165.
ⓐ $8\sqrt{2} - 5\sqrt{2}$
ⓑ $5\sqrt[3]{m} + 2\sqrt[3]{m}$
ⓒ $8\sqrt[4]{m} - 2\sqrt[4]{n}$

166.
ⓐ $7\sqrt{2} - 3\sqrt{2}$
ⓑ $7\sqrt[3]{p} + 2\sqrt[3]{p}$
ⓒ $5\sqrt[3]{x} - 3\sqrt[3]{x}$

167.
ⓐ $3\sqrt{5} + 6\sqrt{5}$
ⓑ $9\sqrt[3]{a} + 3\sqrt[3]{a}$
ⓒ $5\sqrt[4]{2z} + \sqrt[4]{2z}$

168.
ⓐ $4\sqrt{5} + 8\sqrt{5}$
ⓑ $\sqrt[3]{m} - 4\sqrt[3]{m}$
ⓒ $\sqrt{n} + 3\sqrt{n}$

169.
ⓐ $3\sqrt{2a} - 4\sqrt{2a} + 5\sqrt{2a}$
ⓑ $5\sqrt[4]{3ab} - 3\sqrt[4]{3ab} - 2\sqrt[4]{3ab}$

170.
ⓐ $\sqrt{11b} - 5\sqrt{11b} + 3\sqrt{11b}$
ⓑ $8\sqrt[4]{11cd} + 5\sqrt[4]{11cd} - 9\sqrt[4]{11cd}$

171.
ⓐ $8\sqrt{3c} + 2\sqrt{3c} - 9\sqrt{3c}$
ⓑ $2\sqrt[3]{4pq} - 5\sqrt[3]{4pq} + 4\sqrt[3]{4pq}$

172.
ⓐ $3\sqrt{5d} + 8\sqrt{5d} - 11\sqrt{5d}$
ⓑ $11\sqrt[3]{2rs} - 9\sqrt[3]{2rs} + 3\sqrt[3]{2rs}$

173.
ⓐ $\sqrt{27} - \sqrt{75}$
ⓑ $\sqrt[3]{40} - \sqrt[3]{320}$
ⓒ $\frac{1}{2}\sqrt[4]{32} + \frac{2}{3}\sqrt[4]{162}$

174.
ⓐ $\sqrt{72} - \sqrt{98}$
ⓑ $\sqrt[3]{24} + \sqrt[3]{81}$
ⓒ $\frac{1}{2}\sqrt[4]{80} - \frac{2}{3}\sqrt[4]{405}$

175.
ⓐ $\sqrt{48} + \sqrt{27}$
ⓑ $\sqrt[3]{54} + \sqrt[3]{128}$
ⓒ $6\sqrt[4]{5} - \frac{3}{2}\sqrt[4]{320}$

176.
ⓐ $\sqrt{45} + \sqrt{80}$
ⓑ $\sqrt[3]{81} - \sqrt[3]{192}$
ⓒ $\frac{5}{2}\sqrt[4]{80} + \frac{7}{3}\sqrt[4]{405}$

177.
ⓐ $\sqrt{72a^5} - \sqrt{50a^5}$
ⓑ $9\sqrt[4]{80p^4} - 6\sqrt[4]{405p^4}$

178.
ⓐ $\sqrt{48b^5} - \sqrt{75b^5}$
ⓑ $8\sqrt[3]{64q^6} - 3\sqrt[3]{125q^6}$

179.
ⓐ $\sqrt{80c^7} - \sqrt{20c^7}$
ⓑ $2\sqrt[4]{162r^{10}} + 4\sqrt[4]{32r^{10}}$

180.
ⓐ $\sqrt{96d^9} - \sqrt{24d^9}$
ⓑ $5\sqrt[4]{243s^6} + 2\sqrt[4]{3s^6}$

181.
$3\sqrt{128y^2} + 4y\sqrt{162} - 8\sqrt{98y^2}$

182. $3\sqrt{75y^2} + 8y\sqrt{48} - \sqrt{300y^2}$

Multiply Radical Expressions

In the following exercises, simplify.

183.
ⓐ $(-2\sqrt{3})(3\sqrt{18})$
ⓑ $\left(8\sqrt[3]{4}\right)\left(-4\sqrt[3]{18}\right)$

184.
ⓐ $(-4\sqrt{5})(5\sqrt{10})$
ⓑ $\left(-2\sqrt[3]{9}\right)\left(7\sqrt[3]{9}\right)$

185.
ⓐ $(5\sqrt{6})(-\sqrt{12})$
ⓑ $\left(-2\sqrt[4]{18}\right)\left(-\sqrt[4]{9}\right)$

186.
ⓐ $(-2\sqrt{7})(-2\sqrt{14})$

ⓑ $\left(-3\sqrt[4]{8}\right)\left(-5\sqrt[4]{6}\right)$

187.
ⓐ $\left(4\sqrt{12z^3}\right)(3\sqrt{9z})$

ⓑ $\left(5\sqrt[3]{3x^3}\right)\left(3\sqrt[3]{18x^3}\right)$

188.
ⓐ $\left(3\sqrt{2x^3}\right)\left(7\sqrt{18x^2}\right)$

ⓑ $\left(-6\sqrt[3]{20a^2}\right)\left(-2\sqrt[3]{16a^3}\right)$

189.
ⓐ $\left(-2\sqrt{7z^3}\right)\left(3\sqrt{14z^8}\right)$

ⓑ $\left(2\sqrt[4]{8y^2}\right)\left(-2\sqrt[4]{12y^3}\right)$

190.
ⓐ $\left(4\sqrt{2k^5}\right)\left(-3\sqrt{32k^6}\right)$

ⓑ $\left(-\sqrt[4]{6b^3}\right)\left(3\sqrt[4]{8b^3}\right)$

Use Polynomial Multiplication to Multiply Radical Expressions

In the following exercises, multiply.

191.
ⓐ $\sqrt{7}(5 + 2\sqrt{7})$

ⓑ $\sqrt[3]{6}\left(4 + \sqrt[3]{18}\right)$

192.
ⓐ $\sqrt{11}(8 + 4\sqrt{11})$

ⓑ $\sqrt[3]{3}\left(\sqrt[3]{9} + \sqrt[3]{18}\right)$

193.
ⓐ $\sqrt{11}(-3 + 4\sqrt{11})$

ⓑ $\sqrt[4]{3}\left(\sqrt[4]{54} + \sqrt[4]{18}\right)$

194.
ⓐ $\sqrt{2}(-5 + 9\sqrt{2})$

ⓑ $\sqrt[4]{2}\left(\sqrt[4]{12} + \sqrt[4]{24}\right)$

195. $(7 + \sqrt{3})(9 - \sqrt{3})$

196. $(8 - \sqrt{2})(3 + \sqrt{2})$

197.
ⓐ $(9 - 3\sqrt{2})(6 + 4\sqrt{2})$

ⓑ $\left(\sqrt[3]{x} - 3\right)\left(\sqrt[3]{x} + 1\right)$

198.
ⓐ $(3 - 2\sqrt{7})(5 - 4\sqrt{7})$

ⓑ $\left(\sqrt[3]{x} - 5\right)\left(\sqrt[3]{x} - 3\right)$

199.
ⓐ $(1 + 3\sqrt{10})(5 - 2\sqrt{10})$

ⓑ $\left(2\sqrt[3]{x} + 6\right)\left(\sqrt[3]{x} + 1\right)$

200.
ⓐ $(7 - 2\sqrt{5})(4 + 9\sqrt{5})$

ⓑ $\left(3\sqrt[3]{x} + 2\right)\left(\sqrt[3]{x} - 2\right)$

201. $(\sqrt{3} + \sqrt{10})(\sqrt{3} + 2\sqrt{10})$

202. $(\sqrt{11} + \sqrt{5})(\sqrt{11} + 6\sqrt{5})$

203. $(2\sqrt{7} - 5\sqrt{11})(4\sqrt{7} + 9\sqrt{11})$

204. $\left(4\sqrt{6} + 7\sqrt{13}\right)(8\sqrt{6} - 3\sqrt{13})$

205. ⓐ $(3 + \sqrt{5})^2$ ⓑ $(2 - 5\sqrt{3})^2$

206. ⓐ $(4 + \sqrt{11})^2$ ⓑ $(3 - 2\sqrt{5})^2$

207. ⓐ $(9 - \sqrt{6})^2$ ⓑ $(10 + 3\sqrt{7})^2$

208. ⓐ $(5 - \sqrt{10})^2$ ⓑ $(8 + 3\sqrt{2})^2$

209. $(4 + \sqrt{2})(4 - \sqrt{2})$

210. $(7 + \sqrt{10})(7 - \sqrt{10})$

211. $(4 + 9\sqrt{3})(4 - 9\sqrt{3})$

212. $(1 + 8\sqrt{2})(1 - 8\sqrt{2})$

213. $(12 - 5\sqrt{5})(12 + 5\sqrt{5})$

214. $(9 - 4\sqrt{3})(9 + 4\sqrt{3})$

215. $\left(\sqrt[3]{3x} + 2\right)\left(\sqrt[3]{3x} - 2\right)$

216. $\left(\sqrt[3]{4x} + 3\right)\left(\sqrt[3]{4x} - 3\right)$

Mixed Practice

217. $\frac{2}{3}\sqrt{27} + \frac{3}{4}\sqrt{48}$

218. $\sqrt{175k^4} - \sqrt{63k^4}$

219. $\frac{5}{6}\sqrt{162} + \frac{3}{16}\sqrt{128}$

220. $\sqrt[3]{24} + \sqrt[3]{81}$

221. $\frac{1}{2}\sqrt[4]{80} - \frac{2}{3}\sqrt[4]{405}$

222. $8\sqrt[4]{13} - 4\sqrt[4]{13} - 3\sqrt[4]{13}$

223. $5\sqrt{12c^4} - 3\sqrt{27c^6}$

224. $\sqrt{80a^5} - \sqrt{45a^5}$

225. $\frac{3}{5}\sqrt{75} - \frac{1}{4}\sqrt{48}$

226. $21\sqrt[3]{9} - 2\sqrt[3]{9}$

227. $8\sqrt[3]{64q^6} - 3\sqrt[3]{125q^6}$

228. $11\sqrt{11} - 10\sqrt{11}$

229. $\sqrt{3} \cdot \sqrt{21}$

230. $\left(4\sqrt{6}\right)\left(-\sqrt{18}\right)$

231. $\left(7\sqrt[3]{4}\right)\left(-3\sqrt[3]{18}\right)$

232. $\left(4\sqrt{12x^5}\right)\left(2\sqrt{6x^3}\right)$

233. $\left(\sqrt{29}\right)^2$

234. $(-4\sqrt{17})(-3\sqrt{17})$

235. $(-4 + \sqrt{17})(-3 + \sqrt{17})$

236. $\left(3\sqrt[4]{8a^2}\right)\left(\sqrt[4]{12a^3}\right)$

237. $\left(6 - 3\sqrt{2}\right)^2$

238. $\sqrt{3}(4 - 3\sqrt{3})$

239. $\sqrt[3]{3}\left(2\sqrt[3]{9} + \sqrt[3]{18}\right)$

240. $\left(\sqrt{6} + \sqrt{3}\right)\left(\sqrt{6} + 6\sqrt{3}\right)$

Writing Exercises

241. Explain the when a radical expression is in simplest form.

242. Explain the process for determining whether two radicals are like or unlike. Make sure your answer makes sense for radicals containing both numbers and variables.

243.

ⓐ Explain why $(-\sqrt{n})^2$ is always non-negative, for $n \geq 0$.

ⓑ Explain why $-(\sqrt{n})^2$ is always non-positive, for $n \geq 0$.

244. Use the binomial square pattern to simplify $(3 + \sqrt{2})^2$. Explain all your steps.

Self Check

ⓐ After completing the exercises, use this checklist to evaluate your mastery of the objectives of this section.

I can...	Confidently	With some help	No-I don't get it!
add and subtract radical expressions.			
multiply radical expressions.			
use polynomial multiplication to multiply radical expressions.			

ⓑ On a scale of 1-10, how would you rate your mastery of this section in light of your responses on the checklist? How can you improve this?

Divide Radical Expressions

Learning Objectives

By the end of this section, you will be able to:

> Divide radical expressions
> Rationalize a one term denominator
> Rationalize a two term denominator

Be Prepared!

Before you get started, take this readiness quiz.

1. Simplify: $\dfrac{30}{48}$.

    If you missed this problem, review **Example 1.24**.

2. Simplify: $x^2 \cdot x^4$.

    If you missed this problem, review **Example 5.12**.

3. Multiply: $(7 + 3x)(7 - 3x)$.

    If you missed this problem, review **Example 5.32**.

Divide Radical Expressions

We have used the Quotient Property of Radical Expressions to simplify roots of fractions. We will need to use this property 'in reverse' to simplify a fraction with radicals.

We give the Quotient Property of Radical Expressions again for easy reference. Remember, we assume all variables are greater than or equal to zero so that no absolute value bars re needed.

Quotient Property of Radical Expressions

If $\sqrt[n]{a}$ and $\sqrt[n]{b}$ are real numbers, $b \neq 0$, and for any integer $n \geq 2$ then,

$$\sqrt[n]{\dfrac{a}{b}} = \dfrac{\sqrt[n]{a}}{\sqrt[n]{b}} \quad \text{and} \quad \dfrac{\sqrt[n]{a}}{\sqrt[n]{b}} = \sqrt[n]{\dfrac{a}{b}}$$

We will use the Quotient Property of Radical Expressions when the fraction we start with is the quotient of two radicals, and neither radicand is a perfect power of the index. When we write the fraction in a single radical, we may find common factors in the numerator and denominator.

EXAMPLE 8.47

Simplify: ⓐ $\dfrac{\sqrt{72x^3}}{\sqrt{162x}}$ ⓑ $\dfrac{\sqrt[3]{32x^2}}{\sqrt[3]{4x^5}}$.

 **Solution**

$$\frac{\sqrt{72x^3}}{\sqrt{162x}}$$

Rewrite using the quotient property, $\frac{\sqrt[n]{a}}{\sqrt[n]{b}} = \sqrt[n]{\frac{a}{b}}$.

$$\sqrt{\frac{72x^3}{162x}}$$

Remove common factors.

$$\sqrt{\frac{\cancel{18} \cdot 4 \cdot x^2 \cdot \cancel{x}}{\cancel{18} \cdot 9 \cdot \cancel{x}}}$$

Simplify.

$$\sqrt{\frac{4x^2}{9}}$$

Simplify the radical.

$$\frac{2x}{3}$$

ⓑ

$$\frac{\sqrt[3]{32x^2}}{\sqrt[3]{4x^5}}$$

Rewrite using the quotient property, $\frac{\sqrt[n]{a}}{\sqrt[n]{b}} = \sqrt[n]{\frac{a}{b}}$.

$$\sqrt[3]{\frac{32x^2}{4x^5}}$$

Simplify the fraction under the radical.

$$\sqrt[3]{\frac{8}{x^3}}$$

Simplify the radical.

$$\frac{2}{x}$$

> **TRY IT ::** 8.93

Simplify: ⓐ $\frac{\sqrt{50s^3}}{\sqrt{128s}}$ ⓑ $\frac{\sqrt[3]{56a}}{\sqrt[3]{7a^4}}$.

> **TRY IT ::** 8.94

Simplify: ⓐ $\frac{\sqrt{75q^5}}{\sqrt{108q}}$ ⓑ $\frac{\sqrt[3]{72b^2}}{\sqrt[3]{9b^5}}$.

EXAMPLE 8.48

Simplify: ⓐ $\frac{\sqrt{147ab^8}}{\sqrt{3a^3b^4}}$ ⓑ $\frac{\sqrt[3]{-250m\,n^{-2}}}{\sqrt[3]{2m^{-2}n^4}}$.

⊘ **Solution**

ⓐ

$$\frac{\sqrt{147ab^8}}{\sqrt{3a^3b^4}}$$

Rewrite using the quotient property.

$$\sqrt{\frac{147ab^8}{3a^3b^4}}$$

Remove common factors in the fraction.

$$\sqrt{\frac{49b^4}{a^2}}$$

Simplify the radical.

$$\frac{7b^2}{a}$$

ⓑ

$$\frac{\sqrt[3]{-250m\,n^{-2}}}{\sqrt[3]{2m^{-2}n^4}}$$

Rewrite using the quotient property.

$$\sqrt[3]{\frac{-250m\,n^{-2}}{2m^{-2}n^4}}$$

Simplify the fraction under the radical.

$$\sqrt[3]{\frac{-125m^3}{n^6}}$$

Simplify the radical.

$$-\frac{5m}{n^2}$$

> **TRY IT : : 8.95**

Simplify: ⓐ $\dfrac{\sqrt{162x^{10}y^2}}{\sqrt{2x^6y^6}}$ ⓑ $\dfrac{\sqrt[3]{-128x^2y^{-1}}}{\sqrt[3]{2x^{-1}y^2}}$.

> **TRY IT : : 8.96**

Simplify: ⓐ $\dfrac{\sqrt{300m^3n^7}}{\sqrt{3m^5n}}$ ⓑ $\dfrac{\sqrt[3]{-81pq^{-1}}}{\sqrt[3]{3p^{-2}q^5}}$.

EXAMPLE 8.49

Simplify: $\dfrac{\sqrt{54x^5y^3}}{\sqrt{3x^2y}}$.

⊘ **Solution**

$$\frac{\sqrt{54x^5y^3}}{\sqrt{3x^2y}}$$

Rewrite using the quotient property.

$$\sqrt{\frac{54x^5y^3}{3x^2y}}$$

Remove common factors in the fraction.
Rewrite the radicand as a product
using the largest perfect square factor.

$$\sqrt{18x^3y^2}$$

$$\sqrt{9x^2y^2 \cdot 2x}$$

Rewrite the radical as the product of two
radicals.
Simplify.

$$\sqrt{9x^2y^2} \cdot \sqrt{2x}$$

$$3xy\sqrt{2x}$$

> **TRY IT : : 8.97**

Simplify: $\dfrac{\sqrt{64x^4y^5}}{\sqrt{2xy^3}}$.

> **TRY IT : : 8.98**

Simplify: $\dfrac{\sqrt{96a^5b^4}}{\sqrt{2a^3b}}$.

Rationalize a One Term Denominator

Before the calculator became a tool of everyday life, approximating the value of a fraction with a radical in the denominator was a very cumbersome process!

For this reason, a process called **rationalizing the denominator** was developed. A fraction with a radical in the denominator is converted to an equivalent fraction whose denominator is an integer. Square roots of numbers that are not perfect squares are irrational numbers. When we rationalize the denominator, we write an equivalent fraction with a rational number in the denominator.

This process is still used today, and is useful in other areas of mathematics, too.

Rationalizing the Denominator

Rationalizing the denominator is the process of converting a fraction with a radical in the denominator to an equivalent fraction whose denominator is an integer.

Even though we have calculators available nearly everywhere, a fraction with a radical in the denominator still must be rationalized. It is not considered simplified if the denominator contains a radical.

Similarly, a radical expression is not considered simplified if the radicand contains a fraction.

Simplified Radical Expressions

A radical expression is considered simplified if there are

- no factors in the radicand have perfect powers of the index
- no fractions in the radicand
- no radicals in the denominator of a fraction

To rationalize a denominator with a square root, we use the property that $(\sqrt{a})^2 = a$. If we square an irrational square root, we get a rational number.

We will use this property to rationalize the denominator in the next example.

EXAMPLE 8.50

Simplify: ⓐ $\dfrac{4}{\sqrt{3}}$ ⓑ $\sqrt{\dfrac{3}{20}}$ ⓒ $\dfrac{3}{\sqrt{6x}}$.

✓ **Solution**

To rationalize a denominator with one term, we can multiply a square root by itself. To keep the fraction equivalent, we multiply both the numerator and denominator by the same factor.

ⓐ

$$\dfrac{4}{\sqrt{3}}$$

Multiply both the numerator and denominator by $\sqrt{3}$.	$\dfrac{4 \cdot \sqrt{3}}{\sqrt{3} \cdot \sqrt{3}}$
Simplify.	$\dfrac{4\sqrt{3}}{3}$

ⓑ We always simplify the radical in the denominator first, before we rationalize it. This way the numbers stay smaller and easier to work with.

$$\sqrt{\dfrac{3}{20}}$$

The fraction is not a perfect square, so rewrite using the Quotient Property.	$\dfrac{\sqrt{3}}{\sqrt{20}}$
Simplify the denominator.	$\dfrac{\sqrt{3}}{2\sqrt{5}}$
Multiply the numerator and denominator by $\sqrt{5}$.	$\dfrac{\sqrt{3} \cdot \sqrt{5}}{2\sqrt{5} \cdot \sqrt{5}}$

Simplify.	$\dfrac{\sqrt{15}}{2 \cdot 5}$
Simplify.	$\dfrac{\sqrt{15}}{10}$

ⓒ

	$\dfrac{3}{\sqrt{6x}}$
Multiply the numerator and denominator by $\sqrt{6x}$.	$\dfrac{3 \cdot \sqrt{6x}}{\sqrt{6x} \cdot \sqrt{6x}}$
Simplify.	$\dfrac{3\sqrt{6x}}{6x}$
Simplify.	$\dfrac{\sqrt{6x}}{2x}$

> **TRY IT ∷ 8.99**   Simplify: ⓐ $\dfrac{5}{\sqrt{3}}$ ⓑ $\sqrt{\dfrac{3}{32}}$ ⓒ $\dfrac{2}{\sqrt{2x}}$.

> **TRY IT ∷ 8.100**   Simplify: ⓐ $\dfrac{6}{\sqrt{5}}$ ⓑ $\sqrt{\dfrac{7}{18}}$ ⓒ $\dfrac{5}{\sqrt{5x}}$.

When we rationalized a square root, we multiplied the numerator and denominator by a square root that would give us a perfect square under the radical in the denominator. When we took the square root, the denominator no longer had a radical.

We will follow a similar process to rationalize higher roots. To rationalize a denominator with a higher index radical, we multiply the numerator and denominator by a radical that would give us a radicand that is a perfect power of the index. When we simplify the new radical, the denominator will no longer have a radical.

For example,

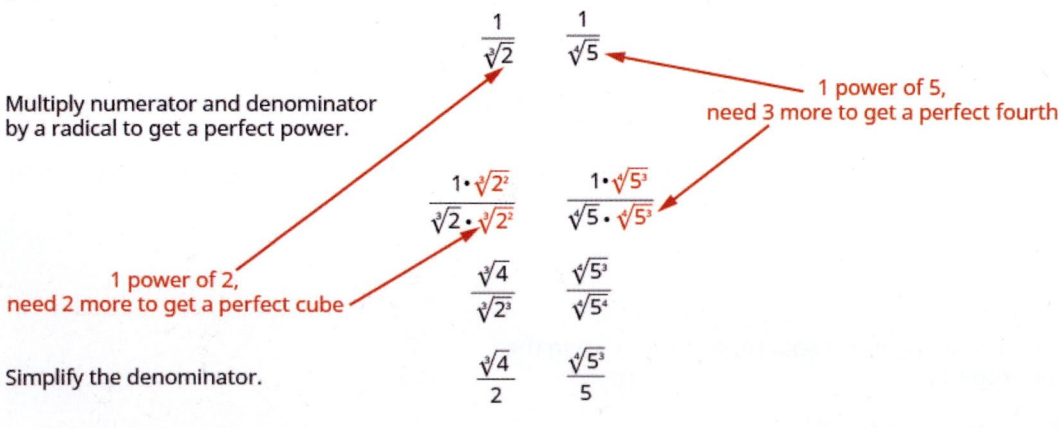

We will use this technique in the next examples.

EXAMPLE 8.51

Simplify ⓐ $\dfrac{1}{\sqrt[3]{6}}$ ⓑ $\sqrt[3]{\dfrac{7}{24}}$ ⓒ $\dfrac{3}{\sqrt[3]{4x}}$.

✓ Solution

To rationalize a denominator with a cube root, we can multiply by a cube root that will give us a perfect cube in the radicand in the denominator. To keep the fraction equivalent, we multiply both the numerator and denominator by the same factor.

ⓐ

$$\frac{1}{\sqrt[3]{6}}$$

The radical in the denominator has one factor of 6. Multiply both the numerator and denominator by $\sqrt[3]{6^2}$, which gives us 2 more factors of 6.	$\dfrac{1 \cdot \sqrt[3]{6^2}}{\sqrt[3]{6} \cdot \sqrt[3]{6^2}}$
Multiply. Notice the radicand in the denominator has 3 powers of 6.	$\dfrac{\sqrt[3]{6^2}}{\sqrt[3]{6^3}}$
Simplify the cube root in the denominator.	$\dfrac{\sqrt[3]{36}}{6}$

ⓑ We always simplify the radical in the denominator first, before we rationalize it. This way the numbers stay smaller and easier to work with.

$$\sqrt[3]{\frac{7}{24}}$$

The fraction is not a perfect cube, so rewrite using the Quotient Property.	$\dfrac{\sqrt[3]{7}}{\sqrt[3]{24}}$
Simplify the denominator.	$\dfrac{\sqrt[3]{7}}{2\sqrt[3]{3}}$
Multiply the numerator and denominator by $\sqrt[3]{3^2}$. This will give us 3 factors of 3.	$\dfrac{\sqrt[3]{7} \cdot \sqrt[3]{3^2}}{2\sqrt[3]{3} \cdot \sqrt[3]{3^2}}$
Simplify.	$\dfrac{\sqrt[3]{63}}{2\sqrt[3]{3^3}}$
Remember, $\sqrt[3]{3^3} = 3$.	$\dfrac{\sqrt[3]{63}}{2 \cdot 3}$
Simplify.	$\dfrac{\sqrt[3]{63}}{6}$

ⓒ

$$\frac{3}{\sqrt[3]{4x}}$$

Rewrite the radicand to show the factors.	$\frac{3}{\sqrt[3]{2^2 \cdot x}}$
Multiply the numerator and denominator by $\sqrt[3]{2 \cdot x^2}$. This will get us 3 factors of 2 and 3 factors of x.	$\frac{3 \cdot \sqrt[3]{2 \cdot x^2}}{\sqrt[3]{2^2 x} \cdot \sqrt[3]{2 \cdot x^2}}$
Simplify.	$\frac{3\sqrt[3]{2x^2}}{\sqrt[3]{2^3 x^3}}$
Simplify the radical in the denominator.	$\frac{3\sqrt[3]{2x^2}}{2x}$

> **TRY IT : : 8.101** Simplify: ⓐ $\frac{1}{\sqrt[3]{7}}$ ⓑ $\sqrt[3]{\frac{5}{12}}$ ⓒ $\frac{5}{\sqrt[3]{9y}}$.

> **TRY IT : : 8.102** Simplify: ⓐ $\frac{1}{\sqrt[3]{2}}$ ⓑ $\sqrt[3]{\frac{3}{20}}$ ⓒ $\frac{2}{\sqrt[3]{25n}}$.

EXAMPLE 8.52

Simplify: ⓐ $\frac{1}{\sqrt[4]{2}}$ ⓑ $\sqrt[4]{\frac{5}{64}}$ ⓒ $\frac{2}{\sqrt[4]{8x}}$.

⊘ **Solution**

To rationalize a denominator with a fourth root, we can multiply by a fourth root that will give us a perfect fourth power in the radicand in the denominator. To keep the fraction equivalent, we multiply both the numerator and denominator by the same factor.

ⓐ

$$\frac{1}{\sqrt[4]{2}}$$

The radical in the denominator has one factor of 2.	
Multiply both the numerator and denominator by $\sqrt[4]{2^3}$, which gives us 3 more factors of 2.	$\frac{1 \cdot \sqrt[4]{2^3}}{\sqrt[4]{2} \cdot \sqrt[4]{2^3}}$
Multiply. Notice the radicand in the denominator has 4 powers of 2.	$\frac{\sqrt[4]{8}}{\sqrt[4]{2^4}}$
Simplify the fourth root in the denominator.	$\frac{\sqrt[4]{8}}{2}$

ⓑ We always simplify the radical in the denominator first, before we rationalize it. This way the numbers stay smaller and easier to work with.

$$\sqrt[4]{\frac{5}{64}}$$

The fraction is not a perfect fourth power, so rewrite using the Quotient Property.	$\dfrac{\sqrt[4]{5}}{\sqrt[4]{64}}$
Rewrite the radicand in the denominator to show the factors.	$\dfrac{\sqrt[4]{5}}{\sqrt[4]{2^6}}$
Simplify the denominator.	$\dfrac{\sqrt[4]{5}}{2\sqrt[4]{2^2}}$
Multiply the numerator and denominator by $\sqrt[4]{2^2}$. This will give us 4 factors of 2.	$\dfrac{\sqrt[4]{5}\cdot\sqrt[4]{2^2}}{2\sqrt[4]{2^2}\cdot\sqrt[4]{2^2}}$
Simplify.	$\dfrac{\sqrt[4]{5}\cdot\sqrt[4]{4}}{2\sqrt[4]{2^4}}$
Remember, $\sqrt[4]{2^4}=2$.	$\dfrac{\sqrt[4]{20}}{2\cdot2}$
Simplify.	$\dfrac{\sqrt[4]{20}}{4}$

ⓒ

	$\dfrac{2}{\sqrt[4]{8x}}$
Rewrite the radicand to show the factors.	$\dfrac{2}{\sqrt[4]{2^3\cdot x}}$
Multiply the numerator and denominator by $\sqrt[4]{2\cdot x^3}$. This will get us 4 factors of 2 and 4 factors of x.	$\dfrac{2\cdot\sqrt[4]{2\cdot x^3}}{\sqrt[4]{2^3x}\cdot\sqrt[4]{2\cdot x^3}}$
Simplify.	$\dfrac{2\sqrt[4]{2x^3}}{\sqrt[4]{2^4x^4}}$
Simplify the radical in the denominator.	$\dfrac{2\sqrt[4]{2x^3}}{2x}$
Simplify the fraction.	$\dfrac{\sqrt[4]{2x^3}}{x}$

> **TRY IT :: 8.103** Simplify: ⓐ $\dfrac{1}{\sqrt[4]{3}}$ ⓑ $\sqrt[4]{\dfrac{3}{64}}$ ⓒ $\dfrac{3}{\sqrt[4]{125x}}$.

> **TRY IT :: 8.104** Simplify: ⓐ $\dfrac{1}{\sqrt[4]{5}}$ ⓑ $\sqrt[4]{\dfrac{7}{128}}$ ⓒ $\dfrac{4}{\sqrt[4]{4x}}$.

Rationalize a Two Term Denominator

When the denominator of a fraction is a sum or difference with square roots, we use the Product of Conjugates Pattern to rationalize the denominator.

$$(a - b)(a + b) \qquad\qquad (2 - \sqrt{5})(2 + \sqrt{5})$$
$$a^2 - b^2 \qquad\qquad 2^2 - (\sqrt{5})^2$$
$$4 - 5$$
$$-1$$

When we multiply a binomial that includes a square root by its conjugate, the product has no square roots.

EXAMPLE 8.53

Simplify: $\dfrac{5}{2 - \sqrt{3}}$.

✓ **Solution**

$$\dfrac{5}{2 - \sqrt{3}}$$

Multiply the numerator and denominator by the conjugate of the denominator.	$\dfrac{5(2 + \sqrt{3})}{(2 - \sqrt{3})(2 + \sqrt{3})}$
Multiply the conjugates in the denominator.	$\dfrac{5(2 + \sqrt{3})}{2^2 - (\sqrt{3})^2}$
Simplify the denominator.	$\dfrac{5(2 + \sqrt{3})}{4 - 3}$
Simplify the denominator.	$\dfrac{5(2 + \sqrt{3})}{1}$
Simplify.	$5(2 + \sqrt{3})$

> **TRY IT :: 8.105** Simplify: $\dfrac{3}{1 - \sqrt{5}}$.

> **TRY IT :: 8.106** Simplify: $\dfrac{2}{4 - \sqrt{6}}$.

Notice we did not distribute the 5 in the answer of the last example. By leaving the result factored we can see if there are any factors that may be common to both the numerator and denominator.

EXAMPLE 8.54

Simplify: $\dfrac{\sqrt{3}}{\sqrt{u} - \sqrt{6}}$.

✓ **Solution**

$$\dfrac{\sqrt{3}}{\sqrt{u} - \sqrt{6}}$$

Multiply the numerator and denominator by the conjugate of the denominator.	$\dfrac{\sqrt{3}(\sqrt{u} + \sqrt{6})}{(\sqrt{u} - \sqrt{6})(\sqrt{u} + \sqrt{6})}$
Multiply the conjugates in the denominator.	$\dfrac{\sqrt{3}(\sqrt{u} + \sqrt{6})}{(\sqrt{u})^2 - (\sqrt{6})^2}$

Simplify the denominator.

$$\frac{\sqrt{3}(\sqrt{u} + \sqrt{6})}{u - 6}$$

> **TRY IT : : 8.107** Simplify: $\dfrac{\sqrt{5}}{\sqrt{x} + \sqrt{2}}$.

> **TRY IT : : 8.108** Simplify: $\dfrac{\sqrt{10}}{\sqrt{y} - \sqrt{3}}$.

Be careful of the signs when multiplying. The numerator and denominator look very similar when you multiply by the conjugate.

EXAMPLE 8.55

Simplify: $\dfrac{\sqrt{x} + \sqrt{7}}{\sqrt{x} - \sqrt{7}}$.

⊘ **Solution**

$$\frac{\sqrt{x} + \sqrt{7}}{\sqrt{x} - \sqrt{7}}$$

Multiply the numerator and denominator by the conjugate of the denominator.	$\dfrac{(\sqrt{x} + \sqrt{7})(\sqrt{x} + \sqrt{7})}{(\sqrt{x} - \sqrt{7})(\sqrt{x} + \sqrt{7})}$
Multiply the conjugates in the denominator.	$\dfrac{(\sqrt{x} + \sqrt{7})(\sqrt{x} + \sqrt{7})}{(\sqrt{x})^2 - (\sqrt{7})^2}$
Simplify the denominator.	$\dfrac{(\sqrt{x} + \sqrt{7})^2}{x - 7}$

We do not square the numerator. Leaving it in factored form, we can see there are no common factors to remove from the numerator and denominator.

> **TRY IT : : 8.109** Simplify: $\dfrac{\sqrt{p} + \sqrt{2}}{\sqrt{p} - \sqrt{2}}$.

> **TRY IT : : 8.110** Simplify: $\dfrac{\sqrt{q} - \sqrt{10}}{\sqrt{q} + \sqrt{10}}$.

▶ **MEDIA : :**
Access these online resources for additional instruction and practice with dividing radical expressions.

- **Rationalize the Denominator (https://openstax.org/l/37RatDenom1)**
- **Dividing Radical Expressions and Rationalizing the Denominator (https://openstax.org/l/37RatDenom2)**
- **Simplifying a Radical Expression with a Conjugate (https://openstax.org/l/37RatDenom3)**
- **Rationalize the Denominator of a Radical Expression (https://openstax.org/l/37RatDenom4)**

8.5 EXERCISES

Practice Makes Perfect

Divide Square Roots

In the following exercises, simplify.

245. ⓐ $\dfrac{\sqrt{128}}{\sqrt{72}}$ ⓑ $\dfrac{\sqrt[3]{128}}{\sqrt[3]{54}}$

246. ⓐ $\dfrac{\sqrt{48}}{\sqrt{75}}$ ⓑ $\dfrac{\sqrt[3]{81}}{\sqrt[3]{24}}$

247. ⓐ $\dfrac{\sqrt{200m^5}}{\sqrt{98m}}$ ⓑ $\dfrac{\sqrt[3]{54y^2}}{\sqrt[3]{2y^5}}$

248. ⓐ $\dfrac{\sqrt{108n^7}}{\sqrt{243n^3}}$ ⓑ $\dfrac{\sqrt[3]{54y}}{\sqrt[3]{16y^4}}$

249. ⓐ $\dfrac{\sqrt{75r^3}}{\sqrt{108r^7}}$ ⓑ $\dfrac{\sqrt[3]{24x^7}}{\sqrt[3]{81x^4}}$

250. ⓐ $\dfrac{\sqrt{196q}}{\sqrt{484q^5}}$ ⓑ $\dfrac{\sqrt[3]{16m^4}}{\sqrt[3]{54m}}$

251. ⓐ $\dfrac{\sqrt{108p^5q^2}}{\sqrt{3p^3q^6}}$ ⓑ $\dfrac{\sqrt[3]{-16a^4b^{-2}}}{\sqrt[3]{2a^{-2}b}}$

252. ⓐ $\dfrac{\sqrt{98rs^{10}}}{\sqrt{2r^3s^4}}$ ⓑ $\dfrac{\sqrt[3]{-375y^4z^{-2}}}{\sqrt[3]{3y^{-2}z^4}}$

253. ⓐ $\dfrac{\sqrt{320mn^{-5}}}{\sqrt{45m^{-7}n^3}}$ ⓑ $\dfrac{\sqrt[3]{16x^4y^{-2}}}{\sqrt[3]{-54x^{-2}y^4}}$

254. ⓐ $\dfrac{\sqrt{810c^{-3}d^7}}{\sqrt{1000c\,d^{-1}}}$ ⓑ $\dfrac{\sqrt[3]{24a^7b^{-1}}}{\sqrt[3]{-81a^{-2}b^2}}$

255. $\dfrac{\sqrt{56x^5y^4}}{\sqrt{2xy^3}}$

256. $\dfrac{\sqrt{72a^3b^6}}{\sqrt{3ab^3}}$

257. $\dfrac{\sqrt[3]{48a^3b^6}}{\sqrt[3]{3a^{-1}b^3}}$

258. $\dfrac{\sqrt[3]{162x^{-3}y^6}}{\sqrt[3]{2x^3y^{-2}}}$

Rationalize a One Term Denominator

In the following exercises, rationalize the denominator.

259. ⓐ $\dfrac{10}{\sqrt{6}}$ ⓑ $\sqrt{\dfrac{4}{27}}$ ⓒ $\dfrac{10}{\sqrt{5x}}$

260. ⓐ $\dfrac{8}{\sqrt{3}}$ ⓑ $\sqrt{\dfrac{7}{40}}$ ⓒ $\dfrac{8}{\sqrt{2y}}$

261. ⓐ $\dfrac{6}{\sqrt{7}}$ ⓑ $\sqrt{\dfrac{8}{45}}$ ⓒ $\dfrac{12}{\sqrt{3p}}$

262. ⓐ $\dfrac{4}{\sqrt{5}}$ ⓑ $\sqrt{\dfrac{27}{80}}$ ⓒ $\dfrac{18}{\sqrt{6q}}$

263. ⓐ $\dfrac{1}{\sqrt[3]{5}}$ ⓑ $\sqrt[3]{\dfrac{5}{24}}$ ⓒ $\dfrac{4}{\sqrt[3]{36a}}$

264. ⓐ $\dfrac{1}{\sqrt[3]{3}}$ ⓑ $\sqrt[3]{\dfrac{5}{32}}$ ⓒ $\dfrac{7}{\sqrt[3]{49b}}$

265. ⓐ $\dfrac{1}{\sqrt[3]{11}}$ ⓑ $\sqrt[3]{\dfrac{7}{54}}$ ⓒ $\dfrac{3}{\sqrt[3]{3x^2}}$

266. ⓐ $\dfrac{1}{\sqrt[3]{13}}$ ⓑ $\sqrt[3]{\dfrac{3}{128}}$ ⓒ $\dfrac{3}{\sqrt[3]{6y^2}}$

267. ⓐ $\dfrac{1}{\sqrt[4]{7}}$ ⓑ $\sqrt[4]{\dfrac{5}{32}}$ ⓒ $\dfrac{4}{\sqrt[4]{4x^2}}$

268. ⓐ $\dfrac{1}{\sqrt[4]{4}}$ ⓑ $\sqrt[4]{\dfrac{9}{32}}$ ⓒ $\dfrac{6}{\sqrt[4]{9x^3}}$

269. ⓐ $\dfrac{1}{\sqrt[4]{9}}$ ⓑ $\sqrt[4]{\dfrac{25}{128}}$ ⓒ $\dfrac{6}{\sqrt[4]{27a}}$

270. ⓐ $\dfrac{1}{\sqrt[4]{8}}$ ⓑ $\sqrt[4]{\dfrac{27}{128}}$ ⓒ $\dfrac{16}{\sqrt[4]{64b^2}}$

Rationalize a Two Term Denominator

In the following exercises, simplify.

271. $\dfrac{8}{1-\sqrt{5}}$

272. $\dfrac{7}{2-\sqrt{6}}$

273. $\dfrac{6}{3-\sqrt{7}}$

274. $\dfrac{5}{4-\sqrt{11}}$

275. $\dfrac{\sqrt{3}}{\sqrt{m}-\sqrt{5}}$

276. $\dfrac{\sqrt{5}}{\sqrt{n}-\sqrt{7}}$

277. $\dfrac{\sqrt{2}}{\sqrt{x}-\sqrt{6}}$

278. $\dfrac{\sqrt{7}}{\sqrt{y}+\sqrt{3}}$

279. $\dfrac{\sqrt{r}+\sqrt{5}}{\sqrt{r}-\sqrt{5}}$

280. $\dfrac{\sqrt{s}-\sqrt{6}}{\sqrt{s}+\sqrt{6}}$

281. $\dfrac{\sqrt{x}+\sqrt{8}}{\sqrt{x}-\sqrt{8}}$

282. $\dfrac{\sqrt{m}-\sqrt{3}}{\sqrt{m}+\sqrt{3}}$

Writing Exercises

283.

ⓐ Simplify $\sqrt{\dfrac{27}{3}}$ and explain all your steps.

ⓑ Simplify $\sqrt{\dfrac{27}{5}}$ and explain all your steps.

ⓒ Why are the two methods of simplifying square roots different?

284. Explain what is meant by the word rationalize in the phrase, "rationalize a denominator."

285. Explain why multiplying $\sqrt{2x}-3$ by its conjugate results in an expression with no radicals.

286. Explain why multiplying $\dfrac{7}{\sqrt[3]{x}}$ by $\dfrac{\sqrt[3]{x}}{\sqrt[3]{x}}$ does not rationalize the denominator.

Self Check

ⓐ After completing the exercises, use this checklist to evaluate your mastery of the objectives of this section.

I can...	Confidently	With some help	No-I don't get it!
divide radical expressions.			
rationalize a one-term denominator.			
rationalize a two-term denominator.			

ⓑ After looking at the checklist, do you think you are well-prepared for the next section? Why or why not?

Solve Radical Equations

Learning Objectives

By the end of this section, you will be able to:

› Solve radical equations
› Solve radical equations with two radicals
› Use radicals in applications

Be Prepared!

Before you get started, take this readiness quiz.

1. Simplify: $(y - 3)^2$.
   If you missed this problem, review **Example 5.31**.

2. Solve: $2x - 5 = 0$.
   If you missed this problem, review **Example 2.2**.

3. Solve $n^2 - 6n + 8 = 0$.
   If you missed this problem, review **Example 6.45**.

Solve Radical Equations

In this section we will solve equations that have a variable in the radicand of a radical expression. An equation of this type is called a **radical equation**.

Radical Equation

An equation in which a variable is in the radicand of a radical expression is called a **radical equation**.

As usual, when solving these equations, what we do to one side of an equation we must do to the other side as well. Once we isolate the radical, our strategy will be to raise both sides of the equation to the power of the index. This will eliminate the radical.

Solving radical equations containing an even index by raising both sides to the power of the index may introduce an algebraic solution that would not be a solution to the original radical equation. Again, we call this an extraneous solution as we did when we solved rational equations.

In the next example, we will see how to solve a radical equation. Our strategy is based on raising a radical with index n to the nth power. This will eliminate the radical.

$$\text{For } a \geq 0, \ \left(\sqrt[n]{a}\right)^n = a.$$

EXAMPLE 8.56   HOW TO SOLVE A RADICAL EQUATION

Solve: $\sqrt{5n - 4} - 9 = 0$.

 Solution

Step 1. Isolate the radical on one side of the equation.	To isolate the radical, add 9 to both sides. Simplify.	$\sqrt{5n - 4} - 9 = 0$ $\sqrt{5n - 4} - 9 + 9 = 0 + 9$ $\sqrt{5n - 4} = 9$
Step 2. Raise both sides of the equation to the power of the index.	Since the index of a square root is 2, we square both sides.	$\left(\sqrt{5n - 4}\right)^2 = (9)^2$
Step 3. Solve the new equation.	Remember, $\left(\sqrt{a}\right)^2 = a$.	$5n - 4 = 81$ $5n = 85$ $n = 17$

Step 4. Check the answer in the original equation.		Check the answer.
		$\sqrt{5n-4}-9=0$
		$\sqrt{5(17)-4}-9\overset{?}{=}0$
		$\sqrt{85-4}-9\overset{?}{=}0$
		$\sqrt{81}-9\overset{?}{=}0$
		$9-9\overset{?}{=}0$
		$0=0\checkmark$
		The solution is $n=17$.

> **TRY IT : : 8.111**   Solve: $\sqrt{3m+2}-5=0$.

> **TRY IT : : 8.112**   Solve: $\sqrt{10z+1}-2=0$.

 HOW TO : : SOLVE A RADICAL EQUATION WITH ONE RADICAL.

Step 1.   Isolate the radical on one side of the equation.

Step 2.   Raise both sides of the equation to the power of the index.

Step 3.   Solve the new equation.

Step 4.   Check the answer in the original equation.

When we use a radical sign, it indicates the principal or positive root. If an equation has a radical with an even index equal to a negative number, that equation will have no solution.

EXAMPLE 8.57

Solve: $\sqrt{9k-2}+1=0$.

⊘ Solution

$$\sqrt{9k-2}+1=0$$

To isolate the radical, subtract 1 to both sides.	$\sqrt{9k-2}+1-1=0-1$
Simplify.	$\sqrt{9k-2}=-1$

Because the square root is equal to a negative number, the equation has no solution.

> **TRY IT : : 8.113**   Solve: $\sqrt{2r-3}+5=0$.

> **TRY IT : : 8.114**   Solve: $\sqrt{7s-3}+2=0$.

If one side of an equation with a square root is a binomial, we use the Product of Binomial Squares Pattern when we square it.

Binomial Squares

$$(a+b)^2 = a^2 + 2ab + b^2$$
$$(a-b)^2 = a^2 - 2ab + b^2$$

Don't forget the middle term!

EXAMPLE 8.58

Solve: $\sqrt{p-1} + 1 = p$.

Solution

$$\sqrt{p-1} + 1 = p$$

To isolate the radical, subtract 1 from both sides.	$\sqrt{p-1} + 1 - 1 = p - 1$
Simplify.	$\sqrt{p-1} = p - 1$
Square both sides of the equation.	$\left(\sqrt{p-1}\right)^2 = (p-1)^2$
Simplify, using the Product of Binomial Squares Pattern on the right. Then solve the new equation.	$p - 1 = p^2 - 2p + 1$
It is a quadratic equation, so get zero on one side.	$0 = p^2 - 3p + 2$
Factor the right side.	$0 = (p-1)(p-2)$
Use the Zero Product Property.	$0 = p - 1 \quad 0 = p - 2$
Solve each equation.	$p = 1 \quad p = 2$
Check the answers.	

$p = 1 \quad \sqrt{p-1} + 1 = p$
$\quad\quad \sqrt{1-1} + 1 \stackrel{?}{=} 1$
$\quad\quad \sqrt{0} + 1 \stackrel{?}{=} 1$
$\quad\quad 1 = 1 ✓$

$p = 2 \quad \sqrt{p-1} + 1 = p$
$\quad\quad \sqrt{2-1} + 1 \stackrel{?}{=} 2$
$\quad\quad \sqrt{1} + 1 \stackrel{?}{=} 2$
$\quad\quad 2 = 2 ✓$

The solutions are $p = 1, \quad p = 2$.

> **TRY IT :: 8.115** Solve: $\sqrt{x-2} + 2 = x$.

> **TRY IT :: 8.116** Solve: $\sqrt{y-5} + 5 = y$.

When the index of the radical is 3, we cube both sides to remove the radical.

$$\left(\sqrt[3]{a}\right)^3 = a$$

EXAMPLE 8.59

Solve: $\sqrt[3]{5x+1} + 8 = 4$.

⊘ Solution

$$\sqrt[3]{5x + 1} + 8 = 4$$

To isolate the radical, subtract 8 from both sides.	$\sqrt[3]{5x + 1} = -4$
Cube both sides of the equation.	$\left(\sqrt[3]{5x + 1}\right)^3 = (-4)^3$
Simplify.	$5x + 1 = -64$
Solve the equation.	$5x = -65$
	$x = -13$

Check the answer.

$x = -13 \quad \sqrt[3]{5x + 1} + 8 = 4$

$\sqrt[3]{5(-13) + 1} + 8 \overset{?}{=} 4$

$\sqrt[3]{-64} + 8 \overset{?}{=} 4$

$-4 + 8 \overset{?}{=} 4$

$4 = 4 \checkmark$

The solution is $x = -13$.

> **TRY IT :: 8.117**   Solve: $\sqrt[3]{4x - 3} + 8 = 5$

> **TRY IT :: 8.118**   Solve: $\sqrt[3]{6x - 10} + 1 = -3$

Sometimes an equation will contain rational exponents instead of a radical. We use the same techniques to solve the equation as when we have a radical. We raise each side of the equation to the power of the denominator of the rational exponent. Since $(a^m)^n = a^{m \cdot n}$, we have for example,

$$\left(x^{\frac{1}{2}}\right)^2 = x, \quad \left(x^{\frac{1}{3}}\right)^3 = x$$

Remember, $x^{\frac{1}{2}} = \sqrt{x}$ and $x^{\frac{1}{3}} = \sqrt[3]{x}$.

EXAMPLE 8.60

Solve: $(3x - 2)^{\frac{1}{4}} + 3 = 5$.

⊘ Solution

$$(3x - 2)^{\frac{1}{4}} + 3 = 5$$

To isolate the term with the rational exponent, subtract 3 from both sides.	$(3x-2)^{\frac{1}{4}} = 2$
Raise each side of the equation to the fourth power.	$\left((3x-2)^{\frac{1}{4}}\right)^4 = (2)^4$
Simplify.	$3x - 2 = 16$
Solve the equation.	$3x = 18$
	$x = 6$

Check the answer.

$$x = 6 \qquad (3x-2)^{\frac{1}{4}} + 3 = 5$$
$$(3 \cdot 6 - 2)^{\frac{1}{4}} + 3 \overset{?}{=} 5$$
$$(16)^{\frac{1}{4}} + 3 \overset{?}{=} 5$$
$$2 + 3 \overset{?}{=} 5$$
$$5 = 5 \checkmark$$

| | The solution is $x = 6$. |

> **TRY IT : : 8.119**
>
> Solve: $(9x+9)^{\frac{1}{4}} - 2 = 1$.

> **TRY IT : : 8.120**
>
> Solve: $(4x-8)^{\frac{1}{4}} + 5 = 7$.

Sometimes the solution of a radical equation results in two algebraic solutions, but one of them may be an extraneous solution!

EXAMPLE 8.61

Solve: $\sqrt{r+4} - r + 2 = 0$.

⊘ **Solution**

	$\sqrt{r+4} - r + 2 = 0$
Isolate the radical.	$\sqrt{r+4} = r - 2$
Square both sides of the equation.	$\left(\sqrt{r+4}\right)^2 = (r-2)^2$
Simplify and then solve the equation	$r + 4 = r^2 - 4r + 4$
It is a quadratic equation, so get zero on one side.	$0 = r^2 - 5r$
Factor the right side.	$0 = r(r-5)$
Use the Zero Product Property.	$0 = r \qquad 0 = r - 5$

Solve the equation.

$$r = 0 \quad r = 5$$

Check your answer.

$r = 0, \quad \sqrt{r+4} - r + 2 = 0$

$\sqrt{0+4} - 0 + 2 \overset{?}{=} 0$

$\sqrt{4} + 2 \overset{?}{=} 0$

$4 \neq 0$

$r = 5, \quad \sqrt{r+4} - r + 2 = 0$

$\sqrt{5+4} - 5 + 2 \overset{?}{=} 0$

$\sqrt{9} - 3 \overset{?}{=} 0$

$0 = 0 \checkmark$ The solution is $r = 5$.

$r = 0$ is an extraneous solution.

> **TRY IT :: 8.121**   Solve: $\sqrt{m+9} - m + 3 = 0$.

> **TRY IT :: 8.122**   Solve: $\sqrt{n+1} - n + 1 = 0$.

When there is a coefficient in front of the radical, we must raise it to the power of the index, too.

EXAMPLE 8.62

Solve: $3\sqrt{3x-5} - 8 = 4$.

⊘ **Solution**

$$3\sqrt{3x-5} - 8 = 4$$

Isolate the radical term.	$3\sqrt{3x-5} = 12$
Isolate the radical by dividing both sides by 3.	$\sqrt{3x-5} = 4$
Square both sides of the equation.	$\left(\sqrt{3x-5}\right)^2 = (4)^2$
Simplify, then solve the new equation.	$3x - 5 = 16$
	$3x = 21$
Solve the equation.	$x = 7$

Check the answer.

$x = 7$ $3\sqrt{3x-5} - 8 = 4$

$3\sqrt{3(7)-5} - 8 \overset{?}{=} 4$

$3\sqrt{21-5} - 8 \overset{?}{=} 4$

$3\sqrt{16} - 8 \overset{?}{=} 4$

$3(4) - 8 \overset{?}{=} 4$

$4 = 4 \checkmark$

The solution is $x = 7$.

> **TRY IT :: 8.123**   Solve: $2\sqrt{4a+4} - 16 = 16$.

> **TRY IT :: 8.124**   Solve: $3\sqrt{2b+3} - 25 = 50$.

Solve Radical Equations with Two Radicals

If the radical equation has two radicals, we start out by isolating one of them. It often works out easiest to isolate the more complicated radical first.

In the next example, when one radical is isolated, the second radical is also isolated.

EXAMPLE 8.63

Solve: $\sqrt[3]{4x-3} = \sqrt[3]{3x+2}$.

Solution

The radical terms are isolated.	$\sqrt[3]{4x-3} = \sqrt[3]{3x+2}$
Since the index is 3, cube both sides of the equation.	$\left(\sqrt[3]{4x-3}\right)^3 = \left(\sqrt[3]{3x+2}\right)^3$
Simplify, then solve the new equation.	$4x-3 = 3x+2$
	$x-3 = 2$
	$x = 5$

The solution is $x = 5$.

Check the answer.
We leave it to you to show that 5 checks!

> **TRY IT :: 8.125**   Solve: $\sqrt[3]{5x-4} = \sqrt[3]{2x+5}$.

> **TRY IT :: 8.126**   Solve: $\sqrt[3]{7x+1} = \sqrt[3]{2x-5}$.

Sometimes after raising both sides of an equation to a power, we still have a variable inside a radical. When that happens, we repeat Step 1 and Step 2 of our procedure. We isolate the radical and raise both sides of the equation to the power of the index again.

EXAMPLE 8.64   HOW TO SOLVE A RADICAL EQUATION

Solve: $\sqrt{m} + 1 = \sqrt{m+9}$.

Solution

Step 1. Isolate one of the radical terms on one side of the equation.	The radical on the right is isolated.	$\sqrt{m} + 1 = \sqrt{m+9}$
Step 2. Raise both sides of the equation to the power of the index.	We square both sides. Simplify—be very careful as you multiply!	$\left(\sqrt{m}+1\right)^2 = \left(\sqrt{m+9}\right)^2$
Step 3. Are there any more radicals? If yes, repeat Step 1 and Step 2 again. If no, solve the new equation.	There is still a radical in the equation. So we must repeat the previous steps. Isolate the radical term. Here, we can easily isolate the radical by dividing both sides by 2. Square both sides.	$m + 2\sqrt{m} + 1 = m+9$ $2\sqrt{m} = 8$ $\sqrt{m} = 4$ $\left(\sqrt{m}\right)^2 = (4)^2$ $m = 16$

Step 4. Check the answer in the original equation.		$\sqrt{m}+1=\sqrt{m+9}$
		$\sqrt{16}+1\stackrel{?}{=}\sqrt{16+9}$
		$4+1\stackrel{?}{=}5$
		$5=5\checkmark$
		The solution is $m=16$.

> **TRY IT ::** 8.127   Solve: $3-\sqrt{x}=\sqrt{x-3}$.

> **TRY IT ::** 8.128   Solve: $\sqrt{x}+2=\sqrt{x+16}$.

We summarize the steps here. We have adjusted our previous steps to include more than one radical in the equation This procedure will now work for any radical equations.

HOW TO :: SOLVE A RADICAL EQUATION.

Step 1.   Isolate one of the radical terms on one side of the equation.

Step 2.   Raise both sides of the equation to the power of the index.

Step 3.   Are there any more radicals?
If yes, repeat Step 1 and Step 2 again.
If no, solve the new equation.

Step 4.   Check the answer in the original equation.

Be careful as you square binomials in the next example. Remember the pattern is $(a+b)^2=a^2+2ab+b^2$ or $(a-b)^2=a^2-2ab+b^2$.

EXAMPLE 8.65

Solve: $\sqrt{q-2}+3=\sqrt{4q+1}$.

⊘ **Solution**

$$\sqrt{q-2}+3=\sqrt{4q+1}$$

The radical on the right is isolated. Square both sides.	$\left(\sqrt{q-2}+3\right)^2=\left(\sqrt{4q+1}\right)^2$
Simplify.	$q-2+6\sqrt{q-2}+9=4q+1$
There is still a radical in the equation so we must repeat the previous steps. Isolate the radical.	$6\sqrt{q-2}=3q-6$
Square both sides. It would not help to divide both sides by 6. Remember to square both the 6 and the $\sqrt{q-2}$.	$\left(6\sqrt{q-2}\right)^2=\left(3q-6\right)^2$ $6^2\left(\sqrt{q-2}\right)^2=(3q)^2-2\cdot 3q\cdot 6+6^2$
Simplify, then solve the new equation.	$36(q-2)=9q^2-36q+36$

Distribute.	$36q - 72 = 9q^2 - 36q + 36$
It is a quadratic equation, so get zero on one side.	$0 = 9q^2 - 72q + 108$
Factor the right side.	$0 = 9(q^2 - 8q + 12)$ $0 = 9(q - 6)(q - 2)$
Use the Zero Product Property.	$q - 6 = 0 \qquad q - 2 = 0$ $q = 6 \qquad\quad q = 2$
The checks are left to you.	The solutions are $q = 6$ and $q = 2$.

> **TRY IT :: 8.129**   Solve: $\sqrt{x - 1} + 2 = \sqrt{2x + 6}$

> **TRY IT :: 8.130**   Solve: $\sqrt{x} + 2 = \sqrt{3x + 4}$

Use Radicals in Applications

As you progress through your college courses, you'll encounter formulas that include radicals in many disciplines. We will modify our Problem Solving Strategy for Geometry Applications slightly to give us a plan for solving applications with formulas from any discipline.

HOW TO :: USE A PROBLEM SOLVING STRATEGY FOR APPLICATIONS WITH FORMULAS.

Step 1.   **Read** the problem and make sure all the words and ideas are understood. When appropriate, draw a figure and label it with the given information.

Step 2.   **Identify** what we are looking for.

Step 3.   **Name** what we are looking for by choosing a variable to represent it.

Step 4.   **Translate** into an equation by writing the appropriate formula or model for the situation. Substitute in the given information.

Step 5.   **Solve the equation** using good algebra techniques.

Step 6.   **Check** the answer in the problem and make sure it makes sense.

Step 7.   **Answer** the question with a complete sentence.

One application of radicals has to do with the effect of gravity on falling objects. The formula allows us to determine how long it will take a fallen object to hit the gound.

Falling Objects

On Earth, if an object is dropped from a height of h feet, the time in seconds it will take to reach the ground is found by using the formula

$$t = \frac{\sqrt{h}}{4}.$$

For example, if an object is dropped from a height of 64 feet, we can find the time it takes to reach the ground by substituting $h = 64$ into the formula.

$$t = \frac{\sqrt{h}}{4}$$

$$t = \frac{\sqrt{64}}{4}$$

Take the square root of 64.

$$t = \frac{8}{4}$$

Simplify the fraction.

$$t = 2$$

It would take 2 seconds for an object dropped from a height of 64 feet to reach the ground.

EXAMPLE 8.66

Marissa dropped her sunglasses from a bridge 400 feet above a river. Use the formula $t = \frac{\sqrt{h}}{4}$ to find how many seconds it took for the sunglasses to reach the river.

✓ **Solution**

Step 1. Read the problem.

Step 2. Identify what we are looking for.	the time it takes for the sunglasses to reach the river
Step 3. Name what we are looking.	Let $t =$ time.
Step 4. Translate into an equation by writing the appropriate formula. Substitute in the given information.	$t = \frac{\sqrt{h}}{4}$, and $h = 400$ $t = \frac{\sqrt{400}}{4}$
Step 5. Solve the equation.	$t = \frac{20}{4}$ $t = 5$
Step 6. Check the answer in the problem and make sure it makes sense.	$5 \stackrel{?}{=} \frac{\sqrt{400}}{4}$ $5 \stackrel{?}{=} \frac{20}{4}$ $5 = 5 \checkmark$
Does 5 seconds seem like a reasonable length of time?	Yes.
Step 7. Answer the question.	It will take 5 seconds for the sunglasses to reach the river.

> **TRY IT : :** 8.131

A helicopter dropped a rescue package from a height of 1,296 feet. Use the formula $t = \frac{\sqrt{h}}{4}$ to find how many seconds it took for the package to reach the ground.

> **TRY IT : :** 8.132

A window washer dropped a squeegee from a platform 196 feet above the sidewalk Use the formula $t = \frac{\sqrt{h}}{4}$ to find how many seconds it took for the squeegee to reach the sidewalk.

Police officers investigating car accidents measure the length of the skid marks on the pavement. Then they use square roots to determine the speed, in miles per hour, a car was going before applying the brakes.

Skid Marks and Speed of a Car

If the length of the skid marks is *d* feet, then the speed, *s*, of the car before the brakes were applied can be found by using the formula

$$s = \sqrt{24d}$$

EXAMPLE 8.67

After a car accident, the skid marks for one car measured 190 feet. Use the formula $s = \sqrt{24d}$ to find the speed of the car before the brakes were applied. Round your answer to the nearest tenth.

 **Solution**

Step 1. Read the problem	
Step 2. Identify what we are looking for.	the speed of a car
Step 3. Name what we are looking for,	Let $s =$ the speed.
Step 4. Translate into an equation by writing the appropriate formula. Substitute in the given information.	$s = \sqrt{24d}$, and $d = 190$ $s = \sqrt{24(190)}$
Step 5. Solve the equation.	$s = \sqrt{4,560}$
	$s = 67.52777...$
Round to 1 decimal place.	$s \approx 67.5$
	$67.5 \overset{?}{\approx} \sqrt{24(190)}$ $67.5 \overset{?}{\approx} \sqrt{4560}$ $67.5 \approx 67.5277... \checkmark$
	The speed of the car before the brakes were applied was 67.5 miles per hour.

> **TRY IT : :** 8.133

An accident investigator measured the skid marks of the car. The length of the skid marks was 76 feet. Use the formula $s = \sqrt{24d}$ to find the speed of the car before the brakes were applied. Round your answer to the nearest tenth.

> **TRY IT : :** 8.134

The skid marks of a vehicle involved in an accident were 122 feet long. Use the formula $s = \sqrt{24d}$ to find the speed of the vehicle before the brakes were applied. Round your answer to the nearest tenth.

▶ | **MEDIA : :**

Access these online resources for additional instruction and practice with solving radical equations.

- **Solving an Equation Involving a Single Radical (https://openstax.org/l/37RadEquat1)**
- **Solving Equations with Radicals and Rational Exponents (https://openstax.org/l/37RadEquat2)**
- **Solving Radical Equations (https://openstax.org/l/37RadEquat3)**
- **Solve Radical Equations (https://openstax.org/l/37RadEquat4)**
- **Radical Equation Application (https://openstax.org/l/37RadEquat5)**

8.6 EXERCISES

Practice Makes Perfect

Solve Radical Equations

In the following exercises, solve.

287. $\sqrt{5x - 6} = 8$

288. $\sqrt{4x - 3} = 7$

289. $\sqrt{5x + 1} = -3$

290. $\sqrt{3y - 4} = -2$

291. $\sqrt[3]{2x} = -2$

292. $\sqrt[3]{4x - 1} = 3$

293. $\sqrt{2m - 3} - 5 = 0$

294. $\sqrt{2n - 1} - 3 = 0$

295. $\sqrt{6v - 2} - 10 = 0$

296. $\sqrt{12u + 1} - 11 = 0$

297. $\sqrt{4m + 2} + 2 = 6$

298. $\sqrt{6n + 1} + 4 = 8$

299. $\sqrt{2u - 3} + 2 = 0$

300. $\sqrt{5v - 2} + 5 = 0$

301. $\sqrt{u - 3} - 3 = u$

302. $\sqrt{v - 10} + 10 = v$

303. $\sqrt{r - 1} = r - 1$

304. $\sqrt{s - 8} = s - 8$

305. $\sqrt[3]{6x + 4} = 4$

306. $\sqrt[3]{11x + 4} = 5$

307. $\sqrt[3]{4x + 5} - 2 = -5$

308. $\sqrt[3]{9x - 1} - 1 = -5$

309. $(6x + 1)^{\frac{1}{2}} - 3 = 4$

310. $(3x - 2)^{\frac{1}{2}} + 1 = 6$

311. $(8x + 5)^{\frac{1}{3}} + 2 = -1$

312. $(12x - 5)^{\frac{1}{3}} + 8 = 3$

313. $(12x - 3)^{\frac{1}{4}} - 5 = -2$

314. $(5x - 4)^{\frac{1}{4}} + 7 = 9$

315. $\sqrt{x + 1} - x + 1 = 0$

316. $\sqrt{y + 4} - y + 2 = 0$

317. $\sqrt{z + 100} - z = -10$

318. $\sqrt{w + 25} - w = -5$

319. $3\sqrt{2x - 3} - 20 = 7$

320. $2\sqrt{5x + 1} - 8 = 0$

321. $2\sqrt{8r + 1} - 8 = 2$

322. $3\sqrt{7y + 1} - 10 = 8$

Solve Radical Equations with Two Radicals

In the following exercises, solve.

323. $\sqrt{3u + 7} = \sqrt{5u + 1}$

324. $\sqrt{4v + 1} = \sqrt{3v + 3}$

325. $\sqrt{8 + 2r} = \sqrt{3r + 10}$

326. $\sqrt{10 + 2c} = \sqrt{4c + 16}$

327. $\sqrt[3]{5x - 1} = \sqrt[3]{x + 3}$

328. $\sqrt[3]{8x - 5} = \sqrt[3]{3x + 5}$

329.
$\sqrt[3]{2x^2 + 9x - 18} = \sqrt[3]{x^2 + 3x - 2}$

330.
$\sqrt[3]{x^2 - x + 18} = \sqrt[3]{2x^2 - 3x - 6}$

331. $\sqrt{a} + 2 = \sqrt{a + 4}$

332. $\sqrt{r} + 6 = \sqrt{r + 8}$

333. $\sqrt{u} + 1 = \sqrt{u + 4}$

334. $\sqrt{x} + 1 = \sqrt{x + 2}$

335. $\sqrt{a + 5} - \sqrt{a} = 1$

336. $-2 = \sqrt{d - 20} - \sqrt{d}$

337. $\sqrt{2x + 1} = 1 + \sqrt{x}$

338. $\sqrt{3x + 1} = 1 + \sqrt{2x - 1}$

339. $\sqrt{2x - 1} - \sqrt{x - 1} = 1$

340. $\sqrt{x + 1} - \sqrt{x - 2} = 1$

341. $\sqrt{x+7} - \sqrt{x-5} = 2$ **342.** $\sqrt{x+5} - \sqrt{x-3} = 2$

Use Radicals in Applications

In the following exercises, solve. Round approximations to one decimal place.

343. Landscaping Reed wants to have a square garden plot in his backyard. He has enough compost to cover an area of 75 square feet. Use the formula $s = \sqrt{A}$ to find the length of each side of his garden. Round your answer to the nearest tenth of a foot.

344. Landscaping Vince wants to make a square patio in his yard. He has enough concrete to pave an area of 130 square feet. Use the formula $s = \sqrt{A}$ to find the length of each side of his patio. Round your answer to the nearest tenth of a foot.

345. Gravity A hang glider dropped his cell phone from a height of 350 feet. Use the formula $t = \dfrac{\sqrt{h}}{4}$ to find how many seconds it took for the cell phone to reach the ground.

346. Gravity A construction worker dropped a hammer while building the Grand Canyon skywalk, 4000 feet above the Colorado River. Use the formula $t = \dfrac{\sqrt{h}}{4}$ to find how many seconds it took for the hammer to reach the river.

347. Accident investigation The skid marks for a car involved in an accident measured 216 feet. Use the formula $s = \sqrt{24d}$ to find the speed of the car before the brakes were applied. Round your answer to the nearest tenth.

348. Accident investigation An accident investigator measured the skid marks of one of the vehicles involved in an accident. The length of the skid marks was 175 feet. Use the formula $s = \sqrt{24d}$ to find the speed of the vehicle before the brakes were applied. Round your answer to the nearest tenth.

Writing Exercises

349. Explain why an equation of the form $\sqrt{x} + 1 = 0$ has no solution.

350.

ⓐ Solve the equation $\sqrt{r+4} - r + 2 = 0$.

ⓑ Explain why one of the "solutions" that was found was not actually a solution to the equation.

Self Check

ⓐ *After completing the exercises, use this checklist to evaluate your mastery of the objectives of this section.*

I can...	Confidently	With some help	No-I don't get it!
solve radical equations.			
solve radical equations with two radicals.			
use radicals in applications.			

ⓑ *After reviewing this checklist, what will you do to become confident for all objectives?*

8.7 Use Radicals in Functions

Learning Objectives

By the end of this section, you will be able to:

> Evaluate a radical function
> Find the domain of a radical function
> Graph radical functions

Be Prepared!

Before you get started, take this readiness quiz.

1. Solve: $1 - 2x \geq 0$.
   If you missed this problem, review **Example 2.50**.
2. For $f(x) = 3x - 4$, evaluate $f(2)$, $f(-1)$, $f(0)$.
   If you missed this problem, review **Example 3.48**.
3. Graph $f(x) = \sqrt{x}$. State the domain and range of the function in interval notation.
   If you missed this problem, review **Example 3.56**.

Evaluate a Radical Function

In this section we will extend our previous work with functions to include radicals. If a function is defined by a radical expression, we call it a **radical function**.

The square root function is $f(x) = \sqrt{x}$.

The cube root function is $f(x) = \sqrt[3]{x}$.

Radical Function

A **radical function** is a function that is defined by a radical expression.

To evaluate a radical function, we find the value of $f(x)$ for a given value of x just as we did in our previous work with functions.

EXAMPLE 8.68

For the function $f(x) = \sqrt{2x - 1}$, find ⓐ $f(5)$ ⓑ $f(-2)$.

 Solution

ⓐ

$$f(x) = \sqrt{2x - 1}$$

To evaluate $f(5)$, substitute 5 for x. $\qquad f(5) = \sqrt{2 \cdot 5 - 1}$

Simplify. $\qquad\qquad\qquad\qquad\qquad\qquad f(5) = \sqrt{9}$

Take the square root. $\qquad\qquad\qquad\quad f(5) = 3$

ⓑ

$$f(x) = \sqrt{2x - 1}$$

To evaluate $f(-2)$, substitute -2 for x. $\quad f(-2) = \sqrt{2(-2) - 1}$

Simplify. $\qquad\qquad\qquad\qquad\qquad\qquad f(-2) = \sqrt{-5}$

Since the square root of a negative number is not a real number, the function does not have a value at $x = -2$.

> **TRY IT :: 8.135**   For the function $f(x) = \sqrt{3x - 2}$, find ⓐ $f(6)$ ⓑ $f(0)$.

> **TRY IT ::** 8.136   For the function $g(x) = \sqrt{5x + 5}$, find ⓐ $g(4)$ ⓑ $g(-3)$.

We follow the same procedure to evaluate cube roots.

EXAMPLE 8.69

For the function $g(x) = \sqrt[3]{x - 6}$, find ⓐ $g(14)$ ⓑ $g(-2)$.

✓ **Solution**

ⓐ

$$g(x) = \sqrt[3]{x - 6}$$

To evaluate $g(14)$, substitute 14 for x. $g(14) = \sqrt[3]{14 - 6}$

Simplify. $g(14) = \sqrt[3]{8}$

Take the cube root. $g(14) = 2$

ⓑ

$$g(x) = \sqrt[3]{x - 6}$$

To evaluate $g(-2)$, substitute -2 for x. $g(-2) = \sqrt[3]{-2 - 6}$

Simplify. $g(-2) = \sqrt[3]{-8}$

Take the cube root. $g(-2) = -2$

> **TRY IT ::** 8.137   For the function $g(x) = \sqrt[3]{3x - 4}$, find ⓐ $g(4)$ ⓑ $g(1)$.

> **TRY IT ::** 8.138   For the function $h(x) = \sqrt[3]{5x - 2}$, find ⓐ $h(2)$ ⓑ $h(-5)$.

The next example has fourth roots.

EXAMPLE 8.70

For the function $f(x) = \sqrt[4]{5x - 4}$, find ⓐ $f(4)$ ⓑ $f(-12)$

✓ **Solution**

ⓐ

$$f(x) = \sqrt[4]{5x - 4}$$

To evaluate $f(4)$, substitute 4 for x. $f(4) = \sqrt[4]{5 \cdot 4 - 4}$

Simplify. $f(4) = \sqrt[4]{16}$

Take the fourth root. $f(4) = 2$

ⓑ

$$f(x) = \sqrt[4]{5x - 4}$$

To evaluate $f(-12)$, substitute -12 for x. $f(-12) = \sqrt[4]{5(-12) - 4}$

Simplify. $f(-12) = \sqrt[4]{-64}$

Since the fourth root of a negative number is not a real number, the function does not have a value at $x = -12$.

> **TRY IT :: 8.139**   For the function $f(x) = \sqrt[4]{3x + 4}$, find ⓐ $f(4)$ ⓑ $f(-1)$.

> **TRY IT :: 8.140**   For the function $g(x) = \sqrt[4]{5x + 1}$, find ⓐ $g(16)$ ⓑ $g(3)$.

Find the Domain of a Radical Function

To find the domain and range of radical functions, we use our properties of radicals. For a radical with an even index, we said the radicand had to be greater than or equal to zero as even roots of negative numbers are not real numbers. For an odd index, the radicand can be any real number. We restate the properties here for reference.

Properties of $\sqrt[n]{a}$

When n is an **even** number and:

- $a \geq 0$, then $\sqrt[n]{a}$ is a real number.

- $a < 0$, then $\sqrt[n]{a}$ is not a real number.

When n is an **odd** number, $\sqrt[n]{a}$ is a real number for all values of a.

So, to find the domain of a radical function with even index, we set the radicand to be greater than or equal to zero. For an odd index radical, the radicand can be any real number.

Domain of a Radical Function

When the **index** of the radical is **even**, the radicand must be greater than or equal to zero.

When the **index** of the radical is **odd**, the radicand can be any real number.

EXAMPLE 8.71

Find the domain of the function, $f(x) = \sqrt{3x - 4}$. Write the domain in interval notation.

⊘ Solution

Since the function, $f(x) = \sqrt{3x - 4}$ has a radical with an index of 2, which is even, we know the radicand must be greater than or equal to 0. We set the radicand to be greater than or equal to 0 and then solve to find the domain.

$$3x - 4 \geq 0$$

Solve.

$$3x \geq 4$$

$$x \geq \frac{4}{3}$$

The domain of $f(x) = \sqrt{3x - 4}$ is all values $x \geq \frac{4}{3}$ and we write it in interval notation as $\left[\frac{4}{3}, \infty\right)$.

> **TRY IT :: 8.141**   Find the domain of the function, $f(x) = \sqrt{6x - 5}$. Write the domain in interval notation.

> **TRY IT :: 8.142**   Find the domain of the function, $f(x) = \sqrt{4 - 5x}$. Write the domain in interval notation.

EXAMPLE 8.72

Find the domain of the function, $g(x) = \sqrt{\dfrac{6}{x - 1}}$. Write the domain in interval notation.

⊘ Solution

Since the function, $g(x) = \sqrt{\dfrac{6}{x - 1}}$ has a radical with an index of 2, which is even, we know the radicand must be greater than or equal to 0.

The radicand cannot be zero since the numerator is not zero.

For $\dfrac{6}{x-1}$ to be greater than zero, the denominator must be positive since the numerator is positive. We know a positive divided by a positive is positive.

We set $x - 1 > 0$ and solve.

$$x - 1 \; > \; 0$$

Solve. $\qquad x \; > \; 1$

Also, since the radicand is a fraction, we must realize that the denominator cannot be zero.

We solve $x - 1 = 0$ to find the value that must be eliminated from the domain.

$$x - 1 \; = \; 0$$

Solve. $\qquad x \; = \; 1$ so $x \neq 1$ in the domain.

Putting this together we get the domain is $x > 1$ and we write it as $(1, \infty)$.

> **TRY IT : : 8.143**
>
> Find the domain of the function, $f(x) = \sqrt{\dfrac{4}{x+3}}$. Write the domain in interval notation.

> **TRY IT : : 8.144**
>
> Find the domain of the function, $h(x) = \sqrt{\dfrac{9}{x-5}}$. Write the domain in interval notation.

The next example involves a cube root and so will require different thinking.

EXAMPLE 8.73

Find the domain of the function, $f(x) = \sqrt[3]{2x^2 + 3}$. Write the domain in interval notation.

⊘ **Solution**

Since the function, $f(x) = \sqrt[3]{2x^2 + 3}$ has a radical with an index of 3, which is odd, we know the radicand can be any real number. This tells us the domain is any real number. In interval notation, we write $(-\infty, \infty)$.

The domain of $f(x) = \sqrt[3]{2x^2 + 3}$ is all real numbers and we write it in interval notation as $(-\infty, \infty)$.

> **TRY IT : : 8.145**
>
> Find the domain of the function, $f(x) = \sqrt[3]{3x^2 - 1}$. Write the domain in interval notation.

> **TRY IT : : 8.146**
>
> Find the domain of the function, $g(x) = \sqrt[3]{5x - 4}$. Write the domain in interval notation.

Graph Radical Functions

Before we graph any radical function, we first find the domain of the function. For the function, $f(x) = \sqrt{x}$, the index is even, and so the radicand must be greater than or equal to 0.

This tells us the domain is $x \geq 0$ and we write this in interval notation as $[0, \infty)$.

Previously we used point plotting to graph the function, $f(x) = \sqrt{x}$. We chose x-values, substituted them in and then created a chart. Notice we chose points that are perfect squares in order to make taking the square root easier.

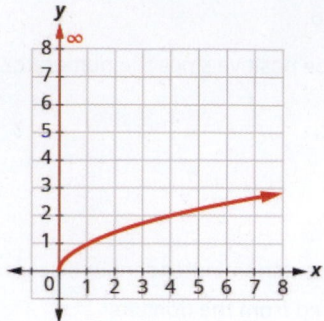

x	$f(x) = \sqrt{x}$	$(x, f(x))$
0	0	(0, 0)
1	1	(1, 1)
4	2	(4, 2)
9	3	(9, 3)

Once we see the graph, we can find the range of the function. The y-values of the function are greater than or equal to zero. The range then is $[0, \infty)$.

EXAMPLE 8.74

For the function $f(x) = \sqrt{x + 3}$,

ⓐ find the domain ⓑ graph the function ⓒ use the graph to determine the range.

✓ Solution

ⓐ Since the radical has index 2, we know the radicand must be greater than or equal to zero. If $x + 3 \geq 0$, then $x \geq -3$. This tells us the domain is all values $x \geq -3$ and written in interval notation as $[-3, \infty)$.

ⓑ To graph the function, we choose points in the interval $[-3, \infty)$ that will also give us a radicand which will be easy to take the square root.

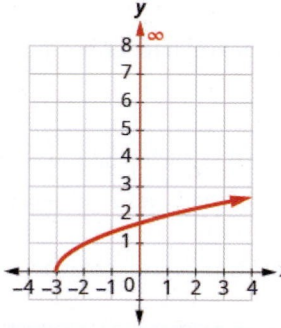

x	$f(x) = \sqrt{x + 3}$	$(x, f(x))$
–3	0	(–3, 0)
–2	1	(–2, 1)
1	2	(1, 2)
6	3	(6, 3)

ⓒ Looking at the graph, we see the y-values of the function are greater than or equal to zero. The range then is $[0, \infty)$.

> **TRY IT ::** 8.147

For the function $f(x) = \sqrt{x + 2}$, ⓐ find the domain ⓑ graph the function ⓒ use the graph to determine the range.

> **TRY IT ::** 8.148

For the function $f(x) = \sqrt{x - 2}$, ⓐ find the domain ⓑ graph the function ⓒ use the graph to determine the range.

In our previous work graphing functions, we graphed $f(x) = x^3$ but we did not graph the function $f(x) = \sqrt[3]{x}$. We will do this now in the next example.

EXAMPLE 8.75

For the function $f(x) = \sqrt[3]{x}$, ⓐ find the domain ⓑ graph the function ⓒ use the graph to determine the range.

✓ **Solution**

ⓐ Since the radical has index 3, we know the radicand can be any real number. This tells us the domain is all real numbers and written in interval notation as $(-\infty, \infty)$

ⓑ To graph the function, we choose points in the interval $(-\infty, \infty)$ that will also give us a radicand which will be easy to take the cube root.

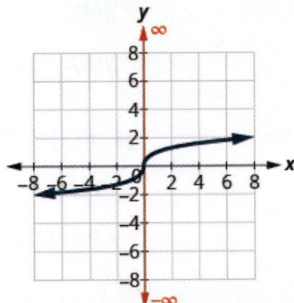

x	$f(x) = \sqrt[3]{x}$	(x, f(x))
−8	−2	(−8, −2)
−1	−1	(−1, −1)
0	0	(0, 0)
1	1	(1, 1)
8	2	(8, 2)

ⓒ Looking at the graph, we see the y-values of the function are all real numbers. The range then is $(-\infty, \infty)$.

> **TRY IT :: 8.149**
>
> For the function $f(x) = -\sqrt[3]{x}$,
>
> ⓐ find the domain ⓑ graph the function ⓒ use the graph to determine the range.

> **TRY IT :: 8.150**
>
> For the function $f(x) = \sqrt[3]{x - 2}$,
>
> ⓐ find the domain ⓑ graph the function ⓒ use the graph to determine the range.

▶ **MEDIA ::**

Access these online resources for additional instruction and practice with radical functions.

- **Domain of a Radical Function (https://openstax.org/l/37RadFuncDom1)**
- **Domain of a Radical Function 2 (https://openstax.org/l/37RadFuncDom2)**
- **Finding Domain of a Radical Function (https://openstax.org/l/37RadFuncDom3)**

8.7 EXERCISES

Practice Makes Perfect

Evaluate a Radical Function

In the following exercises, evaluate each function.

351. $f(x) = \sqrt{4x - 4}$, find

ⓐ $f(5)$

ⓑ $f(0)$.

352. $f(x) = \sqrt{6x - 5}$, find

ⓐ $f(5)$

ⓑ $f(-1)$.

353. $g(x) = \sqrt{6x + 1}$, find

ⓐ $g(4)$

ⓑ $g(8)$.

354. $g(x) = \sqrt{3x + 1}$, find

ⓐ $g(8)$

ⓑ $g(5)$.

355. $F(x) = \sqrt{3 - 2x}$, find

ⓐ $F(1)$

ⓑ $F(-11)$.

356. $F(x) = \sqrt{8 - 4x}$, find

ⓐ $F(1)$

ⓑ $F(-2)$.

357. $G(x) = \sqrt{5x - 1}$, find

ⓐ $G(5)$

ⓑ $G(2)$.

358. $G(x) = \sqrt{4x + 1}$, find

ⓐ $G(11)$

ⓑ $G(2)$.

359. $g(x) = \sqrt[3]{2x - 4}$, find

ⓐ $g(6)$

ⓑ $g(-2)$.

360. $g(x) = \sqrt[3]{7x - 1}$, find

ⓐ $g(4)$

ⓑ $g(-1)$.

361. $h(x) = \sqrt[3]{x^2 - 4}$, find

ⓐ $h(-2)$

ⓑ $h(6)$.

362. $h(x) = \sqrt[3]{x^2 + 4}$, find

ⓐ $h(-2)$

ⓑ $h(6)$.

363. For the function
$f(x) = \sqrt[4]{2x^3}$, find

ⓐ $f(0)$

ⓑ $f(2)$.

364. For the function
$f(x) = \sqrt[4]{3x^3}$, find

ⓐ $f(0)$

ⓑ $f(3)$.

365. For the function
$g(x) = \sqrt[4]{4 - 4x}$, find

ⓐ $g(1)$

ⓑ $g(-3)$.

366. For the function
$g(x) = \sqrt[4]{8 - 4x}$, find

ⓐ $g(-6)$

ⓑ $g(2)$.

Find the Domain of a Radical Function

In the following exercises, find the domain of the function and write the domain in interval notation.

367. $f(x) = \sqrt{3x - 1}$

368. $f(x) = \sqrt{4x - 2}$

369. $g(x) = \sqrt{2 - 3x}$

370. $g(x) = \sqrt{8 - x}$

371. $h(x) = \sqrt{\dfrac{5}{x - 2}}$

372. $h(x) = \sqrt{\dfrac{6}{x + 3}}$

373. $f(x) = \sqrt{\dfrac{x + 3}{x - 2}}$

374. $f(x) = \sqrt{\dfrac{x - 1}{x + 4}}$

375. $g(x) = \sqrt[3]{8x - 1}$

376. $g(x) = \sqrt[3]{6x + 5}$

377. $f(x) = \sqrt[3]{4x^2 - 16}$

378. $f(x) = \sqrt[3]{6x^2 - 25}$

379. $F(x) = \sqrt[4]{8x + 3}$

380. $F(x) = \sqrt[4]{10 - 7x}$

381. $G(x) = \sqrt[5]{2x - 1}$

382. $G(x) = \sqrt[5]{6x - 3}$

Graph Radical Functions

In the following exercises, ⓐ find the domain of the function ⓑ graph the function ⓒ use the graph to determine the range.

383. $f(x) = \sqrt{x + 1}$

384. $f(x) = \sqrt{x - 1}$

385. $g(x) = \sqrt{x + 4}$

386. $g(x) = \sqrt{x - 4}$

387. $f(x) = \sqrt{x} + 2$

388. $f(x) = \sqrt{x} - 2$

389. $g(x) = 2\sqrt{x}$

390. $g(x) = 3\sqrt{x}$

391. $f(x) = \sqrt{3 - x}$

392. $f(x) = \sqrt{4 - x}$

393. $g(x) = -\sqrt{x}$

394. $g(x) = -\sqrt{x} + 1$

395. $f(x) = \sqrt[3]{x + 1}$

396. $f(x) = \sqrt[3]{x - 1}$

397. $g(x) = \sqrt[3]{x + 2}$

398. $g(x) = \sqrt[3]{x - 2}$

399. $f(x) = \sqrt[3]{x} + 3$

400. $f(x) = \sqrt[3]{x} - 3$

401. $g(x) = \sqrt[3]{x}$

402. $g(x) = -\sqrt[3]{x}$

403. $f(x) = 2\sqrt[3]{x}$

404. $f(x) = -2\sqrt[3]{x}$

Writing Exercises

405. Explain how to find the domain of a fourth root function.

406. Explain how to find the domain of a fifth root function.

407. Explain why $y = \sqrt[3]{x}$ is a function.

408. Explain why the process of finding the domain of a radical function with an even index is different from the process when the index is odd.

Self Check

ⓐ *After completing the exercises, use this checklist to evaluate your mastery of the objectives of this section.*

I can...	Confidently	With some help	No-I don't get it!
evaluate a radical function.			
find the domain of a radical function.			
graph a radical function.			

ⓑ *What does this checklist tell you about your mastery of this section? What steps will you take to improve?*

8.8 Use the Complex Number System

Learning Objectives

By the end of this section, you will be able to:

> Evaluate the square root of a negative number
> Add and subtract complex numbers
> Multiply complex numbers
> Divide complex numbers
> Simplify powers of i

Be Prepared!

Before you get started, take this readiness quiz.

1. Given the numbers $-4, \ -\sqrt{7}, 0.\overline{5}, \frac{7}{3}, 3, \sqrt{81},$ list the ⓐ rational numbers, ⓑ irrational numbers, ⓒ real numbers.
   If you missed this problem, review **Example 1.42**.

2. Multiply: $(x-3)(2x+5)$.
   If you missed this problem, review **Example 5.28**.

3. Rationalize the denominator: $\dfrac{\sqrt{5}}{\sqrt{5}-\sqrt{3}}$.
   If you missed this problem, review **Example 5.32**.

Evaluate the Square Root of a Negative Number

Whenever we have a situation where we have a square root of a negative number we say there is no real number that equals that square root. For example, to simplify $\sqrt{-1}$, we are looking for a real number x so that $x^2 = -1$. Since all real numbers squared are positive numbers, there is no real number that equals –1 when squared.

Mathematicians have often expanded their numbers systems as needed. They added 0 to the counting numbers to get the whole numbers. When they needed negative balances, they added negative numbers to get the integers. When they needed the idea of parts of a whole they added fractions and got the rational numbers. Adding the irrational numbers allowed numbers like $\sqrt{5}$. All of these together gave us the real numbers and so far in your study of mathematics, that has been sufficient.

But now we will expand the real numbers to include the square roots of negative numbers. We start by defining the **imaginary unit** i as the number whose square is –1.

Imaginary Unit

The **imaginary unit** i is the number whose square is –1.

$$i^2 = -1 \text{ or } i = \sqrt{-1}$$

We will use the imaginary unit to simplify the square roots of negative numbers.

Square Root of a Negative Number

If b is a positive real number, then

$$\sqrt{-b} = \sqrt{b}\, i$$

We will use this definition in the next example. Be careful that it is clear that the i is not under the radical. Sometimes you will see this written as $\sqrt{-b} = i\sqrt{b}$ to emphasize the i is not under the radical. But the $\sqrt{-b} = \sqrt{b}\, i$ is considered standard form.

EXAMPLE 8.76

Write each expression in terms of i and simplify if possible:

ⓐ $\sqrt{-25}$ ⓑ $\sqrt{-7}$ ⓒ $\sqrt{-12}$.

✓ Solution

ⓐ

	$\sqrt{-25}$
Use the definition of the square root of negative numbers.	$\sqrt{25}\,i$
Simplify.	$5i$

ⓑ

	$\sqrt{-7}$
Use the definition of the square root of negative numbers.	$\sqrt{7}i$
Simplify.	Be careful that it is clear that i is not under the radical sign.

ⓒ

	$\sqrt{-12}$
Use the definition of the square root of negative numbers.	$\sqrt{12}\,i$
Simplify $\sqrt{12}$.	$2\sqrt{3}\,i$

> **TRY IT :: 8.151**   Write each expression in terms of i and simplify if possible:
>
> ⓐ $\sqrt{-81}$ ⓑ $\sqrt{-5}$ ⓒ $\sqrt{-18}$.

> **TRY IT :: 8.152**   Write each expression in terms of i and simplify if possible:
>
> ⓐ $\sqrt{-36}$ ⓑ $\sqrt{-3}$ ⓒ $\sqrt{-27}$.

Now that we are familiar with the imaginary number i, we can expand the real numbers to include imaginary numbers. The **complex number system** includes the real numbers and the imaginary numbers. A **complex number** is of the form $a + bi$, where a, b are real numbers. We call a the real part and b the imaginary part.

Complex Number

A **complex number** is of the form $a + bi$, where a and b are real numbers.

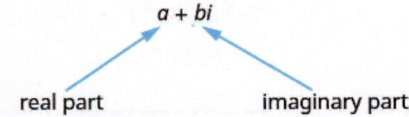

A complex number is in standard form when written as $a + bi$, where a and b are real numbers.

If $b = 0$, then $a + bi$ becomes $a + 0 \cdot i = a$, and is a real number.

If $b \neq 0$, then $a + bi$ is an imaginary number.

If $a = 0$, then $a + bi$ becomes $0 + bi = bi$, and is called a pure imaginary number.

We summarize this here.

	$a + bi$	
$b = 0$	$a + 0 \cdot i$ a	Real number
$b \neq 0$	$a + bi$	Imaginary number
$a = 0$	$0 + bi$ bi	Pure imaginary number

The standard form of a complex number is $a + bi$, so this explains why the preferred form is $\sqrt{-b} = \sqrt{b}\,i$ when $b > 0$.

The diagram helps us visualize the complex number system. It is made up of both the real numbers and the imaginary numbers.

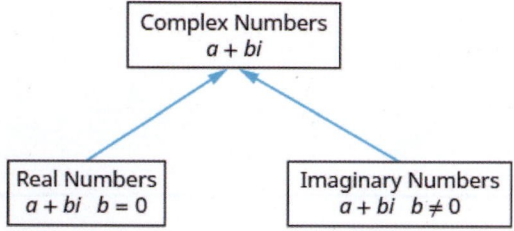

Add or Subtract Complex Numbers

We are now ready to perform the operations of addition, subtraction, multiplication and division on the complex numbers—just as we did with the real numbers.

Adding and subtracting complex numbers is much like adding or subtracting like terms. We add or subtract the real parts and then add or subtract the imaginary parts. Our final result should be in standard form.

EXAMPLE 8.77

Add: $\sqrt{-12} + \sqrt{-27}$.

⊘ Solution

$$\sqrt{-12} + \sqrt{-27}$$

Use the definition of the square root of negative numbers.

$$\sqrt{12}\,i + \sqrt{27}\,i$$

Simplify the square roots.

$$2\sqrt{3}\,i + 3\sqrt{3}\,i$$

Add.

$$5\sqrt{3}\,i$$

> **TRY IT ::** 8.153   Add: $\sqrt{-8} + \sqrt{-32}$.

> **TRY IT ::** 8.154   Add: $\sqrt{-27} + \sqrt{-48}$.

Remember to add both the real parts and the imaginary parts in this next example.

EXAMPLE 8.78

Simplify: ⓐ $(4 - 3i) + (5 + 6i)$ ⓑ $(2 - 5i) - (5 - 2i)$.

⊘ Solution

ⓐ

	$(4 - 3i) + (5 + 6i)$
Use the Associative Property to put the real parts and the imaginary parts together.	$(4 + 5) + (-3i + 6i)$
Simplify.	$9 + 3i$

	$(2 - 5i) - (5 - 2i)$
Distribute.	$2 - 5i - 5 + 2i$
Use the Associative Property to put the real parts and the imaginary parts together.	$2 - 5 - 5i + 2i$
Simplify.	$-3 - 3i$

> **TRY IT ::** 8.155   Simplify: ⓐ $(2 + 7i) + (4 - 2i)$ ⓑ $(8 - 4i) - (2 - i)$.

> **TRY IT ::** 8.156   Simplify: ⓐ $(3 - 2i) + (-5 - 4i)$ ⓑ $(4 + 3i) - (2 - 6i)$.

Multiply Complex Numbers

Multiplying complex numbers is also much like multiplying expressions with coefficients and variables. There is only one special case we need to consider. We will look at that after we practice in the next two examples.

EXAMPLE 8.79

Multiply: $2i(7 - 5i)$.

⊘ Solution

	$2i(7 - 5i)$
Distribute.	$14i - 10i^2$
Simplify i^2.	$14i - 10(-1)$
Multiply.	$14i + 10$
Write in standard form.	$10 + 14i$

> **TRY IT ::** 8.157   Multiply: $4i(5 - 3i)$.

> **TRY IT ::** 8.158   Multiply: $-3i(2 + 4i)$.

In the next example, we multiply the binomials using the Distributive Property or FOIL.

EXAMPLE 8.80

Multiply: $(3 + 2i)(4 - 3i)$.

⊘ Solution

	$(3 + 2i)(4 - 3i)$
Use FOIL.	$12 - 9i + 8i - 6i^2$
Simplify i^2 and combine like terms.	$12 - i - 6(-1)$
Multiply.	$12 - i + 6$
Combine the real parts.	$18 - i$

> **TRY IT ::** 8.159   Multiply: $(5 - 3i)(-1 - 2i)$.

> **TRY IT : :** 8.160   Multiply: $(-4-3i)(2+i)$.

In the next example, we could use FOIL or the Product of Binomial Squares Pattern.

EXAMPLE 8.81

Multiply: $(3+2i)^2$

✓ **Solution**

	$\begin{pmatrix} a+b \\ 3+2i \end{pmatrix}^2$
Use the Product of Binomial Squares Pattern, $(a+b)^2 = a^2 + 2ab + b^2$.	$a^2 + 2 \quad a \quad b + b^2$ $3^2 + 2 \cdot 3 \cdot 2i + (2i)^2$
Simplify.	$9 + 12i + 4i^2$
Simplify i^2.	$9 + 12i + 4(-1)$
Simplify.	$5 + 12i$

> **TRY IT : :** 8.161   Multiply using the Binomial Squares pattern: $(-2-5i)^2$.

> **TRY IT : :** 8.162   Multiply using the Binomial Squares pattern: $(-5+4i)^2$.

Since the square root of a negative number is not a real number, we cannot use the Product Property for Radicals. In order to multiply square roots of negative numbers we should first write them as complex numbers, using $\sqrt{-b} = \sqrt{b}i$. This is one place students tend to make errors, so be careful when you see multiplying with a negative square root.

EXAMPLE 8.82

Multiply: $\sqrt{-36} \cdot \sqrt{-4}$.

✓ **Solution**

To multiply square roots of negative numbers, we first write them as complex numbers.

	$\sqrt{-36} \cdot \sqrt{-4}$
Write as complex numbers using $\sqrt{-b} = \sqrt{b}i$.	$\sqrt{36}\,i \cdot \sqrt{4}\,i$
Simplify.	$6i \cdot 2i$
Multiply.	$12i^2$
Simplify i^2 and multiply.	-12

> **TRY IT : :** 8.163   Multiply: $\sqrt{-49} \cdot \sqrt{-4}$.

> **TRY IT : :** 8.164   Multiply: $\sqrt{-36} \cdot \sqrt{-81}$.

In the next example, each binomial has a square root of a negative number. Before multiplying, each square root of a negative number must be written as a complex number.

EXAMPLE 8.83

Multiply: $(3 - \sqrt{-12})(5 + \sqrt{-27})$.

⊘ Solution

To multiply square roots of negative numbers, we first write them as complex numbers.

$$(3 - \sqrt{-12})(5 + \sqrt{-27})$$

Write as complex numbers using $\sqrt{-b} = \sqrt{b}i$.
$$(3 - 2\sqrt{3}\,i)(5 + 3\sqrt{3}\,i)$$

Use FOIL.
$$15 + 9\sqrt{3}\,i - 10\sqrt{3}\,i - 6 \cdot 3i^2$$

Combine like terms and simplify i^2.
$$15 - \sqrt{3}\,i - 6 \cdot (-3)$$

Multiply and combine like terms.
$$33 - \sqrt{3}\,i$$

> **TRY IT : :** 8.165   Multiply: $(4 - \sqrt{-12})(3 - \sqrt{-48})$.

> **TRY IT : :** 8.166   Multiply: $(-2 + \sqrt{-8})(3 - \sqrt{-18})$.

We first looked at conjugate pairs when we studied polynomials. We said that a pair of binomials that each have the same first term and the same last term, but one is a sum and one is a difference is called a *conjugate pair* and is of the form $(a - b), (a + b)$.

A **complex conjugate pair** is very similar. For a complex number of the form $a + bi$, its conjugate is $a - bi$. Notice they have the same first term and the same last term, but one is a sum and one is a difference.

> **Complex Conjugate Pair**
>
> A **complex conjugate pair** is of the form $a + bi,\ \ a - bi$.

We will multiply a complex conjugate pair in the next example.

EXAMPLE 8.84

Multiply: $(3 - 2i)(3 + 2i)$.

⊘ Solution

$$(3 - 2i)(3 + 2i)$$

Use FOIL.
$$9 + 6i - 6i - 4i^2$$

Combine like terms and simplify i^2.
$$9 - 4(-1)$$

Multiply and combine like terms.
$$13$$

> **TRY IT : :** 8.167   Multiply: $(4 - 3i) \cdot (4 + 3i)$.

> **TRY IT : :** 8.168   Multiply: $(-2 + 5i) \cdot (-2 - 5i)$.

From our study of polynomials, we know the product of conjugates is always of the form $(a - b)(a + b) = a^2 - b^2$. The result is called a difference of squares. We can multiply a complex conjugate pair using this pattern.

The last example we used FOIL. Now we will use the Product of Conjugates Pattern.

$$\left(\overset{a}{3} - \overset{b}{2i}\right)\left(\overset{a}{3} + \overset{b}{2i}\right)$$

$$\overset{a^2}{(3)^2} - \overset{b^2}{(2i)^2}$$

$$9 - 4i^2$$

$$9 - 4(-1)$$

$$13$$

Notice this is the same result we found in **Example 8.84**.

When we multiply complex conjugates, the product of the last terms will always have an i^2 which simplifies to -1.

$$(a - bi)(a + bi)$$
$$a^2 - (bi)^2$$
$$a^2 - b^2 i^2$$
$$a^2 - b^2(-1)$$
$$a^2 + b^2$$

This leads us to the Product of Complex Conjugates Pattern: $(a - bi)(a + bi) = a^2 + b^2$

Product of Complex Conjugates

If a and b are real numbers, then

$$(a - bi)(a + bi) = a^2 + b^2$$

EXAMPLE 8.85

Multiply using the Product of Complex Conjugates Pattern: $(8 - 2i)(8 + 2i)$.

⊘ **Solution**

	$\overset{(a - b)(a + b)}{(8 - 2i)(8 + 2i)}$
Use the Product of Complex Conjugates Pattern, $(a - bi)(a + bi) = a^2 + b^2$.	$\overset{a^2 + b^2}{8^2 + 2^2}$
Simplify the squares.	$64 + 4$
Add.	68

> **TRY IT : :** 8.169   Multiply using the Product of Complex Conjugates Pattern: $(3 - 10i)(3 + 10i)$.

> **TRY IT : :** 8.170   Multiply using the Product of Complex Conjugates Pattern: $(-5 + 4i)(-5 - 4i)$.

Divide Complex Numbers

Dividing complex numbers is much like rationalizing a denominator. We want our result to be in standard form with no imaginary numbers in the denominator.

EXAMPLE 8.86   HOW TO DIVIDE COMPLEX NUMBERS

Divide: $\dfrac{4 + 3i}{3 - 4i}$.

✅ Solution

Step 1. Write both the numerator and denominator in standard form.	They are both in standard form.	$\dfrac{4 + 3i}{3 - 4i}$
Step 2. Multiply the numerator and denominator by the complex conjugate of the denominator.	The complex conjugate of $3 - 4i$ is $3 + 4i$.	$\dfrac{(4 + 3i)(3 + 4i)}{(3 - 4i)(3 + 4i)}$
Step 3. Simplify and write the result in standard form.	Use the pattern $(a - bi)(a + bi) = a^2 + b^2$ in the denominator.	$\dfrac{12 + 16i + 9i + 12i^2}{9 + 16}$
	Combine like terms.	$\dfrac{12 + 25i - 12}{25}$
	Simplify.	$\dfrac{25i}{25}$
	Write the result in standard form.	i

> **TRY IT ::** 8.171   Divide: $\dfrac{2 + 5i}{5 - 2i}$.

> **TRY IT ::** 8.172   Divide: $\dfrac{1 + 6i}{6 - i}$.

We summarize the steps here.

HOW TO :: HOW TO DIVIDE COMPLEX NUMBERS.

Step 1.   Write both the numerator and denominator in standard form.

Step 2.   Multiply the numerator and denominator by the complex conjugate of the denominator.

Step 3.   Simplify and write the result in standard form.

EXAMPLE 8.87

Divide, writing the answer in standard form: $\dfrac{-3}{5 + 2i}$.

✅ Solution

$$\dfrac{-3}{5 + 2i}$$

Multiply the numerator and denominator by the complex conjugate of the denominator.

$$\dfrac{-3(5 - 2i)}{(5 + 2i)(5 - 2i)}$$

Multiply in the numerator and use the Product of Complex Conjugates Pattern in the denominator.

$$\dfrac{-15 + 6i}{5^2 + 2^2}$$

Simplify.

$$\dfrac{-15 + 6i}{29}$$

Write in standard form.

$$-\dfrac{15}{29} + \dfrac{6}{29}i$$

> **TRY IT : : 8.173**   Divide, writing the answer in standard form: $\dfrac{4}{1-4i}$.

> **TRY IT : : 8.174**   Divide, writing the answer in standard form: $\dfrac{-2}{-1+2i}$.

Be careful as you find the conjugate of the denominator.

EXAMPLE 8.88

Divide: $\dfrac{5+3i}{4i}$.

⊘ **Solution**

$$\dfrac{5+3i}{4i}$$

Write the denominator in standard form. $\dfrac{5+3i}{0+4i}$

Multiply the numerator and denominator by $\dfrac{(5+3i)(0-4i)}{(0+4i)(0-4i)}$
the complex conjugate of the denominator.

Simplify. $\dfrac{(5+3i)(-4i)}{(4i)(-4i)}$

Multiply. $\dfrac{-20i-12i^2}{-16i^2}$

Simplify the i^2. $\dfrac{-20i+12}{16}$

Rewrite in standard form. $\dfrac{12}{16}-\dfrac{20}{16}i$

Simplify the fractions. $\dfrac{3}{4}-\dfrac{5}{4}i$

> **TRY IT : : 8.175**   Divide: $\dfrac{3+3i}{2i}$.

> **TRY IT : : 8.176**   Divide: $\dfrac{2+4i}{5i}$.

Simplify Powers of *i*

The powers of i make an interesting pattern that will help us simplify higher powers of *i*. Let's evaluate the powers of i to see the pattern.

i^1	i^2	i^3	i^4
i	-1	$i^2 \cdot i$	$i^2 \cdot i^2$
		$-1 \cdot i$	$(-1)(-1)$
		$-i$	1

i^5	i^6	i^7	i^8
$i^4 \cdot i$	$i^4 \cdot i^2$	$i^4 \cdot i^3$	$i^4 \cdot i^4$
$1 \cdot i$	$1 \cdot i^2$	$1 \cdot i^3$	$1 \cdot 1$
i	i^2	i^3	1
	-1	$-i$	

We summarize this now.

$$
\begin{aligned}
i^1 &= i & i^5 &= i \\
i^2 &= -1 & i^6 &= -1 \\
i^3 &= -i & i^7 &= -i \\
i^4 &= 1 & i^8 &= 1
\end{aligned}
$$

If we continued, the pattern would keep repeating in blocks of four. We can use this pattern to help us simplify powers of i. Since $i^4 = 1$, we rewrite each power, i^n, as a product using i^4 to a power and another power of i.

We rewrite it in the form $i^n = \left(i^4\right)^q \cdot i^r$, where the exponent, q, is the quotient of n divided by 4 and the exponent, r, is the remainder from this division. For example, to simplify i^{57}, we divide 57 by 4 and we get 14 with a remainder of 1. In other words, $57 = 4 \cdot 14 + 1$. So we write $i^{57} = \left(1^4\right)^{14} \cdot i^1$ and then simplify from there.

$$
\begin{array}{cc}
\begin{array}{r}
14 \\
4\overline{)57} \\
4 \\
\overline{17} \\
16 \\
\overline{1}
\end{array}
&
\begin{array}{c}
i^{57} \\
\left(1^4\right)^{14} \cdot i^1 \\
1 \cdot i \\
i
\end{array}
\end{array}
$$

EXAMPLE 8.89

Simplify: i^{86}.

✓ Solution

$$i^{86}$$

Divide 86 by 4 and rewrite i^{86} in the $i^n = \left(i^4\right)^q \cdot i^r$ form.

$$\left(1^4\right)^{21} \cdot i^2$$

$$
\begin{array}{r}
21 \\
4\overline{)86} \\
8 \\
\overline{6} \\
4 \\
\overline{2}
\end{array}
$$

Simplify.

$$(1)^{21} \cdot (-1)$$

Simplify.

$$-1$$

> **TRY IT :: 8.177** Simplify: i^{75}.

> **TRY IT :: 8.178** Simplify: i^{92}.

▶ **MEDIA ::**

Access these online resources for additional instruction and practice with the complex number system.

- **Expressing Square Roots of Negative Numbers with i (https://openstax.org/l/37CompNumb1)**
- **Subtract and Multiply Complex Numbers (https://openstax.org/l/37CompNumb2)**
- **Dividing Complex Numbers (https://openstax.org/l/37CompNumb3)**
- **Rewriting Powers of i (https://openstax.org/l/37CompNumb4)**

8.8 EXERCISES

Practice Makes Perfect

Evaluate the Square Root of a Negative Number

In the following exercises, write each expression in terms of i and simplify if possible.

409.
ⓐ $\sqrt{-16}$

ⓑ $\sqrt{-11}$

ⓒ $\sqrt{-8}$

410.
ⓐ $\sqrt{-121}$

ⓑ $\sqrt{-1}$

ⓒ $\sqrt{-20}$

411.
ⓐ $\sqrt{-100}$

ⓑ $\sqrt{-13}$

ⓒ $\sqrt{-45}$

412.
ⓐ $\sqrt{-49}$

ⓑ $\sqrt{-15}$

ⓒ $\sqrt{-75}$

Add or Subtract Complex Numbers *In the following exercises, add or subtract.*

413. $\sqrt{-75} + \sqrt{-48}$

414. $\sqrt{-12} + \sqrt{-75}$

415. $\sqrt{-50} + \sqrt{-18}$

416. $\sqrt{-72} + \sqrt{-8}$

417. $(1 + 3i) + (7 + 4i)$

418. $(6 + 2i) + (3 - 4i)$

419. $(8 - i) + (6 + 3i)$

420. $(7 - 4i) + (-2 - 6i)$

421. $(1 - 4i) - (3 - 6i)$

422. $(8 - 4i) - (3 + 7i)$

423. $(6 + i) - (-2 - 4i)$

424. $(-2 + 5i) - (-5 + 6i)$

425. $\left(5 - \sqrt{-36}\right) + \left(2 - \sqrt{-49}\right)$

426. $\left(-3 + \sqrt{-64}\right) + \left(5 - \sqrt{-16}\right)$

427.
$\left(-7 - \sqrt{-50}\right) - \left(-32 - \sqrt{-18}\right)$

428. $\left(-5 + \sqrt{-27}\right) - \left(-4 - \sqrt{-48}\right)$

Multiply Complex Numbers

In the following exercises, multiply.

429. $4i(5 - 3i)$

430. $2i(-3 + 4i)$

431. $-6i(-3 - 2i)$

432. $-i(6 + 5i)$

433. $(4 + 3i)(-5 + 6i)$

434. $(-2 - 5i)(-4 + 3i)$

435. $(-3 + 3i)(-2 - 7i)$

436. $(-6 - 2i)(-3 - 5i)$

In the following exercises, multiply using the Product of Binomial Squares Pattern.

437. $(3 + 4i)^2$

438. $(-1 + 5i)^2$

439. $(-2 - 3i)^2$

440. $(-6 - 5i)^2$

In the following exercises, multiply.

441. $\sqrt{-25} \cdot \sqrt{-36}$

442. $\sqrt{-4} \cdot \sqrt{-16}$

443. $\sqrt{-9} \cdot \sqrt{-100}$

444. $\sqrt{-64} \cdot \sqrt{-9}$

445. $(-2 - \sqrt{-27})(4 - \sqrt{-48})$

446. $(5 - \sqrt{-12})(-3 + \sqrt{-75})$

447. $(2 + \sqrt{-8})(-4 + \sqrt{-18})$

448. $(5 + \sqrt{-18})(-2 - \sqrt{-50})$

449. $(2 - i)(2 + i)$

450. $(4 - 5i)(4 + 5i)$

451. $(7 - 2i)(7 + 2i)$

452. $(-3 - 8i)(-3 + 8i)$

In the following exercises, multiply using the Product of Complex Conjugates Pattern.

453. $(7 - i)(7 + i)$

454. $(6 - 5i)(6 + 5i)$

455. $(9 - 2i)(9 + 2i)$

456. $(-3 - 4i)(-3 + 4i)$

Divide Complex Numbers

In the following exercises, divide.

457. $\dfrac{3 + 4i}{4 - 3i}$

458. $\dfrac{5 - 2i}{2 + 5i}$

459. $\dfrac{2 + i}{3 - 4i}$

460. $\dfrac{3 - 2i}{6 + i}$

461. $\dfrac{3}{2 - 3i}$

462. $\dfrac{2}{4 - 5i}$

463. $\dfrac{-4}{3 - 2i}$

464. $\dfrac{-1}{3 + 2i}$

465. $\dfrac{1 + 4i}{3i}$

466. $\dfrac{4 + 3i}{7i}$

467. $\dfrac{-2 - 3i}{4i}$

468. $\dfrac{-3 - 5i}{2i}$

Simplify Powers of *i*

In the following exercises, simplify.

469. i^{41}

470. i^{39}

471. i^{66}

472. i^{48}

473. i^{128}

474. i^{162}

475. i^{137}

476. i^{255}

Writing Exercises

477. Explain the relationship between real numbers and complex numbers.

478. Aniket multiplied as follows and he got the wrong answer. What is wrong with his reasoning?

$$\sqrt{-7} \cdot \sqrt{-7}$$
$$\sqrt{49}$$
$$7$$

479. Why is $\sqrt{-64} = 8i$ but $\sqrt[3]{-64} = -4$.

480. Explain how dividing complex numbers is similar to rationalizing a denominator.

Self Check

ⓐ *After completing the exercises, use this checklist to evaluate your mastery of the objectives of this section.*

I can...	Confidently	With some help	No-I don't get it!
evaluate the square root of a negative number.			
add or subtract complex numbers.			
multiply complex numbers.			
divide complex numbers.			
simplify powers of *i*.			

ⓑ *On a scale of* $1 - 10$, *how would you rate your mastery of this section in light of your responses on the checklist? How can you improve this?*

CHAPTER 8 REVIEW

KEY TERMS

complex conjugate pair A complex conjugate pair is of the form $a + bi$, $a - bi$.

complex number A complex number is of the form $a + bi$, where a and b are real numbers. We call a the real part and b the imaginary part.

complex number system The complex number system is made up of both the real numbers and the imaginary numbers.

imaginary unit The imaginary unit i is the number whose square is –1. $i^2 = -1$ or $i = \sqrt{-1}$.

like radicals Like radicals are radical expressions with the same index and the same radicand.

radical equation An equation in which a variable is in the radicand of a radical expression is called a radical equation.

radical function A radical function is a function that is defined by a radical expression.

rationalizing the denominator Rationalizing the denominator is the process of converting a fraction with a radical in the denominator to an equivalent fraction whose denominator is an integer.

square of a number If $n^2 = m$, then m is the square of n.

square root of a number If $n^2 = m$, then n is a square root of m.

standard form A complex number is in standard form when written as $a + bi$, where a, b are real numbers.

KEY CONCEPTS

8.1 Simplify Expressions with Roots

- **Square Root Notation**
  - $\sqrt{m}$ is read 'the square root of m'
  - If $n^2 = m$, then $n = \sqrt{m}$, for $n \geq 0$.

    radical sign $\longrightarrow \sqrt{m} \longleftarrow$ radicand

  - The square root of m, $\sqrt{m}$, is a positive number whose square is m.

- **n^{th} Root of a Number**
  - If $b^n = a$, then b is an nth root of a.
  - The principal n^{th} root of a is written $\sqrt[n]{a}$.
  - n is called the *index* of the radical.

- **Properties of $\sqrt[n]{a}$**
  - When n is an even number and
    - $a \geq 0$, then $\sqrt[n]{a}$ is a real number
    - $a < 0$, then $\sqrt[n]{a}$ is not a real number
  - When n is an odd number, $\sqrt[n]{a}$ is a real number for all values of a.

- **Simplifying Odd and Even Roots**
  - For any integer $n \geq 2$,
    - when n is odd $\sqrt[n]{a^n} = a$
    - when n is even $\sqrt[n]{a^n} = |a|$
  - We must use the absolute value signs when we take an even root of an expression with a variable in the radical.

8.2 Simplify Radical Expressions

- **Simplified Radical Expression**

- For real numbers a, m and $n \geq 2$

  $\sqrt[n]{a}$ is considered simplified if a has no factors of m^n

- **Product Property of nth Roots**

  - For any real numbers, $\sqrt[n]{a}$ and $\sqrt[n]{b}$, and for any integer $n \geq 2$

    $\sqrt[n]{ab} = \sqrt[n]{a} \cdot \sqrt[n]{b}$ and $\sqrt[n]{a} \cdot \sqrt[n]{b} = \sqrt[n]{ab}$

- **How to simplify a radical expression using the Product Property**

  Step 1. Find the largest factor in the radicand that is a perfect power of the index. Rewrite the radicand as a product of two factors, using that factor.

  Step 2. Use the product rule to rewrite the radical as the product of two radicals.

  Step 3. Simplify the root of the perfect power.

- **Quotient Property of Radical Expressions**

  - If $\sqrt[n]{a}$ and $\sqrt[n]{b}$ are real numbers, $b \neq 0$, and for any integer $n \geq 2$ then,

    $\sqrt[n]{\dfrac{a}{b}} = \dfrac{\sqrt[n]{a}}{\sqrt[n]{b}}$ and $\dfrac{\sqrt[n]{a}}{\sqrt[n]{b}} = \sqrt[n]{\dfrac{a}{b}}$

- **How to simplify a radical expression using the Quotient Property.**

  Step 1. Simplify the fraction in the radicand, if possible.

  Step 2. Use the Quotient Property to rewrite the radical as the quotient of two radicals.

  Step 3. Simplify the radicals in the numerator and the denominator.

8.3 Simplify Rational Exponents

- **Rational Exponent $a^{\frac{1}{n}}$**

  - If $\sqrt[n]{a}$ is a real number and $n \geq 2$, then $a^{\frac{1}{n}} = \sqrt[n]{a}$.

- **Rational Exponent $a^{\frac{m}{n}}$**

  - For any positive integers m and n,

    $a^{\frac{m}{n}} = (\sqrt[n]{a})^m$ and $a^{\frac{m}{n}} = \sqrt[n]{a^m}$

- **Properties of Exponents**

  - If a, b are real numbers and m, n are rational numbers, then

    - **Product Property** $a^m \cdot a^n = a^{m+n}$

    - **Power Property** $(a^m)^n = a^{m \cdot n}$

    - **Product to a Power** $(ab)^m = a^m b^m$

    - **Quotient Property** $\dfrac{a^m}{a^n} = a^{m-n},\ a \neq 0$

    - **Zero Exponent Definition** $a^0 = 1,\ a \neq 0$

    - **Quotient to a Power Property** $\left(\dfrac{a}{b}\right)^m = \dfrac{a^m}{b^m},\ b \neq 0$

    - **Negative Exponent Property** $a^{-n} = \dfrac{1}{a^n},\ a \neq 0$

8.4 Add, Subtract, and Multiply Radical Expressions

- **Product Property of Roots**

  - For any real numbers, $\sqrt[n]{a}$ and $\sqrt[n]{b}$, and for any integer $n \geq 2$

$$\sqrt[n]{ab} = \sqrt[n]{a} \cdot \sqrt[n]{b} \quad \text{and} \quad \sqrt[n]{a} \cdot \sqrt[n]{b} = \sqrt[n]{ab}$$

- **Special Products**

Binomial Squares	Product of Conjugates
$(a+b)^2 = a^2 + 2ab + b^2$	$(a+b)(a-b) = a^2 - b^2$
$(a-b)^2 = a^2 - 2ab + b^2$	

8.5 Divide Radical Expressions

- **Quotient Property of Radical Expressions**

  ◦ If $\sqrt[n]{a}$ and $\sqrt[n]{b}$ are real numbers, $b \neq 0$, and for any integer $n \geq 2$ then,

$$\sqrt[n]{\frac{a}{b}} = \frac{\sqrt[n]{a}}{\sqrt[n]{b}} \quad \text{and} \quad \frac{\sqrt[n]{a}}{\sqrt[n]{b}} = \sqrt[n]{\frac{a}{b}}$$

- **Simplified Radical Expressions**

  ◦ A radical expression is considered simplified if there are:

    ▪ no factors in the radicand that have perfect powers of the index

    ▪ no fractions in the radicand

    ▪ no radicals in the denominator of a fraction

8.6 Solve Radical Equations

- **Binomial Squares**

$$(a+b)^2 = a^2 + 2ab + b^2$$
$$(a-b)^2 = a^2 - 2ab + b^2$$

- **Solve a Radical Equation**

  Step 1. Isolate one of the radical terms on one side of the equation.

  Step 2. Raise both sides of the equation to the power of the index.

  Step 3. Are there any more radicals?
  If yes, repeat Step 1 and Step 2 again.
  If no, solve the new equation.

  Step 4. Check the answer in the original equation.

- **Problem Solving Strategy for Applications with Formulas**

  Step 1. Read the problem and make sure all the words and ideas are understood. When appropriate, draw a figure and label it with the given information.

  Step 2. Identify what we are looking for.

  Step 3. Name what we are looking for by choosing a variable to represent it.

  Step 4. Translate into an equation by writing the appropriate formula or model for the situation. Substitute in the given information.

  Step 5. Solve the equation using good algebra techniques.

  Step 6. Check the answer in the problem and make sure it makes sense.

  Step 7. Answer the question with a complete sentence.

- **Falling Objects**

  ◦ On Earth, if an object is dropped from a height of h feet, the time in seconds it will take to reach the ground is found by using the formula $t = \frac{\sqrt{h}}{4}$.

- **Skid Marks and Speed of a Car**

  ◦ If the length of the skid marks is d feet, then the speed, s, of the car before the brakes were applied can be found by using the formula $s = \sqrt{24d}$.

8.7 Use Radicals in Functions

- **Properties of $\sqrt[n]{a}$**
  - When n is an **even** number and:

    $a \geq 0$, then $\sqrt[n]{a}$ is a real number.

    $a < 0$, then $\sqrt[n]{a}$ is not a real number.

  - When n is an **odd** number, $\sqrt[n]{a}$ is a real number for all values of a.

- **Domain of a Radical Function**
  - When the **index** of the radical is **even**, the radicand must be greater than or equal to zero.
  - When the **index** of the radical is **odd**, the radicand can be any real number.

8.8 Use the Complex Number System

- **Square Root of a Negative Number**
  - If b is a positive real number, then $\sqrt{-b} = \sqrt{b}i$

	$a + bi$	
$b = 0$	$a + 0 \cdot i$ a	Real number
$b \neq 0$	$a + bi$	Imaginary number
$a = 0$	$0 + bi$ bi	Pure imaginary number

Table 8.32

- A complex number is in **standard form** when written as $a + bi$, where a, b are real numbers.

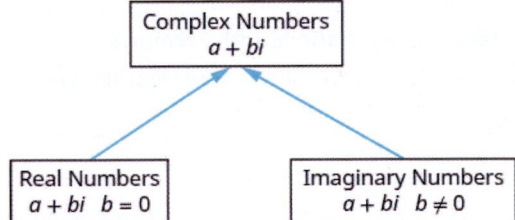

- **Product of Complex Conjugates**
  - If a, b are real numbers, then

    $(a - bi)(a + bi) = a^2 + b^2$

- **How to Divide Complex Numbers**

  Step 1. Write both the numerator and denominator in standard form.

  Step 2. Multiply the numerator and denominator by the complex conjugate of the denominator.

  Step 3. Simplify and write the result in standard form.

REVIEW EXERCISES

8.1 Simplify Expressions with Roots

Simplify Expressions with Roots

In the following exercises, simplify.

481. ⓐ $\sqrt{225}$ ⓑ $-\sqrt{16}$

482. ⓐ $-\sqrt{169}$ ⓑ $\sqrt{-8}$

483. ⓐ $\sqrt[3]{8}$ ⓑ $\sqrt[4]{81}$ ⓒ $\sqrt[5]{243}$

484. ⓐ $\sqrt[3]{-512}$ ⓑ $\sqrt[4]{-81}$ ⓒ $\sqrt[5]{-1}$

Estimate and Approximate Roots

In the following exercises, estimate each root between two consecutive whole numbers.

485. ⓐ $\sqrt{68}$ ⓑ $\sqrt[3]{84}$

In the following exercises, approximate each root and round to two decimal places.

486. ⓐ $\sqrt{37}$ ⓑ $\sqrt[3]{84}$ ⓒ $\sqrt[4]{125}$

Simplify Variable Expressions with Roots

In the following exercises, simplify using absolute values as necessary.

487.
ⓐ $\sqrt[3]{a^3}$
ⓑ $\sqrt[7]{b^7}$

488.
ⓐ $\sqrt{a^{14}}$
ⓑ $\sqrt{w^{24}}$

489.
ⓐ $\sqrt[4]{m^8}$
ⓑ $\sqrt[5]{n^{20}}$

490.
ⓐ $\sqrt{121m^{20}}$
ⓑ $-\sqrt{64a^2}$

491.
ⓐ $\sqrt[3]{216a^6}$
ⓑ $\sqrt[5]{32b^{20}}$

492.
ⓐ $\sqrt{144x^2y^2}$
ⓑ $\sqrt{169w^8y^{10}}$
ⓒ $\sqrt[3]{8a^{51}b^6}$

8.2 Simplify Radical Expressions

Use the Product Property to Simplify Radical Expressions

In the following exercises, use the Product Property to simplify radical expressions.

493. $\sqrt{125}$

494. $\sqrt{675}$

495. ⓐ $\sqrt[3]{625}$ ⓑ $\sqrt[6]{128}$

In the following exercises, simplify using absolute value signs as needed.

496.
ⓐ $\sqrt{a^{23}}$
ⓑ $\sqrt[3]{b^8}$
ⓒ $\sqrt[8]{c^{13}}$

497.
ⓐ $\sqrt{80s^{15}}$
ⓑ $\sqrt[5]{96a^7}$
ⓒ $\sqrt[6]{128b^7}$

498.
ⓐ $\sqrt{96r^3s^3}$
ⓑ $\sqrt[3]{80x^7y^6}$
ⓒ $\sqrt[4]{80x^8y^9}$

499.
ⓐ $\sqrt[5]{-32}$
ⓑ $\sqrt[8]{-1}$

500.
ⓐ $8 + \sqrt{96}$
ⓑ $\dfrac{2 + \sqrt{40}}{2}$

Use the Quotient Property to Simplify Radical Expressions

In the following exercises, use the Quotient Property to simplify square roots.

501. ⓐ $\sqrt{\dfrac{72}{98}}$ ⓑ $\sqrt[3]{\dfrac{24}{81}}$ ⓒ $\sqrt[4]{\dfrac{6}{96}}$

502. ⓐ $\sqrt{\dfrac{y^4}{y^8}}$ ⓑ $\sqrt[5]{\dfrac{u^{21}}{u^{11}}}$ ⓒ $\sqrt[6]{\dfrac{v^{30}}{v^{12}}}$

503. $\sqrt{\dfrac{300m^5}{64}}$

504.

ⓐ $\sqrt{\dfrac{28p^7}{q^2}}$

ⓑ $\sqrt[3]{\dfrac{81s^8}{t^3}}$

ⓒ $\sqrt[4]{\dfrac{64p^{15}}{q^{12}}}$

505.

ⓐ $\sqrt{\dfrac{27p^2q}{108p^4q^3}}$

ⓑ $\sqrt[3]{\dfrac{16c^5d^7}{250c^2d^2}}$

ⓒ $\sqrt[6]{\dfrac{2m^9n^7}{128m^3n}}$

506.

ⓐ $\dfrac{\sqrt{80q^5}}{\sqrt{5q}}$

ⓑ $\dfrac{\sqrt[3]{-625}}{\sqrt[3]{5}}$

ⓒ $\dfrac{\sqrt[4]{80m^7}}{\sqrt[4]{5m}}$

8.3 Simplify Rational Exponents

Simplify expressions with $a^{\frac{1}{n}}$

In the following exercises, write as a radical expression.

507. ⓐ $r^{\frac{1}{2}}$ ⓑ $s^{\frac{1}{3}}$ ⓒ $t^{\frac{1}{4}}$

In the following exercises, write with a rational exponent.

508. ⓐ $\sqrt{21p}$ ⓑ $\sqrt[4]{8q}$ ⓒ $4\sqrt[6]{36r}$

In the following exercises, simplify.

509.

ⓐ $625^{\frac{1}{4}}$

ⓑ $243^{\frac{1}{5}}$

ⓒ $32^{\frac{1}{5}}$

510.

ⓐ $(-1{,}000)^{\frac{1}{3}}$

ⓑ $-1{,}000^{\frac{1}{3}}$

ⓒ $(1{,}000)^{-\frac{1}{3}}$

511.

ⓐ $(-32)^{\frac{1}{5}}$

ⓑ $(243)^{-\frac{1}{5}}$

ⓒ $-125^{\frac{1}{3}}$

Simplify Expressions with $a^{\frac{m}{n}}$

In the following exercises, write with a rational exponent.

512.

ⓐ $\sqrt[4]{r^7}$

ⓑ $\left(\sqrt[5]{2pq}\right)^3$

ⓒ $\sqrt[4]{\left(\dfrac{12m}{7n}\right)^3}$

In the following exercises, simplify.

513.

ⓐ $25^{\frac{3}{2}}$

ⓑ $9^{-\frac{3}{2}}$

ⓒ $(-64)^{\frac{2}{3}}$

514.

ⓐ $-64^{\frac{3}{2}}$

ⓑ $-64^{-\frac{3}{2}}$

ⓒ $(-64)^{\frac{3}{2}}$

Use the Laws of Exponents to Simplify Expressions with Rational Exponents

In the following exercises, simplify.

515.

ⓐ $6^{\frac{5}{2}} \cdot 6^{\frac{1}{2}}$

ⓑ $\left(b^{15}\right)^{\frac{3}{5}}$

ⓒ $\dfrac{w^{\frac{2}{7}}}{w^{\frac{9}{7}}}$

516.

ⓐ $\dfrac{a^{\frac{3}{4}} \cdot a^{-\frac{1}{4}}}{a^{-\frac{10}{4}}}$

ⓑ $\left(\dfrac{27b^{\frac{2}{3}}c^{-\frac{5}{2}}}{b^{-\frac{7}{3}}c^{\frac{1}{2}}}\right)^{\frac{1}{3}}$

8.4 Add, Subtract and Multiply Radical Expressions

Add and Subtract Radical Expressions

In the following exercises, simplify.

517.

ⓐ $7\sqrt{2} - 3\sqrt{2}$

ⓑ $7\sqrt[3]{p} + 2\sqrt[3]{p}$

ⓒ $5\sqrt[3]{x} - 3\sqrt[3]{x}$

518.

ⓐ $\sqrt{11b} - 5\sqrt{11b} + 3\sqrt{11b}$

ⓑ $8\sqrt[4]{11cd} + 5\sqrt[4]{11cd} - 9\sqrt[4]{11cd}$

519.

ⓐ $\sqrt{48} + \sqrt{27}$

ⓑ $\sqrt[3]{54} + \sqrt[3]{128}$

ⓒ $6\sqrt[4]{5} - \dfrac{3}{2}\sqrt[4]{320}$

520.

ⓐ $\sqrt{80c^7} - \sqrt{20c^7}$

ⓑ $2\sqrt[4]{162r^{10}} + 4\sqrt[4]{32r^{10}}$

521. $3\sqrt{75y^2} + 8y\sqrt{48} - \sqrt{300y^2}$

Multiply Radical Expressions

In the following exercises, simplify.

522.

ⓐ $\left(5\sqrt{6}\right)\left(-\sqrt{12}\right)$

ⓑ $\left(-2\sqrt[4]{18}\right)\left(-\sqrt[4]{9}\right)$

523.

ⓐ $\left(3\sqrt{2x^3}\right)\left(7\sqrt{18x^2}\right)$

ⓑ $\left(-6\sqrt[3]{20a^2}\right)\left(-2\sqrt[3]{16a^3}\right)$

Use Polynomial Multiplication to Multiply Radical Expressions

In the following exercises, multiply.

524.

ⓐ $\sqrt{11}\left(8 + 4\sqrt{11}\right)$

ⓑ $\sqrt[3]{3}\left(\sqrt[3]{9} + \sqrt[3]{18}\right)$

525.

ⓐ $(3 - 2\sqrt{7})(5 - 4\sqrt{7})$

ⓑ $\left(\sqrt[3]{x} - 5\right)\left(\sqrt[3]{x} - 3\right)$

526. $\left(2\sqrt{7} - 5\sqrt{11}\right)\left(4\sqrt{7} + 9\sqrt{11}\right)$

527.

ⓐ $(4 + \sqrt{11})^2$

ⓑ $(3 - 2\sqrt{5})^2$

528. $(7 + \sqrt{10})(7 - \sqrt{10})$

529. $(\sqrt[3]{3x} + 2)(\sqrt[3]{3x} - 2)$

8.5 Divide Radical Expressions

Divide Square Roots

In the following exercises, simplify.

530.

ⓐ $\dfrac{\sqrt{48}}{\sqrt{75}}$

ⓑ $\dfrac{\sqrt[3]{81}}{\sqrt[3]{24}}$

531.

ⓐ $\dfrac{\sqrt{320mn^{-5}}}{\sqrt{45m^{-7}n^3}}$

ⓑ $\dfrac{\sqrt[3]{16x^4y^{-2}}}{\sqrt[3]{-54x^{-2}y^4}}$

Rationalize a One Term Denominator

In the following exercises, rationalize the denominator.

532. ⓐ $\dfrac{8}{\sqrt{3}}$ ⓑ $\sqrt{\dfrac{7}{40}}$ ⓒ $\dfrac{8}{\sqrt{2y}}$

533. ⓐ $\dfrac{1}{\sqrt[3]{11}}$ ⓑ $\sqrt[3]{\dfrac{7}{54}}$ ⓒ $\dfrac{3}{\sqrt[3]{3x^2}}$

534. ⓐ $\dfrac{1}{\sqrt[4]{4}}$ ⓑ $\sqrt[4]{\dfrac{9}{32}}$ ⓒ $\dfrac{6}{\sqrt[4]{9x^3}}$

Rationalize a Two Term Denominator

In the following exercises, simplify.

535. $\dfrac{7}{2 - \sqrt{6}}$

536. $\dfrac{\sqrt{5}}{\sqrt{n} - \sqrt{7}}$

537. $\dfrac{\sqrt{x} + \sqrt{8}}{\sqrt{x} - \sqrt{8}}$

8.6 Solve Radical Equations

Solve Radical Equations

In the following exercises, solve.

538. $\sqrt{4x - 3} = 7$

539. $\sqrt{5x + 1} = -3$

540. $\sqrt[3]{4x - 1} = 3$

541. $\sqrt{u - 3} + 3 = u$

542. $\sqrt[3]{4x + 5} - 2 = -5$

543. $(8x + 5)^{\frac{1}{3}} + 2 = -1$

544. $\sqrt{y + 4} - y + 2 = 0$

545. $2\sqrt{8r + 1} - 8 = 2$

Solve Radical Equations with Two Radicals

In the following exercises, solve.

546. $\sqrt{10 + 2c} = \sqrt{4c + 16}$

547.

$\sqrt[3]{2x^2 + 9x - 18} = \sqrt[3]{x^2 + 3x - 2}$

548. $\sqrt{r} + 6 = \sqrt{r + 8}$

549. $\sqrt{x + 1} - \sqrt{x - 2} = 1$

Use Radicals in Applications

In the following exercises, solve. Round approximations to one decimal place.

550. Landscaping Reed wants to have a square garden plot in his backyard. He has enough compost to cover an area of 75 square feet. Use the formula $s = \sqrt{A}$ to find the length of each side of his garden. Round your answer to the nearest tenth of a foot.

551. Accident investigation An accident investigator measured the skid marks of one of the vehicles involved in an accident. The length of the skid marks was 175 feet. Use the formula $s = \sqrt{24d}$ to find the speed of the vehicle before the brakes were applied. Round your answer to the nearest tenth.

8.7 Use Radicals in Functions

Evaluate a Radical Function

In the following exercises, evaluate each function.

552. $g(x) = \sqrt{6x + 1}$, find

ⓐ $g(4)$

ⓑ $g(8)$

553. $G(x) = \sqrt{5x - 1}$, find

ⓐ $G(5)$

ⓑ $G(2)$

554. $h(x) = \sqrt[3]{x^2 - 4}$, find

ⓐ $h(-2)$

ⓑ $h(6)$

555. For the function $g(x) = \sqrt[4]{4 - 4x}$, find

ⓐ $g(1)$

ⓑ $g(-3)$

Find the Domain of a Radical Function

In the following exercises, find the domain of the function and write the domain in interval notation.

556. $g(x) = \sqrt{2 - 3x}$

557. $F(x) = \sqrt{\dfrac{x + 3}{x - 2}}$

558. $f(x) = \sqrt[3]{4x^2 - 16}$

559. $F(x) = \sqrt[4]{10 - 7x}$

Graph Radical Functions

In the following exercises, ⓐ find the domain of the function ⓑ graph the function ⓒ use the graph to determine the range.

560. $g(x) = \sqrt{x + 4}$

561. $g(x) = 2\sqrt{x}$

562. $f(x) = \sqrt[3]{x - 1}$

563. $f(x) = \sqrt[3]{x} + 3$

8.8 Use the Complex Number System

Evaluate the Square Root of a Negative Number

In the following exercises, write each expression in terms of i and simplify if possible.

564.

ⓐ $\sqrt{-100}$

ⓑ $\sqrt{-13}$

ⓒ $\sqrt{-45}$

Add or Subtract Complex Numbers

In the following exercises, add or subtract.

565. $\sqrt{-50} + \sqrt{-18}$

566. $(8 - i) + (6 + 3i)$

567. $(6 + i) - (-2 - 4i)$

568.
$\left(-7 - \sqrt{-50}\right) - \left(-32 - \sqrt{-18}\right)$

Multiply Complex Numbers

In the following exercises, multiply.

569. $(-2 - 5i)(-4 + 3i)$

570. $-6i(-3 - 2i)$

571. $\sqrt{-4} \cdot \sqrt{-16}$

572. $\left(5 - \sqrt{-12}\right)\left(-3 + \sqrt{-75}\right)$

In the following exercises, multiply using the Product of Binomial Squares Pattern.

573. $(-2 - 3i)^2$

In the following exercises, multiply using the Product of Complex Conjugates Pattern.

574. $(9 - 2i)(9 + 2i)$

Divide Complex Numbers

In the following exercises, divide.

575. $\dfrac{2 + i}{3 - 4i}$

576. $\dfrac{-4}{3 - 2i}$

Simplify Powers of *i*

In the following exercises, simplify.

577. i^{48}

578. i^{255}

PRACTICE TEST

In the following exercises, simplify using absolute values as necessary.

579. $\sqrt[3]{125x^9}$

580. $\sqrt{169x^8 y^6}$

581. $\sqrt[3]{72x^8 y^4}$

582. $\sqrt{\dfrac{45x^3 y^4}{180x^5 y^2}}$

In the following exercises, simplify. Assume all variables are positive.

583. ⓐ $216^{-\frac{1}{4}}$ ⓑ $-49^{\frac{3}{2}}$

584. $\sqrt{-45}$

585. $\dfrac{x^{-\frac{1}{4}} \cdot x^{\frac{5}{4}}}{x^{-\frac{3}{4}}}$

586. $\left(\dfrac{8x^{\frac{2}{3}} y^{-\frac{5}{2}}}{x^{-\frac{7}{3}} y^{\frac{1}{2}}}\right)^{\frac{1}{3}}$

587. $\sqrt{48x^5} - \sqrt{75x^5}$

588. $\sqrt{27x^2} - 4x\sqrt{12} + \sqrt{108x^2}$

589. $2\sqrt{12x^5} \cdot 3\sqrt{6x^3}$

590. $\sqrt[3]{4}\left(\sqrt[3]{16} - \sqrt[3]{6}\right)$

591. $\left(4 - 3\sqrt{3}\right)\left(5 + 2\sqrt{3}\right)$

592. $\dfrac{\sqrt[3]{128}}{\sqrt[3]{54}}$

593. $\dfrac{\sqrt{245xy^{-4}}}{\sqrt{45x^{-4}y^3}}$

594. $\dfrac{1}{\sqrt[3]{5}}$

595. $\dfrac{3}{2 + \sqrt{3}}$

596. $\sqrt{-4} \cdot \sqrt{-9}$

597. $-4i(-2 - 3i)$

598. $\dfrac{4 + i}{3 - 2i}$

599. i^{172}

In the following exercises, solve.

600. $\sqrt{2x + 5} + 8 = 6$

601. $\sqrt{x + 5} + 1 = x$

602.
$$\sqrt[3]{2x^2 - 6x - 23} = \sqrt[3]{x^2 - 3x + 5}$$

In the following exercise, ⓐ find the domain of the function ⓑ graph the function ⓒ use the graph to determine the range.

603. $g(x) = \sqrt{x + 2}$

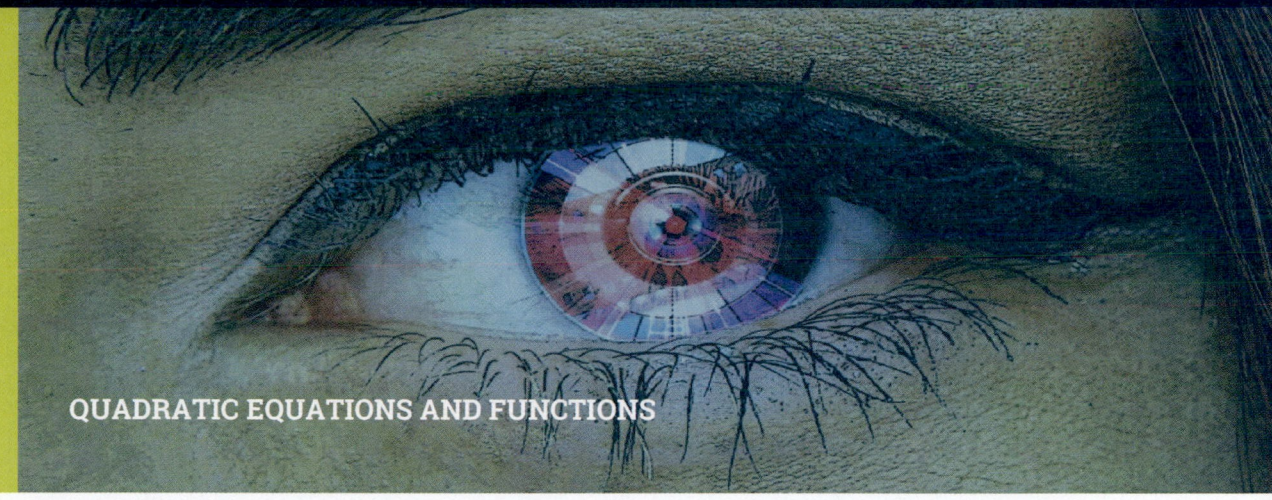

| 9 | QUADRATIC EQUATIONS AND FUNCTIONS |

Figure 9.1 Several companies have patented contact lenses equipped with cameras, suggesting that they may be the future of wearable camera technology. (credit: "intographics"/Pixabay)

Chapter Outline

☑ Introduction

Blink your eyes. You've taken a photo. That's what will happen if you are wearing a contact lens with a built-in camera. Some of the same technology used to help doctors see inside the eye may someday be used to make cameras and other devices. These technologies are being developed by biomedical engineers using many mathematical principles, including an understanding of quadratic equations and functions. In this chapter, you will explore these kinds of equations and learn to solve them in different ways. Then you will solve applications modeled by quadratics, graph them, and extend your understanding to quadratic inequalities.

9.1 Solve Quadratic Equations Using the Square Root Property

Learning Objectives

By the end of this section, you will be able to:

› Solve quadratic equations of the form $ax^2 = k$ using the Square Root Property

› Solve quadratic equations of the form $a(x - h)^2 = k$ using the Square Root Property

Be Prepared!

Before you get started, take this readiness quiz.

1. Simplify: $\sqrt{128}$.
   If you missed this problem, review **Example 8.13**.

2. Simplify: $\sqrt{\dfrac{32}{5}}$.

If you missed this problem, review **Example 8.50**.

3. Factor: $9x^2 - 12x + 4$.

    If you missed this problem, review **Example 6.23**.

A quadratic equation is an equation of the form $ax^2 + bx + c = 0$, where $a \neq 0$. Quadratic equations differ from linear equations by including a quadratic term with the variable raised to the second power of the form ax^2. We use different methods to solve quadratic equations than linear equations, because just adding, subtracting, multiplying, and dividing terms will not isolate the variable.

We have seen that some quadratic equations can be solved by factoring. In this chapter, we will learn three other methods to use in case a quadratic equation cannot be factored.

Solve Quadratic Equations of the form $ax^2 = k$ using the Square Root Property

We have already solved some quadratic equations by factoring. Let's review how we used factoring to solve the quadratic equation $x^2 = 9$.

$$x^2 = 9$$

Put the equation in standard form.	$x^2 - 9 = 0$
Factor the difference of squares.	$(x-3)(x+3) = 0$
Use the Zero Product Property.	$x - 3 = 0 \quad x - 3 = 0$
Solve each equation.	$x = 3 \quad\quad x = -3$

We can easily use factoring to find the solutions of similar equations, like $x^2 = 16$ and $x^2 = 25$, because 16 and 25 are perfect squares. In each case, we would get two solutions, $x = 4$, $x = -4$ and $x = 5$, $x = -5$.

But what happens when we have an equation like $x^2 = 7$? Since 7 is not a perfect square, we cannot solve the equation by factoring.

Previously we learned that since 169 is the square of 13, we can also say that 13 is a *square root* of 169. Also, $(-13)^2 = 169$, so -13 is also a square root of 169. Therefore, both 13 and -13 are square roots of 169. So, every positive number has two square roots—one positive and one negative. We earlier defined the square root of a number in this way:

$$\text{If } n^2 = m, \text{ then } n \text{ is a square root of } m.$$

Since these equations are all of the form $x^2 = k$, the square root definition tells us the solutions are the two square roots of k. This leads to the **Square Root Property**.

Square Root Property

If $x^2 = k$, then

$$x = \sqrt{k} \quad \text{or} \quad x = -\sqrt{k} \quad \text{or} \quad x = \pm\sqrt{k}.$$

Notice that the Square Root Property gives two solutions to an equation of the form $x^2 = k$, the principal square root of k and its opposite. We could also write the solution as $x = \pm\sqrt{k}$. We read this as x equals positive or negative the square root of k.

Now we will solve the equation $x^2 = 9$ again, this time using the Square Root Property.

$$x^2 = 9$$

Use the Square Root Property.	$x = \pm\sqrt{9}$
	$x = \pm 3$

$$\text{So } x = 3 \text{ or } x = -3.$$

What happens when the constant is not a perfect square? Let's use the Square Root Property to solve the equation $x^2 = 7$.

$$x^2 = 7$$

Use the Square Root Property.	$x = \sqrt{7}, \quad x = -\sqrt{7}$

We cannot simplify $\sqrt{7}$, so we leave the answer as a radical.

EXAMPLE 9.1   HOW TO SOLVE A QUADRATIC EQUATION OF THE FORM $AX^2 = K$ USING THE SQUARE ROOT PROPERTY

Solve: $x^2 - 50 = 0$.

✅ Solution

Step 1. Isolate the quadratic term and make its coefficient one.	Add 50 to both sides to get x^2 by itself.	$x^2 - 50 = 0$ $x^2 = 50$
Step 2. Use Square Root Property.	Remember to write the $\pm$ symbol.	$x = \pm\sqrt{50}$
Step 3. Simplify the radical.	 Rewrite to show two solutions.	$x = \pm\sqrt{25} \cdot \sqrt{2}$ $x = \pm5\sqrt{2}$ $x = 5\sqrt{2},\ x = -5\sqrt{2}$
Step 4. Check the solutions.	Substitute in $x = 5\sqrt{2}$ and $x = -5\sqrt{2}$	$x^2 - 50 = 0$ $\left(5\sqrt{2}\right)^2 - 50 \overset{?}{=} 0$ $25 \cdot 2 - 50 \overset{?}{=} 0$ $0 = 0 ✓$ $x^2 - 50 = 0$ $\left(-5\sqrt{2}\right)^2 - 50 \overset{?}{=} 0$ $25 \cdot 2 - 50 \overset{?}{=} 0$ $0 = 0 ✓$

> **TRY IT ::** 9.1   Solve: $x^2 - 48 = 0$.

> **TRY IT ::** 9.2   Solve: $y^2 - 27 = 0$.

The steps to take to use the Square Root Property to solve a quadratic equation are listed here.

HOW TO :: SOLVE A QUADRATIC EQUATION USING THE SQUARE ROOT PROPERTY.

Step 1.   Isolate the quadratic term and make its coefficient one.

Step 2.   Use Square Root Property.

Step 3.   Simplify the radical.

Step 4.   Check the solutions.

In order to use the Square Root Property, the coefficient of the variable term must equal one. In the next example, we must divide both sides of the equation by the coefficient 3 before using the Square Root Property.

EXAMPLE 9.2

Solve: $3z^2 = 108$.

Solution

$$3z^2 = 108$$

The quadratic term is isolated. Divide by 3 to make its coefficient 1.	$\dfrac{3z^2}{3} = \dfrac{108}{3}$
Simplify.	$z^2 = 36$
Use the Square Root Property.	$z = \pm\sqrt{36}$
Simplify the radical.	$z = \pm 6$
Rewrite to show two solutions.	$z = 6, \quad z = -6$

Check the solutions:

$$3z^2 = 108 \qquad\qquad 3z^2 = 108$$
$$3(6)^2 \overset{?}{=} 108 \qquad 3(-6)^2 \overset{?}{=} 108$$
$$3(36) \overset{?}{=} 108 \qquad 3(36) \overset{?}{=} 108$$
$$108 = 108 \checkmark \qquad 108 = 108 \checkmark$$

> **TRY IT :: 9.3**   Solve: $2x^2 = 98$.

> **TRY IT :: 9.4**   Solve: $5m^2 = 80$.

The Square Root Property states 'If $x^2 = k$,' What will happen if $k < 0$? This will be the case in the next example.

EXAMPLE 9.3

Solve: $x^2 + 72 = 0$.

Solution

$$x^2 + 72 = 0$$

Isolate the quadratic term.	$x^2 = -72$
Use the Square Root Property.	$x = \pm\sqrt{-72}$
Simplify using complex numbers.	$x = \pm\sqrt{72}\,i$
Simplify the radical.	$x = \pm 6\sqrt{2}\,i$
Rewrite to show two solutions.	$x = 6\sqrt{2}\,i, \quad x = -6\sqrt{2}\,i$

Check the solutions:

$$x^2 + 72 = 0 \qquad\qquad x^2 + 72 = 0$$

$$\left(6\sqrt{2}\,i\right)^2 + 72 \overset{?}{=} 0 \qquad \left(6\sqrt{2}\,i\right)^2 + 72 \overset{?}{=} 0$$

$$6^2\left(\sqrt{2}\right)^2 i^2 + 72 \overset{?}{=} 0 \qquad (-6)^2\left(\sqrt{2}\right)^2 i^2 + 72 \overset{?}{=} 0$$

$$36 \cdot 2 \cdot (-1) + 72 \overset{?}{=} 0 \qquad 36 \cdot 2 \cdot (-1) + 72 \overset{?}{=} 0$$

$$0 = 0 \checkmark \qquad\qquad 0 = 0 \checkmark$$

> **TRY IT :: 9.5**   Solve: $c^2 + 12 = 0$.

> **TRY IT :: 9.6**   Solve: $q^2 + 24 = 0$.

Our method also works when fractions occur in the equation, we solve as any equation with fractions. In the next example, we first isolate the quadratic term, and then make the coefficient equal to one.

EXAMPLE 9.4

Solve: $\frac{2}{3}u^2 + 5 = 17$.

✓ Solution

$$\frac{2}{3}u^2 + 5 = 17$$

Isolate the quadratic term.	$\frac{2}{3}u^2 = 12$
Multiply by $\frac{3}{2}$ to make the coefficient 1.	$\frac{3}{2} \cdot \frac{2}{3}u^2 = \frac{3}{2} \cdot 12$
Simplify.	$u^2 = 18$
Use the Square Root Property.	$u = \pm\sqrt{18}$
Simplify the radical.	$u = \pm\sqrt{9 \cdot 2}$
Simplify.	$u = \pm 3\sqrt{2}$
Rewrite to show two solutions.	$u = 3\sqrt{2}, \quad u = -3\sqrt{2}$

Check:

$$\frac{2}{3}u^2 + 5 = 17 \qquad\qquad \frac{2}{3}u^2 + 5 = 17$$

$$\frac{2}{3}\left(3\sqrt{2}\right)^2 + 5 \overset{?}{=} 17 \qquad \frac{2}{3}\left(-3\sqrt{2}\right)^2 + 5 \overset{?}{=} 17$$

$$\frac{2}{3} \cdot 18 + 5 \overset{?}{=} 17 \qquad \frac{2}{3} \cdot 18 + 5 \overset{?}{=} 17$$

$$12 + 5 \overset{?}{=} 17 \qquad\qquad 12 + 5 \overset{?}{=} 17$$

$$17 = 17 \checkmark \qquad\qquad 17 = 17 \checkmark$$

> **TRY IT :: 9.7**   Solve: $\frac{1}{2}x^2 + 4 = 24$.

> **TRY IT : : 9.8**   Solve: $\frac{3}{4}y^2 - 3 = 18$.

The solutions to some equations may have fractions inside the radicals. When this happens, we must rationalize the denominator.

EXAMPLE 9.5

Solve: $2x^2 - 8 = 41$.

⊘ **Solution**

	$2x^2 - 8 = 41$
Isolate the quadratic term.	$2x^2 = 49$
Divide by 2 to make the coefficient 1.	$\frac{2x^2}{2} = \frac{49}{2}$
Simplify.	$x^2 = \frac{49}{2}$
Use the Square Root Property.	$x = \pm\sqrt{\frac{49}{2}}$
Rewrite the radical as a fraction of square roots.	$x = \pm\frac{\sqrt{49}}{\sqrt{2}}$
Rationalize the denominator.	$x = \pm\frac{\sqrt{49}\cdot\sqrt{2}}{\sqrt{2}\cdot\sqrt{2}}$
Simplify.	$x = \pm\frac{7\sqrt{2}}{2}$
Rewrite to show two solutions.	$x = \frac{7\sqrt{2}}{2},\quad x = -\frac{7\sqrt{2}}{2}$

Check:
We leave the check for you.

> **TRY IT : : 9.9**   Solve: $5r^2 - 2 = 34$.

> **TRY IT : : 9.10**   Solve: $3t^2 + 6 = 70$.

Solve Quadratic Equations of the Form $a(x - h)^2 = k$ Using the Square Root Property

We can use the Square Root Property to solve an equation of the form $a(x - h)^2 = k$ as well. Notice that the quadratic term, x, in the original form $ax^2 = k$ is replaced with $(x - h)$.

$$ax^2 = k \qquad a(x - h)^2 = k$$

The first step, like before, is to isolate the term that has the variable squared. In this case, a binomial is being squared. Once the binomial is isolated, by dividing each side by the coefficient of a, then the Square Root Property can be used on $(x - h)^2$.

EXAMPLE 9.6

Solve: $4(y - 7)^2 = 48$.

✓ Solution

$$4(y - 7)^2 = 48$$

Divide both sides by the coefficient 4.	$(y - 7)^2 = 12$
Use the Square Root Property on the binomial	$y - 7 = \pm\sqrt{12}$
Simplify the radical.	$y - 7 = \pm 2\sqrt{3}$
Solve for y.	$y = 7 \pm 2\sqrt{3}$
Rewrite to show two solutions.	$y = 7 + 2\sqrt{3}, \quad y = 7 - 2\sqrt{3}$

Check:

$$4(y - 7)^2 = 48 \qquad\qquad 4(y - 7)^2 = 48$$

$$4(7 + 2\sqrt{3} - 7)^2 \overset{?}{=} 48 \qquad 4(7 - 2\sqrt{3} - 7)^2 \overset{?}{=} 48$$

$$4(2\sqrt{3})^2 \overset{?}{=} 48 \qquad\qquad 4(-2\sqrt{3})^2 \overset{?}{=} 48$$

$$4(12) \overset{?}{=} 48 \qquad\qquad 4(12) \overset{?}{=} 48$$

$$48 = 48 \checkmark \qquad\qquad 48 = 48 \checkmark$$

> **TRY IT : : 9.11**   Solve: $3(a - 3)^2 = 54$.

> **TRY IT : : 9.12**   Solve: $2(b + 2)^2 = 80$.

Remember when we take the square root of a fraction, we can take the square root of the numerator and denominator separately.

EXAMPLE 9.7

Solve: $\left(x - \dfrac{1}{3}\right)^2 = \dfrac{5}{9}$.

⊘ Solution

$$\left(x - \frac{1}{3}\right)^2 = \frac{5}{9}$$

Use the Square Root Property.	$x - \frac{1}{3} = \pm\sqrt{\frac{5}{9}}$
Rewrite the radical as a fraction of square roots.	$x - \frac{1}{3} = \pm\frac{\sqrt{5}}{\sqrt{9}}$
Simplify the radical.	$x - \frac{1}{3} = \pm\frac{\sqrt{5}}{3}$
Solve for x.	$x = \frac{1}{3} \pm \frac{\sqrt{5}}{3}$
Rewrite to show two solutions.	$x = \frac{1}{3} + \frac{\sqrt{5}}{3}, \quad x = \frac{1}{3} - \frac{\sqrt{5}}{3}$

Check:
We leave the check for you.

> **TRY IT :: 9.13**   Solve: $\left(x - \frac{1}{2}\right)^2 = \frac{5}{4}$.

> **TRY IT :: 9.14**   Solve: $\left(y + \frac{3}{4}\right)^2 = \frac{7}{16}$.

We will start the solution to the next example by isolating the binomial term.

EXAMPLE 9.8

Solve: $2(x - 2)^2 + 3 = 57$.

⊘ Solution

$$2(x - 2)^2 + 3 = 57$$

Subtract 3 from both sides to isolate the binomial term.	$2(x - 2)^2 = 54$
Divide both sides by 2.	$(x - 2)^2 = 27$
Use the Square Root Property.	$x - 2 = \pm\sqrt{27}$
Simplify the radical.	$x - 2 = \pm 3\sqrt{3}$
Solve for x.	$x = 2 \pm 3\sqrt{3}$
Rewrite to show two solutions.	$x = 2 + 3\sqrt{3}, \quad x = 2 - 3\sqrt{3}$

Check:
We leave the check for you.

> **TRY IT :: 9.15**   Solve: $5(a - 5)^2 + 4 = 104$.

> **TRY IT :: 9.16**   Solve: $3(b + 3)^2 - 8 = 88$.

Sometimes the solutions are complex numbers.

EXAMPLE 9.9

Solve: $(2x - 3)^2 = -12$.

⊘ Solution

$$(2x - 3)^2 = -12$$

Use the Square Root Property. $2x - 3 = \pm\sqrt{-12}$

Simplify the radical. $2x - 3 = \pm 2\sqrt{3}\, i$

Add 3 to both sides. $2x = 3 \pm 2\sqrt{3}\, i$

Divide both sides by 2. $x = \dfrac{3 \pm 2\sqrt{3}\, i}{2}$

Rewrite in standard form.

Simplify. $x = \dfrac{3}{2} \pm \dfrac{2\sqrt{3}\, i}{2}$

$x = \dfrac{3}{2} \pm \sqrt{3}\, i$

Rewrite to show two solutions. $x = \dfrac{3}{2} + \sqrt{3}\, i, \quad x = \dfrac{3}{2} - \sqrt{3}\, i$

Check:

We leave the check for you.

> **TRY IT : : 9.17**   Solve: $(3r + 4)^2 = -8$.

> **TRY IT : : 9.18**   Solve: $(2t - 8)^2 = -10$.

The left sides of the equations in the next two examples do not seem to be of the form $a(x - h)^2$. But they are perfect square trinomials, so we will factor to put them in the form we need.

EXAMPLE 9.10

Solve: $4n^2 + 4n + 1 = 16$.

⊘ Solution

We notice the left side of the equation is a perfect square trinomial. We will factor it first.

$$4n^2 + 4n + 1 = 16$$

Factor the perfect square trinomial.	$(2n + 1)^2 = 16$
Use the Square Root Property.	$2n + 1 = \pm\sqrt{16}$
Simplify the radical.	$2n + 1 = \pm 4$
Solve for n.	$2n = -1 \pm 4$
Divide each side by 2.	$\dfrac{2n}{2} = \dfrac{-1 \pm 4}{2}$
	$n = \dfrac{-1 \pm 4}{2}$
Rewrite to show two solutions.	$n = \dfrac{-1 + 4}{2}, \; n = \dfrac{-1 - 4}{2}$
Simplify each equation.	$n = \dfrac{3}{2}, \qquad n = -\dfrac{5}{2}$

Check:

$$4n^2 + 4n + 1 = 16 \qquad\qquad 4n^2 + 4n + 1 = 16$$

$$4\left(\frac{3}{2}\right)^2 + 4\left(\frac{3}{2}\right) + 1 \overset{?}{=} 16 \qquad 4\left(-\frac{5}{2}\right)^2 + 4\left(-\frac{5}{2}\right) + 1 \overset{?}{=} 16$$

$$4\left(\frac{9}{4}\right) + 4\left(\frac{3}{2}\right) + 1 \overset{?}{=} 16 \qquad 4\left(\frac{25}{4}\right) + 4\left(-\frac{5}{2}\right) + 1 \overset{?}{=} 16$$

$$9 + 6 + 1 \overset{?}{=} 16 \qquad\qquad 25 - 10 + 1 \overset{?}{=} 16$$

$$16 = 16 \checkmark \qquad\qquad 16 = 16 \checkmark$$

> **TRY IT :: 9.19**   Solve: $9m^2 - 12m + 4 = 25$.

> **TRY IT :: 9.20**   Solve: $16n^2 + 40n + 25 = 4$.

▶ **MEDIA ::**

Access this online resource for additional instruction and practice with using the Square Root Property to solve quadratic equations.

- **Solving Quadratic Equations: The Square Root Property (https://openstax.org/l/37SqRtProp1)**
- **Using the Square Root Property to Solve Quadratic Equations (https://openstax.org/l/37SqRtProp2)**

9.1 EXERCISES

Practice Makes Perfect

Solve Quadratic Equations of the Form $ax^2 = k$ Using the Square Root Property

In the following exercises, solve each equation.

1. $a^2 = 49$

2. $b^2 = 144$

3. $r^2 - 24 = 0$

4. $t^2 - 75 = 0$

5. $u^2 - 300 = 0$

6. $v^2 - 80 = 0$

7. $4m^2 = 36$

8. $3n^2 = 48$

9. $\frac{4}{3}x^2 = 48$

10. $\frac{5}{3}y^2 = 60$

11. $x^2 + 25 = 0$

12. $y^2 + 64 = 0$

13. $x^2 + 63 = 0$

14. $y^2 + 45 = 0$

15. $\frac{4}{3}x^2 + 2 = 110$

16. $\frac{2}{3}y^2 - 8 = -2$

17. $\frac{2}{5}a^2 + 3 = 11$

18. $\frac{3}{2}b^2 - 7 = 41$

19. $7p^2 + 10 = 26$

20. $2q^2 + 5 = 30$

21. $5y^2 - 7 = 25$

22. $3x^2 - 8 = 46$

Solve Quadratic Equations of the Form $a(x - h)^2 = k$ Using the Square Root Property

In the following exercises, solve each equation.

23. $(u - 6)^2 = 64$

24. $(v + 10)^2 = 121$

25. $(m - 6)^2 = 20$

26. $(n + 5)^2 = 32$

27. $\left(r - \frac{1}{2}\right)^2 = \frac{3}{4}$

28. $\left(x + \frac{1}{5}\right)^2 = \frac{7}{25}$

29. $\left(y + \frac{2}{3}\right)^2 = \frac{8}{81}$

30. $\left(t - \frac{5}{6}\right)^2 = \frac{11}{25}$

31. $(a - 7)^2 + 5 = 55$

32. $(b - 1)^2 - 9 = 39$

33. $4(x + 3)^2 - 5 = 27$

34. $5(x + 3)^2 - 7 = 68$

35. $(5c + 1)^2 = -27$

36. $(8d - 6)^2 = -24$

37. $(4x - 3)^2 + 11 = -17$

38. $(2y + 1)^2 - 5 = -23$

39. $m^2 - 4m + 4 = 8$

40. $n^2 + 8n + 16 = 27$

41. $x^2 - 6x + 9 = 12$

42. $y^2 + 12y + 36 = 32$

43. $25x^2 - 30x + 9 = 36$

44. $9y^2 + 12y + 4 = 9$

45. $36x^2 - 24x + 4 = 81$

46. $64x^2 + 144x + 81 = 25$

Mixed Practice

In the following exercises, solve using the Square Root Property.

47. $2r^2 = 32$

48. $4t^2 = 16$

49. $(a - 4)^2 = 28$

50. $(b + 7)^2 = 8$

51. $9w^2 - 24w + 16 = 1$

52. $4z^2 + 4z + 1 = 49$

53. $a^2 - 18 = 0$

54. $b^2 - 108 = 0$

55. $\left(p - \frac{1}{3}\right)^2 = \frac{7}{9}$

56. $\left(q - \frac{3}{5}\right)^2 = \frac{3}{4}$

57. $m^2 + 12 = 0$

58. $n^2 + 48 = 0.$

59. $u^2 - 14u + 49 = 72$

60. $v^2 + 18v + 81 = 50$

61. $(m - 4)^2 + 3 = 15$

62. $(n - 7)^2 - 8 = 64$

63. $(x + 5)^2 = 4$

64. $(y - 4)^2 = 64$

65. $6c^2 + 4 = 29$

66. $2d^2 - 4 = 77$

67. $(x - 6)^2 + 7 = 3$

68. $(y - 4)^2 + 10 = 9$

Writing Exercises

69. In your own words, explain the Square Root Property.

70. In your own words, explain how to use the Square Root Property to solve the quadratic equation $(x + 2)^2 = 16$.

Self Check

ⓐ *After completing the exercises, use this checklist to evaluate your mastery of the objectives of this section.*

I can...	Confidently	With some help	No-I don't get it!
solve quadratic equations of the form $ax^2 = k$ using the square root property.			
solve quadratic equations of the form $a(x - h)^2 = k$ using the square root property.			

Choose how would you respond to the statement "I can solve quadratic equations of the form a times the square of x minus h equals k using the Square Root Property." "Confidently," "with some help," or "No, I don't get it."

ⓑ *If most of your checks were:*

...confidently. *Congratulations! You have achieved the objectives in this section. Reflect on the study skills you used so that you can continue to use them. What did you do to become confident of your ability to do these things? Be specific.*

...with some help. *This must be addressed quickly because topics you do not master become potholes in your road to success. In math every topic builds upon previous work. It is important to make sure you have a strong foundation before you move on. Who can you ask for help? Your fellow classmates and instructor are good resources. Is there a place on campus where math tutors are available? Can your study skills be improved?*

...no - I don't get it! *This is a warning sign and you must not ignore it. You should get help right away or you will quickly be* *overwhelmed. See your instructor as soon as you can to discuss your situation. Together you can come up with a plan to get you the help you need.*

9.2   Solve Quadratic Equations by Completing the Square

Learning Objectives

By the end of this section, you will be able to:

> Complete the square of a binomial expression

> Solve quadratic equations of the form $x^2 + bx + c = 0$ by completing the square

> Solve quadratic equations of the form $ax^2 + bx + c = 0$ by completing the square

---

Be Prepared!

Before you get started, take this readiness quiz.

1. Expand: $(x + 9)^2$.
   If you missed this problem, review **Example 5.32**.

2. Factor $y^2 - 14y + 49$.
   If you missed this problem, review **Example 6.9**.

3. Factor $5n^2 + 40n + 80$.
   If you missed this problem, review **Example 6.14**.

So far we have solved quadratic equations by factoring and using the Square Root Property. In this section, we will solve quadratic equations by a process called **completing the square**, which is important for our work on conics later.

Complete the Square of a Binomial Expression

In the last section, we were able to use the Square Root Property to solve the equation $(y - 7)^2 = 12$ because the left side was a perfect square.

$$
\begin{aligned}
(y - 7)^2 &= 12 \\
y - 7 &= \pm\sqrt{12} \\
y - 7 &= \pm 2\sqrt{3} \\
y &= 7 \pm 2\sqrt{3}
\end{aligned}
$$

We also solved an equation in which the left side was a perfect square trinomial, but we had to rewrite it the form $(x - k)^2$ in order to use the Square Root Property.

$$
\begin{aligned}
x^2 - 10x + 25 &= 18 \\
(x - 5)^2 &= 18
\end{aligned}
$$

What happens if the variable is not part of a perfect square? Can we use algebra to make a perfect square?

Let's look at two examples to help us recognize the patterns.

$$
\begin{array}{cc}
(x + 9)^2 & (y - 7)^2 \\
(x + 9)(x + 9) & (y - 7)(y - 7) \\
x^2 + 9x + 9x + 81 & y^2 - 7y - 7y + 49 \\
x^2 + 18x + 81 & y^2 - 14y + 49
\end{array}
$$

We restate the patterns here for reference.

Binomial Squares Pattern

If a and b are real numbers,

$$(a+b)^2 = a^2 + 2ab + b^2$$

$(a+b)^2$	=	a^2	+	$2ab$	+	b^2
(binomial)^2		(first term)^2		2·(product of terms)		(second term)^2

$$(a-b)^2 = a^2 - 2ab + b^2$$

$(a-b)^2$	=	a^2	−	$2ab$	+	b^2
(binomial)^2		(first term)^2		2·(product of terms)		(second term)^2

We can use this pattern to "make" a perfect square.

We will start with the expression $x^2 + 6x$. Since there is a plus sign between the two terms, we will use the $(a+b)^2$ pattern, $a^2 + 2ab + b^2 = (a+b)^2$.

$$\overset{a^2 + 2ab + b^2}{x^2 + 6x + \underline{\quad}}$$

We ultimately need to find the last term of this trinomial that will make it a perfect square trinomial. To do that we will need to find b. But first we start with determining a. Notice that the first term of $x^2 + 6x$ is a square, x^2. This tells us that $a = x$.

$$\overset{a^2 + \quad 2ab \quad + b^2}{x^2 + 2 \cdot x \cdot b + b^2}$$

What number, b, when multiplied with $2x$ gives $6x$? It would have to be 3, which is $\frac{1}{2}(6)$. So $b = 3$.

$$\overset{a^2 + \quad 2ab \quad + b^2}{x^2 + 2 \cdot 3 \cdot x + \underline{\quad}}$$

Now to complete the perfect square trinomial, we will find the last term by squaring b, which is $3^2 = 9$.

$$\overset{a^2 + 2ab + b^2}{x^2 + 6x + 9}$$

We can now factor.

$$\overset{(a+b)^2}{(x+3)^2}$$

So we found that adding 9 to $x^2 + 6x$ 'completes the square', and we write it as $(x+3)^2$.

HOW TO :: COMPLETE A SQUARE OF $x^2 + bx$.

Step 1. Identify b, the coefficient of x.

Step 2. Find $\left(\frac{1}{2}b\right)^2$, the number to complete the square.

Step 3. Add the $\left(\frac{1}{2}b\right)^2$ to $x^2 + bx$.

Step 4. Factor the perfect square trinomial, writing it as a binomial squared.

EXAMPLE 9.11

Complete the square to make a perfect square trinomial. Then write the result as a binomial squared.

ⓐ $x^2 - 26x$ ⓑ $y^2 - 9y$ ⓒ $n^2 + \frac{1}{2}n$

✓ **Solution**

ⓐ

$$x^2 - bx$$
$$x^2 - 26x$$

The coefficient of x is -26.

Find $\left(\frac{1}{2}b\right)^2$.

$$\left(\frac{1}{2} \cdot (-26)\right)^2$$

$$(13)^2$$

$$169$$

| Add 169 to the binomial to complete the square. | $x^2 - 26x + 169$ |

Factor the perfect square trinomial, writing it as a binomial squared. $(x - 13)^2$

ⓑ

$$x^2 - bx$$
$$y^2 - 9y$$

The coefficient of y is -9.

Find $\left(\frac{1}{2}b\right)^2$.

$$\left(\frac{1}{2} \cdot (-9)\right)^2$$

$$\left(-\frac{9}{2}\right)^2$$

$$\frac{81}{4}$$

Add $\frac{81}{4}$ to the binomial to complete the square. $y^2 - 9y + \frac{81}{4}$

Factor the perfect square trinomial, writing it as a binomial squared. $\left(y - \frac{9}{2}\right)^2$

ⓒ

$$x^2 + bx$$
$$n^2 + \frac{1}{2}n$$

The coefficient of n is $\frac{1}{2}$.

Find $\left(\frac{1}{2}b\right)^2$.

$$\left(\frac{1}{2}\cdot\frac{1}{2}\right)^2$$

$$\left(\frac{1}{4}\right)^2$$

$$\frac{1}{16}$$

Add $\frac{1}{16}$ to the binomial to complete the square.	$n^2 + \frac{1}{2}n + \frac{1}{16}$
Rewrite as a binomial square.	$\left(n + \frac{1}{4}\right)^2$

> **TRY IT :: 9.21**
>
> Complete the square to make a perfect square trinomial. Then write the result as a binomial squared.
>
> ⓐ $a^2 - 20a$ ⓑ $m^2 - 5m$ ⓒ $p^2 + \frac{1}{4}p$

> **TRY IT :: 9.22**
>
> Complete the square to make a perfect square trinomial. Then write the result as a binomial squared.
>
> ⓐ $b^2 - 4b$ ⓑ $n^2 + 13n$ ⓒ $q^2 - \frac{2}{3}q$

Solve Quadratic Equations of the Form $x^2 + bx + c = 0$ by Completing the Square

In solving equations, we must always do the same thing to both sides of the equation. This is true, of course, when we solve a quadratic equation by completing the square too. When we add a term to one side of the equation to make a perfect square trinomial, we must also add the same term to the other side of the equation.

For example, if we start with the equation $x^2 + 6x = 40$, and we want to complete the square on the left, we will add 9 to both sides of the equation.

$$x^2 + 6x = 40$$

$$x^2 + 6x + \underline{\quad} = 40 + \underline{\quad}$$

$$x^2 + 6x + 9 = 40 + 9$$

Add 9 to both sides to complete the square. $\qquad (x + 3)^2 = 49$

Now the equation is in the form to solve using the Square Root Property! Completing the square is a way to transform an equation into the form we need to be able to use the Square Root Property.

EXAMPLE 9.12 HOW TO SOLVE A QUADRATIC EQUATION OF THE FORM $x^2 + bx + c = 0$ BY COMPLETING THE SQUARE

Solve by completing the square: $x^2 + 8x = 48$.

✓ Solution

Step 1. Isolate the variable terms on one side and the constant terms on the other.	This equation has all the variables on the left.	$x^2 + bx \quad c$ $x^2 + 8x = 48$
Step 2. Find $\left(\frac{1}{2} \cdot b\right)^2$, the number to complete the square. Add it to both sides of the equation.	Take half of 8 and square it. $4^2 = 16$ Add 16 to BOTH sides of the equation.	$x^2 + 8x + \underset{\left(\frac{1}{2} \cdot 8\right)^2}{\underline{\quad\quad}} = 48$ $x^2 + 8x + 16 = 48 + 16$
Step 3. Factor the perfect square trinomial as a binomial square.	$x^2 + 8x + 16 = (x+4)^2$ Add the terms on the right.	$(x+4)^2 = 64$
Step 4. Use the Square Root Property.		$x + 4 = \pm\sqrt{64}$
Step 5. Simplify the radical and then solve the two resulting equations.		$x + 4 = \pm 8$ $x + 4 = 8 \quad x + 4 = -8$ $x = 4 \quad\quad x = -12$
Step 6. Check the solutions.	Put each answer in the original equation to check. Substitute $x = 4$. Substitute $x = -12$.	$x^2 + 8x = 48$ $(4)^2 + 8(4) \overset{?}{=} 48$ $16 + 32 \overset{?}{=} 48$ $48 = 48 \checkmark$ $x^2 + 8x = 48$ $(-12)^2 + 8(-12) \overset{?}{=} 48$ $144 - 96 \overset{?}{=} 48$ $48 = 48 \checkmark$

> **TRY IT :: 9.23**   Solve by completing the square: $x^2 + 4x = 5$.

> **TRY IT :: 9.24**   Solve by completing the square: $y^2 - 10y = -9$.

The steps to solve a quadratic equation by completing the square are listed here.

HOW TO :: SOLVE A QUADRATIC EQUATION OF THE FORM $x^2 + bx + c = 0$ BY COMPLETING THE SQUARE.

Step 1.   Isolate the variable terms on one side and the constant terms on the other.

Step 2.   Find $\left(\frac{1}{2} \cdot b\right)^2$, the number needed to complete the square. Add it to both sides of the equation.

Step 3.   Factor the perfect square trinomial, writing it as a binomial squared on the left and simplify by adding the terms on the right

Step 4.   Use the Square Root Property.

Step 5.   Simplify the radical and then solve the two resulting equations.

Step 6.   Check the solutions.

When we solve an equation by completing the square, the answers will not always be integers.

EXAMPLE 9.13

Solve by completing the square: $x^2 + 4x = -21$.

✓ Solution

$$x^2 + bx \quad c$$
$$x^2 + 4x = -21$$

The variable terms are on the left side. Take half of 4 and square it.	$x^2 + 4x + \underline{\quad} = -21$ $\left(\frac{1}{2} \cdot 4\right)^2$
$\left(\frac{1}{2}(4)\right)^2 = 4$	
Add 4 to both sides.	$x^2 + 4x + 4 = -21 + 4$
Factor the perfect square trinomial, writing it as a binomial squared.	$(x + 2)^2 = -17$
Use the Square Root Property.	$x + 2 = \pm\sqrt{-17}$
Simplify using complex numbers.	$x + 2 = \pm\sqrt{17}\, i$
Subtract 2 from each side.	$x = -2 \pm \sqrt{17}\, i$
Rewrite to show two solutions.	$x = -2 + \sqrt{17}\, i, \quad x = -2 - \sqrt{17}\, i$
We leave the check to you.	

> **TRY IT :: 9.25**   Solve by completing the square: $y^2 - 10y = -35$.

> **TRY IT :: 9.26**   Solve by completing the square: $z^2 + 8z = -19$.

In the previous example, our solutions were complex numbers. In the next example, the solutions will be irrational numbers.

EXAMPLE 9.14

Solve by completing the square: $y^2 - 18y = -6$.

✓ Solution

$$x^2 - bx \quad c$$
$$y^2 - 18y = -6$$

The variable terms are on the left side. Take half of -18 and square it.	
$\left(\frac{1}{2}(-18)\right)^2 = 81$	$y^2 - 18y + \underline{\quad} = -6$ $\left(\frac{1}{2} \cdot (-18)\right)^2$
Add 81 to both sides.	$y^2 - 18y + 81 = -6 + 81$
Factor the perfect square trinomial, writing it as a binomial squared.	$(y - 9)^2 = 75$

Use the Square Root Property.	$y - 9 = \pm\sqrt{75}$
Simplify the radical.	$y - 9 = \pm 5\sqrt{3}$
Solve for y.	$y = 9 \pm 5\sqrt{3}$

Check.

$$y^2 - 18y = -6 \qquad\qquad y^2 - 18y = -6$$

$$(9 + 5\sqrt{3})^2 - 18(9 + 5\sqrt{3}) \overset{?}{=} -6 \qquad (9 - 5\sqrt{3})^2 - 18(9 - 5\sqrt{3}) \overset{?}{=} -6$$

$$81 + 90\sqrt{3} + 75 - 162 + 90\sqrt{3} \overset{?}{=} -6 \qquad 81 + 90\sqrt{3} + 75 - 162 + 90\sqrt{3} \overset{?}{=} -6$$

$$-6 = -6 \checkmark \qquad\qquad -6 = -6 \checkmark$$

Another way to check this would be to use a calculator. Evaluate $y^2 - 18y$ for both of the solutions. The answer should be -6.

> **TRY IT :: 9.27** Solve by completing the square: $x^2 - 16x = -16$.

> **TRY IT :: 9.28** Solve by completing the square: $y^2 + 8y = 11$.

We will start the next example by isolating the variable terms on the left side of the equation.

EXAMPLE 9.15

Solve by completing the square: $x^2 + 10x + 4 = 15$.

⊘ **Solution**

	$x^2 + 10x + 4 = 15$
Isolate the variable terms on the left side. Subtract 4 to get the constant terms on the right side.	$x^2 + 10x = 11$
Take half of 10 and square it.	
$\left(\frac{1}{2}(10)\right)^2 = 25$	$x^2 - 10x + \underline{} = 11$ $\left(\frac{1}{2}\cdot(10)\right)^2$
Add 25 to both sides.	$x^2 + 10x + 25 = 11 + 25$
Factor the perfect square trinomial, writing it as a binomial squared.	$(x + 5)^2 = 36$
Use the Square Root Property.	$x + 5 = \pm\sqrt{36}$
Simplify the radical.	$x + 5 = \pm 6$
Solve for x.	$x = -5 \pm 6$
Rewrite to show two solutions.	$x = -5 + 6, \qquad x = -5 - 6$
Solve the equations.	$x = 1, \qquad\qquad x = -11$

Check:

$x^2 + 10x + 4 = 15$

$(1)^2 + 10(1) + 4 \overset{?}{=} 15$

$1 + 10 + 4 \overset{?}{=} 15$

$15 = 15 \checkmark$

$x^2 + 10x + 4 = 15$

$(-11)^2 + 10(-11) + 4 \overset{?}{=} 15$

$121 + 110 + 4 \overset{?}{=} 15$

$15 = 15 \checkmark$

> **TRY IT :: 9.29** Solve by completing the square: $a^2 + 4a + 9 = 30$.

> **TRY IT :: 9.30** Solve by completing the square: $b^2 + 8b - 4 = 16$.

To solve the next equation, we must first collect all the variable terms on the left side of the equation. Then we proceed as we did in the previous examples.

EXAMPLE 9.16

Solve by completing the square: $n^2 = 3n + 11$.

✓ **Solution**

$$n^2 = 3n + 11$$

Subtract $3n$ to get the variable terms on the left side.

$$n^2 - 3n = 11$$

Take half of -3 and square it.

$$\left(\tfrac{1}{2}(-3)\right)^2 = \tfrac{9}{4}$$

$$n^2 - 3n + \underline{}_{\left(\tfrac{1}{2}\cdot(-3)\right)^2} = 11$$

Add $\tfrac{9}{4}$ to both sides.

$$n^2 - 3n + \tfrac{9}{4} = 11 + \tfrac{9}{4}$$

Factor the perfect square trinomial, writing it as a binomial squared.

$$\left(n - \tfrac{3}{2}\right)^2 = \tfrac{44}{4} + \tfrac{9}{4}$$

Add the fractions on the right side.

$$\left(n - \tfrac{3}{2}\right)^2 = \tfrac{53}{4}$$

Use the Square Root Property.

$$n - \tfrac{3}{2} = \pm\sqrt{\tfrac{53}{4}}$$

Simplify the radical.

$$n - \tfrac{3}{2} = \pm\tfrac{\sqrt{53}}{2}$$

Solve for n.

$$n = \tfrac{3}{2} \pm \tfrac{\sqrt{53}}{2}$$

Rewrite to show two solutions.

$$n = \tfrac{3}{2} + \tfrac{\sqrt{53}}{2}, \quad n = \tfrac{3}{2} - \tfrac{\sqrt{53}}{2}$$

Check:
We leave the check for you!

> **TRY IT :: 9.31** Solve by completing the square: $p^2 = 5p + 9$.

> **TRY IT ::** 9.32   Solve by completing the square: $q^2 = 7q - 3$.

Notice that the left side of the next equation is in factored form. But the right side is not zero. So, we cannot use the Zero Product Property since it says "If $a \cdot b = 0$, then $a = 0$ or $b = 0$." Instead, we multiply the factors and then put the equation into standard form to solve by completing the square.

EXAMPLE 9.17

Solve by completing the square: $(x - 3)(x + 5) = 9$.

✓ **Solution**

	$(x - 3)(x + 5) = 9$
We multiply the binomials on the left.	$x^2 + 2x - 15 = 9$
Add 15 to isolate the constant terms on the right.	$x^2 + 2x \quad\ = 24$
Take half of 2 and square it.	
$\left(\frac{1}{2} \cdot (2)\right)^2 = 1$	$x^2 + 2x + \underset{\left(\frac{1}{2} \cdot (2)\right)^2}{\underline{\qquad}} = 24$
Add 1 to both sides.	$x^2 + 2x + 1 = 24 + 1$
Factor the perfect square trinomial, writing it as a binomial squared.	$(x + 1)^2 = 25$
Use the Square Root Property.	$x + 1 = \pm\sqrt{25}$
Solve for x.	$x = -1 \pm 5$
Rewrite to show two solutions.	$x = -1 + 5, \; x = -1 - 5$
Simplify.	$x = 4, \qquad x = -6$
Check: We leave the check for you!	

> **TRY IT ::** 9.33   Solve by completing the square: $(c - 2)(c + 8) = 11$.

> **TRY IT ::** 9.34   Solve by completing the square: $(d - 7)(d + 3) = 56$.

Solve Quadratic Equations of the Form $ax^2 + bx + c = 0$ by Completing the Square

The process of completing the square works best when the coefficient of x^2 is 1, so the left side of the equation is of the form $x^2 + bx + c$. If the x^2 term has a coefficient other than 1, we take some preliminary steps to make the coefficient equal to 1.

Sometimes the coefficient can be factored from all three terms of the trinomial. This will be our strategy in the next example.

EXAMPLE 9.18

Solve by completing the square: $3x^2 - 12x - 15 = 0$.

✓ **Solution**

To complete the square, we need the coefficient of x^2 to be one. If we factor out the coefficient of x^2 as a common factor, we can continue with solving the equation by completing the square.

	$3x^2 - 12x - 15 = 0$
Factor out the greatest common factor.	$3(x^2 - 4x - 5) = 0$
Divide both sides by 3 to isolate the trinomial with coefficient 1.	$\dfrac{3(x^2 - 4x - 5)}{3} = \dfrac{0}{3}$
Simplify.	$x^2 - 4x - 5 = 0$
Add 5 to get the constant terms on the right side.	$x^2 - 4x \quad\quad = 5$
Take half of 4 and square it. $\left(\tfrac{1}{2}(-4)\right)^2 = 4$	$x^2 - 4x + \underset{\left(\frac{1}{2}\,\cdot\,(4)\right)^2}{\underline{\quad\quad}} = 5$
Add 4 to both sides.	$x^2 - 4x + 4 = 5 + 4$
Factor the perfect square trinomial, writing it as a binomial squared.	$(x - 2)^2 = 9$
Use the Square Root Property.	$x - 2 = \pm\sqrt{9}$
Solve for x.	$x - 2 = \pm 3$
Rewrite to show two solutions.	$x = 2 + 3,\ x = 2 - 3$
Simplify.	$x = 5, \quad\quad x = -1$

Check:

$$x = 5 \qquad\qquad\qquad\qquad x = -1$$
$$3x^2 - 12x - 15 = 0 \qquad\qquad 3x^2 - 12x - 15 = 0$$
$$3(5)^2 - 12(5) - 15 \overset{?}{=} 0 \qquad 3(-1)^2 - 12(-1) - 15 \overset{?}{=} 0$$
$$75 - 60 - 15 \overset{?}{=} 0 \qquad\qquad 3 + 12 - 15 \overset{?}{=} 0$$
$$0 = 0\ \checkmark \qquad\qquad\qquad\qquad 0 = 0\ \checkmark$$

> **TRY IT :: 9.35**   Solve by completing the square: $2m^2 + 16m + 14 = 0$.

> **TRY IT :: 9.36**   Solve by completing the square: $4n^2 - 24n - 56 = 8$.

To complete the square, the coefficient of the x^2 must be 1. When the leading coefficient is not a factor of all the terms, we will divide both sides of the equation by the leading coefficient! This will give us a fraction for the second coefficient. We have already seen how to complete the square with fractions in this section.

EXAMPLE 9.19

Solve by completing the square: $2x^2 - 3x = 20$.

⊘ **Solution**

To complete the square we need the coefficient of x^2 to be one. We will divide both sides of the equation by the coefficient of x^2. Then we can continue with solving the equation by completing the square.

$$2x^2 - 3x = 20$$

Divide both sides by 2 to get the coefficient of x^2 to be 1.	$\dfrac{2x^2 - 3x}{2} = \dfrac{20}{2}$
Simplify.	$x^2 - \dfrac{3}{2}x = 10$
Take half of $-\dfrac{3}{2}$ and square it.	
$\left(\dfrac{1}{2}\left(-\dfrac{3}{2}\right)\right)^2 = \dfrac{9}{16}$	$x^2 - \dfrac{3}{2}x + \underbrace{\phantom{\dfrac{9}{16}}}_{\left(\frac{1}{2}\cdot\left(-\frac{3}{2}\right)\right)^2} = 10$
Add $\dfrac{9}{16}$ to both sides.	$x^2 - \dfrac{3}{2}x + \dfrac{9}{16} = 10 + \dfrac{9}{16}$
Factor the perfect square trinomial, writing it as a binomial squared.	$\left(x - \dfrac{3}{4}\right)^2 = \dfrac{160}{16} + \dfrac{9}{16}$
Add the fractions on the right side.	$\left(x - \dfrac{3}{4}\right)^2 = \dfrac{169}{16}$
Use the Square Root Property.	$x - \dfrac{3}{4} = \pm\sqrt{\dfrac{169}{16}}$
Simplify the radical.	$x - \dfrac{3}{4} = \pm\dfrac{13}{4}$
Solve for x.	$x = \dfrac{3}{4} \pm \dfrac{13}{4}$
Rewrite to show two solutions.	$x = \dfrac{3}{4} + \dfrac{13}{4}, \quad x = \dfrac{3}{4} - \dfrac{13}{4}$
Simplify.	$x = 4, \qquad x = -\dfrac{5}{2}$
Check: We leave the check for you!	

> **TRY IT :: 9.37**   Solve by completing the square: $3r^2 - 2r = 21$.

> **TRY IT :: 9.38**   Solve by completing the square: $4t^2 + 2t = 20$.

Now that we have seen that the coefficient of x^2 must be 1 for us to complete the square, we update our procedure for solving a quadratic equation by completing the square to include equations of the form $ax^2 + bx + c = 0$.

HOW TO :: SOLVE A QUADRATIC EQUATION OF THE FORM $ax^2 + bx + c = 0$ **BY COMPLETING THE SQUARE.**

Step 1. Divide by a to make the coefficient of x^2 term 1.

Step 2. Isolate the variable terms on one side and the constant terms on the other.

Step 3. Find $\left(\frac{1}{2} \cdot b\right)^2$, the number needed to complete the square. Add it to both sides of the equation.

Step 4. Factor the perfect square trinomial, writing it as a binomial squared on the left and simplify by adding the terms on the right

Step 5. Use the Square Root Property.

Step 6. Simplify the radical and then solve the two resulting equations.

Step 7. Check the solutions.

EXAMPLE 9.20

Solve by completing the square: $3x^2 + 2x = 4$.

✓ **Solution**

Again, our first step will be to make the coefficient of x^2 one. By dividing both sides of the equation by the coefficient of x^2, we can then continue with solving the equation by completing the square.

$$3x^2 + 2x = 4$$

Divide both sides by 3 to make the coefficient of x^2 equal 1.	$\dfrac{3x^2 + 2x}{3} = \dfrac{4}{3}$
Simplify.	$x^2 + \dfrac{2}{3}x = \dfrac{4}{3}$
Take half of $\dfrac{2}{3}$ and square it.	
$\left(\dfrac{1}{2} \cdot \dfrac{2}{3}\right)^2 = \dfrac{1}{9}$	$x^2 + \dfrac{2}{3}x + \underset{\left(\frac{1}{2} \cdot \frac{2}{3}\right)^2}{\underline{}} = \dfrac{4}{3}$
Add $\dfrac{1}{9}$ to both sides.	$x^2 + \dfrac{2}{3}x + \dfrac{1}{9} = \dfrac{4}{3} + \dfrac{1}{9}$
Factor the perfect square trinomial, writing it as a binomial squared.	$\left(x + \dfrac{1}{3}\right)^2 = \dfrac{12}{9} + \dfrac{1}{9}$
Use the Square Root Property.	$x + \dfrac{1}{3} = \pm\sqrt{\dfrac{13}{9}}$
Simplify the radical.	$x + \dfrac{1}{3} = \pm\dfrac{\sqrt{13}}{3}$
Solve for x.	$x = -\dfrac{1}{3} \pm \dfrac{\sqrt{13}}{3}$
Rewrite to show two solutions.	$x = -\dfrac{1}{3} + \dfrac{\sqrt{13}}{3}, \quad x = -\dfrac{1}{3} - \dfrac{\sqrt{13}}{3}$
Check: We leave the check for you!	

> **TRY IT : :** 9.39   Solve by completing the square: $4x^2 + 3x = 2$.

> **TRY IT : :** 9.40   Solve by completing the square: $3y^2 - 10y = -5$.

▶ **MEDIA : :**

Access these online resources for additional instruction and practice with completing the square.

- **Completing Perfect Square Trinomials (https://openstax.org/l/37CompTheSq1)**
- **Completing the Square 1 (https://openstax.org/l/37CompTheSq2)**
- **Completing the Square to Solve Quadratic Equations (https://openstax.org/l/37CompTheSq3)**
- **Completing the Square to Solve Quadratic Equations: More Examples (https://openstax.org/l/37CompTheSq4)**
- **Completing the Square 4 (https://openstax.org/l/37CompTheSq5)**

9.2 EXERCISES

Practice Makes Perfect

Complete the Square of a Binomial Expression

In the following exercises, complete the square to make a perfect square trinomial. Then write the result as a binomial squared.

71.
ⓐ $m^2 - 24m$
ⓑ $x^2 - 11x$
ⓒ $p^2 - \frac{1}{3}p$

72.
ⓐ $n^2 - 16n$
ⓑ $y^2 + 15y$
ⓒ $q^2 + \frac{3}{4}q$

73.
ⓐ $p^2 - 22p$
ⓑ $y^2 + 5y$
ⓒ $m^2 + \frac{2}{5}m$

74.
ⓐ $q^2 - 6q$
ⓑ $x^2 - 7x$
ⓒ $n^2 - \frac{2}{3}n$

Solve Quadratic Equations of the form $x^2 + bx + c = 0$ by Completing the Square

In the following exercises, solve by completing the square.

75. 5. $u^2 + 2u = 3$

76. $z^2 + 12z = -11$

77. $x^2 - 20x = 21$

78. $y^2 - 2y = 8$

79. $m^2 + 4m = -44$

80. $n^2 - 2n = -3$

81. $r^2 + 6r = -11$

82. $t^2 - 14t = -50$

83. $a^2 - 10a = -5$

84. $b^2 + 6b = 41$

85. $x^2 + 5x = 2$

86. $y^2 - 3y = 2$

87. $u^2 - 14u + 12 = 1$

88. $z^2 + 2z - 5 = 2$

89. $r^2 - 4r - 3 = 9$

90. $t^2 - 10t - 6 = 5$

91. $v^2 = 9v + 2$

92. $w^2 = 5w - 1$

93. $x^2 - 5 = 10x$

94. $y^2 - 14 = 6y$

95. $(x + 6)(x - 2) = 9$

96. $(y + 9)(y + 7) = 80$

97. $(x + 2)(x + 4) = 3$

98. $(x - 2)(x - 6) = 5$

Solve Quadratic Equations of the form $ax^2 + bx + c = 0$ by Completing the Square

In the following exercises, solve by completing the square.

99. $3m^2 + 30m - 27 = 6$

100. $2x^2 - 14x + 12 = 0$

101. $2n^2 + 4n = 26$

102. $5x^2 + 20x = 15$

103. $2c^2 + c = 6$

104. $3d^2 - 4d = 15$

105. $2x^2 + 7x - 15 = 0$

106. $3x^2 - 14x + 8 = 0$

107. $2p^2 + 7p = 14$

108. $3q^2 - 5q = 9$

109. $5x^2 - 3x = -10$

110. $7x^2 + 4x = -3$

Writing Exercises

111. Solve the equation $x^2 + 10x = -25$

ⓐ by using the Square Root Property

ⓑ by Completing the Square

ⓒ Which method do you prefer? Why?

112. Solve the equation $y^2 + 8y = 48$ by completing the square and explain all your steps.

Self Check

ⓐ *After completing the exercises, use this checklist to evaluate your mastery of the objectives of this section.*

I can...	Confidently	With some help	No-I don't get it!
complete the square of a binomial expression.			
solve quadratic equations of the form $x^2 + bx + c = 0$ by completing the square.			
solve quadratic equations of the form $ax^2 + bx + c = 0$ by completing the square.			

ⓑ *After reviewing this checklist, what will you do to become confident for all objectives?*

$\boxed{9.3}$ Solve Quadratic Equations Using the Quadratic Formula

Learning Objectives

By the end of this section, you will be able to:

> Solve quadratic equations using the Quadratic Formula
> Use the discriminant to predict the number and type of solutions of a quadratic equation
> Identify the most appropriate method to use to solve a quadratic equation

Be Prepared!

Before you get started, take this readiness quiz.

1. Evaluate $b^2 - 4ab$ when $a = 3$ and $b = -2$.
   If you missed this problem, review **Example 1.21**.

2. Simplify: $\sqrt{108}$.
   If you missed this problem, review **Example 8.13**.

3. Simplify: $\sqrt{50}$.
   If you missed this problem, review **Example 8.76**.

Solve Quadratic Equations Using the Quadratic Formula

When we solved quadratic equations in the last section by completing the square, we took the same steps every time. By the end of the exercise set, you may have been wondering 'isn't there an easier way to do this?' The answer is 'yes'. Mathematicians look for patterns when they do things over and over in order to make their work easier. In this section we will derive and use a formula to find the solution of a quadratic equation.

We have already seen how to solve a formula for a specific variable 'in general', so that we would do the algebraic steps only once, and then use the new formula to find the value of the specific variable. Now we will go through the steps of completing the square using the general form of a quadratic equation to solve a quadratic equation for x.

We start with the standard form of a quadratic equation and solve it for x by completing the square.

$$ax^2 + bx + c = 0 \qquad a \neq 0$$

Isolate the variable terms on one side.	$ax^2 + bx = -c$
Make the coefficient of x^2 equal to 1, by dividing by a.	$\dfrac{ax^2}{a} + \dfrac{b}{a}x = -\dfrac{c}{a}$
Simplify.	$x^2 + \dfrac{b}{a}x = -\dfrac{c}{a}$
To complete the square, find $\left(\dfrac{1}{2} \cdot \dfrac{b}{a}\right)^2$ and add it to both sides of the equation. $\left(\dfrac{1}{2}\dfrac{b}{a}\right)^2 = \dfrac{b^2}{4a^2}$	$x^2 + \dfrac{b}{a}x + \dfrac{b^2}{4a^2} + = -\dfrac{c}{a} + \dfrac{b^2}{4a^2}$
The left side is a perfect square, factor it.	$\left(x + \dfrac{b}{2a}\right)^2 = -\dfrac{c}{a} + \dfrac{b^2}{4a^2}$
Find the common denominator of the right side and write equivalent fractions with the common denominator.	$\left(x + \dfrac{b}{2a}\right)^2 = \dfrac{b^2}{4a^2} - \dfrac{c \cdot 4a}{a \cdot 4a}$

Simplify.	$\left(x + \dfrac{b}{2a}\right)^2 = \dfrac{b^2}{4a^2} - \dfrac{4ac}{4a^2}$
Combine to one fraction.	$\left(x + \dfrac{b}{2a}\right)^2 = \dfrac{b^2 - 4ac}{4a^2}$
Use the square root property.	$x + \dfrac{b}{2a} = \pm\sqrt{\dfrac{b^2 - 4ac}{4a^2}}$
Simplify the radical.	$x + \dfrac{b}{2a} = \pm\dfrac{\sqrt{b^2 - 4ac}}{2a}$
Add $-\dfrac{b}{2a}$ to both sides of the equation.	$x = -\dfrac{b}{2a} \pm \dfrac{\sqrt{b^2 - 4ac}}{2a}$
Combine the terms on the right side.	$x = \dfrac{-b \pm \sqrt{b^2 - 4ac}}{2a}$
	This equation is the Quadratic Formula.

Quadratic Formula

The solutions to a quadratic equation of the form $ax^2 + bx + c = 0$, where $a \neq 0$ are given by the formula:

$$x = \frac{-b \pm \sqrt{b^2 - 4ac}}{2a}$$

To use the Quadratic Formula, we substitute the values of a, b, and c from the standard form into the expression on the right side of the formula. Then we simplify the expression. The result is the pair of solutions to the quadratic equation.

Notice the formula is an equation. Make sure you use both sides of the equation.

EXAMPLE 9.21   HOW TO SOLVE A QUADRATIC EQUATION USING THE QUADRATIC FORMULA

Solve by using the Quadratic Formula: $2x^2 + 9x - 5 = 0$.

⊘ **Solution**

Step 1. Write the quadratic equation in standard form. Identify the a, b, c values.	This equation is in standard form.	$ax^2 + bx + c = 0$ $2x^2 + 9x - 5 = 0$ $a = 2, b = 9, c = -5$
Step 2. Write the quadratic formula. Then substitute in the values of a, b, c.	Substitute in $a = 2, b = 9, c = -5$	$x = \dfrac{-b \pm \sqrt{b^2 - 4ac}}{2a}$ $x = \dfrac{-9 \pm \sqrt{9^2 - 4 \cdot 2 \cdot (-5)}}{2 \cdot 2}$

Step 3. Simplify the fraction, and solve for x.		$x = \dfrac{-9 \pm \sqrt{81 - (-40)}}{4}$
		$x = \dfrac{-9 \pm \sqrt{121}}{4}$
		$x = \dfrac{-9 \pm 11}{4}$
		$x = \dfrac{-9 + 11}{4} \qquad x = \dfrac{-9 - 11}{4}$
		$x = \dfrac{2}{4} \qquad x = \dfrac{-20}{4}$
		$x = \dfrac{1}{2} \qquad x = -5$
Step 4. Check the solutions.	Put each answer in the original equation to check. Substitute $x = \frac{1}{2}$.	$2x^2 + 9x - 5 = 0$
		$2\left(\dfrac{1}{2}\right)^2 + 9 \cdot \dfrac{1}{2} - 5 \overset{?}{=} 0$
		$2 \cdot \dfrac{1}{4} + 9 \cdot \dfrac{1}{2} - 5 \overset{?}{=} 0$
		$2 \cdot \dfrac{1}{4} + 9 \cdot \dfrac{1}{2} - 5 \overset{?}{=} 0$
		$\dfrac{1}{2} + \dfrac{9}{2} - 5 \overset{?}{=} 0$
		$\dfrac{10}{2} - 5 \overset{?}{=} 0$
		$5 - 5 \overset{?}{=} 0$
		$0 = 0 \checkmark$
	Substitute $x = -5$.	$2x^2 + 9x - 5 = 0$
		$2(-5)^2 + 9(-5) - 5 \overset{?}{=} 0$
		$2 \cdot 25 - 45 - 5 \overset{?}{=} 0$
		$50 - 45 - 5 \overset{?}{=} 0$
		$0 = 0 \checkmark$

> **TRY IT :: 9.41**   Solve by using the Quadratic Formula: $3y^2 - 5y + 2 = 0$.

> **TRY IT :: 9.42**   Solve by using the Quadratic Formula: $4z^2 + 2z - 6 = 0$.

HOW TO :: SOLVE A QUADRATIC EQUATION USING THE QUADRATIC FORMULA.

Step 1. Write the quadratic equation in standard form, $ax^2 + bx + c = 0$. Identify the values of a, b, and c.

Step 2. Write the Quadratic Formula. Then substitute in the values of a, b, and c.

Step 3. Simplify.

Step 4. Check the solutions.

If you say the formula as you write it in each problem, you'll have it memorized in no time! And remember, the Quadratic

Formula is an EQUATION. Be sure you start with "$x =$".

EXAMPLE 9.22

Solve by using the Quadratic Formula: $x^2 - 6x = -5$.

⊘ Solution

	$x^2 - 6x = -5$
Write the equation in standard form by adding 5 to each side.	$x^2 - 6x + 5 = 0$
This equation is now in standard form.	$ax^2 + bx + c = 0$ $x^2 - 6x + 5 = 0$
Identify the values of a, b, c.	$a = 1, b = -6, c = 5$
Write the Quadratic Formula.	$x = \dfrac{-b \pm \sqrt{b^2 - 4ac}}{2a}$
Then substitute in the values of a, b, c.	$x = \dfrac{-(-6) \pm \sqrt{(-6)^2 - 4 \cdot 1 \cdot (5)}}{2 \cdot 1}$
Simplify.	$x = \dfrac{6 \pm \sqrt{36 - 20}}{2}$
	$x = \dfrac{6 \pm \sqrt{16}}{2}$
	$x = \dfrac{6 \pm 4}{2}$
Rewrite to show two solutions.	$x = \dfrac{6 + 4}{2}, \quad x = \dfrac{6 - 4}{2}$
Simplify.	$x = \dfrac{10}{2}, \quad x = \dfrac{2}{2}$
	$x = 5, \quad\quad x = 1$

Check:

$$x^2 - 6x + 5 = 0 \qquad\qquad x^2 - 6x + 5 = 0$$
$$5^2 - 6 \cdot 5 + 5 \overset{?}{=} 0 \qquad\quad 1^2 - 6 \cdot 1 + 5 \overset{?}{=} 0$$
$$25 - 30 + 5 \overset{?}{=} 0 \qquad\qquad 1 - 6 + 5 \overset{?}{=} 0$$
$$0 = 0 \checkmark \qquad\qquad\qquad 0 = 0 \checkmark$$

> **TRY IT :: 9.43**   Solve by using the Quadratic Formula: $a^2 - 2a = 15$.

> **TRY IT :: 9.44**   Solve by using the Quadratic Formula: $b^2 + 24 = -10b$.

When we solved quadratic equations by using the Square Root Property, we sometimes got answers that had radicals. That can happen, too, when using the Quadratic Formula. If we get a radical as a solution, the final answer must have the radical in its simplified form.

EXAMPLE 9.23

Solve by using the Quadratic Formula: $2x^2 + 10x + 11 = 0$.

✓ Solution

$$2x^2 + 10x + 11 = 0$$

This equation is in standard form.	$ax^2 + bx + c = 0$ $2x^2 + 10x + 11 = 0$
Identify the values of a, b, and c.	$a = 2, b = 10, c = 11$
Write the Quadratic Formula.	$x = \dfrac{-b \pm \sqrt{b^2 - 4ac}}{2a}$
Then substitute in the values of a, b, and c.	$x = \dfrac{-(10) \pm \sqrt{(10)^2 - 4 \cdot 2 \cdot (11)}}{2 \cdot 2}$
Simplify.	$x = \dfrac{-10 \pm \sqrt{100 - 88}}{4}$
	$x = \dfrac{-10 \pm \sqrt{12}}{4}$
Simplify the radical.	$x = \dfrac{-10 \pm 2\sqrt{3}}{4}$
Factor out the common factor in the numerator.	$x = \dfrac{2(-5 \pm \sqrt{3})}{4}$
Remove the common factors.	$x = \dfrac{-5 \pm \sqrt{3}}{2}$
Rewrite to show two solutions.	$x = \dfrac{-5 + \sqrt{3}}{2}, \quad x = \dfrac{-5 - \sqrt{3}}{2}$

Check:
We leave the check for you!

> **TRY IT :: 9.45** Solve by using the Quadratic Formula: $3m^2 + 12m + 7 = 0$.

> **TRY IT :: 9.46** Solve by using the Quadratic Formula: $5n^2 + 4n - 4 = 0$.

When we substitute a, b, and c into the Quadratic Formula and the radicand is negative, the quadratic equation will have imaginary or complex solutions. We will see this in the next example.

EXAMPLE 9.24

Solve by using the Quadratic Formula: $3p^2 + 2p + 9 = 0$.

✓ Solution

$$3p^2 + 2p + 9 = 0$$

This equation is in standard form	$ax^2 + bx + c = 0$ $3p^2 + 2p + 9 = 0$
Identify the values of a, b, c.	$a = 3, b = 2, c = 9$
Write the Quadratic Formula.	$p = \dfrac{-b \pm \sqrt{b^2 - 4ac}}{2a}$
Then substitute in the values of a, b, c.	$p = \dfrac{-(2) \pm \sqrt{(2)^2 - 4 \cdot 3 \cdot (9)}}{2 \cdot 3}$

Simplify.	$p = \dfrac{-2 \pm \sqrt{4 - 108}}{6}$
	$p = \dfrac{-2 \pm \sqrt{-104}}{6}$
Simplify the radical using complex numbers.	$p = \dfrac{-2 \pm \sqrt{104}\,i}{6}$
Simplify the radical.	$p = \dfrac{-2 \pm 2\sqrt{26}\,i}{6}$
Factor the common factor in the numerator.	$p = \dfrac{2(-1 \pm \sqrt{26}\,i)}{6}$
Remove the common factors.	$p = \dfrac{-1 \pm \sqrt{26}\,i}{3}$
Rewrite in standard $a + bi$ form.	$p = -\dfrac{1}{3} \pm \dfrac{\sqrt{26}\,i}{3}$
Write as two solutions.	$p = -\dfrac{1}{3} + \dfrac{\sqrt{26}\,i}{3}, \quad p = -\dfrac{1}{3} - \dfrac{\sqrt{26}\,i}{3}$

> **TRY IT :: 9.47**   Solve by using the Quadratic Formula: $4a^2 - 2a + 8 = 0$.

> **TRY IT :: 9.48**   Solve by using the Quadratic Formula: $5b^2 + 2b + 4 = 0$.

Remember, to use the Quadratic Formula, the equation must be written in standard form, $ax^2 + bx + c = 0$. Sometimes, we will need to do some algebra to get the equation into standard form before we can use the Quadratic Formula.

EXAMPLE 9.25

Solve by using the Quadratic Formula: $x(x + 6) + 4 = 0$.

⊘ **Solution**

Our first step is to get the equation in standard form.

	$x(x + 6) + 4 = 0$
Distribute to get the equation in standard form.	$x^2 + 6x + 4 = 0$
This equation is now in standard form	$ax^2 + bx + c = 0$ $x^2 + 6x + 4 = 0$
Identify the values of a, b, c.	$a = 1, b = 6, c = 4$
Write the Quadratic Formula.	$x = \dfrac{-b \pm \sqrt{b^2 - 4ac}}{2a}$
Then substitute in the values of a, b, c.	$x = \dfrac{-(6) \pm \sqrt{(6)^2 - 4 \cdot 1 \cdot (4)}}{2 \cdot 1}$
Simplify.	$x = \dfrac{-6 \pm \sqrt{36 - 16}}{2}$
	$x = \dfrac{-6 \pm \sqrt{20}}{2}$
Simplify the radical.	$x = \dfrac{-6 \pm 2\sqrt{5}}{2}$

Factor the common factor in the numerator.	$x = \dfrac{2(-3 \pm 2\sqrt{5})}{2}$
Remove the common factors.	$x = -3 \pm 2\sqrt{5}$
Write as two solutions.	$x = -3 + 2\sqrt{5}, \quad x = -3 - 2\sqrt{5}$
Check: We leave the check for you!	

> **TRY IT :: 9.49**   Solve by using the Quadratic Formula: $x(x+2) - 5 = 0$.

> **TRY IT :: 9.50**   Solve by using the Quadratic Formula: $3y(y-2) - 3 = 0$.

When we solved linear equations, if an equation had too many fractions we cleared the fractions by multiplying both sides of the equation by the LCD. This gave us an equivalent equation—without fractions— to solve. We can use the same strategy with quadratic equations.

EXAMPLE 9.26

Solve by using the Quadratic Formula: $\frac{1}{2}u^2 + \frac{2}{3}u = \frac{1}{3}$.

⊘ Solution

Our first step is to clear the fractions.

	$\frac{1}{2}u^2 + \frac{2}{3}u = \frac{1}{3}$
Multiply both sides by the LCD, 6, to clear the fractions.	$6\left(\frac{1}{2}u^2 + \frac{2}{3}u\right) = 6\left(\frac{1}{3}\right)$
Multiply.	$3u^2 + 4u = 2$
Subtract 2 to get the equation in standard form.	$ax^2 + bx + c = 0$ $3u^2 + 4u - 2 = 0$
Identify the values of a, b, and c.	$a = 3,\ b = 4,\ c = -2$
Write the Quadratic Formula.	$u = \dfrac{-b \pm \sqrt{b^2 - 4ac}}{2a}$
Then substitute in the values of a, b, and c.	$u = \dfrac{-(4) \pm \sqrt{(4)^2 - 4 \cdot 3 \cdot (-2)}}{2 \cdot 3}$
Simplify.	$u = \dfrac{-4 \pm \sqrt{16 + 24}}{6}$
	$u = \dfrac{-4 \pm \sqrt{40}}{6}$
Simplify the radical.	$u = \dfrac{-4 \pm 2\sqrt{10}}{6}$
Factor the common factor in the numerator.	$u = \dfrac{2(-2 \pm \sqrt{10})}{6}$
Remove the common factors.	$u = \dfrac{-2 \pm \sqrt{10}}{3}$
Rewrite to show two solutions.	$u = \dfrac{-2 + \sqrt{10}}{3}, \quad u = \dfrac{-2 - \sqrt{10}}{3}$

Check:
We leave the check for you!

> **TRY IT :: 9.51**   Solve by using the Quadratic Formula: $\frac{1}{4}c^2 - \frac{1}{3}c = \frac{1}{12}$.

> **TRY IT :: 9.52**   Solve by using the Quadratic Formula: $\frac{1}{9}d^2 - \frac{1}{2}d = -\frac{1}{3}$.

Think about the equation $(x - 3)^2 = 0$. We know from the Zero Product Property that this equation has only one solution, $x = 3$.

We will see in the next example how using the Quadratic Formula to solve an equation whose standard form is a perfect square trinomial equal to 0 gives just one solution. Notice that once the radicand is simplified it becomes 0 , which leads to only one solution.

EXAMPLE 9.27

Solve by using the Quadratic Formula: $4x^2 - 20x = -25$.

⊘ **Solution**

$$4x^2 - 20x = -25$$

Add 25 to get the equation in standard form.	$ax^2 + bx + c = 0$ $4x^2 - 20x + 25 = 0$
Identify the values of a, b, and c.	$a = 4$, $b = -20$, $c = 25$
Write the quadratic formula.	$x = \dfrac{-b \pm \sqrt{b^2 - 4ac}}{2a}$
Then substitute in the values of a, b, and c.	$x = \dfrac{-(-20) \pm \sqrt{(-20)^2 - 4 \cdot 4 \cdot (25)}}{2 \cdot 4}$
Simplify.	$x = \dfrac{20 \pm \sqrt{400 - 400}}{8}$
	$x = \dfrac{20 \pm \sqrt{0}}{8}$
Simplify the radical.	$x = \dfrac{20}{8}$
Simplify the fraction.	$x = \dfrac{5}{2}$

Check:
We leave the check for you!

Did you recognize that $4x^2 - 20x + 25$ is a perfect square trinomial. It is equivalent to $(2x - 5)^2$? If you solve $4x^2 - 20x + 25 = 0$ by factoring and then using the Square Root Property, do you get the same result?

> **TRY IT :: 9.53**   Solve by using the Quadratic Formula: $r^2 + 10r + 25 = 0$.

> **TRY IT :: 9.54**   Solve by using the Quadratic Formula: $25t^2 - 40t = -16$.

Use the Discriminant to Predict the Number and Type of Solutions of a Quadratic Equation

When we solved the quadratic equations in the previous examples, sometimes we got two real solutions, one real solution, and sometimes two complex solutions. Is there a way to predict the number and type of solutions to a quadratic equation without actually solving the equation?

Yes, the expression under the radical of the Quadratic Formula makes it easy for us to determine the number and type of solutions. This expression is called the **discriminant**.

Discriminant

In the Quadratic Formula, $x = \dfrac{-b \pm \sqrt{b^2 - 4ac}}{2a}$,

the quantity $b^2 - 4ac$ is called the discriminant.

Let's look at the discriminant of the equations in some of the examples and the number and type of solutions to those quadratic equations.

Quadratic Equation (in standard form)	Discriminant $b^2 - 4ac$	Value of the Discriminant	Number and Type of solutions
$2x^2 + 9x - 5 = 0$	$9^2 - 4 \cdot 2(-5)$ 121	$+$	2 real
$4x^2 - 20x + 25 = 0$	$(-20)^2 - 4 \cdot 4 \cdot 25$ 0	0	1 real
$3p^2 + 2p + 9 = 0$	$2^2 - 4 \cdot 3 \cdot 9$ -104	$-$	2 complex

When the discriminant is positive, the quadratic equation has **2 real solutions**.

$$x = \frac{-b \pm \sqrt{+}}{2a}$$

When the discriminant is zero, the quadratic equation has **1 real solution**.

$$x = \frac{-b \pm \sqrt{0}}{2a}$$

When the discriminant is negative, the quadratic equation has **2 complex solutions**.

$$x = \frac{-b \pm \sqrt{-}}{2a}$$

Using the Discriminant, $b^2 - 4ac$, to Determine the Number and Type of Solutions of a Quadratic Equation

For a quadratic equation of the form $ax^2 + bx + c = 0$, $a \neq 0$,

- If $b^2 - 4ac > 0$, the equation has 2 real solutions.
- if $b^2 - 4ac = 0$, the equation has 1 real solution.
- if $b^2 - 4ac < 0$, the equation has 2 complex solutions.

EXAMPLE 9.28

Determine the number of solutions to each quadratic equation.

ⓐ $3x^2 + 7x - 9 = 0$ ⓑ $5n^2 + n + 4 = 0$ ⓒ $9y^2 - 6y + 1 = 0$.

✓ **Solution**

To determine the number of solutions of each quadratic equation, we will look at its discriminant.

ⓐ

$$3x^2 + 7x - 9 = 0$$

The equation is in standard form, identify a, b, and c.

$$a = 3, \quad b = 7, \quad c = -9$$

Write the discriminant.

$$b^2 - 4ac$$

Substitute in the values of a, b, and c.

$$(7)^2 - 4 \cdot 3 \cdot (-9)$$

Simplify.

$$49 + 108$$
$$157$$

Since the discriminant is positive, there are 2 real solutions to the equation.

$$5n^2 + n + 4 = 0$$

The equation is in standard form, identify a, b, and c.

$$a = 5, \quad b = 1, \quad c = 4$$

Write the discriminant.

$$b^2 - 4ac$$

Substitute in the values of a, b, and c.

$$(1)^2 - 4 \cdot 5 \cdot 4$$

Simplify.

$$1 - 80$$
$$-79$$

Since the discriminant is negative, there are 2 complex solutions to the equation.

$$9y^2 - 6y + 1 = 0$$

The equation is in standard form, identify a, b, and c.

$$a = 9, \quad b = -6, \quad c = 1$$

Write the discriminant.

$$b^2 - 4ac$$

Substitute in the values of a, b, and c.

$$(-6)^2 - 4 \cdot 9 \cdot 1$$

Simplify.

$$36 - 36$$
$$0$$

Since the discriminant is 0, there is 1 real solution to the equation.

> **TRY IT :: 9.55**   Determine the number and type of solutions to each quadratic equation.
>
> ⓐ $8m^2 - 3m + 6 = 0$ ⓑ $5z^2 + 6z - 2 = 0$ ⓒ $9w^2 + 24w + 16 = 0$.

> **TRY IT :: 9.56**   Determine the number and type of solutions to each quadratic equation.
>
> ⓐ $b^2 + 7b - 13 = 0$ ⓑ $5a^2 - 6a + 10 = 0$ ⓒ $4r^2 - 20r + 25 = 0$.

Identify the Most Appropriate Method to Use to Solve a Quadratic Equation

We summarize the four methods that we have used to solve quadratic equations below.

Methods for Solving Quadratic Equations

1. Factoring
2. Square Root Property
3. Completing the Square
4. Quadratic Formula

Given that we have four methods to use to solve a quadratic equation, how do you decide which one to use? Factoring is often the quickest method and so we try it first. If the equation is $ax^2 = k$ or $a(x - h)^2 = k$ we use the Square Root Property. For any other equation, it is probably best to use the Quadratic Formula. Remember, you can solve any quadratic equation by using the Quadratic Formula, but that is not always the easiest method.

What about the method of Completing the Square? Most people find that method cumbersome and prefer not to use it. We needed to include it in the list of methods because we completed the square in general to derive the Quadratic Formula. You will also use the process of Completing the Square in other areas of algebra.

HOW TO :: IDENTIFY THE MOST APPROPRIATE METHOD TO SOLVE A QUADRATIC EQUATION.

Step 1. Try **Factoring** first. If the quadratic factors easily, this method is very quick.

Step 2. Try the **Square Root Property** next. If the equation fits the form $ax^2 = k$ or $a(x-h)^2 = k$, it can easily be solved by using the Square Root Property.

Step 3. Use the **Quadratic Formula**. Any other quadratic equation is best solved by using the Quadratic Formula.

The next example uses this strategy to decide how to solve each quadratic equation.

EXAMPLE 9.29

Identify the most appropriate method to use to solve each quadratic equation.

ⓐ $5z^2 = 17$ ⓑ $4x^2 - 12x + 9 = 0$ ⓒ $8u^2 + 6u = 11$.

✓ Solution

ⓐ

$$5z^2 = 17$$

Since the equation is in the $ax^2 = k$, the most appropriate method is to use the Square Root Property.

ⓑ

$$4x^2 - 12x + 9 = 0$$

We recognize that the left side of the equation is a perfect square trinomial, and so factoring will be the most appropriate method.

ⓒ

$$8u^2 + 6u = 11$$

Put the equation in standard form. $$8u^2 + 6u - 11 = 0$$

While our first thought may be to try factoring, thinking about all the possibilities for trial and error method leads us to choose the Quadratic Formula as the most appropriate method.

> **TRY IT :: 9.57**   Identify the most appropriate method to use to solve each quadratic equation.

ⓐ $x^2 + 6x + 8 = 0$ ⓑ $(n-3)^2 = 16$ ⓒ $5p^2 - 6p = 9$.

> **TRY IT :: 9.58**   Identify the most appropriate method to use to solve each quadratic equation.

ⓐ $8a^2 + 3a - 9 = 0$ ⓑ $4b^2 + 4b + 1 = 0$ ⓒ $5c^2 = 125$.

▶ **MEDIA ::**

Access these online resources for additional instruction and practice with using the Quadratic Formula.

- **Using the Quadratic Formula (https://openstax.org/l/37QuadForm1)**
- **Solve a Quadratic Equation Using the Quadratic Formula with Complex Solutions (https://openstax.org/l/37QuadForm2)**
- **Discriminant in Quadratic Formula (https://openstax.org/l/37QuadForm3)**

9.3 EXERCISES
Practice Makes Perfect

Solve Quadratic Equations Using the Quadratic Formula
In the following exercises, solve by using the Quadratic Formula.

113. $4m^2 + m - 3 = 0$

114. $4n^2 - 9n + 5 = 0$

115. $2p^2 - 7p + 3 = 0$

116. $3q^2 + 8q - 3 = 0$

117. $p^2 + 7p + 12 = 0$

118. $q^2 + 3q - 18 = 0$

119. $r^2 - 8r = 33$

120. $t^2 + 13t = -40$

121. $3u^2 + 7u - 2 = 0$

122. $2p^2 + 8p + 5 = 0$

123. $2a^2 - 6a + 3 = 0$

124. $5b^2 + 2b - 4 = 0$

125. $x^2 + 8x - 4 = 0$

126. $y^2 + 4y - 4 = 0$

127. $3y^2 + 5y - 2 = 0$

128. $6x^2 + 2x - 20 = 0$

129. $2x^2 + 3x + 3 = 0$

130. $2x^2 - x + 1 = 0$

131. $8x^2 - 6x + 2 = 0$

132. $8x^2 - 4x + 1 = 0$

133. $(v + 1)(v - 5) - 4 = 0$

134. $(x + 1)(x - 3) = 2$

135. $(y + 4)(y - 7) = 18$

136. $(x + 2)(x + 6) = 21$

137. $\frac{1}{3}m^2 + \frac{1}{12}m = \frac{1}{4}$

138. $\frac{1}{3}n^2 + n = -\frac{1}{2}$

139. $\frac{3}{4}b^2 + \frac{1}{2}b = \frac{3}{8}$

140. $\frac{1}{9}c^2 + \frac{2}{3}c = 3$

141. $16c^2 + 24c + 9 = 0$

142. $25d^2 - 60d + 36 = 0$

143. $25q^2 + 30q + 9 = 0$

144. $16y^2 + 8y + 1 = 0$

Use the Discriminant to Predict the Number of Solutions of a Quadratic Equation
In the following exercises, determine the number of solutions for each quadratic equation.

145.
ⓐ $4x^2 - 5x + 16 = 0$
ⓑ $36y^2 + 36y + 9 = 0$
ⓒ $6m^2 + 3m - 5 = 0$

146.
ⓐ $9v^2 - 15v + 25 = 0$
ⓑ $100w^2 + 60w + 9 = 0$
ⓒ $5c^2 + 7c - 10 = 0$

147.
ⓐ $r^2 + 12r + 36 = 0$
ⓑ $8t^2 - 11t + 5 = 0$
ⓒ $3v^2 - 5v - 1 = 0$

148.
ⓐ $25p^2 + 10p + 1 = 0$
ⓑ $7q^2 - 3q - 6 = 0$
ⓒ $7y^2 + 2y + 8 = 0$

Identify the Most Appropriate Method to Use to Solve a Quadratic Equation

In the following exercises, identify the most appropriate method (Factoring, Square Root, or Quadratic Formula) to use to solve each quadratic equation. Do not solve.

149.
ⓐ $x^2 - 5x - 24 = 0$
ⓑ $(y + 5)^2 = 12$
ⓒ $14m^2 + 3m = 11$

150.
ⓐ $(8v + 3)^2 = 81$
ⓑ $w^2 - 9w - 22 = 0$
ⓒ $4n^2 - 10 = 6$

151.
ⓐ $6a^2 + 14 = 20$
ⓑ $\left(x - \frac{1}{4}\right)^2 = \frac{5}{16}$
ⓒ $y^2 - 2y = 8$

152.
ⓐ $8b^2 + 15b = 4$
ⓑ $\frac{5}{9}v^2 - \frac{2}{3}v = 1$
ⓒ $\left(w + \frac{4}{3}\right)^2 = \frac{2}{9}$

Writing Exercises

153. Solve the equation $x^2 + 10x = 120$

ⓐ by completing the square

ⓑ using the Quadratic Formula

ⓒ Which method do you prefer? Why?

154. Solve the equation $12y^2 + 23y = 24$

ⓐ by completing the square

ⓑ using the Quadratic Formula

ⓒ Which method do you prefer? Why?

Self Check

ⓐ *After completing the exercises, use this checklist to evaluate your mastery of the objectives of this section.*

I can...	Confidently	With some help	No-I don't get it!
solve quadratic equations using the quadratic formula.			
use the discriminant to predict the number of solutions of a quadratic equation.			
identify the most appropriate method to use to solve a quadratic equation.			

ⓑ *What does this checklist tell you about your mastery of this section? What steps will you take to improve?*

9.4 # Solve Quadratic Equations in Quadratic Form

Learning Objectives

By the end of this section, you will be able to:

> Solve equations in quadratic form

Be Prepared!

Before you get started, take this readiness quiz.

1. Factor by substitution: $y^4 - y^2 - 20$.
   If you missed this problem, review **Example 6.21**.

2. Factor by substitution: $(y - 4)^2 + 8(y - 4) + 15$.
   If you missed this problem, review **Example 6.22**.

3. Simplify: ⓐ $x^{\frac{1}{2}} \cdot x^{\frac{1}{4}}$ ⓑ $\left(x^{\frac{1}{3}}\right)^2$ ⓒ $\left(x^{-1}\right)^2$.

   If you missed this problem, review **Example 8.33**.

Solve Equations in Quadratic Form

Sometimes when we factored trinomials, the trinomial did not appear to be in the $ax^2 + bx + c$ form. So we factored by substitution allowing us to make it fit the $ax^2 + bx + c$ form. We used the standard u for the substitution.

To factor the expression $x^4 - 4x^2 - 5$, we noticed the variable part of the middle term is x^2 and its square, x^4, is the variable part of the first term. (We know $\left(x^2\right)^2 = x^4$.) So we let $u = x^2$ and factored.

$$x^4 - 4x^2 - 5$$

$$(x^2)^2 - 4(x^2) - 5$$

Let $u = x^2$ and substitute.	$u^2 - 4u - 5$
Factor the trinomial.	$(u + 1)(u - 5)$
Replace u with x^2.	$(x^2 + 1)(x^2 - 5)$

Similarly, sometimes an equation is not in the $ax^2 + bx + c = 0$ form but looks much like a quadratic equation. Then, we can often make a thoughtful substitution that will allow us to make it fit the $ax^2 + bx + c = 0$ form. If we can make it fit the form, we can then use all of our methods to solve quadratic equations.

Notice that in the quadratic equation $ax^2 + bx + c = 0$, the middle term has a variable, x, and its square, x^2, is the variable part of the first term. Look for this relationship as you try to find a substitution.

Again, we will use the standard u to make a substitution that will put the equation in quadratic form. If the substitution gives us an equation of the form $ax^2 + bx + c = 0$, we say the original equation was of **quadratic form**.

The next example shows the steps for solving an equation in quadratic form.

EXAMPLE 9.30   HOW TO SOLVE EQUATIONS IN QUADRATIC FORM

Solve: $6x^4 - 7x^2 + 2 = 0$

✓ **Solution**

Step 1. Identify a substitution that will put the equation in quadratic form.	Since $(x^2)^2 = x^4$, we let $u = x^2$.	$6x^4 - 7x^2 + 2 = 0$
Step 2. Rewrite the equation with the substitution to put it in quadratic form.	Rewrite to prepare for the substitution. Substitute $u = x^2$.	$6(x^2)^2 - 7x^2 + 2 = 0$ $6u^2 - 7u + 2 = 0$
Step 3. Solve the quadratic equation for u.	We can solve by factoring. Use the Zero Product Property.	$(2u - 1)(3u - 2) = 0$ $2u - 1 = 0,\ 3u - 2 = 0$ $2u = 1,\ 3u = 2$ $u = \dfrac{1}{2}\quad u = \dfrac{2}{3}$
Step 4. Substitute the original variable back into the results, using the substitution.	Replace u with x^2.	$x^2 = \dfrac{1}{2}\qquad x^2 = \dfrac{2}{3}$
Step 5. Solve for the original variable.	Solve for x, using the Square Root Property.	$x = \pm\sqrt{\dfrac{1}{2}}\qquad x = \pm\sqrt{\dfrac{2}{3}}$ $x = \pm\dfrac{\sqrt{2}}{2}\qquad x = \pm\dfrac{\sqrt{6}}{3}$ There are four solutions. $x = \dfrac{\sqrt{2}}{2}\qquad x = \dfrac{\sqrt{6}}{3}$ $x = -\dfrac{\sqrt{2}}{2}\qquad x = -\dfrac{\sqrt{6}}{3}$
Step 6. Check the solutions.	Check all four solutions. We will show one check here.	$x = \dfrac{\sqrt{2}}{2}$ $6x^4 - 7x^2 + 2 = 0$ $6\left(\dfrac{\sqrt{2}}{2}\right)^4 - 7\left(\dfrac{\sqrt{2}}{2}\right)^2 + 2 \overset{?}{=} 0$ $6\left(\dfrac{4}{16}\right) - 7\left(\dfrac{2}{4}\right) + 2 \overset{?}{=} 0$ $\dfrac{3}{2} - \dfrac{7}{2} + \dfrac{4}{2} \overset{?}{=} 0$ $0 = 0\ \checkmark$ We leave the other checks to you!

> **TRY IT ::** 9.59   Solve: $x^4 - 6x^2 + 8 = 0$.

> **TRY IT ::** 9.60   Solve: $x^4 - 11x^2 + 28 = 0$.

We summarize the steps to solve an equation in quadratic form.

HOW TO :: SOLVE EQUATIONS IN QUADRATIC FORM.

Step 1.   Identify a substitution that will put the equation in quadratic form.

Step 2.   Rewrite the equation with the substitution to put it in quadratic form.

Step 3.   Solve the quadratic equation for u.

Step 4.   Substitute the original variable back into the results, using the substitution.

Step 5.   Solve for the original variable.

Step 6.   Check the solutions.

In the next example, the binomial in the middle term, $(x-2)$ is squared in the first term. If we let $u = x - 2$ and substitute, our trinomial will be in $ax^2 + bx + c$ form.

EXAMPLE 9.31

Solve: $(x-2)^2 + 7(x-2) + 12 = 0$.

 **Solution**

	$(x-2)^2 + 7(x-2) + 12 = 0$
Prepare for the substitution.	$(x-2)^2 + 7(x-2) + 12 = 0$
Let $u = x - 2$ and substitute.	$u^2 + 7u + 12 = 0$
Solve by factoring.	$(u+3)(u+4) = 0$
	$u + 3 = 0, \quad u + 4 = 0$
	$u = -3, \qquad u = -4$
Replace u with $x - 2$.	$x - 2 = -3, \ x - 2 = -4$
Solve for x.	$x = -1, \qquad x = -2$

Check:

$$x = -1$$
$$(x-2)^2 + 7(x-2) + 12 = 0$$
$$(-1-2)^2 + 7(-1-2) + 12 \overset{?}{=} 0$$
$$(-3)^2 + 7(-3) + 12 \overset{?}{=} 0$$
$$9 - 21 + 12 \overset{?}{=} 0$$
$$0 = 0 \ \checkmark$$

$$x = -2$$
$$(x-2)^2 + 7(x-2) + 12 = 0$$
$$(-2-2)^2 + 7(-2-2) + 12 \overset{?}{=} 0$$
$$(-4)^2 + 7(-4) + 12 \overset{?}{=} 0$$
$$16 - 28 + 12 \overset{?}{=} 0$$
$$0 = 0 \ \checkmark$$

> **TRY IT :: 9.61**   Solve: $(x-5)^2 + 6(x-5) + 8 = 0$.

> **TRY IT :: 9.62**   Solve: $(y-4)^2 + 8(y-4) + 15 = 0$.

In the next example, we notice that $(\sqrt{x})^2 = x$. Also, remember that when we square both sides of an equation, we may introduce extraneous roots. Be sure to check your answers!

EXAMPLE 9.32

Solve: $x - 3\sqrt{x} + 2 = 0$.

✓ Solution

The $\sqrt{x}$ in the middle term, is squared in the first term $(\sqrt{x})^2 = x$. If we let $u = \sqrt{x}$ and substitute, our trinomial will be in $ax^2 + bx + c = 0$ form.

$$x - 3\sqrt{x} + 2 = 0$$

Rewrite the trinomial to prepare for the substitution.	$(\sqrt{x})^2 - 3\sqrt{x} + 2 = 0$
Let $u = \sqrt{x}$ and substitute.	$u^2 - 3u + 2 = 0$
Solve by factoring.	$(u - 2)(u - 1) = 0$
	$u - 2 = 0, \quad u - 1 = 0$
	$u = 2, \qquad u = 1$
Replace u with $\sqrt{x}$.	$\sqrt{x} = 2, \qquad \sqrt{x} = 1$
Solve for x, by squaring both sides.	$x = 4, \qquad x = 1$

Check:

$$
\begin{array}{ll}
x = 4 & x = 1 \\
x - 3\sqrt{x} + 2 = 0 & x - 3\sqrt{x} + 2 = 0 \\
4 - 3\sqrt{4} + 2 \stackrel{?}{=} 0 & 1 - 3\sqrt{1} + 2 \stackrel{?}{=} 0 \\
4 - 6 + 2 \stackrel{?}{=} 0 & 1 - 3 + 2 \stackrel{?}{=} 0 \\
0 = 0 \checkmark & 0 = 0 \checkmark
\end{array}
$$

> **TRY IT :: 9.63**   Solve: $x - 7\sqrt{x} + 12 = 0$.

> **TRY IT :: 9.64**   Solve: $x - 6\sqrt{x} + 8 = 0$.

Substitutions for rational exponents can also help us solve an equation in quadratic form. Think of the properties of exponents as you begin the next example.

EXAMPLE 9.33

Solve: $x^{\frac{2}{3}} - 2x^{\frac{1}{3}} - 24 = 0$.

✓ Solution

The $x^{\frac{1}{3}}$ in the middle term is squared in the first term $\left(x^{\frac{1}{3}}\right)^2 = x^{\frac{2}{3}}$. If we let $u = x^{\frac{1}{3}}$ and substitute, our trinomial will be in $ax^2 + bx + c = 0$ form.

$$x^{\frac{2}{3}} - 2x^{\frac{1}{3}} - 24 = 0$$

Rewrite the trinomial to prepare for the substitution.	$\left(x^{\frac{1}{3}}\right)^2 - 2\left(x^{\frac{1}{3}}\right) - 24 = 0$
	$u^2 - 2u - 24 = 0$
Let $u = x^{\frac{1}{3}}$ and substitute.	

Solve by factoring.	$(u-6)(u+4)=0$
	$u-6=0, \quad u+4=0$
	$u=6, \qquad u=-4$
Replace u with $x^{\frac{1}{3}}$.	$x^{\frac{1}{3}}=6, \qquad x^{\frac{1}{3}}=-4$
Solve for x by cubing both sides.	$\left(x^{\frac{1}{3}}\right)^3=(6)^3, \quad \left(x^{\frac{1}{3}}\right)^3=(-4)^3$
	$x=216, \qquad x=-64$

Check:

$x=216$	$x=-64$
$x^{\frac{2}{3}}-2x^{\frac{1}{3}}-24=0$	$x^{\frac{2}{3}}-2x^{\frac{1}{3}}-24=0$
$(216)^{\frac{2}{3}}-2(216)^{\frac{1}{3}}-24\overset{?}{=}0$	$(-64)^{\frac{2}{3}}-2(-64)^{\frac{1}{3}}-24\overset{?}{=}0$
$36-12-24\overset{?}{=}0$	$16+8-24\overset{?}{=}0$
$0=0\checkmark$	$0=0\checkmark$

> **TRY IT :: 9.65**
> Solve: $x^{\frac{2}{3}}-5x^{\frac{1}{3}}-14=0$.

> **TRY IT :: 9.66**
> Solve: $x^{\frac{1}{2}}+8x^{\frac{1}{4}}+15=0$.

In the next example, we need to keep in mind the definition of a negative exponent as well as the properties of exponents.

EXAMPLE 9.34

Solve: $3x^{-2}-7x^{-1}+2=0$.

⊘ **Solution**

The x^{-1} in the middle term is squared in the first term $\left(x^{-1}\right)^2=x^{-2}$. If we let $u=x^{-1}$ and substitute, our trinomial will be in $ax^2+bx+c=0$ form.

	$3x^{-2}-7x^{-1}+2=0$
Rewrite the trinomial to prepare for the substitution.	$3(x^{-1})^2-7(x^{-1})+2=0$
Let $u=x^{-1}$ and substitute.	$3u^2-7u+2=0$
Solve by factoring.	$(3u-1)(u-2)=0$
	$3u-1=0, \quad u-2=0$
	$u=\dfrac{1}{3}, \qquad u=2$
Replace u with x^{-1}.	$x^{-1}=\dfrac{1}{3}, \qquad x^{-1}=2$
Solve for x by taking the reciprocal since $x^{-1}=\dfrac{1}{x}$.	$x=3, \qquad x=\dfrac{1}{2}$

Check:

$$x = 3 \qquad\qquad x = \tfrac{1}{2}$$

$$3x^{-2} - 7x^{-1} + 2 = 0 \qquad\qquad 3x^{-2} - 7x^{-1} + 2 = 0$$

$$3(3)^{-2} - 7(3)^{-1} + 2 \overset{?}{=} 0 \qquad\qquad 3\left(\tfrac{1}{2}\right)^{-2} - 7\left(\tfrac{1}{2}\right)^{-1} + 2 \overset{?}{=} 0$$

$$3\left(\tfrac{1}{9}\right) - 7\left(\tfrac{1}{3}\right) + 2 \overset{?}{=} 0 \qquad\qquad 3(4) - 7(2) + 2 \overset{?}{=} 0$$

$$\tfrac{1}{3} - \left(\tfrac{7}{3}\right) + \tfrac{6}{3} \overset{?}{=} 0 \qquad\qquad 12 - 14 + 2 \overset{?}{=} 0$$

$$0 = 0 \checkmark \qquad\qquad 0 = 0 \checkmark$$

> **TRY IT ::** 9.67

Solve: $8x^{-2} - 10x^{-1} + 3 = 0$.

> **TRY IT ::** 9.68

Solve: $6x^{-2} - 23x^{-1} + 20 = 0$.

▶ **MEDIA ::**

Access this online resource for additional instruction and practice with solving quadratic equations.

- **Solving Equations in Quadratic Form (https://openstax.org/l/37QuadForm4)**

9.4 EXERCISES

Practice Makes Perfect

Solve Equations in Quadratic Form

In the following exercises, solve.

155. $x^4 - 7x^2 + 12 = 0$

156. $x^4 - 9x^2 + 18 = 0$

157. $x^4 - 13x^2 - 30 = 0$

158. $x^4 + 5x^2 - 36 = 0$

159. $2x^4 - 5x^2 + 3 = 0$

160. $4x^4 - 5x^2 + 1 = 0$

161. $2x^4 - 7x^2 + 3 = 0$

162. $3x^4 - 14x^2 + 8 = 0$

163.
$(x - 3)^2 - 5(x - 3) - 36 = 0$

164.
$(x + 2)^2 - 3(x + 2) - 54 = 0$

165. $(3y + 2)^2 + (3y + 2) - 6 = 0$

166.
$(5y - 1)^2 + 3(5y - 1) - 28 = 0$

167.
$(x^2 + 1)^2 - 5(x^2 + 1) + 4 = 0$

168.
$(x^2 - 4)^2 - 4(x^2 - 4) + 3 = 0$

169.
$2(x^2 - 5)^2 - 5(x^2 - 5) + 2 = 0$

170.
$2(x^2 - 5)^2 - 7(x^2 - 5) + 6 = 0$

171. $x - \sqrt{x} - 20 = 0$

172. $x - 8\sqrt{x} + 15 = 0$

173. $x + 6\sqrt{x} - 16 = 0$

174. $x + 4\sqrt{x} - 21 = 0$

175. $6x + \sqrt{x} - 2 = 0$

176. $6x + \sqrt{x} - 1 = 0$

177. $10x - 17\sqrt{x} + 3 = 0$

178. $12x + 5\sqrt{x} - 3 = 0$

179. $x^{\frac{2}{3}} + 9x^{\frac{1}{3}} + 8 = 0$

180. $x^{\frac{2}{3}} - 3x^{\frac{1}{3}} = 28$

181. $x^{\frac{2}{3}} + 4x^{\frac{1}{3}} = 12$

182. $x^{\frac{2}{3}} - 11x^{\frac{1}{3}} + 30 = 0$

183. $6x^{\frac{2}{3}} - x^{\frac{1}{3}} = 12$

184. $3x^{\frac{2}{3}} - 10x^{\frac{1}{3}} = 8$

185. $8x^{\frac{2}{3}} - 43x^{\frac{1}{3}} + 15 = 0$

186. $20x^{\frac{2}{3}} - 23x^{\frac{1}{3}} + 6 = 0$

187. $x + 8x^{\frac{1}{2}} + 7 = 0$

188. $2x - 7x^{\frac{1}{2}} = 15$

189. $6x^{-2} + 13x^{-1} + 5 = 0$

190. $15x^{-2} - 26x^{-1} + 8 = 0$

191. $8x^{-2} - 2x^{-1} - 3 = 0$

192. $15x^{-2} - 4x^{-1} - 4 = 0$

Writing Exercises

193. Explain how to recognize an equation in quadratic form.

194. Explain the procedure for solving an equation in quadratic form.

Self Check

ⓐ *After completing the exercises, use this checklist to evaluate your mastery of the objectives of this section.*

I can...	Confidently	With some help	No-I don't get it!
solve equations in quadratic form.			

ⓑ *On a scale of 1-10, how would you rate your mastery of this section in light of your responses on the checklist? How can you improve this?*

9.5 | Solve Applications of Quadratic Equations

Learning Objectives

By the end of this section, you will be able to:

› Solve applications modeled by quadratic equations

Be Prepared!

Before you get started, take this readiness quiz.

1. The sum of two consecutive odd numbers is –100. Find the numbers.
   If you missed this problem, review **Example 2.18**.

2. Solve: $\dfrac{2}{x+1} + \dfrac{1}{x-1} = \dfrac{1}{x^2-1}$.

   If you missed this problem, review **Example 7.35**.

3. Find the length of the hypotenuse of a right triangle with legs 5 inches and 12 inches.
   If you missed this problem, review **Example 2.34**.

Solve Applications Modeled by Quadratic Equations

We solved some applications that are modeled by quadratic equations earlier, when the only method we had to solve them was factoring. Now that we have more methods to solve quadratic equations, we will take another look at applications.

Let's first summarize the methods we now have to solve quadratic equations.

Methods to Solve Quadratic Equations

1. Factoring
2. Square Root Property
3. Completing the Square
4. Quadratic Formula

As you solve each equation, choose the method that is most convenient for you to work the problem. As a reminder, we will copy our usual Problem-Solving Strategy here so we can follow the steps.

HOW TO : : USE A PROBLEM-SOLVING STRATEGY.

Step 1.   **Read** the problem. Make sure all the words and ideas are understood.

Step 2.   **Identify** what we are looking for.

Step 3.   **Name** what we are looking for. Choose a variable to represent that quantity.

Step 4.   **Translate** into an equation. It may be helpful to restate the problem in one sentence with all the important information. Then, translate the English sentence into an algebraic equation.

Step 5.   **Solve** the equation using algebra techniques.

Step 6.   **Check** the answer in the problem and make sure it makes sense.

Step 7.   **Answer** the question with a complete sentence

We have solved number applications that involved consecutive even and odd integers, by modeling the situation with linear equations. Remember, we noticed each even integer is 2 more than the number preceding it. If we call the first one *n*, then the next one is *n* + 2. The next one would be *n* + 2 + 2 or *n* + 4. This is also true when we use odd integers. One set of even integers and one set of odd integers are shown below.

Consecutive even integers		Consecutive odd integers	
64, 66, 68		77, 79, 81	
n	1st even integer	n	1st odd integer
$n + 2$	2nd consecutive even integer	$n + 2$	2nd consecutive odd integer
$n + 4$	3rd consecutive even integer	$n + 4$	3rd consecutive odd integer

Some applications of odd or even consecutive integers are modeled by quadratic equations. The notation above will be helpful as you name the variables.

EXAMPLE 9.35

The product of two consecutive odd integers is 195. Find the integers.

 **Solution**

Step 1. Read the problem.

Step 2. Identify what we are looking for.   We are looking for two consecutive odd integers.

Step 3. Name what we are looking for.   Let $n =$ the first odd integer.

$n + 2 =$ the next odd integer

Step 4. Translate into an equation. State the problem in one sentence.

"The product of two consecutive odd integers is 195."

The product of the first odd integer and the second odd integer is 195.

Translate into an equation.

$$n(n + 2) = 195$$

Step 5. Solve the equation. Distribute.
Write the equation in standard form.
Factor.

$$n^2 + 2n = 195$$
$$n^2 + 2n - 195 = 0$$
$$(n + 15)(n - 13) = 0$$

Use the Zero Product Property.
Solve each equation.

$$n + 15 = 0 \quad n - 13 = 0$$
$$n = -15, \quad n = 13$$

There are two values of n that are solutions. This will give us two pairs of consecutive odd integers for our solution.

First odd integer $n = 13$ First odd integer $n = -15$

next odd integer $n + 2$ next odd integer $n + 2$

$13 + 2$ $-15 + 2$

15 -13

Step 6. Check the answer.

Do these pairs work?

Are they consecutive odd integers?

    13, 15   yes
$-13, -15$ yes

Is their product 195?

    $13 \cdot 15 = 195$ yes
$-13(-15) = 195$ yes

Step 7. Answer the question.   Two consecutive odd integers whose product is

195 are 13, 15 and $-13, -15$.

> **TRY IT : : 9.69**   The product of two consecutive odd integers is 99. Find the integers.

> **TRY IT : :** 9.70   The product of two consecutive even integers is 168. Find the integers.

We will use the formula for the area of a triangle to solve the next example.

Area of a Triangle

For a triangle with **base**, b, and height, h, the area, A, is given by the formula $A = \frac{1}{2}bh$.

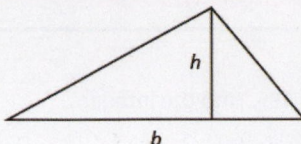

Recall that when we solve geometric applications, it is helpful to draw the figure.

EXAMPLE 9.36

An architect is designing the entryway of a restaurant. She wants to put a triangular window above the doorway. Due to energy restrictions, the window can only have an area of 120 square feet and the architect wants the base to be 4 feet more than twice the height. Find the base and height of the window.

⊘ **Solution**

Step 1. Read the problem. Draw a picture.	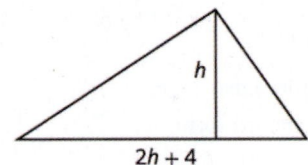
Step 2. Identify what we are looking for.	We are looking for the base and height.
Step 3. Name what we are looking for.	Let h = the height of the triangle. $2h + 4$ = the base of the triangle
Step 4. Translate into an equation. We know the area. Write the formula for the area of a triangle.	$A = \frac{1}{2}bh$
Step 5. Solve the equation. Substitute in the values.	$120 = \frac{1}{2}(2h + 4)h$
Distribute.	$120 = h^2 + 2h$
This is a quadratic equation, rewrite it in standard form.	$h^2 + 2h - 120 = 0$
Factor.	$(h - 10)(h + 12) = 0$
Use the Zero Product Property.	$h - 10 = 0 \quad h + 12 = 0$
Simplify.	$h = 10, \quad \cancel{h = -12}$

Since h is the height of a window, a value of $h = -12$ does not make sense.

The height of the triangle $h = 10$.

$$\text{The base of the triangle } 2h + 4.$$
$$2 \cdot 10 + 4$$
$$24$$

Step 6. Check the answer.
Does a triangle with height 10 and base 24
have area 120? Yes.

Step 7. Answer the question.

The height of the triangular window is 10 feet and
the base is 24 feet.

 **TRY IT : :** 9.71

Find the base and height of a triangle whose base is four inches more than six times its height and has an area of
456 square inches.

 **TRY IT : :** 9.72

If a triangle that has an area of 110 square feet has a base that is two feet less than twice the height, what is the
length of its base and height?

In the two preceding examples, the number in the radical in the Quadratic Formula was a perfect square and so the
solutions were rational numbers. If we get an irrational number as a solution to an application problem, we will use a
calculator to get an approximate value.

We will use the formula for the area of a rectangle to solve the next example.

Area of a Rectangle

For a rectangle with length, L, and width, W, the area, A, is given by the formula $A = LW$.

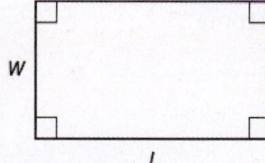

EXAMPLE 9.37

Mike wants to put 150 square feet of artificial turf in his front yard. This is the maximum area of artificial turf allowed by
his homeowners association. He wants to have a rectangular area of turf with length one foot less than 3 times the width.
Find the length and width. Round to the nearest tenth of a foot.

⊘ **Solution**

Step 1. Read the problem.
Draw a picture.

w ⬚ $3w - 1$

Step 2. Identify what we are looking for.

We are looking for the length and width.

Step 3. Name what we are looking for.

Let $w =$ the width of the rectangle.
$3w - 1 =$ the length of the rectangle

Step 4. **Translate** into an equation. We know the area. Write the formula for the area of a rectangle.	$A = L \cdot W$
Step 5. **Solve** the equation. Substitute in the values.	$150 = (3w - 1)w$
Distribute.	$150 = 3w^2 - w$
This is a quadratic equation; rewrite it in standard form. Solve the equation using the Quadratic Formula.	$ax^2 + bx + c = 0$ $3w^2 - w - 150 = 0$
Identify the a, b, c values.	$a = 3, b = -1, c = -150$
Write the Quadratic Formula.	$w = \dfrac{-b \pm \sqrt{b^2 - 4ac}}{2a}$
Then substitute in the values of a, b, c.	$w = \dfrac{-(-1) \pm \sqrt{(-1)^2 - 4 \cdot 3 \cdot (-150)}}{2 \cdot 3}$
Simplify.	$w = \dfrac{1 \pm \sqrt{1 + 1800}}{6}$ $w = \dfrac{1 \pm \sqrt{1801}}{6}$
Rewrite to show two solutions.	$w = \dfrac{1 + \sqrt{1801}}{6}, \quad w = \dfrac{1 - \sqrt{1801}}{6}$
Approximate the answers using a calculator. We eliminate the negative solution for the width.	$w \approx 7.2, \qquad \cancel{w \approx -6.9}$ Width $w \approx 7.2$ Length $\approx 3w - 1$ $\approx 3(7.2) - 1$ ≈ 20.6
Step 6. **Check** the answer. Make sure that the answers make sense. Since the answers are approximate, the area will not come out exactly to 150.	
Step 7. **Answer** the question.	The width of the rectangle is approximately 7.2 feet and the length is approximately 20.6 feet.

> **TRY IT : : 9.73**

The length of a 200 square foot rectangular vegetable garden is four feet less than twice the width. Find the length and width of the garden, to the nearest tenth of a foot.

> **TRY IT : : 9.74**

A rectangular tablecloth has an area of 80 square feet. The width is 5 feet shorter than the length. What are the length and width of the tablecloth to the nearest tenth of a foot.?

The Pythagorean Theorem gives the relation between the legs and hypotenuse of a right triangle. We will use the Pythagorean Theorem to solve the next example.

Pythagorean Theorem

In any right triangle, where a and b are the lengths of the legs, and c is the length of the hypotenuse, $a^2 + b^2 = c^2$.

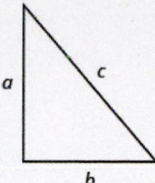

EXAMPLE 9.38

Rene is setting up a holiday light display. He wants to make a 'tree' in the shape of two right triangles, as shown below, and has two 10-foot strings of lights to use for the sides. He will attach the lights to the top of a pole and to two stakes on the ground. He wants the height of the pole to be the same as the distance from the base of the pole to each stake. How tall should the pole be?

✓ **Solution**

Step 1. Read the problem. Draw a picture.	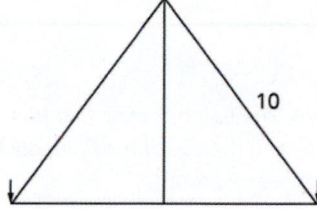

Step 2. Identify what we are looking for.	We are looking for the height of the pole.

Step 3. Name what we are looking for.	The distance from the base of the pole to either stake is the same as the height of the pole.

Let $x =$ the height of the pole.
$x =$ the distance from pole to stake

Each side is a right triangle. We draw a picture of one of them.

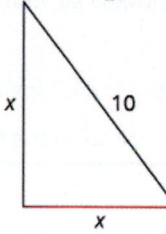

Step 4. Translate into an equation. We can use the Pythagorean Theorem to solve for x. Write the Pythagorean Theorem.	$a^2 + b^2 = c^2$

Step 5. Solve the equation. Substitute.	$x^2 + x^2 = 10^2$

Simplify.	$2x^2 = 100$

Divide by 2 to isolate the variable.	$\dfrac{2x^2}{2} = \dfrac{100}{2}$

Simplify.	$x^2 = 50$

Use the Square Root Property.	$x = \pm\sqrt{50}$
Simplify the radical.	$x = \pm 5\sqrt{2}$
Rewrite to show two solutions.	$x = 5\sqrt{2}, \quad \cancel{x = -5\sqrt{2}}$
	If we approximate this number to the nearest tenth with a calculator, we find $x \approx 7.1$.
Step 6. Check the answer. Check on your own in the Pythagorean Theorem.	
Step 7. Answer the question.	The pole should be about 7.1 feet tall.

> **TRY IT :: 9.75**
>
> The sun casts a shadow from a flag pole. The height of the flag pole is three times the length of its shadow. The distance between the end of the shadow and the top of the flag pole is 20 feet. Find the length of the shadow and the length of the flag pole. Round to the nearest tenth.

> **TRY IT :: 9.76**
>
> The distance between opposite corners of a rectangular field is four more than the width of the field. The length of the field is twice its width. Find the distance between the opposite corners. Round to the nearest tenth.

The height of a projectile shot upward from the ground is modeled by a quadratic equation. The initial velocity, v_0, propels the object up until gravity causes the object to fall back down.

Projectile motion

The height in feet, h, of an object shot upwards into the air with initial velocity, v_0, after t seconds is given by the formula

$$h = -16t^2 + v_0 t$$

We can use this formula to find how many seconds it will take for a firework to reach a specific height.

EXAMPLE 9.39

A firework is shot upwards with initial velocity 130 feet per second. How many seconds will it take to reach a height of 260 feet? Round to the nearest tenth of a second.

 **Solution**

Step 1. Read the problem.	
Step 2. Identify what we are looking for.	We are looking for the number of seconds, which is time.
Step 3. Name what we are looking for.	Let $t =$ the number of seconds.
Step 4. Translate into an equation. Use the formula.	$h = -16t^2 + v_0 t$
Step 5. Solve the equation. We know the velocity v_0 is 130 feet per second. The height is 260 feet. Substitute the values.	$260 = -16t^2 + 130t$

This is a quadratic equation, rewrite it in standard form. Solve the equation using the Quadratic Formula.	$ax^2 + bx + c = 0$ $16t^2 - 130t + 260 = 0$
Identify the values of a, b, c.	$a = 16$, $b = -130$, $c = 260$
Write the Quadratic Formula.	$t = \dfrac{-b \pm \sqrt{b^2 - 4ac}}{2a}$
Then substitute in the values of a, b, c.	$t = \dfrac{-(-130) \pm \sqrt{(-130)^2 - 4 \cdot 16 \cdot (260)}}{2 \cdot 16}$
Simplify.	$t = \dfrac{130 \pm \sqrt{16,900 - 16,640}}{32}$ $t = \dfrac{130 \pm \sqrt{260}}{32}$
Rewrite to show two solutions.	$t = \dfrac{130 + \sqrt{260}}{32}$, $\quad t = \dfrac{130 - \sqrt{260}}{32}$
Approximate the answer with a calculator.	$t \approx 4.6$ seconds, $\quad t \approx 3.6$ seconds

Step 6. Check the answer.
The check is left to you.

Step 7. Answer the question.

The firework will go up and then fall back down. As the firework goes up, it will reach 260 feet after approximately 3.6 seconds. It will also pass that height on the way down at 4.6 seconds.

> **TRY IT : :** 9.77

An arrow is shot from the ground into the air at an initial speed of 108 ft/s. Use the formula $h = -16t^2 + v_0 t$ to determine when the arrow will be 180 feet from the ground. Round the nearest tenth.

> **TRY IT : :** 9.78

A man throws a ball into the air with a velocity of 96 ft/s. Use the formula $h = -16t^2 + v_0 t$ to determine when the height of the ball will be 48 feet. Round to the nearest tenth.

We have solved uniform motion problems using the formula $D = rt$ in previous chapters. We used a table like the one below to organize the information and lead us to the equation.

	Rate	•	Time	=	Distance

The formula $D = rt$ assumes we know r and t and use them to find D. If we know D and r and need to find t, we would solve the equation for t and get the formula $t = \dfrac{D}{r}$.

Some uniform motion problems are also modeled by quadratic equations.

EXAMPLE 9.40

Professor Smith just returned from a conference that was 2,000 miles east of his home. His total time in the airplane for the round trip was 9 hours. If the plane was flying at a rate of 450 miles per hour, what was the speed of the jet stream?

⊘ **Solution**

This is a uniform motion situation. A diagram will help us visualize the situation.

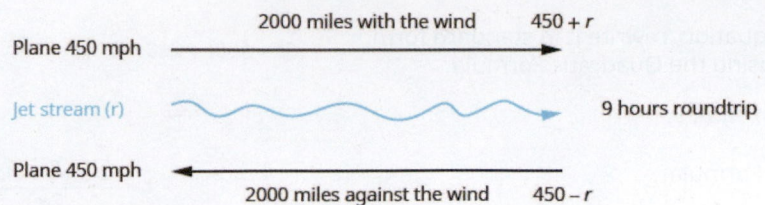

We fill in the chart to organize the information.

We are looking for the speed of the jet stream.   Let $r =$ the speed of the jet stream.

When the plane flies with the wind, the wind increases its speed and so the rate is $450 + r$.

When the plane flies against the wind, the wind decreases its speed and the rate is $450 - r$.

Write in the rates.
Write in the distances.
Since $D = r \cdot t$, we solve for
t and get $t = \dfrac{D}{r}$.

We divide the distance by
the rate in each row, and
place the expression in the
time column.

Type	Rate	·	Time	=	Distance
Headwind	$450 - r$		$\dfrac{2000}{450 - r}$		2000
Tailwind	$450 + r$		$\dfrac{2000}{450 + r}$		2000
			9		

We know the times add to 9
and so we write our equation.

$$\frac{2000}{450 - r} + \frac{2000}{450 + r} = 9$$

We multiply both sides by the LCD.

$$(450 - r)(450 + r)\left(\frac{2000}{450 - r} + \frac{2000}{450 + r}\right) = 9(450 - r)(450 + r)$$

Simplify.

$$2000(450 + r) + 2000(450 - r) = 9(450 - r)(450 + r)$$

Factor the 2,000.

$$2000(450 + r + 450 - r) = 9\left(450^2 - r^2\right)$$

Solve.

$$2000(900) = 9\left(450^2 - r^2\right)$$

Divide by 9.

$$2000(100) = 450^2 - r^2$$

Simplify.

$$200000 = 202500 - r^2$$
$$-2500 = -r^2$$
$$50 = r \quad \text{The speed of the jet stream.}$$

Check:
Is 50 mph a reasonable speed for the
jet stream? Yes.
If the plane is traveling 450 mph and
the wind is 50 mph,
Tailwind

$450 + 50 = 500$ mph $\dfrac{2000}{500} = 4$ hours

Headwind

$450 - 50 = 400$ mph $\dfrac{2000}{400} = 5$ hours

The times add to 9 hours, so it checks.

The speed of the jet stream was 50 mph.

> **TRY IT ::9.79**

MaryAnne just returned from a visit with her grandchildren back east . The trip was 2400 miles from her home and her total time in the airplane for the round trip was 10 hours. If the plane was flying at a rate of 500 miles per hour, what was the speed of the jet stream?

> **TRY IT ::9.80**

Gerry just returned from a cross country trip. The trip was 3000 miles from his home and his total time in the airplane for the round trip was 11 hours. If the plane was flying at a rate of 550 miles per hour, what was the speed of the jet stream?

Work applications can also be modeled by quadratic equations. We will set them up using the same methods we used when we solved them with rational equations.We'll use a similar scenario now.

EXAMPLE 9.41

The weekly gossip magazine has a big story about the presidential election and the editor wants the magazine to be printed as soon as possible. She has asked the printer to run an extra printing press to get the printing done more quickly. Press #1 takes 12 hours more than Press #2 to do the job and when both presses are running they can print the job in 8 hours. How long does it take for each press to print the job alone?

⊘ **Solution**

This is a work problem. A chart will help us organize the information.

We are looking for how many hours it would take each press separately to complete the job.

Let $x =$ the number of hours for Press #2 to complete the job.
Enter the hours per job for Press #1, Press #2, and when they work together.

	Number of hours needed to complete the job.	Part of job completed/hour
Press #1	$x + 12$	$\frac{1}{x + 12}$
Press #2	x	$\frac{1}{x}$
Together	8	$\frac{1}{8}$

The part completed by Press #1 plus the part completed by Press #2 equals the amount completed together. Translate to an equation.	Work completed by Press #1 + Press #2 = Together $$\frac{1}{x + 12} + \frac{1}{x} = \frac{1}{8}$$
Solve.	$$\frac{1}{x + 12} + \frac{1}{x} = \frac{1}{8}$$
Multiply by the LCD, $8x(x + 12)$.	$$8x(x + 12)\left(\frac{1}{x + 12} + \frac{1}{x}\right) = \left(\frac{1}{8}\right)8x(x + 12)$$
Simplify.	$8x + 8(x + 12) = x(x + 12)$ $8x + 8x + 96 = x^2 + 12x$ $0 = x^2 - 4x - 96$
Solve.	$0 = (x - 12)(x + 8)$ $x - 12 = 0, x + 8 = 0$ $x = 12, x = -8$ hours
Since the idea of negative hours does not make sense, we use the value $x = 12$.	$12 + 12$ 12 24 hours 12 hours
Write our sentence answer.	Press #1 would take 24 hours and Press #2 would take 12 hours to do the job alone.

> **TRY IT : :** 9.81

The weekly news magazine has a big story naming the Person of the Year and the editor wants the magazine to be printed as soon as possible. She has asked the printer to run an extra printing press to get the printing done more quickly. Press #1 takes 6 hours more than Press #2 to do the job and when both presses are running they can print the job in 4 hours. How long does it take for each press to print the job alone?

> **TRY IT : :** 9.82

Erlinda is having a party and wants to fill her hot tub. If she only uses the red hose it takes 3 hours more than if she only uses the green hose. If she uses both hoses together, the hot tub fills in 2 hours. How long does it take for each hose to fill the hot tub?

▶ **MEDIA : :**

Access these online resources for additional instruction and practice with solving applications modeled by quadratic equations.

- **Word Problems Involving Quadratic Equations (https://openstax.org/l/37QuadForm5)**
- **Quadratic Equation Word Problems (https://openstax.org/l/37QuadForm6)**
- **Applying the Quadratic Formula (https://openstax.org/l/37QuadForm7)**

9.5 EXERCISES

Practice Makes Pefect

Solve Applications Modeled by Quadratic Equations

In the following exercises, solve using any method.

195. The product of two consecutive odd numbers is 255. Find the numbers.

196. The product of two consecutive even numbers is 360. Find the numbers.

197. The product of two consecutive even numbers is 624. Find the numbers.

198. The product of two consecutive odd numbers is 1,023. Find the numbers.

199. The product of two consecutive odd numbers is 483. Find the numbers.

200. The product of two consecutive even numbers is 528. Find the numbers.

In the following exercises, solve using any method. Round your answers to the nearest tenth, if needed.

201. A triangle with area 45 square inches has a height that is two less than four times the base Find the base and height of the triangle.

202. The base of a triangle is six more than twice the height. The area of the triangle is 88 square yards. Find the base and height of the triangle.

203. The area of a triangular flower bed in the park has an area of 120 square feet. The base is 4 feet longer that twice the height. What are the base and height of the triangle?

204. A triangular banner for the basketball championship hangs in the gym. It has an area of 75 square feet. What is the length of the base and height , if the base is two-thirds of the height?

205. The length of a rectangular driveway is five feet more than three times the width. The area is 50 square feet. Find the length and width of the driveway.

206. A rectangular lawn has area 140 square yards. Its width that is six less than twice the length. What are the length and width of the lawn?

207. A rectangular table for the dining room has a surface area of 24 square feet. The length is two more feet than twice the width of the table. Find the length and width of the table.

208. The new computer has a surface area of 168 square inches. If the the width is 5.5 inches less that the length, what are the dimensions of the computer?

209. The hypotenuse of a right triangle is twice the length of one of its legs. The length of the other leg is three feet. Find the lengths of the three sides of the triangle.

210. The hypotenuse of a right triangle is 10 cm long. One of the triangle's legs is three times as the length of the other leg . Round to the nearest tenth. Find the lengths of the three sides of the triangle.

211. A rectangular garden will be divided into two plots by fencing it on the diagonal. The diagonal distance from one corner of the garden to the opposite corner is five yards longer than the width of the garden. The length of the garden is three times the width. Find the length of the diagonal of the garden.

212. Nautical flags are used to represent letters of the alphabet. The flag for the letter, O consists of a yellow right triangle and a red right triangle which are sewn together along their hypotenuse to form a square. The hypotenuse of the two triangles is three inches longer than a side of the flag. Find the length of the side of the flag.

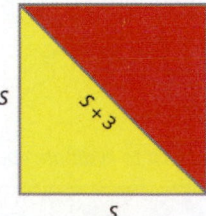

213. Gerry plans to place a 25-foot ladder against the side of his house to clean his gutters. The bottom of the ladder will be 5 feet from the house.How for up the side of the house will the ladder reach?

214. John has a 10-foot piece of rope that he wants to use to support his 8-foot tree. How far from the base of the tree should he secure the rope?

215. A firework rocket is shot upward at a rate of 640 ft/sec. Use the projectile formula $h = -16t^2 + v_0t$ to determine when the height of the firework rocket will be 1200 feet.

216. An arrow is shot vertically upward at a rate of 220 feet per second. Use the projectile formula $h = -16t^2 + v_0t$, to determine when height of the arrow will be 400 feet.

217. A bullet is fired straight up from a BB gun with initial velocity 1120 feet per second at an initial height of 8 feet. Use the formula $h = -16t^2 + v_0t + 8$ to determine how many seconds it will take for the bullet to hit the ground. (That is, when will $h = 0$?)

218. A stone is dropped from a 196-foot platform. Use the formula $h = -16t^2 + v_0t + 196$ to determine how many seconds it will take for the stone to hit the ground. (Since the stone is dropped, $v_0 = 0$.)

219. The businessman took a small airplane for a quick flight up the coast for a lunch meeting and then returned home. The plane flew a total of 4 hours and each way the trip was 200 miles. What was the speed of the wind that affected the plane which was flying at a speed of 120 mph?

220. The couple took a small airplane for a quick flight up to the wine country for a romantic dinner and then returned home. The plane flew a total of 5 hours and each way the trip was 300 miles. If the plane was flying at 125 mph, what was the speed of the wind that affected the plane?

221. Roy kayaked up the river and then back in a total time of 6 hours. The trip was 4 miles each way and the current was difficult. If Roy kayaked at a speed of 5 mph, what was the speed of the current?

222. Rick paddled up the river, spent the night camping, and and then paddled back. He spent 10 hours paddling and the campground was 24 miles away. If Rick kayaked at a speed of 5 mph, what was the speed of the current?

223. Two painters can paint a room in 2 hours if they work together. The less experienced painter takes 3 hours more than the more experienced painter to finish the job. How long does it take for each painter to paint the room individually?

224. Two gardeners can do the weekly yard maintenance in 8 minutes if they work together. The older gardener takes 12 minutes more than the younger gardener to finish the job by himself. How long does it take for each gardener to do the weekly yard maintenence individually?

225. It takes two hours for two machines to manufacture 10,000 parts. If Machine #1 can do the job alone in one hour less than Machine #2 can do the job, how long does it take for each machine to manufacture 10,000 parts alone?

226. Sully is having a party and wants to fill his swimming pool. If he only uses his hose it takes 2 hours more than if he only uses his neighbor's hose. If he uses both hoses together, the pool fills in 4 hours. How long does it take for each hose to fill the hot tub?

Writing Exercises

227. Make up a problem involving the product of two consecutive odd integers.

ⓐ Start by choosing two consecutive odd integers. What are your integers?

ⓑ What is the product of your integers?

ⓒ Solve the equation $n(n + 2) = p$, where p is the product you found in part (b).

ⓓ Did you get the numbers you started with?

228. Make up a problem involving the product of two consecutive even integers.

ⓐ Start by choosing two consecutive even integers. What are your integers?

ⓑ What is the product of your integers?

ⓒ Solve the equation $n(n + 2) = p$, where p is the product you found in part (b).

ⓓ Did you get the numbers you started with?

Self Check

ⓐ After completing the exercises, use this checklist to evaluate your mastery of the objectives of this section.

I can...	Confidently	With some help	No-I don't get it!
solve applications of the quadratic formula.			

ⓑ After looking at the checklist, do you think you are well-prepared for the next section? Why or why not?

9.6 # Graph Quadratic Functions Using Properties

Learning Objectives

By the end of this section, you will be able to:

> Recognize the graph of a quadratic function
> Find the axis of symmetry and vertex of a parabola
> Find the intercepts of a parabola
> Graph quadratic functions using properties
> Solve maximum and minimum applications

Be Prepared!

Before you get started, take this readiness quiz.

1. Graph the function $f(x) = x^2$ by plotting points.
   If you missed this problem, review **Example 3.54**.

2. Solve: $2x^2 + 3x - 2 = 0$.
   If you missed this problem, review **Example 6.45**.

3. Evaluate $-\dfrac{b}{2a}$ when $a = 3$ and $b = -6$.
   If you missed this problem, review **Example 1.21**.

Recognize the Graph of a Quadratic Function

Previously we very briefly looked at the function $f(x) = x^2$, which we called the square function. It was one of the first non-linear functions we looked at. Now we will graph functions of the form $f(x) = ax^2 + bx + c$ if $a \neq 0$. We call this kind of function a quadratic function.

Quadratic Function

A **quadratic function**, where a, b, and c are real numbers and $a \neq 0$, is a function of the form

$$f(x) = ax^2 + bx + c$$

We graphed the quadratic function $f(x) = x^2$ by plotting points.

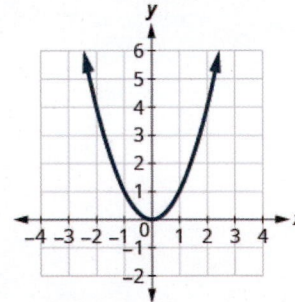

x	$f(x) = x^2$	$(x, f(x))$
-3	9	(-3, 9)
-2	4	(-2, 4)
-1	1	(-1, 1)
0	0	(0, 0)
1	1	(1, 1)
2	4	(2, 4)
3	9	(3, 9)

Every quadratic function has a graph that looks like this. We call this figure a **parabola**. Let's practice graphing a parabola by plotting a few points.

EXAMPLE 9.42

Graph $f(x) = x^2 - 1$.

 Solution

We will graph the function by plotting points.

Choose integer values for x, substitute them into the equation and simplify to find $f(x)$.

Record the values of the ordered pairs in the chart.

$f(x) = x^2 - 1$	
x	$f(x)$
0	−1
1	0
−1	0
2	3
−2	3

Plot the points, and then connect them with a smooth curve. The result will be the graph of the function $f(x) = x^2 - 1$.

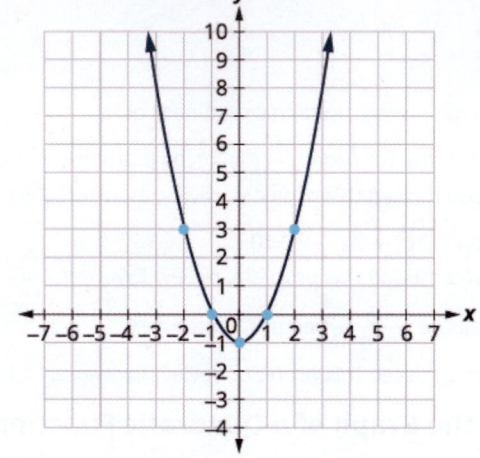

> **TRY IT ::** 9.83

Graph $f(x) = -x^2$..

> **TRY IT ::** 9.84

Graph $f(x) = x^2 + 1$.

All graphs of quadratic functions of the form $f(x) = ax^2 + bx + c$ are parabolas that open upward or downward. See **Figure 9.2**.

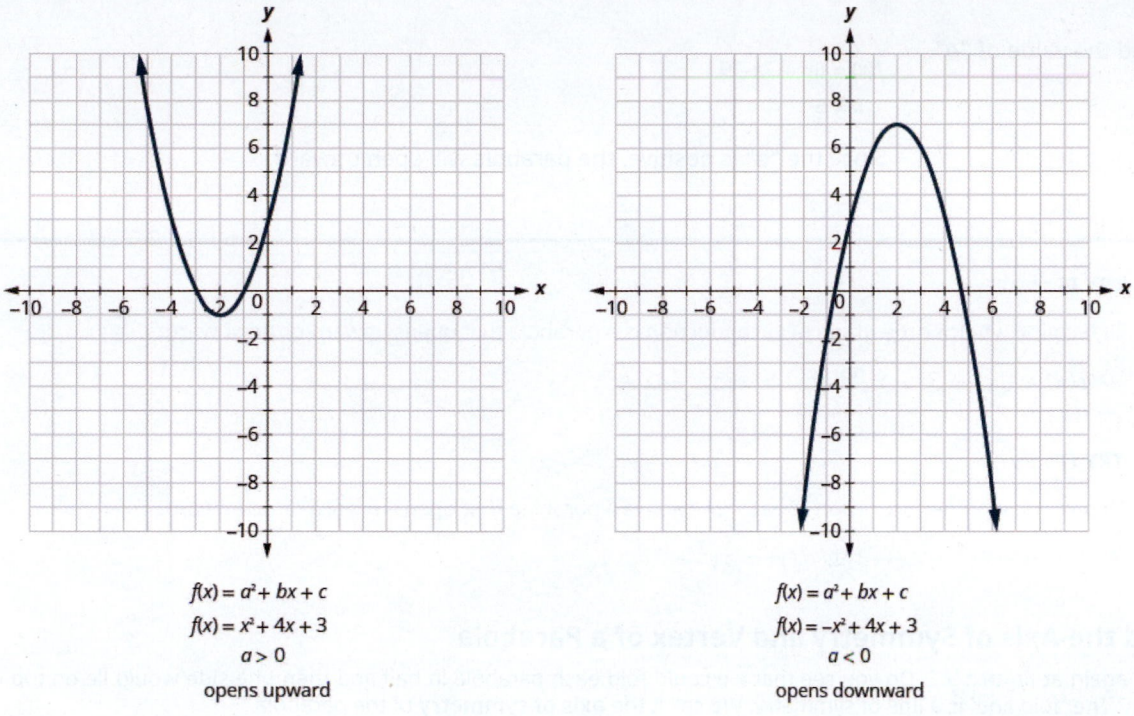

$f(x) = a^2 + bx + c$

$f(x) = x^2 + 4x + 3$

$a > 0$

opens upward

$f(x) = a^2 + bx + c$

$f(x) = -x^2 + 4x + 3$

$a < 0$

opens downward

Figure 9.2

Notice that the only difference in the two functions is the negative sign before the quadratic term (x^2 in the equation of the graph in **Figure 9.2**). When the quadratic term, is positive, the parabola opens upward, and when the quadratic term is negative, the parabola opens downward.

Parabola Orientation

For the graph of the quadratic function $f(x) = ax^2 + bx + c$, if

- $a > 0$, the parabola opens upward

- $a < 0$, the parabola opens downward

EXAMPLE 9.43

Determine whether each parabola opens upward or downward:

ⓐ $f(x) = -3x^2 + 2x - 4$ ⓑ $f(x) = 6x^2 + 7x - 9$.

✓ **Solution**

ⓐ

Find the value of "a".

$f(x) = ax^2 + bx + c$

$f(x) = -3x^2 + 2x - 4$

$a = -3$

Since the "a" is negative, the parabola will open downward.

ⓑ

Find the value of "a". $f(x) = ax^2 + bx + c$
                             $f(x) = 6x^2 + 7x - 9$

                             $a = 6$

Since the "a" is positive, the parabola will open upward.

> **TRY IT : : 9.85**
>
> Determine whether the graph of each function is a parabola that opens upward or downward:
>
> ⓐ $f(x) = 2x^2 + 5x - 2$ ⓑ $f(x) = -3x^2 - 4x + 7$.

> **TRY IT : : 9.86**
>
> Determine whether the graph of each function is a parabola that opens upward or downward:
>
> ⓐ $f(x) = -2x^2 - 2x - 3$ ⓑ $f(x) = 5x^2 - 2x - 1$.

Find the Axis of Symmetry and Vertex of a Parabola

Look again at **Figure 9.2**. Do you see that we could fold each parabola in half and then one side would lie on top of the other? The 'fold line' is a line of symmetry. We call it the **axis of symmetry** of the parabola.

We show the same two graphs again with the axis of symmetry. See **Figure 9.3**.

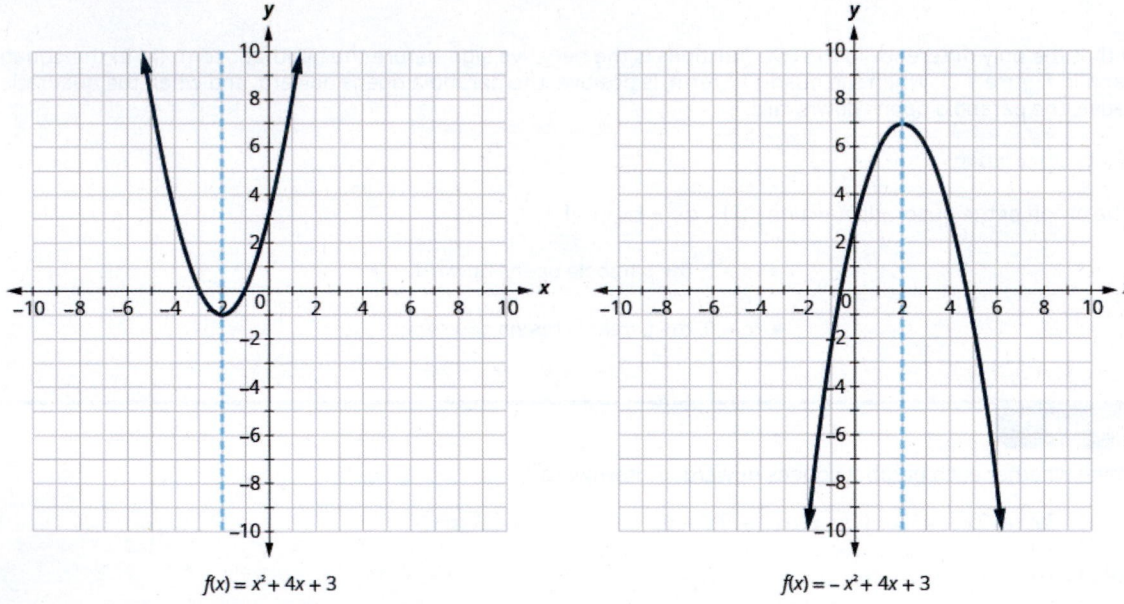

$f(x) = x^2 + 4x + 3$ $f(x) = -x^2 + 4x + 3$

Figure 9.3

The equation of the axis of symmetry can be derived by using the Quadratic Formula. We will omit the derivation here and proceed directly to using the result. The equation of the axis of symmetry of the graph of $f(x) = ax^2 + bx + c$ is $x = -\dfrac{b}{2a}$.

So to find the equation of symmetry of each of the parabolas we graphed above, we will substitute into the formula $x = -\dfrac{b}{2a}$.

$$f(x) = ax^2 + bx + c$$
$$f(x) = x^2 + 4x + 3$$

axis of symmetry

$$x = -\frac{b}{2a}$$
$$x = -\frac{4}{2 \cdot 1}$$
$$x = -2$$

$$f(x) = ax^2 + bx + c$$
$$f(x) = -x^2 + 4x + 3$$

axis of symmetry

$$x = -\frac{b}{2a}$$
$$x = -\frac{4}{2(-1)}$$
$$x = 2$$

Notice that these are the equations of the dashed blue lines on the graphs.

The point on the parabola that is the lowest (parabola opens up), or the highest (parabola opens down), lies on the axis of symmetry. This point is called the **vertex** of the parabola.

We can easily find the coordinates of the vertex, because we know it is on the axis of symmetry. This means its x-coordinate is $-\frac{b}{2a}$. To find the y-coordinate of the vertex we substitute the value of the x-coordinate into the quadratic function.

$$f(x) = x^2 + 4x + 3$$

axis of symmetry is $x = -2$

vertex is $(-2, \underline{\hspace{1em}})$

$$f(x) = -x^2 + 4x + 3$$

axis of symmetry is $x = 2$

vertex is $(2, \underline{\hspace{1em}})$

$$f(x) = x^2 + 4x + 3$$
$$f(x) = (-2)^2 + 4(-2) + 3$$
$$f(x) = -1$$

vertex is $(-2, -1)$

$$f(x) = -x^2 + 4x + 3$$
$$f(x) = -(2)^2 + 4(2) + 3$$
$$f(x) = 7$$

vertex is $(2, 7)$

Axis of Symmetry and Vertex of a Parabola

The graph of the function $f(x) = ax^2 + bx + c$ is a parabola where:

- the axis of symmetry is the vertical line $x = -\frac{b}{2a}$.

- the vertex is a point on the axis of symmetry, so its x-coordinate is $-\frac{b}{2a}$.

- the y-coordinate of the vertex is found by substituting $x = -\frac{b}{2a}$ into the quadratic equation.

EXAMPLE 9.44

For the graph of $f(x) = 3x^2 - 6x + 2$ find:

ⓐ the axis of symmetry ⓑ the vertex.

⊘ Solution

ⓐ

$$f(x) = ax^2 + bx + c$$
$$f(x) = 3x^2 - 6x + 2$$

The axis of symmetry is the vertical line $x = -\frac{b}{2a}$.	
Substitute the values of a, b into the equation.	$x = -\frac{-6}{2 \cdot 3}$
Simplify.	$x = 1$

The axis of symmetry is the line $x = 1$.

$$f(x) = 3x^2 - 6x + 2$$

The vertex is a point on the line of symmetry, so its *x*-coordinate will be $x = 1$. Find $f(1)$.	$f(1) = 3(1)^2 - 6(1) + 2$
Simplify.	$f(1) = 3 \cdot 1 - 6 + 2$
The result is the *y*-coordinate.	$f(1) = -1$
	The vertex is $(1, -1)$.

> **TRY IT : : 9.87**   For the graph of $f(x) = 2x^2 - 8x + 1$ find:
>
> ⓐ the axis of symmetry ⓑ the vertex.

> **TRY IT : : 9.88**   For the graph of $f(x) = 2x^2 - 4x - 3$ find:
>
> ⓐ the axis of symmetry ⓑ the vertex.

Find the Intercepts of a Parabola

When we graphed linear equations, we often used the *x*- and *y*-intercepts to help us graph the lines. Finding the coordinates of the intercepts will help us to graph parabolas, too.

Remember, at the *y*-intercept the value of *x* is zero. So to find the *y*-intercept, we substitute *x* = 0 into the function.

Let's find the *y*-intercepts of the two parabolas shown in **Figure 9.4**.

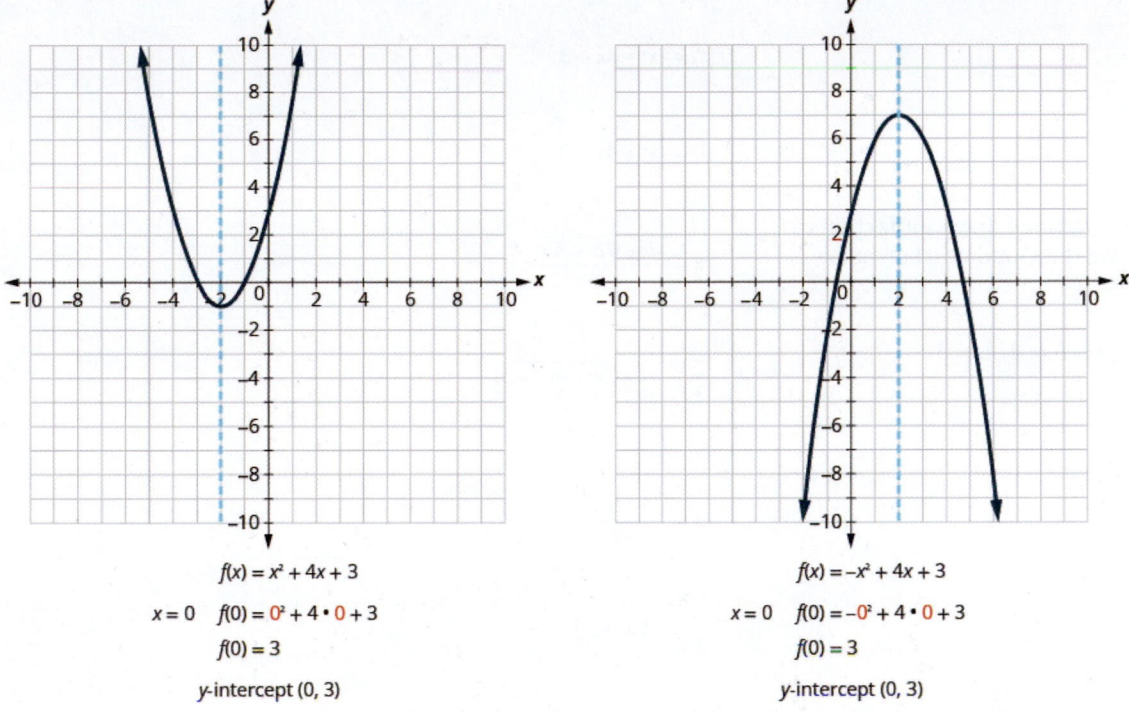

$f(x) = x^2 + 4x + 3$

$x = 0 \quad f(0) = 0^2 + 4 \cdot 0 + 3$

$f(0) = 3$

y-intercept $(0, 3)$

$f(x) = -x^2 + 4x + 3$

$x = 0 \quad f(0) = -0^2 + 4 \cdot 0 + 3$

$f(0) = 3$

y-intercept $(0, 3)$

Figure 9.4

An x-intercept results when the value of $f(x)$ is zero. To find an x-intercept, we let $f(x) = 0$. In other words, we will need to solve the equation $0 = ax^2 + bx + c$ for x.

$$f(x) = ax^2 + bx + c$$
$$0 = ax^2 + bx + c$$

Solving quadratic equations like this is exactly what we have done earlier in this chapter!

We can now find the x-intercepts of the two parabolas we looked at. First we will find the x-intercepts of the parabola whose function is $f(x) = x^2 + 4x + 3$.

$$f(x) = x^2 + 4x + 3$$

Let $f(x) = 0$.	$0 = x^2 + 4x + 3$
Factor.	$0 = (x + 1)(x + 3)$
Use the Zero Product Property.	$x + 1 = 0 \quad x + 3 = 0$
Solve.	$x = -1 \qquad x = -3$
	The x-intercepts are $(-1, 0)$ and $(-3, 0)$.

Now we will find the x-intercepts of the parabola whose function is $f(x) = -x^2 + 4x + 3$.

$$f(x) = -x^2 + 4x + 3$$

Let $f(x) = 0$.	$0 = -x^2 + 4x + 3$
This quadratic does not factor, so we use the Quadratic Formula.	$x = \dfrac{-b \pm \sqrt{b^2 - 4ac}}{2a}$
$a = -1,\ b = 4,\ c = 3$	$x = \dfrac{-4 \pm \sqrt{4^2 - 4(-1)(3)}}{2(-1)}$
Simplify.	$x = \dfrac{-4 \pm \sqrt{28}}{-2}$
	$x = \dfrac{-4 \pm 2\sqrt{7}}{-2}$
	$x = \dfrac{-2(2 \pm \sqrt{7})}{-2}$
	$x = 2 \pm \sqrt{7}$
	The x-intercepts are $\left(2 + \sqrt{7},\ 0\right)$ and $\left(2 - \sqrt{7},\ 0\right)$.

We will use the decimal approximations of the x-intercepts, so that we can locate these points on the graph,

$$\left(2 + \sqrt{7},\ 0\right) \approx (4.6,\ 0) \qquad \left(2 - \sqrt{7},\ 0\right) \approx (-0.6,\ 0)$$

Do these results agree with our graphs? See **Figure 9.5**.

Find the intercepts of the parabola for the function $f(x) = 5x^2 + x + 4$.

✓ **Solution**

$$f(x) = 5x^2 + x + 4$$

To find the *y*-intercept, let $x = 0$ and solve for $f(x)$.	$f(0) = 5 \cdot 0^2 + 0 + 4$
	$f(0) = 4$
	When $x = 0$, then $f(0) = 4$. The *y*-intercept is the point $(0, 4)$.
To find the *x*-intercept, let $f(x) = 0$ and solve for x.	$f(x) = 5x^2 + x + 4$
	$0 = 5x^2 + x + 4$

Find the value of the discriminant to predict the number of solutions which is also the number of *x*-intercepts.

$b^2 - 4ac$
$1^2 - 4 \cdot 5 \cdot 4$
$1 - 80$
-79

Since the value of the discriminant is negative, there is no real solution to the equation.
There are no *x*-intercepts.

> **TRY IT :: 9.91**   Find the intercepts of the parabola whose function is $f(x) = 3x^2 + 4x + 4$.

> **TRY IT :: 9.92**   Find the intercepts of the parabola whose function is $f(x) = x^2 - 4x - 5$.

Graph Quadratic Functions Using Properties

Now we have all the pieces we need in order to graph a quadratic function. We just need to put them together. In the next example we will see how to do this.

EXAMPLE 9.47   HOW TO GRAPH A QUADRATIC FUNCTION USING PROPERTIES

Graph $f(x) = x^2 - 6x + 8$ by using its properties.

✓ **Solution**

Step 1. Determine whether the parabola opens upward or downward.	Look at *a* in the equation. $f(x) = x^2 - 6x + 8$ Since *a* is positive, the parabola opens upward.	$f(x) = x^2 - 6x + 8$ $a = 1, b = -6, c = 8$ **The parabola opens upward.**

Step 2. Find the axis of symmetry.	$f(x) = x^2 - 6x + 8$ The axis of symmetry is the line $x = -\dfrac{b}{2a}$.	Axis of Symmetry $x = -\dfrac{b}{2a}$ $x = -\dfrac{(-6)}{2 \cdot 1}$ $x = 3$ **The axis of symmetry is the line $x = 3$.**
Step 3. Find the vertex.	The vertex is on the axis of symmetry. Substitute $x = 3$ into the function.	Vertex $f(x) = x^2 - 6x + 8$ $f(3) = (3)^2 - 6(3) + 8$ $f(3) = -1$ **The vertex is (3, –1).**
Step 4. Find the y-intercept. Find the point symmetric to the y-intercept across the axis of symmetry.	We find $f(0)$. We use the axis of symmetry to find a point symmetric to the y-intercept. The y-intercept is 3 units left of the axis of symmetry, $x = 3$. A point 3 units to the right of the axis of symmetry has $x = 6$.	y-intercept $f(x) = x^2 - 6x + 8$ $f(0) = (0)^2 - 6(0) + 8$ $f(0) = 8$ **The y-intercept is (0, 8).** Point symmetric to y-intercept: **The point is (6, 8).**
Step 5. Find the x-intercepts. Find additional points if needed.	We solve $f(x) = 0$. We can solve this quadratic equation by factoring.	x-intercepts $f(x) = x^2 - 6x + 8$ $0 = x^2 - 6x + 8$ $0 = (x - 2)(x - 4)$ $x = 2$ or $x = 4$ **The x-intercepts are (2, 0) and (4, 0).**
Step 6. Graph the parabola.	We graph the vertex, intercepts, and the point symmetric to the y-intercept. We connect these 5 points to sketch the parabola.	

> **TRY IT : : 9.93**   Graph $f(x) = x^2 + 2x - 8$ by using its properties.

> | **TRY IT : :** 9.94 Graph $f(x) = x^2 - 8x + 12$ by using its properties.

We list the steps to take in order to graph a quadratic function here.

> **HOW TO : :** TO GRAPH A QUADRATIC FUNCTION USING PROPERTIES.
>
> Step 1.   Determine whether the parabola opens upward or downward.
>
> Step 2.   Find the equation of the axis of symmetry.
>
> Step 3.   Find the vertex.
>
> Step 4.   Find the y-intercept. Find the point symmetric to the y-intercept across the axis of symmetry.
>
> Step 5.   Find the x-intercepts. Find additional points if needed.
>
> Step 6.   Graph the parabola.

We were able to find the x-intercepts in the last example by factoring. We find the x-intercepts in the next example by factoring, too.

EXAMPLE 9.48

Graph $f(x) = x^2 + 6x - 9$ by using its properties.

 Solution

$$f(x) = ax^2 + bx + c$$
$$f(x) = -x^2 + 6x - 9$$

Since a is -1, the parabola opens downward.

To find the equation of the axis of symmetry, use $x = -\dfrac{b}{2a}$.

$$x = -\frac{b}{2a}$$

$$x = -\frac{6}{2(-1)}$$

$$x = 3$$

The axis of symmetry is $x = 3$.
The vertex is on the line $x = 3$.

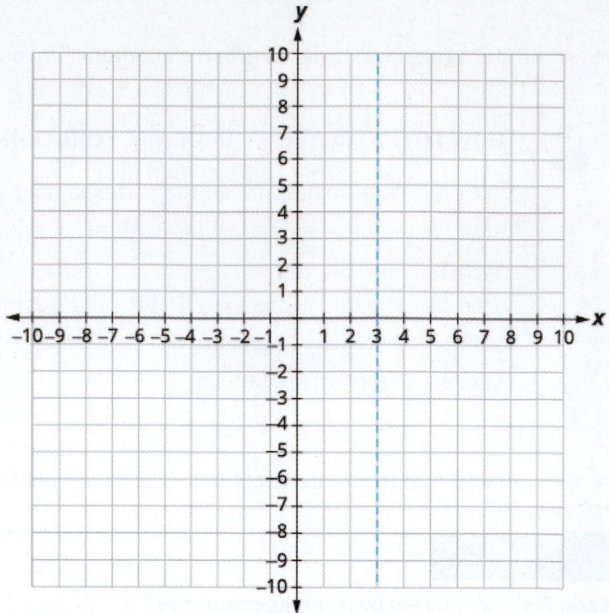

Find $f(3)$.	$f(x) = -x^2 + 6x - 9$
	$f(3) = -3^2 + 6 \cdot 3 - 9$
	$f(3) = -9 + 18 - 9$
	$f(3) = 0$

The vertex is $(3, 0)$.

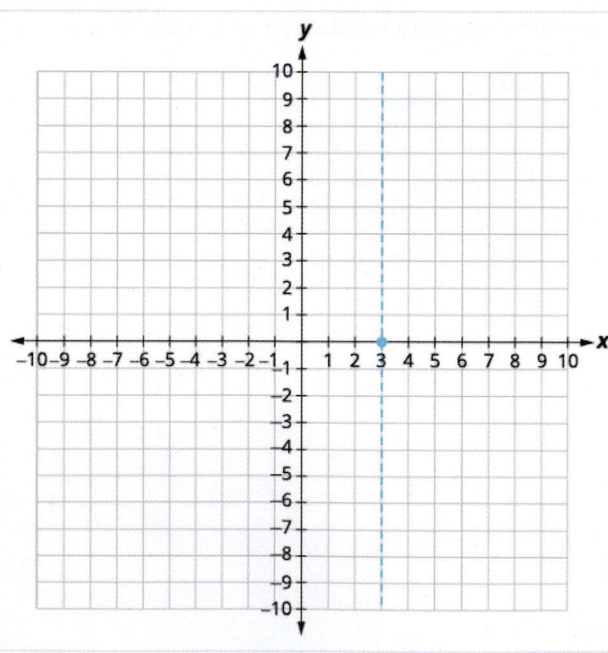

The *y*-intercept occurs when $x = 0$. Find $f(0)$.	$f(x) = -x^2 + 6x - 9$
Substitute $x = 0$.	$f(0) = -0^2 + 6 \cdot 0 - 9$
Simplify.	$f(0) = -9$

The *y*-intercept is $(0, -9)$.

The point $(0, -9)$ is three units to the left of the line of symmetry. The point three units to the right of the line of symmetry is $(6, -9)$.

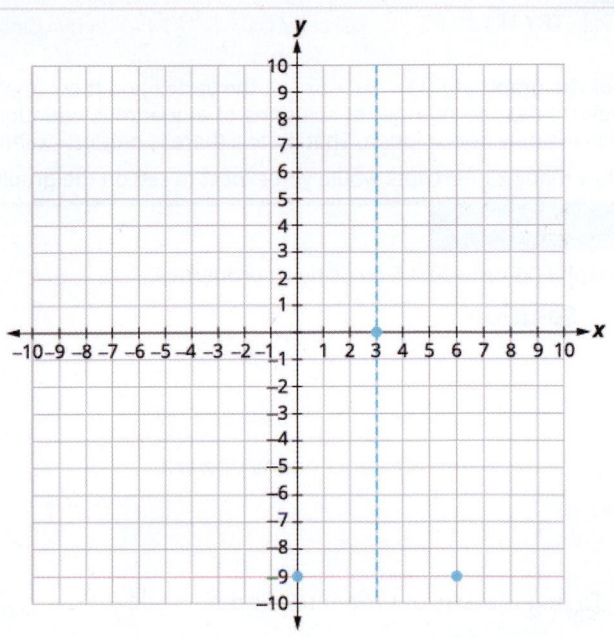

Point symmetric to the *y*-intercept is $(6, -9)$

The *x*-intercept occurs when $f(x) = 0$.	$f(x) = -x^2 + 6x - 9$
Find $f(x) = 0$.	$0 = -x^2 + 6x - 9$
Factor the GCF.	$0 = -(x^2 - 6x + 9)$
Factor the trinomial.	$0 = -(x - 3)^2$
Solve for *x*.	$0 = 3$
Connect the points to graph the parabola.	

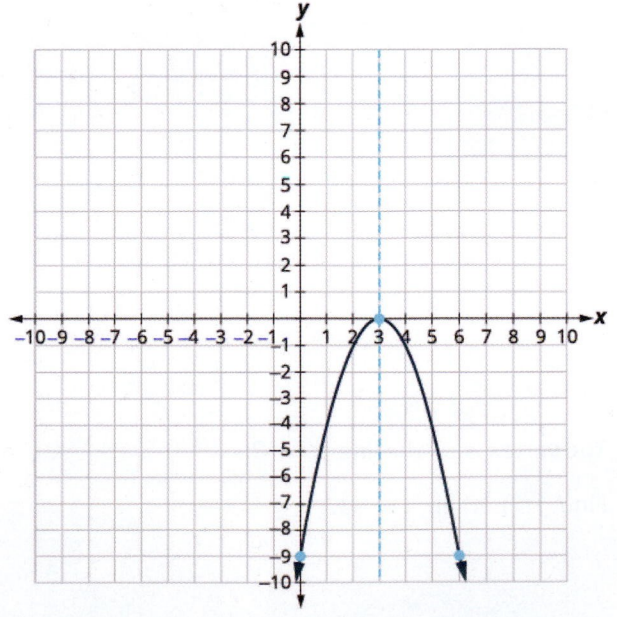

> **TRY IT : : 9.95**   Graph $f(x) = 3x^2 + 12x - 12$ by using its properties.

> **TRY IT : : 9.96**   Graph $f(x) = 4x^2 + 24x + 36$ by using its properties.

For the graph of $f(x) = -x^2 + 6x - 9$, the vertex and the x-intercept were the same point. Remember how the discriminant determines the number of solutions of a quadratic equation? The discriminant of the equation $0 = -x^2 + 6x - 9$ is 0, so there is only one solution. That means there is only one x-intercept, and it is the vertex of the parabola.

How many x-intercepts would you expect to see on the graph of $f(x) = x^2 + 4x + 5$?

EXAMPLE 9.49

Graph $f(x) = x^2 + 4x + 5$ by using its properties.

⊘ **Solution**

$$f(x) = ax^2 + bx + c$$
$$f(x) = x^2 + 4x + 5$$

Since a is 1, the parabola opens upward.

To find the axis of symmetry, find $x = -\dfrac{b}{2a}$.	$x = -\dfrac{b}{2a}$
	$x = -\dfrac{4}{(2)1}$
	$x = -2$
	The equation of the axis of symmetry is $x = -2$.

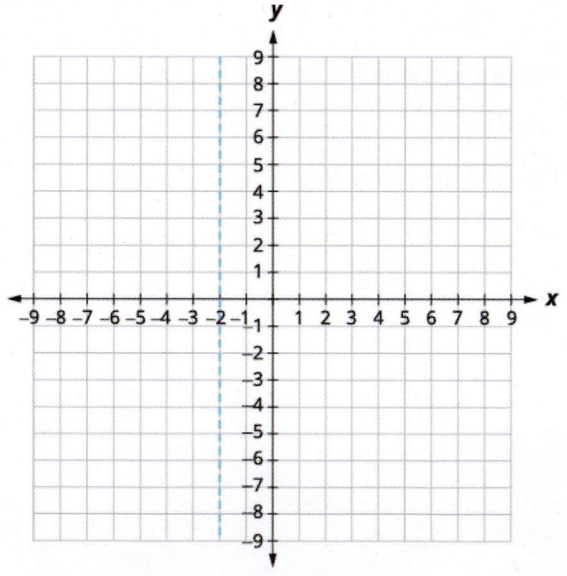

The vertex is on the line $x = -2$.

Find $f(x)$ when $x = -2$. $f(x) = x^2 + 4x + 5$

$$f(-2) = (-2)^2 + 4(-2) + 5$$
$$f(-2) = 4 - 8 + 5$$

$$f(-2) = 1$$

The vertex is $(-2, 1)$.

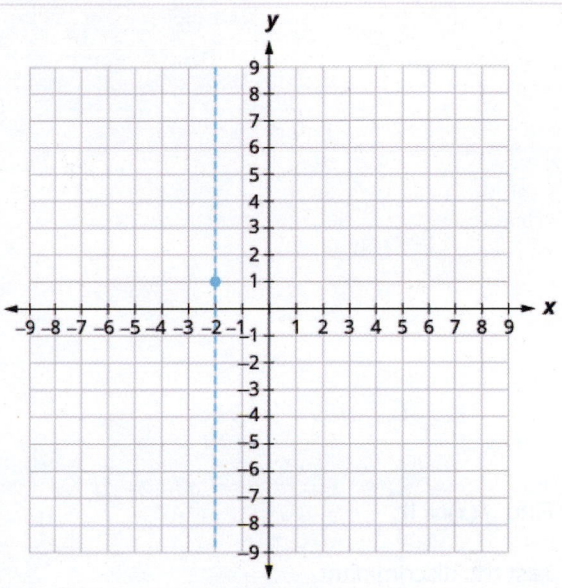

The y-intercept occurs when $x = 0$.

$$f(x) = x^2 + 4x - 5$$

Find $f(0)$.

$$f(0) = 5$$

Simplify.

$$f(0) = 5$$

The y-intercept is $(0, 5)$.

The point $(-4, 5)$ is two units to the left of the line of symmetry.
The point two units to the right of the line of symmetry is $(0, 5)$.

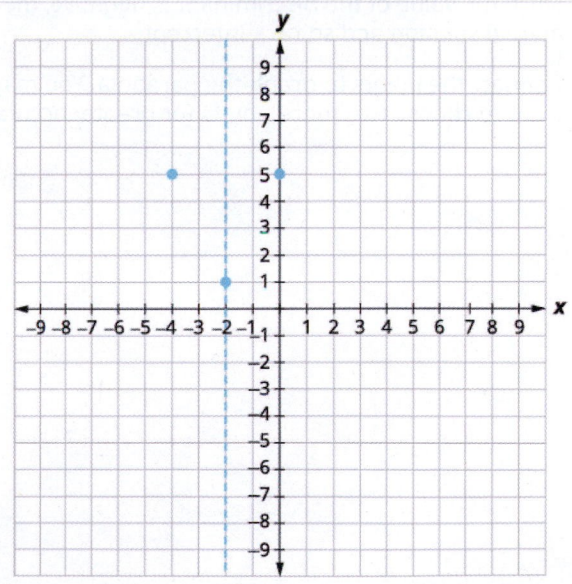

Point symmetric to the y-intercept is $(-4, 5)$.

The *x*-intercept occurs when $f(x) = 0$.

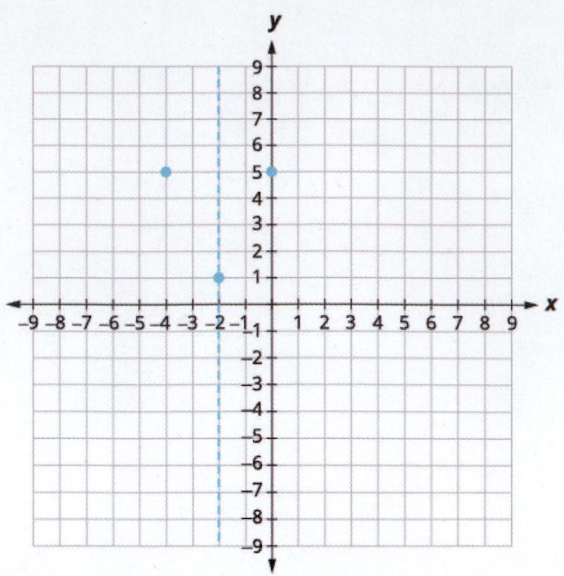

Find $f(x) = 0$.	$0 = x^2 + 4x + 5$

Test the discriminant.

$$b^2 - 4ac$$

$$4^2 - 4 \cdot 1 \cdot 5$$

$$16 - 20$$

$$-4$$

Since the value of the discriminant is negative, there is no real solution and so no *x*-intercept.

Connect the points to graph the parabola. You may want to choose two more points for greater accuracy.

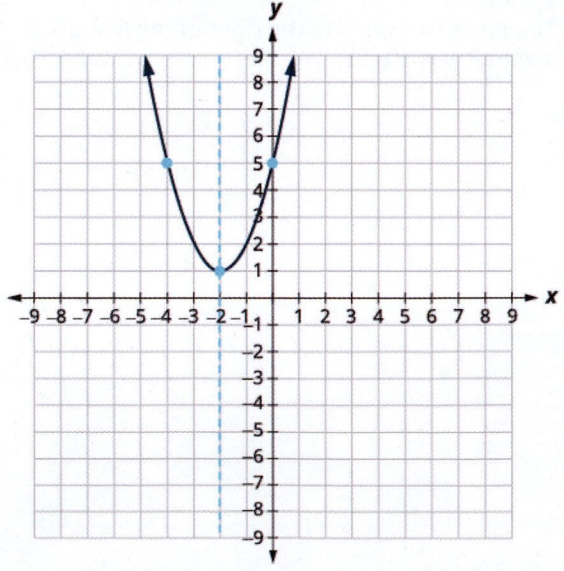

> **TRY IT :: 9.97**   Graph $f(x) = x^2 - 2x + 3$ by using its properties.

> **TRY IT :: 9.98**   Graph $f(x) = -3x^2 - 6x - 4$ by using its properties.

Finding the y-intercept by finding $f(0)$ is easy, isn't it? Sometimes we need to use the Quadratic Formula to find the x-intercepts.

EXAMPLE 9.50

Graph $f(x) = 2x^2 - 4x - 3$ by using its properties.

✓ **Solution**

$$f(x) = ax^2 + bx + c$$
$$f(x) = 2x^2 - 4x - 3$$

Since a is 2, the parabola opens upward. ⌣↗	
To find the equation of the axis of symmetry, use $x = -\dfrac{b}{2a}$.	$x = -\dfrac{b}{2a}$
	$x = -\dfrac{-4}{2 \cdot 2}$
	$x = 1$
	The equation of the axis of symmetry is $x = 1$.
The vertex is on the line $x = 1$.	$f(x) = 2x^2 - 4x - 3$
Find $f(1)$.	$f(x) = 2(1)^2 - 4(1) - 3$
	$f(1) = 2 - 4 - 3$
	$f(1) = -5$
	The vertex is $(1, -5)$.
The y-intercept occurs when $x = 0$.	$f(x) = 2x^2 - 4x - 3$
Find $f(0)$.	$f(0) = 2(0)^2 - 4(0) - 3$
Simplify.	$f(0) = -3$
	The y-intercept is $(0, -3)$.
The point $(0, -3)$ is one unit to the left of the line of symmetry. The point one unit to the right of the line of symmetry is $(2, -3)$.	**Point symmetric to the y-intercept is $(2, -3)$**
The x-intercept occurs when $y = 0$.	$f(x) = 2x^2 - 4x - 3$
Find $f(x) = 0$.	$0 = 2x^2 - 4x - 3$
Use the Quadratic Formula.	$x = \dfrac{-b \pm \sqrt{b^2 - 4ac}}{2a}$
Substitute in the values of a, b, and c.	$x = \dfrac{-(-4) \pm \sqrt{(-4)^2 - 4(2)(3)}}{2(2)}$
Simplify.	$x = \dfrac{-4 \pm \sqrt{16 + 24}}{4}$

Simplify inside the radical.	$x = \dfrac{4 \pm \sqrt{40}}{4}$
Simplify the radical.	$x = \dfrac{4 \pm 2\sqrt{10}}{4}$
Factor the GCF.	$x = \dfrac{2\left(2 \pm \sqrt{10}\right)}{4}$
Remove common factors.	$x = \dfrac{2 \pm \sqrt{10}}{2}$
Write as two equations.	$x = \dfrac{2 + \sqrt{10}}{2}, \quad x = \dfrac{2 - \sqrt{10}}{2}$
Approximate the values.	$x \approx 2.5, \quad x \approx -0.6$

The approximate values of the x-intercepts are $(2.5, 0)$ and $(-0.6, 0)$.

Graph the parabola using the points found.

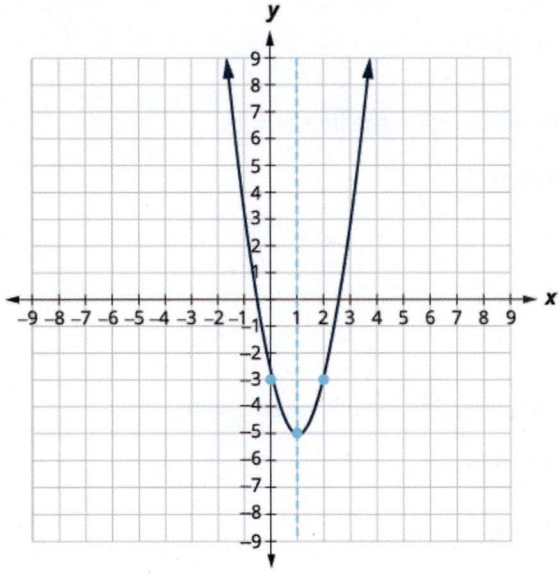

> **TRY IT :: 9.99**   Graph $f(x) = 5x^2 + 10x + 3$ by using its properties.

> **TRY IT :: 9.100**   Graph $f(x) = -3x^2 - 6x + 5$ by using its properties.

Solve Maximum and Minimum Applications

Knowing that the vertex of a parabola is the lowest or highest point of the parabola gives us an easy way to determine the minimum or maximum value of a quadratic function. The y-coordinate of the vertex is the minimum value of a parabola that opens upward. It is the maximum value of a parabola that opens downward. See **Figure 9.6**.

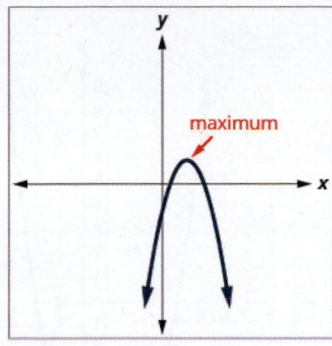 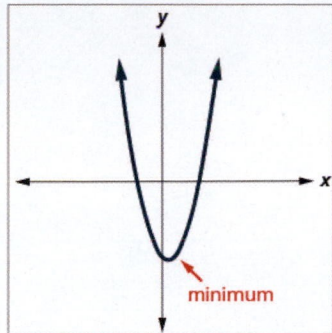

Figure 9.6

Minimum or Maximum Values of a Quadratic Function

The **y-coordinate of the vertex** of the graph of a quadratic function is the

- *minimum* value of the quadratic equation if the parabola opens *upward*.
- *maximum* value of the quadratic equation if the parabola opens *downward*.

EXAMPLE 9.51

Find the minimum or maximum value of the quadratic function $f(x) = x^2 + 2x - 8$.

✓ **Solution**

$$f(x) = x^2 + 2x - 8$$

Since a is positive, the parabola opens upward. The quadratic equation has a minimum.	
Find the equation of the axis of symmetry.	$x = -\dfrac{b}{2a}$
	$x = -\dfrac{2}{2 \times 1}$
	$x = -1$
	The equation of the axis of symmetry is $x = -1$.
The vertex is on the line $x = -1$.	$f(x) = x^2 + 2x - 8$
Find $f(-1)$.	$f(-1) = (-1)^2 + 2(-1) - 8$
	$f(-1) = 1 - 2 - 8$
	$f(-1) = -9$
	The vertex is $(-1, -9)$.

Since the parabola has a minimum, the *y*-coordinate of the vertex is the minimum *y*-value of the quadratic equation.
The minimum value of the quadratic is -9 and it occurs when $x = -1$.

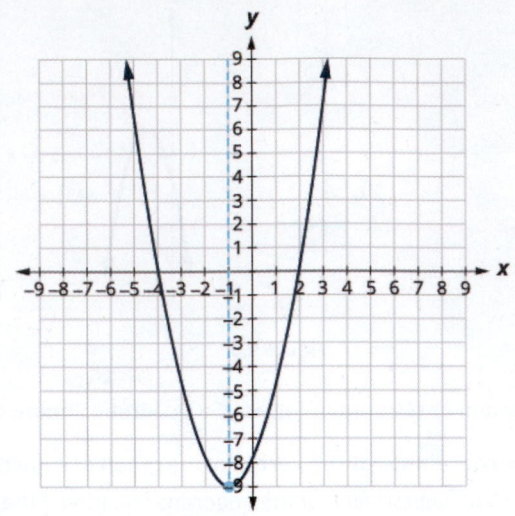

Show the graph to verify the result.

> **TRY IT ::** 9.101
Find the maximum or minimum value of the quadratic function $f(x) = x^2 - 8x + 12$.

> **TRY IT ::** 9.102
Find the maximum or minimum value of the quadratic function $f(x) = -4x^2 + 16x - 11$.

We have used the formula

$$h(t) = -16t^2 + v_0 t + h_0$$

to calculate the height in feet, h, of an object shot upwards into the air with initial velocity, v_0, after t seconds .

This formula is a quadratic function, so its graph is a parabola. By solving for the coordinates of the vertex (t, h), we can find how long it will take the object to reach its maximum height. Then we can calculate the maximum height.

EXAMPLE 9.52

The quadratic equation $h(x) = -16t^2 + 176t + 4$ models the height of a volleyball hit straight upwards with velocity 176 feet per second from a height of 4 feet.

ⓐ How many seconds will it take the volleyball to reach its maximum height? ⓑ Find the maximum height of the volleyball.

✓ **Solution**

$$h(t) = -16t^2 + 176t + 4$$

Since a is negative, the parabola opens downward.
The quadratic function has a maximum.

ⓐ

Find the equation of the axis of symmetry.

$$t = -\frac{b}{2a}$$

$$t = -\frac{176}{2(-16)}$$

$$t = 5.5$$

The equation of the axis of symmetry is $t = 5.5$.

The vertex is on the line $t = 5.5$.

The maximum occurs when $t = 5.5$ seconds.

ⓑ

Find $h(5.5)$.

$$h(t) = -16t^2 + 176t + 4$$

$$h(t) = -16(5.5)^2 + 176(5.5) + 4$$

Use a calculator to simplify.

$$h(t) = 488$$

The vertex is $(5.5,\ 488)$.

Since the parabola has a maximum, the h-coordinate of the vertex is the maximum value of the quadratic function. The maximum value of the quadratic is 488 feet and it occurs when $t = 5.5$ seconds.

After 5.5 seconds, the volleyball will reach its maximum height of 488 feet.

 TRY IT : : 9.103

Solve, rounding answers to the nearest tenth.

The quadratic function $h(x) = -16t^2 + 128t + 32$ is used to find the height of a stone thrown upward from a height of 32 feet at a rate of 128 ft/sec. How long will it take for the stone to reach its maximum height? What is the maximum height?

 TRY IT : : 9.104

A path of a toy rocket thrown upward from the ground at a rate of 208 ft/sec is modeled by the quadratic function of. $h(x) = -16t^2 + 208t$. When will the rocket reach its maximum height? What will be the maximum height?

▶ **MEDIA : :**

Access these online resources for additional instruction and practice with graphing quadratic functions using properties.

- **Quadratic Functions: Axis of Symmetry and Vertex (https://openstax.org/l/37QuadFunct1)**
- **Finding x- and y-intercepts of a Quadratic Function (https://openstax.org/l/37QuadFunct2)**
- **Graphing Quadratic Functions (https://openstax.org/l/37QuadFunct3)**
- **Solve Maxiumum or Minimum Applications (https://openstax.org/l/37QuadFunct4)**
- **Quadratic Applications: Minimum and Maximum (https://openstax.org/l/37QuadFunct5)**

9.6 EXERCISES

Practice Makes Perfect

Recognize the Graph of a Quadratic Function

In the following exercises, graph the functions by plotting points.

229. $f(x) = x^2 + 3$

230. $f(x) = x^2 - 3$

231. $y = -x^2 + 1$

232. $f(x) = -x^2 - 1$

For each of the following exercises, determine if the parabola opens up or down.

233.
ⓐ $f(x) = -2x^2 - 6x - 7$
ⓑ $f(x) = 6x^2 + 2x + 3$

234.
ⓐ $f(x) = 4x^2 + x - 4$
ⓑ $f(x) = -9x^2 - 24x - 16$

235.
ⓐ $f(x) = -3x^2 + 5x - 1$
ⓑ $f(x) = 2x^2 - 4x + 5$

236.
ⓐ $f(x) = x^2 + 3x - 4$
ⓑ $f(x) = -4x^2 - 12x - 9$

Find the Axis of Symmetry and Vertex of a Parabola

In the following functions, find ⓐ *the equation of the axis of symmetry and* ⓑ *the vertex of its graph.*

237. $f(x) = x^2 + 8x - 1$

238. $f(x) = x^2 + 10x + 25$

239. $f(x) = -x^2 + 2x + 5$

240. $f(x) = -2x^2 - 8x - 3$

Find the Intercepts of a Parabola

In the following exercises, find the intercepts of the parabola whose function is given.

241. $f(x) = x^2 + 7x + 6$

242. $f(x) = x^2 + 10x - 11$

243. $f(x) = x^2 + 8x + 12$

244. $f(x) = x^2 + 5x + 6$

245. $f(x) = -x^2 + 8x - 19$

246. $f(x) = -3x^2 + x - 1$

247. $f(x) = x^2 + 6x + 13$

248. $f(x) = x^2 + 8x + 12$

249. $f(x) = 4x^2 - 20x + 25$

250. $f(x) = -x^2 - 14x - 49$

251. $f(x) = -x^2 - 6x - 9$

252. $f(x) = 4x^2 + 4x + 1$

Graph Quadratic Functions Using Properties

In the following exercises, graph the function by using its properties.

253. $f(x) = x^2 + 6x + 5$

254. $f(x) = x^2 + 4x - 12$

255. $f(x) = x^2 + 4x + 3$

256. $f(x) = x^2 - 6x + 8$

257. $f(x) = 9x^2 + 12x + 4$

258. $f(x) = -x^2 + 8x - 16$

259. $f(x) = -x^2 + 2x - 7$

260. $f(x) = 5x^2 + 2$

261. $f(x) = 2x^2 - 4x + 1$

262. $f(x) = 3x^2 - 6x - 1$

263. $f(x) = 2x^2 - 4x + 2$

264. $f(x) = -4x^2 - 6x - 2$

265. $f(x) = -x^2 - 4x + 2$

266. $f(x) = x^2 + 6x + 8$

267. $f(x) = 5x^2 - 10x + 8$

268. $f(x) = -16x^2 + 24x - 9$

269. $f(x) = 3x^2 + 18x + 20$

270. $f(x) = -2x^2 + 8x - 10$

Solve Maximum and Minimum Applications

In the following exercises, find the maximum or minimum value of each function.

271. $f(x) = 2x^2 + x - 1$

272. $y = -4x^2 + 12x - 5$

273. $y = x^2 - 6x + 15$

274. $y = -x^2 + 4x - 5$

275. $y = -9x^2 + 16$

276. $y = 4x^2 - 49$

In the following exercises, solve. Round answers to the nearest tenth.

277. An arrow is shot vertically upward from a platform 45 feet high at a rate of 168 ft/sec. Use the quadratic function $h(t) = -16t^2 + 168t + 45$ find how long it will take the arrow to reach its maximum height, and then find the maximum height.

278. A stone is thrown vertically upward from a platform that is 20 feet height at a rate of 160 ft/sec. Use the quadratic function $h(t) = -16t^2 + 160t + 20$ to find how long it will take the stone to reach its maximum height, and then find the maximum height.

279. A ball is thrown vertically upward from the ground with an initial velocity of 109 ft/sec. Use the quadratic function $h(t) = -16t^2 + 109t + 0$ to find how long it will take for the ball to reach its maximum height, and then find the maximum height.

280. A ball is thrown vertically upward from the ground with an initial velocity of 122 ft/sec. Use the quadratic function $h(t) = -16t^2 + 122t + 0$ to find how long it will take for the ball to reach its maxiumum height, and then find the maximum height.

281. A computer store owner estimates that by charging x dollars each for a certain computer, he can sell $40 - x$ computers each week. The quadratic function $R(x) = -x^2 + 40x$ is used to find the revenue, R, received when the selling price of a computer is x, Find the selling price that will give him the maximum revenue, and then find the amount of the maximum revenue.

282. A retailer who sells backpacks estimates that by selling them for x dollars each, he will be able to sell $100 - x$ backpacks a month. The quadratic function $R(x) = -x^2 + 100x$ is used to find the R, received when the selling price of a backpack is x. Find the selling price that will give him the maximum revenue, and then find the amount of the maximum revenue.

283. A retailer who sells fashion boots estimates that by selling them for x dollars each, he will be able to sell $70 - x$ boots a week. Use the quadratic function $R(x) = -x^2 + 70x$ to find the revenue received when the average selling price of a pair of fashion boots is x. Find the selling price that will give him the maximum revenue, and then find the amount of the maximum revenue.

284. A cell phone company estimates that by charging x dollars each for a certain cell phone, they can sell $8 - x$ cell phones per day. Use the quadratic function $R(x) = -x^2 + 8x$ to find the revenue received when the selling price of a cell phone is x. Find the selling price that will give them the maximum revenue, and then find the amount of the maximum revenue.

285. A rancher is going to fence three sides of a corral next to a river. He needs to maximize the corral area using 240 feet of fencing. The quadratic equation $A(x) = x(240 - 2x)$ gives the area of the corral, A, for the length, x, of the corral along the river. Find the length of the corral along the river that will give the maximum area, and then find the maximum area of the corral.

286. A veterinarian is enclosing a rectangular outdoor running area against his building for the dogs he cares for. He needs to maximize the area using 100 feet of fencing. The quadratic function $A(x) = x(100 - 2x)$ gives the area, A, of the dog run for the length, x, of the building that will border the dog run. Find the length of the building that should border the dog run to give the maximum area, and then find the maximum area of the dog run.

287. A land owner is planning to build a fenced in rectangular patio behind his garage, using his garage as one of the "walls." He wants to maximize the area using 80 feet of fencing. The quadratic function $A(x) = x(80 - 2x)$ gives the area of the patio, where x is the width of one side. Find the maximum area of the patio.

288. A family of three young children just moved into a house with a yard that is not fenced in. The previous owner gave them 300 feet of fencing to use to enclose part of their backyard. Use the quadratic function $A(x) = x(300 - 2x)$ to determine the maximum area of the fenced in yard.

Writing Exercise

289. How do the graphs of the functions $f(x) = x^2$ and $f(x) = x^2 - 1$ differ? We graphed them at the start of this section. What is the difference between their graphs? How are their graphs the same?

290. Explain the process of finding the vertex of a parabola.

291. Explain how to find the intercepts of a parabola.

292. How can you use the discriminant when you are graphing a quadratic function?

Self Check

ⓐ *After completing the exercises, use this checklist to evaluate your mastery of the objectives of this section.*

I can...	Confidently	With some help	No-I don't get it!
recognize the graph of a quadratic equation.			
find the axis of symmetry and vertex of a parabola.			
find the intercepts of a parabola.			
graph quadratic equations in two variables.			
solve maximum and minimum applications.			

ⓑ *After looking at the checklist, do you think you are well-prepared for the next section? Why or why not?*

9.7 | Graph Quadratic Functions Using Transformations

Learning Objectives

By the end of this section, you will be able to:

> Graph quadratic functions of the form $f(x) = x^2 + k$

> Graph quadratic functions of the form $f(x) = (x - h)^2$

> Graph quadratic functions of the form $f(x) = ax^2$

> Graph quadratic functions using transformations

> Find a quadratic function from its graph

Be Prepared!

Before you get started, take this readiness quiz.

1. Graph the function $f(x) = x^2$ by plotting points.
   If you missed this problem, review **Example 3.54**.

2. Factor completely: $y^2 - 14y + 49$.
   If you missed this problem, review **Example 6.24**.

3. Factor completely: $2x^2 - 16x + 32$.
   If you missed this problem, review **Example 6.26**.

Graph Quadratic Functions of the form $f(x) = x^2 + k$

In the last section, we learned how to graph quadratic functions using their properties. Another method involves starting with the basic graph of $f(x) = x^2$ and 'moving' it according to information given in the function equation. We call this graphing quadratic functions using transformations.

In the first example, we will graph the quadratic function $f(x) = x^2$ by plotting points. Then we will see what effect adding a constant, k, to the equation will have on the graph of the new function $f(x) = x^2 + k$.

EXAMPLE 9.53

Graph $f(x) = x^2$, $g(x) = x^2 + 2$, and $h(x) = x^2 - 2$ on the same rectangular coordinate system. Describe what effect adding a constant to the function has on the basic parabola.

⊘ Solution

Plotting points will help us see the effect of the constants on the basic $f(x) = x^2$ graph. We fill in the chart for all three functions.

x	$f(x) = x^2$	$(x, f(x))$	$g(x) = x^2 + 2$	$(x, g(x))$	$h(x) = x^2 - 2$	$(x, h(x))$
–3	9	(–3, 9)	9 + 2	(–3, 11)	9 – 2	(–3, 7)
–2	4	(–2, 4)	4 + 2	(–2, 6)	4 – 2	(–2, 2)
–1	1	(–1, 1)	1 + 2	(–1, 3)	1 – 2	(–1, 1)
0	0	(0, 0)	0 + 2	(0, 2)	0 – 2	(0, –2)
1	1	(1, 1)	1 + 2	(1, 3)	1 – 2	(1, –1)
2	4	(2, 4)	4 + 2	(2, 6)	4 – 2	(2, 2)
3	9	(3, 9)	9 + 2	(3, 11)	9 – 2	(3, 7)

The $g(x)$ values are two more than the $f(x)$ values. Also, the $h(x)$ values are two less than the $f(x)$ values. Now we will graph all three functions on the same rectangular coordinate system.

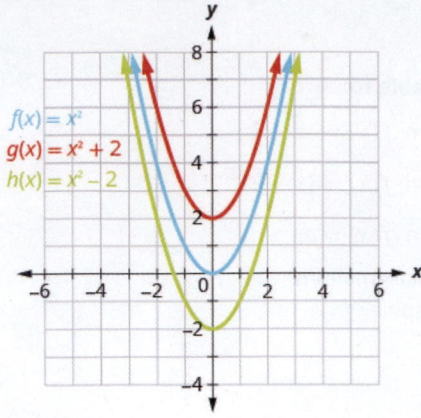

The graph of $g(x) = x^2 + 2$ is the same as the graph of $f(x) = x^2$ but shifted up 2 units.

The graph of $h(x) = x^2 - 2$ is the same as the graph of $f(x) = x^2$ but shifted down 2 units.

The graph of $g(x) = x^2 + 2$ is the same as the graph of $f(x) = x^2$ but shifted up 2 units.

The graph of $h(x) = x^2 - 2$ is the same as the graph of $f(x) = x^2$ but shifted down 2 units.

> **TRY IT : : 9.105**
>
> ⓐ Graph $f(x) = x^2$, $g(x) = x^2 + 1$, and $h(x) = x^2 - 1$ on the same rectangular coordinate system.
>
> ⓑ Describe what effect adding a constant to the function has on the basic parabola.

> **TRY IT : : 9.106**
>
> ⓐ Graph $f(x) = x^2$, $g(x) = x^2 + 6$, and $h(x) = x^2 - 6$ on the same rectangular coordinate system.
>
> ⓑ Describe what effect adding a constant to the function has on the basic parabola.

The last example shows us that to graph a quadratic function of the form $f(x) = x^2 + k$, we take the basic parabola graph of $f(x) = x^2$ and vertically shift it up $(k > 0)$ or shift it down $(k < 0)$.

This transformation is called a vertical shift.

Graph a Quadratic Function of the form $f(x) = x^2 + k$ Using a Vertical Shift

The graph of $f(x) = x^2 + k$ shifts the graph of $f(x) = x^2$ vertically k units.

- If $k > 0$, shift the parabola vertically up k units.
- If $k < 0$, shift the parabola vertically down $|k|$ units.

Now that we have seen the effect of the constant, k, it is easy to graph functions of the form $f(x) = x^2 + k$. We just start with the basic parabola of $f(x) = x^2$ and then shift it up or down.

It may be helpful to practice sketching $f(x) = x^2$ quickly. We know the values and can sketch the graph from there.

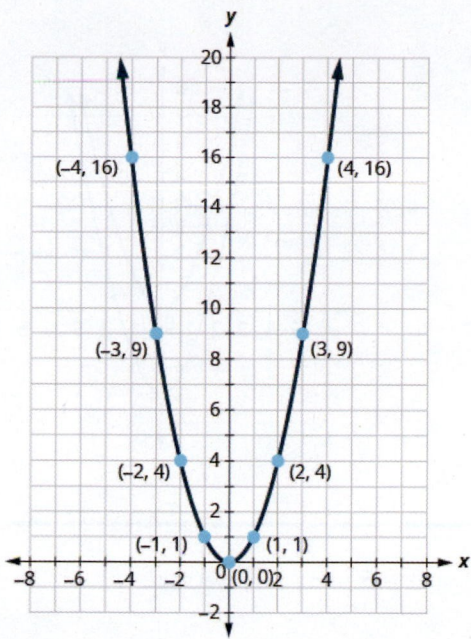

Once we know this parabola, it will be easy to apply the transformations. The next example will require a vertical shift.

EXAMPLE 9.54

Graph $f(x) = x^2 - 3$ using a vertical shift.

✓ **Solution**

We first draw the graph of $f(x) = x^2$ on the grid.

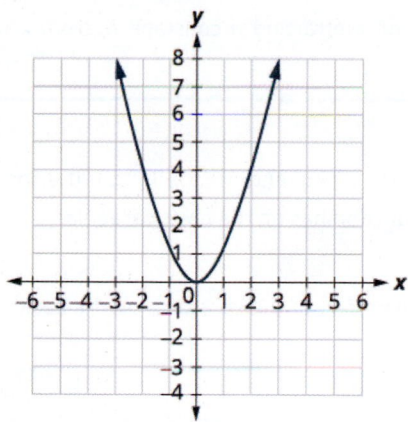

Determine k.

$$f(x) = x^2 + k$$
$$f(x) = x^2 - 3$$

$$k = -3$$

Shift the graph $f(x) = x^2$ down 3.

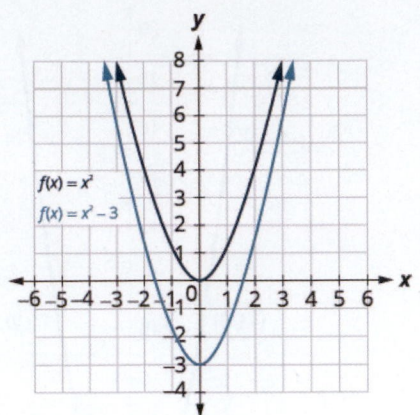

> **TRY IT :: 9.107**   Graph $f(x) = x^2 - 5$ using a vertical shift.

> **TRY IT :: 9.108**   Graph $f(x) = x^2 + 7$ using a vertical shift.

Graph Quadratic Functions of the form $f(x) = (x - h)^2$

In the first example, we graphed the quadratic function $f(x) = x^2$ by plotting points and then saw the effect of adding a constant k to the function had on the resulting graph of the new function $f(x) = x^2 + k$.

We will now explore the effect of subtracting a constant, h, from x has on the resulting graph of the new function $f(x) = (x - h)^2$.

EXAMPLE 9.55

Graph $f(x) = x^2$, $g(x) = (x - 1)^2$, and $h(x) = (x + 1)^2$ on the same rectangular coordinate system. Describe what effect adding a constant to the function has on the basic parabola.

⊘ **Solution**

Plotting points will help us see the effect of the constants on the basic $f(x) = x^2$ graph. We fill in the chart for all three functions.

x	$f(x) = x^2$	$(x, f(x))$	$g(x) = (x - 1)^2$	$(x, g(x))$	$h(x) = (x + 1)^2$	$(x, h(x))$
−3	9	(−3, 9)	16	(−3, 16)	4	(−3, 4)
−2	4	(−2, 4)	9	(−2, 9)	1	(−2, 1)
−1	1	(−1, 1)	4	(−1, 4)	0	(−1, 0)
0	0	(0, 0)	1	(0, 1)	1	(0, 1)
1	1	(1, 1)	0	(1, 0)	4	(1, 4)
2	4	(2, 4)	1	(2, 1)	9	(2, 9)
3	9	(3, 9)	4	(3, 4)	16	(3, 16)

The $g(x)$ values and the $h(x)$ values share the common numbers 0, 1, 4, 9, and 16, but are shifted.

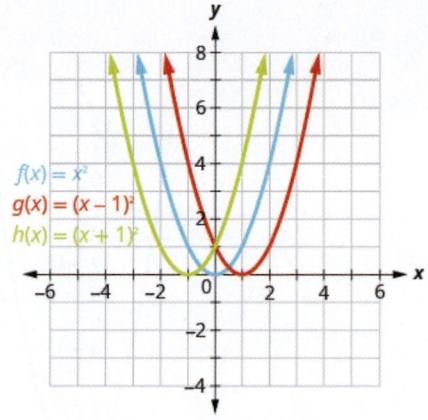

The graph of $g(x) = (x - 1)^2$ is the same as the graph of $f(x) = x^2$ but shifted right 1 unit.

The graph of $h(x) = (x + 1)^2$ is the same as the graph of $f(x) = x^2$ but shifted left 1 unit.

$g(x) = (x - 1)^2$ $h(x) = (x + 1)^2$

⟶ 1 unit   ⟵ 1 unit

> **TRY IT :: 9.109**

ⓐ Graph $f(x) = x^2$, $g(x) = (x + 2)^2$, and $h(x) = (x - 2)^2$ on the same rectangular coordinate system.

ⓑ Describe what effect adding a constant to the function has on the basic parabola.

> **TRY IT :: 9.110**

ⓐ Graph $f(x) = x^2$, $g(x) = x^2 + 5$, and $h(x) = x^2 - 5$ on the same rectangular coordinate system.

ⓑ Describe what effect adding a constant to the function has on the basic parabola.

The last example shows us that to graph a quadratic function of the form $f(x) = (x - h)^2$, we take the basic parabola graph of $f(x) = x^2$ and shift it left ($h > 0$) or shift it right ($h < 0$).

This transformation is called a horizontal shift.

Graph a Quadratic Function of the form $f(x) = (x - h)^2$ Using a Horizontal Shift

The graph of $f(x) = (x - h)^2$ shifts the graph of $f(x) = x^2$ horizontally h units.

- If $h > 0$, shift the parabola horizontally left h units.
- If $h < 0$, shift the parabola horizontally right $|h|$ units.

Now that we have seen the effect of the constant, h, it is easy to graph functions of the form $f(x) = (x - h)^2$. We just start with the basic parabola of $f(x) = x^2$ and then shift it left or right.

The next example will require a horizontal shift.

EXAMPLE 9.56

Graph $f(x) = (x - 6)^2$ using a horizontal shift.

✓ Solution

We first draw the graph of $f(x) = x^2$ on the grid.

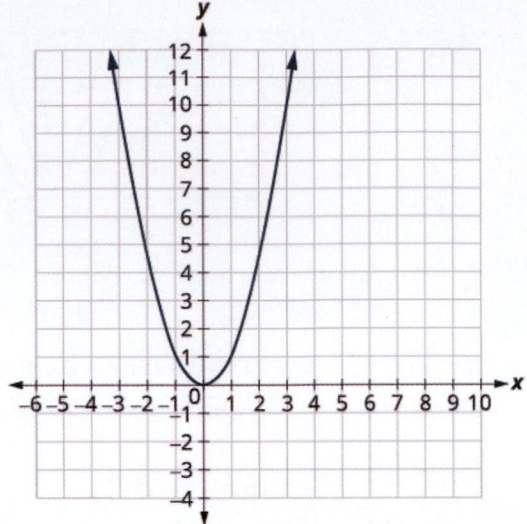

| Determine h. | $f(x) = (x-h)^2$ |
| | $f(x) = (x-6)^2$ |

$$h = 6$$

Shift the graph $f(x) = x^2$ to the right 6 units.

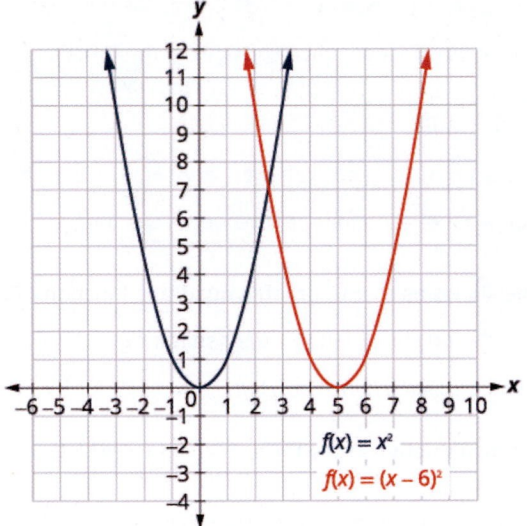

> **TRY IT ::: 9.111**   Graph $f(x) = (x-4)^2$ using a horizontal shift.

> **TRY IT ::: 9.112**   Graph $f(x) = (x+6)^2$ using a horizontal shift.

Now that we know the effect of the constants h and k, we will graph a quadratic function of the form $f(x) = (x-h)^2 + k$ by first drawing the basic parabola and then making a horizontal shift followed by a vertical shift. We could do the vertical shift followed by the horizontal shift, but most students prefer the horizontal shift followed by the vertical.

EXAMPLE 9.57

Graph $f(x) = (x + 1)^2 - 2$ using transformations.

⊘ Solution

This function will involve two transformations and we need a plan.

Let's first identify the constants h, k.

$$f(x) = (x + 1)^2 - 2$$

$$f(x) = \quad (x - h)^2 \quad + \quad k$$
$$f(x) = (x - (-1))^2 + (-2)$$

$$h = -1 \quad k = -2$$

The h constant gives us a horizontal shift and the k gives us a vertical shift.

$$f(x) = x^2 \quad \longrightarrow \quad f(x) = (x + 1)^2 \quad \longrightarrow \quad f(x) = (x + 1)^2 - 2$$

$$h = -1 \qquad\qquad k = -2$$

Shift left 1 unit   Shift down 2 units

We first draw the graph of $f(x) = x^2$ on the grid.

To graph $f(x) = (x + 1)^2$, shift the graph $f(x) = x^2$ to the left 1 unit.

To graph $f(x) = (x + 1)^2 - 2$, shift the graph $f(x) = (x + 1)^2$ down 2 units.

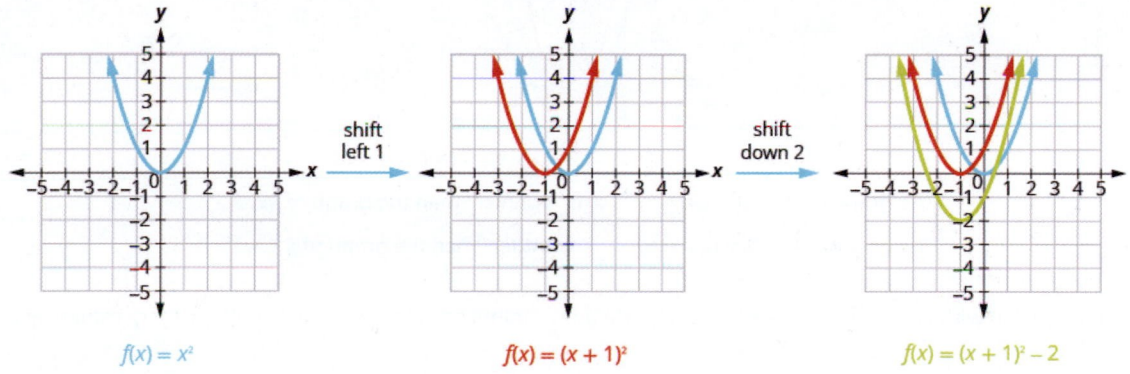

$f(x) = x^2$ $f(x) = (x + 1)^2$ $f(x) = (x + 1)^2 - 2$

> **TRY IT :: 9.113**   Graph $f(x) = (x + 2)^2 - 3$ using transformations.

> **TRY IT :: 9.114**   Graph $f(x) = (x - 3)^2 + 1$ using transformations.

Graph Quadratic Functions of the Form $f(x) = ax^2$

So far we graphed the quadratic function $f(x) = x^2$ and then saw the effect of including a constant h or k in the equation had on the resulting graph of the new function. We will now explore the effect of the coefficient a on the resulting graph of the new function $f(x) = ax^2$.

Let's look at the quadratic functions $f(x) = x^2$, $g(x) = 2x^2$ and $h(x) = \frac{1}{2}x^2$.

x	$f(x) = x^2$	$(x, f(x))$	$g(x) = 2x^2$	$(x, g(x))$	$h(x) = \frac{1}{2}x^2$	$(x, h(x))$
-2	4	$(-2, 4)$	$2 \cdot 4$	$(-2, 8)$	$\frac{1}{2} \cdot 4$	$(-2, 2)$
-1	1	$(-1, 1)$	$2 \cdot 1$	$(-1, 2)$	$\frac{1}{2} \cdot 1$	$\left(-1, \frac{1}{2}\right)$
0	0	$(0, 0)$	$2 \cdot 0$	$(0, 0)$	$\frac{1}{2} \cdot 0$	$(0, 0)$
1	1	$(1, 1)$	$2 \cdot 1$	$(1, 2)$	$\frac{1}{2} \cdot 1$	$\left(1, \frac{1}{2}\right)$
2	4	$(2, 4)$	$2 \cdot 4$	$(2, 8)$	$\frac{1}{2} \cdot 4$	$(2, 2)$

If we graph these functions, we can see the effect of the constant a, assuming $a > 0$.

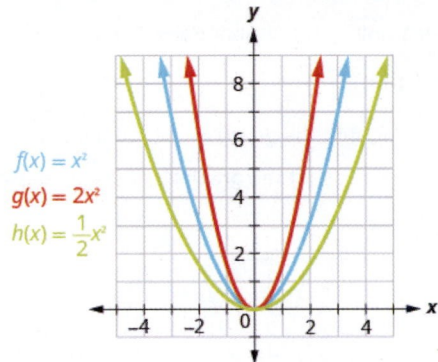

$f(x) = x^2$

$g(x) = 2x^2$

$h(x) = \frac{1}{2}x^2$

The graph of the function $g(x) = 2x^2$ is "skinnier" than the graph of $f(x) = x^2$.

The graph of the function $h(x) = \frac{1}{2}x^2$ is "wider" than the graph of $f(x) = x^2$.

To graph a function with constant a it is easiest to choose a few points on $f(x) = x^2$ and multiply the y-values by a.

Graph of a Quadratic Function of the form $f(x) = ax^2$

The coefficient a in the function $f(x) = ax^2$ affects the graph of $f(x) = x^2$ by stretching or compressing it.

- If $0 < |a| < 1$, the graph of $f(x) = ax^2$ will be "wider" than the graph of $f(x) = x^2$.

- If $|a| > 1$, the graph of $f(x) = ax^2$ will be "skinnier" than the graph of $f(x) = x^2$.

EXAMPLE 9.58

Graph $f(x) = 3x^2$.

✓ **Solution**

We will graph the functions $f(x) = x^2$ and $g(x) = 3x^2$ on the same grid. We will choose a few points on $f(x) = x^2$ and then multiply the y-values by 3 to get the points for $g(x) = 3x^2$.

	$f(x) = x^2$	$g(x) = 3x^2$	
x	**(x, f(x))**	**(x, g(x))**	
−2	(−2, 4)	(−2, 12)	3 · 4 = 12
−1	(−1, 1)	(−1, 3)	3 · 1 = 3
0	(0, 0)	(0, 0)	3 · 0 = 0
1	(1, 1)	(1, 3)	3 · 1 = 3
2	(2, 4)	(2, 12)	3 · 4 = 12

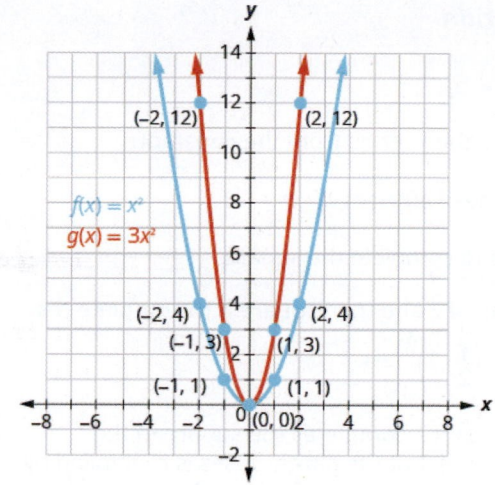

> **TRY IT : : 9.115**   Graph $f(x) = -3x^2$.

> **TRY IT : : 9.116**   Graph $f(x) = 2x^2$.

Graph Quadratic Functions Using Transformations

We have learned how the constants a, h, and k in the functions, $f(x) = x^2 + k$, $f(x) = (x - h)^2$, and $f(x) = ax^2$ affect their graphs. We can now put this together and graph quadratic functions $f(x) = ax^2 + bx + c$ by first putting them into the form $f(x) = a(x - h)^2 + k$ by completing the square. This form is sometimes known as the vertex form or standard form.

We must be careful to both add and subtract the number to the SAME side of the function to complete the square. We cannot add the number to both sides as we did when we completed the square with quadratic equations.

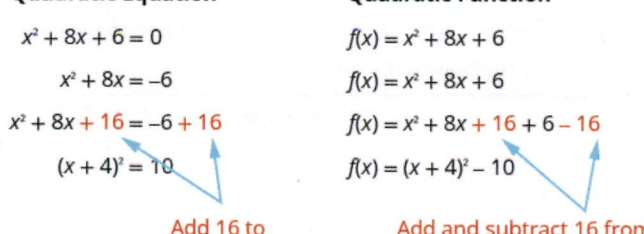

When we complete the square in a function with a coefficient of x^2 that is not one, we have to factor that coefficient from just the x-terms. We do not factor it from the constant term. It is often helpful to move the constant term a bit to the right to make it easier to focus only on the x-terms.

Once we get the constant we want to complete the square, we must remember to multiply it by that coefficient before we then subtract it.

Rewrite $f(x) = -3x^2 - 6x - 1$ in the $f(x) = a(x - h)^2 + k$ form by completing the square.

✓ Solution

	$f(x) = -3x^2 - 6x - 1$
Separate the x terms from the constant.	$f(x) = -3x^2 - 6x \quad -1$
Factor the coefficient of x^2, -3.	$f(x) = -3(x^2 + 2x) \quad -1$
Prepare to complete the square.	$f(x) = -3(x^2 + 2x \quad) -1$
Take half of 2 and then square it to complete the square. $\left(\frac{1}{2} \cdot 2\right)^2 = 1$	
The constant 1 completes the square in the parentheses, but the parentheses is multiplied by -3. So we are really adding -3 We must then add 3 to not change the value of the function.	$f(x) = -3(x^2 + 2x + 1) - 1 + 3$ $-3 \cdot 1 = -3$ so add 3
Rewrite the trinomial as a square and subtract the constants.	$f(x) = -3(x + 1) + 2$
The function is now in the $f(x) = a(x - h)^2 + k$ form.	$f(x) = \quad a(x - h)^2 \quad + k$ $f(x) = -3(x + 1)^2 + 2$

> **TRY IT :: 9.117**
>
> Rewrite $f(x) = -4x^2 - 8x + 1$ in the $f(x) = a(x - h)^2 + k$ form by completing the square.

> **TRY IT :: 9.118**   Rewrite $f(x) = 2x^2 - 8x + 3$ in the $f(x) = a(x - h)^2 + k$ form by completing the square.

Once we put the function into the $f(x) = (x - h)^2 + k$ form, we can then use the transformations as we did in the last few problems. The next example will show us how to do this.

EXAMPLE 9.60

Graph $f(x) = x^2 + 6x + 5$ by using transformations.

✓ Solution

Step 1. Rewrite the function in $f(x) = a(x - h)^2 + k$ vertex form by completing the square.

	$f(x) = x^2 + 6x + 5$
Separate the x terms from the constant.	$f(x) = x^2 + 6x + 5$
Take half of 6 and then square it to complete the square. $\left(\frac{1}{2} \cdot 6\right)^2 = 9$	
We both add 9 and subtract 9 to not change the value of the function.	$f(x) = x^2 + 6x + 9 + 5 - 9$
Rewrite the trinomial as a square and subtract the constants.	$f(x) = (x + 3)^2 - 4$
The function is now in the $f(x) = (x - h)^2 + k$ form.	$f(x) = (x - h)^2 + k$ $f(x) = (x + 3)^2 - 4$

Step 2: Graph the function using transformations.

Looking at the h, k values, we see the graph will take the graph of $f(x) = x^2$ and shift it to the left 3 units and down 4 units.

$f(x) = x^2 \longrightarrow f(x) = (x + 3)^2 \longrightarrow f(x) = (x + 3)^2 - 4$

$h = -3$ $k = -4$

Shift left 3 units Shift down 4 units

We first draw the graph of $f(x) = x^2$ on the grid.

To graph $f(x) = (x + 3)^2$, shift the graph $f(x) = x^2$ to the left 3 units.

To graph $f(x) = (x + 3)^2 - 4$, shift the graph $f(x) = (x + 3)^2$ down 4 units.

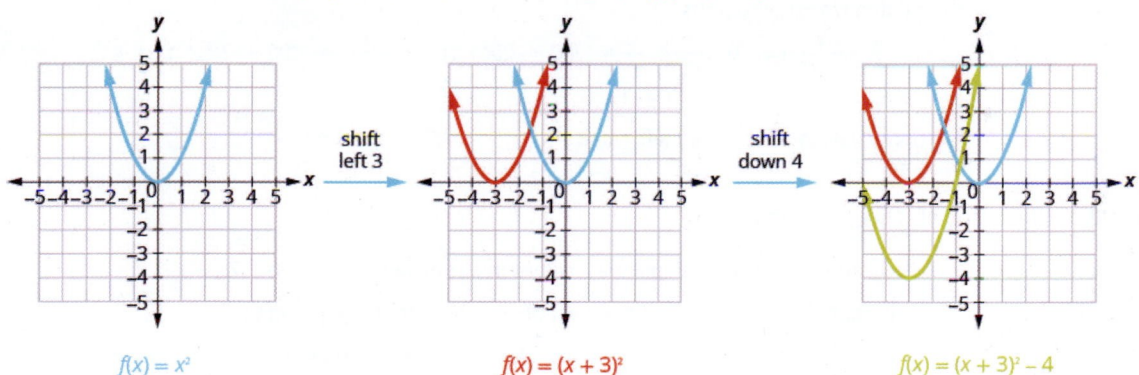

$f(x) = x^2$ $f(x) = (x + 3)^2$ $f(x) = (x + 3)^2 - 4$

 TRY IT :: 9.119 Graph $f(x) = x^2 + 2x - 3$ by using transformations.

> **TRY IT :: 9.120** Graph $f(x) = x^2 - 8x + 12$ by using transformations.

We list the steps to take to graph a quadratic function using transformations here.

HOW TO :: GRAPH A QUADRATIC FUNCTION USING TRANSFORMATIONS.

Step 1. Rewrite the function in $f(x) = a(x - h)^2 + k$ form by completing the square.

Step 2. Graph the function using transformations.

EXAMPLE 9.61

Graph $f(x) = -2x^2 - 4x + 2$ by using transformations.

✓ Solution

Step 1. Rewrite the function in $f(x) = a(x - h)^2 + k$ vertex form by completing the square.

	$f(x) = -2x^2 - 4x + 2$
Separate the x terms from the constant.	$f(x) = -2x^2 - 4x + 2$
We need the coefficient of x^2 to be one. We factor -2 from the x-terms.	$f(x) = -2(x^2 + 2x) + 2$
Take half of 2 and then square it to complete the square. $\left(\frac{1}{2} \cdot 2\right)^2 = 1$	
We add 1 to complete the square in the parentheses, but the parentheses is multiplied by -2. Se we are really adding -2. To not change the value of the function we add 2.	$f(x) = -2(x^2 + 2x + 1) + 2 + 2$
Rewrite the trinomial as a square and subtract the constants.	$f(x) = -2(x + 1)^2 + 4$
The function is now in the $f(x) = a(x - h)^2 + k$ form.	$f(x) = \quad a(x - h)^2 \quad + k$ $f(x) = -2(x + 1)^2 + 4$

Step 2. Graph the function using transformations.

$f(x) = x^2$ ⟶ $f(x) = -2x^2$ ⟶ $f(x) = -2(x + 1)^2$ ⟶ $f(x) = -2(x + 1)^2 + 4$

$a = -2$ $h = -1$ $k = 4$

Multiply y-values by -2 Shift left 1 unit   Shift up 4 units

We first draw the graph of $f(x) = x^2$ on the grid.

To graph $f(x) = -2x^2$, multiply the y-values in parabola of $f(x) = x^2$ by -2.

To graph $f(x) = -2(x + 1)^2$, shift the graph $f(x) = -2x^2$ to the left 1 unit.

To graph $f(x) = -2(x + 1)^2 + 4$, shift the graph $f(x) = (x + 1)^2$ up 4 units.

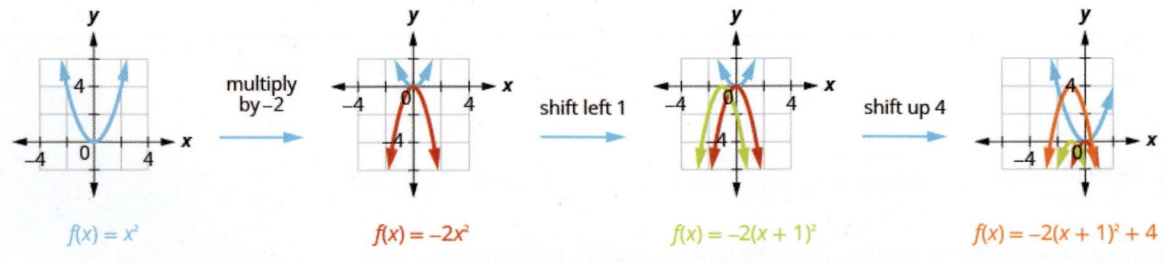

$f(x) = x^2$ multiply by -2 $f(x) = -2x^2$ shift left 1 $f(x) = -2(x + 1)^2$ shift up 4 $f(x) = -2(x + 1)^2 + 4$

> **TRY IT : : 9.121**   Graph $f(x) = -3x^2 + 12x - 4$ by using transformations.

> **TRY IT : : 9.122**   Graph $f(x) = -2x^2 + 12x - 9$ by using transformations.

Now that we have completed the square to put a quadratic function into $f(x) = a(x - h)^2 + k$ form, we can also use this technique to graph the function using its properties as in the previous section.

If we look back at the last few examples, we see that the vertex is related to the constants *h* and *k*.

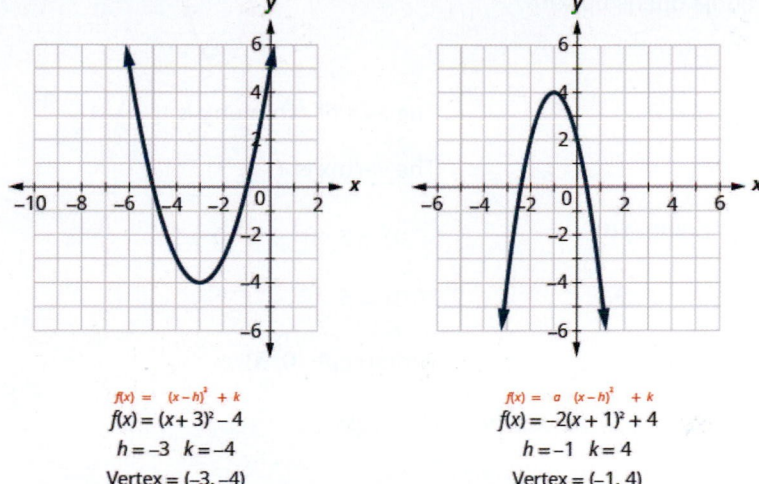

$$f(x) = (x - h)^2 + k$$
$$f(x) = (x + 3)^2 - 4$$
$$h = -3 \quad k = -4$$
Vertex = (-3, -4)

$$f(x) = a(x - h)^2 + k$$
$$f(x) = -2(x + 1)^2 + 4$$
$$h = -1 \quad k = 4$$
Vertex = (-1, 4)

In each case, the vertex is (*h*, *k*). Also the axis of symmetry is the line *x* = *h*.

We rewrite our steps for graphing a quadratic function using properties for when the function is in $f(x) = a(x - h)^2 + k$ form.

HOW TO :: GRAPH A QUADRATIC FUNCTION IN THE FORM $f(x) = a(x - h)^2 + k$ USING PROPERTIES.

Step 1.   Rewrite the function in $f(x) = a(x - h)^2 + k$ form.

Step 2.   Determine whether the parabola opens upward, *a* > 0, or downward, *a* < 0.

Step 3.   Find the axis of symmetry, *x* = *h*.

Step 4.   Find the vertex, (*h*, *k*).

Step 5.   Find the *y*-intercept. Find the point symmetric to the *y*-intercept across the axis of symmetry.

Step 6.   Find the *x*-intercepts.

Step 7.   Graph the parabola.

ⓐ Rewrite $f(x) = 2x^2 + 4x + 5$ in $f(x) = a(x - h)^2 + k$ form and ⓑ graph the function using properties.

✓ **Solution**

Rewrite the function in $f(x) = a(x - h)^2 + k$ form by completing the square. $f(x) = 2x^2 + 4x + 5$

$$f(x) = 2(x^2 + 2x) + 5$$

$$f(x) = 2(x^2 + 2x + 1) + 5 - 2$$

$$f(x) = 2(x + 1)^2 + 3$$

Identify the constants *a*, *h*, *k*. $a = 2 \quad h = -1 \quad k = 3$

Since $a = 2$, the parabola opens upward.

The axis of symmetry is $x = h$.	The axis of symmetry is $x = -1$.
The vertex is (h, k).	The vertex is $(-1, 3)$.
Find the y-intercept by finding $f(0)$.	$f(0) = 2 \cdot 0^2 + 4 \cdot 0 + 5$
	$f(0) = 5$
	y-intercept $(0, 5)$
Find the point symmetric to $(0, 5)$ across the axis of symmetry.	$(-2, 5)$
Find the x-intercepts.	The discriminant negative, so there are no x-intercepts. Graph the parabola.

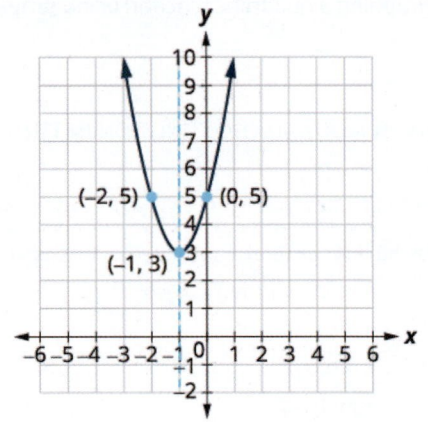

> **TRY IT : :** 9.123

ⓐ Rewrite $f(x) = 3x^2 - 6x + 5$ in $f(x) = a(x - h)^2 + k$ form and ⓑ graph the function using properties.

> **TRY IT : :** 9.124

ⓐ Rewrite $f(x) = -2x^2 + 8x - 7$ in $f(x) = a(x - h)^2 + k$ form and ⓑ graph the function using properties.

Find a Quadratic Function from its Graph

So far we have started with a function and then found its graph.

Now we are going to reverse the process. Starting with the graph, we will find the function.

EXAMPLE 9.63

Determine the quadratic function whose graph is shown.

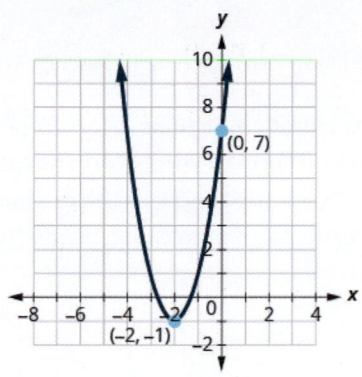

✓ Solution

Since it is quadratic, we start with the
$f(x) = a(x - h)^2 + k$ form.

The vertex, (h, k), is $(-2, -1)$ so $h = -2$ and $k = -1$. $f(x) = a(x - (-2))^2 - 1$
To find a, we use the y-intercept, $(0, 7)$.

So $f(0) = 7$. $7 = a(0 + 2)^2 - 1$
Solve for a. $7 = 4a - 1$
                                            $8 = 4a$
                                            $2 = a$

Write the function. $f(x) = a(x - h)^2 + k$
Substitute in $h = -2$, $k = -1$ and $a = 2$. $f(x) = 2(x + 2)^2 - 1$

> **TRY IT ::** 9.125   Write the quadratic function in $f(x) = a(x - h)^2 + k$ form whose graph is shown.

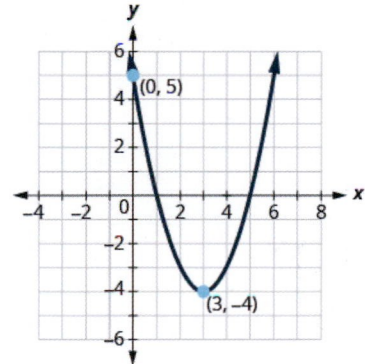

> **TRY IT ::** 9.126   Determine the quadratic function whose graph is shown.

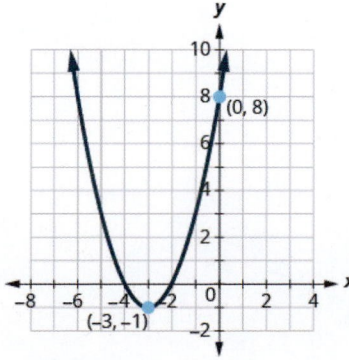

▶ **MEDIA** : :

Access these online resources for additional instruction and practice with graphing quadratic functions using transformations.

- **Function Shift Rules Applied to Quadratic Functions (https://openstax.org/l/37QuadFuncTran1)**
- **Changing a Quadratic from Standard Form to Vertex Form (https://openstax.org/l/37QuadFuncTran2)**
- **Using Transformations to Graph Quadratic Functions (https://openstax.org/l/37QuadFuncTran3)**
- **Finding Quadratic Equation in Vertex Form from Graph (https://openstax.org/l/37QuadFuncTran4)**

9.7 EXERCISES

Practice Makes Perfect

Graph Quadratic Functions of the form $f(x) = x^2 + k$

In the following exercises, ⓐ *graph the quadratic functions on the same rectangular coordinate system and* ⓑ *describe what effect adding a constant, k, to the function has on the basic parabola.*

293. $f(x) = x^2$, $g(x) = x^2 + 4$, and $h(x) = x^2 - 4$.

294. $f(x) = x^2$, $g(x) = x^2 + 7$, and $h(x) = x^2 - 7$.

In the following exercises, graph each function using a vertical shift.

295. $f(x) = x^2 + 3$

296. $f(x) = x^2 - 7$

297. $g(x) = x^2 + 2$

298. $g(x) = x^2 + 5$

299. $h(x) = x^2 - 4$

300. $h(x) = x^2 - 5$

Graph Quadratic Functions of the form $f(x) = (x - h)^2$

In the following exercises, ⓐ *graph the quadratic functions on the same rectangular coordinate system and* ⓑ *describe what effect adding a constant, h, to the function has on the basic parabola.*

301. $f(x) = x^2$, $g(x) = (x - 3)^2$, and $h(x) = (x + 3)^2$.

302. $f(x) = x^2$, $g(x) = (x + 4)^2$, and $h(x) = (x - 4)^2$.

In the following exercises, graph each function using a horizontal shift.

303. $f(x) = (x - 2)^2$

304. $f(x) = (x - 1)^2$

305. $f(x) = (x + 5)^2$

306. $f(x) = (x + 3)^2$

307. $f(x) = (x - 5)^2$

308. $f(x) = (x + 2)^2$

In the following exercises, graph each function using transformations.

309. $f(x) = (x + 2)^2 + 1$

310. $f(x) = (x + 4)^2 + 2$

311. $f(x) = (x - 1)^2 + 5$

312. $f(x) = (x - 3)^2 + 4$

313. $f(x) = (x + 3)^2 - 1$

314. $f(x) = (x + 5)^2 - 2$

315. $f(x) = (x - 4)^2 - 3$

316. $f(x) = (x - 6)^2 - 2$

Graph Quadratic Functions of the form $f(x) = ax^2$

In the following exercises, graph each function.

317. $f(x) = -2x^2$

318. $f(x) = 4x^2$

319. $f(x) = -4x^2$

320. $f(x) = -x^2$

321. $f(x) = \frac{1}{2}x^2$

322. $f(x) = \frac{1}{3}x^2$

323. $f(x) = \frac{1}{4}x^2$

324. $f(x) = -\frac{1}{2}x^2$

Graph Quadratic Functions Using Transformations

In the following exercises, rewrite each function in the $f(x) = a(x-h)^2 + k$ *form by completing the square.*

325. $f(x) = -3x^2 - 12x - 5$

326. $f(x) = 2x^2 - 12x + 7$

327. $f(x) = 3x^2 + 6x - 1$

328. $f(x) = -4x^2 - 16x - 9$

In the following exercises, ⓐ *rewrite each function in* $f(x) = a(x-h)^2 + k$ *form and* ⓑ *graph it by using transformations.*

329. $f(x) = x^2 + 6x + 5$

330. $f(x) = x^2 + 4x - 12$

331. $f(x) = x^2 + 4x - 12$

332. $f(x) = x^2 - 6x + 8$

333. $f(x) = x^2 - 6x + 15$

334. $f(x) = x^2 + 8x + 10$

335. $f(x) = -x^2 + 8x - 16$

336. $f(x) = -x^2 + 2x - 7$

337. $f(x) = -x^2 - 4x + 2$

338. $f(x) = -x^2 + 4x - 5$

339. $f(x) = 5x^2 - 10x + 8$

340. $f(x) = 3x^2 + 18x + 20$

341. $f(x) = 2x^2 - 4x + 1$

342. $f(x) = 3x^2 - 6x - 1$

343. $f(x) = -2x^2 + 8x - 10$

344. $f(x) = -3x^2 + 6x + 1$

In the following exercises, ⓐ *rewrite each function in* $f(x) = a(x-h)^2 + k$ *form and* ⓑ *graph it using properties.*

345. $f(x) = 2x^2 + 4x + 6$

346. $f(x) = 3x^2 - 12x + 7$

347. $f(x) = -x^2 + 2x - 4$

348. $f(x) = -2x^2 - 4x - 5$

Matching

In the following exercises, match the graphs to one of the following functions: ⓐ $f(x) = x^2 + 4$ ⓑ $f(x) = x^2 - 4$ ⓒ $f(x) = (x+4)^2$ ⓓ $f(x) = (x-4)^2$ ⓔ $f(x) = (x+4)^2 - 4$ ⓕ $f(x) = (x+4)^2 + 4$ ⓖ $f(x) = (x-4)^2 - 4$ ⓗ $f(x) = (x-4)^2 + 4$

349.

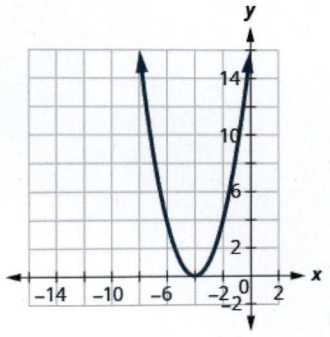

350.

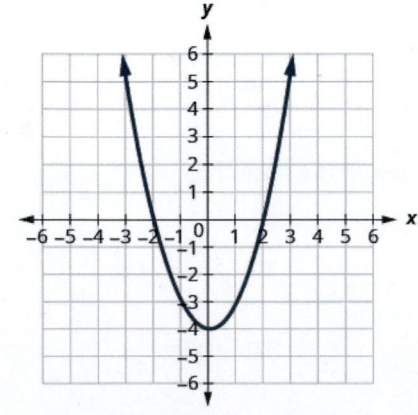

351.

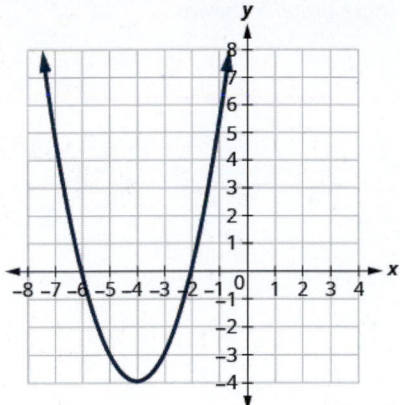

352.

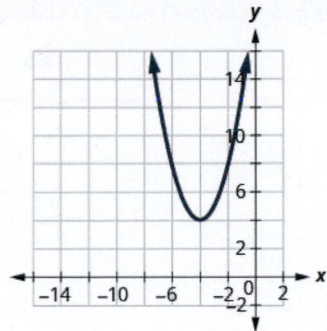

353.

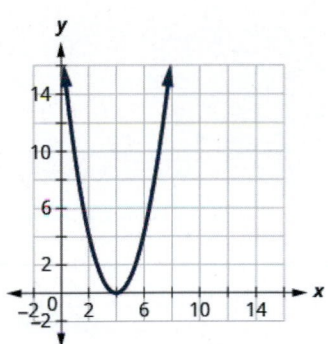

354.

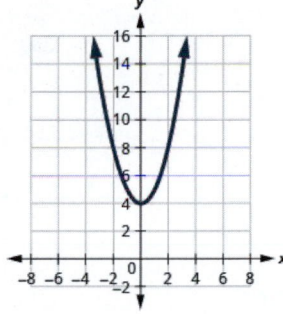

355.

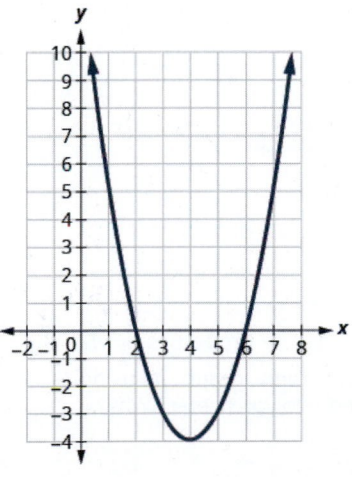

356.

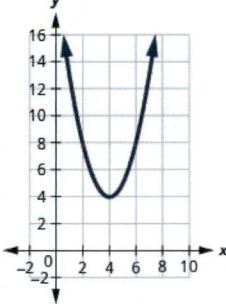

Find a Quadratic Function from its Graph

In the following exercises, write the quadratic function in $f(x) = a(x - h)^2 + k$ *form whose graph is shown.*

357.

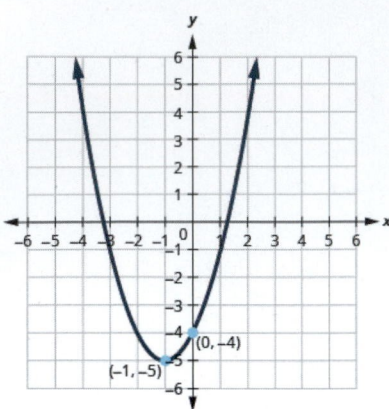

358.

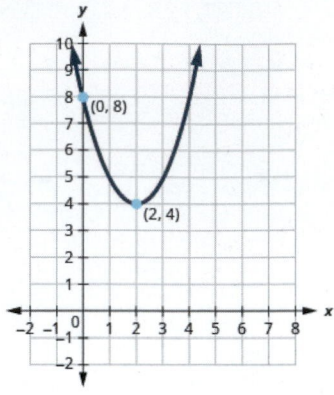

359.

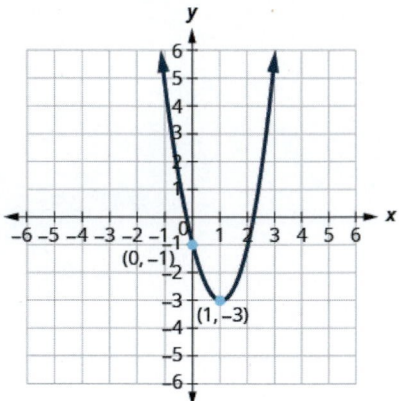

360.

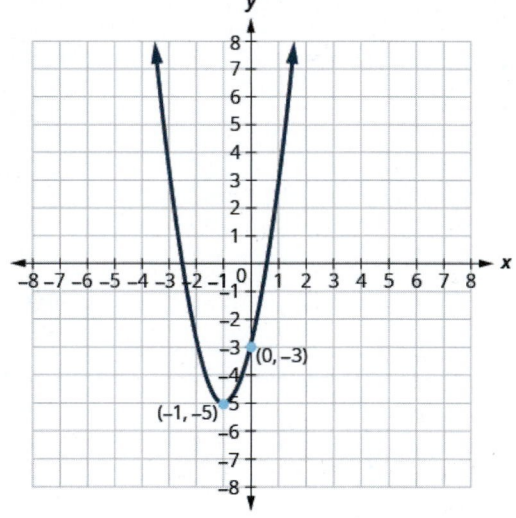

Writing Exercise

361. Graph the quadratic function $f(x) = x^2 + 4x + 5$ first using the properties as we did in the last section and then graph it using transformations. Which method do you prefer? Why?

362. Graph the quadratic function $f(x) = 2x^2 - 4x - 3$ first using the properties as we did in the last section and then graph it using transformations. Which method do you prefer? Why?

Self Check

ⓐ After completing the exercises, use this checklist to evaluate your mastery of the objectives of this section.

I can...	Confidently	With some help	No-I don't get it!
graph Quadratic Functions of the form $f(x) = x^2 + k$.			
graph Quadratic Functions of the form $f(x) = (x - h)^2$.			
graph Quadratic Functions of the form $f(x) = ax^2$.			
graph Quadratic Functions Using Transformations.			
find a Quadratic Function from its Graph.			

ⓑ After looking at the checklist, do you think you are well-prepared for the next section? Why or why not?

9.8 **Solve Quadratic Inequalities**

Learning Objectives

By the end of this section, you will be able to:
  › Solve quadratic inequalities graphically
  › Solve quadratic inequalities algebraically

Be Prepared!

Before you get started, take this readiness quiz.

1. Solve: $2x - 3 = 0$.
   If you missed this problem, review **Example 2.2**.

2. Solve: $2y^2 + y = 15$.
   If you missed this problem, review **Example 6.45**.

3. Solve $\dfrac{1}{x^2 + 2x - 8} > 0$
   If you missed this problem, review **Example 7.56**.

We have learned how to solve linear inequalities and rational inequalities previously. Some of the techniques we used to solve them were the same and some were different.

We will now learn to solve inequalities that have a quadratic expression. We will use some of the techniques from solving linear and rational inequalities as well as quadratic equations.

We will solve quadratic inequalities two ways—both graphically and algebraically.

Solve Quadratic Inequalities Graphically

A quadratic equation is in standard form when written as $ax^2 + bx + c = 0$. If we replace the equal sign with an inequality sign, we have a **quadratic inequality** in standard form.

Quadratic Inequality

A **quadratic inequality** is an inequality that contains a quadratic expression.

The standard form of a quadratic inequality is written:

$$ax^2 + bx + c < 0 \qquad ax^2 + bx + c \le 0$$

$$ax^2 + bx + c > 0 \qquad ax^2 + bx + c \ge 0$$

The graph of a quadratic function $f(x) = ax^2 + bx + c = 0$ is a parabola. When we ask when is $ax^2 + bx + c < 0$, we are asking when is f(x) < 0. We want to know when the parabola is below the x-axis.

When we ask when is $ax^2 + bx + c > 0$, we are asking when is $f(x) > 0$. We want to know when the parabola is above the y-axis.

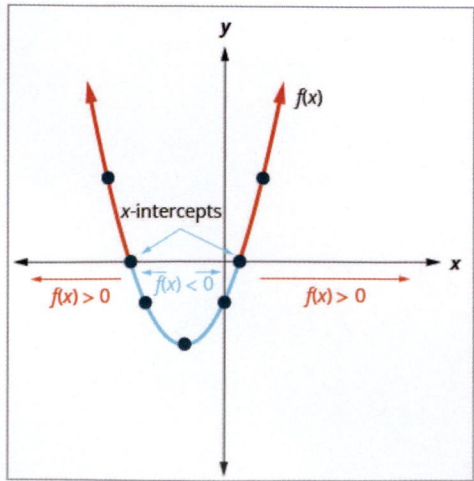
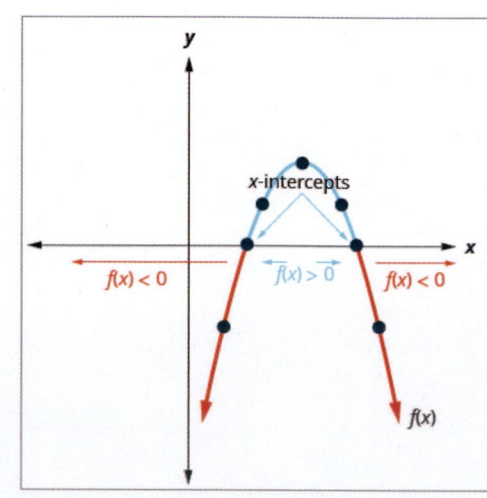

EXAMPLE 9.64 HOW TO SOLVE A QUADRATIC INEQUALITY GRAPHICALLY

Solve $x^2 - 6x + 8 < 0$ graphically. Write the solution in interval notation.

✓ **Solution**

| **Step 1.** Write the quadratic inequality in standard form. | The inequality is in standard form | $x^2 - 6x + 8 < 0$ |

Step 2. Graph the function $f(x) = ax^2 + bx + c$ using properties or transformations.	We will graph using the properties.	$f(x) = x^2 - 6x + 8$ $a = 1, b = -6, c = 8$
	Look at a in the equation. $f(x) = x^2 - 6x + 8$ Since a is positive, the parabola opens upward. 	**The parabola opens upward.**
	$f(x) = x^2 - 6x + 8$ The axis of symmetry is the line $x = -\dfrac{b}{2a}$.	Axis of Symmetry $x = -\dfrac{b}{2a}$ $x = -\dfrac{(-6)}{2 \cdot 1}$ $x = 3$ **The axis of symmetry is the line $x = 3$.**
	The vertex is on the axis of symmetry. Substitute $x = 3$ into the function.	Vertex $f(x) = x^2 - 6x + 8$ $f(3) = (3)^2 - 6(3) + 8$ $f(3) = -1$ **The vertex is (3, –1).**
	We find $f(0)$	y-intercept $f(x) = x^2 - 6x + 8$ $f(0) = (0)^2 - 6(0) + 8$ $f(0) = 8$ **The y-intercept is (0, 8).**
	We use the axis of symmetry to find a point symmetric to the y-intercept. The y-intercept is 3 units left of the axis of symmetry, $x = 3$. A point 3 units to the right of the axis of symmetry has $x = 6$.	Point symmetric to y-intercept **The point is (6, 8).**
	We solve $f(x) = 0$. We can solve this quadratic equation by factoring.	x-intercepts $f(x) = x^2 - 6x + 8$ $0 = x^2 - 6x + 8$ $0 = (x - 2)(x - 4)$ $x = 2$ or $x = 4$ **The x-intercepts are (2, 0) and (4, 0).**
	We graph the vertex, intercepts, and the point symmetric to the y-intercept. We connect these 5 points to sketch the parabola.	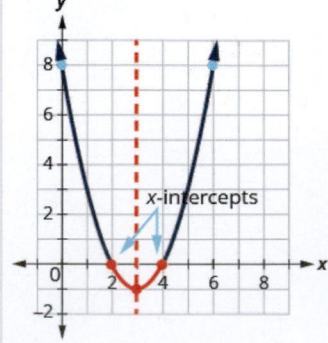

Step 3. Determine the solution from the graph.	$x^2 - 6x + 8 < 0$ The inequality asks for the values of x which make the function less than 0. Which values of x make the parabola below the x-axis. We do not include the values 2, 4 as the inequality is less than only.	The solution, in interval notation, is (2, 4).

> **TRY IT : : 9.127**   ⓐ Solve $x^2 + 2x - 8 < 0$ graphically and ⓑ write the solution in interval notation.

> **TRY IT : : 9.128**   ⓐ Solve $x^2 - 8x + 12 \geq 0$ graphically and ⓑ write the solution in interval notation.

We list the steps to take to solve a quadratic inequality graphically.

HOW TO : : SOLVE A QUADRATIC INEQUALITY GRAPHICALLY.

Step 1.   Write the quadratic inequality in standard form.

Step 2.   Graph the function $f(x) = ax^2 + bx + c$.

Step 3.   Determine the solution from the graph.

In the last example, the parabola opened upward and in the next example, it opens downward. In both cases, we are looking for the part of the parabola that is below the x-axis but note how the position of the parabola affects the solution.

EXAMPLE 9.65

Solve $-x^2 - 8x - 12 \leq 0$ graphically. Write the solution in interval notation.

⊘ **Solution**

The quadratic inequality in standard form.	$-x^2 - 8x - 12 \leq 0$

Graph the function $f(x) = -x^2 - 8x - 12$.	The parabola opens downward.

Find the line of symmetry.	$x = -\dfrac{b}{2a}$ $x = -\dfrac{-8}{2(-1)}$ $x = -4$

Find the vertex.	$f(x) = -x^2 - 8x - 12$ $f(-4) = -(-4)^2 - 8(-4) - 12$ $f(-4) = -16 + 32 - 12$ $f(-4) = 4$ Vertex $(-4, 4)$

Find the x-intercepts. Let $f(x) = 0$.	$f(x) = -x^2 - 8x - 12$
	$0 = -x^2 - 8x - 12$

Factor.	$0 = -1(x + 6)(x + 2)$
Use the Zero Product Property.	$x = -6 \qquad x = -2$

Graph the parabola.	x-intercepts $(-6, 0), (-2, 0)$

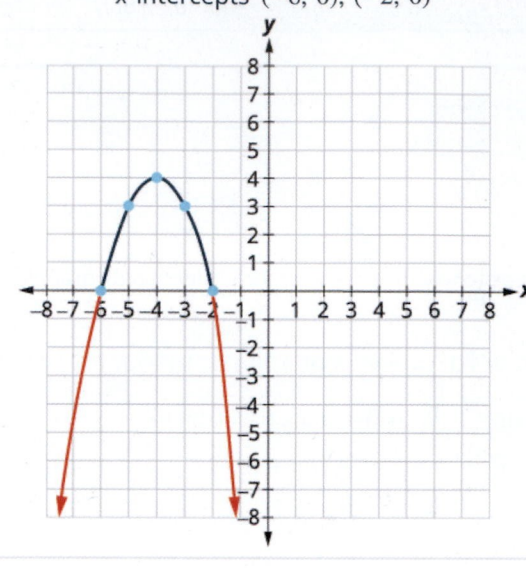

Determine the solution from the graph. We include the x-intercepts as the inequality is "less than or equal to."	$(-\infty, -6] \cup [-2, \infty)$

> **TRY IT :: 9.129**   ⓐ Solve $-x^2 - 6x - 5 > 0$ graphically and ⓑ write the solution in interval notation.

> **TRY IT :: 9.130**   ⓐ Solve $-x^2 + 10x - 16 \leq 0$ graphically and ⓑ write the solution in interval notation.

Solve Quadratic Inequalities Algebraically

The algebraic method we will use is very similar to the method we used to solve rational inequalities. We will find the critical points for the inequality, which will be the solutions to the related quadratic equation. Remember a polynomial expression can change signs only where the expression is zero.

We will use the critical points to divide the number line into intervals and then determine whether the quadratic expression willl be postive or negative in the interval. We then determine the solution for the inequality.

EXAMPLE 9.66   HOW TO SOLVE QUADRATIC INEQUALITIES ALGEBRAICALLY

Solve $x^2 - x - 12 \geq 0$ algebraically. Write the solution in interval notation.

✓ **Solution**

Step 1. Write the quadratic inequality in standard form.	The inequality is in standard form	$x^2 - x - 12 \geq 0$

Step 2. Determine the critical points—the solutions to the related quadratic equation.	Change the inequality sign to an equal sign and then solve the equation.	$x^2 - x - 12 = 0$ $(x + 3)(x - 4) = 0$ $x + 3 = 0 \quad x - 4 = 0$ $x = -3 \quad x = 4$
Step 3. Use the critical points to divide the number line into intervals.	Use –3 and 4 to divide the number line into intervals	(number line from –4 to 4 with marks at –3 and 4)
Step 4. Above the number line show the sign of each quadratic expression using test points from each interval substituted into the original inequality.	Test: $x = -5$ $x = 0$ $x = 5$	$x^2 - x - 12$ $x^2 - x - 12$ $x^2 - x - 12$ $(-5)^2 - (-5) - 12$ $0^2 - 0 - 12$ $5^2 - 5 - 12$ 18 –12 8 + – + (number line from –4 to 4 with marks at –3 and 4)
Step 5. Determine the intervals where the inequality is correct. Write the solution in interval notation.	$x^2 - x - 12 \geq 0$ The inequality is positive in the first and last intervals and equals 0 at the points –4, 3.	The solution, in interval notation, is $(-\infty, -3] \cup [4, \infty)$.

> **TRY IT :: 9.131**

Solve $x^2 + 2x - 8 \geq 0$ algebraically. Write the solution in interval notation.

> **TRY IT :: 9.132**

Solve $x^2 - 2x - 15 \leq 0$ algebraically. Write the solution in interval notation.

In this example, since the expression $x^2 - x - 12$ factors nicely, we can also find the sign in each interval much like we did when we solved rational inequalities. We find the sign of each of the factors, and then the sign of the product. Our number line would like this:

$x^2 - x - 12$ $(x + 3)$ –   +   +

$(x + 3)(x - 4)$ $(x - 4)$ –   –   +

(number line with critical points –3 and 4)

$(x + 3)(x - 4)$ +   –   +

The result is the same as we found using the other method.

We summarize the steps here.

HOW TO :: SOLVE A QUADRATIC INEQUALITY ALGEBRAICALLY.

Step 1.   Write the quadratic inequality in standard form.

Step 2.   Determine the critical points—the solutions to the related quadratic equation.

Step 3.   Use the critical points to divide the number line into intervals.

Step 4.   Above the number line show the sign of each quadratic expression using test points from each interval substituted into the original inequality.

Step 5.   Determine the intervals where the inequality is correct. Write the solution in interval notation.

EXAMPLE 9.67

Solve $x^2 + 6x - 7 \geq 0$ algebraically. Write the solution in interval notation.

✓ **Solution**

Write the quadratic inequality in standard form.	$-x^2 + 6x - 7 \geq 0$
Multiply both sides of the inequality by -1. Remember to reverse the inequality sign.	$x^2 - 6x + 7 \leq 0$
Determine the critical points by solving the related quadratic equation.	$x^2 - 6x + 7 = 0$
Write the Quadratic Formula.	$x = \dfrac{-b \pm \sqrt{b^2 - 4ac}}{2a}$
Then substitute in the values of a, b, c.	$x = \dfrac{-(-6) \pm \sqrt{(-6)^2 - 4 \cdot 1 \cdot (7)}}{2 \cdot 1}$
Simplify.	$x = \dfrac{6 \pm \sqrt{8}}{2}$
Simplify the radical.	$x = \dfrac{6 \pm 2\sqrt{2}}{2}$
Remove the common factor, 2.	$x = \dfrac{2(3 \pm \sqrt{2})}{2}$ $x = 3 \pm \sqrt{2}$ $x = 3 + \sqrt{2} \qquad x = 3 - \sqrt{2}$ $x \approx 1.6 \qquad x \approx 4.4$
Use the critical points to divide the number line into intervals. Test numbers from each interval in the original inequality.	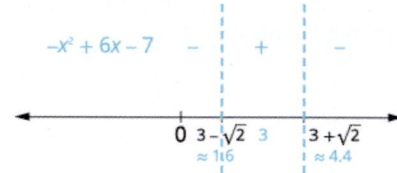
Determine the intervals where the inequality is correct. Write the solution in interval notation.	$-x^2 + 6x - 7 \geq 0$ in the middle interval $\left[3 - \sqrt{2}, \ 3 + \sqrt{2}\right]$

> **TRY IT :: 9.133**   Solve $-x^2 + 2x + 1 \geq 0$ algebraically. Write the solution in interval notation.

> **TRY IT :: 9.134**   Solve $-x^2 + 8x - 14 < 0$ algebraically. Write the solution in interval notation.

The solutions of the quadratic inequalities in each of the previous examples, were either an interval or the union of two intervals. This resulted from the fact that, in each case we found two solutions to the corresponding quadratic equation $ax^2 + bx + c = 0$. These two solutions then gave us either the two x-intercepts for the graph or the two critical points to divide the number line into intervals.

This correlates to our previous discussion of the number and type of solutions to a quadratic equation using the discriminant.

For a quadratic equation of the form $ax^2 + bx + c = 0$, $a \neq 0$.

Discriminant	Number/Type of solution	Typical Graph
$b^2 - 4ac > 0$	2 real solutions 2 x-intercepts on graph	
$b^2 - 4ac = 0$	1 real solution 1 x-intercept on graph	
$b^2 - 4ac < 0$	2 complex solutions No x-intercept	

The last row of the table shows us when the parabolas never intersect the x-axis. Using the Quadratic Formula to solve the quadratic equation, the radicand is a negative. We get two complex solutions.

In the next example, the quadratic inequality solutions will result from the solution of the quadratic equation being complex.

EXAMPLE 9.68

Solve, writing any solution in interval notation:

ⓐ $x^2 - 3x + 4 > 0$ ⓑ $x^2 - 3x + 4 \leq 0$

✓ **Solution**

ⓐ

Write the quadratic inequality in standard form.	$-x^2 - 3x + 4 > 0$
Determine the critical points by solving the related quadratic equation.	$x^2 - 3x + 4 = 0$
Write the Quadratic Formula.	$x = \dfrac{-b \pm \sqrt{b^2 - 4ac}}{2a}$
Then substitute in the values of a, b, c.	$x = \dfrac{-(-3) \pm \sqrt{(-3)^2 - 4 \cdot 1 \cdot (4)}}{2 \cdot 1}$
Simplify.	$x = \dfrac{3 \pm \sqrt{-7}}{2}$
Simplify the radicand.	$x = \dfrac{3 \pm \sqrt{7}i}{2}$

The complex solutions tell us the
parabola does not intercept the x-axis.
Also, the parabola opens upward. This
tells us that the parabola is completely above the x-axis.

Complex solutions

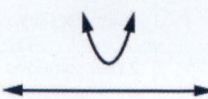

We are to find the solution to $x^2 - 3x + 4 > 0$. Since for all values of x the graph is above the x-axis, all values of x make the inequality true. In interval notation we write $(-\infty, \infty)$.

ⓑ

Write the quadratic inequality in standard form. $x^2 - 3x + 4 \leq 0$

Determine the critical points by solving
the related quadratic equation $x^2 - 3x + 4 = 0$

Since the corresponding quadratic equation is the same as in part (a), the parabola will be the same. The parabola opens upward and is completely above the x-axis—no part of it is below the x-axis.

We are to find the solution to $x^2 - 3x + 4 \leq 0$. Since for all values of x the graph is never below the x-axis, no values of x make the inequality true. There is no solution to the inequality.

> **TRY IT : : 9.135**   Solve and write any solution in interval notation:

ⓐ $-x^2 + 2x - 4 \leq 0$ ⓑ $-x^2 + 2x - 4 \geq 0$

> **TRY IT : : 9.136**   Solve and write any solution in interval notation:

ⓐ $x^2 + 3x + 3 < 0$ ⓑ $x^2 + 3x + 3 > 0$

9.8 EXERCISES

Practice Makes Perfect

Solve Quadratic Inequalities Graphically

In the following exercises, ⓐ solve graphically and ⓑ write the solution in interval notation.

363. $x^2 + 6x + 5 > 0$

364. $x^2 + 4x - 12 < 0$

365. $x^2 + 4x + 3 \leq 0$

366. $x^2 - 6x + 8 \geq 0$

367. $-x^2 - 3x + 18 \leq 0$

368. $-x^2 + 2x + 24 < 0$

369. $-x^2 + x + 12 \geq 0$

370. $-x^2 + 2x + 15 > 0$

In the following exercises, solve each inequality algebraically and write any solution in interval notation.

371. $x^2 + 3x - 4 \geq 0$

372. $x^2 + x - 6 \leq 0$

373. $x^2 - 7x + 10 < 0$

374. $x^2 - 4x + 3 > 0$

375. $x^2 + 8x > -15$

376. $x^2 + 8x < -12$

377. $x^2 - 4x + 2 \leq 0$

378. $-x^2 + 8x - 11 < 0$

379. $x^2 - 10x > -19$

380. $x^2 + 6x < -3$

381. $-6x^2 + 19x - 10 \geq 0$

382. $-3x^2 - 4x + 4 \leq 0$

383. $-2x^2 + 7x + 4 \geq 0$

384. $2x^2 + 5x - 12 > 0$

385. $x^2 + 3x + 5 > 0$

386. $x^2 - 3x + 6 \leq 0$

387. $-x^2 + x - 7 > 0$

388. $-x^2 - 4x - 5 < 0$

389. $-2x^2 + 8x - 10 < 0$

390. $-x^2 + 2x - 7 \geq 0$

Writing Exercises

391. Explain critical points and how they are used to solve quadratic inequalities algebraically.

392. Solve $x^2 + 2x \geq 8$ both graphically and algebraically. Which method do you prefer, and why?

393. Describe the steps needed to solve a quadratic inequality graphically.

394. Describe the steps needed to solve a quadratic inequality algebraically.

Self Check

ⓐ After completing the exercises, use this checklist to evaluate your mastery of the objectives of this section.

I can...	Confidently	With some help	No–I don't get it!
solve quadratic inequalities graphically.			
solve quadratic inequalities algebraically.			

ⓑ On a scale of 1-10, how would you rate your mastery of this section in light of your responses on the checklist? How can you improve this?

CHAPTER 9 REVIEW

KEY TERMS

discriminant
In the Quadratic Formula, $x = \dfrac{-b \pm \sqrt{b^2 - 4ac}}{2a}$, the quantity $b^2 - 4ac$ is called the discriminant.

quadratic function A quadratic function, where a, b, and c are real numbers and $a \neq 0$, is a function of the form
$f(x) = ax^2 + bx + c.$

quadratic inequality A quadratic inequality is an inequality that contains a quadratic expression.

KEY CONCEPTS

9.1 Solve Quadratic Equations Using the Square Root Property

- Square Root Property
  - If $x^2 = k$, then $x = \sqrt{k}$ or $x = -\sqrt{k}$ or $x = \pm\sqrt{k}$

  How to solve a quadratic equation using the square root property.
  Step 1. Isolate the quadratic term and make its coefficient one.
  Step 2. Use Square Root Property.
  Step 3. Simplify the radical.
  Step 4. Check the solutions.

9.2 Solve Quadratic Equations by Completing the Square

- Binomial Squares Pattern
  If a and b are real numbers,

$(a+b)^2 = a^2 + 2ab + b^2$	$\underbrace{(a+b)^2}_{\text{(binomial)}^2}$	$=$	$\underbrace{a^2}_{\text{(first term)}^2}$	$+$	$\underbrace{2ab}_{2\,\cdot\,\text{(product of terms)}}$	$+$	$\underbrace{b^2}_{\text{(second term)}^2}$

$(a-b)^2 = a^2 - 2ab + b^2$	$\underbrace{(a-b)^2}_{\text{(binomial)}^2}$	$=$	$\underbrace{a^2}_{\text{(first term)}^2}$	$-$	$\underbrace{2ab}_{2\,\cdot\,\text{(product of terms)}}$	$+$	$\underbrace{b^2}_{\text{(second term)}^2}$

- How to Complete a Square
  Step 1. Identify b, the coefficient of x.
  Step 2. Find $\left(\frac{1}{2}b\right)^2$, the number to complete the square.
  Step 3. Add the $\left(\frac{1}{2}b\right)^2$ to $x^2 + bx$
  Step 4. Rewrite the trinomial as a binomial square
- How to solve a quadratic equation of the form $ax^2 + bx + c = 0$ by completing the square.
  Step 1. Divide by a to make the coefficient of x^2 term 1.
  Step 2. Isolate the variable terms on one side and the constant terms on the other.
  Step 3. Find $\left(\frac{1}{2} \cdot b\right)^2$, the number needed to complete the square. Add it to both sides of the equation.
  Step 4. Factor the perfect square trinomial, writing it as a binomial squared on the left and simplify by adding the terms on the right.
  Step 5. Use the Square Root Property.
  Step 6. Simplify the radical and then solve the two resulting equations.

Step 7. Check the solutions.

9.3 Solve Quadratic Equations Using the Quadratic Formula

- Quadratic Formula
  - The solutions to a quadratic equation of the form $ax^2 + bx + c = 0$, $a \neq 0$ are given by the formula:

$$x = \frac{-b \pm \sqrt{b^2 - 4ac}}{2a}$$

- How to solve a quadratic equation using the Quadratic Formula.

  Step 1. Write the quadratic equation in standard form, $ax^2 + bx + c = 0$. Identify the values of a, b, c.

  Step 2. Write the Quadratic Formula. Then substitute in the values of a, b, c.

  Step 3. Simplify.

  Step 4. Check the solutions.

- Using the Discriminant, $b^2 - 4ac$, to Determine the Number and Type of Solutions of a Quadratic Equation
  - For a quadratic equation of the form $ax^2 + bx + c = 0$, $a \neq 0$,

    - If $b^2 - 4ac > 0$, the equation has 2 real solutions.
    - if $b^2 - 4ac = 0$, the equation has 1 real solution.
    - if $b^2 - 4ac < 0$, the equation has 2 complex solutions.

- Methods to Solve Quadratic Equations:
  - Factoring
  - Square Root Property
  - Completing the Square
  - Quadratic Formula

- How to identify the most appropriate method to solve a quadratic equation.

  Step 1. Try Factoring first. If the quadratic factors easily, this method is very quick.

  Step 2. Try the **Square Root Property** next. If the equation fits the form $ax^2 = k$ or $a(x - h)^2 = k$, it can easily be solved by using the Square Root Property.

  Step 3. Use the **Quadratic Formula.** Any other quadratic equation is best solved by using the Quadratic Formula.

9.4 Solve Quadratic Equations in Quadratic Form

- How to solve equations in quadratic form.

  Step 1. Identify a substitution that will put the equation in quadratic form.

  Step 2. Rewrite the equation with the substitution to put it in quadratic form.

  Step 3. Solve the quadratic equation for u.

  Step 4. Substitute the original variable back into the results, using the substitution.

  Step 5. Solve for the original variable.

  Step 6. Check the solutions.

9.5 Solve Applications of Quadratic Equations

- Methods to Solve Quadratic Equations
  - Factoring
  - Square Root Property
  - Completing the Square
  - Quadratic Formula

- How to use a Problem-Solving Strategy.

  Step 1. **Read** the problem. Make sure all the words and ideas are understood.

  Step 2. **Identify** what we are looking for.

  Step 3. **Name** what we are looking for. Choose a variable to represent that quantity.

Step 4. **Translate** into an equation. It may be helpful to restate the problem in one sentence with all the important information. Then, translate the English sentence into an algebra equation.

Step 5. **Solve** the equation using good algebra techniques.

Step 6. **Check** the answer in the problem and make sure it makes sense.

Step 7. **Answer** the question with a complete sentence.

- Area of a Triangle

  ◦ For a triangle with base, b, and height, h, the area, A, is given by the formula $A = \frac{1}{2}bh$.

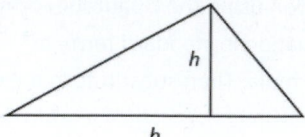

- Area of a Rectangle

  ◦ For a rectangle with length, L, and width, W, the area, A, is given by the formula $A = LW$.

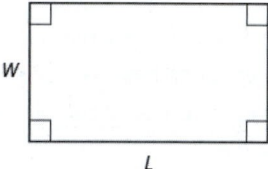

- Pythagorean Theorem

  ◦ In any right triangle, where a and b are the lengths of the legs, and c is the length of the hypotenuse, $a^2 + b^2 = c^2$.

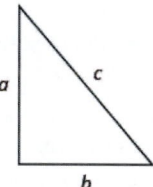

- Projectile motion

  ◦ The height in feet, h, of an object shot upwards into the air with initial velocity, v_0, after t seconds is given by the formula $h = -16t^2 + v_0 t$.

9.6 Graph Quadratic Functions Using Properties

- Parabola Orientation

  ◦ For the graph of the quadratic function $f(x) = ax^2 + bx + c,$ if

    ▪ $a > 0$, the parabola opens upward.
    ▪ $a < 0$, the parabola opens downward.

- Axis of Symmetry and Vertex of a Parabola The graph of the function $f(x) = ax^2 + bx + c$ is a parabola where:

  ◦ the axis of symmetry is the vertical line $x = -\dfrac{b}{2a}$.

  ◦ the vertex is a point on the axis of symmetry, so its x-coordinate is $-\dfrac{b}{2a}$.

  ◦ the y-coordinate of the vertex is found by substituting $x = -\dfrac{b}{2a}$ into the quadratic equation.

- Find the Intercepts of a Parabola

  ◦ To find the intercepts of a parabola whose function is $f(x) = ax^2 + bx + c$:

y-intercept	**x-intercepts**
Let $x = 0$ and solve for $f(x)$.	Let $f(x) = 0$ and solve for x.

- How to graph a quadratic function using properties.

  Step 1. Determine whether the parabola opens upward or downward.

  Step 2. Find the equation of the axis of symmetry.

  Step 3. Find the vertex.

  Step 4. Find the y-intercept. Find the point symmetric to the y-intercept across the axis of symmetry.

  Step 5. Find the x-intercepts. Find additional points if needed.

  Step 6. Graph the parabola.

- Minimum or Maximum Values of a Quadratic Equation

  - The y-coordinate of the vertex of the graph of a quadratic equation is the

  - *minimum* value of the quadratic equation if the parabola opens *upward*.

  - *maximum* value of the quadratic equation if the parabola opens *downward*.

9.7 Graph Quadratic Functions Using Transformations

- Graph a Quadratic Function of the form $f(x) = x^2 + k$ Using a Vertical Shift

  - The graph of $f(x) = x^2 + k$ shifts the graph of $f(x) = x^2$ vertically k units.

    - If $k > 0$, shift the parabola vertically up k units.
    - If $k < 0$, shift the parabola vertically down $|k|$ units.

- Graph a Quadratic Function of the form $f(x) = (x - h)^2$ Using a Horizontal Shift

  - The graph of $f(x) = (x - h)^2$ shifts the graph of $f(x) = x^2$ horizontally h units.

    - If $h > 0$, shift the parabola horizontally left h units.
    - If $h < 0$, shift the parabola horizontally right $|h|$ units.

- Graph of a Quadratic Function of the form $f(x) = ax^2$

  - The coefficient a in the function $f(x) = ax^2$ affects the graph of $f(x) = x^2$ by stretching or compressing it.

    If $0 < |a| < 1$, then the graph of $f(x) = ax^2$ will be "wider" than the graph of $f(x) = x^2$.

    If $|a| > 1$, then the graph of $f(x) = ax^2$ will be "skinnier" than the graph of $f(x) = x^2$.

- How to graph a quadratic function using transformations

  Step 1. Rewrite the function in $f(x) = a(x - h)^2 + k$ form by completing the square.

  Step 2. Graph the function using transformations.

- Graph a quadratic function in the vertex form $f(x) = a(x - h)^2 + k$ using properties

  Step 1. Rewrite the function in $f(x) = a(x - h)^2 + k$ form.

  Step 2. Determine whether the parabola opens upward, $a > 0$, or downward, a < 0.

  Step 3. Find the axis of symmetry, $x = h$.

  Step 4. Find the vertex, (h, k).

  Step 5. Find the y-intercept. Find the point symmetric to the y-intercept across the axis of symmetry.

  Step 6. Find the x-intercepts, if possible.

  Step 7. Graph the parabola.

9.8 Solve Quadratic Inequalities

- Solve a Quadratic Inequality Graphically

  Step 1. Write the quadratic inequality in standard form.

  Step 2. Graph the function $f(x) = ax^2 + bx + c$ using properties or transformations.

  Step 3. Determine the solution from the graph.

- How to Solve a Quadratic Inequality Algebraically

  Step 1. Write the quadratic inequality in standard form.

  Step 2. Determine the critical points -- the solutions to the related quadratic equation.

  Step 3. Use the critical points to divide the number line into intervals.

  Step 4. Above the number line show the sign of each quadratic expression using test points from each interval substituted into the original inequality.

  Step 5. Determine the intervals where the inequality is correct. Write the solution in interval notation.

REVIEW EXERCISES

9.1 Solve Quadratic Equations Using the Square Root Property

Solve Quadratic Equations of the form $ax^2 = k$ Using the Square Root Property

In the following exercises, solve using the Square Root Property.

395. $y^2 = 144$

396. $n^2 - 80 = 0$

397. $4a^2 = 100$

398. $2b^2 = 72$

399. $r^2 + 32 = 0$

400. $t^2 + 18 = 0$

401. $\frac{2}{3}w^2 - 20 = 30$

402. 11. $5c^2 + 3 = 19$

Solve Quadratic Equations of the Form $a(x - h)^2 = k$ Using the Square Root Property

In the following exercises, solve using the Square Root Property.

403. $(p - 5)^2 + 3 = 19$

404. $(u + 1)^2 = 45$

405. $\left(x - \frac{1}{4}\right)^2 = \frac{3}{16}$

406. $\left(y - \frac{2}{3}\right)^2 = \frac{2}{9}$

407. $(n - 4)^2 - 50 = 150$

408. $(4c - 1)^2 = -18$

409. $n^2 + 10n + 25 = 12$

410. $64a^2 + 48a + 9 = 81$

9.2 Solve Quadratic Equations by Completing the Square

Solve Quadratic Equations Using Completing the Square

In the following exercises, complete the square to make a perfect square trinomial. Then write the result as a binomial squared.

411. $x^2 + 22x$

412. $m^2 - 8m$

413. $a^2 - 3a$

414. $b^2 + 13b$

In the following exercises, solve by completing the square.

415. $d^2 + 14d = -13$

416. $y^2 - 6y = 36$

417. $m^2 + 6m = -109$

418. $t^2 - 12t = -40$

419. $v^2 - 14v = -31$

420. $w^2 - 20w = 100$

421. $m^2 + 10m - 4 = -13$ **422.** $n^2 - 6n + 11 = 34$ **423.** $a^2 = 3a + 8$

424. $b^2 = 11b - 5$ **425.** $(u + 8)(u + 4) = 14$ **426.** $(z - 10)(z + 2) = 28$

Solve Quadratic Equations of the form $ax^2 + bx + c = 0$ by Completing the Square

In the following exercises, solve by completing the square.

427. $3p^2 - 18p + 15 = 15$ **428.** $5q^2 + 70q + 20 = 0$ **429.** $4y^2 - 6y = 4$

430. $2x^2 + 2x = 4$ **431.** $3c^2 + 2c = 9$ **432.** $4d^2 - 2d = 8$

433. $2x^2 + 6x = -5$ **434.** $2x^2 + 4x = -5$

9.3 Solve Quadratic Equations Using the Quadratic Formula

In the following exercises, solve by using the Quadratic Formula.

435. $4x^2 - 5x + 1 = 0$ **436.** $7y^2 + 4y - 3 = 0$ **437.** $r^2 - r - 42 = 0$

438. $t^2 + 13t + 22 = 0$ **439.** $4v^2 + v - 5 = 0$ **440.** $2w^2 + 9w + 2 = 0$

441. $3m^2 + 8m + 2 = 0$ **442.** $5n^2 + 2n - 1 = 0$ **443.** $6a^2 - 5a + 2 = 0$

444. $4b^2 - b + 8 = 0$ **445.** $u(u - 10) + 3 = 0$ **446.** $5z(z - 2) = 3$

447. $\frac{1}{8}p^2 - \frac{1}{5}p = -\frac{1}{20}$ **448.** $\frac{2}{5}q^2 + \frac{3}{10}q = \frac{1}{10}$ **449.** $4c^2 + 4c + 1 = 0$

450. $9d^2 - 12d = -4$

Use the Discriminant to Predict the Number of Solutions of a Quadratic Equation

In the following exercises, determine the number of solutions for each quadratic equation.

451.
ⓐ $9x^2 - 6x + 1 = 0$
ⓑ $3y^2 - 8y + 1 = 0$
ⓒ $7m^2 + 12m + 4 = 0$
ⓓ $5n^2 - n + 1 = 0$

452.
ⓐ $5x^2 - 7x - 8 = 0$
ⓑ $7x^2 - 10x + 5 = 0$
ⓒ $25x^2 - 90x + 81 = 0$
ⓓ $15x^2 - 8x + 4 = 0$

Identify the Most Appropriate Method to Use to Solve a Quadratic Equation

In the following exercises, identify the most appropriate method (Factoring, Square Root, or Quadratic Formula) to use to solve each quadratic equation. Do not solve.

453.
ⓐ $16r^2 - 8r + 1 = 0$
ⓑ $5t^2 - 8t + 3 = 9$
ⓒ $3(c + 2)^2 = 15$

454.
ⓐ $4d^2 + 10d - 5 = 21$
ⓑ $25x^2 - 60x + 36 = 0$
ⓒ $6(5v - 7)^2 = 150$

9.4 Solve Equations in Quadratic Form

Solve Equations in Quadratic Form

In the following exercises, solve.

455. $x^4 - 14x^2 + 24 = 0$

456. $x^4 + 4x^2 - 32 = 0$

457. $4x^4 - 5x^2 + 1 = 0$

458.
$(2y + 3)^2 + 3(2y + 3) - 28 = 0$

459. $x + 3\sqrt{x} - 28 = 0$

460. $6x + 5\sqrt{x} - 6 = 0$

461. $x^{\frac{2}{3}} - 10x^{\frac{1}{3}} + 24 = 0$

462. $x + 7x^{\frac{1}{2}} + 6 = 0$

463. $8x^{-2} - 2x^{-1} - 3 = 0$

9.5 Solve Applications of Quadratic Equations

Solve Applications Modeled by Quadratic Equations

In the following exercises, solve by using the method of factoring, the square root principle, or the Quadratic Formula. Round your answers to the nearest tenth, if needed.

464. Find two consecutive odd numbers whose product is 323.

465. Find two consecutive even numbers whose product is 624.

466. A triangular banner has an area of 351 square centimeters. The length of the base is two centimeters longer than four times the height. Find the height and length of the base.

467. Julius built a triangular display case for his coin collection. The height of the display case is six inches less than twice the width of the base. The area of the of the back of the case is 70 square inches. Find the height and width of the case.

468. A tile mosaic in the shape of a right triangle is used as the corner of a rectangular pathway. The hypotenuse of the mosaic is 5 feet. One side of the mosaic is twice as long as the other side. What are the lengths of the sides? Round to the nearest tenth.

469. A rectangular piece of plywood has a diagonal which measures two feet more than the width. The length of the plywood is twice the width. What is the length of the plywood's diagonal? Round to the nearest tenth.

470. The front walk from the street to Pam's house has an area of 250 square feet. Its length is two less than four times its width. Find the length and width of the sidewalk. Round to the nearest tenth.

471. For Sophia's graduation party, several tables of the same width will be arranged end to end to give serving table with a total area of 75 square feet. The total length of the tables will be two more than three times the width. Find the length and width of the serving table so Sophia can purchase the correct size tablecloth . Round answer to the nearest tenth.

472. A ball is thrown vertically in the air with a velocity of 160 ft/sec. Use the formula $h = -16t^2 + v_0 t$ to determine when the ball will be 384 feet from the ground. Round to the nearest tenth.

473. The couple took a small airplane for a quick flight up to the wine country for a romantic dinner and then returned home. The plane flew a total of 5 hours and each way the trip was 360 miles. If the plane was flying at 150 mph, what was the speed of the wind that affected the plane?

474. Ezra kayaked up the river and then back in a total time of 6 hours. The trip was 4 miles each way and the current was difficult. If Roy kayaked at a speed of 5 mph, what was the speed of the current?

475. Two handymen can do a home repair in 2 hours if they work together. One of the men takes 3 hours more than the other man to finish the job by himself. How long does it take for each handyman to do the home repair individually?

9.6 Graph Quadratic Functions Using Properties

Recognize the Graph of a Quadratic Function

In the following exercises, graph by plotting point.

476. Graph $y = x^2 - 2$

477. Graph $y = -x^2 + 3$

In the following exercises, determine if the following parabolas open up or down.

478.

ⓐ $y = -3x^2 + 3x - 1$

ⓑ $y = 5x^2 + 6x + 3$

479.

ⓐ $y = x^2 + 8x - 1$

ⓑ $y = -4x^2 - 7x + 1$

Find the Axis of Symmetry and Vertex of a Parabola

In the following exercises, find ⓐ the equation of the axis of symmetry and ⓑ the vertex.

480. $y = -x^2 + 6x + 8$

481. $y = 2x^2 - 8x + 1$

Find the Intercepts of a Parabola

In the following exercises, find the x- and y-intercepts.

482. $y = x^2 - 4x + 5$

483. $y = x^2 - 8x + 15$

484. $y = x^2 - 4x + 10$

485. $y = -5x^2 - 30x - 46$

486. $y = 16x^2 - 8x + 1$

487. $y = x^2 + 16x + 64$

Graph Quadratic Functions Using Properties

In the following exercises, graph by using its properties.

488. $y = x^2 + 8x + 15$

489. $y = x^2 - 2x - 3$

490. $y = -x^2 + 8x - 16$

491. $y = 4x^2 - 4x + 1$

492. $y = x^2 + 6x + 13$

493. $y = -2x^2 - 8x - 12$

Solve Maximum and Minimum Applications

In the following exercises, find the minimum or maximum value.

494. $y = 7x^2 + 14x + 6$

495. $y = -3x^2 + 12x - 10$

In the following exercises, solve. Rounding answers to the nearest tenth.

496. A ball is thrown upward from the ground with an initial velocity of 112 ft/sec. Use the quadratic equation $h = -16t^2 + 112t$ to find how long it will take the ball to reach maximum height, and then find the maximum height.

497. A daycare facility is enclosing a rectangular area along the side of their building for the children to play outdoors. They need to maximize the area using 180 feet of fencing on three sides of the yard. The quadratic equation $A = -2x^2 + 180x$ gives the area, A, of the yard for the length, x, of the building that will border the yard. Find the length of the building that should border the yard to maximize the area, and then find the maximum area.

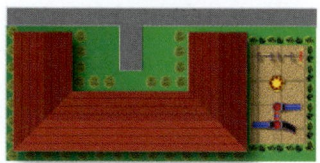

9.7 Graph Quadratic Functions Using Transformations

Graph Quadratic Functions of the form $f(x) = x^2 + k$

In the following exercises, graph each function using a vertical shift.

498. $g(x) = x^2 + 4$

499. $h(x) = x^2 - 3$

In the following exercises, graph each function using a horizontal shift.

500. $f(x) = (x + 1)^2$

501. $g(x) = (x - 3)^2$

In the following exercises, graph each function using transformations.

502. $f(x) = (x + 2)^2 + 3$

503. $f(x) = (x + 3)^2 - 2$

504. $f(x) = (x - 1)^2 + 4$

505. $f(x) = (x - 4)^2 - 3$

Graph Quadratic Functions of the form $f(x) = ax^2$

In the following exercises, graph each function.

506. $f(x) = 2x^2$

507. $f(x) = -x^2$

508. $f(x) = \frac{1}{2}x^2$

Graph Quadratic Functions Using Transformations

In the following exercises, rewrite each function in the $f(x) = a(x - h)^2 + k$ form by completing the square.

509. $f(x) = 2x^2 - 4x - 4$

510. $f(x) = 3x^2 + 12x + 8$

In the following exercises, ⓐ *rewrite each function in* $f(x) = a(x - h)^2 + k$ *form and* ⓑ *graph it by using transformations.*

511. $f(x) = 3x^2 - 6x - 1$

512. $f(x) = -2x^2 - 12x - 5$

513. $f(x) = 2x^2 + 4x + 6$

514. $f(x) = 3x^2 - 12x + 7$

In the following exercises, ⓐ *rewrite each function in* $f(x) = a(x - h)^2 + k$ *form and* ⓑ *graph it using properties.*

515. $f(x) = -3x^2 - 12x - 5$

516. $f(x) = 2x^2 - 12x + 7$

Find a Quadratic Function from its Graph

In the following exercises, write the quadratic function in $f(x) = a(x - h)^2 + k$ *form.*

517.

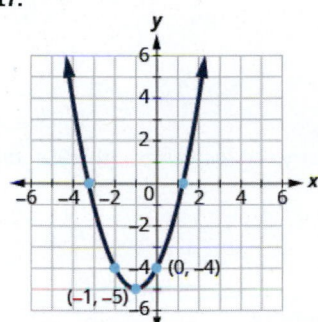

518.

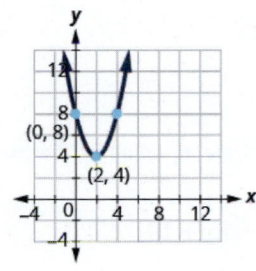

9.8 Solve Quadratic Inequalities

Solve Quadratic Inequalities Graphically

In the following exercises, solve graphically and write the solution in interval notation.

519. $x^2 - x - 6 > 0$

520. $x^2 + 4x + 3 \leq 0$

521. $-x^2 - x + 2 \geq 0$

522. $-x^2 + 2x + 3 < 0$

In the following exercises, solve each inequality algebraically and write any solution in interval notation.

523. $x^2 - 6x + 8 < 0$

524. $x^2 + x > 12$

525. $x^2 - 6x + 4 \leq 0$

526. $2x^2 + 7x - 4 > 0$

527. $-x^2 + x - 6 > 0$

528. $x^2 - 2x + 4 \geq 0$

PRACTICE TEST

529. Use the Square Root Property to solve the quadratic equation $3(w + 5)^2 = 27$.

530. Use Completing the Square to solve the quadratic equation $a^2 - 8a + 7 = 23$.

531. Use the Quadratic Formula to solve the quadratic equation $2m^2 - 5m + 3 = 0$.

Solve the following quadratic equations. Use any method.

532. $2x(3x - 2) - 1 = 0$

533. $\frac{9}{4}y^2 - 3y + 1 = 0$

Use the discriminant to determine the number and type of solutions of each quadratic equation.

534. $6p^2 - 13p + 7 = 0$

535. $3q^2 - 10q + 12 = 0$

Solve each equation.

536. $4x^4 - 17x^2 + 4 = 0$

537. $y^{\frac{2}{3}} + 2y^{\frac{1}{3}} - 3 = 0$

For each parabola, find ⓐ which direction it opens, ⓑ the equation of the axis of symmetry, ⓒ the vertex, ⓓ the x- and y-intercepts, and e) the maximum or minimum value.

538. $y = 3x^2 + 6x + 8$

539. $y = -x^2 - 8x + 16$

Graph each quadratic function using intercepts, the vertex, and the equation of the axis of symmetry.

540. $f(x) = x^2 + 6x + 9$

541. $f(x) = -2x^2 + 8x + 4$

In the following exercises, graph each function using transformations.

542. $f(x) = (x + 3)^2 + 2$

543. $f(x) = x^2 - 4x - 1$

In the following exercises, solve each inequality algebraically and write any solution in interval notation.

544. $x^2 - 6x - 8 \le 0$

545. $2x^2 + x - 10 > 0$

Model the situation with a quadratic equation and solve by any method.

546. Find two consecutive even numbers whose product is 360.

547. The length of a diagonal of a rectangle is three more than the width. The length of the rectangle is three times the width. Find the length of the diagonal. (Round to the nearest tenth.)

548. A water balloon is launched upward at the rate of 86 ft/sec. Using the formula $h = -16t^2 + 86t$ find how long it will take the balloon to reach the maximum height, and then find the maximum height. Round to the nearest tenth.

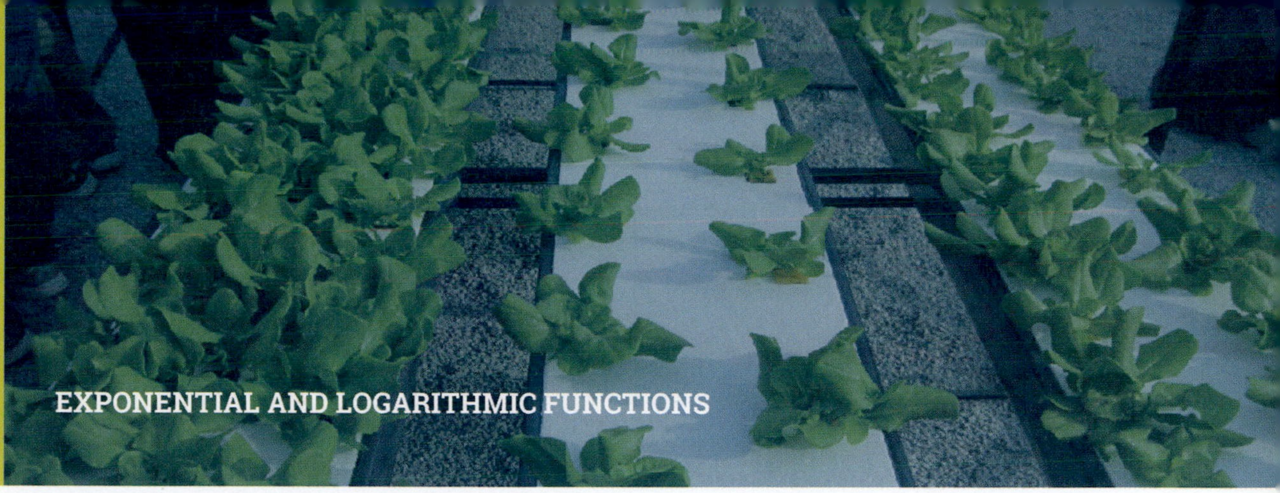

10 EXPONENTIAL AND LOGARITHMIC FUNCTIONS

Figure 10.1 Hydroponic systems allow botanists to grow crops without land. (credit: "Izhamwong"/Wikimedia Commons)

Chapter Outline

Introduction

As the world population continues to grow, food supplies are becoming less able to meet the increasing demand. At the same time, available resources of fertile soil for growing plants is dwindling. One possible solution—grow plants without soil. Botanists around the world are expanding the potential of hydroponics, which is the process of growing plants without soil. To provide the plants with the nutrients they need, the botanists keep careful growth records. Some growth is described by the types of functions you will explore in this chapter—exponential and logarithmic. You will evaluate and graph these functions, and solve equations using them.

10.1 Finding Composite and Inverse Functions

Learning Objectives

By the end of this section, you will be able to:

- Find and evaluate composite functions
- Determine whether a function is one-to-one
- Find the inverse of a function

Be Prepared!

Before you get started, take this readiness quiz.

1. If $f(x) = 2x - 3$ and $g(x) = x^2 + 2x - 3$, find $f(4)$.
   If you missed this problem, review Example 3.48.
2. Solve for x, $3x + 2y = 12$.
   If you missed this problem, review Example 2.31.
3. Simplify: $5\frac{(x+4)}{5} - 4$.
   If you missed this problem, review Example 1.25.

In this chapter, we will introduce two new types of functions, exponential functions and logarithmic functions. These functions are used extensively in business and the sciences as we will see.

Find and Evaluate Composite Functions

Before we introduce the functions, we need to look at another operation on functions called composition. In composition, the output of one function is the input of a second function. For functions f and g, the composition is written $f \circ g$ and is defined by $(f \circ g)(x) = f(g(x))$.

We read $f(g(x))$ as "f of g of x."

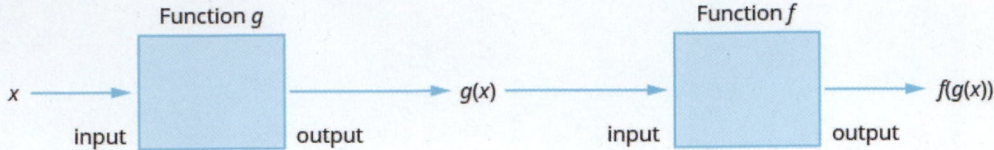

To do a composition, the output of the first function, $g(x)$, becomes the input of the second function, f, and so we must be sure that it is part of the domain of f.

Composition of Functions

The composition of functions f and g is written $f \cdot g$ and is defined by

$$(f \circ g)(x) = f(g(x))$$

We read $f(g(x))$ as f of g of x.

We have actually used composition without using the notation many times before. When we graphed quadratic functions using translations, we were composing functions. For example, if we first graphed $g(x) = x^2$ as a parabola and then shifted it down vertically four units, we were using the composition defined by $(f \circ g)(x) = f(g(x))$ where $f(x) = x - 4$.

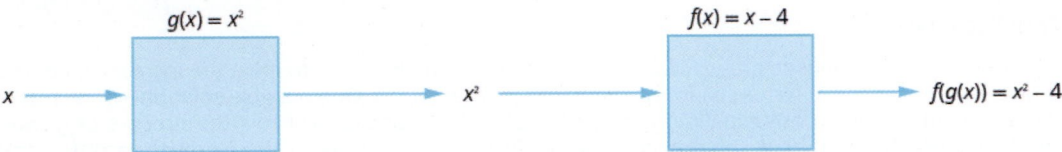

The next example will demonstrate that $(f \circ g)(x)$, $(g \circ f)(x)$ and $(f \cdot g)(x)$ usually result in different outputs.

EXAMPLE 10.1

For functions $f(x) = 4x - 5$ and $g(x) = 2x + 3$, find: ⓐ $(f \circ g)(x)$, ⓑ $(g \circ f)(x)$, and ⓒ $(f \cdot g)(x)$.

✓ Solution

ⓐ

Use the definition of $(f \circ g)(x)$.	$(f \circ g)(x) = f(g(x))$
Substitute $2x + 3$ for $g(x)$.	$(f \circ g)(x) = f(2x + 3)$
Find $f(2x + 3)$ where $f(x) = 4x - 5$.	$(f \circ g)(x) = 4(2x + 3) - 5$
Distribute.	$(f \circ g)(x) = 8x + 12 - 5$
Simplify.	$(f \circ g)(x) = 8x + 7$

ⓑ

Use the definition of $(f \circ g)(x)$.	$(g \circ f)(x) = g(f(x))$
Substitute $4x - 5$ for $f(x)$.	$(g \circ f)(x) = g(4x - 5)$
Find $g(4x - 5)$ where $g(x) = 2x + 3$.	$(g \circ f)(x) = 2(4x - 5) + 3$
Distribute.	$(g \circ f)(x) = 8x - 10 + 3$
Simplify.	$(g \circ f)(x) = 8x - 7$

Notice the difference in the result in part ⓐ and part ⓑ.

ⓒ Notice that $(f \cdot g)(x)$ is different than $(f \circ g)(x)$. In part ⓐ we did the composition of the functions. Now in part ⓒ we are not composing them, we are multiplying them.

Use the definition of $(f \cdot g)(x)$. $(f \cdot g)(x) = f(x) \cdot g(x)$

Substitute $f(x) = 4x - 5$ and $g(x) = 2x + 3$. $(f \cdot g)(x) = (4x - 5) \cdot (2x + 3)$

Multiply. $(f \cdot g)(x) = 8x^2 + 2x - 15$

> **TRY IT :: 10.1**   For functions $f(x) = 3x - 2$ and $g(x) = 5x + 1$, find ⓐ $(f \circ g)(x)$ ⓑ $(g \circ f)(x)$ ⓒ $(f \cdot g)(x)$.

> **TRY IT :: 10.2**

   For functions $f(x) = 4x - 3$, and $g(x) = 6x - 5$, find ⓐ $(f \circ g)(x)$, ⓑ $(g \circ f)(x)$, and ⓒ $(f \cdot g)(x)$.

In the next example we will evaluate a composition for a specific value.

EXAMPLE 10.2

For functions $f(x) = x^2 - 4$, and $g(x) = 3x + 2$, find: ⓐ $(f \circ g)(-3)$, ⓑ $(g \circ f)(-1)$, and ⓒ $(f \circ f)(2)$.

⊘ **Solution**

ⓐ

Use the definition of $(f \circ g)(-3)$.	$(f \circ g)(-3) = f(g(-3))$
Find $g(-3)$ where $g(x) = 3x + 2$.	$(f \circ g)(-3) = f(3 \cdot (-3) + 2)$
Simplify.	$(f \circ g)(-3) = f(-7)$
Find $f(-7)$ where $f(x) = x^2 - 4$.	$(f \circ g)(-3) = (-7)^2 - 4$
Simplify.	$(f \circ g)(-3) = 45$

ⓑ

Use the definition of $(g \circ f)(-1)$.	$(g \circ f)(-1) = g(f(-1))$
Find $f(-1)$ where $f(x) = x^2 - 4$.	$(g \circ f)(-1) = g((-1)^2 - 4)$
Simplify.	$(g \circ f)(-1) = g(-3)$
Find $g(-3)$ where $g(x) = 3x + 2$.	$(g \circ f)(-1) = 3(-3) + 2$
Simplify.	$(g \circ f)(-1) = -7$

Use the definition of $(f \circ f)(2)$.	$(f \circ f)(2) = f(f(2))$
Find $f(2)$ where $f(x) = x^2 - 4$.	$(f \circ f)(2) = f(2^2 - 4)$
Simplify.	$(f \circ f)(2) = f(0)$
Find $f(0)$ where $f(x) = x^2 - 4$.	$(f \circ f)(2) = 0^2 - 4$
Simplify.	$(f \circ f)(2) = -4$

> **TRY IT : :** 10.3

For functions $f(x) = x^2 - 9$, and $g(x) = 2x + 5$, find ⓐ $(f \circ g)(-2)$, ⓑ $(g \circ f)(-3)$, and ⓒ $(f \circ f)(4)$.

> **TRY IT : :** 10.4

For functions $f(x) = x^2 + 1$, and $g(x) = 3x - 5$, find ⓐ $(f \circ g)(-1)$, ⓑ $(g \circ f)(2)$, and ⓒ $(f \circ f)(-1)$.

Determine Whether a Function is One-to-One

When we first introduced functions, we said a function is a relation that assigns to each element in its domain exactly one element in the range. For each ordered pair in the relation, each x-value is matched with only one y-value.

We used the birthday example to help us understand the definition. Every person has a birthday, but no one has two birthdays and it is okay for two people to share a birthday. Since each person has exactly one birthday, that relation is a function.

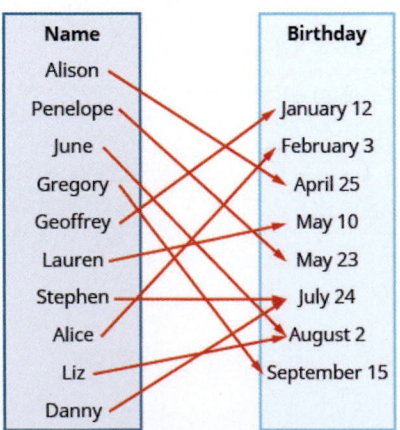

A function is **one-to-one** if each value in the range has exactly one element in the domain. For each ordered pair in the function, each y-value is matched with only one x-value.

Our example of the birthday relation is not a one-to-one function. Two people can share the same birthday. The range value August 2 is the birthday of Liz and June, and so one range value has two domain values. Therefore, the function is not one-to-one.

One-to-One Function

A function is **one-to-one** if each value in the range corresponds to one element in the domain. For each ordered pair in the function, each y-value is matched with only one x-value. There are no repeated y-values.

EXAMPLE 10.3

For each set of ordered pairs, determine if it represents a function and, if so, if the function is one-to-one.

ⓐ $\{(-3, 27), (-2, 8), (-1, 1), (0, 0), (1, 1), (2, 8), (3, 27)\}$ and ⓑ $\{(0, 0), (1, 1), (4, 2), (9, 3), (16, 4)\}$.

⊘ **Solution**

ⓐ

$$\{(-3, 27), (-2, 8), (-1, 1), (0, 0), (1, 1), (2, 8), (3, 27)\}$$

Each x-value is matched with only one y-value. So this relation is a function.

But each y-value is not paired with only one x-value, $(-3, 27)$ and $(3, 27),$ for example. So this function is not one-to-one.

ⓑ

$$\{(0, 0), (1, 1), (4, 2), (9, 3), (16, 4)\}$$

Each x-value is matched with only one y-value. So this relation is a function.

Since each y-value is paired with only one x-value, this function is one-to-one.

> **TRY IT : :** 10.5

> For each set of ordered pairs, determine if it represents a function and if so, is the function one-to-one.
>
> ⓐ $\{(-3, -6), (-2, -4), (-1, -2), (0, 0), (1, 2), (2, 4), (3, 6)\}$
>
> ⓑ $\{(-4, 8), (-2, 4), (-1, 2), (0, 0), (1, 2), (2, 4), (4, 8)\}$

> **TRY IT : :** 10.6

> For each set of ordered pairs, determine if it represents a function and if so, is the function one-to-one.
>
> ⓐ $\{(27, -3), (8, -2), (1, -1), (0, 0), (1, 1), (8, 2), (27, 3)\}$
>
> ⓑ $\{(7, -3), (-5, -4), (8, 0), (0, 0), (-6, 4), (-2, 2), (-1, 3)\}$

To help us determine whether a relation is a function, we use the vertical line test. A set of points in a rectangular coordinate system is the graph of a function if every vertical line intersects the graph in at most one point. Also, if any vertical line intersects the graph in more than one point, the graph does not represent a function.

The vertical line is representing an x-value and we check that it intersects the graph in only one y-value. Then it is a function.

To check if a function is one-to-one, we use a similar process. We use a horizontal line and check that each horizontal line intersects the graph in only one point. The horizontal line is representing a y-value and we check that it intersects the graph in only one x-value. If every horizontal line intersects the graph of a function in at most one point, it is a one-to-one function. This is the **horizontal line test**.

Horizontal Line Test

If every horizontal line intersects the graph of a function in at most one point, it is a one-to-one function.

We can test whether a graph of a relation is a function by using the vertical line test. We can then tell if the function is one-to-one by applying the horizontal line test.

EXAMPLE 10.4

Determine ⓐ whether each graph is the graph of a function and, if so, ⓑ whether it is one-to-one.

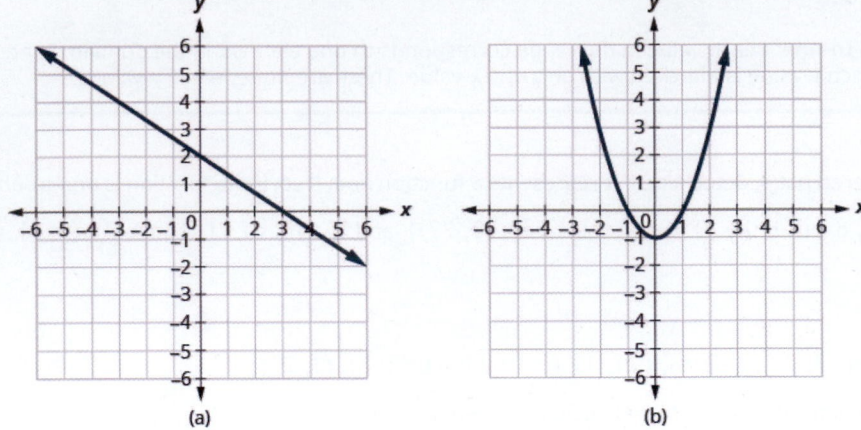

(a)   (b)

✓ **Solution**

ⓐ

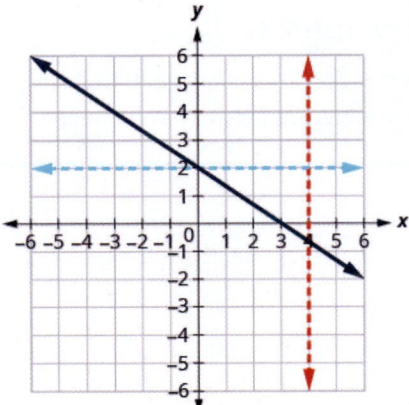

Since any vertical line intersects the graph in at most one point, the graph is the graph of a function. Since any horizontal line intersects the graph in at most one point, the graph is the graph of a one-to-one function.

ⓑ

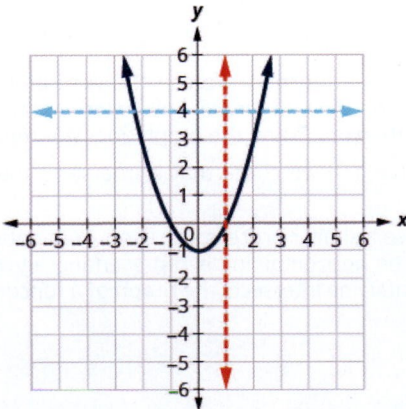

Since any vertical line intersects the graph in at most one point, the graph is the graph of a function. The horizontal line shown on the graph intersects it in two points. This graph does not represent a one-to-one function.

> **TRY IT : : 10.7**

Determine ⓐ whether each graph is the graph of a function and, if so, ⓑ whether it is one-to-one.

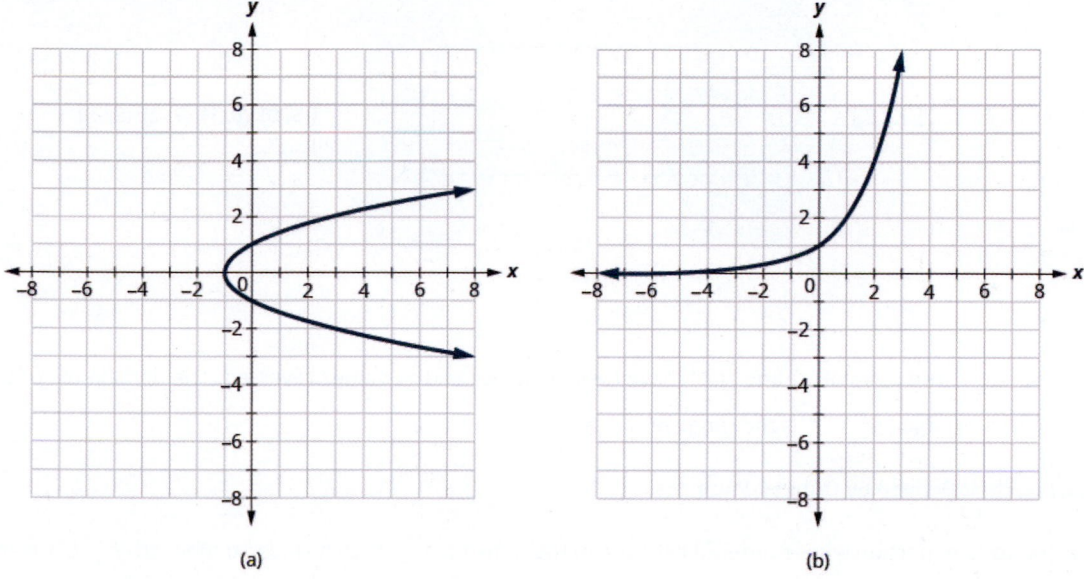

(a)   (b)

> **TRY IT : : 10.8**

Determine ⓐ whether each graph is the graph of a function and, if so, ⓑ whether it is one-to-one.

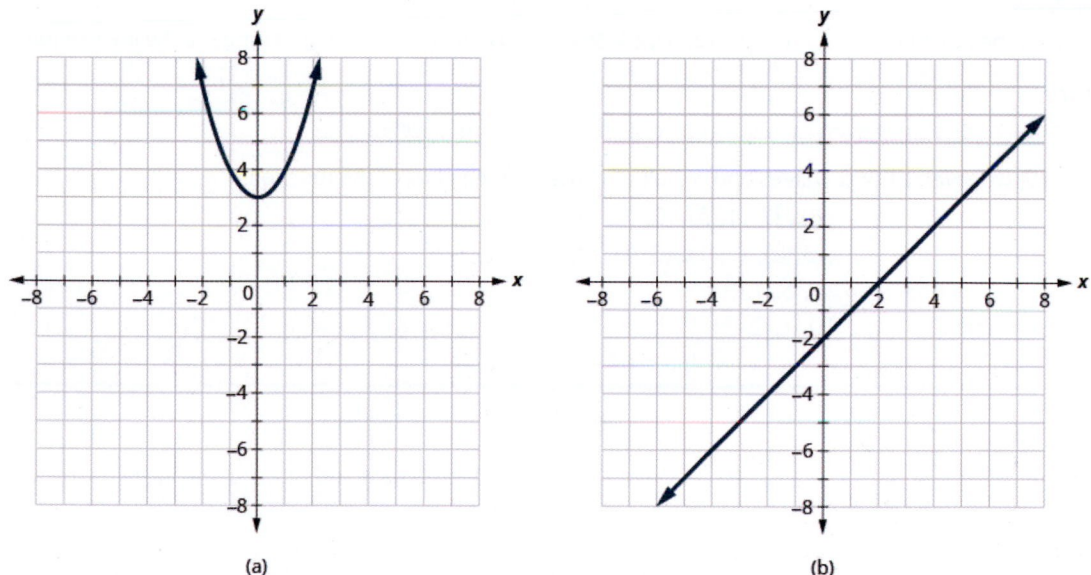

(a)   (b)

Find the Inverse of a Function

Let's look at a one-to one function, f, represented by the ordered pairs $\{(0, 5), (1, 6), (2, 7), (3, 8)\}$. For each x-value, f adds 5 to get the y-value. To 'undo' the addition of 5, we subtract 5 from each y-value and get back to the original x-value. We can call this "taking the inverse of f" and name the function f^{-1}.

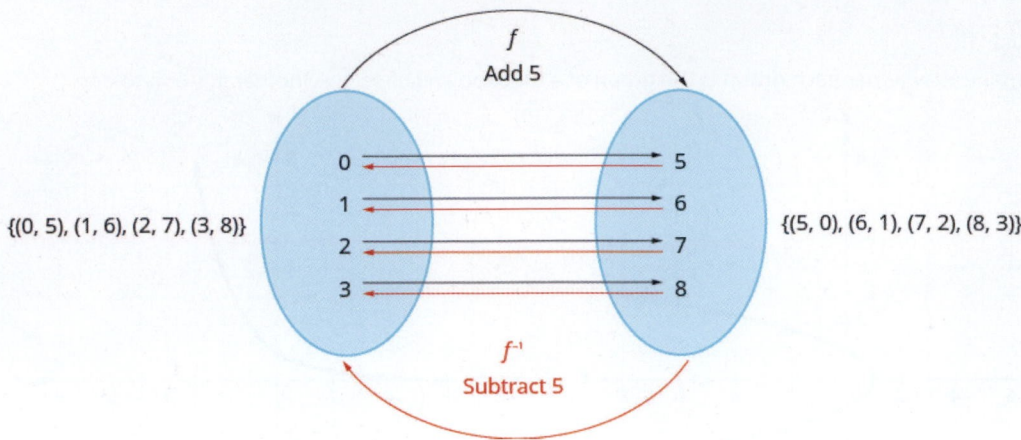

Notice that that the ordered pairs of f and f^{-1} have their x-values and y-values reversed. The domain of f is the range of f^{-1} and the domain of f^{-1} is the range of f.

Inverse of a Function Defined by Ordered Pairs

If $f(x)$ is a one-to-one function whose ordered pairs are of the form (x, y), then its inverse function $f^{-1}(x)$ is the set of ordered pairs (y, x).

In the next example we will find the inverse of a function defined by ordered pairs.

EXAMPLE 10.5

Find the inverse of the function $\{(0, 3), (1, 5), (2, 7), (3, 9)\}$. Determine the domain and range of the inverse function.

⊘ Solution

This function is one-to-one since every x-value is paired with exactly one y-value.

To find the inverse we reverse the x-values and y-values in the ordered pairs of the function.

Function	$\{(0, 3), (1, 5), (2, 7), (3, 9)\}$
Inverse Function	$\{(3, 0), (5, 1), (7, 2), (9, 3)\}$
Domain of Inverse Function	$\{3, 5, 7, 9\}$
Range of Inverse Function	$\{0, 1, 2, 3\}$

> **TRY IT ::** 10.9
>
> Find the inverse of $\{(0, 4), (1, 7), (2, 10), (3, 13)\}$. Determine the domain and range of the inverse function.

> **TRY IT ::** 10.10
>
> Find the inverse of $\{(-1, 4), (-2, 1), (-3, 0), (-4, 2)\}$. Determine the domain and range of the inverse function.

We just noted that if $f(x)$ is a one-to-one function whose ordered pairs are of the form (x, y), then its inverse function $f^{-1}(x)$ is the set of ordered pairs (y, x).

So if a point (a, b) is on the graph of a function $f(x)$, then the ordered pair (b, a) is on the graph of $f^{-1}(x)$. See **Figure 10.2**.

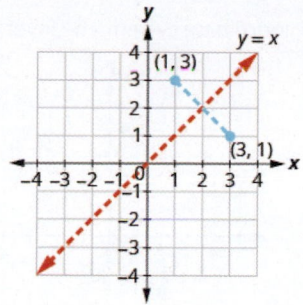

Figure 10.2

The distance between any two pairs (a, b) and (b, a) is cut in half by the line $y = x$. So we say the points are mirror images of each other through the line $y = x$.

Since every point on the graph of a function $f(x)$ is a mirror image of a point on the graph of $f^{-1}(x)$, we say the graphs are mirror images of each other through the line $y = x$. We will use this concept to graph the inverse of a function in the next example.

EXAMPLE 10.6

Graph, on the same coordinate system, the inverse of the one-to one function shown.

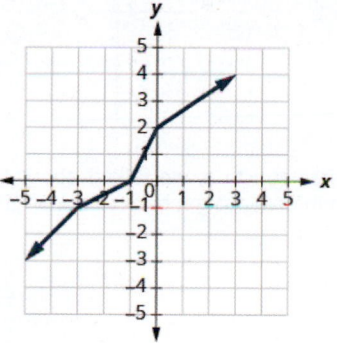

⊘ **Solution**

We can use points on the graph to find points on the inverse graph. Some points on the graph are: $(-5, -3), (-3, -1), (-1, 0), (0, 2), (3, 4)$.

So, the inverse function will contain the points: $(-3, -5), (-1, -3), (0, -1), (2, 0), (4, 3)$.

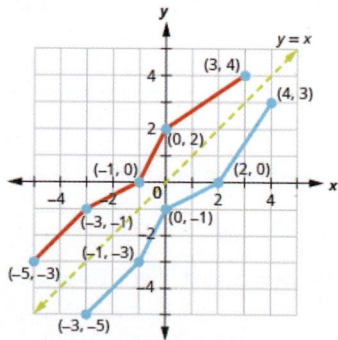

Notice how the graph of the original function and the graph of the inverse functions are mirror images through the line $y = x$.

> **TRY IT :: 10.11**   Graph, on the same coordinate system, the inverse of the one-to one function.

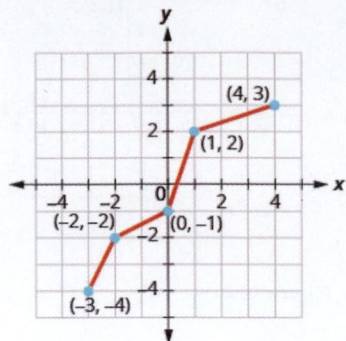

> **TRY IT :: 10.12**   Graph, on the same coordinate system, the inverse of the one-to one function.

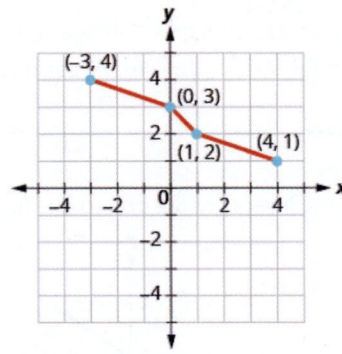

When we began our discussion of an inverse function, we talked about how the inverse function 'undoes' what the original function did to a value in its domain in order to get back to the original x-value.

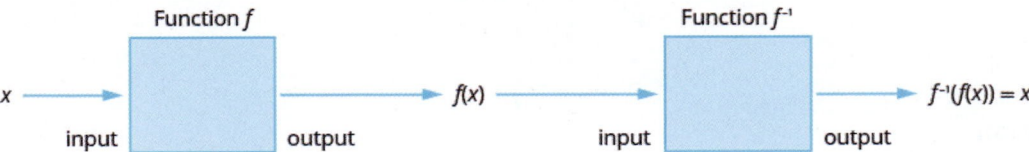

Inverse Functions

$$f^{-1}(f(x)) \;=\; x, \quad \text{for all } x \text{ in the domain of } f$$
$$f(f^{-1}(x)) \;=\; x, \quad \text{for all } x \text{ in the domain of } f^{-1}$$

We can use this property to verify that two functions are inverses of each other.

EXAMPLE 10.7

Verify that $f(x) = 5x - 1$ and $g(x) = \dfrac{x+1}{5}$ are inverse functions.

⊘ Solution

The functions are inverses of each other if $g(f(x)) = x$ and $f(g(x)) = x$.

	$g(f(x)) \stackrel{?}{=} x$
Substitute $5x - 1$ for $f(x)$.	$g(5x - 1) \stackrel{?}{=} x$
Find $g(5x - 1)$ where $g(x) = \frac{x + 1}{5}$.	$\frac{(5x - 1) + 1}{5} \stackrel{?}{=} x$
Simplify.	$\frac{5x}{5} \stackrel{?}{=} x$
Simplify.	$x = x ✓$

	$f(g(x)) \stackrel{?}{=} x$
Substitute $\frac{x + 1}{5}$ for $g(x)$.	$f\left(\frac{x + 1}{5}\right) \stackrel{?}{=} x$
Find $f\left(\frac{x + 1}{5}\right)$ where $f(x) = 5x - 1$.	$5\left(\frac{x + 1}{5}\right) - 1 \stackrel{?}{=} x$
Simplify.	$x + 1 - 1 \stackrel{?}{=} x$
Simplify.	$x = x ✓$

Since both $g(f(x)) = x$ and $f(g(x)) = x$ are true, the functions $f(x) = 5x - 1$ and $g(x) = \frac{x + 1}{5}$ are inverse functions. That is, they are inverses of each other.

> **TRY IT : : 10.13**   Verify that the functions are inverse functions.

$$f(x) = 4x - 3 \text{ and } g(x) = \frac{x + 3}{4}.$$

> **TRY IT : : 10.14**   Verify that the functions are inverse functions.

$$f(x) = 2x + 6 \text{ and } g(x) = \frac{x - 6}{2}.$$

We have found inverses of function defined by ordered pairs and from a graph. We will now look at how to find an inverse using an algebraic equation. The method uses the idea that if $f(x)$ is a one-to-one function with ordered pairs (x, y), then its inverse function $f^{-1}(x)$ is the set of ordered pairs (y, x).

If we reverse the x and y in the function and then solve for y, we get our inverse function.

EXAMPLE 10.8   HOW TO FIND THE INVERSE OF A ONE-TO-ONE FUNCTION

Find the inverse of $f(x) = 4x + 7$.

⊘ **Solution**

Step 1. Substitute y for $f(x)$.	Replace $f(x)$ with y.	$f(x) = 4x + 7$
		$y = 4x + 7$
Step 2. Interchange the variables x and y.	Replace x with y and then y with x.	$x = 4y + 7$

Step 3. Solve for y.	Subtract 7 from each side.	$x - 7 = 4y$
	Divide by 4.	$\dfrac{x - 7}{4} = y$
Step 4. Substitute $f^{-1}(x)$ for y.	Replace y with $f^{-1}(x)$.	$\dfrac{x - 7}{4} = f^{-1}(x)$
Step 5. Verify that the functions are inverses.	Show $f^{-1}(f(x)) = x$ and $f(f^{-1}(x)) = x$	$f^{-1}(f(x)) \overset{?}{=} x$ $f^{-1}(4x + 7) \overset{?}{=} x$ $\dfrac{(4x + 7) - 7}{4} \overset{?}{=} x$ $\dfrac{4x}{4} \overset{?}{=} x$ $x = x \checkmark$ $f(f^{-1}(x)) \overset{?}{=} x$ $f\left(\dfrac{x - 7}{4}\right) \overset{?}{=} x$ $4\left(\dfrac{x - 7}{4}\right) + 7 \overset{?}{=} x$ $x - 7 + 7 \overset{?}{=} x$ $x = x \checkmark$

> **TRY IT ∷ 10.15**   Find the inverse of the function $f(x) = 5x - 3$.

> **TRY IT ∷ 10.16**   Find the inverse of the function $f(x) = 8x + 5$.

We summarize the steps below.

HOW TO ∷ HOW TO FIND THE INVERSE OF A ONE-TO-ONE FUNCTION

Step 1.   Substitute y for $f(x)$.

Step 2.   Interchange the variables x and y.

Step 3.   Solve for y.

Step 4.   Substitute $f^{-1}(x)$ for y.

Step 5.   Verify that the functions are inverses.

EXAMPLE 10.9   HOW TO FIND THE INVERSE OF A ONE-TO-ONE FUNCTION

Find the inverse of $f(x) = \sqrt[5]{2x - 3}$.

⊘ Solution

$$f(x) = \sqrt[5]{2x - 3}$$

Substitute y for $f(x)$. $\qquad\qquad\qquad\qquad y = \sqrt[5]{2x - 3}$

Interchange the variables x and y. $\qquad\qquad x = \sqrt[5]{2y - 3}$

Solve for y. $\qquad\qquad\qquad\qquad\qquad (x)^5 = \left(\sqrt[5]{2y - 3}\right)^5$

$$x^5 = 2y - 3$$

$$x^5 + 3 = 2y$$

$$\frac{x^5 + 3}{2} = y$$

Substitute $f^{-1}(x)$ for y. $\qquad\qquad f^{-1}(x) = \frac{x^5 + 3}{2}$

Verify that the functions are inverses.

$$f^{-1}(f(x)) \stackrel{?}{=} x \qquad\qquad\qquad f(f^{-1}(x)) \stackrel{?}{=} x$$

$$f^{-1}\left(\sqrt[5]{2x - 3}\right) \stackrel{?}{=} x \qquad\qquad f\left(\frac{x^5 + 3}{2}\right) \stackrel{?}{=} x$$

$$\frac{\left(\sqrt[5]{2x - 3}\right)^5 + 3}{2} \stackrel{?}{=} x \qquad\qquad \sqrt[5]{2\left(\frac{x^5 + 3}{2}\right) - 3} \stackrel{?}{=} x$$

$$\frac{2x - 3 + 3}{2} \stackrel{?}{=} x \qquad\qquad\qquad \sqrt[5]{x^5 + 3 - 3} \stackrel{?}{=} x$$

$$\frac{2x}{2} \stackrel{?}{=} x \qquad\qquad\qquad\qquad \sqrt[5]{x^5} \stackrel{?}{=} x$$

$$x = x \checkmark \qquad\qquad\qquad\qquad x = x \checkmark$$

> **TRY IT :: 10.17**

Find the inverse of the function $f(x) = \sqrt[5]{3x - 2}$.

> **TRY IT :: 10.18**

Find the inverse of the function $f(x) = \sqrt[4]{6x - 7}$.

10.1 EXERCISES

Practice Makes Perfect

Find and Evaluate Composite Functions

In the following exercises, find ⓐ (f ∘ g)(x), ⓑ (g ∘ f)(x), and ⓒ (f · g)(x).

1. $f(x) = 4x + 3$ and $g(x) = 2x + 5$

2. $f(x) = 3x - 1$ and $g(x) = 5x - 3$

3. $f(x) = 6x - 5$ and $g(x) = 4x + 1$

4. $f(x) = 2x + 7$ and $g(x) = 3x - 4$

5. $f(x) = 3x$ and $g(x) = 2x^2 - 3x$

6. $f(x) = 2x$ and $g(x) = 3x^2 - 1$

7. $f(x) = 2x - 1$ and $g(x) = x^2 + 2$

8. $f(x) = 4x + 3$ and $g(x) = x^2 - 4$

In the following exercises, find the values described.

9. For functions $f(x) = 2x^2 + 3$ and $g(x) = 5x - 1$, find
ⓐ $(f \circ g)(-2)$
ⓑ $(g \circ f)(-3)$
ⓒ $(f \circ f)(-1)$

10. For functions $f(x) = 5x^2 - 1$ and $g(x) = 4x - 1$, find
ⓐ $(f \circ g)(1)$
ⓑ $(g \circ f)(-1)$
ⓒ $(f \circ f)(2)$

11. For functions $f(x) = 2x^3$ and $g(x) = 3x^2 + 2$, find
ⓐ $(f \circ g)(-1)$
ⓑ $(g \circ f)(1)$
ⓒ $(g \circ g)(1)$

12. For functions $f(x) = 3x^3 + 1$ and $g(x) = 2x^2 - 3$, find
ⓐ $(f \circ g)(-2)$
ⓑ $(g \circ f)(-1)$
ⓒ $(g \circ g)(1)$

Determine Whether a Function is One-to-One

In the following exercises, determine if the set of ordered pairs represents a function and if so, is the function one-to-one.

13. {(−3, 9), (−2, 4), (−1, 1), (0, 0), (1, 1), (2, 4), (3, 9)}

14. {(9, −3), (4, −2), (1, −1), (0, 0), (1, 1), (4, 2), (9, 3)}

15. {(−3, −5), (−2, −3), (−1, −1), (0, 1), (1, 3), (2, 5), (3, 7)}

16. {(5, 3), (4, 2), (3, 1), (2, 0), (1, −1), (0, −2), (−1, −3)}

In the following exercises, determine whether each graph is the graph of a function and if so, is it one-to-one.

17. (a)

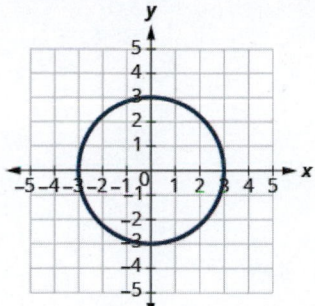

(b)

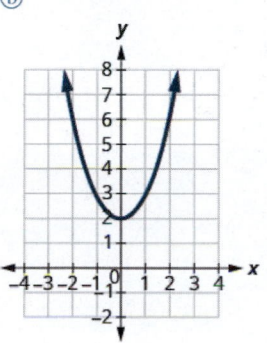

18. (a)

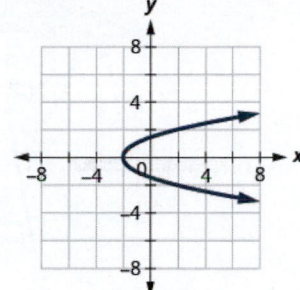

(b)

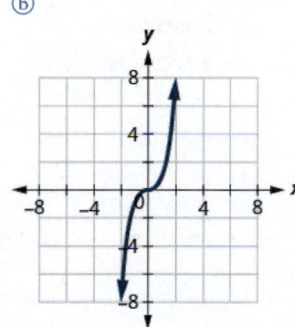

19. (a)

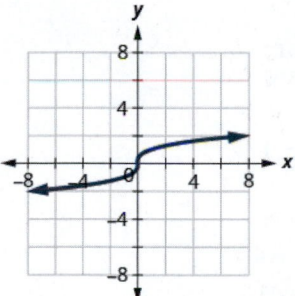

(b)

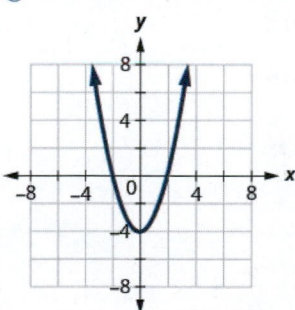

20. (a)

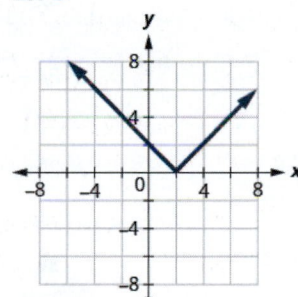

(b)

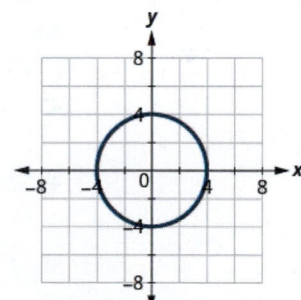

In the following exercises, find the inverse of each function. Determine the domain and range of the inverse function.

21. $\{(2, 1), (4, 2), (6, 3), (8, 4)\}$

22. $\{(6, 2), (9, 5), (12, 8), (15, 11)\}$

23. $\{(0, -2), (1, 3), (2, 7), (3, 12)\}$

24. $\{(0, 0), (1, 1), (2, 4), (3, 9)\}$

25. $\{(-2, -3), (-1, -1), (0, 1), (1, 3)\}$

26. $\{(5, 3), (4, 2), (3, 1), (2, 0)\}$

In the following exercises, graph, on the same coordinate system, the inverse of the one-to-one function shown.

27.

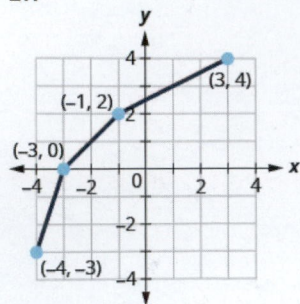

28.

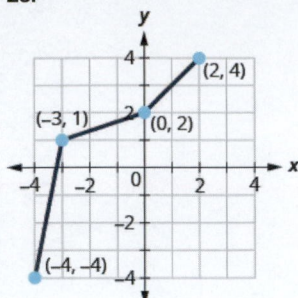

29.

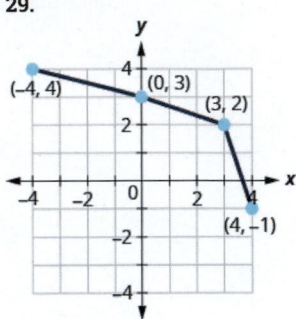

30.

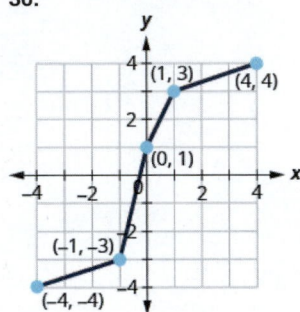

In the following exercises, determine whether or not the given functions are inverses.

31. $f(x) = x + 8$ and $g(x) = x - 8$

32. $f(x) = x - 9$ and $g(x) = x + 9$

33. $f(x) = 7x$ and $g(x) = \frac{x}{7}$

34. $f(x) = \frac{x}{11}$ and $g(x) = 11x$

35. $f(x) = 7x + 3$ and $g(x) = \frac{x-3}{7}$

36. $f(x) = 5x - 4$ and $g(x) = \frac{x-4}{5}$

37. $f(x) = \sqrt{x+2}$ and $g(x) = x^2 - 2$

38. $f(x) = \sqrt[3]{x-4}$ and $g(x) = x^3 + 4$

In the following exercises, find the inverse of each function.

39. $f(x) = x - 12$

40. $f(x) = x + 17$

41. $f(x) = 9x$

42. $f(x) = 8x$

43. $f(x) = \frac{x}{6}$

44. $f(x) = \frac{x}{4}$

45. $f(x) = 6x - 7$

46. $f(x) = 7x - 1$

47. $f(x) = -2x + 5$

48. $f(x) = -5x - 4$

49. $f(x) = x^2 + 6, \quad x \geq 0$

50. $f(x) = x^2 - 9, \quad x \geq 0$

51. $f(x) = x^3 - 4$

52. $f(x) = x^3 + 6$

53. $f(x) = \dfrac{1}{x+2}$

54. $f(x) = \dfrac{1}{x-6}$

55. $f(x) = \sqrt{x-2}, \quad x \geq 2$

56. $f(x) = \sqrt{x+8}, \quad x \geq -8$

57. $f(x) = \sqrt[3]{x-3}$

58. $f(x) = \sqrt[3]{x+5}$

59. $f(x) = \sqrt[4]{9x-5}, \quad x \geq \dfrac{5}{9}$

60. $f(x) = \sqrt[4]{8x-3}, \quad x \geq \dfrac{3}{8}$

61. $f(x) = \sqrt[5]{-3x+5}$

62. $f(x) = \sqrt[5]{-4x-3}$

Writing Exercises

63. Explain how the graph of the inverse of a function is related to the graph of the function.

64. Explain how to find the inverse of a function from its equation. Use an example to demonstrate the steps.

Self Check

ⓐ *After completing the exercises, use this checklist to evaluate your mastery of the objectives of this section.*

I can...	Confidently	With some help	No-I don't get it!
find and evaluate composite functions.			
determine whether a function is one-to-one.			
find the inverse of a function.			

ⓑ *If most of your checks were:*

...confidently. Congratulations! You have achieved the objectives in this section. Reflect on the study skills you used so that you can continue to use them. What did you do to become confident of your ability to do these things? Be specific.

...with some help. This must be addressed quickly because topics you do not master become potholes in your road to success. In math every topic builds upon previous work. It is important to make sure you have a strong foundation before you move on. Who can you ask for help? Your fellow classmates and instructor are good resources. Is there a place on campus where math tutors are available? Can your study skills be improved?

...no—I don't get it! This is a warning sign and you must not ignore it. You should get help right away or you will quickly be overwhelmed. See your instructor as soon as you can to discuss your situation. Together you can come up with a plan to get you the help you need.

10.2 Evaluate and Graph Exponential Functions

Learning Objectives

By the end of this section, you will be able to:

> Graph exponential functions
> Solve Exponential equations
> Use exponential models in applications

Be Prepared!

Before you get started, take this readiness quiz.

1. Simplify: $\left(\dfrac{x^3}{x^2}\right)$.

   If you missed this problem, review **Example 5.13**.

2. Evaluate: ⓐ 2^0 ⓑ $\left(\dfrac{1}{3}\right)^0$.

   If you missed this problem, review **Example 5.14**.

3. Evaluate: ⓐ 2^{-1} ⓑ $\left(\dfrac{1}{3}\right)^{-1}$.

   If you missed this problem, review **Example 5.15**.

Graph Exponential Functions

The functions we have studied so far do not give us a model for many naturally occurring phenomena. From the growth of populations and the spread of viruses to radioactive decay and compounding interest, the models are very different from what we have studied so far. These models involve exponential functions.

An **exponential function** is a function of the form $f(x) = a^x$ where $a > 0$ and $a \neq 1$.

Exponential Function

An exponential function, where $a > 0$ and $a \neq 1$, is a function of the form

$$f(x) = a^x$$

Notice that in this function, the variable is the exponent. In our functions so far, the variables were the base.

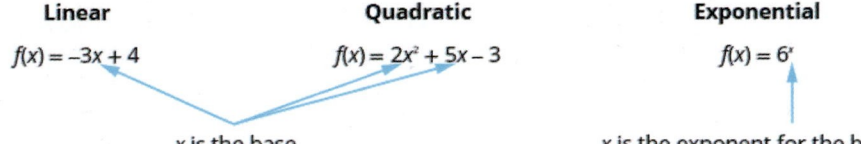

Linear	Quadratic	Exponential
$f(x) = -3x + 4$	$f(x) = 2x^2 + 5x - 3$	$f(x) = 6^x$

x is the base   *x* is the exponent for the base 6

Our definition says $a \neq 1$. If we let $a = 1$, then $f(x) = a^x$ becomes $f(x) = 1^x$. Since $1^x = 1$ for all real numbers, $f(x) = 1$. This is the constant function.

Our definition also says $a > 0$. If we let a base be negative, say -4, then $f(x) = (-4)^x$ is not a real number when $x = \dfrac{1}{2}$.

$$
\begin{aligned}
f(x) &= (-4)^x \\
f\left(\tfrac{1}{2}\right) &= (-4)^{\frac{1}{2}} \\
f\left(\tfrac{1}{2}\right) &= \sqrt{-4} \quad \text{not a real number}
\end{aligned}
$$

In fact, $f(x) = (-4)^x$ would not be a real number any time x is a fraction with an even denominator. So our definition requires $a > 0$.

By graphing a few exponential functions, we will be able to see their unique properties.

EXAMPLE 10.10

On the same coordinate system graph $f(x) = 2^x$ and $g(x) = 3^x$.

⊘ Solution

We will use point plotting to graph the functions.

x	$f(x) = 2^x$	$(x, f(x))$	$g(x) = 3^x$	$(x, g(x))$
-2	$2^{-2} = \frac{1}{2^2} = \frac{1}{4}$	$\left(-2, \frac{1}{4}\right)$	$3^{-2} = \frac{1}{3^2} = \frac{1}{9}$	$\left(-2, \frac{1}{9}\right)$
-1	$2^{-1} = \frac{1}{2^1} = \frac{1}{2}$	$\left(-1, \frac{1}{2}\right)$	$3^{-1} = \frac{1}{3^1} = \frac{1}{3}$	$\left(-1, \frac{1}{3}\right)$
0	$2^0 = 1$	$(0, 1)$	$3^0 = 1$	$(0, 1)$
1	$2^1 = 2$	$(1, 2)$	$3^1 = 3$	$(1, 3)$
2	$2^2 = 4$	$(2, 4)$	$3^2 = 9$	$(2, 9)$
3	$2^3 = 8$	$(3, 8)$	$3^3 = 27$	$(3, 27)$

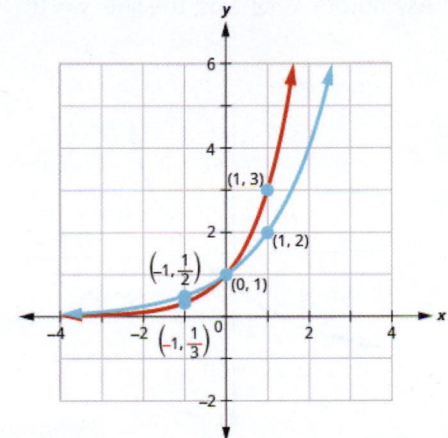

> **TRY IT :: 10.19**   Graph: $f(x) = 4^x$.

> **TRY IT :: 10.20**   Graph: $g(x) = 5^x$.

If we look at the graphs from the previous Example and Try Its, we can identify some of the properties of exponential functions.

The graphs of $f(x) = 2^x$ and $g(x) = 3^x$, as well as the graphs of $f(x) = 4^x$ and $g(x) = 5^x$, all have the same basic shape. This is the shape we expect from an exponential function where $a > 1$.

We notice, that for each function, the graph contains the point $(0, 1)$. This make sense because $a^0 = 1$ for any a.

The graph of each function, $f(x) = a^x$ also contains the point $(1, a)$. The graph of $f(x) = 2^x$ contained $(1, 2)$ and the graph of $g(x) = 3^x$ contained $(1, 3)$. This makes sense as $a^1 = a$.

Notice too, the graph of each function $f(x) = a^x$ also contains the point $\left(-1, \frac{1}{a}\right)$. The graph of $f(x) = 2^x$ contained $\left(-1, \frac{1}{2}\right)$ and the graph of $g(x) = 3^x$ contained $\left(-1, \frac{1}{3}\right)$. This makes sense as $a^{-1} = \frac{1}{a}$.

What is the domain for each function? From the graphs we can see that the domain is the set of all real numbers. There is no restriction on the domain. We write the domain in interval notation as $(-\infty, \infty)$.

Look at each graph. What is the range of the function? The graph never hits the x-axis. The range is all positive numbers.

We write the range in interval notation as $(0, \infty)$.

Whenever a graph of a function approaches a line but never touches it, we call that line an **asymptote**. For the exponential functions we are looking at, the graph approaches the x-axis very closely but will never cross it, we call the line $y = 0$, the x-axis, a horizontal asymptote.

Properties of the Graph of $f(x) = a^x$ when $a > 1$

Domain	$(-\infty, \infty)$
Range	$(0, \infty)$
x-intercept	None
y-intercept	$(0, 1)$
Contains	$(1, a), \left(-1, \frac{1}{a}\right)$
Asymptote	x-axis, the line $y = 0$

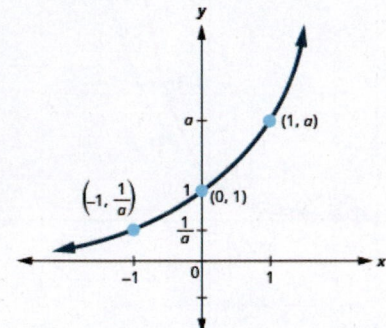

Our definition of an exponential function $f(x) = a^x$ says $a > 0$, but the examples and discussion so far has been about functions where $a > 1$. What happens when $0 < a < 1$? The next example will explore this possibility.

EXAMPLE 10.11

On the same coordinate system, graph $f(x) = \left(\frac{1}{2}\right)^x$ and $g(x) = \left(\frac{1}{3}\right)^x$.

⊘ **Solution**

We will use point plotting to graph the functions.

x	$f(x) = \left(\frac{1}{2}\right)^x$	$(x, f(x))$	$g(x) = \left(\frac{1}{3}\right)^x$	$(x, g(x))$
-2	$\left(\frac{1}{2}\right)^{-2} = 2^2 = 4$	$(-2, 4)$	$\left(\frac{1}{3}\right)^{-2} = 3^2 = 9$	$(-2, 9)$
-1	$\left(\frac{1}{2}\right)^{-1} = 2^1 = 2$	$(-1, 2)$	$\left(\frac{1}{3}\right)^{-1} = 3^1 = 3$	$(-1, 3)$
0	$\left(\frac{1}{2}\right)^0 = 1$	$(0, 1)$	$\left(\frac{1}{3}\right)^0 = 1$	$(0, 1)$
1	$\left(\frac{1}{2}\right)^1 = \frac{1}{2}$	$\left(1, \frac{1}{2}\right)$	$\left(\frac{1}{3}\right)^1 = \frac{1}{3}$	$\left(1, \frac{1}{3}\right)$
2	$\left(\frac{1}{2}\right)^2 = \frac{1}{4}$	$\left(2, \frac{1}{4}\right)$	$\left(\frac{1}{3}\right)^2 = \frac{1}{9}$	$\left(2, \frac{1}{9}\right)$
3	$\left(\frac{1}{2}\right)^3 = \frac{1}{8}$	$\left(3, \frac{1}{8}\right)$	$\left(\frac{1}{3}\right)^3 = \frac{1}{27}$	$\left(3, \frac{1}{27}\right)$

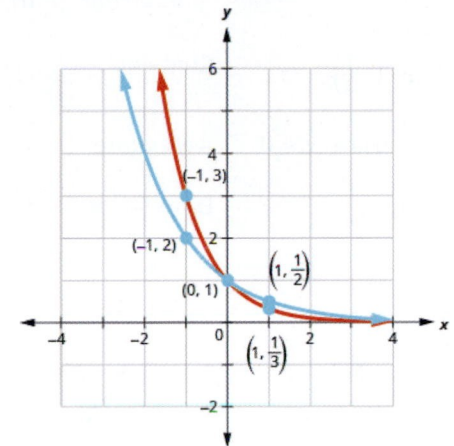

> **TRY IT : : 10.21**   Graph: $f(x) = \left(\frac{1}{4}\right)^x$.

> **TRY IT : : 10.22**   Graph: $g(x) = \left(\frac{1}{5}\right)^x$.

Now let's look at the graphs from the previous Example and Try Its so we can now identify some of the properties of exponential functions where $0 < a < 1$.

The graphs of $f(x) = \left(\frac{1}{2}\right)^x$ and $g(x) = \left(\frac{1}{3}\right)^x$ as well as the graphs of $f(x) = \left(\frac{1}{4}\right)^x$ and $g(x) = \left(\frac{1}{5}\right)^x$ all have the same basic shape. While this is the shape we expect from an exponential function where $0 < a < 1$, the graphs go down from left to right while the previous graphs, when $a > 1$, went from up from left to right.

We notice that for each function, the graph still contains the point (0, 1). This make sense because $a^0 = 1$ for any a.

As before, the graph of each function, $f(x) = a^x$, also contains the point $(1, a)$. The graph of $f(x) = \left(\frac{1}{2}\right)^x$ contained $\left(1, \frac{1}{2}\right)$ and the graph of $g(x) = \left(\frac{1}{3}\right)^x$ contained $\left(1, \frac{1}{3}\right)$. This makes sense as $a^1 = a$.

Notice too that the graph of each function, $f(x) = a^x$, also contains the point $\left(-1, \frac{1}{a}\right)$. The graph of $f(x) = \left(\frac{1}{2}\right)^x$ contained $(-1, 2)$ and the graph of $g(x) = \left(\frac{1}{3}\right)^x$ contained $(-1, 3)$. This makes sense as $a^{-1} = \frac{1}{a}$.

What is the domain and range for each function? From the graphs we can see that the domain is the set of all real numbers and we write the domain in interval notation as $(-\infty, \infty)$. Again, the graph never hits the x-axis. The range is all positive numbers. We write the range in interval notation as $(0, \infty)$.

We will summarize these properties in the chart below. Which also include when $a > 1$.

Properties of the Graph of $f(x) = a^x$

when $a > 1$		when $0 < a < 1$	
Domain	$(-\infty, \infty)$	Domain	$(-\infty, \infty)$
Range	$(0, \infty)$	Range	$(0, \infty)$
x-intercept	none	x-intercept	none
y-intercept	$(0, 1)$	y-intercept	$(0, 1)$
Contains	$(1, a), \left(-1, \frac{1}{a}\right)$	Contains	$(1, a), \left(-1, \frac{1}{a}\right)$
Asymptote	x-axis, the line $y = 0$	Asymptote	x-axis, the line $y = 0$
Basic shape	increasing	Basic shape	decreasing

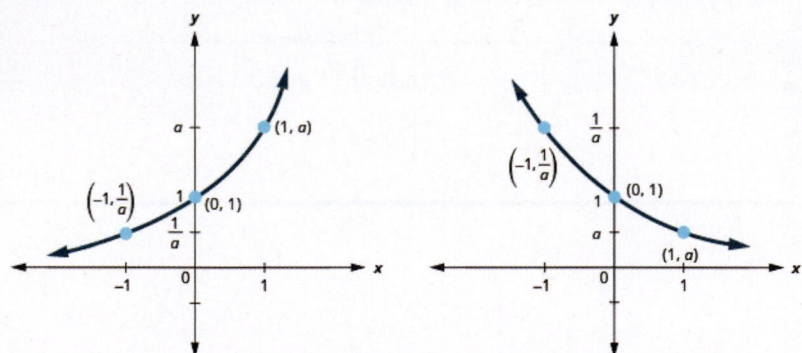

It is important for us to notice that both of these graphs are one-to-one, as they both pass the horizontal line test. This means the exponential function will have an inverse. We will look at this later.

When we graphed quadratic functions, we were able to graph using translation rather than just plotting points. Will that work in graphing exponential functions?

EXAMPLE 10.12

On the same coordinate system graph $f(x) = 2^x$ and $g(x) = 2^{x+1}$.

⊘ **Solution**

We will use point plotting to graph the functions.

x	$f(x) = 2^x$	$(x, f(x))$	$g(x) = 2^{x+1}$	$(x, g(x))$
–2	$2^{-2} = \frac{1}{2^2} = \frac{1}{4}$	$\left(-2, \frac{1}{4}\right)$	$2^{-2+1} = \frac{1}{2^1} = \frac{1}{2}$	$\left(-2, \frac{1}{2}\right)$
–1	$2^{-1} = \frac{1}{2^1} = \frac{1}{2}$	$\left(-1, \frac{1}{2}\right)$	$2^{-1+1} = 2^0 = 1$	$(-1, 1)$
0	$2^0 = 1$	$(0, 1)$	$2^{0+1} = 2^1 = 2$	$(0, 2)$
1	$2^1 = 2$	$(1, 2)$	$2^{1+1} = 2^2 = 4$	$(1, 4)$
2	$2^2 = 4$	$(2, 4)$	$2^{2+1} = 2^3 = 8$	$(2, 8)$
3	$2^3 = 8$	$(3, 8)$	$2^{3+1} = 2^4 = 16$	$(3, 16)$

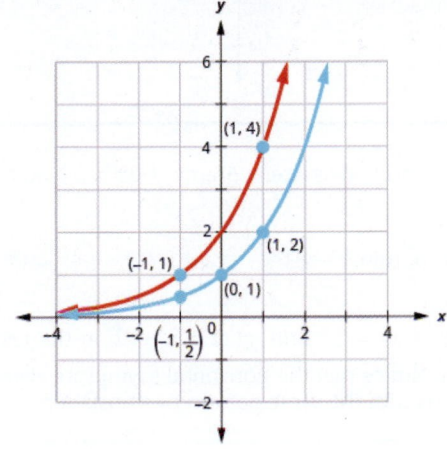

> **TRY IT : :** 10.23
On the same coordinate system, graph: $f(x) = 2^x$ and $g(x) = 2^{x-1}$.

> **TRY IT : :** 10.24
On the same coordinate system, graph: $f(x) = 3^x$ and $g(x) = 3^{x+1}$.

Looking at the graphs of the functions $f(x) = 2^x$ and $g(x) = 2^{x+1}$ in the last example, we see that adding one in the exponent caused a horizontal shift of one unit to the left. Recognizing this pattern allows us to graph other functions with the same pattern by translation.

Let's now consider another situation that might be graphed more easily by translation, once we recognize the pattern.

EXAMPLE 10.13

On the same coordinate system graph $f(x) = 3^x$ and $g(x) = 3^x - 2$.

⊘ **Solution**

We will use point plotting to graph the functions.

x	$f(x) = 3^x$	$(x, g(x))$	$g(x) = 3^x - 2$	$(x, g(x))$
–2	$3^{-2} = \frac{1}{9}$	$\left(-2, \frac{1}{9}\right)$	$3^{-2} - 2 = \frac{1}{9} - 2 = -\frac{17}{9}$	$\left(-2, -\frac{17}{9}\right)$
–1	$3^{-1} = \frac{1}{3}$	$\left(-1, \frac{1}{3}\right)$	$3^{-1} - 2 = \frac{1}{3} - 2 = -\frac{5}{3}$	$\left(-1, -\frac{5}{3}\right)$
0	$3^0 = 1$	$(0, 1)$	$3^0 - 2 = 1 - 2 = -1$	$(0, -1)$
1	$3^1 = 3$	$(1, 3)$	$3^1 - 2 = 3 - 2 = 1$	$(1, 1)$
2	$3^2 = 9$	$(2, 9)$	$3^2 - 2 = 9 - 2 = 7$	$(2, 8)$

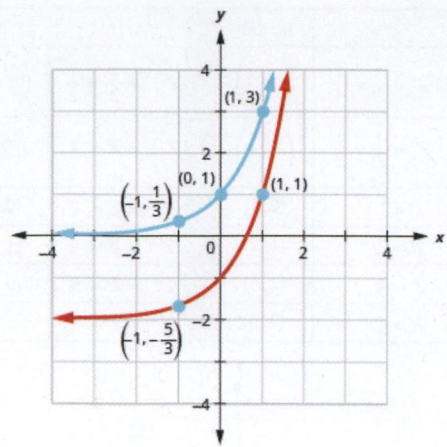

> **TRY IT : : 10.25**   On the same coordinate system, graph: $f(x) = 3^x$ and $g(x) = 3^x + 2$.

> **TRY IT : : 10.26**   On the same coordinate system, graph: $f(x) = 4^x$ and $g(x) = 4^x - 2$.

Looking at the graphs of the functions $f(x) = 3^x$ and $g(x) = 3^x - 2$ in the last example, we see that subtracting 2 caused a vertical shift of down two units. Notice that the horizontal asymptote also shifted down 2 units. Recognizing this pattern allows us to graph other functions with the same pattern by translation.

All of our exponential functions have had either an integer or a rational number as the base. We will now look at an exponential function with an irrational number as the base.

Before we can look at this exponential function, we need to define the irrational number, e. This number is used as a base in many applications in the sciences and business that are modeled by exponential functions. The number is defined as the value of $\left(1 + \frac{1}{n}\right)^n$ as n gets larger and larger. We say, as n approaches infinity, or increases without bound. The table shows the value of $\left(1 + \frac{1}{n}\right)^n$ for several values of n.

n	$\left(1 + \frac{1}{n}\right)^n$
1	2
2	2.25
5	2.48832
10	2.59374246
100	2.704813829...
1,000	2.716923932...
10,000	2.718145927...
100,000	2.718268237...
1,000,000	2.718280469...
1,000,000,000	2.718281827...

Table 10.7

$$e \approx 2.718281827$$

The number e is like the number π in that we use a symbol to represent it because its decimal representation never stops or repeats. The irrational number e is called the **natural base**.

Natural Base e

The number e is defined as the value of $\left(1 + \frac{1}{n}\right)^n$, as n increases without bound. We say, as n approaches infinity,

$$e \approx 2.718281827...$$

The exponential function whose base is e, $f(x) = e^x$ is called the **natural exponential function**.

Natural Exponential Function

The natural exponential function is an exponential function whose base is e

$$f(x) = e^x$$

The domain is $(-\infty, \infty)$ and the range is $(0, \infty)$.

Let's graph the function $f(x) = e^x$ on the same coordinate system as $g(x) = 2^x$ and $h(x) = 3^x$.

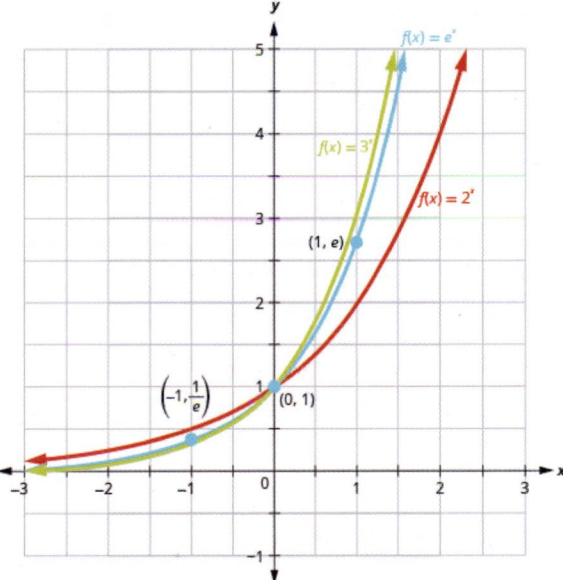

Notice that the graph of $f(x) = e^x$ is "between" the graphs of $g(x) = 2^x$ and $h(x) = 3^x$. Does this make sense as $2 < e < 3$?

Solve Exponential Equations

Equations that include an exponential expression a^x are called exponential equations. To solve them we use a property that says as long as $a > 0$ and $a \neq 1$, if $a^x = a^y$ then it is true that $x = y$. In other words, in an exponential equation, if the bases are equal then the exponents are equal.

One-to-One Property of Exponential Equations

For $a > 0$ and $a \neq 1$,

$$\text{If } a^x = a^y, \text{ then } x = y.$$

To use this property, we must be certain that both sides of the equation are written with the same base.

EXAMPLE 10.14   HOW TO SOLVE AN EXPONENTIAL EQUATION

Solve: $3^{2x-5} = 27$.

⊘ Solution

Step 1. Write both sides of the equation with the same base.	Since the left side has base 3, we write the right side with base 3. $27 = 3^3$	$3^{2x-5} = 27$ $3^{2x-5} = 3^3$
Step 2. Write a new equation by setting the exponents equal.	Since the bases are the same, the exponents must be equal.	$2x - 5 = 3$
Step 3. Solve the equation.	Add 5 to each side. Divide by 2.	$2x = 8$ $x = 4$
Step 4. Check the solution.	Substitute $x = 4$ into the original equation.	$3^{2x-5} = 27$ $3^{2 \cdot 4 - 5} \overset{?}{=} 27$ $3^3 \overset{?}{=} 27$ $27 = 27$ ✓

> **TRY IT : : 10.27**   Solve: $3^{3x-2} = 81$.

> **TRY IT : : 10.28**   Solve: $7^{x-3} = 7$.

The steps are summarized below.

HOW TO : : HOW TO SOLVE AN EXPONENTIAL EQUATION

Step 1.   Write both sides of the equation with the same base, if possible.

Step 2.   Write a new equation by setting the exponents equal.

Step 3.   Solve the equation.

Step 4.   Check the solution.

In the next example, we will use our properties on exponents.

EXAMPLE 10.15

Solve $\dfrac{e^{x^2}}{e^3} = e^{2x}$.

⊘ Solution

$$\frac{e^{x^2}}{e^3} = e^{2x}$$

Use the Property of Exponents: $\dfrac{a^m}{a^n} = a^{m-n}$.	$e^{x^2 - 3} = e^{2x}$
Write a new equation by setting the exponents equal.	$x^2 - 3 = 2x$
Solve the equation.	$x^2 - 2x - 3 = 0$

$$(x-3)(x+1) = 0$$

$$x = 3, \quad x = -1$$

Check the solutions.

$x = 3$	$x = -1$
$\dfrac{e^{x^2}}{e^3} \overset{?}{=} e^{2x}$	$\dfrac{e^{x^2}}{e^3} \overset{?}{=} e^{2x}$
$\dfrac{e^{3^2}}{e^3} \overset{?}{=} e^{2 \cdot 3}$	$\dfrac{e^{(-1)^2}}{e^3} \overset{?}{=} e^{2 \cdot (-1)}$
$\dfrac{e^9}{e^3} \overset{?}{=} e^6$	$\dfrac{e^1}{e^3} \overset{?}{=} e^{-2}$
$e^6 = e^6 \checkmark$	$e^{-2} = e^{-2} \checkmark$

> **TRY IT :: 10.29**
>
> Solve: $\dfrac{e^{x^2}}{e^x} = e^2$.

> **TRY IT :: 10.30**
>
> Solve: $\dfrac{e^{x^2}}{e^x} = e^6$.

Use Exponential Models in Applications

Exponential functions model many situations. If you own a bank account, you have experienced the use of an exponential function. There are two formulas that are used to determine the balance in the account when interest is earned. If a principal, P, is invested at an interest rate, r, for t years, the new balance, A, will depend on how often the interest is compounded. If the interest is compounded n times a year we use the formula $A = P\left(1 + \frac{r}{n}\right)^{nt}$. If the interest is compounded continuously, we use the formula $A = Pe^{rt}$. These are the formulas for **compound interest**.

Compound Interest

For a principal, P, invested at an interest rate, r, for t years, the new balance, A, is:

$$A = P\left(1 + \frac{r}{n}\right)^{nt} \qquad \text{when compounded } n \text{ times a year.}$$

$$A = Pe^{rt} \qquad \text{when compounded continuously.}$$

As you work with the Interest formulas, it is often helpful to identify the values of the variables first and then substitute them into the formula.

EXAMPLE 10.16

A total of $\$10,000$ was invested in a college fund for a new grandchild. If the interest rate is 5%, how much will be in the account in 18 years by each method of compounding?

ⓐ compound quarterly

ⓑ compound monthly

ⓒ compound continuously

✓ Solution

A	$=$	?
P	$=$	\$10,000
r	$=$	0.05
t	$=$	18 years

Identify the values of each variable in the formulas. Remember to express the percent as a decimal.

ⓐ

For quarterly compounding, $n = 4$. There are 4 quarters in a year.

$$A = P\left(1 + \tfrac{r}{n}\right)^{nt}$$

Substitute the values in the formula.

$$A = 10{,}000\left(1 + \tfrac{0.05}{4}\right)^{4 \cdot 18}$$

Compute the amount. Be careful to consider the order of operations as you enter the expression into your calculator.

$$A = \$24{,}459.20$$

ⓑ

For monthly compounding, $n = 12$. There are 12 months in a year.

$$A = P\left(1 + \tfrac{r}{n}\right)^{nt}$$

Substitute the values in the formula.

$$A = 10{,}000\left(1 + \tfrac{0.05}{12}\right)^{12 \cdot 18}$$

Compute the amount.

$$A = \$24{,}550.08$$

ⓒ

For compounding continuously,

$$A = Pe^{rt}$$

Substitute the values in the formula.

$$A = 10{,}000e^{0.05 \cdot 18}$$

Compute the amount.

$$A = \$24{,}596.03$$

> **TRY IT : : 10.31**
>
> Angela invested \$15,000 in a savings account. If the interest rate is 4%, how much will be in the account in 10 years by each method of compounding?
>
> ⓐ compound quarterly
>
> ⓑ compound monthly
>
> ⓒ compound continuously

> **TRY IT : : 10.32**
>
> Allan invested \$10,000 in a mutual fund. If the interest rate is 5%, how much will be in the account in 15 years by each method of compounding?
>
> ⓐ compound quarterly
>
> ⓑ compound monthly
>
> ⓒ compound continuously

Other topics that are modeled by exponential functions involve growth and decay. Both also use the formula $A = Pe^{rt}$ we used for the growth of money. For growth and decay, generally we use A_0, as the original amount instead of calling it P, the principal. We see that **exponential growth** has a positive rate of growth and **exponential decay** has a negative rate of growth.

Exponential Growth and Decay

For an original amount, A_0, that grows or decays at a rate, r, for a certain time, t, the final amount, A, is:

$$A = A_0 e^{rt}$$

Exponential growth is typically seen in the growth of populations of humans or animals or bacteria. Our next example looks at the growth of a virus.

EXAMPLE 10.17

Chris is a researcher at the Center for Disease Control and Prevention and he is trying to understand the behavior of a new and dangerous virus. He starts his experiment with 100 of the virus that grows at a rate of 25% per hour. He will check on the virus in 24 hours. How many viruses will he find?

⊘ **Solution**

Identify the values of each variable in the formulas.	$A = ?$
Be sure to put the percent in decimal form.	$A_0 = 100$
Be sure the units match—the rate is per hour and	$r = 0.25/\text{hour}$
the time is in hours.	$t = 24 \text{ hours}$
Substitute the values in the formula: $A = A_0 e^{rt}$.	$A = 100e^{0.25 \cdot 24}$
Compute the amount.	$A = 40{,}342.88$
Round to the nearest whole virus.	$A = 40{,}343$
	The researcher will find 40,343 viruses.

> **TRY IT :: 10.33**
>
> Another researcher at the Center for Disease Control and Prevention, Lisa, is studying the growth of a bacteria. She starts his experiment with 50 of the bacteria that grows at a rate of 15% per hour. He will check on the bacteria every 8 hours. How many bacteria will he find in 8 hours?

> **TRY IT :: 10.34**
>
> Maria, a biologist is observing the growth pattern of a virus. She starts with 100 of the virus that grows at a rate of 10% per hour. She will check on the virus in 24 hours. How many viruses will she find?

▶ **MEDIA ::**

Access these online resources for additional instruction and practice with evaluating and graphing exponential functions.

- **Graphing Exponential Functions (https://openstax.org/l/37Graphexponent)**
- **Solving Exponential Equations (https://openstax.org/l/37Solvelikebase)**
- **Applications of Exponential Functions (https://openstax.org/l/37Exponentapp)**
- **Continuously Compound Interest (https://openstax.org/l/37Compoundint)**
- **Radioactive Decay and Exponential Growth (https://openstax.org/l/37Exponentdecay)**

10.2 EXERCISES

Practice Makes Perfect

Graph Exponential Functions

In the following exercises, graph each exponential function.

65. $f(x) = 2^x$

66. $g(x) = 3^x$

67. $f(x) = 6^x$

68. $g(x) = 7^x$

69. $f(x) = (1.5)^x$

70. $g(x) = (2.5)^x$

71. $f(x) = \left(\frac{1}{2}\right)^x$

72. $g(x) = \left(\frac{1}{3}\right)^x$

73. $f(x) = \left(\frac{1}{6}\right)^x$

74. $g(x) = \left(\frac{1}{7}\right)^x$

75. $f(x) = (0.4)^x$

76. $g(x) = (0.6)^x$

In the following exercises, graph each function in the same coordinate system.

77. $f(x) = 4^x,\ g(x) = 4^{x-1}$

78. $f(x) = 3^x,\ g(x) = 3^{x-1}$

79. $f(x) = 2^x,\ g(x) = 2^{x-2}$

80. $f(x) = 2^x,\ g(x) = 2^{x+2}$

81. $f(x) = 3^x,\ g(x) = 3^x + 2$

82. $f(x) = 4^x,\ g(x) = 4^x + 2$

83. $f(x) = 2^x,\ g(x) = 2^x + 1$

84. $f(x) = 2^x,\ g(x) = 2^x - 1$

In the following exercises, graph each exponential function.

85. $f(x) = 3^{x+2}$

86. $f(x) = 3^{x-2}$

87. $f(x) = 2^x + 3$

88. $f(x) = 2^x - 3$

89. $f(x) = \left(\frac{1}{2}\right)^{x-4}$

90. $f(x) = \left(\frac{1}{2}\right)^x - 3$

91. $f(x) = e^x + 1$

92. $f(x) = e^{x-2}$

93. $f(x) = -2^x$

94. $f(x) = 3^x$

Solve Exponential Equations

In the following exercises, solve each equation.

95. $2^{3x-8} = 16$

96. $2^{2x-3} = 32$

97. $3^{x+3} = 9$

98. $3^{x^2} = 81$

99. $4^{x^2} = 4$

100. $4^x = 32$

101. $4^{x+2} = 64$

102. $4^{x+3} = 16$

103. $2^{x^2+2x} = \frac{1}{2}$

104. $3^{x^2-2x} = \frac{1}{3}$

105. $e^{3x} \cdot e^4 = e^{10}$

106. $e^{2x} \cdot e^3 = e^9$

107. $\dfrac{e^{x^2}}{e^2} = e^x$

108. $\dfrac{e^{x^2}}{e^3} = e^{2x}$

In the following exercises, match the graphs to one of the following functions: ⓐ 2^x ⓑ 2^{x+1} ⓒ 2^{x-1} ⓓ $2^x + 2$ ⓔ $2^x - 2$ ⓕ 3^x

109.

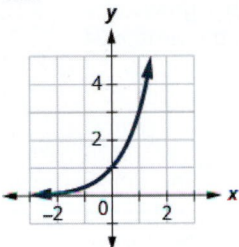

110.

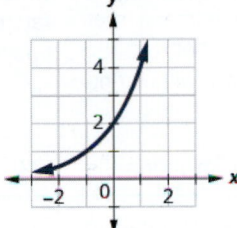

111.

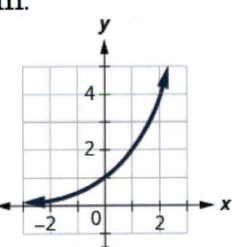

112.

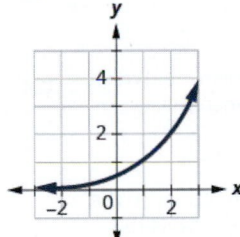

113.

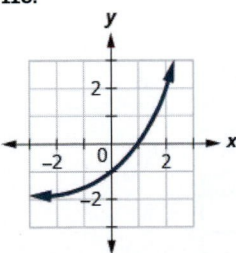

114.

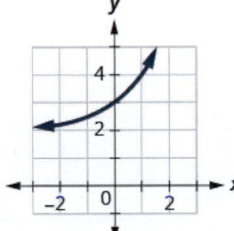

Use exponential models in applications

In the following exercises, use an exponential model to solve.

115. Edgar accumulated $5,000 in credit card debt. If the interest rate is 20% per year, and he does not make any payments for 2 years, how much will he owe on this debt in 2 years by each method of compounding?

ⓐ compound quarterly

ⓑ compound monthly

ⓒ compound continuously

116. Cynthia invested $12,000 in a savings account. If the interest rate is 6%, how much will be in the account in 10 years by each method of compounding?

ⓐ compound quarterly

ⓑ compound monthly

ⓒ compound continuously

117. Rochelle deposits $5,000 in an IRA. What will be the value of her investment in 25 years if the investment is earning 8% per year and is compounded continuously?

118. Nazerhy deposits $8,000 in a certificate of deposit. The annual interest rate is 6% and the interest will be compounded quarterly. How much will the certificate be worth in 10 years?

119. A researcher at the Center for Disease Control and Prevention is studying the growth of a bacteria. He starts his experiment with 100 of the bacteria that grows at a rate of 6% per hour. He will check on the bacteria every 8 hours. How many bacteria will he find in 8 hours?

120. A biologist is observing the growth pattern of a virus. She starts with 50 of the virus that grows at a rate of 20% per hour. She will check on the virus in 24 hours. How many viruses will she find?

121. In the last ten years the population of Indonesia has grown at a rate of 1.12% per year to 258,316,051. If this rate continues, what will be the population in 10 more years?

122. In the last ten years the population of Brazil has grown at a rate of 0.9% per year to 205,823,665. If this rate continues, what will be the population in 10 more years?

Writing Exercises

123. Explain how you can distinguish between exponential functions and polynomial functions.

124. Compare and contrast the graphs of $y = x^2$ and $y = 2^x$.

125. What happens to an exponential function as the values of x decreases? Will the graph ever cross the y-axis? Explain.

Self Check

ⓐ *After completing the exercises, use this checklist to evaluate your mastery of the objectives of this section.*

I can...	Confidently	With some help	No-I don't get it!
graph exponential eunctions.			
solve exponential equations.			
use exponential models in applications.			

ⓑ *After reviewing this checklist, what will you do to become confident for all objectives?*

10.3 Evaluate and Graph Logarithmic Functions

Learning Objectives

By the end of this section, you will be able to:

> Convert between exponential and logarithmic form
> Evaluate logarithmic functions
> Graph Logarithmic functions
> Solve logarithmic equations
> Use logarithmic models in applications

Be Prepared!

Before you get started, take this readiness quiz.

1. Solve: $x^2 = 81$.
   If you missed this problem, review **Example 6.46**.

2. Evaluate: 3^{-2}.
   If you missed this problem, review **Example 5.15**.

3. Solve: $2^4 = 3x - 5$.
   If you missed this problem, review **Example 2.2**.

We have spent some time finding the inverse of many functions. It works well to 'undo' an operation with another operation. Subtracting 'undoes' addition, multiplication 'undoes' division, taking the square root 'undoes' squaring.

As we studied the exponential function, we saw that it is one-to-one as its graphs pass the horizontal line test. This means an exponential function does have an inverse. If we try our algebraic method for finding an inverse, we run into a problem.

$$f(x) = a^x$$

Rewrite with $y = f(x)$. $y = a^x$

Interchange the variables x and y. $x = a^y$

Solve for y.   Oops! We have no way to solve for y!

To deal with this we define the logarithm function with base a to be the inverse of the exponential function $f(x) = a^x$.

We use the notation $f^{-1}(x) = \log_a x$ and say the inverse function of the exponential function is the logarithmic function.

Logarithmic Function

The function $f(x) = \log_a x$ is the **logarithmic function** with base a, where $a > 0$, $x > 0$, and $a \neq 1$.

$$y = \log_a x \text{ is equivalent to } x = a^y$$

Convert Between Exponential and Logarithmic Form

Since the equations $y = \log_a x$ and $x = a^y$ are equivalent, we can go back and forth between them. This will often be the method to solve some exponential and logarithmic equations. To help with converting back and forth let's take a close look at the equations. See **Figure 10.3**. Notice the positions of the exponent and base.

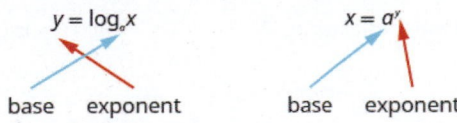

Figure 10.3

If we realize the logarithm is the exponent it makes the conversion easier. You may want to repeat, "base to the exponent give us the number."

EXAMPLE 10.18

Convert to logarithmic form: ⓐ $2^3 = 8$, ⓑ $5^{\frac{1}{2}} = \sqrt{5}$, and ⓒ $\left(\frac{1}{2}\right)^x = \frac{1}{16}$.

✓ Solution

Identify the **base** and the exponent.

(a)	(b)	(c)
$2^3 = 8$	$5^{\frac{1}{2}} = \sqrt{5}$	$\left(\frac{1}{2}\right)^4 = \frac{1}{16}$
$y = \log_a x$	$y = \log_a x$	$y = \log_a x$
$3 = \log_2 8$	$\frac{1}{2} = \log_5 \sqrt{5}$	$4 = \log_{\frac{1}{2}} \frac{1}{16}$
If $2^3 = 8$, then $3 = \log_2 8$.	If $5^{\frac{1}{2}} = \sqrt{5}$, then $\frac{1}{2} = \log_5 \sqrt{5}$.	If $\left(\frac{1}{2}\right)^4 = \frac{1}{16}$, then $4 = \log_{\frac{1}{2}} \frac{1}{16}$.

> **TRY IT :: 10.35**
> Convert to logarithmic form: ⓐ $3^2 = 9$ ⓑ $7^{\frac{1}{2}} = \sqrt{7}$ ⓒ $\left(\frac{1}{3}\right)^x = \frac{1}{27}$

> **TRY IT :: 10.36**
> Convert to logarithmic form: ⓐ $4^3 = 64$ ⓑ $4^{\frac{1}{3}} = \sqrt[3]{4}$ ⓒ $\left(\frac{1}{2}\right)^x = \frac{1}{32}$

In the next example we do the reverse—convert logarithmic form to exponential form.

EXAMPLE 10.19

Convert to exponential form: ⓐ $2 = \log_8 64$, ⓑ $0 = \log_4 1$, and ⓒ $-3 = \log_{10} \frac{1}{1000}$.

✓ Solution

Identify the **base** and the exponent.

(a)	(b)	(c)
$2 = \log_8 64$	$0 = \log_4 1$	$-3 = \log_{10} \frac{1}{1000}$
$x = a^y$	$x = a^y$	$x = a^y$
$64 = 8^2$	$1 = 4^0$	$\frac{1}{1000} = 10^{-3}$
If $2 = \log_8 64$, then $64 = 8^2$.	If $0 = \log_4 1$, then $1 = 4^0$.	If $-3 = \log_{10} \frac{1}{1000}$, then $\frac{1}{1000} = 10^{-3}$.

> **TRY IT :: 10.37**
> Convert to exponential form: ⓐ $3 = \log_4 64$ ⓑ $0 = \log_x 1$ ⓒ $-2 = \log_{10} \frac{1}{100}$

> **TRY IT :: 10.38**
> Convert to exponential form: ⓐ $3 = \log_3 27$ ⓑ $0 = \log_x 1$ ⓒ $-1 = \log_{10} \frac{1}{10}$

Evaluate Logarithmic Functions

We can solve and evaluate logarithmic equations by using the technique of converting the equation to its equivalent exponential equation.

EXAMPLE 10.20

Find the value of x: ⓐ $\log_x 36 = 2$, ⓑ $\log_4 x = 3$, and ⓒ $\log_{\frac{1}{2}} \frac{1}{8} = x$.

✓ **Solution**

ⓐ

$$\log_x 36 = 2$$

Convert to exponential form. $\qquad x^2 = 36$

Solve the quadratic. $\qquad x = 6, \qquad \cancel{x = -6}$

The base of a logarithmic function must be

positive, so we eliminate $x = -6$. $\qquad\qquad x = 6 \qquad$ Therefore, $\log_6 36 = 2$.

ⓑ

$$\log_4 x = 3$$

Convert to exponential form. $\qquad 4^3 = x$

Simplify. $\qquad\qquad x = 64 \qquad$ Therefore, $\log_4 64 = 3$.

ⓒ

$$\log_{\frac{1}{2}} \frac{1}{8} = x$$

Convert to exponential form. $\qquad \left(\frac{1}{2}\right)^x = \frac{1}{8}$

Rewrite $\frac{1}{8}$ as $\left(\frac{1}{2}\right)^3$. $\qquad \left(\frac{1}{2}\right)^x = \left(\frac{1}{2}\right)^3$

With the same base, the exponents must be equal. $\qquad x = 3 \qquad$ Therefore, $\log_{\frac{1}{2}} \frac{1}{8} = 3$

> **TRY IT :: 10.39**   Find the value of x: ⓐ $\log_x 64 = 2$ ⓑ $\log_5 x = 3$ ⓒ $\log_{\frac{1}{2}} \frac{1}{4} = x$

> **TRY IT :: 10.40**   Find the value of x: ⓐ $\log_x 81 = 2$ ⓑ $\log_3 x = 5$ ⓒ $\log_{\frac{1}{3}} \frac{1}{27} = x$

When see an expression such as $\log_3 27$, we can find its exact value two ways. By inspection we realize it means "3 to what power will be 27"? Since $3^3 = 27$, we know $\log_3 27 = 3$. An alternate way is to set the expression equal to x and then convert it into an exponential equation.

EXAMPLE 10.21

Find the exact value of each logarithm without using a calculator:

ⓐ $\log_5 25$,

ⓑ $\log_9 3$, and ⓒ $\log_2 \frac{1}{16}$.

✓ **Solution**

ⓐ

	$\log_5 25$
5 to what power will be 25 ?	$\log_5 25 \;=\; 2$
Or	
Set the expression equal to x.	$\log_5 25 \;=\; x$
Change to exponential form.	$5^x \;=\; 25$
Rewrite 25 as 5^2.	$5^x \;=\; 5^2$
With the same base the exponents must be equal.	$x \;=\; 2$ Therefore, $\log_5 25 = 2$.

ⓑ

	$\log_9 3$
Set the expression equal to x.	$\log_9 3 \;=\; x$
Change to exponential form.	$9^x \;=\; 3$
Rewrite 9 as 3^2.	$\left(3^2\right)^x \;=\; 3^1$
Simplify the exponents.	$3^{2x} \;=\; 3^1$
With the same base the exponents must be equal.	$2x \;=\; 1$
Solve the equation.	$x \;=\; \frac{1}{2}$ Therefore, $\log_9 3 = \frac{1}{2}$.

ⓒ

	$\log_2 \frac{1}{16}$
Set the expression equal to x.	$\log_2 \frac{1}{16} \;=\; x$
Change to exponential form.	$2^x \;=\; \frac{1}{16}$
Rewrite 16 as 2^4.	$2^x \;=\; \frac{1}{2^4}$
	$2^x \;=\; 2^{-4}$
With the same base the exponents must be equal.	$x \;=\; -4$ Therefore, $\log_2 \frac{1}{16} = -4$.

> **TRY IT : : 10.41**   Find the exact value of each logarithm without using a calculator:

  ⓐ $\log_{12} 144$

  ⓑ $\log_4 2$

  ⓒ $\log_2 \frac{1}{32}$

> **TRY IT : : 10.42**   Find the exact value of each logarithm without using a calculator:

  ⓐ $\log_9 81$

  ⓑ $\log_8 2$

  ⓒ $\log_3 \frac{1}{9}$

Graph Logarithmic Functions

To graph a logarithmic function $y = \log_a x$,   it is easiest to convert the equation to its exponential form, $x = a^y$.

Generally, when we look for ordered pairs for the graph of a function, we usually choose an x-value and then determine its corresponding y-value. In this case you may find it easier to choose y-values and then determine its corresponding x-value.

EXAMPLE 10.22

Graph $y = \log_2 x$.

⊘ Solution

To graph the function, we will first rewrite the logarithmic equation, $y = \log_2 x$, in exponential form, $2^y = x$.

We will use point plotting to graph the function. It will be easier to start with values of y and then get x.

y	$2^y = x$	(x, y)
-2	$2^{-2} = \frac{1}{2^2} = \frac{1}{4}$	$\left(\frac{1}{4}, 2\right)$
-1	$2^{-1} = \frac{1}{2^1} = \frac{1}{2}$	$\left(\frac{1}{2}, -1\right)$
0	$2^0 = 1$	$(1, 0)$
1	$2^1 = 2$	$(2, 1)$
2	$2^2 = 4$	$(4, 2)$
3	$2^3 = 8$	$(8, 3)$

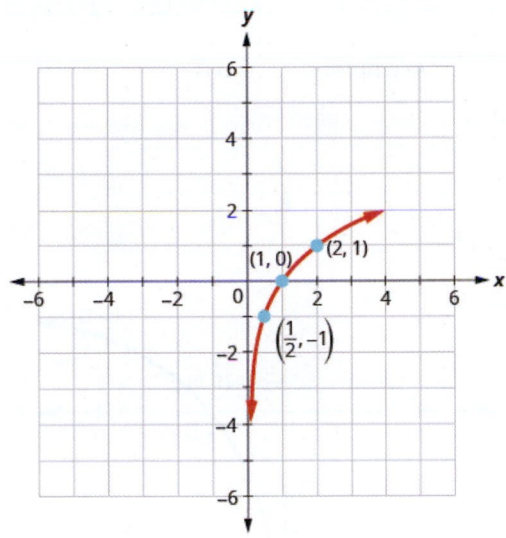

> **TRY IT :: 10.43** Graph: $y = \log_3 x$.

> **TRY IT :: 10.44** Graph: $y = \log_5 x$.

The graphs of $y = \log_2 x$, $y = \log_3 x$, and $y = \log_5 x$ are the shape we expect from a logarithmic function where $a > 1$.

We notice that for each function the graph contains the point $(1, 0)$. This make sense because $0 = \log_a 1$ means $a^0 = 1$ which is true for any a.

The graph of each function, also contains the point $(a, 1)$. This makes sense as $1 = \log_a a$ means $a^1 = a$. which is true for any a.

Notice too, the graph of each function $y = \log_a x$ also contains the point $\left(\frac{1}{a}, -1\right)$. This makes sense as $-1 = \log_a \frac{1}{a}$

means $a^{-1} = \frac{1}{a}$, which is true for any a.

Look at each graph again. Now we will see that many characteristics of the logarithm function are simply 'mirror images' of the characteristics of the corresponding exponential function.

What is the domain of the function? The graph never hits the y-axis. The domain is all positive numbers. We write the domain in interval notation as $(0, \infty)$.

What is the range for each function? From the graphs we can see that the range is the set of all real numbers. There is no restriction on the range. We write the range in interval notation as $(-\infty, \infty)$.

When the graph approaches the y-axis so very closely but will never cross it, we call the line $x = 0$, the y-axis, a vertical asymptote.

Properties of the Graph of $y = \log_a x$ when $a > 1$

Domain	$(0, \infty)$
Range	$(-\infty, \infty)$
x-intercept	$(1, 0)$
y-intercept	None
Contains	$(a, 1)$, $\left(\frac{1}{a}, -1\right)$
Asymptote	y-axis

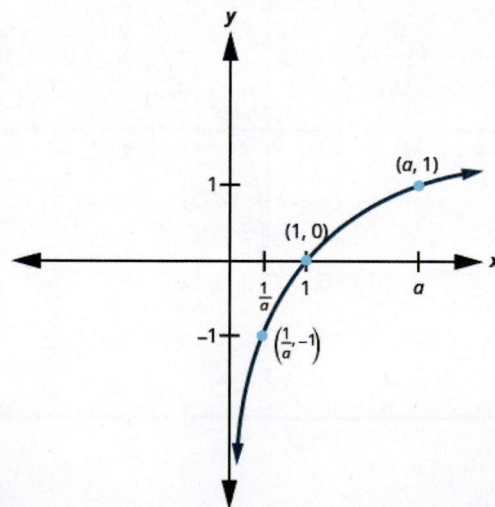

Our next example looks at the graph of $y = \log_a x$ when $0 < a < 1$.

EXAMPLE 10.23

Graph $y = \log_{\frac{1}{3}} x$.

✓ Solution

To graph the function, we will first rewrite the logarithmic equation, $y = \log_{\frac{1}{3}} x$, in exponential form, $\left(\frac{1}{3}\right)^y = x$.

We will use point plotting to graph the function. It will be easier to start with values of y and then get x.

y	$\left(\frac{1}{3}\right)^y = x$	(x, y)
-2	$\left(\frac{1}{3}\right)^{-2} = 3^2 = 9$	$(9, -2)$
-1	$\left(\frac{1}{3}\right)^{-1} = 3^1 = 3$	$(3, -1)$
0	$\left(\frac{1}{3}\right)^{0} = 1$	$(1, 0)$
1	$\left(\frac{1}{3}\right)^{1} = \frac{1}{3}$	$\left(\frac{1}{3}, 1\right)$
2	$\left(\frac{1}{3}\right)^{2} = \frac{1}{9}$	$\left(\frac{1}{9}, 2\right)$
3	$\left(\frac{1}{3}\right)^{3} = \frac{1}{27}$	$\left(\frac{1}{27}, 3\right)$

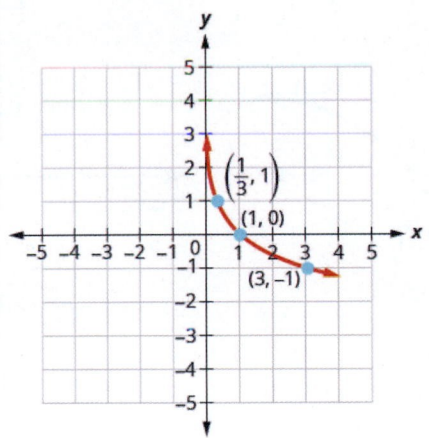

> **TRY IT : : 10.45**   Graph: $y = \log_{\frac{1}{2}} x$.

> **TRY IT : : 10.46**   Graph: $y = \log_{\frac{1}{4}} x$.

Now, let's look at the graphs $y = \log_{\frac{1}{2}} x$, $y = \log_{\frac{1}{3}} x$ and $y = \log_{\frac{1}{4}} x$, so we can identify some of the properties of logarithmic functions where $0 < a < 1$.

The graphs of all have the same basic shape. While this is the shape we expect from a logarithmic function where $0 < a < 1$.

We notice, that for each function again, the graph contains the points, $(1, 0)$, $(a, 1)$, $\left(\frac{1}{a}, -1\right)$. This make sense for the same reasons we argued above.

We notice the domain and range are also the same—the domain is $(0, \infty)$ and the range is $(-\infty, \infty)$. The y-axis is again the vertical asymptote.

We will summarize these properties in the chart below. Which also include when $a > 1$.

Properties of the Graph of $y = \log_a x$

when $a > 1$		when $0 < a < 1$	
Domain	$(0, \infty)$	Domain	$(0, \infty)$
Range	$(-\infty, \infty)$	Range	$(-\infty, \infty)$
x-intercept	$(1, 0)$	x-intercept	$(1, 0)$
y-intercept	none	y-intercept	None
Contains	$(a, 1)$, $\left(\frac{1}{a}, -1\right)$	Contains	$(a, 1)$, $\left(\frac{1}{a}, -1\right)$
Asymptote	y-axis	Asymptote	y-axis
Basic shape	increasing	Basic shape	Decreasing

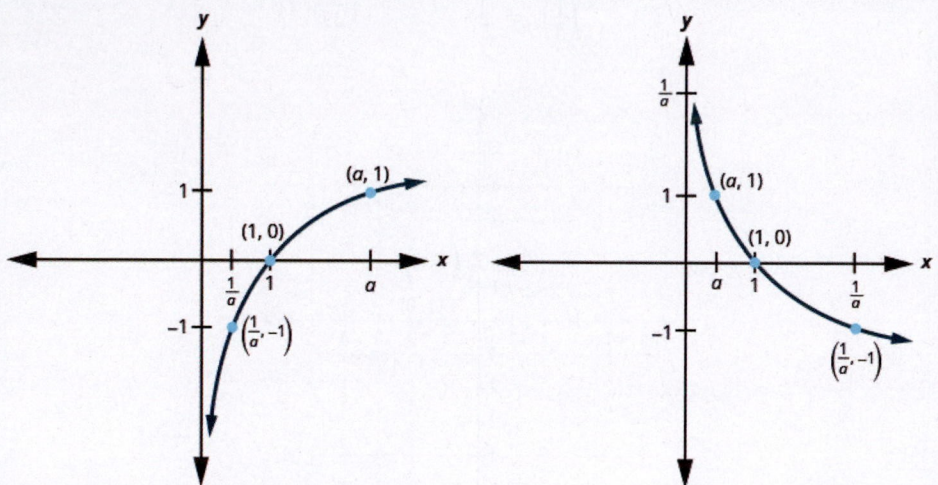

We talked earlier about how the logarithmic function $f^{-1}(x) = \log_a x$ is the inverse of the exponential function $f(x) = a^x$. The graphs in **Figure 10.4** show both the exponential (blue) and logarithmic (red) functions on the same graph for both $a > 1$ and $0 < a < 1$.

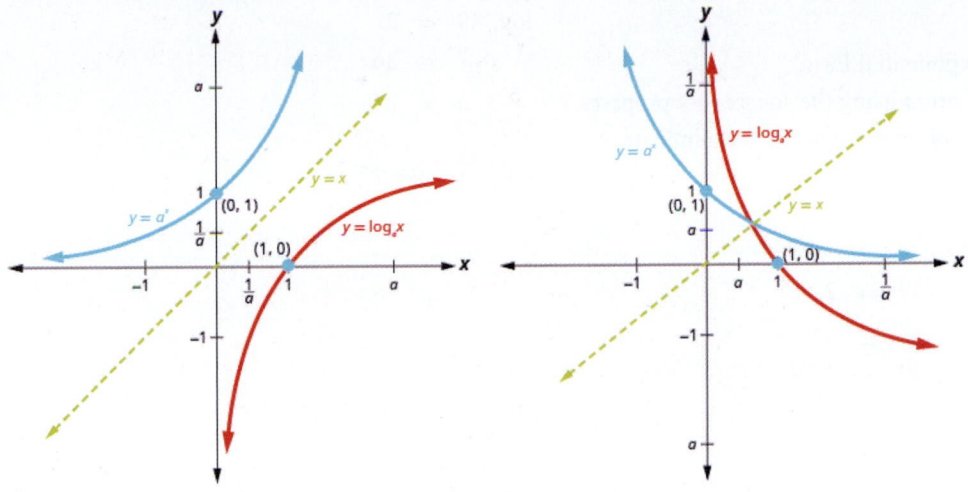

Figure 10.4

Notice how the graphs are reflections of each other through the line $y = x$. We know this is true of inverse functions.

Keeping a visual in your mind of these graphs will help you remember the domain and range of each function. Notice the x-axis is the horizontal asymptote for the exponential functions and the y-axis is the vertical asymptote for the logarithmic functions.

Solve Logarithmic Equations

When we talked about exponential functions, we introduced the number e. Just as e was a base for an exponential function, it can be used a base for logarithmic functions too. The logarithmic function with base e is called the **natural logarithmic function**. The function $f(x) = \log_e x$ is generally written $f(x) = \ln x$ and we read it as "el en of x."

Natural Logarithmic Function

The function $f(x) = \ln x$ is the **natural logarithmic function** with base e, where $x > 0$.

$$y = \ln x \text{ is equivalent to } x = e^y$$

When the base of the logarithm function is 10, we call it the **common logarithmic function** and the base is not shown. If the base a of a logarithm is not shown, we assume it is 10.

Common Logarithmic Function

The function $f(x) = \log x$ is the **common logarithmic function** with base 10, where $x > 0$.

$$y = \log x \text{ is equivalent to } x = 10^y$$

It will be important for you to use your calculator to evaluate both common and natural logarithms.

Look for the log and ln keys on your calculator.

To solve logarithmic equations, one strategy is to change the equation to exponential form and then solve the exponential equation as we did before. As we solve logarithmic equations, $y = \log_a x$, we need to remember that for the base a, $a > 0$ and $a \neq 1$. Also, the domain is $x > 0$. Just as with radical equations, we must check our solutions to eliminate any extraneous solutions.

EXAMPLE 10.24

Solve: ⓐ $\log_a 49 = 2$ and ⓑ $\ln x = 3$.

✓ **Solution**

ⓐ

$$\log_a 49 = 2$$

Rewrite in exponential form. $$a^2 = 49$$
Solve the equation using the square root property. $$a = \pm 7$$
The base cannot be negative, so we eliminate
$a = -7$. $a = 7, \quad \cancel{a = -7}$

Check.
$a = 7 \qquad \log_a 49 = 2$

$$\log_7 49 \overset{?}{=} 2$$

$$7^2 \overset{?}{=} 49$$

$$49 = 49 \checkmark$$

ⓑ

$$\ln x = 3$$

Rewrite in exponential form. $$e^3 = x$$
Check.
$x = e^3 \qquad \ln x = 3$

$$\ln e^3 \overset{?}{=} 3$$

$$e^3 = e^3 \checkmark$$

> **TRY IT : :** 10.47   Solve: ⓐ $\log_a 121 = 2$ ⓑ $\ln x = 7$

> **TRY IT : :** 10.48   Solve: ⓐ $\log_a 64 = 3$ ⓑ $\ln x = 9$

EXAMPLE 10.25

Solve: ⓐ $\log_2 (3x - 5) = 4$ and ⓑ $\ln e^{2x} = 4$.

✓ **Solution**

ⓐ

$$\log_2 (3x - 5) = 4$$

Rewrite in exponential form. $$2^4 = 3x - 5$$
Simplify. $$16 = 3x - 5$$
Solve the equation. $$21 = 3x$$
                                                                    $$7 = x$$

Check.
$x = 7 \qquad \log_2 (3x - 5) = 4$

$$\log_2 (3 \cdot 7 - 5) \overset{?}{=} 4$$

$$\log_2 (16) \overset{?}{=} 4$$

$$2^4 \overset{?}{=} 16$$

$$16 = 16 \checkmark$$

ⓑ

$$\ln e^{2x} = 4$$

Rewrite in exponential form. $\qquad e^4 = e^{2x}$

Since the bases are the same the exponents are equal. $\qquad 4 = 2x$

Solve the equation. $\qquad 2 = x$

Check.

$x = 2 \qquad \ln e^{2x} = 4$

$\qquad\qquad \ln e^{2 \cdot 2} \overset{?}{=} 4$

$\qquad\qquad\quad \ln e^4 \overset{?}{=} 4$

$\qquad\qquad\qquad e^4 = e^4 \checkmark$

> **TRY IT :: 10.49** Solve: ⓐ $\log_2(5x - 1) = 6$ ⓑ $\ln e^{3x} = 6$

> **TRY IT :: 10.50** Solve: ⓐ $\log_3(4x + 3) = 3$ ⓑ $\ln e^{4x} = 4$

Use Logarithmic Models in Applications

There are many applications that are modeled by logarithmic equations. We will first look at the logarithmic equation that gives the decibel (dB) level of sound. Decibels range from 0, which is barely audible to 160, which can rupture an eardrum. The 10^{-12} in the formula represents the intensity of sound that is barely audible.

Decibel Level of Sound

The loudness level, D, measured in decibels, of a sound of intensity, I, measured in watts per square inch is

$$D = 10\log\left(\frac{I}{10^{-12}}\right)$$

EXAMPLE 10.26

Extended exposure to noise that measures 85 dB can cause permanent damage to the inner ear which will result in hearing loss. What is the decibel level of music coming through ear phones with intensity 10^{-2} watts per square inch?

✓ **Solution**

$$D = 10\log\left(\frac{I}{10^{-12}}\right)$$

Substitute in the intensity level, I.	$D = 10\log\left(\frac{10^{-2}}{10^{-12}}\right)$
Simplify.	$D = 10\log(10^{10})$
Since $\log 10^{10} = 10$.	$D = 10 \cdot 10$
Multiply.	$D = 100$
	The decibel level of music coming through earphones is 100 dB.

> **TRY IT :: 10.51** What is the decibel level of one of the new quiet dishwashers with intensity 10^{-7} watts per square inch?

> **TRY IT :: 10.52** What is the decibel level heavy city traffic with intensity 10^{-3} watts per square inch?

The magnitude R of an earthquake is measured by a logarithmic scale called the Richter scale. The model is $R = \log I$, where I is the intensity of the shock wave. This model provides a way to measure earthquake intensity.

Earthquake Intensity

The magnitude R of an earthquake is measured by $R = \log I$,   where I is the intensity of its shock wave.

EXAMPLE 10.27

In 1906, San Francisco experienced an intense earthquake with a magnitude of 7.8 on the Richter scale. Over 80% of the city was destroyed by the resulting fires. In 2014, Los Angeles experienced a moderate earthquake that measured 5.1 on the Richter scale and caused $108 million dollars of damage. Compare the intensities of the two earthquakes.

⊘ **Solution**

To compare the intensities, we first need to convert the magnitudes to intensities using the log formula. Then we will set up a ratio to compare the intensities.

Convert the magnitudes to intensities.	$R = \log I$
1906 earthquake	$7.8 = \log I$
Convert to exponential form.	$I = 10^{7.8}$
2014 earthquake	$5.1 = \log I$
Convert to exponential form.	$I = 10^{5.1}$
Form a ratio of the intensities.	$\dfrac{\text{Intensity for 1906}}{\text{Intensity for 2014}}$
Substitute in the values.	$\dfrac{10^{7.8}}{10^{5.1}}$
Divide by subtracting the exponents.	$10^{2.7}$
Evaluate.	501

The intensity of the 1906 earthquake was about 501 times the intensity of the 2014 earthquake.

> **TRY IT ::** 10.53

In 1906, San Francisco experienced an intense earthquake with a magnitude of 7.8 on the Richter scale. In 1989, the Loma Prieta earthquake also affected the San Francisco area, and measured 6.9 on the Richter scale. Compare the intensities of the two earthquakes.

> **TRY IT ::** 10.54

In 2014, Chile experienced an intense earthquake with a magnitude of 8.2 on the Richter scale. In 2014, Los Angeles also experienced an earthquake which measured 5.1 on the Richter scale. Compare the intensities of the two earthquakes.

▶ **MEDIA ::**

Access these online resources for additional instruction and practice with evaluating and graphing logarithmic functions.

- Re-writing logarithmic equations in exponential form (https://openstax.org/l/37logasexponent)
- Simplifying Logarithmic Expressions (https://openstax.org/l/37Simplifylog)
- Graphing logarithmic functions (https://openstax.org/l/37Graphlog)
- Using logarithms to calculate decibel levels (https://openstax.org/l/37Finddecibel)

10.3 EXERCISES

Practice Makes Perfect

Convert Between Exponential and Logarithmic Form

In the following exercises, convert from exponential to logarithmic form.

126. $4^2 = 16$

127. $2^5 = 32$

128. $3^3 = 27$

129. $5^3 = 125$

130. $10^3 = 1000$

131. $10^{-2} = \frac{1}{100}$

132. $x^{\frac{1}{2}} = \sqrt{3}$

133. $x^{\frac{1}{3}} = \sqrt[3]{6}$

134. $32^x = \sqrt[4]{32}$

135. $17^x = \sqrt[5]{17}$

136. $\left(\frac{1}{4}\right)^2 = \frac{1}{16}$

137. $\left(\frac{1}{3}\right)^4 = \frac{1}{81}$

138. $3^{-2} = \frac{1}{9}$

139. $4^{-3} = \frac{1}{64}$

140. $e^x = 6$

141. $e^3 = x$

In the following exercises, convert each logarithmic equation to exponential form.

142. $3 = \log_4 64$

143. $6 = \log_2 64$

144. $4 = \log_x 81$

145. $5 = \log_x 32$

146. $0 = \log_{12} 1$

147. $0 = \log_7 1$

148. $1 = \log_3 3$

149. $1 = \log_9 9$

150. $-4 = \log_{10} \frac{1}{10,000}$

151. $3 = \log_{10} 1,000$

152. $5 = \log_e x$

153. $x = \log_e 43$

Evaluate Logarithmic Functions

In the following exercises, find the value of x in each logarithmic equation.

154. $\log_x 49 = 2$

155. $\log_x 121 = 2$

156. $\log_x 27 = 3$

157. $\log_x 64 = 3$

158. $\log_3 x = 4$

159. $\log_5 x = 3$

160. $\log_2 x = -6$

161. $\log_3 x = -5$

162. $\log_{\frac{1}{4}} \frac{1}{16} = x$

163. $\log_{\frac{1}{3}} \frac{1}{9} = x$

164. $\log_{\frac{1}{4}} 64 = x$

165. $\log_{\frac{1}{9}} 81 = x$

In the following exercises, find the exact value of each logarithm without using a calculator.

166. $\log_7 49$

167. $\log_6 36$

168. $\log_4 1$

169. $\log_5 1$

170. $\log_{16} 4$

171. $\log_{27} 3$

172. $\log_{\frac{1}{2}} 2$

173. $\log_{\frac{1}{2}} 4$

174. $\log_2 \frac{1}{16}$

175. $\log_3 \frac{1}{27}$

176. $\log_4 \frac{1}{16}$

177. $\log_9 \frac{1}{81}$

Graph Logarithmic Functions

In the following exercises, graph each logarithmic function.

178. $y = \log_2 x$

179. $y = \log_4 x$

180. $y = \log_6 x$

181. $y = \log_7 x$

182. $y = \log_{1.5} x$

183. $y = \log_{2.5} x$

184. $y = \log_{\frac{1}{3}} x$

185. $y = \log_{\frac{1}{5}} x$

186. $y = \log_{0.4} x$

187. $y = \log_{0.6} x$

Solve Logarithmic Equations

In the following exercises, solve each logarithmic equation.

188. $\log_a 16 = 2$

189. $\log_a 81 = 2$

190. $\log_a 8 = 3$

191. $\log_a 27 = 3$

192. $\log_a 32 = 2$

193. $\log_a 24 = 3$

194. $\ln x = 5$

195. $\ln x = 4$

196. $\log_2 (5x + 1) = 4$

197. $\log_2 (6x + 2) = 5$

198. $\log_3 (4x - 3) = 2$

199. $\log_3 (5x - 4) = 4$

200. $\log_4 (5x + 6) = 3$

201. $\log_4 (3x - 2) = 2$

202. $\ln e^{4x} = 8$

203. $\ln e^{2x} = 6$

204. $\log x^2 = 2$

205. $\log(x^2 - 25) = 2$

206. $\log_2\left(x^2 - 4\right) = 5$

207. $\log_3\left(x^2 + 2\right) = 3$

Use Logarithmic Models in Applications

In the following exercises, use a logarithmic model to solve.

208. What is the decibel level of normal conversation with intensity 10^{-6} watts per square inch?

209. What is the decibel level of a whisper with intensity 10^{-10} watts per square inch?

210. What is the decibel level of the noise from a motorcycle with intensity 10^{-2} watts per square inch?

211. What is the decibel level of the sound of a garbage disposal with intensity 10^{-2} watts per square inch?

212. In 2014, Chile experienced an intense earthquake with a magnitude of 8.2 on the Richter scale. In 2010, Haiti also experienced an intense earthquake which measured 7.0 on the Richter scale. Compare the intensities of the two earthquakes.

213. The Los Angeles area experiences many earthquakes. In 1994, the Northridge earthquake measured magnitude of 6.7 on the Richter scale. In 2014, Los Angeles also experienced an earthquake which measured 5.1 on the Richter scale. Compare the intensities of the two earthquakes.

Writing Exercises

214. Explain how to change an equation from logarithmic form to exponential form.

215. Explain the difference between common logarithms and natural logarithms.

216. Explain why $\log_a a^x = x$.

217. Explain how to find the $\log_7 32$ on your calculator.

Self Check

ⓐ

After completing the exercises, use this checklist to evaluate your mastery of the objectives of this section.

I can...	Confidently	With some help	No-I don't get it!
convert between exponential and logarithmic form.			
evaluate logarithmic functions.			
graph logarithmic functions.			
solve logarithmic functions.			
use logarithmic models in applications.			

ⓑ *After reviewing this checklist, what will you do to become confident for all objectives?*

10.4 Use the Properties of Logarithms

Learning Objectives

By the end of this section, you will be able to:

> Use the properties of logarithms
> Use the Change of Base Formula

Be Prepared!

Before you get started, take this readiness quiz.

1. Evaluate: ⓐ a^0 ⓑ a^1.
    If you missed this problem, review **Example 5.14**.

2. Write with a rational exponent: $\sqrt[3]{x^2 y}$.
    If you missed this problem, review **Example 8.27**.

3. Round to three decimal places: 2.5646415.
    If you missed this problem, review **Example 1.34**.

Use the Properties of Logarithms

Now that we have learned about exponential and logarithmic functions, we can introduce some of the properties of logarithms. These will be very helpful as we continue to solve both exponential and logarithmic equations.

The first two properties derive from the definition of logarithms. Since $a^0 = 1,$ we can convert this to logarithmic form and get $\log_a 1 = 0.$ Also, since $a^1 = a,$ we get $\log_a a = 1.$

Properties of Logarithms

$$\log_a 1 = 0 \qquad\qquad \log_a a = 1$$

In the next example we could evaluate the logarithm by converting to exponential form, as we have done previously, but recognizing and then applying the properties saves time.

EXAMPLE 10.28

Evaluate using the properties of logarithms: ⓐ $\log_8 1$ and ⓑ $\log_6 6$.

 Solution

ⓐ

$$\log_8 1$$

Use the property, $\log_a 1 = 0$. 0 $\log_8 1 = 0$

ⓑ

$$\log_6 6$$

Use the property, $\log_a a = 1$. 1 $\log_6 6 = 1$

> **TRY IT :: 10.55**   Evaluate using the properties of logarithms: ⓐ $\log_{13} 1$ ⓑ $\log_9 9$.

> **TRY IT :: 10.56**   Evaluate using the properties of logarithms: ⓐ $\log_5 1$ ⓑ $\log_7 7$.

The next two properties can also be verified by converting them from exponential form to logarithmic form, or the reverse.

The exponential equation $a^{\log_a x} = x$ converts to the logarithmic equation $\log_a x = \log_a x,$ which is a true statement for positive values for x only.

The logarithmic equation $\log_a a^x = x$ converts to the exponential equation $a^x = a^x$, which is also a true statement.

These two properties are called inverse properties because, when we have the same base, raising to a power "undoes" the log and taking the log "undoes" raising to a power. These two properties show the composition of functions. Both ended up with the identity function which shows again that the exponential and logarithmic functions are inverse functions.

Inverse Properties of Logarithms

For $a > 0$, $x > 0$ and $a \neq 1$,

$$a^{\log_a x} = x \qquad\qquad \log_a a^x = x$$

In the next example, apply the inverse properties of logarithms.

EXAMPLE 10.29

Evaluate using the properties of logarithms: ⓐ $4^{\log_4 9}$ and ⓑ $\log_3 3^5$.

✓ **Solution**

ⓐ

$$4^{\log_4 9}$$

Use the property, $a^{\log_a x} = x$. 9 $4^{\log_4 9} = 9$

ⓑ

$$\log_3 3^5$$

Use the property, $a^{\log_a x} = x$. 5 $\log_3 3^5 = 5$

> **TRY IT :: 10.57**   Evaluate using the properties of logarithms: ⓐ $5^{\log_5 15}$ ⓑ $\log_7 7^4$.

> **TRY IT :: 10.58**   Evaluate using the properties of logarithms: ⓐ $2^{\log_2 8}$ ⓑ $\log_2 2^{15}$.

There are three more properties of logarithms that will be useful in our work. We know exponential functions and logarithmic function are very interrelated. Our definition of logarithm shows us that a logarithm is the exponent of the equivalent exponential. The properties of exponents have related properties for exponents.

In the Product Property of Exponents, $a^m \cdot a^n = a^{m+n}$, we see that to multiply the same base, we add the exponents. The **Product Property of Logarithms**, $\log_a M \cdot N = \log_a M + \log_a N$ tells us to take the log of a product, we add the log of the factors.

Product Property of Logarithms

If $M > 0, N > 0, a > 0$ and $a \neq 1$, then,

$$\log_a (M \cdot N) = \log_a M + \log_a N$$

The logarithm of a product is the sum of the logarithms.

We use this property to write the log of a product as a sum of the logs of each factor.

EXAMPLE 10.30

Use the Product Property of Logarithms to write each logarithm as a sum of logarithms. Simplify, if possible: ⓐ $\log_3 7x$ and ⓑ $\log_4 64xy$.

✓ Solution

ⓐ

	$\log_3 7x$
Use the Product Property, $\log_a (M \cdot N) = \log_a M + \log_a N$.	$\log_3 7 + \log_3 x$
	$\log_3 7x = \log_3 7 + \log_3 x$

ⓑ

	$\log_4 64xy$
Use the Product Property, $\log_a (M \cdot N) = \log_a M + \log_a N$.	$\log_4 64 + \log_4 x + \log_4 y$
Simplify by evaluating $\log_4 64$.	$3 + \log_4 x + \log_4 y$
	$\log_4 64xy = 3 + \log_4 x + \log_4 y$

> **TRY IT ::** 10.59

Use the Product Property of Logarithms to write each logarithm as a sum of logarithms. Simplify, if possible.

ⓐ $\log_3 3x$ ⓑ $\log_2 8xy$

> **TRY IT ::** 10.60

Use the Product Property of Logarithms to write each logarithm as a sum of logarithms. Simplify, if possible.

ⓐ $\log_9 9x$ ⓑ $\log_3 27xy$

Similarly, in the Quotient Property of Exponents, $\frac{a^m}{a^n} = a^{m-n}$, we see that to divide the same base, we subtract the exponents. The **Quotient Property of Logarithms**, $\log_a \frac{M}{N} = \log_a M - \log_a N$ tells us to take the log of a quotient, we subtract the log of the numerator and denominator.

Quotient Property of Logarithms

If $M > 0$, $N > 0$, $a > 0$ and $a \neq 1$, then,

$$\log_a \frac{M}{N} = \log_a M - \log_a N$$

The logarithm of a quotient is the difference of the logarithms.

Note that $\log_a M - \log_a N \neq \log_a(M - N)$.

We use this property to write the log of a quotient as a difference of the logs of each factor.

EXAMPLE 10.31

Use the Quotient Property of Logarithms to write each logarithm as a difference of logarithms. Simplify, if possible.
ⓐ $\log_5 \frac{5}{7}$ and ⓑ $\log \frac{x}{100}$

✓ Solution

ⓐ

Use the Quotient Property, $\log_a \dfrac{M}{N} = \log_a M - \log_a N$.

Simplify.

$\log_5 \dfrac{5}{7}$

$\log_5 5 - \log_5 7$

$1 - \log_5 7$

$\log_5 \dfrac{5}{7} = 1 - \log_5 7$

ⓑ

Use the Quotient Property, $\log_a \dfrac{M}{N} = \log_a M - \log_a N$.

Simplify.

$\log \dfrac{x}{100}$

$\log x - \log 100$

$\log x - 2$

$\log \dfrac{x}{100} = \log x - 2$

> **TRY IT** :: 10.61

Use the Quotient Property of Logarithms to write each logarithm as a difference of logarithms. Simplify, if possible.

ⓐ $\log_4 \dfrac{3}{4}$ ⓑ $\log \dfrac{x}{1000}$

> **TRY IT** :: 10.62

Use the Quotient Property of Logarithms to write each logarithm as a difference of logarithms. Simplify, if possible.

ⓐ $\log_2 \dfrac{5}{4}$ ⓑ $\log \dfrac{10}{y}$

The third property of logarithms is related to the Power Property of Exponents, $(a^m)^n = a^{m \cdot n}$, we see that to raise a power to a power, we multiply the exponents. The **Power Property of Logarithms**, $\log_a M^p = p \log_a M$ tells us to take the log of a number raised to a power, we multiply the power times the log of the number.

Power Property of Logarithms

If $M > 0$, $a > 0$, $a \neq 1$ and p is any real number then,

$$\log_a M^p = p \log_a M$$

The log of a number raised to a power as the product product of the power times the log of the number.

We use this property to write the log of a number raised to a power as the product of the power times the log of the number. We essentially take the exponent and throw it in front of the logarithm.

EXAMPLE 10.32

Use the Power Property of Logarithms to write each logarithm as a product of logarithms. Simplify, if possible.
ⓐ $\log_5 4^3$ and ⓑ $\log x^{10}$

✓ **Solution**

ⓐ

Use the Power Property, $\log_a M^p = p \log_a M$.

$\log_5 4^3$

$3\log_5 4$

$\log_5 4^3 = 3\log_5 4$

ⓑ

Use the Power Property, $\log_a M^p = p \log_a M$.

$\log x^{10}$

$10 \log x$

$\log x^{10} = 10\log x$

> **TRY IT : :** 10.63

Use the Power Property of Logarithms to write each logarithm as a product of logarithms. Simplify, if possible.

ⓐ $\log_7 5^4$ ⓑ $\log x^{100}$

> **TRY IT : :** 10.64

Use the Power Property of Logarithms to write each logarithm as a product of logarithms. Simplify, if possible.

ⓐ $\log_2 3^7$ ⓑ $\log x^{20}$

We summarize the Properties of Logarithms here for easy reference. While the natural logarithms are a special case of these properties, it is often helpful to also show the natural logarithm version of each property.

Properties of Logarithms

If $M > 0$, $a > 0$, $a \neq 1$ and p is any real number then,

Property	Base a	Base e
	$\log_a 1 = 0$	$\ln 1 = 0$
	$\log_a a = 1$	$\ln e = 1$
Inverse Properties	$a^{\log_a x} = x$ $\log_a a^x = x$	$e^{\ln x} = x$ $\ln e^x = x$
Product Property of Logarithms	$\log_a (M \cdot N) = \log_a M + \log_a N$	$\ln(M \cdot N) = \ln M + \ln N$
Quotient Property of Logarithms	$\log_a \frac{M}{N} = \log_a M - \log_a N$	$\ln \frac{M}{N} = \ln M - \ln N$
Power Property of Logarithms	$\log_a M^p = p \log_a M$	$\ln M^p = p \ln M$

Now that we have the properties we can use them to "expand" a logarithmic expression. This means to write the logarithm as a sum or difference and without any powers.

We generally apply the Product and Quotient Properties before we apply the Power Property.

EXAMPLE 10.33

Use the Properties of Logarithms to expand the logarithm $\log_4 \left(2x^3 y^2\right)$. Simplify, if possible.

Solution

$$\log_4\left(2x^3 y^2\right)$$

Use the Product Property, $\log_a M \cdot N = \log_a M + \log_a N$.

$$\log_4 2 + \log_4 x^3 + \log_4 y^2$$

Use the Power Property, $\log_a M^p = p\log_a M$, on the last two terms.

$$\log_4 2 + 3\log_4 x + 2\log_4 y$$

Simplify.

$$\tfrac{1}{2} + 3\log_4 x + 2\log_4 y$$

$$\log_4\left(2x^3 y^2\right) = \tfrac{1}{2} + 3\log_4 x + 2\log_4 y$$

> **TRY IT :: 10.65**

Use the Properties of Logarithms to expand the logarithm $\log_2\left(5x^4 y^2\right)$. Simplify, if possible.

> **TRY IT :: 10.66**

Use the Properties of Logarithms to expand the logarithm $\log_3\left(7x^5 y^3\right)$. Simplify, if possible.

When we have a radical in the logarithmic expression, it is helpful to first write its radicand as a rational exponent.

EXAMPLE 10.34

Use the Properties of Logarithms to expand the logarithm $\log_2 \sqrt[4]{\dfrac{x^3}{3y^2 z}}$. Simplify, if possible.

Solution

$$\log_2 \sqrt[4]{\dfrac{x^3}{3y^2 z}}$$

Rewrite the radical with a rational exponent.

$$\log_2\left(\dfrac{x^3}{3y^2 z}\right)^{\frac{1}{4}}$$

Use the Power Property, $\log_a M^p = p\log_a M$.

$$\tfrac{1}{4}\log_2\left(\dfrac{x^3}{3y^2 z}\right)$$

Use the Quotient Property, $\log_a M \cdot N = \log_a M - \log_a N$.

$$\tfrac{1}{4}\left(\log_2\left(x^3\right) - \log_2\left(3y^2 z\right)\right)$$

Use the Product Property, $\log_a M \cdot N = \log_a M + \log_a N$, in the second term.

$$\tfrac{1}{4}\left(\log_2\left(x^3\right) - \left(\log_2 3 + \log_2 y^2 + \log_2 z\right)\right)$$

Use the Power Property, $\log_a M^p = p\log_a M$, inside the parentheses.

$$\tfrac{1}{4}\left(3\log_2 x - \left(\log_2 3 + 2\log_2 y + \log_2 z\right)\right)$$

Simplify by distributing.

$$\tfrac{1}{4}\left(3\log_2 x - \log_2 3 - 2\log_2 y - \log_2 z\right)$$

$$\log_2 \sqrt[4]{\dfrac{x^3}{3y^2 z}} = \tfrac{1}{4}\left(3\log_2 x - \log_2 3 - 2\log_2 y - \log_2 z\right)$$

> **TRY IT : : 10.67**

Use the Properties of Logarithms to expand the logarithm $\log_4 \sqrt[5]{\dfrac{x^4}{2y^3 z^2}}$. Simplify, if possible.

> **TRY IT : : 10.68**

Use the Properties of Logarithms to expand the logarithm $\log_3 \sqrt[3]{\dfrac{x^2}{5y\,z}}$. Simplify, if possible.

The opposite of expanding a logarithm is to condense a sum or difference of logarithms that have the same base into a single logarithm. We again use the properties of logarithms to help us, but in reverse.

To condense logarithmic expressions with the same base into one logarithm, we start by using the Power Property to get the coefficients of the log terms to be one and then the Product and Quotient Properties as needed.

EXAMPLE 10.35

Use the Properties of Logarithms to condense the logarithm $\log_4 3 + \log_4 x - \log_4 y$. Simplify, if possible.

⊘ Solution

The log expressions all have the same base, 4.

$\log_4 3 + \log_4 x - \log_4 y$

The first two terms are added, so we use the Product Property,
$\log_a M + \log_a N = \log_a M \cdot N$.

$\log_4 3x - \log_4 y$

Since the logs are subtracted, we use the Quotient Property,
$\log_a M - \log_a N = \log_a \dfrac{M}{N}$.

$\log_4 \dfrac{3x}{y}$

$\log_4 3 + \log_4 x - \log_4 y = \log_4 \dfrac{3x}{y}$

> **TRY IT : : 10.69**

Use the Properties of Logarithms to condense the logarithm $\log_2 5 + \log_2 x - \log_2 y$. Simplify, if possible.

> **TRY IT : : 10.70**

Use the Properties of Logarithms to condense the logarithm $\log_3 6 - \log_3 x - \log_3 y$. Simplify, if possible.

EXAMPLE 10.36

Use the Properties of Logarithms to condense the logarithm $2\log_3 x + 4\log_3 (x+1)$. Simplify, if possible.

⊘ Solution

The log expressions have the same base, 3.

$2\log_3 x + 4\log_3 (x+1)$

Use the Power Property, $\log_a M + \log_a N = \log_a M \cdot N$.

$\log_3 x^2 + \log_3 (x+1)^4$

The terms are added, so we use the Product Property, $\log_a M + \log_a N = \log_a M \cdot N$.

$\log_3 x^2 (x+1)^4$

$2\log_3 x + 4\log_3 (x+1) = \log_3 x^2 (x+1)^4$

> **TRY IT : : 10.71**

Use the Properties of Logarithms to condense the logarithm $3\log_2 x + 2\log_2 (x-1)$. Simplify, if possible.

> **TRY IT : : 10.72**

Use the Properties of Logarithms to condense the logarithm $2\log x + 2\log(x+1)$. Simplify, if possible.

Use the Change-of-Base Formula

To evaluate a logarithm with any other base, we can use the Change-of-Base Formula. We will show how this is derived.

Suppose we want to evaluate $\log_a M$.	$\log_a M$
Let $y = \log_a M$.	$y = \log_a M$
Rewrite the expression in exponential form.	$a^y = M$
Take the $\log_b$ of each side.	$\log_b a^y = \log_b M$
Use the Power Property.	$y\log_b a = \log_b M$
Solve for y.	$y = \dfrac{\log_b M}{\log_b a}$
Substitute $y = \log_a M$.	$\log_a M = \dfrac{\log_b M}{\log_b a}$

The Change-of-Base Formula introduces a new base b. This can be any base b we want where $b > 0$, $b \neq 1$. Because our calculators have keys for logarithms base 10 and base e, we will rewrite the Change-of-Base Formula with the new base as 10 or e.

Change-of-Base Formula

For any logarithmic bases a, b and $M > 0$,

$$\log_a M = \frac{\log_b M}{\log_b a} \qquad \log_a M = \frac{\log M}{\log a} \qquad \log_a M = \frac{\ln M}{\ln a}$$

$$\text{new base } b \qquad\qquad \text{new base } 10 \qquad\qquad \text{new base } e$$

When we use a calculator to find the logarithm value, we usually round to three decimal places. This gives us an approximate value and so we use the approximately equal symbol ($\approx$).

EXAMPLE 10.37

Rounding to three decimal places, approximate $\log_4 35$.

⊘ Solution

	$\log_4 35$
Use the Change-of-Base Formula.	$\log_c M = \dfrac{\log_b M}{\log_b a}$
Identify a and M. Choose 10 for b.	$\log_4 35 = \dfrac{\log 35}{\log 4}$
Enter the expression $\dfrac{\log 35}{\log 4}$ in the calculator using the log button for base 10. Round to three decimal places.	$\log_4 35 \approx 2.565$

> **TRY IT : : 10.73**   Rounding to three decimal places, approximate $\log_3 42$.

> **TRY IT : : 10.74**   Rounding to three decimal places, approximate $\log_5 46$.

▶ **MEDIA : :**

Access these online resources for additional instruction and practice with using the properties of logarithms.

- **Using Properties of Logarithms to Expand Logs (https://openstax.org/l/37Logproperties)**
- **Using Properties of Logarithms to Condense Logs (https://openstax.org/l/37Condenselogs)**
- **Change of Base (https://openstax.org/l/37Changeofbase)**

10.4 EXERCISES
Practice Makes Perfect

Use the Properties of Logarithms
In the following exercises, use the properties of logarithms to evaluate.

218. ⓐ $\log_4 1$ ⓑ $\log_8 8$

219. ⓐ $\log_{12} 1$ ⓑ $\ln e$

220. ⓐ $3^{\log_3 6}$ ⓑ $\log_2 2^7$

221. ⓐ $5^{\log_5 10}$ ⓑ $\log_4 4^{10}$

222. ⓐ $8^{\log_8 7}$ ⓑ $\log_6 6^{-2}$

223. ⓐ $6^{\log_6 15}$ ⓑ $\log_8 8^{-4}$

224. ⓐ $10^{\log\sqrt{5}}$ ⓑ $\log 10^{-2}$

225. ⓐ $10^{\log\sqrt{3}}$ ⓑ $\log 10^{-1}$

226. ⓐ $e^{\ln 4}$ ⓑ $\ln e^2$

227. ⓐ $e^{\ln 3}$ ⓑ $\ln e^7$

In the following exercises, use the Product Property of Logarithms to write each logarithm as a sum of logarithms. Simplify if possible.

228. $\log_4 6x$

229. $\log_5 8y$

230. $\log_2 32xy$

231. $\log_3 81xy$

232. $\log 100x$

233. $\log 1000y$

In the following exercises, use the Quotient Property of Logarithms to write each logarithm as a sum of logarithms. Simplify if possible.

234. $\log_3 \frac{3}{8}$

235. $\log_6 \frac{5}{6}$

236. $\log_4 \frac{16}{y}$

237. $\log_5 \frac{125}{x}$

238. $\log \frac{x}{10}$

239. $\log \frac{10,000}{y}$

240. $\ln \frac{e^3}{3}$

241. $\ln \frac{e^4}{16}$

In the following exercises, use the Power Property of Logarithms to expand each. Simplify if possible.

242. $\log_3 x^2$

243. $\log_2 x^5$

244. $\log x^{-2}$

245. $\log x^{-3}$

246. $\log_4 \sqrt{x}$

247. $\log_5 \sqrt[3]{x}$

248. $\ln x^{\sqrt{3}}$

249. $\ln x^{\sqrt[3]{4}}$

In the following exercises, use the Properties of Logarithms to expand the logarithm. Simplify if possible.

250. $\log_5\left(4x^6 y^4\right)$

251. $\log_2\left(3x^5 y^3\right)$

252. $\log_3\left(\sqrt{2}x^2\right)$

253. $\log_5\left(\sqrt[4]{21}y^3\right)$

254. $\log_3 \dfrac{xy^2}{z^2}$

255. $\log_5 \dfrac{4ab^3 c^4}{d^2}$

256. $\log_4 \dfrac{\sqrt{x}}{16y^4}$

257. $\log_3 \dfrac{\sqrt[3]{x^2}}{27y^4}$

258. $\log_2 \dfrac{\sqrt{2x+y^2}}{z^2}$

259. $\log_3 \dfrac{\sqrt{3x+2y^2}}{5z^2}$

260. $\log_2 \sqrt[4]{\dfrac{5x^3}{2y^2 z^4}}$

261. $\log_5 \sqrt[3]{\dfrac{3x^2}{4y^3 z}}$

In the following exercises, use the Properties of Logarithms to condense the logarithm. Simplify if possible.

262. $\log_6 4 + \log_6 9$

263. $\log 4 + \log 25$

264. $\log_2 80 - \log_2 5$

265. $\log_3 36 - \log_3 4$

266. $\log_3 4 + \log_3 (x+1)$

267. $\log_2 5 - \log_2 (x-1)$

268. $\log_7 3 + \log_7 x - \log_7 y$

269. $\log_5 2 - \log_5 x - \log_5 y$

270. $4\log_2 x + 6\log_2 y$

271. $6\log_3 x + 9\log_3 y$

272. $\log_3\left(x^2 - 1\right) - 2\log_3 (x-1)$

273. $\log\left(x^2 + 2x + 1\right) - 2\log(x+1)$

274. $4\log x - 2\log y - 3\log z$

275. $3\ln x + 4\ln y - 2\ln z$

276. $\frac{1}{3}\log x - 3\log(x+1)$

277. $2\log(2x+3) + \frac{1}{2}\log(x+1)$

Use the Change-of-Base Formula

In the following exercises, use the Change-of-Base Formula, rounding to three decimal places, to approximate each logarithm.

278. $\log_3 42$

279. $\log_5 46$

280. $\log_{12} 87$

281. $\log_{15} 93$

282. $\log_{\sqrt{2}} 17$

283. $\log_{\sqrt{3}} 21$

Writing Exercises

284. Write the Product Property in your own words. Does it apply to each of the following? $\log_a 5x$, $\log_a (5 + x)$. Why or why not?

285. Write the Power Property in your own words. Does it apply to each of the following? $\log_a x^p$, $(\log_a x)^r$. Why or why not?

286. Use an example to show that $\log(a + b) \neq \log a + \log b$?

287. Explain how to find the value of $\log_7 15$ using your calculator.

Self Check

ⓐ *After completing the exercises, use this checklist to evaluate your mastery of the objectives of this section.*

I can...	Confidently	With some help	No-I don't get it!
use the properties of logarithms.			
use the Change of Base Formula.			

ⓑ *On a scale of 1 – 10, how would you rate your mastery of this section in light of your responses on the checklist? How can you improve this?*

10.5   Solve Exponential and Logarithmic Equations

Learning Objectives

By the end of this section, you will be able to:

› Solve logarithmic equations using the properties of logarithms
› Solve exponential equations using logarithms
› Use exponential models in applications

Be Prepared!

Before you get started, take this readiness quiz.

1. Solve: $x^2 = 16$.
   If you missed this problem, review Example 6.46.

2. Solve: $x^2 - 5x + 6 = 0$.
   If you missed this problem, review Example 6.45.

3. Solve: $x(x + 6) = 2x + 5$.
   If you missed this problem, review Example 6.47.

Solve Logarithmic Equations Using the Properties of Logarithms

In the section on logarithmic functions, we solved some equations by rewriting the equation in exponential form. Now that we have the properties of logarithms, we have additional methods we can use to solve logarithmic equations.

If our equation has two logarithms we can use a property that says that if $\log_a M = \log_a N$ then it is true that $M = N$. This is the **One-to-One Property of Logarithmic Equations**.

One-to-One Property of Logarithmic Equations

For $M > 0$, $N > 0$, $a > 0$, and $a \neq 1$ is any real number:

$$\text{If } \log_a M = \log_a N, \text{ then } M = N.$$

To use this property, we must be certain that both sides of the equation are written with the same base.

Remember that logarithms are defined only for positive real numbers. Check your results in the original equation. You may have obtained a result that gives a logarithm of zero or a negative number.

EXAMPLE 10.38

Solve: $2\log_5 x = \log_5 81$.

⊘ Solution

$$2\log_5 x = \log_5 81$$

Use the Power Property. $$\log_5 x^2 = \log_5 81$$

Use the One-to-One Property, if $\log_a M = \log_a N$, $$x^2 = 81$$
then $M = N$.
Solve using the Square Root Property. $$x = \pm 9$$
We eliminate $x = -9$ as we cannot take the logarithm $$x = 9, \quad \cancel{x = -9}$$
of a negative number.
Check.

$x = 9$ $$2\log_5 x = \log_5 81$$

$$2\log_5 9 \overset{?}{=} \log_5 81$$

$$\log_5 9^2 \overset{?}{=} \log_5 81$$

$$\log_5 81 = \log_5 81 \checkmark$$

> **TRY IT :: 10.75**   Solve: $2\log_3 x = \log_3 36$

> **TRY IT :: 10.76**   Solve: $3\log x = \log 64$

Another strategy to use to solve logarithmic equations is to condense sums or differences into a single logarithm.

EXAMPLE 10.39

Solve: $\log_3 x + \log_3 (x - 8) = 2$.

⊘ Solution

$$\log_3 x + \log_3 (x - 8) = 2$$

Use the Product Property, $\log_a M + \log_a N = \log_a M \cdot N$. $$\log_3 x(x - 8) = 2$$

Rewrite in exponential form. $$3^2 = x(x - 8)$$

Simplify. $$9 = x^2 - 8x$$

Subtract 9 from each side. $$0 = x^2 - 8x - 9$$
Factor. $$0 = (x - 9)(x + 1)$$
Use the Zero-Product Property. $$x - 9 = 0, \quad x + 1 = 0$$
Solve each equation. $$x = 9, \quad\quad \cancel{x = -1}$$
Check.

$x = -1$ $$\log_3 x + \log_3 (x - 8) = 2$$

$$\log_3 (-1) + \log_3 (-1 - 8) \overset{?}{=} 2$$

We cannot take the log of a negative number.

$x = 9$ $$\log_3 x + \log_3 (x - 8) = 2$$

$$\log_3 9 + \log_3 (9 - 8) \overset{?}{=} 2$$

$$2 + 0 \overset{?}{=} 2$$

$$2 = 2 \checkmark$$

> **TRY IT :: 10.77**   Solve: $\log_2 x + \log_2 (x - 2) = 3$

> **TRY IT :: 10.78**   Solve: $\log_2 x + \log_2 (x - 6) = 4$

When there are logarithms on both sides, we condense each side into a single logarithm. Remember to use the Power Property as needed.

EXAMPLE 10.40

Solve: $\log_4 (x + 6) - \log_4 (2x + 5) = -\log_4 x$.

⊘ Solution

$$\log_4 (x + 6) - \log_4 (2x + 5) = -\log_4 x$$

Use the Quotient Property on the left side and the Power Property on the right.

$$\log_4 \left(\frac{x+6}{2x+5} \right) = \log_4 x^{-1}$$

Rewrite $x^{-1} = \frac{1}{x}$.

$$\log_4 \left(\frac{x+6}{2x+5} \right) = \log_4 \frac{1}{x}$$

Use the One-to-One Property, if $\log_a M = \log_a N$, then $M = N$.

$$\frac{x+6}{2x+5} = \frac{1}{x}$$

Solve the rational equation.

$$x(x + 6) = 2x + 5$$

Distribute.

$$x^2 + 6x = 2x + 5$$

Write in standard form.

$$x^2 + 4x - 5 = 0$$

Factor.

$$(x + 5)(x - 1) = 0$$

Use the Zero-Product Property.

$$x + 5 = 0, \qquad x - 1 = 0$$

Solve each equation.

$$x = -5, \qquad x = 1$$

Check.
We leave the check for you.

> **TRY IT :: 10.79**   Solve: $\log(x + 2) - \log(4x + 3) = -\log x$.

> **TRY IT :: 10.80**   Solve: $\log(x - 2) - \log(4x + 16) = \log \frac{1}{x}$.

Solve Exponential Equations Using Logarithms

In the section on exponential functions, we solved some equations by writing both sides of the equation with the same base. Next we wrote a new equation by setting the exponents equal.

It is not always possible or convenient to write the expressions with the same base. In that case we often take the common logarithm or natural logarithm of both sides once the exponential is isolated.

EXAMPLE 10.41

Solve $5^x = 11$. Find the exact answer and then approximate it to three decimal places.

⊘ Solution

$$5^x = 11$$

Since the exponential is isolated, take the logarithm of both sides.

$$\log 5^x = \log 11$$

Use the Power Property to get the x as a factor, not an exponent.

$$x \log 5 = \log 11$$

Solve for x. Find the exact answer.

$$x = \frac{\log 11}{\log 5}$$

Approximate the answer.

$$x \approx 1.490$$

Since $5^1 = 5$ and $5^2 = 25$, does it makes sense that $5^{1.490} \approx 11$?

> **TRY IT :: 10.81**   Solve $7^x = 43$. Find the exact answer and then approximate it to three decimal places.

> **TRY IT : : 10.82**   Solve $8^x = 98$. Find the exact answer and then approximate it to three decimal places.

When we take the logarithm of both sides we will get the same result whether we use the common or the natural logarithm (try using the natural log in the last example. Did you get the same result?) When the exponential has base e, we use the natural logarithm.

EXAMPLE 10.42

Solve $3e^{x+2} = 24$. Find the exact answer and then approximate it to three decimal places.

⊘ **Solution**

	$3e^{x+2} = 24$
Isolate the exponential by dividing both sides by 3.	$e^{x+2} = 8$
Take the natural logarithm of both sides.	$\ln e^{x+2} = \ln 8$
Use the Power Property to get the x as a factor, not an exponent.	$(x+2)\ln e = \ln 8$
Use the property $\ln e = 1$ to simplify.	$x + 2 = \ln 8$
Solve the equation. Find the exact answer.	$x = \ln 8 - 2$
Approximate the answer.	$x \approx 0.079$

> **TRY IT : : 10.83**   Solve $2e^{x-2} = 18$. Find the exact answer and then approximate it to three decimal places.

> **TRY IT : : 10.84**   Solve $5e^{2x} = 25$. Find the exact answer and then approximate it to three decimal places.

Use Exponential Models in Applications

In previous sections we were able to solve some applications that were modeled with exponential equations. Now that we have so many more options to solve these equations, we are able to solve more applications.

We will again use the Compound Interest Formulas and so we list them here for reference.

> ### Compound Interest
>
> For a principal, P, invested at an interest rate, r, for t years, the new balance, A is:
>
> $$A = P\left(1 + \frac{r}{n}\right)^{nt} \qquad \text{when compounded } n \text{ times a year.}$$
>
> $$A = Pe^{rt} \qquad \text{when compounded continuously.}$$

EXAMPLE 10.43

Jermael's parents put $10,000 in investments for his college expenses on his first birthday. They hope the investments will be worth $50,000 when he turns 18. If the interest compounds continuously, approximately what rate of growth will they need to achieve their goal?

⊘ Solution

Identify the variables in the formula.	$A = \$50,000$
	$P = \$10,000$
	$r = ?$
	$t = 17 \text{ years}$
	$A = Pe^{rt}$
Substitute the values into the formula.	$50,000 = 10,000e^{r \cdot 17}$
Solve for r. Divide each side by 10,000.	$5 = e^{17r}$
Take the natural log of each side.	$\ln 5 = \ln e^{17r}$
Use the Power Property.	$\ln 5 = 17r \ln e$
Simplify.	$\ln 5 = 17r$
Divide each side by 17.	$\dfrac{\ln 5}{17} = r$
Approximate the answer.	$r \approx 0.095$
Convert to a percentage.	$r \approx 9.5\%$

They need the rate of growth to be approximately 9.5%.

> **TRY IT : : 10.85**
>
> Hector invests $\$10,000$ at age 21. He hopes the investments will be worth $\$150,000$ when he turns 50. If the interest compounds continuously, approximately what rate of growth will he need to achieve his goal?

> **TRY IT : : 10.86**
>
> Rachel invests $\$15,000$ at age 25. She hopes the investments will be worth $\$90,000$ when she turns 40. If the interest compounds continuously, approximately what rate of growth will she need to achieve her goal?

We have seen that growth and decay are modeled by exponential functions. For growth and decay we use the formula $A = A_0 e^{kt}$. Exponential growth has a positive rate of growth or growth constant, k, and exponential decay has a negative rate of growth or decay constant, k.

Exponential Growth and Decay

For an original amount, A_0, that grows or decays at a rate, k, for a certain time, t, the final amount, A, is:

$$A = A_0 e^{kt}$$

We can now solve applications that give us enough information to determine the rate of growth. We can then use that rate of growth to predict other situations.

EXAMPLE 10.44

Researchers recorded that a certain bacteria population grew from 100 to 300 in 3 hours. At this rate of growth, how many bacteria will there be 24 hours from the start of the experiment?

⊘ Solution

This problem requires two main steps. First we must find the unknown rate, k. Then we use that value of k to help us find the unknown number of bacteria.

Identify the variables in the formula.	$A = 300$ $A_0 = 100$ $k = ?$ $t = 3 \text{ hours}$ $A = A_0 e^{kt}$
Substitute the values in the formula.	$300 = 100e^{k \cdot 3}$
Solve for k. Divide each side by 100.	$3 = e^{3k}$
Take the natural log of each side.	$\ln 3 = \ln e^{3k}$
Use the Power Property.	$\ln 3 = 3k \ln e$
Simplify.	$\ln 3 = 3k$
Divide each side by 3.	$\dfrac{\ln 3}{3} = k$
Approximate the answer.	$k \approx 0.366$
We use this rate of growth to predict the number of bacteria there will be in 24 hours.	$A = ?$ $A_0 = 100$ $k = \dfrac{\ln 3}{3}$ $t = 24 \text{ hours}$ $A = A_0 e^{kt}$
Substitute in the values. Evaluate.	$A = 100e^{\frac{\ln 3}{3} \cdot 24}$ $A \approx 656{,}100$ At this rate of growth, they can expect 656,100 bacteria.

> **TRY IT :: 10.87**
>
> Researchers recorded that a certain bacteria population grew from 100 to 500 in 6 hours. At this rate of growth, how many bacteria will there be 24 hours from the start of the experiment?

> **TRY IT :: 10.88**
>
> Researchers recorded that a certain bacteria population declined from 700,000 to 400,000 in 5 hours after the administration of medication. At this rate of decay, how many bacteria will there be 24 hours from the start of the experiment?

Radioactive substances decay or decompose according to the exponential decay formula. The amount of time it takes for the substance to decay to half of its original amount is called the half-life of the substance.

Similar to the previous example, we can use the given information to determine the constant of decay, and then use that constant to answer other questions.

EXAMPLE 10.45

The half-life of radium-226 is 1,590 years. How much of a 100 mg sample will be left in 500 years?

⊘ **Solution**

This problem requires two main steps. First we must find the decay constant k. If we start with 100-mg, at the half-life there will be 50-mg remaining. We will use this information to find k. Then we use that value of k to help us find the amount of sample that will be left in 500 years.

Identify the variables in the formula.	$A = 50$
	$A_0 = 100$
	$k = ?$
	$t = 1590 \text{years}$
	$A = A_0 e^{kt}$
Substitute the values in the formula.	$50 = 100 e^{k \cdot 1590}$
Solve for k. Divide each side by 100.	$0.5 = e^{1590k}$
Take the natural log of each side.	$\ln 0.5 = \ln e^{1590k}$
Use the Power Property.	$\ln 0.5 = 1590k \ln e$
Simplify.	$\ln 0.5 = 1590k$
Divide each side by 1590.	$\dfrac{\ln 0.5}{1590} = k \text{ exact answer}$

	$A = ?$
	$A_0 = 100$
We use this rate of growth to predict the amount that will be left in 500 years.	$k = \dfrac{\ln 0.5}{1590}$
	$t = 500 \text{years}$
	$A = A_0 e^{kt}$
Substitute in the values.	$A = 100 e^{\frac{\ln 0.5}{1590} \cdot 500}$
Evaluate.	$A \approx 80.4 \, \text{mg}$

In 500 years there would be approximately 80.4 mg remaining.

> **TRY IT : : 10.89**

The half-life of magnesium-27 is 9.45 minutes. How much of a 10-mg sample will be left in 6 minutes?

> **TRY IT : : 10.90**

The half-life of radioactive iodine is 60 days. How much of a 50-mg sample will be left in 40 days?

▶ **MEDIA : :**

Access these online resources for additional instruction and practice with solving exponential and logarithmic equations.

- **Solving Logarithmic Equations (https://openstax.org/l/37Solvelog)**
- **Solving Logarithm Equations (https://openstax.org/l/37sSolvelogeqs2)**
- **Finding the rate or time in a word problem on exponential growth or decay (https://openstax.org/l/37Solveforrate)**
- **Finding the rate or time in a word problem on exponential growth or decay (https://openstax.org/l/37Solveforrate2)**

10.5 EXERCISES

Practice Makes Perfect

Solve Logarithmic Equations Using the Properties of Logarithms

In the following exercises, solve for x.

288. $\log_4 64 = 2\log_4 x$

289. $\log 49 = 2\log x$

290. $3\log_3 x = \log_3 27$

291. $3\log_6 x = \log_6 64$

292. $\log_5 (4x - 2) = \log_5 10$

293. $\log_3 (x^2 + 3) = \log_3 4x$

294. $\log_3 x + \log_3 x = 2$

295. $\log_4 x + \log_4 x = 3$

296. $\log_2 x + \log_2 (x - 3) = 2$

297. $\log_3 x + \log_3 (x + 6) = 3$

298. $\log x + \log(x + 3) = 1$

299. $\log x + \log(x - 15) = 2$

300. $\log(x + 4) - \log(5x + 12) = -\log x$

301. $\log(x - 1) - \log(x + 3) = \log\frac{1}{x}$

302. $\log_5 (x + 3) + \log_5 (x - 6) = \log_5 10$

303. $\log_5 (x + 1) + \log_5 (x - 5) = \log_5 7$

304. $\log_3 (2x - 1) = \log_3 (x + 3) + \log_3 3$

305. $\log(5x + 1) = \log(x + 3) + \log 2$

Solve Exponential Equations Using Logarithms

In the following exercises, solve each exponential equation. Find the exact answer and then approximate it to three decimal places.

306. $3^x = 89$

307. $2^x = 74$

308. $5^x = 110$

309. $4^x = 112$

310. $e^x = 16$

311. $e^x = 8$

312. $\left(\frac{1}{2}\right)^x = 6$

313. $\left(\frac{1}{3}\right)^x = 8$

314. $4e^{x+1} = 16$

315. $3e^{x+2} = 9$

316. $6e^{2x} = 24$

317. $2e^{3x} = 32$

318. $\frac{1}{4}e^x = 3$

319. $\frac{1}{3}e^x = 2$

320. $e^{x+1} + 2 = 16$

321. $e^{x-1} + 4 = 12$

In the following exercises, solve each equation.

322. $3^{3x+1} = 81$

323. $6^{4x-17} = 216$

324. $\dfrac{e^{x^2}}{e^{14}} = e^{5x}$

325. $\dfrac{e^{x^2}}{e^x} = e^{20}$

326. $\log_a 64 = 2$

327. $\log_a 81 = 4$

328. $\ln x = -8$

329. $\ln x = 9$

330. $\log_5(3x - 8) = 2$

331. $\log_4(7x + 15) = 3$

332. $\ln e^{5x} = 30$

333. $\ln e^{6x} = 18$

334. $3\log x = \log 125$

335. $7\log_3 x = \log_3 128$

336. $\log_6 x + \log_6(x - 5) = 24$

337. $\log_9 x + \log_9(x - 4) = 12$

338. $\log_2(x + 2) - \log_2(2x + 9) = -\log_2 x$

339. $\log_6(x + 1) - \log_6(4x + 10) = \log_6 \frac{1}{x}$

In the following exercises, solve for x, giving an exact answer as well as an approximation to three decimal places.

340. $6^x = 91$

341. $\left(\frac{1}{2}\right)^x = 10$

342. $7e^{x-3} = 35$

343. $8e^{x+5} = 56$

Use Exponential Models in Applications

In the following exercises, solve.

344. Sung Lee invests $\$5,000$ at age 18. He hopes the investments will be worth $\$10,000$ when he turns 25. If the interest compounds continuously, approximately what rate of growth will he need to achieve his goal? Is that a reasonable expectation?

345. Alice invests $\$15,000$ at age 30 from the signing bonus of her new job. She hopes the investments will be worth $\$30,000$ when she turns 40. If the interest compounds continuously, approximately what rate of growth will she need to achieve her goal?

346. Coralee invests $\$5,000$ in an account that compounds interest monthly and earns 7%. How long will it take for her money to double?

347. Simone invests $\$8,000$ in an account that compounds interest quarterly and earns 5%. How long will it take for his money to double?

348. Researchers recorded that a certain bacteria population declined from 100,000 to 100 in 24 hours. At this rate of decay, how many bacteria will there be in 16 hours?

349. Researchers recorded that a certain bacteria population declined from 800,000 to 500,000 in 6 hours after the administration of medication. At this rate of decay, how many bacteria will there be in 24 hours?

350. A virus takes 6 days to double its original population $(A = 2A_0)$. How long will it take to triple its population?

351. A bacteria doubles its original population in 24 hours $(A = 2A_0)$. How big will its population be in 72 hours?

352. Carbon-14 is used for archeological carbon dating. Its half-life is 5,730 years. How much of a 100-gram sample of Carbon-14 will be left in 1000 years?

353. Radioactive technetium-99m is often used in diagnostic medicine as it has a relatively short half-life but lasts long enough to get the needed testing done on the patient. If its half-life is 6 hours, how much of the radioactive material form a 0.5 ml injection will be in the body in 24 hours?

Writing Exercises

354. Explain the method you would use to solve these equations: $3^{x+1} = 81$, $3^{x+1} = 75$. Does your method require logarithms for both equations? Why or why not?

355. What is the difference between the equation for exponential growth versus the equation for exponential decay?

Self Check

ⓐ *After completing the exercises, use this checklist to evaluate your mastery of the objectives of this section.*

I can...	Confidently	With some help	No-I don't get it!
solve logarithmic equations using the properties of logarithms.			
solve exponential equations using logarithms.			
use exponential models in applications.			

ⓑ *After looking at the checklist, do you think you are well-prepared for the next section? Why or why not?*

CHAPTER 10 REVIEW

KEY TERMS

asymptote A line which a graph of a function approaches closely but never touches.

common logarithmic function The function $f(x) = \log x$ is the common logarithmic function with base 10, where $x > 0$.

$$y = \log x \text{ is equivalent to } x = 10^y$$

exponential function An exponential function, where $a > 0$ and $a \neq 1$, is a function of the form $f(x) = a^x$.

logarithmic function The function $f(x) = \log_a x$ is the logarithmic function with base a, where $a > 0$, $x > 0$, and $a \neq 1$.

$$y = \log_a x \text{ is equivalent to } x = a^y$$

natural base The number e is defined as the value of $\left(1 + \frac{1}{n}\right)^n$, as n gets larger and larger. We say, as n increases without bound, $e \approx 2.718281827...$

natural exponential function The natural exponential function is an exponential function whose base is e: $f(x) = e^x$. The domain is $(-\infty, \infty)$ and the range is $(0, \infty)$.

natural logarithmic function The function $f(x) = \ln x$ is the natural logarithmic function with base e, where $x > 0$.

$$y = \ln x \text{ is equivalent to } x = e^y$$

one-to-one function A function is one-to-one if each value in the range has exactly one element in the domain. For each ordered pair in the function, each *y*-value is matched with only one *x*-value.

KEY CONCEPTS

10.1 Finding Composite and Inverse Functions

- **Composition of Functions:** The composition of functions f and g, is written $f \circ g$ and is defined by

$$(f \circ g)(x) = f(g(x))$$

  We read $f(g(x))$ as f of g of x.

- **Horizontal Line Test:** If every horizontal line, intersects the graph of a function in at most one point, it is a one-to-one function.

- **Inverse of a Function Defined by Ordered Pairs:** If $f(x)$ is a one-to-one function whose ordered pairs are of the form (x, y), then its inverse function $f^{-1}(x)$ is the set of ordered pairs (y, x).

- **Inverse Functions:** For every x in the domain of one-to-one function f and f^{-1},

$$f^{-1}(f(x)) = x$$
$$f(f^{-1}(x)) = x$$

- **How to Find the Inverse of a One-to-One Function:**

  Step 1. Substitute y for $f(x)$.

  Step 2. Interchange the variables x and y.

  Step 3. Solve for y.

  Step 4. Substitute $f^{-1}(x)$ for y.

  Step 5. Verify that the functions are inverses.

10.2 Evaluate and Graph Exponential Functions

- **Properties of the Graph of** $f(x) = a^x$:

when $a > 1$		when $0 < a < 1$	
Domain	$(-\infty, \infty)$	Domain	$(-\infty, \infty)$
Range	$(0, \infty)$	Range	$(0, \infty)$
x-intercept	none	x-intercept	none
y-intercept	$(0, 1)$	y-intercept	$(0, 1)$
Contains	$(1, a), \left(-1, \frac{1}{a}\right)$	Contains	$(1, a), \left(-1, \frac{1}{a}\right)$
Asymptote	x-axis, the line $y = 0$	Asymptote	x-axis, the line $y = 0$
Basic shape	increasing	Basic shape	decreasing

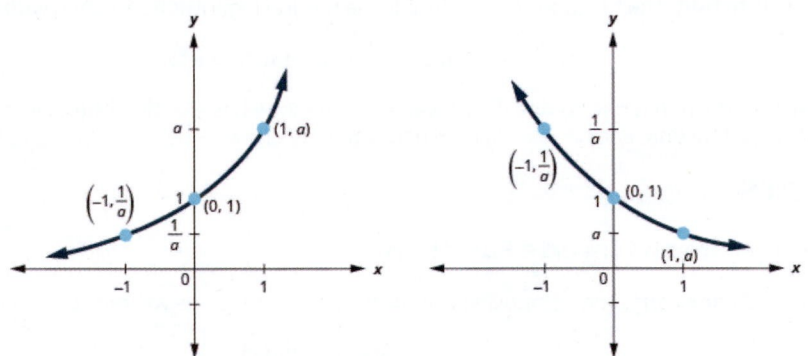

- **One-to-One Property of Exponential Equations:**
  For $a > 0$ and $a \neq 1$,

$$A = A_0 e^{rt}$$

- **How to Solve an Exponential Equation**

  Step 1. Write both sides of the equation with the same base, if possible.

  Step 2. Write a new equation by setting the exponents equal.

  Step 3. Solve the equation.

  Step 4. Check the solution.

- **Compound Interest:** For a principal, P, invested at an interest rate, r, for t years, the new balance, A, is

  $A = P\left(1 + \frac{r}{n}\right)^{nt}$ when compounded n times a year.

  $A = Pe^{rt}$ when compounded continuously.

- **Exponential Growth and Decay:** For an original amount, A_0 that grows or decays at a rate, r, for a certain time t, the final amount, A, is $A = A_0 e^{rt}$.

10.3 Evaluate and Graph Logarithmic Functions

- **Properties of the Graph of** $y = \log_a x$:

when $a > 1$		when $0 < a < 1$	
Domain	$(0, \infty)$	Domain	$(0, \infty)$
Range	$(-\infty, \infty)$	Range	$(-\infty, \infty)$
x-intercept	$(1, 0)$	x-intercept	$(1, 0)$
y-intercept	none	y-intercept	none
Contains	$(a, 1)$, $\left(\frac{1}{a}, -1\right)$	Contains	$(a, 1)$, $\left(\frac{1}{a}, -1\right)$
Asymptote	y-axis	Asymptote	y-axis
Basic shape	increasing	Basic shape	decreasing

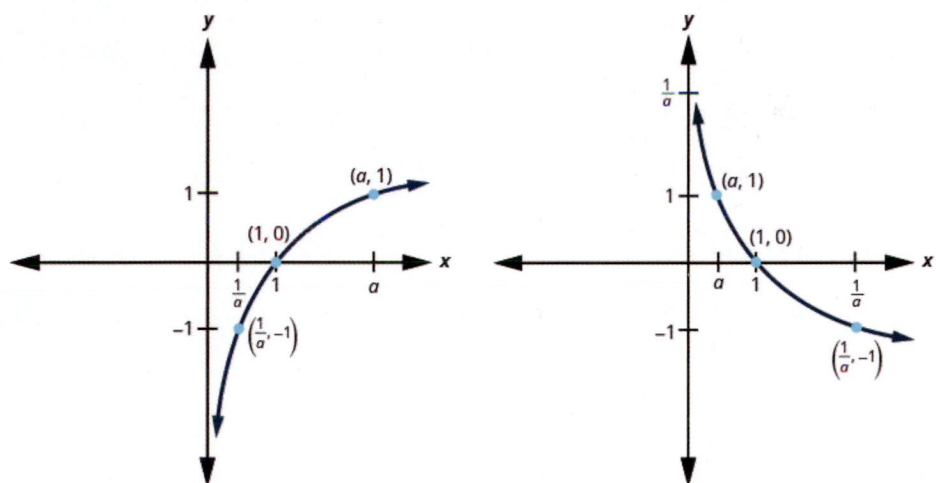

- **Decibel Level of Sound:** The loudness level, D, measured in decibels, of a sound of intensity, I, measured in watts per square inch is $D = 10\log\left(\dfrac{I}{10^{-12}}\right)$.

- **Earthquake Intensity:** The magnitude R of an earthquake is measured by $R = \log I$, where I is the intensity of its shock wave.

10.4 Use the Properties of Logarithms

- **Properties of Logarithms**

$$\log_a 1 = 0 \qquad\qquad \log_a a = 1$$

- **Inverse Properties of Logarithms**
  - For $a > 0$, $x > 0$ and $a \neq 1$

$$a^{\log_a x} = x \qquad\qquad \log_a a^x = x$$

- **Product Property of Logarithms**
  - If $M > 0, N > 0, a > 0$ and $a \neq 1$, then,

$$\log_a M \cdot N = \log_a M + \log_a N$$

The logarithm of a product is the sum of the logarithms.

- **Quotient Property of Logarithms**

◦ If $M > 0$, $N > 0$, $a > 0$ and $a \neq 1$, then,

$$\log_a \frac{M}{N} = \log_a M - \log_a N$$

The logarithm of a quotient is the difference of the logarithms.

- **Power Property of Logarithms**
  ◦ If $M > 0$, $a > 0$, $a \neq 1$ and p is any real number then,

$$\log_a M^p = p \log_a M$$

The log of a number raised to a power is the product of the power times the log of the number.

- **Properties of Logarithms Summary**
  If $M > 0$, $a > 0$, $a \neq 1$ and p is any real number then,

Property	Base a	Base e
	$\log_a 1 = 0$	$\ln 1 = 0$
	$\log_a a = 1$	$\ln e = 1$
Inverse Properties	$a^{\log_a x} = x$ $\log_a a^x = x$	$e^{\ln x} = x$ $\ln e^x = x$
Product Property of Logarithms	$\log_a (M \cdot N) = \log_a M + \log_a N$	$\ln(M \cdot N) = \ln M + \ln N$
Quotient Property of Logarithms	$\log_a \frac{M}{N} = \log_a M - \log_a N$	$\ln\frac{M}{N} = \ln M - \ln N$
Power Property of Logarithms	$\log_a M^p = p \log_a M$	$\ln M^p = p\ln M$

- **Change-of-Base Formula**
  For any logarithmic bases a and b, and $M > 0$,

$$\log_a M = \frac{\log_b M}{\log_b a} \qquad \log_a M = \frac{\log M}{\log a} \qquad \log_a M = \frac{\ln M}{\ln a}$$

$$\text{new base } b \qquad\qquad \text{new base } 10 \qquad\qquad \text{new base } e$$

10.5 Solve Exponential and Logarithmic Equations

- **One-to-One Property of Logarithmic Equations:** For $M > 0$, $N > 0$, $a > 0$, and $a \neq 1$ is any real number:

$$\text{If } \log_a M = \log_a N, \text{ then } M = N.$$

- **Compound Interest:**
  For a principal, P, invested at an interest rate, r, for t years, the new balance, A, is:

$$A = P\left(1 + \frac{r}{n}\right)^{nt} \qquad \text{when compounded } n \text{ times a year.}$$

$$A = Pe^{rt} \qquad \text{when compounded continuously.}$$

- **Exponential Growth and Decay:** For an original amount, A_0 that grows or decays at a rate, r, for a certain time t, the final amount, A, is $A = A_0 e^{rt}$.

REVIEW EXERCISES

10.1 Finding Composite and Inverse Functions

Find and Evaluate Composite Functions

In the following exercises, for each pair of functions, find ⓐ $(f \circ g)(x)$, ⓑ $(g \circ f)(x)$, and ⓒ $(f \cdot g)(x)$.

356. $f(x) = 7x - 2$ and

$g(x) = 5x + 1$

357. $f(x) = 4x$ and

$g(x) = x^2 + 3x$

In the following exercises, evaluate the composition.

358. For functions

$f(x) = 3x^2 + 2$ and

$g(x) = 4x - 3$, find

ⓐ $(f \circ g)(-3)$

ⓑ $(g \circ f)(-2)$

ⓒ $(f \circ f)(-1)$

359. For functions

$f(x) = 2x^3 + 5$ and

$g(x) = 3x^2 - 7$, find

ⓐ $(f \circ g)(-1)$

ⓑ $(g \circ f)(-2)$

ⓒ $(g \circ g)(1)$

Determine Whether a Function is One-to-One

In the following exercises, for each set of ordered pairs, determine if it represents a function and if so, is the function one-to-one.

360. $\{(-3, -5), (-2, -4), (-1, -3), (0, -2),$

$(-1, -1), (-2, 0), (-3, 1)\}$

361. $\{(-3, 0), (-2, -2), (-1, 0), (0, 1),$

$(1, 2), (2, 1), (3, -1)\}$

362. $\{(-3, 3), (-2, 1), (-1, -1), (0, -3),$

$(1, -5), (2, -4), (3, -2)\}$

In the following exercises, determine whether each graph is the graph of a function and if so, is it one-to-one.

363. ⓐ

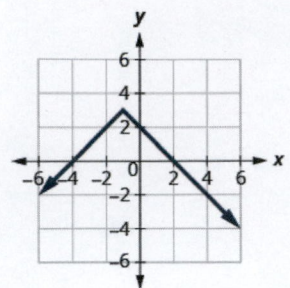

364. ⓐ

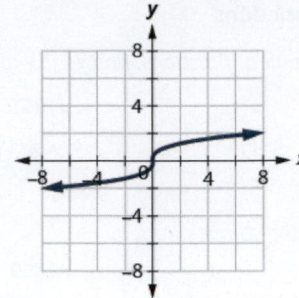

ⓑ

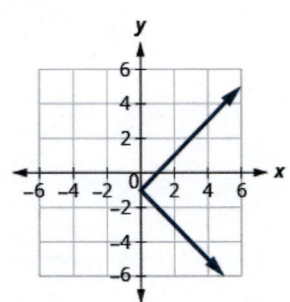

ⓑ

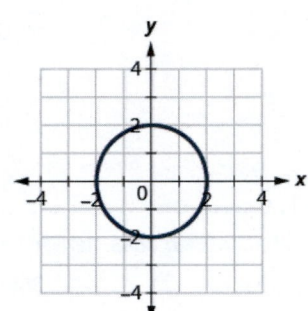

Find the Inverse of a Function

In the following exercise, find the inverse of the function. Determine the domain and range of the inverse function.

365. $\{(-3, 10), (-2, 5), (-1, 2), (0, 1)\}$

In the following exercise, graph the inverse of the one-to-one function shown.

366.

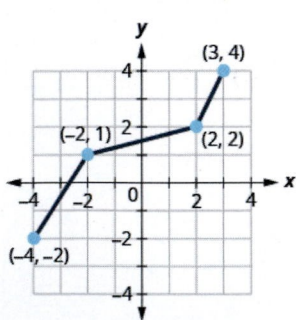

In the following exercises, verify that the functions are inverse functions.

367. $f(x) = 3x + 7$ and
$g(x) = \dfrac{x-7}{3}$

368. $f(x) = 2x + 9$ and
$g(x) = \dfrac{x+9}{2}$

In the following exercises, find the inverse of each function.

369. $f(x) = 6x - 11$

370. $f(x) = x^3 + 13$

371. $f(x) = \dfrac{1}{x+5}$

372. $f(x) = \sqrt[5]{x-1}$

10.2 Evaluate and Graph Exponential Functions

Graph Exponential Functions

In the following exercises, graph each of the following functions.

373. $f(x) = 4^x$

374. $f(x) = \left(\dfrac{1}{5}\right)^x$

375. $g(x) = (0.75)^x$

376. $g(x) = 3^{x+2}$

377. $f(x) = (2.3)^x - 3$

378. $f(x) = e^x + 5$

379. $f(x) = -e^x$

Solve Exponential Equations

In the following exercises, solve each equation.

380. $3^{5x-6} = 81$

381. $2^{x^2} = 16$

382. $9^x = 27$

383. $5^{x^2+2x} = \dfrac{1}{5}$

384. $e^{4x} \cdot e^7 = e^{19}$

385. $\dfrac{e^{x^2}}{e^{15}} = e^{2x}$

Use Exponential Models in Applications

In the following exercises, solve.

386. Felix invested $12,000 in a savings account. If the interest rate is 4% how much will be in the account in 12 years by each method of compounding?

ⓐ compound quarterly

ⓑ compound monthly

ⓒ compound continuously.

387. Sayed deposits $20,000 in an investment account. What will be the value of his investment in 30 years if the investment is earning 7% per year and is compounded continuously?

388. A researcher at the Center for Disease Control and Prevention is studying the growth of a bacteria. She starts her experiment with 150 of the bacteria that grows at a rate of 15% per hour. She will check on the bacteria every 24 hours. How many bacteria will he find in 24 hours?

389. In the last five years the population of the United States has grown at a rate of 0.7% per year to about 318,900,000. If this rate continues, what will be the population in 5 more years?

10.3 Evaluate and Graph Logarithmic Functions

Convert Between Exponential and Logarithmic Form

In the following exercises, convert from exponential to logarithmic form.

390. $5^4 = 625$

391. $10^{-3} = \dfrac{1}{1,000}$

392. $63^{\frac{1}{5}} = \sqrt[5]{63}$

393. $e^y = 16$

In the following exercises, convert each logarithmic equation to exponential form.

394. $7 = \log_2 128$

395. $5 = \log 100{,}000$

396. $4 = \ln x$

Evaluate Logarithmic Functions

In the following exercises, solve for x.

397. $\log_x 125 = 3$

398. $\log_7 x = -2$

399. $\log_{\frac{1}{2}} \frac{1}{16} = x$

In the following exercises, find the exact value of each logarithm without using a calculator.

400. $\log_2 32$

401. $\log_8 1$

402. $\log_3 \frac{1}{9}$

Graph Logarithmic Functions

In the following exercises, graph each logarithmic function.

403. $y = \log_5 x$

404. $y = \log_{\frac{1}{4}} x$

405. $y = \log_{0.8} x$

Solve Logarithmic Equations

In the following exercises, solve each logarithmic equation.

406. $\log_a 36 = 5$

407. $\ln x = -3$

408. $\log_2 (5x - 7) = 3$

409. $\ln e^{3x} = 24$

410. $\log(x^2 - 21) = 2$

Use Logarithmic Models in Applications

411. What is the decibel level of a train whistle with intensity 10^{-3} watts per square inch?

10.4 Use the Properties of Logarithms

Use the Properties of Logarithms

In the following exercises, use the properties of logarithms to evaluate.

412. ⓐ $\log_7 1$ ⓑ $\log_{12} 12$

413. ⓐ $5^{\log_5 13}$ ⓑ $\log_3 3^{-9}$

414. ⓐ $10^{\log \sqrt{5}}$ ⓑ $\log 10^{-3}$

415. ⓐ $e^{\ln 8}$ ⓑ $\ln e^5$

In the following exercises, use the Product Property of Logarithms to write each logarithm as a sum of logarithms. Simplify if possible.

416. $\log_4(64xy)$

417. $\log 10{,}000m$

In the following exercises, use the Quotient Property of Logarithms to write each logarithm as a sum of logarithms. Simplify, if possible.

418. $\log_7 \dfrac{49}{y}$

419. $\ln \dfrac{e^5}{2}$

In the following exercises, use the Power Property of Logarithms to expand each logarithm. Simplify, if possible.

420. $\log x^{-9}$

421. $\log_4 \sqrt[7]{z}$

In the following exercises, use properties of logarithms to write each logarithm as a sum of logarithms. Simplify if possible.

422. $\log_3\left(\sqrt{4x^7 y^8}\right)$

423. $\log_5 \dfrac{8a^2 b^6 c}{d^3}$

424. $\ln \dfrac{\sqrt{3x^2 - y^2}}{z^4}$

425. $\log_6 \sqrt[3]{\dfrac{7x^2}{6y^3 z^5}}$

In the following exercises, use the Properties of Logarithms to condense the logarithm. Simplify if possible.

426. $\log_2 56 - \log_2 7$

427. $3\log_3 x + 7\log_3 y$

428. $\log_5(x^2 - 16) - 2\log_5(x + 4)$

429. $\frac{1}{4}\log y - 2\log(y - 3)$

Use the Change-of-Base Formula

In the following exercises, rounding to three decimal places, approximate each logarithm.

430. $\log_5 97$

431. $\log_{\sqrt{3}} 16$

10.5 Solve Exponential and Logarithmic Equations

Solve Logarithmic Equations Using the Properties of Logarithms

In the following exercises, solve for x.

432. $3\log_5 x = \log_5 216$

433. $\log_2 x + \log_2(x - 2) = 3$

434. $\log(x - 1) - \log(3x + 5) = -\log x$

435. $\log_4(x - 2) + \log_4(x + 5) = \log_4 8$

436. $\ln(3x - 2) = \ln(x + 4) + \ln 2$

Solve Exponential Equations Using Logarithms

In the following exercises, solve each exponential equation. Find the exact answer and then approximate it to three decimal places.

437. $2^x = 101$

438. $e^x = 23$

439. $\left(\frac{1}{3}\right)^x = 7$

440. $7e^{x+3} = 28$

441. $e^{x-4} + 8 = 23$

Use Exponential Models in Applications

442. Jerome invests $18,000 at age 17. He hopes the investments will be worth $30,000 when he turns 26. If the interest compounds continuously, approximately what rate of growth will he need to achieve his goal? Is that a reasonable expectation?

443. Elise invests $4500 in an account that compounds interest monthly and earns 6%. How long will it take for her money to double?

444. Researchers recorded that a certain bacteria population grew from 100 to 300 in 8 hours. At this rate of growth, how many bacteria will there be in 24 hours?

445. Mouse populations can double in 8 months $(A = 2A_0)$. How long will it take for a mouse population to triple?

446. The half-life of radioactive iodine is 60 days. How much of a 50 mg sample will be left in 40 days?

PRACTICE TEST

447. For the functions, $f(x) = 6x + 1$ and $g(x) = 8x - 3$, find ⓐ $(f \circ g)(x)$, ⓑ $(g \circ f)(x)$, and ⓒ $(f \cdot g)(x)$.

448. Determine if the following set of ordered pairs represents a function and if so, is the function one-to-one. $\{(-2, 2), (-1, -3), (0, 1), (1, -2), (2, -3)\}$

449. Determine whether each graph is the graph of a function and if so, is it one-to-one.

ⓐ

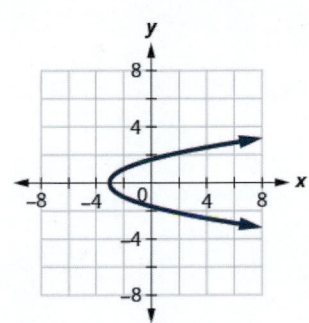

ⓑ

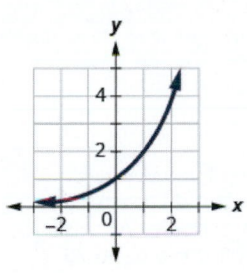

450. Graph, on the same coordinate system, the inverse of the one-to-one function shown.

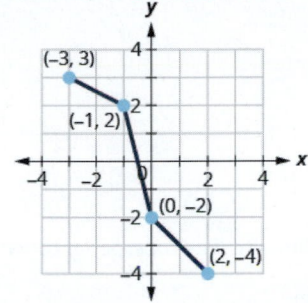

451. Find the inverse of the function $f(x) = x^5 - 9$.

452. Graph the function $g(x) = 2^{x-3}$.

453. Solve the equation $2^{2x-4} = 64$.

454. Solve the equation $\dfrac{e^{x^2}}{e^4} = e^{3x}$.

455. Megan invested $21,000 in a savings account. If the interest rate is 5%, how much will be in the account in 8 years by each method of compounding?
ⓐ compound quarterly
ⓑ compound monthly
ⓒ compound continuously.

456. Convert the equation from exponential to logarithmic form: $10^{-2} = \dfrac{1}{100}$.

457. Convert the equation from logarithmic equation to exponential form: $3 = \log_7 343$

458. Solve for x: $\log_5 x = -3$

459. Evaluate $\log_{11} 1$.

460. Evaluate $\log_4 \dfrac{1}{64}$.

461. Graph the function
$y = \log_3 x$.

462. Solve for x:
$\log(x^2 - 39) = 1$

463. What is the decibel level of a small fan with intensity 10^{-8} watts per square inch?

464. Evaluate each. ⓐ $6^{\log_6 17}$
ⓑ $\log_9 9^{-3}$

In the following exercises, use properties of logarithms to write each expression as a sum of logarithms, simplifying if possible.

465. $\log_5 25ab$

466. $\ln\dfrac{e^{12}}{8}$

467. $\log_2 \sqrt[4]{\dfrac{5x^3}{16y^2 z^7}}$

In the following exercises, use the Properties of Logarithms to condense the logarithm, simplifying if possible.

468. $5\log_4 x + 3\log_4 y$

469. $\dfrac{1}{6}\log x - 3\log(x+5)$

470. Rounding to three decimal places, approximate $\log_4 73$.

471. Solve for x:
$\log_7(x+2) + \log_7(x-3) = \log_7 24$

In the following exercises, solve each exponential equation. Find the exact answer and then approximate it to three decimal places.

472. $\left(\dfrac{1}{5}\right)^x = 9$

473. $5e^{x-4} = 40$

474. Jacob invests $14,000 in an account that compounds interest quarterly and earns 4%. How long will it take for his money to double?

475. Researchers recorded that a certain bacteria population grew from 500 to 700 in 5 hours. At this rate of growth, how many bacteria will there be in 20 hours?

476. A certain beetle population can double in 3 months $(A = 2A_0)$. How long will it take for that beetle population to triple?

Figure 11.1 Aerospace engineers use rockets such as this one to launch people and objects into space. (credit: WikiImages/Pixabay)

Chapter Outline

11.1 Distance and Midpoint Formulas; Circles

11.2 Parabolas

11.3 Ellipses

11.4 Hyperbolas

11.5 Solve Systems of Nonlinear Equations

 **Introduction**

Five, Four. Three. Two. One. Lift off. The rocket launches off the ground headed toward space. Unmanned spaceships, and spaceships in general, are designed by aerospace engineers. These engineers are investigating reusable rockets that return safely to Earth to be used again. Someday, rockets may carry passengers to the International Space Station and beyond. One essential math concept for aerospace engineers is that of conics. In this chapter, you will learn about conics, including circles, parabolas, ellipses, and hyperbolas. Then you will use what you learn to investigate systems of nonlinear equations.

 Distance and Midpoint Formulas; Circles

Learning Objectives

By the end of this section, you will be able to:

> Use the Distance Formula
> Use the Midpoint Formula
> Write the equation of a circle in standard form
> Graph a circle

Be Prepared!

Before you get started, take this readiness quiz.

1. Find the length of the hypotenuse of a right triangle whose legs are 12 and 16 inches.
   If you missed this problem, review **Example 2.34**.

2. Factor: $x^2 - 18x + 81$.
   If you missed this problem, review **Example 6.24**.

3. Solve by completing the square: $x^2 - 12x - 12 = 0$.
   If you missed this problem, review **Example 9.22**.

In this chapter we will be looking at the conic sections, usually called the conics, and their properties. The conics are curves that result from a plane intersecting a double cone—two cones placed point-to-point. Each half of a double cone is called a nappe.

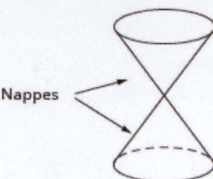

Nappes

There are four conics—the **circle**, **parabola**, **ellipse**, and **hyperbola**. The next figure shows how the plane intersecting the double cone results in each curve.

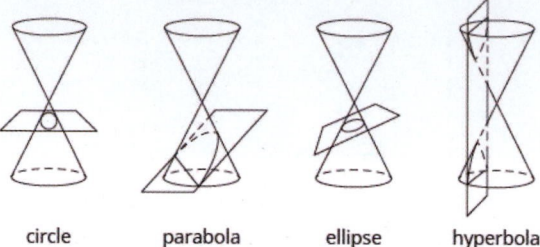

circle   parabola   ellipse   hyperbola

Each of the curves has many applications that affect your daily life, from your cell phone to acoustics and navigation systems. In this section we will look at the properties of a circle.

Use the Distance Formula

We have used the Pythagorean Theorem to find the lengths of the sides of a right triangle. Here we will use this theorem again to find distances on the rectangular coordinate system. By finding distance on the rectangular coordinate system, we can make a connection between the geometry of a conic and algebra—which opens up a world of opportunities for application.

Our first step is to develop a formula to find distances between points on the rectangular coordinate system. We will plot the points and create a right triangle much as we did when we found slope in Graphs and Functions. We then take it one step further and use the Pythagorean Theorem to find the length of the hypotenuse of the triangle—which is the distance between the points.

EXAMPLE 11.1

Use the rectangular coordinate system to find the distance between the points $(6, 4)$ and $(2, 1)$.

⊘ Solution

Plot the two points. Connect the two points with a line. Draw a right triangle as if you were going to find slope.	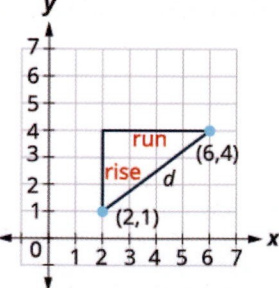
Find the length of each leg.	The rise is 3. The run is 4. 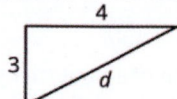
Use the Pythagorean Theorem to find d, the distance between the two points.	$$a^2 + b^2 = c^2$$
Substitute in the values.	$$3^2 + 4^2 = d^2$$

Simplify.	$9 + 16 = d^2$
	$25 = d^2$
Use the Square Root Property.	$d = 5 \quad \cancel{d = -5}$
Since distance, d is positive, we can eliminate $d = -5$.	The distance between the points $(6,\ 4)$ and $(2,\ 1)$ is 5.

> **TRY IT : : 11.1**
>
> Use the rectangular coordinate system to find the distance between the points $(6,\ 1)$ and $(2,\ -2)$.

> **TRY IT : : 11.2**
>
> Use the rectangular coordinate system to find the distance between the points $(5,\ 3)$ and $(-3,\ -3)$.

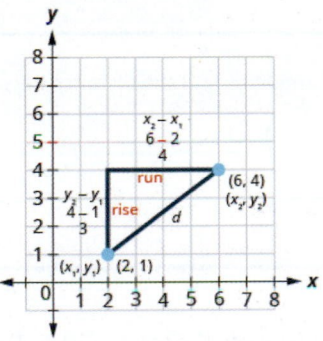

The method we used in the last example leads us to the formula to find the distance between the two points $(x_1,\ y_1)$ and $(x_2,\ y_2)$.

When we found the length of the horizontal leg we subtracted $6 - 2$ which is $x_2 - x_1$.

When we found the length of the vertical leg we subtracted $4 - 1$ which is $y_2 - y_1$.

If the triangle had been in a different position, we may have subtracted $x_1 - x_2$ or $y_1 - y_2$. The expressions $x_2 - x_1$ and $x_1 - x_2$ vary only in the sign of the resulting number. To get the positive value-since distance is positive- we can use absolute value. So to generalize we will say $|x_2 - x_1|$ and $|y_2 - y_1|$.

In the Pythagorean Theorem, we substitute the general expressions $|x_2 - x_1|$ and $|y_2 - y_1|$ rather than the numbers.

$$a^2 + b^2 = c^2$$

Substitute in the values.	$(|x_2 - x_1|)^2 + (|y_2 - y_1|)^2 = d^2$
Squaring the expressions makes them positive, so we eliminate the absolute value bars.	$(x_2 - x_1)^2 + (y_2 - y_1)^2 = d^2$
Use the Square Root Property.	$d = \pm\sqrt{(x_2 - x_1)^2 + (y_2 - y_1)^2}$
Distance is positive, so eliminate the negative value.	$d = \sqrt{(x_2 - x_1)^2 + (y_2 - y_1)^2}$

This is the Distance Formula we use to find the distance d between the two points $(x_1,\ y_1)$ and $(x_2,\ y_2)$.

Distance Formula

The distance d between the two points (x_1, y_1) and (x_2, y_2) is

$$d = \sqrt{(x_2 - x_1)^2 + (y_2 - y_1)^2}$$

EXAMPLE 11.2

Use the Distance Formula to find the distance between the points $(-5, -3)$ and $(7, 2)$.

⊘ Solution

Write the Distance Formula.

$$d = \sqrt{(x_2 - x_1)^2 + (y_2 - y_1)^2}$$

Label the points, $\left(\overset{x_1,\ y_1}{-5,\ -3}\right), \left(\overset{x_2,\ y_2}{7,\ 2}\right)$ and substitute.

$$d = \sqrt{(7 - (-5))^2 + (2 - (-3))^2}$$

Simplify.

$$d = \sqrt{12^2 + 5^2}$$
$$d = \sqrt{144 + 25}$$
$$d = \sqrt{169}$$
$$d = 13$$

> **TRY IT ::** 11.3 Use the Distance Formula to find the distance between the points $(-4, -5)$ and $(5, 7)$.

> **TRY IT ::** 11.4
>
> Use the Distance Formula to find the distance between the points $(-2, -5)$ and $(-14, -10)$.

EXAMPLE 11.3

Use the Distance Formula to find the distance between the points $(10, -4)$ and $(-1, 5)$. Write the answer in exact form and then find the decimal approximation, rounded to the nearest tenth if needed.

⊘ Solution

Write the Distance Formula.

$$d = \sqrt{(x_2 - x_1)^2 + (y_2 - y_1)^2}$$

Label the points, $\left(\overset{x_1,\ y_1}{10,\ -4}\right), \left(\overset{x_2,\ y_2}{-1,\ 5}\right)$ and substitute.

$$d = \sqrt{(-1 - 10)^2 + (5 - (-4))^2}$$

Simplify.

$$d = \sqrt{(-11)^2 + 9^2}$$
$$d = \sqrt{121 + 81}$$
$$d = \sqrt{202}$$

Since 202 is not a perfect square, we can leave the answer in exact form or find a decimal approximation.

$$d = \sqrt{202}$$
or
$$d \approx 14.2$$

> **TRY IT ::** 11.5
>
> Use the Distance Formula to find the distance between the points $(-4, -5)$ and $(3, 4)$. Write the answer in exact form and then find the decimal approximation, rounded to the nearest tenth if needed.

TRY IT : : 11.6

Use the Distance Formula to find the distance between the points $(-2, -5)$ and $(-3, -4)$. Write the answer in exact form and then find the decimal approximation, rounded to the nearest tenth if needed.

Use the Midpoint Formula

It is often useful to be able to find the midpoint of a segment. For example, if you have the endpoints of the diameter of a circle, you may want to find the center of the circle which is the midpoint of the diameter. To find the midpoint of a line segment, we find the average of the *x*-coordinates and the average of the *y*-coordinates of the endpoints.

Midpoint Formula

The midpoint of the line segment whose endpoints are the two points (x_1, y_1) and (x_2, y_2) is

$$\left(\frac{x_1 + x_2}{2}, \frac{y_1 + y_2}{2}\right)$$

To find the midpoint of a line segment, we find the average of the *x*-coordinates and the average of the *y*-coordinates of the endpoints.

EXAMPLE 11.4

Use the Midpoint Formula to find the midpoint of the line segments whose endpoints are $(-5, -4)$ and $(7, 2)$. Plot the endpoints and the midpoint on a rectangular coordinate system.

⊘ Solution

Write the Midpoint Formula.	$\left(\frac{x_1 + x_2}{2}, \frac{y_1 + y_2}{2}\right)$
Label the points, $\left(\overset{x_1, y_1}{-5, -4}\right), \left(\overset{x_2, y_2}{7, 2}\right)$ and substitute.	$\left(\frac{-5 + 7}{2}, \frac{-4 + 2}{2}\right)$
Simplify.	$\left(\frac{2}{2}, \frac{-2}{2}\right)$

$$(1, -1)$$

The midpoint of the segment is the point $(1, -1)$.

Plot the endpoints and midpoint.

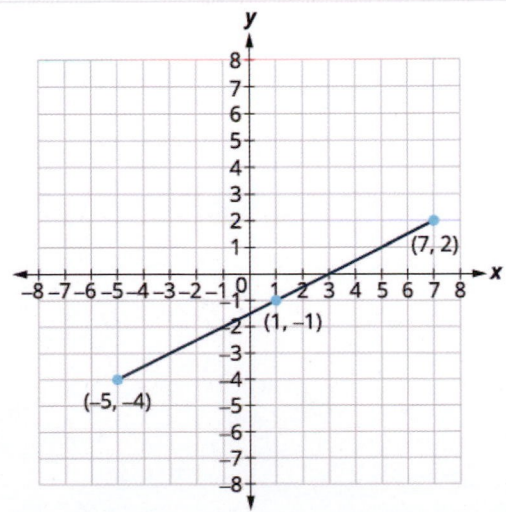

> **TRY IT : : 11.7**

Use the Midpoint Formula to find the midpoint of the line segments whose endpoints are $(-3, -5)$ and $(5, 7)$.

Plot the endpoints and the midpoint on a rectangular coordinate system.

> **TRY IT : : 11.8**

Use the Midpoint Formula to find the midpoint of the line segments whose endpoints are $(-2, -5)$ and $(6, -1)$.

Plot the endpoints and the midpoint on a rectangular coordinate system.

Both the Distance Formula and the Midpoint Formula depend on two points, (x_1, y_1) and (x_2, y_2). It is easy to confuse which formula requires addition and which subtraction of the coordinates. If we remember where the formulas come from, is may be easier to remember the formulas.

<table>
<tr><td align="center">**Distance Formula**</td><td align="center">**Midpoint Formula**</td></tr>
<tr><td align="center">$d = \sqrt{(x_2 - x_1)^2 + (y_2 - y_1)^2}$</td><td align="center">$\left(\dfrac{x_1 + x_2}{2}, \dfrac{y_1 + y_2}{2} \right)$</td></tr>
<tr><td align="center">Subtract the coordinates.</td><td align="center">Add the coordinates.</td></tr>
</table>

Write the Equation of a Circle in Standard Form

As we mentioned, our goal is to connect the geometry of a conic with algebra. By using the coordinate plane, we are able to do this easily.

circle

We define a circle as all points in a plane that are a fixed distance from a given point in the plane. The given point is called the *center*, (h, k), and the fixed distance is called the *radius*, r, of the circle.

Circle

A circle is all points in a plane that are a fixed distance from a given point in the plane. The given point is called the **center**, (h, k), and the fixed distance is called the **radius**, r, of the circle.

We look at a circle in the rectangular coordinate system. The radius is the distance from the center, (h, k), to a point on the circle, (x, y).

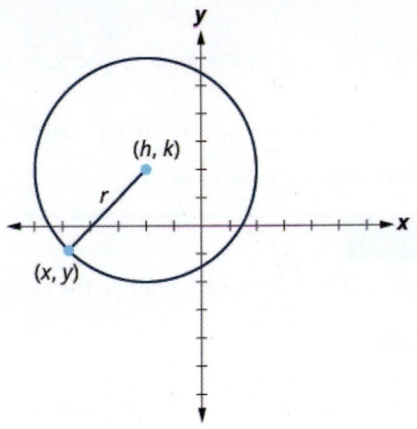

To derive the equation of a circle, we can use the distance formula with the points (h, k), (x, y) and the distance, r.	$d = \sqrt{(x_2 - x_1)^2 + (y_2 - y_1)^2}$
Substitute the values.	$r = \sqrt{(x - h)^2 + (y - k)^2}$
Square both sides.	$r^2 = (x - h)^2 + (y - k)^2$

This is the standard form of the equation of a circle with center, (h, k), and radius, r.

Standard Form of the Equation a Circle

The standard form of the equation of a circle with center, (h, k), and radius, r, is

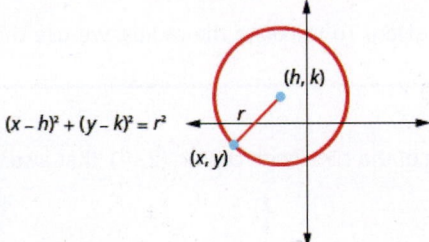

EXAMPLE 11.5

Write the standard form of the equation of the circle with radius 3 and center $(0, 0)$.

✓ **Solution**

Use the standard form of the equation of a circle	$(x - h)^2 + (y - k)^2 = r^2$
Substitute in the values $r = 3$, $h = 0$, and $k = 0$.	$(x - 0)^2 + (y - 0)^2 = 3^2$
Center: $\binom{h,\ k}{0,\ 0}$	
Simplify.	$x^2 + y^2 = 9$

> **TRY IT : : 11.9**   Write the standard form of the equation of the circle with a radius of 6 and center $(0, 0)$.

> **TRY IT : : 11.10**   Write the standard form of the equation of the circle with a radius of 8 and center $(0, 0)$.

In the last example, the center was $(0, 0)$. Notice what happened to the equation. Whenever the center is $(0, 0)$, the standard form becomes $x^2 + y^2 = r^2$.

EXAMPLE 11.6

Write the standard form of the equation of the circle with radius 2 and center $(-1, 3)$.

⊘ Solution

Use the standard form of the equation of a circle.	$(x - h)^2 + (y - k)^2 = r^2$
Substitute in the values.	$(x - (-1))^2 + (y - 3)^2 = 2^2$
Center: $\left(\overset{h,\ k}{-1,\ 3}\right)$	
Simplify.	$(x + 1)^2 + (y - 3)^2 = 4$

> **TRY IT : : 11.11**   Write the standard form of the equation of the circle with a radius of 7 and center $(2, -4)$.

> **TRY IT : : 11.12**
>
> Write the standard form of the equation of the circle with a radius of 9 and center $(-3, -5)$.

In the next example, the radius is not given. To calculate the radius, we use the Distance Formula with the two given points.

EXAMPLE 11.7

Write the standard form of the equation of the circle with center $(2, 4)$ that also contains the point $(-2, 1)$.

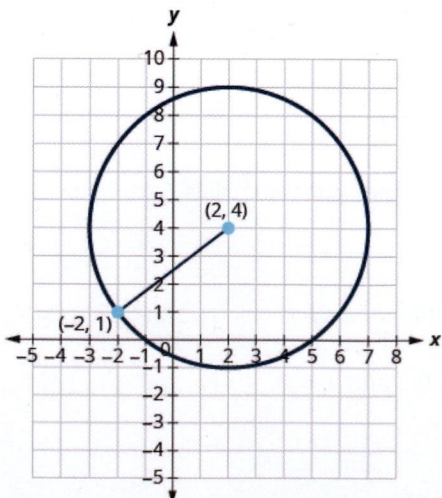

⊘ Solution

The radius is the distance from the center to any point on the circle so we can use the distance formula to calculate it. We will use the center $(2, 4)$ and point $(-2, 1)$

Use the Distance Formula to find the radius.	$r = \sqrt{(x_2 - x_1)^2 + (y_2 - y_1)^2}$
Substitute the values. $\left(\overset{x_1,\ y_1}{2,\ 4}\right), \left(\overset{x_2,\ y_2}{-2,\ 1}\right)$	$r = \sqrt{(-2 - 2)^2 + (1 - 4)^2}$

Simplify.

$$r = \sqrt{(-4)^2 + (-3)^2}$$
$$r = \sqrt{16 + 9}$$
$$r = \sqrt{25}$$
$$r = 5$$

Now that we know the radius, $r = 5$, and the center, $(2, 4)$, we can use the standard form of the equation of a circle to find the equation.

Use the standard form of the equation of a circle.	$(x - h)^2 + (y - k)^2 = r^2$
Substitute in the values.	$(x - 2)^2 + (y - 4)^2 = 5^2$
Simplify.	$(x - 2)^2 + (y - 4)^2 = 25$

> **TRY IT : : 11.13**

  Write the standard form of the equation of the circle with center $(2, 1)$ that also contains the point $(-2, -2)$.

> **TRY IT : : 11.14**

  Write the standard form of the equation of the circle with center $(7, 1)$ that also contains the point $(-1, -5)$.

Graph a Circle

Any equation of the form $(x - h)^2 + (y - k)^2 = r^2$ is the standard form of the equation of a circle with center, (h, k), and radius, r. We can then graph the circle on a rectangular coordinate system.

Note that the standard form calls for subtraction from x and y. In the next example, the equation has $x + 2$, so we need to rewrite the addition as subtraction of a negative.

EXAMPLE 11.8

Find the center and radius, then graph the circle: $(x + 2)^2 + (y - 1)^2 = 9$.

✓ **Solution**

$$(x + 2)^2 + (y - 1)^2 = 9$$

Use the standard form of the equation of a circle. Identify the center, (h, k) and radius, r.	$(x - h)^2 + (y - k)^2 = r^2$ $(x - (-2))^2 + (y - 1)^2 = 3^2$ Center: $(-2, 1)$ radius: 3
Graph the circle.	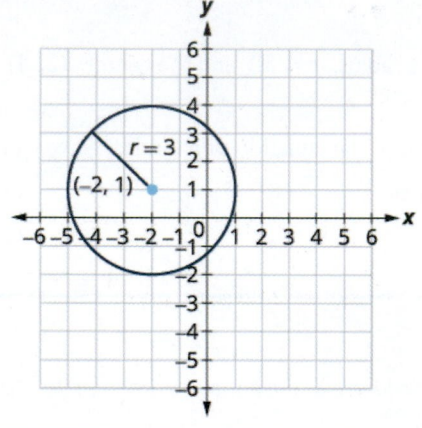

> **TRY IT :: 11.15**   ⓐ Find the center and radius, then ⓑ graph the circle: $(x - 3)^2 + (y + 4)^2 = 4$.

> **TRY IT :: 11.16**   ⓐ Find the center and radius, then ⓑ graph the circle: $(x - 3)^2 + (y - 1)^2 = 16$.

To find the center and radius, we must write the equation in standard form. In the next example, we must first get the coefficient of x^2, y^2 to be one.

EXAMPLE 11.9

Find the center and radius and then graph the circle, $4x^2 + 4y^2 = 64$.

✓ **Solution**

$$4x^2 + 4y^2 = 64$$

Divide each side by 4.	$x^2 + y^2 = 16$
Use the standard form of the equation of a circle. Identify the center, (h, k) and radius, r.	$(x - h)^2 + (y - k)^2 = r^2$ $(x - 0)^2 + (y - 0)^2 = 4^2$ Center: $(0, 0)$ radius: 4

Graph the circle.

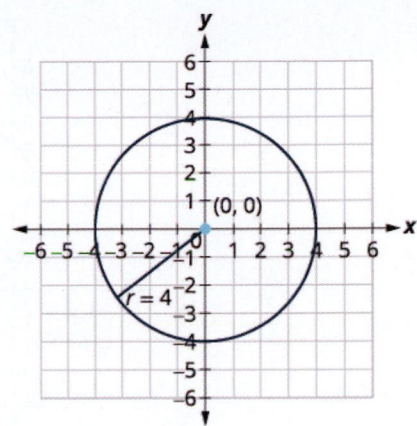

> **TRY IT : : 11.17**

ⓐ Find the center and radius, then ⓑ graph the circle: $3x^2 + 3y^2 = 27$

> **TRY IT : : 11.18**

ⓐ Find the center and radius, then ⓑ graph the circle: $5x^2 + 5y^2 = 125$

If we expand the equation from **Example 11.8**, $(x + 2)^2 + (y - 1)^2 = 9$, the equation of the circle looks very different.

$$(x + 2)^2 + (y - 1)^2 = 9$$

Square the binomials.

$$x^2 + 4x + 4 + y^2 - 2y + 1 = 9$$

Arrange the terms in descending degree order, and get zero on the right

$$x^2 + y^2 + 4x - 2y - 4 = 0$$

This form of the equation is called the general form of the equation of the circle.

General Form of the Equation of a Circle

The general form of the equation of a circle is

$$x^2 + y^2 + ax + by + c = 0$$

If we are given an equation in general form, we can change it to standard form by completing the squares in both x and y. Then we can graph the circle using its center and radius.

EXAMPLE 11.10

ⓐ Find the center and radius, then ⓑ graph the circle: $x^2 + y^2 - 4x - 6y + 4 = 0$.

⊘ **Solution**

We need to rewrite this general form into standard form in order to find the center and radius.

$$x^2 + y^2 - 4x - 6y + 4 = 0$$

Group the x-terms and y-terms. Collect the constants on the right side.	$x^2 - 4x + y^2 - 6y = -4$
Complete the squares.	$x^2 - 4x + 4 + y^2 - 6y + 9 = -4 + 4 + 9$
Rewrite as binomial squares.	$(x - 2)^2 + (y - 3)^2 = 9$
Identify the center and radius.	Center: $(2, 3)$ radius: 3

Graph the circle.

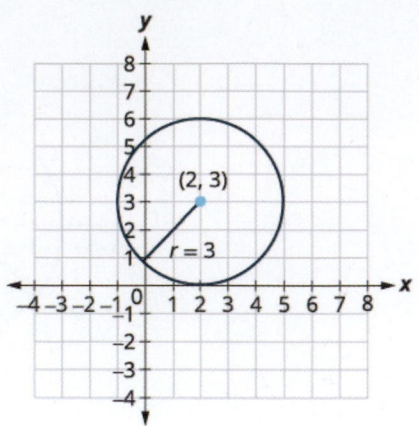

> **TRY IT : :** 11.19   ⓐ Find the center and radius, then ⓑ graph the circle: $x^2 + y^2 - 6x - 8y + 9 = 0$.

> **TRY IT : :** 11.20   ⓐ Find the center and radius, then ⓑ graph the circle: $x^2 + y^2 + 6x - 2y + 1 = 0$.

In the next example, there is a y-term and a y^2-term. But notice that there is no x-term, only an x^2-term. We have seen this before and know that it means h is 0. We will need to complete the square for the y terms, but not for the x terms.

EXAMPLE 11.11

ⓐ Find the center and radius, then ⓑ graph the circle: $x^2 + y^2 + 8y = 0$.

✓ **Solution**

We need to rewrite this general form into standard form in order to find the center and radius.

	$x^2 + y^2 + 8y = 0$
Group the x-terms and y-terms.	$x^2 + y^2 + 8y = 0$
There are no constants to collect on the right side.	
Complete the square for $y^2 + 8y$.	$x^2 + y^2 + 8y + 16 = 0 + 16$
Rewrite as binomial squares.	$(x - 0)^2 + (y + 4)^2 = 16$
Identify the center and radius.	Center: $(0, -4)$ radius: 4

Graph the circle.

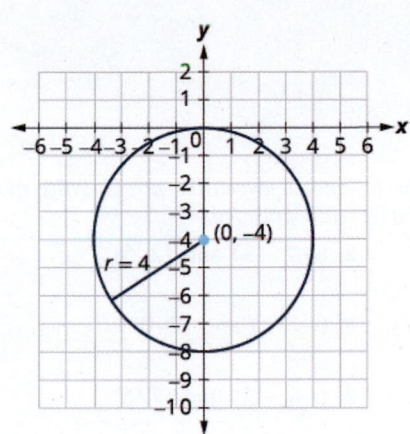

> **TRY IT :: 11.21**   ⓐ Find the center and radius, then ⓑ graph the circle: $x^2 + y^2 - 2x - 3 = 0$.

> **TRY IT :: 11.22**   ⓐ Find the center and radius, then ⓑ graph the circle: $x^2 + y^2 - 12y + 11 = 0$.

▶ **MEDIA ::**

Access these online resources for additional instructions and practice with using the distance and midpoint formulas, and graphing circles.

- **Distance-Midpoint Formulas and Circles (https://openstax.org/l/37distmidcircle)**
- **Finding the Distance and Midpoint Between Two Points (https://openstax.org/l/37distmid2pts)**
- **Completing the Square to Write Equation in Standard Form of a Circle (https://openstax.org/l/37stformcircle)**

11.1 EXERCISES

Practice Makes Perfect

Use the Distance Formula

In the following exercises, find the distance between the points. Write the answer in exact form and then find the decimal approximation, rounded to the nearest tenth if needed.

1. $(2, 0)$ and $(5, 4)$

2. $(-4, -3)$ and $(2, 5)$

3. $(-4, -3)$ and $(8, 2)$

4. $(-7, -3)$ and $(8, 5)$

5. $(-1, 4)$ and $(2, 0)$

6. $(-1, 3)$ and $(5, -5)$

7. $(1, -4)$ and $(6, 8)$

8. $(-8, -2)$ and $(7, 6)$

9. $(-3, -5)$ and $(0, 1)$

10. $(-1, -2)$ and $(-3, 4)$

11. $(3, -1)$ and $(1, 7)$

12. $(-4, -5)$ and $(7, 4)$

Use the Midpoint Formula

In the following exercises, ⓐ find the midpoint of the line segments whose endpoints are given and ⓑ plot the endpoints and the midpoint on a rectangular coordinate system.

13. $(0, -5)$ and $(4, -3)$

14. $(-2, -6)$ and $(6, -2)$

15. $(3, -1)$ and $(4, -2)$

16. $(-3, -3)$ and $(6, -1)$

Write the Equation of a Circle in Standard Form

In the following exercises, write the standard form of the equation of the circle with the given radius and center $(0, 0)$.

17. Radius: 7

18. Radius: 9

19. Radius: $\sqrt{2}$

20. Radius: $\sqrt{5}$

In the following exercises, write the standard form of the equation of the circle with the given radius and center

21. Radius: 1, center: $(3, 5)$

22. Radius: 10, center: $(-2, 6)$

23. Radius: 2.5, center: $(1.5, -3.5)$

24. Radius: 1.5, center: $(-5.5, -6.5)$

For the following exercises, write the standard form of the equation of the circle with the given center with point on the circle.

25. Center $(3, -2)$ with point $(3, 6)$

26. Center $(6, -6)$ with point $(2, -3)$

27. Center $(4, 4)$ with point $(2, 2)$

28. Center $(-5, 6)$ with point $(-2, 3)$

Graph a Circle

In the following exercises, ⓐ find the center and radius, then ⓑ graph each circle.

29. $(x + 5)^2 + (y + 3)^2 = 1$

30. $(x - 2)^2 + (y - 3)^2 = 9$

31. $(x - 4)^2 + (y + 2)^2 = 16$

32. $(x + 2)^2 + (y - 5)^2 = 4$

33. $x^2 + (y + 2)^2 = 25$

34. $(x - 1)^2 + y^2 = 36$

35. $(x - 1.5)^2 + (y + 2.5)^2 = 0.25$

36. $(x - 1)^2 + (y - 3)^2 = \frac{9}{4}$

37. $x^2 + y^2 = 64$

38. $x^2 + y^2 = 49$ **39.** $2x^2 + 2y^2 = 8$ **40.** $6x^2 + 6y^2 = 216$

In the following exercises, ⓐ identify the center and radius and ⓑ graph.

41. $x^2 + y^2 + 2x + 6y + 9 = 0$ **42.** $x^2 + y^2 - 6x - 8y = 0$ **43.** $x^2 + y^2 - 4x + 10y - 7 = 0$

44.
$x^2 + y^2 + 12x - 14y + 21 = 0$ **45.** $x^2 + y^2 + 6y + 5 = 0$ **46.** $x^2 + y^2 - 10y = 0$

47. $x^2 + y^2 + 4x = 0$ **48.** $x^2 + y^2 - 14x + 13 = 0$

Writing Exercises

49. Explain the relationship between the distance formula and the equation of a circle.

50. Is a circle a function? Explain why or why not.

51. In your own words, state the definition of a circle.

52. In your own words, explain the steps you would take to change the general form of the equation of a circle to the standard form.

Self Check

ⓐ *After completing the exercises, use this checklist to evaluate your mastery of the objectives of this section.*

I can...	Confidently	With some help	No-I don't get it!
use the distance formula.			
use the midpoint formula.			
write the equation of a circle in standard form.			
graph a circle.			

ⓑ *If most of your checks were:*

...confidently. Congratulations! You have achieved the objectives in this section. Reflect on the study skills you used so that you can continue to use them. What did you do to become confident of your ability to do these things? Be specific.

...with some help. This must be addressed quickly because topics you do not master become potholes in your road to success. In math every topic builds upon previous work. It is important to make sure you have a strong foundation before you move on. Who can you ask for help? Your fellow classmates and instructor are good resources. Is there a place on campus where math tutors are available? Can your study skills be improved?

...no - I don't get it! This is a warning sign and you must not ignore it. You should get help right away or you will quickly be overwhelmed. See your instructor as soon as you can to discuss your situation. Together you can come up with a plan to get you the help you need.

11.2 | Parabolas

Learning Objectives

By the end of this section, you will be able to:

> Graph vertical parabolas
> Graph horizontal parabolas
> Solve applications with parabolas

Be Prepared!

Before you get started, take this readiness quiz.

1. Graph: $y = -3x^2 + 12x - 12$.
   If you missed this problem, review **Example 9.47**.

2. Solve by completing the square: $x^2 - 6x + 6 = 0$.
   If you missed this problem, review **Example 9.12**.

3. Write in standard form: $y = 3x^2 - 6x + 5$.
   If you missed this problem, review **Example 9.59**.

Graph Vertical Parabolas

The next conic section we will look at is a **parabola**. We define a parabola as all points in a plane that are the same distance from a fixed point and a fixed line. The fixed point is called the **focus,** and the fixed line is called the **directrix** of the parabola.

parabola

Parabola

A **parabola** is all points in a plane that are the same distance from a fixed point and a fixed line. The fixed point is called the **focus,** and the fixed line is called the **directrix** of the parabola.

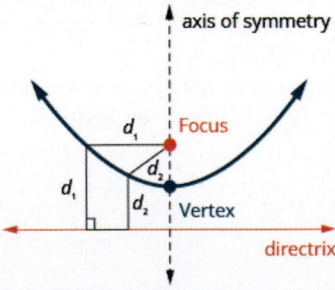

Previously, we learned to graph vertical parabolas from the general form or the standard form using properties. Those methods will also work here. We will summarize the properties here.

Vertical Parabolas		
	General form $y = ax^2 + bx + c$	**Standard form** $y = a(x - h)^2 + k$
Orientation	$a > 0$ up; $a < 0$ down	$a > 0$ up; $a < 0$ down
Axis of symmetry	$x = -\dfrac{b}{2a}$	$x = h$
Vertex	Substitute $x = -\dfrac{b}{2a}$ and solve for y.	(h, k)
y-intercept	Let $x = 0$	Let $x = 0$
x-intercepts	Let $y = 0$	Let $y = 0$

The graphs show what the parabolas look like when they open up or down. Their position in relation to the *x*- or *y*-axis is merely an example.

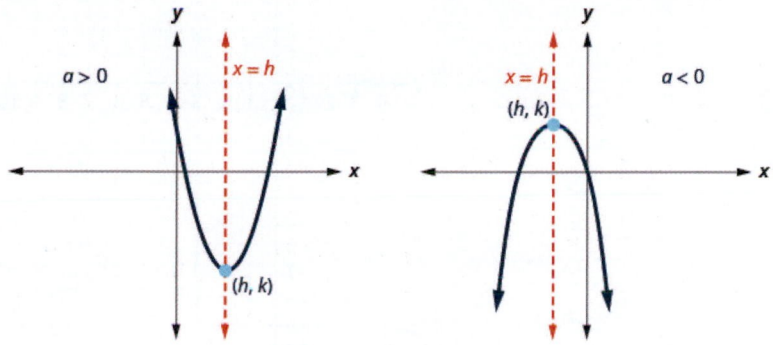

To graph a parabola from these forms, we used the following steps.

 HOW TO :: GRAPH VERTICAL PARABOLAS $\left(y = ax^2 + bx + c \text{ or } f(x) = a(x - h)^2 + k \right)$ USING PROPERTIES.

Step 1. Determine whether the parabola opens upward or downward.

Step 2. Find the axis of symmetry.

Step 3. Find the vertex.

Step 4. Find the *y*-intercept. Find the point symmetric to the *y*-intercept across the axis of symmetry.

Step 5. Find the *x*-intercepts.

Step 6. Graph the parabola.

The next example reviews the method of graphing a parabola from the general form of its equation.

EXAMPLE 11.12

Graph $y = -x^2 + 6x - 8$ by using properties.

Solution

$$y = ax^2 + bx + c$$
$$y = -x^2 + 6x - 8$$

Since a is -1, the parabola opens downward.

To find the axis of symmetry, find $x = -\dfrac{b}{2a}$.

$$x = -\dfrac{b}{2a}$$

$$x = -\dfrac{6}{2(-1)}$$

$$x = 3$$

The axis of symmetry is $x = 3$.

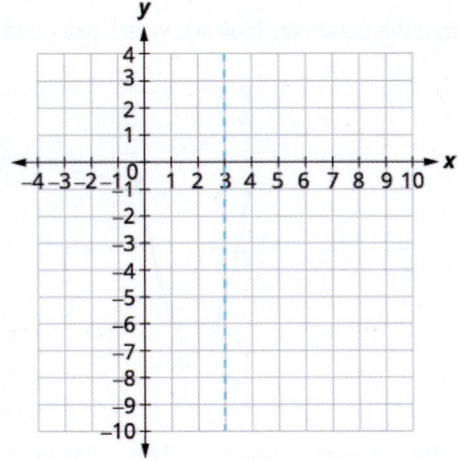

The vertex is on the line $x = 3$.

$$y = -x^2 + 6x - 8$$

Let $x = 3$.

$$y = -3^2 + 6 \cdot 3 - 8$$

$$y = -9 + 18 - 8$$

$$y = 1$$

The vertex is $(3, 1)$.

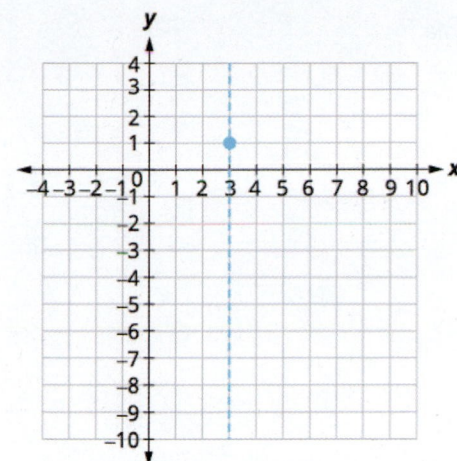

The *y*-intercept occurs when $x = 0$.	$y = -x^2 + 6x - 8$
Substitute $x = 0$.	$y = -0^2 + 6 \cdot 0 - 8$
Simplify.	$y = -8$
	The *y*-intercept is $(0, -8)$.

The point $(0, -8)$ is three units to the left of the line of symmetry. The point three units to the right of the line of symmetry is $(6, -8)$.	Point symmetric to the *y*-intercept is $(6, -8)$.

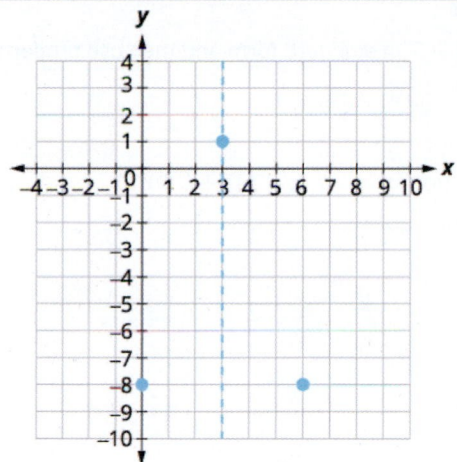

The *x*-intercept occurs when $y = 0$.	$y = -x^2 + 6x - 8$
Let $y = 0$.	$0 = -x^2 + 6x - 8$
Factor the GCF.	$0 = -(x^2 - 6x + 8)$
Factor the trinomial.	$0 = -(x - 4)(x - 2)$
Solve for *x*.	$x = 4, \quad x = 2$
	The *x*-intercepts are $(4, 0), (2, 0)$.

Graph the parabola.

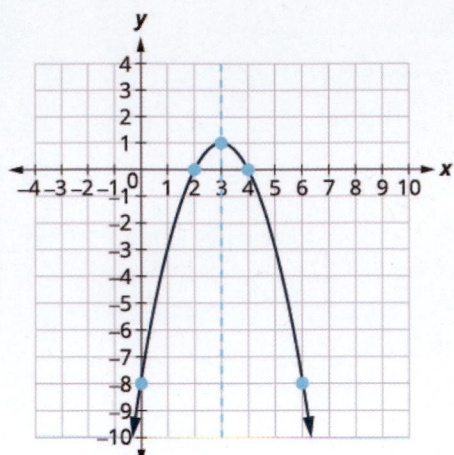

> **TRY IT : : 11.23**   Graph $y = -x^2 + 5x - 6$ by using properties.

> **TRY IT : : 11.24**   Graph $y = -x^2 + 8x - 12$ by using properties.

The next example reviews the method of graphing a parabola from the standard form of its equation, $y = a(x - h)^2 + k$.

EXAMPLE 11.13

Write $y = 3x^2 - 6x + 5$ in standard form and then use properties of standard form to graph the equation.

 Solution

Rewrite the function in $y = a(x - h)^2 + k$ form by completing the square.	$y = 3x^2 - 6x + 5$
	$y = 3(x^2 - 2x \quad) + 5$
	$y = 3(x^2 - 2x + 1) + 5 - 3$
	$y = 3(x - 1)^2 + 2$
Identify the constants a, h, k.	$a = 3$, $h = 1$, $k = 2$
Since $a = 2$, the parabola opens upward.	

$$\bigcup$$

The axis of symmetry is $x = h$.	The axis of symmetry is $x = 1$.
The vertex is (h, k).	The vertex is $(1, 2)$.

Find the *y*-intercept by substituting $x = 0$.	$y = 3(x-1)^2 + 2$ $y = 3 \cdot 0^2 - 6 \cdot 0 + 5$
	$y = 5$
	y-intercept $(0, 5)$
Find the point symmetric to $(0, 5)$ across the axis of symmetry.	$(2, 5)$
Find the *x*-intercepts.	$\begin{aligned} y &= 3(x-1)^2 + 2 \\ 0 &= 3(x-1)^2 + 2 \\ -2 &= 3(x-1)^2 \\ -\frac{2}{3} &= (x-1)^2 \\ \pm\sqrt{-\frac{2}{3}} &= x - 1 \end{aligned}$
	The square root of a negative number tells us the solutions are complex numbers. So there are no *x*-intercepts.
Graph the parabola.	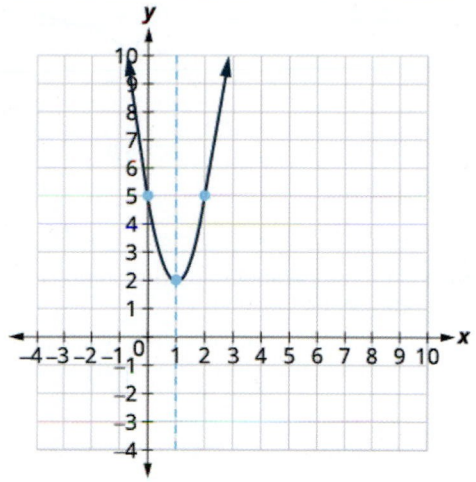

> **TRY IT : : 11.25**

⒜ Write $y = 2x^2 + 4x + 5$ in standard form and ⒝ use properties of standard form to graph the equation.

> **TRY IT : : 11.26**

⒜ Write $y = -2x^2 + 8x - 7$ in standard form and ⒝ use properties of standard form to graph the equation.

Graph Horizontal Parabolas

Our work so far has only dealt with parabolas that open up or down. We are now going to look at horizontal parabolas. These parabolas open either to the left or to the right. If we interchange the *x* and *y* in our previous equations for parabolas, we get the equations for the parabolas that open to the left or to the right.

Horizontal Parabolas		
	General form $x = ay^2 + by + c$	**Standard form** $x = a(y - k)^2 + h$
Orientation	$a > 0$ right; $a < 0$ left	$a > 0$ right; $a < 0$ left
Axis of symmetry	$y = -\dfrac{b}{2a}$	$y = k$
Vertex	Substitute $y = -\dfrac{b}{2a}$ and solve for x.	(h, k)
y-intercepts	Let $x = 0$	Let $x = 0$
x-intercept	Let $y = 0$	Let $y = 0$

Table 11.11

The graphs show what the parabolas look like when they to the left or to the right. Their position in relation to the x- or y-axis is merely an example.

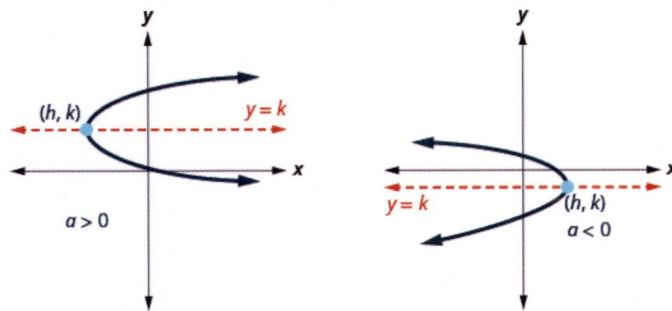

Looking at these parabolas, do their graphs represent a function? Since both graphs would fail the vertical line test, they do not represent a function.

To graph a parabola that opens to the left or to the right is basically the same as what we did for parabolas that open up or down, with the reversal of the x and y variables.

 HOW TO :: GRAPH HORIZONTAL PARABOLAS $\left(x = ay^2 + by + c \text{ or } x = a(y - k)^2 + h\right)$ **USING PROPERTIES.**

Step 1. Determine whether the parabola opens to the left or to the right.

Step 2. Find the axis of symmetry.

Step 3. Find the vertex.

Step 4. Find the x-intercept. Find the point symmetric to the x-intercept across the axis of symmetry.

Step 5. Find the y-intercepts.

Step 6. Graph the parabola.

EXAMPLE 11.14

Graph $x = 2y^2$ by using properties.

⊘ Solution

$$x = ay^2 + by + c$$
$$x = 2y^2$$

Since $a = 2$, the parabola opens to the right.

To find the axis of symmetry, find $y = -\dfrac{b}{2a}$.	$y = -\dfrac{b}{2a}$
	$y = -\dfrac{0}{2(2)}$
	$y = 0$
	The axis of symmetry is $y = 0$.
The vertex is on the line $y = 0$.	$x = 2y^2$
Let $y = 0$.	$x = 2 \cdot 0^2$
	$x = 0$
	The vertex is $(0, 0)$.

Since the vertex is $(0, 0)$, both the x- and y-intercepts are the point $(0, 0)$. To graph the parabola we need more points. In this case it is easiest to choose values of y.

$$x = 2y^2 \qquad\qquad x = 2y^2$$

			x	y	
$y = 1$	$x = 2 \cdot 1^2$	$y = 2$ $x = 2 \cdot 2^2$	2	1	(2, 1)
	$x = 2$	$x = 8$	8	2	(8, 2)

We also plot the points symmetric to $(2, 1)$ and $(8, 2)$ across the y-axis, the points $(2, -1)$, $(8, -2)$.

Graph the parabola.

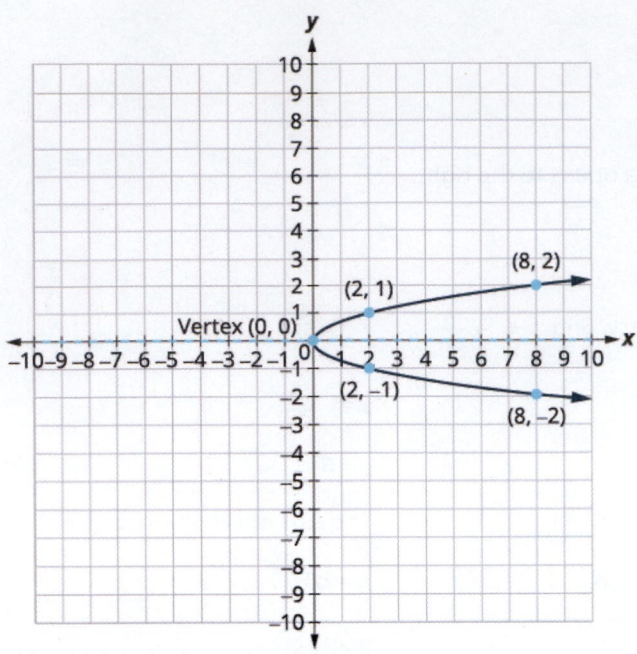

> TRY IT :: 11.27   Graph $x = y^2$ by using properties.

> TRY IT :: 11.28   Graph $x = -y^2$ by using properties.

In the next example, the vertex is not the origin.

EXAMPLE 11.15

Graph $x = -y^2 + 2y + 8$ by using properties.

⊘ **Solution**

$$x = ay^2 + by + c$$
$$x = -y^2 + 2y + 8$$

Since $a = -1$, the parabola opens to the left.

To find the axis of symmetry, find $y = -\dfrac{b}{2a}$.

$$y = -\dfrac{b}{2a}$$

$$y = -\dfrac{2}{2(-1)}$$

$$y = 1$$

The axis of symmetry is $y = 1$.

The vertex is on the line $y = 1$.

$$x = -y^2 + 2y + 8$$

Let $y = 1$.

$$x = -1^2 + 2 \cdot 1 + 8$$

	$x = 9$
	The vertex is $(9,\ 1)$.
The x-intercept occurs when $y = 0$.	$x = -y^2 + 2y\ \ +8$
	$x = -0^2 + 2 \cdot 0 + 8$
	$x = 8$
	The x-intercept is $(8,\ 0)$.
The point $(8,\ 0)$ is one unit below the line of symmetry. The symmetric point one unit above the line of symmetry is $(8,\ 2)$	Symmetric point is $(8,\ 2)$.
The y-intercept occurs when $x = 0$.	$x = -y^2 + 2y\ \ +8$
Substitute $x = 0$.	$0 = -y^2 + 2y\ \ +8$
Solve.	$y^2 - 2y - 8 = 0$
	$(y-4)(y+2) = 0$
	$y = 4,\ y = -2$
	The y-intercepts are $(0,\ 4)$ and $(0,\ -2)$.

Connect the points to graph the parabola.

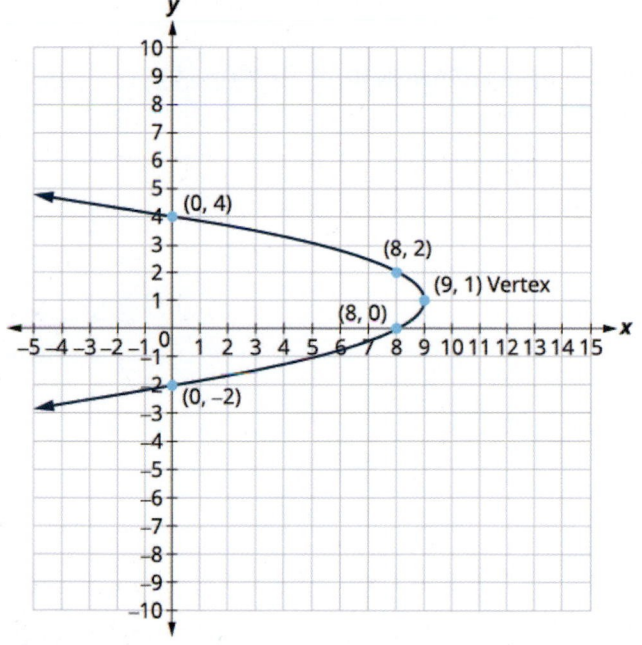

> **TRY IT : : 11.29**   Graph $x = -y^2 - 4y + 12$ by using properties.

> **TRY IT : : 11.30**   Graph $x = -y^2 + 2y - 3$ by using properties.

In **Table 11.11**, we see the relationship between the equation in standard form and the properties of the parabola. The

How To box lists the steps for graphing a parabola in the standard form $x = a(y - k)^2 + h.$ We will use this procedure in the next example.

EXAMPLE 11.16

Graph $x = 2(y - 2)^2 + 1$ using properties.

⊘ **Solution**

$$x = a(y - k)^2 + h$$
$$x = 2(y - 2)^2 + 1$$

Identify the constants a, h, k.	$a = 2, \quad h = 1, \quad k = 2$
Since $a = 2$, the parabola opens to the right.	
The axis of symmetry is $y = k$.	The axis of symmetry is $y = 2$.
The vertex is (h, k).	The vertex is $(1, 2)$.
Find the x-intercept by substituting $y = 0$.	$x = 2(y - 2)^2 + 1$ $x = 2(0 - 2)^2 + 1$ $x = 9$
	The x-intercept is $(9, 0)$.
Find the point symmetric to $(9, 0)$ across the axis of symmetry.	$(9, 4)$
Find the y-intercepts. Let $x = 0$.	$x = 2(y - 2)^2 + 1$ $0 = 2(y - 2)^2 + 1$ $-1 = 2(y - 2)^2$
	A square cannot be negative, so there is no real solution. So there are no y-intercepts.

Graph the parabola.

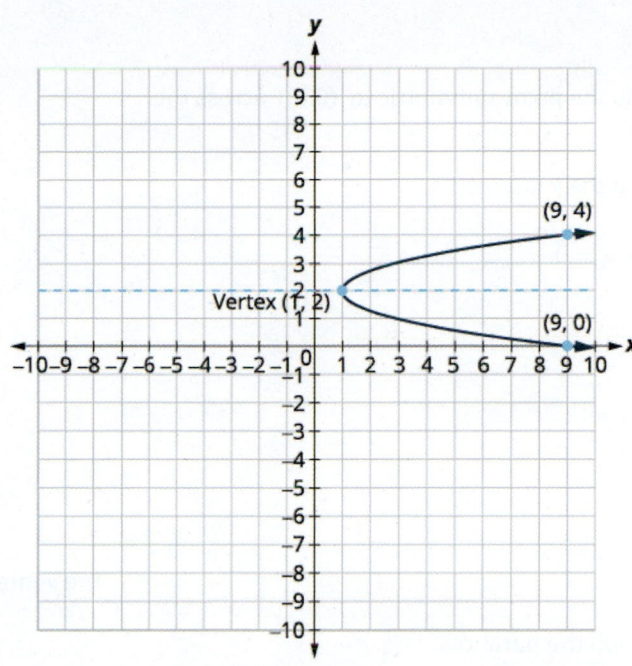

> **TRY IT ::** 11.31

Graph $x = 3(y - 1)^2 + 2$ using properties.

> **TRY IT ::** 11.32

Graph $x = 2(y - 3)^2 + 2$ using properties.

In the next example, we notice the a is negative and so the parabola opens to the left.

EXAMPLE 11.17

Graph $x = -4(y + 1)^2 + 4$ using properties.

⊘ **Solution**

$$x = a(y - k)^2 + h$$
$$x = -4(y + 1)^2 + 4$$

Identify the constants a, h, k.	$a = -4, \quad h = 4, \quad k = -1$
Since $a = -4$, the parabola opens to the left.	

The axis of symmetry is $y = k$.	The axis of symmetry is $y = -1$.
The vertex is (h, k).	The vertex is $(4, -1)$.
Find the *x*-intercept by substituting $y = 0$.	$x = -4(y + 1)^2 + 4$
	$x = -4(0 + 1)^2 + 4$
	$x = 0$

The *x*-intercept is $(0, 0)$.

Find the point symmetric to $(0, 0)$ across the axis of symmetry.	$(0, -2)$
Find the *y*-intercepts.	$x = -4(y + 1)^2 + 4$
Let $x = 0$.	$\begin{aligned} 0 &= -4(y + 1)^2 + 4 \\ -4 &= -4(y + 1)^2 \\ 1 &= (y + 1)^2 \\ y + 1 &= \pm 1 \end{aligned}$
	$y = -1 + 1 \qquad y = -1 - 1$
	$y = 0 \qquad\qquad y = -2$

The *y*-intercepts are $(0, 0)$ and $(0, -2)$.

Graph the parabola.

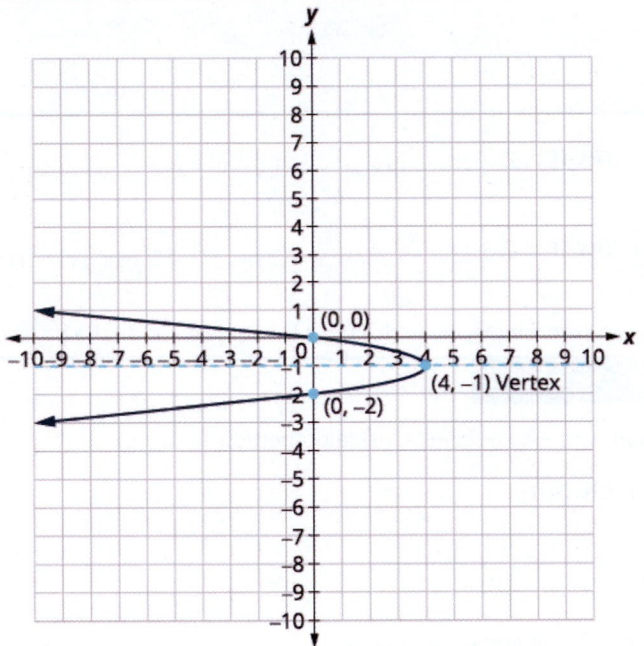

> **TRY IT : : 11.33**   Graph $x = -4(y + 2)^2 + 4$ using properties.

> **TRY IT : : 11.34**   Graph $x = -2(y + 3)^2 + 2$ using properties.

The next example requires that we first put the equation in standard form and then use the properties.

EXAMPLE 11.18

Write $x = 2y^2 + 12y + 17$ in standard form and then use the properties of the standard form to graph the equation.

⊘ Solution

$$x = 2y^2 + 12y + 17$$

Rewrite the function in $x = a(y - k)^2 + h$ form by completing the square.	$x = 2(y^2 + 6y) + 17$
	$x = 2(y^2 + 6y + 9) + 17 - 18$
	$x = 2(y + 3)^2 - 1$
	$x = a(y - k)^2 + h$ $x = 2(y + 3)^2 - 1$
Identify the constants a, h, k.	$a = 2, \quad h = -1, \quad k = -3$
Since $a = 2$, the parabola opens to the right.	

The axis of symmetry is $y = k$.	The axis of symmetry is $y = -3$.
The vertex is (h, k).	The vertex is $(-1, -3)$.
Find the x-intercept by substituting $y = 0$.	$\begin{aligned} x &= 2(y + 3)^2 - 1 \\ x &= 2(0 + 3)^2 - 1 \\ x &= 17 \end{aligned}$
	The x-intercept is $(17, 0)$.
Find the point symmetric to $(17, 0)$ across the axis of symmetry.	$(17, -6)$
Find the y-intercepts. Let $x = 0$.	$\begin{aligned} x &= 2(y + 3)^2 - 1 \\ 0 &= 2(y + 3)^2 - 1 \\ 1 &= 2(y + 3)^2 \\ \tfrac{1}{2} &= (y + 3)^2 \\ y + 3 &= \pm\sqrt{\tfrac{1}{2}} \\ y &= -3 \pm \tfrac{\sqrt{2}}{2} \end{aligned}$
	$y = -3 + \dfrac{\sqrt{2}}{2} \quad y = -3 - \dfrac{\sqrt{2}}{2}$
	$y \approx -2.3 \qquad y \approx -3.7$
	The y-intercepts are $\left(0, -3 + \dfrac{\sqrt{2}}{2}\right), \left(0, -3 - \dfrac{\sqrt{2}}{2}\right)$.

Graph the parabola.

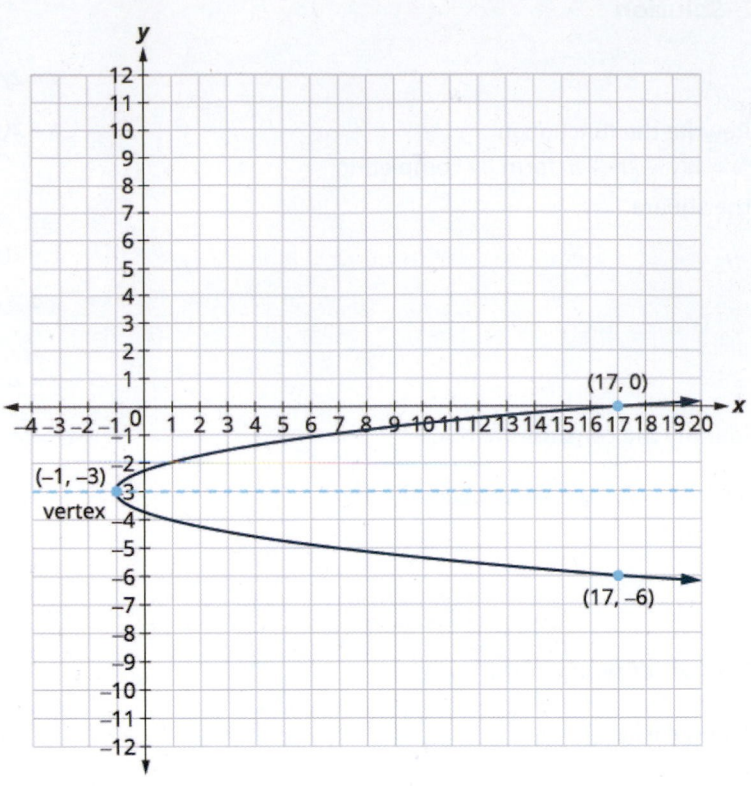

> **TRY IT : :** 11.35

ⓐ Write $x = 3y^2 + 6y + 7$ in standard form and ⓑ use properties of the standard form to graph the equation.

> **TRY IT : :** 11.36

ⓐ Write $x = -4y^2 - 16y - 12$ in standard form and ⓑ use properties of the standard form to graph the equation.

Solve Applications with Parabolas

Many architectural designs incorporate parabolas. It is not uncommon for bridges to be constructed using parabolas as we will see in the next example.

EXAMPLE 11.19

Find the equation of the parabolic arch formed in the foundation of the bridge shown. Write the equation in standard form.

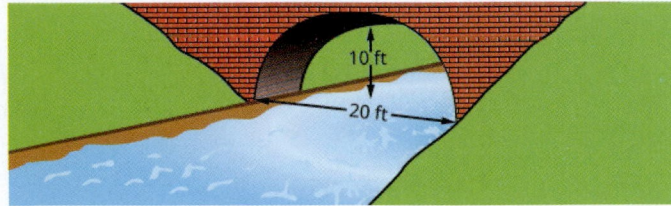

✓ Solution

We will first set up a coordinate system and draw the parabola. The graph will give us the information we need to write the equation of the graph in the standard form $y = a(x - h)^2 + k$.

Let the lower left side of the bridge be the origin of the coordinate grid at the point $(0, 0)$.

Since the base is 20 feet wide the point $(20, 0)$ represents the lower right side.

The bridge is 10 feet high at the highest point. The highest point is the vertex of the parabola so the y-coordinate of the vertex will be 10.

Since the bridge is symmetric, the vertex must fall halfway between the left most point, $(0, 0)$, and the rightmost point $(20, 0)$. From this we know that the x-coordinate of the vertex will also be 10.

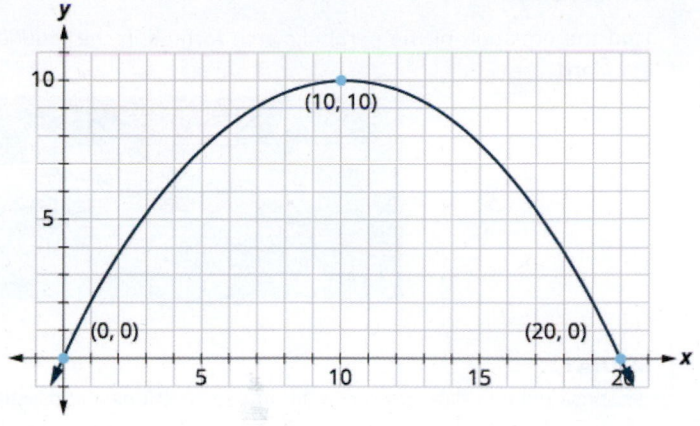

Identify the vertex, (h, k).	$(h, k) = (10, 10)$
	$h = 10, \quad k = 10$
Substitute the values into the standard form.	$y = a(x - h)^2 + k$ $\quad\quad\quad\quad y = a(x - 10)^2 + 10$
The value of a is still unknown. To find the value of a use one of the other points on the parabola.	$(x, y) = (0, 0)$
Substitute the values of the other point into the equation.	$y = a(x - 10)^2 + 10$ $\quad\quad\quad\quad 0 = a(0 - 10)^2 + 10$
Solve for a.	$0 = a(0 - 10)^2 + 10$ $-10 = a(-10)^2$ $-10 = 100a$ $\dfrac{-10}{100} = a$ $a = -\dfrac{1}{10}$
	$y = a(x - 10)^2 + 10$
Substitute the value for a into the equation.	$y = -\dfrac{1}{10}(x - 10)^2 + 10$

 TRY IT : : 11.37

Find the equation of the parabolic arch formed in the foundation of the bridge shown. Write the equation in standard form.

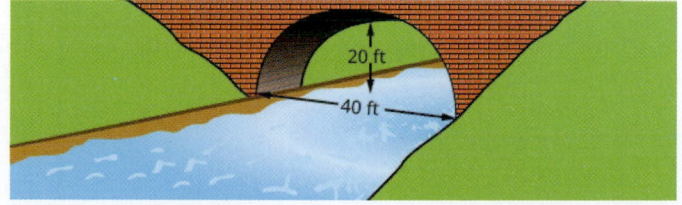

> **TRY IT : :** 11.38

Find the equation of the parabolic arch formed in the foundation of the bridge shown. Write the equation in standard form.

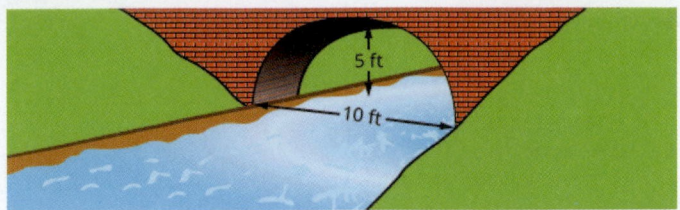

▶ **MEDIA : :**

Access these online resources for additional instructions and practice with quadratic functions and parabolas.

- **Quadratic Functions (https://openstax.org/l/37quadfunc)**
- **Introduction to Conics and Graphing Horizontal Parabolas (https://openstax.org/l/37conhorizpbola)**

11.2 EXERCISES

Practice Makes Perfect

Graph Vertical Parabolas

In the following exercises, graph each equation by using properties.

53. $y = -x^2 + 4x - 3$

54. $y = -x^2 + 8x - 15$

55. $y = 6x^2 + 2x - 1$

56. $y = 8x^2 - 10x + 3$

In the following exercises, ⓐ write the equation in standard form and ⓑ use properties of the standard form to graph the equation.

57. $y = -x^2 + 2x - 4$

58. $y = 2x^2 + 4x + 6$

59. $y = -2x^2 - 4x - 5$

60. $y = 3x^2 - 12x + 7$

Graph Horizontal Parabolas

In the following exercises, graph each equation by using properties.

61. $x = -2y^2$

62. $x = 3y^2$

63. $x = 4y^2$

64. $x = -4y^2$

65. $x = -y^2 - 2y + 3$

66. $x = -y^2 - 4y + 5$

67. $x = y^2 + 6y + 8$

68. $x = y^2 - 4y - 12$

69. $x = (y - 2)^2 + 3$

70. $x = (y - 1)^2 + 4$

71. $x = -(y - 1)^2 + 2$

72. $x = -(y - 4)^2 + 3$

73. $x = (y + 2)^2 + 1$

74. $x = (y + 1)^2 + 2$

75. $x = -(y + 3)^2 + 2$

76. $x = -(y + 4)^2 + 3$

77. $x = -3(y - 2)^2 + 3$

78. $x = -2(y - 1)^2 + 2$

79. $x = 4(y + 1)^2 - 4$

80. $x = 2(y + 4)^2 - 2$

In the following exercises, ⓐ write the equation in standard form and ⓑ use properties of the standard form to graph the equation.

81. $x = y^2 + 4y - 5$

82. $x = y^2 + 2y - 3$

83. $x = -2y^2 - 12y - 16$

84. $x = -3y^2 - 6y - 5$

Mixed Practice

In the following exercises, match each graph to one of the following equations: ⓐ $x^2 + y^2 = 64$ ⓑ $x^2 + y^2 = 49$
ⓒ $(x + 5)^2 + (y + 2)^2 = 4$ ⓓ $(x - 2)^2 + (y - 3)^2 = 9$ ⓔ $y = -x^2 + 8x - 15$ ⓕ $y = 6x^2 + 2x - 1$

85.

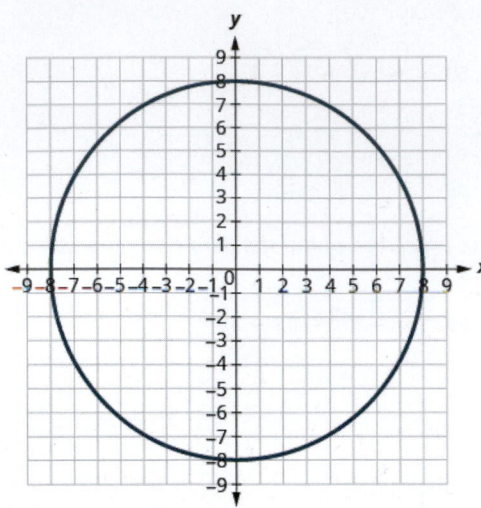

86.

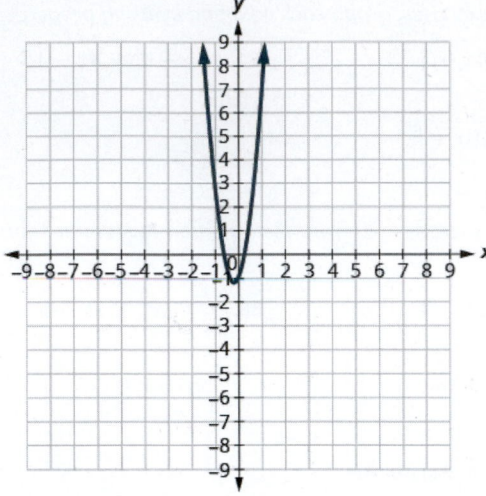

87.

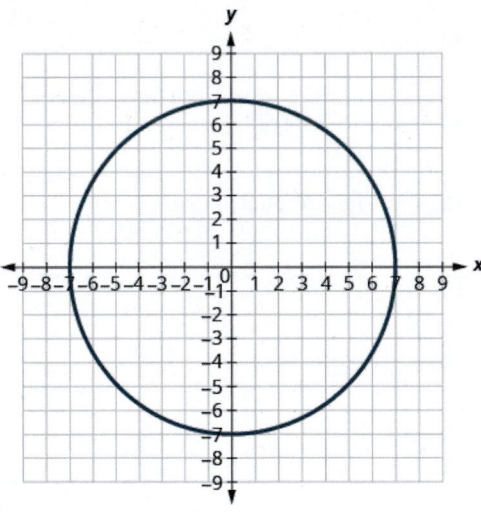

88.

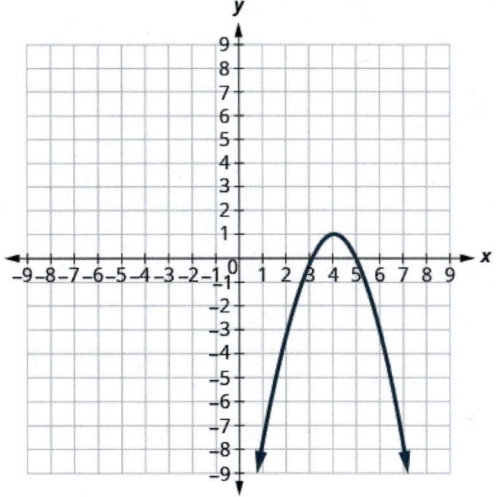

89.

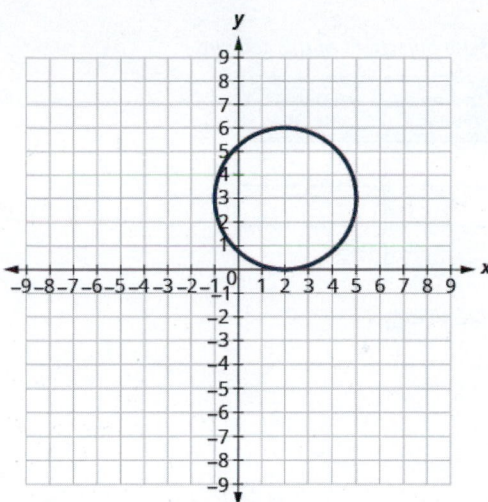

90.

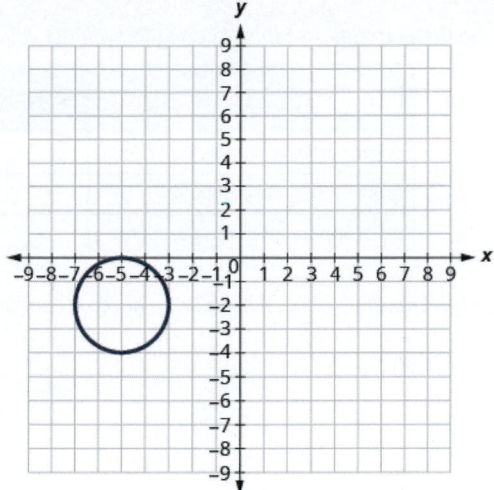

Solve Applications with Parabolas

91. Write the equation of the parabolic arch formed in the foundation of the bridge shown. Write the equation in standard form.

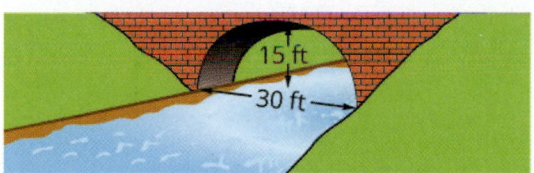

92. Write the equation of the parabolic arch formed in the foundation of the bridge shown. Write the equation in standard form.

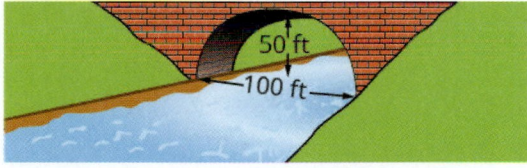

93. Write the equation of the parabolic arch formed in the foundation of the bridge shown. Write the equation in standard form.

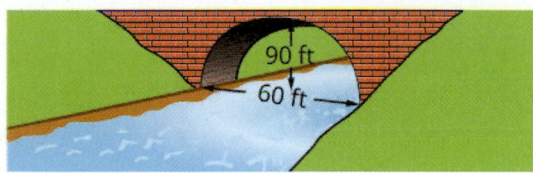

94. Write the equation of the parabolic arch formed in the foundation of the bridge shown. Write the equation in standard form.

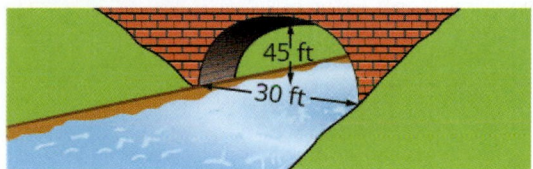

Writing Exercises

95. In your own words, define a parabola.

96. Is the parabola $y = x^2$ a function? Is the parabola $x = y^2$ a function? Explain why or why not.

97. Write the equation of a parabola that opens up or down in standard form and the equation of a parabola that opens left or right in standard form. Provide a sketch of the parabola for each one, label the vertex and axis of symmetry.

98. Explain in your own words, how you can tell from its equation whether a parabola opens up, down, left or right.

Self Check

ⓐ After completing the exercises, use this checklist to evaluate your mastery of the objectives of this section.

I can...	Confidently	With some help	No-I don't get it!
graph vertical parabolas.			
graph horizontal parabolas.			
solve applications with parabolas.			

ⓑ After reviewing this checklist, what will you do to become confident for all objectives?

Ellipses

Learning Objectives

By the end of this section, you will be able to:

› Graph an ellipse with center at the origin
› Find the equation of an ellipse with center at the origin
› Graph an ellipse with center not at the origin
› Solve application with ellipses

Be Prepared!

Before you get started, take this readiness quiz.

1. Graph $y = (x - 1)^2 - 2$ using transformations.
   If you missed this problem, review **Example 9.57**.

2. Complete the square: $x^2 - 8x = 8$.
   If you missed this problem, review **Example 9.12**.

3. Write in standard form. $y = 2x^2 - 12x + 14$
   If you missed this problem, review **Example 9.59**.

Graph an Ellipse with Center at the Origin

The next conic section we will look at is an **ellipse**. We define an ellipse as all points in a plane where the sum of the distances from two fixed points is constant. Each of the given points is called a **focus** of the ellipse.

Ellipse

An **ellipse** is all points in a plane where the sum of the distances from two fixed points is constant. Each of the fixed points is called a **focus** of the ellipse.

ellipse

We can draw an ellipse by taking some fixed length of flexible string and attaching the ends to two thumbtacks. We use a pen to pull the string taut and rotate it around the two thumbtacks. The figure that results is an ellipse.

A line drawn through the foci intersect the ellipse in two points. Each point is called a **vertex** of the ellipse. The segment connecting the vertices is called the **major axis**. The midpoint of the segment is called the **center** of the ellipse. A segment perpendicular to the major axis that passes through the center and intersects the ellipse in two points is called the **minor axis**.

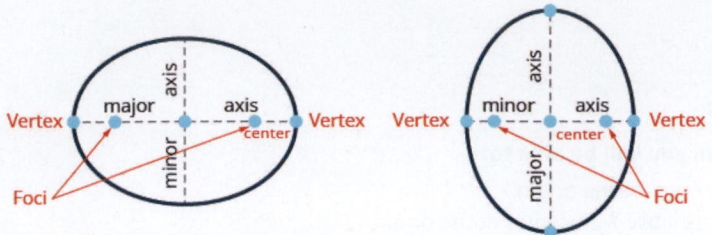

We mentioned earlier that our goal is to connect the geometry of a conic with algebra. Placing the ellipse on a rectangular coordinate system gives us that opportunity. In the figure, we placed the ellipse so the foci $((-c, 0), (c, 0))$ are on the x-axis and the center is the origin.

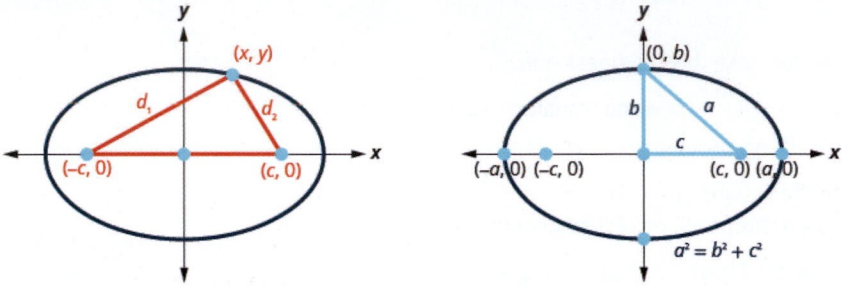

The definition states the sum of the distance from the foci to a point (x, y) is constant. So $d_1 + d_2$ is a constant that we will call $2a$ so, $d_1 + d_2 = 2a$. We will use the distance formula to lead us to an algebraic formula for an ellipse.

$$d_1 \quad + \quad d_2 \quad = \quad 2a$$

Use the distance formula to find d_1, d_2.

$$\sqrt{(x-(-c))^2 + (y-0)^2} + \sqrt{(x-c)^2 + (y-0)^2} = 2a$$

After eliminating radicals and simplifying, we get:

$$\frac{x^2}{a^2} + \frac{y^2}{a^2 - c^2} = 1$$

To simplify the equation of the ellipse, we let $a^2 - c^2 = b^2$.

So, the equation of an ellipse centered at the origin in standard form is:

$$\frac{x^2}{a^2} + \frac{y^2}{b^2} = 1$$

To graph the ellipse, it will be helpful to know the intercepts. We will find the x-intercepts and y-intercepts using the formula.

y-intercepts

$$\frac{x^2}{a^2} + \frac{y^2}{b^2} = 1$$

Let $x = 0$. $\quad \dfrac{0^2}{a^2} + \dfrac{y^2}{b^2} = 1$

$$\frac{y^2}{b^2} = 1$$

$$y^2 = b^2$$

$$y = \pm b$$

The y-intercepts are $(0, b)$ and $(0, -b)$.

x-intercepts

$$\frac{x^2}{a^2} + \frac{y^2}{b^2} = 1$$

Let $y = 0$. $\quad \dfrac{x^2}{a^2} + \dfrac{0^2}{b^2} = 1$

$$\frac{x^2}{a^2} = 1$$

$$x^2 = a^2$$

$$x = \pm a$$

The x-intercepts are $(a, 0)$ and $(-a, 0)$.

Standard Form of the Equation an Ellipse with Center $(0, 0)$

The standard form of the equation of an ellipse with center $(0, 0)$, is

$$\frac{x^2}{a^2} + \frac{y^2}{b^2} = 1$$

The *x*-intercepts are $(a, 0)$ and $(-a, 0)$.

The *y*-intercepts are $(0, b)$ and $(0, -b)$.

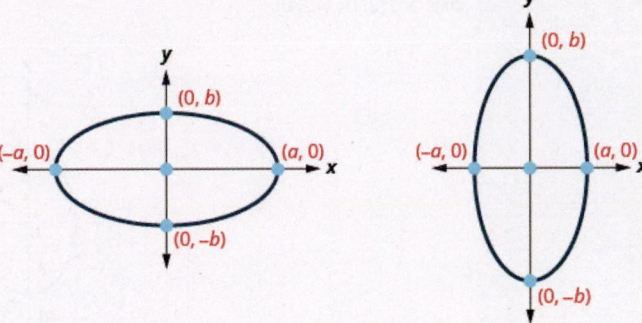

Notice that when the major axis is horizontal, the value of *a* will be greater than the value of *b* and when the major axis is vertical, the value of *b* will be greater than the value of *a*. We will use this information to graph an ellipse that is centered at the origin.

Ellipse with Center $(0, 0)$		
$\dfrac{x^2}{a^2} + \dfrac{y^2}{b^2} = 1$	$a > b$	$b > a$
Major axis	on the *x*- axis.	on the *y*-axis.
x-intercepts	$(-a, 0),\ (a, 0)$	
y-intercepts	$(0, -b),\ (0, b)$	

Table 11.18

EXAMPLE 11.20   HOW TO GRAPH AN ELLIPSE WITH CENTER (0, 0)

Graph: $\dfrac{x^2}{4} + \dfrac{y^2}{9} = 1$.

✓ **Solution**

Step 1. Write the equation in standard form.	It is in standard form.	$\dfrac{x^2}{4} + \dfrac{y^2}{9} = 1$
Step 2. Determine whether the major axis is horizontal or vertical.	Since $9 > 4$ and 9 is in the y^2 term, the major axis is vertical.	Major axis is vertical.
Step 3. Find the endpoints of the major axis.	The endpoints will be the *y*-intercepts. Since $b^2 = 9$, then $b = \pm3$. The endpoints of the major axis are $(0, 3), (0, -3)$.	The endpoints of the major axis are $(0, 3), (0, -3)$.

Step 4. Find the endpoints of the minor axis.	The endpoints will be the x-intercepts. Since $a^2 = 4$, then $a = \pm 2$. The endpoints of the minor axis are (2, 0), (–2, 0).	The endpoints of the major axis are (2, 0), (–2, 0).
Step 5. Sketch the ellipse.		

 TRY IT : : 11.39

Graph: $\dfrac{x^2}{4} + \dfrac{y^2}{16} = 1$.

 TRY IT : : 11.40

Graph: $\dfrac{x^2}{9} + \dfrac{y^2}{16} = 1$.

We summarize the steps for reference.

> **HOW TO : :** HOW TO GRAPH AN ELLIPSE WITH CENTER $(0, 0)$.
>
> Step 1. Write the equation in standard form.
> Step 2. Determine whether the major axis is horizontal or vertical.
> Step 3. Find the endpoints of the major axis.
> Step 4. Find the endpoints of the minor axis
> Step 5. Sketch the ellipse.

Sometimes our equation will first need to be put in standard form.

EXAMPLE 11.21

Graph $x^2 + 4y^2 = 16$.

⊘ **Solution**

We recognize this as the equation of an ellipse since both the x and y terms are squared and have different coefficients.	$x^2 + 4y^2 = 16$
To get the equation in standard form, divide both sides by 16 so that the equation is equal to 1.	$\dfrac{x^2}{16} + \dfrac{4y^2}{16} = \dfrac{16}{16}$

Simplify.

$$\frac{x^2}{16} + \frac{y^2}{4} = 1$$

The equation is in standard form. The ellipse is centered at the origin.

The center is $(0, 0)$.

Since $16 > 4$ and 16 is in the x^2 term, the major axis is horizontal.

$a^2 = 16,\ a = \pm 4$

$b^2 = 4,\ \ b = \pm 2$

The vertices are $(4, 0)$, $(-4, 0)$.

The endpoints of the minor axis are $(0, 2)$, $(0, -2)$.

Sketch the parabola.

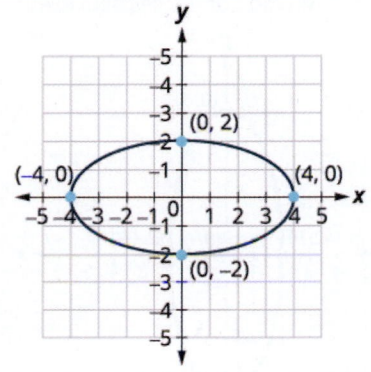

> **TRY IT : : 11.41**   Graph $9x^2 + 16y^2 = 144$.

> **TRY IT : : 11.42**   Graph $16x^2 + 25y^2 = 400$.

Find the Equation of an Ellipse with Center at the Origin

If we are given the graph of an ellipse, we can find the equation of the ellipse.

EXAMPLE 11.22

Find the equation of the ellipse shown.

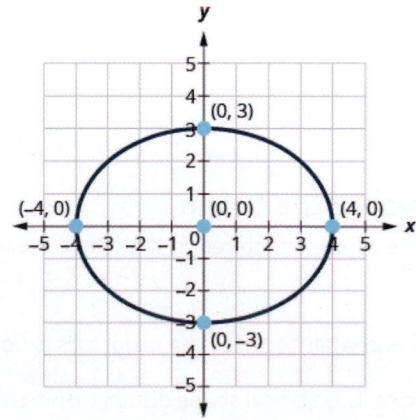

⊘ Solution

We recognize this as an ellipse that is centered at the origin.	$\dfrac{x^2}{a^2} + \dfrac{y^2}{b^2} = 1$
Since the major axis is horizontal and the distance from the center to the vertex is 4, we know $a = 4$ and so $a^2 = 16$.	$\dfrac{x^2}{16} + \dfrac{y^2}{b^2} = 1$
The minor axis is vertical and the distance from the center to the ellipse is 3, we know $b = 3$ and so $b^2 = 9$.	$\dfrac{x^2}{16} + \dfrac{y^2}{9} = 1$

> **TRY IT : : 11.43**   Find the equation of the ellipse shown.

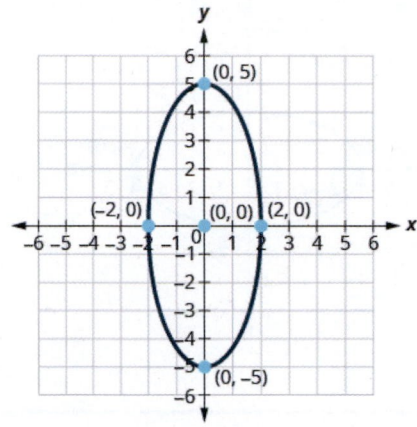

> **TRY IT : : 11.44**   Find the equation of the ellipse shown.

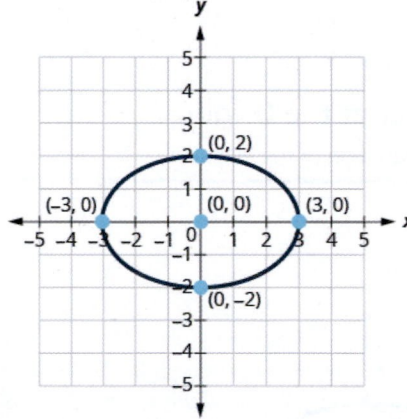

Graph an Ellipse with Center Not at the Origin

The ellipses we have looked at so far have all been centered at the origin. We will now look at ellipses whose center is (h, k).

The equation is $\dfrac{(x - h)^2}{a^2} + \dfrac{(y - k)^2}{b^2} = 1$ and when $a > b$, the major axis is horizontal so the distance from the center to the vertex is a. When $b > a$, the major axis is vertical so the distance from the center to the vertex is b.

> **Standard Form of the Equation an Ellipse with Center (h, k)**
>
> The standard form of the equation of an ellipse with center (h, k), is

$$\frac{(x-h)^2}{a^2} + \frac{(y-k)^2}{b^2} = 1$$

When $a > b$, the major axis is horizontal so the distance from the center to the vertex is a.

When $b > a$, the major axis is vertical so the distance from the center to the vertex is b.

EXAMPLE 11.23

Graph: $\frac{(x-3)^2}{9} + \frac{(y-1)^2}{4} = 1$.

⊘ **Solution**

The equation is in standard form,
$\frac{(x-h)^2}{a^2} + \frac{(y-k)^2}{b^2} = 1$.

$$\frac{(x-3)^2}{9} + \frac{(y-1)^2}{4} = 1$$

The ellipse is centered at (h, k).

The center is $(3, 1)$.

Since $9 > 4$ and 9 is in the x^2 term, the major axis is horizontal.

$a^2 = 9,\ a = \pm 3$

$b^2 = 4,\ b = \pm 2$

The distance from the center to the vertices is 3.
The distance from the center to the endpoints of the minor axis is 2.

Sketch the ellipse.

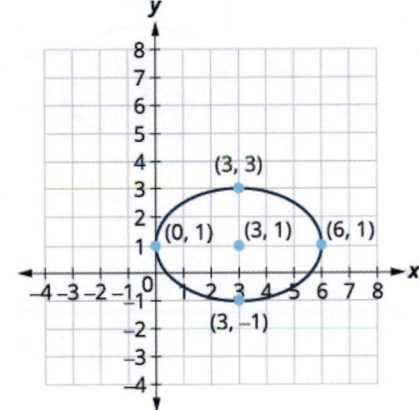

TRY IT : : 11.45

Graph: $\frac{(x+3)^2}{4} + \frac{(y-5)^2}{16} = 1$.

TRY IT : : 11.46

Graph: $\frac{(x-1)^2}{25} + \frac{(y+3)^2}{16} = 1$.

If we look at the equations of $\frac{x^2}{9} + \frac{y^2}{4} = 1$ and $\frac{(x-3)^2}{9} + \frac{(y-1)^2}{4} = 1$, we see that they are both ellipses with $a = 3$ and $b = 2$. So they will have the same size and shape. They are different in that they do not have the same center.

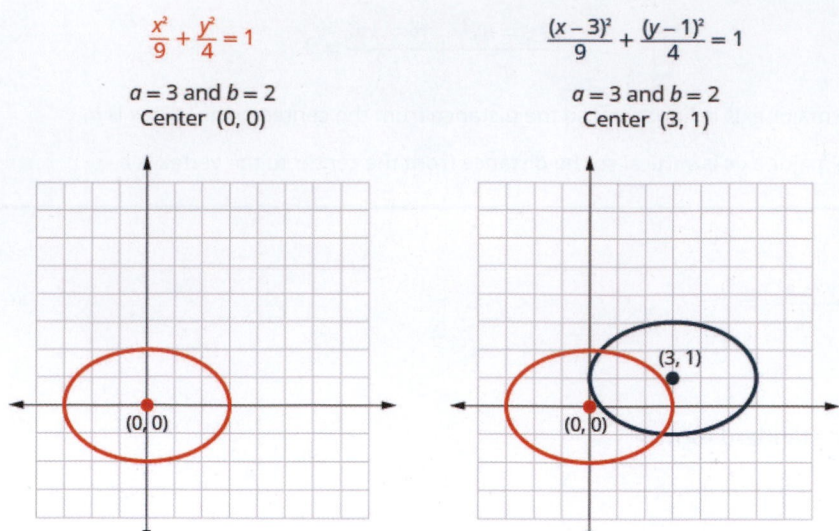

$$\frac{x^2}{9} + \frac{y^2}{4} = 1$$

$a = 3$ and $b = 2$
Center $(0, 0)$

$$\frac{(x-3)^2}{9} + \frac{(y-1)^2}{4} = 1$$

$a = 3$ and $b = 2$
Center $(3, 1)$

Notice in the graph above that we could have graphed $\dfrac{(x-3)^2}{9} + \dfrac{(y-1)^2}{4} = 1$ by translations. We moved the original ellipse to the right 3 units and then up 1 unit.

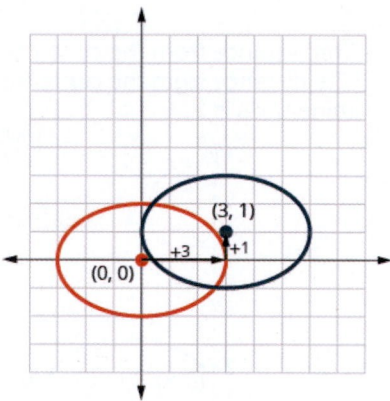

In the next example we will use the translation method to graph the ellipse.

EXAMPLE 11.24

Graph $\dfrac{(x+4)^2}{16} + \dfrac{(y-6)^2}{9} = 1$ by translation.

⊘ **Solution**

This ellipse will have the same size and shape as $\dfrac{x^2}{16} + \dfrac{y^2}{9} = 1$ whose center is $(0, 0)$. We graph this ellipse first.

The center is $(0, 0)$.	Center $(0, 0)$
Since $16 > 9$, the major axis is horizontal.	
$a^2 = 16$, $a = \pm 4$ $b^2 = 9$, $b = \pm 3$	The vertices are $(4, 0)$, $(-4, 0)$. The endpoints of the minor axis are $(0, 3)$, $(0, -3)$.

Sketch the ellipse.

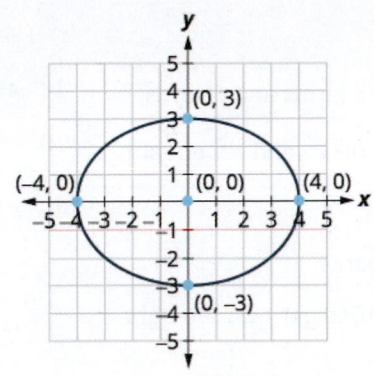

The original equation is in standard form, $\dfrac{(x-h)^2}{a^2} + \dfrac{(y-k)^2}{b^2} = 1$.	$\dfrac{(x-(-4))^2}{16} + \dfrac{(y-6)^2}{9} = 1$
The ellipse is centered at (h, k).	The center is $(-4, 6)$.

We translate the graph of $\dfrac{x^2}{16} + \dfrac{y^2}{9} = 1$ four

units to the left and then up 6 units.
Verify that the center is $(-4, 6)$.

The new ellipse is the ellipse whose equation
is

$$\dfrac{(x+4)^2}{16} + \dfrac{(y-6)^2}{9} = 1.$$

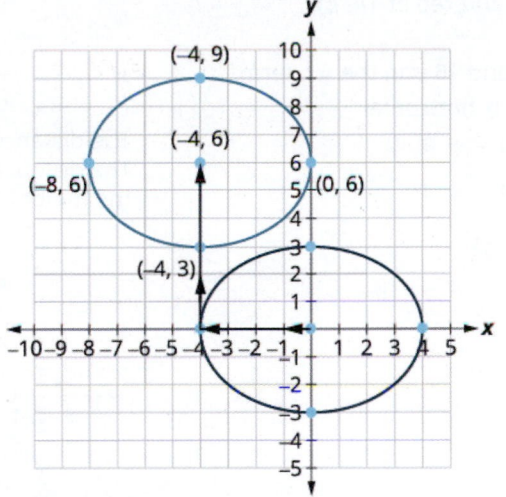

> **TRY IT : : 11.47**

Graph $\dfrac{(x-5)^2}{9} + \dfrac{(y+4)^2}{4} = 1$ by translation.

> **TRY IT : : 11.48**

Graph $\dfrac{(x+6)^2}{16} + \dfrac{(y+2)^2}{25} = 1$ by translation.

When an equation has both an x^2 and a y^2 with different coefficients, we verify that it is an ellipsis by putting it in standard form. We will then be able to graph the equation.

EXAMPLE 11.25

Write the equation $x^2 + 4y^2 - 4x + 24y + 24 = 0$ in standard form and graph.

✓ **Solution**

We put the equation in standard form by completing the squares in both x and y.

$$x^2 + 4y^2 - 4x + 24y + 24 = 0$$

Rewrite grouping the *x* terms and *y* terms.	$(x^2 - 4x +\ ) + (4y^2 + 24y +\ ) = -24$
Make the coefficients of x^2 and y^2 equal 1.	$(x^2 - 4x +\ ) + 4(y^2 + 6y +\ ) = -24$
Complete the squares.	$(x^2 - 4x + 4) + 4(y^2 + 6y + 9) = -24 + 4 + 36$
Write as binomial squares.	$(x - 2)^2 + 4(y + 3)^2 = 16$
Divide both sides by 16 to get 1 on the right.	$\dfrac{(x-2)^2}{16} + \dfrac{4(y+3)^2}{16} = \dfrac{16}{16}$
Simplify.	$\dfrac{(x-2)^2}{16} + \dfrac{(y+3)^2}{4} = 1$
The equation is in standard form, $\dfrac{(x-h)^2}{a^2} + \dfrac{(y-k)^2}{b^2} = 1$	$\dfrac{(x-2)^2}{16} + \dfrac{(y+3)^2}{4} = 1$
The ellipse is centered at (h, k).	The center is $(2, -3)$.
Since $16 > 4$ and 16 is in the x^2 term, the major axis is horizontal. $\quad a^2 = 16,\ a = \pm 4$ $\quad b^2 = 4,\quad b = \pm 2$	The distance from the center to the vertices is 4. The distance from the center to the endpoints of the minor axis is 2.
Sketch the ellipse.	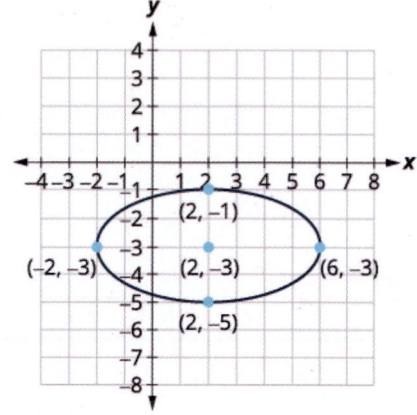

> **TRY IT : : 11.49**
ⓐ Write the equation $6x^2 + 4y^2 + 12x - 32y + 34 = 0$ in standard form and ⓑ graph.

> **TRY IT : : 11.50**
ⓐ Write the equation $4x^2 + y^2 - 16x - 6y + 9 = 0$ in standard form and ⓑ graph.

Solve Application with Ellipses

The orbits of the planets around the sun follow elliptical paths.

EXAMPLE 11.26

Pluto (a dwarf planet) moves in an elliptical orbit around the Sun. The closest Pluto gets to the Sun is approximately 30 astronomical units (AU) and the furthest is approximately 50 AU. The Sun is one of the foci of the elliptical orbit. Letting the ellipse center at the origin and labeling the axes in AU, the orbit will look like the figure below. Use the graph to write an equation for the elliptical orbit of Pluto.

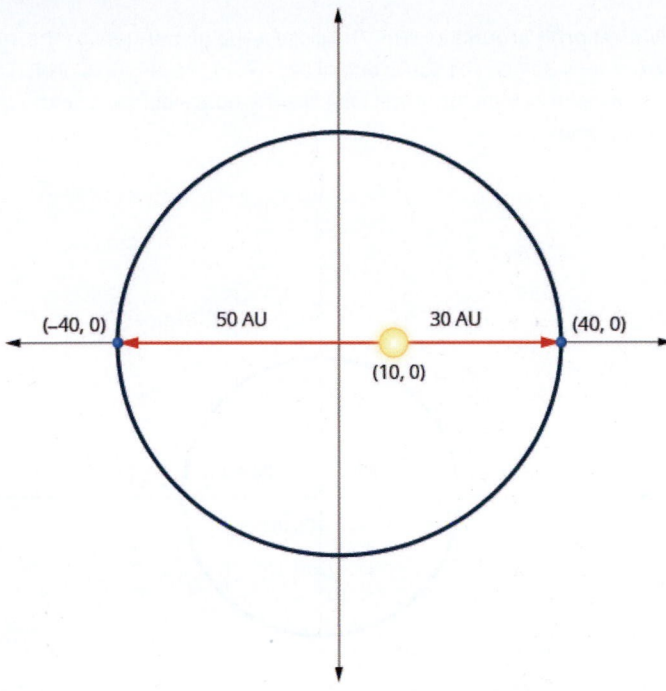

(–40, 0)   50 AU   30 AU   (40, 0)

(10, 0)

✓ **Solution**

We recognize this as an ellipse that is centered at the origin.

$$\frac{x^2}{a^2} + \frac{y^2}{b^2} = 1$$

Since the major axis is horizontal and the distance from the center to the vertex is 40, we know $a = 40$ and so $a^2 = 1600$.

$$\frac{x^2}{1600} + \frac{y^2}{b^2} = 1$$

The minor axis is vertical but the end points aren't given. To find b we will use the location of the Sun. Since the Sun is a focus of the ellipse at the point $(10, 0)$, we know $c = 10$. Use this to solve for b^2.

$$b^2 = a^2 - c^2$$
$$b^2 = 40^2 - 10^2$$
$$b^2 = 1600 - 100$$
$$b^2 = 1500$$

Substitute a^2 and b^2 into the standard form of the ellipse.

$$\frac{x^2}{1600} + \frac{y^2}{1500} = 1$$

> **TRY IT : :** 11.51

A planet moves in an elliptical orbit around its sun. The closest the planet gets to the sun is approximately 20 AU and the furthest is approximately 30 AU. The sun is one of the foci of the elliptical orbit. Letting the ellipse center at the origin and labeling the axes in AU, the orbit will look like the figure below. Use the graph to write an equation for the elliptical orbit of the planet.

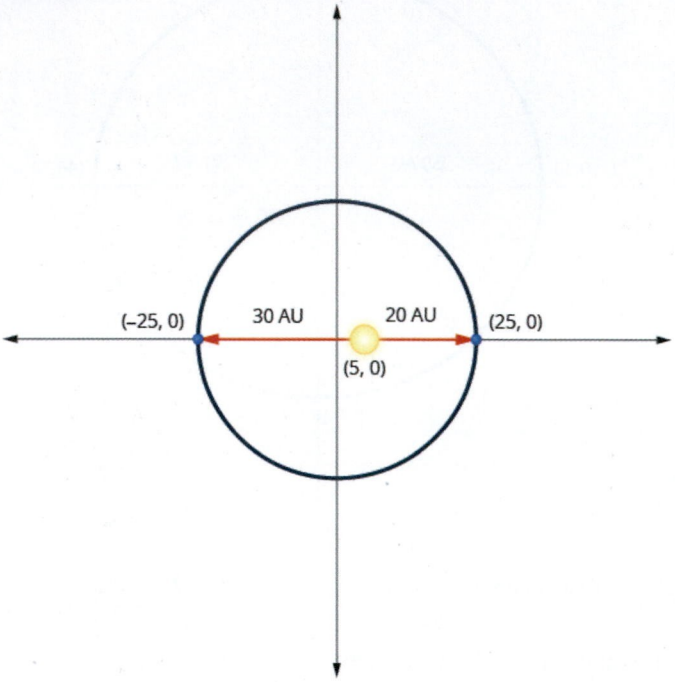

> **TRY IT : :** 11.52

A planet moves in an elliptical orbit around its sun. The closest the planet gets to the sun is approximately 20 AU and the furthest is approximately 50 AU. The sun is one of the foci of the elliptical orbit. Letting the ellipse center at the origin and labeling the axes in AU, the orbit will look like the figure below. Use the graph to write an equation for the elliptical orbit of the planet.

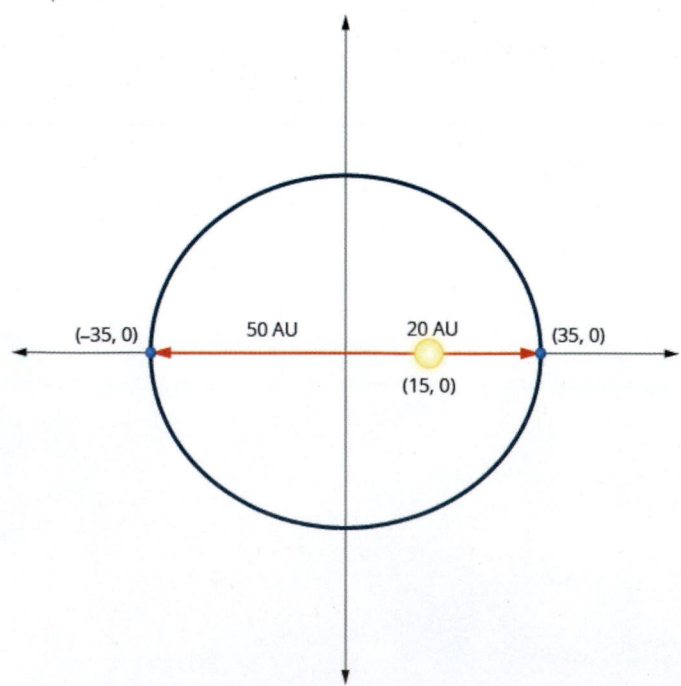

▶ | **MEDIA : :**

Access these online resources for additional instructions and practice with ellipses.

- **Conic Sections: Graphing Ellipses Part 1 (https://openstax.org/l/37graphellipse1)**
- **Conic Sections: Graphing Ellipses Part 2 (https://openstax.org/l/37graphellipse2)**
- **Equation for Ellipse From Graph (https://openstax.org/l/37eqellipse)**

11.3 EXERCISES

Practice Makes Perfect

Graph an Ellipse with Center at the Origin

In the following exercises, graph each ellipse.

99. $\dfrac{x^2}{4} + \dfrac{y^2}{25} = 1$

100. $\dfrac{x^2}{9} + \dfrac{y^2}{25} = 1$

101. $\dfrac{x^2}{25} + \dfrac{y^2}{36} = 1$

102. $\dfrac{x^2}{16} + \dfrac{y^2}{36} = 1$

103. $\dfrac{x^2}{36} + \dfrac{y^2}{16} = 1$

104. $\dfrac{x^2}{25} + \dfrac{y^2}{9} = 1$

105. $x^2 + \dfrac{y^2}{4} = 1$

106. $\dfrac{x^2}{9} + y^2 = 1$

107. $4x^2 + 25y^2 = 100$

108. $16x^2 + 9y^2 = 144$

109. $16x^2 + 36y^2 = 576$

110. $9x^2 + 25y^2 = 225$

Find the Equation of an Ellipse with Center at the Origin

In the following exercises, find the equation of the ellipse shown in the graph.

111.

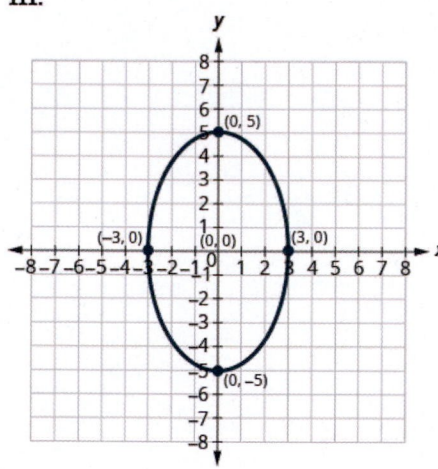

112.

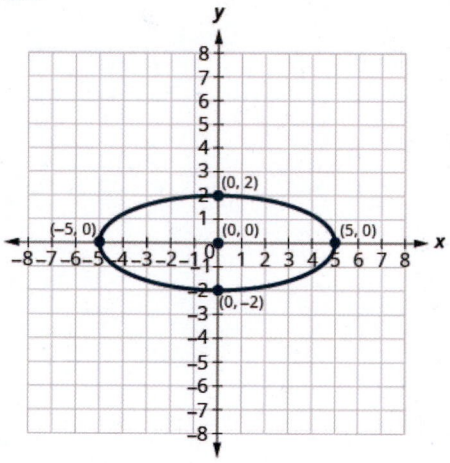

113.

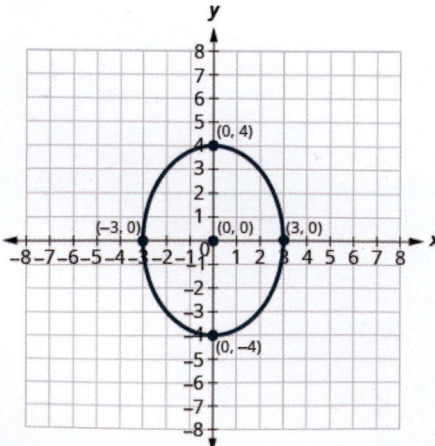

114.

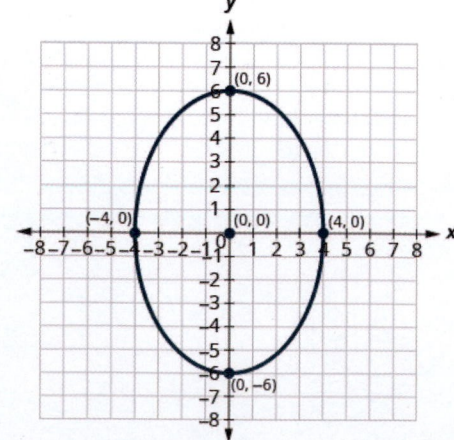

Graph an Ellipse with Center Not at the Origin

In the following exercises, graph each ellipse.

115. $\dfrac{(x+1)^2}{4} + \dfrac{(y+6)^2}{25} = 1$

116. $\dfrac{(x-3)^2}{25} + \dfrac{(y+2)^2}{9} = 1$

117. $\dfrac{(x+4)^2}{4} + \dfrac{(y-2)^2}{9} = 1$

118. $\dfrac{(x-4)^2}{9} + \dfrac{(y-1)^2}{16} = 1$

In the following exercises, graph each equation by translation.

119. $\dfrac{(x-3)^2}{4} + \dfrac{(y-7)^2}{25} = 1$

120. $\dfrac{(x+6)^2}{16} + \dfrac{(y+5)^2}{4} = 1$

121. $\dfrac{(x-5)^2}{9} + \dfrac{(y+4)^2}{25} = 1$

122. $\dfrac{(x+5)^2}{36} + \dfrac{(y-3)^2}{16} = 1$

In the following exercises, ⓐ write the equation in standard form and ⓑ graph.

123. $25x^2 + 9y^2 - 100x - 54y - 44 = 0$

124. $4x^2 + 25y^2 + 8x + 100y + 4 = 0$

125. $4x^2 + 25y^2 - 24x - 64 = 0$

126. $9x^2 + 4y^2 + 56y + 160 = 0$

In the following exercises, graph the equation.

127. $x = -2(y-1)^2 + 2$

128. $x^2 + y^2 = 49$

129. $(x+5)^2 + (y+2)^2 = 4$

130. $y = -x^2 + 8x - 15$

131. $\dfrac{(x+3)^2}{16} + \dfrac{(y+1)^2}{4} = 1$

132. $(x-2)^2 + (y-3)^2 = 9$

133. $\dfrac{x^2}{25} + \dfrac{y^2}{36} = 1$

134. $x = 4(y+1)^2 - 4$

135. $x^2 + y^2 = 64$

136. $\dfrac{x^2}{9} + \dfrac{y^2}{25} = 1$

137. $y = 6x^2 + 2x - 1$

138. $\dfrac{(x-2)^2}{9} + \dfrac{(y+3)^2}{25} = 1$

Solve Application with Ellipses

139. A planet moves in an elliptical orbit around its sun. The closest the planet gets to the sun is approximately 10 AU and the furthest is approximately 30 AU. The sun is one of the foci of the elliptical orbit. Letting the ellipse center at the origin and labeling the axes in AU, the orbit will look like the figure below. Use the graph to write an equation for the elliptical orbit of the planet.

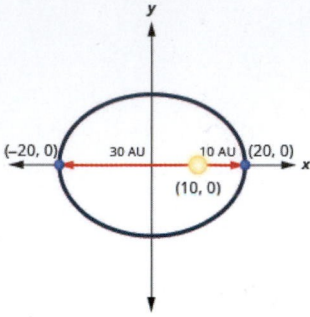

140. A planet moves in an elliptical orbit around its sun. The closest the planet gets to the sun is approximately 10 AU and the furthest is approximately 70 AU. The sun is one of the foci of the elliptical orbit. Letting the ellipse center at the origin and labeling the axes in AU, the orbit will look like the figure below. Use the graph to write an equation for the elliptical orbit of the planet.

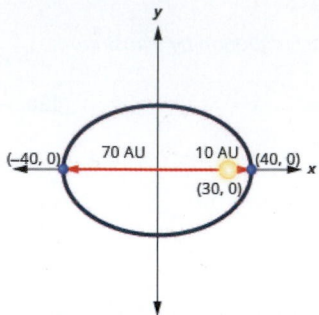

141. A comet moves in an elliptical orbit around a sun. The closest the comet gets to the sun is approximately 15 AU and the furthest is approximately 85 AU. The sun is one of the foci of the elliptical orbit. Letting the ellipse center at the origin and labeling the axes in AU, the orbit will look like the figure below. Use the graph to write an equation for the elliptical orbit of the comet.

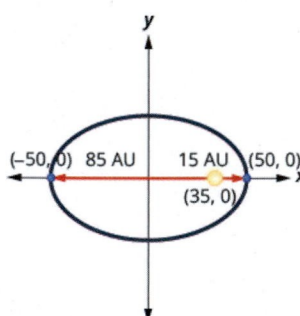

142. A comet moves in an elliptical orbit around a sun. The closest the comet gets to the sun is approximately 15 AU and the furthest is approximately 95 AU. The sun is one of the foci of the elliptical orbit. Letting the ellipse center at the origin and labeling the axes in AU, the orbit will look like the figure below. Use the graph to write an equation for the elliptical orbit of the comet.

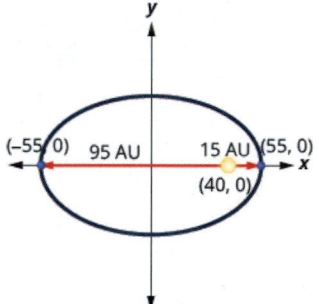

Writing Exercises

143. In your own words, define an ellipse and write the equation of an ellipse centered at the origin in standard form. Draw a sketch of the ellipse labeling the center, vertices and major and minor axes.

144. Explain in your own words how to get the axes from the equation in standard form.

145. Compare and contrast the graphs of the equations $\frac{x^2}{4} + \frac{y^2}{9} = 1$ and $\frac{x^2}{9} + \frac{y^2}{4} = 1$.

146. Explain in your own words, the difference between a vertex and a focus of the ellipse.

Self Check

ⓐ *After completing the exercises, use this checklist to evaluate your mastery of the objectives of this section.*

I can...	Confidently	With some help	No-I don't get it!
graph an ellipse with center at the origin.			
find the equation of an ellipse with center at the origin.			
graph an ellipse with center not at the origin.			
solve applications with ellipses.			

ⓑ *What does this checklist tell you about your mastery of this section? What steps will you take to improve?*

Hyperbolas

Learning Objectives

By the end of this section, you will be able to:

> Graph a hyperbola with center at $(0, 0)$

> Graph a hyperbola with center at (h, k)

> Identify conic sections by their equations

Be Prepared!

Before you get started, take this readiness quiz.

1. Solve: $x^2 = 12$.
   If you missed this problem, review **Example 9.1**.

2. Expand: $(x - 4)^2$.
   If you missed this problem, review **Example 5.32**.

3. Graph $y = -\frac{2}{3}x$.
   If you missed this problem, review **Example 3.4**.

Graph a Hyperbola with Center at (0, 0)

The last conic section we will look at is called a **hyperbola**. We will see that the equation of a hyperbola looks the same as the equation of an ellipse, except it is a difference rather than a sum. While the equations of an ellipse and a hyperbola are very similar, their graphs are very different.

We define a **hyperbola** as all points in a plane where the difference of their distances from two fixed points is constant. Each of the fixed points is called a **focus** of the hyperbola.

Hyperbola

A **hyperbola** is all points in a plane where the difference of their distances from two fixed points is constant. Each of the fixed points is called a **focus** of the hyperbola.

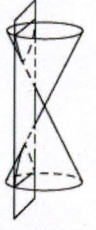

hyperbola

The line through the foci, is called the **transverse axis**. The two points where the transverse axis intersects the hyperbola are each a **vertex** of the hyperbola. The midpoint of the segment joining the foci is called the **center** of the hyperbola. The line perpendicular to the transverse axis that passes through the center is called the **conjugate axis**. Each piece of the graph is called a **branch** of the hyperbola.

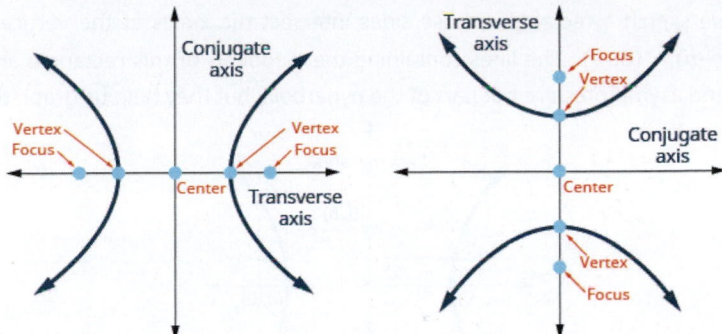

Again our goal is to connect the geometry of a conic with algebra. Placing the hyperbola on a rectangular coordinate system gives us that opportunity. In the figure, we placed the hyperbola so the foci $((-c, 0), (c, 0))$ are on the x-axis and the center is the origin.

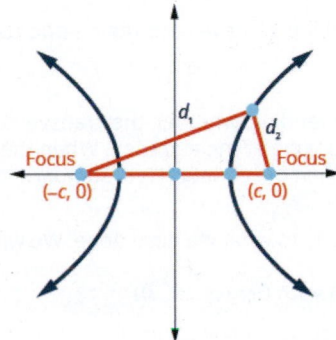

The definition states the difference of the distance from the foci to a point (x, y) is constant. So $|d_1 - d_2|$ is a constant that we will call $2a$ so $|d_1 - d_2| = 2a$. We will use the distance formula to lead us to an algebraic formula for an ellipse.

$$|d_1 \quad - \quad d_2| \quad = 2a$$

Use the distance formula to find d_1, d_2
$$\left| \sqrt{(x-(-c))^2 + (y-0)^2} - \sqrt{(x-c)^2 + (y-0)^2} \right| = 2a$$

Eliminate the radicals.
To simplify the equation of the ellipse, we let $c^2 - a^2 = b^2$.
$$\frac{x^2}{a^2} + \frac{y^2}{c^2 - a^2} = 1$$

So, the equation of a hyperbola centered at the origin in standard form is:
$$\frac{x^2}{a^2} - \frac{y^2}{b^2} = 1$$

To graph the hyperbola, it will be helpful to know about the intercepts. We will find the x-intercepts and y-intercepts using the formula.

x-intercepts	**y-intercepts**
$\dfrac{x^2}{a^2} - \dfrac{y^2}{b^2} = 1$	$\dfrac{x^2}{a^2} - \dfrac{y^2}{b^2} = 1$
Let $y = 0$. $\quad \dfrac{x^2}{a^2} - \dfrac{0^2}{b^2} = 1$	Let $x = 0$. $\quad \dfrac{0^2}{a^2} - \dfrac{y^2}{b^2} = 1$
$\dfrac{x^2}{a^2} = 1$	$-\dfrac{y^2}{b^2} = 1$
$x^2 = a^2$	$y^2 = -b^2$
$x = \pm a$	$y = \pm\sqrt{-b^2}$
The x-intercepts are $(a, 0)$ and $(-a, 0)$.	There are no y-intercepts.

The a, b values in the equation also help us find the asymptotes of the hyperbola. The asymptotes are intersecting straight lines that the branches of the graph approach but never intersect as the x, y values get larger and larger.

To find the asymptotes, we sketch a rectangle whose sides intersect the x-axis at the vertices $(-a, 0)$, $(a, 0)$ and intersect the y-axis at $(0, -b)$, $(0, b)$. The lines containing the diagonals of this rectangle are the asymptotes of the hyperbola. The rectangle and asymptotes are not part of the hyperbola, but they help us graph the hyperbola.

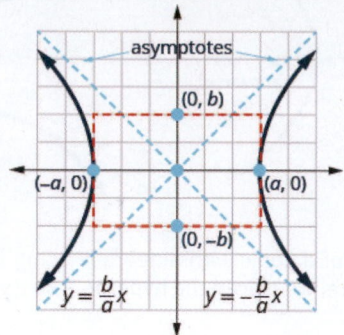

The asymptotes pass through the origin and we can evaluate their slope using the rectangle we sketched. They have equations $y = \frac{b}{a}x$ and $y = -\frac{b}{a}x$.

There are two equations for hyperbolas, depending whether the transverse axis is vertical or horizontal. We can tell whether the transverse axis is horizontal by looking at the equation. When the equation is in standard form, if the x^2-term is positive, the transverse axis is horizontal. When the equation is in standard form, if the y^2-term is positive, the transverse axis is vertical.

The second equations could be derived similarly to what we have done. We will summarize the results here.

Standard Form of the Equation a Hyperbola with Center $(0, 0)$

The standard form of the equation of a hyperbola with center $(0, 0)$, is

$$\frac{x^2}{a^2} - \frac{y^2}{b^2} = 1 \qquad \text{or} \qquad \frac{y^2}{a^2} - \frac{x^2}{b^2} = 1$$

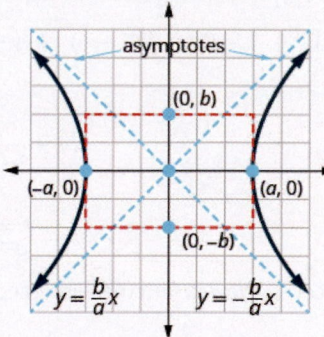
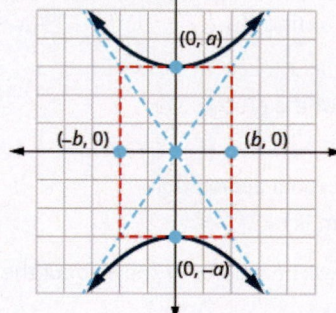

Notice that, unlike the equation of an ellipse, the denominator of x^2 is not always a^2 and the denominator of y^2 is not always b^2.

Notice that when the x^2-term is positive, the transverse axis is on the x-axis. When the y^2-term is positive, the transverse axis is on the y-axis.

Standard Forms of the Equation a Hyperbola with Center $(0, 0)$		
	$\dfrac{x^2}{a^2} - \dfrac{y^2}{b^2} = 1$	$\dfrac{y^2}{a^2} - \dfrac{x^2}{b^2} = 1$
Orientation	Transverse axis on the x-axis. Opens left and right	Transverse axis on the y-axis. Opens up and down
Vertices	$(-a, 0), \quad (a, 0)$	$(0, -a), \quad (0, a)$
x-intercepts	$(-a, 0), \quad (a, 0)$	none
y-intercepts	none	$(0, -a), \quad (0, a)$
Rectangle	Use $(\pm a, 0) \; (0, \pm b)$	Use $(0, \pm a) \; (\pm b, 0)$
asymptotes	$y = \dfrac{b}{a}x, \quad y = -\dfrac{b}{a}x$	$y = \dfrac{a}{b}x, \quad y = -\dfrac{a}{b}x$

We will use these properties to graph hyperbolas.

EXAMPLE 11.27 HOW TO GRAPH A HYPERBOLA WITH CENTER $(0, 0)$

Graph $\dfrac{x^2}{25} - \dfrac{y^2}{4} = 1$.

✓ **Solution**

Step 1. Write the equation in standard form.	The equation is in standard form.	$\dfrac{x^2}{25} - \dfrac{y^2}{4} = 1$
Step 2. Determine whether the transverse axis is horizontal or vertical.	Since the x^2-term is positive, the transverse axis is horizontal.	The transverse axis is horizontal.
Step 3. Find the vertices.	Since $a^2 = 25$ then $a = \pm 5$. The vertices are on the x-axis.	$(-5, 0), (5, 0)$
Step 4. Sketch the rectangle centered at the origin intersecting one axis at $\pm a$ and the other at $\pm b$.	Since $a = \pm 5$, the rectangle will intersect the x-axis at the vertices. Since $b = \pm 2$, the rectangle will intersect the y-axis at $(0, -2)$ and $(0, 2)$.	

Step 5. Sketch the asymptotes—the lines through the diagonals of the rectangle.	The asymptotes have the equations $y = \frac{5}{2}x$, $y = -\frac{5}{2}x$	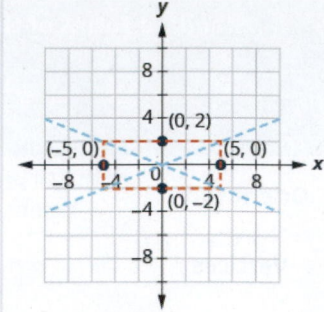
Step 6. Draw the two branches of the hyperbola.	Start at each vertex and use the asymptotes as a guide.	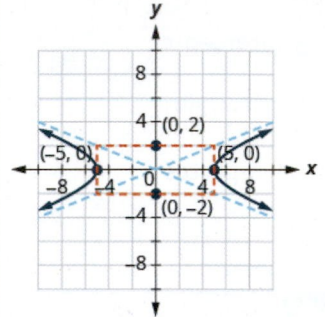

> **TRY IT : :** 11.53
>
> Graph $\frac{x^2}{16} - \frac{y^2}{4} = 1$.

> **TRY IT : :** 11.54
>
> Graph $\frac{x^2}{9} - \frac{y^2}{16} = 1$.

We summarize the steps for reference.

 HOW TO : : GRAPH A HYPERBOLA CENTERED AT $(0, 0)$.

Step 1. Write the equation in standard form.

Step 2. Determine whether the transverse axis is horizontal or vertical.

Step 3. Find the vertices.

Step 4. Sketch the rectangle centered at the origin intersecting one axis at $\pm a$ and the other at $\pm b$.

Step 5. Sketch the asymptotes—the lines through the diagonals of the rectangle.

Step 6. Draw the two branches of the hyperbola.

Sometimes the equation for a hyperbola needs to be first placed in standard form before we graph it.

EXAMPLE 11.28

Graph $4y^2 - 16x^2 = 64$.

✓ **Solution**

$$4y^2 - 16x^2 = 64$$

To write the equation in standard form, divide each term by 64 to make the equation equal to 1.	$\dfrac{4y^2}{64} - \dfrac{16x^2}{64} = \dfrac{64}{64}$
Simplify.	$\dfrac{y^2}{16} - \dfrac{x^2}{4} = 1$
Since the y^2-term is positive, the transverse axis is vertical. Since $a^2 = 16$ then $a = \pm 4$.	
The vertices are on the y-axis, $(0, -a)$, $(0, a)$. Since $b^2 = 4$ then $b = \pm 2$.	$(0, -4)$, $(0, 4)$
Sketch the rectangle intersecting the x-axis at $(-2, 0)$, $(2, 0)$ and the y-axis at the vertices. Sketch the asymptotes through the diagonals of the rectangle. Draw the two branches of the hyperbola.	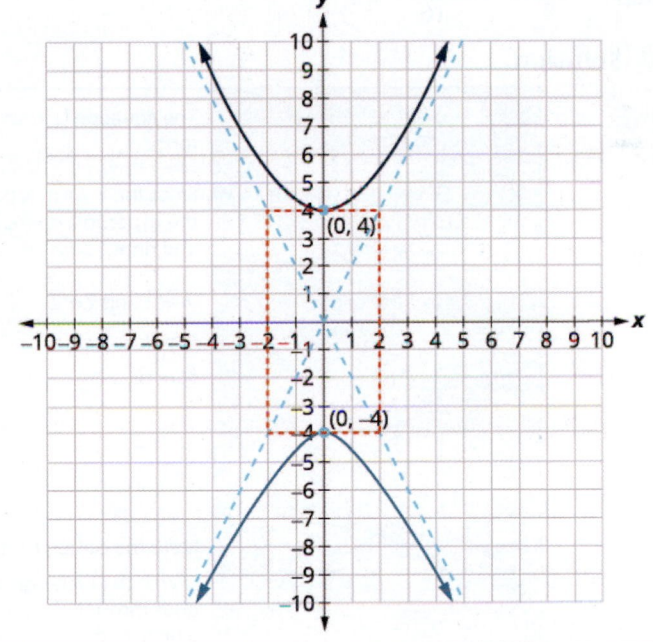

> **TRY IT : : 11.55**   Graph $4y^2 - 25x^2 = 100$.

> **TRY IT : : 11.56**   Graph $25y^2 - 9x^2 = 225$.

Graph a Hyperbola with Center at (h, k)

Hyperbolas are not always centered at the origin. When a hyperbola is centered at (h, k) the equations changes a bit as reflected in the table.

Standard Forms of the Equation a Hyperbola with Center (h, k)		
	$\dfrac{(x-h)^2}{a^2} - \dfrac{(y-k)^2}{b^2} = 1$	$\dfrac{(y-k)^2}{a^2} - \dfrac{(x-h)^2}{b^2} = 1$
Orientation	Transverse axis is horizontal. Opens left and right	Transverse axis is vertical. Opens up and down
Center	(h, k)	(h, k)
Vertices	a units to the left and right of the center	a units above and below the center
Rectangle	Use a units left/right of center b units above/ below the center	Use a units above/below the center b units left/right of center

EXAMPLE 11.29   HOW TO GRAPH A HYPERBOLA WITH CENTER (h, k)

Graph $\dfrac{(x-1)^2}{9} - \dfrac{(y-2)^2}{16} = 1$

✓ Solution

Step 1. Write the equation in standard form.	The equation is in standard form.	$\dfrac{(x-1)^2}{9} - \dfrac{(y-2)^2}{16} = 1$
Step 2. Determine whether the transverse axis is horizontal or vertical.	Since the x^2-term is positive, the hyperbola opens left and right.	The transverse axis is horizontal. The hyperbola opens left and right.
Step 3. Find the center and a, b.	$h = 1$ and $k = 2$ $a^2 = 9$ $b^2 = 16$	$\dfrac{(\overset{x-h}{x-1})^2}{9} - \dfrac{(\overset{y-k}{y-2})^2}{16} = 1$ Center: (1, 2) $a = 3$ $b = 4$
Step 4. Sketch the rectangle centered at (h, k) using a, b.	Mark the center, (1, 2). Sketch the rectangle that goes through the points 3 units to the left/ right of the center and 4 units above and below the center.	

Step 5. Sketch the asymptotes—the lines through the diagonals of the rectangle. Mark the vertices.	Sketch the diagonals. Mark the vertices, which are on the rectangle 3 units to the left and right of the center.	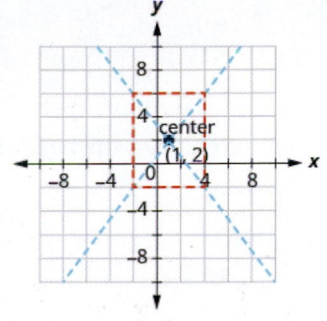
Step 6. Draw the two branches of the hyperbola.	Start at each vertex and use the asymptotes as a guide.	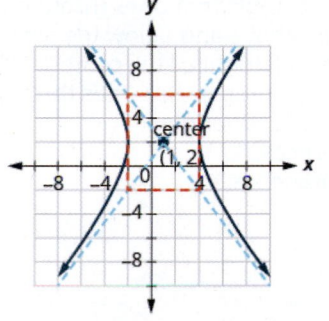

> **TRY IT :: 11.57**
> Graph $\dfrac{(x-3)^2}{25} - \dfrac{(y-1)^2}{9} = 1$.

> **TRY IT :: 11.58**
> Graph $\dfrac{(x-2)^2}{4} - \dfrac{(y-2)^2}{9} = 1$.

We summarize the steps for easy reference.

 HOW TO :: GRAPH A HYPERBOLA CENTERED AT (h, k).

Step 1.   Write the equation in standard form.

Step 2.   Determine whether the transverse axis is horizontal or vertical.

Step 3.   Find the center and a, b.

Step 4.   Sketch the rectangle centered at (h, k) using a, b.

Step 5.   Sketch the asymptotes—the lines through the diagonals of the rectangle. Mark the vertices.

Step 6.   Draw the two branches of the hyperbola.

Be careful as you identify the center. The standard equation has $x - h$ and $y - k$ with the center as (h, k).

EXAMPLE 11.30

Graph $\dfrac{(y+2)^2}{9} - \dfrac{(x+1)^2}{4} = 1$.

✓ Solution

$$\frac{(y+2)^2}{9} - \frac{(x+1)^2}{4} = 1$$

Since the y^2-term is positive, the hyperbola opens up and down.	$$\frac{\left(y - (-2)\right)^2}{9} - \frac{\left(x - (-1)\right)^2}{4} = 1$$
Find the center, (h, k).	Center: $(-1, -2)$
Find a, b.	$a = 3 \quad b = 2$
Sketch the rectangle that goes through the points 3 units above and below the center and 2 units to the left/right of the center. Sketch the asymptotes—the lines through the diagonals of the rectangle. Mark the vertices. Graph the branches.	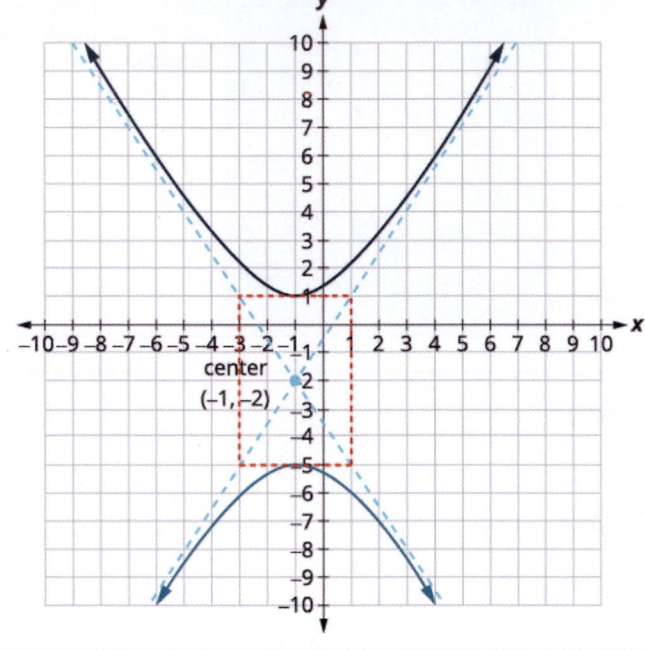

> **TRY IT : : 11.59**   Graph $\dfrac{(y+3)^2}{16} - \dfrac{(x+2)^2}{9} = 1$.

> **TRY IT : : 11.60**   Graph $\dfrac{(y+2)^2}{9} - \dfrac{(x+2)^2}{9} = 1$.

Again, sometimes we have to put the equation in standard form as our first step.

EXAMPLE 11.31

Write the equation in standard form and graph $4x^2 - 9y^2 - 24x - 36y - 36 = 0$.

✓ Solution

$$4x^2 - 9y^2 - 24x - 36y - 36 = 0$$

To get to standard form, complete the squares.	$$4(x^2 - 6x) - 9(y^2 + 4y) = 36$$
	$$4(x^2 - 6x + 9) - 9(y^2 + 4y + 4) = 36 + 36 - 36$$

$$4(x-3)^2 - 9(y+2)^2 = 36$$

Divide each term by 36 to get the constant to be 1.	$$\dfrac{4(x-3)^2}{36} - \dfrac{9(y+2)^2}{36} = \dfrac{36}{36}$$
	$$\dfrac{(x-3)^2}{9} - \dfrac{(y+2)^2}{4} = 1$$
Since the x^2-term is positive, the hyperbola opens left and right.	
Find the center, (h, k).	Center: $(3, -2)$
Find a, b.	$a = 3$ $b = 4$

Sketch the rectangle that goes through the points 3 units to the left/right of the center and 2 units above and below the center.
Sketch the asymptotes—the lines through the diagonals of the rectangle.
Mark the vertices.
Graph the branches.

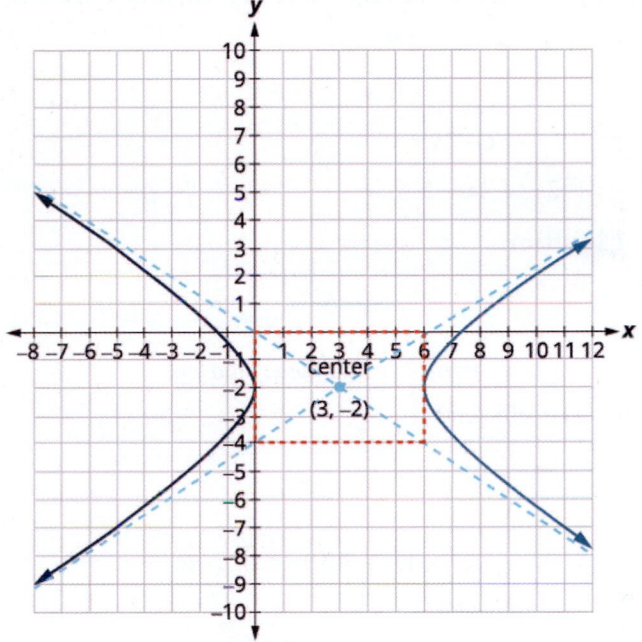

> **TRY IT : : 11.61** ⓐ Write the equation in standard form and ⓑ graph $9x^2 - 16y^2 + 18x + 64y - 199 = 0$.

> **TRY IT : : 11.62** ⓐ Write the equation in standard form and ⓑ graph $16x^2 - 25y^2 + 96x - 50y - 281 = 0$.

Identify Conic Sections by their Equations

Now that we have completed our study of the conic sections, we will take a look at the different equations and recognize some ways to identify a conic by its equation. When we are given an equation to graph, it is helpful to identify the conic so we know what next steps to take.

To identify a conic from its equation, it is easier if we put the variable terms on one side of the equation and the constants on the other.

Conic	Characteristics of x^2- and y^2- terms	Example
Parabola	Either x^2 OR y^2. Only one variable is squared.	$x = 3y^2 - 2y + 1$
Circle	x^2- and y^2- terms have the same coefficients	$x^2 + y^2 = 49$
Ellipse	x^2- and y^2- terms have the **same** sign, different coefficients	$4x^2 + 25y^2 = 100$
Hyperbola	x^2- and y^2- terms have **different** signs, different coefficients	$25y^2 - 4x^2 = 100$

EXAMPLE 11.32

Identify the graph of each equation as a circle, parabola, ellipse, or hyperbola.

ⓐ $9x^2 + 4y^2 + 56y + 160 = 0$

ⓑ $9x^2 - 16y^2 + 18x + 64y - 199 = 0$

ⓒ $x^2 + y^2 - 6x - 8y = 0$

ⓓ $y = -2x^2 - 4x - 5$

⊘ **Solution**

ⓐ

The x^2- and y^2-terms have the same sign and different coefficients.

$$9x^2 + 4y^2 + 56y + 160 = 0$$

Ellipse

ⓑ

The x^2- and y^2-terms have different signs and different coefficients.

$$9x^2 - 16y^2 + 18x + 64y - 199 = 0$$

Hyperbola

ⓒ

The x^2- and y^2-terms have the same coefficients.

$$x^2 + y^2 - 6x - 8y = 0$$

Circle

ⓓ

Only one variable, x, is squared.

$$y = -2x^2 - 4x - 5$$

Parabola

> **TRY IT :: 11.63**   Identify the graph of each equation as a circle, parabola, ellipse, or hyperbola.

ⓐ $x^2 + y^2 - 8x - 6y = 0$

ⓑ $4x^2 + 25y^2 = 100$

ⓒ $y = 6x^2 + 2x - 1$

ⓓ $16y^2 - 9x^2 = 144$

> **TRY IT : : 11.64**   Identify the graph of each equation as a circle, parabola, ellipse, or hyperbola.

        ⓐ $16x^2 + 9y^2 = 144$

        ⓑ $y = 2x^2 + 4x + 6$

        ⓒ $x^2 + y^2 + 2x + 6y + 9 = 0$

        ⓓ $4x^2 - 16y^2 = 64$

▶ **MEDIA : :**

Access these online resources for additional instructions and practice with hyperbolas.

- **Graph a Hyperbola with Center at the Origin (https://openstax.org/l/37ghyperborig)**
- **Graph a Hyperbola with Center not at the Origin (https://openstax.org/l/37ghyperbnorig)**
- **Graph a Hyperbola in General Form (https://openstax.org/l/37ghyperbgen)**
- **Identifying Conic Sections in General Form (https://openstax.org/l/37conicsgen)**

11.4 EXERCISES

Practice Makes Perfect

Graph a Hyperbola with Center at $(0, 0)$

In the following exercises, graph.

147. $\dfrac{x^2}{9} - \dfrac{y^2}{4} = 1$

148. $\dfrac{x^2}{25} - \dfrac{y^2}{9} = 1$

149. $\dfrac{x^2}{16} - \dfrac{y^2}{25} = 1$

150. $\dfrac{x^2}{9} - \dfrac{y^2}{36} = 1$

151. $\dfrac{y^2}{25} - \dfrac{x^2}{4} = 1$

152. $\dfrac{y^2}{36} - \dfrac{x^2}{16} = 1$

153. $16y^2 - 9x^2 = 144$

154. $25y^2 - 9x^2 = 225$

155. $4y^2 - 9x^2 = 36$

156. $16y^2 - 25x^2 = 400$

157. $4x^2 - 16y^2 = 64$

158. $9x^2 - 4y^2 = 36$

Graph a Hyperbola with Center at (h, k)

In the following exercises, graph.

159. $\dfrac{(x-1)^2}{16} - \dfrac{(y-3)^2}{4} = 1$

160. $\dfrac{(x-2)^2}{4} - \dfrac{(y-3)^2}{16} = 1$

161. $\dfrac{(y-4)^2}{9} - \dfrac{(x-2)^2}{25} = 1$

162. $\dfrac{(y-1)^2}{25} - \dfrac{(x-4)^2}{16} = 1$

163. $\dfrac{(y+4)^2}{25} - \dfrac{(x+1)^2}{36} = 1$

164. $\dfrac{(y+1)^2}{16} - \dfrac{(x+1)^2}{4} = 1$

165. $\dfrac{(y-4)^2}{16} - \dfrac{(x+1)^2}{25} = 1$

166. $\dfrac{(y+3)^2}{16} - \dfrac{(x-3)^2}{36} = 1$

167. $\dfrac{(x-3)^2}{25} - \dfrac{(y+2)^2}{9} = 1$

168. $\dfrac{(x+2)^2}{4} - \dfrac{(y-1)^2}{9} = 1$

In the following exercises, ⓐ write the equation in standard form and ⓑ graph.

169. $9x^2 - 4y^2 - 18x + 8y - 31 = 0$

170. $16x^2 - 4y^2 + 64x - 24y - 36 = 0$

171. $y^2 - x^2 - 4y + 2x - 6 = 0$

172. $4y^2 - 16x^2 - 24y + 96x - 172 = 0$

173.
$9y^2 - x^2 + 18y - 4x - 4 = 0$

Identify the Graph of each Equation as a Circle, Parabola, Ellipse, or Hyperbola

In the following exercises, identify the type of graph.

174.

ⓐ $x = -y^2 - 2y + 3$

ⓑ $9y^2 - x^2 + 18y - 4x - 4 = 0$

ⓒ $9x^2 + 25y^2 = 225$

ⓓ $x^2 + y^2 - 4x + 10y - 7 = 0$

175.

ⓐ $x = -2y^2 - 12y - 16$

ⓑ $x^2 + y^2 = 9$

ⓒ $16x^2 - 4y^2 + 64x - 24y - 36 = 0$

ⓓ $16x^2 + 36y^2 = 576$

Mixed Practice

In the following exercises, graph each equation.

176. $\dfrac{(y-3)^2}{9} - \dfrac{(x+2)^2}{16} = 1$

177. $x^2 + y^2 - 4x + 10y - 7 = 0$

178. $y = (x-1)^2 + 2$

179. $\dfrac{x^2}{9} + \dfrac{y^2}{25} = 1$

180. $(x+2)^2 + (y-5)^2 = 4$

181. $y^2 - x^2 - 4y + 2x - 6 = 0$

182. $x = -y^2 - 2y + 3$

183. $16x^2 + 9y^2 = 144$

Writing Exercises

184. In your own words, define a hyperbola and write the equation of a hyperbola centered at the origin in standard form. Draw a sketch of the hyperbola labeling the center, vertices, and asymptotes.

185. Explain in your own words how to create and use the rectangle that helps graph a hyperbola.

186. Compare and contrast the graphs of the equations $\dfrac{x^2}{4} - \dfrac{y^2}{9} = 1$ and $\dfrac{y^2}{9} - \dfrac{x^2}{4} = 1$.

187. Explain in your own words, how to distinguish the equation of an ellipse with the equation of a hyperbola.

Self Check

ⓐ *After completing the exercises, use this checklist to evaluate your mastery of the objectives of this section.*

I can...	Confidently	With some help	No-I don't get it!
graph a hyperbola with center at (0, 0).			
graph a hyperbola with center at (h, k).			
identify conic sections by their equations.			

ⓑ *On a scale of 1-10, how would you rate your mastery of this section in light of your responses on the checklist? How can you improve this?*

11.5 Solve Systems of Nonlinear Equations

Learning Objectives

By the end of this section, you will be able to:

› Solve a system of nonlinear equations using graphing
› Solve a system of nonlinear equations using substitution
› Solve a system of nonlinear equations using elimination
› Use a system of nonlinear equations to solve applications

Be Prepared!

1. Solve the system by graphing: $\begin{cases} x - 3y = -3 \\ x + y = 5 \end{cases}$.

   If you missed this problem, review Example 4.2.

2. Solve the system by substitution: $\begin{cases} x - 4y = -4 \\ -3x + 4y = 0 \end{cases}$.

   If you missed this problem, review Example 4.7.

3. Solve the system by elimination: $\begin{cases} 3x - 4y = -9 \\ 5x + 3y = 14 \end{cases}$.

   If you missed this problem, review Example 4.9.

Solve a System of Nonlinear Equations Using Graphing

We learned how to solve systems of linear equations with two variables by graphing, substitution and elimination. We will be using these same methods as we look at nonlinear systems of equations with two equations and two variables. A **system of nonlinear equations** is a system where at least one of the equations is not linear.

For example each of the following systems is a **system of nonlinear equations**.

$$\begin{cases} x^2 + y^2 = 9 \\ x^2 - y = 9 \end{cases} \qquad \begin{cases} 9x^2 + y^2 = 9 \\ y = 3x - 3 \end{cases} \qquad \begin{cases} x + y = 4 \\ y = x^2 + 2 \end{cases}$$

System of Nonlinear Equations

A **system of nonlinear equations** is a system where at least one of the equations is not linear.

Just as with systems of linear equations, a solution of a nonlinear system is an ordered pair that makes both equations true. In a nonlinear system, there may be more than one solution. We will see this as we solve a system of nonlinear equations by graphing.

When we solved systems of linear equations, the solution of the system was the point of intersection of the two lines. With systems of nonlinear equations, the graphs may be circles, parabolas or hyperbolas and there may be several points of intersection, and so several solutions. Once you identify the graphs, visualize the different ways the graphs could intersect and so how many solutions there might be.

To solve systems of nonlinear equations by graphing, we use basically the same steps as with systems of linear equations modified slightly for nonlinear equations. The steps are listed below for reference.

HOW TO :: SOLVE A SYSTEM OF NONLINEAR EQUATIONS BY GRAPHING.

Step 1. Identify the graph of each equation. Sketch the possible options for intersection.
Step 2. Graph the first equation.
Step 3. Graph the second equation on the same rectangular coordinate system.
Step 4. Determine whether the graphs intersect.
Step 5. Identify the points of intersection.
Step 6. Check that each ordered pair is a solution to both original equations.

EXAMPLE 11.33

Solve the system by graphing: $\begin{cases} x - y = -2 \\ y = x^2 \end{cases}$.

✓ Solution

Identify each graph.

$\begin{cases} x - y = -2 & \text{line} \\ y = x^2 & \text{parabola} \end{cases}$

Sketch the possible options for intersection of a parabola and a line.

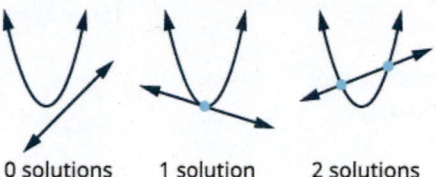

0 solutions   1 solution   2 solutions

Graph the line, $x - y = -2$.

Slope-intercept form $y = x + 2$.

Graph the parabola, $y = x^2$.

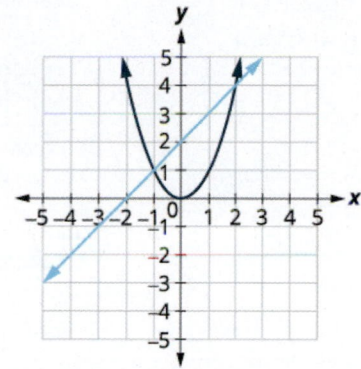

Identify the points of intersection.

The points of intersection appear to be $(2, 4)$ and $(-1, 1)$.

Check to make sure each solution makes both equations true.

$(2, 4)$

$$x - y = -2 \qquad y = x^2$$
$$2 - 4 \stackrel{?}{=} -2 \qquad 4 \stackrel{?}{=} 2^2$$
$$-2 = -2 \checkmark \qquad 4 = 4 \checkmark$$

$(-1, 1)$

$$x - y = -2 \qquad y = x^2$$
$$-1 - 1 \stackrel{?}{=} -2 \qquad 1 \stackrel{?}{=} (-1)^2$$
$$-2 = -2 \checkmark \qquad 1 = 1 \checkmark$$

The solutions are $(2, 4)$ and $(-1, 1)$.

> **TRY IT : : 11.65**

Solve the system by graphing: $\begin{cases} x + y = 4 \\ y = x^2 + 2 \end{cases}$.

> **TRY IT : : 11.66**

Solve the system by graphing: $\begin{cases} x - y = -1 \\ y = -x^2 + 3 \end{cases}$.

To identify the graph of each equation, keep in mind the characteristics of the x^2 and y^2 terms of each conic.

EXAMPLE 11.34

Solve the system by graphing: $\begin{cases} y = -1 \\ (x-2)^2 + (y+3)^2 = 4 \end{cases}$

⊘ **Solution**

Identify each graph.

$\begin{cases} y = -1 & \text{line} \\ (x-2)^2 + (y+3)^2 = 4 & \text{circle} \end{cases}$

Sketch the possible options for the intersection of a circle and a line.

0 solutions   1 solution   2 solutions

Graph the circle, $(x-2)^2 + (y+3)^2 = 4$

Center: $(2, -3)$ radius: 2

Graph the line, $y = -1$.

It is a horizontal line.

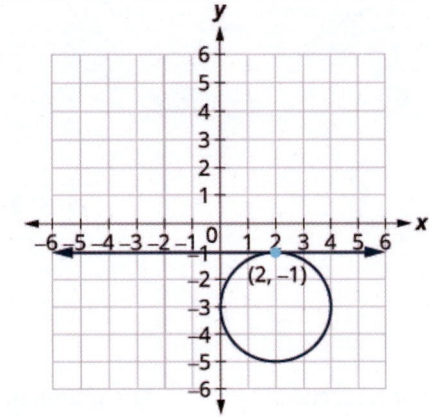

Identify the points of intersection.

The point of intersection appears to be $(2, -1)$.

Check to make sure the solution makes both equations true.
$(2, -1)$

$$(x-2)^2 + (y+3)^2 = 4 \qquad\qquad y = -1$$
$$(2-2)^2 + (-1+3)^2 \overset{?}{=} 4 \qquad\qquad -1 = -1 \checkmark$$
$$(0)^2 + (2)^2 \overset{?}{=} 4$$
$$4 = 4 \checkmark$$

The solution is $(2, -1)$.

> **TRY IT :: 11.67**
>
> Solve the system by graphing: $\begin{cases} x = -6 \\ (x+3)^2 + (y-1)^2 = 9 \end{cases}$

> **TRY IT :: 11.68**
>
> Solve the system by graphing: $\begin{cases} y = 4 \\ (x-2)^2 + (y+3)^2 = 4 \end{cases}$

Solve a System of Nonlinear Equations Using Substitution

The graphing method works well when the points of intersection are integers and so easy to read off the graph. But more often it is difficult to read the coordinates of the points of intersection. The substitution method is an algebraic method that will work well in many situations. It works especially well when it is easy to solve one of the equations for one of the variables.

The substitution method is very similar to the substitution method that we used for systems of linear equations. The steps are listed below for reference.

HOW TO :: SOLVE A SYSTEM OF NONLINEAR EQUATIONS BY SUBSTITUTION.

Step 1.   Identify the graph of each equation. Sketch the possible options for intersection.

Step 2.   Solve one of the equations for either variable.

Step 3.   Substitute the expression from Step 2 into the other equation.

Step 4.   Solve the resulting equation.

Step 5.   Substitute each solution in Step 4 into one of the original equations to find the other variable.

Step 6.   Write each solution as an ordered pair.

Step 7.   Check that each ordered pair is a solution to **both** original equations.

EXAMPLE 11.35

Solve the system by using substitution: $\begin{cases} 9x^2 + y^2 = 9 \\ y = 3x - 3 \end{cases}$.

⊘ **Solution**

Identify each graph.	$\begin{cases} 9x^2 + y^2 = 9 & \text{ellipse} \\ y = 3x - 3 & \text{line} \end{cases}$

Sketch the possible options for intersection of an ellipse and a line.

0 solutions   1 solution   2 solutions

The equation $y = 3x - 3$ is solved for y.	$y = 3x - 3$
	$9x^2 + y^2 = 9$
Substitute $3x - 3$ for y in the first equation.	$9x^2 + (3x - 3)^2 = 9$
Solve the equation for x.	$9x^2 + 9x^2 - 18x + 9 = 9$
	$18x^2 - 18x = 0$
	$18x(x - 1) = 0$
	$x = 0 \quad x = 1$
Substitute $x = 0$ and $x = 1$ into $y = 3x - 3$ to find y.	$y = 3x - 3 \qquad y = 3x - 3$
	$y = 3 \cdot 0 - 3 \qquad y = 3 \cdot 1 - 3$
	$y = -3 \qquad\qquad y = 0$
	The ordered pairs are $(0, -3), \quad (1, 0)$.

Check **both** ordered pairs in **both** equations.

(0, −3)

$$9x^2 + y^2 = 9 \qquad\qquad y = 3x - 3$$
$$9 \cdot 0^2 + (-3)^2 \stackrel{?}{=} 9 \qquad\qquad -3 \stackrel{?}{=} 3 \cdot 0 - 3$$
$$0 + 9 \stackrel{?}{=} 9 \qquad\qquad -3 \stackrel{?}{=} 0 - 3$$
$$9 = 9 \checkmark \qquad\qquad -3 = -3 \checkmark$$

(1, 0)

$$9x^2 + y^2 = 9 \qquad\qquad y = 3x - 3$$
$$9 \cdot 1^2 + 0^2 \stackrel{?}{=} 9 \qquad\qquad 0 \stackrel{?}{=} 3 \cdot 1 - 3$$
$$9 + 0 \stackrel{?}{=} 9 \qquad\qquad 0 \stackrel{?}{=} 3 - 3$$
$$9 = 9 \checkmark \qquad\qquad 0 = 0 \checkmark$$

The solutions are $(0, -3)$, $(1, 0)$.

> **TRY IT : : 11.69**

Solve the system by using substitution: $\begin{cases} x^2 + 9y^2 = 9 \\ y = \frac{1}{3}x - 3 \end{cases}$.

> **TRY IT : : 11.70**

Solve the system by using substitution: $\begin{cases} 4x^2 + y^2 = 4 \\ y = x + 2 \end{cases}$.

So far, each system of nonlinear equations has had at least one solution. The next example will show another option.

EXAMPLE 11.36

Solve the system by using substitution: $\begin{cases} x^2 - y = 0 \\ y = x - 2 \end{cases}$.

⊘ **Solution**

Identify each graph.

$$\begin{cases} x^2 - y = 0 & \text{parabola} \\ y = x - 2 & \text{line} \end{cases}$$

Sketch the possible options for intersection of a parabola and a line

| 0 solutions | 1 solution | 2 solutions |

The equation $y = x - 2$ is solved for y.

$$y = x - 2$$

$$x^2 - y = 0$$

Substitute $x - 2$ for y in the first equation.

$$x^2 - (x - 2) = 0$$

Solve the equation for x.

$$x^2 - x + 2 = 0$$

This doesn't factor easily, so we can check the discriminant.

$$b^2 - 4ac$$
$$(-1)^2 - 4 \cdot 1 \cdot 2$$
$$-7$$

The discriminant is negative, so there is no real solution. The system has no solution.

> **TRY IT :: 11.71**

Solve the system by using substitution: $\begin{cases} x^2 - y = 0 \\ y = 2x - 3 \end{cases}$.

> **TRY IT :: 11.72**

Solve the system by using substitution: $\begin{cases} y^2 - x = 0 \\ y = 3x - 2 \end{cases}$.

Solve a System of Nonlinear Equations Using Elimination

When we studied systems of linear equations, we used the method of elimination to solve the system. We can also use elimination to solve systems of nonlinear equations. It works well when the equations have both variables squared. When using elimination, we try to make the coefficients of one variable to be opposites, so when we add the equations together, that variable is eliminated.

The elimination method is very similar to the elimination method that we used for systems of linear equations. The steps are listed for reference.

HOW TO :: SOLVE A SYSTEM OF EQUATIONS BY ELIMINATION.

Step 1. Identify the graph of each equation. Sketch the possible options for intersection.

Step 2. Write both equations in standard form.

Step 3. Make the coefficients of one variable opposites.
Decide which variable you will eliminate.
Multiply one or both equations so that the coefficients of that variable are opposites.

Step 4. Add the equations resulting from Step 3 to eliminate one variable.

Step 5. Solve for the remaining variable.

Step 6. Substitute each solution from Step 5 into one of the original equations. Then solve for the other variable.

Step 7. Write each solution as an ordered pair.

Step 8. Check that each ordered pair is a solution to **both** original equations.

EXAMPLE 11.37

Solve the system by elimination: $\begin{cases} x^2 + y^2 = 4 \\ x^2 - y = 4 \end{cases}$.

⊘ **Solution**

Identify each graph.

$\begin{cases} x^2 + y^2 = 4 & \text{circle} \\ x^2 - y = 4 & \text{parabola} \end{cases}$

Sketch the possible options for intersection of a circle and a parabola.

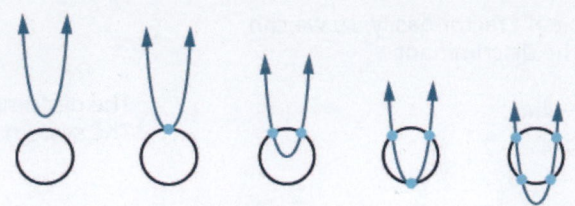

0 solutions 1 solution 2 solutions 3 solutions 4 solutions

Both equations are in standard form.	$\begin{cases} x^2 + y^2 = 4 \\ x^2 - y = 4 \end{cases}$
To get opposite coefficients of x^2, we will multiply the second equation by -1.	$\begin{cases} x^2 + y^2 = 4 \\ -1(x^2 - y) = -1(4) \end{cases}$
Simplify.	$\begin{cases} x^2 + y^2 = 4 \\ -x^2 + y = 4 \end{cases}$
Add the two equations to eliminate x^2.	$\begin{array}{r} x^2 + y^2 = 4 \\ \underline{-x^2 + y = 4} \\ y^2 + y = 0 \end{array}$
Solve for y.	$y(y + 1) = 0$
	$y = 0 \quad y + 1 = 0$ $y = -1$
Substitute $y = 0$ and $y = -1$ into one of the original equations. Then solve for x.	$y = 0 \quad y = -1$
	$\begin{array}{ll} x^2 - y = 4 & x^2 - y = 4 \\ x^2 - 0 = 4 & x^2 - (-1) = 4 \\ x^2 = 4 & x^2 = 3 \\ x = \pm 2 & x = \pm\sqrt{3} \end{array}$
Write each solution as an ordered pair.	The ordered pairs are $(-2, 0) \quad (2, 0)$. $(\sqrt{3}, -1)(-\sqrt{3}, -1)$
Check that each ordered pair is a solution to **both** original equations.	
We will leave the checks for each of the four solutions to you.	The solutions are $(-2, 0)$, $(2, 0)$, $(\sqrt{3}, -1)$, and $(-\sqrt{3}, -1)$.

> **TRY IT :: 11.73**
>
> Solve the system by elimination: $\begin{cases} x^2 + y^2 = 9 \\ x^2 - y = 9 \end{cases}$.

> **TRY IT :: 11.74**
>
> Solve the system by elimination: $\begin{cases} x^2 + y^2 = 1 \\ -x + y^2 = 1 \end{cases}$.

There are also four options when we consider a circle and a hyperbola.

EXAMPLE 11.38

Solve the system by elimination: $\begin{cases} x^2 + y^2 = 7 \\ x^2 - y^2 = 1 \end{cases}$.

✓ Solution

Identify each graph.	$\begin{cases} x^2 + y^2 = 7 & \text{circle} \\ x^2 - y^2 = 1 & \text{hyperbola} \end{cases}$

Sketch the possible options for intersection of a circle and hyperbola.

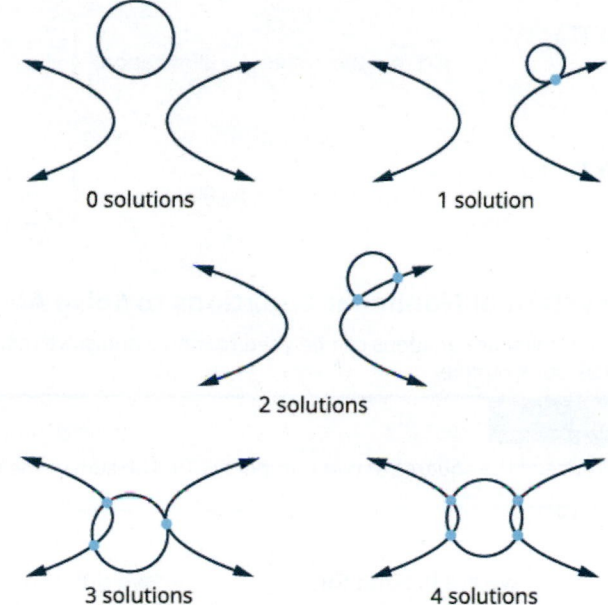

0 solutions   1 solution

2 solutions

3 solutions   4 solutions

Both equations are in standard form.	$\begin{cases} x^2 + y^2 = 7 \\ x^2 - y^2 = 1 \end{cases}$
The coefficients of y^2 are opposite, so we will add the equations.	$\begin{cases} x^2 + y^2 = 7 \\ x^2 - y^2 = 1 \end{cases}$ $2x^2 \qquad = 8$
Simplify.	$x^2 = 4$ $x = \pm 2$ $x = 2 \qquad x = -2$

Substitute $x = 2$ and $x = -2$ into one of the original equations. Then solve for y.

$$\begin{aligned} x^2 + y^2 &= 7 & x^2 + y^2 &= 7 \\ 2^2 + y^2 &= 7 & (-2)^2 + y^2 &= 7 \\ 4 + y^2 &= 7 & 4 + y^2 &= 7 \\ y^2 &= 3 & y^2 &= 3 \\ y &= \pm\sqrt{3} & y &= \pm\sqrt{3} \end{aligned}$$

Write each solution as an ordered pair.

The ordered pairs are $(-2, \sqrt{3})$, $(-2, -\sqrt{3})$, $(2, \sqrt{3})$, and $(2, -\sqrt{3})$.

Check that the ordered pair is a solution to **both** original equations.

We will leave the checks for each of the four solutions to you.	The solutions are $(-2, \sqrt{3})$, $(-2, -\sqrt{3})$, $(2, \sqrt{3})$, and $(2, -\sqrt{3})$.

> **TRY IT : : 11.75**

Solve the system by elimination: $\begin{cases} x^2 + y^2 = 25 \\ y^2 - x^2 = 7 \end{cases}$.

> **TRY IT : : 11.76**

Solve the system by elimination: $\begin{cases} x^2 + y^2 = 4 \\ x^2 - y^2 = 4 \end{cases}$.

Use a System of Nonlinear Equations to Solve Applications

Systems of nonlinear equations can be used to model and solve many applications. We will look at an everyday geometric situation as our example.

EXAMPLE 11.39

The difference of the squares of two numbers is 15. The sum of the numbers is 5. Find the numbers.

⊘ **Solution**

Identify what we are looking for.	Two different numbers.
Define the variables.	$x =$ first number $\quad y =$ second number
Translate the information into a system of equations.	
First sentence.	The difference of the squares of two numbers is 15.
	$x^2 - y^2 = 15$
Second sentence.	The sum of the numbers is 5.
	$x + y = 5$
Solve the system by substitution	$\begin{cases} x^2 - y^2 = 15 \\ x + y = 5 \end{cases}$
Solve the second equation for x.	$x = 5 - y$
Substitute x into the first equation.	$x^2 - y^2 = 15$
	$(5 - y)^2 - y^2 = 15$
Expand and simplify.	$(25 - 10y + y^2) - y^2 = 15$
	$25 - 10y + y^2 - y^2 = 15$
	$25 - 10y = 15$
Solve for y.	$-10y = -10$
	$y = 1$

Substitute back into the second equation.	$x + y = 5$
	$x + (1) = 5$
	$x = 4$
	The numbers are 1 and 4.

> **TRY IT : : 11.77**

The difference of the squares of two numbers is $-20.$ The sum of the numbers is 10. Find the numbers.

> **TRY IT : : 11.78**

The difference of the squares of two numbers is 35. The sum of the numbers is $-1.$ Find the numbers.

EXAMPLE 11.40

Myra purchased a small 25" TV for her kitchen. The size of a TV is measured on the diagonal of the screen. The screen also has an area of 300 square inches. What are the length and width of the TV screen?

⊘ **Solution**

Identify what we are looking for.	The length and width of the rectangle
Define the variables.	Let $x =$ width of the rectangle $y =$ length of the rectangle
Draw a diagram to help visualize the situation.	
	Area is 300 square inches.
Translate the information into a system of equations.	The diagonal of the right triangle is 25 inches.
	$x^2 + y^2 = 25^2$ $x^2 + y^2 = 625$
	The area of the rectangle is 300 square inches.
	$x \cdot y = 300$ $\begin{cases} x^2 + y^2 = 625 \\ x \cdot y = 300 \end{cases}$
Solve the system using substitution.	$x \cdot y = 300$
Solve the second equation for x.	$x = \dfrac{300}{y}$
Substitute x into the first equation.	$x^2 + y^2 = 625$
	$\left(\dfrac{300}{y}\right)^2 + y^2 = 625$

Simplify.	$\frac{90000}{y^2} + y^2 = 625$
Multiply by y^2 to clear the fractions.	$90000 + y^4 = 625y^2$
Put in standard form.	$y^4 - 625y^2 + 90000 = 0$
Solve by factoring.	$(y^2 - 225)(y^2 - 400) = 0$
	$y^2 - 225 = 0 \quad y^2 - 400 = 0$
	$y^2 = 225 \qquad y^2 = 400$ $y = \pm 15 \qquad y = \pm 20$
Since y is a side of the rectangle, we discard the negative values.	$y = 15 \qquad y = 20$
Substitute back into the second equation.	$x \cdot y = 300 \quad x \cdot y = 300$
	$x \cdot 15 = 300 \quad x \cdot 20 = 300$ $x = 20 \qquad x = 15$
	If the length is 15 inches, the width is 20 inches.
	If the length is 20 inches, the width is 15 inches.

> **TRY IT : : 11.79**
>
> Edgar purchased a small 20″ TV for his garage. The size of a TV is measured on the diagonal of the screen. The screen also has an area of 192 square inches. What are the length and width of the TV screen?

> **TRY IT : : 11.80**
>
> The Harper family purchased a small microwave for their family room. The diagonal of the door measures 15 inches. The door also has an area of 108 square inches. What are the length and width of the microwave door?

▶ **MEDIA : :**

Access these online resources for additional instructions and practice with solving nonlinear equations.

- **Nonlinear Systems of Equations (https://openstax.org/l/37nonsyseq)**
- **Solve a System of Nonlinear Equations (https://openstax.org/l/37nonsyseq2)**
- **Solve a System of Nonlinear Equations by Elimination (https://openstax.org/l/37nonsyselim)**
- **System of Nonlinear Equations – Area and Perimeter Application (https://openstax.org/l/37nonsysapps)**

11.5 EXERCISES

Practice Makes Perfect

Solve a System of Nonlinear Equations Using Graphing

In the following exercises, solve the system of equations by using graphing.

188. $\begin{cases} y = 2x + 2 \\ y = -x^2 + 2 \end{cases}$

189. $\begin{cases} y = 6x - 4 \\ y = 2x^2 \end{cases}$

190. $\begin{cases} x + y = 2 \\ x = y^2 \end{cases}$

191. $\begin{cases} x - y = -2 \\ x = y^2 \end{cases}$

192. $\begin{cases} y = \frac{3}{2}x + 3 \\ y = -x^2 + 2 \end{cases}$

193. $\begin{cases} y = x - 1 \\ y = x^2 + 1 \end{cases}$

194. $\begin{cases} x = -2 \\ x^2 + y^2 = 4 \end{cases}$

195. $\begin{cases} y = -4 \\ x^2 + y^2 = 16 \end{cases}$

196. $\begin{cases} x = 2 \\ (x + 2)^2 + (y + 3)^2 = 16 \end{cases}$

197. $\begin{cases} y = -1 \\ (x - 2)^2 + (y - 4)^2 = 25 \end{cases}$

198. $\begin{cases} y = -2x + 4 \\ y = \sqrt{x} + 1 \end{cases}$

199. $\begin{cases} y = -\frac{1}{2}x + 2 \\ y = \sqrt{x} - 2 \end{cases}$

Solve a System of Nonlinear Equations Using Substitution

In the following exercises, solve the system of equations by using substitution.

200. $\begin{cases} x^2 + 4y^2 = 4 \\ y = \frac{1}{2}x - 1 \end{cases}$

201. $\begin{cases} 9x^2 + y^2 = 9 \\ y = 3x + 3 \end{cases}$

202. $\begin{cases} 9x^2 + y^2 = 9 \\ y = x + 3 \end{cases}$

203. $\begin{cases} 9x^2 + 4y^2 = 36 \\ x = 2 \end{cases}$

204. $\begin{cases} 4x^2 + y^2 = 4 \\ y = 4 \end{cases}$

205. $\begin{cases} x^2 + y^2 = 169 \\ x = 12 \end{cases}$

206. $\begin{cases} 3x^2 - y = 0 \\ y = 2x - 1 \end{cases}$

207. $\begin{cases} 2y^2 - x = 0 \\ y = x + 1 \end{cases}$

208. $\begin{cases} y = x^2 + 3 \\ y = x + 3 \end{cases}$

209. $\begin{cases} y = x^2 - 4 \\ y = x - 4 \end{cases}$

210. $\begin{cases} x^2 + y^2 = 25 \\ x - y = 1 \end{cases}$

211. $\begin{cases} x^2 + y^2 = 25 \\ 2x + y = 10 \end{cases}$

Solve a System of Nonlinear Equations Using Elimination

In the following exercises, solve the system of equations by using elimination.

212. $\begin{cases} x^2 + y^2 = 16 \\ x^2 - 2y = 8 \end{cases}$

213. $\begin{cases} x^2 + y^2 = 16 \\ x^2 - y = 4 \end{cases}$

214. $\begin{cases} x^2 + y^2 = 4 \\ x^2 + 2y = 1 \end{cases}$

215. $\begin{cases} x^2 + y^2 = 4 \\ x^2 - y = 2 \end{cases}$

216. $\begin{cases} x^2 + y^2 = 9 \\ x^2 - y = 3 \end{cases}$

217. $\begin{cases} x^2 + y^2 = 4 \\ y^2 - x = 2 \end{cases}$

218. $\begin{cases} x^2 + y^2 = 25 \\ 2x^2 - 3y^2 = 5 \end{cases}$

219. $\begin{cases} x^2 + y^2 = 20 \\ x^2 - y^2 = -12 \end{cases}$

220. $\begin{cases} x^2 + y^2 = 13 \\ x^2 - y^2 = 5 \end{cases}$

221. $\begin{cases} x^2 + y^2 = 16 \\ x^2 - y^2 = 16 \end{cases}$

222. $\begin{cases} 4x^2 + 9y^2 = 36 \\ 2x^2 - 9y^2 = 18 \end{cases}$

223. $\begin{cases} x^2 - y^2 = 3 \\ 2x^2 + y^2 = 6 \end{cases}$

224. $\begin{cases} 4x^2 - y^2 = 4 \\ 4x^2 + y^2 = 4 \end{cases}$

225. $\begin{cases} x^2 - y^2 = -5 \\ 3x^2 + 2y^2 = 30 \end{cases}$

226. $\begin{cases} x^2 - y^2 = 1 \\ x^2 - 2y = 4 \end{cases}$

227. $\begin{cases} 2x^2 + y^2 = 11 \\ x^2 + 3y^2 = 28 \end{cases}$

Use a System of Nonlinear Equations to Solve Applications

In the following exercises, solve the problem using a system of equations.

228. The sum of two numbers is -6 and the product is 8. Find the numbers.

229. The sum of two numbers is 11 and the product is -42. Find the numbers.

230. The sum of the squares of two numbers is 65. The difference of the number is 3. Find the numbers.

231. The sum of the squares of two numbers is 113. The difference of the number is 1. Find the numbers.

232. The difference of the squares of two numbers is 15. The difference of twice the square of the first number and the square of the second number is 30. Find the numbers.

233. The difference of the squares of two numbers is 20. The difference of the square of the first number and twice the square of the second number is 4. Find the numbers.

234. The perimeter of a rectangle is 32 inches and its area is 63 square inches. Find the length and width of the rectangle.

235. The perimeter of a rectangle is 52 cm and its area is 165 cm^2. Find the length and width of the rectangle.

236. Dion purchased a new microwave. The diagonal of the door measures 17 inches. The door also has an area of 120 square inches. What are the length and width of the microwave door?

237. Jules purchased a microwave for his kitchen. The diagonal of the front of the microwave measures 26 inches. The front also has an area of 240 square inches. What are the length and width of the microwave?

238. Roman found a widescreen TV on sale, but isn't sure if it will fit his entertainment center. The TV is 60". The size of a TV is measured on the diagonal of the screen and a widescreen has a length that is larger than the width. The screen also has an area of 1728 square inches. His entertainment center has an insert for the TV with a length of 50 inches and width of 40 inches. What are the length and width of the TV screen and will it fit Roman's entertainment center?

239. Donnette found a widescreen TV at a garage sale, but isn't sure if it will fit her entertainment center. The TV is 50". The size of a TV is measured on the diagonal of the screen and a widescreen has a length that is larger than the width. The screen also has an area of 1200 square inches. Her entertainment center has an insert for the TV with a length of 38 inches and width of 27 inches. What are the length and width of the TV screen and will it fit Donnette's entertainment center?

Writing Exercises

240. In your own words, explain the advantages and disadvantages of solving a system of equations by graphing.

241. Explain in your own words how to solve a system of equations using substitution.

242. Explain in your own words how to solve a system of equations using elimination.

243. A circle and a parabola can intersect in ways that would result in 0, 1, 2, 3, or 4 solutions. Draw a sketch of each of the possibilities.

Self Check

ⓐ *After completing the exercises, use this checklist to evaluate your mastery of the objectives of this section.*

I can...	Confidently	With some help	No-I don't get it!
solve a system of nonlinear equations using graphing.			
solve a system of nonlinear equations using substitution.			
solve a system of nonlinear equations using elimination.			
use a system of nonlinear equations to solve applications.			

ⓑ *After looking at the checklist, do you think you are well-prepared for the next section? Why or why not?*

CHAPTER 11 REVIEW

KEY TERMS

circle A circle is all points in a plane that are a fixed distance from a fixed point in the plane.

ellipse An ellipse is all points in a plane where the sum of the distances from two fixed points is constant.

hyperbola A hyperbola is defined as all points in a plane where the difference of their distances from two fixed points is constant.

parabola A parabola is all points in a plane that are the same distance from a fixed point and a fixed line.

system of nonlinear equations A system of nonlinear equations is a system where at least one of the equations is not linear.

KEY CONCEPTS

11.1 Distance and Midpoint Formulas; Circles

- **Distance Formula:** The distance d between the two points (x_1, y_1) and (x_2, y_2) is

$$d = \sqrt{(x_2 - x_1)^2 + (y_2 - y_1)^2}$$

- **Midpoint Formula:** The midpoint of the line segment whose endpoints are the two points (x_1, y_1) and (x_2, y_2) is

$$\left(\frac{x_1 + x_2}{2}, \frac{y_1 + y_2}{2}\right)$$

  To find the midpoint of a line segment, we find the average of the x-coordinates and the average of the y-coordinates of the endpoints.

- **Circle:** A circle is all points in a plane that are a fixed distance from a fixed point in the plane. The given point is called the *center,* $(h, k),$ and the fixed distance is called the *radius, r,* of the circle.

- **Standard Form of the Equation a Circle:** The standard form of the equation of a circle with center, $(h, k),$ and radius, r, is

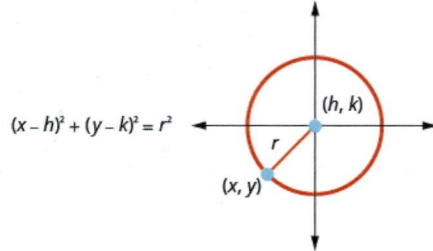

$(x - h)^2 + (y - k)^2 = r^2$

- **General Form of the Equation of a Circle:** The general form of the equation of a circle is

$$x^2 + y^2 + ax + by + c = 0$$

11.2 Parabolas

- **Parabola:** A **parabola** is all points in a plane that are the same distance from a fixed point and a fixed line. The fixed point is called the **focus,** and the fixed line is called the **directrix** of the parabola.

Vertical Parabolas		
	General form $y = ax^2 + bx + c$	**Standard form** $y = a(x - h)^2 + k$
Orientation	$a > 0$ up; $a < 0$ down	$a > 0$ up; $a < 0$ down
Axis of symmetry	$x = -\dfrac{b}{2a}$	$x = h$
Vertex	Substitute $x = -\dfrac{b}{2a}$ and solve for y.	(h, k)
y- intercept	Let $x = 0$	Let $x = 0$
x-intercepts	Let $y = 0$	Let $y = 0$

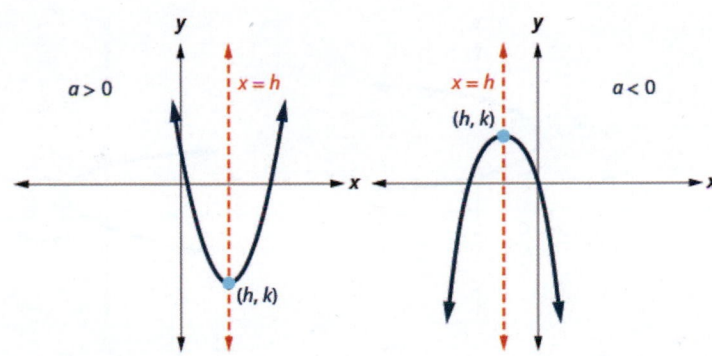

- **How to graph vertical parabolas** $(y = ax^2 + bx + c$ **or** $f(x) = a(x - h)^2 + k)$ **using properties.**

  Step 1. Determine whether the parabola opens upward or downward.

  Step 2. Find the axis of symmetry.

  Step 3. Find the vertex.

  Step 4. Find the *y*-intercept. Find the point symmetric to the *y*-intercept across the axis of symmetry.

  Step 5. Find the *x*-intercepts.

  Step 6. Graph the parabola.

Horizontal Parabolas		
	General form $x = ay^2 + by + c$	**Standard form** $x = a(y - k)^2 + h$
Orientation	$a > 0$ right; $a < 0$ left	$a > 0$ right; $a < 0$ left
Axis of symmetry	$y = -\dfrac{b}{2a}$	$y = k$
Vertex	Substitute $y = -\dfrac{b}{2a}$ and solve for x.	(h, k)
y-intercepts	Let $x = 0$	Let $x = 0$
x-intercept	Let $y = 0$	Let $y = 0$

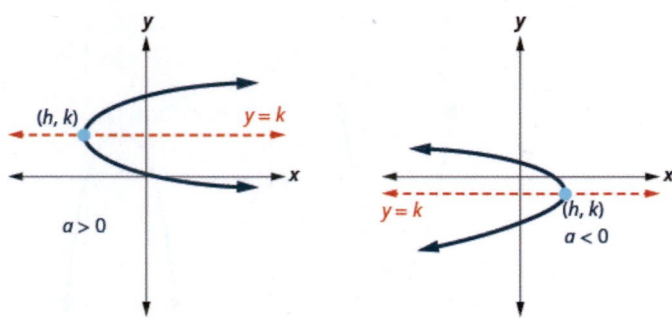

- **How to graph horizontal parabolas** $(x = ay^2 + by + c$ or $x = a(y - k)^2 + h)$ **using properties.**

  Step 1. Determine whether the parabola opens to the left or to the right.

  Step 2. Find the axis of symmetry.

  Step 3. Find the vertex.

  Step 4. Find the x-intercept. Find the point symmetric to the x-intercept across the axis of symmetry.

  Step 5. Find the y-intercepts.

  Step 6. Graph the parabola.

11.3 Ellipses

- **Ellipse:** An **ellipse** is all points in a plane where the sum of the distances from two fixed points is constant. Each of the fixed points is called a **focus** of the ellipse.

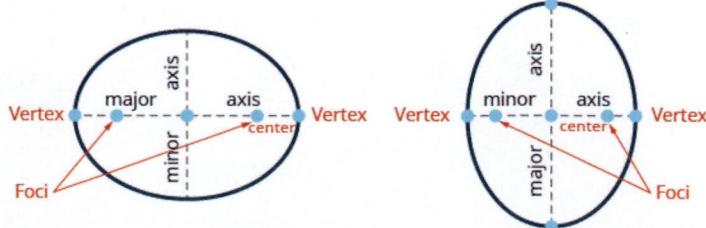

If we draw a line through the foci intersects the ellipse in two points—each is called a **vertex** of the ellipse.
The segment connecting the vertices is called the **major axis**.
The midpoint of the segment is called the **center** of the ellipse.
A segment perpendicular to the major axis that passes through the center and intersects the ellipse in two points is called the **minor axis**.

- **Standard Form of the Equation an Ellipse with Center** $(0, 0)$: The standard form of the equation of an ellipse with center $(0, 0),$ is

$$\frac{x^2}{a^2} + \frac{y^2}{b^2} = 1$$

The x-intercepts are $(a, 0)$ and $(-a, 0)$.
The y-intercepts are $(0, b)$ and $(0, -b)$.

- **How to an Ellipse with Center** $(0, 0)$

Step 1. Write the equation in standard form.
Step 2. Determine whether the major axis is horizontal or vertical.
Step 3. Find the endpoints of the major axis.
Step 4. Find the endpoints of the minor axis
Step 5. Sketch the ellipse.

- **Standard Form of the Equation an Ellipse with Center** (h, k) : The standard form of the equation of an ellipse with center $(h, k),$ is

$$\frac{(x - h)^2}{a^2} + \frac{(y - k)^2}{b^2} = 1$$

When $a > b,$ the major axis is horizontal so the distance from the center to the vertex is a.
When $b > a,$ the major axis is vertical so the distance from the center to the vertex is b.

11.4 Hyperbolas

- **Hyperbola:** A **hyperbola** is all points in a plane where the difference of their distances from two fixed points is constant.

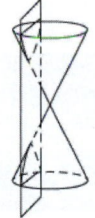

hyperbola

Each of the fixed points is called a **focus** of the hyperbola.
The line through the foci, is called the **transverse axis**.
The two points where the transverse axis intersects the hyperbola are each a **vertex** of the hyperbola.
The midpoint of the segment joining the foci is called the **center** of the hyperbola.
The line perpendicular to the transverse axis that passes through the center is called the **conjugate axis**.
Each piece of the graph is called a **branch** of the hyperbola.

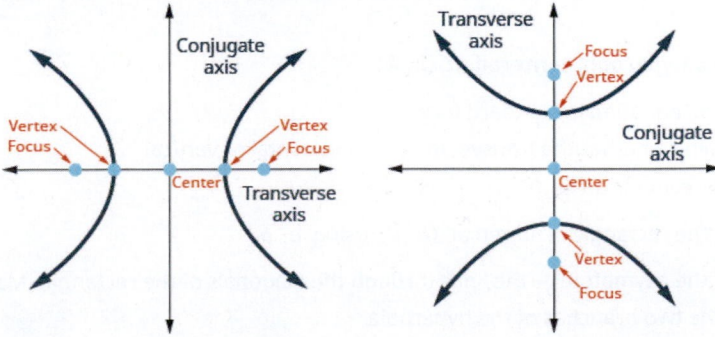

Standard Forms of the Equation a Hyperbola with Center $(0, 0)$		
	$\dfrac{x^2}{a^2} - \dfrac{y^2}{b^2} = 1$	$\dfrac{y^2}{a^2} - \dfrac{x^2}{b^2} = 1$
Orientation	Transverse axis on the *x*-axis. Opens left and right	Transverse axis on the *y*-axis. Opens up and down
Vertices	$(-a, 0), \quad (a, 0)$	$(0, -a), \quad (0, a)$
***x*-intercepts**	$(-a, 0), \quad (a, 0)$	none
***y*-intercepts**	none	$(0, -a), (0, a)$
Rectangle	Use $(\pm a, 0) \ (0, \pm b)$	Use $(0, \pm a) \ (\pm b, 0)$
asymptotes	$y = \dfrac{b}{a}x, \quad y = -\dfrac{b}{a}x$	$y = \dfrac{a}{b}x, \quad y = -\dfrac{a}{b}x$

- **How to graph a hyperbola centered at $(0, 0)$.**

  Step 1. Write the equation in standard form.

  Step 2. Determine whether the transverse axis is horizontal or vertical.

  Step 3. Find the vertices.

  Step 4. Sketch the rectangle centered at the origin intersecting one axis at $\pm a$ and the other at $\pm b$.

  Step 5. Sketch the asymptotes—the lines through the diagonals of the rectangle.

  Step 6. Draw the two branches of the hyperbola.

Standard Forms of the Equation a Hyperbola with Center (h, k)		
	$\dfrac{(x-h)^2}{a^2} - \dfrac{(y-k)^2}{b^2} = 1$	$\dfrac{(y-k)^2}{a^2} - \dfrac{(x-h)^2}{b^2} = 1$
Orientation	Transverse axis is horizontal. Opens left and right	Transverse axis is vertical. Opens up and down
Center	(h, k)	(h, k)
Vertices	*a* units to the left and right of the center	*a* units above and below the center
Rectangle	Use *a* units left/right of center *b* units above/below the center	Use *a* units above/below the center *b* units left/right of center

- **How to graph a hyperbola centered at (h, k).**

  Step 1. Write the equation in standard form.

  Step 2. Determine whether the transverse axis is horizontal or vertical.

  Step 3. Find the center and a, b.

  Step 4. Sketch the rectangle centered at (h, k) using a, b.

  Step 5. Sketch the asymptotes—the lines through the diagonals of the rectangle. Mark the vertices.

  Step 6. Draw the two branches of the hyperbola.

Conic	Characteristics of x^2- and y^2- terms	Example
Parabola	Either x^2 OR y^2. Only one variable is squared.	$x = 3y^2 - 2y + 1$
Circle	x^2- and y^2- terms have the same coefficients	$x^2 + y^2 = 49$
Ellipse	x^2- and y^2- terms have the **same** sign, different coefficients	$4x^2 + 25y^2 = 100$
Hyperbola	x^2- and y^2- terms have **different** signs, different coefficients	$25y^2 - 4x^2 = 100$

11.5 Solve Systems of Nonlinear Equations

- **How to solve a system of nonlinear equations by graphing.**

  Step 1. Identify the graph of each equation. Sketch the possible options for intersection.

  Step 2. Graph the first equation.

  Step 3. Graph the second equation on the same rectangular coordinate system.

  Step 4. Determine whether the graphs intersect.

  Step 5. Identify the points of intersection.

  Step 6. Check that each ordered pair is a solution to both original equations.

- **How to solve a system of nonlinear equations by substitution.**

  Step 1. Identify the graph of each equation. Sketch the possible options for intersection.

  Step 2. Solve one of the equations for either variable.

  Step 3. Substitute the expression from Step 2 into the other equation.

  Step 4. Solve the resulting equation.

  Step 5. Substitute each solution in Step 4 into one of the original equations to find the other variable.

  Step 6. Write each solution as an ordered pair.

  Step 7. Check that each ordered pair is a solution to **both** original equations.

- **How to solve a system of equations by elimination.**

  Step 1. Identify the graph of each equation. Sketch the possible options for intersection.

  Step 2. Write both equations in standard form.

  Step 3. Make the coefficients of one variable opposites.
  Decide which variable you will eliminate.
  Multiply one or both equations so that the coefficients of that variable are opposites.

  Step 4. Add the equations resulting from Step 3 to eliminate one variable.

  Step 5. Solve for the remaining variable.

  Step 6. Substitute each solution from Step 5 into one of the original equations. Then solve for the other variable.

  Step 7. Write each solution as an ordered pair.

  Step 8. Check that each ordered pair is a solution to **both** original equations.

REVIEW EXERCISES

11.1 Distance and Midpoint Formulas; Circles

Use the Distance Formula

In the following exercises, find the distance between the points. Round to the nearest tenth if needed.

244. $(-5, 1)$ and $(-1, 4)$ **245.** $(-2, 5)$ and $(1, 5)$ **246.** $(8, 2)$ and $(-7, -3)$

247. $(1, -4)$ and $(5, -5)$

Use the Midpoint Formula
In the following exercises, find the midpoint of the line segments whose endpoints are given.

248. $(-2, -6)$ and $(-4, -2)$ **249.** $(3, 7)$ and $(5, 1)$ **250.** $(-8, -10)$ and $(9, 5)$

251. $(-3, 2)$ and $(6, -9)$

Write the Equation of a Circle in Standard Form
In the following exercises, write the standard form of the equation of the circle with the given information.

252. radius is 15 and center is $(0, 0)$ **253.** radius is $\sqrt{7}$ and center is $(0, 0)$ **254.** radius is 9 and center is $(-3, 5)$

255. radius is 7 and center is $(-2, -5)$ **256.** center is $(3, 6)$ and a point on the circle is $(3, -2)$ **257.** center is $(2, 2)$ and a point on the circle is $(4, 4)$

Graph a Circle

In the following exercises, ⓐ find the center and radius, then ⓑ graph each circle.

258. $2x^2 + 2y^2 = 450$ **259.** $3x^2 + 3y^2 = 432$ **260.** $(x + 3)^2 + (y - 5)^2 = 81$

261. $(x + 2)^2 + (y + 5)^2 = 49$ **262.** $x^2 + y^2 - 6x - 12y - 19 = 0$ **263.** $x^2 + y^2 - 4y - 60 = 0$

11.2 Parabolas
Graph Vertical Parabolas
In the following exercises, graph each equation by using its properties.

264. $y = x^2 + 4x - 3$ **265.** $y = 2x^2 + 10x + 7$

266. $y = -6x^2 + 12x - 1$ **267.** $y = -x^2 + 10x$

In the following exercises, ⓐ write the equation in standard form, then ⓑ use properties of the standard form to graph the equation.

268. $y = x^2 + 4x + 7$ **269.** $y = 2x^2 - 4x - 2$

270. $y = -3x^2 - 18x - 29$ **271.** $y = -x^2 + 12x - 35$

Graph Horizontal Parabolas
In the following exercises, graph each equation by using its properties.

272. $x = 2y^2$ **273.** $x = 2y^2 + 4y + 6$

274. $x = -y^2 + 2y - 4$ **275.** $x = -3y^2$

In the following exercises, ⓐ write the equation in standard form, then ⓑ use properties of the standard form to graph the equation.

276. $x = 4y^2 + 8y$

277. $x = y^2 + 4y + 5$

278. $x = -y^2 - 6y - 7$

279. $x = -2y^2 + 4y$

Solve Applications with Parabolas

In the following exercises, create the equation of the parabolic arch formed in the foundation of the bridge shown. Give the answer in standard form.

280.

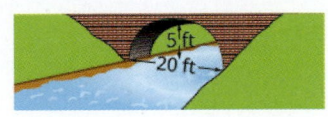

281.

11.3 Ellipses

Graph an Ellipse with Center at the Origin

In the following exercises, graph each ellipse.

282. $\dfrac{x^2}{36} + \dfrac{y^2}{25} = 1$

283. $\dfrac{x^2}{4} + \dfrac{y^2}{81} = 1$

284. $49x^2 + 64y^2 = 3136$

285. $9x^2 + y^2 = 9$

Find the Equation of an Ellipse with Center at the Origin

In the following exercises, find the equation of the ellipse shown in the graph.

286.

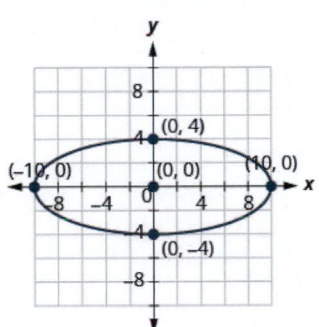

287.

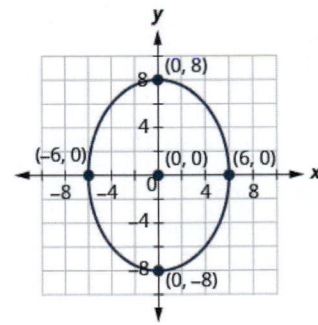

Graph an Ellipse with Center Not at the Origin

In the following exercises, graph each ellipse.

288. $\dfrac{(x-1)^2}{25} + \dfrac{(y-6)^2}{4} = 1$

289. $\dfrac{(x+4)^2}{16} + \dfrac{(y+1)^2}{9} = 1$

290. $\dfrac{(x-5)^2}{16} + \dfrac{(y+3)^2}{36} = 1$

291. $\dfrac{(x+3)^2}{9} + \dfrac{(y-2)^2}{25} = 1$

In the following exercises, ⓐ write the equation in standard form and ⓑ graph.

292. $x^2 + y^2 + 12x + 40y + 120 = 0$

293. $25x^2 + 4y^2 - 150x - 56y + 321 = 0$

294. $25x^2 + 4y^2 + 150x + 125 = 0$

295. $4x^2 + 9y^2 - 126x + 405 = 0$

Solve Applications with Ellipses

In the following exercises, write the equation of the ellipse described.

296. A comet moves in an elliptical orbit around a sun. The closest the comet gets to the sun is approximately 10 AU and the furthest is approximately 90 AU. The sun is one of the foci of the elliptical orbit. Letting the ellipse center at the origin and labeling the axes in AU, the orbit will look like the figure below. Use the graph to write an equation for the elliptical orbit of the comet.

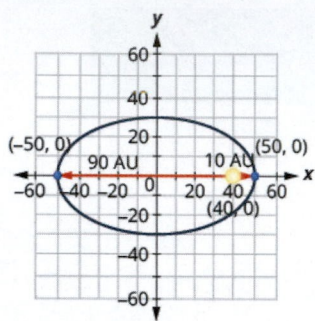

11.4 Hyperbolas

Graph a Hyperbola with Center at $(0, 0)$

In the following exercises, graph.

297. $\dfrac{x^2}{25} - \dfrac{y^2}{9} = 1$

298. $\dfrac{y^2}{49} - \dfrac{x^2}{16} = 1$

299. $9y^2 - 16x^2 = 144$

300. $16x^2 - 4y^2 = 64$

Graph a Hyperbola with Center at (h, k)

In the following exercises, graph.

301. $\dfrac{(x+1)^2}{4} - \dfrac{(y+1)^2}{9} = 1$

302. $\dfrac{(x-2)^2}{4} - \dfrac{(y-3)^2}{16} = 1$

303. $\dfrac{(y+2)^2}{9} - \dfrac{(x+1)^2}{9} = 1$

304. $\dfrac{(y-1)^2}{25} - \dfrac{(x-2)^2}{9} = 1$

In the following exercises, ⓐ write the equation in standard form and ⓑ graph.

305. $4x^2 - 16y^2 + 8x + 96y - 204 = 0$

306. $16x^2 - 4y^2 - 64x - 24y - 36 = 0$

307. $4y^2 - 16x^2 + 32x - 8y - 76 = 0$

308. $36y^2 - 16x^2 - 96x + 216y - 396 = 0$

Identify the Graph of each Equation as a Circle, Parabola, Ellipse, or Hyperbola

In the following exercises, identify the type of graph.

309.

ⓐ $16y^2 - 9x^2 - 36x - 96y - 36 = 0$

ⓑ $x^2 + y^2 - 4x + 10y - 7 = 0$

ⓒ $y = x^2 - 2x + 3$

ⓓ $25x^2 + 9y^2 = 225$

310.

ⓐ $x^2 + y^2 + 4x - 10y + 25 = 0$

ⓑ $y^2 - x^2 - 4y + 2x - 6 = 0$

ⓒ $x = -y^2 - 2y + 3$

ⓓ $16x^2 + 9y^2 = 144$

11.5 Solve Systems of Nonlinear Equations

Solve a System of Nonlinear Equations Using Graphing

In the following exercises, solve the system of equations by using graphing.

311. $\begin{cases} 3x^2 - y = 0 \\ y = 2x - 1 \end{cases}$

312. $\begin{cases} y = x^2 - 4 \\ y = x - 4 \end{cases}$

313. $\begin{cases} x^2 + y^2 = 169 \\ x = 12 \end{cases}$

314. $\begin{cases} x^2 + y^2 = 25 \\ y = -5 \end{cases}$

Solve a System of Nonlinear Equations Using Substitution

In the following exercises, solve the system of equations by using substitution.

315. $\begin{cases} y = x^2 + 3 \\ y = -2x + 2 \end{cases}$

316. $\begin{cases} x^2 + y^2 = 4 \\ x - y = 4 \end{cases}$

317. $\begin{cases} 9x^2 + 4y^2 = 36 \\ y - x = 5 \end{cases}$

318. $\begin{cases} x^2 + 4y^2 = 4 \\ 2x - y = 1 \end{cases}$

Solve a System of Nonlinear Equations Using Elimination

In the following exercises, solve the system of equations by using elimination.

319. $\begin{cases} x^2 + y^2 = 16 \\ x^2 - 2y - 1 = 0 \end{cases}$

320. $\begin{cases} x^2 - y^2 = 5 \\ -2x^2 - 3y^2 = -30 \end{cases}$

321. $\begin{cases} 4x^2 + 9y^2 = 36 \\ 3y^2 - 4x = 12 \end{cases}$

322. $\begin{cases} x^2 + y^2 = 14 \\ x^2 - y^2 = 16 \end{cases}$

Use a System of Nonlinear Equations to Solve Applications

In the following exercises, solve the problem using a system of equations.

323. The sum of the squares of two numbers is 25. The difference of the numbers is 1. Find the numbers.

324. The difference of the squares of two numbers is 45. The difference of the square of the first number and twice the square of the second number is 9. Find the numbers.

325. The perimeter of a rectangle is 58 meters and its area is 210 square meters. Find the length and width of the rectangle.

326. Colton purchased a larger microwave for his kitchen. The diagonal of the front of the microwave measures 34 inches. The front also has an area of 480 square inches. What are the length and width of the microwave?

PRACTICE TEST

In the following exercises, find the distance between the points and the midpoint of the line segment with the given endpoints. Round to the nearest tenth as needed.

327. $(-4, -3)$ and $(-10, -11)$

328. $(6, 8)$ and $(-5, -3)$

In the following exercises, write the standard form of the equation of the circle with the given information.

329. radius is 11 and center is $(0, 0)$

330. radius is 12 and center is $(10, -2)$

331. center is $(-2, 3)$ and a point on the circle is $(2, -3)$

332. Find the equation of the ellipse shown in the graph.

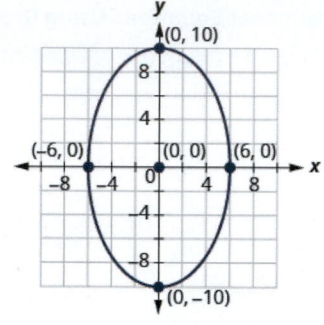

In the following exercises, ⓐ identify the type of graph of each equation as a circle, parabola, ellipse, or hyperbola, and ⓑ graph the equation.

333. $4x^2 + 49y^2 = 196$

334. $y = 3(x - 2)^2 - 2$

335. $3x^2 + 3y^2 = 27$

336. $\dfrac{y^2}{100} - \dfrac{x^2}{36} = 1$

337. $\dfrac{x^2}{16} + \dfrac{y^2}{81} = 1$

338. $x = 2y^2 + 10y + 7$

339. $64x^2 - 9y^2 = 576$

In the following exercises, ⓐ identify the type of graph of each equation as a circle, parabola, ellipse, or hyperbola, ⓑ write the equation in standard form, and ⓒ graph the equation.

340. $25x^2 + 64y^2 + 200x - 256y - 944 = 0$

341. $x^2 + y^2 + 10x + 6y + 30 = 0$

342. $x = -y^2 + 2y - 4$

343. $9x^2 - 25y^2 - 36x - 50y - 214 = 0$

344. $y = x^2 + 6x + 8$

345. Solve the nonlinear system of equations by graphing:
$$\begin{cases} 3y^2 - x = 0 \\ y = -2x - 1 \end{cases}$$

346. Solve the nonlinear system of equations using substitution:
$$\begin{cases} x^2 + y^2 = 8 \\ y = -x - 4 \end{cases}$$

347. Solve the nonlinear system of equations using elimination:
$$\begin{cases} x^2 + 9y^2 = 9 \\ 2x^2 - 9y^2 = 18 \end{cases}$$

348. Create the equation of the parabolic arch formed in the foundation of the bridge shown. Give the answer in $y = ax^2 + bx + c$ form.

349. A comet moves in an elliptical orbit around a sun. The closest the comet gets to the sun is approximately 20 AU and the furthest is approximately 70 AU. The sun is one of the foci of the elliptical orbit. Letting the ellipse center at the origin and labeling the axes in AU, the orbit will look like the figure below. Use the graph to write an equation for the elliptical orbit of the comet.

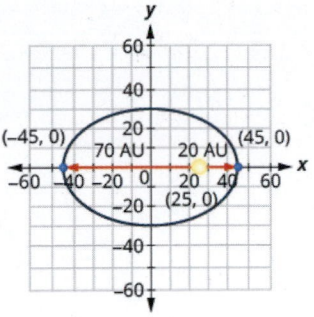

350. The sum of two numbers is 22 and the product is −240. Find the numbers.

351. For her birthday, Olive's grandparents bought her a new widescreen TV. Before opening it she wants to make sure it will fit her entertainment center. The TV is 55". The size of a TV is measured on the diagonal of the screen and a widescreen has a length that is larger than the width. The screen also has an area of 1452 square inches. Her entertainment center has an insert for the TV with a length of 50 inches and width of 40 inches. What are the length and width of the TV screen and will it fit Olive's entertainment center?

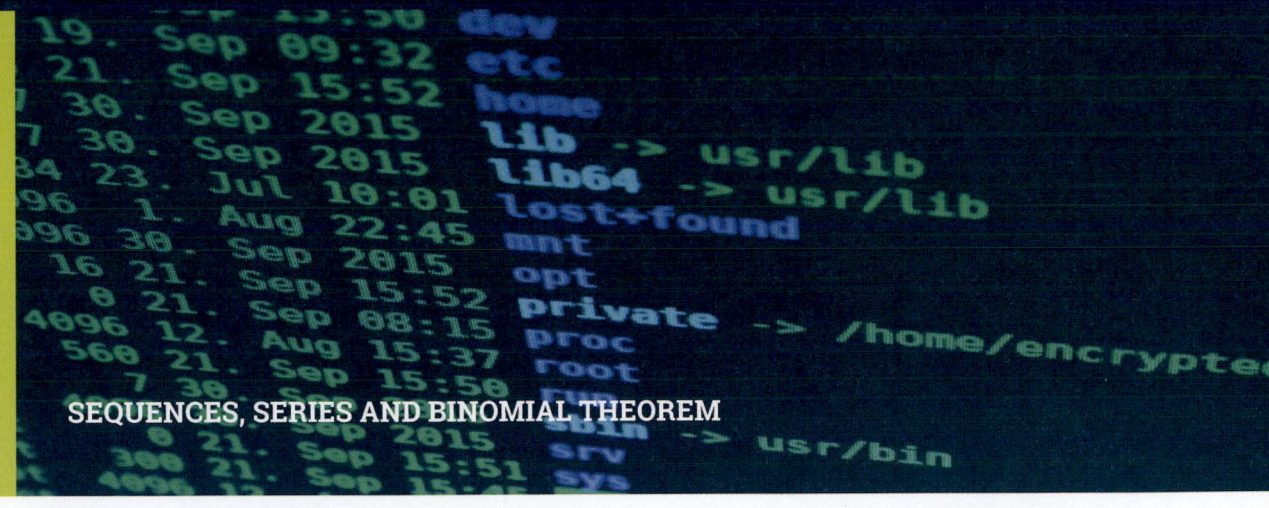

Figure 12.1 Cryptographers protect private data by encrypting it; this means they convert the data into a code that hackers and thieves cannot easily break. (credit: "joffi"/pixabay)

Chapter Outline

12.1 Sequences

12.2 Arithmetic Sequences

12.3 Geometric Sequences and Series

12.4 Binomial Theorem

Introduction

A strange charge suddenly appears on your credit card. But your card is in your wallet—it's not even lost or stolen. Sadly, you may have been a victim of cyber crime. In this day and age, most transactions take advantage of the benefit of computers in some way. Cyber crime is any type of crime that uses a computer or computer network. Thankfully, many people are working to prevent cyber crime. Sometimes known as cryptographers, these people develop complex patterns in computer codes that block access to would-be thieves as well as write codes to intercept and decode information from them so that they may be identified. In this chapter, you will explore basic sequences and series related to those used by computer programmers to prevent cyber crime.

12.1 Sequences

Learning Objectives

By the end of this section, you will be able to:

> Write the first few terms of a sequence
> Find a formula for the general term (nth term) of a sequence
> Use factorial notation
> Find the partial sum
> Use summation notation to write a sum

Be Prepared!

Before you get started, take this readiness quiz.

1. Evaluate $2n + 3$ for the integers 1, 2, 3, and 4.
   If you missed this problem, review **Example 1.6**.

2. Evaluate $(-1)^n$ for the integers 1, 2, 3, and 4.
   If you missed this problem, review **Example 1.19**.

3. If $f(n) = n^2 + 2$, find $f(1) + f(2) + f(3)$.

If you missed this problem, review **Example 3.49**.

Write the First Few Terms of a Sequence

Let's look at the function $f(x) = 2x$ and evaluate it for just the counting numbers.

$f(x) = 2x$	
x	$2x$
1	2
2	4
3	6
4	8
5	10
...	...

If we list the function values in order as 2, 4, 6, 8, and 10, ... we have a sequence. A **sequence** is a function whose domain is the counting numbers.

Sequences

A **sequence** is a function whose domain is the counting numbers.

A sequence can also be seen as an ordered list of numbers and each number in the list is a *term*. A sequence may have an infinite number of terms or a finite number of terms. Our sequence has three dots (ellipsis) at the end which indicates the list never ends. If the domain is the set of all counting numbers, then the sequence is an **infinite sequence**. Its domain is all counting numbers and there is an infinite number of counting numbers.

$$2, 4, 6, 8, 10, \ldots,$$

If we limit the domain to a finite number of counting numbers, then the sequence is a **finite sequence**. If we use only the first four counting numbers, 1, 2, 3, 4 our sequence would be the finite sequence,

$$2, 4, 6, 8$$

Often when working with sequences we do not want to write out all the terms. We want more compact way to show how each term is defined. When we worked with functions, we wrote $f(x) = 2x$ and we said the expression $2x$ was the rule that defined values in the range. While a sequence is a function, we do not use the usual function notation. Instead of writing the function as $f(x) = 2x$, we would write it as $a_n = 2n$. The a_n is the nth term of the sequence, the term in the nth position where n is a value in the domain. The formula for writing the nth term of the sequence is called the **general term** or formula of the sequence.

General Term of a Sequence

The **general term** of the sequence is found from the formula for writing the nth term of the sequence. The nth term of the sequence, a_n, is the term in the nth position where n is a value in the domain.

When we are given the general term of the sequence, we can find the terms by replacing n with the counting numbers in order. For $a_n = 2n$,

n	1	2	3	4	5	a_n
a_n	$2 \cdot 1$ 2	$2 \cdot 2$ 4	$2 \cdot 3$ 6	$2 \cdot 4$ 8	$2 \cdot 5$ 10	$2n$

$$a_1, \quad a_2, \quad a_3, \quad a_4, \quad a_5, \ldots, a_n, \ldots$$
$$2, \quad 4, \quad 6, \quad 8, \quad 10, \ldots$$

To find the values of a sequence, we substitute in the counting numbers in order into the general term of the sequence.

EXAMPLE 12.1

Write the first five terms of the sequence whose general term is $a_n = 4n - 3$.

✓ Solution

We substitute the values 1, 2, 3, 4, and 5 into the formula, $a_n = 4n - 3$, in order.

$a_n = 4n - 3$	$a_n = 4n - 3$	$a_n = 4n - 3$	$a_n = 4n - 3$	$a_n = 4n - 3$
$a_1 = 4 \cdot 1 - 3$	$a_2 = 4 \cdot 2 - 3$	$a_3 = 4 \cdot 3 - 3$	$a_4 = 4 \cdot 4 - 3$	$a_5 = 4 \cdot 5 - 3$
$a_1 = 1$	$a_2 = 5$	$a_3 = 9$	$a_4 = 13$	$a_5 = 17$

The first five terms of the sequence are 1, 5, 9, 13, and 17.

> **TRY IT :: 12.1**   Write the first five terms of the sequence whose general term is $a_n = 3n - 4$.

> **TRY IT :: 12.2**   Write the first five terms of the sequence whose general term is $a_n = 2n - 5$.

For some sequences, the variable is an exponent.

EXAMPLE 12.2

Write the first five terms of the sequence whose general term is $a_n = 2^n + 1$.

✓ Solution

We substitute the values 1, 2, 3, 4, and 5 into the formula, $a_n = 2^n + 1$, in order.

$a_n = 2^n + 1$	$a_n = 2^n + 1$	$a_n = 2^n + 1$	$a_n = 2^n + 1$	$a_n = 2^n + 1$
$a_1 = 2^1 + 1$	$a_2 = 2^2 + 1$	$a_3 = 2^3 + 1$	$a_4 = 2^4 + 1$	$a_5 = 2^5 + 1$
$a_1 = 3$	$a_2 = 5$	$a_3 = 9$	$a_4 = 17$	$a_5 = 33$

The first five terms of the sequence are 3, 5, 9, 17, and 33.

> **TRY IT :: 12.3**   Write the first five terms of the sequence whose general term is $a_n = 3^n + 4$.

> **TRY IT :: 12.4**   Write the first five terms of the sequence whose general term is $a_n = 2^n - 5$.

It is not uncommon to see the expressions $(-1)^n$ or $(-1)^{n+1}$ in the general term for a sequence. If we evaluate each of these expressions for a few values, we see that this expression alternates the sign for the terms.

n	1	2	3	4	5
$(-1)^n$	$(-1)^1$	$(-1)^2$	$(-1)^3$	$(-1)^4$	$(-1)^5$
	-1	1	-1	1	-1
$(-1)^{n+1}$	$(-1)^{1+1}$	$(-1)^{2+1}$	$(-1)^{3+1}$	$(-1)^{4+1}$	$(-1)^{5+1}$
	1	-1	1	-1	1

$$a_1, \quad a_2, \quad a_3, \quad a_4, \quad a_5, \ldots, \quad a_n, \ldots$$
$$-1, \quad 1, \quad -1, \quad 1, \quad -1 \ldots$$
$$1, \quad -1, \quad 1, \quad -1, \quad 1 \ldots$$

The terms in the next example will alternate signs as a result of the powers of -1.

EXAMPLE 12.3

Write the first five terms of the sequence whose general term is $a_n = (-1)^n n^3$.

✓ **Solution**

We substitute the values 1, 2, 3, 4, and 5 into the formula, $a_n = (-1)^n n^3$, in order.

$a_n = (-1)^n n^3$	$a_n = (-1)^n n^3$	$a_n = (-1)^n n^3$	$a_n = (-1)^n n^3$	$a_n = (-1)^n n^3$
$a_1 = (-1)^1 \cdot 1^3$	$a_2 = (-1)^2 \cdot 2^3$	$a_3 = (-1)^3 \cdot 3^3$	$a_4 = (-1)^4 \cdot 4^3$	$a_5 = (-1)^5 \cdot 5^3$
$a_1 = -1$	$a_2 = 8$	$a_3 = -27$	$a_4 = 64$	$a_5 = -125$

The first five terms of the sequence are $-1, 8, -27, 64,$ and -125.

> **TRY IT :: 12.5**   Write the first five terms of the sequence whose general term is $a_n = (-1)^n n^2$.

> **TRY IT :: 12.6**   Write the first five terms of the sequence whose general term is $a_n = (-1)^{n+1} n^3$.

Find a Formula for the General Term (*n*th Term) of a Sequence

Sometimes we have a few terms of a sequence and it would be helpful to know the general term or *n*th term. To find the general term, we look for patterns in the terms. Often the patterns involve multiples or powers. We also look for a pattern in the signs of the terms.

EXAMPLE 12.4

Find a general term for the sequence whose first five terms are shown.

$$4, 8, 12, 16, 20, \ldots$$

✓ **Solution**

	4,	8,	12,	16,	20, ...
n:	1,	2,	3,	4,	5, ...*n*
We look for a pattern in the terms.	Terms:	4,	8,	12,	16,	20, ...
The numbers are all multiples of 4.	Pattern: $4 \cdot 1, 4 \cdot 2, 4 \cdot 3, 4 \cdot 4, 4 \cdot 5, \ldots, 4 \cdot n$

The general term of the sequence is $a_n = 4n$.

> **TRY IT :: 12.7**   Find a general term for the sequence whose first five terms are shown.
>
> $3, 6, 9, 12, 15, \ldots$

> **TRY IT :: 12.8**   Find a general term for the sequence whose first five terms are shown.
>
> $5, 10, 15, 20, 25, \ldots$

EXAMPLE 12.5

Find a general term for the sequence whose first five terms are shown.

$$2, -4, 8, -16, 32, \ldots$$

⊘ Solution

	2,	−4,	8,	−16,	32, …

n:	1,	2, 3,	4,	5, …n

We look for a pattern in the terms.	Terms:	2, −4, 8, −16, 32, …
The numbers are powers of 2. The signs are alternating, with even n negative.	Pattern: $(-1)^{1+1}\,2^1, (-1)^{2+1}\,2^2, (-1)^{3+1}\,2^3, (-1)^{4+1}\,2^4, (-1)^{5+1}\,2^5, …, (-1)^{n+1}\,2^n$
	The general term of the sequence is $a_n = (-1)^{n+1}\,2^n$.

> **TRY IT : : 12.9**   Find a general term for the sequence whose first five terms are shown.
>
> $$-3, 9, -27, 81, -243, …$$

> **TRY IT : : 12.10**   Find a general term for the sequence whose first five terms are shown
>
> $$1, -4, 9, -16, 25, …$$

EXAMPLE 12.6

Find a general term for the sequence whose first five terms are shown.

$$\frac{1}{3}, \frac{1}{9}, \frac{1}{27}, \frac{1}{81}, \frac{1}{243}, \cdots$$

⊘ Solution

	$\dfrac{1}{3}, \dfrac{1}{9}, \dfrac{1}{27}, \dfrac{1}{81}, \dfrac{1}{243}, \cdots$

n:	1,	2,	3,	4,	5, …n

We look for a pattern in the terms.	Terms:	$\dfrac{1}{3}, \dfrac{1}{9}, \dfrac{1}{27}, \dfrac{1}{81}, \dfrac{1}{243}, \cdots$
The numerators are all 1.	Pattern:	$\dfrac{1}{3^1}, \dfrac{1}{3^2}, \dfrac{1}{3^3}, \dfrac{1}{3^4}, \dfrac{1}{3^5}, \cdots \dfrac{1}{3^n}$
The denominators are powers of 3.	The general term of the sequence is $a_n = \dfrac{1}{3^n}$.

> **TRY IT : : 12.11**   Find a general term for the sequence whose first five terms are shown.
>
> $$\frac{1}{2}, \frac{1}{4}, \frac{1}{8}, \frac{1}{16}, \frac{1}{32}, \cdots$$

> **TRY IT : : 12.12**   Find a general term for the sequence whose first five terms are shown.
>
> $$\frac{1}{1}, \frac{1}{4}, \frac{1}{9}, \frac{1}{16}, \frac{1}{25}, \cdots$$

Use Factorial Notation

Sequences often have terms that are products of consecutive integers. We indicate these products with a special notation called *factorial notation*. For example, $5!$, read 5 factorial, means $5 \cdot 4 \cdot 3 \cdot 2 \cdot 1$. The exclamation point is not punctuation here; it indicates the **factorial notation**.

Factorial Notation

If n is a positive integer, then $n!$ is

$$n! = n(n-1)(n-2)\dots$$

We define $0!$ as 1, so $0! = 1$.

The values of $n!$ for the first 5 positive integers are shown.

$1!$	$2!$	$3!$	$4!$	$5!$
1	$2 \cdot 1$	$3 \cdot 2 \cdot 1$	$4 \cdot 3 \cdot 2 \cdot 1$	$5 \cdot 4 \cdot 3 \cdot 2 \cdot 1$
1	2	6	24	120

EXAMPLE 12.7

Write the first five terms of the sequence whose general term is $a_n = \dfrac{1}{n!}$.

✓ **Solution**

We substitute the values 1, 2, 3, 4, 5 into the formula, $a_n = \dfrac{1}{n!}$, in order.

$$a_n = \frac{1}{n!} \qquad a_n = \frac{1}{n!} \qquad a_n = \frac{1}{n!} \qquad a_n = \frac{1}{n!} \qquad a_n = \frac{1}{n!}$$

$$a_1 = \frac{1}{1!} \qquad a_2 = \frac{1}{2!} \qquad a_3 = \frac{1}{3!} \qquad a_4 = \frac{1}{4!} \qquad a_5 = \frac{1}{5!}$$

$$a_1 = \frac{1}{1} \qquad a_2 = \frac{1}{2 \cdot 1} \qquad a_3 = \frac{1}{3 \cdot 2 \cdot 1} \qquad a_4 = \frac{1}{4 \cdot 3 \cdot 2 \cdot 1} \qquad a_5 = \frac{1}{5 \cdot 4 \cdot 3 \cdot 2 \cdot 1}$$

$$a_1 = 1 \qquad a_2 = \frac{1}{2} \qquad a_3 = \frac{1}{6} \qquad a_4 = \frac{1}{24} \qquad a_5 = \frac{1}{120}$$

The first five terms of the sequence are $1, \dfrac{1}{2}, \dfrac{1}{6}, \dfrac{1}{24}, \dfrac{1}{120}$.

> **TRY IT : : 12.13**   Write the first five terms of the sequence whose general term is $a_n = \dfrac{2}{n!}$.

> **TRY IT : : 12.14**   Write the first five terms of the sequence whose general term is $a_n = \dfrac{3}{n!}$.

When there is a fraction with factorials in the numerator and denominator, we line up the factors vertically to make our calculations easier.

EXAMPLE 12.8

Write the first five terms of the sequence whose general term is $a_n = \dfrac{(n+1)!}{(n-1)!}$.

✓ **Solution**

We substitute the values 1, 2, 3, 4, 5 into the formula, $a_n = \dfrac{(n+1)!}{(n-1)!}$, in order.

$$a_n = \frac{(n+1)!}{(n-1)!} \qquad a_n = \frac{(n+1)!}{(n-1)!} \qquad a_n = \frac{(n+1)!}{(n-1)!} \qquad a_n = \frac{(n+1)!}{(n-1)!} \qquad a_n = \frac{(n+1)!}{(n-1)!}$$

$$a_1 = \frac{(1+1)!}{(1-1)!} \qquad a_2 = \frac{(2+1)!}{(2-1)!} \qquad a_3 = \frac{(3+1)!}{(3-1)!} \qquad a_4 = \frac{(4+1)!}{(4-1)!} \qquad a_5 = \frac{(5+1)!}{(5-1)!}$$

$$a_1 = \frac{(2)!}{(0)!} \qquad a_2 = \frac{(3)!}{(1)!} \qquad a_3 = \frac{(4)!}{(2)!} \qquad a_4 = \frac{(5)!}{(3)!} \qquad a_5 = \frac{(6)!}{(4)!}$$

$$a_1 = \frac{2 \cdot 1}{1} \qquad a_2 = \frac{3 \cdot 2 \cdot 1}{1} \qquad a_3 = \frac{4 \cdot 3 \cdot 2 \cdot 1}{2 \cdot 1} \qquad a_4 = \frac{5 \cdot 4 \cdot 3 \cdot 2 \cdot 1}{3 \cdot 2 \cdot 1} \qquad a_5 = \frac{6 \cdot 5 \cdot 4 \cdot 3 \cdot 2 \cdot 1}{4 \cdot 3 \cdot 2 \cdot 1}$$

$$a_1 = 2 \qquad a_2 = 6 \qquad a_3 = 12 \qquad a_4 = 20 \qquad a_5 = 30$$

The first five terms of the sequence are 2, 6, 12, 20, and 30.

> **TRY IT :: 12.15**
>
> Write the first five terms of the sequence whose general term is $a_n = \frac{(n-1)!}{(n+1)!}$.

> **TRY IT :: 12.16**
>
> Write the first five terms of the sequence whose general term is $a_n = \frac{n!}{(n+1)!}$.

Find the Partial Sum

Sometimes in applications, rather than just list the terms, it is important for us to add the terms of a sequence. Rather than just connect the terms with plus signs, we can use **summation notation**.

For example, $a_1 + a_2 + a_3 + a_4 + a_5$ can be written as $\displaystyle\sum_{i=1}^{5} a_i$. We read this as "the sum of a sub i from i equals one to five." The symbol $\sum$ means to add and the i is the index of summation. The 1 tells us where to start (initial value) and the 5 tells us where to end (terminal value).

Summation Notation

The sum of the first n terms of a sequence whose nth term is a_n is written in summation notation as:

$$\sum_{i=1}^{n} a_i = a_1 + a_2 + a_3 + a_4 + a_5 + \ldots + a_n$$

The i is the index of summation and the 1 tells us where to start and the n tells us where to end.

When we add a finite number of terms, we call the sum a **partial sum**.

EXAMPLE 12.9

Expand the partial sum and find its value: $\displaystyle\sum_{i=1}^{5} 2i$.

Solution

$$\sum_{i=1}^{5} 2i$$

We substitute the values 1, 2, 3, 4, 5 in order. $2 \cdot 1 + 2 \cdot 2 + 2 \cdot 3 + 2 \cdot 4 + 2 \cdot 5$

Simplify. $2 + 4 + 6 + 8 + 10$

Add. 30

$$\sum_{i=1}^{5} 2i = 30$$

> **TRY IT ::** 12.17

Expand the partial sum and find its value: $\displaystyle\sum_{i=1}^{5} 3i$.

> **TRY IT ::** 12.18

Expand the partial sum and find its value: $\displaystyle\sum_{i=1}^{5} 4i$.

The index does not always have to be i we can use any letter, but i and k are commonly used. The index does not have to start with 1 either—it can start and end with any positive integer.

EXAMPLE 12.10

Expand the partial sum and find its value: $\displaystyle\sum_{k=0}^{3} \frac{1}{k!}$.

Solution

$$\sum_{k=0}^{3} \frac{1}{k!}$$

We substitute the values 0, 1, 2, 3, in order. $\frac{1}{0!} + \frac{1}{1!} + \frac{1}{2!} + \frac{1}{3!}$

Evaluate the factorials. $\frac{1}{1} + \frac{1}{1} + \frac{1}{2} + \frac{1}{6}$

Simplify. $1 + 1 + \frac{3}{6} + \frac{1}{6}$

Simplify. $\frac{16}{6}$

Simplify. $\frac{8}{3}$

$$\sum_{k=0}^{3} \frac{1}{k!} = \frac{8}{3}$$

> **TRY IT ::** 12.19

Expand the partial sum and find its value: $\displaystyle\sum_{k=0}^{3} \frac{2}{k!}$.

> **TRY IT ::** 12.20

Expand the partial sum and find its value: $\displaystyle\sum_{k=0}^{3} \frac{3}{k!}$.

Use Summation Notation to Write a Sum

In the last two examples, we went from summation notation to writing out the sum. Now we will start with a sum and change it to summation notation. This is very similar to finding the general term of a sequence. We will need to look at the terms and find a pattern. Often the patterns involve multiples or powers.

EXAMPLE 12.11

Write the sum using summation notation: $1 + \frac{1}{2} + \frac{1}{3} + \frac{1}{4} + \frac{1}{5}$.

⊘ Solution

$$1 + \frac{1}{2} + \frac{1}{3} + \frac{1}{4} + \frac{1}{5}$$

	n: 1, 2, 3, 4, 5
We look for a pattern in the terms.	Terms: $1, \frac{1}{2}, \frac{1}{3}, \frac{1}{4}, \frac{1}{5}$
The numerators are all one.	Pattern: $\frac{1}{1}, \frac{1}{2}, \frac{1}{3}, \frac{1}{4}, \frac{1}{5}, \cdots \frac{1}{n}$
The denominators are the counting numbers from one to five.	The sum written in summation notation is

$$1 + \frac{1}{2} + \frac{1}{3} + \frac{1}{4} + \frac{1}{5} = \sum_{n=1}^{5} \frac{1}{n}.$$

> **TRY IT ::** 12.21   Write the sum using summation notation: $\frac{1}{2} + \frac{1}{4} + \frac{1}{8} + \frac{1}{16} + \frac{1}{32}$.

> **TRY IT ::** 12.22   Write the sum using summation notation: $1 + \frac{1}{4} + \frac{1}{9} + \frac{1}{16} + \frac{1}{25}$.

When the terms of a sum have negative coefficients, we must carefully analyze the pattern of the signs.

EXAMPLE 12.12

Write the sum using summation notation: $-1 + 8 - 27 + 64 - 125$.

⊘ Solution

$$-1 + 8 - 27 + 64 - 125$$

	n:	1,	2,	3,	4, 5
We look for a pattern in the terms.	Terms:	-1,	8,	-27,	64 -125
The signs of the terms alternate, and the odd terms are negative.	Pattern: $(-1)^1 \cdot 1$, $(-1)^2 \cdot 8$, $(-1)^3 \cdot 27$, $(-1)^4 \cdot 64$, $(-1)^5 \cdot 125$				
The numbers are the cubes of the counting numbers from one to five.	Pattern: $(-1)^1 \cdot 1^3$, $(-1)^2 \cdot 2^3$, $(-1)^3 \cdot 3^3$, $(-1)^4 \cdot 4^3$, $(-1)^5 \cdot 5^3$				
	Pattern: $(-1)^n \cdot n^3$				
	The sum written in summation notation is				

$$-1 + 8 - 27 + 64 - 125 = \sum_{n=1}^{5} (-1)^n \cdot n^3$$

> **TRY IT ::** 12.23   Write each sum using summation notation: $1 - 4 + 9 - 16 + 25$.

> **TRY IT ::** 12.24   Write each sum using summation notation: $-2 + 4 - 6 + 8 - 10$.

▶ **MEDIA ::**

Access this online resource for additional instruction and practice with sequences.

- **Series and Sequences-Finding Patterns (https://openstax.org/l/37serseqfindpat)**

12.1 EXERCISES

Practice Makes Perfect

Write the First Few Terms of a Sequence

In the following exercises, write the first five terms of the sequence whose general term is given.

1. $a_n = 2n - 7$

2. $a_n = 5n - 1$

3. $a_n = 3n + 1$

4. $a_n = 4n + 2$

5. $a_n = 2^n + 3$

6. $a_n = 3^n - 1$

7. $a_n = 3^n - 2n$

8. $a_n = 2^n - 3n$

9. $a_n = \dfrac{2^n}{n^2}$

10. $a_n = \dfrac{3^n}{n^3}$

11. $a_n = \dfrac{4n - 2}{2^n}$

12. $a_n = \dfrac{3n + 3}{3^n}$

13. $a_n = (-1)^n \cdot 2n$

14. $a_n = (-1)^n \cdot 3n$

15. $a_n = (-1)^{n+1} n^2$

16. $a_n = (-1)^{n+1} n^4$

17. $a_n = \dfrac{(-1)^{n+1}}{n^2}$

18. $a_n = \dfrac{(-1)^{n+1}}{2n}$

Find a Formula for the General Term (*n*th Term) of a Sequence

In the following exercises, find a general term for the sequence whose first five terms are shown.

19. 8, 16, 24, 32, 40, …

20. 7, 14, 21, 28, 35, …

21. 6, 7, 8, 9, 10, …

22. −3, −2, −1, 0, 1, …

23. $e^3, e^4, e^5, e^6, e^7, \ldots$

24. $\dfrac{1}{e^2}, \dfrac{1}{e}, 1, e, e^2, \ldots$

25. −5, 10, −15, 20, −25, …

26. −6, 11, −16, 21, −26, …

27. −1, 8, −27, 64, −125, …

28. 2, −5, 10, −17, 26, …

29. −2, 4, −6, 8, −10, …

30. 1, −3, 5, −7, 9, …

31. $\dfrac{1}{4}, \dfrac{1}{16}, \dfrac{1}{64}, \dfrac{1}{256}, \dfrac{1}{1{,}024}, \cdots$

32. $\dfrac{1}{1}, \dfrac{1}{8}, \dfrac{1}{27}, \dfrac{1}{64}, \dfrac{1}{125}, \cdots$

33. $-\dfrac{1}{2}, -\dfrac{2}{3}, -\dfrac{3}{4}, -\dfrac{4}{5}, -\dfrac{5}{6}, \cdots$

34. $-2, -\dfrac{3}{2}, -\dfrac{4}{3}, -\dfrac{5}{4}, -\dfrac{6}{5}, \cdots$

35. $-\dfrac{5}{2}, -\dfrac{5}{4}, -\dfrac{5}{8}, -\dfrac{5}{16}, -\dfrac{5}{32}, \cdots$

36. $4, \dfrac{1}{2}, \dfrac{4}{27}, \dfrac{4}{64}, \dfrac{4}{125}, \cdots$

Use Factorial Notation

In the following exercises, using factorial notation, write the first five terms of the sequence whose general term is given.

37. $a_n = \dfrac{4}{n!}$

38. $a_n = \dfrac{5}{n!}$

39. $a_n = 3n!$

40. $a_n = 2n!$

41. $a_n = (2n)!$

42. $a_n = (3n)!$

43. $a_n = \dfrac{(n-1)!}{(n)!}$

44. $a_n = \dfrac{n!}{(n+1)!}$

45. $a_n = \dfrac{n!}{n^2}$

46. $a_n = \dfrac{n^2}{n!}$

47. $a_n = \dfrac{(n+1)!}{n^2}$

48. $a_n = \dfrac{(n+1)!}{2n}$

Find the Partial Sum

In the following exercises, expand the partial sum and find its value.

49. $\displaystyle\sum_{i=1}^{5} i^2$

50. $\displaystyle\sum_{i=1}^{5} i^3$

51. $\displaystyle\sum_{i=1}^{6} (2i+3)$

52. $\displaystyle\sum_{i=1}^{6} (3i-2)$

53. $\displaystyle\sum_{i=1}^{4} 2^i$

54. $\displaystyle\sum_{i=1}^{4} 3^i$

55. $\displaystyle\sum_{k=0}^{3} \dfrac{4}{k!}$

56. $\displaystyle\sum_{k=0}^{4} -\dfrac{1}{k!}$

57. $\displaystyle\sum_{k=1}^{5} k(k+1)$

58. $\displaystyle\sum_{k=1}^{5} k(2k-3)$

59. $\displaystyle\sum_{n=1}^{5} \dfrac{n}{n+1}$

60. $\displaystyle\sum_{n=1}^{4} \dfrac{n}{n+2}$

Use Summation Notation to write a Sum

In the following exercises, write each sum using summation notation.

61. $\dfrac{1}{3} + \dfrac{1}{9} + \dfrac{1}{27} + \dfrac{1}{81} + \dfrac{1}{243}$

62. $\dfrac{1}{4} + \dfrac{1}{16} + \dfrac{1}{64} + \dfrac{1}{256}$

63. $1 + \dfrac{1}{8} + \dfrac{1}{27} + \dfrac{1}{64} + \dfrac{1}{125}$

64. $\dfrac{1}{5} + \dfrac{1}{25} + \dfrac{1}{125} + \dfrac{1}{625}$

65. $2 + 1 + \dfrac{2}{3} + \dfrac{1}{2} + \dfrac{2}{5}$

66. $3 + \dfrac{3}{2} + 1 + \dfrac{3}{4} + \dfrac{3}{5} + \dfrac{1}{2}$

67. $3 - 6 + 9 - 12 + 15$

68. $-5 + 10 - 15 + 20 - 25$

69. $-2 + 4 - 6 + 8 - 10 + \ldots + 20$

70. $1 - 3 + 5 - 7 + 9 + \ldots + 21$

71. $14 + 16 + 18 + 20 + 22 + 24 + 26$

72. $9 + 11 + 13 + 15 + 17 + 19 + 21$

Writing Exercises

73. In your own words, explain how to write the terms of a sequence when you know the formula. Show an example to illustrate your explanation.

74. Which terms of the sequence are negative when the nth term of the sequence is $a_n = (-1)^n (n + 2)$?

75. In your own words, explain what is meant by $n!$ Show some examples to illustrate your explanation.

76. Explain what each part of the notation $\displaystyle\sum_{k=1}^{12} 2k$ means.

Self Check

ⓐ After completing the exercises, use this checklist to evaluate your mastery of the objectives of this section.

I can...	Confidently	With some help	No-I don't get it!
write the first few terms of a sequence.			
find a formula for the nth term of a sequence.			
use factorial notation.			
find the partial sum.			
use summation notation to write a sum.			

ⓑ If most of your checks were:

...confidently. Congratulations! You have achieved the objectives in this section. Reflect on the study skills you used so that you can continue to use them. What did you do to become confident of your ability to do these things? Be specific.

...with some help. This must be addressed quickly because topics you do not master become potholes in your road to success. In math, every topic builds upon previous work. It is important to make sure you have a strong foundation before you move on. Who can you ask for help? Your fellow classmates and instructor are good resources. Is there a place on campus where math tutors are available? Can your study skills be improved?

...no - I don't get it! This is a warning sign and you must not ignore it. You should get help right away or you will quickly be overwhelmed. See your instructor as soon as you can to discuss your situation. Together you can come up with a plan to get you the help you need.

12.2 Arithmetic Sequences

Learning Objectives

By the end of this section, you will be able to:
> Determine if a sequence is arithmetic
> Find the general term (n th term) of an arithmetic sequence
> Find the sum of the first n terms of an arithmetic sequence

> **Be Prepared!**
>
> Before you get started, take this readiness quiz.
>
> 1. Evaluate $4n - 1$ for the integers 1, 2, 3, and 4.
>   If you missed this problem, review **Example 1.6**.
>
> 2. Solve the system of equations: $\begin{cases} x + y = 7 \\ 3x + 4y = 23 \end{cases}$.
>   If you missed this problem, review **Example 4.9**.
>
> 3. If $f(n) = \frac{n}{2}(3n + 5)$, find $f(1) + f(20)$.
>   If you missed this problem, review **Example 3.49**.

Determine if a Sequence is Arithmetic

The last section introduced sequences and now we will look at two specific types of sequences that each have special properties. In this section we will look at arithmetic sequences and in the next section, geometric sequences.

An **arithmetic sequence** is a sequence where the difference between consecutive terms is constant. The difference between consecutive terms in an arithmetic sequence, $a_n - a_{n-1}$, is d, the **common difference**, for n greater than or equal to two.

> **Arithmetic Sequence**
>
> An **arithmetic sequence** is a sequence where the difference between consecutive terms is always the same.
>
> The difference between consecutive terms, $a_n - a_{n-1}$, is d, the **common difference**, for n greater than or equal to two.

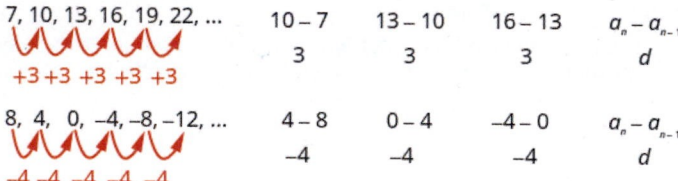

In each of these sequences, the difference between consecutive terms is constant, and so the sequence is arithmetic.

EXAMPLE 12.13

Determine if each sequence is arithmetic. If so, indicate the common difference.

ⓐ 5, 9, 13, 17, 21, 25, ...

ⓑ 4, 9, 12, 17, 20, 25, ...

ⓒ 10, 3, −4, −11, −18, −25, ...

⊘ Solution

To determine if the sequence is arithmetic, we find the difference of the consecutive terms shown.

ⓐ

	5,	9,	13,	17	21,	25, ...

Find the difference of the consecutive terms.

	$9-5$	$13-9$	$17-13$	$21-17$	$25-21$
	4	4	4	4	4

The sequence is arithmetic. The common difference is $d = 4$.

ⓑ

	4,	9,	12,	17	20,	25, ...

Find the difference of the consecutive terms.

	$9-4$	$12-9$	$17-12$	$20-17$	$25-20$
	2	3	5	3	5

The sequence is not arithmetic as all the differences between the consecutive terms are not the same. There is no common difference.

ⓒ

	10,	3,	−4,	−11	−18,	−25, ...

Find the difference of the consecutive terms.

	$3-10$	$-4-3$	$-11-(-4)$	$-18-(-11)$	$-25-(-18)$
	−7	−7	−7	−7	−7

The sequence is arithmetic. The common difference is $d = -7$.

> **TRY IT : : 12.25**   Determine if each sequence is arithmetic. If so, indicate the common difference.

ⓐ 9, 20, 31, 42, 53, 64, ... ⓑ 12, 6, 0, −6, −12, −18, ... ⓒ 7, 1, 10, 4, 13, 7, ...

> **TRY IT : : 12.26**   Determine if each sequence is arithmetic. If so, indicate the common difference.

ⓐ −4, 4, 2, 10, 8, 16, ... ⓑ −3, −1, 1, 3, 5, 7, ... ⓒ 7, 2, −3, −8, −13, −18, ...

If we know the first term, a_1, and the common difference, d, we can list a finite number of terms of the sequence.

EXAMPLE 12.14

Write the first five terms of the sequence where the first term is 5 and the common difference is $d = -6$.

⊘ **Solution**

We start with the first term and add the common difference. Then we add the common difference to that result to get the next term, and so on.

a_1	a_2	a_3	a_4	a_5
5	$5+(-6)$	$-1+(-6)$	$-7+(-6)$	$-13+(-6)$
	−1	−7	−13	−19

The sequence is $5, -1, -7, -13, -19, \ldots$

> **TRY IT : : 12.27**

Write the first five terms of the sequence where the first term is 7 and the common difference is $d = -4$.

> **TRY IT :: 12.28**

Write the first five terms of the sequence where the first term is 11 and the common difference is $d = -8$.

Find the General Term (*n*th Term) of an Arithmetic Sequence

Just as we found a formula for the general term of a sequence, we can also find a formula for the general term of an arithmetic sequence.

Let's write the first few terms of a sequence where the first term is a_1 and the common difference is d. We will then look for a pattern.

As we look for a pattern we see that each term starts with a_1.

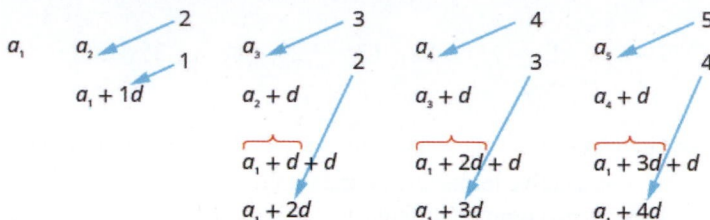

The first term adds $0d$ to the a_1, the second term adds $1d$, the third term adds $2d$, the fourth term adds $3d$, and the fifth term adds $4d$. The number of ds that were added to a_1 is one less than the number of the term. This leads us to the following

$$a_n = a_1 + (n - 1)d$$

General Term (*n*th term) of an Arithmetic Sequence

The general term of an arithmetic sequence with first term a_1 and the common difference d is

$$a_n = a_1 + (n - 1)d$$

We will use this formula in the next example to find the 15th term of a sequence.

EXAMPLE 12.15

Find the fifteenth term of a sequence where the first term is 3 and the common difference is 6.

✓ **Solution**

To find the fifteenth term, a_{15}, use the formula with $a_1 = 3$ and $d = 6$.

$$a_n = a_1 + (n - 1)d$$

Substitute in the values.

$$a_{15} = 3 + (15 - 1)6$$

Simplify.

$$a_{15} = 3 + (14)6$$

$$a_{15} = 87$$

> **TRY IT :: 12.29**

Find the twenty-seventh term of a sequence where the first term is 7 and the common difference is 9.

> **TRY IT :: 12.30**

Find the eighteenth term of a sequence where the first term is 13 and the common difference is -7.

Sometimes we do not know the first term and we must use other given information to find it before we find the requested term.

EXAMPLE 12.16

Find the twelfth term of a sequence where the seventh term is 10 and the common difference is -2. Give the formula for the general term.

⊘ Solution

To first find the first term, a_1, use the formula with $a_7 = 10$, $n = 7$, and $d = -2$.

$$a_n = a_1 + (n-1)d$$

Substitute in the values.

$$10 = a_1 + (7-1)(-2)$$

Simplify.

$$10 = a_1 + (6)(-2)$$
$$10 = a_1 - 12$$
$$a_1 = 22$$

Find the twelfth term, a_{12}, using the formula with $a_1 = 22$, $n = 12$, and $d = -2$.

$$a_n = a_1 + (n-1)d$$

Substitute in the values.

$$a_{12} = 22 + (12-1)(-2)$$

Simplify.

$$a_{12} = 22 + (11)(-2)$$
$$a_{12} = 0$$

The twelfth term of the sequence is $0, a_{12} = 0$.

To find the general term, substitute the values into the formula.

$$a_n = a_1 + (n-1)d$$

$$a_n = 22 + (n-1)(-2)$$
$$a_n = 22 - 2n + 2$$

The general term is $a_n = -2n + 24$.

> **TRY IT :: 12.31**
>
> Find the eleventh term of a sequence where the ninth term is 8 and the common difference is -3. Give the formula for the general term.

> **TRY IT :: 12.32**
>
> Find the nineteenth term of a sequence where the fifth term is 1 and the common difference is -4. Give the formula for the general term.

Sometimes the information given leads us to two equations in two unknowns. We then use our methods for solving systems of equations to find the values needed.

EXAMPLE 12.17

Find the first term and common difference of a sequence where the fifth term is 19 and the eleventh term is 37. Give the formula for the general term.

⊘ Solution

Since we know two terms, we can make a system of equations using the formula for the general term.

$$a_n = a_1 + (n-1)d$$

We know the value of a_5 and a_{11}, so we will use $n=5$ and $n=11$.	$\begin{cases} a_5 = a_1 + (5-1)d \\ a_{11} = a_1 + (11-1)d \end{cases}$
Substitute in the values, $a_5 = 19$ and $a_{11} = 37$.	$\begin{cases} 19 = a_1 + (5-1)d \\ 37 = a_1 + (11-1)d \end{cases}$
Simplify.	$\begin{cases} 19 = a_1 + 4d \\ 37 = a_1 + 10d \end{cases}$
Prepare to eliminate the a_1 term by multiplying the top equation by -1. Add the equations.	$\begin{array}{r} -19 = -a_1 - 4d \\ 37 = a_1 + 10d \\ \hline 18 = 6d \\ 3 = d \end{array}$
Substituting $d=3$ back into the first equation.	$19 = a_1 + 4 \cdot 3$
Solve for a_1.	$19 = a_1 + 12$ $7 = a_1$
Use the formula with $a_1 = 7$ and $d = 3$.	$a_n = a_1 + (n-1)d$
Substitute in the values.	$a_n = 7 + (n-1)3$
Simplify.	$a_n = 7 + 3n - 3$ $a_n = 3n + 4$

The first term is $a_1 = 7$.

The common difference is $d = 3$.

The general term of the sequence is $a_n = 3n + 4$.

> **TRY IT : : 12.33**

Find the first term and common difference of a sequence where the fourth term is 17 and the thirteenth term is 53. Give the formula for the general term.

> **TRY IT : : 12.34**

Find the first term and common difference of a sequence where the third term is 2 and the twelfth term is -25. Give the formula for the general term.

Find the Sum of the First n Terms of an Arithmetic Sequence

As with the general sequences, it is often useful to find the sum of an arithmetic sequence. The sum, S_n, of the first n terms of any arithmetic sequence is written as $S_n = a_1 + a_2 + a_3 + ... + a_n$. To find the sum by merely adding all the terms can be tedious. So we can also develop a formula to find the sum of a sequence using the first and last term of the sequence.

We can develop this new formula by first writing the sum by starting with the first term, a_1, and keep adding a d to get the next term as:

$$S_n = a_1 + (a_1 + d) + (a_1 + 2d) + ... + a_n.$$

We can also reverse the order of the terms and write the sum by starting with a_n and keep subtracting d to get the next term as

$$S_n = a_n + (a_n - d) + (a_n - 2d) + \ ... \ + a_1.$$

If we add these two expressions for the sum of the first n terms of an arithmetic sequence, we can derive a formula for the sum of the first n terms of any arithmetic series.

$$
\begin{array}{rcl}
S_n &=& a_1 \quad\ + (a_1 + d) + (a_1 + 2d) + \ ... \ + a_n \\
+S_n &=& a_n \quad\ + (a_n - d) + (a_n - 2d) + \ ... \ + a_1 \\
\hline
2S_n &=& (a_1 + a_n) + (a_1 + a_n) + (a_1 + a_n) + \ ... \ + (a_1 + a_n)
\end{array}
$$

Because there are n sums of $(a_1 + a_n)$ on the right side of the equation, we rewrite the right side as $n(a_1 + a_n)$.

$$2S_n = n(a_1 + a_n)$$

We divide by two to solve for S_n.

$$S_n = \tfrac{n}{2}(a_1 + a_n)$$

This give us a general formula for the sum of the first n terms of an arithmetic sequence.

Sum of the First n Terms of an Arithmetic Sequence

The sum, S_n, of the first n terms of an arithmetic sequence is

$$S_n = \tfrac{n}{2}(a_1 + a_n)$$

where a_1 is the first term and a_n is the nth term.

We apply this formula in the next example where the first few terms of the sequence are given.

EXAMPLE 12.18

Find the sum of the first 30 terms of the arithmetic sequence: 8, 13, 18, 23, 28, ...

✓ Solution

To find the sum, we will use the formula $S_n = \tfrac{n}{2}(a_1 + a_n)$. We know $a_1 = 8$, $d = 5$ and $n = 30$, but we need to find a_n in order to use the sum formula.

$$
\begin{array}{rcl}
a_n &=& a_1 + (n-1)d
\end{array}
$$

Find a_n where $a_1 = 8$, $d = 5$ and $n = 30$.

$$
\begin{array}{rcl}
a_{30} &=& 8 + (30 - 1)5
\end{array}
$$

Simplify.

$$
\begin{array}{rcl}
a_{30} &=& 8 + (29)5 \\
a_{30} &=& 153
\end{array}
$$

Knowing $a_1 = 8$, $n = 30$, and $a_{30} = 153$, use the sum formula.

$$
S_n = \tfrac{n}{2}(a_1 + a_n)
$$

Substitute in the values.

$$
S_{30} = \tfrac{30}{2}(8 + 153)
$$

Simplify.

$$
S_{30} = 15(161)
$$

Simplify.

$$
S_{30} = 2{,}415
$$

> **TRY IT :: 12.35**   Find the sum of the first 30 terms of the arithmetic sequence: 5, 9, 13, 17, 21, ...

> **TRY IT :: 12.36**   Find the sum of the first 30 terms of the arithmetic sequence: 7, 10, 13, 16, 19, ...

In the next example, we are given the general term for the sequence and are asked to find the sum of the first 50 terms.

EXAMPLE 12.19

Find the sum of the first 50 terms of the arithmetic sequence whose general term is $a_n = 3n - 4$.

✓ Solution

To find the sum, we will use the formula $S_n = \frac{n}{2}(a_1 + a_n)$. We know $n = 50$, but we need to find a_1 and a_n in order to use the sum formula.

	$a_n = 3n - 4$
Find a_1, by substituting $n = 1$.	$a_1 = 3 \cdot 1 - 4$ $a_1 = -1$
Find a_n by substituting $n = 50$.	$a_n = 3n - 4$ $a_{50} = 3 \cdot 50 - 4$
Simplify.	$a_{50} = 146$
Knowing $n = 50$, $a_1 = -1$, and $a_{50} = 146$ use the sum formula.	$S_n = \frac{n}{2}(a_1 + a_n)$
Substitute in the values.	$S_n = \frac{50}{2}(-1 + 146)$
Simplify.	$S_{50} = 25(145)$
Simplify.	$S_{50} = 3,625$

> **TRY IT : : 12.37**
>
> Find the sum of the first 50 terms of the arithmetic sequence whose general term is $a_n = 2n - 5$.

> **TRY IT : : 12.38**
>
> Find the sum of the first 50 terms of the arithmetic sequence whose general term is $a_n = 4n + 3$.

In the next example we are given the sum in summation notation. To add all the terms would be tedious, so we extract the information needed to use the formula to find the sum of the first *n* terms.

EXAMPLE 12.20

Find the sum: $\displaystyle\sum_{i=1}^{25} (4i + 7)$.

✓ Solution

To find the sum, we will use the formula $S_n = \frac{n}{2}(a_1 + a_n)$. We know $n = 25$, but we need to find a_1 and a_n in order to use the sum formula.

Expand the summation notation.	$\sum_{i=1}^{25} 4i + 7 = (4 \cdot 1 + 7) + (4 \cdot 2 + 7) + (4 \cdot 3 + 7) + \ldots + (4 \cdot 25 + 7)$
Simplify.	$\sum_{i=1}^{25} 4i + 7 = \quad 11 \quad + \quad 15 \quad + \quad 19 \quad + \ldots + \quad 107$
Identify a_1.	$a_1 = 11$
Identify a_{25}.	$a_{25} = 107$
Knowing $n = 25$, $a_1 = 11$, and $a_{25} = 107$ use the sum formula.	$S_n = \frac{n}{2}(a_1 + a_n)$
Substitute in the values.	$S_{25} = \frac{25}{2}(11 + 107)$
Simplify.	$S_{25} = \frac{25}{2}(118)$
Simplify.	$S_{25} = 1{,}475$

> **TRY IT : : 12.39**
>
> Find the sum: $\displaystyle\sum_{i=1}^{30} (6i - 4)$.

> **TRY IT : : 12.40**
>
> Find the sum: $\displaystyle\sum_{i=1}^{35} (5i - 3)$.

▶ **MEDIA : :**

Access these online resources for additional instruction and practice with arithmetic sequences

- **Arithmetic Sequences (https://openstax.org/l/37ArithSequenc)**
- **Arithmetic Sequences: A Formula for the 'n-th' Term (https://openstax.org/l/37AritSeqnthter)**
- **Arithmetic Series (https://openstax.org/l/37ArithSeries)**

12.2 EXERCISES
Practice Makes Perfect

Determine if a Sequence is Arithmetic

In the following exercises, determine if each sequence is arithmetic, and if so, indicate the common difference.

77. 4, 12, 20, 28, 36, 44, …

78. $-7, -2, 3, 8, 13, 18, …$

79. $-15, -16, 3, 12, 21, 30, …$

80. $11, 5, -1, -7 - 13, -19, …$

81. $8, 5, 2, -1, -4, -7, …$

82. $15, 5, -5, -15, -25, -35, …$

In the following exercises, write the first five terms of each sequence with the given first term and common difference.

83. $a_1 = 11$ and $d = 7$

84. $a_1 = 18$ and $d = 9$

85. $a_1 = -7$ and $d = 4$

86. $a_1 = -8$ and $d = 5$

87. $a_1 = 14$ and $d = -9$

88. $a_1 = -3$ and $d = -3$

Find the General Term (*n*th Term) of an Arithmetic Sequence

In the following exercises, find the term described using the information provided.

89. Find the twenty-first term of a sequence where the first term is three and the common difference is eight.

90. Find the twenty-third term of a sequence where the first term is six and the common difference is four.

91. Find the thirtieth term of a sequence where the first term is -14 and the common difference is five.

92. Find the fortieth term of a sequence where the first term is -19 and the common difference is seven.

93. Find the sixteenth term of a sequence where the first term is 11 and the common difference is -6.

94. Find the fourteenth term of a sequence where the first term is eight and the common difference is -3.

95. Find the twentieth term of a sequence where the fifth term is -4 and the common difference is -2. Give the formula for the general term.

96. Find the thirteenth term of a sequence where the sixth term is -1 and the common difference is -4. Give the formula for the general term.

97. Find the eleventh term of a sequence where the third term is 19 and the common difference is five. Give the formula for the general term.

98. Find the fifteenth term of a sequence where the tenth term is 17 and the common difference is seven. Give the formula for the general term.

99. Find the eighth term of a sequence where the seventh term is -8 and the common difference is -5. Give the formula for the general term.

100. Find the fifteenth term of a sequence where the tenth term is -11 and the common difference is -3. Give the formula for the general term.

In the following exercises, find the first term and common difference of the sequence with the given terms. Give the formula for the general term.

101. The second term is 14 and the thirteenth term is 47.

102. The third term is 18 and the fourteenth term is 73.

103. The second term is 13 and the tenth term is -51.

104. The third term is four and the tenth term is -38.

105. The fourth term is -6 and the fifteenth term is 27.

106. The third term is -13 and the seventeenth term is 15.

Find the Sum of the First n Terms of an Arithmetic Sequence

In the following exercises, find the sum of the first 30 terms of each arithmetic sequence.

107. 11, 14, 17, 20, 23, ...

108. 12, 18, 24, 30, 36, ...

109. 8, 5, 2, −1, −4, ...

110. 16, 10, 4, −2, −8, ...

111. −17, −15, −13, −11, −9, ...

112. −15, −12, −9, −6, −3, ...

In the following exercises, find the sum of the first 50 terms of the arithmetic sequence whose general term is given.

113. $a_n = 5n - 1$

114. $a_n = 2n + 7$

115. $a_n = -3n + 5$

116. $a_n = -4n + 3$

In the following exercises, find each sum.

117. $\displaystyle\sum_{i=1}^{40} (8i - 7)$

118. $\displaystyle\sum_{i=1}^{45} (7i - 5)$

119. $\displaystyle\sum_{i=1}^{50} (3i + 6)$

120. $\displaystyle\sum_{i=1}^{25} (4i + 3)$

121. $\displaystyle\sum_{i=1}^{35} (-6i - 2)$

122. $\displaystyle\sum_{i=1}^{30} (-5i + 1)$

Writing Exercises

123. In your own words, explain how to determine whether a sequence is arithmetic.

124. In your own words, explain how the first two terms are used to find the tenth term. Show an example to illustrate your explanation.

125. In your own words, explain how to find the general term of an arithmetic sequence.

126. In your own words, explain how to find the sum of the first n terms of an arithmetic sequence without adding all the terms.

Self Check

ⓐ *After completing the exercises, use this checklist to evaluate your mastery of the objectives of this section.*

I can...	Confidently	With some help	No-I don't get it!
determine if a sequence is arithmetic.			
find the general term (nth term) of an arithmetic sequence.			
find the sum of the first n terms of an arithmetic sequence			

ⓑ *After reviewing this checklist, what will you do to become confident for all objectives?*

12.3 Geometric Sequences and Series

Learning Objectives

By the end of this section, you will be able to:

› Determine if a sequence is geometric
› Find the general term (nth term) of a geometric sequence
› Find the sum of the first n terms of a geometric sequence
› Find the sum of an infinite geometric series
› Apply geometric sequences and series in the real world

Be Prepared!

Before you get started, take this readiness quiz.

1. Simplify: $\frac{24}{32}$.

   If you missed this problem, review **Example 1.24**.

2. Evaluate: ⓐ 3^4 ⓑ $\left(\frac{1}{2}\right)^4$.

   If you missed this problem, review **Example 1.19**.

3. If $f(x) = 4 \cdot 3^x$, find ⓐ $f(1)$ ⓑ $f(2)$ ⓒ $f(3)$.

   If you missed this problem, review **Example 3.49**.

Determine if a Sequence is Geometric

We are now ready to look at the second special type of sequence, the geometric sequence.

A sequence is called a **geometric sequence** if the ratio between consecutive terms is always the same. The ratio between consecutive terms in a geometric sequence is r, the **common ratio**, where r is greater than or equal to two.

Geometric Sequence

A **geometric sequence** is a sequence where the ratio between consecutive terms is always the same.

The ratio between consecutive terms, $\frac{a_n}{a_{n-1}}$, is r, the **common ratio**. r is greater than or equal to two.

Consider these sequences.

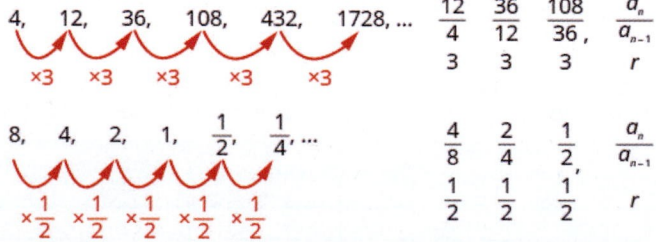

EXAMPLE 12.21

Determine if each sequence is geometric. If so, indicate the common ratio.

ⓐ 4, 8, 16, 32, 64, 128, …

ⓑ −2, 6, −12, 36, −72, 216, …

ⓒ 27, 9, 3, 1, $\frac{1}{3}$, $\frac{1}{9}$, …

✓ Solution

To determine if the sequence is geometric, we find the ratio of the consecutive terms shown.

ⓐ

$$4, \quad 8, \quad 16, \quad 32, \quad 64, \quad 128, \ldots$$

Find the ratio of
the consecutive terms.

$$\frac{8}{4} \quad \frac{16}{8} \quad \frac{32}{16} \quad \frac{64}{32} \quad \frac{128}{64}$$
$$2 \quad \ 2 \quad \ \ 2 \quad \ \ 2 \quad \ \ 2$$

The sequence is geometric. The common ratio is $r = 2$.

ⓑ

$$-2, \quad 6, \quad -12, \quad 36, \quad -72, \quad 216, \ldots$$

Find the ratio of
the consecutive terms.

$$\frac{6}{-2} \quad \frac{-12}{6} \quad \frac{36}{-12} \quad \frac{-72}{36} \quad \frac{216}{-72}$$
$$-3 \quad \ -2 \quad \ \ -3 \quad \ \ -2 \quad \ \ -3$$

The sequence is not geometric. There is no common ratio.

ⓒ

$$27, 9, 3, 1, \frac{1}{3}, \frac{1}{9}, \ldots$$

Find the ratio of
the consecutive terms.

$$\frac{9}{27} \quad \frac{3}{9} \quad \frac{1}{3} \quad \frac{\frac{1}{3}}{1} \quad \frac{\frac{1}{9}}{\frac{1}{3}}$$

$$\frac{1}{3} \quad \frac{1}{3} \quad \frac{1}{3} \quad \frac{1}{3} \quad \frac{1}{3}$$

The sequence is geometric. The common ratio is $r = \frac{1}{3}$.

> **TRY IT :: 12.41**   Determine if each sequence is geometric. If so indicate the common ratio.
>
> ⓐ $7, 21, 63, 189, 567, 1{,}701, \ldots$
>
> ⓑ $64, 16, 4, 1, \frac{1}{4}, \frac{1}{16}, \ldots$
>
> ⓒ $2, 4, 12, 48, 240, 1{,}440, \ldots$

> **TRY IT :: 12.42**   Determine if each sequence is geometric. If so indicate the common ratio.
>
> ⓐ $-150, -30, -15, -5, -\frac{5}{2}, 0, \ldots$
>
> ⓑ $5, 10, 20, 40, 80, 160, \ldots$
>
> ⓒ $8, 4, 2, 1, \frac{1}{2}, \frac{1}{4}, \ldots$

If we know the first term, a_1, and the common ratio, r, we can list a finite number of terms of the sequence.

EXAMPLE 12.22

Write the first five terms of the sequence where the first term is 3 and the common ratio is $r = -2$.

✓ **Solution**

We start with the first term and multiply it by the common ratio. Then we multiply that result by the common ratio to get the next term, and so on.

$$
\begin{array}{ccccc}
a_1 & a_2 & a_3 & a_4 & a_5 \\
3 & 3 \cdot (-2) & -6 \cdot (-2) & 12 \cdot (-2) & -24 \cdot (-2) \\
 & -6 & 12 & -24 & 48
\end{array}
$$

The sequence is $3, -6, 12, -24, 48, \ldots$

> **TRY IT :: 12.43**

Write the first five terms of the sequence where the first term is 7 and the common ratio is $r = -3$.

> **TRY IT :: 12.44**

Write the first five terms of the sequence where the first term is 6 and the common ratio is $r = -4$.

Find the General Term (*n*th Term) of a Geometric Sequence

Just as we found a formula for the general term of a sequence and an arithmetic sequence, we can also find a formula for the general term of a geometric sequence.

Let's write the first few terms of the sequence where the first term is a_1 and the common ratio is r. We will then look for a pattern.

As we look for a pattern in the five terms above, we see that each of the terms starts with a_1.

The first term, a_1, is not multiplied by any r. In the second term, the a_1 is multiplied by r. In the third term, the a_1 is multiplied by r two times ($r \cdot r$ or r^2). In the fourth term, the a_1 is multiplied by r three times ($r \cdot r \cdot r$ or r^3) and in the fifth term, the a_1 is multiplied by r four times. In each term, the number of times a_1 is multiplied by r is one less than the number of the term. This leads us to the following

$$a_n = a_1 r^{n-1}$$

General Term (*n*th term) of a Geometric Sequence

The general term of a geometric sequence with first term a_1 and the common ratio r is

$$a_n = a_1 r^{n-1}$$

We will use this formula in the next example to find the fourteenth term of a sequence.

EXAMPLE 12.23

Find the fourteenth term of a sequence where the first term is 64 and the common ratio is $r = \frac{1}{2}$.

⊘ Solution

To find the fourteenth term, a_{14},
use the formula with $a_1 = 64$ and $r = \frac{1}{2}$.

$$a_n = a_1 r^{n-1}$$

Substitute in the values.

$$a_{14} = 64\left(\frac{1}{2}\right)^{14-1}$$

Simplify.

$$a_{14} = 64\left(\frac{1}{2}\right)^{13}$$

$$a_{14} = \frac{1}{128}$$

> **TRY IT : :** 12.45

Find the thirteenth term of a sequence where the first term is 81 and the common ratio is $r = \frac{1}{3}$.

> **TRY IT : :** 12.46

Find the twelfth term of a sequence where the first term is 256 and the common ratio is $r = \frac{1}{4}$.

Sometimes we do not know the common ratio and we must use the given information to find it before we find the requested term.

EXAMPLE 12.24

Find the twelfth term of the sequence 3, 6, 12, 24, 48, 96, ... Find the general term for the sequence.

⊘ Solution

To find the twelfth term, we use the formula, $a_n = a_1 r^{n-1}$, and so we need to first determine a_1 and the common ratio r.

3, 6, 12, 24, 48, 96, …

The first term is three.

$$a_1 = 3$$

Find the common ratio.

$$\frac{6}{3} \quad \frac{12}{6} \quad \frac{24}{12} \quad \frac{48}{24} \quad \frac{96}{48}$$
$$2 \quad 2 \quad 2 \quad 2 \quad 2$$
The common ratio is $r = 2$.

To find the twelfth term, a_{12}, use the formula with $a_1 = 3$ and $r = 2$.

$$a_n = a_1 r^{n-1}$$

Substitute in the values.

$$a_{12} = 3 \cdot 2^{12-1}$$

Simplify.

$$a_{12} = 3 \cdot 2^{11}$$
$$a_{12} = 6,144$$

Find the general term.
We use the formula with $a_1 = 3$ and $r = 2$.

$$a_n = a_1 r^{n-1}$$
$$a_n = 3(2)^{n-1}$$

> **TRY IT : :** 12.47

Find the ninth term of the sequence 6, 18, 54, 162, 486, 1,458, ... Then find the general term for the sequence.

> TRY IT :: 12.48

Find the eleventh term of the sequence 7, 14, 28, 56, 112, 224, … Then find the general term for the sequence.

Find the Sum of the First n Terms of a Geometric Sequence

We found the sum of both general sequences and arithmetic sequence. We will now do the same for geometric sequences. The sum, S_n, of the first n terms of a geometric sequence is written as $S_n = a_1 + a_2 + a_3 + \dots + a_n$. We can write this sum by starting with the first term, a_1, and keep multiplying by r to get the next term as:

$$S_n = a_1 + a_1 r + a_1 r^2 + \dots + a_1 r^{n-1}$$

Let's also multiply both sides of the equation by r.

$$r S_n = a_1 r + a_1 r^2 + a_1 r^3 + \dots + a_1 r^n$$

Next, we subtract these equations. We will see that when we subtract, all but the first term of the top equation and the last term of the bottom equation subtract to zero.

$$
\begin{aligned}
S_n &= a_1 + a_1 r + a_1 r^2 + a_1 r^3 + \dots + a_1 r^{n-1} \\
r S_n &= a_1 r + a_1 r^2 + a_1 r^3 + \dots + a_1 r^{n-1} + a_1 r^n \\
\hline
S_n - r S_n &= a_1 \phantom{+ a_1 r + a_1 r^2 + a_1 r^3 + \dots + a_1 r^{n-1}} - a_1 r^n
\end{aligned}
$$

We factor both sides.

$$S_n(1 - r) = a_1(1 - r^n)$$

To obtain the formula for S_n, divide both sides by $(1 - r)$.

$$S_n = \frac{a_1(1 - r^n)}{1 - r}$$

Sum of the First n Terms of a Geometric Series

The sum, S_n, of the first n terms of a geometric sequence is

$$S_n = \frac{a_1(1 - r^n)}{1 - r}$$

where a_1 is the first term and r is the common ratio, and r is not equal to one.

We apply this formula in the next example where the first few terms of the sequence are given. Notice the sum of a geometric sequence typically gets very large when the common ratio is greater than one.

EXAMPLE 12.25

Find the sum of the first 20 terms of the geometric sequence 7, 14, 28, 56, 112, 224, …

 **Solution**

To find the sum, we will use the formula $S_n = \frac{a_1(1 - r^n)}{1 - r}$. We know $a_1 = 7$, $r = 2$, and $n = 20$.

Knowing $a_1 = 7$, $r = 2$, and $n = 20$, use the sum formula.

$$S_n = \frac{a_1(1 - r^n)}{1 - r}$$

Substitute in the values.

$$S_{20} = \frac{7\left(1 - 2^{20}\right)}{1 - 2}$$

Simplify.

$$S_{20} = 7{,}340{,}025$$

> TRY IT :: 12.49   Find the sum of the first 20 terms of the geometric sequence 3, 6, 12, 24, 48, 96, …

> TRY IT :: 12.50   Find the sum of the first 20 terms of the geometric sequence 6, 18, 54, 162, 486, 1,458, …

In the next example, we are given the sum in summation notation. While adding all the terms might be possible, most often it is easiest to use the formula to find the sum of the first n terms.

To use the formula, we need r. We can find it by writing out the first few terms of the sequence and find their ratio. Another option is to realize that in summation notation, a sequence is written in the form $\sum_{i=1}^{k} a(r)^i$, where r is the common ratio.

EXAMPLE 12.26

Find the sum: $\sum_{i=1}^{15} 2(3)^i$.

✓ **Solution**

To find the sum, we will use the formula $S_n = \dfrac{a_1(1-r^n)}{1-r}$, which requires a_1 and r. We will write out a few of the terms, so we can get the needed information.

$$\sum_{i=1}^{15} 2(3)^i$$

Write out the first few terms.	$2 \cdot 3^1, 2 \cdot 3^2, 2 \cdot 3^3, \dots$ $6, \quad 18, \quad 54,$
Identify a_1.	$a_1 = 6$
Find the common ratio.	$\dfrac{18}{6} \quad \dfrac{54}{18}$ or $\sum_{i=1}^{15} 2(3)^i$ $\quad 3 \qquad 3$ The common ratio is $r = 3$.
Knowing $a_1 = 6$, $r = 3$, and $n = 15$, use the sum formula.	$S_n = \dfrac{a_1(1-r^n)}{1-r}$
Substitute in the values.	$S_{15} = \dfrac{6(1-3^{15})}{1-3}$
Simplify.	$S_{15} = 43{,}046{,}718$

> **TRY IT :: 12.51**   Find the sum: $\sum_{i=1}^{15} 6(2)^i$.

> **TRY IT :: 12.52**   Find the sum: $\sum_{i=1}^{10} 5(2)^i$.

Find the Sum of an Infinite Geometric Series

If we take a geometric sequence and add the terms, we have a sum that is called a geometric series. An **infinite geometric series** is an infinite sum whose first term is a_1 and common ratio is r and is written

$$a_1 + a_1 r + a_1 r^2 + \dots + a_1 r^{n-1} + \dots$$

Infinite Geometric Series

An **infinite geometric series** is an infinite sum whose first term is a_1 and common ratio is r and is written

$$a_1 + a_1 r + a_1 r^2 + \dots + a_1 r^{n-1} + \dots$$

We know how to find the sum of the first n terms of a geometric series using the formula, $S_n = \dfrac{a_1(1-r^n)}{1-r}$. But how do we find the sum of an infinite sum?

Let's look at the infinite geometric series $3 + 6 + 12 + 24 + 48 + 96 + \ldots$. Each term gets larger and larger so it makes sense that the sum of the infinite number of terms gets larger. Let's look at a few partial sums for this series. We see $a_1 = 3$ and $r = 2$

$$S_n = \dfrac{a_1(1-r^n)}{1-r} \qquad S_n = \dfrac{a_1(1-r^n)}{1-r} \qquad S_n = \dfrac{a_1(1-r^n)}{1-r}$$

$$S_{10} = \dfrac{3(1-2^{10})}{1-2} \qquad S_{30} = \dfrac{3(1-2^{30})}{1-2} \qquad S_{50} = \dfrac{3(1-2^{50})}{1-2}$$

$$S_{10} = 3{,}069 \qquad S_{30} = 3{,}221{,}225{,}469 \qquad S_{50} \approx 3.38 \times 10^{15}$$

As n gets larger and larger, the sum gets larger and larger. This is true when $|r| \geq 1$ and we call the series divergent. We cannot find a sum of an infinite geometric series when $|r| \geq 1$.

Let's look at an infinite geometric series whose common ratio is a fraction less than one, $\frac{1}{2} + \frac{1}{4} + \frac{1}{8} + \frac{1}{16} + \frac{1}{32} + \frac{1}{64} + \ldots$. Here the terms get smaller and smaller as n gets larger. Let's look at a few finite sums for this series. We see $a_1 = \frac{1}{2}$ and $r = \frac{1}{2}$.

$$S_n = \dfrac{a_1(1-r^n)}{1-r} \qquad S_n = \dfrac{a_1(1-r^n)}{1-r} \qquad S_n = \dfrac{a_1(1-r^n)}{1-r}$$

$$S_{10} = \dfrac{\frac{1}{2}\left(1-\left(\frac{1}{2}\right)^{10}\right)}{1-\frac{1}{2}} \qquad S_{20} = \dfrac{\frac{1}{2}\left(1-\left(\frac{1}{2}\right)^{20}\right)}{1-\frac{1}{2}} \qquad S_{30} = \dfrac{\frac{1}{2}\left(1-\left(\frac{1}{2}\right)^{30}\right)}{1-\frac{1}{2}}$$

$$S_{10} \approx .9990234375 \qquad S_{20} \approx 0.9999990463 \qquad S_{30} \approx 0.9999999991$$

Notice the sum gets larger and larger but also gets closer and closer to one. When $|r| < 1$, the expression r^n gets smaller and smaller. In this case, we call the series convergent. As n approaches infinity, (gets infinitely large), r^n gets closer and closer to zero. In our sum formula, we can replace the r^n with zero and then we get a formula for the sum, S, for an infinite geometric series when $|r| < 1$.

$$S_n = \dfrac{a_1(1-r^n)}{1-r}$$
$$S = \dfrac{a_1(1-0)}{1-r}$$
$$S = \dfrac{a_1}{1-r}$$

This formula gives us the sum of the infinite geometric sequence. Notice the S does not have the subscript n as in S_n as we are not adding a finite number of terms.

Sum of an Infinite Geometric Series

For an infinite geometric series whose first term is a_1 and common ratio r,

If $|r| < 1$, the sum is

$$S = \dfrac{a_1}{1-r}$$

If $|r| \geq 1$, the infinite geometric series does not have a sum. We say the series diverges.

EXAMPLE 12.27

Find the sum of the infinite geometric series $54 + 18 + 6 + 2 + \frac{2}{3} + \frac{2}{9} + \ldots$

⊘ Solution

To find the sum, we first have to verify that the common ratio $|r| < 1$ and then we can use the sum formula $S = \frac{a_1}{1-r}$.

Find the common ratio.	$r = \frac{18}{54} \qquad r = \frac{6}{18} \ldots$
	$r = \frac{1}{3} \qquad r = \frac{1}{3} \qquad |r| < 1$
Identify a_1.	$a_1 = 54$
Knowing $a_1 = 54$, $r = \frac{1}{3}$, use the sum formula.	$S = \frac{a_1}{1-r}$
Substitute in the values.	$S = \frac{54}{1 - \frac{1}{3}}$
Simplify.	$S = 81$

> **TRY IT :: 12.53**   Find the sum of the infinite geometric series $48 + 24 + 12 + 6 + 3 + \frac{3}{2} + \ldots$

> **TRY IT :: 12.54**   Find the sum of the infinite geometric series $64 + 16 + 4 + 1 + \frac{1}{4} + \frac{1}{16} + \ldots$

An interesting use of infinite geometric series is to write a repeating decimal as a fraction.

EXAMPLE 12.28

Write the repeating decimal $0.\overline{5}$ as a fraction.

⊘ **Solution**

Rewrite the $0.\overline{5}$ showing the repeating five.	$0.5555555555555...$
Use place value to rewrite this as a sum. This is an infinite geometric series.	$0.5 + 0.05 + 0.005 + 0.0005 + ...$

Find the common ratio.
$$r = \frac{0.05}{0.5} \quad r = \frac{0.005}{0.05} \quad ...$$
$$r = 0.1 \quad r = 0.1 \quad |r| < 1$$

Identify a_1.
$$a_1 = 0.5$$

Knowing $a_1 = 0.5$, $r = 0.1$, use the sum formula.
$$S = \frac{a_1}{1-r}$$

Substitute in the values.
$$S = \frac{0.5}{1-0.1}$$

Simplify.
$$S = \frac{0.5}{0.9}$$

Multiply numerator and denominator by 10.
$$S = \frac{5}{9}$$

We are asked to find the fraction form.
$$0.\overline{5} = \frac{5}{9}$$

> **TRY IT :: 12.55** Write the repeating decimal $0.\overline{4}$ as a fraction.

> **TRY IT :: 12.56** Write the repeating decimal $0.\overline{8}$ as a fraction.

Apply Geometric Sequences and Series in the Real World

One application of geometric sequences has to do with consumer spending. If a tax rebate is given to each household, the effect on the economy is many times the amount of the individual rebate.

EXAMPLE 12.29

The government has decided to give a $1,000 tax rebate to each household in order to stimulate the economy. The government statistics say that each household will spend 80% of the rebate in goods and services. The businesses and individuals who benefitted from that 80% will then spend 80% of what they received and so on. The result is called the multiplier effect. What is the total effect of the rebate on the economy?

⊘ **Solution**

Every time money goes into the economy, 80% of it is spent and is then in the economy to be spent. Again, 80% of this money is spent in the economy again. This situation continues and so leads us to an infinite geometric series.
$$1000 + 1000(0.8) + 1000(0.8)^2 + ...$$

Here the first term is 1,000, $a_1 = 1000$. The common ratio is 0.8, $r = 0.8$. We can evaluate this sum since $0.8 < 1$.

We use the formula for the sum on an infinite geometric series.
$$S = \frac{a_1}{1-r}$$

Substitute in the values, $a_1 = 1,000$ and $r = 0.8$.
$$S = \frac{1,000}{1-0.8}$$

Evaluate.
$$S = 5,000$$

The total effect of the $1,000 received by each household will be a $5,000 growth in the economy.

> **TRY IT : : 12.57**

What is the total effect on the economy of a government tax rebate of $1,000 to each household in order to stimulate the economy if each household will spend 90% of the rebate in goods and services?

> **TRY IT : : 12.58**

What is the total effect on the economy of a government tax rebate of $500 to each household in order to stimulate the economy if each household will spend 85% of the rebate in goods and services?

We have looked at a compound interest formula where a principal, P, is invested at an interest rate, r, for t years. The new balance, A, is $A = P\left(1 + \frac{r}{n}\right)^{nt}$ when interest is compounded n times a year. This formula applies when a lump sum was invested upfront and tells us the value after a certain time period.

An **annuity** is an investment that is a sequence of equal periodic deposits. We will be looking at annuities that pay the interest at the time of the deposits. As we develop the formula for the value of an annuity, we are going to let $n = 1$. That means there is one deposit per year.

$$A = P\left(1 + \frac{r}{n}\right)^{nt}$$

Let $n = 1$.
$$A = P\left(1 + \frac{r}{1}\right)^{1t}$$

Simplify.
$$A = P(1 + r)^t$$

Suppose P dollars is invested at the end of each year. One year later that deposit is worth $P(1 + r)^1$ dollars, and another year later it is worth $P(1 + r)^2$ dollars. After t years, it will be worth $A = P(1 + r)^t$ dollars.

	End of year 1	End of year 2	End of year 3
First Deposit P @ end of year 1	P	Amount 1 year later $P(1 + r)^1$	Amount 2 years later $P(1 + r)^2$
2nd Deposit P @ end of year 2		P	Amount 1 year later $P(1 + r)^1$
3rd Deposit P @ end of year 3			P

After three years, the value of the annuity is

$$\underbrace{P}_{\substack{\text{money dep} \\ \text{end of year 3}}} + \underbrace{P(1 + r)^1}_{\substack{\text{money dep} \\ \text{end of year 2}}} + \underbrace{P(1 + r)^2}_{\substack{\text{money dep} \\ \text{end of year 1}}}$$

This a sum of the terms of a geometric sequence where the first term is P and the common ratio is $1 + r$. We substitute these values into the sum formula. Be careful, we have two different uses of r. The r in the sum formula is the common ratio of the sequence. In this case, that is $1 + r$ where r is the interest rate.

$$S_t = \frac{a_1(1 - r^t)}{1 - r}$$

Substitute in the values.
$$S_t = \frac{P\left(1 - (1 + r)^t\right)}{1 - (1 + r)}$$

Simplify.
$$S_t = \frac{P\left(1 - (1 + r)^t\right)}{-r}$$

$$S_t = \frac{P\left((1 + r)^t - 1\right)}{r}$$

Remember our premise was that one deposit was made at the end of each year.

We can adapt this formula for n deposits made per year and the interest is compounded n times a year.

Value of an Annuity with Interest Compounded n Times a Year

For a principal, P, invested at the end of a compounding period, with an interest rate, r, which is compounded n times a year, the new balance, A, after t years, is

$$A_t = \frac{P\left(\left(1 + \frac{r}{n}\right)^{nt} - 1\right)}{\frac{r}{n}}$$

EXAMPLE 12.30

New parents decide to invest $100 per month in an annuity for their baby daughter. The account will pay 5% interest per year which is compounded monthly. How much will be in the child's account at her eighteenth birthday?

⊘ **Solution**

To find the Annuity formula, $A_t = \frac{P\left(\left(1 + \frac{r}{n}\right)^{nt} - 1\right)}{\frac{r}{n}}$, we need to identify P, r, n, and t.

Identify P, the amount invested each month. $P = 100$

Identify r, the annual interest rate, in decimal form. $r = 0.05$

Identify n, the number of times the deposit will be made and the interest compounded each year. $n = 12$

Identify t, the number of years. $t = 18$

Knowing $P = 100$, $r = 0.05$, $n = 12$ and $t = 18$, use the sum formula. $A_t = \frac{P\left(\left(1 + \frac{r}{n}\right)^{nt} - 1\right)}{\frac{r}{n}}$

Substitute in the values. $A_t = \frac{100\left(\left(1 + \frac{0.05}{12}\right)^{12 \cdot 18} - 1\right)}{\frac{0.05}{12}}$

Use the calculator to evaluate. Be sure to use parentheses as needed. $A_t = 34{,}920.20$

The child will have $34,920.20 when she turns 18.

> **TRY IT : : 12.59**
>
> New grandparents decide to invest $200 per month in an annuity for their grandson. The account will pay 5% interest per year which is compounded monthly. How much will be in the child's account at his twenty-first birthday?

> **TRY IT : : 12.60**
>
> Arturo just got his first full-time job after graduating from college at age 27. He decided to invest $200 per month in an IRA (an annuity). The interest on the annuity is 8%, which is compounded monthly. How much will be in the Arturo's account when he retires at his sixty-seventh birthday?

MEDIA : :

Access these online resources for additional instruction and practice with sequences.

- **Geometric Sequences (https://openstax.org/l/37GeomSequence)**
- **Geometric Series (https://openstax.org/l/37GeometricSer)**
- **Future Value Annuities and Geometric Series (https://openstax.org/l/37FutValAnnGeoS)**
- **Application of a Geometric Series: Tax Rebate (https://openstax.org/l/37AppGeomSerTax)**

12.3 EXERCISES

Practice Makes Perfect

Determine if a Sequence is Geometric

In the following exercises, determine if the sequence is geometric, and if so, indicate the common ratio.

127. 3, 12, 48, 192, 768, 3072, ...

128. 2, 10, 50, 250, 1250, 6250, ...

129. 48, 24, 12, 6, 3, $\frac{3}{2}$, ...

130. 54, 18, 6, 2, $\frac{2}{3}$, $\frac{2}{9}$, ...

131. −3, 6, −12, 24, −48, 96, ...

132. 2, −6, 18, −54, 162, −486, ...

In the following exercises, determine if each sequence is arithmetic, geometric or neither. If arithmetic, indicate the common difference. If geometric, indicate the common ratio.

133. 48, 24, 12, 6, 3, $\frac{3}{2}$, ...

134. 12, 6, 0, −6, −12, −18, ...

135. −7, −2, 3, 8, 13, 18, ...

136. 5, 9, 13, 17, 21, ...

137. $\frac{1}{2}$, $\frac{1}{4}$, $\frac{1}{8}$, $\frac{1}{16}$, $\frac{1}{32}$, $\frac{1}{64}$, ...

138. 4, 8, 12, 24, 48, 96, ...

In the following exercises, write the first five terms of each geometric sequence with the given first term and common ratio.

139. $a_1 = 4$ and $r = 3$

140. $a_1 = 9$ and $r = 2$

141. $a_1 = -4$ and $r = -2$

142. $a_1 = -5$ and $r = -3$

143. $a_1 = 27$ and $r = \frac{1}{3}$

144. $a_1 = 64$ and $r = \frac{1}{4}$

Find the General Term (*n*th Term) of a Geometric Sequence

In the following exercises, find the indicated term of a sequence where the first term and the common ratio is given.

145. Find a_{11} given $a_1 = 8$ and $r = 3$.

146. Find a_{13} given $a_1 = 7$ and $r = 2$.

147. Find a_{10} given $a_1 = -6$ and $r = -2$.

148. Find a_{15} given $a_1 = -4$ and $r = -3$.

149. Find a_{10} given $a_1 = 100,000$ and $r = 0.1$.

150. Find a_8 given $a_1 = 1,000,000$ and $r = 0.01$.

In the following exercises, find the indicated term of the given sequence. Find the general term for the sequence.

151. Find a_9 of the sequence, 9, 18, 36, 72, 144, 288, ...

152. Find a_{12} of the sequence, 5, 15, 45, 135, 405, 1215, ...

153. Find a_{15} of the sequence, −486, 162, −54, 18, −6, 2, ...

154. Find a_{16} of the sequence, 224, −112, 56, −28, 14, −7, ...

155. Find a_{10} of the sequence, 1, 0.1, 0.01, 0.001, 0.0001, 0.00001, ...

156. Find a_9 of the sequence, 1000, 100, 10, 1, 0.1, 0.01, ...

Find the Sum of the First *n* terms of a Geometric Sequence

In the following exercises, find the sum of the first fifteen terms of each geometric sequence.

157. 8, 24, 72, 216, 648, 1944, ...

158. 7, 14, 28, 56, 112, 224, ...

159. −6, 12, −24, 48, −96, 192, ...

160. −4, 12, −36, 108, −324, 972, ...

161. 81, 27, 9, 3, 1, $\frac{1}{3}$, ...

162. 256, 64, 16, 4, 1, $\frac{1}{4}$, $\frac{1}{16}$, ...

In the following exercises, find the sum of the geometric sequence.

163. $\sum\limits_{i=1}^{15} (2)^i$

164. $\sum\limits_{i=1}^{10} (3)^i$

165. $\sum\limits_{i=1}^{9} 4(2)^i$

166. $\sum\limits_{i=1}^{8} 5(3)^i$

167. $\sum\limits_{i=1}^{10} 9\left(\frac{1}{3}\right)^i$

168. $\sum\limits_{i=1}^{15} 4\left(\frac{1}{2}\right)^i$

Find the Sum of an Infinite Geometric Series

In the following exercises, find the sum of each infinite geometric series.

169. $1 + \frac{1}{3} + \frac{1}{9} + \frac{1}{27} + \frac{1}{81} + \frac{1}{243} + \frac{1}{729} + \ ...$

170. $1 + \frac{1}{2} + \frac{1}{4} + \frac{1}{8} + \frac{1}{16} + \frac{1}{32} + \frac{1}{64} + \ ...$

171. $6 - 2 + \frac{2}{3} - \frac{2}{9} + \frac{2}{27} - \frac{2}{81} + \ ...$

172. $-4 + 2 - 1 + \frac{1}{2} - \frac{1}{4} + \frac{1}{8} - \ ...$

173. $6 + 12 + 24 + 48 + 96 + 192 + \ ...$

174. $5 + 15 + 45 + 135 + 405 + 1215 + \ ...$

175. $1,024 + 512 + 256 + 128 + 64 + 32 + \ ...$

176. $6,561 + 2187 + 729 + 243 + 81 + 27 + \ ...$

In the following exercises, write each repeating decimal as a fraction.

177. $0.\overline{3}$

178. $0.\overline{6}$

179. $0.\overline{7}$

180. $0.\overline{2}$

181. $0.\overline{45}$

182. $0.\overline{27}$

Apply Geometric Sequences and Series in the Real World

In the following exercises, solve the problem.

183. Find the total effect on the economy of each government tax rebate to each household in order to stimulate the economy if each household will spend the indicated percent of the rebate in goods and services.

	Tax rebate to each household	Percent spent on goods and services	Total Effect on the economy
ⓐ	$1,000	85%	
ⓑ	$1,000	85%	
ⓒ	$1,500	90%	
ⓓ	$1,500	80%	

184. New grandparents decide to invest $100 per month in an annuity for their grandchild. The account will pay 6% interest per year which is compounded monthly (12 times a year). How much will be in the child's account at their twenty-first birthday?

185. Berenice just got her first full-time job after graduating from college at age 30. She decided to invest $500 per quarter in an IRA (an annuity). The interest on the annuity is 7% which is compounded quarterly (4 times a year). How much will be in the Berenice's account when she retires at age 65?

186. Alice wants to purchase a home in about five years. She is depositing $500 a month into an annuity that earns 5% per year that is compounded monthly (12 times a year). How much will Alice have for her down payment in five years?

187. Myra just got her first full-time job after graduating from college. She plans to get a master's degree, and so is depositing $2,500 a year from her year-end bonus into an annuity. The annuity pays 6.5% per year and is compounded yearly. How much will she have saved in five years to pursue her master's degree?

Writing Exercises

188. In your own words, explain how to determine whether a sequence is geometric.

189. In your own words, explain how to find the general term of a geometric sequence.

190. In your own words, explain the difference between a geometric sequence and a geometric series.

191. In your own words, explain how to determine if an infinite geometric series has a sum and how to find it.

Self Check

I can...	Confidently	With some help	No-I don't get it!
determine if a sequence is geometric.			
find the general term (*n*th term) of a geometric sequence.			
find the sum of the first *n* terms of a geometric sequence.			
find the sum of an infinite geometric series.			
use geometric sequences to solve applications.			

ⓐ After completing the exercises, use this checklist to evaluate your mastery of the objectives of this section.

ⓑ What does this checklist tell you about your mastery of this section? What steps will you take to improve?

12.4 ## Binomial Theorem

Learning Objectives

By the end of this section, you will be able to:

> ⟩ Use Pascal's Triangle to expand a binomial
> ⟩ Evaluate a binomial coefficient
> ⟩ Use the Binomial Theorem to expand a binomial

Be Prepared!

Before you get started, take this readiness quiz.

1. Simplify: $\dfrac{7 \cdot 6 \cdot 5 \cdot 4}{4 \cdot 3 \cdot 2 \cdot 1}$.

   If you missed this problem, review **Example 1.25**.

2. Expand: $(3x + 5)^2$.

   If you missed this problem, review **Example 5.32**.

3. Expand: $(x - y)^2$.

   If you missed this problem, review **Example 5.32**.

Use Pascal's Triangle to Expand a Binomial

In our previous work, we have squared binomials either by using FOIL or by using the Binomial Squares Pattern. We can also say that we expanded $(a + b)^2$.

$$(a + b)^2 = a^2 + 2ab + b^2$$

To expand $(a + b)^3$, we recognize that this is $(a + b)^2 (a + b)$ and multiply.

$$(a + b)^3$$
$$(a + b)^2 (a + b)$$
$$\left(a^2 + 2ab + b^2\right)(a + b)$$
$$a^3 + 2a^2 b + ab^2 + a^2 b + 2ab^2 + b^3$$
$$a^3 + 3a^2 b + 3ab^2 + b^3$$
$$(a + b)^3 = a^3 + 3a^2 b + 3ab^2 + b^3$$

To find a method that is less tedious that will work for higher expansions like $(a + b)^7$, we again look for patterns in some expansions.

	Number of terms	First term	Last term
$(a + b)^1 = a + b$	2	a^1	b^1
$(a + b)^2 = a^2 + 2ab + b^2$	3	a^2	b^2
$(a + b)^3 = a^3 + 3a^2 b + 3ab^2 + b^3$	4	a^3	b^3
$(a + b)^4 = a^4 + 4a^3 b + 6a^2 b^2 + 4ab^3 + b^4$	5	a^4	b^4
$(a + b)^5 = a^5 + 5a^4 b + 10a^3 b^2 + 10a^2 b^3 + 5ab^4 + b^5$	6	a^5	b^5
$(a + b)^n$	n	a^n	b^n

Notice the first and last terms show only one variable. Recall that $a^0 = 1,$ so we could rewrite the first and last terms to include both variables. For example, we could expand $(a + b)^3$ to show each term with both variables.

$$(a + b)^3 = a^3 b^0 + 3a^2 b + 3ab^2 + a^0 b^3$$

Generally, we don't show the zero exponents, just as we usually write x rather than $1x$.

Patterns in the expansion of $(a + b)^n$

- The number of terms is $n + 1$.
- The first term is a^n and the last term is b^n.
- The exponents on a decrease by one on each term going left to right.
- The exponents on b increase by one on each term going left to right.
- The sum of the exponents on any term is n.

Let's look at an example to highlight the last three patterns.

$$(a + b)^5 = a^5 + 5a^4 b^1 + 10a^3 b^2 + 10a^2 b^3 + 5a^1 b^4 + b^5$$

The exponents of a go from $5 \rightarrow 4 \rightarrow 3 \rightarrow 2 \rightarrow 1$

The exponents of b go from $1 \rightarrow 2 \rightarrow 3 \rightarrow 4 \rightarrow 5$

In each term, the sum of the exponents is 5.

From the patterns we identified, we see the variables in the expansion of $(a + b)^n,$ would be

$$(a + b)^n = a^n + \underline{\quad}a^{n-1}b^1 + \underline{\quad}a^{n-2}b^2 + ... + \underline{\quad}a^1 b^{n-1} + b^n.$$

To find the coefficients of the terms, we write our expansions again focusing on the coefficients. We rewrite the coefficients to the right forming an array of coefficients.

$(a + b)^0 = 1$	1
$(a + b)^1 = 1a + 1b$	1 1
$(a + b)^2 = 1a^2 + 2ab + 1b^2$	1 2 1
$(a + b)^3 = 1a^3 + 3a^2 b + 3ab^2 + 1b^3$	1 3 3 1
$(a + b)^4 = 1a^4 + 4a^3 b + 6a^2 b^2 + 4ab^3 + 1b^4$	1 4 6 4 1
$(a + b)^5 = 1a^5 + 5a^4 b + 10a^3 b^2 + 10a^2 b^3 + 5ab^4 + 1b^5$	1 5 10 10 5 1

The array to the right is called **Pascal's Triangle**. Notice each number in the array is the sum of the two closest numbers in the row above. We can find the next row by starting and ending with one and then adding two adjacent numbers.

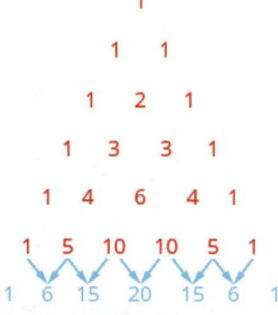

This triangle gives the coefficients of the terms when we expand binomials.

Pascal's Triangle

$$
\begin{array}{c}
1 \\
1 \quad 1 \\
1 \quad 2 \quad 1 \\
1 \quad 3 \quad 3 \quad 1 \\
1 \quad 4 \quad 6 \quad 4 \quad 1 \\
1 \quad 5 \quad 10 \quad 10 \quad 5 \quad 1 \\
1 \quad 6 \quad 15 \quad 20 \quad 15 \quad 6 \quad 1
\end{array}
$$

In the next example, we will use this triangle and the patterns we recognized to expand the binomial.

EXAMPLE 12.31

Use Pascal's Triangle to expand $(x + y)^6$.

⊘ Solution

We know the variables for this expansion will follow the pattern we identified. The nonzero exponents of x will start at six and decrease to one. The nonzero exponents of y will start at one and increase to six. The sum of the exponents in each term will be six. In our pattern, $a = x$ and $b = y$.

$$(a+b)^n = a^n + \underline{\quad}a^{n-1}b^1 + \underline{\quad}a^{n-2}b^2 + ... + \underline{\quad}a^1 b^{n-1} + b^n$$

$$(x+y)^6 = x^6 + \underline{\quad}x^5 y^1 + \underline{\quad}x^4 y^2 + \underline{\quad}x^3 y^3 + \underline{\quad}x^2 y^4 + \underline{\quad}x^1 y^5 + y^6$$

To find the coefficients, we go to Pascal's Triangle and read off the coefficients from the row whose second entry is n, in this case, 6.

$$
\begin{array}{c}
1 \\
1 \quad 1 \\
1 \quad 2 \quad 1 \\
1 \quad 3 \quad 3 \quad 1 \\
1 \quad 4 \quad 6 \quad 4 \quad 1 \\
1 \quad 5 \quad 10 \quad 10 \quad 5 \quad 1 \\
1 \quad 6 \quad 15 \quad 20 \quad 15 \quad 6 \quad 1
\end{array}
$$

$$(x+y)^6 = x^6 + \underline{\quad}x^5 y^1 + \underline{\quad}x^4 y^2 + \underline{\quad}x^3 y^3 + \underline{\quad}x^2 y^4 + \underline{\quad}x^1 y^5 + y^6$$

$$(x+y)^6 = 1x^6 + 6x^5 y^1 + 15x^4 y^2 + 20x^3 y^3 + 15x^2 y^4 + 6x^1 y^5 + 1y^6$$

$$(x+y)^6 = x^6 + 6x^5 y^1 + 15x^4 y^2 + 20x^3 y^3 + 15x^2 y^4 + 6x^1 y^5 + y^6$$

> **TRY IT : : 12.61**   Use Pascal's Triangle to expand $(x + y)^5$.

> **TRY IT : : 12.62**   Use Pascal's Triangle to expand $(p + q)^7$.

In the next example we want to expand a binomial with one variable and one constant. We need to identify the a and b to carefully apply the pattern.

EXAMPLE 12.32

Use Pascal's Triangle to expand $(x + 3)^5$.

⊘ Solution

We identify the a and b of the pattern.

$$(a + b)^n$$
$$(x + 3)^5$$

In our pattern, $a = x$ and $b = 3$.

We know the variables for this expansion will follow the pattern we identified. The sum of the exponents in each term will be five.

$$(a + b)^n = a^n + \underline{\quad}a^{n-1}b^1 + \underline{\quad}a^{n-2}b^2 + \ldots + \underline{\quad}a^1 b^{n-1} + b^n$$
$$(x + 3)^5 = x^5 + \underline{\quad}x^4 \cdot 3^1 + \underline{\quad}x^3 \cdot 3^2 + \underline{\quad}x^2 \cdot 3^3 + \underline{\quad}x^1 \cdot 3^4 + 3^5$$

To find the coefficients, we go to Pascal's Triangle and read off the coefficients from the row whose second entry is n, in this case, 5.

```
              1
            1   1
          1   2   1
        1   3   3   1
      1   4   6   4   1
    1   5   10  10  5   1
  1   6   15  20  15  6   1
```

$$(x + 3)^5 = 1x^5 + 5 \cdot 3x^4 + 10 \cdot 9x^3 + 10 \cdot 27x^2 + 5 \cdot 81x^1 + 1 \cdot 243$$

$$(x + 3)^5 = x^5 + 15x^4 + 90x^3 + 270x^2 + 405x^1 + 243$$

> **TRY IT :: 12.63**   Use Pascal's Triangle to expand $(x + 2)^4$.

> **TRY IT :: 12.64**   Use Pascal's Triangle to expand $(x + 1)^6$.

In the next example, the binomial is a difference and the first term has a constant times the variable. Once we identify the a and b of the pattern, we must once again carefully apply the pattern.

EXAMPLE 12.33

Use Pascal's Triangle to expand $(3x - 2)^4$.

⊘ Solution

We identify the a and b of the pattern.

$$(a + b)^n$$
$$(3x - 2)^4$$

In our pattern, $a = 3x$ and $b = -2$.

To find the coefficients, we go to Pascal's Triangle and read off the coefficients from the row whose second entry is n, in this case, 4.

```
              1
            1   1
          1   2   1
        1   3   3   1
      1   4   6   4   1
    1   5   10  10  5   1
  1   6   15  20  15  6   1
```

$$(a + b)^n = a^n + \underline{\quad}a^{n-1}b^1 + \underline{\quad}a^{n-2}b^2 + \ldots + \underline{\quad}a^1 b^{n-1} + b^n$$

$$(3x - 2)^4 = 1 \cdot (^3 + 4(3x)^3 (-2)^1 + 6(3x)^2 (-2)^2 + 4(3x)^1 (-2)^3 + 1 \cdot (-2)^4$$

$$(3x - 2)^4 = 81x^4 + 4(27x^3)(-2) + 6(9x^2)(4) + 4(3x)(-8) + 1 \cdot 16$$

$$(3x - 2)^4 = 81x^4 - 216x^3 + 216x^2 - 96x + 16$$

> **TRY IT : : 12.65**   Use Pascal's Triangle to expand $(2x - 3)^4$.

> **TRY IT : : 12.66**   Use Pascal's Triangle to expand $(2x - 1)^6$.

Evaluate a Binomial Coefficient

While Pascal's Triangle is one method to expand a binomial, we will also look at another method. Before we get to that, we need to introduce some more factorial notation. This notation is not only used to expand binomials, but also in the study and use of probability.

To find the coefficients of the terms of expanded binomials, we will need to be able to evaluate the notation $\binom{n}{r}$ which is called a binomial coefficient. We read $\binom{n}{r}$ as "n choose r" or "n taken r at a time".

Binomial Coefficient $\binom{n}{r}$

A binomial coefficient $\binom{n}{r}$, where r and n are integers with $0 \le r \le n$, is defined as

$$\binom{n}{r} = \frac{n!}{r!(n-r)!}$$

We read $\binom{n}{r}$ as "n choose r" or "n taken r at a time".

EXAMPLE 12.34

Evaluate: ⓐ $\binom{5}{1}$ ⓑ $\binom{7}{7}$ ⓒ $\binom{4}{0}$ ⓓ $\binom{8}{5}$.

⊘ **Solution**

ⓐ We will use the definition of a binomial coefficient, $\binom{n}{r} = \frac{n!}{r!(n-r)!}$.

$$\binom{5}{1}$$

Use the definition, $\binom{n}{r} = \dfrac{n!}{r!(n-r)!}$, where $n = 5$, $r = 1$.

$$\dfrac{5!}{1!(5-1)!}$$

Simplify.

$$\dfrac{5!}{1!(4)!}$$

Rewrite $5!$ as $5 \cdot 4!$

$$\dfrac{5 \cdot 4!}{1! \cdot 4!}$$

Simplify, by removing common factors.

$$\dfrac{5 \cdot \cancel{4!}}{1 \cdot \cancel{4!}}$$

Simplify.

$$5$$

$$\binom{5}{1} = 5$$

ⓑ

$$\binom{7}{7}$$

Use the definition, $\binom{n}{r} = \dfrac{n!}{r!(n-r)!}$, where $n = 7$, $r = 7$.

$$\dfrac{7!}{7!(7-7)!}$$

Simplify.

$$\dfrac{7!}{7!(0)!}$$

Simplify. Remember $0! = 1$.

$$1$$

$$\binom{7}{7} = 1$$

ⓒ

$$\binom{4}{0}$$

Use the definition, $\binom{n}{r} = \dfrac{n!}{r!(n-r)!}$, where $n = 4$, $r = 0$.

$$\dfrac{4!}{0!(4-0)!}$$

Simplify.

$$\dfrac{4!}{0!(4)!}$$

Simplify.

$$1$$

$$\binom{4}{0} = 1$$

ⓓ

$$\binom{8}{5}$$

Use the definition, $\binom{n}{r} = \dfrac{n!}{r!(n-r)!}$, where

$n = 8$, $r = 5$.

$$\frac{8!}{5!(8-5)!}$$

Simplify.

$$\frac{8!}{5!(3)!}$$

Rewrite $8!$ as $8 \cdot 7 \cdot 6 \cdot 5!$ and remove common factors.

$$\frac{8 \cdot 7 \cdot \cancel{6} \cdot \cancel{5!}}{\cancel{5!} \cdot \cancel{3} \cdot \cancel{2} \cdot 1}$$

Simplify.

$$56$$

$$\binom{8}{5} = 56$$

> **TRY IT :: 12.67**   Evaluate each binomial coefficient:

ⓐ $\binom{6}{1}$ ⓑ $\binom{8}{8}$ ⓒ $\binom{5}{0}$ ⓓ $\binom{7}{3}$.

> **TRY IT :: 12.68**   Evaluate each binomial coefficient:

ⓐ $\binom{2}{1}$ ⓑ $\binom{11}{11}$ ⓒ $\binom{9}{0}$ ⓓ $\binom{6}{5}$.

In the previous example, parts (a), (b), (c) demonstrate some special properties of binomial coefficients.

Properties of Binomial Coefficients

$$\binom{n}{1} = n \qquad \binom{n}{n} = 1 \qquad \binom{n}{0} = 1$$

Use the Binomial Theorem to Expand a Binomial

We are now ready to use the alternate method of expanding binomials. The **Binomial Theorem** uses the same pattern for the variables, but uses the binomial coefficient for the coefficient of each term.

Binomial Theorem

For any real numbers a and b, and positive integer n,

$$(a+b)^n = \binom{n}{0}a^n + \binom{n}{1}a^{n-1}b^1 + \binom{n}{2}a^{n-2}b^2 + \dots + \binom{n}{r}a^{n-r}b^r + \dots + \binom{n}{n}b^n$$

EXAMPLE 12.35

Use the Binomial Theorem to expand $(p+q)^4$.

⊘ **Solution**

We identify the a and b of the pattern.

$$(a+b)^n$$
$$(p+q)^4$$

In our pattern, $a = p$ and $b = q$.

We use the Binomial Theorem.

$$(a+b)^n = \binom{n}{0}a^n + \binom{n}{1}a^{n-1}b^1 + \binom{n}{2}a^{n-2}b^2 + \dots + \binom{n}{r}a^{n-r}b^r + \dots + \binom{n}{n}b^n$$

Substitute in the values $a = p$, $b = q$ and $n = 4$.

$$(p+q)^4 = \binom{4}{0}p^4 + \binom{4}{1}p^{4-1}q^1 + \binom{4}{2}p^{4-2}q^2 + \binom{4}{3}p^{4-3}q^3 + \binom{4}{4}q^4$$

Simplify the exponents.

$$(p+q)^4 = \binom{4}{0}p^4 + \binom{4}{1}p^3 q + \binom{4}{2}p^2 q^2 + \binom{4}{3}p q^3 + \binom{4}{4}q^4$$

Evaluate the coefficients. Remember, $\binom{n}{1} = n$, $\binom{n}{n} = 1$, $\binom{n}{0} = 1$.

$$(p+q)^4 = 1p^4 + 4p^3 q^1 + \frac{4!}{2!(2)!}p^2 q^2 + \frac{4!}{3!(4-3)!}p^1 q^3 + 1q^4$$

$$(p+q)^4 = p^4 + 4p^3 q + 6p^2 q^2 + 4p q^3 + q^4$$

> **TRY IT :: 12.69**   Use the Binomial Theorem to expand $(x+y)^5$.

> **TRY IT :: 12.70**   Use the Binomial Theorem to expand $(m+n)^6$.

Notice that when we expanded $(p+q)^4$ in the last example, using the Binomial Theorem, we got the same coefficients we would get from using Pascal's Triangle.

$$(p+q)^4 = \binom{4}{0}p^4 + \binom{4}{1}p^3 q + \binom{4}{2}p^2 q^2 + \binom{4}{3}p q^3 + \binom{4}{4}q^4$$

$$(p+q)^4 = p^4 + 4p^3 q + 6p^2 q^2 + 4p q^3 + q^4$$

```
                1
              1   1
            1   2   1
          1   3   3   1
        1   4   6   4   1
      1   5   10  10   5   1
    1   6   15  20  15   6   1
```

The next example, the binomial is a difference. When the binomial is a difference, we must be careful in identifying the values we will use in the pattern.

EXAMPLE 12.36

Use the Binomial Theorem to expand $(x-2)^5$.

⊘ **Solution**

We identify the a and b of the pattern.

$$(a+b)^n$$
$$(x-2)^5$$

In our pattern, $a = x$ and $b = -2$.

We use the Binomial Theorem.

$$(a+b)^n = \binom{n}{0}a^n + \binom{n}{1}a^{n-1}b^1 + \binom{n}{2}a^{n-2}b^2 + \ldots + \binom{n}{r}a^{n-r}b^r + \ldots + \binom{n}{n}b^n$$

Substitute in the values $a = x$, $b = -2$, and $n = 5$.

$$(x-2)^5 = \binom{5}{0}x^5 + \binom{5}{1}x^{5-1}(-2)^1 + \binom{5}{2}x^{5-2}(-2)^2 + \binom{5}{3}x^{5-3}(-2)^3 + \binom{5}{4}x^{5-4}(-2)^4 + \binom{5}{5}(-2)^5$$

Simplify the exponents and evaluate the coefficients. Remember, $\binom{n}{1} = n$, $\binom{n}{n} = 1$, $\binom{n}{0} = 1$.

$$(x-2)^5 = \binom{5}{0}x^5 + \binom{5}{1}x^4(-2) + \binom{5}{2}x^3(-2)^2 + \binom{5}{3}x^2(-2)^3 + \binom{5}{4}x(-2)^4 + \binom{5}{5}(-2)^5$$

$$(x-2)^5 = 1x^5 + 5(-2)x^4 + \frac{5!}{2! \cdot 3!}(-2)^2 x^3 + \frac{5!}{3! \cdot 2!}(-2)^3 x^2 + \frac{5!}{4! \cdot 1!}(-2)^4 x + 1(-2)^5$$

$$(x-2)^5 = x^5 + 5(-2)x^4 + 10 \cdot 4 \cdot x^3 + 10(-8)x^2 + 5 \cdot 16 \cdot x + 1(-32)$$

$$(x-2)^5 = x^5 - 10x^4 + 40x^3 - 80x^2 + 80x - 32$$

> **TRY IT :: 12.71**   Use the Binomial Theorem to expand $(x-3)^5$.

> **TRY IT :: 12.72**   Use the Binomial Theorem to expand $(y-1)^6$.

Things can get messy when both terms have a coefficient and a variable.

EXAMPLE 12.37

Use the Binomial Theorem to expand $(2x - 3y)^4$.

⊘ **Solution**

We identify the a and b of the pattern.

$$(a+b)^n$$
$$(2x-3y)^4$$

In our pattern, $a = 2x$ and $b = -3y$.

We use the Binomial Theorem.

$$(a+b)^n = \binom{n}{0}a^n + \binom{n}{1}a^{n-1}b^1 + \binom{n}{2}a^{n-2}b^2 + \ldots + \binom{n}{r}a^{n-r}b^r + \ldots + \binom{n}{n}b^n$$

Substitute in the values $a = 2x$, $b = -3y$ and $n = 4$.

$$(2x-3y)^4 = \binom{4}{0}(2x)^4 + \binom{4}{1}(2x)^{4-1}(-3y)^1 + \binom{4}{2}(2x)^{4-2}(-3y)^2 + \binom{4}{3}(2x)^{4-3}(-3y)^3 + \binom{4}{4}(-3y)^4$$

Simplify the exponents.

$$(2x-3y)^4 = \binom{4}{0}(2x)^4 + \binom{4}{1}(2x)^3(-3y)^1 + \binom{4}{2}(2x)^2(-3y)^2 + \binom{4}{3}(2x)^1(-3y)^3 + \binom{4}{4}(-3y)^4$$

Evaluate the coefficients. Remember, $\binom{n}{1} = n$, $\binom{n}{n} = 1$, $\binom{n}{0} = 1$.

$$(2x-3y)^4 = 1(2x)^4 + 4(2x)^3(-3y)^1 + \frac{4!}{2!(2)!}(2x)^2(-3y)^2 + \frac{4!}{3!(4-3)!}(2x)^1(-3y)^3 + 1(-3y)^4$$

$$(2x-3y)^4 = 16x^4 + 4 \cdot 8x^3(-3y) + 6(4x^2)(9y^2) + 4(2x)(-27y^3) + 81y^4$$

$$(2x-3y)^4 = 16x^4 - 96x^3 y + 216x^2 y^2 - 216xy^3 + 81y^4$$

> **TRY IT :: 12.73**   Use the Binomial Theorem to expand $(3x - 2y)^5$.

> **TRY IT :: 12.74**   Use the Binomial Theorem to expand $(4x - 3y)^4$.

The real beauty of the Binomial Theorem is that it gives a formula for any particular term of the expansion without having to compute the whole sum. Let's look for a pattern in the Binomial Theorem.

$$(a + b)^n = \binom{n}{0}a^n b^0 + \binom{n}{1}a^{n-1}b^1 + \binom{n}{2}a^{n-2}b^2 + \ldots + \binom{n}{r}a^{n-r}b^r + \ldots + \binom{n}{n}b^n$$

| 1st term | 2nd term | 3rd term | (r+1) st term |

Notice, that in each case the exponent on the b is one less than the number of the term. The $(r + 1)st$ term is the term where the exponent of b is r. So we can use the format of the $(r + 1)st$ term to find the value of a specific term.

Find a Specific Term in a Binomial Expansion

The $(r + 1)^{st}$ term in the expansion of $(a + b)^n$ is

$$\binom{n}{r}a^{n-r}b^r$$

EXAMPLE 12.38

Find the fourth term of $(x + y)^7$.

✓ Solution

In our pattern, $n = 7$, $a = x$ and $b = y$.	$(a + b)^n$ $(x + y)^7$
We are looking for the fourth term. $\quad$ Since $r + 1 \;=\; 4$, $\qquad$ then $r \;=\; 3$.	
Write the formula.	$\binom{n}{r}a^{n-r}b^r$
Substitute in the values, $n = 7$, $r = 3$, $a = x$, and $b = y$.	$\binom{7}{3}a^{7-3}b^3$
Use $\binom{n}{r} = \dfrac{n!}{r!(n-r)!}$.	$\dfrac{7!}{3!4!}a^{7-3}b^3$
Simplify.	$\dfrac{7 \cdot 6 \cdot 5 \cdot \cancel{4!}}{\cancel{4!} \cdot 3 \cdot 2 \cdot 1}a^4 b^3$
Simplify.	$35a^4 b^3$

> **TRY IT : : 12.75** $\qquad$ Find the third term of $(x + y)^6$.

> **TRY IT : : 12.76** $\qquad$ Find the fifth term of $(a + b)^8$.

EXAMPLE 12.39

Find the coefficient of the x^6 term of $(x + 3)^9$.

⊘ Solution

In our pattern, then $n = 9$, $a = x$, and $b = 3$.	$(a + b)^n$ $(x + 3)^9$
We are looking for the coefficient of the x^6 term. Since $a = x$, and $x^{9-r} = x^6$, we know $r = 3$.	
Write the formula.	$\binom{n}{r} a^{n-r} b^r$
Substitute in the values, $n = 9$, $r = 3$, $a = x$, and $b = 3$.	$\binom{9}{3} x^{9-3} \cdot 3^3$
Use $\binom{n}{r} = \dfrac{n!}{r!(n-r)!}$.	$\dfrac{9!}{3!6!} x^{9-3} \cdot 3^3$
Simplify.	$\dfrac{9 \cdot \cancel{7} \cdot \cancel{6!}}{\cancel{3!} \cdot \cancel{2} \cdot 1 \cdot \cancel{6!}} x^6 \cdot 27$
Simplify.	$9 \cdot 27 \cdot x^6$
Simplify.	$243x^6$
	The coefficient of the x^6 term is 243.

> **TRY IT : : 12.77**   Find the coefficient of the x^5 term of $(x + 4)^8$.

> **TRY IT : : 12.78**   Find the coefficient of the x^4 term of $(x + 2)^7$.

▶ **MEDIA : :**

Access these online resources for additional instruction and practice with sequences.

- **Binomial Expansion Using Pascal's Triangle (https://openstax.org/l/37binexpanpastr)**
- **Binomial Coefficients (https://openstax.org/l/37binocoeffic)**

12.4 EXERCISES

Practice Makes Perfect

Use Pascal's Triangle to Expand a Binomial

In the following exercises, expand each binomial using Pascal's Triangle.

192. $(x + y)^4$

193. $(a + b)^8$

194. $(m + n)^{10}$

195. $(p + q)^9$

196. $(x - y)^5$

197. $(a - b)^6$

198. $(x + 4)^4$

199. $(x + 5)^3$

200. $(y + 2)^5$

201. $(y + 1)^7$

202. $(z - 3)^5$

203. $(z - 2)^6$

204. $(4x - 1)^3$

205. $(3x - 1)^5$

206. $(3x - 4)^4$

207. $(3x - 5)^3$

208. $(2x + 3y)^3$

209. $(3x + 5y)^3$

Evaluate a Binomial Coefficient

In the following exercises, evaluate.

210. ⓐ $\binom{8}{1}$ ⓑ $\binom{10}{10}$ ⓒ $\binom{6}{0}$ ⓓ $\binom{9}{3}$

211. ⓐ $\binom{7}{1}$ ⓑ $\binom{4}{4}$ ⓒ $\binom{3}{0}$ ⓓ $\binom{10}{8}$

212. ⓐ $\binom{3}{1}$ ⓑ $\binom{9}{9}$ ⓒ $\binom{7}{0}$ ⓓ $\binom{5}{3}$

213. ⓐ $\binom{4}{1}$ ⓑ $\binom{5}{5}$ ⓒ $\binom{8}{0}$ ⓓ $\binom{11}{9}$

Use the Binomial Theorem to Expand a Binomial

In the following exercises, expand each binomial.

214. $(x + y)^3$

215. $(m + n)^5$

216. $(a + b)^6$

217. $(s + t)^7$

218. $(x - 2)^4$

219. $(y - 3)^4$

220. $(p - 1)^5$

221. $(q - 4)^3$

222. $(3x - y)^5$

223. $(5x - 2y)^4$

224. $(2x + 5y)^4$

225. $(3x + 4y)^5$

In the following exercises, find the indicated term in the expansion of the binomial.

226. Sixth term of $(x + y)^{10}$

227. Fifth term of $(a + b)^9$

228. Fourth term of $(x - y)^8$

229. Seventh term of $(x - y)^{11}$

In the following exercises, find the coefficient of the indicated term in the expansion of the binomial.

230. y^3 term of $(y + 5)^4$

231. x^6 term of $(x + 2)^8$

232. x^5 term of $(x - 4)^6$

233. x^7 term of $(x - 3)^9$

234. $a^4 b^2$ term of $(2a + b)^6$

235. $p^5 q^4$ term of $(3p + q)^9$

Writing Exercises

236. In your own words explain how to find the rows of the Pascal's Triangle. Write the first five rows of Pascal's Triangle.

237. In your own words, explain the pattern of exponents for each variable in the expansion of.

238. In your own words, explain the difference between $(a + b)^n$ and $(a - b)^n$.

239. In your own words, explain how to find a specific term in the expansion of a binomial without expanding the whole thing. Use an example to help explain.

Self Check

ⓐ *After completing the exercises, use this checklist to evaluate your mastery of the objectives of this section.*

I can...	Confidently	With some help	No-I don't get it!
use Pascal's Triangle to expand a binomial.			
evaluate a binomial coefficient.			
use the binomial theorem to expand a binomial.			

ⓑ *On a scale of 1-10, how would you rate your mastery of this section in light of your responses on the checklist? How can you improve this?*

CHAPTER 12 REVIEW

KEY TERMS

annuity An annuity is an investment that is a sequence of equal periodic deposits.

arithmetic sequence An arithmetic sequence is a sequence where the difference between consecutive terms is constant.

common difference The difference between consecutive terms in an arithmetic sequence, $a_n - a_{n-1}$, is d, the common difference, for n greater than or equal to two.

common ratio The ratio between consecutive terms in a geometric sequence, $\frac{a_n}{a_{n-1}}$, is r, the common ratio, where r greater than or equal to two.

finite sequence A sequence with a domain that is limited to a finite number of counting numbers.

general term of a sequence The general term of the sequence is the formula for writing the nth term of the sequence. The nth term of the sequence, a_n, is the term in the nth position where n is a value in the domain.

geometric sequence A geometric sequence is a sequence where the ratio between consecutive terms is always the same

infinite geometric series An infinite geometric series is an infinite sum infinite geometric sequence.

infinite sequence A sequence whose domain is all counting numbers and there is an infinite number of counting numbers.

partial sum When we add a finite number of terms of a sequence, we call the sum a partial sum.

sequence A sequence is a function whose domain is the counting numbers.

KEY CONCEPTS

12.1 Sequences

- **Factorial Notation**
  If n is a positive integer, then $n!$ is

  $$n! = n(n-1)(n-2) \ldots (3)(2)(1).$$

  We define $0!$ as 1, so $0! = 1$

- **Summation Notation**
  The sum of the first n terms of a sequence whose nth term a_n is written in summation notation as:

  $$\sum_{i=1}^{n} a_i = a_1 + a_2 + a_3 + a_4 + a_5 + \ldots + a_n$$

  The i is the index of summation and the 1 tells us where to start and the n tells us where to end.

12.2 Arithmetic Sequences

- **General Term (nth term) of an Arithmetic Sequence**
  The general term of an arithmetic sequence with first term a_1 and the common difference d is

  $$a_n = a_1 + (n-1)d$$

- **Sum of the First n Terms of an Arithmetic Sequence**
  The sum, S_n, of the first n terms of an arithmetic sequence, where a_1 is the first term and a_n is the nth term is

  $$S_n = \frac{n}{2}(a_1 + a_n)$$

12.3 Geometric Sequences and Series

- **General Term (nth term) of a Geometric Sequence:** The general term of a geometric sequence with first term a_1 and the common ratio r is

$$a_n = a_1 r^{n-1}$$

- **Sum of the First n Terms of a Geometric Series:** The sum, S_n, of the n terms of a geometric sequence is

$$S_n = \frac{a_1(1 - r^n)}{1 - r}$$

where a_1 is the first term and r is the common ratio.

- **Infinite Geometric Series:** An infinite geometric series is an infinite sum whose first term is a_1 and common ratio is r and is written

$$a_1 + a_1 r + a_1 r^2 + \ldots + a_1 r^{n-1} + \ldots$$

- **Sum of an Infinite Geometric Series:** For an infinite geometric series whose first term is a_1 and common ratio r,

    If $|r| < 1$, the sum is

$$S = \frac{a_1}{1 - r}$$

    We say the series converges.

    If $|r| \geq 1$, the infinite geometric series does not have a sum. We say the series diverges.

- **Value of an Annuity with Interest Compounded n Times a Year:** For a principal, P, invested at the end of a compounding period, with an interest rate, r, which is compounded n times a year, the new balance, A, after t years, is

$$A_t = \frac{P\left(\left(1 + \frac{r}{n}\right)^{nt} - 1\right)}{\frac{r}{n}}$$

12.4 Binomial Theorem

- **Patterns in the expansion of $(a + b)^n$**
    - The number of terms is $n + 1$.
    - The first term is a^n and the last term is b^n.
    - The exponents on a decrease by one on each term going left to right.
    - The exponents on b increase by one on each term going left to right.
    - The sum of the exponents on any term is n.

- **Pascal's Triangle**

```
              1
            1   1
          1   2   1
        1   3   3   1
      1   4   6   4   1
    1   5  10  10   5   1
  1   6  15  20  15   6   1
```

- **Binomial Coefficient $\binom{n}{r}$:** A binomial coefficient $\binom{n}{r}$, where r and n are integers with $0 \leq r \leq n$, is defined as

$$\binom{n}{r} = \frac{n!}{r!(n-r)!}$$

We read $\binom{n}{r}$ as "n choose r" or "n taken r at a time".

- **Properties of Binomial Coefficients**

$$\binom{n}{1} = n \qquad \binom{n}{n} = 1 \qquad \binom{n}{0} = 1$$

- **Binomial Theorem:** For any real numbers a, b, and positive integer n,

$$(a+b)^n = \binom{n}{0}a^n + \binom{n}{1}a^{n-1}b^1 + \binom{n}{2}a^{n-2}b^2 + \ldots + \binom{n}{r}a^{n-r}b^r + \ldots + \binom{n}{n}b^n$$

REVIEW EXERCISES

12.1 Sequences

Write the First Few Terms of a Sequence

In the following exercises, write the first five terms of the sequence whose general term is given.

240. $a_n = 7n - 5$

241. $a_n = 3^n + 4$

242. $a_n = 2^n + n$

243. $a_n = \dfrac{2n+1}{4^n}$

244. $a_n = \dfrac{(-1)^n}{n^2}$

Find a Formula for the General Term (*n*th Term) of a Sequence

In the following exercises, find a general term for the sequence whose first five terms are shown.

245. 9, 18, 27, 36, 45, …

246. −5, −4, −3, −2, −1, …

247. $\dfrac{1}{e^3}, \dfrac{1}{e^2}, \dfrac{1}{e}, 1, e, \ldots$

248. 1, −8, 27, −64, 125, …

249. $-\dfrac{1}{3}, -\dfrac{1}{2}, -\dfrac{3}{5}, -\dfrac{2}{3}, -\dfrac{5}{7}, \ldots$

Use Factorial Notation

In the following exercises, using factorial notation, write the first five terms of the sequence whose general term is given.

250. $a_n = 4n!$

251. $a_n = \dfrac{n!}{(n+2)!}$

252. $a_n = \dfrac{(n-1)!}{(n+1)^2}$

Find the Partial Sum

In the following exercises, expand the partial sum and find its value.

253. $\displaystyle\sum_{i=1}^{7} (2i - 5)$

254. $\displaystyle\sum_{i=1}^{3} 5^i$

255. $\displaystyle\sum_{k=0}^{4} \dfrac{4}{k!}$

256. $\displaystyle\sum_{k=1}^{4} (k+1)(2k+1)$

Use Summation Notation to write a Sum

In the following exercises, write each sum using summation notation.

257. $-\frac{1}{3}+\frac{1}{9}-\frac{1}{27}+\frac{1}{81}-\frac{1}{243}$

258. $4-8+12-16+20-24$

259. $4+2+\frac{4}{3}+1+\frac{4}{5}$

12.2 Arithmetic Sequences

Determine if a Sequence is Arithmetic

In the following exercises, determine if each sequence is arithmetic, and if so, indicate the common difference.

260. $1, 2, 4, 8, 16, 32, \ldots$

261. $-7, -1, 5, 11, 17, 23, \ldots$

262. $13, 9, 5, 1, -3, -7, \ldots$

In the following exercises, write the first five terms of each arithmetic sequence with the given first term and common difference.

263. $a_1 = 5$ and $d = 3$

264. $a_1 = 8$ and $d = -2$

265. $a_1 = -13$ and $d = 6$

Find the General Term (*n*th Term) of an Arithmetic Sequence

In the following exercises, find the term described using the information provided.

266. Find the twenty-fifth term of a sequence where the first term is five and the common difference is three.

267. Find the thirtieth term of a sequence where the first term is 16 and the common difference is -5.

268. Find the seventeenth term of a sequence where the first term is -21 and the common difference is two.

In the following exercises, find the indicated term and give the formula for the general term.

269. Find the eighteenth term of a sequence where the fifth term is 12 and the common difference is seven.

270. Find the twenty-first term of a sequence where the seventh term is 14 and the common difference is -3.

In the following exercises, find the first term and common difference of the sequence with the given terms. Give the formula for the general term.

271. The fifth term is 17 and the fourteenth term is 53.

272. The third term is -26 and the sixteenth term is $-91.$.

Find the Sum of the First *n* Terms of an Arithmetic Sequence

In the following exercises, find the sum of the first 30 terms of each arithmetic sequence.

273. $7, 4, 1, -2, -5, \ldots$

274. $1, 6, 11, 16, 21, \ldots$

In the following exercises, find the sum of the first fifteen terms of the arithmetic sequence whose general term is given.

275. $a_n = 4n + 7$

276. $a_n = -2n + 19$

In the following exercises, find each sum.

277. $\displaystyle\sum_{i=1}^{50} (4i - 5)$

278. $\displaystyle\sum_{i=1}^{30} (-3i - 7)$

279. $\displaystyle\sum_{i=1}^{35} (i + 10)$

12.3 Geometric Sequences and Series

Determine if a Sequence is Geometric

In the following exercises, determine if the sequence is geometric, and if so, indicate the common ratio.

280. 3, 12, 48, 192, 768, 3072, …

281. 5, 10, 15, 20, 25, 30, …

282. 112, 56, 28, 14, 7, $\frac{7}{2}$, …

283. 9, −18, 36, −72, 144, −288, …

In the following exercises, write the first five terms of each geometric sequence with the given first term and common ratio.

284. $a_1 = -3$ and $r = 5$

285. $a_1 = 128$ and $r = \frac{1}{4}$

286. $a_1 = 5$ and $r = -3$

Find the General Term (*n*th Term) of a Geometric Sequence

In the following exercises, find the indicated term of a sequence where the first term and the common ratio is given.

287. Find a_9 given $a_1 = 6$ and $r = 2$.

288. Find a_{11} given $a_1 = 10,000,000$ and $r = 0.1$.

In the following exercises, find the indicated term of the given sequence. Find the general term of the sequence.

289. Find a_{12} of the sequence, 6, −24, 96, −384, 1536, −6144, …

290. Find a_9 of the sequence, 4374, 1458, 486, 162, 54, 18, …

Find the Sum of the First *n* terms of a Geometric Sequence

In the following exercises, find the sum of the first fifteen terms of each geometric sequence.

291. −4, 8, −16, 32, −64, 128…

292. 3, 12, 48, 192, 768, 3072…

293. 3125, 625, 125, 25, 5, 1…

In the following exercises, find the sum

294. $\displaystyle\sum_{i=1}^{8} 7(3)^i$

295. $\displaystyle\sum_{i=1}^{6} 24\left(\frac{1}{2}\right)^i$

Find the Sum of an Infinite Geometric Series

In the following exercises, find the sum of each infinite geometric series.

296. $1 - \frac{1}{3} + \frac{1}{9} - \frac{1}{27} + \frac{1}{81} - \frac{1}{243} + \frac{1}{729} - \cdots$

297. $49 + 7 + 1 + \frac{1}{7} + \frac{1}{49} + \frac{1}{343} + \cdots$

In the following exercises, write each repeating decimal as a fraction.

298. $0.\overline{8}$

299. $0.\overline{36}$

Apply Geometric Sequences and Series in the Real World

In the following exercises, solve the problem.

300. What is the total effect on the economy of a government tax rebate of 360 to each household in order to stimulate the economy if each household will spend 60% of the rebate in goods and services?

301. Adam just got his first full-time job after graduating from high school at age 17. He decided to invest 300 per month in an IRA (an annuity). The interest on the annuity is 7% which is compounded monthly. How much will be in Adam's account when he retires at his sixty-seventh birthday?

12.4 Binomial Theorem

Use Pascal's Triangle to Expand a Binomial

In the following exercises, expand each binomial using Pascal's Triangle.

302. $(a+b)^7$

303. $(x-y)^4$

304. $(x+6)^3$

305. $(2y-3)^5$

306. $(7x+2y)^3$

Evaluate a Binomial Coefficient

In the following exercises, evaluate.

307.

ⓐ $\binom{11}{1}$

ⓑ $\binom{12}{12}$

ⓒ $\binom{13}{0}$

ⓓ $\binom{8}{3}$

308.

ⓐ $\binom{7}{1}$

ⓑ $\binom{5}{5}$

ⓒ $\binom{9}{0}$

ⓓ $\binom{9}{5}$

309.

ⓐ $\binom{1}{1}$

ⓑ $\binom{15}{15}$

ⓒ $\binom{4}{0}$

ⓓ $\binom{11}{2}$

Use the Binomial Theorem to Expand a Binomial

In the following exercises, expand each binomial, using the Binomial Theorem.

310. $(p+q)^6$

311. $(t-1)^9$

312. $(2x+1)^4$

313. $(4x+3y)^4$

314. $(x-3y)^5$

In the following exercises, find the indicated term in the expansion of the binomial.

315. Seventh term of $(a + b)^9$

316. Third term of $(x - y)^7$

In the following exercises, find the coefficient of the indicated term in the expansion of the binomial.

317. y^4 term of $(y + 3)^6$

318. x^5 term of $(x - 2)^8$

319. $a^3 b^4$ term of $(2a + b)^7$

PRACTICE TEST

In the following exercises, write the first five terms of the sequence whose general term is given.

320. $a_n = \dfrac{5n - 3}{3^n}$

321. $a_n = \dfrac{(n + 2)!}{(n + 3)!}$

322. Find a general term for the sequence, $-\dfrac{2}{3}, -\dfrac{4}{5}, -\dfrac{6}{7}, -\dfrac{8}{9}, -\dfrac{10}{11}, \ldots$

323. Expand the partial sum and find its value.
$$\sum_{i=1}^{4} (-4)^i$$

324. Write the following using summation notation. $-1 + \dfrac{1}{4} - \dfrac{1}{9} + \dfrac{1}{16} - \dfrac{1}{25}$

325. Write the first five terms of the arithmetic sequence with the given first term and common difference. $a_1 = -13$ and $d = 3$

326. Find the twentieth term of an arithmetic sequence where the first term is two and the common difference is -7.

327. Find the twenty-third term of an arithmetic sequence whose seventh term is 11 and common difference is three. Then find a formula for the general term.

328. Find the first term and common difference of an arithmetic sequence whose ninth term is -1 and the sixteenth term is -15. Then find a formula for the general term.

329. Find the sum of the first 25 terms of the arithmetic sequence, $5, 9, 13, 17, 21, \ldots$

330. Find the sum of the first 50 terms of the arithmetic sequence whose general term is $a_n = -3n + 100$.

331. Find the sum. $\displaystyle\sum_{i=1}^{40} (5i - 21)$

In the following exercises, determine if the sequence is arithmetic, geometric, or neither. If arithmetic, then find the common difference. If geometric, then find the common ratio.

332. $14, 3, -8, -19, -30, -41, \ldots$

333. $324, 108, 36, 12, 4, \dfrac{4}{3}, \ldots$

334. Write the first five terms of the geometric sequence with the given first term and common ratio. $a_1 = 6$ and $r = -2$

335. In the geometric sequence whose first term and common ratio are $a_1 = 5$ and $r = 4$, find a_{11}.

336. Find a_{10} of the geometric sequence, $1250, 250, 50, 10, 2, \dfrac{2}{5}, \ldots$ Then find a formula for the general term.

337. Find the sum of the first thirteen terms of the geometric sequence, $2, -6, 18, -54, 162, -486\ldots$

In the following exercises, find the sum.

338. $\displaystyle\sum_{i=1}^{9} 5(2)^i$

339. $1 - \dfrac{1}{5} + \dfrac{1}{25} - \dfrac{1}{125} + \dfrac{1}{625} - \dfrac{1}{3125} + \ldots$

340. Write the repeating decimal as a fraction. $0.\overline{81}$

341. Dave just got his first full-time job after graduating from high school at age 18. He decided to invest \$450 per month in an IRA (an annuity). The interest on the annuity is 6% which is compounded monthly. How much will be in Adam's account when he retires at his sixty-fifth birthday?

342. Expand the binomial using Pascal's Triangle. $(m - 2n)^5$

343. Evaluate each binomial coefficient. ⓐ $\binom{8}{1}$ ⓑ $\binom{16}{16}$ ⓒ $\binom{12}{0}$ ⓓ $\binom{10}{6}$

344. Expand the binomial using the Binomial Theorem. $(4x + 5y)^3$

ANSWER KEY
Chapter 1
Try It

1.1. ⓐ yes ⓑ yes ⓒ no ⓓ yes ⓔ no

1.2. ⓐ no ⓑ yes ⓒ yes ⓓ no ⓔ no

1.3. $2 \cdot 2 \cdot 2 \cdot 2 \cdot 5$

1.4. $2 \cdot 2 \cdot 3 \cdot 5$

1.5. 36

1.6. 72

1.7. 16

1.8. 23

1.9. 86

1.10. 1

1.11. ⓐ 9 ⓑ 64 ⓒ 40

1.12. ⓐ 216 ⓑ 64 ⓒ 185

1.13. $10x^2 + 16x + 17$

1.14. $12y^2 + 9y + 7$

1.15. ⓐ $14x^2 - 13$ ⓑ $12x \div 2$ ⓒ $z + 13$ ⓓ $8x - 18$

1.16. ⓐ $17y^2 + 19$ ⓑ $7y$ ⓒ $x + 11$ ⓓ $11a - 14$

1.17. ⓐ $4(p + q)$ ⓑ $4p + q$

1.18. ⓐ $2x - 8$ ⓑ $2(x - 8)$

1.19. $w - 7$

1.20. $l - 6$

1.21. $4q - 8$

1.22. $7n + 3$

1.23. ⓐ > ⓑ > ⓒ < ⓓ =

1.24. ⓐ > ⓑ = ⓒ > ⓓ <

1.25. 16

1.26. 9

1.27. ⓐ -6 ⓑ 2 ⓒ -2

1.28. ⓐ -7 ⓑ 3 ⓒ -3

1.29. ⓐ 2 ⓑ -2 ⓒ -10 ⓓ 10

1.30. ⓐ 3 ⓑ -3 ⓒ -11 ⓓ 11

1.31. ⓐ 8, 8 ⓑ -18, -18 ⓒ 19, 19 ⓓ -4, -4

1.32. ⓐ 8, 8 ⓑ -22, -22 ⓒ 23, 23 ⓓ 3, 3

1.33. 3

1.34. 13

1.35. ⓐ 23 ⓑ 60 ⓒ -63 ⓓ -9

1.36. ⓐ 39 ⓑ 39 ⓒ -28 ⓓ -7

1.37. ⓐ 81 ⓑ -81

1.38. ⓐ 49 ⓑ -49

1.39. ⓐ 4 ⓑ 21

1.40. ⓐ 9 ⓑ 6

1.41. 31

1.42. 67

1.43. $(9 + (-16)) + 4; \; -3$

1.44. $(-8 + (-12)) + 7; \; -13$

1.45. The difference in temperatures was 45 degrees Fahrenheit.

1.46. The difference in temperatures was 9 degrees.

1.47. $-\frac{23}{40}$

1.48. $-\frac{5}{8}$

1.49. $-33a$

1.50. $-26b$

1.51. $\frac{4}{15}$

1.52. $\frac{2}{3}$

1.53. $\frac{3}{4b}$

1.54. $\frac{4}{q}$

1.55. $\frac{79}{60}$

1.56. $\frac{103}{60}$

1.57. ⓐ $\frac{27a - 32}{36}$ ⓑ $\frac{2a}{3}$

1.58. ⓐ $\frac{24k - 5}{30}$ ⓑ $\frac{2k}{15}$

1.59. 4

1.60. 2

1.61. $\frac{1}{90}$

1.62. 272

1.63. 2

1.64. $\frac{2}{7}$

1.65. $-\frac{1}{2}$

1.66. $\frac{2}{3}$

1.67. ⓐ 6.58 ⓑ 6.6 ⓒ 7

1.68. ⓐ 15.218 ⓑ 15.22 ⓒ 15.2

1.69. ⓐ -16.49 ⓑ -0.42

1.70. ⓐ -23.593 ⓑ -12.58

1.71. -27.4815

1.72. -87.6148

1.73. ⓐ 25.8 ⓑ 258 ⓒ 2,580

1.74. ⓐ 142 ⓑ 1,420 ⓒ 14,200

1.75. 587.3

1.76. 34.25

1.77. ⓐ $\frac{117}{500}$ ⓑ -0.875

1.78. ⓐ $\frac{3}{125}$ ⓑ -0.375

1.79. ⓐ 0.09, 0.87, 0.039 ⓑ 17%, 175%, 8.25%

1.80. ⓐ 0.03, 0.91, 0.083 ⓑ 41%, 225%, 9.25%

1.81. ⓐ 6 ⓑ 13 ⓒ −15

1.82. ⓐ 4 ⓑ 14 ⓒ −10

1.83. ⓐ 4, $\sqrt{49}$ ⓑ −3, 4, $\sqrt{49}$ ⓒ −3, $0.\overline{3}$, $\frac{9}{5}$, 4, $\sqrt{49}$ ⓓ $-\sqrt{2}$ ⓔ −3, $-\sqrt{2}$, $0.\overline{3}$, $\frac{9}{5}$, 4, $\sqrt{49}$

1.84. ⓐ 6, $\sqrt{121}$

ⓑ $-\sqrt{25}$, −1, 6, $\sqrt{121}$

ⓒ $-\sqrt{25}$, $-\frac{3}{8}$, −1, 6, $\sqrt{121}$

ⓓ 2.041975...

ⓔ

$-\sqrt{25}$, $-\frac{3}{8}$, −1, 6, $\sqrt{121}$, 2.041975...

1.85.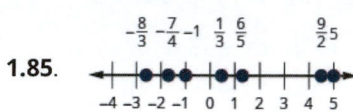

1.86.
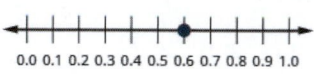

1.87. ⓐ

ⓑ
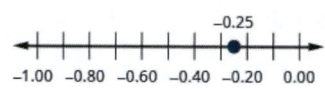

1.88. ⓐ
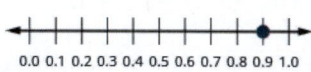

ⓑ
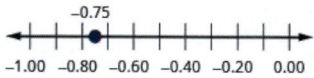

1.89. $32r + 29s$

1.90. $41m + 6n$

1.91. $1\frac{7}{15}$

1.92. $1\frac{2}{9}$

1.93. $-48a$

1.94. $-92x$

1.95. $\frac{5}{49}$

1.96. $\frac{11}{25}$

1.97. ⓐ 0 ⓑ undefined

1.98. ⓐ 0 ⓑ undefined

1.99. $4x + 8$

1.100. $6x + 42$

1.101. $5y + 3$

1.102. $4n + 9$

1.103. $70 + 15p$

1.104. $4 + 35d$

1.105. $-10 + 15a$

1.106. $-56 + 105y$

1.107. $-z + 11$

1.108. $-x + 4$

1.109. $3 - 3x$

1.110. $2x - 20$

1.111. $5x - 66$

1.112. $7x - 13$

Section Exercises

1. Divisible by 2, 3, 6
7. $2 \cdot 43$
13. 24
19. 5
25. 149
31. ⓐ 64 ⓑ 16 ⓒ 7
37. $10x + 6$

3. Divisible by 2
9. $5 \cdot 7 \cdot 13$
15. 420
21. 58
27. 50
33. 21
39. $22a + 1$

5. Divisible by 3, 5
11. $2 \cdot 2 \cdot 2 \cdot 2 \cdot 3 \cdot 3 \cdot 3$
17. 440
23. 29
29. 5
35. 9
41. $17x^2 + 20x + 16$

43. ⓐ $5x^2 - 6xy$ ⓑ $\dfrac{6y^2}{5x}$
ⓒ $y^2 + 21$ ⓓ $81x^2 - 6x$

45. ⓐ $4ab^2 + 3a^2 b$ ⓑ $20xy^2$
ⓒ $m + 15$ ⓓ $121x^2 - 9x$
$9x < 121x^2$

47. ⓐ $8(y - 9)$ ⓑ $8y - 9$

49. ⓐ $5(3x + y)$ ⓑ $15x + y$

51. $14 > 2c$

53. $3n - 7$

55. Answers will vary.

57. Answers will vary.

59. ⓐ $>$ ⓑ $>$ ⓒ $>$ ⓓ $>$

61. ⓐ $=$ ⓑ $=$ ⓒ $>$ ⓓ $=$

63. 0

65. 8

67. 15

69. 1

71. ⓐ -11 ⓑ -3 ⓒ 3

73. 32

75. -22

77. 29

79. ⓐ 6 ⓑ -6 ⓒ -20 ⓓ 20

81. -59

83. 22

85. ⓐ 16 ⓑ16

87. ⓐ 45 ⓑ45

89. 0

91. 4

93. 6

95. ⓐ -32 ⓑ -65 ⓒ -4
ⓓ 13

97. ⓐ -4 ⓑ -12 ⓒ -39
ⓓ 14

99. ⓐ 64 ⓑ -64

101. -47

103. 90

105. 9

107. 41

109. -9

111. -29

113. 1

115. -3

117. 5

119. ⓐ -47 ⓑ 16

121. 121

123. -56

125. 6

127. $(3 + (-15)) + 7; \; -5$

129. ⓐ $10 - (-18); 28$
ⓑ $-25 - 11; \; -36$

131. $\dfrac{-6}{a + b}$

133. $96°$

135. 21

137. $-\$28$

139. Answers will vary.

141. Answers will vary.

143. $-\dfrac{12}{7}$

145. $\dfrac{10}{21}$

147. $\dfrac{2x^2}{3y}$

149. $\dfrac{-21a^2}{11b^2}$

151. $\dfrac{1}{3}$

153. $-\dfrac{21}{50}$

155. $\dfrac{11}{30}$

157. $9n$

159. $\dfrac{33}{4x}$

161. $-\dfrac{4}{9}$

163. $\dfrac{10u}{9v}$

165. $-\dfrac{1}{16}$

167. $-\dfrac{10}{9}$

169. $-\dfrac{2}{5}$

171. $\dfrac{2m}{3n}$

173. $\dfrac{29}{24}$

175. $\dfrac{1}{48}$

177. $\dfrac{17}{105}$

179. $-\dfrac{53}{40}$

181. $\dfrac{1}{12}$

183. $\dfrac{4x + 3}{12}$

185. ⓐ $\dfrac{5}{6}$ ⓑ 4

187. ⓐ $\dfrac{25n}{16}$ ⓑ $\dfrac{25n - 16}{30}$

189. ⓐ $\dfrac{-8x - 15}{18}$ ⓑ $-\dfrac{10k}{27}$

191. ⓐ $\dfrac{-5(a + 1)}{3}$ ⓑ a

193. $\dfrac{9}{7}$

195. -8

197. $\dfrac{11}{6}$

199. $\dfrac{5}{2}$

201. 54

203. $\dfrac{49}{25}$

205. $\dfrac{15}{4}$

207. $\dfrac{5}{21}$

209. $\dfrac{5}{4}$

211. $\dfrac{1}{24}$

213. $\dfrac{-28 - 15y}{60}$

215. $\dfrac{33}{64}$

217. $\dfrac{7}{9}$

219. -5

221. $\dfrac{23}{24}$

223. $\dfrac{11}{5}$

225. 1

227. $\dfrac{13}{3}$

229. ⓐ $\dfrac{1}{5}$ ⓑ $\dfrac{6}{5}$

231. $-\dfrac{1}{9}$

233. $-\dfrac{5}{11}$

235. Answers will vary.

237. Answers will vary.

239. ⓐ 5.78 ⓑ 5.8 ⓒ 6

241. ⓐ 0.30 ⓑ 0.3 ⓒ 0

243. ⓐ 63.48 ⓑ 63.5 ⓒ 63

245. −40.91

247. −7.22

249. −27.5

251. 02.212

253. 51.31

255. −4.89

257. −11.653

259. 337.8914

261. 1.305

263. 92.4

265. 2.5

267. 55200

269. $2.44

271. −4.8

273. 2.08

275. $\dfrac{1}{25}$

277. $\dfrac{19}{200}$

279. 0.85

281. −12.4

283. 0.71

285. 0.393

287. 156%

289. 6.25%

291. 8

293. 12

295. −10

297. ⓐ 0, $\sqrt{36}$, 9 ⓑ −8, 0, $\sqrt{36}$, 9

ⓒ −8, 0, $\sqrt{36}$, 9 ⓓ 1.95286...,

ⓔ −8, 0, 1.95286..., $\dfrac{12}{5}$, $\sqrt{36}$, 9

299. ⓐ none ⓑ $-\sqrt{100}$, −7, −1

ⓒ $-\sqrt{100}$, −7, $-\dfrac{8}{3}$, −1, 0.77, $3\dfrac{1}{4}$

ⓓ none

ⓔ

$-\sqrt{100}$, −7, $-\dfrac{8}{3}$, −1, 0.77, $3\dfrac{1}{4}$

301.

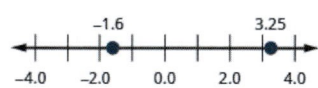

303.

305.

307.

309. Answers will vary.

311. Answers will vary.

313. $27m + (-21n)$

315. $\dfrac{5}{4}g + \dfrac{1}{2}h$

317. $2.43p + 8.26q$

319. −63

321. $1\dfrac{5}{6}$

323. 17

325. 14.88

327. $10p$

329. 44

331. $\dfrac{7}{8}$

333. d

335. $\dfrac{49}{11}$

337. 0

339. undefined

341. undefined

343. $32y + 72$

345. $6c - 78$

347. $\dfrac{3}{4}q + 3$

349. $5y - 3$

351. $3 + 8r$

353. $36d + 90$

355. $rs - 18r$

357. $yp + 4p$

359. $-28p - 7$

361. $-3x + 18$

363. $-3x + 7$

365. $-3y - 8$

367. $-33c + 26$

369. $-a + 19$

371. $4m - 10$

373. $72x - 25$

375. $22n + 9$

377. $6c + 34$

379. $12y + 63$

381. Answers will vary.

383. Answers will vary.

Review Exercises

385. Divisible by 2, 3, 5, 6

387. 120

389. 4

391. 3

393. $6x^2 - x + 5$

395. ⓐ $11(y-2)$ ⓑ $11y - 2$

397. ⓐ $=$ ⓑ $=$ ⓒ $>$ ⓓ $=$

399. -9

401. ⓐ 8 ⓑ -8 ⓒ -22 ⓓ 22

403. ⓐ 40 ⓑ 40

405. ⓐ -3 ⓑ -15 ⓒ -56 ⓓ 17

407. 16

409. -12

411. 9

413. $(-4 + (-9)) + 23;\ 10$

415. $46°$

417. $-\dfrac{15x^3}{11y^2}$

419. $\dfrac{8x}{15y}$

421. $\dfrac{31}{36}$

423. ⓐ $\dfrac{11}{8}$ ⓑ $\dfrac{5}{6}$

425. $-\dfrac{1}{6}$

427. 75

429. $-\dfrac{1}{5}$

431. 6.17

433. 96.978

435. 488.1813

437. 56.3

439. -23

441. -9.6

443. $1.\overline{27}$

445. 4.75%

447. no real number

449.

451. $\dfrac{3}{4}x + y$

453. $1\dfrac{11}{15}$

455. $\dfrac{9}{17}$

457. undefined

459. $8b + 10$

461. $xp - 5p$

463. $-6x - 6$

465. $y + 16$

Practice Test

467. $7n + 7$

469. $-8 - 11;\ -19$
$(-8 - (-3)) + 5;\ 0$

471. ⓐ 28.15 ⓑ 28.146

473.

475. 1

477. -8

479. $\dfrac{15}{17}$

481. $-\dfrac{5}{3}$

483. 3

485. $-\dfrac{7}{6}$

487. -65.4

489. 160

491. $1\dfrac{8}{13}$

493. ⓐ 0 ⓑ undefined

495. $13y - 3$

Chapter 2

Try It

2.1. ⓐ no ⓑ yes

2.2. ⓐ yes ⓑ no

2.3. $m = 2$

2.4. $a = 0$

2.5. $u = 2$

2.6. $x = 4$

2.7. $p = -2$

2.8. $q = -8$

2.9. $y = -\dfrac{17}{5}$

2.10. $z = 0$

2.11. identity; all real numbers

2.12. identity; all real numbers

2.13. conditional equation;
$q = -\dfrac{9}{11}$

2.14. conditional equation;
$k = \dfrac{201}{14}$

2.15. contradiction; no solution

2.16. contradiction; no solution

2.17. $x = \dfrac{1}{2}$

2.18. $x = -2$

2.19. $x = 12$

2.20. $u = -12$

2.21. $n = 2$

2.22. $m = -1$ **2.23.** $r = 1$ **2.24.** $s = -8$

2.25. $n = 9$ **2.26.** $d = 16$ **2.27.** He bought two notebooks.

2.28. He did seven crosswords puzzles.   **2.29.** 3   **2.30.** 6

2.31. $-15, -8$ **2.32.** $-29, 11$ **2.33.** $-33, -32, -31$

2.34. $-13, -12, -11$ **2.35.** 32, 34, 36   **2.36.** $-10, -8, -6$

2.37. The average cost was $5,000.   **2.38.** The median price was $19,300.   **2.39.** ⓐ 36 ⓑ $26 ⓒ 125%

2.40. ⓐ 33 ⓑ $36 ⓐ 175%   **2.41.** 25 grams   **2.42.** 2,375 mg

2.43. 50%   **2.44.** 40%   **2.45.** 8.8%

2.46. 50%   **2.47.** ⓐ $600 ⓑ $1,800   **2.48.** ⓐ $2,975 ⓑ $11,475

2.49. He will earn $2,500.   **2.50.** She earned $7,020.   **2.51.** The rate of simple interest was 6%.

2.52. The rate of simple interest was 5.5%.   **2.53.** He paid $17,590.   **2.54.** She deposited $9,600.

2.55. $b = \dfrac{2A}{h}$ **2.56.** $h = \dfrac{2A}{b}$ **2.57.** $C = \dfrac{5}{9}(F - 32)$

2.58. $b = \dfrac{2A - Bh}{h}$ **2.59.** $t = \dfrac{A - P}{Pr}$ **2.60.** $r = \dfrac{A - P}{Pt}$

2.61. $y = \dfrac{9 - 4x}{7}$ **2.62.** $y = \dfrac{1 - 5x}{8}$ **2.63.** The window's height is 12 meters.

2.64. The length of the base is 6 feet.   **2.65.** The measures of the angles are 20°, 70°, and 90°.   **2.66.** The measures of the angles are 30°, 60°, and 90°.

2.67. The length of the leg is 8.   **2.68.** The length of the leg is 12.   **2.69.** The length is 16 inches and the width is 39 inches.

2.70. The length is 17 yards and the width is 26 yards.   **2.71.** The lengths of the sides of the triangle are 5, 11 and 12 inches.   **2.72.** The lengths of the sides of the triangle are 4, 7 and 9 feet.

2.73. The length of the swimming pool is 70 feet and the width is 30 feet.   **2.74.** The length of the garden is 90 yards and the width is 60 yards.   **2.75.** The ladder reaches 12 feet.

2.76. He should attach the lights 8 feet from the base of the mast.   **2.77.** Jess has 41 nickels and 18 quarters.   **2.78.** Elane has 22 nickels and 59 dimes.

2.79. Eric bought thirty-two 49-cent stamps and twelve 35-cent stamps.   **2.80.** Kailee bought twenty-six 49-cent stamps and ten 20-cent stamps.   **2.81.** 84 adult tickets, 31 student tickets

2.82. 615 children's tickets and 195 adult tickets   **2.83.** Orlando mixed five pounds of cereal squares and 25 pounds of nuts.   **2.84.** Becca mixed 21 gallons of fruit punch and seven gallons of soda.

2.85. The speed of the local train is 48 mph and the speed of the express train is 60 mph.   **2.86.** Jeromy drove at a speed of 80 mph and his mother drove 60 mph.   **2.87.** Christopher's speed was 50 mph and his parents' speed was 40 mph.

2.88. Ashley's parents drove 55 mph and Ashley drove 62 mph.   **2.89.** Pierre and Monique will be 429 miles apart in 3 hours.   **2.90.** Thanh and Nhat will be 330 miles apart in 2.2 hours.

2.91. Suzy's speed uphill is 1.8 mph and downhill is three mph.   **2.92.** The boat's speed upstream is eight mph and downstream is 12 mph.   **2.93.** Hamilton drove 40 mph in the city and 70 mph in the desert.

2.94. Phuong rode uphill at a speed of 12 mph and on the flat street at 20 mph.

2.95. ⓐ

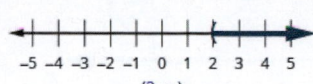

(2, ∞)

ⓑ

−1.5

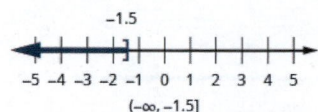

(−∞, −1.5]

ⓒ

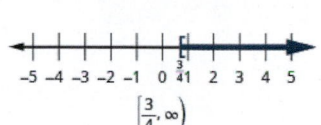

$\left[\frac{3}{4}, \infty\right)$

2.96. ⓐ

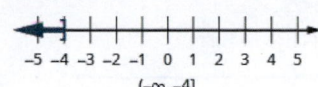

(−∞, −4]

ⓑ

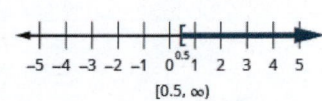

[0.5, ∞)

ⓒ

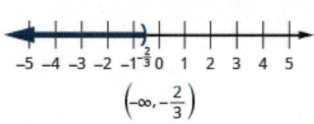

$\left(-\infty, -\frac{2}{3}\right)$

2.97. ⓐ

−2 < x < 1

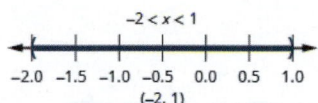

(−2, 1)

ⓑ

−5 ≤ x < −4

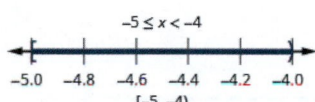

[−5, −4)

ⓒ

1 ≤ x ≤ 4.25

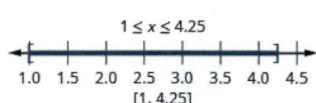

[1, 4.25]

2.98. ⓐ

−6 < x < 2

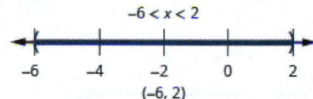

(−6, 2)

ⓑ

−3 ≤ x < −1

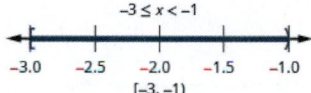

[−3, −1)

ⓒ

2.5 ≤ x ≤ 6

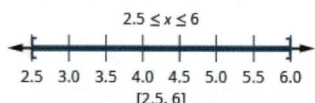

[2.5, 6]

2.99. ⓐ

$p \geq \frac{11}{12}$

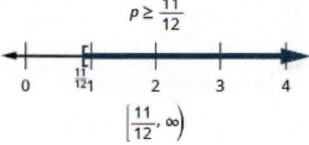

$\left[\frac{11}{12}, \infty\right)$

ⓑ

c > 8

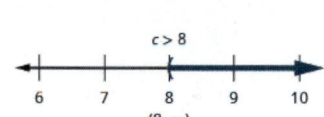

(8, ∞)

ⓒ

m ≥ 64

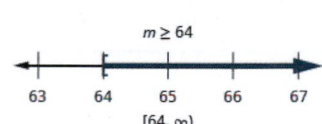

[64, ∞)

2.100. ⓐ

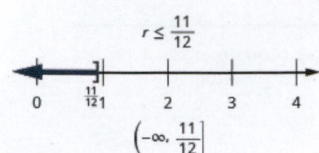

$$\left(-\infty, \frac{11}{12}\right]$$

ⓑ

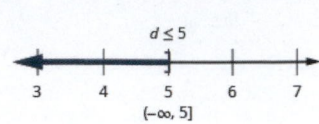

$(-\infty, 5]$

ⓒ

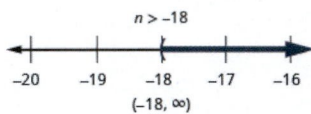

$(-18, \infty)$

2.101. ⓐ

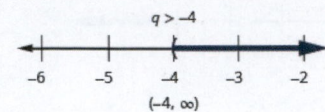

$(-4, \infty)$

ⓑ

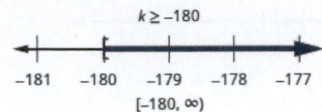

$[-180, \infty)$

2.102. ⓐ

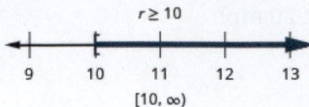

$[10, \infty)$

ⓑ

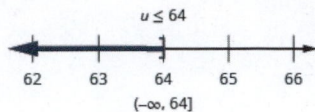

$(-\infty, 64]$

2.103.

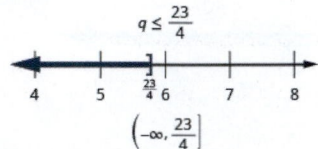

$$\left(-\infty, \frac{23}{4}\right]$$

2.104.

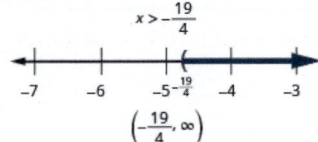

$$\left(-\frac{19}{4}, \infty\right)$$

2.105.

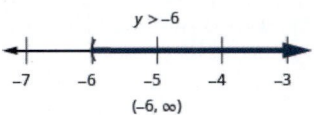

$(-6, \infty)$

2.106.

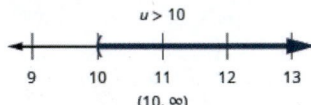

$(10, \infty)$

2.107.

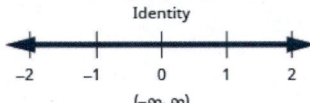

$(-\infty, \infty)$

2.108.

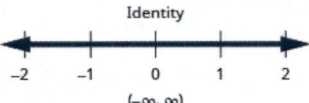

$(-\infty, \infty)$

2.109.

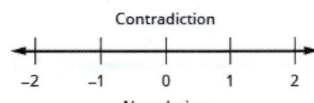

No solution

2.110.

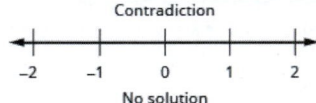

No solution

2.111.

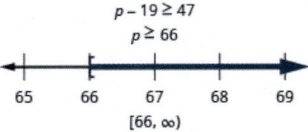

$[66, \infty)$

2.112.

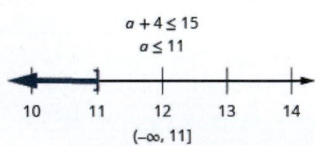

$(-\infty, 11]$

2.113. Angie can buy 7 packs of juice.

2.114. Daniel can have 11 people at the party.

2.115. Sergio and Lizeth can travel no more than 500 miles.
2.118. Elliot must work at least 85 jobs.

2.116. Rameen can use no more than 76 therms.
2.119. Brenda must babysit at least 27 hours.

2.117. Caleb must work at least 96 hours.
2.120. Josue must shovel at least 20 driveways.

2.121.

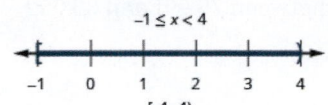

$-1 \le x < 4$

$[-1, 4)$

2.122.

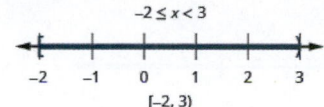

$-2 \le x < 3$

$[-2, 3)$

2.123.

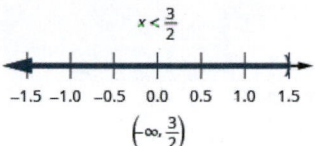

$x < \frac{3}{2}$

$\left(-\infty, \frac{3}{2}\right)$

2.124.

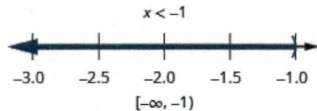

$x < -1$

$[-\infty, -1)$

2.125.

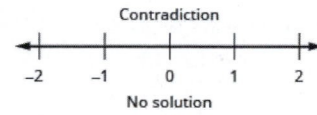

Contradiction

No solution

2.126.

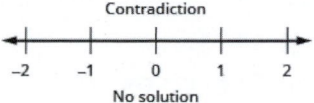

Contradiction

No solution

2.127.

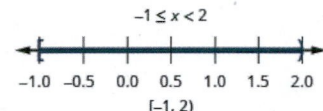

$-1 \le x < 2$

$[-1, 2)$

2.128.

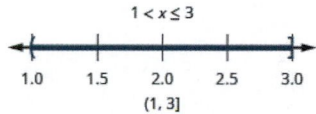

$1 < x \le 3$

$(1, 3]$

2.129.

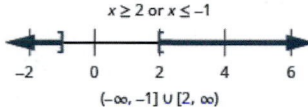

$x \ge 2 \text{ or } x \le -1$

$(-\infty, -1] \cup [2, \infty)$

2.130.

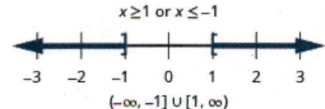

$x \ge 1 \text{ or } x \le -1$

$(-\infty, -1] \cup [1, \infty)$

2.131.

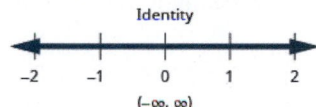

Identity

$(-\infty, \infty)$

2.132.

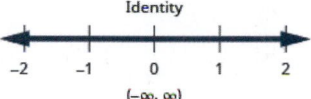

Identity

$(-\infty, \infty)$

2.133. The homeowner can use 5−20 hcf and still fall within the "conservation usage" billing range.

2.134. The homeowner can use 16−40 hcf and still fall within the "normal usage" billing range.

2.135. ⓐ ± 2 ⓑ no solution ⓒ 0

2.136. ⓐ ± 11 ⓑ no solution ⓒ 0

2.137. $x = 4, x = -\frac{2}{3}$

2.138. $x = -1, x = \frac{5}{2}$

2.139. $x = 8, x = 0$

2.140. $x = 8, x = 2$

2.141. No solution

2.142. No solution

2.143. $x = -\frac{2}{5}, \quad x = \frac{5}{2}$

2.144. $x = 3, \quad x = \frac{1}{9}$

2.145.

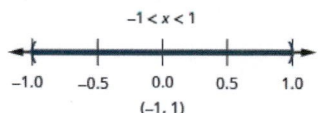

$-9 < x < 9$

$(-9, 9)$

2.146.

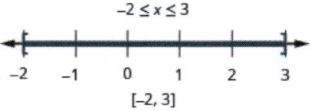

$-1 < x < 1$

$(-1, 1)$

2.147.

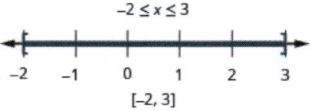

$-2 \le x \le 3$

$[-2, 3]$

2.148.

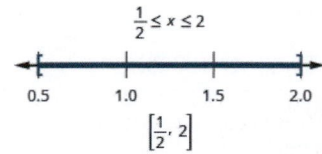

$\frac{1}{2} \le x \le 2$

$\left[\frac{1}{2}, 2\right]$

2.149.

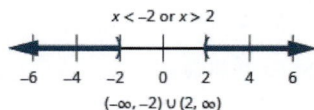

$x < -2 \text{ or } x > 2$

$(-\infty, -2) \cup (2, \infty)$

2.150.

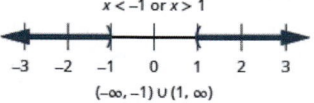

$x < -1 \text{ or } x > 1$

$(-\infty, -1) \cup (1, \infty)$

2.151.

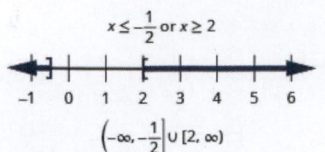

$$\left(-\infty, -\frac{1}{2}\right] \cup [2, \infty)$$

2.152.

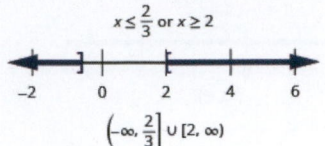

$$\left(-\infty, \frac{2}{3}\right] \cup [2, \infty)$$

2.153. The diameter of the rod can be between 79.991 and 80.009 mm.

2.154. The diameter of the rod can be between 74.95 and 75.05 mm.

Section Exercises

1. ⓐ yes ⓑ no

3. ⓐ no ⓑ yes

5. $y = 5$

7. $w = -18$

9. $q = 3$

11. $x = 14$

13. $c = 4$

15. $c = \frac{5}{2}$

17. $n = 2$

19. $y = -5$

21. $s = 10$

23. $p = -4$

25. $c = -4$

27. $h = \frac{3}{4}$

29. $m = 6$

31. identity; all real numbers

33. conditional equation; $j = \frac{2}{5}$

35. conditional equation; $m = \frac{16}{5}$

37. contradiction; no solution

39. contradiction; no solution

41. identity; all real numbers

43. $x = -1$

45. $y = -1$

47. $a = \frac{3}{4}$

49. $x = 4$

51. $w = \frac{9}{4}$

53. $b = 12$

55. $p = -41$

57. $x = -\frac{5}{2}$

59. $n = -3$

61. $x = -2$

63. $u = 3$

65. $x = 18$

67. $x = 20$

69. $n = 9$

71. $d = 8$

73. $L = 19.75$ feet

75. Answers will vary.

77. Answers will vary.

79. Answers will vary.

81. Answers will vary.

83. six boys

85. 58 hardback books

87. 15

89. 5

91. 12

93. -5

95. 18, 24

97. 8, 12

99. 32, 46

101. 4, 10

103. 38, 39

105. 25, 26, 27

107. $-11, -12, -13$

109. 84, 86, 88

111. $-69, -71, -73$

113. $750

115. $13,500

117. ⓐ 54 ⓑ 108 ⓐ 30%

119. ⓐ 162.5 ⓑ $35 ⓐ 150%

121. $11.88

123. 24.2 g

125. 2407 mg

127. 45%

129. 25%

131. 12%

133. 200%

135. -2.5%

137. -11%

139. ⓐ $26.97 ⓑ $17.98

141. ⓐ $576 ⓑ 30%

143. ⓐ $7.20 ⓑ $23.20

145. ⓐ $0.20 ⓑ $0.80

147. $116

149. $4836

151. 3%

153. 3.75%

155. $35,000

157. $3345

159. 17.5%

161. Answers will vary.

163. Answers will vary.

165. $d = \frac{C}{\pi}$

167. $L = \frac{V}{WH}$

169. $b = \frac{2A}{h}$

171. $d_1 = \frac{2A}{d_2}$

173. $b_1 = \frac{2A}{h} - b_2$

175. $a = \frac{2h - 108t}{t^2}$

177. $a = 180 - b - c$

179. $p = \frac{2A - 2B}{l}$

181. $L = \dfrac{P - 2W}{2}$

183. $y = 15 - 8x$

185. $y = -6 + 4x$

187. $y = 4 + x$

189. $y = \dfrac{7 - 4x}{3}$

191. $y = \dfrac{12 - 2x}{3}$

193. $y = \dfrac{18 - 3x}{-2}$

195. 1 foot

197. 23 inches

199. $45°, 45°, 90°$

201. $30°, 60°, 90°$

203. 15

205. 25

207. 8

209. 12

211. 10.2

213. 9.8

215. 18 meters, 11 meters

217. 13.5 m, 12.8 m

219. 25 ft, 50 ft

221. 7 m, 11 m

223. 12 ft, 13 ft, 14 ft

225. 3 ft, 6 ft, 8 ft

227. 120 yd, 160 yd

229. 40 ft, 85 ft

231. 5 feet

233. 14.1 feet

235. $104°$ F

237. 24 ft

239. Answers will vary.

241. ⓐ Answers will vary. ⓑ The areas are the same. The 2×8 rectangle has a larger perimeter than the 4×4 square.

ⓒ Answers will vary.

243. nine nickels, 16 dimes

245. ten $10 bills, five $5 bills

247. 63 dimes, 20 quarters

249. 16 nickels, 12 dimes, seven quarters

251. 330 day passes, 367 tournament passes

253. 40 postcards, 100 stamps

255. 15 $10 shares, five $12 shares

257. 31 general, 61 youth

259. 114 general, 246 student

261. Four pounds of macadamia nuts, eight pounds almonds

263. 3.6 lbs Bermuda seed, 5.4 lbs Fescue seed

265. $33,000 in Fund A, $22,000 in Fund B

267. 5.9%

269. Kathy 5 mph, Cheryl 3 mph

271. commercial 540 mph, private plane 330 mph

273. Violet 65 mph, Charlie 55 mph

275. Ethan 22 mph, Leo 16 mph

277. DaMarcus 16 mph, Fabian 22 mph

279. four hours

281. 4.5 hours

283. uphill 1.6 mph, downhill 4.8 mph

285. light traffic 54 mph, heavy traffic 30 mph

287. freeway 72 mph, mountain road 24 mph

289. running eight mph, walking three mph

291. ⓐ 15 minutes ⓑ 20 minutes

ⓒ one hour (d) 1:25

293. Answers will vary.

295. Answers will vary.

297.

ⓐ

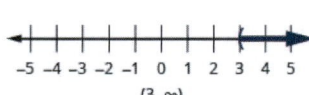

$(3, \infty)$

ⓑ

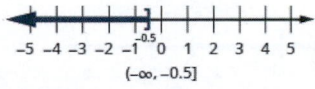

$(-\infty, -0.5]$

ⓒ

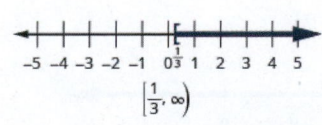

$\left[\dfrac{1}{3}, \infty\right)$

299.

ⓐ

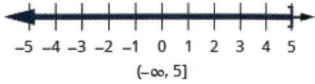

$(-\infty, 5]$

ⓑ

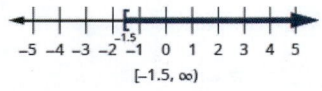

$[-1.5, \infty)$

ⓒ

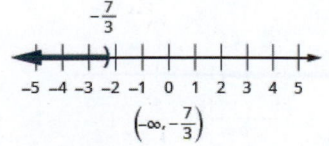

$\left(-\infty, -\dfrac{7}{3}\right)$

301.

ⓐ

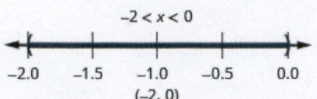

$-2 < x < 0$
$(-2, 0)$

ⓑ

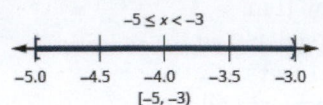

$-5 \le x < -3$
$[-5, -3)$

ⓒ

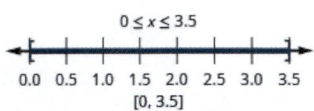

$0 \le x \le 3.5$
$[0, 3.5]$

303.

ⓐ

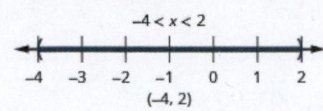

$-4 < x < 2$
$(-4, 2)$

ⓑ

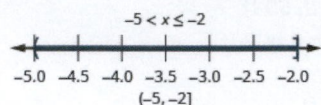

$-5 < x \le -2$
$(-5, -2]$

ⓒ

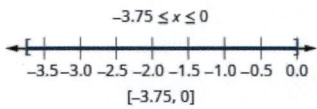

$-3.75 \le x \le 0$
$[-3.75, 0]$

305.

ⓐ

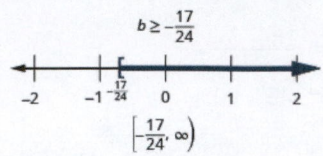

$b \ge -\frac{17}{24}$
$\left[-\frac{17}{24}, \infty\right)$

ⓑ

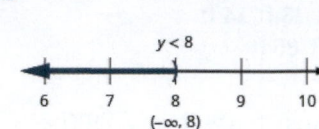

$y < 8$
$(-\infty, 8)$

ⓒ

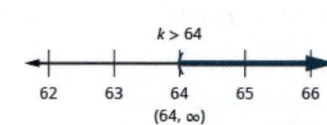

$k > 64$
$(64, \infty)$

307.

ⓐ

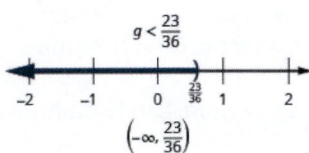

$g < \frac{23}{36}$
$\left(-\infty, \frac{23}{36}\right)$

ⓑ

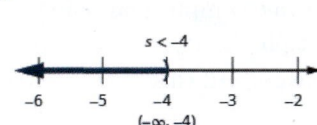

$s < -4$
$(-\infty, -4)$

ⓒ

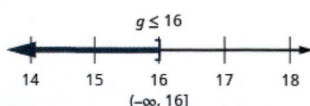

$g \le 16$
$(-\infty, 16]$

309.

ⓐ

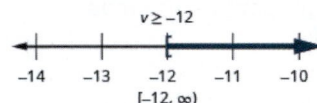

$v \ge -12$
$[-12, \infty)$

ⓑ

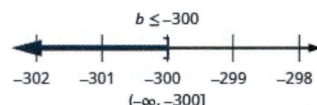

$b \le -300$
$(-\infty, -300]$

311.

ⓐ

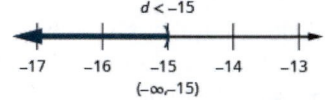

$d < -15$
$(-\infty, -15)$

ⓑ

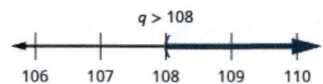

$q > 108$

313.

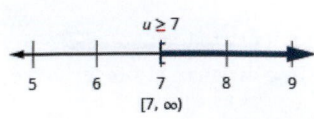

$u \ge 7$
$[7, \infty)$

319.

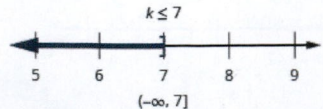

$k \le 7$
$(-\infty, 7]$

315.

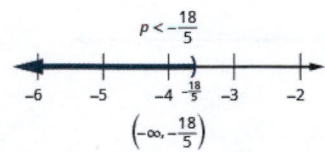

$p < -\frac{18}{5}$
$\left(-\infty, -\frac{18}{5}\right)$

321.

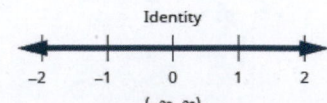

Identity
$(-\infty, \infty)$

317.

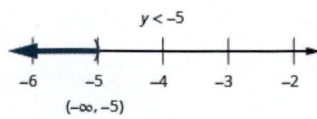

$y < -5$
$(-\infty, -5)$

323.

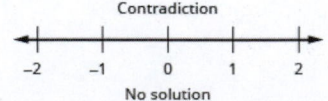

Contradiction
No solution

325.

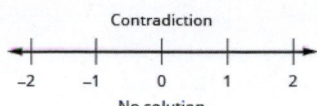

327.

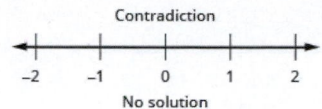

329.

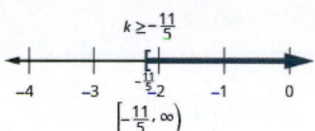

331.

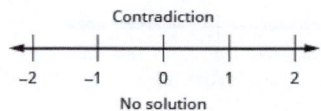

333.

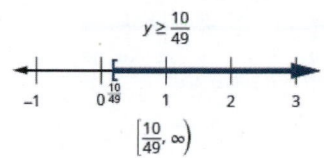

335.

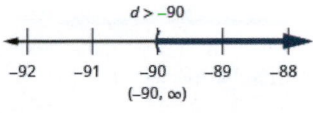

337.

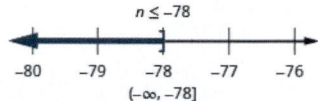

339.

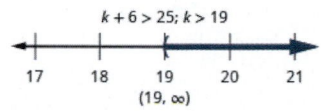

341.

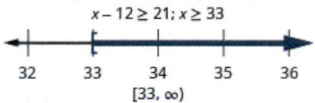

343.

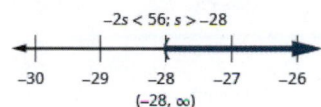

345.

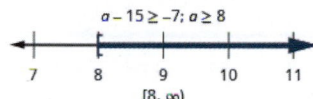

347. A maximum of 14 people can safely ride in the elevator.

349. five drinks

355. $110,000

361. 32 jobs

367. 20 hours

373. Answers will vary.

351. 86 cars

357. 260 messages

363. 62 necklaces

369. 20 people

375. Answers will vary.

353. $16,875

359. 35 people

365. seven lawns

371. 42 guests

377.

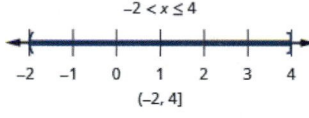

379.

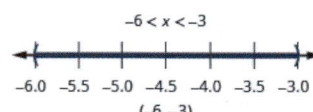

381.

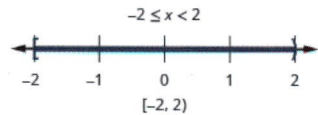

383.

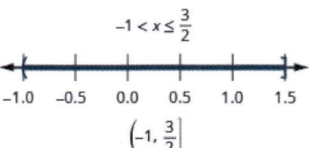

385.

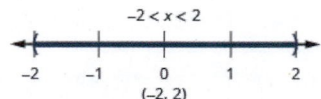

387.

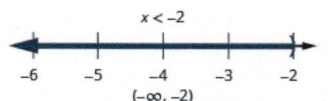

389.

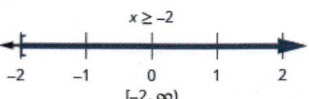

391.

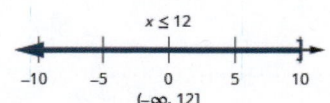

393.

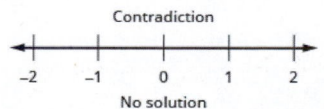

395.

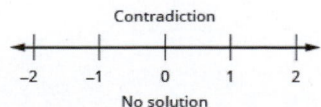

397.

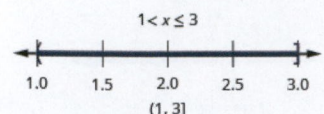

$$1 < x \le 3$$

(1, 3]

399.

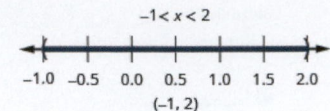

$$-1 < x < 2$$

(−1, 2)

401.

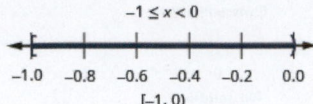

$$-1 \le x < 0$$

[−1, 0)

403.

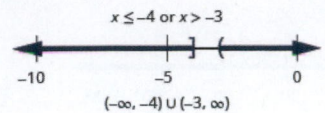

$$x \le -4 \text{ or } x > -3$$

(−∞, −4) ∪ (−3, ∞)

405.

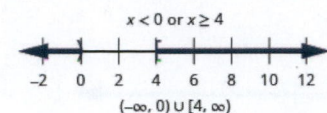

$$x < 0 \text{ or } x \ge 4$$

(−∞, 0) ∪ [4, ∞)

407.

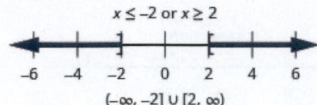

$$x \le -2 \text{ or } x \ge 2$$

(−∞, −2] ∪ [2, ∞)

409.

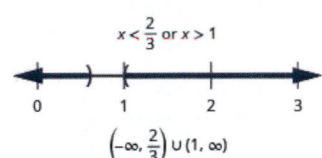

$$x < \frac{2}{3} \text{ or } x > 1$$

$\left(-\infty, \frac{2}{3}\right) \cup (1, \infty)$

411.

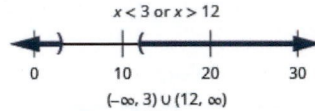

$$x < 3 \text{ or } x > 12$$

(−∞, 3) ∪ (12, ∞)

413.

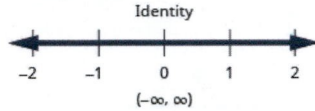

Identity

(−∞, ∞)

415.

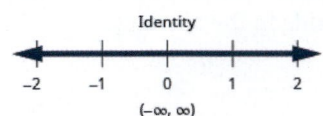

Identity

(−∞, ∞)

417.

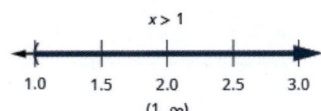

$$x > 1$$

(1, ∞)

419.

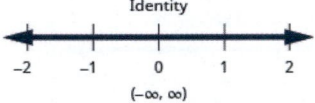

Identity

(−∞, ∞)

421.

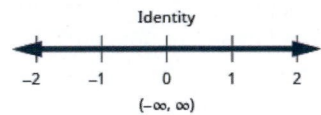

Identity

(−∞, ∞)

423.

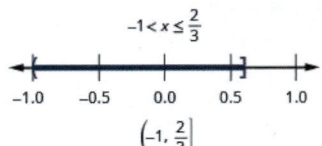

$$-1 < x \le \frac{2}{3}$$

$\left(-1, \frac{2}{3}\right]$

425.

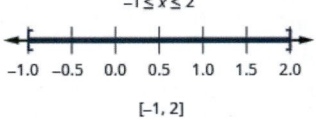

$$-1 \le x \le 2$$

[−1, 2]

427. $5 \le n \le 24$

429. $6 \le w \le 12$

431. ⓐ answers vary ⓑ answers vary

433. Answers will vary.

435. ⓐ $x = 4, x = -4$ ⓑ no solution ⓒ $z = 0$

437. ⓐ $x = 3, x = -3$ ⓑ no solution ⓒ $z = 0$

439. $x = 1, x = -\frac{1}{2}$

441. $x = -1, x = -\frac{5}{2}$

443. $x = 7, x = 1$

445. $x = 4, x = -1$

447. $x = 7, x = 1$

449. no solution

451. no solution

453. $x = -1, x = \frac{2}{3}$

455. $x = -3, x = 3$

457.

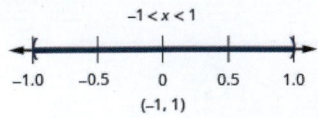

$$-1 < x < 1$$

(−1, 1)

459.

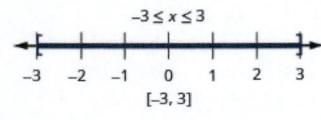

$$-3 \le x \le 3$$

[−3, 3]

461.

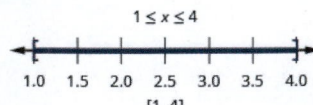

$$1 \le x \le 4$$

[1, 4]

463.

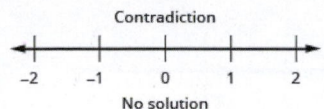

465.

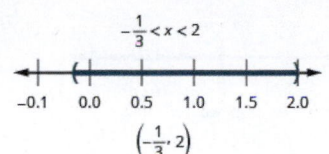

$$\left(-\frac{1}{3}, 2\right)$$

467.

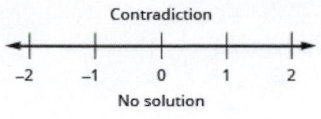

469.

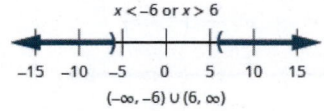

$(-\infty, -6) \cup (6, \infty)$

471.

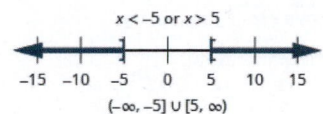

$(-\infty, -5] \cup [5, \infty)$

473.

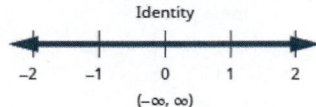

$(-\infty, \infty)$

475.

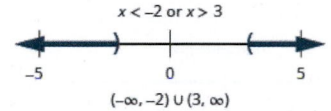

$(-\infty, -2) \cup (3, \infty)$

477.

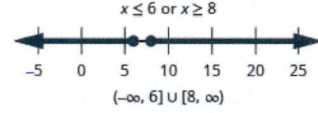

$(-\infty, 6] \cup [8, \infty)$

479.

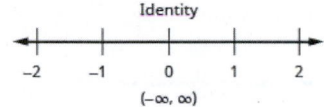

$(-\infty, \infty)$

481. $x = 4, x = \frac{2}{7}$

483. $x = 3, x = 2$

485. $x = 3, x = -\frac{11}{3}$

487. $x = \frac{3}{2}, x = -\frac{1}{2}$

489.

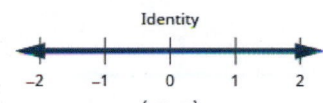

$(-\infty, \infty)$

491. The minimum to maximum expected production is 207,500 to 2,225,000 bottles

493. The acceptable weight is 22.5 to 25.5 ounces.

495. Answers will vary.

Review Exercises

497. no

499. $s = -22$

501. $m = -14$

503. $q = 18$

505. $x = -1$

507. $k = \frac{3}{4}$

509. contradiction; no solution

511. $n = 2$

513. $k = 23$

515. $x = 5$

517. There are 116 people.

519. 38

521. $-3, -10$

523. 76, 78, 80

525. $922

527. 160

529. $3.89

531. 35%

533. 32%

535. ⓐ $105 ⓑ 52.5%

537. $428.22

539. $1,900

541. $d_2 = \frac{2A}{d_1}$

543. $y = \frac{4x}{3} - 4$

545. $22.5°, 67.5°, 90°$

547. 26

549. 6 feet

551. 24.5 cm, 12.5 cm

553. 9 ft, 14 ft, 12 ft

555. nine pennies, six dimes, 12 quarters

557. 57 students, 68 adults

559. 2.2 lbs of raisins, 7.8 lbs of nuts

561. 9.7%

563. Louellen 65 mph, Tracy 66 mph

565. upstream 3 mph, downstream 5 mph

567. heavy traffic 32 mph, light traffic 66 mph

569.

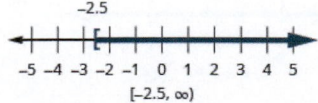

$[-2.5, \infty)$

571.

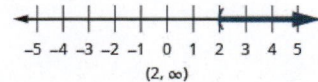

$(2, \infty)$

573.

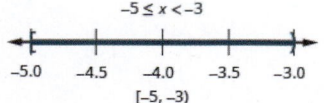

$[-5, -3)$

575.

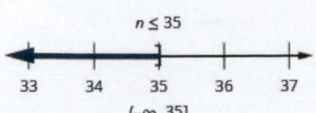

$n \le 35$

$(-\infty, 35]$

577.

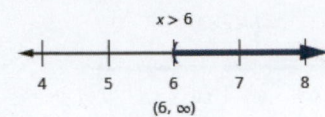

$x > 6$

$(6, \infty)$

579.

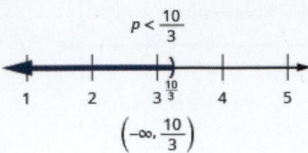

$p < \dfrac{10}{3}$

$\left(-\infty, \dfrac{10}{3}\right)$

581.

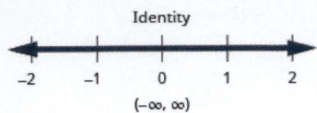

Identity

$(-\infty, \infty)$

583.

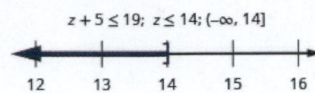

$z + 5 \le 19; \ z \le 14; \ (-\infty, 14]$

585.

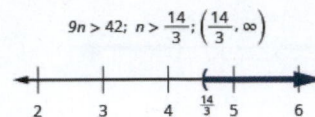

$9n > 42; \ n > \dfrac{14}{3}; \ \left(\dfrac{14}{3}, \infty\right)$

587. $33 per day

589. at least $300,000

591. at least 112 jobs

593.

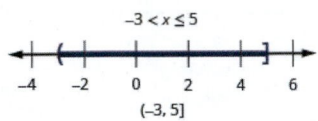

$-3 < x \le 5$

$(-3, 5]$

595.

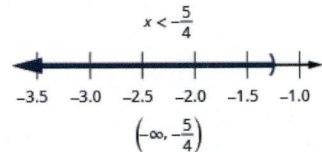

$x < -\dfrac{5}{4}$

$\left(-\infty, -\dfrac{5}{4}\right)$

597.

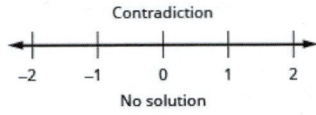

Contradiction

No solution

599.

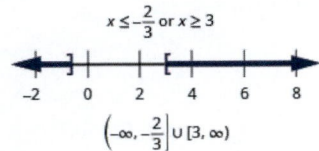

$x \le -\dfrac{2}{3}$ or $x \ge 3$

$\left(-\infty, -\dfrac{2}{3}\right] \cup [3, \infty)$

601.

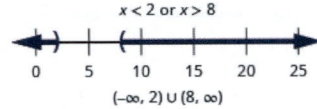

$x < 2$ or $x > 8$

$(-\infty, 2) \cup (8, \infty)$

603.

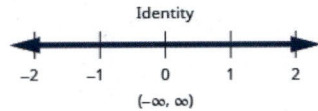

Identity

$(-\infty, \infty)$

605. $6 \le w \le 12$

607. no solution

609. $x = 2, \ x = \dfrac{2}{3}$

611. $x = 9, \ x = -3$

613. $x = 2, \ x = \dfrac{1}{4}$

615.

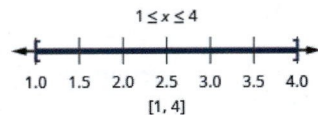

$1 \le x \le 4$

$[1, 4]$

617.

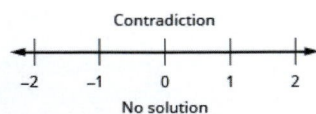

Contradiction

No solution

619.

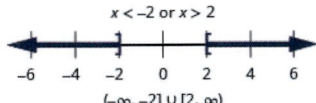

$x < -2$ or $x > 2$

$(-\infty, -2] \cup [2, \infty)$

621.

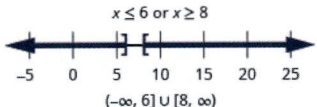

$x \le 6$ or $x \ge 8$

$(-\infty, 6] \cup [8, \infty)$

623. The minimum to maximum expected usage is 210,000 to 220,000 bottles

Practice Test

625. $x = -5$

627. $a = 41$

629. contradiction; no solution

631. $x = 6$

633. $x = -2, \ x = -\dfrac{1}{3}$

635.

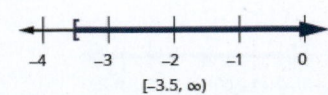

$[-3.5, \infty)$

637.

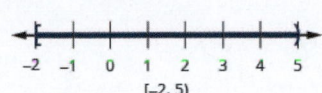

[−2, 5)

639.

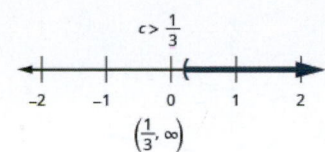

$\left(\frac{1}{3}, \infty\right)$

641.

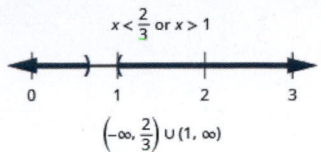

$\left(-\infty, \frac{2}{3}\right) \cup (1, \infty)$

643.

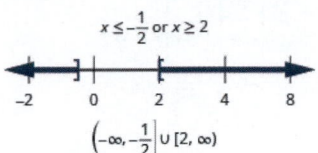

$\left(-\infty, -\frac{1}{2}\right] \cup [2, \infty)$

645. 10.8

647. −57, −55

649. 12 dimes, seven quarters

655. At most $55.56 per costume.

651. 30°, 60°, 90°

653. 2.5 hours

Chapter 3

Try It

3.1.

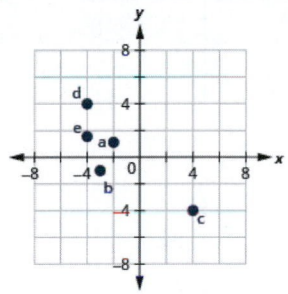

3.2.

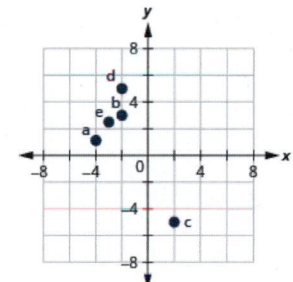

3.3. ⓐ yes, yes ⓑ yes, yes

3.4. ⓐ no, no ⓑ yes, yes

3.5.

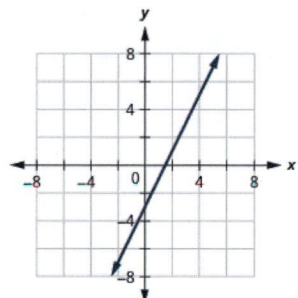

3.6.

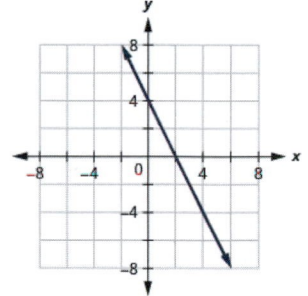

3.7.

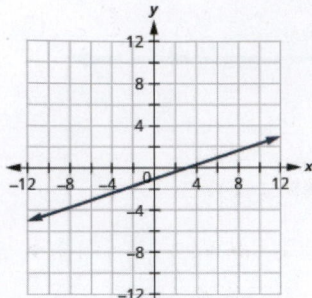

3.8.

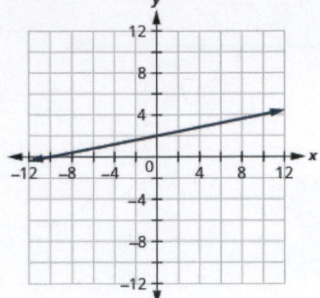

3.9. ⓐ

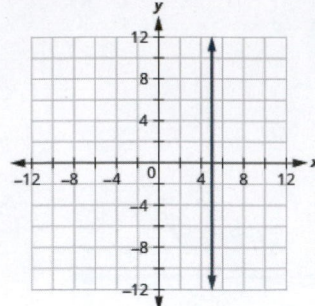

ⓑ

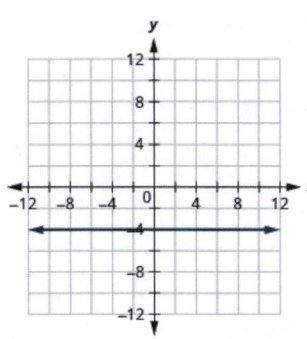

3.10. ⓐ

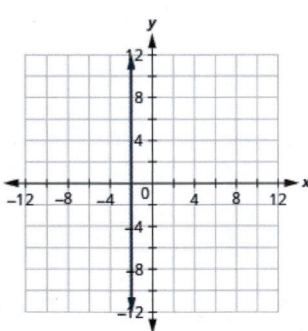

3.11.

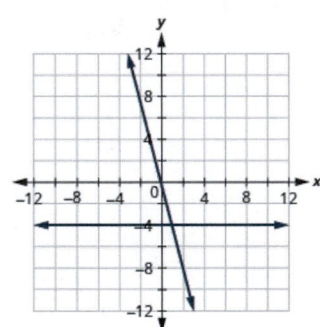

3.12.

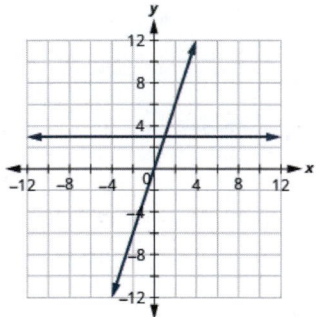

ⓑ

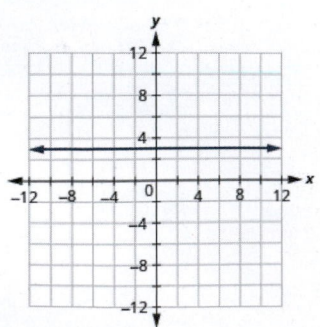

3.13. *x*-intercept: (2, 0),
y-intercept: (0, −2)

3.14. *x*-intercept: (3, 0),
y-intercept: (0, 2)

3.15. *x*-intercept: (4, 0),
y-intercept: (0, 12)

3.16. *x*-intercept: (8, 0),
y-intercept: (0, 2)

3.17.

3.18.

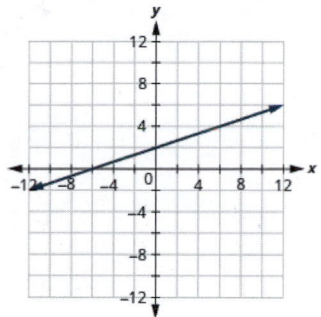

3.19.

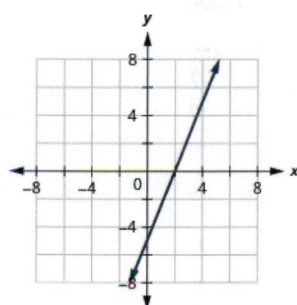

3.20.

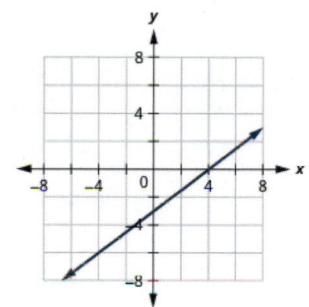

3.21.

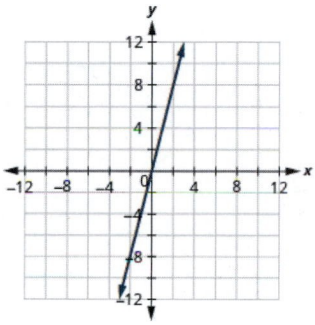

3.22.

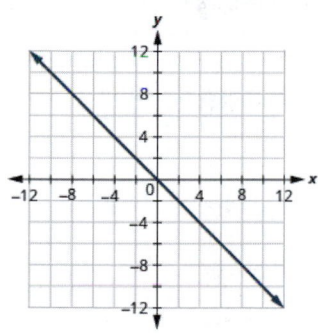

3.23. $-\dfrac{4}{3}$

3.24. $-\dfrac{3}{5}$

3.25. undefined

3.26. 0

3.27. −1

3.28. 10

3.29.

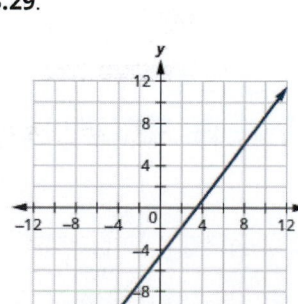

3.30.

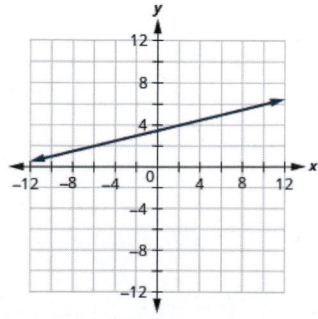

3.31. ⓐ $m = \frac{2}{5}$; $(0, -1)$

ⓑ $m = -\frac{1}{4}$; $(0, 2)$

3.32. ⓐ $m = -\frac{4}{3}$; $(0, 1)$

ⓑ $m = -\frac{3}{2}$; $(0, 6)$

3.33.

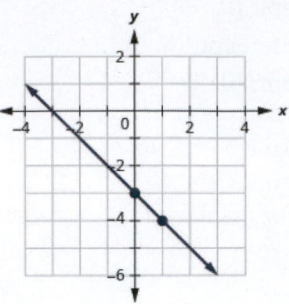

3.34.

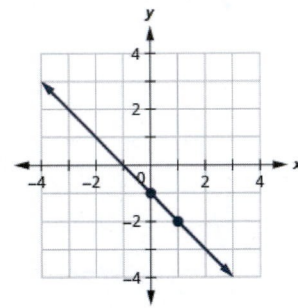

3.35. ⓐ intercepts ⓑ horizontal line ⓒ slope-intercept ⓓ vertical line

3.36. ⓐ vertical line ⓑ slope-intercept ⓒ horizontal line ⓓ intercepts

3.37. ⓐ 50 inches

ⓑ 66 inches

ⓒ The slope, 2, means that the height, h, increases by 2 inches when the shoe size, s, increases by 1. The h-intercept means that when the shoe size is 0, the height is 50 inches.

ⓓ

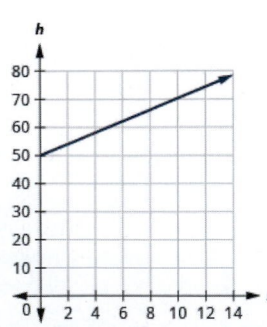

3.38. ⓐ 40 degrees

ⓑ 65 degrees

ⓒ The slope, $\frac{1}{4}$, means that the temperature Fahrenheit (F) increases 1 degree when the number of chirps, n, increases by 4. The T-intercept means that when the number of chirps is 0, the temperature is 40°.

ⓓ

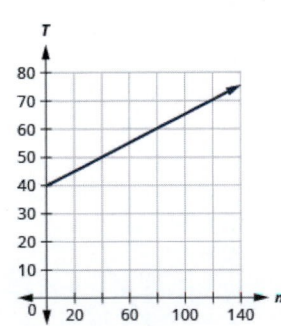

3.39. ⓐ $25

ⓑ $85

ⓒ The slope, 4, means that the weekly cost, C, increases by $4 when the number of pizzas sold, p, increases by 1. The C-intercept means that when the number of pizzas sold is 0, the weekly cost is $25.

ⓓ

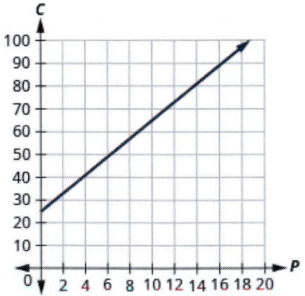

3.40. ⓐ $35

ⓑ $170

ⓒ The slope, 1.8, means that the weekly cost, C, increases by 1.80 when the number of invitations, n, increases by 1. The C-intercept means that when the number of invitations is 0, the weekly cost is $35.

ⓓ

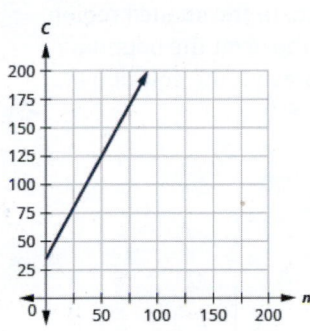

3.41. ⓐ parallel ⓑ not parallel; same line

3.42. ⓐ parallel ⓑ not parallel; same line

3.43. ⓐ parallel ⓑ parallel

3.44. ⓐ parallel ⓑ parallel

3.45. ⓐ perpendicular ⓑ not perpendicular

3.46. ⓐ perpendicular ⓑ not perpendicular

3.47. $y = \frac{2}{5}x + 4$

3.48. $y = -x - 3$

3.49. $y = \frac{3}{5}x + 1$

3.50. $y = \frac{4}{3}x - 5$

3.51. $y = -\frac{2}{5}x - 1$

3.52. $y = -\frac{3}{4}x - 4$

3.53. $y = 8$

3.54. $y = 4$

3.55. $y = \frac{1}{3}x - \frac{10}{3}$

3.56. $y = -\frac{2}{5}x - \frac{23}{5}$

3.57. $x = 5$

3.58. $x = -4$

3.59. $y = 3x - 10$

3.60. $y = \frac{1}{2}x + 1$

3.61. $y = -\frac{1}{3}x + \frac{10}{3}$

3.62. $y = -2x + 16$

3.63. $y = -5$

3.64. $y = -1$

3.65. $x = -5$

3.66. $x = -4$

3.67. ⓐ yes ⓑ yes ⓒ yes ⓓ yes ⓔ no

3.68. ⓐ yes ⓑ yes ⓒ no ⓓ no ⓔ yes

3.69. $y \geq -2x + 3$

3.70. $y \leq \frac{1}{2}x - 4$

3.71. $x - 4y \leq 8$

3.72. $3x - y \geq 6$

3.73.

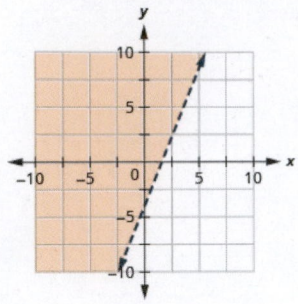

All points in the shaded region and on the boundary line, represent the solutions to $y > \frac{5}{2}x - 4$.

3.74.

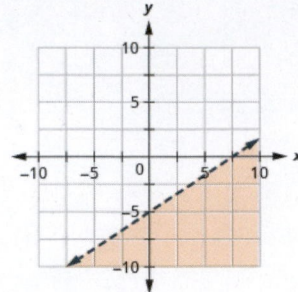

All points in the shaded region, but not those on the boundary line, represent the solutions to $y < \frac{2}{3}x - 5$.

3.75.

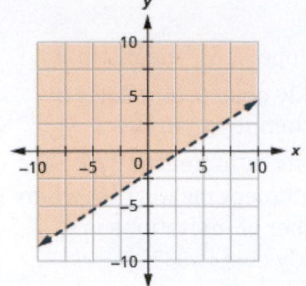

All points in the shaded region, but not those on the boundary line, represent the solutions to $2x - 3y < 6$.

3.76.

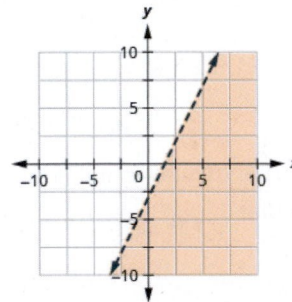

All points in the shaded region, but not those on the boundary line, represent the solutions to $2x - y > 3$.

3.77.

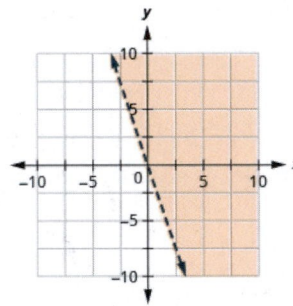

All points in the shaded region, but not those on the boundary line, represent the solutions to $y > -3x$.

3.78.

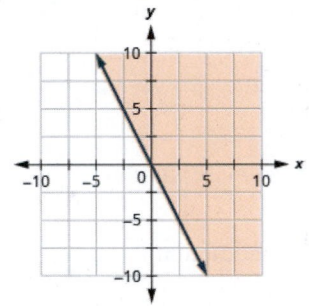

All points in the shaded region and on the boundary line, represent the solutions to $y \geq -2x$.

3.79.

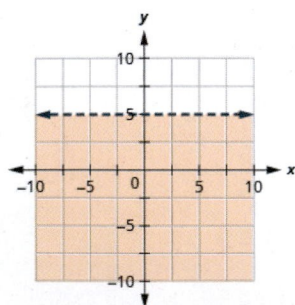

All points in the shaded region, but not those on the boundary line, represent the solutions to $y < 5$.

3.80.

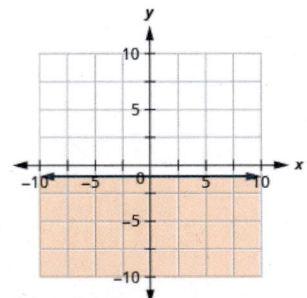

All points in the shaded region and on the boundary line represent the solutions to $y \leq -1$.

3.81. ⓐ $10x + 13y \geq 260$

ⓑ

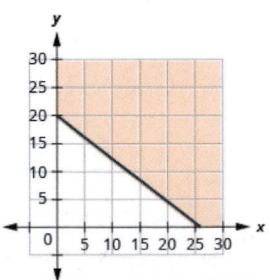

ⓒ Answers will vary.

3.82. ⓐ $10x + 17.5y \geq 280$

ⓑ

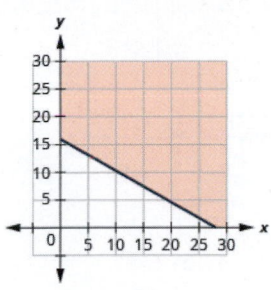

ⓒ Answers will vary.

3.85. ⓐ (Khanh Nguyen, kn68413), (Abigail Brown, ab56781), (Sumantha Mishal, sm32479), (Jose Hern and ez, jh47983) ⓑ {Khanh Nguyen, Abigail Brown, Sumantha Mishal, Jose Hern and ez} ⓒ {kn68413, ab56781, sm32479, jh47983}

3.88. ⓐ
$(-3, 0), (-3, 5), (-3, -6),$
$(-1, -2), (1, 2), (4, -4)$

ⓑ {−3, −1, 1, 4}

ⓒ {−6, 0, 5, −2, 2, −4}

3.91. ⓐ no ⓑ {NBC, HGTV, HBO}

ⓒ {Ellen Degeneres Show, Law and Order, Tonight Show, Property Brothers, House Hunters, Love it or List it, Game of Thrones, True Detective, Sesame Street}

3.94. ⓐ no ⓑ yes ⓒ yes

3.97. ⓐ $4m^2 - 7$ ⓑ $4x - 19$

ⓒ $x - 12$

3.100. ⓐ t IND; N DEP ⓑ 460; the number of unread emails in Anthony's account on the fourteenth day

3.83. ⓐ {1, 2, 3, 4, 5}

ⓑ {1, 8, 27, 64, 125}

3.86. ⓐ (Maria, November 6), (Arm and o, January 18), (Cynthia, December 8), (Kelly, March 15), (Rachel, November 6) ⓑ {Maria, Arm and o, Cynthia, Kelly, Rachel} ⓒ {November 6, January 18, December 8, March 15}

3.89. ⓐ Yes;
{−3, −2, −1, 0, 1, 2, 3};
{−6, −4, −2, 0, 2, 4, 6}

ⓑ No; {0, 2, 4, 8};
{−4, −2, −1, 0, 1, 2, 4}

3.92. ⓐ No ⓑ {Neal, Krystal, Kelvin, George, Christa, Mike} ⓒ {123-567-4839 work, 231-378-5941 cell, 743-469-9731 cell, 567-534-2970 work, 684-369-7231 cell, 798-367-8541 cell, 639-847-6971 cell}

3.95. ⓐ $f(3) = 22$ ⓑ $f(-1) = 6$

ⓒ $f(t) = 3t^2 - 2t - 1$

3.98. ⓐ $2k^2 + 1$ ⓑ $2x + 3$

ⓒ $2x + 4$

3.101. ⓐ yes ⓑ no

3.84. ⓐ {1, 2, 3, 4, 5}

ⓑ {3, 6, 9, 12, 15}

3.87. ⓐ $(-3, 3), (-2, 2), (-1, 0),$
$(0, -1), (2, -2), (4, -4)$

ⓑ {−3, −2, −1, 0, 2, 4}

ⓒ {3, 2, 0, −1, −2, −4}

3.90. ⓐ No; {0, 1, 8, 27};
{−3, −2, −1, 0, 2, 2, 3}

ⓑ Yes;
{7, −5, 8, 0, −6, −2, −1};
{−3, −4, 0, 4, 2, 3}

3.93. ⓐ yes ⓑ no ⓒ yes

3.96. ⓐ (2) = 13 ⓑ $f(-3) = 3$

ⓒ $f(h) = 2h^2 + 4h - 3$

3.99. ⓐ t IND; N DEP ⓑ 205; the number of unread emails in Bryan's account on the seventh day.

3.102. ⓐ no ⓑ yes

3.103.

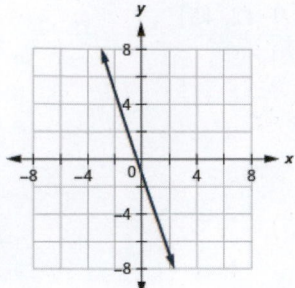

3.104.

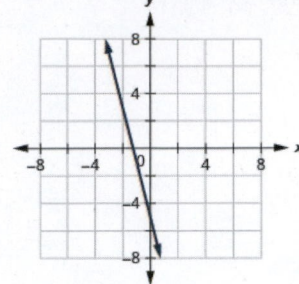

3.105.

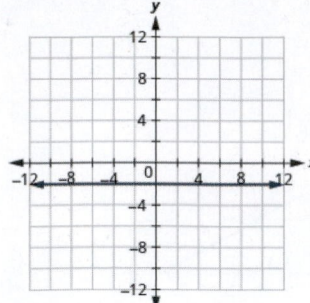

3.106.

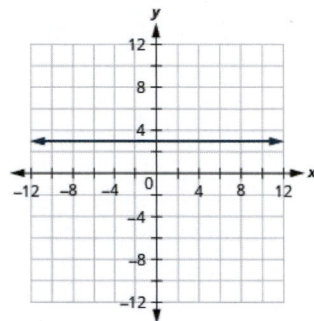

3.107.

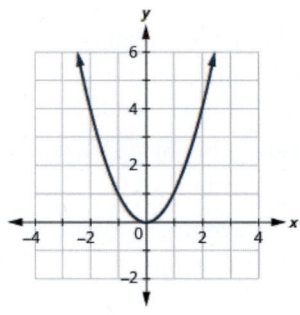

3.108.

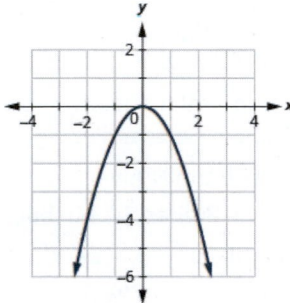

3.109.

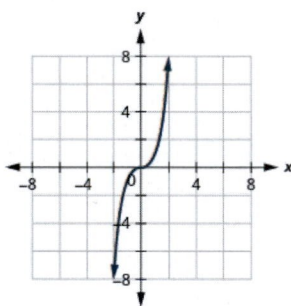

3.110.

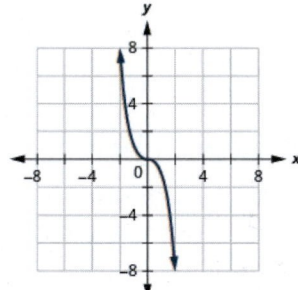

3.111.

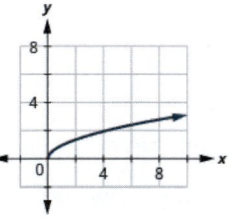

3.112.

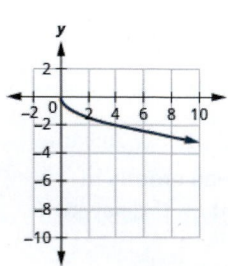

3.113.

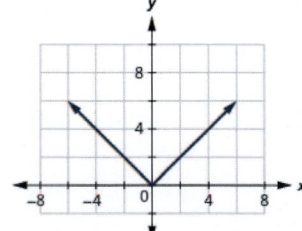

3.114.

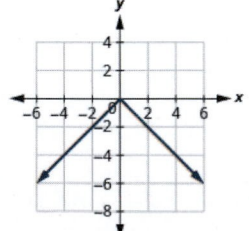

3.115. The domain is $[-5, 1]$. The range is $[-4, 2]$.

3.116. The domain is $[-2, 4]$. The range is $[-5, 3]$.

3.117. ⓐ $f(0) = 0$ ⓑ $f = \left(\frac{\pi}{2}\right) = 2$

ⓒ $f = \left(\frac{-3\pi}{2}\right) = 2$ ⓓ $f(x) = 0$ for

$x = -2\pi, -\pi, 0, \pi, 2\pi$ ⓔ

$(-2\pi, 0), (-\pi, 0), (0, 0), (\pi, 0), (2\pi, 0)$

ⓕ $(0, 0)$ ⓖ $[-2\pi, 2\pi]$ ⓗ

$[-2, 2]$

3.118. ⓐ $f(0) = 1$ ⓑ $f(\pi) = -1$

ⓒ $f(-\pi) = -1$ ⓓ $f(x) = 0$ for

$x = -\frac{3\pi}{2}, -\frac{\pi}{2}, \frac{\pi}{2}, \frac{3\pi}{2}$ ⓔ

$(-2pi, 0), (-pi, 0), (0, 0), (pi, 0), (2pi, 0)$

ⓕ $(0, 1)$ ⓖ $[-2pi, 2pi]$ ⓗ

$[-1, 1]$

Section Exercises

1.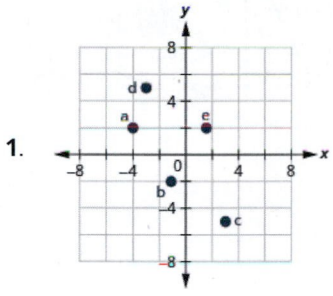

3.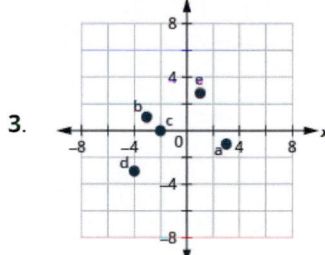

5. ⓐ A: yes, B: no, C: yes, D: yes ⓑ
A: yes, B: no, C: yes, D: yes

7. ⓐ A: yes, B: yes, C: yes, D: no ⓑ
A: yes, B: yes, C: yes, D: no

9.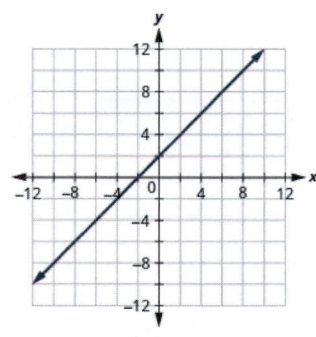

11.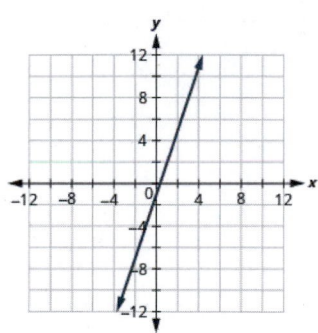

13.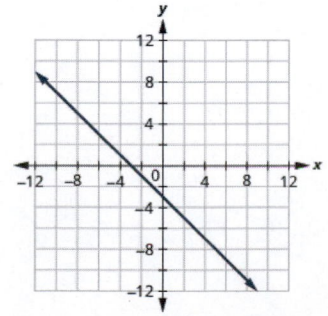

15.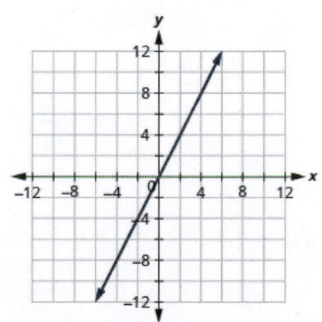

17.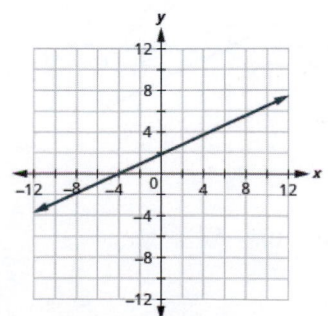

19.

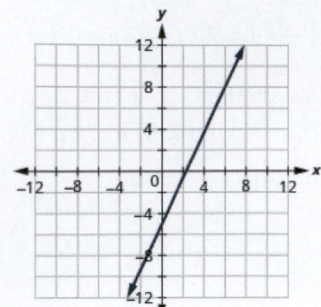

21.

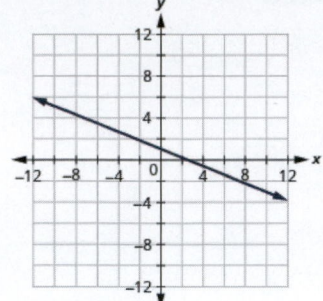

23.

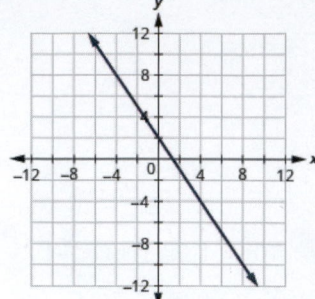

25. (a)

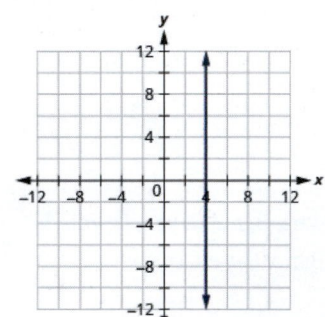

27. (a)

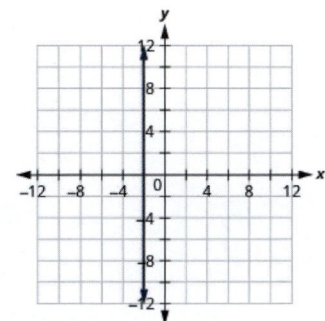

29.

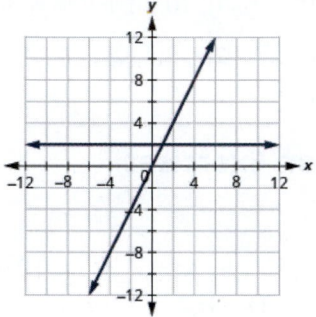

(b)

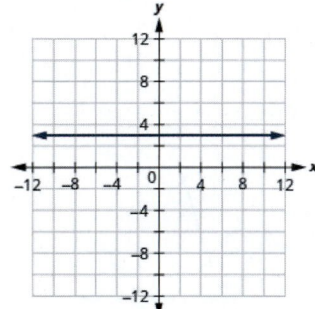

(b)

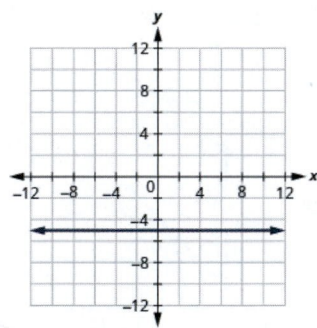

31.

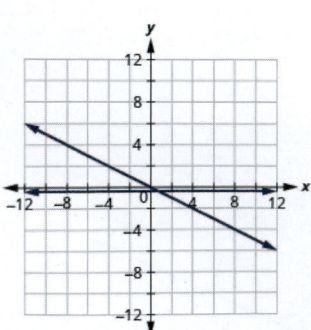

33. (3, 0), (0, 3)

35. (5, 0), (0, −5)

37. (5, 0), (0, −5)

39. (2, 0), (0, 6)

41. (2, 0), (0, −8)

43. (5, 0), (0, 2)

45.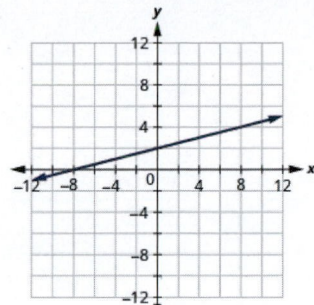

47.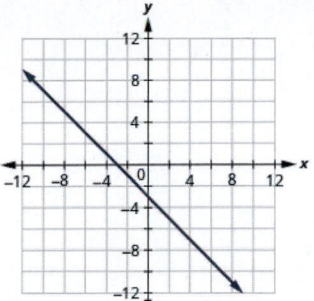

49.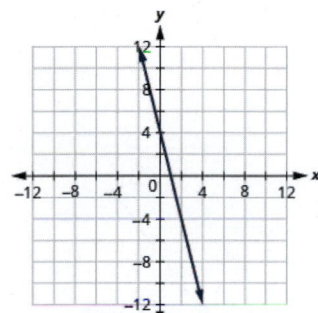

51.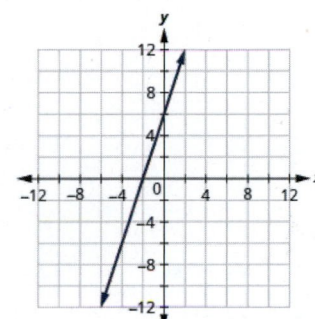

53.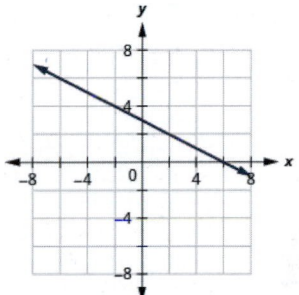

55.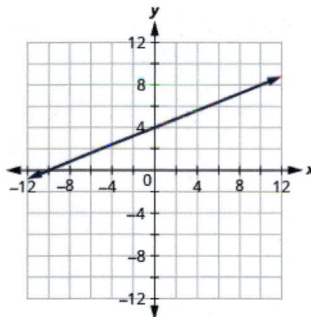

57.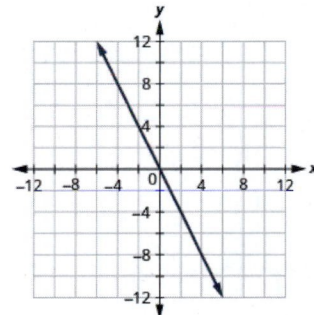

59.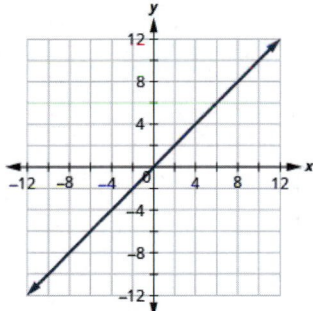

61.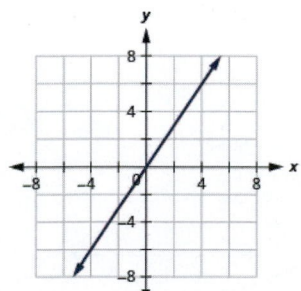

63.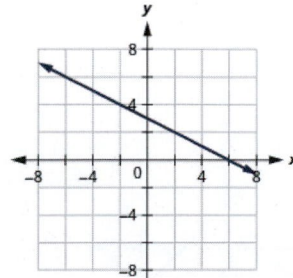

65.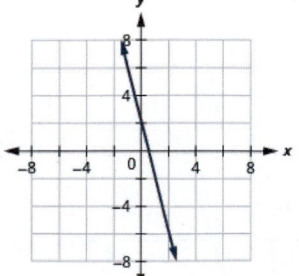

67.

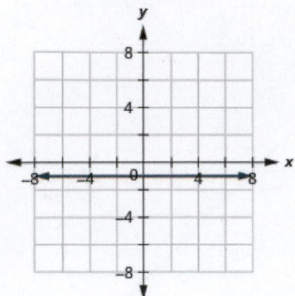

69. Answers will vary.

71. Answers will vary.

73. $\frac{2}{5}$

75. $\frac{5}{4}$

77. $-\frac{1}{3}$

79. $-\frac{5}{2}$

81. 0

83. undefined

85. $-\frac{5}{2}$

87. $-\frac{8}{5}$

89. $\frac{7}{3}$

91. -1

93.

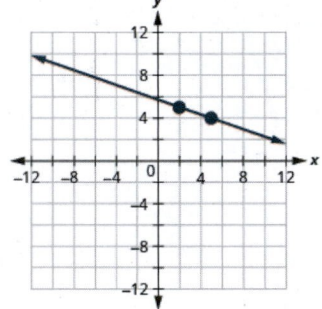

95.

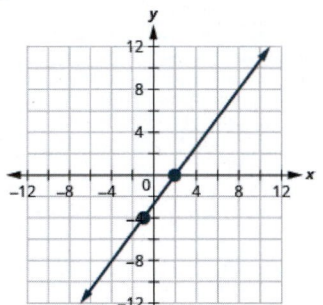

97.

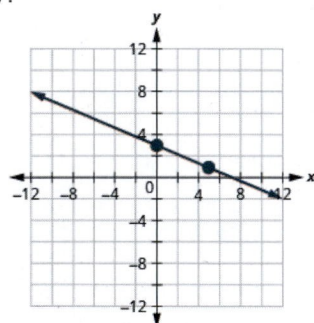

99.

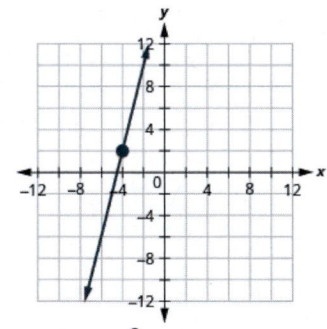

101. $m = -7; \ (0, \ 3)$

103. $m = -3; \ (0, \ 5)$

105. $m = -\frac{3}{2}; \ (0, \ 3)$

107. $m = \frac{5}{2}; \ (0, \ -3)$

109.

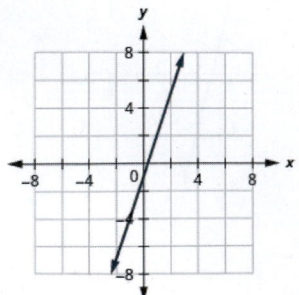

111.

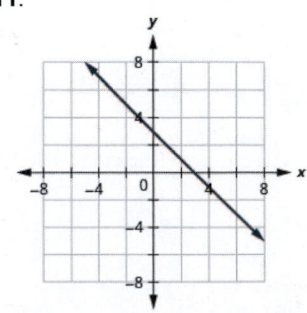

113.

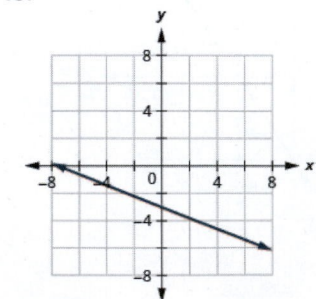

115.

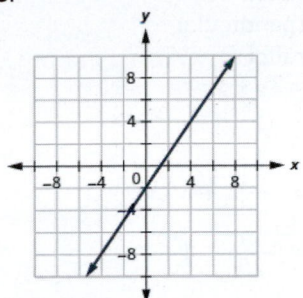

117. vertical line

119. slope-intercept

121. intercepts

123. intercepts

125. ⓐ $31

ⓑ $52

ⓒ The slope, 1.75, means that the payment, P, increases by $1.75 when the number of units of water used, w, increases by 1. The P-intercept means that when the number units of water Tuyet used is 0, the payment is $31.

ⓓ

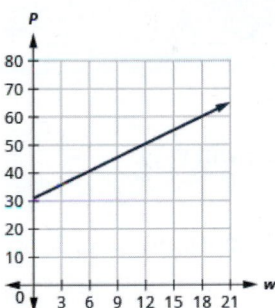

127. ⓐ $42

ⓑ $168.50

ⓒ The slope, 0.575 means that the amount he is reimbursed, R, increases by $0.575 when the number of miles driven, m, increases by 1. The R-intercept means that when the number miles driven is 0, the amount reimbursed is $42.

ⓓ

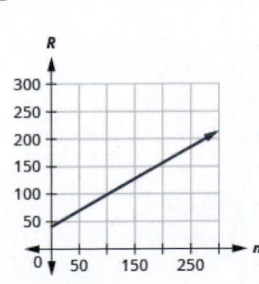

129. ⓐ $400

ⓑ $940

ⓒ The slope, 0.15, means that Cherie's salary, S, increases by $0.15 for every $1 increase in her sales. The S-intercept means that when her sales are $0, her salary is $400.

ⓓ

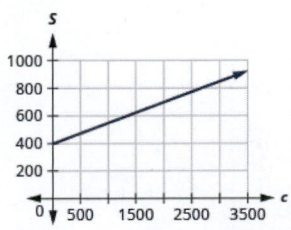

131. ⓐ $1570

ⓑ $5690

ⓒ The slope gives the cost per guest. The slope, 28, means that the cost, C, increases by $28 when the number of guests increases by 1. The C-intercept means that if the number of guests was 0, the cost would be $450.

ⓓ

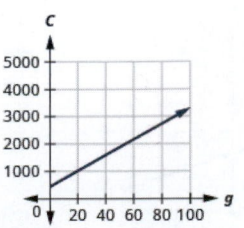

133. parallel
135. neither
137. parallel
139. perpendicular
141. neither
143. perpendicular
145. perpendicular
147. neither
149. parallel
151. Answers will vary.
153. Answers will vary.
155. $y = 3x + 5$
157. $y = -3x - 1$
159. $y = \frac{1}{5}x - 5$
161. $y = -1$
163. $y = 3x - 5$
165. $y = \frac{1}{2}x - 3$
167. $y = -\frac{4}{3}x + 3$
169. $y = -2$
171. $y = \frac{5}{8}x - 2$
173. $y = -\frac{3}{5}x + 1$
175. $y = -\frac{3}{2}x + 9$
177. $y = -7x - 10$
179. $y = 5$
181. $y = -7$
183. $y = -x + 8$
185. $y = \frac{1}{4}x - \frac{13}{4}$
187. $y = 2x + 5$
189. $y = -\frac{7}{2}x + 4$
191. $x = 7$
193. $y = -4$
195. $y = 4x - 2$
197. $y = 2x - 6$
199. $x = -3$
201. $y = -2$
203. $y = \frac{1}{2}x + 1$
205. $y = -\frac{4}{3}x$
207. $y = -\frac{3}{2}x + 5$
209. $y = \frac{5}{2}x$
211. $y = 4$
213. $y = -4$
215. $x = -2$
217. $y = 4$
219. $y = -\frac{1}{2}x + 5$
221. $y = \frac{1}{6}x$
223. $y = -\frac{4}{3}x - 3$
225. $y = -\frac{3}{4}x + 1$
227. $x = -2$
229. $x = -2$
231. $y = -\frac{1}{5}x - \frac{23}{5}$
233. $y = -2x - 2$
235. Answers will vary.
237. ⓐ yes ⓑ yes ⓒ no ⓓ no ⓔ no
239. ⓐ no ⓑ no ⓒ yes ⓓ yes ⓔ no
241. ⓐ yes ⓑ no ⓒ no ⓓ no ⓔ no
243. $y < 3x - 4$
245. $y \leq -\frac{1}{2}x + 1$
247. $x + y \geq 5$
249. $3x - y < 6$
251.

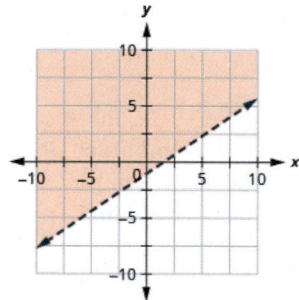

253.

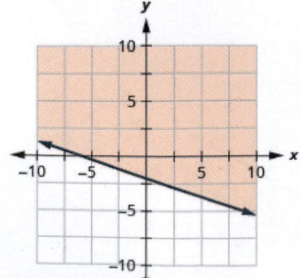

255.

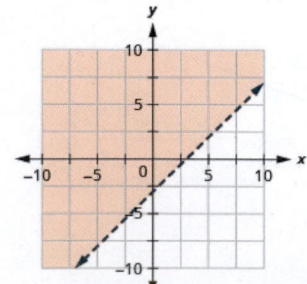

257.

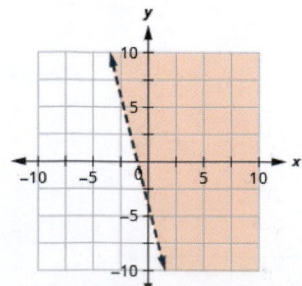

259.

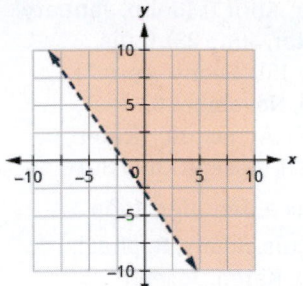

261.

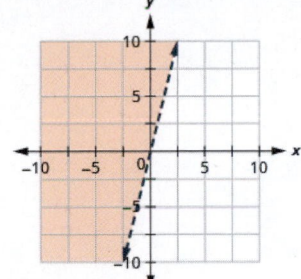

263.

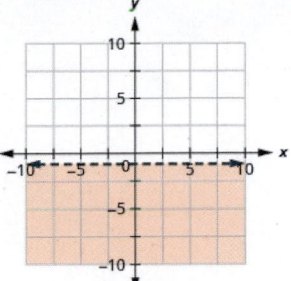

265.

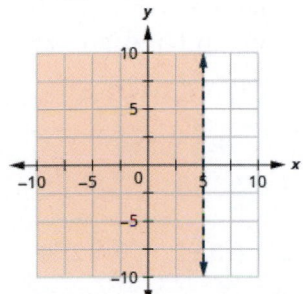

267.

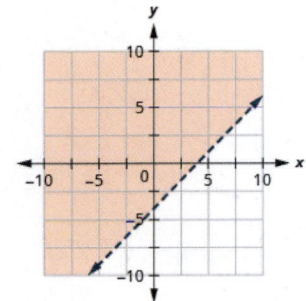

269.

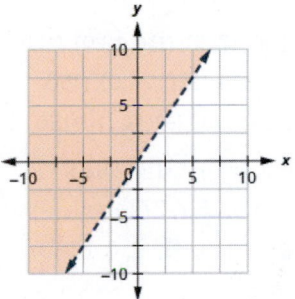

271.

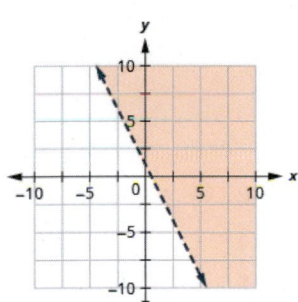

273.

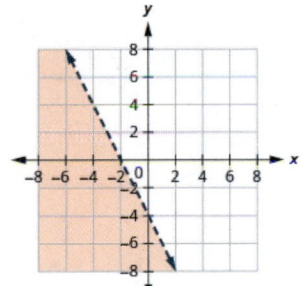

275.

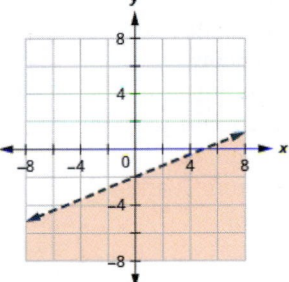

277. ⓐ $11x + 16.5y \geq 330$

ⓑ

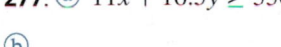

ⓒ Answers will vary.

279. ⓐ $15x + 10y \geq 500$

ⓑ

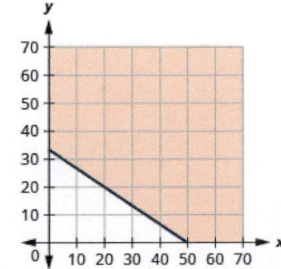

ⓒ Answers will vary.

281. Answers will vary.

283. ⓐ {1, 2, 3, 4, 5} ⓑ {4, 8, 12, 16, 20}

285. ⓐ {1, 5, 7, -2} ⓑ {7, 3, 9, -3, 8}

287. ⓐ (Rebecca, January 18), (Jennifer, April 1), (John, January 18), (Hector, June 23), (Luis, February 15), (Ebony, April 7), (Raphael, November 6), (Meredith, August 19), (Karen, August 19), (Joseph, July 30)

ⓑ {Rebecca, Jennifer, John, Hector, Luis, Ebony, Raphael, Meredith, Karen, Joseph}

ⓒ {January 18, April 1, June 23, February 15, April 7, November 6, August 19, July 30}

289. ⓐ (+100, 17. 2), (110, 18.9), (120, 20.6), (130, 22.3), (140, 24.0), (150, 25.7), (160, 27.5) ⓑ {+100, 110, 120, 130, 140, 150, 160,} ⓒ {17.2, 18.9, 20.6, 22.3, 24.0, 25.7, 27.5}

291. ⓐ (2, 3), (4, -3), (-2, -1), (-3, 4), (4, -1), (0, -3) ⓑ {-3, -2, 0, 2, 4}

ⓒ {-3, -1, 3, 4}

293. ⓐ (1, 4), (1, -4), (-1, 4), (-1, -4), (0, 3), (0, -3) ⓑ {-1, 0, 1} ⓒ {-4, -3, 3,4}

295. ⓐ yes ⓑ {-3, -2, -1, 0, 1, 2, 3}

ⓒ {9, 4, 1, 0}

297. ⓐ yes ⓑ {-3, -2, -1, 0, 1, 2, 3}

ⓒ 0, 1, 8, 27}

299. ⓐ yes ⓑ {-3, -2, -1, 0, 1, 2, 3}

ⓒ {0, 1, 2, 3}

301. ⓐ no ⓑ {Jenny, R and y, Dennis, Emily, Raul} ⓒ {RHern and ez@state.edu, JKim@gmail.com, Raul@gmail.com, ESmith@state.edu, DBroen@aol.com, jenny@aol.cvom, R and y@gmail.com}

303. ⓐ yes ⓑ yes ⓒ no

305. ⓐ yes ⓑ no ⓒ yes

307. ⓐ $f(2) = 7$ ⓑ $f(-1) = -8$

ⓒ $f(a) = 5a - 3$

309. ⓐ $f(2) = -6$ ⓑ $f(-1) = 6$

ⓒ $f(a) = -4a + 2$

311. ⓐ $f(2) = 5$ ⓑ $f(-1) = 5$

ⓒ $f(a) = a^2 - a + 3$

313. ⓐ $f(2) = 9$ ⓑ $f(-1) = 6$

ⓒ $f(a) = 2a^2 - a + 3$

315. ⓐ $g(h^2) = 2h^2 + 1$

ⓑ $g(x + 2) = 4x + 5$

ⓒ $g(x) + g(2) = 2x + 6$

317. ⓐ $g(h^2) = -3h^2 - 2$

ⓑ $g(x + 2) = -3x - 8$

ⓒ $g(x) + g(2) = -3x - 10$

319. ⓐ $g(h^2) = 3 - h^2$

ⓑ $g(x + 2) = 1 - x$

ⓒ $g(x) + g(2) = 4 - x$

321. 2

323. 6

325. 22

327. 4

329. ⓐ t IND; N DEP

ⓑ $N(4) = 165$ the number of unwatched shows in Sylvia's DVR at the fourth week.

331. ⓐ x IND; C DEP

ⓑ $N(0) = 1500$ the daily cost if no books are printed

ⓒ $N(1000) = 4750$ the daily cost of printing 1000 books

337. ⓐ no ⓑ yes

339. ⓐ no ⓑ yes

341. ⓐ

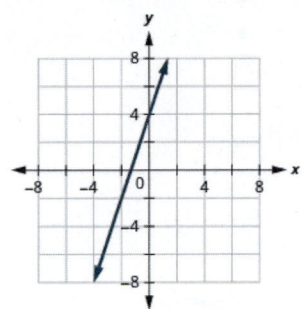

ⓑ D:(-∞,∞), R:(-∞,∞)

343. ⓐ

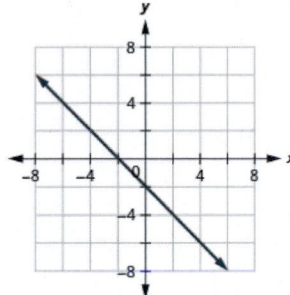

ⓑ D:(-∞,∞), R:(-∞,∞)

345. ⓐ

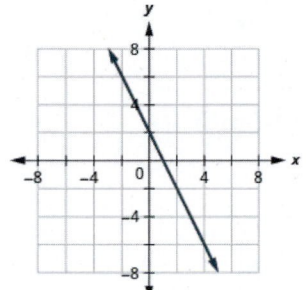

ⓑ D:(-∞,∞), R:(-∞,∞)

347. ⓐ

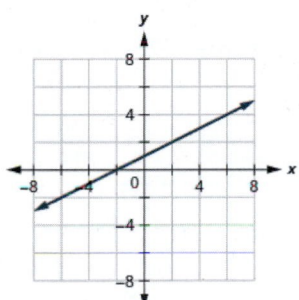

ⓑ D:(-∞,∞), R:(-∞,∞)

349. ⓐ

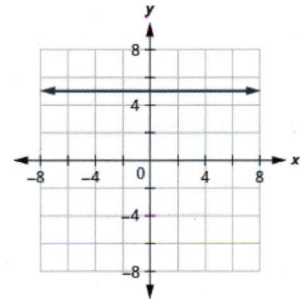

ⓑ D:(-∞,∞), R:{5}

351. ⓐ

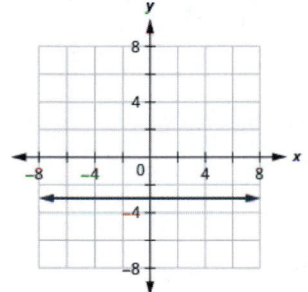

ⓑ D:(-∞,∞), R: $\{-3\}$

353. ⓐ

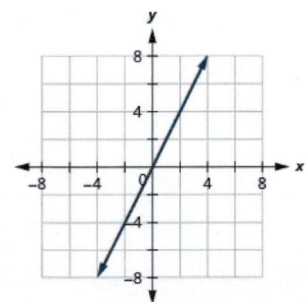

ⓑ D:(-∞,∞), R:(-∞,∞)

355. ⓐ

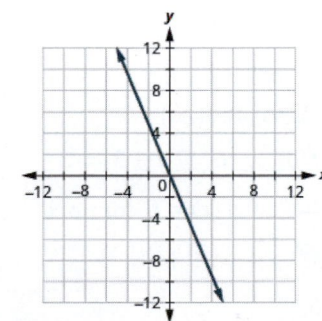

ⓑ D:(-∞,∞), R:(-∞,∞)

357. ⓐ

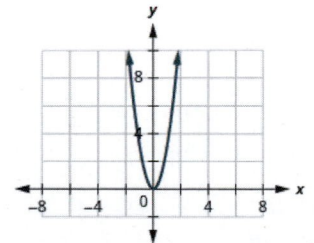

ⓑ D:(-∞,∞), R:(0,∞)

359. ⓐ

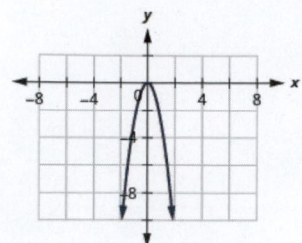

ⓑ (-∞,∞), R:(-∞,0)

365. ⓐ

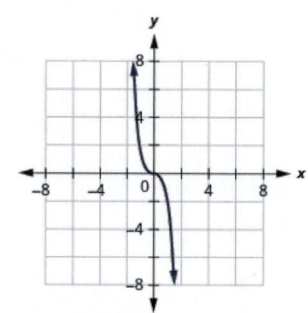

ⓑ D:(-∞,∞), R:(-∞,∞)

371. ⓐ

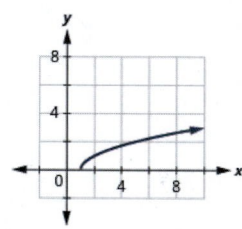

ⓑ D:[1,∞), R:[0,∞)

377. D: [2,∞), R: [0,∞)

383. ⓐ $f(0) = 0$ ⓑ $(pi/2) = -1$

ⓒ $f(-3pi/2) = -1$ ⓓ $f(x) = 0$ for

$x = -2pi, -pi, 0, pi, 2pi$

ⓔ (−2pi, 0), (−pi, 0),

(0, 0), (pi, 0), (2pi, 0) (f)(0, 0)

ⓖ [−2pi, 2pi] ⓗ [−1, 1]

361. ⓐ

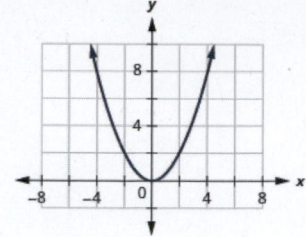

ⓑ (-∞,∞), R:(-∞,0)

367. ⓐ

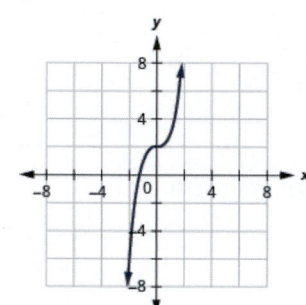

ⓑ D:(-∞,∞), R:(-∞,∞)

373. ⓐ

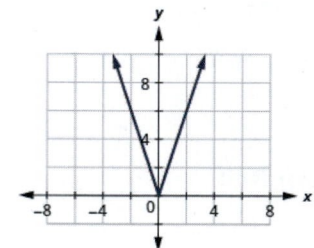

ⓑ D:[−1, ∞), R:[0,∞)

379. D: (-∞,∞), R: [4,∞)

385. ⓐ $f(0) = -6$ ⓑ $f(-3) = 3$

ⓒ $f(3) = 3$ ⓓ $f(x) = 0$ for no x

ⓔ none (f) $y = 6$ ⓖ [−3, 3]

ⓗ [−3, 6]

363. ⓐ

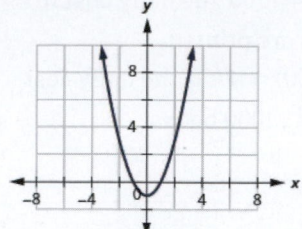

ⓑ (-∞,∞), R:(−1, ∞)

369. ⓐ

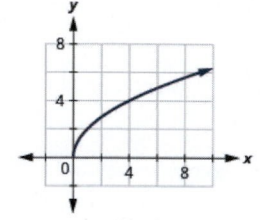

ⓑ D:[0,∞), R:[0,∞)

375. ⓐ

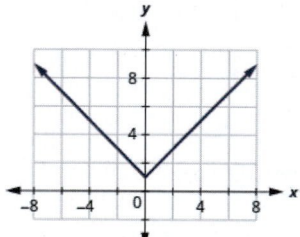

ⓑ D:(-∞,∞), R:[1,∞)

381. D: [−2, 2], R: [0, 2]

Review Exercises

391.

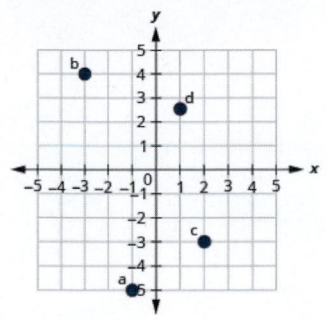

393. ⓑ, ⓒ

395.

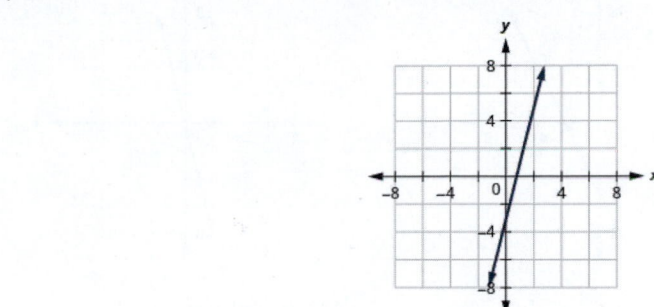

397.

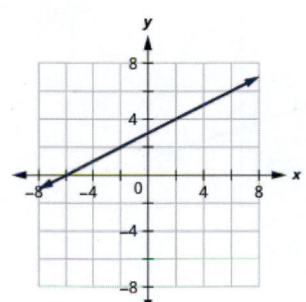

399.

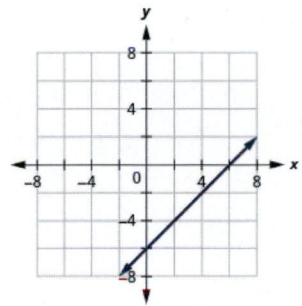

401.

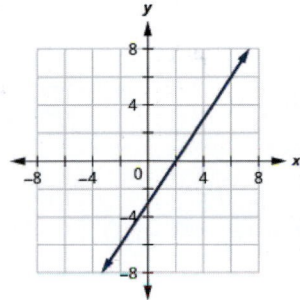

403.

405.

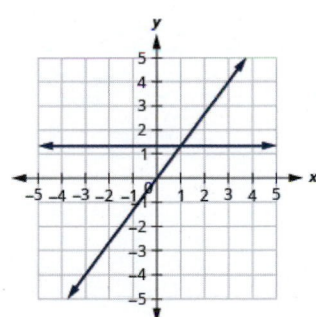

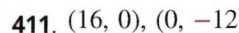

407. $(0, 3)(3, 0)$

409. $(6, 0), (0, 3)$

411. $(16, 0), (0, -12)$

413.

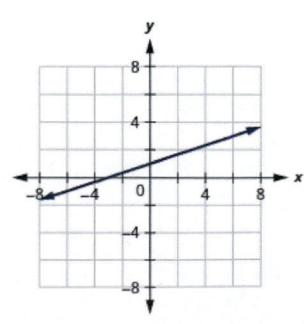

415.

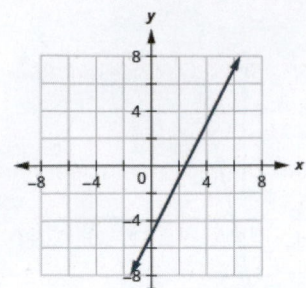

417.

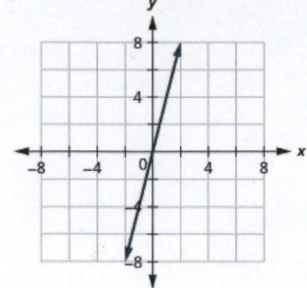

419. 1

421. $-\frac{1}{2}$

427. -6

423. undefined

429. $\frac{5}{2}$

425. 0

431.

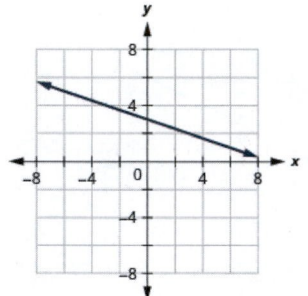

433.

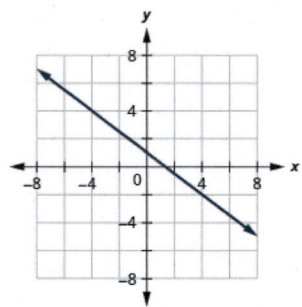

435. $m = \frac{5}{3}; (0, -6)$

437. $m = \frac{4}{5}; \left(0, -\frac{8}{5}\right)$

439.

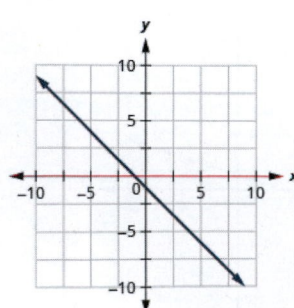

441.

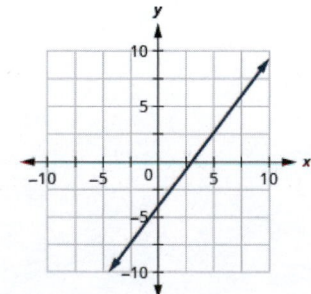

443. horizontal line

445. intercepts

447. plotting points

449. ⓐ −$250

ⓑ $450

ⓒ The slope, 35, means that Marjorie's weekly profit, P, increases by $35 for each additional student lesson she teaches.
The P-intercept means that when the number of lessons is 0, Marjorie loses $250.

ⓓ

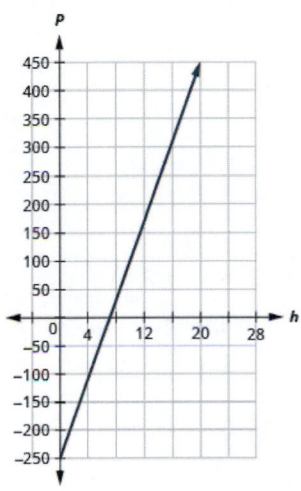

451. neither

457. $y = -2x$

463. $y = \frac{3}{5}x$

469. $y = 2$

475. $y = -\frac{3}{2}x - 6$

481. $y > \frac{2}{3}x - 3$

453. not parallel

459. $y = -3x + 5$

465. $y = -2x - 5$

471. $y = -\frac{2}{5}x + 8$

477. $y = 1$

483. $x - 2y \geq 6$

455. $y = -5x - 3$

461. $y = -4$

467. $y = \frac{1}{2}x - \frac{5}{2}$

473. $y = 3$

479. ⓐ yes ⓑ no ⓒ yes ⓓ yes; ⓔ no

485.

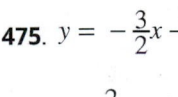

487.

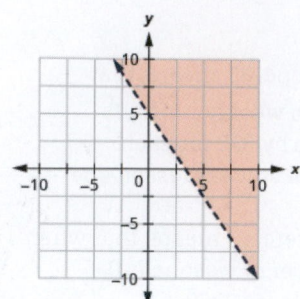

489.

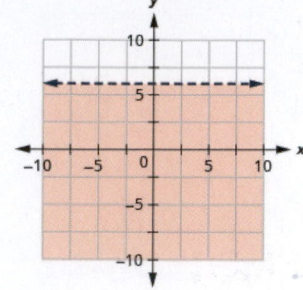

491. ⓐ $20x + 15y \geq 600$

ⓑ

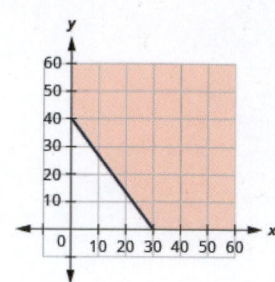

ⓒ Answers will vary.

493. ⓐ D: {-3, -2, -1, 0}

ⓑ R: {7, 3, 9, -3, 8}

495. ⓐ (4, 3), (-2, -3), (-2, -1), (-3, 1), (0, -1), (0, 4),

ⓑ D: {-3, -2, 0, 4}

ⓒ R: {-3, -1, 1, 3, 4}

497. ⓐ yes ⓑ {-3, -2, -1, 0, 1, 2, 3}

ⓒ {0, 1, 8, 27}

499. ⓐ {-3, -2, -1, 0, 1, 2, 3}

ⓑ {-3, -2, -1, 0, 1, 2, 3}

ⓒ {-243, -32, -1, 0, 1, 32, 243}

501. yes

503. yes

505. ⓐ $f(-2) = -10$ ⓑ $f(3) = 5$

ⓒ $f(a) = 3a - 4$

507. ⓐ $f(-2) = 20$ ⓑ $f(3) = 0$

ⓒ $f(a) = a^2 - 5a + 6$

509. 2

511. 18

513. yes

515. no

517. yes

519. no

521. ⓐ

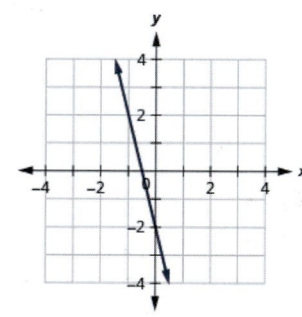

ⓑ D: (-∞,∞), R: (-∞,∞)

523. ⓐ

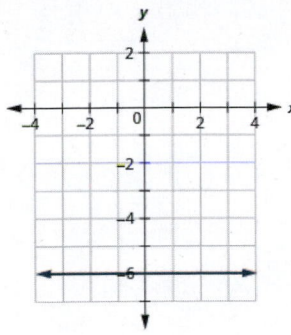

ⓑ D: (-∞,∞), R: (-∞,∞)

525. ⓐ

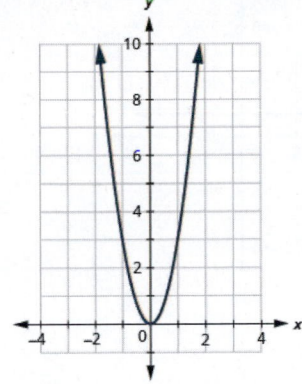

ⓑ D: (-∞,∞), R: (-∞,0]

527. ⓐ

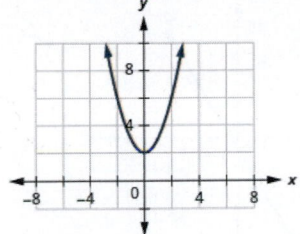

ⓑ D: (-∞,∞), R: (-∞,∞)

533. D: (-∞,∞), R: [2,∞)

529. ⓐ

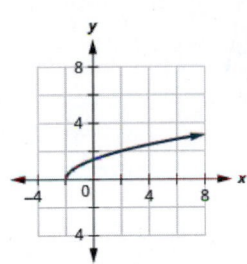

ⓑ D: [−2, ∞), R: [0,∞)

531. ⓐ

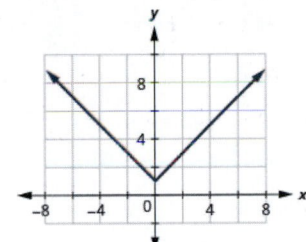

ⓑ D: (-∞,∞), R: [1,∞)

535. ⓐ $f(x) = 0$ ⓑ $f(\pi/2) = 1$
ⓒ $f(-3\pi/2) = 1$ ⓓ $f(x) = 0$ for
$x = -2\pi, -\pi, 0, \pi, 2\pi$
ⓔ $(-2\pi, 0), \quad (-\pi, 0), \quad (0, 0),$
$(\pi, 0), \quad (2\pi, 0)$ $(f)(0, 0)$
ⓖ $[-2\pi, 2\pi]$ ⓗ $[-1, 1]$

Practice Test

537.

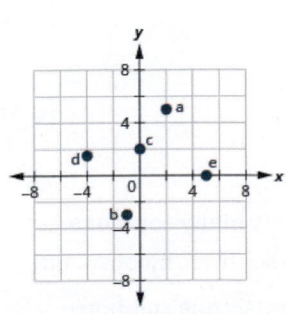

539. ⓐ $-\dfrac{3}{5}$ ⓑ undefined

541.

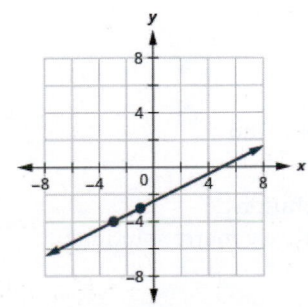

543.

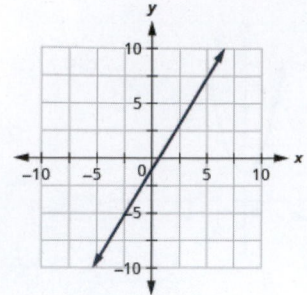

545.

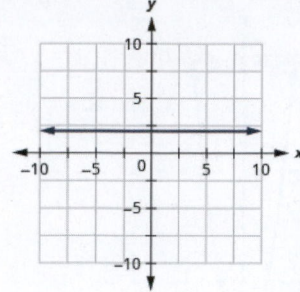

547. $y = 2x + 5$

549. $y = -\frac{4}{5}x - 5$

551.

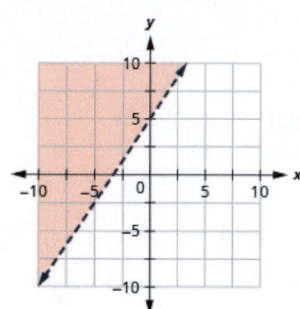

553.

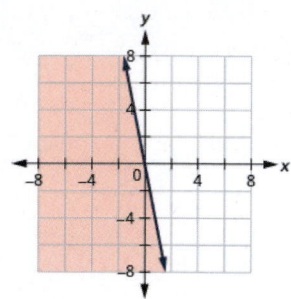

555. ⓐ yes ⓑ
$\{-3, -2, -1, 0, 1, 2, 3\}$ ⓒ $\{0, 1, 8, 27\}$

557. 12

559. ⓐ

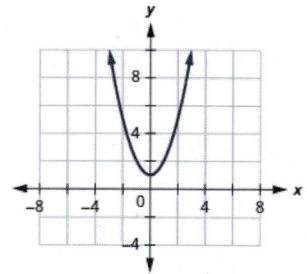

ⓑ D: (-∞,∞), R: [1,∞)

561. ⓐ $x = -2, 2$ ⓑ $y = -4$

ⓒ $f(-1) = -3$ ⓓ $f(1) = -3$

ⓔ D: (-∞,∞) ⓕ R: $[-4, \infty)$

Chapter 4

Try It

4.1. ⓐ yes ⓑ no

4.4. $(2, 3)$

4.7. no solution

4.10. infinitely many solutions

4.2. ⓐ no ⓑ yes

4.5. $(3, 4)$

4.8. no solution

4.11. ⓐ no solution, inconsistent, independent ⓑ one solution, consistent, independent

4.3. $(3, 2)$

4.6. $(5, -4)$

4.9. infinitely many solutions

4.12. ⓐ no solution, inconsistent, independent ⓑ one solution, consistent, independent

4.13. $(6, 1)$

4.14. $(-3, 5)$

4.15. $\left(2, \frac{3}{2}\right)$

4.16. $\left(-\frac{1}{2}, -2\right)$

4.17. $(2, -1)$

4.18. $(-2, 3)$

4.19. $(1, 3)$

4.20. $(4, -3)$

4.21. $(6, 2)$

4.22. $(1, -2)$

4.23. infinitely many solutions

4.24. infinitely many solutions

4.27. 3, 7

4.25. ⓐ Since both equations are in standard form, using elimination will be most convenient. ⓑ Since one equation is already solved for x, using substitution will be most convenient.

4.26. ⓐ Since one equation is already solved for y, using substitution will be most convenient. ⓑ Since both equations are in standard form, using elimination will be most convenient.

4.28. 2, −8

4.29. 160 policies

4.30. 1000 suits

4.31. Mark burned 11 calories for each minute of yoga and 7 calories for each minute of jumping jacks.

4.32. Erin burned 11 calories for each minute on the rowing machine and 5 calories for each minute of weight lifting.

4.33. The angle measures are 55 and 35.

4.34. The angle measures are 5 and 85.

4.35. The angle measures are 42 and 138.

4.36. The angle measures are 66 and 114.

4.37. 22, 68

4.38. 36, 54

4.39. The length is 60 feet and the width is 35 feet.

4.40. The length is 60 feet and the width is 38 feet.

4.41. It will take Clark 4 hours to catch Mitchell.

4.42. It will take Sally $1\frac{1}{2}$ hours to catch up to Charlie.

4.43. The rate of the boat is 11 mph and the rate of the current is 1 mph.

4.44. The speed of the canoe is 7 mph and the speed of the current is 1 mph.

4.45. The speed of the jet is 235 mph and the speed of the wind is 30 mph.

4.46. The speed of the jet is 408 mph and the speed of the wind is 24 mph.

4.47. 206 adults, 347 children

4.48. 42 adults, 105 children

4.49. 13 dimes and 29 quarters

4.50. 19 quarters and 51 nickels

4.51. 3 pounds peanuts and 2 pounds cashews

4.52. 10 pounds of beans, 10 pounds of ground beef

4.53. 120 ml of 25% solution and 30 ml of 50% solution

4.54. 125 ml of 10% solution and 125 ml of 40% solution

4.55. $42,000 in the stock fund and $8000 in the savings account

4.56. $1750 at 11% and $5250 at 13%

4.57. Bank $4,000; Federal $14,000

4.58. $41,200 at 4.5%, $24,000 at 7.2%

4.59. ⓐ $C(x) = 15x + 25,500$ ⓑ $R(x) = 32x$ ⓒ

4.60. ⓐ $C(x) = 120x + 150,000$ ⓑ $R(x) = 170x$ ⓒ

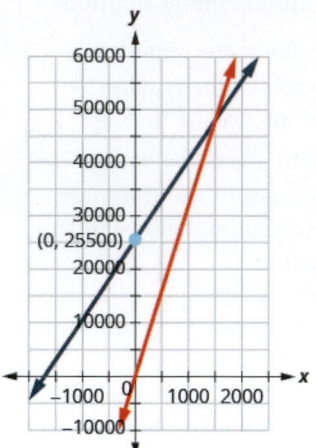

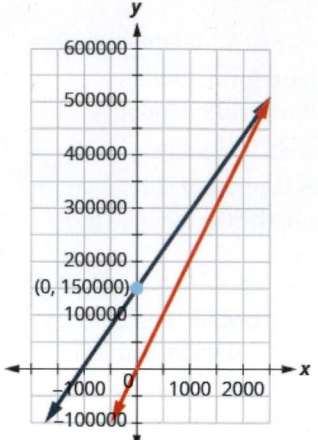

ⓓ 1,500 ; when 1,500 benches are sold, the cost and revenue will be both 48,000

ⓓ 3,000 ; when 3,000 benches are sold, the revenue and costs are both $510,000

4.61. ⓐ yes ⓑ no

4.62. ⓐ no ⓑ yes

4.63. $(2, -1, 3)$

4.64. $(-2, 3, 4)$

4.65. $(-3, 4, -2)$

4.66. $(-2, 3, -1)$

4.67. no solution

4.68. no solution

4.69. infinitely many solutions $(x, 3, z)$ where $x = z - 3$; $y = 3$; z is any real number

4.70. infinitely many solutions (x, y, z) where $x = 5z - 2$; $y = 4z - 3$; z is any real number

4.71. The fine arts department sold 75 adult tickets, 200 student tickets, and 75 child tickets.

4.72. The soccer team sold 200 adult tickets, 300 student tickets, and 100 child tickets.

4.73.
ⓐ $\begin{bmatrix} 3 & 8 & -3 \\ 2 & 5 & -3 \end{bmatrix}$

ⓑ $\begin{bmatrix} 2 & -5 & 3 & 8 \\ 3 & -1 & 4 & 7 \\ 1 & 3 & 2 & -3 \end{bmatrix}$

4.74.
ⓐ $\begin{bmatrix} 11 & 9 & -5 \\ 7 & 5 & -1 \end{bmatrix}$

ⓑ $\begin{bmatrix} 5 & -3 & 2 & -5 \\ 2 & -1 & -1 & 4 \\ 3 & -2 & 2 & -7 \end{bmatrix}$

4.75. $\begin{cases} x - y + 2z = 3 \\ 2x + y - 2z = 1 \\ 4x - y + 2z = 0 \end{cases}$

4.76. $\begin{cases} x + y + z = 4 \\ 2x + 3y - z = 8 \\ x + y - z = 3 \end{cases}$

4.77.
ⓐ $\begin{bmatrix} -2 & 3 & 0 & -2 \\ 4 & -1 & -4 & 4 \\ 5 & -2 & -2 & -2 \end{bmatrix}$

ⓑ $\begin{bmatrix} -2 & 3 & 0 & -2 \\ 4 & -1 & -4 & 4 \\ 15 & -6 & -6 & -6 \end{bmatrix}$

ⓒ $\begin{bmatrix} -2 & 3 & 0 & -2 \\ 3 & 4 & -13 & -16 & -8 \\ 15 & -6 & -6 & -6 \end{bmatrix}$

4.78.
ⓐ $\begin{bmatrix} 4 & 1 & -3 & 2 \\ 2 & -3 & -2 & -4 \\ 5 & 0 & 4 & -1 \end{bmatrix}$

ⓑ $\begin{bmatrix} 8 & 2 & -6 & 4 \\ 2 & -3 & -2 & -4 \\ 5 & 0 & 4 & -1 \end{bmatrix}$

ⓒ $\begin{bmatrix} 14 & -7 & -12 & -8 \\ 2 & -3 & -2 & -4 \\ 5 & 0 & 4 & -1 \end{bmatrix}$

4.79. $\begin{bmatrix} 1 & -1 & 2 \\ 0 & -3 & -4 \end{bmatrix}$

4.80. $\begin{bmatrix} 1 & -1 & 3 \\ 0 & -5 & 8 \end{bmatrix}$

4.81. The solution is $(4, -1)$.

4.82. The solution is $(-2, 0)$.

4.83. $(6, -1, -3)$

4.84. $(5, 7, 4)$

4.85. no solution

4.86. no solution

4.87. infinitely many solutions (x, y, z), where
$x = z - 3$; $y = 3$; z is any real
number.

4.88. infinitely many solutions (x, y, z), where
$x = 5z - 2$; $y = 4z - 3$; z is any
real number.

4.89. ⓐ -14; ⓑ -28

4.90. ⓐ 2 ⓑ -15

4.91. ⓐ 3 ⓑ 11 ⓒ 2

4.92. ⓐ -3 ⓑ 2 ⓒ 3

4.93. 37

4.94. 7

4.95. -11

4.96. 8

4.97. $(-\frac{15}{7}, \frac{24}{7})$

4.98. $(-2, 0)$

4.99. $(-9, 3, -1)$

4.100. $(-6, 3, -2)$

4.101. no solution

4.102. infinite solutions

4.103. yes

4.104. yes

4.105. ⓐ no ⓑ yes

4.106. ⓐ yes ⓑ no

4.107.

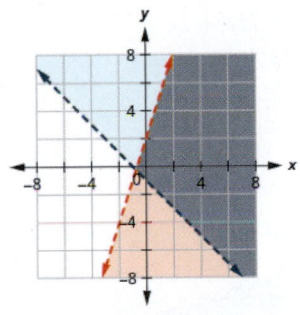

The solution is the grey region.

4.108.

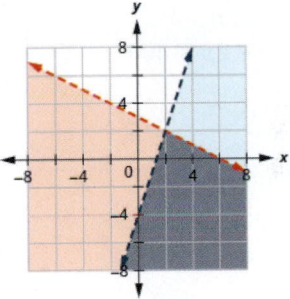

The solution is the grey region.

4.109.

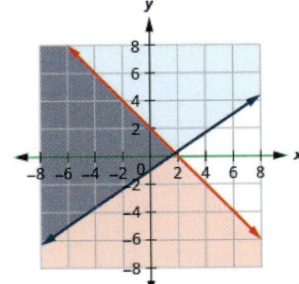

The solution is the grey region.

4.110.

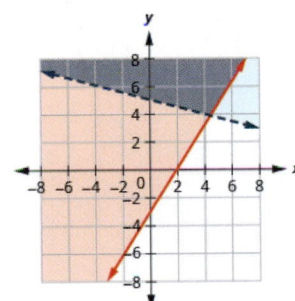

The solution is the grey region.

4.111.

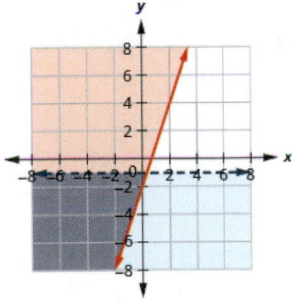

The solution is the grey region.

4.112.

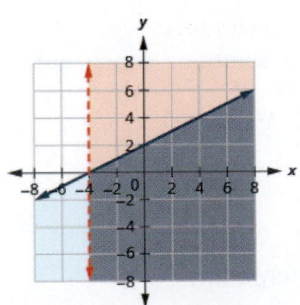

The solution is the grey region.

4.113.

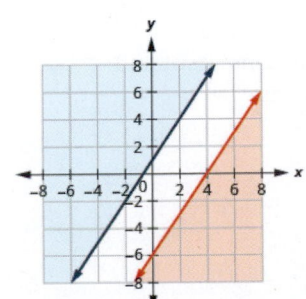

No solution.

4.114.

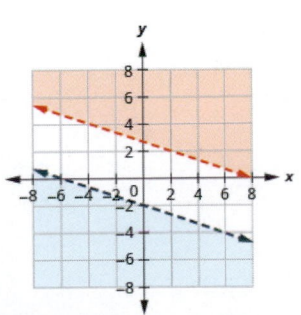

No solution.

4.115.

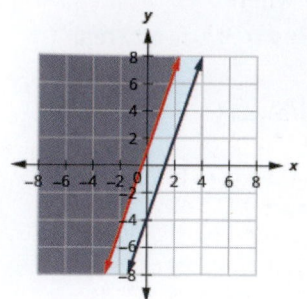

The solution is the grey region.

4.116.

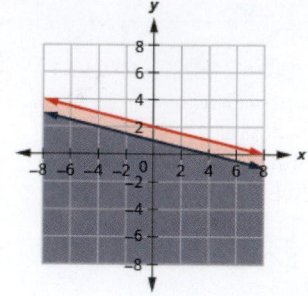

The solution is the grey region.

4.117. ⓐ $\begin{cases} 30m + 20p \le 160 \\ 2m + 3p \le 15 \end{cases}$

ⓑ

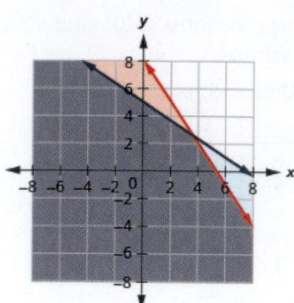

ⓒ yes

ⓓ no

4.118. ⓐ $\begin{cases} a \ge p + 5 \\ a + 2p \le 400 \end{cases}$

ⓑ

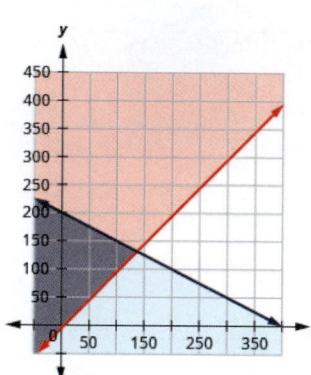

ⓒ no

ⓓ no

4.119. ⓐ $\begin{cases} 0.75d + 2e \le 25 \\ 360d + 110e \ge 1000 \end{cases}$

ⓑ

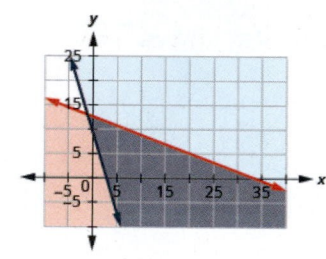

ⓒ yes

ⓓ no

4.120. ⓐ $\begin{cases} 140p + 125j \ge 1000 \\ 1.80p + 1.25j \le 12 \end{cases}$

ⓑ

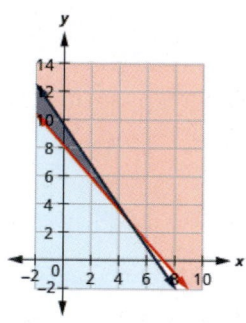

ⓒ yes

ⓓ no

Section Exercises

1. ⓐ yes ⓑ no

7. $(0, 2)$

13. $(3, 3)$

19. no solution

25. 1 point, consistent and independent

31. $(1, -4)$

37. $(-5, 4)$

43. $(4, 0)$

49. $(2, 3)$

55. $(-3, 2)$

61. infinitely many

3. ⓐ yes ⓑ no

9. $(2, 4)$

15. $(6, -4)$

21. infinite solutions

27. 1 point, consistent and independent

33. $(-3, 2)$

39. $(0, 10)$

45. none

51. $(-9, 3)$

57. $(-15/7, 24/7)$

63. infinitely many

5. $(-3, 2)$

11. $(-2, 2)$

17. no solution

23. infinite solutions

29. infinite solutions, consistent, dependent

35. $(-1/2, 5/2)$

41. $(4, -2)$

47. $(6, 1)$

53. $(9, 5)$

59. infinitely many

65. ⓐ substitution ⓑ elimination

67. ⓐ elimination ⓑ substituion
73. 13 and 17
79. 22 and −67

85. 8 and 40 gallons

91. Package of paper $4, stapler $7

97. 53.5 degrees and 36.5 degrees
103. 37 degrees and 143 degrees
109. Width is 41 feet and length is 118 feet.
115. 1.5 hour

121. Jet rate is 265 mph and wind speed is 22 mph.
127. 110 adult tickets, 190 child tickets
133. 13 nickels, 3 dimes
139. 80 pounds nuts and 40 pounds raisins

145. 7.5 liters of each solution

151. $1600 at 8%, 960 at 6%

157. $55,000 on loan at 6% and $30,000 on loan at 4.5%

69. Answers will vary.

75. −7 and −19
81. Eighty cable packages would need to be sold to make the total pay the same.
87. 1000 calories playing basketball and 400 calories canoeing
93. Hot dog 150 calories, cup of cottage cheese 220 calories

99. 16 degrees and 74 degrees

105. 16° and 74°
111. Width is 10 feet and length is 40 feet.
117. Boat rate is 16 mph and current rate is 4 mph.
123. Jet rate is 415 mph and wind speed is 25 mph.
129. 6 good seats, 10 cheap seats

135. 42 dimes, 8 quarters
141. 9 pounds of Chicory coffee, 3 pounds of Jamaican Blue Mountain coffee
147. 80 liters of the 25% solution and 40 liters of the 10% solution

153. $28,000 at 9%, $36,000 at 5.5%

159. ⓐ $C(x) = 5x + 6500$ ⓑ

$R(x) = 10x$ ⓒ

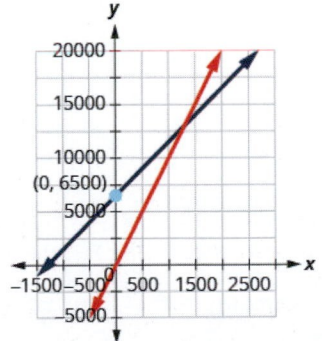

ⓓ 1,500; when 1,500 water bottles are sold, the cost and the revenue equal $15,000

71. Answers will vary.

77. 14 and 23
83. Mitchell would need to sell 120 stoves for the companies to be equal.
89. Oranges cost $2 per pound and bananas cost $1 per pound

95. Owen will need 80 quarts of water and 20 quarts of concentrate to make 100 quarts of lemonade.
101. 134 degrees and 46 degrees

107. 45° and 45°
113. 11 hours

119. Boat rate is 18 mph and current rate is 2 mph.
125. Answers will vary.

131. 92 adult tickets, 220 children tickets
137. 17 $10 bills, 37 $20 bills
143. 10 bags of M&M's, 15 bags of Reese's Pieces

149. 240 liters of the 90% solution and 120 liters of the 75% solution
155. $8500 CD, $1500 savings account
161. Answers will vary.

163. ⓐ no ⓑ yes
169. (7, 12, −2)

165. ⓐ no ⓑ yes
171. (−3, −5, 4)

167. (4, 5, 2)
173. (2, −3, −2)

175. $(6, -9, -3)$

177. $(3, -4, -2)$

179. $(-3, 2, 3)$

181. $(-2, 0, -3)$

183. no solution

185. no solution

187. (x, y, z) where
$x = 5z + 2; y = -3z + 1; z$ is any real number

189. (x, y, z) where
$x = 5z - 2; y = 4z - 3; z$ is any real number

191. $42, 50, 58$

193. $20, 5, 10$

195. Answers will vary.

197.

ⓐ $\begin{bmatrix} 2 & 4 & -5 \\ 3 & -2 & 2 \end{bmatrix}$

ⓑ $\begin{bmatrix} 3 & -2 & -1 & -2 \\ -2 & 1 & 0 & 5 \\ 5 & 4 & 1 & -1 \end{bmatrix}$

199.

ⓐ $\begin{bmatrix} 2 & -5 & -3 \\ 4 & -3 & -1 \end{bmatrix}$

ⓑ $\begin{bmatrix} 4 & 3 & -2 & -3 \\ -2 & 1 & -3 & 4 \\ -1 & -4 & 5 & -2 \end{bmatrix}$

201. $\begin{cases} 2x - 4y = -2 \\ 3x - 3y = -1 \end{cases}$

203. $\begin{cases} 2x - 2y = -1 \\ 2y - z = 2 \\ 3x - z = -2 \end{cases}$

205.

ⓐ $\begin{bmatrix} 3 & 2 & 1 \\ 4 & -6 & -3 \end{bmatrix}$

ⓑ $\begin{bmatrix} 12 & 8 & 4 \\ 4 & -6 & -3 \end{bmatrix}$

ⓒ $\begin{bmatrix} 12 & 8 & 4 \\ 24 & -10 & -5 \end{bmatrix}$

207.

ⓐ $\begin{bmatrix} 2 & 1 & -4 & 5 \\ 6 & -5 & 2 & 3 \\ 3 & -3 & 1 & -1 \end{bmatrix}$

ⓑ $\begin{bmatrix} 2 & 1 & -4 & 5 \\ 6 & -5 & 2 & 3 \\ 3 & -3 & 1 & -1 \end{bmatrix}$

ⓒ $\begin{bmatrix} 2 & 1 & -4 & 5 \\ 6 & -5 & 2 & 3 \\ -4 & 7 & -6 & 7 \end{bmatrix}$

209. $\begin{bmatrix} 1 & -2 & 3 & -4 \\ 0 & 5 & -11 & 17 \\ 0 & 1 & -10 & 7 \end{bmatrix}$

211. $(1, -1)$

213. $(3, 3)$

215. $(-2, 5, 2)$

217. $(-3, -5, 4)$

219. $(-3, 2, 3)$

221. $(-2, 0, -3)$

223. no solution

225. no solution

227. infinitely many solutions (x, y, z) where
$x = \frac{1}{2}z + 4; y = \frac{1}{2}z - 6; z$ is any real number

229. infinitely many solutions (x, y, z) where
$x = 5z + 2; y = -3z + 1; z$ is any real number

231. Answers will vary.

233. 4

235. 10

237. ⓐ 6 ⓑ -14 ⓒ -6

239. ⓐ 9 ⓑ -3 ⓒ 8

241. -77

243. 49

245. -24

247. 25

249. $(7, 6)$

251. $(-2, 0)$

253. $(-3, 2)$

255. $(-9, 3)$

257. $(-3, -5, 4)$

259. $(2, -3, -2)$

261. $(-3, 2, 3)$

263. $(-2, 0, -3)$

265. infinitely many solutions

267. inconsistent

269. inconsistent

271. infinitely many solutions

273. yes

275. no

277. Answers will vary.

279. Answers will vary.

281. ⓐ false ⓑ true

283. ⓐ false ⓑ true

285. ⓐ false ⓑ true

287.

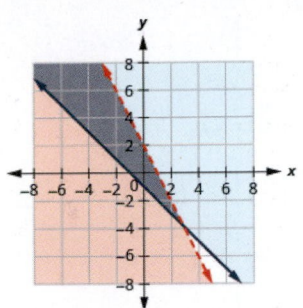

The solution is the grey region.

289.

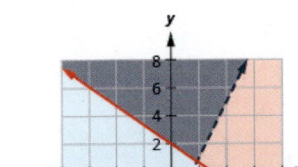

The solution is the grey region.

291.

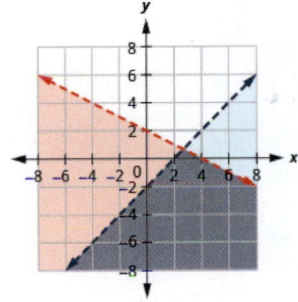

The solution is the grey region.

293.

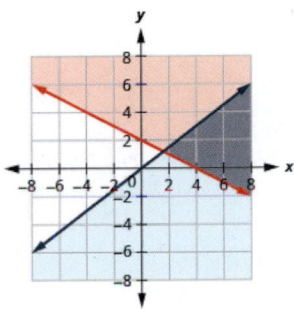

The solution is the grey region.

295.

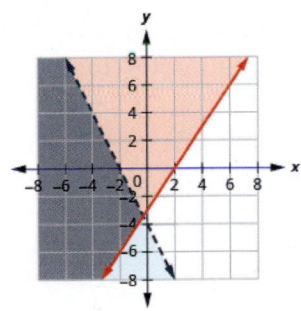

The solution is the grey region.

297.

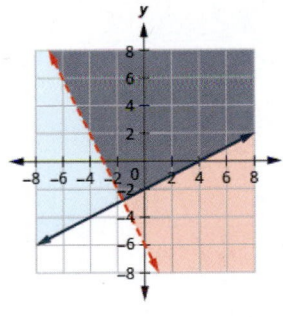

The solution is the grey region.

299.

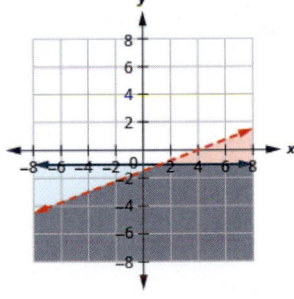

The solution is the grey region.

301.

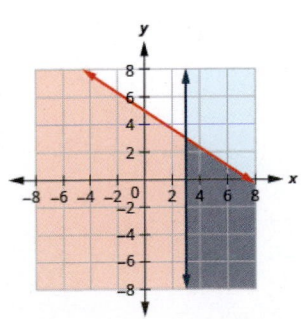

The solution is the grey region.

303.

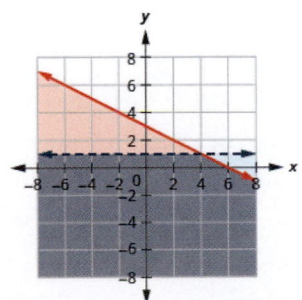

The solution is the grey region.

305.

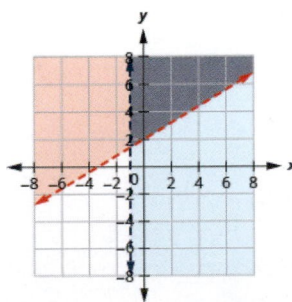

The solution is the grey region.

307.

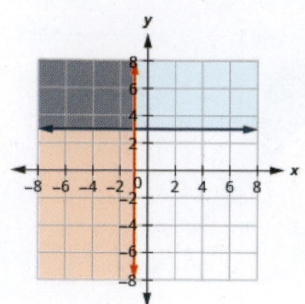

The solution is the grey region.

309.

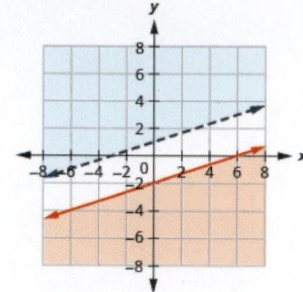

No solution.

311.

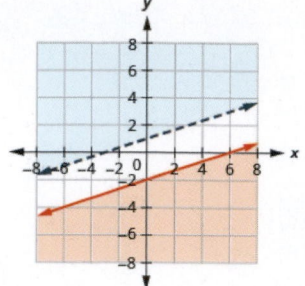

No solution.

313.

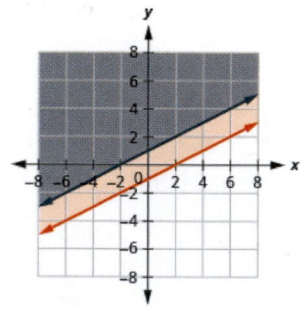

The solution is the grey region.

315.

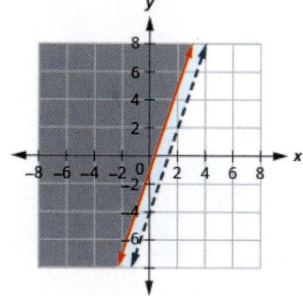

The solution is the grey region.

317.

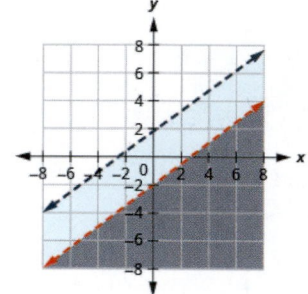

The solution is the grey region.

319. ⓐ
$$\begin{cases} f \geq 0 \\ p \geq 0 \\ f + p \leq 20 \\ 2f + 5p \leq 50 \end{cases}$$

ⓑ

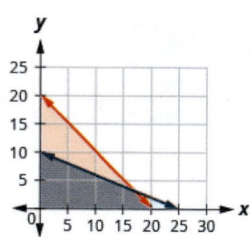

ⓒ yes

ⓓ no

321. ⓐ
$$\begin{cases} c \geq 0 \\ a \geq 0 \\ c + a \leq 24 \\ a \geq 3c \end{cases}$$

ⓑ

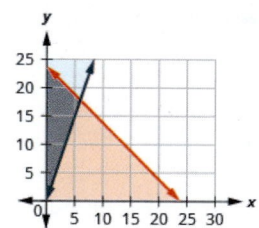

ⓒ yes

ⓓ no

323. ⓐ
$$\begin{cases} w \geq 0 \\ b \geq 0 \\ 27w + 16b > 80 \\ 3.20w + 1.75b \leq 10 \end{cases}$$

ⓑ

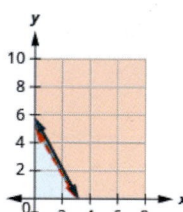

ⓒ no

ⓓ yes

325. ⓐ $\begin{cases} w \geq 0 \\ r \geq 0 \\ w + r \geq 4 \\ 270w + 650r \geq 1500 \end{cases}$

327. Answers will vary.

ⓑ

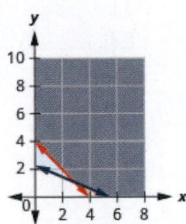

ⓒ no

ⓓ yes

Review Exercises

329. ⓐ yes ⓑ no

331.

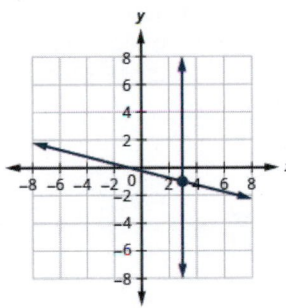

$(3, -1)$

333.

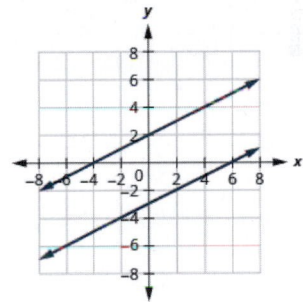

no solution

335. one solution, consistent system, independent equations

337. $(4, 5)$

339. $(3, 1)$

341. infinitely many solutions

343. $(4, -1)$

345. $(6, 2)$

347. elimination

349. 50 irises and 150 tulips

351. 10 calories jogging and 10 calories cycling

353. 119, 61

355. $35°$ and $55°$

357. the length is 450 feet, the width is 264 feet

359. $\frac{1}{2}$ an hour

361. the rate of the jet is 395 mph, the rate of the wind is 7 mph

363. 41 dimes and 11 pennies

365. $46\frac{2}{3}$ liters of 30% solution, $23\frac{1}{3}$ liters of 60% solution

367. $29,000 for the federal loan, $14,000 for the private loan

369. ⓐ no ⓑ yes

371. $(-3, 2, -4)$

373. no solution

375. 25, 20, 15

377. $\begin{bmatrix} 4 & 3 & 0 & -2 \\ 1 & -2 & -3 & 7 \\ 2 & -1 & 2 & -6 \end{bmatrix}$

379. $\begin{cases} x - 3z = -1 \\ x - 2y = -27 \\ -y + 2z = 3 \end{cases}$

381.

(a) $\begin{bmatrix} 1 & -3 & -2 & 4 \\ 4 & -2 & -3 & -1 \\ 2 & 2 & -1 & -3 \end{bmatrix}$

(b) $\begin{bmatrix} 2 & -6 & -4 & 8 \\ 4 & -2 & -3 & -1 \\ 2 & 2 & -1 & -3 \end{bmatrix}$

(c) $\begin{bmatrix} 2 & -6 & -4 & 8 \\ 4 & -2 & -3 & -1 \\ 0 & -6 & -1 & 5 \end{bmatrix}$

383. $(-2, 5, -2)$

385. no solution

387. -4

389. 21

391. $(-3, 2)$

393. $(-3, 2, 3)$

395. inconsistent

397. (a) yes (b) no

399.

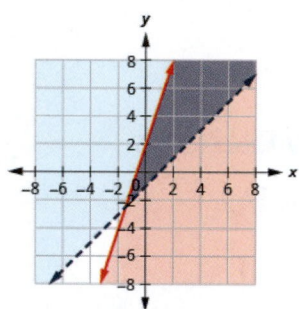

The solution is the grey region.

401.

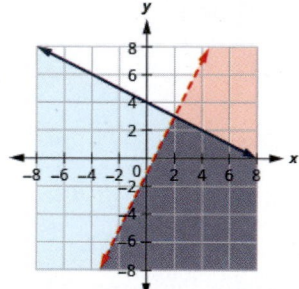

The solution is the grey region.

403.

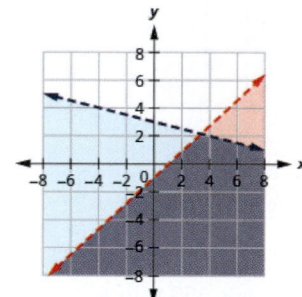

No solution.

405. (a) $\begin{cases} b \geq 0 \\ n \geq 0 \\ b + n \leq 40 \\ 12b + 18n \geq 500 \end{cases}$

(b)

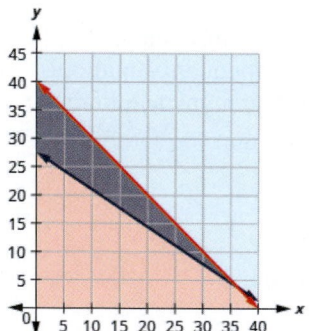

(c) yes

(d) no

Practice Test

407.

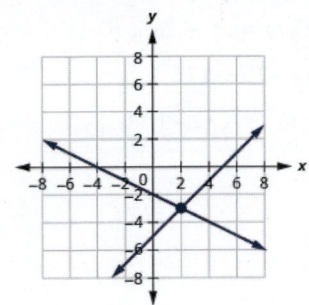

$(2, -3)$

409. $(2, 1)$

411. $(2, -2, 1)$

413. $(5, 7, 4)$

415. 9

417. 15 liters of 1% solution, 5 liters of 5% solution

419. The candy cost $20; the cookies cost $5; and the popcorn cost $10.

421. (a)  $\begin{cases} C \geq 0 \\ L \geq 0 \\ C + 0.5L \leq 50 \\ L \geq 3C \end{cases}$

(b)

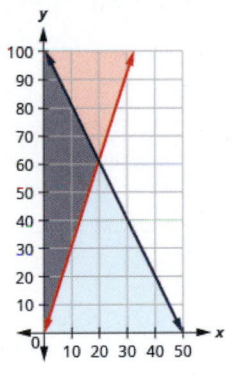

(c) no

(d) yes

Chapter 5

Try It

5.1. (a) monomial, 0 (b) polynomial, 3 (c) trinomial, 3 (d) binomial, 2 (e) monomial, 10

5.2. (a) binomial, 3 (b) trinomial, 3 (c) monomial, 0 (d) polynomial, 4 (e) monomial, 7

5.3. (a) $21q^2$ (b) $13mn^3$

5.4. (a) $-7c^2$ (b) $-10y^2z^3$

5.5. (a) $5y^2 + 3z^2$ (b) $m^2n^2 - 8m^2 + 4n^2$

5.6. (a) $-4m^2 + n^2$ (b) $pq^2 - 6p - 5q^2$

5.7. $8x^2 - 11x + 8$

5.8. $17y^2 + 14y + 1$

5.9. $x^2 + 3x - 5$

5.10. $6b^2 + 3$

5.11. $-5ab + 7b^2$

5.12. $7mn + 4n^2$

5.13. $5x^2 - 5xy + 5y^2$

5.14. $7x^2 - 6xy - 2y^2$

5.15. $x^3 + y^3$

5.16. $p^3 - 3p^2q + q^3$

5.17. (a) 18 (b) 50 (c) -15

5.18. (a) 20 (b) 2 (c) -4

5.19. The height is 150 feet.

5.20. The height is 31 feet.

5.21. ⓐ $(f+g)(x) = 3x^2 - 6x - 3$
ⓑ $(f+g)(3) = 6$
ⓒ $(f-g)(x) = x^2 - 2x + 9$
ⓓ $(f-g)(-2) = 17$

5.22. ⓐ $(f+g)(x) = 6x^2 - x + 7$
ⓑ $(f+g)(3) = 58$
ⓒ $(f-g)(x) = 4x^2 - 7x - 9$
ⓓ $(f-g)(-2) = 21$

5.23. ⓐ b^{17} ⓑ 4^{3x} ⓒ $12p^6$
ⓓ x^{18}

5.24. ⓐ x^{16} ⓑ 10^{x+1} ⓐ $12z^8$
ⓓ b^{19}

5.25. ⓐ x^5 ⓑ 6^9 ⓒ $\dfrac{1}{x^4}$
ⓓ $\dfrac{1}{12^{15}}$

5.26. ⓐ y^6 ⓑ 10^8 ⓒ $\dfrac{1}{m^8}$
ⓓ $\dfrac{1}{9^{11}}$

5.27. ⓐ 1 ⓑ 1

5.28. ⓐ 1 ⓑ 1

5.29. ⓐ $\dfrac{1}{z^3}$ ⓑ $\dfrac{1}{10^7}$ ⓒ p^8 ⓓ 64

5.30. ⓐ $\dfrac{1}{n^2}$ ⓑ $\dfrac{1}{10,000}$ ⓒ q^7
ⓓ 16

5.31. ⓐ $\dfrac{81}{16}$ ⓑ $\dfrac{n^2}{m^2}$

5.32. ⓐ $\dfrac{125}{27}$ ⓑ $\dfrac{b^4}{a^4}$

5.33. ⓐ $\dfrac{1}{z^9}$ ⓑ $\dfrac{1}{p^3 q^3}$ ⓒ $-\dfrac{12v^5}{u}$

5.34. ⓐ $\dfrac{1}{c^{15}}$ ⓑ $\dfrac{1}{r^2 s^8}$ ⓒ $\dfrac{30d^3}{c^8}$

5.35. ⓐ b^{35} ⓑ 5^{12} ⓒ a^{48}

5.36. ⓐ z^{54} ⓑ 3^{49} ⓒ q^{29}

5.37. ⓐ $32w^5 x^5$ ⓑ 1 ⓒ $\dfrac{1}{16b^{12}}$
ⓓ $\dfrac{64}{a^8}$

5.38. ⓐ $-27y^3$ ⓑ 1 ⓒ $\dfrac{1}{16x^8}$
ⓓ $\dfrac{8}{c^{12}}$

5.39. ⓐ $\dfrac{p^4}{10000}$ ⓑ $\dfrac{n^7}{m^7}$
ⓒ $\dfrac{81a^4 b^{12}}{c^8}$ ⓓ $\dfrac{27}{x^6 y^9}$

5.40. ⓐ $\dfrac{-8}{q^3}$ ⓑ $\dfrac{x^4}{w^4}$ ⓒ $\dfrac{x^2 y^6}{9z^4}$
ⓓ $\dfrac{8n^6}{m^6}$

5.41. ⓐ $81c^{24} d^{30}$ ⓑ $\dfrac{1}{a^{18}}$
ⓒ y^{15}

5.42. ⓐ $256a^{22} b^{24}$ ⓑ $\dfrac{1}{p^{39}}$
ⓒ $2x^3 y^{10}$

5.43. ⓐ 9.6×10^4 ⓑ 7.8×10^{-3}

5.44. ⓐ 4.83×10^4
ⓑ 1.29×10^{-2}

5.45. ⓐ 1,300 ⓑ -0.00012

5.46. ⓐ $-950,000$ ⓑ 0.075

5.47. ⓐ -0.006 ⓑ 20,000

5.48. ⓐ -0.009 ⓑ 400,000

5.49. ⓐ $-35y^{11}$ ⓑ $6a^5 b^6$

5.50. ⓐ $54b^9$ ⓑ $8r^{11} s^8$

5.51. ⓐ $-15y^3 - 24y^2 + 21y$
ⓑ $12x^4 y^2 - 20x^3 y^3 + 12x^2 y^4$

5.52. ⓐ $8x^4 - 24x^3 + 20x^2$
ⓑ $-18a^5 b + 12a^4 b^2 - 36a^3 b^3$

5.53. ⓐ $x^2 + 17x + 72$
ⓑ $15c^2 + 14c - 8$

5.54. ⓐ $20x^2 + 51x + 27$
ⓑ $30y^2 - 3y - 6$

5.55. ⓐ $x^2 - 2x - 35$
ⓑ $15x^2 + 29x - 14$

5.56. ⓐ $b^2 + 3b - 18$
ⓑ $16y^2 - 20y - 50$

5.57. ⓐ $x^3 - 8x^2 + 6x - 48$
ⓑ $8a^2 b^2 + 12ab - 20$

5.58. ⓐ $y^3 - 9y^2 + 7y - 63$
ⓑ $8x^2 y^2 + 2xy - 15$

5.59. $15m^2 - 51m + 42$

5.60. $42b^2 - 53b + 15$

5.61. ⓐ $y^3 - 8y^2 + 17y - 6$
ⓑ $y^3 - 8y^2 + 17y - 6$

5.62. ⓐ $2x^3 + 5x^2 - 7x + 20$
ⓑ $y^3 - 8y^2 + 17y - 6$

5.63. ⓐ $x^2 + 18x + 81$
ⓑ $4c^2 - 4cd + d^2$

5.64. ⓐ $y^2 + 22y + 121$
ⓑ $16x^2 - 40xy + 25y^2$

5.65. ⓐ $36x^2 - 25$
ⓑ $16p^2 - 49q^2$

5.66. ⓐ $4x^2 - 49$ ⓑ $9x^2 - y^2$

5.67. ⓐ FOIL; $18b^2 + 77b - 18$
ⓑ Binomial Squares;
$81p^2 - 72p + 16$
ⓒ Binomial Squares;
$49y^2 + 14y + 1$
ⓓ Product of Conjugates;
$16r^2 - 9$

5.68. ⓐ Binomial Squares;
$36x^2 + 84x + 49$ ⓑ Product of
Conjugates; $9x^2 - 16$ ⓒ FOIL;
$10x^2 - 29x + 10$ ⓓ Binomial
Squares; $36n^2 - 12n + 1$

5.69. ⓐ
$(f \cdot g)(x) = x^3 - 7x^2 + 13x - 15$
ⓑ $(f \cdot g)(2) = -9$

5.70. ⓐ
$(f \cdot g)(x) = x^3 + x^2 - 52x - 28$
ⓑ $(f \cdot g)(2) = -120$

5.71. $-\dfrac{9}{a^5 b}$

5.72. $\dfrac{-9d}{c^4}$

5.73. $\dfrac{4y^2}{7x^4}$

5.74. $\dfrac{5}{8m^5 n^3}$

5.75. $-4a + 2b$

5.76. $8a^5 b + 6a^3 b^2$

5.77. $y + 7$

5.78. $m + 5$

5.79. $x^3 - 3x^2 + 2x + 1 + \dfrac{3}{x + 3}$

5.80. $x^3 - 3x^2 - 2x - 1 - \dfrac{3}{x + 3}$

5.81. $x^2 + 4x + 16$

5.82. $25x^2 + 10x + 4$

5.83. $3x^2 + 4x - 2; 2$

5.84. $4x^2 - 3x + 1; 1$

5.85. $x^3 - 4x^2 + 5; 0$

5.86. $x^3 - 3x^2 + 2; 0$

5.87. ⓐ $\left(\dfrac{f}{g}\right)(x) = x - 8$
ⓑ $\left(\dfrac{f}{g}\right)(-3) = -11$

5.88. ⓐ $\left(\dfrac{f}{g}\right)(x) = x - 9$
ⓑ $\left(\dfrac{f}{g}\right)(-5) = -14$

5.89. -1

5.90. 6

5.91. yes

5.92. yes

Section Exercises

1. ⓐ trinomial, 5 ⓑ polynomial, 3
ⓒ binomial, 1 ⓓ monomial, 1
ⓔ binomial, 1

3. ⓐ binomial ⓑ trinomial
ⓒ polynomial ⓓ trinomial
ⓔ monomial

5. ⓐ 2^0 ⓑ 3^0 ⓒ 1^0 ⓓ 3^0
ⓔ 0^0

7. ⓐ 1^0 ⓑ 2^0 ⓒ 3^0 ⓓ 3^0
ⓔ 0^0

9. ⓐ $12x^2$ ⓑ $-5a$

11. ⓐ $6w$ ⓑ $19x^2 y$

13. $12x^2 - 5a$

15. $6w + 19x^2 y$

17. ⓐ $-22b$ ⓑ $16xy$

19. ⓐ $-10a + 5b$
ⓑ $pq^2 - 4p - 3q^2$

21. ⓐ $-4a^2 + b^2$
ⓑ $x^2 y - 3x + 7xy^2$

23. ⓐ $xy^2 - 5x - 5y^2$
ⓑ $19y + 5z$

25. $x - 3y + a^2 b - 4a - 5ab^2$

27. $-u^2 + 4v^2 + 12a + 8b$

29. $-4a - 3b$

31. $-7x^6$

33. $11y^2 + 4y + 11$

35. $-3x^2 + 17x - 1$

37. $11x^2 - 5x + 5$

39. $6a^2 - 4a - 1$

41. $2m^2 - 7m + 4$

43. $5a + 3$

45. $12s^2 - 14s + 9$

47. $3x^2 - x + 4$

49. $w^2 + 3w + 4$

51. $11w - 64$

53. $10x^2 - 7xy + 6y^2$

55. $10m^2 + 3mn - 8n^2$

57. $-3ab + 3b^2$

59. $p^3 - 6p^2q + pq^2 + 4q^3$

61. $x^3 + 2x^2y - 5xy^2 + y^3$

63. ⓐ 187 ⓑ 40 ⓒ 2

65. ⓐ -104 ⓑ 4 ⓒ 40

67. The height is 11 feet.

69. The revenue is $10,800.

71. The cost is $456.

73. ⓐ $(f+g)(x) = 7x^2 + 4x + 4$
ⓑ $(f+g)(2) = 40$
ⓒ $(f-g)(x) = -3x^2 - 12x - 2$
ⓓ $(f-g)(-3) = 7$

75.
ⓐ $(f+g)(x) = 6x^3 - x^2 - 9x + 3$
ⓑ $(f+g)(2) = 29$
ⓒ $(f-g)(x) = -x^2 + 5x + 3$
ⓓ $(f-g)(-3) = -21$

77. Answers will vary.

79. Answers will vary.

81. ⓐ d^9 ⓑ 4^{14x} ⓒ $8y^4$ ⓓ w^6

83. ⓐ n^{31} ⓑ 3^{x+6} ⓒ $56w^6$ ⓓ a^{16}

85. m^{x+3}

87. y^{a+b}

89. ⓐ x^{15} ⓑ 5^9 ⓒ $\frac{1}{q^{18}}$ ⓓ $\frac{1}{10}$

91. ⓐ p^{14} ⓑ 4^{12} ⓒ $\frac{1}{b^8}$ ⓓ $\frac{1}{4^5}$

93. ⓐ 1 ⓑ 1

95. ⓐ -1 ⓑ -1

97. ⓐ $\frac{1}{a^2}$ ⓑ $\frac{1}{1000}$ ⓒ c^5 ⓓ 9

99. ⓐ $\frac{1}{r^3}$ ⓑ $\frac{1}{100,000}$ ⓒ q^{10} ⓓ 1,000

101. ⓐ $\frac{64}{25}$ ⓑ $\frac{a^2}{b^2}$

103. ⓐ $\frac{729}{64}$ ⓑ $-\frac{v^5}{u^5}$

105. ⓐ $\frac{1}{25}$ ⓑ $\frac{1}{25}$ ⓒ 25 ⓓ -25

107. ⓐ $\frac{3}{5}$ ⓑ $\frac{1}{15}$

109. ⓐ $\frac{1}{b^4}$ ⓑ $\frac{w^2}{x^9}$ ⓒ $-12cd^4$

111. ⓐ 1 ⓑ $\frac{1}{u^4v^5}$ ⓒ $-36\frac{r^2}{j^5}$

113. $\frac{1}{p}$

115. ⓐ m^8 ⓑ 10^{18} ⓒ $\frac{1}{x^{12}}$

117. ⓐ y^{3x} ⓑ 5^{xy} ⓒ $\frac{1}{q^{48}}$

119. ⓐ $9x^2y^2$ ⓑ 1 ⓒ $\frac{1}{25x^4}$ ⓓ $\frac{16}{y^6}$

121. ⓐ $-125a^3b^3$ ⓑ 1 ⓒ $\frac{1}{36x^6}$ ⓓ $\frac{9}{y^8}$

123. ⓐ $\frac{p^5}{32}$ ⓑ $\frac{x^6}{y^6}$ ⓒ $\frac{8x^3y^6}{z^3}$ ⓓ $\frac{16}{p^6q^4}$

125. ⓐ $\frac{a^4}{81b^4}$ ⓑ $\frac{16m^2}{25}$ ⓒ $\frac{8x^3y^6}{z^3}$ ⓓ $\frac{16}{p^6q^4}$

127. ⓐ $1125t^8$ ⓑ $\frac{1}{t^{19}}$ ⓒ $\frac{y^4}{3x^2}$

129. ⓐ $16m^8n^{22}$ ⓑ $\frac{4}{p^6}$

131. ⓐ $\frac{7}{n}$ ⓑ $\frac{1}{7n}$ ⓒ $-\frac{1}{7n}$

133. ⓐ $\frac{1}{9p^2}$ ⓑ $\frac{3}{p^2}$ ⓒ $\frac{-3}{p^2}$

135. x^{14}

137. x^{30}

139. $2m^{18}$

141. $1,000x^6y^3$

143. $16a^{12}b^8$

145. $\frac{8}{27}x^6y^3$

147. $1,024a^{10}$

149. $25,000p^{24}$

151. $x^{18}y^{18}$

153. $144m^8n^{22}$

155. ⓐ $45x^3$ ⓑ $48y^4$

157. ⓐ $\frac{1}{2r^4}$ ⓑ $\frac{1}{3}x^{11}$

159. $\frac{1}{j^3}$

161. $-\frac{4000}{n^{12}}$

163. ⓐ 34×10^4 ⓑ 41×10^{-3}

165. ⓐ 1.29×10^6 ⓑ 103×10^{-8}

167. ⓐ -830 ⓑ 0.038

169. ⓐ 16,000,000,000 ⓑ 0.00000843

171. ⓐ 0.02 ⓑ 500,000,000

173. ⓐ 0.0000056 ⓑ 20,000,000

175. Answers will vary.

177. Answers will vary.

179. ⓐ $30x^8$ ⓑ $15x^8y^2$

181. ⓐ $72c^5$ ⓑ $\frac{1}{3}m^5 n^5$

183. ⓐ $-5t^3 - 15t^2 + 90t$

ⓑ $9sr^5 - 27s^2 r^4 + 45s^3 r^3$

185. ⓐ $-5m^3 - 15m^2 + 90m$

ⓑ $-21x^4 y^2 - 30x^3 y^3 + 3x^2 y^4$

187. $y^2 + 12y + 27$

189. $21q^2 - 44q - 32$

191. $y^2 - 8y + 12$

193. $6p^2 + 11p + 5$

195. $m^2 + 7m - 44$

197. $33r^2 - 85r - 8$

199. $y^3 + 3y^2 - 4y - 12$

201. $6x^2 y^2 + 13xy + 6$

203. $y^4 - 11y^2 + 28$

205. $9r^2 s^2 - 33rs + 28$

207. $u^3 + 7u^2 + 14u + 8$

209. $3a^3 + 31a^2 + 5a - 50$

211.
$6a^4 - 13a^3 + 15a^2 + 35a - 50$

213. $p^3 - 10p^2 + 33p - 36$

215. $6r^3 - 41r^2 - 61r - 9$

217. $q^2 + 24q + 144$

219. $4y^2 - 12yz + 9z^2$

221. $x^2 + \frac{4}{3}x + \frac{4}{9}$

223. $\frac{1}{64}x^2 - \frac{1}{36}xy + \frac{1}{81}y^2$

225. $25u^4 + 90u^2 + 81$

227. $64p^6 - 48p^3 + 9$

229. $64j^2 - 16$

231. $81c^2 - 25$

233. $49w^2 - 100x^2$

235. $p^2 - \frac{16}{25}q^2$

237. $x^2 y^2 - 81$

239. $225m^4 - 64n^8$

241. $t^2 - 18t + 81$

243. $2x^2 - 3xy - 2y^2$

245. $9p^2 - 64$

247. $k^2 - 12k + 36$

249. $8x^3 - x^2 y^2 + 64xy - 8y^3$

251. $y^8 + 4y^4 z + 4z^2$

253. $m^6 - 16m^3 n + 64n^2$

255. $r^5 + r^2 s^2 - r^3 s^3 - s^5$

257. $18p - 9$

259. $-10j - 15$

261. $9d^7$

263. $72m^5 n^8$

265. $5q^5 - 10q^4 + 30q^3$

267. $y^3 - y^2 - 2y$

269. $6k^3 - 11k^2 - 26k + 4$

271. $121 - b^2$

273. $4x^4 - 9y^8$

275. $9d^2 + 6d + 1$

277. $9z^2 - \frac{6}{5}z + \frac{1}{25}$

279.
ⓐ $(f \cdot g)(x) = 4x^3 - x^2 - 8x + 5$

ⓑ $(f \cdot g)(-2) = -15$

281. ⓐ $(f \cdot g)(x) = 49x^2 - 64$

ⓑ $(f \cdot g)(-2) = 187$

283.
ⓐ

$(f \cdot g)(x) = x^4 + 6x^3 + 9x^2 + 10x - 12$

ⓑ $(f \cdot g)(1) = 14$

285. Answers will vary.

287. Answers will vary.

289. $\frac{2m^3}{3n^5}$

291. $\frac{-3y^3}{4x^3}$

293. $\frac{-3q^5}{p^5}$

295. $\frac{5v^5}{u^2}$

297. $4x^2 + 3x$

299. $-6y^2 + 3y$

301. $6r^3 + 11r^2 s - 8rs^2$

303. $-5y - 3 + \frac{1}{4y}$

305. $a - 7$

307. $4x + 3$

309. $p + 3 - \frac{8}{p + 8}$

311. $2n^2 - 6n + 8$

313. $m^2 - 10m + 100$

315. $25y^2 + 20x + 16$

317. $x^2 - 5x + 6; 0$

319. $2x^2 - 3x + 4; 4$

321. $x^3 - 2x^2 + 5x - 4; -2$

323. $3x^3 - 2x^2 - 4x - 2; 0$

325. ⓐ $\left(\frac{f}{g}\right)(x) = x - 6$

ⓑ $\left(\frac{f}{g}\right)(-5) = -11$

327. ⓐ $\left(\frac{f}{g}\right)(x) = x^2 + 5x - 4$

ⓑ $\left(\frac{f}{g}\right)(0) = -4$

329.
ⓐ

$(f \cdot g)(x) = x^4 + 6x^3 + 9x^2 + 10x - 12;$

ⓑ $(f \cdot g)(1) = 14$

331. -9

333. -6

335. no

337. yes

339. answer will vary

341. Answers will vary.

Review Exercises

343. binomial

345. other polynomial

347. $-13y^3$

349. $6m^2 + 19m - 4$

351. $5u^2 + 4u + 7$

353. $2x^2 - 2x + 23$

355. $-7b - 3a$

357. $8a^2 - 5a - 2$

359. $-3m + 3$

361. $3y^2 - 8y - 14$

363. $x^3 + 2x^2 y - 4xy^2$

365. (a) 165 (b) 39 (c) 5

367. The height is 64 feet.

369. (a) $(f + g)(x) = 4x^2 - 5x - 2$

(b) $(f + g)(3) = 19$

(c) $(f - g)(x) = -3x - 12$

(d) $(f - g)(-2) = -6$

371. p^{13}

373. a^6

375. y^{a+b}

377. a^5

379. $\dfrac{1}{x^4}$

381. 1

383. -3

385. $-\dfrac{1}{1000}$

387. $\dfrac{1}{8n}$

389. $\dfrac{1}{1000}$

391. 36

393. -125

395. $-\dfrac{1}{125}$

397. $\dfrac{x^3}{27}$

399. 3^{10}

401. x^5

403. $\dfrac{1}{u^3 v^5}$

405. $\dfrac{1}{m^5}$

407. $\dfrac{1}{q^{16}}$

409. n^2

411. 1

413. $\dfrac{9}{y^8}$

415. $\dfrac{81x^4 y^8}{z^4}$

417. $27x^7 y^{17}$

419. $\dfrac{3y^4}{4x^4}$

421. 5.3×10^6

423. 29,000

425. 0.00009413

427. 0.00072

429. 9,000

431. $10c^{10}$

433. $\dfrac{m^7 n^{10}}{9}$

435. $a^4 - 9a^3 - 36a^2$

437. $8n^4 - 10n^3$

439. $y^2 + 8y - 48$

441. $18p^2 - 93p + 110$

443. $k^2 - 3k - 54$

445. $10y^2 - 59y + 63$

447. $x^2 + x - 72$

449. $30a^2 - 33a + 3$

451. $15b^3 - b^2 - 47b + 18$

453. $24y^2 - 54y^2 + 32y - 5$

455. $x^2 + \dfrac{3}{2}x + \dfrac{9}{16}$

457. $25p^2 + 70pq + 49q^2$

459. $36x^2 - y^2$

461. $144x^6 - 49y^4$

463. $9p^9$

465. $-\dfrac{3}{y^4}$

467. $\dfrac{4a^5}{b^2}$

469. $\dfrac{4m^9}{n^4}$

471. $-9y^2 + 4y$

473. $-3x - 1 + \dfrac{3}{4x}$

475. $y - 3 + \dfrac{33}{q + 6}$

477. $a^2 + a + 1$

479. $2x^2 - 5x - 4; 0$

481. (a) $\left(\dfrac{f}{g}\right)(x) = x - 6$

(b) $\left(\dfrac{f}{g}\right)(-2) = -8$

483. -9

485. no

Practice Test

487. (a) trinomial (b) 4

489. $6x^2 - 3x + 11$

491. x

493. 1

495. $\dfrac{1}{8y^3}$

497. x^9

499. $\dfrac{4r}{s^6}$

501. $-48x^5 y^9$

503. $21m^2 - 19m - 6$

505. $16x^2 - 24x + 9$

507. $3y^2 - 7x$

509. yes

511. 4.4×10^3

513. ⓐ 36 ⓑ 21

515. ⓐ $\left(\dfrac{f}{g}\right)(x) = 3x + 4$

ⓑ $\left(\dfrac{f}{g}\right)(3) = 13$

Chapter 6

Try It

6.1. $5m^2$

6.2. $7x$

6.3. $3y^2\left(3x + 2x^2 + 7y\right)$

6.4. $3p\left(p^2 - 2pq + 3q^2\right)$

6.5. $2x^2(x + 6)$

6.6. $3y^2(2y - 5)$

6.7. $3xy\left(5x^2 - xy + 2y^2\right)$

6.8. $2ab\left(4a^2 + ab - 3b^2\right)$

6.9. $-4b\left(b^2 - 4b + 2\right)$

6.10. $-7a\left(a^2 - 3a + 2\right)$

6.11. $(m + 3)(4m - 7)$

6.12. $(n - 4)(8n + 5)$

6.13. $(x + 8)(y + 3)$

6.14. $(a + 7)(b + 8)$

6.15. ⓐ $(x - 5)(x + 2)$

ⓑ $(5x - 4)(4x - 3)$

6.16. ⓐ $(y + 4)(y - 7)$

6.17. $(q + 4)(q + 6)$

6.18. $(t + 2)(t + 12)$

ⓑ $(7m - 3)(6m - 5)$

6.19. $(u - 3)(u - 6)$

6.20. $(y - 7)(y - 9)$

6.21. $(m + 3)(m + 6)$

6.22. $(n - 3)(n - 4)$

6.23. $(a - b)(a - 10b)$

6.24. $(m - n)(m - 12n)$

6.25. prime

6.26. prime

6.27. $5x(x - 1)(x + 4)$

6.28. $6y(y - 2)(y + 5)$

6.29. $(a + 1)(2a + 3)$

6.30. $(b + 1)(4b + 1)$

6.31. $(2x - 3)(4x - 1)$

6.32. $(2y - 7)(5y - 1)$

6.33. $(3x + 2y)(6x - 5y)$

6.34. $(3x + y)(10x - 21y)$

6.35. $5n(n - 4)(3n - 5)$

6.36. $8q(q + 6)(7q - 2)$

6.37. $(x + 2)(6x + 1)$

6.38. $(2y + 1)(2y + 3)$

6.39. $4(2x - 3)(2x - 1)$

6.40. $3(3w - 2)(2w - 3)$

6.41. $\left(h^2 - 2\right)\left(h^2 + 6\right)$

6.42. $\left(y^2 + 4\right)\left(y^2 - 5\right)$

6.43. $(x - 3)(x - 1)$

6.44. $(y - 1)(y + 1)$

6.45. $(2x + 3)^2$

6.46. $(3y + 4)^2$

6.47. $(8y - 5)^2$

6.48. $(4z - 9)^2$

6.49. $(7x + 6y)^2$

6.50. $(8m + 7n)^2$

6.51. $2y(2x - 3)^2$

6.52. $3q(3p + 5)^2$

6.53. $(11m - 1)(11m + 1)$

6.54. $(9y - 1)(9y + 1)$

6.55. $(16m - 5n)(16m + 5n)$

6.56. $(11p - 3q)(11p + 3q)$

6.57. $2y^2(x - 2)(x + 2)\left(x^2 + 4\right)$

6.58. $7c^2(a - b)(a + b)\left(a^2 + b^2\right)$

6.59. $(x - 5 - y)(x - 5 + y)$

6.60. $(x + 3 - 2y)(x + 3 + 2y)$

6.61. $(x + 3)\left(x^2 - 3x + 9\right)$

6.62. $(y + 2)\left(y^2 - 2y + 4\right)$

6.63. $(2x - 3y)\left(4x^2 - 6xy + 9y^2\right)$

6.64. $(10m - 5n)\left(100m^2 - 50mn + 25n^2\right)$

6.65. $4(5p + q)\left(25p^2 - 5pq + q^2\right)$

6.66. $2(6c + 7d)\left(36c^2 - 42cd + 49d^2\right)$

6.67. $(-2y + 1)\left(13y^2 + 5y + 1\right)$

6.68. $(-4n + 3)\left(31n^2 + 21n + 9\right)$

6.69. $8y(y - 1)(y + 3)$

6.70. $5y(y - 9)(y + 6)$

6.71. $4x(2x - 3)(2x + 3)$

6.72. $3(3y - 4)(3y + 4)$

6.73. $(2x + 5y)^2$

6.74. $(3x - 4y)^2$

6.75. $2xy\left(25x^2 + 36\right)$

6.76. $3xy\left(9y^2 + 16\right)$

6.77.
$2(5m + 6n)\left(25m^2 - 30mn + 36n^2\right)$

6.78. $2(p + 3q)\left(p^2 - 3pq + 9q^2\right)$

6.79. $4ab\left(a^2 + 4\right)(a - 2)(a + 2)$

6.80. $7xy\left(y^2 + 1\right)(y - 1)(y + 1)$

6.81. $6(x + b)(x - 2c)$

6.82. $2(4x - 1)(2x + 3y)$

6.83. $4q(p - 3)(p - 1)$

6.84. $3p(2q + 1)(q - 2)$

6.85. $(2x - 3y - 5)(2x - 3y + 5)$

6.86. $(4x - 3y - 8)(4x - 3y + 8)$

6.87. $m = \frac{2}{3}, m = -\frac{1}{2}$

6.88. $p = -\frac{3}{4}, p = \frac{3}{4}$

6.89. $c = 2, c = \frac{4}{3}$

6.90. $d = 3, d = -\frac{1}{2}$

6.91. $p = \frac{7}{5}, p = -\frac{7}{5}$

6.92. $x = \frac{11}{6}, x = -\frac{11}{6}$

6.93. $m = 1, m = \frac{3}{2}$

6.94. $k = 3, k = -3$

6.95. $a = -\frac{5}{2}, a = \frac{2}{3}$

6.96. $b = -2, b = -\frac{1}{20}$

6.97. $x = 0, x = \frac{3}{2}$

6.98. $y = 0, y = \frac{1}{4}$

6.99. ⓐ $x = -3$ or $x = 5$

ⓑ $(-3, 7)$ $(5, 7)$

6.100. ⓐ $x = 1$ or $x = 7$

ⓑ $(1, -4)$ $(7, -4)$

6.101. ⓐ $x = 1$ or $x = \frac{5}{2}$

ⓑ $(1, 0)$, $\left(\frac{5}{2}, 0\right)$ ⓒ $(0, 5)$

6.102. ⓐ $x = -3$ or $x = \frac{5}{6}$

ⓑ $(-3, 0)$, $\left(\frac{5}{6}, 0\right)$ ⓒ $(0, -15)$

6.103. $-15, -17$ and $15, 17$

6.104. $-23, -21$ and $21, 23$

6.105. The width is 5 feet and length is 6 feet.

6.106. The length of the patio is 12 feet and the width 15 feet.

6.107. 5 feet and 12 feet

6.108. 24 feet and 25 feet

6.109. ⓐ 5 ⓑ 0;3 ⓒ 196

6.110. ⓐ 4 ⓑ 0;2 ⓒ 144

Section Exercises

1. $2pq$

3. $6m^2 n^3$

5. $2a$

7. $5x^3 y$

9. $3(2m + 3)$

11. $9(n - 7)$

13. $3(x^2 + 2x - 3)$

15. $2(p^2 + 4p + 1)$

17. $8y^2(y + 2)$

19. $5x(x^2 - 3x + 4)$

21. $3x\left(8x^2 - 4x + 5\right)$

23. $6y^2(2x + 3x^2 - 5y)$

25. $4xy\left(5x^2 - xy + 3y^2\right)$

27. $-2(x + 4)$

29. $-2x\left(x^2 - 9x + 4\right)$

31. $-4pq\left(p^2 + 3pq - 4q\right)$

33. $(x + 1)(5x + 3)$

35. $(b - 2)(3b - 13)$

37. $(b + 5)(a + 3)$

39. $(y + 5)(8y + 1)$

41. $(u + 2)(v - 9)$

43. $(u - 1)(u + 6)$

45. $(3p - 5)(3p + 4)$

47. $(n - 6)(m - 4)$

49. $(x - 7)(2x - 5)$

51. $-9xy(3x + 2y)$

53. $(x^2 + 2)(3x - 7)$

55. $(x + y)(x + 5)$

57. Answers will vary.

59. Answers will vary.

61. $(p + 5)(p + 6)$

63. $(n + 3)(n + 16)$

65. $(a + 5)(a + 20)$

67. $(x - 2)(x - 6)$

69. $(y - 3)(y - 15)$

71. $(x - 1)(x - 7)$

73. $(p - 1)(p + 6)$

75. $(x - 4)(x - 2)$

77. $(x - 12)(x + 1)$

79. $(x + 8y)(x - 10y)$

81. $(m + n)(m - 65n)$

83. $(a + 8b)(a - 3b)$

85. Prime

87. Prime

89. $p(p - 10)(p + 2)$

91. $3m(m - 5)(m - 2)$

93. $5x^2(x - 3)(x + 5)$

95. $(2t + 5)(t + 1)$

97. $(11x + 1)(x + 3)$

99. $(4w - 1)(w - 1)$

101. $(4q + 1)(q - 2)$

103. $(2p - 5q)(3p - 2q)$

105. $(4a - 3b)(a + 5b)$

107. $-16(x - 1)(x - 1)$

109. $10q(3q + 2)(q + 4)$

111. $(5n + 1)(n + 4)$

113. $(2k - 3)(2k - 5)$

115. $(3y + 5)(2y - 3)$

117. $(2n + 3)(n - 15)$

119. $10(6y - 1)(y + 5)$

121. $3z(8z + 3)(2z - 5)$

123. $8(2s + 3)(s + 1)$

125. $12(4y - 3)(y + 1)$

127. $(x^2 + 1)(x^2 - 7)$

129. $(x^2 - 7)(x^2 + 4)$

131. $(x - 12)(x + 1)$

133. $(3y - 4)(3y - 1)$

135. $(u - 6)(u - 6)$

137. $(r - 4s)(r - 16s)$

139. $(4y - 7)(3y - 2)$

141. $(2n - 1)(3n + 4)$

143. $13(z^2 + 3z - 2)$

145. $3p(p + 7)$

147. $6(r + 2)(r + 3)$

149. $4(2n + 1)(3n + 1)$

151. $(x^2 + 2)(x^2 - 6)$

153. $(x - 9)(x + 6)$

155. Answers will vary.

157. Answers will vary.

159. $(4y + 3)^2$

161. $(6s + 7)^2$

163. $(10x - 1)^2$

165. $(5n - 12)^2$

167. $(7x + 2y)^2$

169. $(50y - 1)(2y - 1)$

171. $10j(k + 4)^2$

173. $3u^2(5u - v)^2$

175. $(5v - 1)(5v + 1)$

177. $(7x - 2)(7x + 2)$

179. $6p^2(q - 3)(q + 3)$

181. $6(4p^2 + 9)$

183. $(11x - 12y)(11x + 12y)$

185. $(13c - 6d)(13c + 6d)$

187. $(2z - 1)(2z + 1)(4z^2 + 1)$

189.
$2b^2(3a - 2)(3a + 2)(9a^2 + 4)$

191. $(x - 8 - y)(x - 8 + y)$

193. $(a + 3 - 3b)(a + 3 + 3b)$

195. $(x + 5)(x^2 - 5x + 25)$

197. $(z^2 - 3)(z^4 + 3z^2 + 9)$

199. $(2 - 7t)(4 + 14t + 49t^2)$

201.
$(2y - 5z)(4y^2 + 10yz + 25z^2)$

203.
$(6a + 5b)(36a^2 - 30ab + 25b^2)$

205. $7(k + 2)(k^2 - 2k + 4)$

207. $2x^2(1 - 2y)(1 + 2y + 4y^2)$

209. $-9(x + 1)(x^2 + 3)$

211. $-(3y + 5)(21y^2 - 30y + 25)$

213. $(8a - 5)(8a + 5)$

215. $3(3q - 1)(3q + 1)$

217. $(4x - 9)^2$

219. $2(4p^2 + 1)$

221. $(5 - 2y)(25 + 10y + 4y^2)$

223. $5(3n + 2)^2$

225. $(x + y - 5)(x - y - 5)$

227. $(3x + 1)(3x^2 + 1)$

229. Answers will vary.

231. Answers will vary.

233. $(2n - 1)(n + 7)$

235. $a^3(a^2 + 9)$

237. $(11r - s)(11r + s)$

239. $8(m - 2)(m + 2)$

241. $(5w - 6)^2$

243. $(m + 7n)^2$

245. $7(b + 3)(b - 2)$

247. $3(x - 3)(x^2 + 3x + 9)$

249. $(k - 2)(k + 2)(k^2 + 4)$

251. $5xy^2(x^2 + 4)(x + 2)(x - 2)$

253. $3(5p + 4)(q - 1)$

255. $4(x + 3)(x + 7)$

257. $u^2(u + 1)(u^2 - u + 1)$

259. prime

261. $10(m - 5)(m + 5)(m^2 + 25)$

263. $3y(3x + 2)(4x - 1)$

265. $(2x - 3y)(4x^2 + 6xy + 9y^2)$

267.
$(y + 1)(y - 1)(y^2 - y + 1)(y^2 + y + 1)$

269. $(3x - y + 7)(3x - y - 7)$

271. $(9x^2 - 12x + 4)$

273. Answers will vary.

275. Answers will vary.

277. $a = 10/3, a = 7/2$

279. $m = 0, m = 5/12$

281. $x = 1/2$

283. $a = -5/4, a = 6$

285. $m = 5/4, m = 3$

287. $a = -1, a = 0$

289. $m = 12/7, m = -12/7$

291. $y = -9/4, y = 9/4$

293. $n = -6/11, n = 6/11$

295. $x = 2, x = -5$

297. $x = 3/2, x = -1$

299. $x = 3/2, x = -1$

301. $x = -2/3$

303. $x = 2, x = -4/3$

305. $x = -3/2, x = 1/3$

307. $p = 0, p = ¾$

309. $x = 0, x = 6$

311. $x = 0, x = 1/3$

313. ⓐ $x = 2$ or $x = 6$ ⓑ $(2, -4)$
$(6, -4)$

315. ⓐ $x = \frac{3}{2}$ or $x = \frac{3}{4}$
ⓑ $\left(\frac{3}{2}, -4\right)$ $\left(\frac{3}{4}, -4\right)$

317. ⓐ $x = \frac{2}{3}$ or $x = -\frac{2}{3}$
ⓑ $\left(\frac{2}{3}, 0\right), \left(-\frac{2}{3}, 0\right)$ ⓒ $(0, -4)$

319. ⓐ $x = \frac{5}{3}$ or $x = -\frac{1}{2}$
ⓑ $\left(\frac{5}{3}, 0\right), \left(-\frac{1}{2}, 0\right)$ ⓒ $(0, -5)$

321. $-13, -11$ and 11, 13

323. $-14, -12$ and 12, 14

325. -4 and 7

331. $8, 15, 17$

327. $5, 11$

333. ⓐ $0, 2$ ⓑ 1

329. $6, 8$

335. Answers will vary.

Review Exercises

337. $3ab^2$

343. $3x(6x^2 - 5)$

349. $(a + b)(x - y)$

355. $(a + 3)(a + 11)$

361. $(a + 7b)(a - 3b)$

367. $(5y + 9)(y + 1)$

373. $(3a - 1)(6a - 1)$

379. $3(2a - 7)(3a + 1)$

385. $10(2x + 9)^2$

391. $(3 + 11y)(3 - 11y)$

397. $(2z - 1)(2z + 1)(4z^2 + 1)$

403. $2(m + 3)(m^2 - 3m + 9)$

409. $5u^2(u + 3)(u - 3)$

415. $(2b + 5c)(b - c)$

421. $(4x - 3y + 8)(4x - 3y - 8)$

427. $x = -4, x = -5$

433. $x = 2, x = -5$

339. $3y$

345. $4x(x^2 - 3x + 4)$

351. $(x - 3)(x + 7)$

357. $(m + 9)(m - 6)$

363. Prime

369. $(5y + 1)(2y - 11)$

375. $(3x - 1)(5x + 2)$

381. $(x^2 - 15)(x^2 + 2)$

387. $3u^2(5u - v)^2$

393. $n(13n + 1)(13n - 1)$

399. $(a + 3 - 3b)(a + 3 + 3b)$

405. $4x^2(6x + 11)$

411. prime

417. $5(q + 3)(q - 6)$

423. $b = -1/5, b = -1/6$

429. $p = -\frac{5}{2}, p = 8$

435. $p = 0, p = \frac{3}{4}$

341. $7(5y + 12)$

347. $-3x(x^2 - 9x + 4)$

353. $(m^2 + 1)(m + 1)$

359. $(x + 5y)(x + 7y)$

365. $3y(y - 5)(y - 2)$

371. $-9(9a - 1)(a + 2)$

377. $3(x + 4)(x - 3)$

383. $(5x + 3)^2$

389. $(13m + n)(13m - n)$

395. $6(4p^2 + 9)$

401. $(a - 5)(a^2 + 5a + 25)$

407. $(4n - 7m)^2$

413. $(b - 4)(b^2 + 4b + 16)$

419. $10(m - 5)(m + 5)(m^2 + 25)$

425. $x = 1/2$

431. $m = \frac{5}{12}, m = -\frac{5}{12}$

437. ⓐ $x = -7$ or $x = -4$
ⓑ $(-7, -8)$ $(-4, -8)$

439. ⓐ $x = \frac{7}{8}$ or $x = -\frac{7}{8}$
ⓑ $\left(\frac{7}{8}, 0\right)$, $\left(-\frac{7}{8}, 0\right)$ ⓒ $(0, -49)$

441. The numbers are -21 and -19 or 19 and 21.

443. The lengths are $8, 15,$ and 17 ft.

Practice Test

445. $40a^2(2 + 3a)$

451. $(3s - 2)^2$

457. $(3x^2 - 5)(2x^2 - 3)$

447. $(x + 7)(x + 6)$

453. $3(x + 5y)(x - 5y)$

459. $a = 4/5, a = -6$

449. $(x - 8)(y + 7)$

455. $(x + 5)(x^2 - 5x + 25)$

461. The width is 12 inches and the length is 14 inches.

463. ⓐ $x = 3$ or $x = 4$ ⓑ $(3, -7)$ $(4, -7)$

Chapter 7

Try It

7.1. ⓐ $x = 0$ ⓑ $n = -\frac{1}{3}$
ⓒ $a = -1, a = -3$

7.4. $\frac{x - 5}{x - 1}$, $x \neq -2$, $x \neq 1$

7.2. ⓐ $q = 0$ ⓑ $y = -\frac{2}{3}$
ⓒ $m = 2, m = -3$

7.5. $\frac{2(x - 3y)}{3(x + 3y)}$

7.3. $\frac{x + 1}{x - 1}$, $x \neq 2$, $x \neq 1$

7.6. $\frac{5(x - y)}{2(x + 5y)}$

7.7. $-\dfrac{x+1}{x+5}$

7.8. $-\dfrac{x+2}{x+1}$

7.9. $\dfrac{x-2}{2(x+3)}$

7.10. $\dfrac{3(x-6)}{x+5}$

7.11. $\dfrac{x-4}{x-5}$

7.12. $-\dfrac{(b+2)(b-1)}{(1+b)(b+4)}$

7.13. $\dfrac{2(x^2+2x+4)}{(x+2)(x^2-2x+4)}$

7.14. $\dfrac{2z}{z-1}$

7.15. $\dfrac{x+2}{4}$

7.16. $\dfrac{2}{y+5}$

7.17. $\dfrac{2(m+1)(m+2)}{3(m+4)(m-3)}$

7.18. $\dfrac{(n+5)(n+9)}{2(n+6)(2n+3)}$

7.19. The domain of $R(x)$ is all real numbers where $x \neq 5$ and $x \neq -1$.

7.20. The domain of $R(x)$ is all real numbers where $x \neq 4$ and $x \neq -2$.

7.21. $R(x) = 2$

7.22. $R(x) = \dfrac{1}{3}$

7.23. $R(x) = \dfrac{x-2}{4(x-8)}$

7.24. $R(x) = \dfrac{x(x-2)}{x-1}$

7.25. $x+2$

7.26. $x+3$

7.27. $\dfrac{x-11}{x-2}$

7.28. $\dfrac{x-3}{x+9}$

7.29. $\dfrac{y+3}{y+2}$

7.30. $\dfrac{3n-2}{n-1}$

7.31. ⓐ $(x-4)(x+3)(x+4)$

ⓑ $\dfrac{2x+8}{(x-4)(x+3)(x+4)}$, $\dfrac{x+3}{(x-4)(x+3)(x+4)}$

7.32. ⓐ $(x+2)(x-5)(x+1)$

ⓑ $\dfrac{3x^2+3x}{(x+2)(x-5)(x+1)}$, $\dfrac{5x-25}{(x+2)(x-5)(x+1)}$

7.33. $\dfrac{7x-4}{(x-2)(x+3)}$

7.34. $\dfrac{7m+25}{(m+3)(m+4)}$

7.35. $\dfrac{5m^2-9m+2}{(m+1)(m-2)(m+2)}$

7.36. $\dfrac{2n^2+12n-30}{(n+2)(n-5)(n+3)}$

7.37. $\dfrac{1}{x-2}$

7.38. $\dfrac{-3}{z-3}$

7.39. $\dfrac{5x+1}{(x-6)(x+1)}$

7.40. $\dfrac{y+3}{y+4}$

7.41. $\dfrac{1}{(b+1)(b-1)}$

7.42. $\dfrac{1}{(x+2)(x+1)}$

7.43. $\dfrac{v+3}{v+1}$

7.44. $\dfrac{3w}{w+7}$

7.45. $\dfrac{x-7}{x-4}$

7.46. $\dfrac{x^2-3x+18}{(x+3)(x-3)}$

7.47. $\dfrac{2}{3(x-1)}$

7.48. $\dfrac{1}{2(x-3)}$

7.49. $\dfrac{14}{11}$

7.50. $\dfrac{10}{23}$

7.51. $\dfrac{y+x}{y-x}$

7.52. $\dfrac{ab}{b-a}$

7.53. $\dfrac{b(b+2)(b-5)}{3b-5}$

7.54. $\dfrac{3}{c+3}$

7.55. $\dfrac{7}{3}$

7.56. $\dfrac{10}{3}$

7.57. $\dfrac{b+a}{a^2+b^2}$

7.58. $\dfrac{y-x}{xy}$

7.59. $\dfrac{3(x-2)}{5x+7}$

7.60. $\dfrac{x+21}{6x-43}$

7.61. $\dfrac{3}{5x+22}$

7.62. $\dfrac{2(2y^2+13y+5)}{3y}$

7.63. $\dfrac{x}{x+4}$

7.64. $\dfrac{x(x+1)}{3(x-1)}$

7.65. $y = -\dfrac{7}{15}$

7.66. $x = \dfrac{13}{15}$

7.67. $x = -3,\ x = 5$

7.68. $y = -2,\ y = 6$

7.69. $x = \dfrac{2}{3}$

7.70. $y = 2$

7.71. There is no solution.

7.72. There is no solution.

7.73. $x = 3$

7.74. $y = 7$

7.75. There is no solution.

7.76. There is no solution.

7.77. There is no solution.

7.78. There is no solution.

7.79. ⓐ The domain is all real numbers except $x \neq 3$ and $x \neq 4$. ⓑ $x = 2$, $x = \frac{14}{3}$ ⓒ $(2, 3)$, $\left(\frac{14}{3}, 3\right)$

7.80. ⓐ The domain is all real numbers except $x \neq 1$ and $x \neq 5$. ⓑ $x = \frac{21}{4}$ ⓒ $\left(\frac{21}{4}, 4\right)$

7.81. $y = mx - 4m + 5$

7.82. $y = mx + 5m + 1$

7.83. $a = \frac{b}{cb - 1}$

7.84. $y = \frac{3x}{x + 6}$

7.85. $y = 33$

7.86. $z = 14$

7.87. The pediatrician will prescribe 12 ml of acetaminophen to Emilia.

7.88. The pediatrician will prescribe 180 mg of fever reducer to Isabella.

7.89. The distance is 150 miles.

7.90. The distance is 350 miles.

7.91. The telephone pole is 40 feet tall.

7.92. The pine tree is 60 feet tall.

7.93. Link's biking speed is 15 mph.

7.94. The speed of Danica's boat is 17 mph.

7.95. Dennis's uphill speed was 10 mph and his downhill speed was 5 mph.

7.96. Joon's rate on the country roads is 50 mph.

7.97. Kayla's biking speed was 15 mph.

7.98. Victoria jogged 6 mph on the flat trail.

7.99. When the two gardeners work together it takes 2 hours and 24 minutes.

7.100. When Daria and her mother work together it takes 2 hours and 6 minutes.

7.101. Kristina can paint the room in 12 hours.

7.102. It will take Jordan 6 hours.

7.103. ⓐ $c = 4.8t$ ⓑ He would burn 432 calories.

7.104. ⓐ $d = 50t$ ⓑ It would travel 250 miles.

7.105. ⓐ $h = \frac{130}{t}$ ⓑ $1\frac{2}{3}$ hours

7.106. ⓐ $x = \frac{3500}{p}$ ⓑ 500 units

7.107. $(-\infty, -4) \cup [2, \infty)$

7.108. $(-\infty, -2] \cup (4, \infty)$

7.109. $\left(-\frac{3}{2}, 3\right)$

7.110. $(-8, 4)$

7.111. $(-\infty, -4) \cup (2, \infty)$

7.112. $(-\infty, -4) \cup (3, \infty)$

7.113. $(2, 4)$

7.114. $(3, 6)$

7.115. $(-4, 2]$

7.116. $[-1, 4)$

7.117. ⓐ $c(x) = \frac{20x + 6000}{x}$

ⓑ More than 150 items must be produced to keep the average cost below $60 per item.

7.118. ⓐ $c(x) = \frac{5x + 900}{x}$ ⓑ More than 60 items must be produced to keep the average cost below $20 per item.

Section Exercises

1. ⓐ $z = 0$ ⓑ $p = \frac{5}{6}$ ⓒ $n = -4$, $n = 2$

3. ⓐ $y = 0$ ⓑ $x = -\frac{1}{2}$ ⓒ $u = -4$, $u = 7$

5. $-\frac{4}{5}$

7. $\frac{2m^2}{3n}$

9. $\frac{8}{3}$

11. $\frac{x + 5}{x - 1}$

13. $\frac{a + 2}{a + 8}$

15. $\frac{p^2 + 4}{p - 2}$

17. $\frac{4b(b - 4)}{(b + 5)(b - 8)}$

19. $\frac{3(m + 5n)}{4(m - 5n)}$

21. -1

23. $-\frac{5}{y + 4}$

25. $\dfrac{w^2 - 6w + 36}{w - 6}$

27. $-\dfrac{z - 5}{4 + z}$

29. $\dfrac{3}{10}$

31. $\dfrac{x^3}{8y}$

33. $\dfrac{p(p - 4)}{2(p - 9)}$

35. $\dfrac{y - 5}{3(y + 5)}$

37. $-\dfrac{4(b + 9)}{3(b + 7)}$

39. $\dfrac{(c + 2)(c + 2)}{(c - 2)(c - 3)}$

41. $-\dfrac{(m - 2)(m - 3)}{(3 + m)(m + 4)}$

43. $-\dfrac{1}{v + 5}$

45. $\dfrac{3s}{s + 4}$

47. $\dfrac{4(p^2 - pq + q^2)}{(p - q)(p^2 + pq + q^2)}$

49. $\dfrac{x - 2}{8x}$

51. $\dfrac{2a - 7}{5}$

53. $3(3c - 5)$

55. $\dfrac{4(m + 8)(m + 7)}{3(m - 4)(m + 2)}$

57. $\dfrac{(4p + 1)(p - 7)}{3p(p + 9)(p - 1)}$

59. $x \neq 5$ and $x \neq -5$

61. $x \neq 2$ and $x \neq -3$

63. $R(x) = 2$

65. $R(x) = \dfrac{x + 5}{2x(x + 2)}$

67. $R(x) = \dfrac{3x(x + 7)}{x - 7}$

69. $R(x) = \dfrac{x(x - 5)}{x - 6}$

71. Answers will vary.

73. Answers will vary.

75. $\dfrac{3}{5}$

77. $\dfrac{3c + 5}{4c - 5}$

79. $r + 8$

81. $\dfrac{2w}{w - 4}$

83. $3a + 7$

85. $\dfrac{m - 2}{2}$

87. $\dfrac{p + 3}{p + 5}$

89. $\dfrac{r + 9}{r + 7}$

91. 4

93. $x + 2$

95. $\dfrac{z + 4}{z - 5}$

97. $\dfrac{4b - 3}{b - 7}$

99. ⓐ $(x + 2)(x - 4)(x + 3)$
ⓑ $\dfrac{5x + 15}{(x + 2)(x - 4)(x + 3)}$,
$\dfrac{2x^2 + 4x}{(x + 2)(x - 4)(x + 3)}$

101. ⓐ $(z - 2)(z + 4)(z - 4)$
ⓑ $\dfrac{9z - 36}{(z - 2)(z + 4)(z - 4)}$,
$\dfrac{4z^2 - 8z}{(z - 2)(z + 4)(z - 4)}$

103. ⓐ $(b + 3)(b + 3)(b - 5)$
ⓑ $\dfrac{4b - 20}{(b + 3)(b + 3)(b - 5)}$,
$\dfrac{2b^2 + 6b}{(b + 3)(b + 3)(b - 5)}$

105. ⓐ $(d + 5)(3d - 1)(d - 6)$
ⓑ $\dfrac{2d - 12}{(d + 5)(3d - 1)(d - 6)}$,
$\dfrac{5d^2 + 25d}{(d + 5)(3d - 1)(d - 6)}$

107. $\dfrac{21y + 8x}{30x^2 y^2}$

109. $\dfrac{5r - 7}{(r + 4)(r - 5)}$

111. $\dfrac{11w + 1}{(3w - 2)(w + 1)}$

113. $\dfrac{2y^2 + y + 9}{(y + 3)(y - 1)}$

115. $\dfrac{b(5b + 10 + 2a^2)}{a^2(b - 2)(b + 2)}$

117. $-\dfrac{m}{m + 4}$

119. $\dfrac{3(r^2 + 6r + 18)}{(r + 1)(r + 6)(r + 3)}$

121. $\dfrac{2(7t - 6)}{(t - 6)(t + 6)}$

123. $\dfrac{4a^2 + 25a - 6}{(a + 3)(a + 6)}$

125. $\dfrac{-6}{m - 6}$

127. $\dfrac{p + 2}{p + 3}$

129. $\dfrac{3}{r - 2}$

131. $\dfrac{4(8x + 1)}{10x - 1}$

133. $\dfrac{x - 5}{(x - 4)(x + 1)(x - 1)}$

135. $\dfrac{1}{(x - 1)(x + 1)}$

137. $\dfrac{5a^2 + 7a - 36}{a(a - 2)}$

139. $\dfrac{c - 5}{c + 2}$

141. $\dfrac{3(d + 1)}{d + 2}$

143. ⓐ $R(x) = -\dfrac{(x + 8)(x + 1)}{(x - 2)(x + 3)}$
ⓑ $R(x) = \dfrac{x + 1}{x + 3}$

145. ⓐ $\dfrac{3(3x+8)}{(x-8)(x+8)}$

ⓑ $R(x) = \dfrac{3}{x+8}$

147. Answers will vary.

149. ⓐ Answers will vary.

ⓑ Answers will vary.

ⓒ Answers will vary.

ⓓ $\dfrac{x+y}{xy}$

151. $\dfrac{a-4}{2a}$

153. $\dfrac{1}{2(c-2)}$

155. $\dfrac{12}{13}$

157. $\dfrac{20}{57}$

159. $\dfrac{n^2+m}{m-n^2}$

161. $\dfrac{rt}{t-r}$

163. $\dfrac{(x+1)(x-3)}{2}$

165. $\dfrac{4}{a+1}$

167. $\dfrac{11}{8}$

169. 19

171. $\dfrac{c^2+c}{c-d^2}$

173. $\dfrac{pq}{q-p}$

175. $\dfrac{2x-10}{3x+16}$

177. $\dfrac{3z-19}{3z+8}$

179. $\dfrac{4}{3a-7}$

181. $\dfrac{2c+29}{5c}$

183. $\dfrac{2p-5}{5}$

185. $\dfrac{m(m-5)}{(4m-19)(m+5)}$

187. $\dfrac{13}{24}$

189. $2(a-4)$

191. $\dfrac{3mn}{n-m}$

193. $\dfrac{(x-1)(x-2)}{6}$

195. Answers will vary.

197. $a=10$

199. $v=\dfrac{40}{21}$

201. $m=-2,\ m=4$

203. $p=-5,\ p=-4$

205. $v=14$

207. $x=-\dfrac{4}{5}$

209. $z=-145$

211. $q=-18,\ q=-1$

213. no solution

215. no solution

217. $b=-8$

219. $d=2$

221. $m=1$

223. no solution

225. $s=\dfrac{5}{4}$

227. $x=-\dfrac{4}{3}$

229. no solution

231.
ⓐ The domain is all real numbers except $x\neq-2$ and $x\neq-4$.

ⓑ $x=-3,\ x=-\dfrac{14}{5}$

ⓒ $(-3,5),\left(-\dfrac{14}{5},5\right)$

233.
ⓐ The domain is all real numbers except $x\neq2$ and $x\neq5$.

ⓑ $x=\dfrac{9}{2},$

ⓒ $\left(\dfrac{9}{2},2\right)$

235. $r=\dfrac{C}{2\pi}$

237. $w=2v+7$

239. $c=\dfrac{b+3+2a}{a}$

241. $p=\dfrac{q}{4q-2}$

243. $w=\dfrac{15v}{10+v}$

245. $n=\dfrac{5m+23}{4}$

247. $c=\dfrac{E}{m^2}$

249. $y=\dfrac{20x}{12-x}$

251. Answers will vary.

253. $x=49$

255. $p=-11$

257. $a=16$

259. $m=60$

261. $p=30$

263. ⓐ 162 beats per minute

ⓑ yes

265. 9 ml

267. 159 calories

269. 325 Canadian dollars

271. 3 cups

273. 4 bags

275. ⓐ 6 ⓑ 12

277. 950 miles

279. 680 miles

281. $\dfrac{2}{3}$ foot (8 in.)

283. 247.3 feet

285. 160 mph

287. 29 mph

289. 30 mph

291. 20 mph

293. 4 mph

295. 60 mph
301. 50 mph
307. 2 hours and 44 minutes
313. $y = \frac{14}{3}x$

319. ⓐ $m = 8v$ ⓑ 16 liters
325. $v = \frac{3}{w}$

331. ⓐ $c = \frac{2}{t}$ ⓑ 1 cavity

337. Answers will vary.
343. $(-\infty, 1) \cup (7, \infty)$

349. $(-\infty, -3) \cup (6, \infty)$
355. $(-\infty, -3) \cup (-4, \infty)$

361. $(-\infty, \frac{2}{3}) \cup (\frac{3}{2}, \infty)$

367. $(-4, 4)$
373. $(-\infty, -2) \cup [6, \infty)$

297. 650 mph
303. 4.2 mph
309. 7 hours and 30 minutes
315. $p = 3.2q$

321. ⓐ $L = 3d^2$ ⓑ 300 pounds

327. ⓐ $g = \frac{92,400}{w}$ ⓑ 16.8 mpg

333. ⓐ $c = 2.5m$ ⓑ $55

339. $(-\infty, -4) \cup [3, \infty)$
345. $(-5, 6)$

351. $[-9, 6)$
357. $(1, 4)$

363. $(-\infty, 0) \cup (0, 4) \cup (6, \infty)$

369. $[-10, -1) \cup (2, \infty)$
375. Answers will vary.

299. 50 mph
305. 2 hours
311. 10 min
317. ⓐ $P = 2.5g$ ⓑ $82.50

323. $y = \frac{20}{x}$

329. ⓐ $t = \frac{1000}{r}$ ⓑ 2.5 hours

335. Answers will vary.
341. $[-1, 3)$

347. $\left(-\frac{5}{2}, 5\right)$

353. $(-\infty, -6] \cup (4, \infty)$

359. $(-\infty, -3) \cup (\frac{5}{2}, \infty)$

365. $[-2, 0) \cup (0, 4]$

371. $(2, 5]$

Review Exercises

377. $a \neq \frac{2}{3}$

383. $\frac{x+3}{x+4}$

389. $\frac{3x}{(x+6)(x+6)}$

395. $R(x) = 3$
401. $x + 4$

407. $\frac{3b-2}{b+7}$

413. $\frac{11c-12}{(c-2)(c+3)}$

419. $\frac{2m-7}{m+2}$

425. $R(x) = \frac{2}{x+11}$

431. $\frac{11}{8}$

437. $b = \frac{3}{2}$

379. $y \neq 0$

385. $\frac{1}{6}$

391. $\frac{1}{11+w}$

397. 1

403. $\frac{q-3}{q+5}$

409. $(a+2)(a-5)(a+4)$

415. $\frac{5x^2+26x}{(x+4)(x+4)(x+6)}$

421. $\frac{4}{a-8}$

427. $\frac{x-2}{2x}$

433. $\frac{z-5}{23z+21}$

439. no solution

381. $\frac{3}{4}$

387. $\frac{-3x}{2}$

393. $\frac{5}{c+4}$

399. $y + 5$

405. $\frac{15w+2}{6w-1}$

411. $(3p+1)(p+6)(p+8)$

417. $\frac{2(y^2+10y-2)}{(y+2)(y+8)}$

423. $R(x) = \frac{x+8}{x+5}$

429. $\frac{(x-8)(x-5)}{2}$

435. $x = \frac{6}{7}$

441. ⓐ The domain is all real numbers except $x \neq 2$ and $x \neq 4$. ⓑ $x = 1$, $x = 6$ ⓒ $(1, 1)$, $(6, 1)$

443. $l = \frac{V}{hw}$
449. 15

455. 23 feet
461. $\frac{4}{5}$ hour
467. 301 mph

445. $z = \frac{y+5+7x}{x}$
451. 1161 calories

457. 45 mph
463. 12 days
469. 288 feet

447. $\frac{12}{5}$

453. $b = 9$; $x = 2\frac{1}{3}$

459. 16 mph
465. 7
471. 97 tickets

473. $(-4, 3]$

479. $(-\infty, 2) \cup [5, \infty)$

475. $[-6, 4)$

481.

ⓐ $c(x) = \dfrac{150x + 100000}{x}$

ⓑ More than 10,000 items must be produced to keep the average cost below $160 per item.

477. $(-\infty, -2] \cup [4, \infty)$

Practice Test

483. $\dfrac{a}{3b}$

489. $\dfrac{3n-2}{n-1}$

495. $[-3, 6)$

501. $(2, 5]$

485. $\dfrac{x+3}{3x}$

491. $\dfrac{n-m}{m+n}$

497. $(-\infty, 0) \cup (0, 4] \cup [6, \infty)$

503. $y = \dfrac{81}{16}$

487. $\dfrac{x-3}{x+9}$

493. $z = \dfrac{1}{2}$

499. $R(x) = \dfrac{1}{(x+2)(x+2)}$

505. Oliver's dad would take $4\dfrac{4}{5}$ hours to split the logs himself.

507. The distance between Dayton and Columbus is 64 miles.

Chapter 8

Try It

8.1. ⓐ -8 ⓑ 15

8.4. ⓐ -7 ⓑ not a real number

8.7. ⓐ -3 ⓑ not real ⓒ -2

8.2. ⓐ 10 ⓑ -11

8.5. ⓐ 3 ⓑ 4 ⓒ 3

8.8. ⓐ -6 ⓑ not real ⓒ -4

8.3. ⓐ not a real number ⓑ -9

8.6. ⓐ 10 ⓑ 2 ⓒ 3

8.9. ⓐ $6 < \sqrt{38} < 7$
ⓑ $4 < \sqrt[3]{93} < 5$

8.10. ⓐ $9 < \sqrt{84} < 10$
ⓑ $5 < \sqrt[3]{152} < 6$

8.13. ⓐ $|b|$ ⓑ w ⓒ $|m|$ ⓓ q

8.16. ⓐ m^2 ⓑ $|b^5|$

8.19. ⓐ $8|x|$ ⓑ $-10|p|$

8.22. ⓐ $5p^3$ ⓑ $3q^5$

8.11. ⓐ ≈ 3.32 ⓑ ≈ 4.14
ⓒ ≈ 3.36

8.14. ⓐ $|y|$ ⓑ p ⓒ $|z|$ ⓓ q

8.17. ⓐ $|u^3|$ ⓑ v^5

8.20. ⓐ $13|y|$ ⓑ $-11|y|$

8.23. ⓐ $10|ab|$ ⓑ $12p^6 q^{10}$
ⓒ $2x^{10} y^4$

8.12. ⓐ ≈ 3.61 ⓑ ≈ 4.38
ⓒ ≈ 3.15

8.15. ⓐ $|y^9|$ ⓑ z^6

8.18. ⓐ c^4 ⓑ d^4

8.21. ⓐ $3x^9$ ⓑ $3|q^7|$

8.24. ⓐ $15|mn|$ ⓑ $13|x^5 y^7|$
ⓒ $3w^{12} z^5$

8.25. $4\sqrt{3}$

8.28. ⓐ $12\sqrt{3}$ ⓑ $5\sqrt[3]{5}$ ⓒ $3\sqrt[4]{9}$

8.31. ⓐ $4y^2 \sqrt{2y}$ ⓑ $3p^3 \sqrt[3]{2p}$
ⓒ $2q^2 \sqrt[4]{4q^2}$

8.34. ⓐ $6m^4 |n^5| \sqrt{5mn}$
ⓑ $2x^2 y \sqrt[3]{9y^2}$ ⓒ $2|xy| \sqrt[4]{5x^3}$

8.26. $3\sqrt{5}$

8.29. ⓐ $b^2 \sqrt{b}$ ⓑ $|y| \sqrt[4]{y^2}$ ⓒ $z \sqrt[3]{z^2}$

8.32. ⓐ $5a^4 \sqrt{3a}$ ⓑ $4m^3 \sqrt[3]{2m^2}$
ⓒ $3|n| \sqrt[4]{2n^3}$

8.35. ⓐ -4 ⓑ no real number

8.27. ⓐ $12\sqrt{2}$ ⓑ $3\sqrt[3]{3}$ ⓒ $2\sqrt[4]{4}$

8.30. ⓐ $p^4 \sqrt{p}$ ⓑ $p\sqrt[5]{p^3}$
ⓒ $q^2 \sqrt[6]{q}$

8.33. ⓐ $7|a^3| b^2 \sqrt{2ab}$
ⓑ $2xy \sqrt[3]{7x^2 y}$ ⓒ $2|x| y^2 \sqrt[4]{2x}$

8.36. ⓐ $-5\sqrt[3]{5}$ ⓑ no real number

8.37. ⓐ $5 + 5\sqrt{3}$ ⓑ $2 - \sqrt{3}$

8.38. ⓐ $2 + 7\sqrt{2}$ ⓑ $2 - \sqrt{5}$

8.39. ⓐ $\frac{5}{4}$ ⓑ $\frac{3}{5}$ ⓒ $\frac{2}{3}$

8.40. ⓐ $\frac{7}{9}$ ⓑ $\frac{2}{5}$ ⓒ $\frac{1}{3}$

8.41. ⓐ $|a|$ ⓑ $|x|$ ⓒ y^3

8.42. ⓐ x^2 ⓑ m^2 ⓒ n^2

8.43. $\frac{2|p|\sqrt{6p}}{7}$

8.44. $\frac{2x^2\sqrt{3x}}{5}$

8.45. ⓐ $\frac{4|m|\sqrt{5m}}{|n^3|}$ ⓑ $\frac{3c^3\sqrt[3]{4c}}{d^2}$

ⓒ $\frac{2x^2\sqrt[4]{5x^2}}{|y|}$

8.46. ⓐ $\frac{3u^3\sqrt{6u}}{v^4}$ ⓑ $\frac{2r\sqrt[3]{5}}{s^2}$

ⓒ $\frac{3|m^3|\sqrt[4]{2m^2}}{|n^3|}$

8.47. ⓐ $\frac{5|y|\sqrt{x}}{6}$ ⓑ $\frac{2xy\sqrt[3]{y^2}}{3}$

ⓒ $\frac{|ab|\sqrt[4]{a}}{2}$

8.48. ⓐ $\frac{2|m|\sqrt{3}}{5|n^3|}$ ⓑ $\frac{3xy\sqrt[3]{x^2}}{5}$

ⓒ $\frac{2|ab|\sqrt[4]{a^2}}{3}$

8.49. ⓐ $7z^2$ ⓑ $-5\sqrt[3]{2}$

ⓒ $3|m|\sqrt[4]{2m^2}$

8.50. ⓐ $8m^4$ ⓑ -4 ⓒ $3|n|\sqrt[4]{2}$

8.51. ⓐ $\sqrt{t}$ ⓑ $\sqrt[3]{m}$ ⓒ $\sqrt[4]{r}$

8.52. ⓐ $\sqrt[6]{b}$ ⓑ $\sqrt[5]{z}$ ⓒ $\sqrt[4]{p}$

8.53. ⓐ $(10m)^{\frac{1}{2}}$ ⓑ $(3n)^{\frac{1}{5}}$

ⓒ $3(6y)^{\frac{1}{4}}$

8.54. ⓐ $(3k)^{\frac{1}{7}}$ ⓑ $(5j)^{\frac{1}{4}}$

ⓒ $8(2a)^{\frac{1}{3}}$

8.55. ⓐ 6 ⓑ 2 ⓒ 2

8.56. ⓐ 10 ⓑ 3 ⓒ 3

8.57. ⓐ No real solution ⓑ -8

ⓒ $\frac{1}{8}$

8.58. ⓐ No real solution ⓑ -4

ⓒ $\frac{1}{4}$

8.59. ⓐ $x^{\frac{5}{2}}$ ⓑ $(3y)^{\frac{3}{4}}$ ⓒ $\left(\frac{2m}{3n}\right)^{\frac{5}{2}}$

8.60. ⓐ $a^{\frac{2}{5}}$ ⓑ $(5ab)^{\frac{5}{3}}$

ⓒ $\left(\frac{7xy}{z}\right)^{\frac{3}{2}}$

8.61. ⓐ 9 ⓑ $\frac{1}{729}$ ⓒ $\frac{1}{8}$

8.62. ⓐ 8 ⓑ $\frac{1}{9}$ ⓒ $\frac{1}{125}$

8.63. ⓐ -64 ⓑ $-\frac{1}{64}$ ⓒ not a real number

8.64. ⓐ -729 ⓑ $-\frac{1}{729}$ ⓒ not a real number

8.65. ⓐ $x^{\frac{3}{2}}$ ⓑ x^8 ⓒ $\frac{1}{x}$

8.66. ⓐ $y^{\frac{11}{8}}$ ⓑ m^2 ⓒ $\frac{1}{d}$

8.67. ⓐ $8x^{\frac{1}{5}}$ ⓑ $x^{\frac{1}{2}}y^{\frac{1}{3}}$

8.68. ⓐ $729n^{\frac{3}{5}}$ ⓑ $a^2b^{\frac{2}{3}}$

8.69. ⓐ m^2 ⓑ $\frac{5n}{m^{\frac{1}{4}}}$

8.70. ⓐ u^3 ⓑ $3x^{\frac{1}{5}}y^{\frac{1}{3}}$

8.71. ⓐ $-\sqrt{2}$ ⓑ $11\sqrt[3]{x}$

ⓒ $3\sqrt[4]{x} - 5\sqrt[4]{y}$

8.72. ⓐ $-4\sqrt{3}$ ⓑ $8\sqrt[3]{y}$

ⓒ $5\sqrt[4]{m} - 2\sqrt[3]{m}$

8.73. ⓐ $-2\sqrt{7x}$ ⓑ $-\sqrt[4]{5xy}$

8.74. ⓐ $-\sqrt{3y}$ ⓑ $3\sqrt[3]{7mn}$

8.75. ⓐ $9\sqrt{2}$ ⓑ $2\sqrt[3]{2}$ ⓒ $\sqrt[3]{3}$

8.76. ⓐ $7\sqrt{3}$ ⓑ $-10\sqrt[3]{5}$ ⓒ $-3\sqrt[3]{2}$

8.77. ⓐ $-m^3\sqrt{2m}$ ⓑ $x^2\sqrt[3]{5x}$

8.78. ⓐ $-p\sqrt{3p}$

ⓑ $4y\sqrt[3]{4y^2} - 2n\sqrt[3]{4n^2}$

8.79. ⓐ $12\sqrt{15}$ ⓑ $-18\sqrt[3]{2}$

8.80. ⓐ $27\sqrt{2}$ ⓑ $-36\sqrt[3]{2}$

8.81. ⓐ $36x^3\sqrt{5}$ ⓑ $8y\sqrt[4]{3y^2}$

8.82. ⓐ $144y^2\sqrt{5y}$ ⓑ $-36\sqrt[3]{3a}$

8.83. ⓐ $18 + \sqrt{6}$ ⓑ $-2\sqrt[3]{4} - 2\sqrt[3]{3}$

8.84. ⓐ $-40 + 4\sqrt{2}$ ⓑ $-3 - \sqrt[3]{18}$

8.85. ⓐ $-66 + 15\sqrt{7}$

ⓑ $\sqrt[3]{x^2} - 5\sqrt[3]{x} + 6$

8.86. ⓐ $41 - 14\sqrt{11}$

ⓑ $\sqrt[3]{x^2} + 4\sqrt[3]{x} + 3$

8.87. $1 + 9\sqrt{21}$

8.88. $-12 - 20\sqrt{3}$

8.89. ⓐ $102 + 20\sqrt{2}$ ⓑ $55 + 6\sqrt{6}$

8.90. ⓐ $41 - 12\sqrt{5}$
ⓑ $121 - 36\sqrt{10}$

8.91. -11

8.92. -159

8.93. ⓐ $\frac{5s}{8}$ ⓑ $\frac{2}{a}$

8.94. ⓐ $\frac{5q^2}{6}$ ⓑ $\frac{2}{b}$

8.95. ⓐ $\frac{9x^2}{y^2}$ ⓑ $\frac{-4x}{y}$

8.96. ⓐ $\frac{10n^3}{m}$ ⓑ $\frac{-3p}{q^2}$

8.97. $4xy\sqrt{2x}$

8.98. $4ab\sqrt{3b}$

8.99. ⓐ $\frac{5\sqrt{3}}{3}$ ⓑ $\frac{\sqrt{6}}{8}$ ⓒ $\frac{\sqrt{2x}}{x}$

8.100. ⓐ $\frac{6\sqrt{5}}{5}$ ⓑ $\frac{\sqrt{14}}{6}$ ⓒ $\frac{\sqrt{5x}}{x}$

8.101. ⓐ $\frac{\sqrt[3]{49}}{7}$ ⓑ $\frac{\sqrt[3]{90}}{6}$ ⓒ $\frac{5\sqrt[3]{3y^2}}{3y}$

8.102. ⓐ $\frac{\sqrt[3]{4}}{2}$ ⓑ $\frac{\sqrt[3]{150}}{10}$ ⓒ $\frac{2\sqrt[3]{5n^2}}{5n}$

8.103. ⓐ $\frac{\sqrt[4]{27}}{3}$ ⓑ $\frac{\sqrt[4]{12}}{4}$ ⓒ $\frac{3\sqrt[4]{5x^3}}{5x}$

8.104. ⓐ $\frac{\sqrt[4]{125}}{5}$ ⓑ $\frac{\sqrt[4]{224}}{8}$
ⓒ $\frac{\sqrt[4]{64x^3}}{x}$

8.105. $-\frac{3(1 + \sqrt{5})}{4}$

8.106. $\frac{4 + \sqrt{6}}{5}$

8.107. $\frac{\sqrt{5}(\sqrt{x} - \sqrt{2})}{x - 2}$

8.108. $\frac{\sqrt{10}(\sqrt{y} + \sqrt{3})}{y - 3}$

8.109. $\frac{(\sqrt{p} + \sqrt{2})^2}{p - 2}$

8.110. $\frac{(\sqrt{q} - \sqrt{10})^2}{q - 10}$

8.111. $m = \frac{23}{3}$

8.112. $z = \frac{3}{10}$

8.113. no solution

8.114. no solution

8.115. $x = 2, x = 3$

8.116. $y = 5, y = 6$

8.117. $x = -6$

8.118. $x = -9$

8.119. $x = 8$

8.120. $x = 6$

8.121. $m = 7$

8.122. $n = 3$

8.123. $a = 63$

8.124. $b = 311$

8.125. $x = 3$

8.126. $x = -\frac{6}{5}$

8.127. $x = 4$

8.128. $x = 9$

8.129. $x = 5$

8.130. $x = 0 \ x = 4$

8.131. 9 seconds

8.132. 3.5 seconds

8.133. 42.7 feet

8.134. 54.1 feet

8.135. ⓐ $f(6) = 4$ ⓑ no value at $x = 0$

8.136. ⓐ $g(4) = 5$ ⓑ no value at $f(-3)$

8.137. ⓐ $g(4) = 2$ ⓑ $g(1) = -1$

8.138. ⓐ $h(2) = 2$
ⓑ $h(-5) = -3$

8.139. ⓐ $f(4) = 2$ ⓑ $f(-1) = 1$

8.140. ⓐ $g(16) = 3$ ⓑ $g(3) = 2$

8.141. $\left[\frac{5}{6}, \infty\right)$

8.142. $\left(-\infty, \frac{4}{5}\right]$

8.143. $(-3, \infty)$

8.144. $(5, \infty)$

8.145. $(-\infty, \infty)$

8.146. $(-\infty, \infty)$

8.147. ⓐ domain: $[-2, \infty)$
ⓑ
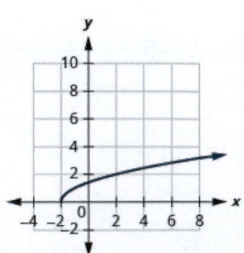
ⓒ range: $[0, \infty)$

8.148. ⓐ domain: $[2, \infty)$

ⓑ

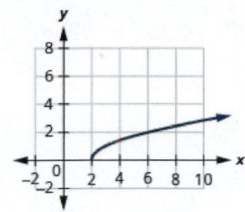

ⓒ range: $[0, \infty)$

8.149. ⓐ domain: $(-\infty, \infty)$

ⓑ

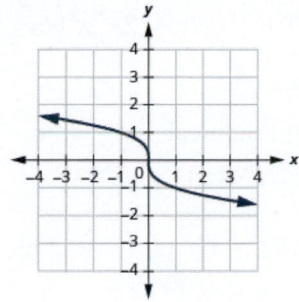

ⓒ range: $(-\infty, \infty)$

8.150. ⓐ domain: $(-\infty, \infty)$

ⓑ

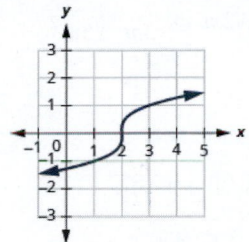

ⓒ range: $(-\infty, \infty)$

8.151. ⓐ $9i$ ⓑ $\sqrt{5}i$ ⓒ $3\sqrt{2}i$

8.154. $7\sqrt{3}i$

8.157. $12 + 20i$

8.160. $-5 - 10i$

8.163. -14

8.166. $6 + 12\sqrt{2}i$

8.169. 109

8.172. i

8.175. $\frac{3}{2} - \frac{3}{2}i$

8.178. 1

8.152. ⓐ $6i$ ⓑ $\sqrt{3}i$ ⓒ $3\sqrt{3}i$

8.155. ⓐ $6 + 5i$ ⓑ $6 - 3i$

8.158. $12 + 6i$

8.161. $-21 - 20i$

8.164. -54

8.167. 25

8.170. 41

8.173. $\frac{4}{17} + \frac{16}{17}i$

8.176. $\frac{4}{5} - \frac{2}{5}i$

8.153. $6\sqrt{2}i$

8.156. ⓐ $-2 - 6i$ ⓑ $2 + 9i$

8.159. $-11 - 7i$

8.162. $9 - 40i$

8.165. $-12 - 22\sqrt{3}i$

8.168. 29

8.171. i

8.174. $\frac{2}{5} + \frac{4}{5}i$

8.177. $-i$

Section Exercises

1. ⓐ 8 ⓑ -9

3. ⓐ 14 ⓑ -1

5. ⓐ $\frac{2}{3}$ ⓑ -0.1

7. ⓐ not real number ⓑ -17

9. ⓐ -15 ⓑ not real number

11. ⓐ 6 ⓑ 4

13. ⓐ 8 ⓑ 3 ⓒ 1

15. ⓐ -2 ⓑ not real ⓒ -2

17. ⓐ -5 ⓑ not real ⓒ -4

19. ⓐ $8 < \sqrt{70} < 9$

ⓑ $4 < \sqrt[3]{71} < 5$

21. ⓐ $14 < \sqrt{200} < 15$

ⓑ $5 < \sqrt[3]{137} < 6$

23. ⓐ 4.36 ⓑ ≈ 4.46

ⓒ ≈ 3.14

25. ⓐ 7.28 ⓑ ≈ 5.28

ⓒ ≈ 4.61

27. ⓐ u ⓑ $|v|$

29. ⓐ $|y|$ ⓑ m

31. ⓐ $\left|x^3\right|$ ⓑ y^8

33. ⓐ x^{12} ⓑ $\left|y^{11}\right|$

35. ⓐ x^3 ⓑ $\left|y^3\right|$

37. ⓐ m^2 ⓑ n^4

39. ⓐ $7|x|$ ⓑ $-9\left|x^9\right|$

41. ⓐ $11m^{10}$ ⓑ $-8|a|$

43. ⓐ $2x^2$ ⓑ $2y^2$

45. ⓐ $6a^2$ ⓑ $2b^4$

47. ⓐ $12|xy|$ ⓑ $13w^4\left|y^5\right|$

ⓒ $2a^{17}b^2$

49. ⓐ $11|ab|$ ⓑ $3c^4d^6$

ⓒ $4x^5y^{22}$

55. $3\sqrt{3}$

61. $20\sqrt{2}$

51. Answers will vary.

57. $5\sqrt{5}$

63. ⓐ $2\sqrt[4]{2}$ ⓑ $2\sqrt[5]{2}$

53. Answers will vary.

59. $7\sqrt{3}$

65. ⓐ $2\sqrt[5]{2}$ ⓑ $4\sqrt[3]{4}$

67. (a) $|y^5|\sqrt{y}$ (b) $r\sqrt[3]{r^2}$ (c) $s^2\sqrt[4]{s^2}$

69. (a) $n^{10}\sqrt{n}$ (b) $q^2\sqrt[3]{q^2}$ (c) $|n|\sqrt[8]{n^2}$

71. (a) $5r^6\sqrt{5r}$ (b) $3x\sqrt[3]{4x^2}$ (c) $2|y|\sqrt[4]{3y^2}$

73. (a) $11|m^{11}|\sqrt{2m}$ (b) $3m^2\sqrt[4]{5m^2}$ (c) $2n\sqrt[5]{5n^3}$

75. (a) $7|m^3n^5|\sqrt{3mn}$ (b) $2x^2y^2\sqrt[3]{6y}$ (c) $2|xy|\sqrt[4]{2x}$

77. (a) $8|qr^3|\sqrt{3qr}$ (b) $3m^3n^3\sqrt[3]{2n}$ (c) $3a^2b^2\sqrt[4]{a}$

79. (a) $-6\sqrt[3]{4}$ (b) not real

81. (a) -2 (b) not real

83. (a) $5+2\sqrt{3}$ (b) $5-\sqrt{6}$

85. (a) $1+3\sqrt{5}$ (b) $1+\sqrt{10}$

87. (a) $\frac{3}{4}$ (b) $\frac{2}{3}$ (c) $\frac{1}{3}$

89. (a) $\frac{5}{3}$ (b) $\frac{3}{5}$ (c) $\frac{1}{4}$

91. (a) x^2 (b) p^3 (c) $|q|$

93. (a) $\frac{1}{y^2}$ (b) u^2 (c) $|v^3|$

95. $\frac{4|x^3|\sqrt{6x}}{11}$

97. $\frac{10m^2\sqrt{3m}}{8}$

99. $\frac{7r^2\sqrt{2r}}{10}$

101. $\frac{2|q^3|\sqrt{7}}{15}$

103. (a) $\frac{5r^4\sqrt{3r}}{s^4}$ (b) $\frac{3a^2\sqrt[3]{2a^2}}{|b|}$ (c) $\frac{2|c|\sqrt[4]{4c}}{|d|}$

105. (a) $\frac{2|p^3|\sqrt{7p}}{|q|}$ (b) $\frac{3s^2\sqrt[3]{3s^2}}{t}$ (c) $\frac{2|p^3|\sqrt[4]{4p^3}}{|q^3|}$

107. (a) $\frac{4|xy|}{3}$ (b) $\frac{y^2\sqrt[3]{x}}{2}$ (c) $\frac{|ab|\sqrt[4]{a}}{4}$

109. (a) $\frac{1}{2|pq|}$ (b) $\frac{2cd\sqrt[5]{2d^2}}{5}$ (c) $\frac{|mn|\sqrt[6]{2}}{2}$

111. (a) $\frac{3p^4\sqrt{p}}{|q|}$ (b) $2\sqrt[4]{2}$ (c) $2x\sqrt[5]{2x}$

113. (a) $5|m^3|$ (b) $5\sqrt[3]{5}$ (c) $3|y|\sqrt[4]{3y^2}$

115. Answers will vary.

117. Answers will vary.

119. (a) $\sqrt{x}$ (b) $\sqrt[3]{y}$ (c) $\sqrt[4]{z}$

121. (a) $\sqrt[5]{u}$ (b) $\sqrt[9]{v}$ (c) $\sqrt[20]{w}$

123. (a) $\frac{1}{x^7}$ (b) $\frac{1}{y^9}$ (c) $f^{\frac{1}{5}}$

125. (a) $(7c)^{\frac{1}{4}}$ (b) $(12d)^{\frac{1}{7}}$ (c) $2(6b)^{\frac{1}{4}}$

127. (a) $(21p)^{\frac{1}{2}}$ (b) $(8q)^{\frac{1}{4}}$ (c) $4(36r)^{\frac{1}{6}}$

129. (a) 9 (b) 5 (c) 8

131. (a) 2 (b) 4 (c) 5

133. (a) -6 (b) -6 (c) $\frac{1}{6}$

135. (a) not real (b) -3 (c) $\frac{1}{3}$

137. (a) not real (b) -6 (c) $\frac{1}{6}$

139. (a) not real (b) -10 (c) $\frac{1}{10}$

141. (a) $m^{\frac{5}{2}}$ (b) $(3y)^{\frac{7}{3}}$ (c) $\left(\frac{4x}{5y}\right)^{\frac{3}{5}}$

143. (a) $u^{\frac{2}{5}}$ (b) $(6x)^{\frac{5}{3}}$ (c) $\left(\frac{18a}{5b}\right)^{\frac{7}{4}}$

145. (a) 32,768 (b) $\frac{1}{729}$ (c) 9

147. (a) 4 (b) $\frac{1}{9}$ (c) not real

149. (a) -27 (b) $-\frac{1}{27}$ (c) not real

151. (a) $c^{\frac{7}{8}}$ (b) p^9 (c) $\frac{1}{r}$

153. (a) $y^{\frac{5}{4}}$ (b) x^8 (c) $\frac{1}{m}$

155. (a) $81q^2$ (b) $a^{\frac{1}{2}}b$

157. (a) $8u^{\frac{1}{4}}$ (b) $8p^{\frac{1}{2}}q^{\frac{3}{4}}$

159. (a) $r^{\frac{7}{2}}$ (b) $\frac{6s}{t}$

161. (a) c^2 (b) $\frac{2x}{3y}$

163. Answers will vary.

165. (a) $3\sqrt{2}$ (b) $7\sqrt[3]{m}$ (c) $6\sqrt[4]{m}$

167. (a) $9\sqrt{5}$ (b) $12\sqrt[3]{a}$ (c) $6\sqrt[4]{2z}$

169. (a) $4\sqrt{2a}$ (b) 0

171. (a) $\sqrt{3c}$ (b) $\sqrt[3]{4pq}$

173. (a) $4\sqrt{3}$ (b) $-2\sqrt[3]{5}$ (c) $3\sqrt[3]{2}$

175. (a) $7\sqrt{3}$ (b) $7\sqrt[3]{2}$ (c) $3\sqrt[4]{5}$

177. (a) $a^2\sqrt{2a}$ (b) 0

179. (a) $2c^3\sqrt{5c}$ (b) $14r^2\sqrt[4]{2r^2}$

181. $4y\sqrt{2}$

183. (a) $-18\sqrt{6}$ (b) $-64\sqrt[3]{9}$

185. (a) $-30\sqrt{2}$) (b) $6\sqrt[4]{2}$

187. (a) $72z^2\sqrt{3}$ (b) $45x^2\sqrt[3]{2}$

189. (a) $-42z^5\sqrt{2z}$ (b) $-8y\sqrt[4]{6y}$

191. (a) $14+5\sqrt{7}$ (b) $4\sqrt[3]{6}+3\sqrt[3]{4}$

193. ⓐ $44 - 3\sqrt{11}$ ⓑ $3\sqrt[4]{2} + \sqrt[4]{54}$

195. $60 + 2\sqrt{3}$

197. ⓐ $30 + 18\sqrt{2}$

ⓑ $\sqrt[3]{x^2} - 2\sqrt[3]{x} - 3$

199. ⓐ $-54 + 13\sqrt{10}$

ⓑ $2\sqrt[3]{x^2} + 8\sqrt[3]{x} + 6$

201. $23 + 3\sqrt{30}$

203. $-439 - 2\sqrt{77}$

205. ⓐ $14 + 6\sqrt{5}$ ⓑ $79 - 20\sqrt{3}$

207. ⓐ $87 - 18\sqrt{6}$

ⓑ $163 + 60\sqrt{7}$

209. 14

211. -227

213. 19

215. $\sqrt[3]{9x^2} - 4$

217. $5\sqrt{3}$

219. $9\sqrt{2}$

221. $-\sqrt[4]{5}$

223. $10c^2\sqrt{3} - 9c^3\sqrt{3}$

225. $2\sqrt{3}$

227. $17q^2$

229. $3\sqrt{7}$

231. $-42\sqrt[3]{9}$

233. 29

235. $29 - 7\sqrt{17}$

237. $72 - 36\sqrt{2}$

239. $6 + 3\sqrt[3]{2}$

241. Answers will vary.

243. Answers will vary.

245. ⓐ $\frac{4}{3}$ ⓑ $\frac{4}{3}$

247. ⓐ $\frac{10m^2}{7}$ ⓑ $\frac{3}{y}$

249. ⓐ $\frac{5}{6r^2}$ ⓑ $\frac{2x}{3}$

251. ⓐ $\frac{6p}{q^2}$ ⓑ $-\frac{2a^2}{b}$

253. ⓐ $\frac{8m^4}{3n^4}$ ⓑ $-\frac{x^2}{2y^2}$

255. $4x^4\sqrt{7y}$

257. $2ab\sqrt[3]{2a}$

259. ⓐ $\frac{5\sqrt{6}}{3}$ ⓑ $\frac{2\sqrt{3}}{9}$ ⓒ $\frac{2\sqrt{5x}}{x}$

261. ⓐ $\frac{6\sqrt{7}}{7}$ ⓑ $\frac{2\sqrt{10}}{15}$ ⓒ $\frac{4\sqrt{3p}}{p}$

263. ⓐ $\frac{\sqrt[3]{25}}{5}$ ⓑ $\frac{\sqrt[3]{45}}{6}$ ⓒ $\frac{2\sqrt[3]{6a^2}}{3a}$

265. ⓐ $\frac{\sqrt[3]{121}}{11}$ ⓑ $\frac{\sqrt[3]{28}}{6}$ ⓒ $\frac{\sqrt[3]{9x}}{x}$

267. ⓐ $\frac{\sqrt[4]{343}}{7}$ ⓑ $\frac{\sqrt[4]{40}}{4}$ ⓒ $\frac{2\sqrt[4]{4x^2}}{x}$

269. ⓐ $\frac{\sqrt[4]{9}}{3}$ ⓑ $\frac{\sqrt[4]{50}}{4}$ ⓒ $\frac{2\sqrt[4]{3a^2}}{a}$

271. $-2(1 + \sqrt{5})$

273. $3(3 + \sqrt{7})$

275. $\frac{\sqrt{3}(\sqrt{m} + \sqrt{5})}{m - 5}$

277. $\frac{\sqrt{2}(\sqrt{x} + \sqrt{6})}{x - 6}$

279. $\frac{(\sqrt{r} + \sqrt{5})^2}{r - 5}$

281. $\frac{(\sqrt{x} + 2\sqrt{2})^2}{x - 8}$

283. Answers will vary.

285. Answers will vary.

287. $m = 14$

289. no solution

291. $x = -4$

293. $m = 14$

295. $v = 17$

297. $m = \frac{7}{2}$

299. no solution

301. $u = 3, u = 4$

303. $r = 1, r = 2$

305. $x = 10$

307. $x = -8$

309. $x = 8$

311. $x = -4$

313. $x = 7$

315. $x = 3$

317. $z = 21$

319. $x = 42$

321. $r = 3$

323. $u = 3$

325. $r = -2$

327. $x = 1$

329. $x = -8, x = 2$

331. $a = 0$

333. $u = \frac{9}{4}$

335. $a = 4$

337. $x = 0\ x = 4$

339. $x = 1\ x = 5$

341. $x = 9$

343. 8.7 feet

345. 4.7 seconds

347. 72 feet

349. Answers will vary.

351. ⓐ $f(5) = 4$ ⓑ no value at $x = 0$

353. ⓐ $g(4) = 5$ ⓑ $g(8) = 7$

355. ⓐ $F(1) = 1$ ⓑ $F(-11) = 5$

357. ⓐ $G(5) = 2\sqrt{6}$ ⓑ $G(2) = 3$

359. ⓐ $g(6) = 2$ ⓑ $g(-2) = -2$

361. ⓐ $h(-2) = 0$ ⓑ $h(6) = 2\sqrt[3]{4}$

363. ⓐ $f(0) = 0$ ⓑ $f(2) = 2$

365. ⓐ $g(1) = 0$ ⓑ $g(-3) = 2$

367. $\left[\frac{1}{3}, \infty\right)$

369. $\left(-\infty, \frac{2}{3}\right]$

371. $(2, \infty)$

373. $(-\infty, -3] \cup (2, \infty)$

375. $(-\infty, \infty)$

377. $(-\infty, \infty)$

379. $\left[-\frac{3}{8}, \infty\right)$

381. $(-\infty, \infty)$

383. (a) domain: $[-1, \infty)$

(b)

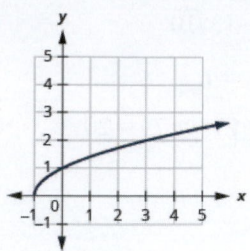

(c) $[0, \infty)$

385. (a) domain: $[-4, \infty)$

(b)

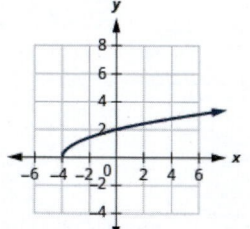

(c) $[0, \infty)$

387. (a) domain: $[0, \infty)$

(b)

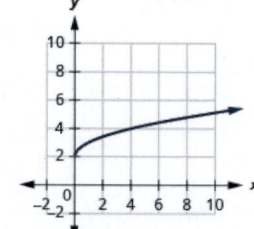

(c) $[2, \infty)$

389. (a) domain: $[0, \infty)$

(b)

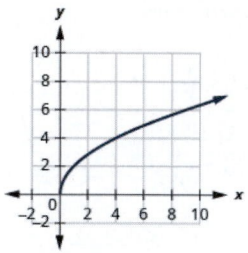

(c) $[0, \infty)$

391. (a) domain: $(-\infty, 3]$

(b)

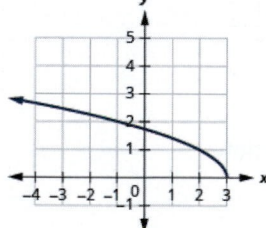

(c) $[0, \infty)$

393. (a) domain: $[0, \infty)$

(b)

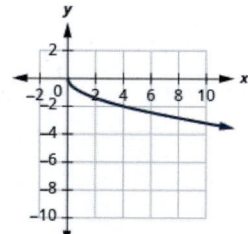

(c) $(-\infty, 0]$

395. (a) domain: $(-\infty, \infty)$

(b)

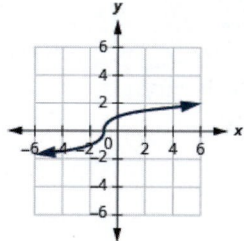

(c) $(-\infty, \infty)$

397. (a) domain: $(-\infty, \infty)$

(b)

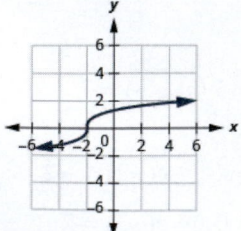

(c) $(-\infty, \infty)$

399. (a) domain: $(-\infty, \infty)$

(b)

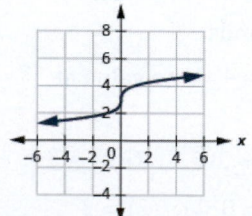

(c) $(-\infty, \infty)$

401. (a) domain: $(-\infty, \infty)$

(b)

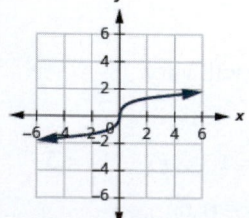

(c) $(-\infty, \infty)$

403. ⓐ domain: $(-\infty, \infty)$

ⓑ

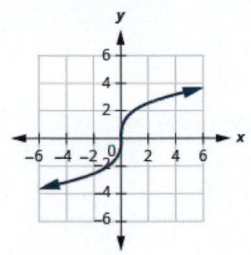

ⓒ $(-\infty, \infty)$

409. ⓐ $4i$ ⓑ $\sqrt{11}i$ ⓒ $2\sqrt{2}i$

411. ⓐ $10i$ ⓑ $\sqrt{13}i$ ⓒ $3\sqrt{5}i$

413. $9\sqrt{3}i$

415. $8\sqrt{2}i$

417. $8 + 7i$

419. $14 + 2i$

421. $-2 + 2i$

423. $8 + 5i$

425. $7 - 13i$

427. $25 - 2\sqrt{2}i$

429. $12 + 20i$

431. $-12 + 18i$

433. $-38 + + 9i$

435. $27 + 15i$

437. $-7 + 24i$

439. $-5 - 12i$

441. -11

443. -30

445. $-44 + 4\sqrt{3}i$

447. $-20 - 2\sqrt{2}i$

449. 5

451. 53

453. 50

455. 85

457. i

459. $\frac{2}{25} + \frac{11}{25}i$

461. $\frac{6}{13} + \frac{9}{13}i$

463. $-\frac{12}{13} - \frac{8}{13}i$

465. $\frac{4}{3} - \frac{1}{3}i$

467. $-\frac{3}{4} + \frac{1}{2}i$

469. i

471. -1

473. 1

475. i

477. Answers will vary.

479. Answers will vary.

Review Exercises

481. ⓐ 15 ⓑ -4

483. ⓐ 2 ⓑ 3 ⓒ 3

485. ⓐ $8 < \sqrt{68} < 9$

ⓑ $4 < \sqrt[3]{84} < 5$

487. ⓐ a ⓑ $|b|$

489. ⓐ m^2 ⓑ n^4

491. ⓐ $6a^2$ ⓑ $2b^4$

493. $5\sqrt{5}$

495. ⓐ $5\sqrt[3]{5}$ ⓑ $2\sqrt[6]{2}$

497. ⓐ $4|s^7|\sqrt{5s}$ ⓑ $2a\sqrt[5]{3a^2}$

ⓒ $2|b|\sqrt[6]{2b}$

499. ⓐ -2 ⓑ not real

501. ⓐ $\frac{6}{7}$ ⓑ $\frac{2}{3}$ ⓒ $\frac{1}{2}$

503. $\frac{10m^2\sqrt{3m}}{8}$

505. ⓐ $\frac{1}{2|pq|}$ ⓑ $\frac{2cd\sqrt[5]{2d^2}}{5}$

507. ⓐ $\sqrt{r}$ ⓑ $\sqrt[3]{s}$ ⓒ $\sqrt[4]{t}$

509. ⓐ 5 ⓑ 3 ⓒ 2

ⓒ $\frac{|mn|\sqrt[6]{2}}{2}$

511. ⓐ -2 ⓑ $\frac{1}{3}$ ⓒ -5

513. ⓐ 125 ⓑ $\frac{1}{27}$ ⓒ 16

515. ⓐ 6^3 ⓑ b^9 ⓒ $\frac{1}{w}$

517. ⓐ $4\sqrt{2}$ ⓑ $9\sqrt[3]{p}$ ⓒ $2\sqrt[3]{x}$

519. ⓐ $7\sqrt{3}$ ⓑ $7\sqrt[3]{2}$ ⓒ $3\sqrt[4]{5}$

521. $37y\sqrt{3}$

523. ⓐ $126x^2\sqrt{2}$ ⓑ $48a\sqrt[3]{a^2}$

525. ⓐ $71 - 22\sqrt{7}$

527. ⓐ $27 + 8\sqrt{11}$ ⓑ $29 - 12\sqrt{5}$

ⓑ $\sqrt[3]{x^2} - 8\sqrt[3]{x} + 15$

529. $\sqrt[3]{9x^2} - 4$

531. ⓐ $\frac{8m^4}{3n^4}$ ⓑ $-\frac{x^2}{2y^2}$

533. ⓐ $\frac{\sqrt[3]{121}}{11}$ ⓑ $\frac{\sqrt[3]{28}}{6}$ ⓒ $\frac{\sqrt[3]{9x}}{x}$

535. $-\dfrac{7(2 + \sqrt{6})}{2}$

537. $\dfrac{\left(\sqrt{x} + 2\sqrt{2}\right)^2}{x - 8}$

539. no solution

541. $u = 3, u = 4$

543. $x = -4$

545. $r = 3$

547. $x = -8, x = 2$

549. $x = 3$

551. 64.8 feet

553. ⓐ $G(5) = 2\sqrt{6}$ ⓑ $G(2) = 3$

555. ⓐ $g(1) = 0$ ⓑ $g(-3) = 2$

557. $(2, \infty)$

559. $\left[\dfrac{7}{10}, \infty\right)$

561. ⓐ domain: $[0, \infty)$
ⓑ

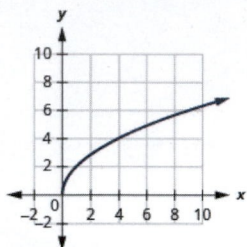

ⓒ range: $[0, \infty)$

563. ⓐ domain: $(-\infty, \infty)$
ⓑ

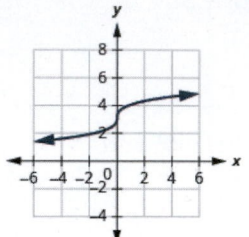

ⓒ range: $(-\infty, \infty)$

565. $8\sqrt{2}i$

567. $8 + 5i$

569. $23 + 14i$

571. -6

573. $-5 - 12i$

575. $\dfrac{2}{25} + \dfrac{11}{25}i$

577. 1

Practice Test

579. $5x^3$

581. $2x^2 y \sqrt[3]{9x^2 y}$

583. ⓐ $\dfrac{1}{4}$ ⓑ -343

585. $x^{\frac{7}{4}}$

587. $-x^2 \sqrt{3x}$

589. $36x^4 \sqrt{2}$

591. $2 - 7\sqrt{3}$

593. $\dfrac{7x^5}{3y^7}$

595. $3(2 - \sqrt{3})$

597. $-12 + 8i$

599. $-i$

601. $x = 4$

603. ⓐ domain: $[-2, \infty)$

ⓑ

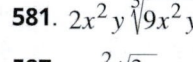

ⓒ range: $[0, \infty)$

Chapter 9
Try It

9.1. $x = 4\sqrt{3}, x = -4\sqrt{3}$

9.2. $y = 3\sqrt{3}, y = -3\sqrt{3}$

9.3. $x = 7, x = -7$

9.4. $m = 4, m = -4$

9.5. $c = 2\sqrt{3}i, \ c = -2\sqrt{3}i$

9.6. $c = 2\sqrt{6}i, \ c = -2\sqrt{6}i$

9.7. $x = 2\sqrt{10}$, $x = -2\sqrt{10}$

9.8. $y = 2\sqrt{7}$, $y = -2\sqrt{7}$

9.9. $r = \frac{6\sqrt{5}}{5}$, $r = -\frac{6\sqrt{5}}{5}$

9.10. $t = \frac{8\sqrt{3}}{3}$, $t = -\frac{8\sqrt{3}}{3}$

9.11. $a = 3 + 3\sqrt{2}$, $a = 3 - 3\sqrt{2}$

9.12.
$b = -2 + 2\sqrt{10}$, $b = -2 - 2\sqrt{10}$

9.13. $x = \frac{1}{2} + \frac{\sqrt{5}}{2}$, $x = \frac{1}{2} - \frac{\sqrt{5}}{2}$

9.14.
$y = -\frac{3}{4} + \frac{\sqrt{7}}{4}$, $y = -\frac{3}{4} - \frac{\sqrt{7}}{4}$

9.15. $a = 5 + 2\sqrt{5}$, $a = 5 - 2\sqrt{5}$

9.16.
$b = -3 + 4\sqrt{2}$, $b = -3 - 4\sqrt{2}$

9.17.
$r = -\frac{4}{3} + \frac{2\sqrt{2}i}{3}$, $r = -\frac{4}{3} - \frac{2\sqrt{2}i}{3}$

9.18. $t = 4 + \frac{\sqrt{10}i}{2}$, $t = 4 - \frac{\sqrt{10}i}{2}$

9.19. $m = \frac{7}{3}$, $m = -1$

9.20. $n = -\frac{3}{4}$, $n = -\frac{7}{4}$

9.21. ⓐ $(a - 10)^2$ ⓑ $\left(b - \frac{5}{2}\right)^2$

ⓒ $\left(p + \frac{1}{8}\right)^2$

9.22. ⓐ $(b - 2)^2$ ⓑ $\left(n + \frac{13}{2}\right)^2$

9.23. $x = -5$, $x = -1$

9.24. $y = 1$, $y = 9$

ⓒ $\left(q - \frac{1}{3}\right)^2$

9.25.
$y = 5 + \sqrt{15}i$, $y = 5 - \sqrt{15}i$

9.26.
$z = -4 + \sqrt{3}i$, $z = -4 - \sqrt{3}i$

9.27. $x = 8 + 4\sqrt{3}$, $x = 8 - 4\sqrt{3}$

9.28.
$y = -4 + 3\sqrt{3}$, $y = -4 - 3\sqrt{3}$

9.29. $a = -7$, $a = 3$

9.30. $b = -10$, $b = 2$

9.31.
$p = \frac{5}{2} + \frac{\sqrt{61}}{2}$, $p = \frac{5}{2} - \frac{\sqrt{61}}{2}$

9.32. $q = \frac{7}{2} + \frac{\sqrt{37}}{2}$, $q = \frac{7}{2} - \frac{\sqrt{37}}{2}$

9.33. $c = -9$, $c = 3$

9.34. $d = 11$, $d = -7$

9.35. $m = -7$, $m = -1$

9.36. $n = -2$, $n = 8$

9.37. $r = -\frac{7}{3}$, $r = 3$

9.38. $t = -\frac{5}{2}$, $t = 2$

9.39.
$x = -\frac{3}{8} + \frac{\sqrt{41}}{8}$, $x = -\frac{3}{8} - \frac{\sqrt{41}}{8}$

9.40. $y = \frac{5}{3} + \frac{\sqrt{10}}{3}$, $y = \frac{5}{3} - \frac{\sqrt{10}}{3}$

9.41. $y = 1$, $y = \frac{2}{3}$

9.42. $z = 1$, $z = -\frac{3}{2}$

9.43. $a = -3$, $a = 5$

9.44. $b = -6$, $b = -4$

9.45.
$m = \frac{-6 + \sqrt{15}}{3}$, $m = \frac{-6 - \sqrt{15}}{3}$

9.46.
$n = \frac{-2 + 2\sqrt{6}}{5}$, $n = \frac{-2 - 2\sqrt{6}}{5}$

9.47.
$a = \frac{1}{4} + \frac{\sqrt{31}}{4}i$, $a = \frac{1}{4} - \frac{\sqrt{31}}{4}i$

9.48.
$b = -\frac{1}{5} + \frac{\sqrt{19}}{5}i$, $b = -\frac{1}{5} - \frac{\sqrt{19}}{5}i$

9.49. $x = -1 + \sqrt{6}$, $x = -1 - \sqrt{6}$

9.50. $y = 1 + \sqrt{2}$, $y = 1 - \sqrt{2}$

9.51. $c = \frac{2 + \sqrt{7}}{3}$, $c = \frac{2 - \sqrt{7}}{3}$

9.52. $d = \frac{9 + \sqrt{33}}{4}$, $d = \frac{9 - \sqrt{33}}{4}$

9.53. $r = -5$

9.54. $t = \frac{4}{5}$

9.55. ⓐ 2 complex solutions; ⓑ 2 real solutions; ⓒ 1 real solution

9.56. ⓐ 2 real solutions; ⓑ 2 complex solutions; ⓒ 1 real solution

9.57. ⓐ factoring; ⓑ Square Root Property; ⓒ Quadratic Formula

9.58. ⓐ Quadratic Forumula; ⓑ Factoring or Square Root Property ⓒ Square Root Property

9.59.
$x = \sqrt{2}$, $x = -\sqrt{2}$, $x = 2$, $x = -2$

9.60.
$x = \sqrt{7}$, $x = -\sqrt{7}$, $x = 2$, $x = -2$

9.61. $x = 3$, $x = 1$

9.62. $y = -1$, $y = 1$

9.63. $x = 9$, $x = 16$

9.64. $x = 4$, $x = 16$

9.65. $x = -8$, $x = 343$

9.66. $x = 81$, $x = 625$

9.67. $x = \frac{4}{3}, x = 2$

9.68. $x = \frac{2}{5}, x = \frac{3}{4}$

9.69. The two consecutive odd integers whose product is 99 are 9, 11, and −9, −11

9.70. The two consecutive even integers whose product is 128 are 12, 14 and −12, −14.

9.71. The height of the triangle is 12 inches and the base is 76 inches.

9.72. The height of the triangle is 11 feet and the base is 20 feet.

9.73. The length of the garden is approximately 18 feet and the width 11 feet.

9.74. The length of the tablecloth is approximatel 11.8 feet and the width 6.8 feet.

9.75. The length of the flag pole's shadow is approximately 6.3 feet and the height of the flag pole is 18.9 feet.

9.76. The distance between the opposite corners is approximately 7.2 feet.

9.77. The arrow will reach 180 feet on its way up after 3 seconds and again on its way down after approximately 3.8 seconds.

9.78. The ball will reach 48 feet on its way up after approximately .6 second and again on its way down after approximately 5.4 seconds.

9.79. The speed of the jet stream was 100 mph.

9.80. The speed of the jet stream was 50 mph.

9.81. Press #1 would take 12 hours, and Press #2 would take 6 hours to do the job alone.

9.82. The red hose take 6 hours and the green hose take 3 hours alone.

9.83.

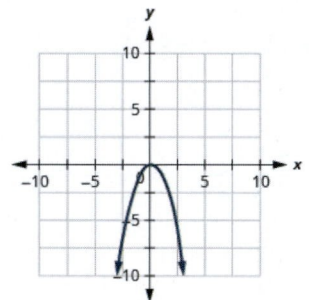

9.84.

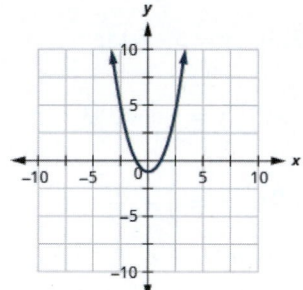

9.85. ⓐ up; ⓑ down

9.86. ⓐ down; ⓑ up

9.87. ⓐ $x = 2$; ⓑ $(2, -7)$

9.88. ⓐ $x = 1$; ⓑ $(1, -5)$

9.89. y-intercept: $(0, -8)$
x-intercepts $(-4, 0), (2, 0)$

9.90. y-intercept: $(0, -12)$
x-intercepts $(-2, 0), (6, 0)$

9.91. y-intercept: $(0, 4)$ no x-intercept

9.92. y-intercept: $(0, -5)$
x-intercepts $(-1, 0), (5, 0)$

9.93.

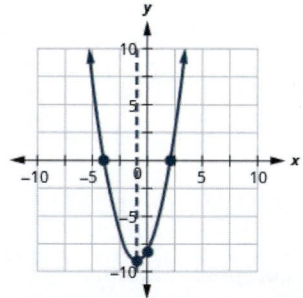

9.94.

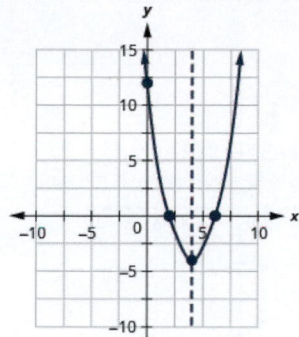

9.95.

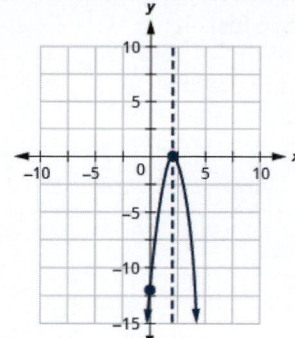

9.96.

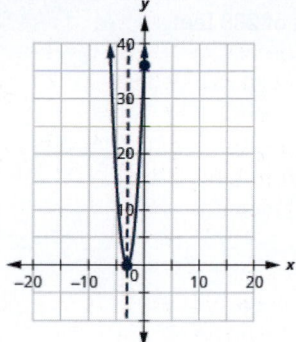

9.97.

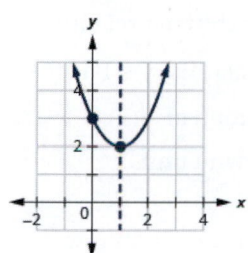

9.98.

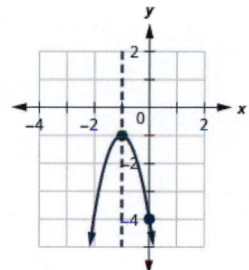

9.99.

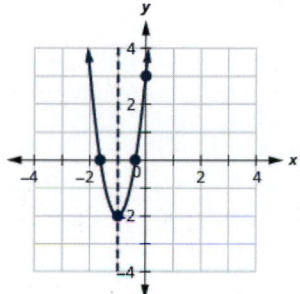

9.100.

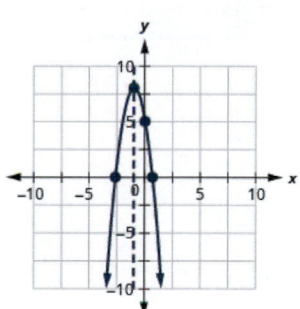

9.101. The minimum value of the quadratic function is –4 and it occurs when $x = 4$.

9.102. The maximum value of the quadratic function is 5 and it occurs when $x = 2$.

9.103. It will take 4 seconds for the stone to reach its maximum height of 288 feet.

9.104. It will 6.5 seconds for the rocket to reach its maximum height of 676 feet.

9.105.

ⓐ

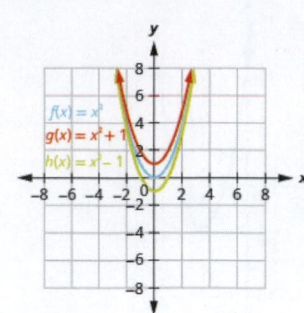

ⓑ The graph of $g(x) = x^2 + 1$ is the same as the graph of $f(x) = x^2$ but shifted up 1 unit. The graph of $h(x) = x^2 - 1$ is the same as the graph of $f(x) = x^2$ but shifted down 1 unit.

9.106.

ⓐ

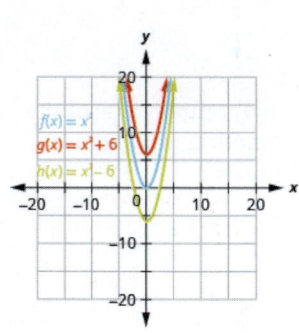

ⓑ The graph of $h(x) = x^2 + 6$ is the same as the graph of $f(x) = x^2$ but shifted up 6 units. The graph of $h(x) = x^2 - 6$ is the same as the graph of $f(x) = x^2$ but shifted down 6 units.

9.107.

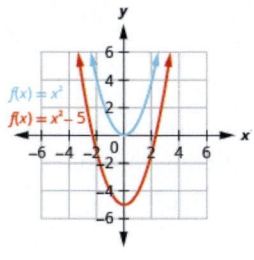

9.108.

9.109.

ⓐ

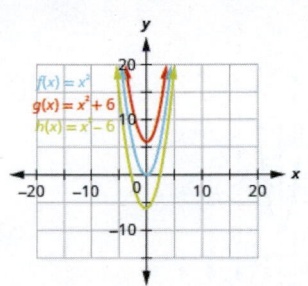

ⓑ The graph of $g(x) = (x+2)^2$ is the same as the graph of $f(x) = x^2$ but shifted left 2 units.

The graph of $h(x) = (x-2)^2$ is the same as the graph of $f(x) = x^2$ but shift right 2 units.

9.110.

ⓐ

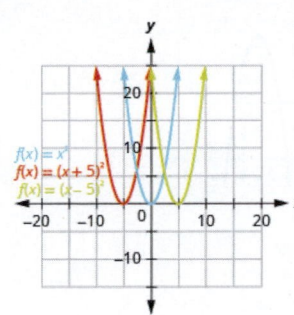

ⓑ The graph of $g(x) = (x+5)^2$ is the same as the graph of $f(x) = x^2$ but shifted left 5 units.

The graph of $h(x) = (x-5)^2$ is the same as the graph of $f(x) = x^2$ but shifted right 5 units.

9.111.

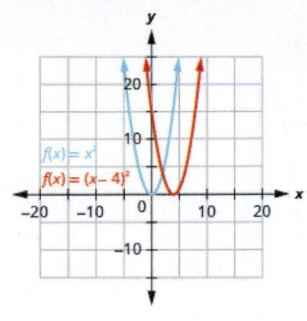

9.112.

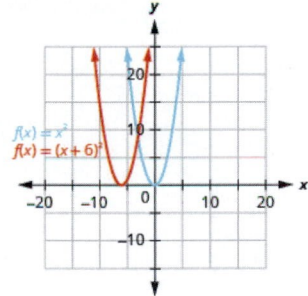

9.113.

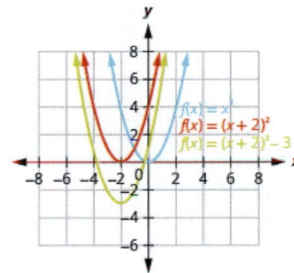

9.114.

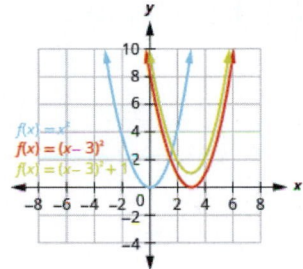

9.115.

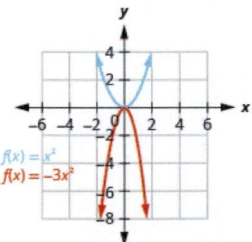

9.116.

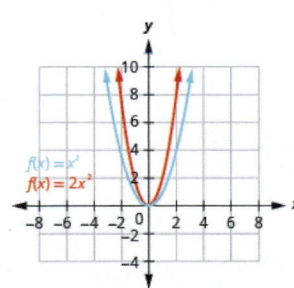

9.117. $f(x) = -4(x+1)^2 + 5$

9.118. $f(x) = 2(x-2)^2 - 5$

9.119.

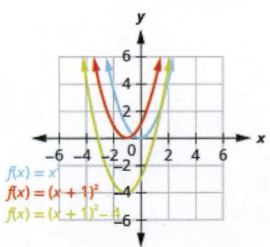

9.120.

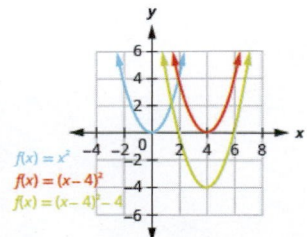

9.121.

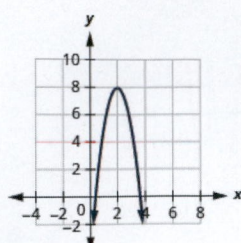

9.122.

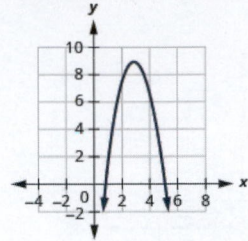

9.123.

ⓐ $f(x) = 3(x-1)^2 + 2$

ⓑ

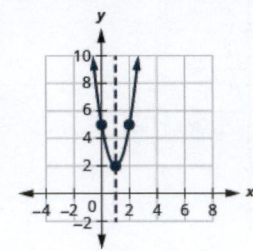

9.124.

ⓐ $f(x) = -2(x-2)^2 + 1$

ⓑ

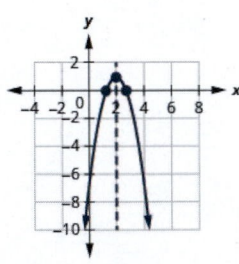

9.125. $f(x) = (x-3)^2 - 4$

9.126. $f(x) = (x+3)^2 - 1$

9.127.

ⓐ

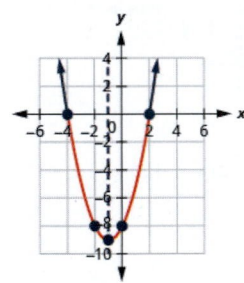

ⓑ $(-4, -2)$

9.128.

ⓐ

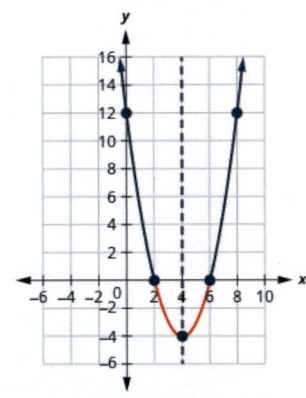

ⓑ $(-\infty, 2] \cup [6, \infty)$

9.129.

ⓐ

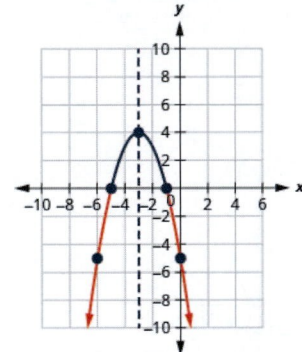

ⓑ $(-1, 5)$

9.130.

ⓐ

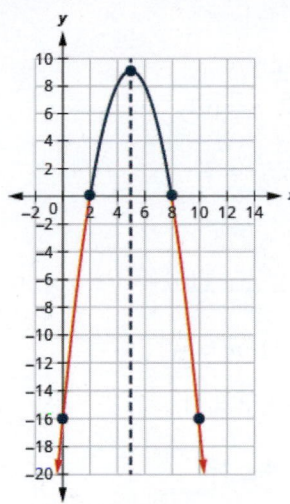

ⓑ $(-\infty, 2] \cup [8, \infty)$

9.131. $(-\infty, -4] \cup [2, \infty)$

9.132. $[-3, 5]$

9.133. $\left[-1 - \sqrt{2}, -1 + \sqrt{2}\right]$

9.134.
$\left(-\infty, 4 - \sqrt{2}\right) \cup \left(4 + \sqrt{2}, \infty\right)$

9.135. ⓐ $(-\infty, \infty)$

ⓑ no solution

9.136. ⓐ no solution

ⓑ $(-\infty, \infty)$

Section Exercises

1. $a = \pm 7$

3. $r = \pm 2\sqrt{6}$

5. $u = \pm 10\sqrt{3}$

7. $m = \pm 3$

9. $x = \pm 6$

11. $x = \pm 5i$

13. $x = \pm 3\sqrt{7}i$

15. $x = \pm 9$

17. $a = \pm 2\sqrt{5}$

19. $p = \pm \dfrac{4\sqrt{7}}{7}$

21. $y = \pm \dfrac{4\sqrt{10}}{5}$

23. $u = 14, u = -2$

25. $m = 6 \pm 2\sqrt{5}$

27. $r = \dfrac{1}{2} \pm \dfrac{\sqrt{3}}{2}$

29. $y = -\dfrac{2}{3} \pm \dfrac{2\sqrt{2}}{9}$

31. $a = 7 \pm 5\sqrt{2}$

33. $x = -3 \pm 2\sqrt{2}$

35. $c = -\dfrac{1}{5} \pm \dfrac{3\sqrt{3}}{5}i$

37. $x = \dfrac{3}{4} \pm \dfrac{\sqrt{7}}{2}i$

39. $m = 2 \pm 2\sqrt{2}$

41. $x = 3 + 2\sqrt{3}, \quad x = 3 - 2\sqrt{3}$

43. $x = -\dfrac{3}{5}, x = \dfrac{9}{5}$

45. $x = -\dfrac{7}{6}, x = \dfrac{11}{6}$

47. $r = \pm 4$

49. $a = 4 \pm 2\sqrt{7}$

51. $w = 1, w = \dfrac{5}{3}$

53. $a = \pm 3\sqrt{2}$

55. $p = \dfrac{1}{3} \pm \dfrac{\sqrt{7}}{3}$

57. no real solution

59. $u = 7 \pm 6\sqrt{2}$

61. $m = 4 \pm 2\sqrt{3}$

63. $x = -3, x = -7$

65. $c = \pm \dfrac{5\sqrt{6}}{6}$

67. no real solution

69. Answers will vary.

71. ⓐ $(m - 12)^2$ ⓑ $\left(x - \dfrac{11}{2}\right)^2$

ⓒ $\left(p - \dfrac{1}{6}\right)^2$

73. ⓐ $(p-11)^2$ ⓑ $\left(y+\frac{5}{2}\right)^2$

ⓒ $\left(m+\frac{1}{5}\right)^2$

75. $u=-3,\ u=1$

77. $x=-1,\ x=21$

79. $m=-2\pm2\sqrt{10}i$

81. $r=-3\pm\sqrt{2}i$

83. $a=5\pm2\sqrt{5}$

85. $x=-\frac{5}{2}\pm\frac{\sqrt{33}}{2}$

87. $u=1,\ u=13$

89. $r=-2,\ r=6$

91. $v=\frac{9}{2}\pm\frac{\sqrt{89}}{2}$

93. $x=5\pm\sqrt{30}$

95. $x=-7,\ x=3$

97. $x=-5,\ x=-1$

99. $m=-11,\ m=1$

101. $n=1\pm2\sqrt{3}$

103. $c=-2,\ c=\frac{3}{2}$

105. $x=-5,\ x=\frac{3}{2}$

107. $p=-\frac{7}{4}\pm\frac{\sqrt{161}}{4}$

109. $x=\frac{3}{10}\pm\frac{\sqrt{191}}{10}i$

111. Answers will vary.

113. $m=-1,\ m=\frac{3}{4}$

115. $p=\frac{1}{3},\ p=2$

117. $p=-4,\ p=-3$

119. $r=-3,\ r=11$

121. $u=\frac{-7\pm\sqrt{73}}{6}$

123. $a=\frac{3\pm\sqrt{3}}{2}$

125. $x=-4\pm2\sqrt{5}$

127. $y=-\frac{2}{3},\ y=-1$

129. $x=-\frac{3}{4}\pm\frac{\sqrt{15}}{4}i$

131. $x=\frac{3}{8}\pm\frac{\sqrt{7}}{8}i$

133. $v=2\pm2\sqrt{2}$

135. $y=-4,\ y=7$

137. $m=-1,\ m=\frac{3}{4}$

139. $b=\frac{-2\pm\sqrt{11}}{6}$

141. $c=-\frac{3}{4}$

143. $q=-\frac{3}{5}$

145. ⓐ no real solutions ⓑ 1

ⓒ 2

147. ⓐ 1 ⓑ no real solutions

ⓒ 2

149. ⓐ factor ⓑ square root

ⓒ Quadratic Formula

151. ⓐ Quadratic Formula

ⓑ square root ⓒ factor

153. Answers will vary.

155. $x=\pm\sqrt{3},\ x=\pm2$

157. $x=\pm\sqrt{15},\ x=\pm\sqrt{2}i$

159. $x=\pm1,\ x=\frac{\pm\sqrt{6}}{2}$

161. $x=\pm\sqrt{3},\ x=\pm\frac{\sqrt{2}}{2}$

163. $x=-1,\ x=12$

165. $x=-\frac{5}{3},\ x=0$

167. $x=0,\ x=\pm\sqrt{3}$

169. $x=\pm\frac{11}{2},\ x=\pm\frac{\sqrt{22}}{2}$

171. $x=25$

173. $x=4$

175. $x=\frac{1}{4}$

177. $x=\frac{1}{25}$

179. $x=-1,\ x=-512$

181. $x=8,\ x=-216$

183. $x=\frac{27}{8},\ x=-\frac{64}{27}$

185. $x=27,\ x=64{,}000$

187. $x=1,\ x=49$

189. $x=-2,\ x=-\frac{3}{5}$

191. $x=-2,\ x=\frac{4}{3}$

193. Answers will vary.

195. Two consecutive odd numbers whose product is 255 are 15 and 17, and –15 and –17.

197. The first and second consecutive odd numbers are 24 and 26, and –26 and –24.

199. Two consecutive odd numbers whose product is 483 are 21 and 23, and –21 and –23.

201. The width of the triangle is 5 inches and the height is 18 inches.

203. The base is 24 feet and the height of the triangle is 10 feet.

205. The length of the driveway is 15.0 feet and the width is 3.3 feet.

207. The length of table is 8 feet and the width is 3 feet.

209. The length of the legs of the right triangle are 3.2 and 9.6 cm.

211. The length of the diagonal fencing is 7.3 yards.

213. The ladder will reach 24.5 feet on the side of the house.

215. The arrow will reach 400 feet on its way up in 2.8 seconds and on the way down in 11 seconds.

217. The bullet will take 70 seconds to hit the ground.

219. The speed of the wind was 49 mph.

221. The speed of the current was 4.3 mph.

223. The less experienced painter takes 6 hours and the experienced painter takes 3 hours to do the job alone.

225. Machine #1 takes 3.6 hours and Machine #2 takes 4.6 hours to do the job alone.

227. Answers will vary.

229.

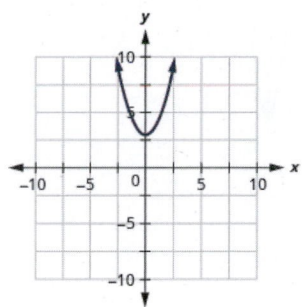

231.

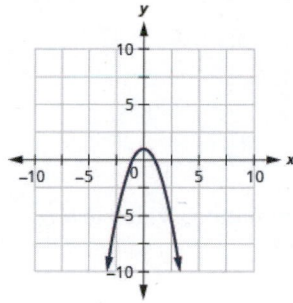

233. ⓐ down ⓑ up

235. ⓐ down ⓑ up

237. ⓐ $x = -4$; ⓑ $(-4, -17)$

239. ⓐ $x = 1$; ⓑ $(1, 2)$

241. y-intercept: $(0, 6)$; x-intercept $(-1, 0), (-6, 0)$

243. y-intercept: $(0, 12)$; x-intercept $(-2, 0), (-6, 0)$

245. y-intercept: $(0, -19)$; x-intercept: none

247. y-intercept: $(0, 13)$; x-intercept: none

249. y-intercept: $(0, -16)$; x-intercept $\left(\frac{5}{2}, 0\right)$

251. y-intercept: $(0, 9)$; x-intercept $(-3, 0)$

253.

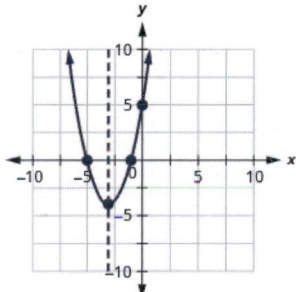

255.

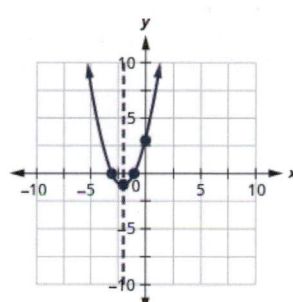

257.

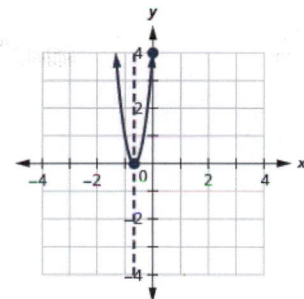

259.

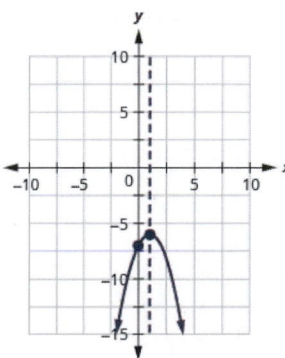

261.

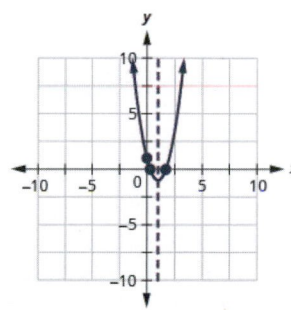

263.

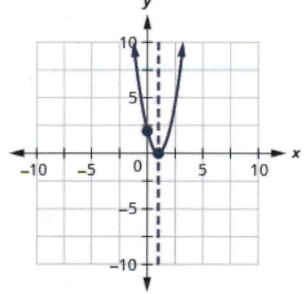

265.

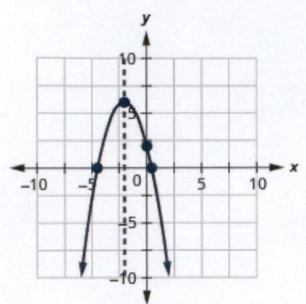

267.

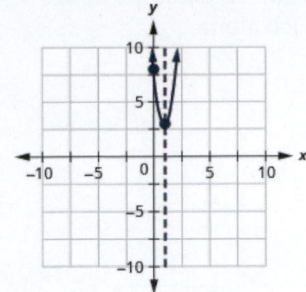

269.

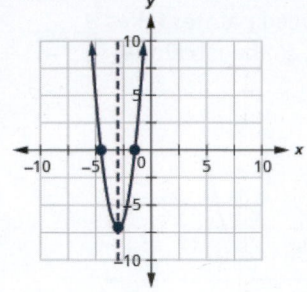

271. The minimum value is $-\frac{9}{8}$ when $x = -\frac{1}{4}$.

277. In 5.3 sec the arrow will reach maximum height of 486 ft.

283. He will be able to sell 35 pairs of boots at the maximum revenue of $1,225.

289. Answers will vary.

273. The maximum value is 6 when $x = 3$.

279. In 3.4 seconds the ball will reach its maximum height of 185.6 feet.

285. The length of the side along the river of the corral is 120 feet and the maximum area is 7,200 square feet.

291. Answers will vary.

275. The maximum value is 16 when $x = 0$.

281. 20 computers will give the maximum of $400 in receipts.

287. The maximum area of the patio is 800 feet.

293.

ⓐ

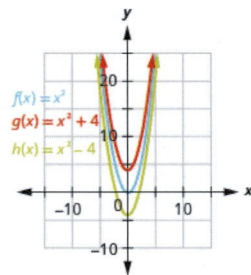

ⓑ The graph of $g(x) = x^2 + 4$ is the same as the graph of $f(x) = x^2$ but shifted up 4 units. The graph of $h(x) = x^2 - 4$ is the same as the graph of $f(x) = x^2$ but shift down 4 units.

295.

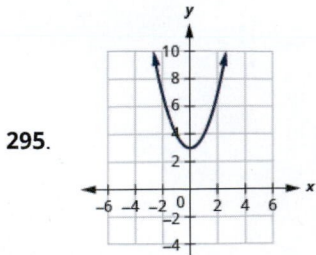

297.

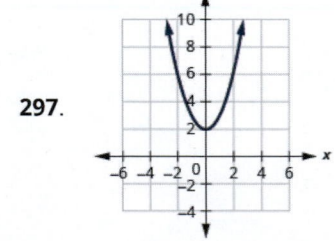

299.

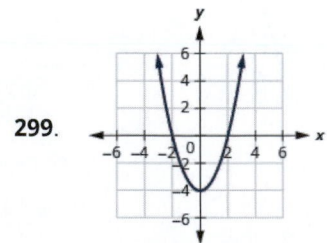

301.

ⓐ

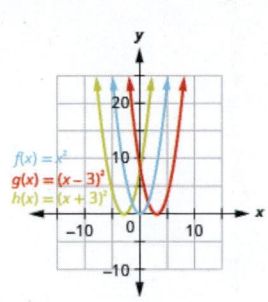

ⓑ The graph of $g(x) = (x - 3)^2$ is the same as the graph of $f(x) = x^2$ but shifted right 3 units. The graph of $h(x) = (x + 3)^2$ is the same as the graph of $f(x) = x^2$ but shifted left 3 units.

303.

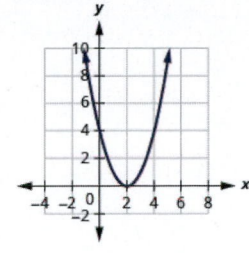

305.

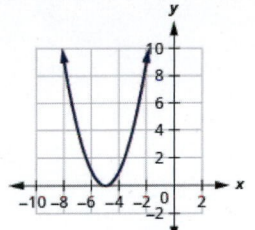

307.

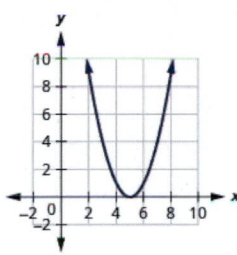

309.

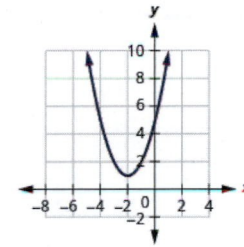

311.

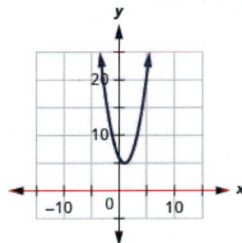

313.

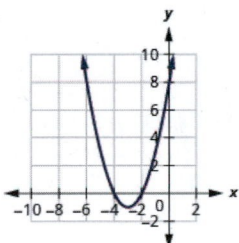

315.

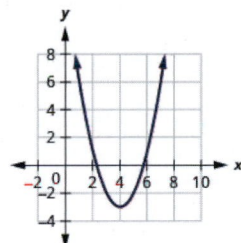

317.

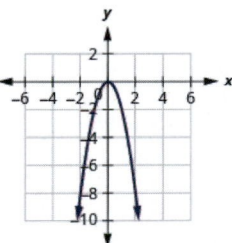

319.

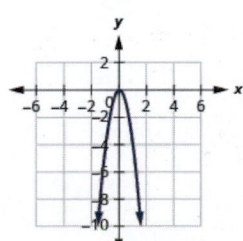

321.

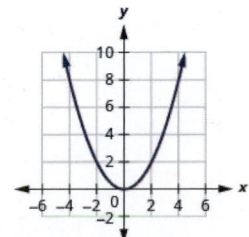

323.

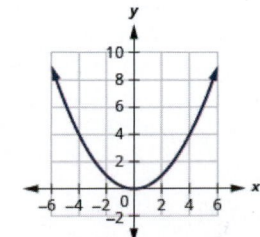

325. $f(x) = -3(x+2)^2 + 7$ **327.** $f(x) = 3(x+1)^2 - 4$ **329.** ⓐ $f(x) = (x+3)^2 - 4$

ⓑ

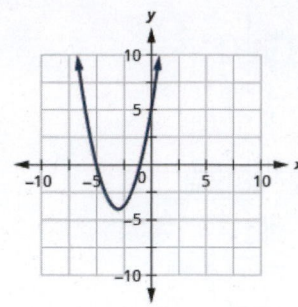

331. ⓐ $f(x) = (x+2)^2 - 1$ **333.** ⓐ $f(x) = (x-3)^2 + 6$ **335.** ⓐ $f(x) = -(x-4)^2 + 0$

ⓑ   ⓑ   ⓑ

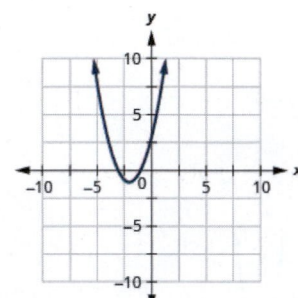   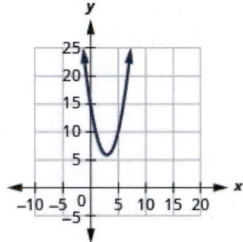   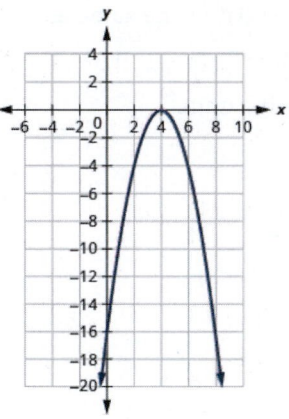

337. ⓐ $f(x) = -(x+2)^2 + 6$ **339.** ⓐ $f(x) = 5(x-1)^2 + 3$ **341.** ⓐ $f(x) = 2(x-1)^2 - 1$

ⓑ   ⓑ   ⓑ

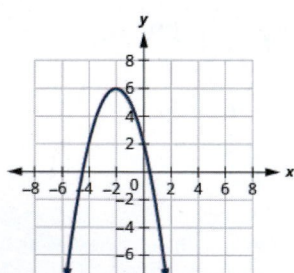   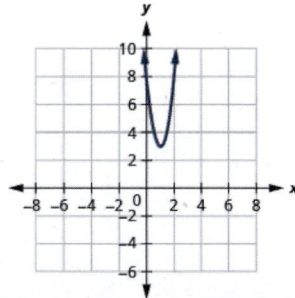   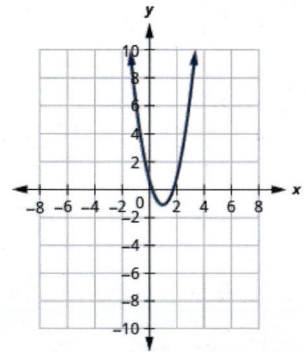

343. ⓐ $f(x) = -2(x - 2)^2 - 2$

ⓑ

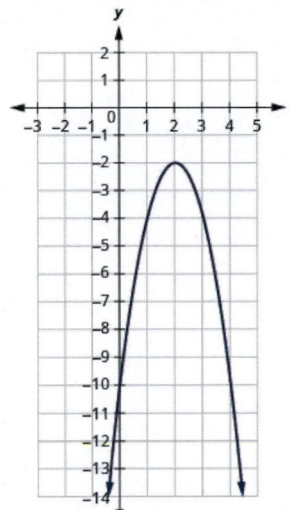

345. ⓐ $f(x) = 2(x + 1)^2 + 4$

ⓑ

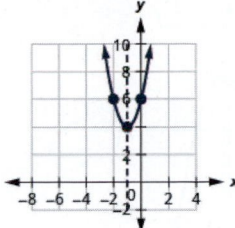

347. ⓐ $f(x) = -(x - 1)^2 - 3$

ⓑ

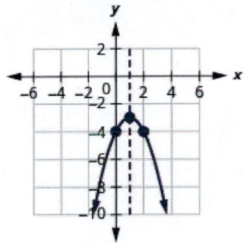

349. ⓒ

355. ⓖ

361. Answers will vary.

351. ⓔ

357. $f(x) = (x + 1)^2 - 5$

363.

ⓐ

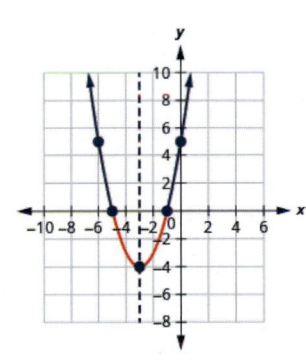

ⓑ $(-\infty, -5) \cup (-1, \infty)$

353. ⓓ

359. $f(x) = 2(x - 1)^2 - 3$

365.

ⓐ

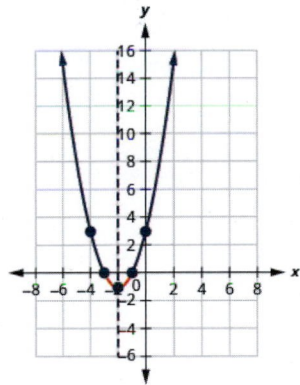

ⓑ $[-3, -1]$

367.

ⓐ

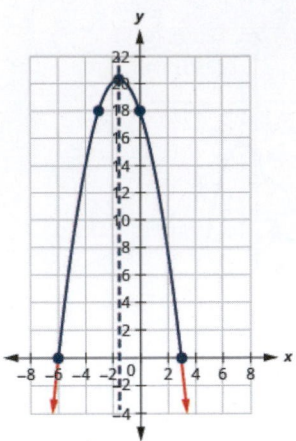

369.

ⓐ

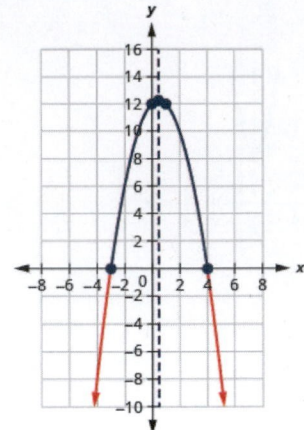

371. $(-\infty, -4] \cup [1, \infty)$

ⓑ $(-\infty, -6] \cup [3, \infty)$

ⓑ $[-3, 4]$

373. $(2, 5)$

375. $(-\infty, -5) \cup (-3, \infty)$

377. $\left[2 - \sqrt{2}, 2 + \sqrt{2}\right]$

379. $\left(-\infty, 5 - \sqrt{6}\right) \cup \left(5 + \sqrt{6}, \infty\right)$

381. $\left(-\infty, -\frac{5}{2}\right] \cup \left[-\frac{2}{3}, \infty\right)$

383. $\left(-\infty, -\frac{1}{2}\right] \cup [4, \infty)$

385. $(-\infty, \infty)$.

387. no solution

389. $(-\infty, \infty)$.

391. Answers will vary.

393. Answers will vary.

Review Exercises

395. $y = \pm 12$

397. $a = \pm 5$

399. $r = \pm 4\sqrt{2}i$

401. $w = \pm 5\sqrt{3}$

403. $p = -1, 9$

405. $x = \frac{1}{4} \pm \frac{\sqrt{3}}{4}$

407. $n = 4 \pm 10\sqrt{2}$

409. $n = -5 \pm 2\sqrt{3}$

411. $(x + 11)^2$

413. $\left(a - \frac{3}{2}\right)^2$

415. $d = -13, -1$

417. $m = -3 \pm 10i$

419. $v = 7 \pm 3\sqrt{2}$

421. $m = -9, -1$

423. $a = \frac{3}{2} \pm \frac{\sqrt{41}}{2}$

425. $u = -6 \pm 2\sqrt{2}$

427. $p = 0, 6$

429. $y = -\frac{1}{2}, 2$

431. $c = -\frac{1}{3} \pm \frac{2\sqrt{7}}{3}$

433. $x = \frac{3}{2} \pm \frac{1}{2}i$

435. $x = \frac{1}{4}, 1$

437. $r = -6, 7$

439. $v = \frac{-1 \pm \sqrt{21}}{8}$

441. $m = \frac{-4 \pm \sqrt{10}}{3}$

443. $a = \frac{5}{12} \pm \frac{\sqrt{23}}{12}i$

445. $u = 5 \pm \sqrt{21}$

447. $p = \frac{4 \pm \sqrt{5}}{5}$

449. $c = -\frac{1}{2}$

451. ⓐ 1 ⓑ 2 ⓒ 2 ⓓ 2

453. ⓐ factor ⓑ Quadratic Formula ⓒ square root

455. $x = \pm \sqrt{2},\ x = \pm 2\sqrt{3}$

457. $x = \pm 1,\ x = \pm \frac{1}{2}$

459. $x = 16$

461. $x = 64, x = 216$

463. $x = -2, x = \frac{4}{3}$

465. Two consecutive even numbers whose product is 624 are 24 and 26, and -24 and -26.

467. The height is 14 inches and the width is 10 inches.

469. The length of the diagonal is 3.6 feet.

471. The width of the serving table is 4.7 feet and the length is 16.1 feet.

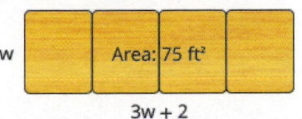

473. The speed of the wind was 30 mph.

475. One man takes 3 hours and the other man 6 hours to finish the repair alone.

477.

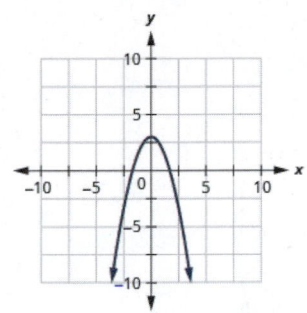

479. ⓐ up ⓑ down

481. $x = 2$; $(2, -7)$

$y : (0, 15)$

483. $x : (3, 0), (5, 0)$

485. $y : (0, -46)$
$x : $ none

487. $y : (0, -64)$
$x : (-8, 0)$

489.

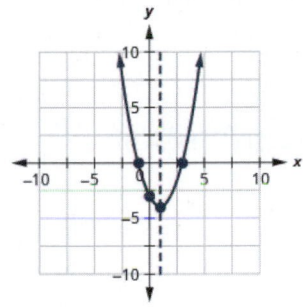

491.

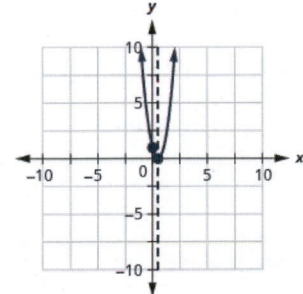

493.

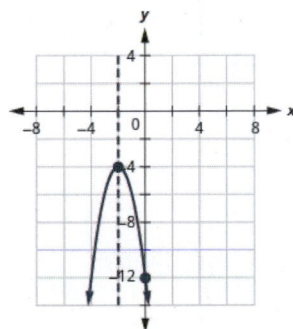

495. The maximum value is 2 when $x = 2$.

497. The length adjacent to the building is 90 feet giving a maximum area of 4,050 square feet.

499.

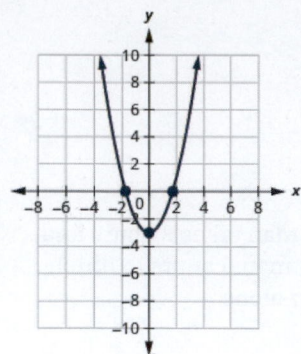

501.

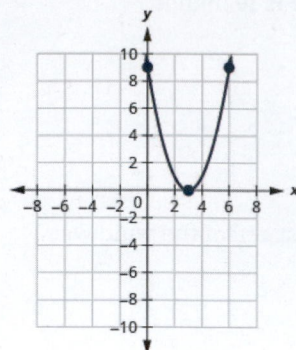

503.

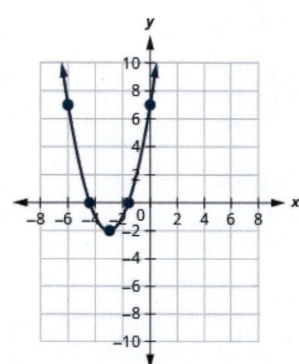

505.

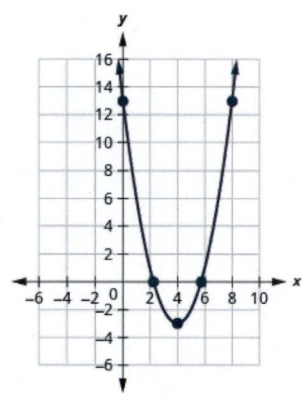

507.

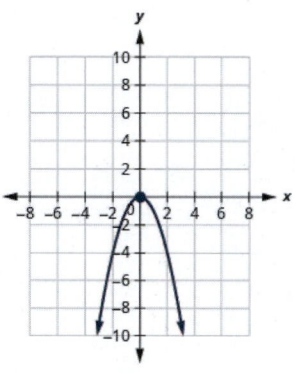

509. $f(x) = 2(x - 1)^2 - 6$

511. ⓐ $f(x) = 3(x - 1)^2 - 4$

ⓑ

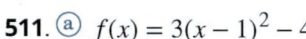

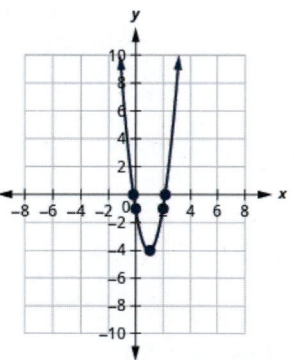

513. ⓐ $f(x) = 2(x + 1)^2 + 4$

ⓑ

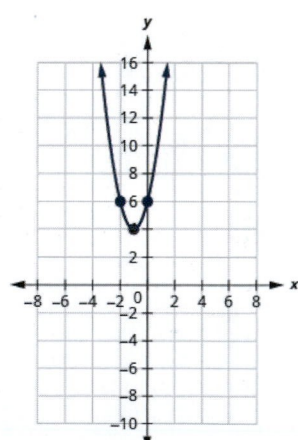

515. ⓐ $f(x) = -3(x+2)^2 + 7$

ⓑ

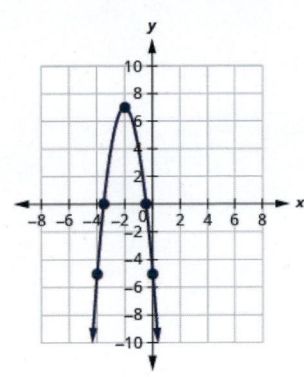

517. $f(x) = (x+1)^2 - 5$

519.

ⓐ

ⓑ $(-\infty, -2) \cup (3, \infty)$

521.

ⓐ

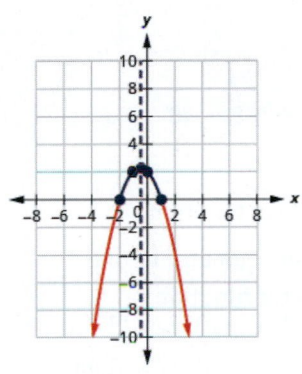

$[-2, 1]$

527. no solution

523. $(2, 4)$

525. $\left[3 - \sqrt{5}, \; 3 + \sqrt{5}\right]$

Practice Test

529. $w = -2, \; w = -8$

531. $m = 1, \; m = \dfrac{3}{2}$

533. $y = \dfrac{2}{3}$

535. 2 complex

537. $y = 1, \; y = -27$

539. ⓐ down ⓑ $x = -4$

ⓒ $(-4, 0)$ ⓓ $y: (0, 16); \; x: (-4, 0)$

ⓔ minimum value of -4 when $x = 0$

541.

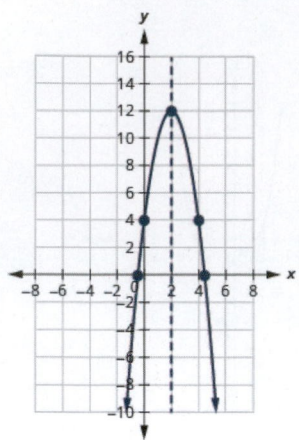

543.

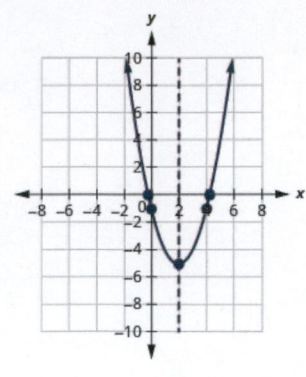

$f(x) = 2(x-1)^2 - 6$

545. $\left(-\infty, -\frac{5}{2}\right) \cup (2, \infty)$

547. The diagonal is 3.8 units long.

Chapter 10

Try It

10.1. ⓐ $15x + 1$ ⓑ $15x - 9$ ⓒ $15x^2 - 7x - 2$

10.2. ⓐ $24x - 23$ ⓑ $24x - 23$ ⓒ $24x^2 - 38x + 15$

10.3. ⓐ -8 ⓑ 5 ⓒ 40

10.4. ⓐ 65 ⓑ 10 ⓒ 5

10.5. ⓐ One-to-one function ⓑ Function; not one-to-one

10.6. ⓐ Not a function ⓑ Function; not one-to-one

10.7. ⓐ Not a function ⓑ One-to-one function

10.8. ⓐ Function; not one-to-one ⓑ One-to-one function

10.9. Inverse function: $\{(4, 0), (7, 1), (10, 2), (13, 3)\}$. Domain: $\{4, 7, 10, 13\}$. Range: $\{0, 1, 2, 3\}$.

10.10. Inverse function: $\{(4, -1), (1, -2), (0, -3), (2, -4)\}$. Domain: $\{0, 1, 2, 4\}$. Range: $\{-4, -3, -2, -1\}$.

10.11.

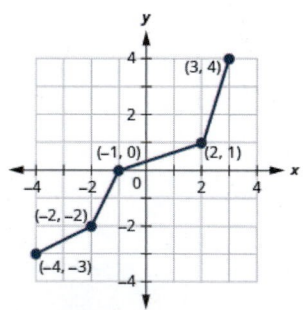

10.12.

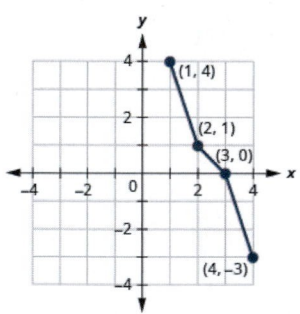

10.13. $g(f(x)) = x$, and $f(g(x)) = x$, so they are inverses.

10.14. $g(f(x)) = x$, and $f(g(x)) = x$, so they are inverses.

10.15. $f^{-1}(x) = \dfrac{x+3}{5}$

10.16. $f^{-1}(x) = \dfrac{x-5}{8}$

10.17. $f^{-1}(x) = \dfrac{x^5 + 2}{3}$

10.18. $f^{-1}(x) = \dfrac{x^4 + 7}{6}$

10.19.

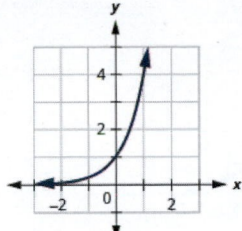

10.20.

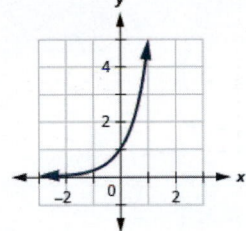

10.21.

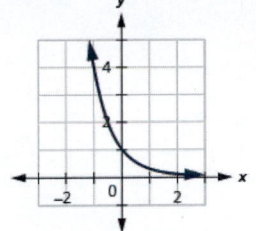

10.22.

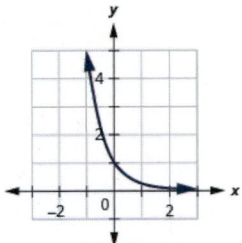

10.23.

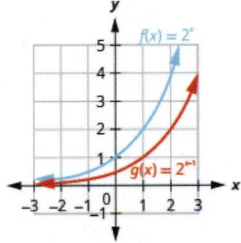

10.24.

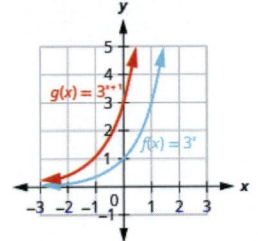

10.25.

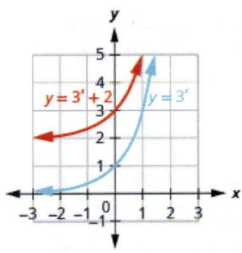

10.26.

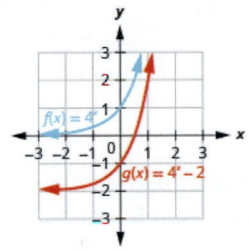

10.27. $x = 2$

10.28. $x = 4$

10.29. $x = -1, x = 2$

10.30. $x = -2, x = 3$

10.31. ⓐ $22,332.96

ⓑ $22,362.49 ⓒ $22,377.37

10.32. ⓐ $21,071.81 ⓑ $21,137.04

ⓒ $21,170.00

10.33. She will find 166 bacteria.

10.34. She will find 1,102 viruses.

10.35. ⓐ $\log_3 9 = 2$

ⓑ $\log_7 \sqrt{7} = \frac{1}{2}$ ⓒ $\log_{\frac{1}{3}} \frac{1}{27} = x$

10.36. ⓐ $\log_4 64 = 3$

ⓑ $\log_4 \sqrt[3]{4} = \frac{1}{3}$ ⓒ $\log_{\frac{1}{2}} \frac{1}{32} = x$

10.37. ⓐ $64 = 4^3$

ⓑ $1 = x^0$ ⓒ $\frac{1}{100} = 10^{-2}$

10.38. ⓐ $27 = 3^3$ ⓑ $1 = x^0$

ⓒ $\frac{1}{10} = 10^{-1}$

10.39.

ⓐ $x = 8$ ⓑ $x = 125$ ⓒ $x = 2$

10.40.

ⓐ

$x = 9$ ⓑ $x = 243$ ⓒ $x = 3$

10.41.

ⓐ

2 ⓑ $\frac{1}{2}$ ⓒ -5

10.42. ⓐ 2 ⓑ $\frac{1}{3}$ ⓒ -2

10.43.

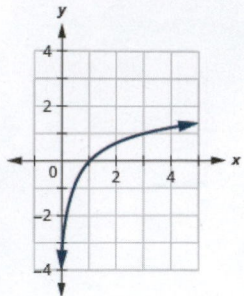

10.44.

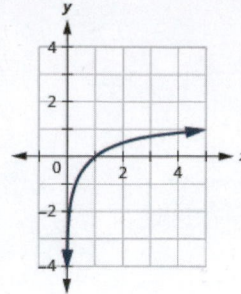

10.45.

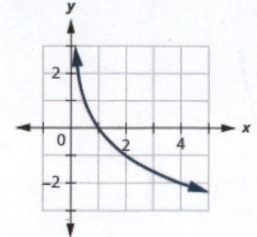

10.46.

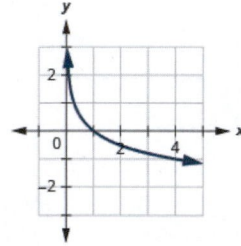

10.47.
ⓐ
$a = 11$
ⓑ $x = e^7$

10.48.
ⓐ
$a = 4$
ⓑ $x = e^9$

10.49.
ⓐ
$x = 13$
ⓑ $x = 2$

10.50.
ⓐ
$x = 6$
ⓑ $x = 1$

10.51. The quiet dishwashers have a decibel level of 50 dB.

10.52. The decibel level of heavy traffic is 90 dB.

10.53. The intensity of the 1906 earthquake was about 8 times the intensity of the 1989 earthquake.

10.54. The intensity of the earthquake in Chile was about 1,259 times the intensity of the earthquake in Los Angeles.

10.55. ⓐ 0 ⓑ 1

10.56. ⓐ 0 ⓑ 1

10.57. ⓐ 15 ⓑ 4

10.58. ⓐ 8 ⓑ 15

10.59. ⓐ $1 + \log_3 x$

ⓑ $3 + \log_2 x + \log_2 y$

10.60. ⓐ $1 + \log_9 x$

ⓑ $3 + \log_3 x + \log_3 y$

10.61. ⓐ $\log_4 3 - 1$ ⓑ $\log x - 3$

10.62. ⓐ $\log_2 5 - 2$ ⓑ $1 - \log y$

10.63. ⓐ $4\log_7 5$ ⓑ $100 \cdot \log x$

10.64. ⓐ $7\log_2 3$ ⓑ $20 \cdot \log x$

10.65. $\log_2 5 + 4\log_2 x + 2\log_2 y$

10.66. $\log_3 7 + 5\log_3 x + 3\log_3 y$

10.67.
$\frac{1}{5}\left(4\log_4 x - \frac{1}{2} - 3\log_4 y - 2\log_4 z\right)$

10.68.
$\frac{1}{3}\left(2\log_3 x - \log_3 5 - \log_3 y - \log_3 z\right)$

10.69. $\log_2 \frac{5x}{y}$

10.70. $\log_3 \frac{6}{xy}$

10.71. $\log_2 x^3 (x-1)^2$

10.72. $\log x^2 (x+1)^2$

10.73. 3.402

10.74. 2.379

10.75. $x = 6$

10.76. $x = 4$

10.77. $x = 4$

10.78. $x = 8$

10.79. $x = 3$

10.80. $x = 8$

10.81. $x = \frac{\log 43}{\log 7} \approx 1.933$

10.82. $x = \frac{\log 98}{\log 8} \approx 2.205$

10.83. $x = \ln 9 + 2 \approx 4.197$

10.84. $x = \frac{\ln 5}{2} \approx 0.805$

10.85. $r \approx 9.3\%$

10.86. $r \approx 11.9\%$

10.87. There will be 62,500 bacteria.

10.88. There will be 5,870,061 bacteria.

10.89. There will be 6.43 mg left.

10.90. There will be 31.5 mg left.

Section Exercises

1. ⓐ $8x + 23$ ⓑ $8x + 11$ ⓒ
$8x^2 + 26x + 15$

3. ⓐ $24x + 1$ ⓑ $24x - 19$
ⓒ $24x^2 + 19x - 5$

5. ⓐ $6x^2 - 9x$ ⓑ $18x^2 - 9x$
ⓒ $6x^3 - 9x^2$

7. ⓐ $2x^2 + 3$ ⓑ $4x^2 - 4x + 3$
ⓒ $2x^3 - x^2 + 4x - 2$

9. ⓐ 245 ⓑ 104 ⓒ 53

11. ⓐ 250 ⓑ 14 ⓒ 77

13. Function; not one-to-one

15. One-to-one function

17. ⓐ Not a function ⓑ Function; not one-to-one

19. ⓐ One-to-one function
ⓑ Function; not one-to-one

21. Inverse function:
$\{(1, 2), (2, 4), (3, 6), (4, 8)\}$.
Domain: $\{1, 2, 3, 4\}$. Range: $\{2, 4, 6, 8\}$.

23. Inverse function:
$\{(-2, 0), (3, 1), (7, 2), (12, 3)\}$.
Domain: $\{-2, 3, 7, 12\}$. Range: $\{0, 1, 2, 3\}$.

25. Inverse function:
$\{(-3, -2), (-1, -1), (1, 0), (3, 1)\}$.
Domain: $\{-3, -1, 1, 3\}$. Range: $\{-2, -1, 0, 1\}$.

27.

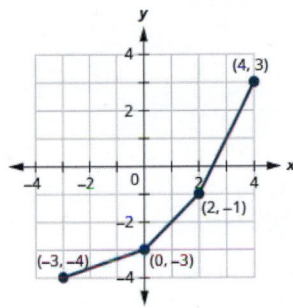

29.

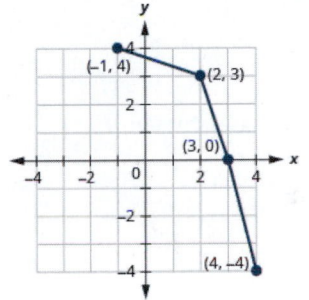

31. $g(f(x)) = x$, and $f(g(x)) = x$, so they are inverses.

33. $g(f(x)) = x$, and $f(g(x)) = x$, so they are inverses.

35. $g(f(x)) = x$, and $f(g(x)) = x$, so they are inverses.

37. $g(f(x)) = x$, and $f(g(x)) = x$, so they are inverses (for nonnegative x).

39. $f^{-1}(x) = x + 12$

41. $f^{-1}(x) = \frac{x}{9}$

43. $f^{-1}(x) = 6x$

45. $f^{-1}(x) = \frac{x + 7}{6}$

47. $f^{-1}(x) = \frac{x - 5}{-2}$

49. $f^{-1}(x) = \sqrt{x - 6}$

51. $f^{-1}(x) = \sqrt[3]{x + 4}$

53. $f^{-1}(x) = \frac{1}{x} - 2$

55. $f^{-1}(x) = x^2 + 2$, $x \geq 0$

57. $f^{-1}(x) = x^3 + 3$

59. $f^{-1}(x) = \frac{x^4 + 5}{9}$, $x \geq 0$

61. $f^{-1}(x) = \frac{x^5 - 5}{-3}$

63. Answers will vary.

65.

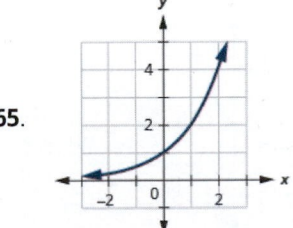

67.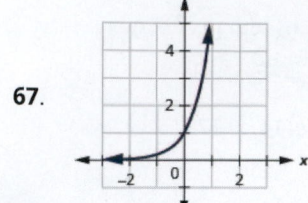

69.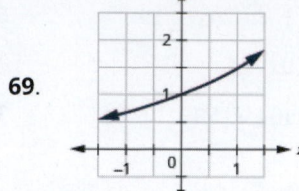

71.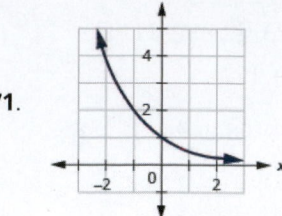

73.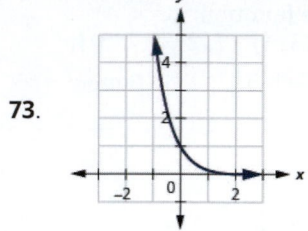

75.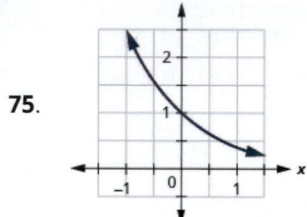

77.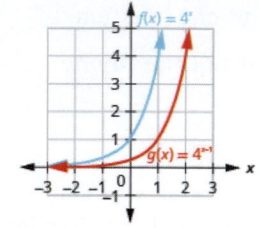

79.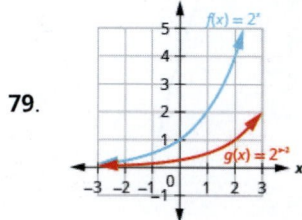

81.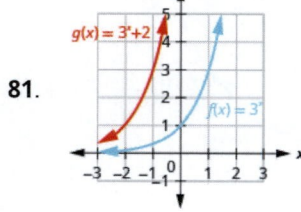

83.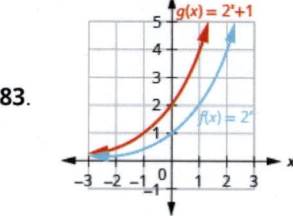

85.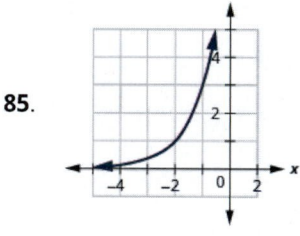

87.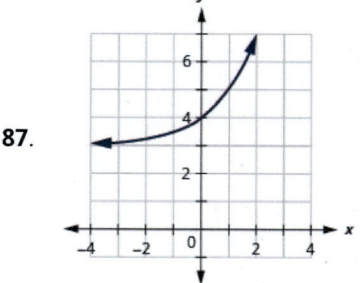

89.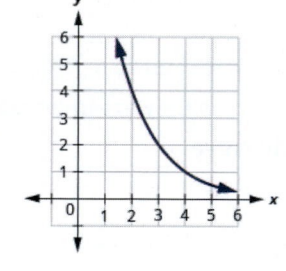

95. $x = 4$

91.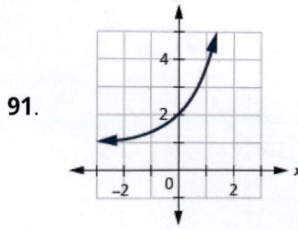

93.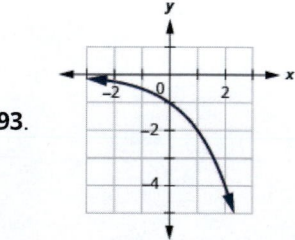

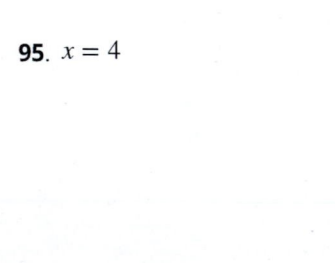

97. $x = -1$

103. $x = -1$

109. (f)

115. (a) $7,387.28 (b) $7,434.57

(c) $7,459.12

99. $x = -1, x = 1$

105. $x = 2$

111. (a)

117. $36,945.28

101. $x = 1$

107. $x = -1, x = 2$

113. (e)

119. 223 bacteria

121. 288,929,825

127. $\log_2 32 = 5$

133. $\log_x \sqrt[3]{6} = \frac{1}{3}$

139. $\log_4 \frac{1}{64} = -3$

145. $32 = x^5$

151. $1,000 = 10^3$

157. $x = 4$

163. $x = 2$

169. 0

175. -3

123. Answers will vary.

129. $\log_5 125 = 3$

135. $\log_{17} \sqrt[5]{17} = x$

141. $\ln x = 3$

147. $1 = 7^0$

153. $43 = e^x$

159. $x = 125$

165. $x = -2$

171. $\frac{1}{3}$

177. -2

125. Answers will vary.

131. $\log \frac{1}{100} = -2$

137. $\log_{\frac{1}{3}} \frac{1}{81} = 4$

143. $64 = 2^6$

149. $9 = 9^1$

155. $x = 11$

161. $x = \frac{1}{243}$

167. 2

173. -2

179.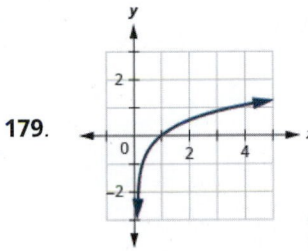

181.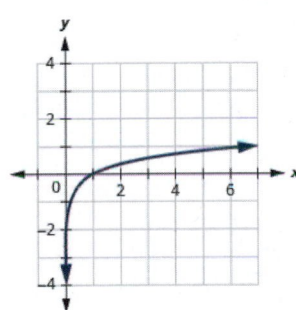

183.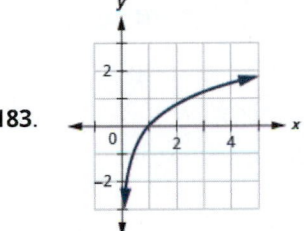

185.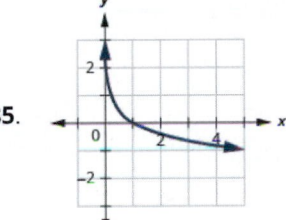

189. $a = 9$

191. $a = 3$

187.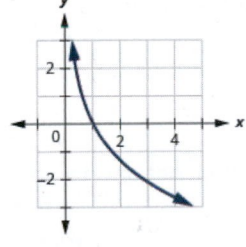

193. $a = \sqrt[3]{24}$

199. $x = 17$

205. $x = -5\sqrt{5}, x = 5\sqrt{5}$

211. The sound of a garbage disposal has a decibel level of 100 dB.

217. Answers will vary.

195. $x = e^4$

201. $x = 6$

207. $x = -5, x = 5$

213. The intensity of the 1994 Northridge earthquake in the Los Angeles area was about 40 times the intensity of the 2014 earthquake.

219. ⓐ 0 ⓑ 1

197. $x = 5$

203. $x = 3$

209. A whisper has a decibel level of 20 dB.

215. Answers will vary.

221. ⓐ 10 ⓑ 10

223. (a) 15 (b) -4

225. (a) $\sqrt{3}$ (b) -1

227. (a) 3 (b) 7

229. $\log_5 8 + \log_5 y$

231. $4 + \log_3 x + \log_3 y$

233. $3 + \log y$

235. $\log_6 5 - 1$

237. $3 - \log_5 x$

239. $4 - \log y$

241. $4 - \ln 16$

243. $5\log_2 x$

245. $-3\log x$

247. $\frac{1}{3}\log_5 x$

249. $\sqrt[3]{4}\ln x$

251. $\log_2 3 + 5\log_2 x + 3\log_2 y$

253. $\frac{1}{4}\log_5 21 + 3\log_5 y$

255. $\log_5 4 + \log_5 a + 3\log_5 b + 4\log_5 c - 2\log_5 d$

257. $\frac{2}{3}\log_3 x - 3 - 4\log_3 y$

259. $\frac{1}{2}\log_3(3x + 2y^2) - \log_3 5 - 2\log_3 z$

261. $\frac{1}{3}(\log_5 3 + 2\log_5 x - \log_5 4 - 3\log_5 y - \log_5 z)$

263. 2

265. 2

267. $\log_2 \frac{5}{x-1}$

269. $\log_5 \frac{2}{xy}$

271. $\log_3 x^6 y^9$

273. 0

275. $\ln\frac{x^3 y^4}{z^2}$

277. $\log(2x + 3)^2 \cdot \sqrt{x+1}$

279. 2.379

281. 1.674

283. 5.542

285. Answers will vary.

287. Answers will vary.

289. $x = 7$

291. $x = 4$

293. $x = 1, \quad x = 3$

295. $x = 8$

297. $x = 3$

299. $x = 20$

301. $x = 3$

303. $x = 6$

305. $x = \frac{5}{3}$

307. $x = \frac{\log 74}{\log 2} \approx 6.209$

309. $x = \frac{\log 112}{\log 4} \approx 3.404$

311. $x = \ln 8 \approx 2.079$

313. $x = \frac{\log 8}{\log\frac{1}{3}} \approx -1.893$

315. $x = \ln 3 - 2 \approx -0.901$

317. $x = \frac{\ln 16}{3} \approx 0.924$

319. $x = \ln 6 \approx 1.792$

321. $x = \ln 8 + 1 \approx 3.079$

323. $x = 5$

325. $x = -4, x = 5$

327. $a = 3$

329. $x = e^9$

331. $x = 7$

333. $x = 3$

335. $x = 2$

337. $x = 6$

339. $x = 5$

341. $x = \frac{\log 10}{\log\frac{1}{2}} \approx -3.322$

343. $x = \ln 7 - 5 \approx -3.054$

345. 6.9%

347. 13.9 years

349. $122{,}070$ bacteria

351. 8 times as large as the original population

353. 0.03 ml

355. Answers will vary.

Review Exercises

357. (a) $4x^2 + 12x$ (b) $16x^2 + 12x$ (c) $4x^3 + 12x^2$

359. (a) -123 (b) 356 (c) 41

361. Function; not one-to-one

363. (a) Function; not one-to-one (b) Not a function

365. Inverse function: $\{(10, -3), (5, -2), (2, -1), (1, 0)\}$. Domain: $\{1, 2, 5, 10\}$. Range: $\{-3, -2, -1, 0\}$.

367. $g(f(x)) = x,$ and $f(g(x)) = x,$ so they are inverses.

369. $f^{-1}(x) = \dfrac{x+11}{6}$

371. $f^{-1}(x) = \dfrac{1}{x} - 5$

373.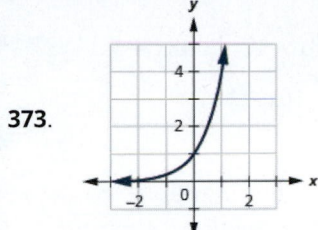

375.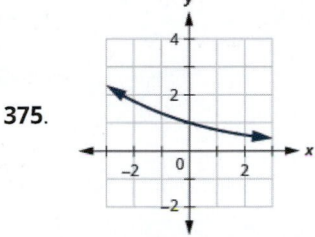

377.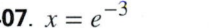

379.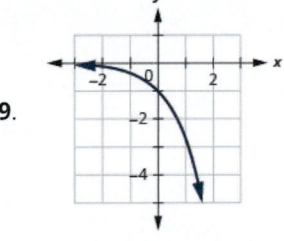

381. $x = -2,\ x = 2$

383. $x = -1$

385. $x = -3,\ x = 5$

387. $163,323.40

389. 330,259,000

391. $\log \dfrac{1}{1,000} = -3$

393. $\ln 16 = y$

395. $100000 = 10^5$

397. $x = 5$

399. $x = 4$

401. 0

403.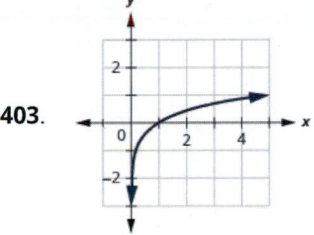

407. $x = e^{-3}$

409. $x = 8$

405.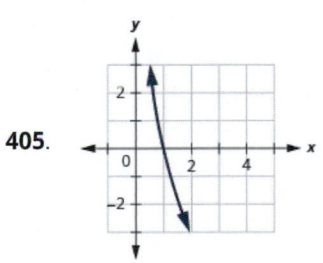

411. 90 dB

413. ⓐ 13 ⓑ −9

415. ⓐ 8 ⓑ 5

417. $4 + \log m$

419. $5 - \ln 2$

421. $\dfrac{1}{7}\log_4 z$

423. $\log_5 8 + 2\log_5 a + 6\log_5 b + \log_5 c - 3\log_5 d$

425. $\dfrac{1}{3}(\log_6 7 + 2\log_6 x - 1 - 3\log_6 y - 5\log_6 z)$

427. $\log_3 x^3 y^7$

429. $\log \dfrac{\sqrt[4]{y}}{(y-3)^2}$

431. 5.047

433. $x = 4$

435. $x = 3$

437. $x = \dfrac{\log 101}{\log 2} \approx 6.658$

439. $x = \dfrac{\log 7}{\log \frac{1}{3}} \approx -1.771$

441. $x = \ln 15 + 4 \approx 6.708$

443. 11.6 years

445. 12.7 months

Practice Test

447. ⓐ $48x - 17$ ⓑ $48x + 5$
ⓒ $48x^2 - 10x - 3$

453. $x = 5$

449. ⓐ Not a function ⓑ One-to-one function

455. ⓐ $31,250.74 ⓑ $31,302.29
ⓒ $31,328.32

451. $f^{-1}(x) = \sqrt[5]{x + 9}$

457. $343 = 7^3$

459. 0

461.

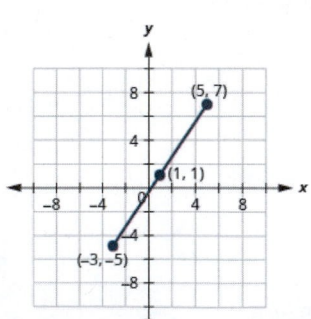

463. 40 dB

465. $2 + \log_5 a + \log_5 b$

467.
$\frac{1}{4}(\log_2 5 + 3\log_2 x - 4 - 2\log_2 y$
$- 7\log_2 z)$

469. $\log\dfrac{\sqrt[6]{x}}{(x + 5)^3}$

471. $x = 6$

473. $x = \ln 8 + 4 \approx 6.079$

475. 1,921 bacteria

Chapter 11

Try It

11.1. $d = 5$

11.2. $d = 10$

11.3. $d = 15$

11.4. $d = 13$

11.5. $d = \sqrt{130},\ d \approx 11.4$

11.6. $d = \sqrt{2},\ d \approx 1.4$

11.7.

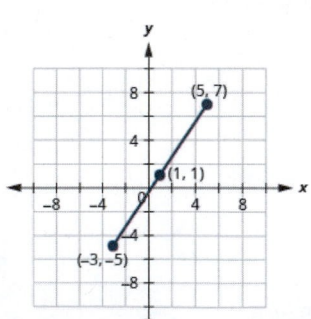

11.8.

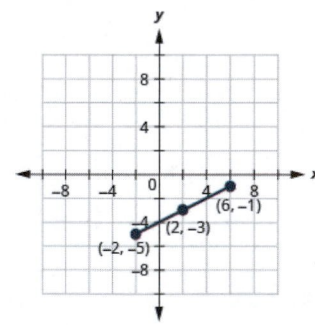

11.9. $x^2 + y^2 = 36$

11.10. $x^2 + y^2 = 64$

11.11. $(x - 2)^2 + (y + 4)^2 = 49$

11.12. $(x + 3)^2 + (y + 5)^2 = 81$

11.13. $(x-2)^2+(y-1)^2=25$

11.14. $(x-7)^2+(y-1)^2=100$

11.15. ⓐ The circle is centered at $(3,-4)$ with a radius of 2.

ⓑ

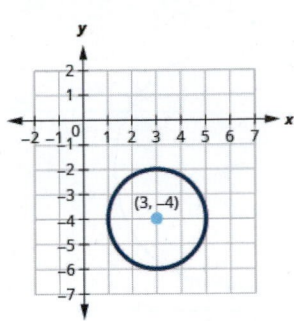

11.16. ⓐ The circle is centered at $(3, 1)$ with a radius of 4.

ⓑ

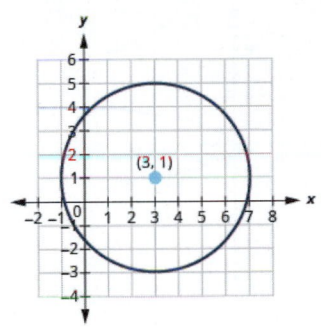

11.17. ⓐ The circle is centered at $(0, 0)$ with a radius of 3.

ⓑ

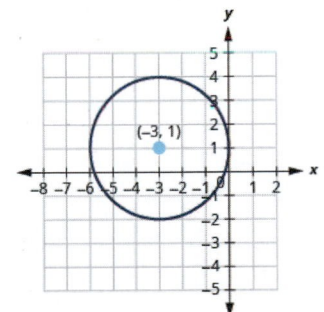

11.18. ⓐ The circle is centered at $(0, 0)$ with a radius of 5.

ⓑ

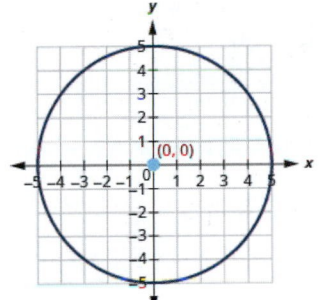

11.19. ⓐ The circle is centered at $(3, 4)$ with a radius of 4.

ⓑ

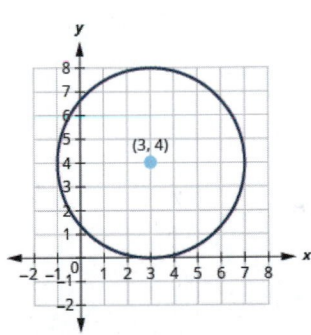

11.20. ⓐ The circle is centered at $(-3, 1)$ with a radius of 3.

ⓑ

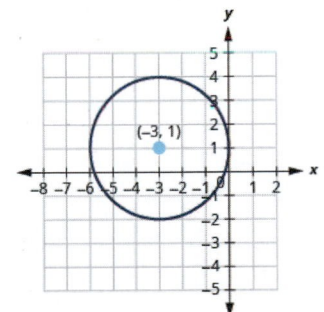

11.21. ⓐ The circle is centered at $(-1, 0)$ with a radius of 2.

ⓐ

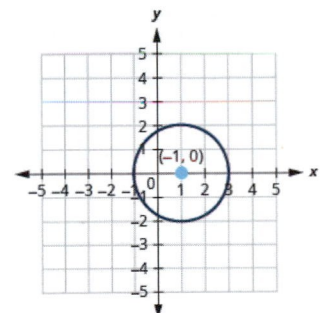

11.22. ⓐ The circle is centered at $(0,\ 6)$ with a radius of 5.

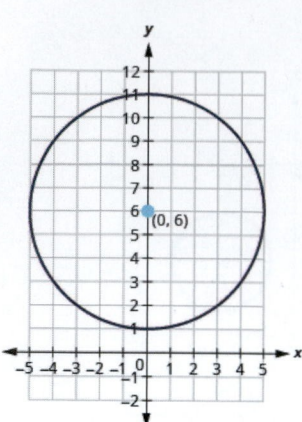

11.23.

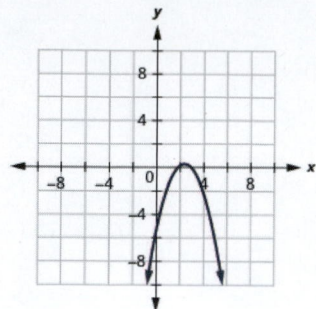

11.24.

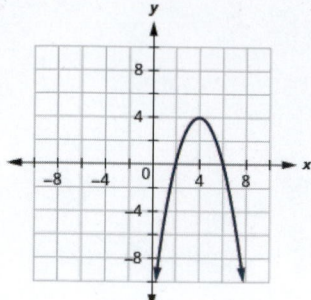

11.25. ⓐ $y = 2(x+1)^2 + 3$

ⓑ

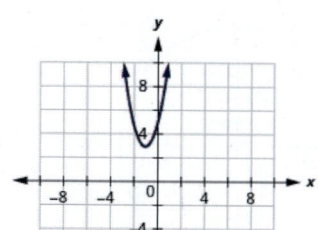

11.26. ⓐ $y = -2(x-2)^2 + 1$

ⓑ

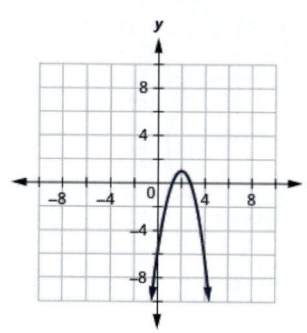

11.27.

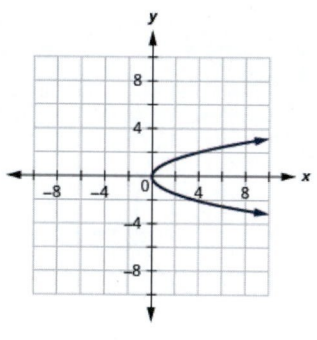

11.28.

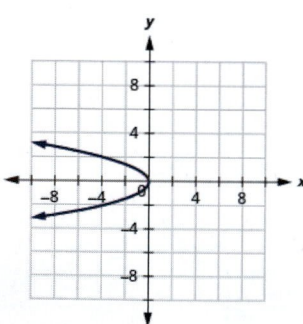

11.29.

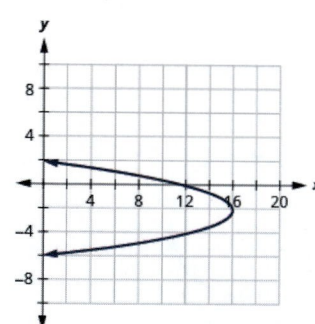

11.30.

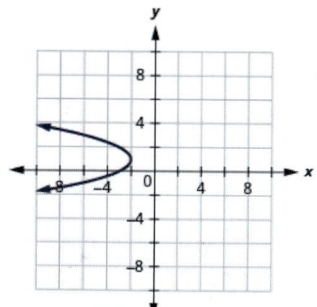

11.31.

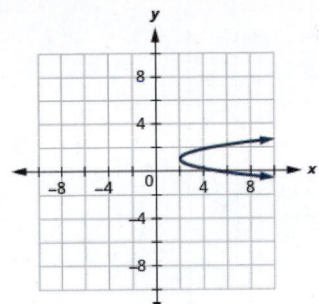

11.32.

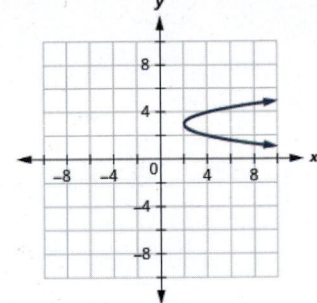

11.33.

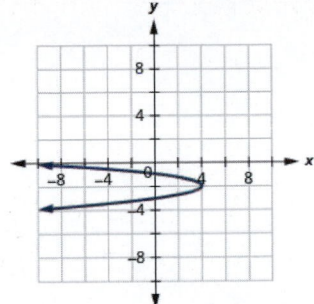

11.34.

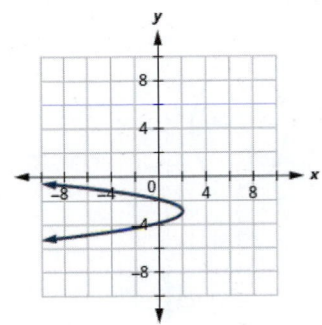

11.35. ⓐ $x = 3(y + 1)^2 + 4$

ⓑ

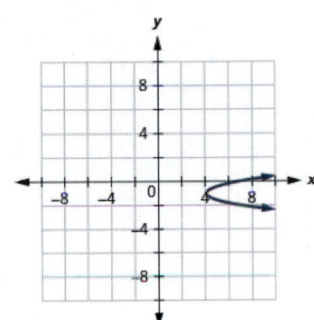

11.36. ⓐ $x = -4(y + 2)^2 + 4$

ⓑ

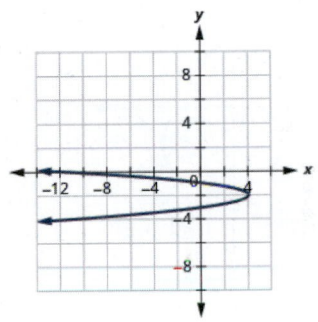

11.37. $y = -\frac{1}{20}(x - 20)^2 + 20$

11.38. $y = -\frac{1}{5}x^2 + 2x$

$y = -\frac{1}{5}(x - 5)^2 + 5$

11.39.

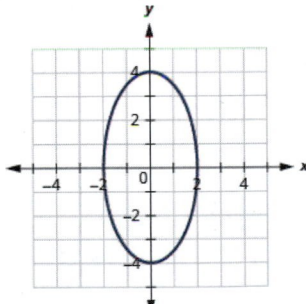

11.40.

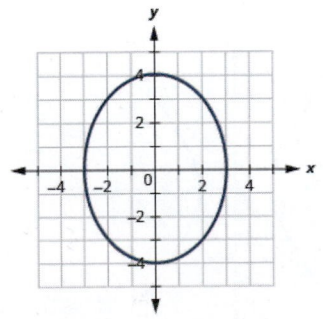

11.41.

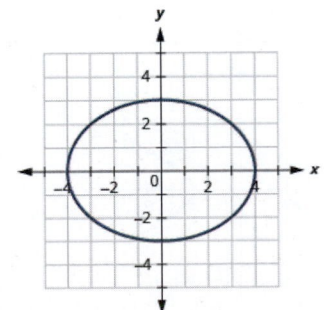

11.42.

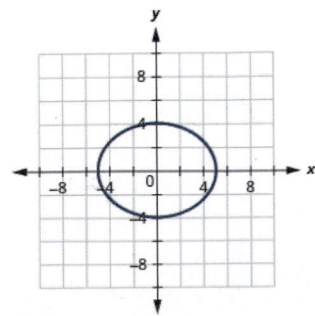

11.43. $\dfrac{x^2}{4} + \dfrac{y^2}{25} = 1$

11.44. $\dfrac{x^2}{9} + \dfrac{y^2}{4} = 1$

11.45.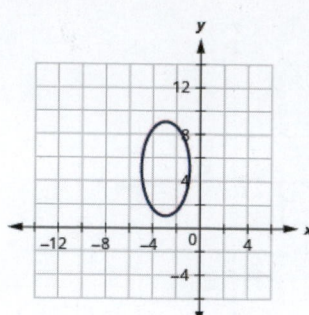

11.46.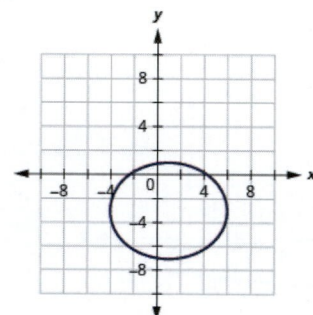

11.47.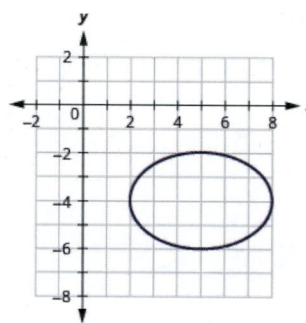

11.48.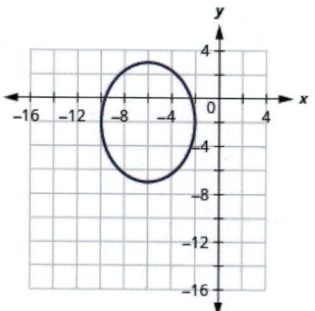

11.49. ⓐ $\dfrac{(x+1)^2}{6} + \dfrac{(y-4)^2}{9} = 1$

ⓑ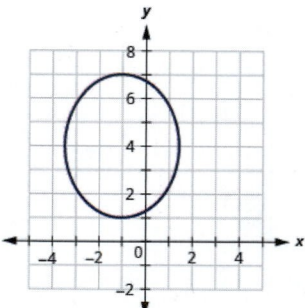

11.50. ⓐ $\dfrac{(x-2)^2}{4} + \dfrac{(y-3)^2}{16} = 1$

ⓑ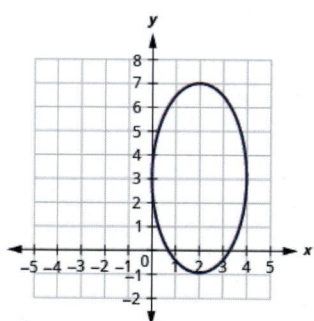

11.51. $\dfrac{x^2}{625} + \dfrac{y^2}{600} = 1$

11.52. $\dfrac{x^2}{1225} + \dfrac{y^2}{1000} = 1$

11.53.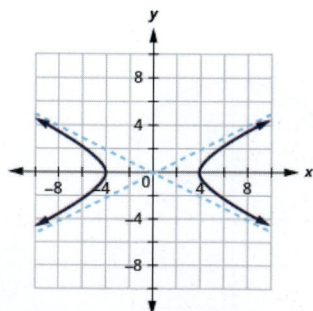

11.54.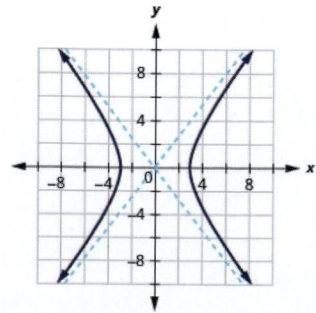

11.55.

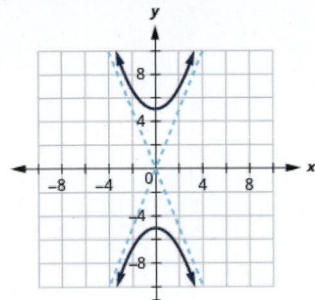

11.56.

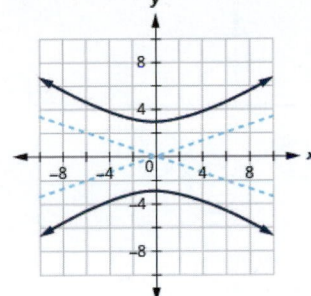

11.57.

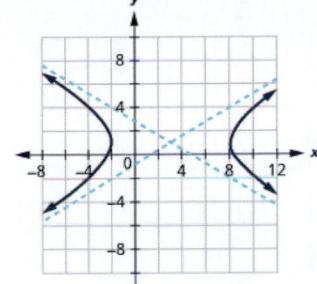

11.58.

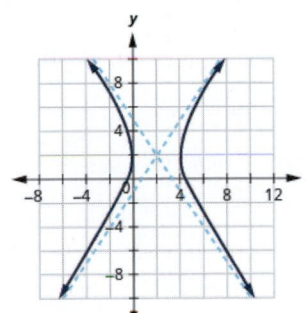

11.59.

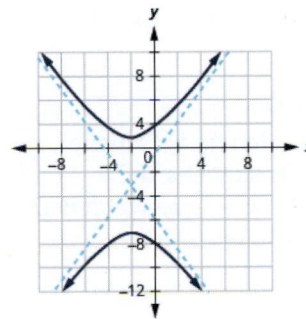

11.60.

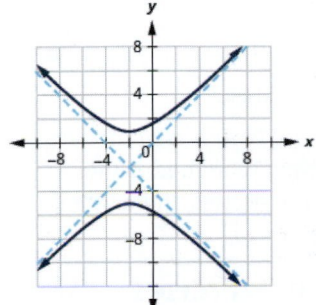

11.61. ⓐ $\dfrac{(x+1)^2}{16} - \dfrac{(y-2)^2}{9} = 1$

ⓑ

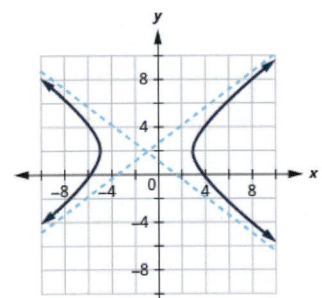

11.62. ⓐ $\dfrac{(x+3)^2}{25} - \dfrac{(y+1)^2}{16} = 1$

ⓑ

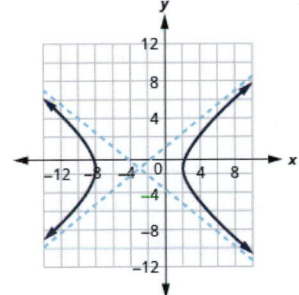

11.63. ⓐ circle ⓑ ellipse ⓒ parabola ⓓ hyperbola

11.64. ⓐ ellipse ⓑ parabola ⓒ circle ⓓ hyperbola

11.65.

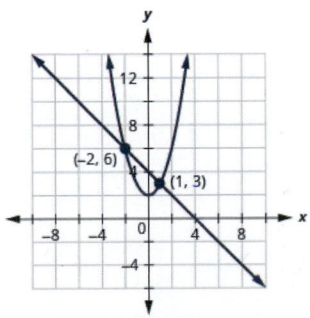

11.66.

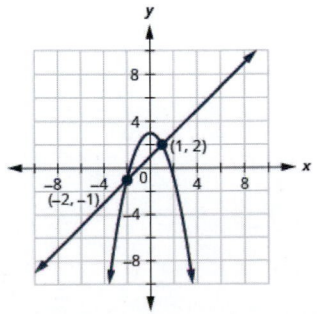

11.67.

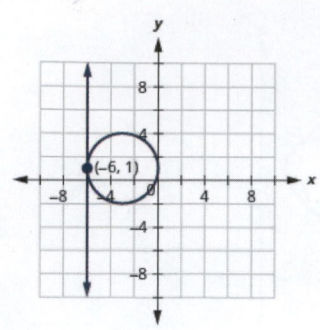

11.68.

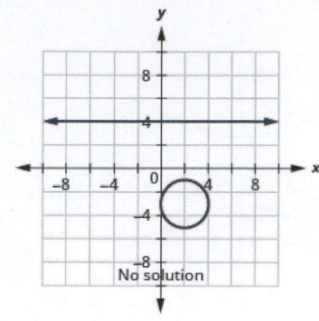

11.69. No solution

11.70. $\left(-\frac{4}{5}, \frac{6}{5}\right)$, $(0, 2)$

11.71. No solution

11.72. $\left(\frac{4}{9}, -\frac{2}{3}\right)$, $(1, 1)$

11.73.
$(-3, 0)$, $(3, 0)$, $\left(-2\sqrt{2}, -1\right)$, $\left(2\sqrt{2}, -1\right)$

11.74. $(-1, 0)$, $(0, 1)$, $(0, -1)$

11.75.
$(-3, -4)$, $(-3, 4)$, $(3, -4)$, $(3, 4)$

11.76. $(-2, 0)$, $(2, 0)$

11.77. 4 and 6

11.78. -18 and 17

11.79. If the length is 12 inches, the width is 16 inches. If the length is 16 inches, the width is 12 inches.

11.80. If the length is 12 inches, the width is 9 inches. If the length is 9 inches, the width is 12 inches.

Section Exercises

1. $d = 5$

3. 13

5. 5

7. 13

9. 76. $d = 3\sqrt{5}$, $d \approx 6.7$

11. $d = \sqrt{68}$, $d \approx 8.2$

13. ⓐ Midpoint: $(2, -4)$

15. ⓐ Midpoint: $\left(3\frac{1}{2}, -1\frac{1}{2}\right)$

17. $x^2 + y^2 = 49$

ⓑ

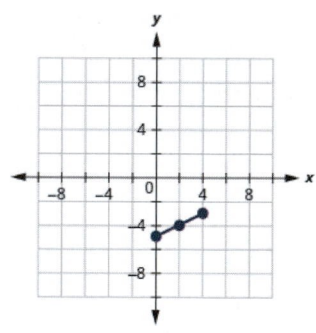

ⓑ

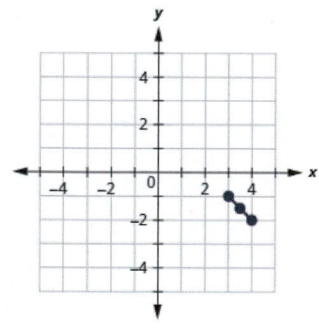

19. $x^2 + y^2 = 2$

21. $(x - 3)^2 + (y - 5)^2 = 1$

23. $(x - 1.5)^2 + (y + 3.5)^2 = 6.25$

25. $(x-3)^2 + (y+2)^2 = 64$

27. $(x-4)^2 + (y-4)^2 = 8$

29. ⓐ The circle is centered at $(-5, -3)$ with a radius of 1.

ⓑ

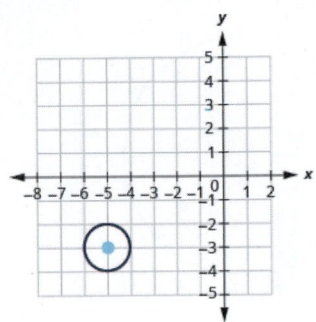

31. ⓐ The circle is centered at $(4, -2)$ with a radius of 4.

ⓑ

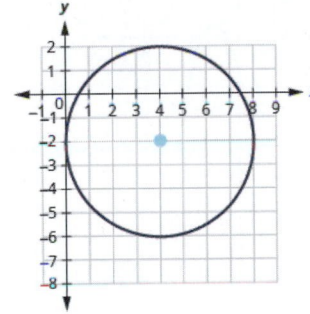

33. ⓐ The circle is centered at $(0, -2)$ with a radius of 5.

ⓑ

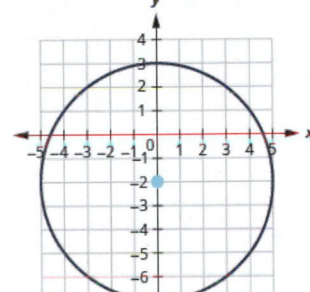

35. ⓐ The circle is centered at $(1.5, 2.5)$ with a radius of 0.5.

ⓑ

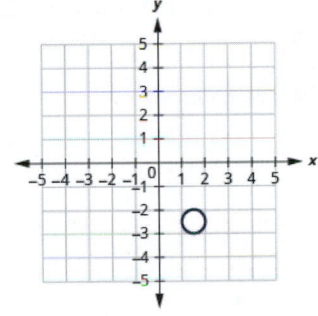

37. ⓐ The circle is centered at $(0, 0)$ with a radius of 8.

ⓑ

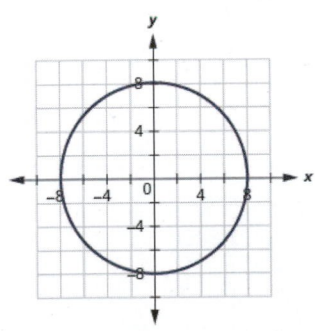

39. ⓐ The circle is centered at $(0, 0)$ with a radius of 2.

ⓑ

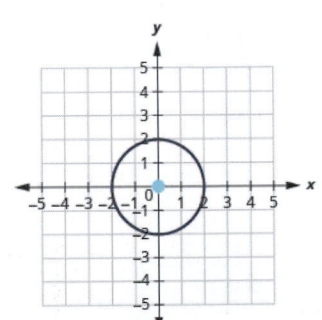

41. ⓐ Center: $(-1, -3)$, radius: 1

ⓑ

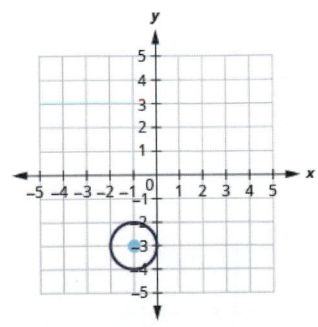

43. ⓐ Center: (2, −5), radius: 6

ⓑ

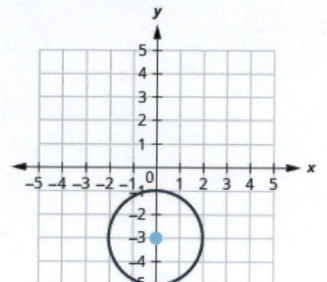

45. ⓐ Center: (0, −3), radius: 2

ⓑ

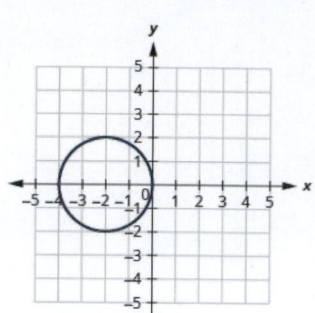

47. ⓐ Center: (−2, 0), radius:

ⓑ

49. Answers will vary.

51. Answers will vary.

53.

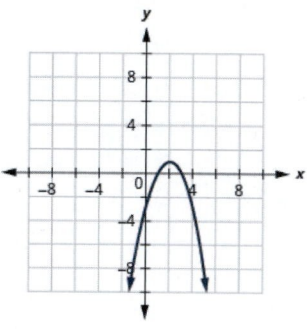

55.

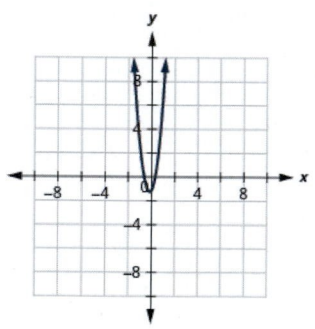

57. ⓐ $y = -(x-1)^2 - 3$

ⓑ

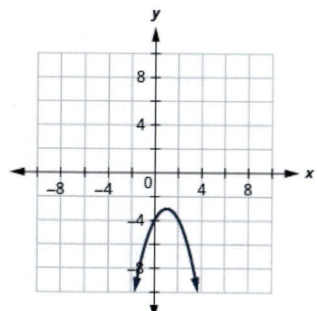

59. ⓐ $y = -2(x+1)^2 - 3$

ⓑ

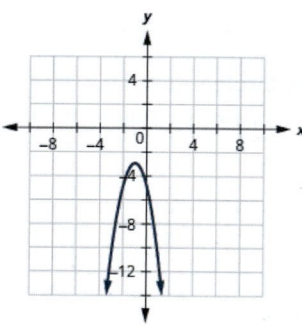

61.

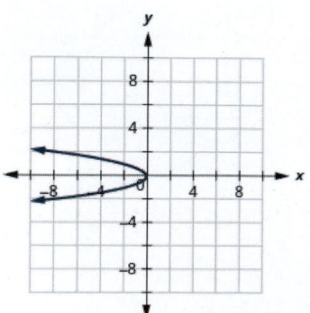

63.

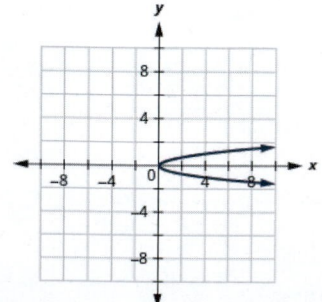

65.

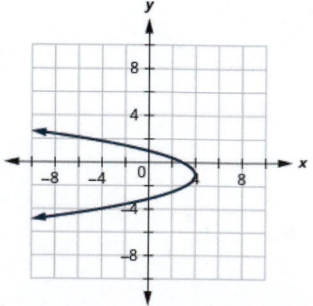

67.

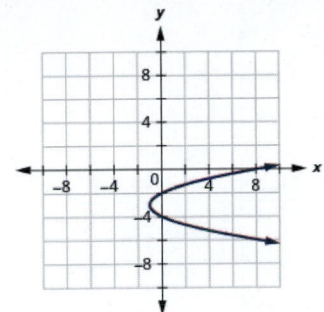

69.

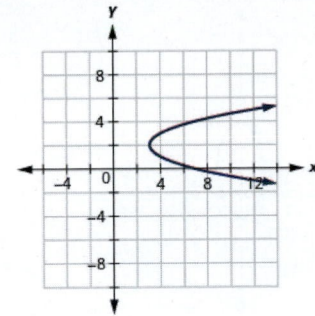

71.

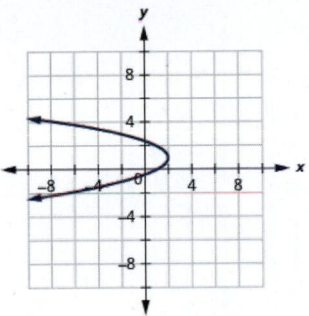

73.

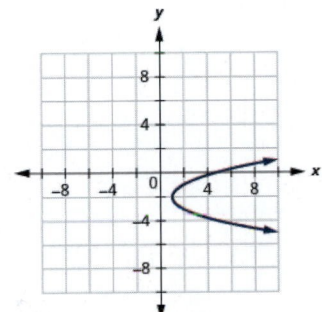

75.

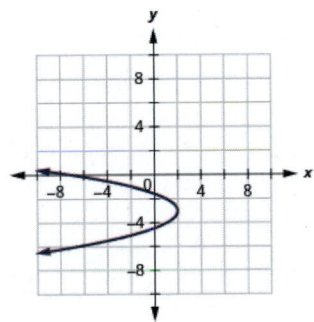

77.

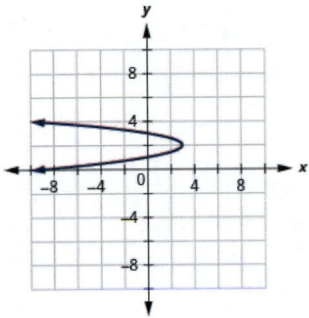

79.

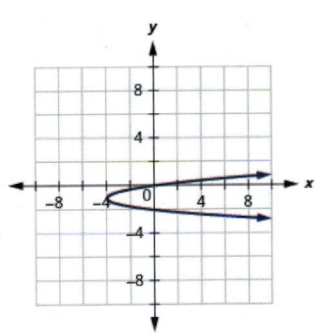

81. (a) $x = (y + 2)^2 - 9$

(b)

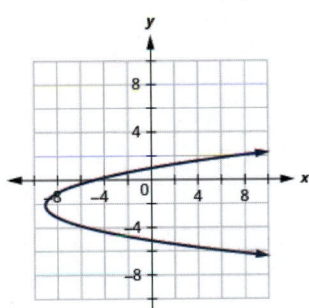

83. (a) $x = -2(y + 3)^2 + 2$

(b)

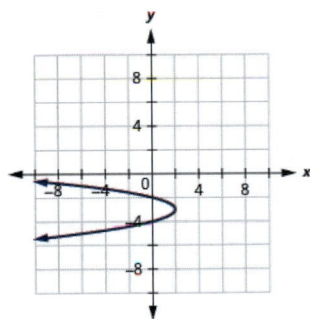

85. (a)

91. $y = -\dfrac{1}{15}(x - 15)^2 + 15$

97. Answers will vary.

87. (b)

93. $y = -\dfrac{1}{10}(x - 30)^2 + 90$

99.

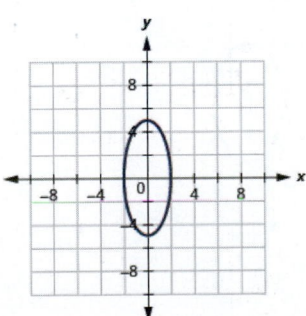

89. (d)

95. Answers will vary.

101.

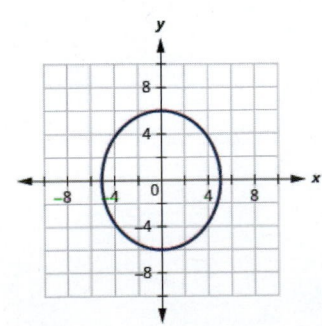

103.

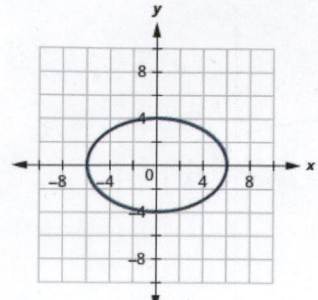

105.

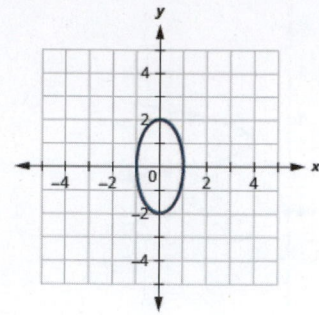

107.

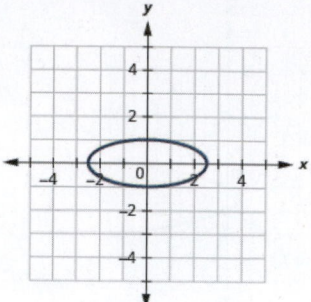

109.

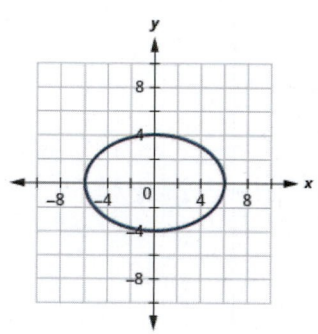

111. $\dfrac{x^2}{9} + \dfrac{y^2}{25} = 1$

113. $\dfrac{x^2}{9} + \dfrac{y^2}{16} = 1$

115.

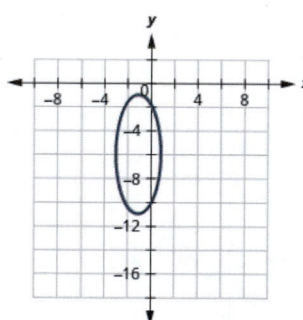

117.

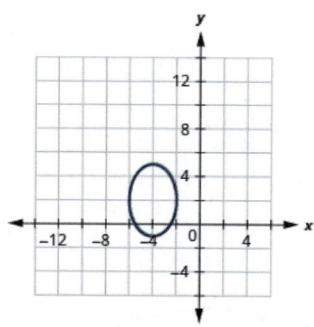

119.

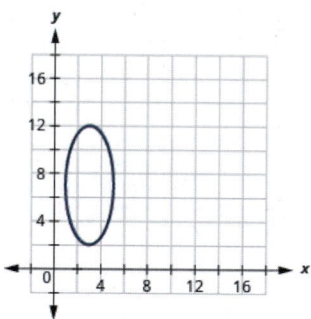

121.

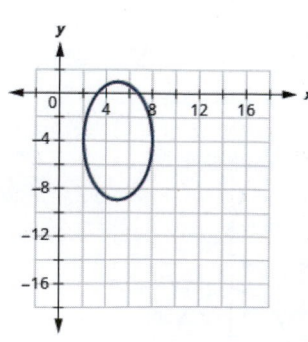

123. ⓐ $\dfrac{(x-2)^2}{9} + \dfrac{(y-3)^2}{25} = 1$

ⓑ

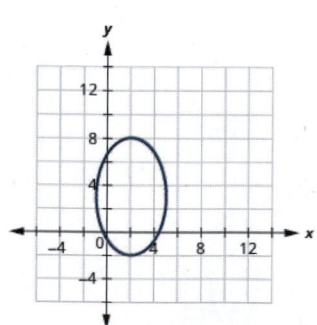

125. ⓐ $\dfrac{y^2}{4} + \dfrac{(x-3)^2}{25} = 1$

ⓑ

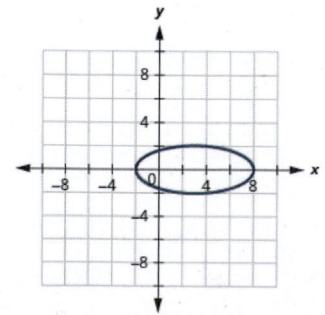

127.

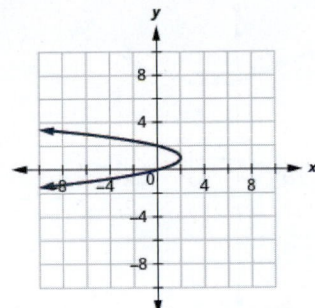

129.

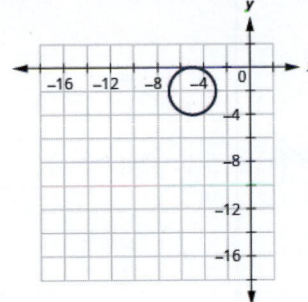

131.

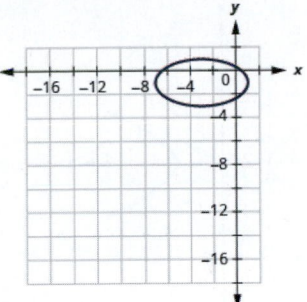

133.

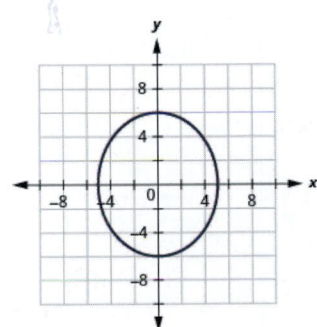

135.

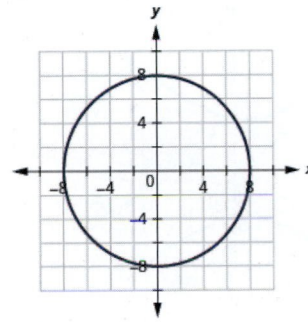

137.

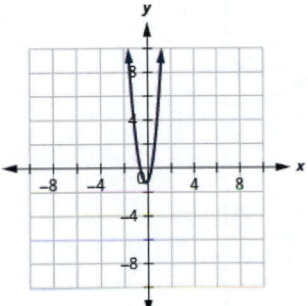

139. $\dfrac{x^2}{400} + \dfrac{y^2}{300} = 1$

141. $\dfrac{x^2}{2500} + \dfrac{y^2}{1275} = 1$

143. Answers will vary.

145. Answers will vary.

147.

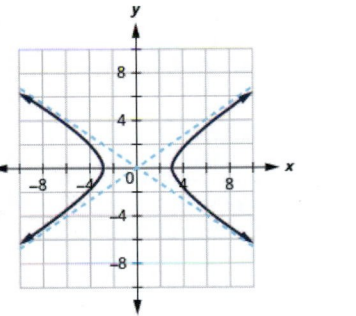

149.

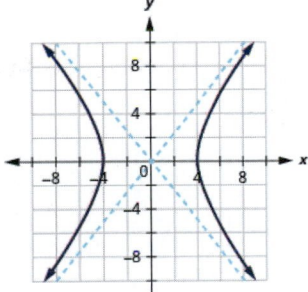

151.

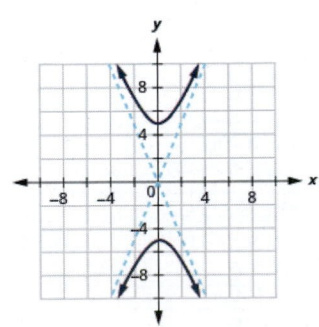

153.

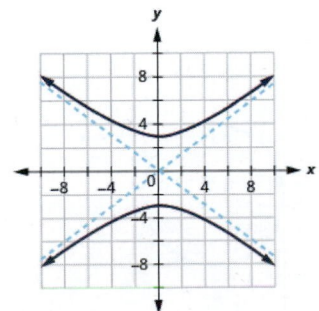

155.

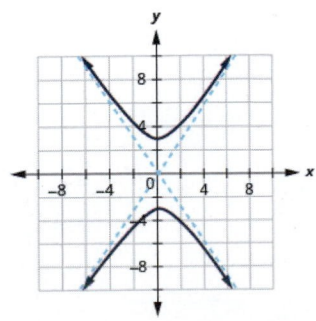

157.

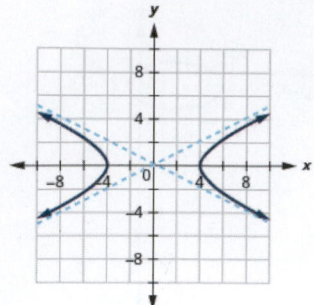

159.

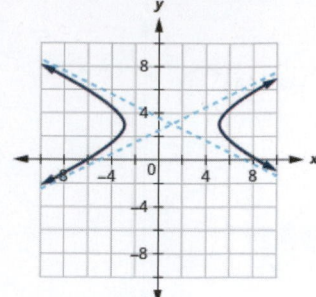

161.

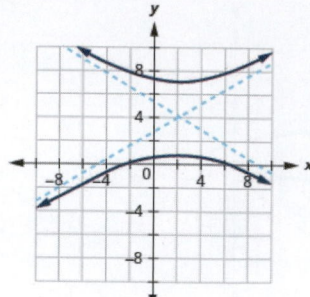

163.

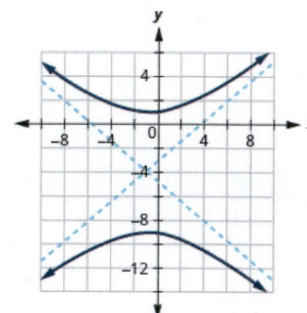

165.

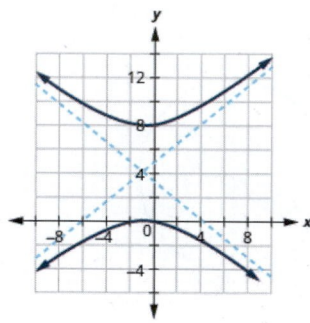

167.

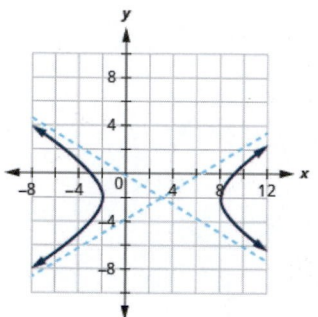

169. ⓐ $\dfrac{(x-1)^2}{4} - \dfrac{(y-1)^2}{9} = 1$

ⓑ

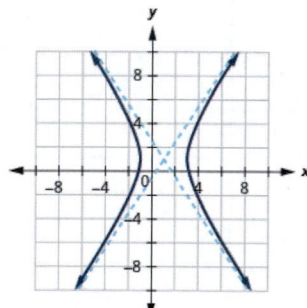

171. ⓐ $\dfrac{(y-2)^2}{9} - \dfrac{(x-1)^2}{9} = 1$

ⓑ

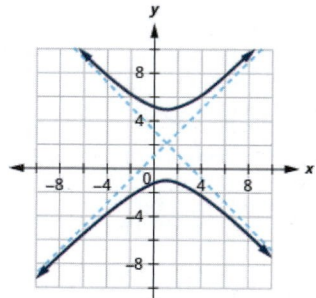

173. ⓐ $\dfrac{(y+1)^2}{1} - \dfrac{(x+2)^2}{9} = 1$

ⓑ

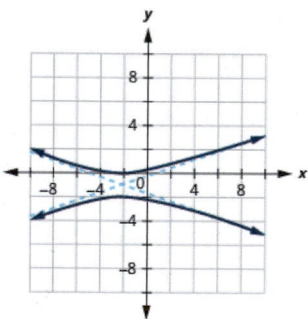

175. ⓐ parabola ⓑ circle ⓒ hyperbola ⓓ ellipse

177.

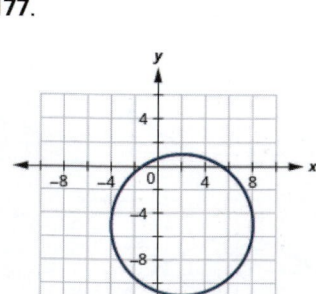

179.

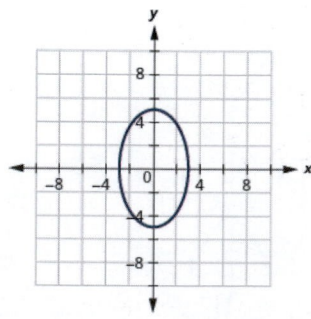

181.

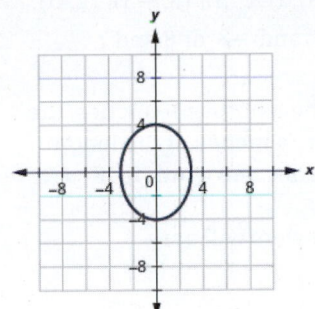

183.

185. Answers will vary.

187. Answers will vary.

189.

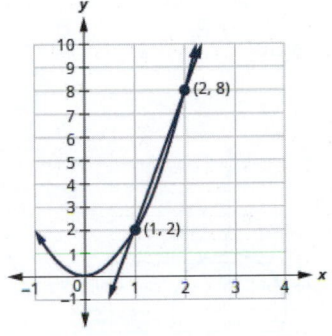

191.

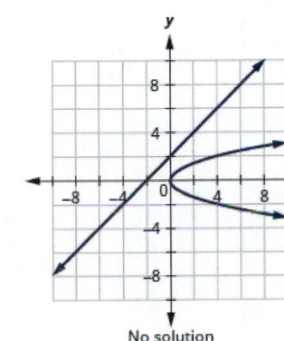

No solution

193.

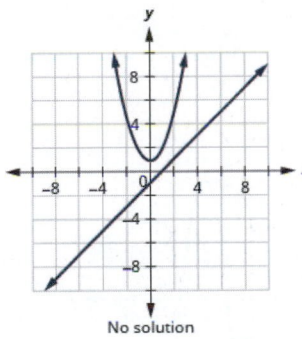

No solution

195.

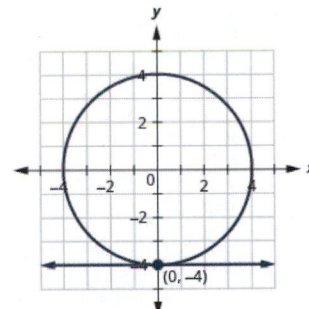

197.

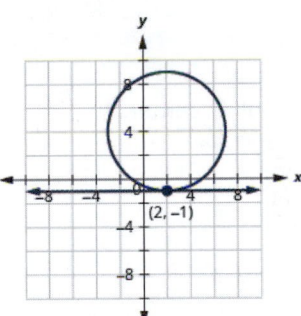

199.

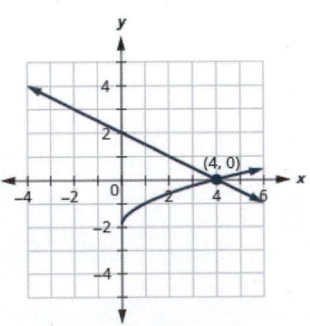

205. $(12, -5), (12, 5)$

211. $(3, 4), (5, 0)$

217. $(-2, 0), (1, -\sqrt{3}), (1, \sqrt{3})$

201. $(-1, 0), (0, 3)$

203. $(2, 0)$

207. No solution

213. $(0, -4), (-\sqrt{7}, 3), (\sqrt{7}, 3)$

219.
$(-2, -4), (-2, 4), (2, -4), (2, 4)$

209. $(0, -4), (1, -3)$

215. $(0, -2), (-\sqrt{3}, 1), (\sqrt{3}, 1)$

221. $(-4, 0), (4, 0)$

223. $\left(-\sqrt{3},\ 0\right),\ \left(\sqrt{3},\ 0\right)$

225.
$(-2,\ -3),\ (-2,\ 3),\ (2,\ -3),\ (2,\ 3)$

227.
$(-1,\ -3),\ (-1,\ 3),\ (1,\ -3),\ (1,\ 3)$

229. -3 and 14

231. -7 and -8 or 8 and 7

233. -6 and -4 or -6 and 4 or 6 and -4 or 6 and 4

235. If the length is 11 cm, the width is 15 cm. If the length is 15 cm, the width is 11 cm.

237. If the length is 10 inches, the width is 24 inches. If the length is 24 inches, the width is 10 inches.

239. The length is 40 inches and the width is 30 inches. The TV will not fit Donnette's entertainment center.

241. Answers will vary.

243. Answers will vary.

Review Exercises

245. $d = 3$

247. $d = \sqrt{17},\ d \approx 4.1$

249. $(4,\ 4)$

251. $\left(\frac{3}{2},\ -\frac{7}{2}\right)$

253. $x^2 + y^2 = 7$

255. $(x + 2)^2 + (y + 5)^2 = 49$

257. $(x - 2)^2 + (y - 2)^2 = 8$

259. ⓐ radius: 12, center: $(0,\ 0)$

ⓑ

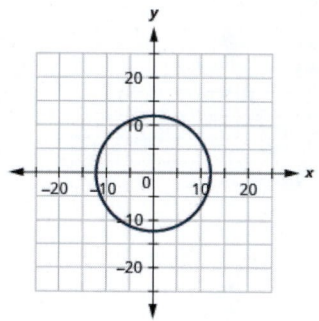

261. ⓐ radius: 7, center: $(-2,\ -5)$

ⓑ

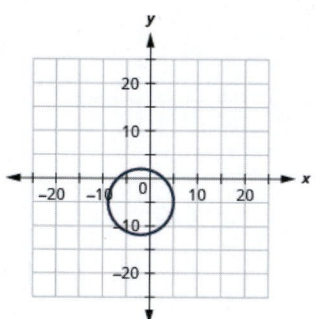

263. ⓐ radius: 8, center: $(0,\ 2)$

ⓑ

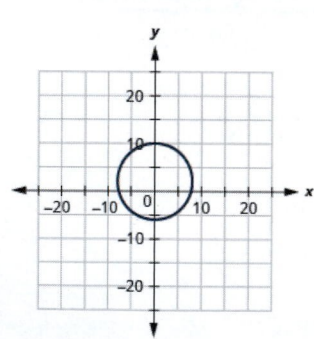

265.

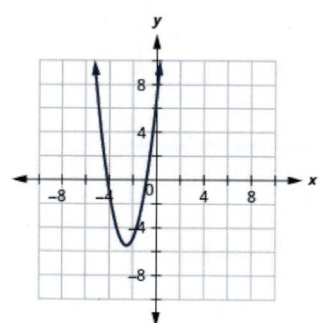

267.

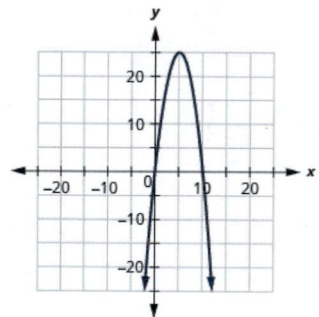

269. ⓐ $y = 2(x - 1)^2 - 4$

ⓑ

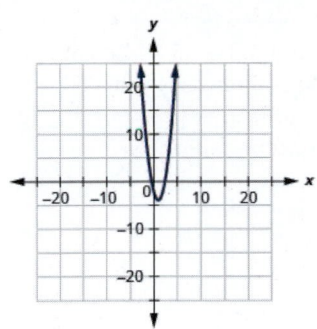

271. ⓐ $y = -(x - 6)^2 + 1$

ⓑ

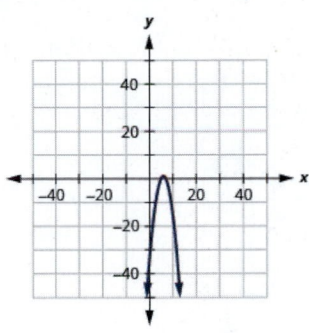

273.

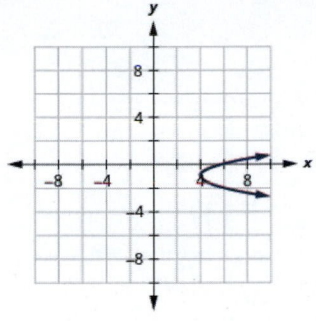

275.

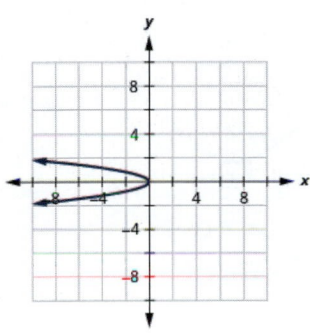

277. ⓐ $x = (y + 2)^2 + 1$

ⓑ

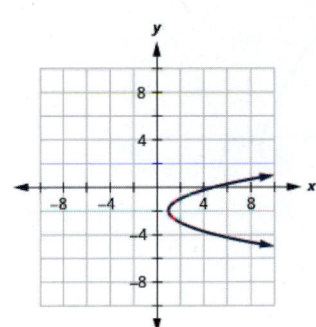

279. ⓐ $x = -2(y - 1)^2 + 2$

ⓑ

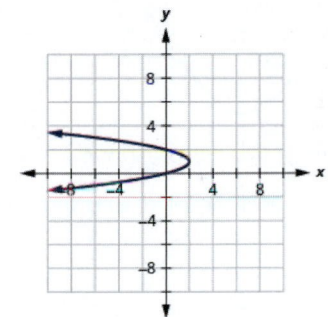

281. $y = -\frac{1}{9}x^2 + \frac{10}{3}x$

283.

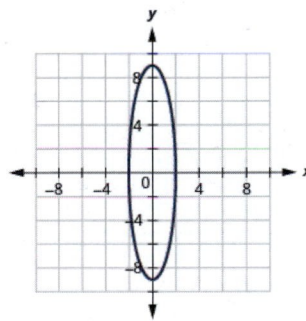

285.

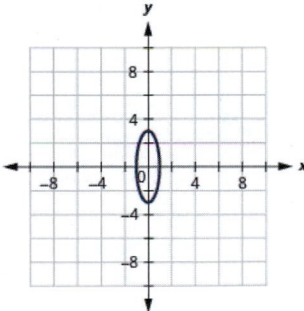

287. $\frac{x^2}{36} + \frac{y^2}{64} = 1$

289.

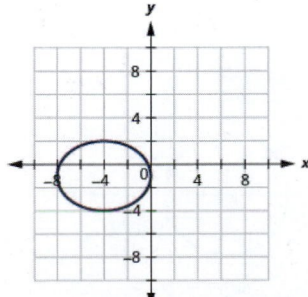

291.

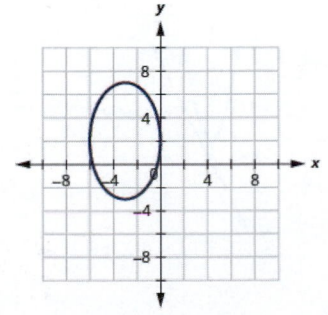

293. ⓐ $\dfrac{(x-3)^2}{4} + \dfrac{(y-7)^2}{25} = 1$

ⓑ

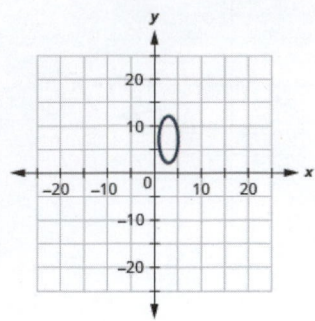

295. ⓐ $\dfrac{x^2}{9} + \dfrac{(y-7)^2}{4} = 1$

ⓑ

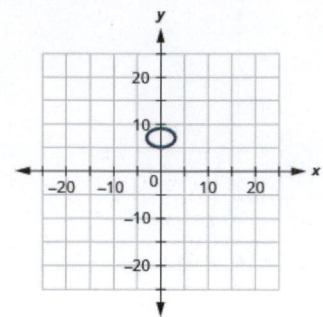

297.

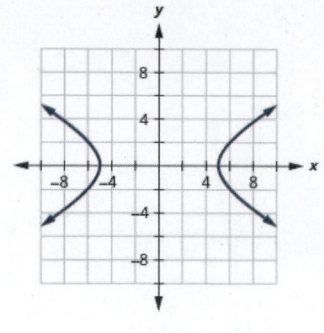

299.

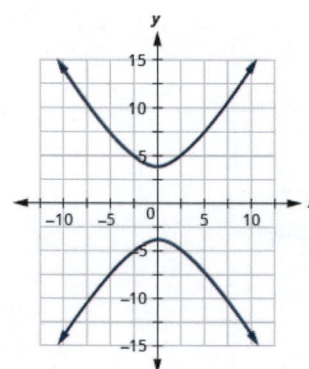

301.

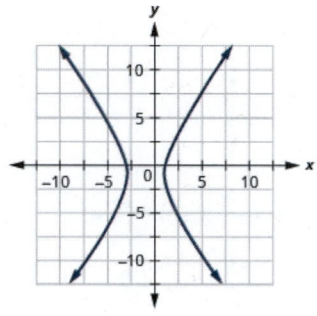

303.

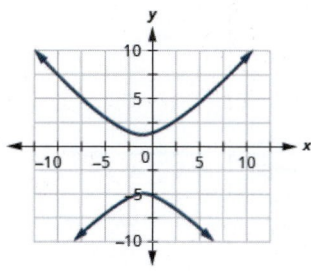

305. ⓐ $\dfrac{(x+1)^2}{16} - \dfrac{(y-3)^2}{4} = 1$

ⓑ

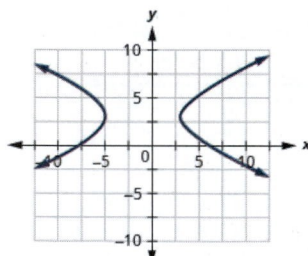

307. ⓐ $\dfrac{(y-1)^2}{16} - \dfrac{(x-1)^2}{4} = 1$

ⓑ

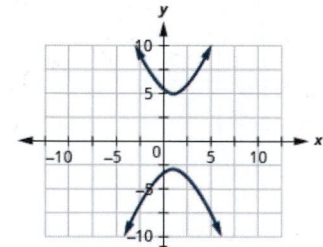

309. ⓐ hyperbola ⓑ circle ⓒ parabola ⓓ ellipse

311.

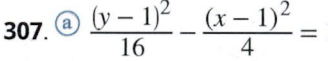

313.

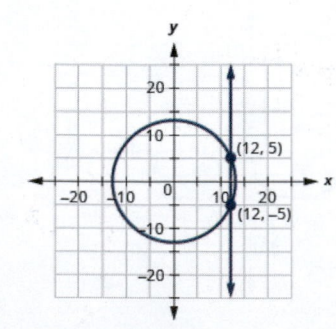

315. $(-1, 4)$

317. No solution

319. $\left(-\sqrt{7},\, 3\right),\, \left(\sqrt{7},\, 3\right)$

321. $(-3, 0),\, (0, -2),\, (0, 2)$

323. -3 and -4 or 4 and 3

325. If the length is 14 inches, the width is 15 inches. If the length is 15 inches, the width is 14 inches.

Practice Test

327. distance: 10, midpoint: $(-7, -7)$

329. $x^2 + y^2 = 121$

331. $(x + 2)^2 + (y - 3)^2 = 52$

333. ⓐ ellipse

ⓑ

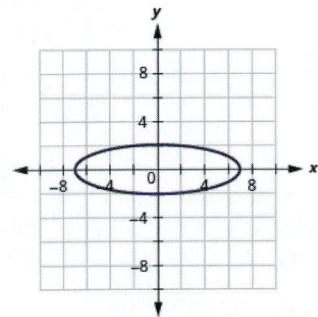

335. ⓐ circle

ⓑ

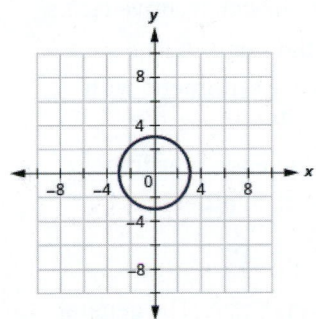

337. ⓐ ellipse

ⓑ

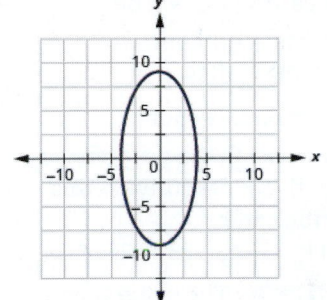

339. ⓐ hyperbola

ⓑ

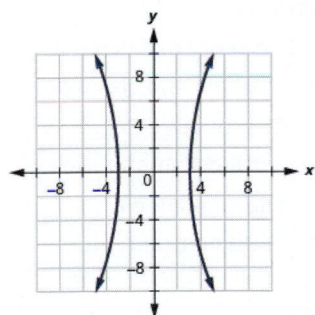

341. ⓐ circle

ⓑ $(x + 5)^2 + (y + 3)^2 = 4$

ⓒ

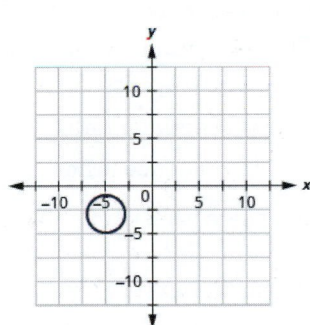

343. ⓐ hyperbola

ⓑ $\dfrac{(x - 2)^2}{25} - \dfrac{(y + 1)^2}{9} = 1$

ⓒ

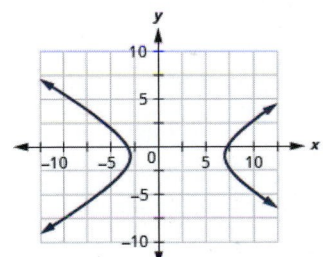

345. No solution

347. $(0, -3), (0, 3)$

349. $\dfrac{x^2}{2025} + \dfrac{y^2}{1400} = 1$

351. The length is 44 inches and the width is 33 inches. The TV will fit Olive's entertainment center.

Chapter 12

Try It

12.1. $-1, 2, 5, 8, 11$
12.4. $-3, -1, 3, 11, 27$
12.7. $a_n = 3n$
12.10. $a_n = (-1)^{n+1} n^2$

12.2. $-3, -1, 1, 3, 5$
12.5. $-1, 4, -9, 16, -25$
12.8. $a_n = 5n$
12.11. $a_n = \dfrac{1}{2^n}$

12.3. $7, 13, 31, 85, 247$
12.6. $1, -8, 27, -64, 125$
12.9. $a_n = (-1)^n 3^n$
12.12. $a_n = \dfrac{1}{n^2}$

12.13. $2, 1, \frac{1}{3}, \frac{1}{12}, \frac{1}{60}$

12.14. $3, \frac{3}{2}, \frac{1}{2}, \frac{1}{8}, \frac{1}{40}$

12.15. $\frac{1}{2}, \frac{1}{6}, \frac{1}{12}, \frac{1}{20}, \frac{1}{30}$

12.16. $\frac{1}{2}, \frac{1}{3}, \frac{1}{4}, \frac{1}{5}, \frac{1}{6}$

12.17. 45

12.18. 60

12.19. $\frac{16}{3}$

12.20. 8

12.21. $\sum_{n=1}^{5} \frac{1}{2^n}$

12.22. $\sum_{n=1}^{5} \frac{1}{n^2}$

12.23. $\sum_{n=1}^{5} (-1)^{n+1} n^2$

12.24. $\sum_{n=1}^{5} (-1)^n 2n$

12.25. ⓐ The sequence is arithmetic with common difference $d = 11$. ⓑ The sequence is arithmetic with common difference $d = -6$. ⓒ The sequence is not arithmetic as all the differences between the consecutive terms are not the same.

12.26. ⓐ The sequence is not arithmetic as all the differences between the consecutive terms are not the same. ⓑ The sequence is arithmetic with common difference $d = 2$. ⓒ The sequence is arithmetic with common difference $d = -5$.

12.27. $7, 3, -1, -5, -9, \ldots$

12.28. $11, 3, -5, -13, -21, \ldots$

12.29. 241

12.30. -106

12.31. $a_{11} = 2$. The general term is $a_n = -3n + 35$.

12.32. $a_{19} = -55$. The general term is $a_n = -4n + 21$.

12.33. $a_1 = 5$, $d = 4$. The general term is $a_n = 4n + 1$.

12.34. $a_1 = 8$, $d = -3$. The general term is $a_n = -3n + 11$.

12.35. 1,890

12.36. 1,515

12.37. 2,300

12.38. 5,250

12.39. 2,670

12.40. 3,045

12.41. ⓐ The sequence is geometric with common ratio $r = 3$. ⓑ The sequence is geometric with common ratio $d = \frac{1}{4}$. ⓒ The sequence is not geometric. There is no common ratio.

12.42. ⓐ The sequence is not geometric. There is no common ratio. ⓑ The sequence is geometric with common ratio $r = 2$. ⓒ The sequence is geometric with common ratio $r = \frac{1}{2}$.

12.43. $7, -21, 63, -189, 567$

12.44. $6, -24, 96, -384, 1536$

12.45. $\frac{1}{6,561}$

12.46. $\frac{1}{16,384}$

12.47. $a_9 = 39,366$. The general term is $a_n = 6(3)^{n-1}$.

12.48. $a_{11} = 7,168$. The general term is $a_n = 7(2)^{n-1}$.

12.49. 3,145,725

12.50. 10,460,353,200

12.51. 393,204

12.52. 10,230

12.53. 96

12.54. $\frac{256}{3}$

12.55. $\frac{4}{9}$

12.56. $\frac{8}{9}$

12.57. \$10,000

12.58. \$3, 333.33

12.59. \$88,868.36

12.60. \$698,201.57

12.61.
$x^5 + 5x^4 y + 10x^3 y^2 + 10x^2 y^3 + 5xy^4 + y^5$

12.62.
$p^7 + 7p^6 q + 21p^5 q^2 + 35p^4 q^3 + 35p^3 q^4 + 21p^2 q^5 + 7pq^6 + q^7$

12.63.
$x^4 + 8x^3 + 24x^2 + 32x + 16$

12.64.
$x^6 + 6x^5 + 15x^4 + 20x^3 + 15x^2 + 6x + 1$

12.65.
$16x^4 - 96x^3 + 216x^2 - 216x + 81$

12.66.
$64x^6 - 192x^5 + 240x^4 - 160x^3 + 60x^2 - 12x + 1$

12.67. (a) 6 (b) 1 (c) 1 (d) 35

12.68. (a) 2 (b) 1 (c) 1 (d) 6

12.69.
$x^5 + 5x^4 y + 10x^3 y^2 + 10x^2 y^3$
$+ 5xy^4 + y^5$

12.70.
$m^6 + 6m^5 n + 15m^4 n^2 + 20m^3 n^3$
$+ 15m^2 n^4 + 6mn^5 + n^6$

12.71. $x^5 - 15x^4 + 90x^3 - 270x^2$
$+ 405x - 243$

12.72.
$y^6 - 6y^5 + 15y^4 - 20y^3 + 15y^2$
$- 6y + 1$

12.73.
$243x^5 - 810x^4 y + 1080x^3 y^2$
$- 720x^2 y^3 + 240xy^4 - 32y^5$

12.74. $256x^4 - 768x^3 y + 864x^2 y^2$
$- 432xy^3 + 81y^4$

12.75. $15x^4 y^2$

12.76. $8ab^7$

12.77. 7,168

12.78. 280

Section Exercises

1. −5, −3, −1, 1, 3

3. 4, 7, 10, 13, 16

5. 5, 7, 11, 19, 35

7. 1, 5, 21, 73, 233

9. $2, 1, \frac{8}{9}, 1, \frac{32}{25}$

11. $1, \frac{3}{2}, \frac{5}{4}, \frac{7}{8}, \frac{9}{16}$

13. −2, 4, −6, 8, −10

15. 1, −4, 9, −16, 25

17. $1, -\frac{1}{4}, \frac{1}{9}, -\frac{1}{16}, \frac{1}{25}$

19. $a_n = 8n$

21. $a_n = n + 5$

23. $a_n = e^{n+2}$

25. $a_n = (-1)^n 5n$

27. $a_n = (-1)^n n^3$

29. $a_n = (-1)^n 2n$

31. $a_n = \frac{1}{4^n}$

33. $a_n = -\frac{n}{n+1}$

35. $-\frac{5}{2^n}$

37. $4, 2, \frac{2}{3}, \frac{1}{6}, \frac{1}{30}$

39. 3, 6, 18, 72, 360

41. 2, 24, 720, 40320, 3628800

43. $1, \frac{1}{2}, \frac{1}{3}, \frac{1}{4}, \frac{1}{5}$

45. $1, \frac{1}{2}, \frac{2}{3}, \frac{3}{2}, \frac{24}{5}$

47. $2, \frac{3}{2}, \frac{8}{3}, \frac{15}{2}, \frac{144}{5}$

49. $1 + 4 + 9 + 16 + 25 = 55$

51.
$5 + 7 + 9 + 11 + 13 + 15 = 60$

53. $2 + 4 + 8 + 16 = 30$

55. $4 + 4 + 3 + \frac{2}{3} = 11\frac{2}{3}$

57. $2 + 6 + 12 + 20 + 30 = 70$

59. $\frac{1}{2} + \frac{2}{3} + \frac{3}{4} + \frac{4}{5} + \frac{5}{6} = \frac{71}{20}$

61. $\sum_{n=1}^{5} \frac{1}{3^n}$

63. $\sum_{n=1}^{5} \frac{1}{n^3}$

65. $\sum_{n=1}^{5} \frac{2}{n}$

67. $\sum_{n=1}^{5} (-1)^{n+1} 3n$

69. $\sum_{n=1}^{10} (-1)^n 2n$

71. $\sum_{n=1}^{7} (2n + 12)$

73. Answers will vary.

75. Answers will vary.

77. The sequence is arithmetic with common difference $d = 8$.

79. The sequence is not arithmetic.

81. The sequence is arithmetic with common difference $d = -3$.

83. 11, 18, 25, 32, 39

85. −7, −3, 1, 5, 9

87. 14, 5, −4, −13, −22

89. 163

91. 131

93. −79

95. $a_{20} = -34$. The general term is $a_n = -2n + 6$.

97. $a_{11} = 59$. The general term is $a_n = 5n + 4$.

99. $a_8 = -13$. The general term is $a_n = -5n + 27$.

101. $a_1 = 11$, $d = 3$. The general term is $a_n = 3n + 8$.

103. $a_1 = 21$, $d = -8$. The general term is $a_n = -8n + 29$.

105. $a_1 = -15$, $d = 3$. The general term is $a_n = 3n - 18$.

107. 1,635

109. −1, 065

111. 360

113. 6,325

115. −3,575

117. 6,280

119. 4,125

121. −3,850

123. Answers will vary.

125. Answers will vary.

127. The sequence is geometric with common ratio $r = 4$.

129. The sequence is geometric with common ratio $r = \frac{1}{2}$.

131. The sequence is not geometric. There is no common ratio.

133. The sequence is geometric with common ratio $r = \frac{1}{2}$.

135. The sequence is arithmetic with common difference $d = 5$.

137. The sequence is geometric with common ratio $r = \frac{1}{2}$.

139. 4, 12, 36, 108, 324

141. $-4, 8, -16, 32, -64$

143. $27, 9, 3, 1, \frac{1}{3}$

145. 472,392

147. 3,072

149. 0.0001

151. $a_9 = 2{,}304$. The general term is $a_n = 9(2)^{n-1}$.

153. $a_{15} = -\frac{2}{19,683}$. The general term is $a_n = -486\left(-\frac{1}{3}\right)^{n-1}$.

155. $a_{10} = 0.000000001$. The general term is $a_n = (0.1)^{n-1}$.

157. 57,395,624

159. $-65{,}538$

161. $\frac{7,174,453}{59,049} \approx 121.5$

163. 65,534

165. 4088

167. $\frac{29,524}{6561} \approx 4.5$

169. $\frac{3}{2}$

171. $\frac{9}{2}$

173. no sum as $r \geq 1$

175. 2,048

177. $\frac{1}{3}$

179. $\frac{7}{9}$

181. $\frac{5}{11}$

183. ⓐ $6666.67 ⓑ $4000 ⓒ $15,000 ⓒ $7500

185. $295,581.88

187. $14,234.10

189. Answers will vary.

191. Answers will vary.

193. $a^8 + 8a^7 b + 28a^6 b^2 + 56a^5 b^3 + 70a^4 b^4 + 56a^3 b^5 + 28a^2 b^6 + 8ab^7 + b^8$

195. $p^9 + 9p^8 q + 36p^7 q^2 + 84p^6 q^3 + 126p^5 q^4 + 126p^4 q^5 + 84p^3 q^6 + 36p^2 q^7 + 9pq^8 + q^9$

197. $a^6 - 6a^5 b + 15a^4 b^2 - 20a^3 b^3 + 15a^2 b^4 - 6ab^5 + b^6$

199. $x^3 + 15x^2 + 75x + 125$

201. $y^7 + 7y^6 + 21y^5 + 35y^4 + 35y^3 + 21y^2 + 7y + 1$

203. $z^6 - 12z^5 + 60z^4 - 160z^3 + 240z^2 - 192z + 64$

205. $243x^5 - 405x^4 + 270x^3 - 90x^2 + 15x - 1$

207. $27x^3 - 135x^2 + 225x - 125$

209. $27x^3 + 135x^2 y + 225xy^2 + 125y^3$

211. ⓐ 7 ⓑ 1 ⓒ 1 ⓓ 45

213. ⓐ 4 ⓑ 1 ⓒ 1 ⓓ 55

215. $m^5 + 5m^4 n + 10m^3 n^2 + 10m^2 n^3 + 5mn^4 + n^5$

217. $s^7 + 7s^6 t + 21s^5 t^2 + 35s^4 t^3 + 35s^3 t^4 + 21s^2 t^5 + 7st^6 + t^7$

219. $y^4 - 12y^3 + 54y^2 - 108y + 81$

221. $q^3 - 12q^2 + 48q - 64$

223. $625x^4 - 1000x^3 y + 600x^2 y^2 - 160xy^3 + 16y^4$

225. $243x^5 + 1620x^4 y + 4320x^3 y^2 + 5760x^2 y^3 + 3840xy^4 + 1024y^5$

227. $126a^5 b^4$

229. $462x^5 y^6$

231. 112

233. 324

235. 30,618

237. Answers will vary.

239. Answers will vary.

Review Exercises

241. 7, 13, 31, 85, 247

243. $\frac{3}{4}, \frac{5}{16}, \frac{7}{64}, \frac{9}{256}, \frac{11}{1024}$

245. $a_n = 9n$

247. $a_n = e^{n-4}$

249. $a_n = -\dfrac{n}{n+2}$

251. $\dfrac{1}{6}, \dfrac{1}{12}, \dfrac{1}{20}, \dfrac{1}{30}, \dfrac{1}{42}$

253. $-3 + (-1) + 1 + 3 + 5$
$+7 + 9 = 21$

255. $4 + 4 + 2 + \dfrac{2}{3} + \dfrac{1}{6} = \dfrac{65}{6}$

257. $\displaystyle\sum_{n=1}^{5} (-1)^n \dfrac{1}{3^n}$

259. $\displaystyle\sum_{n=1}^{5} \dfrac{4}{n}$

261. The sequence is arithmetic with common difference $d = 6$.

263. 5, 8, 11, 14, 17

265. $-13, -7, -1, 5, 11$

267. -129

269. $a_{18} = 103$. The general term is $a_n = 7n - 23$.

271. $a_1 = 1$, $d = 4$. The general term is $a_n = 4n - 3$.

273. -430

275. 585

277. 4850

279. 980

281. The sequence is not geometric.

283. The sequence is geometric with common ratio $r = -2$.

285. $128, 32, 8, 2, \dfrac{1}{2}$

287. 1,536

289. $a_{12} = -25,165,824$. The general term is $a_n = 6(-4)^{n-1}$.

291. 5,460

293. ≈ 3906.25

295. $\dfrac{189}{8} = 23.625$

297. $\dfrac{343}{6} \approx 57.167$

299. $\dfrac{4}{11}$

301. $1,634,421.27

303.
$x^4 - 4x^3 y + 6x^2 y^2 - 4xy^3 + y^4$

305.
$32y^5 - 240y^4 + 720y^3 - 1080y^2$
$+810y - 243$

307. ⓐ 11 ⓑ 1 ⓒ 1 ⓓ 56

309. ⓐ 1 ⓑ 1 ⓒ 1 ⓓ 55

311.
$t^9 - 9t^8 + 36t^7 - 84t^6 + 126t^5$
$-126t^4 + 84t^3 - 36t^2 + 9t - 1$

313. $256x^4 + 768x^3 y + 864x^2 y^2$
$+432xy^3 + 81y^4$

315. $84a^6 b^3$

317. 135

319. 280

Practice Test

321. $\dfrac{1}{4}, \dfrac{1}{5}, \dfrac{1}{6}, \dfrac{1}{7}, \dfrac{1}{8}$

323. $-4 + 16 - 64 + 256 = 204$

325. $-13, -10, -7, -4, -1$

327. $a_{23} = 59$. The general term is $a_n = 3n - 10$.

329. 1,325

331. 3,260

333. The sequence is geometric with common ratio $r = \dfrac{1}{3}$.

335. 5,242,880

337. 797,162

339. $\dfrac{5}{6}$

341. $1,409,344.19

343. ⓐ 8 ⓑ 1 ⓒ 1 ⓓ 210

INDEX